Random variable	Probability density function $f_x(x)$	Mean		ic $\omega)$
Students' $t(n)$	$\frac{\Gamma((n+1)/2)}{\sqrt{\pi n}\,\Gamma(n/2)}(1+x^2/n)^{-(n+1)/2}$, $-\infty < x < \infty$	0		
F distribution	$\frac{\Gamma((m+n)/2)}{\Gamma(m/2)\Gamma(n/2)}\left(\frac{m}{n}\right)^{m/2} x^{m/2-1} \times \left(1+\frac{mx}{n}\right)^{-(m+n)/2}, x > 0$	$\frac{n}{n-2}$, $n > 2$	$\frac{n^2(2m+2n-4)}{m(n-2)^2(n-4)}$, $n > 4$	—
Bernoulli	$P(\mathbf{x}=1) = p, P(\mathbf{x}=0) = 1-p = q$	p	$p(1-p)$	$pe^{j\omega}+q$
Binomial $B(n, p)$	$\binom{n}{k} p^k q^{n-k}$, $k = 0, 1, 2, \ldots, n, p+q = 1$	np	npq	$(pe^{j\omega}+q)^n$
Poisson $P(\lambda)$	$e^{-\lambda}\frac{\lambda^k}{k!}, k = 0, 1, 2, \ldots, \infty$	λ	λ	$e^{-\lambda(1-e^{j\omega})}$
Hypergeometric	$\frac{\binom{M}{k}\binom{N-M}{n-k}}{\binom{N}{n}}$ $\max(0, M+n-N) \le k \le \min(M, n)$	$\frac{nM}{N}$	$n\frac{M}{N}\left(1-\frac{M}{N}\right)\left(1-\frac{n-1}{N-1}\right)$	—
Geometric	pq^k, $k = 0, 1, 2, \ldots, \infty$	$\frac{q}{p}$	$\frac{q}{p^2}$	$\frac{p}{1-qe^{j\omega}}$
	or pq^{k-1}, $k = 1, 2, \ldots, \infty,\ p+q = 1$	$\frac{1}{p}$	$\frac{q}{p^2}$	$\frac{p}{e^{-j\omega}-q}$
Pascal or negative binomial $NB(r, p)$	$\binom{r+k-1}{k} p^r q^k$, $k = 0, 1, 2, \ldots, \infty$	$\frac{rq}{p}$	$\frac{rq}{p^2}$	$\left(\frac{p}{1-qe^{j\omega}}\right)^r$
	or $\binom{k-1}{r-1} p^r q^{k-r}$, $k = r, r+1, \ldots, \infty, p+q = 1$	$\frac{r}{p}$	$\frac{rq}{p^2}$	$\left(\frac{p}{e^{-j\omega}-q}\right)^r$
Discrete uniform	$1/N$, $k = 1, 2, \ldots, N$	$\frac{N+1}{2}$	$\frac{N^2-1}{12}$	$e^{j(N+1)\omega/2}\frac{\sin(N\omega/2)}{\sin(\omega/2)}$
Multivariate Gaussian $X = (x_1, x_2, \ldots, x_n)$ $m = (m_1, m_2, \ldots, m_n)$ $u = (u_1, u_2, \ldots, u_n)$ $C = (C_{ik})$, $i, k = 1, 2, \ldots, n$	$\frac{1}{(2\pi)^{n/2}\det C} e^{-\{(X-m)C^{-1}(X-m)^t/2\}}$ $C_{ik} = E[(\mathbf{x}_i - m_i)(\mathbf{x}_k - m_k)^*]$	m	C (Covariance matrix)	$e^{\{jmu^t - uCu^t/2\}}$

PROBABILITY, RANDOM VARIABLES, AND STOCHASTIC PROCESSES

McGraw-Hill Series in Electrical and Computer Engineering

Stephen W. Director, University of Michigan, Ann Arbor, Senior Consulting Editor

Circuits and Systems
Communications and Signal Processing
Computer Engineering
Control Theory and Robotics
Electromagnetics
Electronics and VLSI Circuits
Introductory
Power
Antennas, Microwaves, and Radar

Previous Consulting Editors

Ronald N. Bracewell, Colin Cherry, James F. Gibbons, Willis W. Harman, Hubert Heffner, Edward W. Herold, John G. Linvill, Simon Ramo, Ronald A. Rohrer, Anthony E. Siegman, Charles Susskind, Frederick E. Terman, John G. Truxal, Ernst Weber, and John R. Whinnery

Related Titles:

Antoniou: *Digital Filters: Analysis and Design*
Carlson/Crilly/Rutledge: *Communication Systems*
Ham/Kostanic: *Principles of Neurocomputing for Science and Engineering*
Keiser: *Local Area Networks*
Keiser: *Optical Fiber Communications*
Kraus: *Antennas*
Leon-Garcia/Widjaja: *Communications Networks*
Lindner: *Introduction to Signals and Systems*
Manolakis/Ingle/Kogon: *Statistical and Adaptive Signal Processing: Spectral Estimation, Signal Modeling, Adaptive Filtering and Array Processing*
Mitra: *Digital Signal Processing: A Computer-Based Approach*
Peebles: *Probability, Random Variables, and Random Signal Principles*
Proakis: *Digital Communications*
Smith: *Modern Communication Circuits*
Taylor: *Hands-On Digital Signal Processing*
Viniotis: *Probability and Random Processes*
Walrand: *Communications Networks*

PROBABILITY, RANDOM VARIABLES, AND STOCHASTIC PROCESSES

FOURTH EDITION

Athanasios Papoulis

University Professor
Polytechnic University

S. Unnikrishna Pillai

Professor of Electrical and Computer Engineering
Polytechnic University

McGraw Hill

Boston Burr Ridge, IL Dubuque, IA Madison, WI New York San Francisco St. Louis
Bangkok Bogotá Caracas Kuala Lumpur Lisbon London Madrid Mexico City
Milan Montreal New Delhi Santiago Seoul Singapore Sydney Taipei Toronto

McGraw-Hill Higher Education

A Division of The **McGraw-Hill** *Companies*

PROBABILITY, RANDOM VARIABLES, AND STOCHASTIC PROCESSES, FOURTH EDITION

Published by McGraw-Hill, a business unit of The McGraw-Hill Companies, Inc., 1221 Avenue of the Americas, New York, NY 10020.

This book is printed on acid-free paper.

International 1 2 3 4 5 6 7 8 9 0 QPF/QPF 0 9 8 7 6 5 4 3 2 1 0
Domestic 1 2 3 4 5 6 7 8 9 0 QPF/QPF 0 9 8 7 6 5 4 3 2 1 0

ISBN 0–07–366011–6
ISBN 0–07–112256–7 (ISE)

General manager: *Thomas E. Casson*
Publisher: *Elizabeth A. Jones*
Sponsoring editor: *Catherine Fields Shultz*
Developmental editor: *Michelle L. Flomenhoft*
Executive marketing manager: *John Wannemacher*
Project manager: *Sheila M. Frank*
Production supervisor: *Sherry L. Kane*
Coordinator of freelance design: *Rick D. Noel*
Cover designer: *So Yon Kim*
Cover image: *©PhotoDisc, Signature Series, Dice, SS10074*
Supplement producer: *Brenda A. Ernzen*
Media technology senior producer: *Phillip Meek*
Compositor: *Interactive Composition Corporation*
Typeface: *10/12 Times Roman*
Printer: *Quebecor World Fairfield, PA*

Library of Congress Cataloging-in-Publication Data

Papoulis, Athanasios, 1921–
Probability, random variables, and stochastic processes / Athanasios Papoulis, S. Unnikrishna Pillai. — 4th ed.
p. cm.
Includes bibliographical references and index.
ISBN 0–07–366011–6 — ISBN 0–07–112256–7 (ISE)
1. Probabilities. 2. Random variables. 3. Stochastic processes. I. Pillai, S. Unnikrishna, 1955 – . II. Title.

QA273 .P2 2002
519.2—dc21 2001044139
CIP

INTERNATIONAL EDITION ISBN 0–07–112256–7

www.mhhe.com

CONTENTS

PREFACE

The fourth edition of this book has been updated significantly from previous editions, and it includes a coauthor. About one-third of the content of this edition is new material, and these additions are incorporated while maintaining the style and spirit of the previous editions that are familiar to many of its readers.

The basic outlook and approach remain the same: To develop the subject of probability theory and stochastic processes as a deductive discipline and to illustrate the theory with basic applications of engineering interest. To this extent, these remarks made in the first edition are still valid: "The book is written neither for the handbook-oriented students nor for the sophisticated few (if any) who can learn the subject from advanced mathematical texts. It is written for the majority of engineers and physicists who have sufficient maturity to appreciate and follow a logical presentation. . . . There is an obvious lack of continuity between the elements of probability as presented in introductory courses, and the sophisticated concepts needed in today's applications. . . . Random variables, transformations, expected values, conditional densities, characteristic functions cannot be mastered with mere exposure. These concepts must be clearly defined and must be developed, one at a time, with sufficient elaboration."

Recognizing these factors, additional examples are added for further clarity, and the new topics include the following.

Chapters 3 and 4 have undergone substantial rewriting. Chapter 3 has a detailed section on *Bernoulli's theorem and games of chance* (Sec. 3-3), and several examples are presented there including the classical *gambler's ruin problem* to stimulate student interest. In Chap. 4 various probability distributions are categorized and illustrated, and two kinds of approximations to the binomial distribution are carried out to illustrate the connections among some of the random variables.

Chapter 5 contains new examples illustrating the usefulness of characteristic functions and moment-generating functions including the proof of the DeMoivre–Laplace theorem.

Chapter 6 has been rewritten with additional examples, and is complete in its description of two random variables and their properties.

Chapter 8 contains a new Sec. 8-3 on *Parameter estimation* that includes key ideas on minimum variance unbiased estimation, the Cramer–Rao bound, the Rao–Blackwell theorem, and the Bhattacharya bound.

In Chaps. 9 and 10, sections on *Poisson processes* are further expanded with additional results. A new detailed section on *random walks* has also been added.

Chapter 12 includes a new subsection describing the parametrization of the class of all admissible spectral extensions given a set of valid autocorrelations.

Because of the importance of *queueing theory,* the old material has undergone complete revision to the extent that two new chapters (15 and 16) are devoted to this topic. Chapter 15 describes *Markov chains,* their properties, characterization, and the long-term (steady state) and transient behavior of the chain and illustrates various theorems through several examples. In particular, Example 15-26 *The Game of Tennis* is an excellent illustration of the theory to analyze practical applications, and the chapter concludes with a detailed study of *branching processes,* which have important applications in queueing theory. Chapter 16 describes *Markov processes and queueing theory* starting with the Chapman–Kolmogorov equations and concentrating on the *birth-death processes* to illustrate markovian queues. The treatment, however, includes non-markovian queues and machine servicing problems, and concludes with an introduction to the network of queues.

The material in this book can be organized for various one semester courses:

- Chapters 1 to 6: *Probability Theory* (for senior and/or first-level graduate students)
- Chapters 7 and 8: *Statistics and Estimation Theory* (as a follow-up course to *Probability Theory*)
- Chapters 9 to 11: *Stochastic Processes* (follow-up course to *Probability Theory*)
- Chapters 12 to 14: *Spectrum Estimation and Filtering* (follow-up course to *Stochastic Processes*)
- Chapters 15 and 16: *Markov Chains and Queueing Theory* (follow-up course to *Probability Theory*)

The authors would like to thank Ms. Catherine Fields Shultz, editor for electrical and computer engineering at McGraw-Hill Publishing Company, Ms. Michelle Flomenhoft and Mr. John Griffin, developmental editors, Ms. Sheila Frank, Project manager and her highly efficient team, and Profs. D. P. Gelopulos, M. Georgiopoulos, A. Haddad, T. Moon, J. Rowland, C. S. Tsang, J. K. Tugnait, and O. C. Ugweje, for their comments, criticism, and guidance throughout the period of this revision. In addition, Dr. Michael Rosse, several colleagues at Polytechnic including Profs. Dante Youla, Henry Bertoni, Leonard Shaw and Ivan Selesnick, as well as students Dr. Hyun Seok Oh, Mr. Jun Ho Jo, and Mr. Seung Hun Cha deserve special credit for their valuable help and encouragement during the preparation of the manuscript. Discussions with Prof. C. Radhakrishna Rao about two of his key theorems in statistics and other items are also gratefully acknowledged.

Athanasios Papoulis
S. Unnikrishna Pillai

PROBABILITY, RANDOM VARIABLES, AND STOCHASTIC PROCESSES

PART I

PROBABILITY AND RANDOM VARIABLES

CHAPTER 1

THE MEANING OF PROBABILITY

1-1 INTRODUCTION

The theory of probability deals with averages of mass phenomena occurring sequentially or simultaneously: electron emission, telephone calls, radar detection, quality control, system failure, games of chance, statistical mechanics, turbulence, noise, birth and death rates, and queueing theory, among many others.

It has been *observed* that in these and other fields certain averages approach a constant value as the number of observations increases and this value remains the same if the averages are evaluated over any subsequence specified before the experiment is performed. In the coin experiment, for example, the percentage of heads approaches 0.5 or some other constant, and the same average is obtained if we consider every fourth, say, toss (no betting system can beat the roulette).

The purpose of the theory is to describe and predict such averages in terms of probabilities of events. The probability of an event A is a number $P(A)$ assigned to this event. This number could be interpreted as:

> If the experiment is performed n times and the event A occurs n_A times, then, *with a high degree of certainty,* the relative frequency n_A/n of the occurrence of A is *close* to $P(A)$:
>
> $$P(A) \simeq n_A/n \tag{1-1}$$
>
> provided that n is *sufficiently large*.

This interpretation is imprecise: The terms "with a high degree of certainty," "close," and "sufficiently large" have no clear meaning. However, this lack of precision cannot be avoided. If we attempt to define in probabilistic terms the "high degree of certainty" we shall only postpone the inevitable conclusion that probability, like any physical theory, is related to physical phenomena only in inexact terms. Nevertheless, the theory is an exact discipline developed logically from clearly defined axioms, and when it is applied to real problems, *it works*.

OBSERVATION, DEDUCTION, PREDICTION. In the applications of probability to real problems, these steps must be clearly distinguished:

Step 1 (physical) We determine by an inexact process the probabilities $P(A_i)$ of certain events A_i.

This process could be based on the relationship (1-1) between probability and observation: The probabilistic data $P(A_i)$ equal the observed ratios n_{A_i}/n. It could also be based on "reasoning" making use of certain symmetries: If, out of a total of N outcomes, there are N_A outcomes favorable to the event A, then $P(A) = N_A/N$.

For example, if a loaded die is rolled 1000 times and *five* shows 200 times, then the probability of *five* equals 0.2. If the die is fair, then, because of its symmetry, the probability of *five* equals $1/6$.

Step 2 (conceptual) We assume that probabilities satisfy certain axioms, and by deductive reasoning we determine from the probabilities $P(A_i)$ of certain events A_i the probabilities $P(B_j)$ of other events B_j.

For example, in the game with a fair die we deduce that the probability of the event *even* equals $3/6$. Our reasoning is of the form:

$$\text{If} \quad P(1) = \cdots = P(6) = \tfrac{1}{6} \qquad \text{then} \quad P(even) = \tfrac{3}{6}$$

Step 3 (physical) We make a physical *prediction* based on the numbers $P(B_j)$ so obtained.

This step could rely on (1-1) applied in reverse: If we perform the experiment n times and an event B occurs n_B times, then $n_B \simeq nP(B)$.

If, for example, we roll a fair die 1000 times, our prediction is that *even* will show about 500 times.

We could not emphasize too strongly the need for separating these three steps in the solution of a problem. We must make a clear distinction between the data that are determined empirically and the results that are deduced logically.

Steps 1 and 3 are based on *inductive reasoning*. Suppose, for example, that we wish to determine the probability of *heads* of a given coin. Should we toss the coin 100 or 1000 times? If we toss it 1000 times and the average number of heads equals 0.48, what kind of prediction can we make on the basis of this observation? Can we deduce that at the next 1000 tosses the number of heads will be about 480? Such questions can be answered only inductively.

In this book, we consider mainly step 2, that is, from certain probabilities we derive *deductively* other probabilities. One might argue that such derivations are mere tautologies because the results are contained in the assumptions. This is true in the same sense that the intricate equations of motion of a satellite are included in Newton's laws.

To conclude, we repeat that the probability $P(A)$ of an event A will be interpreted as a number assigned to this event as mass is assigned to a body or resistance to a resistor. In the development of the theory, we will not be concerned about the "physical meaning" of this number. This is what is done in circuit analysis, in electromagnetic theory, in classical mechanics, or in any other scientific discipline. These theories are, of course, of no value to physics unless they help us solve real problems. We must assign specific, if only approximate, resistances to real resistors and probabilities to real events (step 1); we must also give physical meaning to all conclusions that are derived from the theory (step 3). But this link between concepts and observation must be separated from the purely logical structure of each theory (step 2).

As an illustration, we discuss in Example 1-1 the interpretation of the meaning of resistance in circuit theory.

EXAMPLE 1-1 ▶ A resistor is commonly viewed as a two-terminal device whose voltage is proportional to the current

$$R = \frac{v(t)}{i(t)} \tag{1-2}$$

This, however, is only a convenient abstraction. A real resistor is a complex device with distributed inductance and capacitance having no clearly specified terminals. A relationship of the form (1-2) can, therefore, be claimed only within certain errors, in certain frequency ranges, and with a variety of other qualifications. Nevertheless, in the development of circuit theory we ignore all these uncertainties. We assume that the resistance R is a precise number satisfying (1-2) and we develop a theory based on (1-2) and on Kirchhoff's laws. It would not be wise, we all agree, if at each stage of the development of the theory we were concerned with the *true* meaning of R. ◀

1-2 THE DEFINITIONS

In this section, we discuss various definitions of probability and their roles in our investigation.

Axiomatic Definition

We shall use the following concepts from set theory (for details see Chap. 2): The certain event S is the event that occurs in every trial. The union $A \cup B \equiv A + B$ of two events A and B is the event that occurs when A or B or both occur. The intersection $A \cap B \equiv AB$ of the events A and B is the event that occurs when both events A and B occur. The events A and B are *mutually exclusive* if the occurrence of one of them excludes the occurrence of the other.

We shall illustrate with the die experiment: The certain event is the event that occurs whenever any one of the six faces shows. The union of the events *even* and *less than 3* is the event *1 or 2 or 4 or 6* and their intersection is the event 2. The events *even* and *odd* are mutually exclusive.

The axiomatic approach to probability is based on the following three postulates and on nothing else: The probability $P(A)$ of an event A is a non-negative number assigned to this event:

$$P(A) \geq 0 \tag{1-3}$$

The probability of the certain event equals 1:

$$P(S) = 1 \tag{1-4}$$

If the events A and B are mutually exclusive, then

$$P(A \cup B) = P(A) + P(B) \tag{1-5}$$

This approach to probability is relatively recent (A.N. Kolmogorov,[1] 1933). However, in our view, it is the best way to introduce a probability even in elementary courses. It emphasizes the deductive character of the theory, it avoids conceptual ambiguities, it provides a solid preparation for sophisticated applications, and it offers at least a beginning for a deeper study of this important subject.

The axiomatic development of probability might appear overly mathematical. However, as we hope to show, this is not so. The elements of the theory can be adequately explained with basic calculus.

Relative Frequency Definition

The relative frequency approach is based on the following definition: The probability $P(A)$ of an event A is the limit

$$P(A) = \lim_{n\to\infty} \frac{n_A}{n} \tag{1-6}$$

where n_A is the number of occurrences of A and n is the number of trials.

This definition appears reasonable. Since probabilities are used to describe relative frequencies, it is natural to define them as limits of such frequencies. The problem associated with a priori definitions are eliminated, one might think, and the theory is founded on observation.

However, although the relative frequency concept is fundamental in the applications of probability (steps 1 and 3), its use as the basis of a deductive theory (step 2) must be challenged. Indeed, in a physical experiment, the numbers n_A and n might be large but they are only finite; their ratio cannot, therefore, be equated, even approximately, to a limit. If (1-6) is used to define $P(A)$, the limit must be accepted as a *hypothesis,* not as a number that can be determined experimentally.

Early in the century, Von Mises[2] used (1-6) as the foundation for a new theory. At that time, the prevailing point of view was still the classical, and his work offered a welcome alternative to the a priori concept of probability, challenging its metaphysical implications and demonstrating that it leads to useful conclusions mainly because it makes implicit use of relative frequencies based on our collective experience. The use of (1-6) as the basis for deductive theory has not, however, enjoyed wide acceptance even though (1-6) relates $P(A)$ to observed frequencies. It has generally been recognized that the axiomatic approach (Kolmogorov) is superior.

We shall venture a comparison between the two approaches using as illustration the definition of the resistance R of an ideal resistor. We can define R as a limit

$$R = \lim_{n\to\infty} \frac{e(t)}{i_n(t)}$$

---

[1] A.N. Kolmogorov: Grundbegriffe der Wahrscheinlichkeits Rechnung, *Ergeb. Math und ihrer Grensg.* vol. 2, 1933.

[2] Richard Von Mises: *Probability, Statistics and Truth,* English edition, H. Geiringer, ed., G. Allen and Unwin Ltd., London, 1957.

where $e(t)$ is a voltage source and $i_n(t)$ are the currents of a sequence of real resistors that tend in some sense to an ideal two-terminal element. This definition might show the relationship between real resistors and ideal elements but the resulting theory is complicated. An axiomatic definition of R based on Kirchhoff's laws is, of course, preferable.

Classical Definition

For several centuries, the theory of probability was based on the classical definition. This concept is used today to determine probabilistic data and as a working hypothesis. In the following, we explain its significance.

According to the classical definition, the probability $P(A)$ of an event A is determined a priori without actual experimentation: It is given by the ratio

$$P(A) = \frac{N_A}{N} \tag{1-7}$$

where N is the number of *possible* outcomes and N_A is the number of outcomes that are *favorable* to the event A.

In the die experiment, the possible outcomes are six and the outcomes favorable to the event *even* are three; hence $P(even) = 3/6$.

It is important to note, however, that the significance of the numbers N and N_A is not always clear. We shall demonstrate the underlying ambiguities with Example 1-2.

EXAMPLE 1-2 ▶ We roll two dice and we want to find the probability p that the sum of the numbers that show equals 7.

To solve this problem using (1-7), we must determine the numbers N and N_A. (*a*) We could consider as possible outcomes the 11 sums $2, 3, \ldots, 12$. Of these, only one, namely the sum 7, is favorable; hence $p = 1/11$. This result is of course wrong. (*b*) We could count as possible outcomes all pairs of numbers not distinguishing between the first and the second die. We have now 21 outcomes of which the pairs (3, 4), (5, 2), and (6, 1) are favorable. In this case, $N_A = 3$ and $N = 21$; hence $p = 3/21$. This result is also wrong. (*c*) We now reason that the above solutions are wrong because the outcomes in (*a*) and (*b*) are not *equally likely*. To solve the problem "correctly," we must count all pairs of numbers distinguishing between the first and the second die. The total number of outcomes is now 36 and the favorable outcomes are the six pairs (3, 4), (4, 3), (5, 2), (2, 5), (6, 1), and (1, 6); hence $p = 6/36$. ◀

Example 1-2 shows the need for refining definition (1-7). The improved version reads as follows:

> The probability of an event equals the ratio of its favorable outcomes to the total number of outcomes provided that all outcomes are *equally likely*.

As we shall presently see, this refinement does not eliminate the problems associated with the classical definition.

Notes 1. The classical definition was introduced as a consequence of the *principle of insufficient reason*[3]: "In the absence of any prior knowledge, we *must* assume that the events A_i have equal probabilities." This conclusion is based on the subjective interpretation of probability as a *measure of our state of knowledge* about the events A_i. Indeed, if it were not true that the events A_i have the same probability, then changing their indices we would obtain different probabilities without a change in the state of our knowledge.

2. As we explain in Chap. 14, the principle of insufficient reason is equivalent to the *principle of maximum entropy*.

CRITIQUE. The classical definition can be questioned on several grounds.

A. The term *equally likely* used in the improved version of (1-7) means, actually, *equally probable*. Thus, in the definition, use is made of the concept to be defined. As we have seen in Example 1-2, this often leads to difficulties in determining N and N_A.

B. The definition can be applied only to a limited class of problems. In the die experiment, for example, it is applicable only if the six faces have the same probability. If the die is loaded and the probability of *four* equals 0.2, say, the number 0.2 cannot be derived from (1-7).

C. It appears from (1-7) that the classical definition is a consequence of logical imperatives divorced from experience. This, however, is not so. We accept certain alternatives as equally likely because of our collective experience. The probabilities of the outcomes of a fair die equal $1/6$ not only because the die is symmetrical but also because it was observed in the long history of rolling dice that the ratio n_A/n in (1-1) is close to $1/6$. The next illustration is, perhaps, more convincing:

We wish to determine the probability p that a newborn baby is a boy. It is generally assumed that $p = 1/2$; however, this is not the result of pure reasoning. In the first place, it is only approximately true that $p = 1/2$. Furthermore, without access to long records we would not know that the boy–girl alternatives are equally likely regardless of the sex history of the baby's family, the season or place of its birth, or other conceivable factors. It is only after long accumulation of records that such factors become irrelevant and the two alternatives are accepted as equally likely.

D. If the number of possible outcomes is infinite, then to apply the classical definition we must use length, area, or some other measure of infinity for determining the ratio N_A/N in (1-7). We illustrate the resulting difficulties with the following example known as the *Bertrand paradox*.

EXAMPLE 1-3

BERTRAND PARADOX

▶ We are given a circle C of radius r and we wish to determine the probability p that the length l of a "randomly selected" cord AB is greater than the length $r\sqrt{3}$ of the inscribed equilateral triangle.

---

[3] H. Bernoulli, *Ars Conjectandi,* 1713.

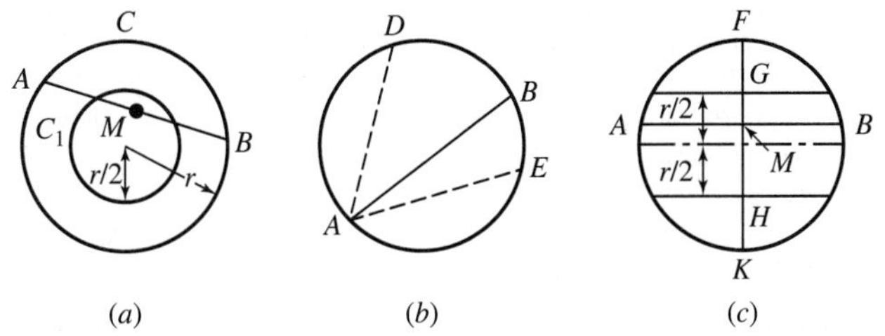

FIGURE 1-1

We shall show that this problem can be given at least three reasonable solutions.

I. If the center M of the cord AB lies inside the circle C_1 of radius $r/2$ shown in Fig. 1-1a, then $l > r\sqrt{3}$. It is reasonable, therefore, to consider as favorable outcomes all points inside the circle C_1 and as possible outcomes all points inside the circle C. Using as measure of their numbers the corresponding areas $\pi r^2/4$ and πr^2, we conclude that

$$p = \frac{\pi r^2/4}{\pi r^2} = \frac{1}{4}$$

II. We now assume that the end A of the cord AB is fixed. This reduces the number of possibilities but it has no effect on the value of p because the number of favorable locations of B is reduced proportionately. If B is on the 120° arc DBE of Fig. 1-1b, then $l > r\sqrt{3}$. The favorable outcomes are now the points on this arc and the total outcomes all points on the circumference of the circle C. Using as their measurements the corresponding lengths $2\pi r/3$ and $2\pi r$, we obtain

$$p = \frac{2\pi r/3}{2\pi r} = \frac{1}{3}$$

III. We assume finally that the direction of AB is perpendicular to the line FK of Fig. 1-1c. As in II this restriction has no effect on the value of p. If the center M of AB is between G and H, then $l > r\sqrt{3}$. Favorable outcomes are now the points on GH and possible outcomes all points on FK. Using as their measures the respective lengths r and $2r$, we obtain

$$p = \frac{r}{2r} = \frac{1}{2}$$ ◀

We have thus found not one but three different solutions for the same problem! One might remark that these solutions correspond to three different experiments. This is true but not obvious and, in any case, it demonstrates the ambiguities associated with the classical definition, and the need for a clear specification of the outcomes of an experiment and the meaning of the terms "possible" and "favorable."

VALIDITY. We shall now discuss the value of the classical definition in the determination of probabilistic data and as a working hypothesis.

A. In many applications, the assumption that there are N equally likely alternatives is well established through long experience. Equation (1-7) is then accepted as self-evident. For example, "If a ball is selected at random from a box containing m black and n white balls, the probability that it is white equals $n/(m+n)$," or, "If a call occurs at random in the time interval $(0, T)$, the probability that it occurs in the interval (t_1, t_2) equals $(t_2 - t_1)/T$."

Such conclusions are of course, valid and useful; however, their validity rests on the meaning of the word *random*. The conclusion of the last example that "the unknown probability equals $(t_2 - t_1)/T$" is not a consequence of the "randomness" of the call. The two statements are merely equivalent and they follow not from a priori reasoning but from past records of telephone calls.

B. In a number of applications it is impossible to determine the probabilities of various events by repeating the underlying experiment a sufficient number of times. In such cases, we have no choice but to *assume* that certain alternatives are equally likely and to determine the desired probabilities from (1-7). This means that we use the classical definition as a *working* hypothesis. The hypothesis is accepted if its observable consequences agree with experience, otherwise it is rejected. We illustrate with an important example from statistical mechanics.

EXAMPLE 1-4

▶ Given n particles and $m > n$ boxes, we place at random each particle in one of the boxes. We wish to find the probability p that in n preselected boxes, one and only one particle will be found.

Since we are interested only in the underlying assumptions, we shall only state the results (the proof is assigned as Prob. 4-34). We also verify the solution for $n = 2$ and $m = 6$. For this special case, the problem can be stated in terms of a pair of dice: The $m = 6$ faces correspond to the m boxes and the $n = 2$ dice to the n particles. We assume that the preselected faces (boxes) are 3 and 4.

The solution to this problem depends on the choice of possible and favorable outcomes We shall consider these three celebrated cases:

MAXWELL–BOLTZMANN STATISTICS

If we accept as outcomes all possible ways of placing n particles in m boxes distinguishing the identity of each particle, then

$$p = \frac{n!}{m^n}$$

For $n = 2$ and $m = 6$ this yields $p = 2/36$. This is the probability for getting 3, 4 in the game of two dice.

BOSE–EINSTEIN STATISTICS

If we assume that the particles are not distinguishable, that is, if all their permutations count as one, then

$$p = \frac{(m-1)!n!}{(n+m-1)!}$$

For $n = 2$ and $m = 6$ this yields $p = 1/21$. Indeed, if we do not distinguish between the two dice, then $N = 21$ and $N_A = 1$ because the outcomes 3, 4 and 4, 3 are counted as one.

FERMI–DIRAC STATISTICS

If we do not distinguish between the particles and also we assume that in each box we are allowed to place at most one particle, then

$$p = \frac{n!(m-n)!}{m!}$$

For $n = 2$ and $m = 6$ we obtain $p = 1/15$. This is the probability for 3, 4 if we do not distinguish between the dice and also we ignore the outcomes in which the two numbers that show are equal.

One might argue, as indeed it was in the early years of statistical mechanics, that only the first of these solutions is logical. The fact is that in the absence of direct or indirect experimental evidence this argument cannot be supported. The three models proposed are actually only *hypotheses* and the physicist accepts the one whose consequences agree with experience. ◀

C. Suppose that we know the probability $P(A)$ of an event A in experiment 1 and the probability $P(B)$ of an event B in experiment 2. In general, from this information we cannot determine the probability $P(AB)$ that both events A and B will occur. However, if we know that the two experiments are *independent*, then

$$P(AB) = P(A)P(B) \tag{1-8}$$

In many cases, this independence can be established a priori by reasoning that the outcomes of experiment 1 have no effect on the outcomes of experiment 2. For example, if in the coin experiment the probability of *heads* equals 1/2 and in the die experiment the probability of *even* equals 1/2, then, we conclude "logically," that if both experiments are performed, the probability that we get *heads* on the coin and *even* on the die equals $1/2 \times 1/2$. Thus, as in (1-7), we accept the validity of (1-8) as a logical necessity without recourse to (1-1) or to any other direct evidence.

D. The classical definition can be used as the basis of a deductive theory if we accept (1-7) as an *assumption*. In this theory, no other assumptions are used and postulates (1-3) to (1-5) become theorems. Indeed, the first two postulates are obvious and the third follows from (1-7) because, if the events A and B are mutually exclusive, then $N_{A+B} = N_A + N_B$; hence

$$P(A \cup B) = \frac{N_{A+B}}{N} = \frac{N_A}{N} + \frac{N_B}{N} = P(A) + P(B)$$

As we show in (2-25), however, this is only a very special case of the axiomatic approach to probability.

1-3 PROBABILITY AND INDUCTION

In the applications of the theory of probability we are faced with the following question: Suppose that we know somehow from past observations the probability $P(A)$ of an event A in a given experiment. What conclusion can we draw about the occurrence of this event in a *single* future performance of this experiment? (See also Sec. 8-1.)

We shall answer this question in two ways depending on the size of $P(A)$: We shall give one kind of an answer if $P(A)$ is a number distinctly different from 0 or 1, for example 0.6, and a different kind of an answer if $P(A)$ is close to 0 or 1, for example 0.999. Although the boundary between these two cases is not sharply defined, the corresponding answers are fundamentally different.

Case 1 Suppose that $P(A) = 0.6$. In this case, the number 0.6 gives us only a "certain degree of confidence that the event A will occur." The known probability is thus used as a "measure of our belief" about the occurrence of A in a single trial. This interpretation of $P(A)$ is subjective in the sense that it cannot be verified experimentally. In a single trial, the event A will either occur or will not occur. If it does not, this will not be a reason for questioning the validity of the assumption that $P(A) = 0.6$.

Case 2 Suppose, however, that $P(A) = 0.999$. We can now state with practical certainty that at the next trial thc cvcnt A will occur. This conclusion is objective in the sense that it can be verified experimentally. At the next trial the event A must occur. If it does not, we must seriously doubt, if not outright reject, the assumption that $P(A) = 0.999$.

The boundary between these two cases, arbitrary though it is (0.9 or 0.99999?), establishes in a sense the line separating "soft" from "hard" scientific conclusions. The theory of probability gives us the analytic tools (step 2) for transforming the "subjective" statements of case 1 to the "objective" statements of case 2. In the following, we explain briefly the underlying reasoning.

As we show in Chap. 3, the information that $P(A) = 0.6$ leads to the conclusion that if the experiment is performed 1000 times, then "almost certainly" the number of times the event A will occur is between 550 and 650. This is shown by considering the repetition of the original experiment 1000 times as a *single* outcome of a new experiment. In this experiment the probability of the event

$$A_1 = \{\text{the number of times } A \text{ occurs is between 550 and 650}\}$$

equals 0.999 (see Prob. 4-25). We must, therefore, conclude that (case 2) the event A_1 will occur with practical certainty.

We have thus succeeded, using the theory of probability, to transform the "subjective" conclusion about A based on the *given* information that $P(A) = 0.6$, to the "objective" conclusion about A_1 based on the *derived* conclusion that $P(A_1) = 0.999$. We should emphasize, however, that both conclusions rely on inductive reasoning. Their difference, although significant, is only quantitative. As in case 1, the "objective" conclusion of case 2 is not a certainty but only an inference. This, however, should not surprise us; after all, no prediction about future events based on past experience can be accepted as logical certainty.

Our inability to make categorical statements about future events is not limited to probability but applies to all sciences. Consider, for example, the development of classical mechanics. It was *observed* that bodies fall according to certain patterns, and on this evidence Newton formulated the laws of mechanics and used them to *predict* future events. His predictions, however, are not logical certainties but only plausible inferences. To "prove" that the future will evolve in the predicted manner we must invoke metaphysical causes.

1-4 CAUSALITY VERSUS RANDOMNESS

We conclude with a brief comment on the apparent controversy between causality and randomness. There is no conflict between causality and randomness or between determinism and probability if we agree, as we must, that scientific theories are not *discoveries* of the laws of nature but rather *inventions* of the human mind. Their consequences are presented in deterministic form if we examine the results of a single trial; they are presented as probabilistic statements if we are interested in averages of many trials. In both cases, all statements are qualified. In the first case, the uncertainties are of the form "with certain errors and in certain ranges of the relevant parameters"; in the second, "with a high degree of certainty if the number of trials is large enough." In the next example, we illustrate these two approaches.

EXAMPLE 1-5

▶ A rocket leaves the ground with an initial velocity v forming an angle θ with the horizontal axis (Fig. 1-2). We shall determine the distance $d = OB$ from the origin to the reentry point B.

From Newton's law it follows that

$$d = \frac{v^2}{g} \sin 2\theta \tag{1-9}$$

This seems to be an unqualified consequence of a causal law; however, this is not so. The result is approximate and it can be given a probabilistic interpretation.

Indeed, (1-9) is not the solution of a real problem but of an idealized model in which we have neglected air friction, air pressure, variation of g, and other uncertainties in the values of v and θ. We must, therefore, accept (1-9) only with qualifications. It holds within an error ε provided that the neglected factors are smaller than δ.

Suppose now that the reentry area consists of numbered holes and we want to find the reentry hole. Because of the uncertainties in v and θ, we are in no position to give a deterministic answer to our problem. We can, however, ask a different question: If many rockets, nominally with the same velocity, are launched, what percentage will enter the nth hole? This question no longer has a causal answer; it can only be given a random interpretation.

Thus the same physical problem can be subjected either to a deterministic or to a probabilistic analysis. One might argue that the problem is inherently deterministic because the rocket has a precise velocity even if we do not know it. If we did, we would know exactly the reentry hole. Probabilistic interpretations are, therefore, necessary because of our ignorance.

Such arguments can be answered with the statement that the physicists are not concerned with what *is true* but only with what *they can observe*. ◀

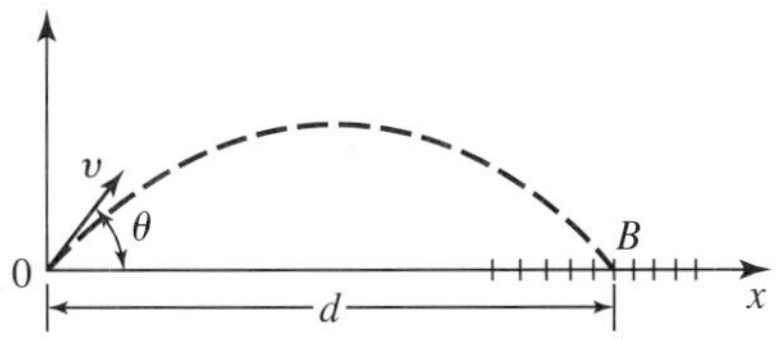

FIGURE 1-2

Historical Perspective

Probability theory has its humble origin in problems related to gambling and games of chance. The origin of the theory of probability goes back to the middle of the 17th century and is connected with the works of Pierre de Fermat (1601–1665), Blaise Pascal (1623–1662), and Christian Huygens (1629–1695). In their works, the concepts of the probability of a stochastic event and the expected or mean value of a random variable can be found. Although their investigations were concerned with problems connected with games of chance, the importance of these new concepts was clear to them, as Huygens points out in the first printed probability text[4] (1657) *On Calculations in Games of Chance:* "The reader will note that we are dealing not only with games, but also that the foundations of a very interesting and profound theory are being laid here." Later, Jacob Bernoulli (1654–1705), Abraham De Moivre (1667–1754), Rev. Thomas Bayes (1702–1761), Marquis Pierre Simon Laplace (1749–1827), Johann Friedrich Carl Gauss (1777–1855), and Siméon Denis Poisson (1781–1840) contributed significantly to the development of probability theory. The notable contributors from the Russian school include P.L. Chebyshev (1821–1894), and his students A. Markov (1856–1922) and A.M. Lyapunov (1857–1918) with important works dealing with the law of large numbers.

The deductive theory based on the axiomatic definition of probability that is popular today is mainly attributed to Andrei Nikolaevich Kolmogorov, who in the 1930s along with Paul Lévy found a close connection between the theory of probability and the mathematical theory of sets and functions of a real variable. Although Emile Borel had arrived at these ideas earlier, putting probability theory on this modern frame work is mainly due to the early 20th century mathematicians.

Concluding Remarks

In this book, we present a deductive theory (step 2) based on the axiomatic definition of probability. Occasionally, we use the classical definition but only to determine probabilistic data (step 1).

To show the link between theory and applications (step 3), we give also a relative frequency interpretation of the important results. This part of the book, written in small print under the title *Frequency interpretation,* does not obey the rules of deductive reasoning on which the theory is based.

---

[4]Although the ecentric scholar (and gambler) Girolamo Cardano (1501–1576) had written *The Book of Games and Chance* around 1520, it was not published until 1663. Cardano had left behind 131 printed works and 111 additional manuscripts.

CHAPTER 2

THE AXIOMS OF PROBABILITY

2-1 SET THEORY

A *set* is a collection of objects called *elements*. For example, "car, apple, pencil" is a set whose elements are a car, an apple, and a pencil. The set "heads, tails" has two elements. The set "1, 2, 3, 5" has four elements.

A *subset* B of a set A is another set whose elements are also elements of A. All sets under consideration will be subsets of a set S, which we shall call *space*.

The elements of a set will be identified mostly by the Greek letter ζ. Thus

$$A = \{\zeta_1, \ldots, \zeta_n\} \tag{2-1}$$

will mean that the set A consists of the elements $\zeta_1, \ldots, \zeta_n$. We shall also identify sets by the properties of their elements. Thus

$$A = \{\text{all positive integers}\} \tag{2-2}$$

will mean the set whose elements are the numbers $1, 2, 3, \ldots$.

The notation

$$\zeta_i \in A \qquad \zeta_i \notin A$$

will mean that ζ_i is or is not an element of A.

The *empty* or *null* set is by definition the set that contains no elements. This set will be denoted by $\{\emptyset\}$.

If a set consists of n elements, then the total number of its subsets equals 2^n.

Note In probability theory, we assign probabilities to the subsets (events) of S and we define various functions (random variables) whose domain consists of the elements of S. We must be careful, therefore, to distinguish between the element ζ and the set $\{\zeta\}$ consisting of the single element ζ.

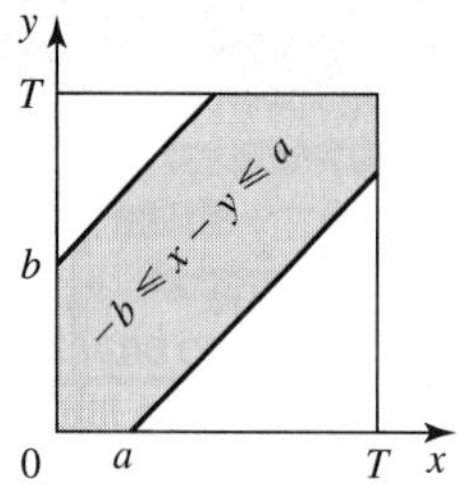

FIGURE 2-1

EXAMPLE 2-1 ▶ We shall denote by f_i the faces of a die. These faces are the elements of the set $S = \{f_1, \ldots, f_6\}$. In this case, $n = 6$; hence S has $2^6 = 64$ subsets:

$$\{\emptyset\}, \{f_1\}, \ldots, \{f_1, f_2\}, \ldots, \{f_1, f_2, f_3\}, \ldots, S$$ ◀

In general, the elements of a set are arbitrary objects. For example, the 64 subsets of the set S in Example 2-1 can be considered as the elements of another set. In Example 2-2, the elements of S are pairs of objects. In Example 2-3, S is the set of points in the square of Fig. 2-1.

EXAMPLE 2-2 ▶ Suppose that a coin is tossed twice. The resulting outcomes are the four objects hh, ht, th, tt forming the set

$$S = \{hh, ht, th, tt\}$$

where hh is an abbreviation for the element "heads–heads." The set S has $2^4 = 16$ subsets. For example,

$$A = \{\text{heads at the first toss}\} = \{hh, ht\}$$
$$B = \{\text{only one head showed}\} = \{ht, th\}$$
$$C = \{\text{heads shows at least once}\} = \{hh, ht, th\}$$

In the first equality, the sets A, B, and C are represented by their properties as in (2-2); in the second, in terms of their elements as in (2-1). ◀

EXAMPLE 2-3 ▶ In this example, S is the set of all points in the square of Fig. 2-1. Its elements are all ordered pairs of numbers (x, y) where

$$0 \leq x \leq T \qquad 0 \leq y \leq T$$

The shaded area is a subset A of S consisting of all points (x, y) such that $-b \leq x - y \leq a$. The notation

$$A = \{-b \leq x - y \leq a\}$$

describes A in terms of the properties of x and y as in (2-2). ◀

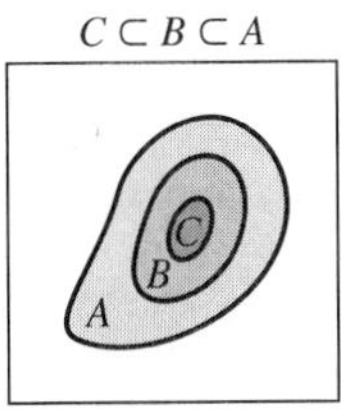

FIGURE 2-2

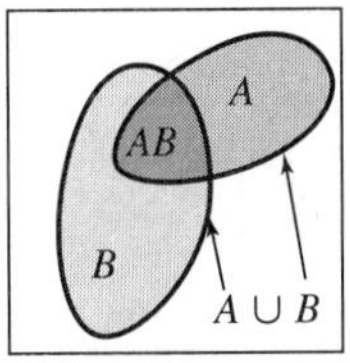

FIGURE 2-3

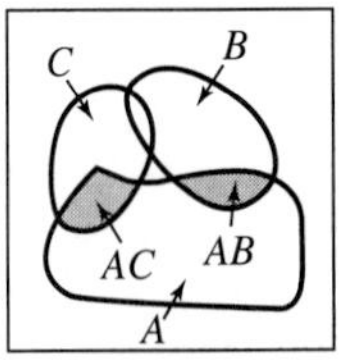

FIGURE 2-4

Set Operations

In the following, we shall represent a set S and its subsets by plane figures as in Fig. 2-2 (*Venn diagrams*).

The notation $B \subset A$ or $A \supset B$ will mean that B is a subset of A (B *belongs to* A), that is, that every element of B is an element of A. Thus, for any A,

$$\{\emptyset\} \subset A \subset A \subset S$$

Transitivity If $C \subset B$ and $B \subset A$ then $C \subset A$
Equality $A = B$ iff[1] $A \subset B$ and $B \subset A$

UNIONS AND INTERSECTIONS. The *sum* or *union* of two sets A and B is a set whose elements are all elements of A or of B or of both (Fig. 2-3). This set will be written in the form

$$A + B \quad \text{or} \quad A \cup B$$

This operation is commutative and associative:

$$A \cup B = B \cup A \qquad (A \cup B) \cup C = A \cup (B \cup C)$$

We note that, if $B \subset A$, then $A \cup B = A$. From this it follows that

$$A \cup A = A \qquad A \cup \{\emptyset\} = A \qquad S \cup A = S$$

The *product* or *intersection* of two sets A and B is a set consisting of all elements that are common to the set A and B (Fig. 2-3). This set is written in the form

$$AB \quad \text{or} \quad A \cap B$$

This operation is commutative, associative, and distributive (Fig. 2-4):

$$AB = BA \qquad (AB)C = A(BC) \qquad A(B \cup C) = AB \cup AC$$

We note that if $A \subset B$, then $AB = A$. Hence

$$AA = A \qquad \{\emptyset\}A = \{\emptyset\} \qquad AS = A$$

---

[1]The term iff is an abbreviation for *if and only if*.

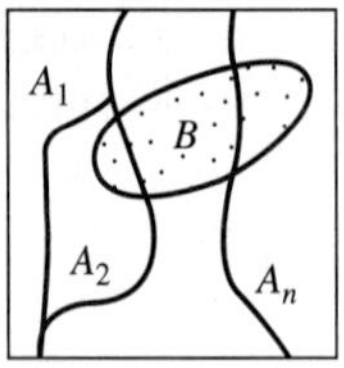

FIGURE 2-5

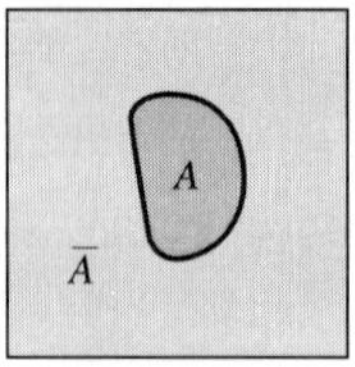

FIGURE 2-6

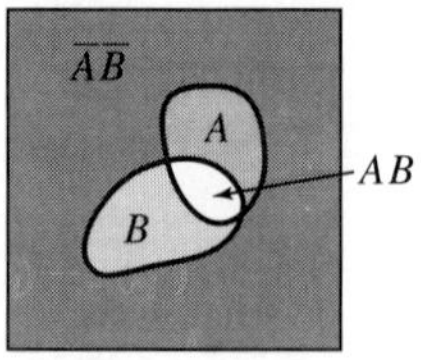

FIGURE 2-7

Note If two sets A and B are described by the properties of their elements as in (2-2), then their intersection AB will be specified by including these properties in braces. For example, if

$$S = \{1, 2, 3, 4, 5, 6\} \qquad A = \{\text{even}\} \qquad B = \{\text{less than } 5\}$$

then[2]

$$AB = \{\text{even, less than } 5\} = \{2, 4\} \tag{2-3}$$

MUTUALLY EXCLUSIVE SETS. Two sets A and B are said to be *mutually exclusive* or *disjoint* if they have no common elements, that is, if

$$AB = \{\emptyset\}$$

Several sets $A_1, A_2, \ldots$ are called mutually exclusive if

$$A_i A_j = \{\emptyset\} \quad \text{for every} \quad i \text{ and } j \neq i$$

PARTITIONS. A partition $\boldsymbol{U}$ of a set S is a collection of mutually exclusive subsets A_i of S whose union equals S (Fig. 2-5).

$$A_1 \cup \cdots \cup A_n = S \qquad A_i A_j = \{\emptyset\} \qquad i \neq j \tag{2-4}$$

Thus

$$\boldsymbol{U} = [A_1, \ldots, A_n]$$

COMPLEMENTS. The complement $\overline{A}$ of a set A is the set consisting of all elements of S that are not in A (Fig. 2-6). From the definition it follows that

$$A \cup \overline{A} = S \qquad A\overline{A} = \{\emptyset\} \qquad \overline{\overline{A}} = A \qquad \overline{S} = \{\emptyset\} \qquad \{\overline{\emptyset}\} = S$$

If $B \subset A$, then $\overline{B} \supset \overline{A}$; if $A = B$, then $\overline{A} = \overline{B}$.

DE MORGAN'S LAW. Clearly (see Fig. 2-7)

$$\overline{A \cup B} = \overline{A}\,\overline{B} \qquad \overline{AB} = \overline{A} \cup \overline{B} \tag{2-5}$$

---

[2]We should stress the difference in the meaning of commas in (2-1) and (2-3). In (2-1) the braces include all elements ζ_i and

$$\{\zeta_1, \ldots, \zeta_n\} = \{\zeta_1\} \cup \cdots \cup \{\zeta_n\}$$

is the union of the sets $\{\zeta_i\}$. In (2-3) the braces include the properties of the sets {even} and {less than 5}, and

$$\{\text{even, less than } 5\} = \{\text{even}\} \cap \{\text{less than } 5\}$$

is the intersection of the sets {even} and {less than 5}.

Repeated application of (2-5) leads to this: If in a set identity we replace all sets by their complements, all unions by intersections, and all intersections by unions, the identity is preserved.

We shall demonstrate this using the identity as an example:

$$A(B \cup C) = AB \cup AC \tag{2-6}$$

From (2-5) it follows that

$$\overline{A(B \cup C)} = \overline{A} \cup \overline{B \cup C} = \overline{A} \cup \overline{B}\,\overline{C}$$

Similarly,

$$\overline{AB \cup AC} = (\overline{AB})(\overline{AC}) = (\overline{A} \cup \overline{B})(\overline{A} \cup \overline{C})$$

and since the two sides of (2-6) are equal, their complements are also equal. Hence

$$\overline{A} \cup \overline{B}\,\overline{C} = (\overline{A} \cup \overline{B})(\overline{A} \cup \overline{C}) \tag{2-7}$$

DUALITY PRINCIPLE. As we know, $\overline{S} = \{\emptyset\}$ and $\overline{\{\emptyset\}} = S$. Furthermore, if in an identity like (2-7) all overbars are removed, the identity is preserved. This leads to the following version of De Morgan's law:

If in a set identity we replace all unions by intersections, all intersections by unions, and the sets S and $\{\emptyset\}$ by the sets $\{\emptyset\}$ and S, the identity is preserved.

Applying these to the identities

$$A(B \cup C) = AB \cup AC \qquad S \cup A = S$$

we obtain the identities

$$A \cup BC = (A \cup B)(A \cup C) \qquad \{\emptyset\}A = \{\emptyset\}$$

2-2 PROBABILITY SPACE

In probability theory, the following set terminology is used: The space, S or Ω is called the *certain event,* its elements *experimental outcomes,* and its subsets *events.* The empty set $\{\emptyset\}$ is the *impossible event,* and the event $\{\zeta_i\}$ consisting of a single element ζ_i is an *elementary event.* All events will be identified by italic letters.

In the applications of probability theory to physical problems, the identification of experimental outcomes is not always unique. We shall illustrate this ambiguity with the die experiment as might be interpreted by players X, Y, and Z.

X says that the outcomes of this experiment are the six faces of the die forming the space $S = \{f_1, \ldots, f_6\}$. This space has $2^6 = 64$ subsets and the event {even} consists of the three outcomes f_2, f_4, and f_6.

Y wants to bet on *even* or *odd* only. He argues, therefore that the experiment has only the two outcomes *even* and *odd* forming the space $S = \{\text{even, odd}\}$. This space has only $2^2 = 4$ subsets and the event {even} consists of a single outcome.

Z bets that *one* will show and the die will rest on the left side of the table. He maintains, therefore, that the experiment has infinitely many outcomes specified by the coordinates of its center and by the six faces. The event {even} consists not of one or of three outcomes but of infinitely many.

In the following, when we talk about an experiment, we shall assume that its outcomes are clearly identified. In the die experiment, for example, S will be the set consisting of the six faces $f_1, \ldots, f_6$.

In the relative frequency interpretation of various results, we shall use the following terminology.

Trial A single performance of an experiment will be called a *trial*. At each trial we observe a single outcome ζ_i. We say that an event A *occurs* during this trial if it contains the element ζ_i. The certain event occurs at every trial and the impossible event never occurs. The event $A \cup B$ occurs when A or B or both occur. The event AB occurs when both events A and B occur. If the events A and B are mutually exclusive and A occurs, then B does not occur. If $A \subset B$ and A occurs, then B occurs. At each trial, either A or $\overline{A}$ occurs.

If, for example, in the die experiment we observe the outcome f_5, then the event $\{f_5\}$, the event $\{\text{odd}\}$, and 30 other events occur.

THE AXIOMS

▶ We assign to each event A a number $P(A)$, which we call *the probability of the event* A. This number is so chosen as to satisfy the following three conditions:

I $$P(A) \geq 0 \tag{2-8}$$

II $$P(S) = 1 \tag{2-9}$$

III $$\text{if} \quad AB = \{\emptyset\} \quad \text{then} \quad P(A \cup B) = P(A) + P(B) \tag{2-10}$$

◀

These conditions are the axioms of the theory of probability. In the development of the theory, all conclusions are based directly or indirectly on the axioms and only on the axioms. Some simple consequences are presented next.

PROPERTIES. The probability of the impossible event is 0:

$$P\{\emptyset\} = 0 \tag{2-11}$$

Indeed, $A\{\emptyset\} = \{\emptyset\}$ and $A \cup \{\emptyset\} = A$; therefore [see (2-10)]

$$P(A) = P(A \cup \emptyset) = P(A) + P\{\emptyset\}$$

For any A,

$$P(A) = 1 - P(\overline{A}) \leq 1 \tag{2-12}$$

because $A \cup \overline{A} = S$ and $A\overline{A} = \{\emptyset\}$; hence

$$1 = P(S) = P(A \cup \overline{A}) = P(A) + P(\overline{A})$$

For any A and B,

$$P(A \cup B) = P(A) + P(B) - P(AB) \leq P(A) + P(B) \tag{2-13}$$

To prove this, we write the events $A \cup B$ and B as unions of two mutually exclusive events:

$$A \cup B = A \cup \overline{A}B \qquad B = AB \cup \overline{A}B$$

Therefore [see (2-10)]

$$P(A \cup B) = P(A) + P(\overline{A}B) \qquad P(B) = P(AB) + P(\overline{A}B)$$

Eliminating $P(\overline{A}B)$, we obtain (2-13).

Finally, if $B \subset A$, then

$$P(A) = P(B) + P(A\overline{B}) \geq P(B) \tag{2-14}$$

because $A = B \cup A\overline{B}$ and $B(A\overline{B}) = \{\emptyset\}$.

Frequency interpretation The axioms of probability are so chosen that the resulting theory gives a satisfactory representation of the physical world. Probabilities as used in real problems must, therefore, be compatible with the axioms. Using the frequency interpretation

$$P(A) \simeq \frac{n_A}{n}$$

of probability, we shall show that they do.

I. Clearly, $P(A) \geq 0$ because $n_A \geq 0$ and $n > 0$.
II. $P(S) = 1$ because S occurs at every trial; hence $n_s = n$.
III. If $AB = \{\emptyset\}$, then $n_{A+B} = n_A + n_B$ because if $A \cup B$ occurs, then A or B occurs but not both. Hence

$$P(A \cup B) \simeq \frac{n_{A \cup B}}{n} = \frac{n_A}{n} + \frac{n_B}{n} \simeq P(A) + P(B)$$

EQUALITY OF EVENTS. Two events A and B are called *equal* if they consist of the same elements. They are called *equal with probability 1* if the set

$$(A \cup B)(\overline{AB}) = A\overline{B} \cup \overline{A}B$$

consisting of all outcomes that are in A or in B but not in AB (shaded area in Fig. 2-8) has zero probability.

From the definition it follows that (see Prob. 2-4) the events A and B are equal with probability 1 iff

$$P(A) = P(B) = P(AB) \tag{2-15}$$

If $P(A) = P(B)$, then we say that A and B are *equal in probability*. In this case, no conclusion can be drawn about the probability of AB. In fact, the events A and B might be mutually exclusive.

From (2-15) it follows that, if an event N equals the impossible event with probability 1 then $P(N) = 0$. This does not, of course, mean that $N = \{\emptyset\}$.

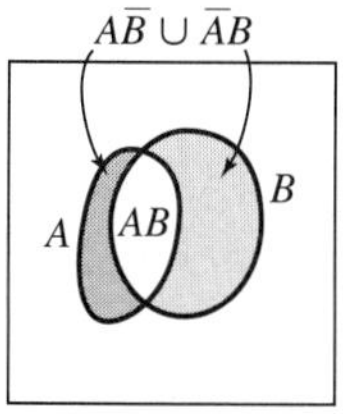

FIGURE 2-8

The Class $\boldsymbol{F}$ of Events

Events are subsets of S to which we have assigned probabilities. As we shall presently explain, we shall not consider as events all subsets of S but only a class $\boldsymbol{F}$ of subsets.

One reason for this might be the nature of the application. In the die experiment, for example, we might want to bet only on *even* or *odd*. In this case, it suffices to consider as events only the four sets $\{\emptyset\}$, $\{\text{even}\}$, $\{\text{odd}\}$, and S.

The main reason, however, for not including all subsets of S in the class $\boldsymbol{F}$ of events is of a mathematical nature: In certain cases involving sets with infinitely many outcomes, it is impossible to assign probabilities to all subsets satisfying all the axioms including the generalized form (2-21) of axiom III.

The class $\boldsymbol{F}$ of events will not be an arbitrary collection of subsets of S. We shall assume that, if A and B are events, then $A \cup B$ and AB are also events. We do so because we will want to know not only the probabilities of various events, but also the probabilities of their unions and intersections. This leads to the concept of a field.

FIELDS

▶ A field $\boldsymbol{F}$ is a nonempty class of sets such that:

$$\text{If} \quad A \in \boldsymbol{F} \quad \text{then} \quad \overline{A} \in \boldsymbol{F} \tag{2-16}$$

$$\text{If} \quad A \in \boldsymbol{F} \quad \text{and} \quad B \in \boldsymbol{F} \quad \text{then} \quad A \cup B \in \boldsymbol{F} \tag{2-17}$$

◀

These two properties give a minimum set of conditions for $\boldsymbol{F}$ to be a field. All other properties follow:

$$\text{If} \quad A \in \boldsymbol{F} \quad \text{and} \quad B \in \boldsymbol{F} \quad \text{then} \quad AB \in \boldsymbol{F} \tag{2-18}$$

Indeed, from (2-16) it follows that $\overline{A} \in \boldsymbol{F}$ and $\overline{B} \in \boldsymbol{F}$. Applying (2-17) and (2-16) to the sets $\overline{A}$ and $\overline{B}$, we conclude that

$$\overline{A} \cup \overline{B} \in \boldsymbol{F} \qquad \overline{\overline{A} \cup \overline{B}} = AB \in \boldsymbol{F}$$

A field contains the certain event and the impossible event:

$$S \in \boldsymbol{F} \qquad \{\emptyset\} \in \boldsymbol{F} \tag{2-19}$$

Indeed, since $\boldsymbol{F}$ is not empty, it contains at least one element A; therefore [see (2-16)] it also contains $\overline{A}$. Hence

$$A \cup \overline{A} = S \in \boldsymbol{F} \qquad A\overline{A} = \{\emptyset\} \in \boldsymbol{F}$$

From this it follows that all sets that can be written as unions or intersections of *finitely many* sets in $\boldsymbol{F}$ are also in $\boldsymbol{F}$. This is not, however, necessarily the case for infinitely many sets.

Borel fields. Suppose that $A_1, \ldots, A_n, \ldots$ is an infinite sequence of sets in $\boldsymbol{F}$. If the union and intersection of these sets also belongs to $\boldsymbol{F}$, then $\boldsymbol{F}$ is called a Borel field.

The class of all subsets of a set S is a Borel field. Suppose that $\boldsymbol{C}$ is a class of subsets of S that is not a field. Attaching to it other subsets of S, all subsets if necessary, we can form a field with $\boldsymbol{C}$ as its subset. It can be shown that there exists a smallest Borel field containing all the elements of $\boldsymbol{C}$.

EXAMPLE 2-4

▶ Suppose that S consists of the four elements $a, b, c,$ and d and C consists of the sets $\{a\}$ and $\{b\}$. Attaching to C the complements of $\{a\}$ and $\{b\}$ and their unions and intersections, we conclude that the smallest field containing $\{a\}$ and $\{b\}$ consists of the sets

$$\{\emptyset\} \quad \{a\} \quad \{b\} \quad \{a, b\} \quad \{c, d\} \quad \{b, c, d\} \quad \{a, c, d\} \quad S$$ ◀

EVENTS. In probability theory, events are certain subsets of S forming a Borel field. This enables us to assign probabilities not only to finite unions and intersections of events, but also to their limits.

For the determination of probabilities of sets that can be expressed as limits, the following extension of axiom III is necessary.

Repeated application of (2-10) leads to the conclusion that, if the events $A_1, \ldots, A_n$ are mutually exclusive, then

$$P(A_1 \cup \cdots \cup A_n) = P(A_1) + \cdots + P(A_n) \tag{2-20}$$

The extension of the preceding to infinitely many sets does not follow from (2-10). It is an additional condition known as the *axiom of infinite additivity:*

AXIOM OF INFINITE ADDITIVITY

▶ **III*a*.** If the events $A_1, A_2, \ldots$ are mutually exclusive, then

$$P(A_1 \cup A_2 \cup \cdots) = P(A_1) + P(A_2) + \cdots \tag{2-21}$$

◀

We shall assume that all probabilities satisfy axioms I, II, III, and III*a*.

Axiomatic Definition of an Experiment

In the theory of probability, an experiment is specified in terms of the following concepts:

1. The set S of all experimental outcomes.
2. The Borel field of all events of S.
3. The probabilities of these events.

The letter S will be used to identify not only the certain event, but also the entire experiment.

We discuss next the determination of probabilities in experiments with finitely many and infinitely many elements.

COUNTABLE SPACES. If the space S consists of N outcomes and N is a finite number, then the probabilities of all events can be expressed in terms of the probabilities

$$P\{\zeta_i\} = p_i$$

of the elementary events $\{\zeta_i\}$. From the axioms it follows, of course, that the numbers p_i must be nonnegative and their sum must equal 1:

$$p_i \geq 0 \qquad p_1 + \cdots + p_N = 1 \tag{2-22}$$

Suppose that A is an event consisting of the r elements ζ_{k_i}. In this case, A can be written as the union of the elementary events $\{\zeta_{k_i}\}$. Hence [see (2-20)]

$$P(A) = P\{\zeta_{k_1}\} + \cdots + P\{\zeta_{k_r}\} = p_{k_1} + \cdots + p_{k_r} \tag{2-23}$$

This is true even if S consists of an infinite but countable number of elements $\zeta_1, \zeta_2, \ldots$ [see (2-21)].

Classical definition If S consists of N outcomes and the probabilities p_i of the elementary events are all equal, then

$$p_i = \frac{1}{N} \tag{2-24}$$

In this case, the probability of an event A consisting of r elements equals r/N:

$$P(A) = \frac{r}{N} \tag{2-25}$$

This very special but important case is equivalent to the classical definition (1-7), with one important difference, however: In the classical definition, (2-25) is deduced as a logical necessity; in the axiomatic development of probability, (2-24), on which (2-25) is based, is a mere assumption.

EXAMPLE 2-5

▶ (*a*) In the coin experiment, the space S consists of the outcomes h and t:

$$S = \{h, t\}$$

and its events are the four sets $\{\emptyset\}, \{t\}, \{h\}, S$. If $P\{h\} = p$ and $P\{t\} = q$, then $p + q = 1$.

(*b*) We consider now the experiment of the toss of a coin three times. The possible outcomes of this experiment are:

$$hhh, hht, hth, htt, thh, tht, tth, ttt$$

We shall assume that all elementary events have the same probability as in (2-24) (fair coin). In this case, the probability of each elementary event equals $1/8$. Thus the probability $P\{hhh\}$ that we get three heads equals $1/8$. The event

$$\{\text{heads at the first two tosses}\} = \{hhh, hht\}$$

consists of the two outcomes hhh and hht; hence its probability equals $2/8$. ◀

THE REAL LINE. If S consists of a noncountable infinity of elements, then its probabilities cannot be determined in terms of the probabilities of the elementary events. This is the case if S is the set of points in an n-dimensional space. In fact, most applications can be presented in terms of events in such a space. We shall discuss the determination of probabilities using as illustration the real line.

Suppose that S is the set of all real numbers. Its subsets can be considered as sets of points on the real line. It can be shown that it is impossible to assign probabilities to all subsets of S so as to satisfy the axioms. To construct a probability space on the real line, we shall consider as events all intervals $x_1 \le x \le x_2$ and their countable unions and intersections. These events form a field $\boldsymbol{F}$ that can be specified as follows:

It is the smallest Borel field that includes all half-lines $x \le x_i$, where x_i is any number.

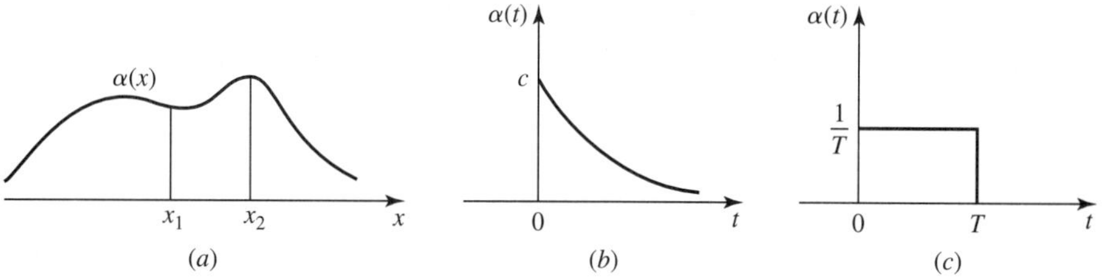

FIGURE 2-9

This field contains all open and closed intervals, all points, and, in fact, every set of points on the real line that is of interest in the applications. One might wonder whether $\boldsymbol{F}$ does not include *all* subsets of S. Actually, it is possible to show that there exist sets of points on the real line that are not countable unions and intersections of intervals. Such sets, however, are of no interest in most applications. To complete the specification of S, it suffices to assign probabilities to the events $\{x \le x_i\}$. All other probabilities can then be determined from the axioms.

Suppose that $\alpha(x)$ is a function such that (Fig. 2-9*a*)

$$\int_{-\infty}^{\infty} \alpha(x)\,dx = 1 \qquad \alpha(x) \ge 0 \tag{2-26}$$

We define the probability of the event $\{x \le x_i\}$ by the integral

$$P\{x \le x_i\} = \int_{-\infty}^{x_i} \alpha(x)\,dx \tag{2-27}$$

This specifies the probabilities of all events of S. We maintain, for example, that the probability of the event $\{x_1 < x \le x_2\}$ consisting of all points in the interval (x_1, x_2) is given by

$$P\{x_1 < x \le x_2\} = \int_{x_1}^{x_2} \alpha(x)\,dx \tag{2-28}$$

Indeed, the events $\{x \le x_1\}$ and $\{x_1 < x \le x_2\}$ are mutually exclusive and their union equals $\{x \le x_2\}$. Hence [see (2-10)]

$$P\{x \le x_1\} + P\{x_1 < x \le x_2\} = P\{x \le x_2\}$$

and (2-28) follows from (2-27).

We note that, if the function $\alpha(x)$ is bounded, then the integral in (2-28) tends to 0 as $x_1 \to x_2$. This leads to the conclusion that the probability of the event $\{x_2\}$ consisting of the single outcome x_2 is 0 for every x_2. In this case, the probability of all elementary events of S equals 0, although the probability of their unions equals 1. This is not in conflict with (2-21) because the total number of elements of S is not countable.

EXAMPLE 2-6

▶ A radioactive substance is selected at $t = 0$ and the time t of emission of a particle is observed. This process defines an experiment whose outcomes are all points on the positive t axis. This experiment can be considered as a special case of the real line experiment if we assume that S is the entire t axis and all events on the negative axis have zero probability.

Suppose then that the function $\alpha(t)$ in (2-26) is given by (Fig. 2-9*b*)

$$\alpha(t) = ce^{-ct}U(t) \qquad U(t) = \begin{cases} 1 & t \geq 0 \\ 0 & t < 0 \end{cases}$$

Inserting into (2-28), we conclude that the probability that a particle will be emitted in the time interval $(0, t_0)$ equals

$$c\int_0^{t_0} e^{-ct}\,dt = 1 - e^{-ct_0}$$ ◀

EXAMPLE 2-7 ▶ A telephone call occurs at *random* in the interval $(0, T)$. This means that the probability that it will occur in the interval $0 \leq t \leq t_0$ equals t_0/T. Thus the outcomes of this experiment are all points in the interval $(0, T)$ and the probability of the event {the call will occur in the interval (t_1, t_2)} equals

$$P\{t_1 \leq t \leq t_2\} = \frac{t_2 - t_1}{T}$$

This is again a special case of (2-28) with $\alpha(t) = 1/T$ for $0 \leq t \leq T$ and 0 otherwise (Fig. 2-9*c*). ◀

PROBABILITY MASSES. The probability $P(A)$ of an event A can be interpreted as the mass of the corresponding figure in its Venn diagram representation. Various identities have similar interpretations. Consider, for example, the identity $P(A \cup B) = P(A) + P(B) - P(AB)$. The left side equals the mass of the event $A \cup B$. In the sum $P(A) + P(B)$, the mass of AB is counted twice (Fig. 2-3). To equate this sum with $P(A \cup B)$, we must, therefore, subtract $P(AB)$.

As Examples 2-8 and 2-9 show, by expressing complicated events as the union of simpler events that are mutually exclusive, their probabilities can be systematically computed.

EXAMPLE 2-8 ▶ A box contains m white balls and n black balls. Balls are drawn at random one at a time without replacement. Find the probability of encountering a white ball by the kth draw.

SOLUTION
Let W_k denote the event

$$W_k = \{\text{a white ball is drawn by the } k\text{th draw}\}$$

The event W_k can occur in the following mutually exclusive ways: a white ball is drawn on the first draw, or a black ball followed by a white ball is drawn, or two black balls followed by a white ball, and so on. Let

$$X_i = \{i \text{ black balls followed by a white ball are drawn}\} \qquad i = 0, 1, 2, \ldots, n$$

Then

$$W_k = X_0 \cup X_1 \cup \cdots \cup X_{k-1}$$

and using (2-20), we obtain

$$P(W_k) = \sum_{i=0}^{k-1} P(X_i)$$

Now

$$P(X_0) = \frac{m}{m+n}$$

$$P(X_1) = \frac{n}{m+n} \cdot \frac{m}{m+n-1}$$

$$\vdots \qquad\qquad \vdots$$

$$P(X_{k-1}) = \frac{n(n-1)\cdots(n-k+1)m}{(m+n)(m+n-1)\cdots(m+n-k+1)}$$

and hence

$$P(W_k) = \frac{m}{m+n}\left(1 + \frac{n}{m+n-1} + \frac{n(n-1)}{(m+n-1)(m+n-2)} + \cdots + \frac{n(n-1)\cdots(n-k+1)}{(m+n-1)(m+n-2)\cdots(m+n-k+1)}\right) \tag{2-29}$$

By the $(n+1)$st draw, we must have a white ball, and hence

$$P(W_{n+1}) = 1$$

and using (2-29) this gives an interesting identity

$$1 + \frac{n}{m+n-1} + \frac{n(n-1)}{(m+n-1)(m+n-2)} + \cdots + \frac{n(n-1)\cdots 2\cdot 1}{(m+n-1)(m+n-2)\cdots(m+1)m} = \frac{m+n}{m} \tag{2-30}$$

◀

EXAMPLE 2-9

▶ Two players A and B draw balls one at a time alternately from a box containing m white balls and n black balls. Suppose the player who picks the first white ball wins the game. What is the probability that the player who starts the game will win?

SOLUTION

Suppose A starts the game. The game can be won by A if he extracts a white ball at the start or if A and B draw a black ball each and then A draws a white one, or if A and B extract two black balls each and then A draws a white one and so on. Let

$$X_k = \{A \text{ and } B \text{ alternately draw } k \text{ black balls each and then } A \text{ draws a white ball}\} \qquad k = 0, 1, 2, \ldots$$

where the X_ks represent mutually exclusive events and moreover the event

$$\{A \text{ wins}\} = X_0 \cup X_1 \cup X_2 \cup \cdots$$

Hence

$$P_A \triangleq P(A \text{ wins}) = P(X_0 \cup X_1 \cup X_2 \cup \cdots) = P(X_0) + P(X_1) + P(X_2) + \cdots$$

where we have made use of the axiom of additivity in (2-20). Now

$$P(X_0) = \frac{m}{m+n}$$

$$P(X_1) = \frac{n}{n+m} \cdot \frac{n-1}{m+n-1} \cdot \frac{m}{m+n-2}$$
$$= \frac{n(n-1)m}{(m+n)(m+n-1)(m+n-2)}$$

and

$$P(X_2) = \frac{n(n-1)(n-2)(n-3)m}{(m+n)(m+n-1)(m+n-2)(m+n-3)}$$

so that

$$P_A = \frac{m}{m+n}\left(1 + \frac{n(n-1)}{(m+n-1)(m+n-2)} + \frac{n(n-1)(n-2)(n-3)}{(m+n-1)(m+n-2)(m+n-3)} + \cdots\right) \tag{2-31}$$

This above sum has a finite number of terms and it ends as soon as a term equals zero. In a similar manner,

$$Q_B = P(B \text{ wins})$$
$$= \frac{m}{m+n}\left(\frac{n}{m+n-1} + \frac{n(n-1)(n-2)}{(m+n-1)(m+n-2)(m+n-3)} + \cdots\right) \tag{2-32}$$

But one of the players must win the game. Hence

$$P_A + Q_B = 1$$

and using (2-31) to (2-32) this leads to the same identity in (2-30). This should not be surprising considering that these two problems are closely related. ◀

2-3 CONDITIONAL PROBABILITY

The *conditional probability* of an event A assuming another event M, denoted by $P(A \mid M)$, is by definition the ratio

$$P(A \mid M) = \frac{P(AM)}{P(M)} \tag{2-33}$$

where we assume that $P(M)$ is not 0.

The following properties follow readily from the definition:

$$\text{If} \quad M \subset A \quad \text{then} \quad P(A \mid M) = 1 \tag{2-34}$$

because then $AM = M$. Similarly,

$$\text{if} \quad A \subset M \quad \text{then} \quad P(A \mid M) = \frac{P(A)}{P(M)} \geq P(A) \tag{2-35}$$

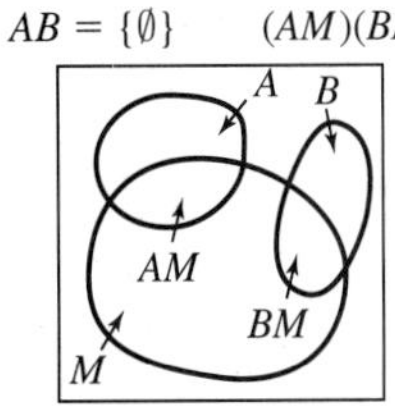

FIGURE 2-10

Frequency interpretation Denoting by n_A, n_M, and n_{AM} the number of occurrences of the events A, M, and AM respectively, we conclude from (1-1) that

$$P(A) \simeq \frac{n_A}{n} \qquad P(M) \simeq \frac{n_M}{n} \qquad P(AM) \simeq \frac{n_{AM}}{n}$$

Hence

$$P(A \mid M) = \frac{P(AM)}{P(M)} \simeq \frac{n_{AM}/n}{n_M/n} = \frac{n_{AM}}{n_M}$$

This result can be phrased as follows: If we discard all trials in which the event M did not occur and we retain only the subsequence of trials in which M occurred, then $P(A \mid M)$ equals the relative frequency of occurrence n_{AM}/n_M of the event A in that subsequence.

FUNDAMENTAL REMARK. We shall show that, for a specific M, the conditional probabilities are indeed probabilities; that is, they satisfy the axioms.

The first axiom is obviously satisfied because $P(AM) \geq 0$ and $P(M) > 0$:

$$P(A \mid M) \geq 0 \tag{2-36}$$

The second follows from (2-34) because $M \subset S$:

$$P(S \mid M) = 1 \tag{2-37}$$

To prove the third, we observe that if the events A and B are mutually exclusive, then (Fig. 2-10) the events AM and BM are also mutually exclusive. Hence

$$P(A \cup B \mid M) = \frac{P[(A \cup B)M]}{P(M)} = \frac{P(AM) + P(BM)}{P(M)}$$

This yields the third axiom:

$$P(A \cup B \mid M) = P(A \mid M) + P(B \mid M) \tag{2-38}$$

From this it follows that all results involving probabilities holds also for conditional probabilities. The significance of this conclusion will be appreciated later (see (2-44)).

EXAMPLE 2-10

▶ In the fair-die experiment, we shall determine the conditional probability of the event $\{f_2\}$ assuming that the event *even* occurred. With

$$A = \{f_2\} \qquad M = \{\text{even}\} = \{f_2, f_4, f_6\}$$

we have $P(A) = 1/6$ and $P(M) = 3/6$. And since $AM = A$, (2-33) yields

$$P\{f_2 \mid \text{even}\} = \frac{P\{f_2\}}{P\{\text{even}\}} = \frac{1}{3}$$

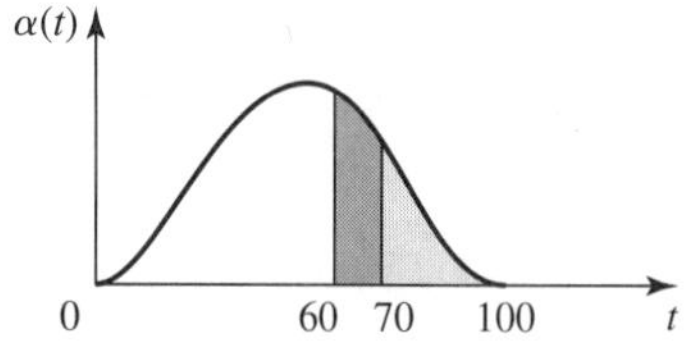

FIGURE 2-11

This equals the relative frequency of the occurrence of the event {two} in the subsequence whose outcomes are even numbers. ◀

EXAMPLE 2-11

▶ We denote by t the age of a person when he dies. The probability that $t \le t_o$ is given by

$$P\{t \le t_o\} = \int_0^{t_o} \alpha(t)\,dt$$

where $\alpha(t)$ is a function determined from mortality records. We shall assume that

$$\alpha(t) = 3 \times 10^{-9} t^2 (100 - t)^2 \qquad 0 \le t \le 100 \text{ years}$$

and 0 otherwise (Fig. 2-11).

From (2-28) it follows that the probability that a person will die between the ages of 60 and 70 equals

$$P\{60 \le t \le 70\} = \int_{60}^{70} \alpha(t)\,dt = 0.154$$

This equals the number of people who die between the ages of 60 and 70 divided by the total population.

With

$$A = \{60 \le t \le 70\} \qquad M = \{t \ge 60\} \qquad AM = A$$

it follows from (2-33) that the probability that a person will die between the ages of 60 and 70 assuming that he was alive at 60 equals

$$P\{60 \le t \le 70 \mid t \ge 60\} = \frac{\int_{60}^{70} \alpha(t)\,dt}{\int_{60}^{100} \alpha(t)\,dt} = 0.486$$

This equals the number of people who die between the ages 60 and 70 divided by the number of people that are alive at age 60. ◀

EXAMPLE 2-12

▶ A box contains three white balls w_1, w_2, and w_3 and two red balls r_1 and r_2. We remove at random two balls in succession. What is the probability that the first removed ball is white and the second is red?

We shall give two solutions to this problem. In the first, we apply (2-25); in the second, we use conditional probabilities.

FIRST SOLUTION
The space of our experiment consists of all *ordered* pairs that we can form with the five balls:

$$w_1w_2 \quad w_1w_3 \quad w_1r_1 \quad w_1r_2 \quad \cdots \quad r_2w_1 \quad r_2w_2 \quad r_2w_3 \quad r_2r_1$$

The number of such pairs equals $5 \times 4 = 20$. The event {white first, red second} consists of the six outcomes

$$w_1r_1 \quad w_1r_2 \quad w_2r_1 \quad w_2r_2 \quad w_3r_1 \quad w_3r_2$$

Hence [see (2-25)] its probability equals 6/20.

SECOND SOLUTION
Because the box contains three white and two red balls, the probability of the event $W_1 = \{\text{white first}\}$ equals 3/5. If a white ball is removed, there remain two white and two red balls; hence the conditional probability $P(R_2 \mid W_1)$ of the event $R_2 = \{\text{red second}\}$ assuming {white first} equals 2/4. From this and (2-33) it follows that

$$P(W_1R_2) = P(R_2 \mid W_1)P(W_1) = \frac{2}{4} \times \frac{3}{5} = \frac{6}{20}$$

where W_1R_2 is the event {white first, red second}. ◀

EXAMPLE 2-13 ▶ A box contains white and black balls. When two balls are drawn without replacement, suppose the probability that both are white is 1/3. (*a*) Find the smallest number of balls in the box. (*b*) How small can the total number of balls be if black balls are even in number?

SOLUTION
(*a*) Let a and b denote the number of white and black balls in the box, and W_k the event

$$W_k = \text{“a white ball is drawn at the } k\text{th draw”}$$

We are given that $P(W_1 \cap W_2) = 1/3$. But

$$P(W_1 \cap W_2) = P(W_2 \cap W_1) = P(W_2 \mid W_1)P(W_1) = \frac{a-1}{a+b-1} \cdot \frac{a}{a+b} = \frac{1}{3} \qquad (2\text{-}39)$$

Because

$$\frac{a}{a+b} > \frac{a-1}{a+b-1} \qquad b > 0$$

we can rewrite (2-39) as

$$\left(\frac{a-1}{a+b-1}\right)^2 < \frac{1}{3} < \left(\frac{a}{a+b}\right)^2$$

This gives the inequalities

$$(\sqrt{3}+1)b/2 < a < 1 + (\sqrt{3}+1)b/2 \qquad (2\text{-}40)$$

For $b = 1$, this gives $1.36 < a < 2.36$, or $a = 2$, and we get

$$P(W_2 \cap W_1) = \frac{2}{3} \cdot \frac{1}{2} = \frac{1}{3}$$

Thus the smallest number of balls required is 3.

TABLE 2-1

b	a from (2-40)	$P(W_2 W_1)$
2	3	$\frac{3}{5}\cdot\frac{2}{4} = \frac{3}{10} \neq \frac{1}{3}$
4	6	$\frac{6}{10}\cdot\frac{5}{9} = \frac{1}{3}$

(*b*) For even value of b, we can use (2-40) with $b = 2, 4, \ldots$ as shown in Table 2-1. From the table, when b is even, 10 is the smallest number of balls ($a = 6, b = 4$) that gives the desired probability. ◀

Total Probability and Bayes' Theorem

If $\boldsymbol{U} = [A_1, \ldots, A_n]$ is a partition of S and B is an arbitrary event (Fig. 2-5), then

$$P(B) = P(B \mid A_1)P(A_1) + \cdots + P(B \mid A_n)P(A_n) \tag{2-41}$$

Proof. Clearly,

$$B = BS = B(A_1 \cup \cdots \cup A_n) = BA_1 \cup \cdots \cup BA_n$$

But the events BA_i and BA_j are mutually exclusive because the events A_i and A_j are mutually exclusive [see (2-4)]. Hence

$$P(B) = P(BA_1) + \cdots + P(BA_n)$$

and (2-41) follows because [see (2-33)]

$$P(BA_i) = P(B \mid A_i)P(A_i) \tag{2-42}$$

This result is known as the *total probability theorem.*

Since $P(BA_i) = P(A_i \mid B)P(B)$ we conclude with (2-42) that

$$P(A_i \mid B) = P(B \mid A_i)\frac{P(A_i)}{P(B)} \tag{2-43}$$

Inserting (2-41) into (2-43), we obtain *Bayes' theorem*[3]:

$$P(A_i \mid B) = \frac{P(B \mid A_i)P(A_i)}{P(B \mid A_1)P(A_1) + \cdots + P(B \mid A_n)P(A_n)} \tag{2-44}$$

Note The terms *a priori* and *a posteriori* are often used for the probabilities $P(A_i)$ and $P(A_i \mid B)$.

EXAMPLE 2-14

▶ Suppose box 1 contains a white balls and b black balls, and box 2 contains c white balls and d black balls. One ball of unknown color is transferred from the first box into the second one and then a ball is drawn from the latter. What is the probability that it will be a white ball?

[3]The main idea of this theorem is due to Rev. Thomas Bayes (*ca.* 1760). However, its final form (2-44) was given by Laplace several years later.

SOLUTION

If no ball is transferred from the first box into the second box, the probability of obtaining a white ball from the second one is simply $c/(c + d)$. In the present case, a ball is first transferred from box 1 to box 2 and there are only two mutually exclusive possibilities for this event—the transferred ball is either a white ball or a black ball. Let

$$W = \{\text{transferred ball is white}\} \qquad B = \{\text{transferred ball is black}\}$$

Note that W together with B form a partition ($W \cup B = S$) and

$$P(W) = \frac{a}{a+b} \qquad P(B) = \frac{b}{a+b}$$

The event of interest

$$A = \{\text{white ball is drawn from the second box}\}$$

can happen only under the two mentioned mutually exclusive possibilities. Hence

$$\begin{aligned} P(A) &= P\{A \cap (W \cup B)\} = P\{(A \cap W) \cup (A \cap B)\} \\ &= P(A \cap W) + P(A \cap B) \\ &= P(A \mid W)P(W) + P(A \mid B)P(B) \end{aligned} \tag{2-45}$$

But

$$P(A \mid W) = \frac{c+1}{c+d+1} \qquad P(A \mid B) = \frac{c}{c+d+1}$$

Hence

$$P(A) = \frac{a(c+1)}{(a+b)(c+d+1)} + \frac{bc}{(a+b)(c+d+1)} = \frac{ac+bc+a}{(a+b)(c+d+1)} \tag{2-46}$$

gives the probability of picking a white ball from box 2 after one ball of unknown color has been transferred from the first box. ◀

The concepts of conditional probability and Bayes theorem can be rather confusing. As Example 2-15 shows, care should be used in interpreting them.

EXAMPLE 2-15

▶ A certain test for a particular cancer is known to be 95% accurate. A person submits to the test and the results are positive. Suppose that the person comes from a population of 100,000, where 2000 people suffer from that disease. What can we conclude about the probability that the person under test has that particular cancer?

SOLUTION

Although it will be tempting to jump to the conclusion that based on the test the probability of having cancer for that person is 95%, the test data simply does not support that. The test is known to be 95% accurate, which means that 95% of all positive tests are correct and 95% of all negative tests are correct. Thus if the events $\{T > 0\}$ stands for the test being positive and $\{T < 0\}$ stands for the test being negative, then with H and C representing the sets of healthy and cancer patients, we have

$$P(T > 0 \mid C) = 0.95 \qquad P(T > 0 \mid H) = 0.05$$
$$P(T < 0 \mid C) = 0.05 \qquad P(T < 0 \mid H) = 0.95$$

The space of this particular experiment consists of 98,000 healthy people and 2000 cancer patients so that in the absence of any other information a person chosen at random is healthy with probability 98,000/100,000 = 0.98 and suffers from cancer with probability 0.02. We denote this by $P(H) = 0.98$, and $P(C) = 0.02$. To interpret the test results properly, we can now use the Bayes' theorem. In this case, from (2-44) the probability that the person suffers from cancer given that the test is positive is

$$P(C \mid T > 0) = \frac{P(T > 0 \mid C)P(C)}{P(T > 0)} = \frac{P(T > 0 \mid C)P(C)}{P(T > 0 \mid C)P(C) + P(T > 0 \mid H)P(H)}$$
$$= \frac{0.95 \times 0.02}{0.95 \times 0.02 + 0.05 \times 0.98} = 0.278 \qquad (2\text{-}47)$$

This result states that if the test is taken by someone from this population *without knowing* whether that person has the disease or not, then even a positive test only suggests that there is a 27.6% chance of having the disease. However, if the person knows that he or she has the disease, then the test is 95% accurate. ◀

EXAMPLE 2-16

▶ We have four boxes. Box 1 contains 2000 components of which 5% are defective. Box 2 contains 500 components of which 40% are defective. Boxes 3 and 4 contain 1000 each with 10% defective. We select *at random* one of the boxes and we remove *at random* a single component.

(*a*) What is the probability that the selected component is defective?

SOLUTION

The space of this experiment consists of 4000 good (g) components and 500 defective (d) components arranged as:

$$\text{Box 1: } 1900g, 100d \qquad \text{Box 2: } 300g, 200d$$
$$\text{Box 3: } 900g, 100d \qquad \text{Box 4: } 900g, 100d$$

We denote by B_i the event consisting of all components in the ith box and by D the event consisting of all defective components. Clearly,

$$P(B_1) = P(B_2) = P(B_3) = P(B_4) = \tfrac{1}{4} \qquad (2\text{-}48)$$

because the boxes are selected at random. The probability that a component taken from a specific box is defective equals the ratio of the defective to the total number of components in that box. This means that

$$P(D \mid B_1) = \frac{100}{2000} = 0.05 \qquad P(D \mid B_2) = \frac{200}{500} = 0.4$$
$$P(D \mid B_3) = \frac{100}{1000} = 0.1 \qquad P(D \mid B_4) = \frac{100}{1000} = 0.1 \qquad (2\text{-}49)$$

And since the events B_1, B_2, B_3, and B_4 form a partition of S, we conclude from (2-41) that

$$P(D) = 0.05 \times \tfrac{1}{4} + 0.4 \times \tfrac{1}{4} + 0.1 \times \tfrac{1}{4} + 0.1 \times \tfrac{1}{4} = 0.1625$$

This is the probability that the selected component is defective.

(*b*) We examine the selected component and we find it defective. On the basis of this evidence, we want to determine the probability that it came from box 2.

We now want the conditional probability $P(B_2 \mid D)$. Since

$$P(D) = 0.1625 \qquad P(D \mid B_2) = 0.4 \qquad P(B_2) = 0.25$$

(2-43) yields

$$P(B_2 \mid D) = 0.4 \times \frac{0.25}{0.1625} = 0.615$$

Thus the a priori probability of selecting box 2 equals 0.25 and the a posteriori probability assuming that the selected component is defective equals 0.615. These probabilities have this frequency interpretation: If the experiment is performed n times, then box 2 is selected $0.25n$ times. If we consider only the n_D experiments in which the removed part is defective, then the number of times the part is taken from box 2 equals $0.615n_D$.

We conclude with a comment on the distinction between assumptions and deductions: Equations (2-48) and (2-49) are not derived; they are merely reasonable *assumptions*. Based on these assumptions and on the axioms, we *deduce* that $P(D) = 0.1625$ and $P(B_2 \mid D) = 0.615$. ◀

Independence

Two events A and B are called *independent* if

$$P(AB) = P(A)P(B) \tag{2-50}$$

The concept of independence is fundamental. In fact, it is this concept that justifies the mathematical development of probability, not merely as a topic in measure theory, but as a separate discipline. The significance of independence will be appreciated later in the context of repeated trials. We discuss here only various simple properties.

Frequency interpretation Denoting by n_A, n_B, and n_{AB} the number of occurrences of the events A, B, and AB, respectively, we have

$$P(A) \simeq \frac{n_A}{n} \qquad P(B) \simeq \frac{n_B}{n} \qquad P(AB) \simeq \frac{n_{AB}}{n}$$

If the events A and B are independent, then

$$\frac{n_A}{n} \simeq P(A) = \frac{P(AB)}{P(B)} \simeq \frac{n_{AB}/n}{n_B/n} = \frac{n_{AB}}{n_B}$$

Thus, if A and B are independent, then the relative frequency n_A/n of the occurrence of A in the original sequence of n trials equals the relative frequency n_{AB}/n_B of the occurrence of A in the subsequence in which B occurs.

We show next that if the events A and B are independent, then the events $\overline{A}$ and B and the events $\overline{A}$ and $\overline{B}$ are also independent.

As we know, the events AB and $\overline{A}B$ are mutually exclusive and

$$B = AB \cup \overline{A}B \qquad P(\overline{A}) = 1 - P(A)$$

From this and (2-50) it follows that

$$P(\overline{A}B) = P(B) - P(AB) = [1 - P(A)]P(B) = P(\overline{A})P(B)$$

This establishes the independence of $\overline{A}$ and B. Repeating the argument, we conclude that $\overline{A}$ and $\overline{B}$ are also independent.

In Examples 2-17 and 2-18, we illustrate the concept of independence. In Example 2-17*a*, we start with a known experiment and we show that two of its events are independent. In Examples 2-17*b* and 2-18 we use the concept of independence to complete the specification of each experiment. This idea is developed further in Chap. 3.

EXAMPLE 2-17

▶ If we toss a coin twice, we generate the four outcomes hh, ht, th, and tt.

(*a*) To construct an experiment with these outcomes, it suffices to assign probabilities to its elementary events. With a and b two positive numbers such that $a + b = 1$, we assume that

$$P\{hh\} = a^2 \qquad P\{ht\} = P\{th\} = ab \qquad P\{tt\} = b^2$$

These probabilities are consistent with the axioms because

$$a^2 + ab + ab + b^2 = (a + b)^2 = 1$$

In the experiment so constructed, the events

$$H_1 = \{\text{heads at first toss}\} = \{hh, ht\}$$

$$H_2 = \{\text{heads at second toss}\} = \{hh, th\}$$

consist of two elements each, and their probabilities are [see (2-23)]

$$P(H_1) = P\{hh\} + P\{ht\} = a^2 + ab = a$$

$$P(H_2) = P\{hh\} + P\{th\} = a^2 + ab = a$$

The intersection H_1H_2 of these two events consists of the single outcome $\{hh\}$. Hence

$$P(H_1H_2) = P\{hh\} = a^2 = P(H_1)P(H_2)$$

This shows that the events H_1 and H_2 are independent.

(*b*) The experiment in part (*a*) of this example can be specified in terms of the probabilities $P(H_1) = P(H_2) = a$ of the events H_1 and H_2, and the information that these events are independent.

Indeed, as we have shown, the events $\overline{H}_1$ and H_2 and the events $\overline{H}_1$ and $\overline{H}_2$ are also independent. Furthermore,

$$H_1H_2 = \{hh\} \qquad H_1H_2 = \{ht\} \qquad \overline{H}_1\overline{H}_2 = \{th\} \qquad \overline{H}_1\overline{H}_2 = \{tt\}$$

and $P(\overline{H}_1) = 1 - P(H_1) = 1 - a$, $P(\overline{H}_2) = 1 - P(H_2) = 1 - a$. Hence

$$P\{hh\} = a^2 \qquad P\{ht\} = a(1 - a) \qquad P\{th\} = (1 - a)a \qquad P\{tt\} = (1 - a)^2$$ ◀

EXAMPLE 2-18

▶ Trains X and Y arrive at a station at random between 8 A.M. and 8.20 A.M. Train X stops for four minutes and train Y stops for five minutes. Assuming that the trains arrive independently of each other, we shall determine various probabilities related to the

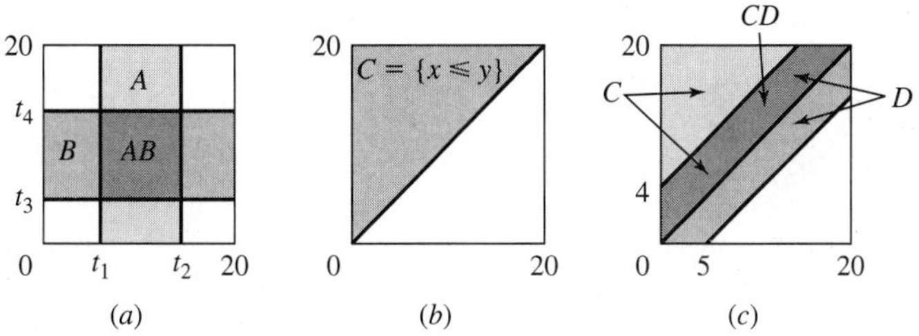

FIGURE 2-12

times x and y of their respective arrivals. To do so, we must first specify the underlying experiment.

The outcomes of this experiment are all points (x, y) in the square of Fig. 2-12. The event

$$A = \{X \text{ arrives in the interval } (t_1, t_2)\} = \{t_1 \leq x \leq t_2\}$$

is a vertical strip as in Fig. 2-12*a* and its probability equals $(t_2 - t_1)/20$. This is our interpretation of the information that the train arrives at random. Similarly, the event

$$B = \{Y \text{ arrives in the interval } (t_3, t_4)\} = \{t_3 \leq y \leq t_4\}$$

is a horizontal strip and its probability equals $(t_4 - t_3)/20$.

Proceeding similarly, we can determine the probabilities of any horizontal or vertical sets of points. To complete the specification of the experiment, we must determine also the probabilities of their intersections. Interpreting the independence of the arrival times as independence of the events A and B, we obtain

$$P(AB) = P(A)P(B) = \frac{(t_2 - t_1)(t_4 - t_3)}{20 \times 20}$$

The event AB is the rectangle shown in the figure. Since the coordinates of this rectangle are arbitrary, we conclude that the probability of any rectangle equals its area divided by 400. In the plane, all events are unions and intersections of rectangles forming a Borel field. This shows that the probability that the point (x, y) will be in an arbitrary region R of the plane equals the area of R divided by 400. This completes the specification of the experiment.

(*a*) We shall determine the probability that train X arrives before train Y. This is the probability of the event

$$C = \{x \leq y\}$$

shown in Fig. 2-12*b*. This event is a triangle with area 200. Hence

$$P(C) = \frac{200}{400}$$

(*b*) We shall determine the probability that the trains meet at the station. For the trains to meet, x must be less than $y + 5$ and y must be less than $x + 4$. This is the event

$$D = \{-4 \leq x - y \leq 5\}$$

of Fig. 2-12*c*. As we see from the figure, the region D consists of two trapezoids with

$AB = BC = AC = ABC$

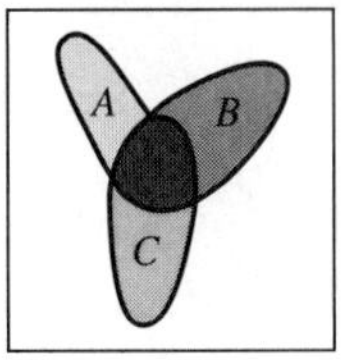

FIGURE 2-13

common base, and its area equals 159.5. Hence

$$P(D) = \frac{159.5}{400}$$

(*c*) Assuming that the trains met, we shall determine the probability that train X arrived before train Y. We wish to find the conditional probability $P(C \mid D)$. The event CD is a trapezoid as shown and its area equals 72. Hence

$$P(C \mid D) = \frac{P(CD)}{P(D)} = \frac{72}{159.5}$$ ◀

INDEPENDENCE OF THREE EVENTS. The events A_1, A_2, and A_3 are called (mutually) independent if they are independent in pairs:

$$P(A_i A_j) = P(A_i)P(A_j) \qquad i \neq j \tag{2-51}$$

and

$$P(A_1 A_2 A_3) = P(A_1)P(A_2)P(A_3) \tag{2-52}$$

We should emphasize that three events might be independent in pairs but not independent. The next example is an illustration.

EXAMPLE 2-19 ▶ Suppose that the events A, B, and C of Fig. 2-13 have the same probability

$$P(A) = P(B) = P(C) = \tfrac{1}{5}$$

and the intersections AB, AC, BC, and ABC also have the same probability

$$p = P(AB) = P(AC) = P(BC) = P(ABC)$$

(*a*) If $p = 1/25$, then these events are independent in pairs but they are not independent because

$$P(ABC) \neq P(A)P(B)P(C)$$

(*b*) If $p = 1/125$, then $P(ABC) = P(A)P(B)P(C)$ but the events are not independent because

$$P(AB) \neq P(A)P(B)$$ ◀

From the independence of the events A, B, and C it follows that:

1. Any one of them is independent of the intersection of the other two.
   Indeed, from (2-51) and (2-52) it follows that

$$P(A_1 A_2 A_3) = P(A_1)P(A_2)P(A_3) = P(A_1)P(A_2 A_3) \tag{2-53}$$

Hence the events A_1 and $A_2 A_3$ are independent.

2. If we replace one or more of these events with their complements, the resulting events are also independent.

Indeed, since

$$A_1A_2 = A_1A_2A_3 \cup A_1A_2\overline{A}_3; \qquad P(\overline{A}_3) = 1 - P(A_3)$$

we conclude with (2-53) that

$$P(A_1A_2\overline{A}_3) = P(A_1A_2) - P(A_1A_2)P(A_3) = P(A_1)P(A_2)P(\overline{A}_3)$$

Hence the events A_1, A_2, and $\overline{A}_3$ are independent because they satisfy (2-52) and, as we have shown earlier in the section, they are also independent in pairs.

3. Any one of them is independent of the union of the other two.

To show that the events A_1 and $A_2 \cup A_3$ are independent, it suffices to show that the events A_1 and $\overline{A_2 \cup A_3} = \overline{A}_2\overline{A}_3$ are independent. This follows from 1 and 2.

Generalization. The independence of n events can be defined inductively: Suppose that we have defined independence of k events for every $k < n$. We then say that the events $A_1, \ldots, A_n$ are independent if any $k < n$ of them are independent and

$$P(A_1 \cdots A_n) = P(A_1) \cdots P(A_n) \tag{2-54}$$

This completes the definition for any n because we have defined independence for $n = 2$.

EXAMPLE 2-20

BIRTHDAY PAIRING

▶ In a group of n people, (a) what is the probability that two or more persons will have the same birthday (month and date)? (b) What is the probability that someone in that group will have birthday that matches yours?

SOLUTION

There are $N = 365$ equally likely ways (number of days in a year) where the birthdays of each person can fall independently. The event of interest $A =$ "two or more persons have the same birthday" is the complement of the simpler event $B =$ "no two persons have the same birthday." To compute the number of ways no matching birthdays can occur among n persons, note that there are N ways for the first person to have a birthday, $N - 1$ ways for the second person without matching the first person, and finally $N - n + 1$ ways for the last person without matching any others. Using the independence assumption this gives $N(N-1)\cdots(N-r+1)$ possible "no matches." Without any such restrictions, there are N choices for each person's birthday and hence there are a total of N^n ways of assigning birthdays to n persons. Using the classical definition of probability in (1-7) this gives

$$P(B) = \frac{N(N-1)\cdots(N-n+1)}{N^n} = \prod_{k=1}^{n-1}\left(1 - \frac{k}{N}\right)$$

and hence the probability of the desired event

$$P(\text{at least one matching pair among } n \text{ persons}) = P(\overline{B}) = 1 - P(B)$$

$$= 1 - \prod_{k=1}^{n-1}\left(1 - \frac{k}{N}\right) \simeq 1 - e^{-\sum_{k=1}^{n-1} k/N} = 1 - e^{-n(n-1)/2N} \tag{2-55}$$

where we have used the approximation $e^{-x} \simeq 1 - x$ that is valid for small x. For example, $n = 23$ gives the probability of at least one match to be 0.5, whereas in a group of 50 persons, the probability of a birthday match is 0.97.

(*b*) To compute the probability for a personal match, once again it is instructive to look at the complement event. In that case there are $N - 1$ "unfavorable days" among N days for each person not to match your birthday. Hence the probability of each person missing your birthday is $(N - 1)/N$. For a group of n persons, this gives the probability that none of them will match your birthday to be $(1 - 1/N)^n \simeq e^{-n/N}$, and hence the probability of at least one match is $1 - e^{-n/N}$. For a modest 50–50 chance in this case, the group size needs to be about 253. In a group of 1000 people, chances are about 93% that there will be someone sharing your birthday. ◀

EXAMPLE 2-21

▶ Three switches connected in parallel operate independently. Each switch remains closed with probability p. (*a*) Find the probability of receiving an input signal at the output. (*b*) Find the probability that switch S_1 is open given that an input signal is received at the output.

SOLUTION

(*a*) Let A_i = "Switch S_i is closed." Then $P(A_i) = p, i = 1, 2, 3$. Since switches operate independently, we have

$$P(A_i A_j) = P(A_i)P(A_j) \qquad P(A_1 A_2 A_3) = P(A_1)P(A_2)P(A_3)$$

Let R represents the event "Input signal is received at the output." For the event R to occur either switch 1 or switch 2 or switch 3 must remain closed (Fig. 2-14), that is,

$$R = A_1 \cup A_2 \cup A_3 \tag{2-56}$$

$$\begin{aligned} P(R) &= 1 - P(\overline{R}) = 1 - P(\overline{A}_1 \overline{A}_2 \overline{A}_3) = 1 - P(\overline{A}_1)P(\overline{A}_2)P(\overline{A}_3) \\ &= 1 - (1 - p)^3 = 3p - 3p^2 + p^3 \end{aligned} \tag{2-57}$$

We can also derive (2-57) in a different manner. Since any event and its compliment form a trivial partition, we can always write

$$P(R) = P(R \mid A_1)P(A_1) + P(R \mid \overline{A}_1)P(\overline{A}_1) \tag{2-58}$$

But $P(R \mid A_1) = 1$, and $P(R \mid \overline{A}_1) = P(A_2 \cup A_3) = 2p - p^2$, and using these in (2-58) we obtain

$$P(R) = p + (2p - p^2)(1 - p) = 3p - 3p^2 + p^3 \tag{2-59}$$

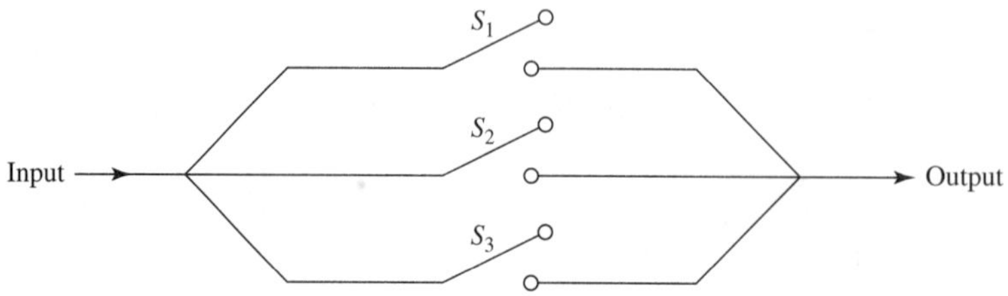

FIGURE 2-14

which agrees with (2-57). Note that the events A_1, A_2, and A_3 do not form a partition, since they are not mutually exclusive. Obviously any two or all three switches can be closed (or open) simultaneously. Moreover, $P(A_1) + P(A_2) + P(A_3) \neq 1$.

(*b*) We need $P(\overline{A}_1 \mid R)$. From Bayes' theorem

$$P(\overline{A}_1 \mid R) = \frac{P(R \mid \overline{A}_1)P(\overline{A}_1)}{P(R)} = \frac{(2p - p^2)(1-p)}{3p - 3p^2 + p^3} = \frac{2 - 3p + p^2}{3 - 3p + p^2}. \tag{2-60}$$

Because of the symmetry of the switches, we also have

$$P(\overline{A}_1 \mid R) = P(\overline{A}_2 \mid R) = P(\overline{A}_3 \mid R).$$

◀

EXAMPLE 2-22

▶ A biased coin is tossed till a head appears for the first time. What is the probability that the number of required tosses is odd?

SOLUTION

Let

$$\begin{aligned} A_i &= \text{“Head appears at the } i\text{th toss for the first time”} \\ &= \{\underbrace{T, T, T, \ldots, T}_{i-1}, H\} \end{aligned}$$

Assuming that each trial is independent of the rest,

$$P(A_i) = P(\{T, T, \ldots, T, H\}) = P(T)P(T)\cdots P(T)P(H) = q^{i-1}q \tag{2-61}$$

where $P(H) = p$, $P(T) = q = 1 - p$. Thus

$$\begin{aligned} &P(\text{“Head appears on an odd toss”}) \\ &\quad = P(A_1 \cup A_3 \cup A_5 \cup \cdots) \\ &\quad = \sum_{i=0}^{\infty} P(A_{2i+1}) = \sum_{i=0}^{\infty} q^{2i} p = p \sum_{i=0}^{\infty} q^{2i} \\ &\quad = \frac{p}{1-q^2} = \frac{p}{(1+q)(1-q)} \\ &\quad = \frac{1}{1+q} = \frac{1}{2-p} \end{aligned} \tag{2-62}$$

because $A_i \cup A_j = \emptyset$, $i \neq j$. Even for a fair coin, the probability of “Head first appears on an odd toss” is 2/3. ◀

As Theorems 2-1 through 2-3 show, a number of important consequences can be derived using the “generalized additive law” in (2-21).

THEOREM 2-1

▶ If $A_1, A_2, \ldots$ is an “increasing sequence” of events, that is, a sequence such that $A_1 \subset A_2 \subset \cdots$, then

$$P\left(\bigcup_k A_k\right) = \lim_{n\to\infty} P(A_n). \tag{2-63}$$

Proof. Clearly, the events

$$B_1 = A_1, \qquad B_2 = A_2\overline{A}_1 \stackrel{\Delta}{=} A_2 - A_1, \ldots, \qquad B_n = A_n - \bigcup_{k=1}^{n-1} B_k, \ldots \tag{2-64}$$

are mutually exclusive and have union $\bigcup_k A_k$. Moreover,

$$\bigcup_{k=1}^{n} B_k = A_n \tag{2-65}$$

Therefore, by (2-21)

$$P\left(\bigcup_k A_k\right) = P\left(\bigcup_k B_k\right) = \sum_k P(B_k) = \lim_{n\to\infty} \sum_{k=1}^{n} P(B_k)$$
$$= \lim_{n\to\infty} P\left(\bigcup_{k=1}^{n} B_k\right) = \lim_{n\to\infty} P(A_n) \tag{2-66}$$

◀

THEOREM 2-2

▶ If $A_1, A_2, \ldots$ is a "decreasing sequence" of events, that is, a sequence such that $A_1 \supset A_2 \supset \cdots$, then

$$P\left(\bigcap_k A_k\right) = \lim_{n\to\infty} P(A_n) \tag{2-67}$$

Proof. Considering the complementary events, we get $\overline{A}_1 \subset \overline{A}_2 \subset \cdots$, and hence, by (2-63)

$$P\left(\bigcap_k A_k\right) = 1 - P\left(\bigcup_k \overline{A}_k\right) = 1 - \lim_{n\to\infty} P(\overline{A}_n) = \lim_{n\to\infty} [1 - P(\overline{A}_n)] = \lim_{n\to\infty} P(A_n)$$

◀

In the case of arbitrary events, we have the result in Theorem 2-3.

THEOREM 2-3

▶ The inequality

$$P\left(\bigcup_k A_k\right) \le \sum_k P(A_k) \tag{2-68}$$

holds for arbitrary events $A_1, A_2, \ldots$.

Proof. Proceeding as in (2-64), $\bigcup_k A_k$ can be expressed as the union of the mutually exclusive events $B_1, B_2, \ldots$, where $B_k \subset A_k$ and hence $P(B_k) < P(A_k)$. Therefore

$$P\left(\bigcup_k A_k\right) = P\left(\bigcup_k B_k\right) = \sum_k P(B_k) \le \sum_k P(A_k)$$

Notice that (2-68) is a direct generalization of (2-21), where the events are not mutually exclusive. We can make use of Theorems 2-1 to 2-3 to prove an important result known as *Borel–Cantelli lemma.*

BOREL–CANTELLI LEMMA Given a sequence of events $A_1, A_2, \ldots,$ with probabilities $p_k = P(A_k), k = 1, 2, \ldots,$ (*i*) suppose

$$\sum_{k=1}^{\infty} p_k < \infty \tag{2-69}$$

that is, the series on the left converges. Then, with probability 1 only finitely many of the events $A_1, A_2, \ldots,$ occur.

(*ii*) Suppose $A_1, A_2, \ldots$ are also *independent* events, and

$$\sum_{k=1}^{\infty} p_k = \infty \tag{2-70}$$

that is, the series on the left diverges. Then, with probability 1 infinitely many of the events $A_1, A_2, \ldots$ occur.

Proof. (*i*) Let B be the event that *"infinitely many of the events $A_1, A_2, \ldots$ occur,"* and let

$$B_n = \bigcup_{k \geq n} A_k \tag{2-71}$$

so that B_n is the event that at least one of the events $A_n, A_{n+1}, \ldots$ occurs. Clearly B occurs if and only if B_n occurs for every $n = 1, 2, \ldots$. To see this, let the outcome ξ belong to an infinite number of events A_i. Then ξ must belong to every B_n, and hence it is contained in their intersection $\bigcap_n B_n$. Conversely if ξ belongs to this intersection, then it belongs to every B_n, which is possible only if ξ belongs to an infinite number of events A_i. Thus

$$B = \bigcap_n B_n = \bigcap_n \left(\bigcup_{k \geq n} A_k \right). \tag{2-72}$$

Further, $B_1 \supset B_2 \supset \cdots$, and hence, by Theorem 2-2,

$$P(B) = \lim_{n \to \infty} P(B_n) \tag{2-73}$$

But, by Theorem 2-3

$$P(B_n) \leq \sum_{k \geq n} P(A_k) = \sum_{k \geq n} p_k \to 0 \quad \text{as} \quad n \to \infty \tag{2-74}$$

because of (2-69). Therefore

$$P(B) = \lim_{n \to \infty} P(B_n) = \lim_{n \to \infty} \sum_{k \geq n} p_k = 0 \tag{2-75}$$

that is, the probability of infinitely many of the events $A_1, A_2, \ldots$ occurring is 0. Equivalently, the probability of only finitely many of the events $A_1, A_2, \ldots$ occurring is 1.

(*ii*) To prove the second part, taking complements of the events B_n and B in (2-71) and (2-72), we get

$$\overline{B}_n = \bigcap_{k \geq n} \overline{A}_k \qquad \overline{B} = \bigcup_n \overline{B}_n \tag{2-76}$$

Further,

$$\overline{B}_n \subset \bigcap_{k=n}^{n+m} \overline{A}_k$$

for every $m = 0, 1, 2, \ldots$. Therefore, by the independence of the events $\overline{A}_1, \overline{A}_2, \ldots,$ we get

$$P(\overline{B}_n) \leq P\left(\bigcap_{k=n}^{n+m} \overline{A}_k\right) = P(\overline{A}_n)\cdots P(\overline{A}_{n+m})$$

$$= (1-p_n)\cdots(1-p_{n+m}) \leq \exp\left(-\sum_{k=n}^{n+m} p_k\right) \qquad (2\text{-}77)$$

where we have made use of the inequality $1 - x \leq e^{-x}, x \geq 0$. Notice that if $A_1, A_2, \ldots$ is a sequence of independent events, then so is the sequence of complementary events $\overline{A}_1, \overline{A}_2, \ldots$. But from (2-70)

$$\sum_{k=n}^{n+m} p_k \to \infty \quad \text{as} \quad m \to \infty \qquad (2\text{-}78)$$

Therefore, passing to the limit $m \to \infty$ in (2-77), we find that $P(\overline{B}_n) = 0$ for every $n = 1, 2, \ldots$. Thus using (2-76)

$$P(\overline{B}) \leq \sum_n P(\overline{B}_n) = 0$$

and hence

$$P(B) = 1 - P(\overline{B}) = 1 \qquad (2\text{-}79)$$

that is, the probability of infinitely many of the events $A_1, A_2, \ldots$ occurring is 1. Notice that the second part of the Borel–Cantelli lemma, which represents a converse to the first part, requires the additional assumption that the events involved be independent of each other. ◀

As an example, consider the event "$HH \cdots H$" occurring in a sequence of Bernoulli trials. To determine the probability that such an *"all success"* sequence of length n appears infinitely often, let A_k stand for the event *"Head appears on the kth toss,"* and define $B_i = A_i \cap A_{i+1} \cap \cdots A_{i+n-1}, i \geq 1$. We have $P(B_i) = p_i = p^n$. The events B_i are not independent, however, the events $B_1, B_{n+1}, B_{2n+1}, \ldots$ are independent, and the series $\sum_{k=0}^{\infty} p_{kn+1}$ diverges. Hence, from the second part of the Borel–Cantelli lemma, it follows that with probability one the pattern "$HH \cdots H$" (as well as any other arbitrary pattern) will occur infinitely often. To summarize, if the sum of the probabilities of an infinite set of independent events diverge, then with probability 1, infinitely many of those events will occur in the long run.

PROBLEMS

2-1 Show that (*a*) $\overline{\overline{A} \cup \overline{B}} \cup \overline{\overline{A} \cup B} = A$; (*b*) $(A \cup B)(\overline{AB}) = A\overline{B} \cup B\overline{A}$.

2-2 If $A = \{2 \leq x \leq 5\}$ and $B = \{3 \leq x \leq 6\}$, find $A \cup B$, AB, and $(A \cup B)(\overline{AB})$.

2-3 Show that if $AB = \{\emptyset\}$, then $P(A) \leq P(\overline{B})$.

2-4 Show that (*a*) if $P(A) = P(B) = P(AB)$, then $P(A\overline{B} \cup B\overline{A}) = 0$; (*b*) if $P(A) = P(B) = 1$, then $P(AB) = 1$.

2-5 Prove and generalize the following identity

$$P(A \cup B \cup C) = P(A) + P(B) + P(C) - P(AB) - P(AC) - P(BC) + P(ABC)$$

2-6 Show that if S consists of a countable number of elements ζ_i and each subset $\{\zeta_i\}$ is an event, then all subsets of S are events.

2-7 If $S = \{1, 2, 3, 4\}$, find the smallest field that contains the sets $\{1\}$ and $\{2, 3\}$.

2-8 If $A \subset B$, $P(A) = 1/4$, and $P(B) = 1/3$, find $P(A \mid B)$ and $P(B \mid A)$.

2-9 Show that $P(AB \mid C) = P(A \mid BC)P(B \mid C)$ and $P(ABC) = P(A \mid BC)P(B \mid C)P(C)$.

2-10 (*Chain rule*) Show that

$$P(A_n \cdots A_1) = P(A_n \mid A_{n-1} \cdots A_1) \cdots P(A_2 \mid A_1)P(A_1)$$

2-11 We select at random m objects from a set S of n objects and we denote by A_m the set of the selected objects. Show that the probability p that a particular element ζ_0 of S is in A_m equals m/n.

Hint: p equals the probability that a randomly selected element of S is in A_m.

2-12 A call occurs at time t, where t is a random point in the interval $(0, 10)$. (*a*) Find $P\{6 \le t \le 8\}$. (*b*) Find $P\{6 \le t \le 8 \mid t > 5\}$.

2-13 The space S is the set of all positive numbers t. Show that if $P\{t_0 \le t \le t_0 + t_1 \mid t \ge t_0\} = P\{t \le t_1\}$ for every t_0 and t_1, then $P\{t \le t_1\} = 1 - e^{-ct_1}$, where c is a constant.

2-14 The events A and B are mutually exclusive. Can they be independent?

2-15 Show that if the events $A_1, \ldots, A_n$ are independent and B_i equals A_i or $\overline{A}_i$ or S, then the events $B_1, \ldots, B_n$ are also independent.

2-16 A box contains n identical balls numbered 1 through n. Suppose k balls are drawn in succession. (*a*) What is the probability that m is the largest number drawn? (*b*) What is the probability that the largest number drawn is less than or equal to m?

2-17 Suppose k identical boxes contain n balls numbered 1 through n. One ball is drawn from each box. What is the probability that m is the largest number drawn?

2-18 Ten passengers get into a train that has three cars. Assuming a random placement of passengers, what is the probability that the first car will contain three of them?

2-19 A box contains m white and n black balls. Suppose k balls are drawn. Find the probability of drawing at least one white ball.

2-20 A player tosses a penny from a distance onto the surface of a square table ruled in 1 in. squares. If the penny is 3/4 in. in diameter, what is the probability that it will fall entirely inside a square (assuming that the penny lands on the table).

2-21 In the New York State lottery, six numbers are drawn from the sequence of numbers 1 through 51. What is the probability that the six numbers drawn will have (*a*) all one digit numbers? (*b*) two one-digit and four two-digit numbers?

2-22 Show that $2^n - (n + 1)$ equations are needed to establish the independence of n events.

2-23 Box 1 contains 1 white and 999 red balls. Box 2 contains 1 red and 999 white balls. A ball is picked from a randomly selected box. If the ball is red what is the probability that it came from box 1?

2-24 Box 1 contains 1000 bulbs of which 10% are defective. Box 2 contains 2000 bulbs of which 5% are defective. Two bulbs are picked from a randomly selected box. (*a*) Find the probability that both bulbs are defective. (*b*) Assuming that both are defective, find the probability that they came from box 1.

2-25 A train and a bus arrive at the station at random between 9 A.M. and 10 A.M. The train stops for 10 minutes and the bus for x minutes. Find x so that the probability that the bus and the train will meet equals 0.5.

2-26 Show that a set S with n elements has

$$\frac{n(n-1)\cdots(n-k+1)}{1 \cdot 2 \cdots k} = \frac{n!}{k!(n-k)!}$$

k-element subsets.

2-27 We have two coins; the first is fair and the second two-headed. We pick one of the coins at random, we toss it twice and heads shows both times. Find the probability that the coin picked is fair.

CHAPTER 3

REPEATED TRIALS

3-1 COMBINED EXPERIMENTS

We are given two experiments: The first experiment is the rolling of a fair die

$$S_1 = \{f_1, \ldots, f_6\} \qquad P_1\{f_i\} = \tfrac{1}{6}$$

The second experiment is the tossing of a fair coin

$$S_2 = \{h, t\} \qquad P_2\{h\} = P_2\{t\} = \tfrac{1}{2}$$

We perform both experiments and we want to find the probability that we get "two" on the die and "heads" on the coin.

If we make the reasonable assumption that the outcomes of the first experiment are independent of the outcomes of the second, we conclude that the unknown probability equals $1/6 \times 1/2$.

This conclusion is reasonable; however, the notion of independence used in its derivation does not agree with the definition given in (2-50). In that definition, the events A and B were subsets of the *same* space. In order to fit this conclusion into our theory, we must, therefore, construct a space S having as subsets the events "two" and "heads." This is done as follows:

The two experiments are viewed as a single experiment whose outcomes are pairs $\zeta_1\zeta_2$, where ζ_1 is one of the six faces of the die and ζ_2 is heads or tails.[1] The resulting space consists of the 12 elements

$$f_1h, \ldots, f_6h, f_1t, \ldots, f_6t$$

[1] In the earlier discussion, the symbol ζ_i represented a *single* element of a set S. From now on, ζ_i will also represent an arbitrary element of a set S_i. We will understand from the context whether ζ_i is one particular element or any element of S_i.

In this space, {two} is not an elementary event but a subset consisting of two elements

$$\{\text{two}\} = \{f_2h, f_2t\}$$

Similarly, {heads} is an event with six elements

$$\{\text{heads}\} = \{f_1h, \ldots, f_6h\}$$

To complete the experiment, we must assign probabilities to all subsets of S. Clearly, the event {two} occurs if the die shows "two" no matter what shows on the coin. Hence

$$P\{\text{two}\} = P_1\{f_2\} = \tfrac{1}{6}$$

Similarly,

$$P\{\text{heads}\} = P_2\{h\} = \tfrac{1}{2}$$

The intersection of the events {two} and {heads} is the elementary event $\{f_2h\}$. Assuming that the events {two} and {heads} are independent in the sense of (2-50), we conclude that $P\{f_2h\} = 1/6 \times 1/2$ in agreement with our earlier conclusion.

Cartesian Products

Given two sets S_1 and S_2 with elements ζ_1 and ζ_2, respectively, we form all *ordered* pairs $\zeta_1\zeta_2$, where ζ_1 is any element of S_1 and ζ_2 is any element of S_2. The *cartesian product* of the sets S_1 and S_2 is a set S whose elements are all such pairs. This set is written in the form

$$S = S_1 \times S_2$$

EXAMPLE 3-1

▶ The cartesian product of the sets

$$S_1 = \{\text{car, apple, bird}\} \qquad S_2 = \{h, t\}$$

has six elements

$$S_1 \times S_2 = \{\text{car-}h, \text{car-}t, \text{apple-}h, \text{apple-}t, \text{bird-}h, \text{bird-}t\}$$ ◀

EXAMPLE 3-2

▶ If $S_1 = \{h, t\}$, $S_2 = \{h, t\}$. Then

$$S_1 \times S_2 = \{hh, ht, th, tt\}$$

In this example, the sets S_1 and S_2 are identical. We note also that the element *ht* is different from the element *th*. ◀

If A is a subset of S_1 and B is a subset of S_2, then the set

$$C = A \times B$$

consisting of all pairs $\zeta_1\zeta_2$, where $\zeta_1 \in A$ and $\zeta_2 \in B$, is a subset of S.

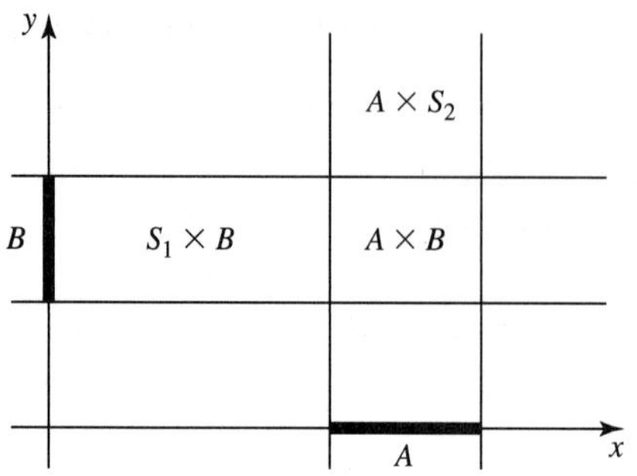

FIGURE 3-1

Forming similarly the sets $A \times S_2$ and $S_1 \times B$, we conclude that their intersection is the set $A \times B$:

$$A \times B = (A \times S_2) \cap (S_1 \times B) \tag{3-1}$$

Note Suppose that S_1 is the x axis, S_2 is the y axis, and A and B are two intervals:

$$A = \{x_1 \le x \le x_2\} \qquad B = \{y_1 \le y \le y_2\}$$

In this case, $A \times B$ is a rectangle, $A \times S_2$ is a vertical strip, and $S_1 \times B$ is a horizontal strip (Fig. 3-1).
We can thus interpret the cartesian product $A \times B$ of two arbitrary sets as a generalized rectangle.

CARTESIAN PRODUCT OF TWO EXPERIMENTS. The cartesian product of two experiments S_1 and S_2 is a new experiment $S = S_1 \times S_2$ whose events are all cartesian products of the form

$$A \times B \tag{3-2}$$

where A is an event of S_1 and B is an event of S_2, and their unions and intersections.

In this experiment, the probabilities of the events $A \times S_2$ and $S_1 \times B$ are such that

$$P(A \times S_2) = P_1(A) \qquad P(S_1 \times B) = P_2(B) \tag{3-3}$$

where $P_1(A)$ is the probability of the event A in the experiments S_1 and $P_2(B)$ is the probability of the event B in the experiments S_2. This fact is motivated by the interpretation of S as a combined experiment. Indeed, the event $A \times S_2$ of the experiment S occurs if the event A of the experiment S_1 occurs no matter what the outcome of S_2 is. Similarly, the event $S_1 \times B$ of the experiment S occurs if the event B of the experiment S_2 occurs no matter what the outcome of S_1 is. This justifies the two equations in (3-3).

These equations determine only the probabilities of the events $A \times S_2$ and $S_1 \times B$. The probabilities of events of the form $A \times B$ and of their unions and intersections cannot in general be expressed in terms of P_1 and P_2. To determine them, we need additional information about the experiments S_1 and S_2.

INDEPENDENT EXPERIMENTS. In many applications, the events $A \times S_2$ and $S_1 \times B$ of the combined experiment S are independent for any A and B. Since the intersection of these events equals $A \times B$ [see (3-1)], we conclude from (2-50) and (3-3) that

$$P(A \times B) = P(A \times S_2)P(S_1 \times B) = P_1(A)P_2(B) \tag{3-4}$$

This completes the specification of the experiment S because all its events are unions and intersections of events of the form $A \times B$.

We note in particular that the elementary event $\{\zeta_1\zeta_2\}$ can be written as a cartesian product $\{\zeta_1\} \times \{\zeta_2\}$ of the elementary events $\{\zeta_1\}$ and $\{\zeta_2\}$ of S_1 and S_2. Hence

$$P\{\zeta_1\zeta_2\} = P_1\{\zeta_1\}P_2\{\zeta_2\} \tag{3-5}$$

EXAMPLE 3-3

▶ A box B_1 contains 10 white and 5 red balls and a box B_2 contains 20 white and 20 red balls. A ball is drawn from each box. What is the probability that the ball from B_1 will be white and the ball from B_2 red?

This operation can be considered as a combined experiment. Experiment S_1 is the drawing from B_1 and experiment S_2 is the drawing from B_2. The space S_1 has 15 elements: 10 white and 5 red balls. The event

$$W_1 = \{\text{a white ball is drawn from } B_1\}$$

has 10 favorable elements and its probability equals 10/15. The space S_2 has 40 elements: 20 white and 20 red balls. The event

$$R_2 = \{\text{a red ball is drawn from } B_2\}$$

has 20 favorable elements and its probability equals 20/40. The space $S_1 \times S_2$ has 40×15 elements: all possible pairs that can be drawn.

We want the probability of the event

$$W_1 \times R_2 = \{\text{white from } B_1 \text{ and red from } B_2\}$$

Assuming independence of the two experiments, we conclude from (3-4) that

$$P(W_1 \times R_2) = P_1(W_1)P_2(R_2) = \frac{10}{15} \times \frac{20}{40}$$ ◀

EXAMPLE 3-4

▶ Consider the coin experiment where the probability of "heads" equals p and the probability of "tails" equals $q = 1 - p$. If we toss the coin twice, we obtain the space

$$S = S_1 \times S_2 \qquad S_1 = S_2 = \{h, t\}$$

Thus S consists of the four outcomes hh, ht, th, and tt. Assuming that the experiments S_1 and S_2 are independent, we obtain

$$P\{hh\} = P_1\{h\}P_2\{h\} = p^2$$

Similarly,

$$P\{ht\} = pq \qquad P\{th\} = qp \qquad P\{tt\} = q^2$$

We shall use this information to find the probability of the event

$$H_1 = \{\text{heads at the first toss}\} = \{hh, ht\}$$

Since H_1 consists of the two outcomes hh and ht, (2-23) yields

$$P(H_1) = P\{hh\} + P\{ht\} = p^2 + pq = p$$

This follows also from (3-4) because $H_1 = \{h\} \times S_2$. ◀

GENERALIZATION. Given n experiments $S_1, \ldots, S_n$, we define as their cartesian product

$$S = S_1 \times \cdots \times S_n \tag{3-6}$$

the experiment whose elements are all ordered n tuplets $\zeta_1 \cdots \zeta_n$ where ζ_i is an element of the set S_i. Events in this space are all sets of the form

$$A_1 \times \cdots \times A_n$$

where $A_i \subset S_i$, and their unions and intersections. If the experiments are independent and $P_i(A_i)$ is the probability of the event A_i in the experiment S_i, then

$$P(A_1 \times \cdots \times A_n) = P_1(A_1) \cdots P_n(A_n) \tag{3-7}$$

EXAMPLE 3-5

▶ If we toss the coin of Example 3-4 n times, we obtain the space $S = S_1 \times \cdots \times S_n$ consisting of the 2^n elements $\zeta_1 \cdots \zeta_n$, where $\zeta_i = h$ or t. Clearly,

$$P\{\zeta_1 \cdots \zeta_n\} = P_1\{\zeta_1\} \cdots P_n\{\zeta_n\} \qquad P_i\{\zeta_i\} = \begin{cases} p & \zeta_i = h \\ q & \zeta_i = t \end{cases} \tag{3-8}$$

If, in particular, $p = q = 1/2$, then

$$P\{\zeta_1 \cdots \zeta_n\} = \frac{1}{2^n}$$

From (3-8) it follows that, if the elementary event $\{\zeta_1 \cdots \zeta_n\}$ consists of k heads and $n - k$ tails (in a specific order), then

$$P\{\zeta_1 \cdots \zeta_n\} = p^k q^{n-k} \tag{3-9}$$

We note that the event $H_1 = \{\text{heads at the first toss}\}$ consists of 2^{n-1} outcomes $\zeta_1 \cdots \zeta_n$, where $\zeta_1 = h$ and $\zeta_i = t$ or h for $i > 1$. The event H_1 can be written as a cartesian product

$$H_1 = \{h\} \times S_2 \times \cdots \times S_n$$

Hence [see (3-7)]

$$P(H_1) = P_1\{h\} P_2(S_2) \cdots P_n(S_n) = p$$

because $P_i(S_i) = 1$. We can similarly show that if

$$H_i = \{\text{heads at the } i\text{th toss}\} \qquad I_i = \{\text{tails at the } i\text{th toss}\}$$

then

$$P(H_i) = p \qquad P(I_i) = q$$

◀

DUAL MEANING OF REPEATED TRIALS. In the theory of probability, the notion of repeated trials has two fundamentally different meanings. The first is the approximate relationship (1-1) between the probability $P(A)$ of an event A in an experiment S and the relative frequency of the occurrence of A. The second is the creation of the experiment $S \times \cdots \times S$.

For example, the repeated tossings of a coin can be given the following two interpretations:

First interpretation (physical) Our experiment is the *single* toss of a fair coin. Its space has two elements and the probability of each elementary event equals $1/2$. A trial is the toss of the coin *once*.

If we toss the coin n times and heads shows n_h times, then almost certainly $n_h/n \approx 1/2$ provided that n is sufficiently large. Thus the first interpretation of repeated trials is the above inprecise statement relating probabilities with observed frequencies.

Second interpretation (conceptual) Our experiment is now the toss of the coin n *times,* where n is any number large or small. Its space has 2^n elements and the probability of each elementary event equals $1/2^n$. A trial is the toss of the coin n *times*. All statements concerning the number of heads are precise and in the form of probabilities.

We can, of course, give a relative frequency interpretation to these statements. However, to do so, we must repeat the n *tosses of the coin* a large number of times.

3-2 BERNOULLI TRIALS

A set of n distinct objects can be placed in several different orders forming *permutations*. Thus, for example, the possible permutations of three objects a, b, c are: abc, bac, bca, acb, cab, cba, (6 different permutations out of 3 objects). In general, given n objects the first spot can be selected n different ways, and for every such choice the next spot the remaining $n-1$ ways, and so on. Thus the number of permutations of n objects equal $n(n-1)(n-2)\cdots 3\cdot 2\cdot 1 = n!$.

Suppose only $k < n$ objects are taken out of n objects at a time, attention being paid to the order of objects in each such group. Once again the first spot can be selected n distinct ways, and for every such selection the next spot can be chosen $(n-1)$ distinct ways, ..., and the kth spot $(n-k+1)$ distinct ways from the remaining objects. Thus the total number of distinct arrangements (permutations) of n objects taken k at a time is given by

$$n(n-1)(n-2)\cdots(n-k+1) = \frac{n!}{(n-k)!} \tag{3-10}$$

For example, taking two objects out of the three objects a, b, c, we get the permutations ab, ba, ac, ca, bc, cb.

Next suppose the k objects are taken out of n objects *without* paying any attention to the order of the objects in each group, thus forming *combinations*. In that case, the $k!$ permutations generated by each group of k objects contribute toward only one combination, and hence using (3-10) the total *combinations* of n objects taken k at a time is given by

$$\frac{n(n-1)(n-2)\cdots(n-k+1)}{k!} = \frac{n!}{(n-k)!k!} = \binom{n}{k}$$

Thus, if a set has n elements, then the total number of its subsets consisting of k elements each equals

$$\binom{n}{k} = \frac{n(n-1)\cdots(n-k+1)}{1\cdot 2\cdots k} = \frac{n!}{k!(n-k)!} \tag{3-11}$$

For example, if $n = 4$ and $k = 2$, then

$$\binom{4}{2} = \frac{4\cdot 3}{1\cdot 2} = 6$$

Indeed, the two-element subsets of the four-element set *abcd* are

$$ab \qquad ac \qquad ad \qquad bc \qquad bd \qquad cd$$

This result will be used to find the probability that an event occurs k times in n independent trials of an experiment S. This problem is essentially the same as the problem of obtaining k heads in n tossings of a coin. We start, therefore, with the coin experiment.

EXAMPLE 3-6

▶ A coin with $P\{h\} = p$ is tossed n times. We maintain that the probability $p_n(k)$ that heads shows k times is given by

$$p_n(k) = \binom{n}{k} p^k q^{n-k} \qquad q = 1 - p \tag{3-12}$$

SOLUTION

The experiment under consideration is the n-tossing of a coin. A single outcome is a particular sequence of heads and tails. The event $\{k$ heads in any order$\}$ consists of all sequences containing k heads and $n-k$ tails. To obtain all distinct arrangements with n objects consisting of k heads and $n-k$ tails, note that if they were all distinct objects there would be $n!$ such arrangements. However since the k heads and $n-k$ tails are identical among themselves, the corresponding $k!$ permutations among the heads and the $(n-k)!$ permutations among the tails together only contribute to one distinct sequence. Thus the total distinct arrangements (combinations) are given by $\frac{n!}{k!(n-k)!} = \binom{n}{k}$. Hence the event $\{k$ heads in any order$\}$ consists of $\binom{n}{k}$ elementary events containing k heads and $n-k$ tails in a specific order. Since the probability of each of these elementary events equals $p^k q^{n-k}$, we conclude that

$$P\{k \text{ heads in any order}\} = \binom{n}{k} p^k q^{n-k}$$

Special Case. If $n = 3$ and $k = 2$, then there are three ways of getting two heads, namely, hht, hth, and thh. Hence $p_3(2) = 3p^2q$ in agreement with (3-12). ◀

Success or Failure of an Event A in n Independent Trials

We consider now our main problem. We are given an experiment S and an event A with

$$P(A) = p \quad P(\overline{A}) = q \quad p + q = 1$$

We repeat the experiment n times and the resulting product space we denote by S^n. Thus

$$S^n = S \times \cdots \times S.$$

We shall determine the probability $p_n(k)$ that the event A occurs exactly k times.

FUNDAMENTAL THEOREM

$$p_n(k) = P\{A \text{ occurs } k \text{ times in any order}\} = \binom{n}{k} p^k q^{n-k} \tag{3-13}$$

Proof. The event $\{A$ occurs k times in a specific order$\}$ is a cartesian product $B_1 \times \cdots \times B_n$, where k of the events B_i equal A and the remaining $n - k$ equal $\overline{A}$. As we know from (3-7), the probability of this event equals

$$P(B_1) \cdots P(B_n) = p^k q^{n-k}$$

because

$$P(B_i) = \begin{cases} p & \text{if} \quad B_i = A \\ q & \text{if} \quad B_i = A \end{cases}$$

In other words,

$$P\{A \text{ occurs } k \text{ times in a specific order}\} = p^k q^{n-k} \tag{3-14}$$

The event $\{A$ occurs k times in any order$\}$ is the union of the $\binom{n}{k}$ events $\{A$ occurs k times in a specific order$\}$ and since these events are mutually exclusive, we conclude from (2-20) that $p_n(k)$ is given by (3-13).

In Fig. 3-2, we plot $p_n(k)$ for $n = 9$. The meaning of the dashed curves will be explained later.

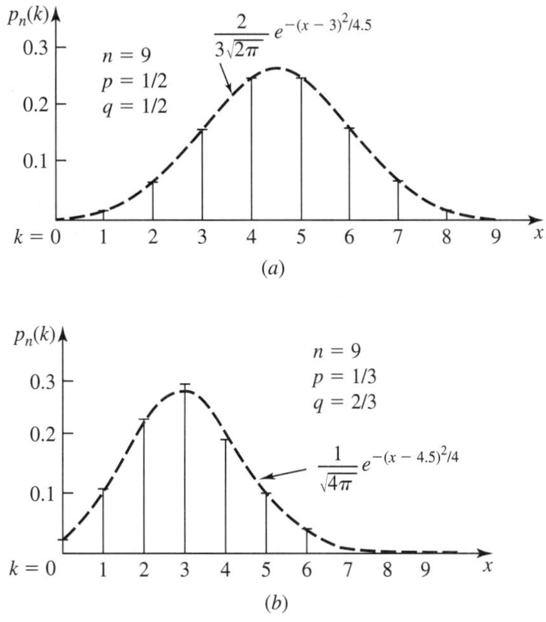

FIGURE 3-2

EXAMPLE 3-7 ▶ A fair die is rolled five times. We shall find the probability $p_5(2)$ that "six" will show twice.

In the single roll of a die, $A = \{\text{six}\}$ is an event with probability $1/6$. Setting

$$P(A) = \tfrac{1}{6} \quad P(\overline{A}) = \tfrac{5}{6} \quad n = 5 \quad k = 2$$

in (3-13), we obtain

$$P_5(2) = \frac{5!}{2!3!}\left(\frac{1}{6}\right)^2\left(\frac{5}{6}\right)^3$$ ◀

The problem in Example 3-8 has an interesting historical content, since part of it was one of the first problems solved by Pascal.

EXAMPLE 3-8 ▶ A pair of dice is rolled n times. (a) Find the probability that "seven" will not show at all. (b) (Pascal) Find the probability of obtaining double six at least once.

SOLUTION

The space of the single roll of two dice consists of the 36 elements $f_i f_j, i, j = 1, 2, \ldots, 6$.

(a) The event $A = \{\text{seven}\}$ consists of the six elements

$$f_1 f_6 \quad f_2 f_5 \quad f_3 f_4 \quad f_4 f_3 \quad f_5 f_2 \quad f_6 f_1$$

Therefore $P(A) = 6/36 = 1/6$ and $P(\overline{A}) = 5/6$. With $k = 0$, (3-13) yields

$$p_n(0) = \left(\frac{5}{6}\right)^n$$

(b) The event $B = \{\text{double six}\}$ consists of the single element $f_6 f_6$. Thus $P(B) = 1/36$, and $P(\overline{B}) = 35/36$. Let

$$X = \{\text{double six at least once in } n \text{ games}\}$$

Then

$$\begin{aligned}\overline{X} &= \{\text{double six will not show in any of the } n \text{ games}\}\\ &= \overline{B}\,\overline{B}\cdots\overline{B}\end{aligned}$$

and this gives

$$P(X) = 1 - P(\overline{X}) = 1 - P(\overline{B})^n = 1 - \left(\frac{35}{36}\right)^n \tag{3-15}$$

where we have made use of the independence of each throw. Similarly, it follows that if one die is rolled in succession n times, the probability of obtaining six at least once would be

$$1 - \left(\frac{5}{6}\right)^n \tag{3-16}$$

Suppose, we are interested in finding the number of throws required to assure a 50% success of obtaining double six at least once. From (3-15), in that case n must satisfy

$$1-\left(\frac{35}{36}\right)^n > \frac{1}{2} \quad \text{or} \quad \left(\frac{35}{36}\right)^n < \frac{1}{2}$$

which gives

$$n > \frac{\log 2}{\log 36 - \log 35} = 24.605$$

Thus in 25 throws one is more likely to get double six at least once than not to get it at all. Also in 24 or less throws, there is a greater chance to fail than to succeed.

In the case of a single die, from (3-16), for 50% success in obtaining six at least once, we must throw a minimum of four times (since $\log 2/(\log 6 - \log 5) = 3.801$).

This problem was suggested to Pascal by Chevalier de Mere, a nobleman well experienced in gambling. He, along with other gamblers, had all along known the advantage of betting for double six in 25 throws or for one six with a single die in 4 throws. The confusion at that time originated from the fact that although there are 36 cases for two dice and 6 cases for one die, yet the above numbers (25 and 4) did not seem to fit into that scheme (36 versus 6). The correct solution given by Pascal removed all apparent "paradoxes;" and in fact he correctly derived the same number 25 that had been observed by gamblers all along. ◀

Example 3-9 is one of the first problems in probability discussed and solved by Fermat and Pascal in their correspondence.

EXAMPLE 3-9

▶ Two players A and B agree to play a series of games on the condition that A wins the series if he succeeds in winning m games before B wins n games. The probability of winning a single game is p for A and $q = 1 - p$ for B. What is the probability that A will win the series?

SOLUTION

Let P_A denote the probability that A will win m games before B wins n games, and let P_B denote the probability that B wins n games before A wins m of them. Clearly by the $(m+n-1)$th game there must be a winner. Thus $P_A + P_B = 1$. To find P_A, notice that A can win in the following mutually exclusive ways. Let

$$X_k = \{A \text{ wins } m \text{ games in exactly } m+k \text{ games}\}, \qquad k = 0, 1, 2, \ldots, n-1.$$

Notice that X_ks are mutually exclusive events, and the event

$$\{A \text{ wins}\} = X_0 \cup X_1 \cup \cdots \cup X_{n-1}$$

so that

$$P_A = P(A \text{ wins}) = P\left(\bigcup_{i=0}^{n-1} X_i\right) = \sum_{i=0}^{n-1} P(X_i) \tag{3-17}$$

To determine $P(X_i)$, we argue as follows: For A to win m games in exactly $m+k$ games, A must win the last game and $(m-1)$ games in any order among the first $(m+k-1)$

games. Since all games are independent of each other, we get

$$\begin{aligned} P(X_i) &= P(A \text{ wins } m-1 \text{ games among the first } (m+k-1) \text{ games}) \\ &\quad \times P(A \text{ wins the last game}) \\ &= \binom{m+k-1}{m-1} p^{m-1} q^k p \\ &= \frac{(m+k-1)!}{(m-1)!k!} p^m q^k, \quad k = 0, 1, 2, \ldots, n-1. \end{aligned} \tag{3-18}$$

Substituting this into (3-17) we get

$$\begin{aligned} P_A &= p^m \sum_{k=0}^{n-1} \frac{(m+k-1)!}{(m-1)!k!} q^k \\ &= p^m \left(1 + \frac{m}{1} q + \frac{m(m+1)}{1 \cdot 2} q^2 + \cdots + \frac{m(m+1)\cdots(m+n-2)}{1 \cdot 2 \cdots (n-1)} q^{n-1}\right) \end{aligned} \tag{3-19}$$

In a similar manner, we obtain the probability that B wins

$$P_B = q^n \left(1 + \frac{n}{1} p + \frac{n(n+1)}{1 \cdot 2} p^2 + \cdots + \frac{n(n+1)\cdots(m+n-2)}{1 \cdot 2 \cdots (m-1)} p^{m-1}\right) \tag{3-20}$$

Since A or B must win by the $(m+n-1)$ game, we have $P_A + P_B = 1$, and substituting (3-19)–(3-20) into this we obtain an interesting identity. See also (2-30). ◀

EXAMPLE 3-10 ▶ We place at random n points in the interval $(0, T)$. What is the probability that k of these points are in the interval (t_1, t_2) (Fig. 3-3)?

This example can be considered as a problem in repeated trials. The experiment S is the placing of a *single* point in the interval $(0, T)$. In this experiment, $A = \{$the point is in the interval $(t_1, t_2)\}$ is an event with probability

$$P(A) = p = \frac{t_2 - t_1}{T}$$

In the space S^n, the event $\{A \text{ occurs } k \text{ times}\}$ means that k of the n points are in the interval (t_1, t_2). Hence [see (3-13)]

$$P\{k \text{ points are in the interval } (t_1, t_2)\} = \binom{n}{k} p^k q^{n-k} \tag{3-21}$$

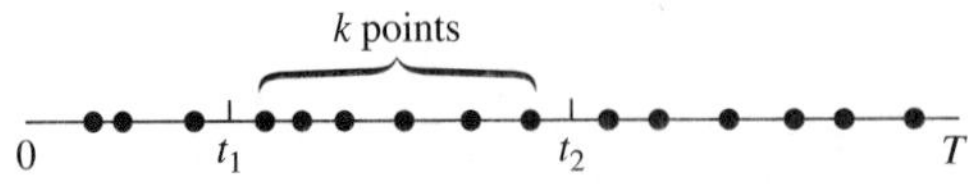

FIGURE 3-3

◀

EXAMPLE 3-11 ▶ A system containing n components is put into operation at $t = 0$. The probability that a particular component will fail in the interval $(0, t)$ equals

$$p = \int_0^t \alpha(\tau)\,d\tau \qquad \text{where} \quad \alpha(t) \ge 0 \qquad \int_0^\infty \alpha(t)\,dt = 1 \tag{3-22}$$

What is the probability that k of these components will fail prior to time t?

This example can also be considered as a problem in repeated trials. Reasoning as before, we conclude that the unknown probability is given by (3-21). ◀

MOST LIKELY NUMBER OF SUCCESSES. We shall now examine the behavior of $p_n(k)$ as a function of k for a fixed n. We maintain that as k increases, $p_n(k)$ increases reaching a maximum for

$$k = k_{\max} = [(n+1)p] \tag{3-23}$$

where the brackets mean the largest integer that does not exceed $(n+1)p$. If $(n+1)p$ is an integer, then $p_n(k)$ is maximum for two consecutive values of k:

$$k = k_1 = (n+1)p \qquad \text{and} \qquad k = k_2 = k_1 - 1 = np - q$$

Proof. We form the ratio

$$\frac{p_n(k-1)}{p_n(k)} = \frac{kq}{(n-k+1)p}$$

If this ratio is less than 1, that is, if $k < (n+1)p$, then $p_n(k-1)$ is less than $p_n(k)$. This shows that as k increases, $p_n(k)$ increases reaching its maximum for $k = [(n+1)p]$. For $k > (n+1)p$, this ratio is greater than 1: hence $p_n(k)$ decreases.

If $k_1 = (n+1)p$ is an integer, then

$$\frac{p_n(k_1-1)}{p_n(k_1)} = \frac{k_1 q}{(n-k_1+1)p} = \frac{(n+1)pq}{[n-(n+1)p+1]p} = 1$$

This shows that $p_n(k)$ is maximum for $k = k_1$ and $k = k_1 - 1$.

EXAMPLE 3-12 ▶ (*a*) If $n = 10$ and $p = 1/3$, then $(n+1)p = 11/3$; hence $k_{\max} = [11/3] = 3$.
(*b*) If $n = 11$ and $p = 1/2$, then $(n+1)p = 6$; hence $k_1 = 6$, $k_2 = 5$. ◀

We shall, finally, find the probability

$$P\{k_1 \le k \le k_2\}$$

that the number k of occurrences of A is between k_1 and k_2. Clearly, the events $\{A$ occurs k times$\}$, where k takes all values from k_1 to k_2, are mutually exclusive and their union is the event $\{k_1 \le k \le k_2\}$. Hence [see (3-13)]

$$P\{k_1 \le k \le k_2\} = \sum_{k=k_1}^{k_2} p_n(k) = \sum_{k=k_1}^{k_2} \binom{n}{k} p^k q^{n-k} \tag{3-24}$$

EXAMPLE 3-13

▶ An order of 10^4 parts is received. The probability that a part is defective equals 0.1. What is the probability that the total number of defective parts does not exceed 1100?

The experiment S is the selection of a single part. The probability of the event $A = \{\text{the part is defective}\}$ equals 0.1. We want the probability that in 10^4 trials, A will occur at most 1100 times. With

$$p = 0.1 \quad n = 10^4 \quad k_1 = 0 \quad k_2 = 1100$$

(3-24) yields

$$P\{0 \le k \le 1100\} = \sum_{k=0}^{1100} \binom{10^4}{k} (0.1)^k (0.9)^{10^4 - k} \tag{3-25}$$

◀

From (3-23)

$$\lim_{n\to\infty} \frac{k_m}{n} = p \tag{3-26}$$

so that as $n \to \infty$, the ratio of the most probable number of successes (A) to the total number of trials in a Bernoulli experiment tends to p, the probability of occurrence of A in a single trial. Notice that (3-26) connects the results of an actual experiment (k_m/n) to the axiomatic definition of p. In this context, as we show below it is possible to obtain a more general result.

3-3 BERNOULLI'S THEOREM AND GAMES OF CHANCE

In this section we shall state and prove one of the most important and beautiful theorems in the theory of probability, discovered and rigorously proved by Jacob Bernoulli (1713). To emphasize its significance to problems of practical importance we shall briefly examine certain games of chance.

THEOREM 3-1

BERNOULLI'S THEOREM

▶ Let A denote an event whose probability of occurrence in a single trial is p. If k denotes the number of occurrences of A in n independent trials, then

$$P\left(\left|\frac{k}{n} - p\right| > \epsilon\right) < \frac{pq}{n\epsilon^2} \tag{3-27}$$

Equation (3-27) states that the frequency definition of probability of an event k/n and its axiomatic definition p can be made compatible to any degree of accuracy with probability 1 or with almost certainty. In other words, given two positive numbers ϵ and δ, the probability of the inequality

$$\left|\frac{k}{n} - p\right| < \epsilon \tag{3-28}$$

will be greater than $1 - \delta$, provided the number of trials is above a certain limit.

Proof. We shall outline a simple proof of Bernoulli's theorem, by Chebyshev (1821–1894), that makes use of certain identities. Note that with $p_n(k)$ as in (3-13), direct computation gives

$$\begin{aligned}\sum_{k=0}^{n} kp_n(k) &= \sum_{k=1}^{n} k\frac{n!}{(n-k)!k!}p^k q^{n-k} = \sum_{k=1}^{n} \frac{n!}{(n-k)!(k-1)!}p^k q^{n-k} \\ &= \sum_{i=0}^{n-1} \frac{n!}{(n-i-1)!i!}p^{i+1} q^{n-i-1} = np\sum_{i=0}^{n-1} \frac{(n-1)!}{(n-i-1)!i!}p^i q^{n-i-1} \\ &= np(p+q)^{n-1} = np \end{aligned} \tag{3-29}$$

Proceeding in a similar manner, it can be shown that

$$\begin{aligned}\sum_{k=0}^{n} k^2 p_n(k) &= \sum_{k=1}^{n} k\frac{n!}{(n-k)!(k-1)!}p^k q^{n-k} \\ &= \sum_{k=2}^{n} \frac{n!}{(n-k)!(k-2)!}p^k q^{n-k} + \sum_{k=1}^{n} \frac{n!}{(n-k)!(k-1)!}p^k q^{n-k} \\ &= n^2p^2 + npq \end{aligned} \tag{3-30}$$

Returning to (3-27), note that

$$\left|\frac{k}{n} - p\right| > \epsilon \text{ is equivalent to } (k-np)^2 > n^2\epsilon^2 \tag{3-31}$$

which in turn is equivalent to

$$\sum_{k=0}^{n} (k-np)^2 p_n(k) > \sum_{k=0}^{n} n^2\epsilon^2 p_n(k) = n^2\epsilon^2 \tag{3-32}$$

Using (3-29) and (3-30), the left side of (3-32) can be expanded to give

$$\begin{aligned}\sum_{k=0}^{n} (k-np)^2 p_n(k) &= \sum_{k=0}^{n} k^2 p_n(k) - 2np\sum_{k=0}^{n} kp_n(k) + n^2p^2 \\ &= n^2p^2 + npq - 2np\cdot np + n^2p^2 = npq \end{aligned} \tag{3-33}$$

Alternatively, the left side of (3-32) can be expressed as

$$\begin{aligned}\sum_{k=0}^{n} (k-np)^2 p_n(k) &= \sum_{|k-np|\le n\epsilon} (k-np)^2 p_n(k) + \sum_{|k-np|>n\epsilon} (k-np)^2 p_n(k) \\ &\ge \sum_{|k-np|>n\epsilon} (k-np)^2 p_n(k) > n^2\epsilon^2 \sum_{|k-np|>n\epsilon} p_n(k) \\ &= n^2\epsilon^2 P\{|k-np| > n\epsilon\} \end{aligned} \tag{3-34}$$

Using (3-33) in (3-34), we get the desired result

$$P\left(\left|\frac{k}{n} - p\right| > \epsilon\right) < \frac{pq}{n\epsilon^2} \tag{3-35}$$

For a given $\epsilon > 0$, $pq/n\epsilon^2$ can be made arbitrarily small by letting n become large. Thus for very large n, we can make the fractional occurrence (relative frequency) k/n of the event A as close to the actual probability p of the event A in a single trial. Thus the theorem states that the probability of event A from the axiomatic framework can be computed from the relative frequency definition quite accurately, provided the number of experiments is large enough. Since $k_{\max}$ is the most likely value of k in n trials, from this discussion, as $n \to \infty$, the plots of $p_n(k)$ tend to concentrate more and more around $k_{\max}$ in (3-23). ◀

Thus Bernoulli's theorem states that with probability approaching 1 or with certainty, we can expect that in a sufficiently long series of independent trials with constant probability, the relative frequency of an event will differ from that probability by less than any specified number, no matter how small. Such an event (with probability approaching 1), although not bound to happen, has a probability of occurring so close to 1 that it may be considered to be a certain event. The immense practical value of Bernoulli's theorem lies in pointing out this advantage in real-life problems where the conditions of the theorem are satisfied.

One case where the conditions of Bernoulli's theorem are satisfied is that of gambling and casino operations. Situations facing the insurance companies are not far from this either. In gambling, one may gain or lose wealth depending on chance. In each game the probabilities of winning and losing are predetermined, and if one continues to play, the interesting question of course concerns the probability of gaining or losing money.

Suppose a player gains an amount a if he wins the game and loses another amount b if he loses the game. Let p and q represent the probability of winning and losing a game. In n games if the player wins k of them, then his net gain is

$$G = ka - (n-k)b \tag{3-36}$$

If n is large, according to Bernoulli's theorem k/n is very close to p, so that the difference or discrepancy $(k - np)$ must be very small. Let us denote this discrepancy value by Δ. Thus

$$\Delta = k - np$$

and by Bernoulli's theorem the probability of $\Delta > -n\epsilon$, where ϵ is any arbitrary positive number, approaches 1 provided n is sufficiently large. Using the discrepancy Δ, the net gain can be rewritten as

$$G = n(pa - qb) + (a+b)\Delta = n\eta + (a+b)\Delta \tag{3-37}$$

where the quantity

$$\eta = pa - qb \tag{3-38}$$

represents the "average gain" in any one game. The average gain η can be positive, zero, or negative. As we shall see, the ultimate fate of the player (gain or loss) depends on the sign of this quantity. Suppose $\eta > 0$ and n is sufficiently large. Then by Bernoulli's theorem the net gain G satisfies the inequality

$$G = n\eta + (a+b)\Delta > n[\eta - \epsilon(a+b)]$$

with probability approaching 1. Thus the net gain will exceed the number $Q = n(\eta - \epsilon(a+b))$, which itself is larger than any specified positive number, if n is sufficiently large (this assumes that ϵ is sufficiently small enough so that $\eta - \epsilon(a+b) > 0$). The conclusion is remarkable: A player whose average gain is positive stands to gain an arbitrarily large amount with probability approaching 1, if he continues to play a sufficiently large number of games.

It immediately follows that if the average gain η is negative, the player is bound to lose a large amount of money with almost certainty in the long run. If $\eta = 0$, then either a huge gain or loss is highly unlikely.

Thus the game favors players with positive average gain over those with negative average gain. All gambling institutions operate on this principle. The average gain of the institution is adjusted to be positive at every game, and consequently the average gain of any gambler turns out to be negative. This agrees with the everyday reality that gambling institutions derive enormous profits at the expense of regular gamblers, who are almost inevitably ruined in the long run.

We shall illustrate this using the profitable business of operating lotteries.

EXAMPLE 3-14

STATE LOTTERY

▶ In the New York State lottery, the player picks 6 numbers from a sequence of 1 through 51. At a lottery drawing, 6 balls are drawn at random from a box containing 51 balls numbered 1 through 51. What is the probability that a player has k matches, $k = 4, 5, 6$?

SOLUTION

Let n represent the total number of balls in the box among which for any player there are m "good ones" (those chosen by the player!). The remaining $(n - m)$ balls are "bad ones." There are in total $\binom{n}{m}$ samples of size m each with equal probability of occurrence. To determine the probability of the event "k matches," we need to determine the number of samples containing exactly k "good" balls (and hence $m - k$ "bad" ones). Since the k good balls must be chosen from m and the $(m - k)$ bad ones from $n - m$, the total number of such samples is

$$\binom{m}{k}\binom{n-m}{m-k}$$

This gives

$$P(k \text{ matches}) = \frac{\binom{m}{k}\binom{n-m}{m-k}}{\binom{n}{m}} \qquad k = 0, 1, 2, \ldots, m \tag{3-39}$$

In particular, with $k = m$, we get a perfect match, and a win. Thus

$$P(\text{winning the lottery}) = \frac{1}{\binom{n}{m}} = \frac{m \cdot (m-1) \cdots 2 \cdot 1}{n(n-1) \cdots (n-m+1)} \tag{3-40}$$

In the New York State lottery, $n = 51$, $m = 6$, so that

$$\begin{aligned} P(\text{winning the lottery}) &= \frac{6 \cdot 5 \cdot 4 \cdot 3 \cdot 2 \cdot 1}{51 \cdot 50 \cdot 49 \cdot 48 \cdot 47 \cdot 46} \\ &= \frac{1}{18{,}009{,}460} \simeq 5.5 \times 10^{-8} \end{aligned} \tag{3-41}$$

Thus the odds for winning the lottery are

$$1 : 18{,}009{,}460. \tag{3-42}$$

Using $k = 5$ and 4 in (3-39), we get the odds for 5 matches and 4 matches in the New York lottery to be 1 : 66,701 and 1 : 1213, respectively.

In a typical game suppose the state lottery pays \$4 million to the winner and \$15,000 for 5 matches and \$200 for 4 matches. Since the ticket costs \$1, this gives the average gain for the player to be

$$\eta_6 = \frac{4{,}000{,}000}{18{,}009{,}460} - 1 \simeq -0.778,$$

$$\eta_5 = \frac{15{,}000}{66{,}701} - 1 \simeq -0.775$$

and

$$\eta_4 = \frac{200}{1213} - 1 \simeq -0.835$$

for winning 5 matches and 4 matches, respectively. Notice that the average gain for the player is always negative. On the other hand, the average gain for the lottery institution is always positive, and because of the large number of participants involved in the lottery, the state stands to gain a very large amount in each game. ◀

The inference from Bernoulli's theorem is that when a large number of games are played under identical conditions between two parties, the one with a positive average gain in a single game stands to gain a fortune, and at the same time the one with negative average gain will almost certainly be ruined. These conclusions assume that the games are played indefinitely to take advantage of Bernoulli's theorem, and the actual account settlement is done only at the very end. Interestingly, the stock market situation does allow the possibility of long-time play without the need to settle accounts intermittently. Hence if one holds onto stocks with positive average gains, in the long run that should turn out to be a much more profitable strategy compared to day-to-day trading[2] (which is equivalent to gambling). The *key* is not to engage in games that call for account settlement quite frequently. In regular gambling, however, payment adjustment is made at the end of each game, and it is quite possible that one may lose all his capital and will have to quit playing long before reaping the advantage that a large number of games would have brought to him.

In this context, next we examine a classic problem involving the ruin of gamblers. Since probability theory had its humble origin in computing chances of players in different games, the important question of the ruin of gamblers was discussed at a very early stage in the historical development of the theory of probability. The gambler's ruin problem has a long history and extensive literature is available on this topic. The simplest problem of its kind was first solved by C. Huygens (1657), followed by J. Bernoulli (1680), and the general case was proved by A. De Moivre in 1711. More important, over the years it has played a significant role as a source of theorems and has contributed to various generalizations including the subject of random walks (see Chapter 10). The

[2]Among others, this strategy worked very well for the late Prof. Donald Othmer of Polytechnic, who together with his wife Mildred had initially invested \$25,000 each in the early 1960s with the legendary investor, Warren Buffett who runs the Berkshire Hathaway company. In 1998, the New York Times reported that the Othmer's net assets in the Berkshire Hathaway stock fund were around \$800,000,000.

underlying principles are used also by casinos, state lotteries, and more respectable institutions such as insurance companies in deciding their operational strategies.

EXAMPLE 3-15

GAMBLER'S RUIN PROBLEM

▶ Two players A and B play a game consecutively till one of them loses all his capital. Suppose A starts with a capital of $\$a$ and B with a capital of $\$b$ and the loser pays \$1 to the winner in each game. Let p represent the probability of winning each game for A and $q = 1 - p$ for player B. Find the probability of ruin for each player if no limit is set for the number of games.[3]

SOLUTION

Let P_n denote the probability of the event X_n = "A's ultimate ruin when his wealth is $\$n$" $(0 \le n \le a+b)$. His ruin can occur in only two mutually exclusive ways: either A can win the next game with probability p and his wealth increases to $\$(n+1)$ so that the probability of being ruined ultimately equals P_{n+1}, or A can lose the next game with probability q and reduce his wealth to $\$(n-1)$, in which case the probability of being ruined later is P_{n-1}. More explicitly, with H = "A succeeds in the next game," by the theorem of total probability we obtain the equation

$$X_n = X_n(H \cup \overline{H}) = X_nH \cup X_n\overline{H}$$

and hence

$$\begin{aligned} P_n = P(X_n) &= P(X_n \mid H)P(H) + P(X_n \mid \overline{H})P(\overline{H}) \\ &= pP_{n+1} + qP_{n-1} \end{aligned} \tag{3-43}$$

with initial conditions

$$P_0 = 1 \qquad P_{a+b} = 0 \tag{3-44}$$

The first initial condition states that A is certainly ruined if he has no money left, and the second one states that if his wealth is $(a+b)$ then B has no money left to play, and the ruin of A is impossible.

To solve the difference equation in (3-43), we first rewrite it as

$$p(P_{n+1} - P_n) = q(P_n - P_{n-1}) \tag{3-45}$$

or

$$P_{n+1} - P_n = \frac{q}{p}(P_n - P_{n-1}) = \left(\frac{q}{p}\right)^n (P_1 - 1)$$

where we have made use of the first initial condition. To exploit the remaining initial condition, consider $P_{a+b} - P_n$. Clearly, for $p \ne q$

$$\begin{aligned} P_{a+b} - P_n &= \sum_{k=n}^{a+b-1} P_{k+1} - P_k = \sum_{k=n}^{a+b-1} \left(\frac{q}{p}\right)^k (P_1 - 1) \\ &= (P_1 - 1)\frac{\left(\frac{q}{p}\right)^n - \left(\frac{q}{p}\right)^{a+b}}{1 - \frac{q}{p}} \end{aligned}$$

[3] Hugyens dealt with the particular case where $a = b = 12$ and $p/q = 5/4$.

Since $P_{a+b} = 0$, it follows that

$$P_n = (1 - P_1)\frac{\left(\frac{q}{p}\right)^n - \left(\frac{q}{p}\right)^{a+b}}{1 - \frac{q}{p}}$$

and since $P_0 = 1$, this expression also gives

$$P_0 = 1 = (1 - P_1)\frac{\left(\frac{q}{p}\right)^0 - \left(\frac{q}{p}\right)^{a+b}}{1 - \frac{q}{p}}$$

Eliminating $(1 - P_1)$ from the last two equations, we get

$$P_n = \frac{\left(\frac{q}{p}\right)^n - \left(\frac{q}{p}\right)^{a+b}}{1 - \left(\frac{q}{p}\right)^{a+b}} \tag{3-46}$$

Substituting $n = a$ into (3-46), we obtain the probability of ruin for player A when his wealth is $\$a$ to be (for $p \neq q$)

$$P_a = \frac{1 - \left(\frac{p}{q}\right)^b}{1 - \left(\frac{p}{q}\right)^{a+b}} \tag{3-47}$$

Proceeding in a similar manner (or interchange p and q as well as a and b) we get the probability of ultimate ruin for player B (when his wealth is $\$b$) to be for $p \neq q$

$$Q_b = \frac{1 - \left(\frac{q}{p}\right)^a}{1 - \left(\frac{q}{p}\right)^{a+b}} \tag{3-48}$$

By direct addition, we also get

$$P_a + Q_b = 1 \tag{3-49}$$

so that the probability that the series of games will continue indefinitely without A or B being ruined is zero. Note that the zero probability does not imply the impossibility of an eternal game. Although an eternal game is not excluded theoretically, for all practical purposes it can be disregarded. From (3-47), $1 - P_a$ represents the probability of A winning the game and from (3-49) it equals his opponents probability of ruin.

Consider the special case where the players are of equal skill. In that case $p = q = 1/2$, and (3-47) and (3-48) simplify to

$$P_a = \frac{b}{a + b} \tag{3-50}$$

and

$$Q_b = \frac{a}{a + b} \tag{3-51}$$

Equations (3-50) and (3-51) state that if both players are of equal skill, then their probabilities of ruin are inversely proportional to the wealth of the players. Thus it is unwise to play indefinitely even against some one of equal skill whose fortune is very large, since the risk of losing all money is practically certain in the long run ($P_a \to 1$, if $b \gg a$). Needless to say if the adversary is also skillful ($q > p$) and wealthy, then as (3-47) shows, A's ruin is certain in the long run ($P_a \to 1$, as $b \to \infty$). All casino games against the house amount to this situation, and a sensible strategy in such cases would be to quit while ahead.

What if odds are in your favor? In that case $p > q$, so that $q/p < 1$, and (3-47) can be rewritten as

$$P_a = \left(\frac{q}{p}\right)^a \frac{1 - \left(\frac{q}{p}\right)^b}{1 - \left(\frac{q}{p}\right)^{a+b}} < \left(\frac{q}{p}\right)^a$$

and P_a converges to $(q/p)^a$ as $b \to \infty$. Thus, while playing a series of advantageous games even against an infinitely rich adversary, the probability of escaping ruin (or gaining wealth) is

$$1 - P_a = 1 - \left(\frac{q}{p}\right)^a \tag{3-52}$$

If a is large enough, this probability can be made as close to 1 as possible. Thus a skillful player who also happens to be reasonably rich, will never be ruined in the course of games, and in fact he will end up even richer in the long run. (Of course, one has to live long enough for all this to happen!)

Casinos and state lotteries work on this principle. They always keep a slight advantage to themselves ($q > p$), and since they also possess large capitals, from (3-48) their ruin is practically impossible ($Q_b \to 0$). This conclusion is also confirmed by experience. It is hard to find a casino that has gone "out of business or doing rather poorly." Interestingly, the same principles underlie the operations of more respectable institutions of great social and public value such as insurance companies. We shall say more about their operational strategies in a later example (see Example 4-29, page 114).

If one must gamble, interestingly (3-47) suggests the following strategy: Suppose a gambler A with initial capital \$$a$ is playing against an adversary who is always willing to play (such as the house), and A has the option of stopping at any time. If A adopts the strategy of playing until either he loses all his capital or increase it to \$$(a + b)$ (with a net gain of \$$b$), then P_a represents his probability of losing and $1 - P_a$ represents his probability of winning. Moreover, the average duration of such a game is given by (see Problem 3-7)

$$N_a = \begin{cases} \dfrac{b}{2p-1} - \dfrac{a+b}{2p-1} \dfrac{1 - \left(\frac{p}{q}\right)^b}{1 - \left(\frac{p}{q}\right)^{a+b}} & p \neq q \\[2ex] ab & p = q = \dfrac{1}{2} \end{cases} \tag{3-53}$$

Table 3-1 illustrates the probability of ruin and average duration for some typical values of a, b, and p.

CHANGING STAKES. Let us now analyze the effect of changing stakes in this situation. Suppose the amount is changed from \$1 to \$$k$ for each play. Notice that its effect is the

TABLE 3-1
Gambler's ruin

p	q	Capital, a	Gain, b	Probability of Ruin, P_a	Probability of Success, $1 - P_a$	Average duration, N_a
0.50	0.50	9	1	0.100	0.900	9
0.50	0.50	90	10	0.100	0.900	900
0.50	0.50	90	5	0.053	0.947	450
0.50	0.50	500	100	0.167	0.833	50,000
0.45	0.55	9	1	0.210	0.790	11
0.45	0.55	50	10	0.866	0.134	419
0.45	0.55	90	5	0.633	0.367	552
0.45	0.55	90	10	0.866	0.134	765
0.45	0.55	100	5	0.633	0.367	615
0.45	0.55	100	10	0.866	0.134	852

same as reducing the capital of each player by a factor of k. Thus the new probability of ruin P_a^* for A, where \$$k$ are staked at each play, is given by (3-47) with a replaced by a/k and b by b/k.

$$P_a^* = \frac{1 - \left(\frac{p}{q}\right)^{b/k}}{1 - \left(\frac{p}{q}\right)^{(a+b)/k}} \tag{3-54}$$

Let $a_0 = a/k$, $b_0 = b/k$, $x = (p/q)^{b_0}$, and $y = (p/q)^{a_0+b_0}$. Then

$$\begin{aligned} P_a &= \frac{1 - x^k}{1 - y^k} = \frac{1 - x}{1 - y} \cdot \frac{1 + x + \cdots + x^{k-1}}{1 + y + \cdots + y^{k-1}} \\ &= P_a^* \frac{1 + x + \cdots + x^{k-1}}{1 + y + \cdots + y^{k-1}} > P_a^* \quad \text{for } p < q, \end{aligned} \tag{3-55}$$

since $x > y$ for $p < q$. Equation (3-55) states that if the stakes are increased while the initial capital remains unchanged, for the disadvantageous player (whose probability of success $p < 1/2$) the probability of ruin decreases and it increases for the adversary (for whom the original game was more advantageous). From Table 3-1, for $a = 90$, $b = 10$, with $p = 0.45$, the probability of ruin for A is founed to be 0.866 for a \$1 stake game. However, if the same game is played for \$10 stakes, the probability of ruin drops down to 0.21. In an unfavorable game of constant stakes, the probability of ruin can be reduced by selecting the stakes to be higher. Thus if the goal is to win \$$b$ starting with capital \$$a$, then the ratio capital/stake must be adjusted in an unfavorable game to fix the overall probability of ruin at the desired level. ◀

Example 3-16 shows that the game of craps is perhaps the most favorable game among those without any strategy (games of chance). The important question in that case is how long one should play to maximize the returns. Interestingly as Example 3-17 shows even that question has an optimum solution.

EXAMPLE 3-16

GAME OF CRAPS

▶ A pair of dice is rolled on every play and the player wins at once if the total for the first throw is 7 or 11, loses at once if 2, 3, or 12 are rolled. Any other throw is called a "carry-over." If the first throw is a carry-over, then the player throws the dice repeatedly until he wins by throwing the same carry-over again, or loses by throwing 7. What is the probability of winning the game?

SOLUTION

A pair of dice when rolled gives rise to 36 equally likely outcomes (refer to Example 3-8). Their combined total T can be any integer from 2 to 12, and for each such outcome the associated probability is shown below in Table 3-2.

The game can be won by throwing a 7 or 11 on the first throw, or by throwing the carry-over on a later throw. Let P_1 and P_2 denote the probabilities of these two mutually exclusive events. Thus the probability of a win on the first throw is given by

$$P_1 = P(T = 7) + P(T = 11) = \frac{6}{36} + \frac{2}{36} = \frac{2}{9} \tag{3-56}$$

Similarly, the probability of loss on the first throw is

$$Q_1 = P(T = 2) + P(T = 3) + P(T = 12) = \frac{1}{36} + \frac{2}{36} + \frac{1}{36} = \frac{1}{9} \tag{3-57}$$

To compute the probability P_2 of winning by throwing a carry-over, we first note that 4, 5, 6, 8, 9, and 10 are the only carry-overs with associated probabilities of occurrence as in Table 3-2. Let B denote the event "winning the game by throwing the carry-over" and let C denote a carry-over. Then using the theorem of total probability

$$P_2 = P(B) = \sum_{k=4, k\neq 7}^{10} P(B \mid C = k)P(C = k) = \sum_{k=4, k\neq 7}^{10} P(B \mid C = k)p_k \tag{3-58}$$

To compute $a_k = P(B \mid C = k)$ note that the player can win by throwing a number of plays that do not count with probability $r_k = 1 - p_k - 1/6$, and then by throwing the carry-over with probability p_k. (The 1/6 in $r_k = 1 - p_k - 1/6$ is the probability of losing by throwing 7 in later plays.) The probability that the player throws the carry-over k on the jth throw (and $j - 1$ do-not-count throws earlier) is $p_k r_k^{j-1}$, $j = 1, 2, 3, \ldots, \infty$. Hence

$$a_k = P(B \mid C = k) = p_k \sum_{j=1}^{\infty} r_k^{j-1} = \frac{p_k}{1 - r_k} = \frac{p_k}{p_k + 1/6}, \tag{3-59}$$

which gives

k	4	5	6	8	9	10
a_k	$\frac{1}{3}$	$\frac{2}{5}$	$\frac{5}{11}$	$\frac{5}{11}$	$\frac{2}{5}$	$\frac{1}{3}$

TABLE 3-2

Total $T = k$	2	3	4	5	6	7	8	9	10	11	12
$p_k = \text{Prob}(T = k)$	$\frac{1}{36}$	$\frac{2}{36}$	$\frac{3}{36}$	$\frac{4}{36}$	$\frac{5}{36}$	$\frac{6}{36}$	$\frac{5}{36}$	$\frac{4}{36}$	$\frac{3}{36}$	$\frac{2}{36}$	$\frac{1}{36}$

Using (3-59) and Table 3-2 in (3-58) we get

$$P_2 = \sum_{k=4,k\neq 7}^{10} a_k p_k = \frac{1}{3}\cdot\frac{3}{36} + \frac{2}{5}\cdot\frac{4}{36} + \frac{5}{11}\cdot\frac{5}{36} + \frac{5}{11}\cdot\frac{5}{36} + \frac{2}{5}\cdot\frac{4}{36} + \frac{1}{3}\cdot\frac{3}{36} = \frac{134}{495} \tag{3-60}$$

Finally, (3-56) and (3-60) give

$$P(\text{winning the game}) = P_1 + P_2 = \frac{2}{9} + \frac{134}{495} = \frac{244}{495} \simeq 0.492929 \tag{3-61}$$

Notice that the game is surprisingly close to even, but as expected slightly to the advantage of the house! ◀

Example 3-17 shows that in games like craps, where the player is only at a slight disadvantage compared to the house, surprisingly it is possible to devise a strategy that works to the player's advantage by restricting the total number of plays to a certain optimum number.

EXAMPLE 3-17

STRATEGY FOR AN UNFAIR GAME[4]

▶ A and B plays a series of games where the probability of winning p in a single play for A is unfairly kept at less than $1/2$. However, A gets to choose in advance the total number of plays. To win the whole game one must score more than half the plays. If the total number of plays is to be even, how many plays should A choose?

SOLUTION

On any play A wins with probability p and B wins with probability $q = 1 - p > p$. Notice that the expected gain for A on any play is $p - q < 0$. At first it appears that since the game is unfairly biased toward A, the best strategy for A is to quit the game as early as possible. If A must play an even number, then perhaps quit after two plays? Indeed if p is extremely small that is the correct strategy. However, if $p = 1/2$, then as $2n$, the total number of plays increases the probability of a tie (the middle binomial term) decreases and the limiting value of A's chances to win tends to $1/2$. In that case, the more plays, the better are the chances for A to succeed. Hence if p is somewhat less that $1/2$, it is reasonable to expect a finite number of plays as the optimum strategy.

To examine this further, let X_k denote the event "A wins k games in a series of $2n$ plays." Then

$$P(X_k) = \binom{2n}{k} p^k q^{2n-k} \qquad k = 0, 1, 2, \ldots, 2n$$

and let P_{2n} denote the probability that A wins in $2n$ games. Then

$$P_{2n} = P\left(\bigcup_{k=n+1}^{2n} X_k\right) = \sum_{k=n+1}^{2n} P(X_k) = \sum_{k=n+1}^{2n} \binom{2n}{k} p^k q^{2n-k} \tag{3-62}$$

where we have used the mutually exclusive nature of the X_is.

[4] "Optimal length of play for a binomial game," *Mathematics Teacher,* Vol. 54, pp. 411–412, 1961.

If $2n$ is indeed the optimum number of plays, then we must have

$$P_{2n-2} \le P_{2n} \ge P_{2n+2} \tag{3-63}$$

where P_{2n+2} denotes the probability that A wins in $2n + 2$ plays. Thus

$$P_{2n+2} = \sum_{k=n+2}^{2n+2} \binom{2n+2}{k} p^k q^{2n+2-k} \tag{3-64}$$

To obtain a relation between the right side expressions in (3-63) and (3-64) we can make use of the binomial expansion

$$\sum_{k=0}^{2n+2} \binom{2n+2}{k} p^k q^{2n+2-k} = (p+q)^{2n+2} = (p+q)^{2n}(p+q)^2$$

$$= \left\{ \sum_{k=0}^{2n} \binom{2n}{k} p^k q^{2n-k} \right\} (p^2 + 2pq + q^2) \tag{3-65}$$

Notice that the later half of the left side expression in (3-65) represents P_{2n+2}. Similarly, the later half of the first term on the right side represents P_{2n}. Equating like powers of terms $p^{n+2}q^n, p^{n+3}q^{n-1}, \ldots, p^{2n+2}$ on both sides of (3-65), after some simple algebra we get the identity

$$P_{2n+2} = P_{2n} + \binom{2n}{n} p^{n+2} q^n - \binom{2n}{n+1} p^{n+1} q^{n+1} \tag{3-66}$$

Equation (3-66) has an interesting interpretation. From (3-66), events involved in winning a game of $2n+2$ plays or winning a game of $2n$ plays differ in only two cases: (i) Having won n games in the first $2n$ plays with probability $\binom{2n}{n} p^n q^n$, A wins the next two plays with probability p^2, thereby increasing the winning probability by $\binom{2n}{n} p^{n+2} q^n$; (ii) Having won $n + 1$ plays in the first $2n$ plays with probability $\binom{2n}{n+1} p^{n+1} q^{n-1}$, A loses the next two plays with probability q^2, thereby decreasing the winning probability by $\binom{2n}{n+1} p^{n+1} q^{n-1} q^2$. Except for these two possibilities, in all other respects they are identical.

If $2n$ is optimum, the right side inequality in (3-63) when applied to (3-66) gives

$$\binom{2n}{n+1} p^{n+1} q^{n+1} \ge \binom{2n}{n} p^{n+2} q^n \tag{3-67}$$

or

$$nq \ge (n+1)p \qquad n(q-p) \ge p \qquad n \ge \frac{p}{1-2p} \tag{3-68}$$

Similarly, the left side inequality in (3-63) gives (replace n by $n - 1$ in (3-66))

$$\binom{2n-2}{n-1} p^{n+1} q^{n-1} \ge \binom{2n-2}{n} p^n q^n \tag{3-69}$$

or

$$np \ge (n-1)q \qquad n(q-p) \le q \qquad n \le \frac{q}{1-2p} \tag{3-70}$$

From (3-68) and (3-70) we get

$$\frac{1}{1-2p} - 1 \le 2n \le \frac{1}{1-2p} + 1 \tag{3-71}$$

which determines $2n$ uniquely as the even integer that is nearest to $1/(1-2p)$. Thus for example, if $p = 0.47$, then $2n = 16$. However, if $1/(1-2p)$ is an odd integer ($p = 0.48$), both adjacent even integers $2n = 1/(1-2p) - 1$ and $2n + 2 = 1/(1-2p) + 1$ give the same probability (show this). Finally if $p \simeq 0$, then (3-71) gives the optimum number of plays to be 2. ◀

Returning to Example 3-16 (game of craps), p was found to be 0.492929 there, which gives $2n = 70$ to be the optimum number of plays. Most people make the mistake of quitting the game long before 70 plays, one reason being the slow progress of the game. (Recall that each play may require many throws because of the do-not-count throws.) However, here is one game where the strategy should be to execute a certain number of plays.

Interestingly, the results from Examples 3-15 and 3-16 can be used to design an optimum strategy for the game of craps involving the amounts of capital, expected return, stakes and probability of success. Table 3-3 lists the probability of success and average duration for some typical values of capital a and gain b. Here P_a represents the probability of ruin computed using (3-47) with $p = 0.492929$, and N_a represents the corresponding expected number of games given by (3-53). Notice that a and b have been chosen here so that the expected number of games is around its optimum value of 70. Thus starting with \$10, in a \$1 stake game of craps the probability of gaining \$7 is 0.529 in about 70 games. Clearly if the capital is increased to \$100, then to maintain the same number of games and risk level, one should raise the stakes to \$10 for an expected gain of \$70. However, if a strategy with reduced risk is preferred, from Table 3-3 one may play the $a = 16$, $b = 4$ game (25% gain) that ensures 75% probability of success in about 67 games. It follows that for a \$100 investment, the stakes in that case should be set at \$6 per game for an expected gain of \$25.

TABLE 3-3
Strategy for a game of craps ($p = 0.492929$)

		Probability of		
Capital, a	**Gain, b**	**Ruin, P_a**	**Success, $1 - P_a$**	**Expected duration, N_a**
9	8	0.5306	0.4694	72.14
10	7	0.4707	0.5293	70.80
11	6	0.4090	0.5910	67.40
12	6	0.3913	0.6087	73.84
13	5	0.3307	0.6693	67.30
14	5	0.3173	0.6827	72.78
15	5	0.3054	0.6946	78.32
16	4	0.2477	0.7523	67.47
17	4	0.2390	0.7610	71.98

PROBLEMS

3-1 Let p represent the probability of an event A. What is the probability that (a) A occurs at least twice in n independent trials; (b) A occurs at least thrice in n independent trials?

3-2 A pair of dice is rolled 50 times. Find the probability of obtaining double six at least three times.

3-3 A pair of fair dice is rolled 10 times. Find the probability that "seven" will show at least once.

3-4 A coin with $p\{h\} = p = 1 - q$ is tossed n times. Show that the probability that the number of heads is even equals $0.5[1 + (q - p)^n]$.

3-5 (*Hypergeometric series*) A shipment contains K good and $N - K$ defective components. We pick at random $n \leq K$ components and test them. Show that the probability p that k of the tested components are good equals (compare with (3-39))

$$p = \binom{K}{k}\binom{N-K}{n-k} \Big/ \binom{N}{n}$$

3-6 Consider the following three events: (a) At least 1 six is obtained when six dice are rolled, (b) at least 2 sixes are obtained when 12 dice are rolled, and (c) at least 3 sixes are obtained when 18 dice are rolled. Which of these events is more likely?

3-7 A player wins \$1 if he throws two heads in succession, otherwise he loses two quarters. If the game is repeated 50 times, what is the probability that the net gain or less exceeds (a) \$1? (b) \$5?

3-8 Suppose there are r successes in n independent Bernoulli trials. Find the conditional probability of a success on the ith trial.

3-9 A standard pack of cards has 52 cards, 13 in each of 4 suits. Suppose 4 players are dealt 13 cards each from a well shuffled pack. What is the probability of dealing a perfect hand (13 of any one suit)?

3-10 Refer to Example 3-15 (Gambler's ruin problem). Let N_a denote the average duration of the game for player A starting with capital a. Show that

$$N_a = \begin{cases} \dfrac{b}{2p-1} - \dfrac{a+b}{2p-1}\dfrac{1-\left(\frac{p}{q}\right)^b}{1-\left(\frac{p}{q}\right)^{a+b}} & p \neq q \\[2ex] ab & p = q = \dfrac{1}{2} \end{cases}$$

(*Hint:* Show that N_k satisfies the iteration $N_k = 1 + pN_{k+1} + qN_{k-1}$ under the initial conditions $N_0 = N_{a+b} = 0$.)

3-11 Refer to Example 3-15. Suppose the stakes of A and B are α and β, and respective capitals are a and b, as before. Find the probabilities for A or B to be ruined.

3-12 Three dice are rolled and the player may bet on any one of the face values 1, 2, 3, 4, 5, and 6. If the player's number appears on one, two, or all three dice, the player receives respectively one, two, or three times his original stake plus his own money back. Determine the expected loss per unit stake for the player.

CHAPTER

4

THE CONCEPT OF A RANDOM VARIABLE

4-1 INTRODUCTION

A random variable is a number $\mathbf{x}(\zeta)$ assigned to every outcome ζ of an experiment. This number could be the gain in a game of chance, the voltage of a random source, the cost of a random component, or any other numerical quantity that is of interest in the performance of the experiment.

EXAMPLE 4-1

▶ (*a*) In the die experiment, we assign to the six outcomes f_i the numbers $\mathbf{x}(f_i) = 10i$. Thus

$$\mathbf{x}(f_1) = 10, \ldots, \mathbf{x}(f_6) = 60$$

(*b*) In the same experiment, instead we can assign the number 1 to every even outcome and the number 0 to every odd outcome. Thus

$$\mathbf{x}(f_1) = \mathbf{x}(f_3) = \mathbf{x}(f_5) = 0 \qquad \mathbf{x}(f_2) = \mathbf{x}(f_4) = \mathbf{x}(f_6) = 1$$ ◀

THE MEANING OF A FUNCTION. A random variable is a function whose domain is the set S of all experimental outcomes. To clarify further this important concept, we review briefly the notion of a function. As we know, a function $x(t)$ is a rule of correspondence between values of t and x. The values of the independent variable t form a set S_t on the t axis called the *domain* of the function and the values of the dependent variable x form a set S_x on the x axis called the *range* of the function. The rule of correspondence between t and x could be a curve, a table, or a formula, for example, $x(t) = t^2$.

The notation $x(t)$ used to represent a function is ambiguous: It might mean either the particular number $x(t)$ corresponding to a specific t, or the function $x(t)$, namely, the rule of correspondence between any t in S_t and the corresponding x in S_x. To distinguish between these two interpretations, we shall denote the latter by x, leaving its dependence on t understood.

The definition of a function can be phrased as: We are given two sets of numbers S_t and S_x. To every $t \in S_t$ we assign a number $x(t)$ belonging to the set S_x. This leads to this generalization: We are given two sets of objects S_α and S_β consisting of the elements α and β, respectively. We say that β is a function of α if to every element of the set S_α we make correspond an element β of the set S_β. The set S_α is the domain of the function and the set S_β its range.

Suppose, for example, that S_α is the set of children in a community and S_β the set of their fathers. The pairing of a child with his or her father is a function.

We note that to a given α there corresponds a single $\beta(\alpha)$. However, more than one element from S_α might be paired with the same β (a child has only one father but a father might have more than one child). In Example 4-1b, the domain of the function consists of the six faces of the die. Its range, however, has only two elements, namely, the numbers 0 and 1.

The Random Variable

We are given an experiment specified by the space S (or Ω), the field of subsets of S called events, and the probability assigned to these events. To every outcome ζ of this experiment, we assign a number $\mathbf{x}(\zeta)$. We have thus created a function $\mathbf{x}$ with domain the set S and range a set of numbers. This function is called random variable if it satisfies certain mild conditions to be soon given.

All random variables will be written in boldface letters. The symbol $\mathbf{x}(\zeta)$ will indicate the number assigned to the specific outcome ζ and the symbol $\mathbf{x}$ will indicate the rule of correspondence between any element of S and the number assigned to it. Example 4-1a, $\mathbf{x}$ is the table pairing the six faces of the die with the six numbers $10, \ldots, 60$. The domain of this function is the set $S = \{f_1, \ldots, f_6\}$ and its range is the set of the above six numbers. The expression $\mathbf{x}(f_2)$ is the number 20.

EVENTS GENERATED BY RANDOM VARIABLES. In the study of random variables, questions of the following form arise: What is the probability that the random variable $\mathbf{x}$ is less than a given number x, or what is the probability that $\mathbf{x}$ is between the numbers x_1 and x_2? If, for example, the random variable is the height of a person, we might want the probability that it will not exceed certain bounds. As we know, probabilities are assigned only to events; therefore, in order to answer such questions, we should be able to express the various conditions imposed on $\mathbf{x}$ as events.

We start with the meaning of the notation

$$\{\mathbf{x} \le x\}$$

This notation represents a subset of S consisting of all outcomes ζ such that $\mathbf{x}(\zeta) \le x$. We elaborate on its meaning: Suppose that the random variable $\mathbf{x}$ is specified by a table. At the left column we list all elements ζ_i of S and at the right the corresponding values (numbers) $\mathbf{x}(\zeta_i)$ of $\mathbf{x}$. Given an arbitrary number x, we find all numbers $\mathbf{x}(\zeta_i)$ that do not exceed x. The corresponding elements ζ_i on the left column form the set $\{\mathbf{x} \le x\}$. Thus $\{\mathbf{x} \le x\}$ is not a set of numbers but *a set of experimental outcomes.*

The meaning of

$$\{x_1 \le \mathbf{x} \le x_2\}$$

is similar. It represents a subset of S consisting of all outcomes ζ such that $x_1 \le \mathbf{x}(\zeta) \le x_2$; where x_1 and x_2 are two given numbers.

The notation

$$\{\mathbf{x} = x\}$$

is a subset of S consisting of all outcomes ζ such that $\mathbf{x}(\zeta) = x$.

Finally, if R is a set of numbers on the x axis, then

$$\{\mathbf{x} \in R\}$$

represents the subset of S consisting of all outcomes ζ such that $\mathbf{x}(\zeta) \in R$.

EXAMPLE 4-2

▶ We shall illustrate the above with the random variable $\mathbf{x}(f_i) = 10i$ of the die experiment (Fig. 4-1).

The set $\{\mathbf{x} \le 35\}$ consists of the elements f_1, f_2, and f_3 because $\mathbf{x}(f_i) \le 35$ only if $i = 1, 2$, or 3.

The set $\{\mathbf{x} \le 5\}$ is empty because there is no outcome such that $\mathbf{x}(f_i) \le 5$.

The set $\{20 \le \mathbf{x} \le 35\}$ consists of the elements f_2 and f_3 because $20 \le \mathbf{x}(f_i) \le 35$ only if $i = 2$ or 3.

The set $\{\mathbf{x} = 40\}$ consists of the element f_4 because $\mathbf{x}(f_i) = 40$ only if $i = 4$.

Finally, $\{\mathbf{x} = 35\}$ is the empty set because there is no experimental outcome such that $\mathbf{x}(f_i) = 35$. ◀

We conclude with a formal definition of a random variable.

DEFINITION

▶ A random variable $\mathbf{x}$ is a process of assigning a number $\mathbf{x}(\zeta)$ to every outcome ζ. The resulting function must satisfy the following two conditions but is otherwise arbitrary:

I. The set $\{\mathbf{x} \le x\}$ is an event for every x.

II. The probabilities of the events $\{\mathbf{x} = \infty\}$ and $\{\mathbf{x} = -\infty\}$ equal 0:

$$P\{\mathbf{x} = \infty\} = 0 \qquad P\{\mathbf{x} = -\infty\} = 0$$ ◀

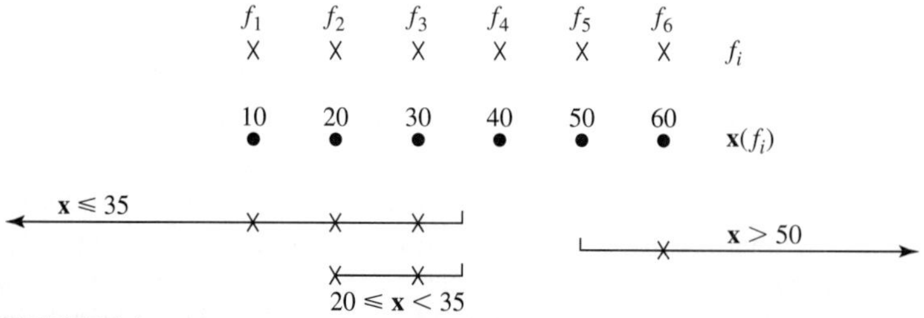

FIGURE 4-1

The second condition states that, although we allow **x** to be $+\infty$ or $-\infty$ for some outcomes, we demand that these outcomes form a set with zero probability.

A *complex* random variable **z** is a sum

$$\mathbf{z} = \mathbf{x} + j\mathbf{y}$$

where **x** and **y** are real random variables. Unless otherwise stated, it will be assumed that all random variables are real.

Note In the applications, we are interested in the probability that a random variable **x** takes values in a certain region R of the x axis. This requires that the set $\{\mathbf{x} \in R\}$ be an event. As we noted in Sec. 2-2, that is not always possible. However, if $\{\mathbf{x} \le x\}$ is an event for every x and R is a countable union and intersection of intervals, then $\{\mathbf{x} \in R\}$ is also an event. In the definition of random variables we shall assume, therefore, that the set $\{\mathbf{x} \le x\}$ is an event. This mild restriction is mainly of mathematical interest.

4-2 DISTRIBUTION AND DENSITY FUNCTIONS

The elements of the set S (or Ω) that are contained in the event $\{\mathbf{x} \le x\}$ change as the number x takes various values. The probability $P\{\mathbf{x} \le x\}$ of the event $\{\mathbf{x} \le x\}$ is, therefore, a number that depends on x. This number is denoted by $F_x(x)$ and is called the *(cumulative) distribution function* of the random variable **x**.

DEFINITION ▶ The distribution function of the random variable **x** is the function

$$F_x(x) = P\{\mathbf{x} \le x\} \tag{4-1}$$

defined for every x from $-\infty$ to ∞.

The distribution functions of the random variables **x**, **y**, and **z** are denoted by $F_x(x)$, $F_y(y)$, and $F_z(z)$, respectively. In this notation, the variables x, y, and z can be identified by any letter. We could, for example, use the notation $F_x(w)$, $F_y(w)$, and $F_z(w)$ to represent these functions. Specifically,

$$F_x(w) = P\{\mathbf{x} \le w\}$$

is the distribution function of the random variable **x**. However, if there is no fear of ambiguity, we shall identify the random variables under consideration by the independent variable in (4-1) omitting the subscripts. Thus the distribution functions of the random variables **x**, **y**, and **z** will be denoted by $F(x)$, $F(y)$, and $F(z)$, respectively. ◀

EXAMPLE 4-3 ▶ In the coin-tossing experiment, the probability of heads equals p and the probability of tails equals q. We define the random variable **x** such that

$$\mathbf{x}(h) = 1 \qquad \mathbf{x}(t) = 0$$

We shall find its distribution function $F(x)$ for every x from $-\infty$ to ∞.

If $x \ge 1$, then $\mathbf{x}(h) = 1 \le x$ and $\mathbf{x}(t) = 0 \le x$. Hence (Fig. 4-2)

$$F(x) = P\{\mathbf{x} \le x\} = P\{h, t\} = 1 \qquad x \ge 1$$

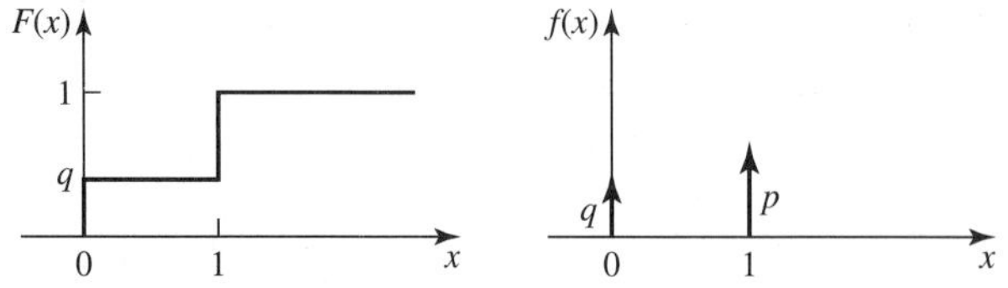

FIGURE 4-2

If $0 \le x < 1$, then $\mathbf{x}(h) = 1 > x$ and $\mathbf{x}(t) = 0 \le x$. Hence

$$F(x) = P\{\mathbf{x} \le x\} = P\{t\} = q \qquad 0 \le x < 1$$

If $x < 0$, then $\mathbf{x}(h) = 1 > x$ and $\mathbf{x}(t) = 0 > x$. Hence

$$F(x) = P\{\mathbf{x} \le x\} = P\{\varnothing\} = 0 \qquad x < 0$$

◀

EXAMPLE 4-4

▶ In the die experiment of Example 4-2, the random variable $\mathbf{x}$ is such that $\mathbf{x}(f_i) = 10i$. If the die is fair, then the distribution function of $\mathbf{x}$ is a staircase function as in Fig. 4-3.

We note, in particular, that

$$F(100) = P\{\mathbf{x} \le 100\} = P(S) = 1$$

$$F(35) = P\{\mathbf{x} \le 35\} = P\{f_1, f_2, f_3\} = \tfrac{3}{6}$$

$$F(30.01) = P\{\mathbf{x} \le 30.01\} = P\{f_1, f_2, f_3\} = \tfrac{3}{6}$$

$$F(30) = P\{\mathbf{x} \le 30\} = P\{f_1, f_2, f_3\} = \tfrac{3}{6}$$

$$F(29.99) = P\{\mathbf{x} \le 29.99\} = P\{f_1, f_2\} = \tfrac{2}{6}$$

◀

EXAMPLE 4-5

▶ A telephone call occurs at random in the interval (0, 1). In this experiment, the outcomes are time distances t between 0 and 1 and the probability that t is between t_1 and t_2 is given by

$$P\{t_1 \le t \le t_2\} = t_2 - t_1$$

We define the random variable $\mathbf{x}$ such that

$$\mathbf{x}(t) = t \qquad 0 \le t \le 1$$

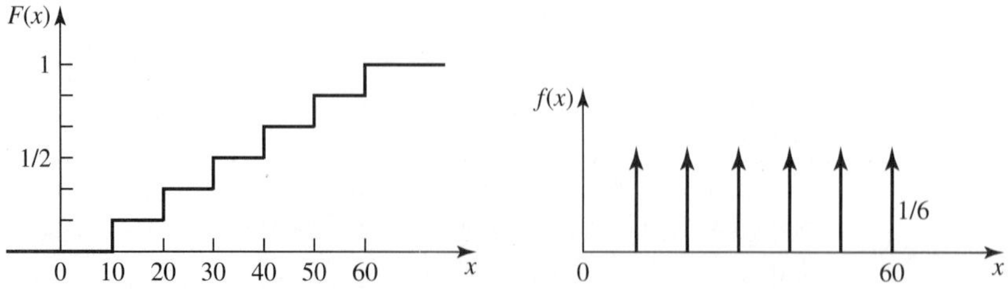

FIGURE 4-3

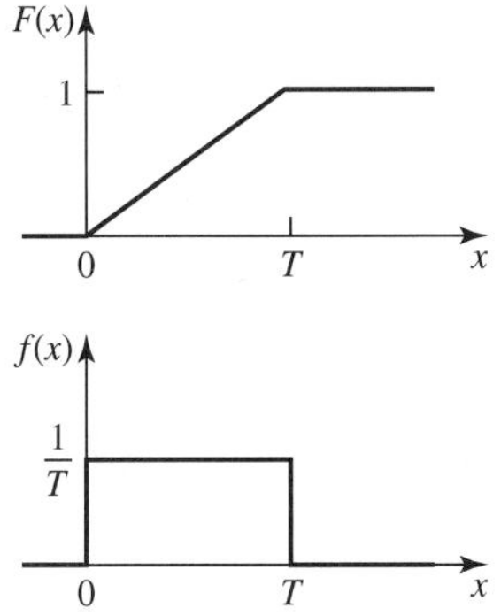

FIGURE 4-4

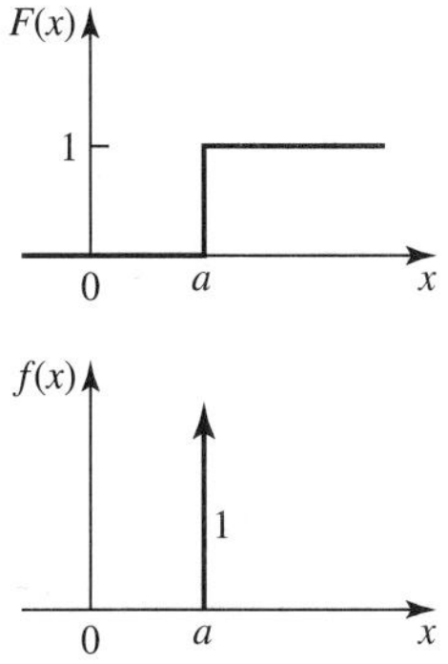

FIGURE 4-5

Thus the variable t has a double meaning: It is the outcome of the experiment and the corresponding value $\mathbf{x}(t)$ of the random variable $\mathbf{x}$. We shall show that the distribution function $F(x)$ of $\mathbf{x}$ is a ramp as in Fig. 4-4.

If $x > 1$, then $\mathbf{x}(t) \leq x$ for every outcome. Hence

$$F(x) = P\{\mathbf{x} \leq x\} = P\{0 \leq t \leq 1\} = P(S) = 1 \qquad x > 1$$

If $0 \leq x \leq 1$, then $\mathbf{x}(t) \leq x$ for every t in the interval $(0, x)$. Hence

$$F(x) = P\{\mathbf{x} \leq x\} = P\{0 \leq t \leq x\} = x \qquad 0 \leq x \leq 1$$

If $x < 0$, then $\{\mathbf{x} \leq x\}$ is the impossible event because $\mathbf{x}(t) \geq 0$ for every t. Hence

$$F(x) = P\{\mathbf{x} \leq x\} = P\{\emptyset\} = 0 \qquad x < 0$$ ◀

EXAMPLE 4-6 ▶ Suppose that a random variable $\mathbf{x}$ is such that $\mathbf{x}(\zeta) = a$ for every ζ in S. We shall find its distribution function.

If $x \geq a$, then $\mathbf{x}(\zeta) = a \leq x$ for every ζ. Hence

$$F(x) = P\{\mathbf{x} \leq x\} = P\{S\} = 1 \qquad x \geq a$$

If $x < a$, then $\{\mathbf{x} \leq x\}$ is the impossible event because $\mathbf{x}(\zeta) = a$. Hence

$$F(x) = P\{\mathbf{x} \leq x\} = P\{\emptyset\} = 0 \qquad x < a$$

Thus a constant can be interpreted as a random variable with distribution function a delayed step $U(x - a)$ as in Fig. 4-5. ◀

Note A complex random variable $\mathbf{z} = \mathbf{x} + j\mathbf{y}$ has no distribution function because the inequality $\mathbf{x} + j\mathbf{y} \leq x + jy$ has no meaning. The statistical properties of $\mathbf{z}$ are specified in terms of the *joint distribution* of the random variables $\mathbf{x}$ and $\mathbf{y}$ (see Chap. 6).

PERCENTILES. The u percentile of a random variable $\mathbf{x}$ is the smallest number x_u such that

$$u = P\{\mathbf{x} \leq x_u\} = F(x_u) \qquad (4\text{-}2)$$

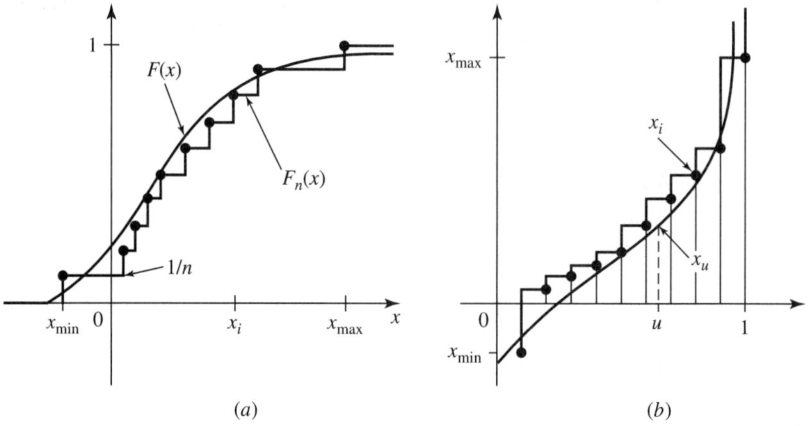

FIGURE 4-6

Thus x_u is the inverse of the function $u = F(x)$. Its domain is the interval $0 \leq u \leq 1$, and its range is the x axis. To find the graph of the function x_u, we interchange the axes of the $F(x)$ curve as in Fig. 4-6. The *median* of $\mathbf{x}$ is the smallest number m such that $F(m) = 0.5$. Thus m is the 0.5 percentile of $\mathbf{x}$.

Frequency interpretation of $F(x)$ and x_u. We perform the experiment n times and we observe n values $x_1, \ldots, x_n$ of the random variable $\mathbf{x}$. We place these numbers on the x axis and we form a staircase function $F_n(x)$ as in Fig 4-6*a*. The steps are located at the points x_i and their height equals $1/n$. They start at the smallest value x_{min} of x_i, and $F_n(x) = 0$ for $x < x_{min}$. The function $F_n(x)$ so constructed is called the *empirical distribution* of the random variable $\mathbf{x}$.

For a specific x, the number of steps of $F_n(x)$ equals the number n_x of $x_i s$ that are smaller than x; thus $F_n(x) = n_x/n$. And since $n_x/n \simeq P\{\mathbf{x} \leq x\}$ for large n, we conclude that

$$F_n(x) = \frac{n_x}{n} \to P\{\mathbf{x} \leq x\} = F(x) \qquad \text{as} \quad n \to \infty \tag{4-3}$$

The empirical interpretation of the u percentile x_u is the *Quetelet curve* defined as: We form n line segments of length x_i and place them vertically in order of increasing length, distance $1/n$ apart. We then form a staircase function with corners at the endpoints of these segments as in Fig. 4-6*b*. The curve so obtained is the empirical interpretation of x_u and it equals the empirical distribution $F_n(x)$ if its axes are interchanged.

Properties of Distribution Functions

In this discussion, the expressions $F(x^+)$ and $F(x^-)$ will mean the limits

$$F(x^+) = \lim F(x + \varepsilon) \qquad F(x^-) = \lim F(x - \varepsilon) \qquad 0 < \varepsilon \to 0$$

The distribution function has the following properties

1. $$F(+\infty) = 1 \qquad F(-\infty) = 0$$

Proof.

$$F(+\infty) = P\{\mathbf{x} \leq +\infty\} = P(S) = 1 \qquad F(-\infty) = P\{\mathbf{x} = -\infty\} = 0$$

2. It is a nondecreasing function of x:

$$\text{if} \quad x_1 < x_2 \qquad \text{then} \quad F(x_1) \leq F(x_2) \tag{4-4}$$

Proof. The event $\{\mathbf{x} \le x_1\}$ is a subset of the event $\{\mathbf{x} \le x_2\}$ because, if $\mathbf{x}(\zeta) \le x_1$ for some ζ, then $\mathbf{x}(\zeta) \le x_2$. Hence [see (2-14)] $P\{\mathbf{x} \le x_1\} \le P\{\mathbf{x} \le x_2\}$ and (4-4) results.

From (4-4) it follows that $F(x)$ increases from 0 to 1 as x increases from $-\infty$ to ∞.

3.
$$\text{If} \quad F(x_0) = 0 \quad \text{then} \quad F(x) = 0 \quad \text{for every} \quad x \le x_0 \tag{4-5}$$

Proof. It follows from (4-4) because $F(-\infty) = 0$. The preceding leads to the conclusion: Suppose that $\mathbf{x}(\zeta) \ge 0$ for every ζ. In this case, $F(0) = P\{\mathbf{x} \le 0\} = 0$ because $\{\mathbf{x} \le 0\}$ is the impossible event. Hence $F(x) = 0$ for every $x \le 0$.

4.
$$P\{\mathbf{x} > x\} = 1 - F(x) \tag{4-6}$$

Proof. The events $\{\mathbf{x} \le x\}$ and $\{\mathbf{x} > x\}$ are mutually exclusive and

$$\{\mathbf{x} \le x\} \cup \{\mathbf{x} > x\} = S$$

Hence $P\{\mathbf{x} \le x\} + P\{\mathbf{x} > x\} = P(S) = 1$ and (4-6) results.

5. The function $F(x)$ is continuous from the right:

$$F(x^+) = F(x) \tag{4-7}$$

Proof. It suffices to show that $P\{\mathbf{x} \le x + \varepsilon\} \to F(x)$ as $\varepsilon \to 0$ because $P\{\mathbf{x} \le x+\varepsilon\} = F(x + \varepsilon)$ and $F(x + \varepsilon) \to F(x^+)$ by definition. To prove the concept in (4-7), we must show that the sets $\{\mathbf{x} \le x + \varepsilon\}$ tend to the set $\{\mathbf{x} \le x\}$ as $\varepsilon \to 0$ and to use the axiom III*a* of finite additivity. We omit, however, the details of the proof because we have not introduced limits of sets.

6.
$$P\{x_1 < \mathbf{x} \le x_2\} = F(x_2) - F(x_1) \tag{4-8}$$

Proof. The events $\{\mathbf{x} \le x_1\}$ and $\{x_1 < \mathbf{x} \le x_2\}$ are mutually exclusive because $\mathbf{x}(\zeta)$ cannot be both less than x_1 and between x_1 and x_2. Furthermore,

$$\{\mathbf{x} \le x_2\} = \{\mathbf{x} \le x_1\} \cup \{x_1 < \mathbf{x} \le x_2\}$$

Hence

$$P\{\mathbf{x} \le x_2\} = P\{\mathbf{x} \le x_1\} + P\{x_1 < \mathbf{x} \le x_2\}$$

and (4-8) results.

7.
$$P\{\mathbf{x} = x\} = F(x) - F(x^-) \tag{4-9}$$

Proof. Setting $x_1 = x - \varepsilon$ and $x_2 = x$ in (4-8), we obtain

$$P\{\mathbf{x} - \varepsilon < \mathbf{x} \le x\} = F(x) - F(x - \varepsilon)$$

and with $\varepsilon \to 0$, (4-9) results.

8.
$$P\{x_1 \le \mathbf{x} \le x_2\} = F(x_2) - F(x_1^-) \tag{4-10}$$

Proof. It follows from (4-8) and (4-9) because

$$\{x_1 \le \mathbf{x} \le x_2\} = \{x_1 < \mathbf{x} \le x_2\} \cup \{\mathbf{x} = x_1\}$$

and the last two events are mutually exclusive.

Statistics. We shall say that the statistics of a random variable **x** are known if we can determine the probability $P\{\mathbf{x} \in R\}$ that **x** is in a set R of the x axis consisting of countable unions or intersections of intervals. From (4-1) and the axioms it follows that the statistics of **x** are determined in terms of its distribution function.

According to (4-7), $F_x(x_0^+)$, the limit of $F_x(x)$ as $x \to x_0$ from the right always exists and equals $F_x(x_0)$. But $F_x(x)$ need not be continuous from the left. At a discontinuity point of the distribution, the left and right limits are different, and from (4-9)

$$P\{\mathbf{x}(\xi) = x_0\} = F_x(x_0) - F_x(x_0^-) > 0 \tag{4-11}$$

Thus the only discontinuities of a distribution function $F_x(x)$ are of the jump type, and occur at points x_0 where (4-11) is satisfied. These points can always be enumerated as a sequence, and moreover they are at most countable in number.

EXAMPLE 4-7

▶ The set of nonnegative real numbers $\{p_i\}$ satisfy $P\{\mathbf{x} = x_i\} = p_i$ for all i, and $\sum_{i=1}^{\infty} p_i = 1$. Determine $F(x)$.

SOLUTION

For $x_i \le x < x_{i+1}$, we have $\{\mathbf{x}(\xi) \le x\} = \bigcup_{x_k \le x} \{\mathbf{x}(\xi) = x_k\} = \bigcup_{k=1}^{i} \{\mathbf{x}(\xi) = x_k\}$ and hence

$$F(x) = P\{\mathbf{x}(\xi) \le x\} = \sum_{k=1}^{i} p_k \qquad x_i \le x < x_{i+1}$$

Here $F(x)$ is a staircase function with an infinite number of steps and the i-th step size equals p_i, $i = 1, 2, \ldots, \infty$ (see Fig 4-7). ◀

EXAMPLE 4-8

▶ Suppose the random variable **x** is such that $\mathbf{x}(\xi) = 1$ if $\xi \in A$ and zero otherwise. Find $F(x)$.

SOLUTION

For $x < 0$, $\{\mathbf{x}(\xi) \le x\} = \{\emptyset\}$, so that $F(x) = 0$. For $0 \le x < 1$, $\{\mathbf{x}(\xi) \le x\} = \{\overline{A}\}$, so that $F(x) = P\{\overline{A}\} = 1 - p = q$, where $p \triangleq P(A)$, and if $x \ge 1$, $\{\mathbf{x}(\xi) \le x\} = \Omega$, so that $F(x) = 1$ (see Fig. 4-2, page 76). Here the event A may refer to *success* and $\overline{A}$ to *failure*. ◀

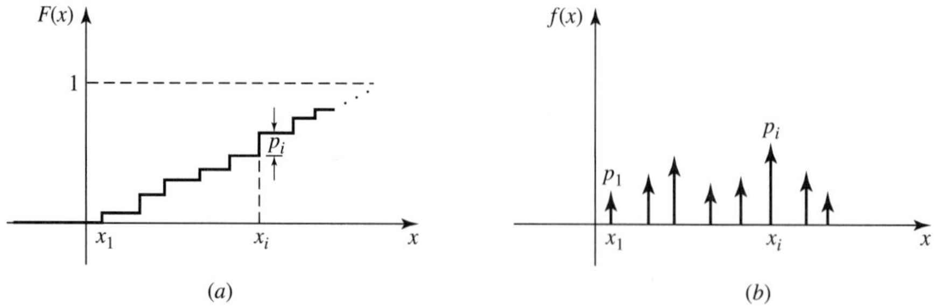

FIGURE 4-7

Continuous, Discrete, and Mixed Types

The random variable **x** is said to be a continuous type if its distribution function $F_x(x)$ is continuous. In that case $F_x(x^-) = F_x(x)$ for all x, and from (4-11) we get $P\{\mathbf{x} = x\} = 0$.

If $F_x(x)$ is constant except for a finite number of jump discontinuities (piecewise constant; step type), then **x** is said to be a discrete-type random variable. If x_i is such a discontinuity point, then from (4-11) (see Fig. 4-9–Fig. 4-10 and also Fig. 4-7 on page 80)

$$P\{\mathbf{x} = x_i\} = F_x(x_i) - F_x(x_i^-) = p_i \tag{4-12}$$

For example, from Fig. 4-5, page 77, at the point of discontinuity we get

$$P\{\mathbf{x} = a\} = F_x(a) - F_x(a^-) = 1 - 0 = 1$$

and from Fig. 4-2, page 76, at such a point

$$P\{\mathbf{x} = 0\} = F_x(0) - F_x(0^-) = q - 0 = q$$

EXAMPLE 4-9

▶ A fair coin is tossed twice, and let the random variable **x** represent the number of heads. Find $F_x(x)$.

SOLUTION

In this case, $\Omega = \{HH, HT, TH, TT\}$, and

$$\mathbf{x}(HH) = 2 \qquad \mathbf{x}(HT) = 1 \qquad \mathbf{x}(TH) = 1 \qquad \mathbf{x}(TT) = 0$$

For $x < 0$, $\{\mathbf{x}(\xi) \le x\} = \phi \Rightarrow F_x(x) = 0$, and for $0 \le x < 1$,

$$\{\mathbf{x}(\xi) \le x\} = \{TT\} \Rightarrow F_x(x) = P\{TT\} = P(T)P(T) = \tfrac{1}{4}$$

Finally for $1 \le x < 2$,

$$\{\mathbf{x}(\xi) \le x\} = \{TT, HT, TH\} \Rightarrow F_x(x) = P\{TT\} + P\{HT\} + P\{TH\} = \tfrac{3}{4}$$

and for $x \ge 2$, $\{\mathbf{x}(\xi) \le x\} = \Omega \Rightarrow F_x(x) = 1$ (see Fig. 4-8). From Fig. 4-8, at a point of discontinuity $P\{\mathbf{x} = 1\} = F_x(1) - F_x(1^-) = 3/4 - 1/4 = 1/2$. ◀

The Probability Density Function (p.d.f.)

The derivative of the probability distribution function $F_x(x)$ is called the probability density function $f_x(x)$ of the random variable **x**. Thus

$$f_x(x) \triangleq \frac{dF_x(x)}{dx} \tag{4-13}$$

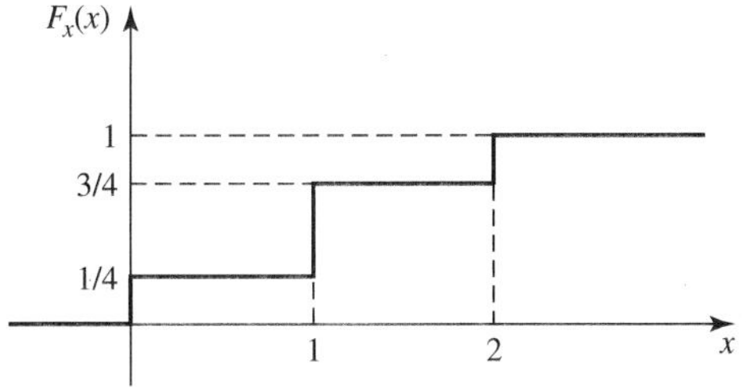

FIGURE 4-8

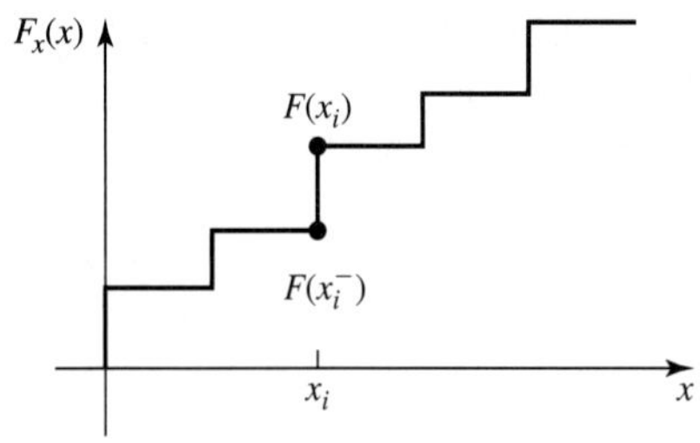

FIGURE 4-9

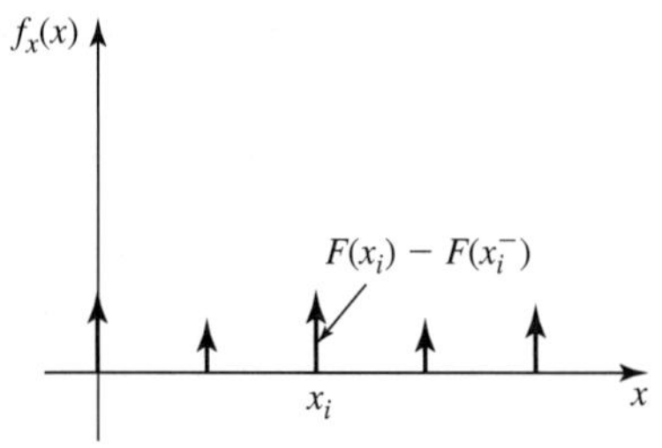

FIGURE 4-10

Since

$$\frac{dF_x(x)}{dx} = \lim_{\Delta x \to 0} \frac{F_x(x + \Delta x) - F_x(x)}{\Delta x} \geq 0 \tag{4-14}$$

from the monotone-nondecreasing nature of $F_x(x)$, it follows that $f_x(x) \geq 0$ for all x. If $\mathbf{x}$ is a continuous-type random variable, $f_x(x)$ will be a continuous function. However, if x is a discrete-type random variable as in Example 4-7, then its p.d.f. has the general form (Figs. 4-7b and 4-10)

$$f_x(x) = \sum_i p_i \delta(x - x_i) \tag{4-15}$$

where x_is represent the jump-discontinuity points in $F_x(x)$. As Fig. 4-10 shows, $f_x(x)$ represents a collection of positive discrete masses in the discrete case, and it is known as the probability mass function (p.m.f.).

From (4-13), we also obtain by integration

$$F_x(x) = \int_{-\infty}^{x} f_x(u)\, du \tag{4-16}$$

Since $F_x(+\infty) = 1$, (4-16) yields

$$\int_{-\infty}^{+\infty} f_x(x)\, dx = 1 \tag{4-17}$$

which justifies its name as the density function. Further, from (4-16), we also get (Fig. 4-11)

$$P\{x_1 < \mathbf{x}(\xi) \leq x_2\} = F_x(x_2) - F_x(x_1) = \int_{x_1}^{x_2} f_x(x)\, dx \tag{4-18}$$

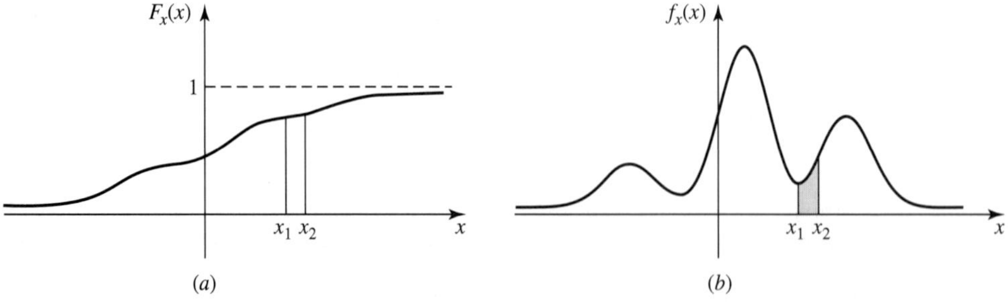

FIGURE 4-11

Thus the area under $f_x(x)$ in the interval (x_1, x_2) represents the probability that the random variable $\mathbf{x}$ lies in the interval (x_1, x_2) as in (4-18).

If the random variable $\mathbf{x}$ is of continuous type, then the set on the left might be replaced by the set $\{x_1 \le \mathbf{x} \le x_2\}$. However, if $F(x)$ is discontinuous at x_1 or x_2, then the integration must include the corresponding impulses of $f(x)$.

With $x_1 = x$ and $x_2 = x + \Delta x$ it follows from (4-18) that, if $\mathbf{x}$ is of continuous type, then

$$P\{x \le \mathbf{x} \le x + \Delta x\} \simeq f(x)\,\Delta x \tag{4-19}$$

provided that Δx is sufficiently small. This shows that $f(x)$ can be defined directly as a limit

$$f(x) = \lim_{\Delta x \to 0} \frac{P\{x \le \mathbf{x} \le x + \Delta x\}}{\Delta x} \tag{4-20}$$

Note As we can see from (4-19), the probability that $\mathbf{x}$ is in a small interval of specified length Δx is proportional to $f(x)$ and it is maximum if that interval contains the point x_m, where $f(x)$ is maximum. This point is called the *mode* or the *most likely value* of $\mathbf{x}$. A random variable is called *unimodal* if it has a single mode.

Frequency interpretation We denote by Δn_x the number of trials such that

$$x \le \mathbf{x}(\zeta) \le x + \Delta \mathbf{x}$$

From (1-1) and (4-19) it follows that

$$f(x)\Delta x \simeq \frac{\Delta n_x}{n} \tag{4-21}$$

4-3 SPECIFIC RANDOM VARIABLES

In Secs. 4-1 and 4-2 we defined random variables starting from known experiments. In this section and throughout the book, we shall often consider random variables having specific distribution or density functions without any reference to a particular probability space.

THEOREM 4-1

EXISTENCE THEOREM

▶ To do so, we must show that given a function $f(x)$ or its integral

$$F(x) = \int_{-\infty}^{x} f(u)\,du$$

we can construct an experiment and a random variable $\mathbf{x}$ with distribution $F(x)$ or density $f(x)$. As we know, these functions must have these properties:

The function $f(x)$ must be non-negative and its area must be 1. The function $F(x)$ must be continuous from the right and, as x increases from $-\infty$ to ∞, it must increase monotonically from 0 to 1.

Proof. We consider as our space S the set of all real numbers, and as its events all intervals on the real line and their unions and intersections. We define the probability of the event $\{x \le x_1\}$ by

$$P\{x \le x_1\} = F(x_1) \tag{4-22}$$

where $F(x)$ is the given function. This specifies the experiment completely (see Sec. 2-2).

The outcomes of our experiment are the real numbers. To define a random variable **x** on this experiment, we must know its value $\mathbf{x}(x)$ for every x. We define **x** such that

$$\mathbf{x}(x) = x \tag{4-23}$$

Thus x is the outcome of the experiment and the corresponding value of the random variable **x** (see also Example 4-5).

We maintain that the distribution function of **x** equals the given $F(x)$. Indeed, the event $\{x \le x_1\}$ consists of all outcomes x such that $\mathbf{x}(x) \le x_1$. Hence

$$P\{\mathbf{x} \le x_1\} = P\{x \le x_1\} = F(x_1) \tag{4-24}$$

and since this is true for every x_1, the theorem is proved. ◀

In the following, we discuss briefly a number of common densities.

Continuous-Type Random Variables

NORMAL (GAUSSIAN) DISTRIBUTION. Normal (Gaussian) distribution is one of the most commonly used distributions. We say that **x** is a normal or Gaussian random variable with parameters μ and σ^2 if its density function is given by

$$f_x(x) = \frac{1}{\sqrt{2\pi\sigma^2}} e^{-(x-\mu)^2/2\sigma^2} \tag{4-25}$$

This is a bell-shaped curve (see Fig. 4-12), symmetric around the parameter μ, and its distribution function is given by

$$F_x(x) = \int_{-\infty}^{x} \frac{1}{\sqrt{2\pi\sigma^2}} e^{-(y-\mu)^2/2\sigma^2}\, dy \triangleq G\left(\frac{x-\mu}{\sigma}\right) \tag{4-26}$$

where the function

$$G(x) \triangleq \int_{-\infty}^{x} \frac{1}{\sqrt{2\pi}} e^{-y^2/2}\, dy \tag{4-27}$$

is often available in tabulated form (see Table 4-1 later in the chapter). Since $f_x(x)$ depends on two parameters μ and σ^2, the notation $\mathbf{x} \sim N(\mu, \sigma^2)$ will be used to represent

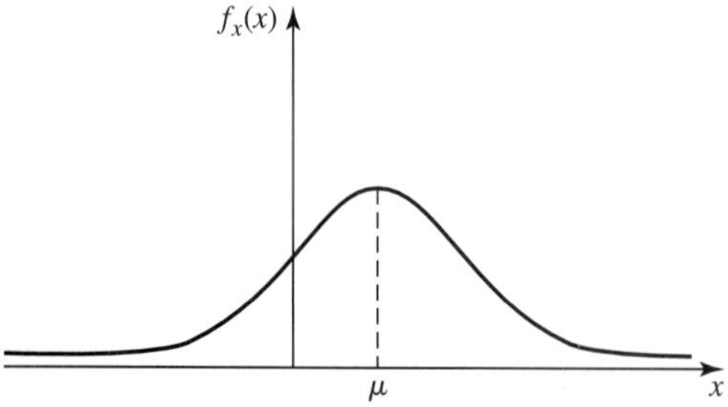

(a) $X \sim N(\mu, \sigma_1^2)$

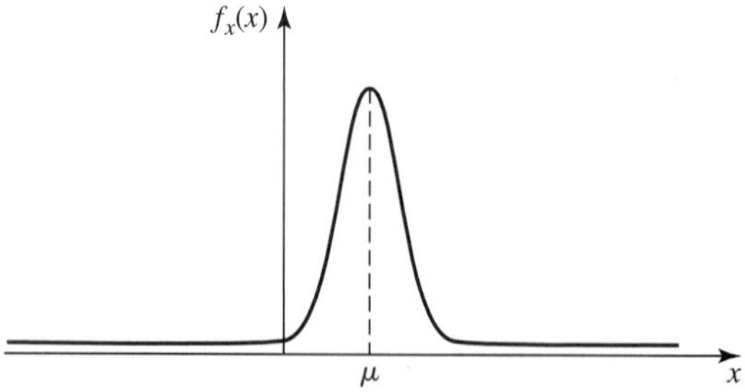

(b) $X \sim N(\mu, \sigma_2^2)$, $\sigma_1^2 > \sigma_2^2$

FIGURE 4-12
Normal density function.

the Gaussian p.d.f. in (4-25). The constant $\sqrt{2\pi\sigma^2}$ in (4-25) is the normalization constant that maintains the area under $f_x(x)$ to be unity.

This follows since if we let

$$Q = \int_{-\infty}^{+\infty} e^{-x^2/2\sigma^2}\, dx \tag{4-28}$$

then

$$\begin{aligned} Q^2 &= \int_{-\infty}^{+\infty}\int_{-\infty}^{+\infty} e^{-(x^2+y^2)/2\sigma^2}\, dx\, dy \\ &= \int_0^{2\pi}\int_0^{+\infty} e^{-r^2/2\sigma^2} r\, dr\, d\theta \\ &= 2\pi\sigma^2 \int_0^{+\infty} e^{-u}\, du = 2\pi\sigma^2 \end{aligned} \tag{4-29}$$

where we have made use of the transformation $x = r\cos\theta$, $y = r\sin\theta$, so that $dx\,dy = r\,dr\,d\theta$ and thus $Q = \sqrt{2\pi\sigma^2}$. The special case $\mathbf{x} \sim N(0, 1)$ is often referred to as the standard normal random variable.

The normal distribution is one of the most important distributions in the study of probability and statistics. Various natural phenomena follow the Gaussian distribution. Maxwell arrived at the normal distribution for the distribution of velocities of molecules, under the assumption that the probability density of molecules with given velocity components is a function of their velocity magnitude and not their directions. Hagen, in developing the theory of errors, showed that under the assumption that the error is the sum of a large number of independent infinitesimal errors due to different causes, all of equal magnitude, the overall error has a normal distribution. This result is a special case of a more general theorem which states that under very general conditions the limiting distribution of the average of any number of independent, identically distributed random variables is normal.

EXPONENTIAL DISTRIBUTION. We say $\mathbf{x}$ is exponential with parameter λ if its density function is given by (see Fig. 4-13)

$$f_x(x) = \begin{cases} \lambda e^{-\lambda x} & x \ge 0 \\ 0 & \text{otherwise} \end{cases} \tag{4-30}$$

If occurrences of events over nonoverlapping intervals are independent, such as arrival times of telephone calls or bus arrival times at a bus stop, then the waiting time distribution of these events can be shown to be exponential. To see this, let $q(t)$ represent

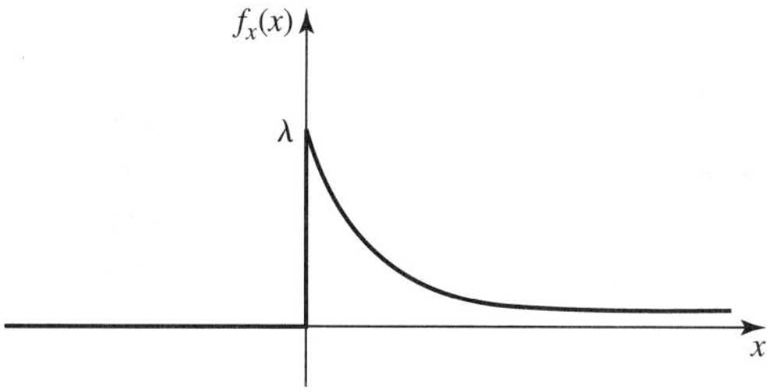

FIGURE 4-13
Exponential density function.

the probability that in a time interval t no event has occurred. If $\mathbf{x}$ represents the waiting time to the first arrival, then by definition $P(\mathbf{x} > t) = q(t)$. If t_1 and t_2 represent two consecutive nonoverlapping intervals, then by the independent assumption we have

$$q(t_1)q(t_2) = q(t_1 + t_2)$$

which has the only nontrivial bounded solution of the form [see also (16-9)–(16-10)]

$$q(t) = e^{-\lambda t}$$

Hence

$$F_x(t) = P(\mathbf{x} \le t) = 1 - q(t) = 1 - e^{-\lambda t} \tag{4-31}$$

and the corresponding density function is exponential as in (4-30).

Memoryless property of exponential distributions. Let $s, t \ge 0$. Consider the events $\{\mathbf{x} > t + s\}$ and $\{\mathbf{x} > s\}$. Then

$$P\{\mathbf{x} > t + s \mid \mathbf{x} > s\} = \frac{P\{\mathbf{x} > t + s\}}{P\{\mathbf{x} > s\}} = \frac{e^{-(t+s)}}{e^{-s}} = e^{-t} = P\{\mathbf{x} > t\} \tag{4-32}$$

since the event $\{\mathbf{x} > t + s\} \subset \{\mathbf{x} > s\}$. If $\mathbf{x}$ represents the lifetime of an equipment, then (4-32) states that if the equipment has been working for time s, then the probability that it will survive an additional time t depends only on t (not on s) and is identical to the probability of survival for time t of a new piece of equipment. In that sense, the equipment does not remember that it has been in use for time s. It follows that for a *continuous non-negative* random variable $\mathbf{x}$, if

$$P\{\mathbf{x} > t + s \mid \mathbf{x} > s\} = P\{\mathbf{x} > t\}$$

holds for all $s, t \ge 0$, then $\mathbf{x}$ must have an exponential distribution.

This memoryless property simplifies many calculations and is mainly the reason for wide applicability of the exponential model. Under this model, an item that has not failed so far is as good as new. This is not the case for other non-negative continuous type random variables. In fact, the conditional probability

$$P\{\mathbf{x} > t + s \mid \mathbf{x} > s\} = \frac{1 - P\{\mathbf{x} \le t + s\}}{1 - P\{\mathbf{x} \le s\}} = \frac{1 - F(t + s)}{1 - F(s)} \tag{4-33}$$

depends on s in general.

EXAMPLE 4-10

LIFE LENGTH OF AN APPLIANCE

▶ Suppose the life length of an appliance has an exponential distribution with $\lambda = 10$ years. A used appliance is bought by someone. What is the probability that it will not fail in the next 5 years?

SOLUTION

Because of the memoryless property, it is irrelevant how many years the appliance has been in service prior to its purchase. Hence if $\mathbf{x}$ is the random variable representing the length of the life time of the appliance and t_0 its actual life duration to the present time instant, then

$$P\{\mathbf{x} > t_0 + 5 \mid \mathbf{x} > t_0\} = P\{\mathbf{x} > 5\} = e^{-5/10} = e^{-1/2} = 0.368$$

As mentioned earlier, for any other lifetime distribution this calculation will depend on the actual life duration time t_0. ◀

EXAMPLE 4-11

WAITING TIME AT A RESTAURANT

▶ Suppose that the amount of waiting time a customer spends at a restaurant has an exponential distribution with a mean value of 5 minutes. Then the probability that a customer will spend more than 10 minutes in the restaurant is given by

$$P(\mathbf{x} > 10) = e^{-10/\lambda} = e^{-10/5} = e^{-2} = 0.1353$$

More interestingly, the (conditional) probability that the customer will spend an additional 10 minutes in the restaurant given that he or she has been there for more that 10 minutes is

$$P\{\mathbf{x} > 10 \mid \mathbf{x} > 10\} = P\{\mathbf{x} > 10\} = e^{-2} = 0.1353$$

In other words, the past does not matter. ◀

A generalization of the exponential distribution leads to the gamma distribution.

GAMMA DISTRIBUTION. $\mathbf{x}$ is said to be a gamma random variable with parameters α, and β, $\alpha > 0$, $\beta > 0$ if

$$f_x(x) = \begin{cases} \dfrac{x^{\alpha-1}}{\Gamma(\alpha)\beta^{\alpha}} e^{-x/\beta} & x \geq 0 \\ 0 & \text{otherwise} \end{cases} \tag{4-34}$$

where $\Gamma(\alpha)$ represents the gamma function defined as

$$\Gamma(\alpha) = \int_0^{\infty} x^{\alpha-1} e^{-x}\, dx \tag{4-35}$$

If α is an integer, integrating (4-35) by parts we get

$$\Gamma(n) = (n-1)\Gamma(n-1) = (n-1)! \tag{4-36}$$

We shall denote the p.d.f. in (4-34) by $G(\alpha, \beta)$.

The gamma density function takes on a wide variety of shapes depending on the values of α and β. For $\alpha < 1$, $f_x(x)$ is strictly decreasing and $f_x(x) \to \infty$ as $x \to 0$, $f_x(x) \to 0$ as $x \to \infty$. For $\alpha > 1$, the density $f_x(x)$ has a unique mode at $x = (\alpha - 1)/\beta$ with maximum value $[(\alpha - 1)e^{-1}]^{\alpha-1}/(\beta\Gamma(\alpha))$. Figure 4.14 gives graphs of some typical gamma probability density functions.

Some special cases of the gamma distribution are widely used and have special names. Notice that the exponential random variable defined in (4-30) is a special case of gamma distribution with $\alpha = 1$. If we let $\alpha = n/2$ and $\beta = 2$, we obtain the χ^2 (chi-square) random variable with n degrees of freedom shown in (4-39).

For $\alpha = n$ in (4-34), we obtain the gamma density function to be (with $\beta = 1/\lambda$)

$$f_x(x) = \begin{cases} \dfrac{\lambda^n x^{n-1}}{(n-1)!} e^{-\lambda x} & x \geq 0, \\ 0 & \text{otherwise} \end{cases} \tag{4-37}$$

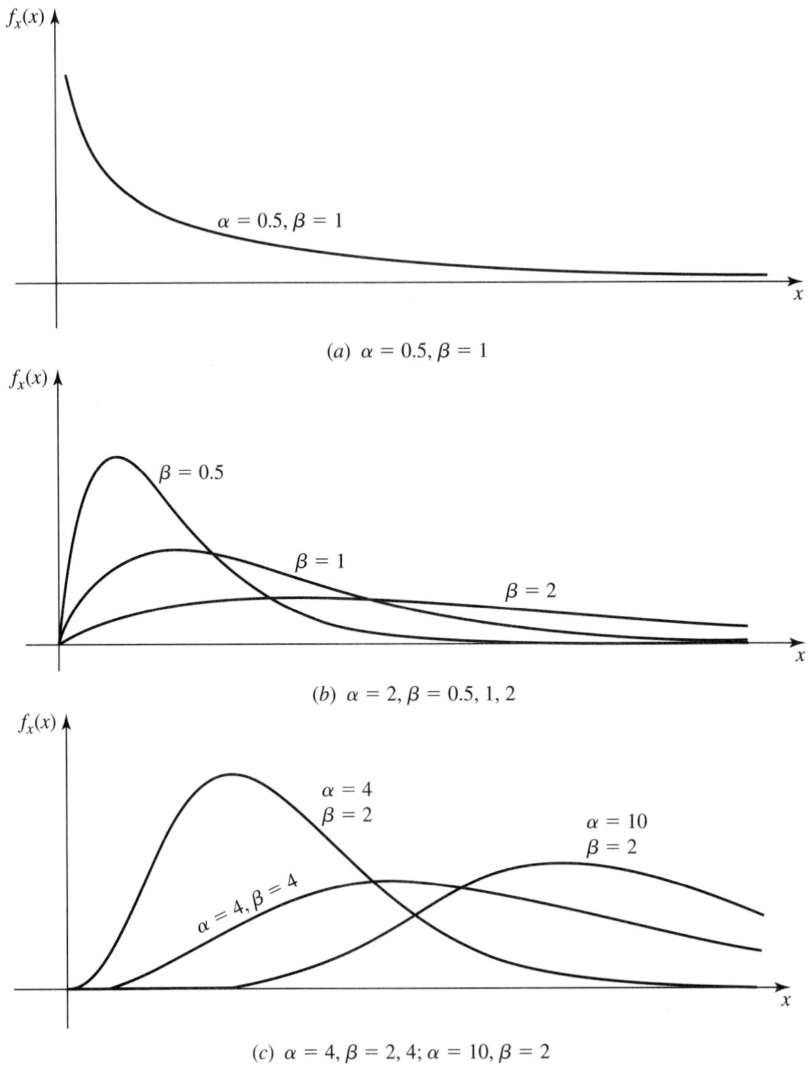

FIGURE 4-14
Gamma density functions.

Integrating (4-37) by parts, we obtain the probability distribution function for the corresponding gamma random variable to be

$$F_x(t) = \int_0^t f_x(x) = 1 - \sum_{k=0}^{n-1} \frac{(\lambda t)^k}{k!} e^{-\lambda t} \tag{4-38}$$

If $\lambda = n\mu$ in (4-37) and (4-38), then it corresponds to an *Erlangian* random variable. Thus $G(n, 1/n\mu)$ corresponds to an *Erlangian* distribution (E_n). In that case, $n = 1$ yields an exponential random variable, and $n \to \infty$ gives a constant distribution ($F_x(t) = 1$, for $t > 1/\mu$ and zero otherwise). Thus randomness to certainty are covered by the Erlangian distribution as n varies between 1 and ∞. Many important distributions occurring in

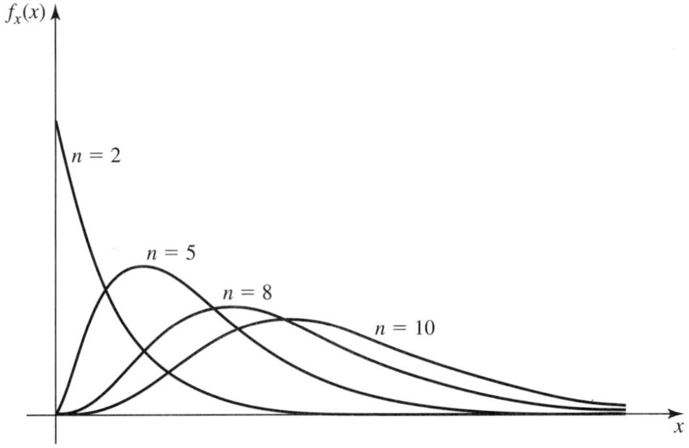

FIGURE 4-15
χ^2 density functions for $n = 2, 5, 8, 10$.

practice lie between these two cases, and they may be approximated by an Erlangian distribution for a proper choice of n.

CHI-SQUARE DISTRIBUTION. $\mathbf{x}$ is said to be $\chi^2(n)$ (chi-square) with n degrees of freedom if

$$f_x(x) = \begin{cases} \dfrac{x^{n/2-1}}{2^{n/2}\Gamma(n/2)} e^{-x/2} & x \geq 0 \\ 0 & \text{otherwise} \end{cases} \tag{4-39}$$

Figure 4.15 shows graphs of $\chi^2(n)$ for various values of n. Note that if we let $n = 2$ in (4-39), we obtain an exponential distribution. It is also possible to generalize the exponential random variable in such a way as to avoid its memoryless property discussed earlier. In reality, most of the appliances deteriorate over time so that an exponential model is inadequate to describe the length of its life duration and its failure rate. In that context, consider the distribution function

$$F_x(x) = 1 - e^{-\int_0^x \lambda(t)\,dt} \qquad x \geq 0 \qquad \lambda(t) \geq 0 \tag{4-40}$$

The associated density function is given by

$$f_x(x) = \lambda(x) e^{-\int_0^x \lambda(t)\,dt} \qquad x \geq 0 \qquad \lambda(t) \geq 0 \tag{4-41}$$

Notice that $\lambda(t) =$ constant, gives rise to the exponential distribution. More generally, consider

$$\lambda(t) = \alpha t^{\beta-1} \tag{4-42}$$

and from (4-41), it corresponds to the p.d.f.

$$f_x(x) = \begin{cases} \alpha x^{\beta-1} e^{-\alpha x^\beta/\beta} & x \geq 0, \\ 0 & \text{otherwise,} \end{cases} \tag{4-43}$$

and it is known as the Weibull distribution (see Fig. 4-16).

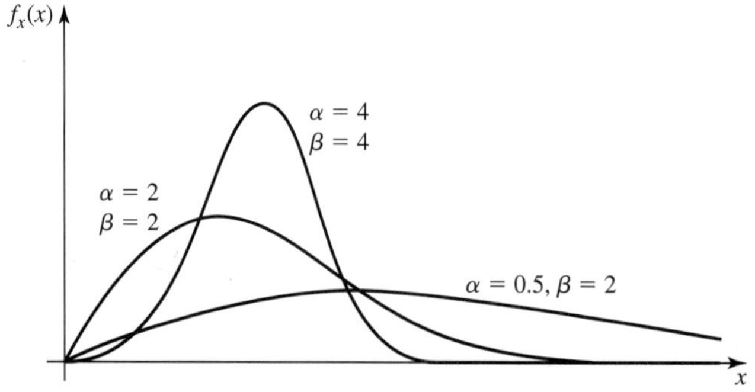

FIGURE 4-16
Weibull density function.

The special case of Weibull with $\alpha = 1/\sigma^2$ and $\beta = 2$ is known as the Rayleigh distribution. Thus Rayleigh has linear rate in (4-42).

RAYLEIGH DISTRIBUTION. The random variable **x** is said to be Rayleigh distribution with parameter σ^2 if

$$f_x(x) = \begin{cases} \dfrac{x}{\sigma^2} e^{-x^2/2\sigma^2} & x \geq 0 \\ 0 & \text{otherwise} \end{cases} \tag{4-44}$$

In communication systems, the signal amplitude values of a randomly received signal usually can be modeled as a Rayleigh distribution.

NAKAGAMI-*m* DISTRIBUTION. A generalization to the Rayleigh distribution (through a parameter m), is given by the Nakagami distribution where

$$f_x(x) = \begin{cases} \dfrac{2}{\Gamma(m)} \left(\dfrac{m}{\Omega}\right)^m x^{2m-1} e^{-mx^2/\Omega} & x > 0 \\ 0 & \text{otherwise} \end{cases} \tag{4-45}$$

Compared to the Rayleigh distribution, Nakagami distribution gives greater flexibility to model randomly fluctuating (fading) channels in communication theory. Notice that in (4-45) $m = 1$ corresponds to the Rayleigh distribution, and the parameter m there can be used to control the tail distribution. As Fig. 4-17 shows, for $m < 1$, the tail distribution decays slowly compared to the Rayleigh distribution, while $m > 1$ corresponds to faster decay.

UNIFORM DISTRIBUTION. **x** is said to be uniformly distributed in the interval (a, b), $-\infty < a < b < \infty$, if

$$f_x(x) = \begin{cases} \dfrac{1}{b-a} & a \leq x \leq b \\ 0 & \text{otherwise} \end{cases} \tag{4-46}$$

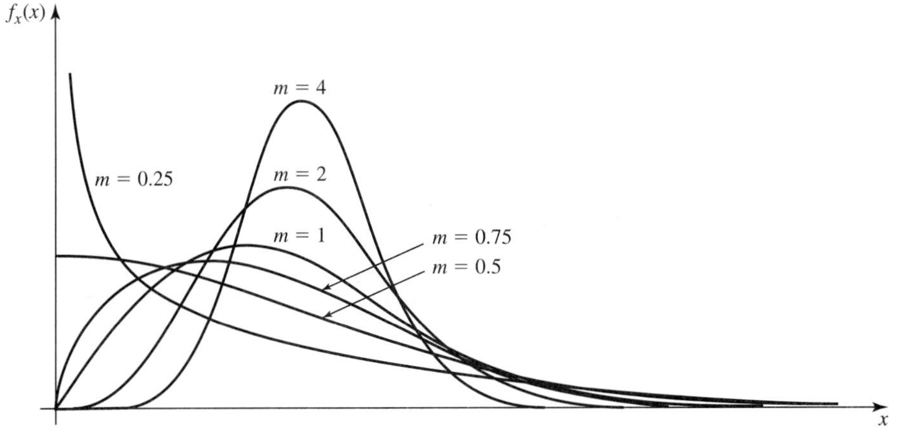

FIGURE 4-17
Nakagami-m density function for $m = 0.25, 0.5, 0.75, 1, 2, 4$.

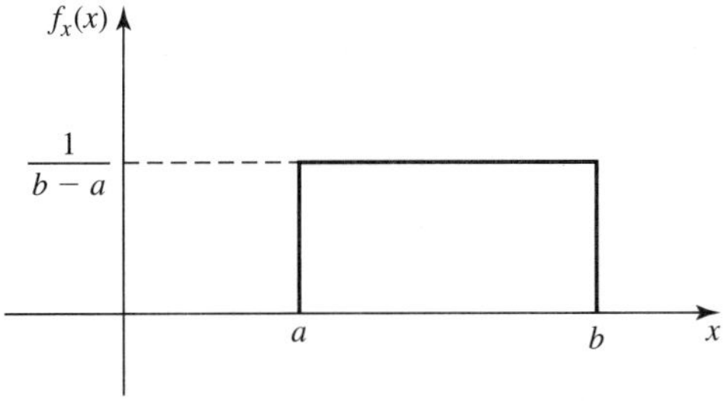

FIGURE 4-18
Uniform density function.

We will write $\mathbf{x} \sim U(a, b)$. The distribution function of $\mathbf{x}$ is given by (see Fig. 4-18)

$$F_x(x) = \begin{cases} 1 & x \geq b \\ \dfrac{x-a}{b-a} & a \leq x < b \\ 0 & x < a \end{cases} \tag{4-47}$$

BETA DISTRIBUTION. The random variable $\mathbf{x}$ is said to have beta distribution with nonnegative parameters α and β if

$$f_x(x) = \begin{cases} \dfrac{1}{B(\alpha, \beta)} x^{\alpha-1}(1-x)^{\beta-1} & 0 < x < b \\ 0 & \text{otherwise} \end{cases} \tag{4-48}$$

where the beta function $B(\alpha, \beta)$ is defined as

$$B(\alpha, \beta) = \int_0^1 x^{\alpha-1}(1-x)^{\beta-1}\,dx = 2\int_0^{2\pi} (\sin\theta)^{2\alpha-1}(\cos\theta)^{2\beta-1}\,d\theta \tag{4-49}$$

The trigonometric form in (4-49) can be obtained by substituting $x = \sin^2\theta$ into the algebraic version there. It is possible to express the beta function in terms of the gamma function defined earlier. If we let $x = y^2$ in (4-35) we obtain

$$\Gamma(\alpha) = 2\int_0^\infty y^{2\alpha-1}e^{-y^2}\,dy \tag{4-50}$$

so that

$$\Gamma(\alpha)\Gamma(\beta) = 4\int_0^\infty\int_0^\infty x^{2\alpha-1}y^{2\beta-1}e^{-(x^2+y^2)}\,dx\,dy$$

Changing to polar coordinates with $x = r\cos\theta$, $y = r\sin\theta$, we obtain

$$\begin{aligned}\Gamma(\alpha)\Gamma(\beta) &= 4\int_0^{\pi/2}\int_0^\infty r^{2(\alpha+\beta)-1}e^{-r^2}(\sin\theta)^{2\alpha-1}(\cos\theta)^{2\beta-1}\,dr\,d\theta \\ &= \left(2\int_0^\infty r^{2(\alpha+\beta)-1}e^{-r^2}\,dr\right)\left(2\int_0^\infty(\sin\theta)^{2\alpha-1}(\cos\theta)^{2\beta-1}d\theta\right) \\ &= \Gamma(\alpha+\beta)B(\alpha,\beta)\end{aligned}$$

or

$$B(\alpha,\beta) = \frac{\Gamma(\alpha)\Gamma(\beta)}{\Gamma(\alpha+\beta)} \tag{4-51}$$

The beta function provides greater flexibility than the uniform distribution on (0, 1), which corresponds to a beta distribution with $\alpha = \beta = 1$. Depending on the values of α and β, the beta distribution takes a variety of shapes. If $\alpha > 1$, $\beta > 1$, then $f_x(x) \to 0$ at both $x = 0$ and $x = 1$, and it has a concave down shape. If $0 < \alpha < 1$, then $f_x(x) \to \infty$ as $x \to 0$, and if $0 < \beta < 1$, then $f_x(x) \to \infty$ as $x \to 1$. If $\alpha < 1$, $\beta < 1$, then $f_x(x)$ is concave up with a unique minimum. When $\alpha = \beta$, the p.d.f. is symmetric about $x = 1/2$ (see Fig. 4-19).

Some other common continuous distributions are listed next.

CAUCHY DISTRIBUTION

$$f_x(x) = \frac{\alpha/\pi}{(x-\mu)^2+\alpha^2} \qquad |x| < \infty \tag{4-52}$$

LAPLACE DISTRIBUTION

$$f_x(x) = \frac{\alpha}{2}e^{-\alpha|x|} \qquad |x| < \infty \tag{4-53}$$

MAXWELL DISTRIBUTION

$$f_x(x) = \begin{cases}\dfrac{4}{\alpha^3\sqrt{\pi}}x^2e^{-x^2/\alpha^2} & x \ge 0 \\ 0 & \text{otherwise}\end{cases} \tag{4-54}$$

Discrete-Type Random Variables

The simplest among the discrete set of random variables is the Bernoulli random variable that corresponds to any experiment with only two possible outcomes—success or failure (head or tail) as in Examples 4-3 and 4-8.

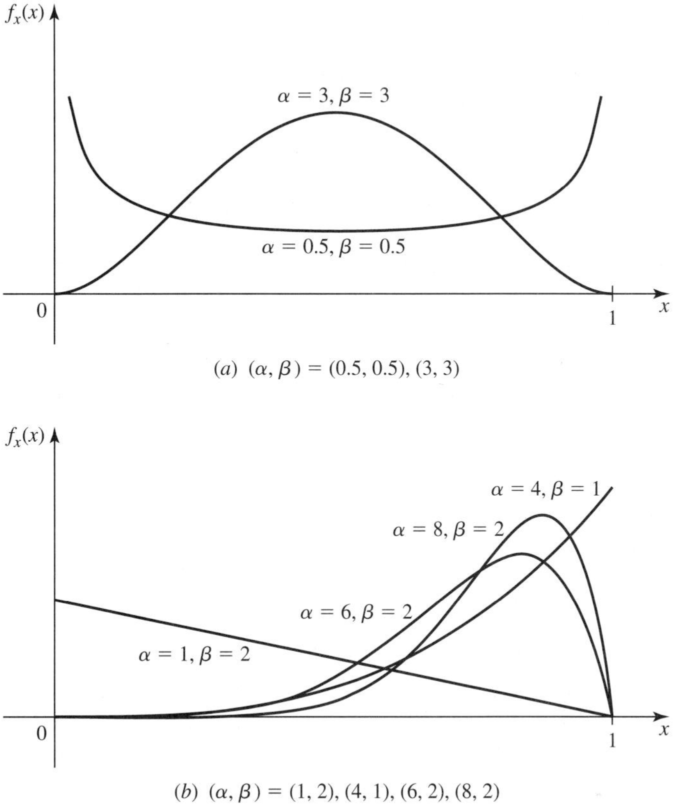

FIGURE 4-19
Beta density function.

BERNOULLI DISTRIBUTION. $\mathbf{x}$ is said to be Bernoulli distributed if $\mathbf{x}$ takes the values 1 and 0 with (Fig. 4-2)

$$P\{\mathbf{x} = 1\} = p \qquad P\{\mathbf{x} = 0\} = q = 1 - p \tag{4-55}$$

In an independent trial of n Bernoulli experiments with p representing the probability of success in each experiment, if $\mathbf{y}$ represents the total number of favorable outcomes, then $\mathbf{y}$ is said to be a Binomial random variable.

BINOMIAL DISTRIBUTION. $\mathbf{y}$ is said to be a Binomial random variable with parameters n and p if $\mathbf{y}$ takes the values $0, 1, 2, \ldots, n$ with

$$P\{\mathbf{y} = k\} = \binom{n}{k} p^k q^{n-k} \qquad p + q = 1 \qquad k = 0, 1, 2, \ldots, n \tag{4-56}$$

The corresponding distribution is a staircase function as in Fig. 4-20. [See also (3-13) and (3-23).]

Another distribution that is closely connected to the binomial distribution is the Poisson distribution, which represents the number of occurrences of a rare event in a large number of trials. Typical examples include the number of telephone calls at an

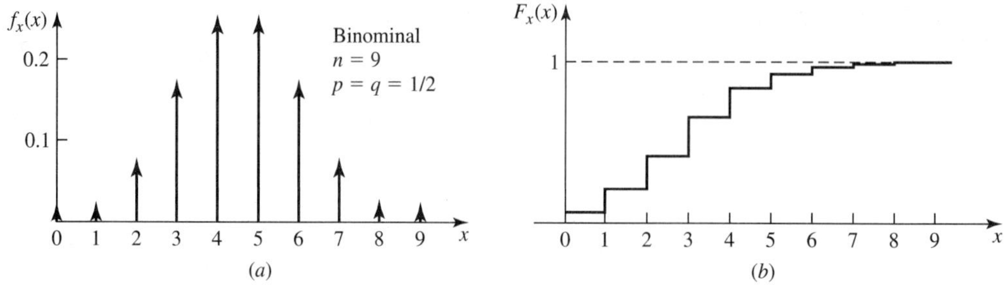

FIGURE 4-20
Binomial distribution ($n = 9$, $p = q = 1/2$).

exchange over a fixed duration, the number of winning tickets among those purchased in a large lottery, and the number of printing errors in a book. Notice that the basic event of interest is a rare one, nevertheless it occurs. The probability distribution of the number of such events is of interest, and it is dictated by the Poisson distribution.

POISSON DISTRIBUTION. $\mathbf{x}$ is said to be a Poisson random variable with parameter λ if $\mathbf{x}$ takes the values $0, 1, 2, \ldots, \infty$, with

$$P\{\mathbf{x} = k\} = e^{-\lambda}\frac{\lambda^k}{k!} \qquad k = 0, 1, 2, \ldots, \infty \tag{4-57}$$

With $p_k = P(\mathbf{x} = k)$, it follows that (see Fig. 4-21)

$$\frac{p_{k-1}}{p_k} = \frac{e^{-\lambda}\lambda^{k-1}/(k-1)!}{e^{-\lambda}\lambda^k/k!} = \frac{k}{\lambda}$$

If $k < \lambda$, then $P(\mathbf{x} = k-1) < P(\mathbf{x} = k)$, but if $k > \lambda$, then $P(\mathbf{x} = k-1) > P(\mathbf{x} = k)$. Finally, if $k = \lambda$, we get $P(\mathbf{x} = k-1) = P(\mathbf{x} = k)$. From this we conclude that $P(\mathbf{x} = k)$ increases with k from 0 till $k \le \lambda$ and falls off beyond λ. If λ is an integer $P(\mathbf{x} = k)$ has two maximal values at $k = \lambda - 1$ and λ.

The corresponding distribution is also a staircase function similar to Fig. 4-20*b* but containing an infinite number of steps.

In summary, if the ratio p_{k-1}/p_k is less than 1, that is, if $k < \lambda$, then as k increases, p_k increases reaching its maximum for $k = [\lambda]$. Hence

if $\lambda < 1$, then p_k is maximum for $k = 0$;

if $\lambda > 1$ but it is not an integer, then p_k increases as k increases, reaching its maximum for $k = [\lambda]$;

if λ is an integer, then p_k is maximum for $k = \lambda - 1$ and $k = \lambda$.

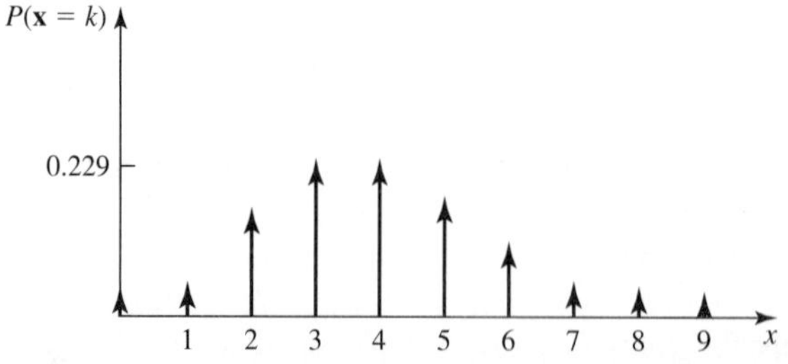

FIGURE 4-21
Poisson distribution ($\lambda = 3$).

EXAMPLE 4-12

POISSON POINTS

▶ In the Poisson points experiment, an outcome ζ is a set of points $\mathbf{t}_i$ on the t axis.

(a) Given a constant t_o, we define the random variable $\mathbf{n}$ such that its value $\mathbf{n}(\zeta)$ equals the number of points $\mathbf{t}_i$ in the interval $(0, t_o)$. Clearly, $\mathbf{n} = k$ means that the number of points in the interval $(0, t_o)$ equals k. Hence [see (4-117) for a proof]

$$P\{\mathbf{n} = k\} = e^{-\lambda t_o} \frac{(\lambda t_o)^k}{k!} \tag{4-58}$$

Thus the number of Poisson points in an interval of length t_o is a Poisson distributed random variable with parameter $a = \lambda t_o$, where λ is the density of the points.

(b) We denote by $\mathbf{t}_1$ the first random point to the right of the fixed point t_o and we define the random variable $\mathbf{x}$ as the distance from t_o to $\mathbf{t}_1$ (Fig. 4-22a). From the definition it follows that $\mathbf{x}(\zeta) \geq 0$ for any ζ. Hence the distribution function of $\mathbf{x}$ is 0 for $x < 0$. We maintain that for $x > 0$ it is given by

$$F(x) = 1 - e^{-\lambda x}$$

Proof. As we know, $F(x)$ equals the probability that $\mathbf{x} \leq x$, where x is a specific number. But $\mathbf{x} \leq x$ means that there is at least one point between t_o and $t_o + x$. Hence $1 - F(x)$ equals the probability p_0 that there are no points in the interval $(t_o, t_o + x)$. And since the length of this interval equals x, (4-58) yields

$$p_0 = e^{-\lambda x} = 1 - F(x)$$

The corresponding density

$$f(x) = \lambda e^{-\lambda x} U(x) \tag{4-59}$$

is *exponential* as in (4-30) (Fig. 4-22b). ◀

As we shall see in the next section, it is possible to establish the Poisson distribution as a limiting case of the binomial distribution under special conditions [see (4-107)].

Recall that the binomial distribution gives the probability of the number of successes in a fixed number of trials. Suppose we are interested in the first success. One might ask how many Bernoulli trials are required to realize the first success. In that case, the number of trials so needed is not fixed in advance, and in fact it is a random number. In a binomial experiment, on the other hand, the number of trials is fixed in advance and the random variable of interest is the total number of successes in n trials.

Let $\mathbf{x}$ denote the number of trials needed to the first success in repeated Bernoulli trials. Then $\mathbf{x}$ is said to be a *geometric* random variable. Thus with A representing the

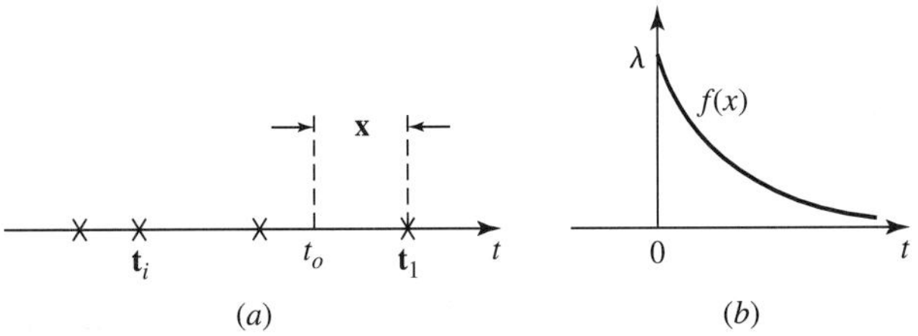

FIGURE 4-22

success event

$$P\{\mathbf{x}=k\} = P(\underbrace{\overline{A}\,\overline{A}\cdots\overline{A}}_{k-1}A) = P(\overline{A})P(\overline{A})\cdots P(\overline{A})P(A)$$
$$= (1-p)^{k-1}p \qquad k = 1, 2, 3, \ldots, \infty$$

GEOMETRIC DISTRIBUTION. **x** is said to be a geometric random variable if

$$P\{\mathbf{x}=k\} = pq^{k-1} \qquad k = 1, 2, 3, \ldots, \infty \tag{4-60}$$

From (4-60), the probability of the event $\{\mathbf{x} > m\}$ is given by

$$P\{\mathbf{x} > m\} = \sum_{k=m+1}^{\infty} P\{\mathbf{x}=k\} = \sum_{k=m+1}^{\infty} pq^{k-1}$$
$$= pq^m(1+q+\cdots) = \frac{pq^m}{1-q} = q^m$$

Thus, for integers $m, n > 1$,

$$P\{\mathbf{x} > m+n \mid \mathbf{x} > m\} = \frac{P\{\mathbf{x} > m+n\}}{P\{\mathbf{x} > m\}} = \frac{q^{m+n}}{q^m} = q^n \tag{4-61}$$

since the event $\{\mathbf{x} > m+n\} \subset \{\mathbf{x} > m\}$. Equation (4-61) states that given that the first m trials had no success, the conditional probability that the first success will appear after an additional n trials depends only on n and not on m (not on the past). Recall that this memoryless property is similar to that exhibited by the exponential random variable.

An obvious generalization to the geometric random variable is to extend it to the number of trials needed for r successes. Let **y** denote the number of Bernoulli trials required to realize r successes. Then **y** is said to be a *negative binomial* random variable. Thus using (4-56) and the independence of trials, we get

$$P\{\mathbf{y}=k\} = P\{r-1 \text{ successes in } k-1 \text{ trials and success at the } k\text{th trial}\}$$
$$= \binom{k-1}{r-1} p^{r-1}q^{k-r}p$$
$$= \binom{k-1}{r-1} p^r q^{k-r} \qquad k = r, r+1, \ldots, \infty \tag{4-62}$$

NEGATIVE BINOMIAL DISTRIBUTION. **y** is said to be negative binomial random variable with parameters r and p if

$$P\{\mathbf{y}=k\} = \binom{k-1}{r-1} p^r q^{k-r} \qquad k = r, r+1, \ldots, \infty \tag{4-63}$$

If n or fewer trials are needed for r successes, then the number of successes in n trials must be at least r. Thus

$$P\{\mathbf{y} \le n\} = P\{\mathbf{x} \ge r\}$$

where $\mathbf{y} \sim NB(r, p)$ as in (4-62) and **x** is a binomial random variable as in (4-56). Since the negative binomial random variable represents the waiting time to the rth success, it is sometimes referred as the waiting-time distribution.

The random variable $\mathbf{z} = \mathbf{y} - r$, that denotes the number of trials (failures) preceding the rth success, has the distribution given by [use (4-62)]

$$P\{\mathbf{z} = k\} = P\{\mathbf{y} = k + r\} = \binom{r+k-1}{r-1} p^r q^k$$

$$= \binom{r+k-1}{k} p^r q^k \qquad k = 0, 1, 2, \ldots, \infty. \tag{4-64}$$

In particular $r = 1$ gives

$$P\{\mathbf{z} = k\} = pq^k \qquad k = 0, 1, 2, \ldots, \infty, \tag{4-65}$$

and sometimes the distribution in (4-65) is referred to also as the *geometric* distribution and that in (4-64) as the *negative binomial* distribution.

EXAMPLE 4-13

▶ Two teams A and B play a series of at most five games. The first team to win three games wins the series. Assume that the outcomes of the games are independent. Let p be the probability for team A to win each game, $0 < p < 1$. Let $\mathbf{x}$ be the number of games needed for A to win. Then $3 \le \mathbf{x} \le 5$. Let the event

$$A_k = \{A \text{ wins on the } k\text{th trial}\} \qquad k = 3, 4, 5$$

We note that $A_k \cap A_l = \phi$, $k \neq l$, so that

$$P(A \text{ wins}) = P\left(\bigcup_{k=3}^{5} A_k\right) = \sum_{k=3}^{5} P(A_k)$$

where

$$P(A_k) = P(3\text{rd success on } k\text{th trial}) = \binom{k-1}{2} p^3 (1-p)^{k-3}$$

Hence

$$P(A \text{ wins}) = \sum_{k=3}^{5} \binom{k-1}{2} p^3 (1-p)^{k-3}$$

If $p = 1/2$, then $P(A \text{ wins}) = 1/2$. The probability that A will win in exactly four games is

$$\binom{3}{2}\left(\frac{1}{2}\right)^4 = \frac{3}{16}$$

The probability that A will win in four games or less is $1/8 + 3/16 = 5/16$.

Given that A has won the first game, the conditional probability of A winning equals

$$\sum_{k=2}^{4} \binom{k-1}{1}\left(\frac{1}{2}\right)^2 \left(\frac{1}{2}\right)^{k-2} = \left(\frac{1}{4} + \frac{2}{8} + \frac{3}{16}\right) = \frac{11}{16}$$ ◀

DISCRETE UNIFORM DISTRIBUTION. The random variable $\mathbf{x}$ is said to be discrete uniform if

$$P\{\mathbf{x} = k\} = \frac{1}{N} \qquad k = 1, 2, \ldots, N \tag{4-66}$$

4-4 CONDITIONAL DISTRIBUTIONS

We recall that the probability of an event A assuming M is given by

$$P(A \mid M) = \frac{P(AM)}{P(M)} \qquad \text{where} \quad P(M) \neq 0$$

The *conditional distribution* $F(x \mid M)$ of a random variable $\mathbf{x}$, assuming M is defined as the conditional probability of the event $\{\mathbf{x} \le x\}$:

$$F(x \mid M) = P\{\mathbf{x} \le x \mid M\} = \frac{P\{\mathbf{x} \le x, M\}}{P(M)} \tag{4-67}$$

In (4-67) $\{\mathbf{x} \le x, M\}$ is the intersection of the events $\{\mathbf{x} \le x\}$ and M, that is, the event consisting of all outcomes ζ such that $\mathbf{x}(\zeta) \le x$ and $\zeta \in M$.

Thus the definition of $F(x \mid M)$ is the same as the definition (4-1) of $F(x)$, provided that all probabilities are replaced by conditional probabilities. From this it follows (see Fundamental remark, Sec. 2-3) that $F(x \mid M)$ has the same properties as $F(x)$. In particular [see (4-3) and (4-8)]

$$F(\infty \mid M) = 1 \qquad F(-\infty \mid M) = 0 \tag{4-68}$$

$$P\{x_1 < \mathbf{x} \le x_2 \mid M\} = F(x_2 \mid M) - F(x_1 \mid M) = \frac{P\{x_1 < \mathbf{x} \le x_2, M\}}{P(M)} \tag{4-69}$$

The *conditional density* $f(x \mid M)$ is the derivative of $F(x \mid M)$:

$$f(x \mid M) = \frac{dF(x \mid M)}{dx} = \lim_{\Delta x \to 0} \frac{P\{x \le \mathbf{x} \le x + \Delta x \mid M\}}{\Delta x} \tag{4-70}$$

This function is nonnegative and its area equals 1.

EXAMPLE 4-14 ▶ We shall determine the conditional $F(x \mid M)$ of the random variable $\mathbf{x}(f_i) = 10i$ of the fair-die experiment (Example 4-4), where $M = \{f_2, f_4, f_6\}$ is the event "even."

If $x \ge 60$, then $\{\mathbf{x} \le x\}$ is the certain event and $\{\mathbf{x} \le x, M\} = M$. Hence (Fig. 4-23)

$$F(x \mid M) = \frac{P(M)}{P(M)} = 1 \qquad x \ge 60$$

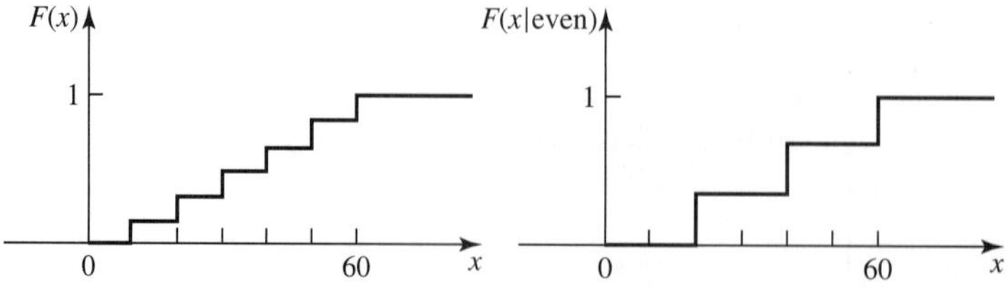

FIGURE 4-23

If $40 \le x < 60$, then $\{\mathbf{x} \le x, M\} = \{f_2, f_4\}$. Hence

$$F(x \mid M) = \frac{P\{f_2, f_4\}}{P(M)} = \frac{2/6}{3/6} \qquad 40 \le x < 60$$

If $20 \le x < 40$, then $\{\mathbf{x} \le x, M\} = \{f_2\}$. Hence

$$F(x \mid M) = \frac{P\{f_2\}}{P(M)} = \frac{1/6}{3/6} \qquad 20 \le x < 60$$

If $x < 20$, then $\{\mathbf{x} \le x, M\} = \{\emptyset\}$. Hence

$$F(x \mid M) = 0 \qquad x < 20$$

◀

To find $F(x \mid M)$, we must, in general, know the underlying experiment. However, if M is an event that can be expressed in terms of the random variable **x**, then, for the determination of $F(x \mid M)$, knowledge of $F(x)$ is sufficient. The two cases presented next are important illustrations.

I. We wish to find the conditional distribution of a random variable **x** assuming that $\mathbf{x} \le a$, where a is number such that $F(a) \ne 0$. This is a special case of (4-67) with

$$M = \{\mathbf{x} \le a\}$$

Thus our problem is to find the function

$$F(x \mid \mathbf{x} \le a) = P\{\mathbf{x} \le x \mid \mathbf{x} \le a\} = \frac{P\{\mathbf{x} \le x, \mathbf{x} \le a\}}{P(\mathbf{x} \le a)}$$

If $x \ge a$, then $\{\mathbf{x} \le x, \mathbf{x} \le a\} = \{\mathbf{x} \le a\}$. Hence (Fig. 4-24)

$$F(x \mid \mathbf{x} \le a) = \frac{P\{\mathbf{x} \le a\}}{P\{\mathbf{x} \le a\}} = 1 \qquad x \ge a$$

If $x < a$, then $\{\mathbf{x} \le x, \mathbf{x} \le a\} = \{\mathbf{x} \le x\}$. Hence

$$F(x \mid \mathbf{x} \le a) = \frac{P\{\mathbf{x} \le x\}}{P\{\mathbf{x} \le a\}} = \frac{F(x)}{F(a)} \qquad x < a$$

Differentiating $F(x \mid \mathbf{x} \le a)$ with respect to x, we obtain the corresponding density: Since $F'(x) = f(x)$, the preceding yields

$$f(x \mid \mathbf{x} \le a) = \frac{f(x)}{F(a)} = \frac{f(x)}{\int_{-\infty}^{a} f(x)\,dx} \qquad \text{for} \quad x < a$$

and it is 0 for $x \ge a$.

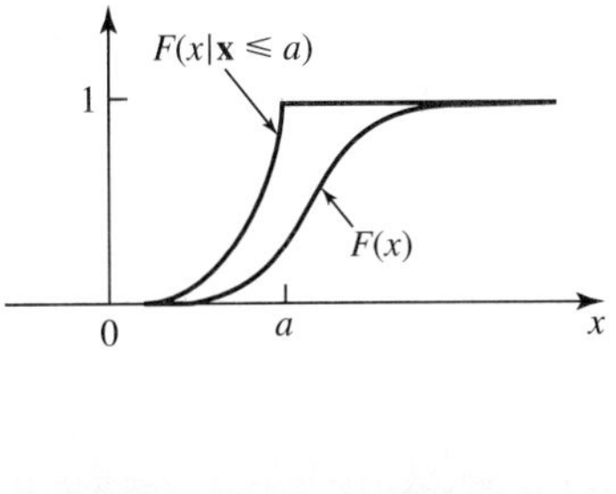

FIGURE 4-24

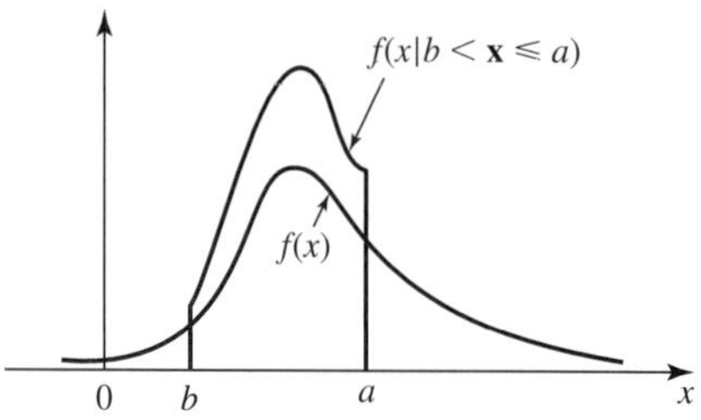

FIGURE 4-25

II. Suppose now that $M = \{b < \mathbf{x} \leq a\}$. In this case, (4-67) yields

$$F(x \mid b < \mathbf{x} \leq a) = \frac{P\{\mathbf{x} \leq x, b < \mathbf{x} \leq a\}}{P\{b < \mathbf{x} \leq a\}}$$

If $x \geq a$, then $\{\mathbf{x} \leq x, b < \mathbf{x} \leq a\} = \{b < \mathbf{x} \leq a\}$. Hence

$$F(x \mid b < \mathbf{x} \leq a) = \frac{F(a) - F(b)}{F(a) - F(b)} = 1 \qquad x \geq a$$

If $b \leq x < a$, then $\{\mathbf{x} \leq x, b < \mathbf{x} \leq a\} = \{b < \mathbf{x} \leq x\}$. Hence

$$F(x \mid b < \mathbf{x} \leq a) = \frac{F(x) - F(b)}{F(a) - F(b)} \qquad b \leq x < a$$

Finally, if $x < b$, then $\{\mathbf{x} \leq x, b < \mathbf{x} \leq a\} = \{\emptyset\}$. Hence

$$F(x \mid b < \mathbf{x} \leq a) = 0 \qquad x < b$$

The corresponding density is given by

$$f(x \mid b < \mathbf{x} \leq a) = \frac{f(x)}{F(a) - F(b)} \qquad \text{for} \quad b \leq x < a$$

and it is 0 otherwise (Fig. 4-25).

EXAMPLE 4-15 ▶ We shall determine the conditional density $f(x \mid |\mathbf{x} - \eta| \leq k\sigma)$ of an $N(\eta; \sigma)$ random variable. Since

$$P\{|\mathbf{x} - \eta| \leq k\sigma\} = P\{\eta - k\sigma \leq \mathbf{x} \leq \eta + k\sigma\} = 2\int_0^k \frac{1}{\sqrt{2\pi}} e^{-x^2/2}\, dx$$

we conclude from (4-72) that

$$f(x \mid |\mathbf{x} - \eta| \leq k\sigma) = \frac{1}{P(|\mathbf{x} - \eta| \leq k\sigma)} \frac{e^{-(x-\eta)^2/2\sigma^2}}{\sigma\sqrt{2\pi}}$$

for $\mathbf{x}$ between $\eta - k\sigma$ and $\eta + k\sigma$ and 0 otherwise. This density is called *truncated normal.* ◀

Frequency interpretation In a sequence of n trials, we reject all outcomes ζ such that $\mathbf{x}(\zeta) \leq b$ or $\mathbf{x}(\zeta) > a$. In the subsequence of the remaining trials, $F(x \mid b < \mathbf{x} \leq a)$ has the same frequency interpretation as $F(x)$ [see (4-3)].

EXAMPLE 4-16 ▶ The memoryless property of a geometric random variable **x** states that [see (4-61)]

$$P\{\mathbf{x} > m + n \mid \mathbf{x} > m\} = P\{\mathbf{x} > n\} \tag{4-71}$$

Show that the converse is also true. (i.e., if **x** is a nonnegative integer valued random variable satisfying (4-71) for any two positive integers m and n, then **x** is a geometric random variable.)

SOLUTION
Let

$$p_k = P\{\mathbf{x} = k\}, \qquad k = 1, 2, 3, \ldots$$

so that

$$P\{\mathbf{x} > n\} = \sum_{k=n+1}^{\infty} p_k = a_n \tag{4-72}$$

and hence using (4-71)

$$\begin{aligned} P\{\mathbf{x} > m + n \mid \mathbf{x} > m\} &= \frac{P\{\mathbf{x} > m + n\}}{P\{\mathbf{x} > m\}} = \frac{a_{m+n}}{a_m} \\ &= P\{\mathbf{x} > n\} = a_n \end{aligned}$$

Hence

$$a_{m+n} = a_m \, a_n$$

or

$$a_{m+1} = a_m \, a_1 = a_1^{m+1}$$

where

$$a_1 = P\{\mathbf{x} > 1\} = 1 - P\{\mathbf{x} = 1\} \triangleq 1 - p$$

Thus

$$a_m = (1 - p)^m$$

and from (4-72)

$$\begin{aligned} P\{\mathbf{x} = n\} &= P\{\mathbf{x} \geq n\} - P\{\mathbf{x} > n\} \\ &= a_{n-1} - a_n = p\,(1 - p)^{n-1} \qquad n = 1, 2, 3, \ldots \end{aligned}$$

comparing with (4-60) the proof is complete. ◀

Total Probability and Bayes' Theorem

We shall now extend the results of Sec. 2-3 to random variables.

1. Setting $B = \{\mathbf{x} \leq x\}$ in (2-41), we obtain

$$P\{\mathbf{x} \leq x\} = P\{\mathbf{x} \leq x \mid A_1\}P(A_1) + \cdots + P\{\mathbf{x} \leq x \mid A_n\}P(A_n)$$

Hence [see (4-67) and (4-70)]

$$F(x) = F(x \mid A_1)P(A_1) + \cdots + F(x \mid A_n)P(A_n) \tag{4-73}$$

$$f(x) = f(x \mid A_1)P(A_1) + \cdots + f(x \mid A_n)P(A_n) \tag{4-74}$$

In the above, the events $A_1, \ldots, A_n$ form a partition of S.

EXAMPLE 4-17

▶ Suppose that the random variable $\mathbf{x}$ is such that $f(x \mid M)$ is $N(\eta_1; \sigma_1)$ and $f(x \mid \overline{M})$ is $N(\eta_2, \sigma_2)$ as in Fig. 4-26. Clearly, the events M and $\overline{M}$ form a partition of S. Setting $A_1 = M$ and $A_2 = \overline{M}$ in (4-74), we conclude that

$$f(x) = pf(x \mid M) + (1-p)f(x \mid \overline{M}) = \frac{p}{\sigma_1}\mathbf{G}\left(\frac{x-\eta_1}{\sigma_1}\right) + \frac{1-p}{\sigma_2}\mathbf{G}\left(\frac{x-\eta_2}{\sigma_2}\right)$$

where $p = P(M)$. ◀

2. From the identity

$$P(A \mid B) = \frac{P(B \mid A)P(A)}{P(B)} \tag{4-75}$$

[see (2-43)] it follows that

$$P(A \mid \mathbf{x} \le x) = \frac{P\{\mathbf{x} \le x \mid A\}}{P\{\mathbf{x} \le x\}}P(A) = \frac{F(x \mid A)}{F(x)}P(A) \tag{4-76}$$

3. Setting $B = \{x_1 < \mathbf{x} \le x_2\}$ in (4-75), we conclude with (4-69) that

$$P\{A \mid x_1 < \mathbf{x} \le x_2\} = \frac{P\{x_1 < \mathbf{x} \le x_2 \mid A\}}{P\{x_1 < \mathbf{x} \le x_2\}}P(A)$$

$$= \frac{F(x_2 \mid A) - F(x_1 \mid A)}{F(x_2) - F(x_1)}P(A) \tag{4-77}$$

4. The conditional probability $P(A \mid \mathbf{x} = x)$ of the event A assuming $\mathbf{x} = x$ cannot be defined as in (2-33) because, in general, $P\{\mathbf{x} = x\} = 0$. We shall define it as a limit

$$P\{A \mid \mathbf{x} = x\} = \lim_{\Delta x \to 0} P\{A \mid x < \mathbf{x} \le x + \Delta x\} \tag{4-78}$$

With $x_1 = x$, $x_2 = x + \Delta x$, we conclude from the above and (4-77) that

$$P\{A \mid \mathbf{x} = x\} = \frac{f(x \mid A)}{f(x)}P(A) \tag{4-79}$$

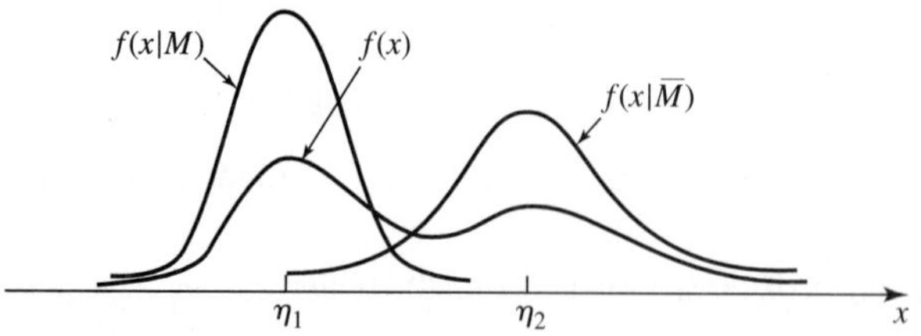

FIGURE 4-26

Total probability theorem. As we know [see (4-68)]

$$F(\infty \mid A) = \int_{-\infty}^{\infty} f(x \mid A)\,dx = 1$$

Multiplying (4-79) by $f(x)$ and integrating, we obtain

$$\int_{-\infty}^{\infty} P(A \mid \mathbf{x} = x) f(x)\,dx = P(A) \tag{4-80}$$

This is the continuous version of the total probability theorem (2-41).

Bayes' theorem. From (4-79) and (4-80) it follows that

$$f(x \mid A) = \frac{P(A \mid \mathbf{x} = x)}{P(A)} f(x) = \frac{P(A \mid \mathbf{x} = x) f(x)}{\int_{-\infty}^{\infty} P(A \mid \mathbf{x} = x) f(x)\,dx} \tag{4-81}$$

This is the continuous version of Bayes' theorem (2-44).

EXAMPLE 4-18

▶ Suppose that the probability of heads in a coin-tossing experiment S is not a number, but a random variable $\mathbf{p}$ with density $f(p)$ defined in some space S_c. The experiment of the toss of a randomly selected coin is a cartesian product $S_c \times S$. In this experiment, the event $H = \{\text{head}\}$ consists of all pairs of the form $\zeta_c h$ where ζ_c is any element of S_c and h is the element heads of the space $S = \{h, t\}$. We shall show that

$$P(H) = \int_0^1 p f(p)\,dp \tag{4-82}$$

SOLUTION

The conditional probability of H assuming $\mathbf{p} = p$ is the probability of heads if the coin with $\mathbf{p} = p$ is tossed. In other words,

$$P\{H \mid \mathbf{p} = p\} = p \tag{4-83}$$

Inserting into (4-80), we obtain (4-81) because $f(p) = 0$ outside the interval (0, 1). ◀

To illustrate the usefulness of this formulation, let us reexamine the coin-tossing problem.

EXAMPLE 4-19

▶ Let $p = P(H)$ represent the probability of obtaining a head in a toss. For a given coin, a priori p can possess any value in the interval (0, 1) and hence we may consider it to be a random variable $\mathbf{p}$. In the absence of any additional information, we may assume the a-priori p.d.f. $f_p(p)$ to be a uniform distribution in that interval (see Fig. 4-27). Now suppose we actually perform an experiment of tossing the coin n times and k heads are observed. This is new information. How do we update $f_p(p)$?

SOLUTION

Let A = "k heads in n specific tosses." Since these tosses result in a specific sequence,

$$P(A \mid \mathbf{p} = p) = p^k q^{n-k} \tag{4-84}$$

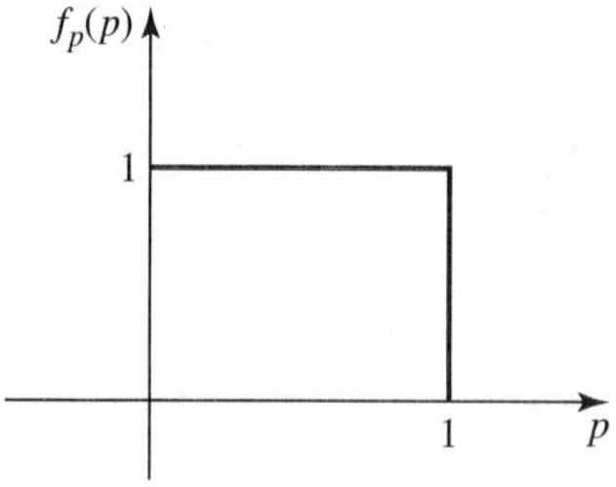

FIGURE 4-27

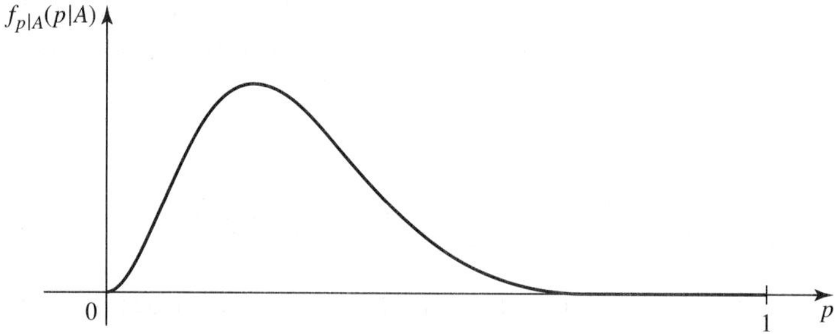

FIGURE 4-28

and using (4-80) we get

$$P(A) = \int_0^1 P(A \mid \mathbf{p} = p) f_p(p)\,dp = \int_0^1 p^k (1-p)^{n-k}\,dp = \frac{(n-k)!k!}{(n+1)!} \tag{4-85}$$

The a posteriori p.d.f. $f_p(p \mid A)$ (see Fig. 4-28) represents the updated information given the event A, and from (4-81)

$$\begin{aligned} f_{p \mid A}(p \mid A) &= \frac{P(A \mid \mathbf{p} = p) f_P(p)}{P(A)} \\ &= \frac{(n+1)!}{(n-k)!k!}\, p^k q^{n-k} \qquad 0 < p < 1 \quad \sim \beta(n, k) \end{aligned} \tag{4-86}$$

Notice that the a posteriori p.d.f. of $\mathbf{p}$ in (4-86) is not a uniform distribution, but a beta distribution. We can use this a posteriori p.d.f. to make further predictions. For example, in the light of this experiment, what can we say about the probability of a head occurring in the next $(n+1)$th toss?

Let B = "head occurring in the $(n+1)$th toss, given that k heads have occurred in n previous tosses." Clearly $P(B \mid \mathbf{p} = p) = p$, and from (4-80)

$$P(B) = \int_0^1 P(B \mid \mathbf{p} = p) f_p(p \mid A)\,dp \tag{4-87}$$

Notice that unlike (4-80), we have used the a posteriori p.d.f. in (4-87) to reflect our knowledge about the experiment already performed. Using (4-86) in (4-87), we get

$$P(B) = \int_0^1 p \cdot \frac{(n+1)!}{(n-k)!k!} p^k q^{n-k}\,dp = \frac{k+1}{n+2} \tag{4-88}$$

Thus, if $n = 10$, and $k = 6$, then

$$P(B) = \frac{7}{12} = 0.58$$

which is better than $p = 0.5$.

To summarize, if the probability density function of a random variable $\mathbf{x}$ is unknown, one should make noncommittal judgment about its a priori p.d.f. $f_x(x)$. Usually, the uniform distribution is a reasonable assumption in the absence of any other information. Then experimental results (A) are obtained, and the knowledge about $\mathbf{x}$ is updated reflecting this new information. Bayes' rule helps to obtain the a posteriori p.d.f. of $\mathbf{x}$ given A. From that point on, this a posteriori p.d.f. $f_{x \mid A}(x \mid A)$ should be used to make further predictions and calculations. ◀

4-5 ASYMPTOTIC APPROXIMATIONS FOR BINOMIAL RANDOM VARIABLE

Let $\mathbf{x}$ represent a binomial random variable as in (4-56). Then from (3-12), (4-15), and (4-18)

$$P\{k_1 \le \mathbf{x} \le k_2\} = \sum_{k=k_1}^{k_2} p_n(k) = \sum_{k=k_1}^{k_2} \binom{n}{k} p^k q^{n-k} \tag{4-89}$$

Since the binomial coefficient

$$\binom{n}{k} = \frac{n!}{(n-k)!k!}$$

grows quite rapidly with n, it is difficult to compute (4-89) for large n. In this context, two approximations—the normal approximation and the Poisson approximation—are extremely useful.

The Normal Approximation (DeMoivre–Laplace Theorem)

Suppose $n \to \infty$ with p held fixed. Then for k in the $\sqrt{npq}$ neighborhood of np, as we shall show in Chap. 5 [see (5-120) and (5-121)], we can approximate

$$\binom{n}{k} p^k q^{n-k} \simeq \frac{1}{\sqrt{2\pi npq}} e^{-(k-np)^2/2npq} \qquad p + q = 1 \tag{4-90}$$

This important approximation, known as the DeMoivre–Laplace theorem, can be stated as an equality in the limit: The ratio of the two sides tends to 1 as $n \to \infty$. Thus if k_1 and k_2 in (4-89) are within or around the neighborhood of the interval $(np - \sqrt{npq},\ np + \sqrt{npq})$, we can approximate the summation in (4-89) by an integration of the normal density function. In that case, (4-89) reduces to

$$P\{k_1 \le \mathbf{x} \le k_2\} = \int_{k_1}^{k_2} \frac{1}{\sqrt{2\pi npq}} e^{-(k-np)^2/2npq}\,dx = \int_{x_1}^{x_2} \frac{1}{2\pi} e^{-y^2/2}\,dy \tag{4-91}$$

TABLE 4-1

$$\operatorname{erf} x = \frac{1}{\sqrt{2\pi}}\int_0^x e^{-y^2/2}\,dy = G(x) - \frac{1}{2},\ x > 0$$

x	erf x	x	erf x	x	erf x	x	erf x
0.05	0.01994	0.80	0.28814	1.55	0.43943	2.30	0.48928
0.10	0.03983	0.85	0.30234	1.60	0.44520	2.35	0.49061
0.15	0.05962	0.90	0.31594	1.65	0.45053	2.40	0.49180
0.20	0.07926	0.95	0.32894	1.70	0.45543	2.45	0.49286
0.25	0.09871	1.00	0.34134	1.75	0.45994	2.50	0.49379
0.30	0.11791	1.05	0.35314	1.80	0.46407	2.55	0.49461
0.35	0.13683	1.10	0.36433	1.85	0.46784	2.60	0.49534
0.40	0.15542	1.15	0.37493	1.90	0.47128	2.65	0.49597
0.45	0.17364	1.20	0.38493	1.95	0.47441	2.70	0.49653
0.50	0.19146	1.25	0.39435	2.00	0.47726	2.75	0.49702
0.55	0.20884	1.30	0.40320	2.05	0.47982	2.80	0.49744
0.60	0.22575	1.35	0.41149	2.10	0.48214	2.85	0.49781
0.65	0.24215	1.40	0.41924	2.15	0.48422	2.90	0.49813
0.70	0.25804	1.45	0.42647	2.20	0.48610	2.95	0.49841
0.75	0.27337	1.50	0.43319	2.25	0.48778	3.00	0.49865

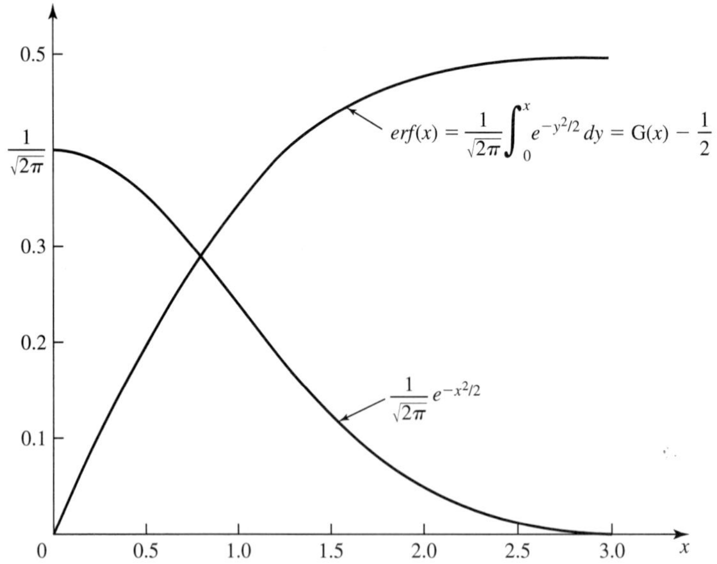

FIGURE 4-29

where

$$x_1 = \frac{k_1 - np}{\sqrt{npq}} \qquad x_2 = \frac{k_2 - np}{\sqrt{npq}}$$

We define (see Fig. 4-29 and Table 4-1)

$$G(x) = \int_{-\infty}^{x} \frac{1}{\sqrt{2\pi}} e^{-y^2/2}\,dy \tag{4-92}$$

as before, and the error function

$$\operatorname{erf} x = \int_0^x \frac{1}{\sqrt{2\pi}} e^{-y^2/2}\, dy = G(x) - \frac{1}{2}, x > 0 \tag{4-93}$$

Note that $G(-x) = 1 - G(x)$, $x > 0$. In terms of $G(x)$, we obtain

$$P(k_1 \le \mathbf{x} \le k_2) = G(x_2) - G(x_1) \tag{4-94}$$

Thus the evaluation of the probability of k successes in n trials, given exactly by (3-13), is reduced to the evaluation of the normal curve

$$\frac{1}{\sqrt{2\pi npq}} e^{-(x-np)^2/2npq} \tag{4-95}$$

for $x = k$.

EXAMPLE 4-20 ▶ A fair coin is tossed 1000 times. Find the probability p_a that heads will show 500 times and the probability p_b that heads will show 510 times.

In this example

$$p = q = 0.5 \qquad n = 1000 \qquad \sqrt{npq} = 5\sqrt{10}$$

(a) If $k = 500$, then $k - np = 0$ and (4-90) yields

$$p_a \simeq \frac{1}{\sqrt{2\pi npq}} = \frac{1}{10\sqrt{5\pi}} = 0.0252$$

(b) If $k = 510$, then $k - np = 10$ and (4-90) yields

$$p_b \simeq \frac{e^{-0.2}}{10\sqrt{5\pi}} = 0.0207$$ ◀

As Example 4-21 indicates, the approximation (4-90) is satisfactory even for moderate values of n.

EXAMPLE 4-21 ▶ We shall determine $p_n(k)$ for $p = 0.5$, $n = 10$, and $k = 5$.

(a) Exactly from (3-13)

$$p_n(k) = \binom{n}{k} p^k q^{n-k} = \frac{10!}{5!5!} \frac{1}{2^{10}} = 0.246$$

(b) Approximately from (4-90)

$$p_n(k) \simeq \frac{1}{\sqrt{2\pi npq}} e^{-(k-np)^2/2npq} = \frac{1}{\sqrt{5\pi}} = 0.252$$ ◀

APPROXIMATE EVALUATION OF $P\{k_1 \le k \le k_2\}$. Using the approximation (4-90), we shall show that

$$\sum_{k=k_1}^{k_2} \binom{n}{k} p^k q^{n-k} \simeq G\left(\frac{k_2 - np}{\sqrt{npq}}\right) - G\left(\frac{k_1 - np}{\sqrt{npq}}\right) \tag{4-96}$$

Thus, to find the probability that in n trials the number of occurrences of an event A is between k_1 and k_2, it suffices to evaluate the tabulated normal function $G(x)$. The approximation is satisfactory if $npq \gg 1$ and the differences $k_1 - np$ and $k_2 - np$ are of the order of $\sqrt{npq}$.

Proof. Inserting (4-90) into (3-24), we obtain

$$\sum_{k=k_1}^{k_2} \binom{n}{k} p^k q^{n-k} \simeq \frac{1}{\sigma\sqrt{2\pi}} \sum_{k=k_1}^{k_2} e^{-(k-np)^2/2\sigma^2} \tag{4-97}$$

The normal curve is nearly constant in any interval of length 1 because $\sigma^2 = npq \gg 1$ by assumption; hence its area in such an interval equals approximately its ordinate (Fig. 4-30). From this it follows that the right side of (4-97) can be approximated by the integral of the normal curve in the interval (k_1, k_2). This yields

$$\frac{1}{\sigma\sqrt{2\pi}} \sum_{k=k_1}^{k_2} e^{-(k-np)^2/2\sigma^2} \simeq \frac{1}{\sigma\sqrt{2\pi}} \int_{k_1}^{k_2} e^{-(x-np)^2/2\sigma^2}\, dx \tag{4-98}$$

and (4-96) results [see (4-94)].

Error correction. The sum on the left of (4-97) consists of $k_2 - k_1 + 1$ terms. The integral in (4-98) is an approximation of the shaded area of Fig. 4-31a, consisting of $k_2 - k_1$ rectangles. If $k_2 - k_1 \gg 1$ the resulting error can be neglected. For moderate values of $k_2 - k_1$, however, the error is no longer negligible. To reduce it, we replace in (4-96) the limits k_1 and k_2 by $k_1 - 1/2$ and $k_2 + 1/2$ respectively (see Fig. 4-31b). This

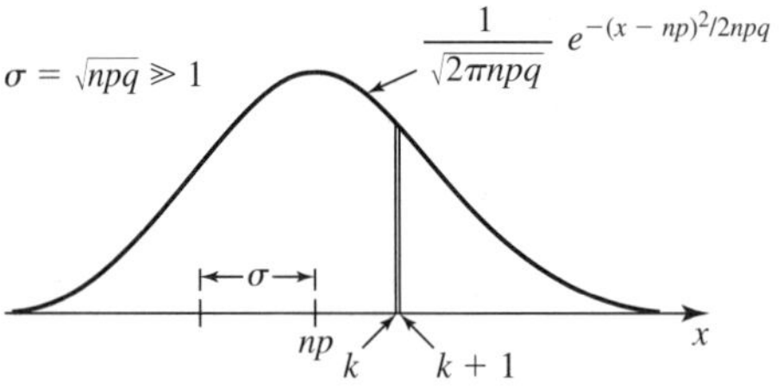

FIGURE 4-30

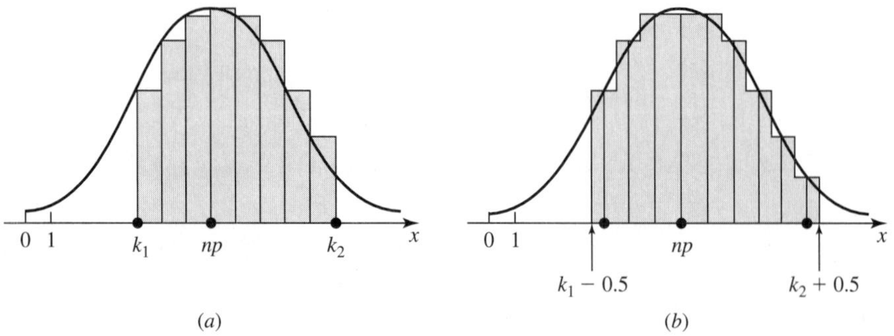

FIGURE 4-31

yields the improved approximation

$$\sum_{k=k_1}^{k_2} \binom{n}{k} p^k q^{n-k} \simeq \mathrm{G}\left(\frac{k_2 + 0.5 - np}{\sqrt{npq}}\right) - \mathrm{G}\left(\frac{k_1 - 0.5 - np}{\sqrt{npq}}\right) \tag{4-99}$$

EXAMPLE 4-22 ▶ A fair coin is tossed 10 000 times. What is the probability that the number of heads is between 4900 and 5100?

In this problem

$$n = 10{,}000 \qquad p = q = 0.5 \qquad k_1 = 4900 \qquad k_2 = 5100$$

Since $(k_2 - np)/\sqrt{npq} = 100/50$ and $(k_1 - np)/\sqrt{npq} = -100/50$, we conclude from (4-96) that the unknown probability equals

$$\mathrm{G}(2) - \mathrm{G}(-2) = 2\mathrm{G}(2) - 1 = 0.9545$$ ◀

EXAMPLE 4-23 ▶ Over a period of 12 hours, 180 calls are made at random. What is the probability that in a four-hour interval the number of calls is between 50 and 70?

This situation can be considered as a problem in repeated trials with $p = 4/12$ the probability that a particular call will occur in the four-hour interval. The probability that k calls will occur in this interval equals [see (4-90)]

$$\binom{180}{k}\left(\frac{1}{3}\right)^k\left(\frac{2}{3}\right)^{180-k} \simeq \frac{1}{4\sqrt{5\pi}} e^{-(k-60)^2/80}$$

and the probability that the number of calls is between 50 and 70 equals [see (4-96)]

$$\sum_{k=50}^{70}\binom{180}{k}\left(\frac{1}{3}\right)^k\left(\frac{2}{3}\right)^{180-k} \simeq \mathrm{G}(\sqrt{2.5}) - \mathrm{G}(-\sqrt{2.5}) \simeq 0.886$$ ◀

Note It seems that we cannot use the approximation (4-96) if $k_1 = 0$ because the sum contains values of k that are not in the $\sqrt{npq}$ vicinity of np. However, the corresponding terms are small compared to the terms with k near np; hence the errors of their estimates are also small. Since

$$\mathrm{G}(-np/\sqrt{npq}) = \mathrm{G}(-\sqrt{np/q}) \simeq 0 \quad \text{for} \quad np/q \gg 1$$

we conclude that if not only $n \gg 1$ but also $np \gg 1$, then

$$\sum_{k=0}^{k_2}\binom{n}{k} p^k q^{n-k} \simeq \mathrm{G}\left(\frac{k_2 - np}{\sqrt{npq}}\right) \tag{4-100}$$

In the sum (3-25) of Example 3-13 (Chap. 3),

$$np = 1000 \qquad npq = 900 \qquad \frac{k_2 - np}{\sqrt{npq}} = \frac{10}{3}$$

Using (4-100), we obtain

$$\sum_{k=0}^{1100}\binom{10^4}{k}(0.1)^k(0.9)^{10^4-k} \simeq \mathrm{G}\left(\frac{10}{3}\right) = 0.99936$$

We note that the sum of the terms of the above sum from 900 to 1100 equals $2\mathrm{G}(10/3) - 1 \simeq 0.99872$.

The Law of Large Numbers

According to the relative frequency interpretation of probability, if an event A with $P(A) = p$ occurs k times in n trials, then $k \simeq np$. In the following, we rephrase this heuristic statement as a limit theorem.

We start with the observation that $k \simeq np$ does not mean that k will be close to np. In fact [see (4-90)]

$$P\{k = np\} \simeq \frac{1}{\sqrt{2\pi npq}} \to 0 \qquad \text{as} \quad n \to \infty \tag{4-101}$$

As we saw in Bernoulli's theorem (Sec. 3-3), the approximation $k \simeq np$ means that the ratio k/n is close to p in the sense that, for any $\varepsilon > 0$, the probability that $|k/n - p| < \varepsilon$ tends to 1 as $n \to \infty$.

EXAMPLE 4-24

▶ Suppose that $p = q = 0.5$ and $\varepsilon = 0.05$. In this case,

$$k_1 = n(p - \varepsilon) = 0.45n \qquad k_2 = n(p + \varepsilon) = 0.55n$$

$$(k_2 - np)/\sqrt{npq} = \varepsilon\sqrt{n/pq} = 0.1\sqrt{n}$$

In the table below we show the probability $2G(0.1\sqrt{n}) - 1$ that k is between $0.45n$ and $0.55n$ for various values of n.

n	100	400	900
$0.1\sqrt{n}$	1	2	3
$2G(0.1\sqrt{n}) - 1$	0.682	0.954	0.997

◀

EXAMPLE 4-25

▶ We now assume that $p = 0.6$ and we wish to find n such that the probability that k is between $0.59n$ and $0.61n$ is at least 0.98.

In this case, $p = 0.6$, $q = 0.4$, and $\varepsilon = 0.01$. Hence

$$P\{0.59n \le k \le 0.61n\} \simeq 2G(0.01\sqrt{n/0.24}) - 1$$

Thus n must be such that

$$2G(0.01\sqrt{n/0.24}) - 1 \ge 0.98$$

From Table 4-1 we see that $G(x) > 0.99$ if $x > 2.35$. Hence $0.01\sqrt{n/0.24} > 2.35$ yielding $n > 13254$. ◀

GENERALIZATION OF BERNOULLI TRIALS. The experiment of repeated trials can be phrased in the following form: The events $A_1 = A$ and $A_2 = \overline{A}$ of the space S form a partition and their respective probabilities equal $p_1 = p$ and $p_2 = 1 - p$. In the space S^n, the probability of the event $\{A_1$ occurs $k_1 = k$ times and A_2 occurs $k_2 = n - k$ times in any order$\}$ equals $p_n(k)$ as in (3-13). We shall now generalize.

Suppose that

$$U = [A_1, \ldots, A_r]$$

is a partition of S consisting of the r events A_i with

$$P(A_i) = p_i \qquad p_1 + \cdots + p_r = 1$$

We repeat the experiment n times and we denote by $p_n(k_1, \ldots, k_r)$ the probability of the event $\{A_1$ occurs k_1 times, ..., A_r occurs k_r times in any order$\}$ where

$$k_1 + \cdots + k_r = n$$

We maintain that

$$p_n(k_1, \ldots, k_r) = \frac{n!}{k_1! \cdots k_r!} p_1^{k_1} \ldots p_r^{k_r} \tag{4-102}$$

Proof. Repeated application of (3-11) leads to the conclusion that the number of events of the form $\{A_1$ occurs k_1 times, ..., A_r occurs k_r times in a specific order$\}$ equals

$$\frac{n!}{k_1! \cdots k_r!}$$

Since the trials are independent, the probability of each such event equals

$$p^{k_1} \ldots p_r^{k_r}$$

and (4-102) results.

EXAMPLE 4-26

▶ A fair die is rolled 10 times. We shall determine the probability that f_1 shows three times, and "even" shows six times.

In this case

$$A_1 = \{f_1\} \qquad A_2 = \{f_2, f_4, f_6\} \qquad A_3 = \{f_3, f_5\}$$

Clearly,

$$p_1 = \tfrac{1}{6} \qquad p_2 = \tfrac{3}{6} \qquad p_3 = \tfrac{2}{6} \qquad k_1 = 3 \qquad k_2 = 6 \qquad k_3 = 1$$

and (4-102) yields

$$p_{10}(3, 6, 1) = \frac{10!}{3!6!1!}\left(\frac{1}{6}\right)^3\left(\frac{1}{2}\right)^6\frac{1}{3} = 0.002$$ ◀

THEOREM 4-2

DEMOIVRE–LAPLACE THEOREM

▶ We can show as in (4-90) that, if k_i is in the $\sqrt{n}$ vicinity of np_i and n is sufficiently large, then

$$\frac{n!}{k_1! \cdots k_r!} p_1^{k_1} \ldots p_1^{k_r} \simeq \frac{\exp\left\{-\frac{1}{2}\left[\frac{(k_1 - np_1)^2}{np_1} + \cdots + \frac{(k_r - np_r)^2}{np_r}\right]\right\}}{\sqrt{(2\pi n)^{r-1} p_1 \ldots p_r}} \tag{4-103}$$

Equation (4-90) is a special case. ◀

The Poisson Approximation

As we have mentioned earlier, for large n, the Gaussian approximation of a binomial random variable is valid only if p is fixed, that is, only if $np \gg 1$ and $npq \gg 1$. From the

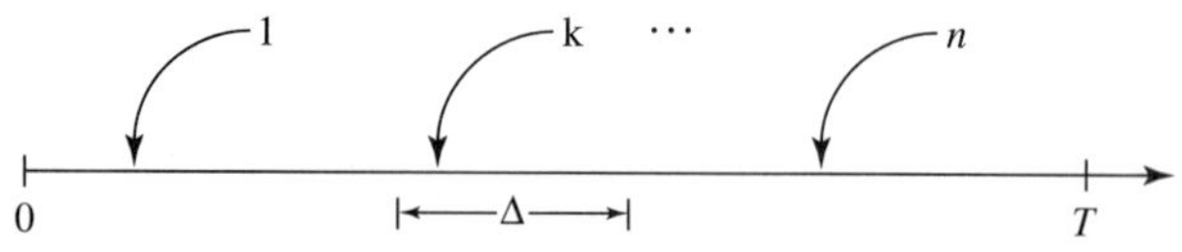

FIGURE 4-32

conditions of the DeMoivre–Laplace theorem, the Gaussian representation deteriorates as $p \to 0$ or $p \to 1$. This representation completely fails for $p = 0$, $q = 1$, or $p = 1$, $q = 0$. However, a variety of problems fall under this category where p (or q) is very small, but the total number of trials is very large, thereby making it necessary to find an asymptotic formula for such rare events. Obviously, that is the case if, for example, $p \to 0$ as $n \to \infty$, such that $np = \lambda$ is a fixed number.

Many random phenomena in nature in fact follow this pattern. The number of calls on a telephone line, claims in an insurance company, and the like tend to follow this type of behavior. Consider random arrivals such as telephone calls over a line. Let n represent the total number of calls in the interval $(0, T)$. From our experience, as $T \to \infty$ we also have $n \to \infty$, so that we may assume $n = \mu T$. Consider a small interval of duration Δ as in Fig. 4-32. Had there been only a single call coming in, the probability p of that single call occurring in that interval must depend on its relative size with respect to T.

Hence as in Example 3-10 we can assume $p = \Delta/T$. Note that $p \to 0$ as $T \to \infty$. However in this case $np = \mu T(\Delta/T) = \mu\Delta = \lambda$ is a constant, and the normal approximation is invalid here.

Suppose the interval Δ in Fig. 4-32 is of interest to us. A call inside that interval is a "success" (H), whereas one outside is a "failure" (T). This is equivalent to the coin tossing situation, and hence the probability $p_n(k)$ of obtaining k calls (in any order) in an interval of duration Δ is given by the binomial p.m.f. Thus from (3-13) and (4-56)

$$p_n(k) = \frac{n!}{(n-k)!k!} p^k (1-p)^{n-k} \tag{4-104}$$

as in (3-21), and as $n \to \infty$ we have $p \to 0$ such that $np = \lambda$. It is easy to obtain an excellent approximation to (4-104) in that situation. To see this, rewrite (4-104) as

$$\begin{aligned} p_n(k) &= \frac{n(n-1)\cdots(n-k+1)}{n^k}\frac{(np)^k}{k!}\left(1-\frac{np}{n}\right)^{n-k} \\ &= \frac{\left(1-\frac{1}{n}\right)\left(1-\frac{2}{n}\right)\cdots\left(1-\frac{k-1}{n}\right)}{\left(1-\frac{\lambda}{n}\right)^k}\frac{\lambda^k}{k!}\left(1-\frac{\lambda}{n}\right)^n \\ &= \left(\prod_{m=0}^{k-1}\frac{\left(1-\frac{m}{n}\right)}{\left(1-\frac{\lambda}{n}\right)}\right)\frac{\lambda^k}{k!}\left(1-\frac{\lambda}{n}\right)^n \\ &= \left(\prod_{m=0}^{k-1}\left(1+\frac{\lambda-m}{n-\lambda}\right)\right)\frac{\lambda^k}{k!}\left(1-\frac{\lambda}{n}\right)^n \end{aligned} \tag{4-105}$$

Thus as $n \to \infty$, $p \to 0$ such that $np = \lambda$

$$p_n(k) \to \frac{\lambda^k}{k!}e^{-\lambda} \tag{4-106}$$

since the finite product

$$\prod_{m=0}^{k-1}\left(1+\frac{\lambda-m}{n-\lambda}\right) \le \prod_{m=0}^{\lfloor\lambda\rfloor}\left(1+\frac{\lambda-m}{n-\lambda}\right)$$

tends to unity as $n \to \infty$, and

$$\lim_{n\to\infty}\left(1-\frac{\lambda}{n}\right)^n = e^{-\lambda}$$

The right side of (4-106) represents the Poisson p.m.f. described in (4-57) and the Poisson approximation to the binomial random variable is valid in situations where the binomial random varible parameters n and p diverge to two extremes ($n \to \infty$, $p \to 0$) such that their product np is a constant. Thus if a phenomenon consists of the sum of several independent and identical Bernoulli random variables, each of which has a small probability of occurrence, then the collective sum tends to be Poisson. We summarize these observations in the following theorem.

POISSON THEOREM. If

$$n \to \infty \qquad p \to 0 \qquad \text{such that} \quad np \to \lambda$$

then

$$\frac{n!}{k!(n-k)!}p^k q^{n-k} \underset{n\to\infty}{\longrightarrow} e^{-\lambda}\frac{\lambda^k}{k!} \qquad k = 0, 1, 2, \ldots \tag{4-107}$$

EXAMPLE 4-27

▶ A system contains 1000 components. Each component fails independently of the others and the probability of its failure in one month equals 10^{-3}. We shall find the probability that the system will function (i.e., no component will fail) at the end of one month.

This can be considered as a problem in repeated trials with $p = 10^{-3}$, $n = 10^3$, and $k = 0$. Hence [see (3-21)]

$$P\{k = 0\} = q^n = 0.999^{1000}$$

Since $np = 1$, the approximation (4-107) yields

$$P\{k = 0\} \simeq e^{-np} = e^{-1} = 0.368$$

◀

Applying (4-107) to the sum in (3-24), we obtain the following approximation for the probability that the number k of occurrences of A is between k_1 and k_2:

$$P\{k_1 \le k \le k_2\} \simeq e^{-np}\sum_{k=k_1}^{k_2}\frac{(np)^k}{k!} \tag{4-108}$$

EXAMPLE 4-28

▶ An order of 3000 parts is received. The probability that a part is defective equals 10^{-3}. We wish to find the probability $P\{k > 5\}$ that there will be more than five defective parts.

Clearly,

$$P\{k > 5\} = 1 - P\{k \le 5\}$$

With $np = 3$, (4-108) yields

$$P\{k \le 5\} = e^{-3} \sum_{k=0}^{5} \frac{3^k}{k!} = 0.916$$

Hence

$$P\{k > 5\} = 0.084$$

◀

EXAMPLE 4-29

▶ An insurance company has issued policies to 100,000 people for a premium of \$500/person. In the event of a causality, the probability of which is assumed to be 0.001, the company pays \$200,000/causality. What is the probability that (*a*) the company will suffer a loss? (*b*) The company will make a profit of at least \$25 million?

SOLUTION

The company collects $\$500 \times 10^5 = \$50$ million in terms of premium from its customers. The probability of each causality is very small ($p = 0.001$) and n is large so that we can take advantage of Poisson's theorem with

$$\lambda = np = 10^5 \times 0.001 = 100$$

(*a*) For the company to suffer a loss, it must make payments to more than

$$n_0 = \frac{\$50 \times 10^6}{\$200{,}000} = 250 \text{ persons}$$

Hence with $\mathbf{x}$ representing the (random) number of causalities, we get

$$P_0 = P\{\text{company suffers a loss}\} = P\{\mathbf{x} > 250\} = \sum_{k=250}^{\infty} e^{-\lambda} \frac{\lambda^k}{k!} \tag{4-109}$$

It is possible to obtain excellent approximations to (4-109) as follows. Clearly

$$e^{-\lambda}\frac{\lambda^n}{n!} < \sum_{k=n}^{\infty} e^{-\lambda}\frac{\lambda^k}{k!} = e^{-\lambda}\frac{\lambda^n}{n!}\left(1 + \frac{\lambda}{n+1} + \frac{\lambda^2}{(n+1)(n+2)} + \cdots\right)$$

$$< e^{-\lambda}\frac{\lambda^n}{n!}\left(1 + \frac{\lambda}{n+1} + \left(\frac{\lambda}{n+1}\right)^2 + \cdots\right)$$

$$= e^{-\lambda}\frac{\lambda^n}{n!}\frac{1}{1 - \lambda/(n+1)} \tag{4-110}$$

We can make use of Stirling's formula

$$n! \simeq \sqrt{2\pi n}\, n^n e^{-n} \qquad n \to \infty \tag{4-111}$$

that is valid for large n to simplify (4-110). Thus

$$\frac{\Delta^n}{\sqrt{2\pi n}} < P(\mathbf{x} > n) < \frac{\Delta^n}{\sqrt{2\pi n}}\frac{1}{1 - \lambda/(n+1)} \tag{4-112}$$

where

$$\Delta = \frac{\lambda}{n} e^{(1-\lambda/n)} \tag{4-113}$$

With $\lambda = 100$, $n_0 = 250$, we get $\Delta = 0.7288$ so that $\Delta^{250} = 0$ and the desired probability is essentially zero.

(*b*) In this case to guarantee a profit of \$25 million, the total number of payments should not exceed n_1, where

$$n_1 = \frac{\$50 \times 10^6 - \$25 \times 10^6}{\$200{,}000} = 125.$$

This gives $\Delta = 0.9771$ so that $\Delta^{n_1} = 0.0554$, and

$$P\{\mathbf{x} \le n_1\} \ge 1 - \frac{\Delta^{n_1}}{\sqrt{2\pi n_1}} \frac{1}{1 - \lambda/(n_1 + 1)} \simeq 0.9904$$

Thus the company is assured a profit of \$25 million with almost certainty.

Notice that so long as the parameter Δ in (4-113) is maintained under 1, the event of interest such as $P(\mathbf{x} \le n_1)$ can be predicted with almost certainty. ◀

EXAMPLE 4-30

▶ The probability of hitting an aircraft is 0.001 for each shot. How many shots should be fired so that the probability of hitting with two or more shots is above 0.95?

SOLUTION

In designing an anti-aircraft gun it is important to know how many rounds should be fired at the incoming aircraft so that the probability of hit is above a certain threshold. The aircraft can be shot down only if it is hit in a vulnerable spot and since the probability of hitting these spots with a single shot is extremely small, it is important to fire at them with a large number of shots. Let $\mathbf{x}$ represent the number of hits when n shots are fired. Using the Poisson approximation with $\lambda = np$ we need

$$P(\mathbf{x} \ge 2) \ge 0.95$$

But

$$P(\mathbf{x} \ge 2) = 1 - [P(X = 0) + P(X = 1)] = 1 - e^{-\lambda}(1 + \lambda)$$

so that

$$(1 + \lambda)e^{-\lambda} < 0.05$$

By trial, $\lambda = 4$ and 5, give $(1 + \lambda)e^{-\lambda}$ to be 0.0916 and 0.0404, respectively, so that we must have $4 \le \lambda \le 5$ or $4000 \le n \le 5000$. If 5000 shots are fired at the aircraft, the probability of miss equals $e^{-5} = 0.00673$. ◀

EXAMPLE 4-31

▶ Suppose one million lottery tickets are issued, with 100 winning tickets among them. (*a*) If a person purchases 100 tickets, what is the probability of his winning the lottery? (*b*) How many tickets should one buy to be 95% confident of having a winning ticket?

SOLUTION

The probability of buying a winning ticket

$$p = \frac{\text{No. of winning tickets}}{\text{Total no. of tickets}} = \frac{100}{10^6} = 10^{-4}.$$

Let $n = 100$ represent the number of purchased tickets, and $\mathbf{x}$ the number of winning tickets in the n purchased tickets. Then $\mathbf{x}$ has an approximate Poisson distribution with parameter $\lambda = np = 100 \times 10^{-4} = 10^{-2}$. Thus

$$P(\mathbf{x} = k) = e^{-\lambda}\frac{\lambda^k}{k!}$$

(a) Probability of winning $= P(\mathbf{x} \geq 1) = 1 - P(\mathbf{x} = 0) = 1 - e^{-\lambda} \approx 0.0099$.
(b) In this case we need $P(\mathbf{x} \geq 1) \geq 0.95$.

$$P\{\mathbf{x} \geq 1\} = 1 - e^{-\lambda} \geq 0.95 \text{ implies } \lambda \geq \ln 20 \simeq 3.$$

But $\lambda = np = n \times 10^{-4} \geq 3$ or $n \geq 30{,}000$. Thus one needs to buy about 30,000 tickets to be 95% confident of having a winning ticket! ◀

EXAMPLE 4-32 ▶ A spacecraft has 20,000 components ($n \to \infty$). The probability of any one component being defective is 10^{-4} ($p \to 0$). The mission will be in danger if five or more components become defective. Find the probability of such an event.

SOLUTION

Here n is large and p is small, and hence Poisson approximation is valid. Thus $np = \lambda = 20{,}000 \times 10^{-4} = 2$, and the desired probability is given by

$$P\{\mathbf{x} \geq 5\} = 1 - P\{\mathbf{x} \leq 4\} = 1 - \sum_{k=0}^{4} e^{-\lambda}\frac{\lambda^k}{k!} = 1 - e^{-2}\sum_{k=0}^{4}\frac{\lambda^k}{k!}$$

$$= 1 - e^{-2}\left(1 + 2 + 2 + \frac{4}{3} + \frac{2}{3}\right) = 0.052$$ ◀

GENERALIZATION OF POISSON THEOREM. Suppose that, $A_1, \ldots, A_{m+1}$ are the $m+1$ events of a partition with $P\{A_i\} = p_i$. Reasoning as in (4-107), we can show that if $np_i \to a_i$ for $i \leq m$, then

$$\frac{n!}{k_1! \cdots k_{m+1}!} p_1^{k_1} \cdots p_{m+1}^{k_{m+1}} \underset{n\to\infty}{\longrightarrow} \frac{e^{-a_1}a_1^{k_1}}{k_1!} \cdots \frac{e^{-a_m}a_n^{k_m}}{k_m!} \tag{4-114}$$

Random Poisson Points

An important application of Poisson's theorem is the approximate evaluation of (3-21) as T and n tend to ∞. We repeat the problem: We place at random n points in the interval $(-T/2, T/2)$ and we denote by $P\{k \text{ in } t_a\}$ the probability that k of these points will lie in an interval (t_1, t_2) of length $t_2 - t_1 = t_a$. As we have shown in (3-21)

$$P\{k \text{ in } t_a\} = \binom{n}{k} p^k q^{n-k} \qquad \text{where} \quad p = \frac{t_a}{T} \tag{4-115}$$

We now assume that $n \gg 1$ and $t_a \ll T$. Applying (4-107), we conclude that

$$P\{k \text{ in } t_a\} \simeq e^{-nt_a/T} \frac{(nt_a/T)^k}{k!} \tag{4-116}$$

for k of the order of nt_a/T.

Suppose, next, that n and T increase indefinitely but the ratio

$$\lambda = n/T$$

remains constant. The result is an infinite set of points covering the entire t axis from $-\infty$ to $+\infty$. As we see from (4-116) the probability that k of these points are in an interval of length t_a is given by

$$P\{k \text{ in } t_a\} = e^{-\lambda t_a} \frac{(\lambda t_a)^k}{k!} \tag{4-117}$$

POINTS IN NONOVERLAPPING INTERVALS. Returning for a moment to the original interval $(-T/2, T/2)$ containing n points, we consider two nonoverlapping subintervals t_a and t_b (Fig. 4-33).

We wish to determine the probability

$$P\{k_a \text{ in } t_a, k_b \text{ in } t_b\}$$

that k_a of the n points are in interval t_a and k_b in the interval t_b. We maintain that

$$P\{k_a \text{ in } t_a, k_b \text{ in } t_b\} = \frac{n!}{k_a!k_b!k_3!} \left(\frac{t_a}{T}\right)^{k_a} \left(\frac{t_b}{T}\right)^{k_b} \left(1 - \frac{t_a}{T} - \frac{t_b}{T}\right)^{k^3} \tag{4-118}$$

where $k_3 = n - k_a - k_b$.

Proof. This material can be considered as a generalized Bernoulli trial. The original experiment S is the random selection of a single point in the interval $(-T/2, T/2)$. In this experiment, the events $A_1 = \{\text{the point is in } t_a\}$, $A_2 = \{\text{the point is in } t_b\}$, and $A_3 = \{\text{the point is outside the intervals } t_a \text{ and } t_a\}$ form a partition and

$$P(A_1) = \frac{t_a}{T} \qquad P(A_2) = \frac{t_b}{T} \qquad P(A_3) = 1 - \frac{t_a}{T} - \frac{t_b}{T}$$

If the experiment S is performed n times, then the event $\{k_a \text{ in } t_a \text{ and } k_b \text{ in } t_b\}$ will equal the event $\{A_1$ occurs $k_1 = k_a$ times, A_2 occurs $k_2 = k_b$ times, and A_3 occurs $k_3 = n - k_1 - k_2$ times$\}$. Hence (4-118) follows from (4-102) with $r = 3$.

We note that the events $\{k_a \text{ in } t_a\}$ and $\{k_b \text{ in } t_b\}$ *are not* independent because the probability (4-118) of their intersection $\{k_a \text{ in } t_a, k_b \text{ in } t_b\}$ does not equal $P\{k_a \text{ in } t_a\}$ $P\{k_b \text{ in } t_b\}$.

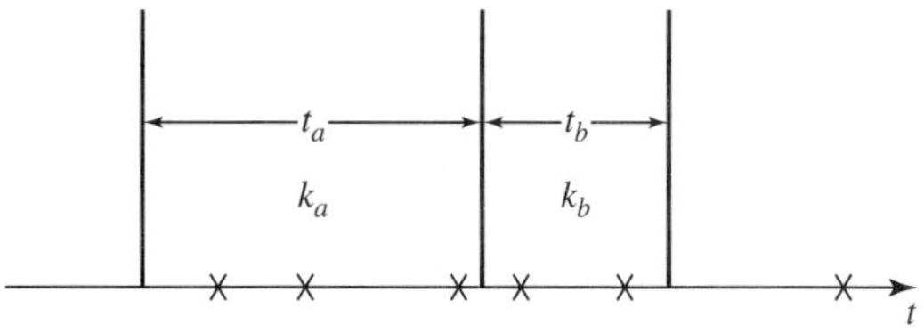

FIGURE 4-33

Suppose now that

$$\frac{n}{T} = \lambda \quad n \to \infty \quad T \to \infty$$

Since $nt_a/T = \lambda t_a$ and $nt_b/T = \lambda t_b$, we conclude from (4-118) and Prob. 4-35 that

$$P\{k_a \text{ in } t_a, k_b \text{ in } t_b\} = e^{-\lambda t_a} \frac{(\lambda t_a)^{k_a}}{k_a!} e^{-\lambda t_b} \frac{(\lambda t_b)^{k_b}}{k_b!} \tag{4-119}$$

From (4-117) and (4-119) it follows that

$$P\{k_a \text{ in } t_a, k_b \text{ in } t_b\} = P\{k_a \text{ in } t_a\} P\{k_b \text{ in } t_b\} \tag{4-120}$$

This shows that the events $\{k_a \text{ in } t_a\}$ and $\{k_b \text{ in } t_b\}$ *are* independent.

We have thus created an experiment whose outcomes are infinite sets of points on the t axis. These outcomes will be called *random Poisson points*. The experiment was formed by a limiting process; however, it is completely specified in terms of the following two properties:

1. The probability $P\{k_a \text{ in } t_a\}$ that the number of points in an interval (t_1, t_2) equals k_a is given by (4-117).
2. If two intervals (t_1, t_2) and (t_3, t_4) are nonoverlapping, then the events $\{k_a \text{ in } (t_1, t_2)\}$ and $\{k_b \text{ in } (t_3, t_4)\}$ are independent.

The experiment of random Poisson points is fundamental in the theory and the applications of probability. As illustrations we mention electron emission, telephone calls, cars crossing a bridge, and shot noise, among many others.

EXAMPLE 4-33

▶ Consider two consecutive intervals (t_1, t_2) and (t_2, t_3) with respective lengths t_a and t_b. Clearly, (t_1, t_3) is an interval with length $t_c = t_a + t_b$. We denote by k_a, k_b, and $k_c = k_a + k_b$ the number of points in these intervals. We assume that the number of points k_c in the interval (t_1, t_3) is specified. We wish to find the probability that k_a of these points are in the interval (t_1, t_2). In other words, we wish to find the conditional probability

$$P\{k_a \text{ in } t_a \mid k_c \text{ in } t_c\}$$

With $k_b = k_c - k_a$, we observe that

$$\{k_a \text{ in } t_a, k_c \text{ in } t_c\} = \{k_a \text{ in } t_a, k_b \text{ in } t_b\}$$

Hence

$$P\{k_a \text{ in } t_a \mid k_c \text{ in } t_c\} = \frac{P\{k_a \text{ in } t_a, k_b \text{ in } t_b\}}{P\{k_c \text{ in } t_c\}}$$

From (4-117) and (4-119) it follows that this fraction equals

$$\frac{e^{-\lambda t_a}\left[(\lambda t_a)^{k_a}/k_a!\right] e^{-\lambda t_b}\left[(\lambda t_b)^{k_b}/k_b!\right]}{e^{-\lambda t_c}\left[(\lambda t_c)^{k_c}/k_c!\right]}$$

Since $t_c = t_a + t_b$ and $k_c = k_a + k_b$, the last equation yields

$$P\{k_a \text{ in } t_a \mid k_c \text{ in } t_c\} = \frac{k_c!}{k_a!k_b!}\left(\frac{t_a}{t_c}\right)^{k_a}\left(\frac{t_b}{t_c}\right)^{k_b} \tag{4-121}$$

This result has the following useful interpretation: Suppose that we place at random k_c points in the interval (t_1, t_3). As we see from (3-21), the probability that k_a of these points are in the interval (t_1, t_2) equals the right side of (4-121). ◀

DENSITY OF POISSON POINTS. The experiment of Poisson points is specified in terms of the parameter λ. We show next that this parameter can be interpreted as the density of the points. Indeed, if the interval $\Delta t = t_2 - t_1$ is sufficiently small, then

$$\lambda \Delta t e^{-\lambda \Delta t} \simeq \lambda \Delta t$$

From this and (4-117) it follows that

$$P\{\text{one point in } (t, t + \Delta t)\} \simeq \lambda \Delta t \tag{4-122}$$

Hence

$$\lambda = \lim_{\Delta t \to 0} \frac{P\{\text{one point in } (t, t + \Delta t)\}}{\Delta t} \tag{4-123}$$

Nonuniform density Using a nonlinear transformation of the t axis, we shall define an experiment whose outcomes are Poisson points specified by a minor modification of property 1 on page 118.

Suppose that $\lambda(t)$ is a function such that $\lambda(t) \geq 0$ but otherwise arbitrary. We define the experiment of the nonuniform Poisson points as follows:

1. The probability that the number of points in the interval (t_1, t_2) equals k is given by

$$P\{k \text{ in } (t_1, t_2)\} = \exp\left[-\int_{t_1}^{t_2} \lambda(t)\, dt\right] \frac{\left[\int_{t_1}^{t_2} \lambda(t)\, dt\right]^k}{k!} \tag{4-124}$$

2. The same as in the uniform case.

The significance of $\lambda(t)$ as density remains the same. Indeed, with $t_2 - t_1 = \Delta t$ and $k = 1$, (4-124) yields

$$P\{\text{one point in } (t, t + \Delta t)\} \simeq \lambda(t)\Delta t \tag{4-125}$$

as in (4-122).

PROBLEMS

4-1 Suppose that x_u is the u percentile of the random variable $\mathbf{x}$, that is, $F(x_u) = u$. Show that if $f(-x) = f(x)$, then $x_{1-u} = -x_u$.

4-2 Show that if $f(x)$ is symmetrical about the point $x = \eta$ and $P\{\eta - a < \mathbf{x} < \eta + a\} = 1 - \alpha$, then $a = \eta - x_{\alpha/2} = x_{1-\alpha/2} - \eta$.

4-3 (*a*) Using Table 3-1 and linear interpolation, find the z_u percentile of the $N(0, 1)$ random variable $\mathbf{z}$ for $u = 0.9, 0.925, 0.95, 0.975$, and 0.99. (*b*) The random variable $\mathbf{x}$ is $N(\eta, \sigma)$. Express its x_u percentiles in terms of z_u.

4-4 The random variable is $\mathbf{x}$ is $N(\eta, \sigma)$ and $P\{\eta - k\sigma < \mathbf{x} < \eta + k\sigma\} = p_k$. (*a*) Find p_k for $k = 1, 2$, and 3. (*b*) Find k for $p_k = 0.9, 0.99$, and 0.999. (*c*) If $P\{\eta - z_u\sigma < \mathbf{x} < \eta + z_u\sigma\} = \gamma$, express z_u in terms of γ.

4-5 Find x_u for $u = 0.1, 0.2, \ldots, 0.9$ (*a*) if $\mathbf{x}$ is uniform in the interval (0, 1); (*b*) if $f(x) = 2e^{-2x}U(x)$.

4-6 We measure for resistance **R** of each resistor in a production line and we accept only the units the resistance of which is between 96 and 104 ohms. Find the percentage of the accepted units (*a*) if **R** is uniform between 95 and 105 ohms; (*b*) if **R** is normal with $\eta = 100$ and $\sigma = 2$ ohms.

4-7 Show that if the random variable $\mathbf{x}$ has an Erlang density with $n = 2$, then $F_x(x) = (1 - e^{-cx} - cxe^{-cx})U(x)$.

4-8 The random variable $\mathbf{x}$ is $N(10; 1)$. Find $f(x \mid (\mathbf{x} - 10)^2 < 4)$.

4-9 Find $f(x)$ if $F(x) = (1 - e^{-\alpha x})U(x - c)$.

4-10 If $\mathbf{x}$ is $N(0, 2)$ find (*a*) $P\{1 \le \mathbf{x} \le 2\}$ and (*b*) $P\{1 \le \mathbf{x} \le 2 \mid \mathbf{x} \ge 1\}$.

4-11 The space S consists of all points t_i in the interval (0, 1) and $P\{0 \le t_i \le y\} = y$ for every $y \le 1$. The function $G(x)$ is increasing from $G(-\infty) = 0$ to $G(\infty) = 1$; hence it has an inverse $G^{(-1)}(y) = H(y)$. The random variable $\mathbf{x}$ is such that $\mathbf{x}(t_i) = H(t_i)$. Show that $F_x(x) = G(x)$.

4-12 If $\mathbf{x}$ is $N(1000; 20)$ find (*a*) $P\{\mathbf{x} < 1024\}$, (*b*) $P\{\mathbf{x} < 1024 \mid \mathbf{x} > 961\}$, and (*c*) $P\{31 < \sqrt{\mathbf{x}} \le 32\}$.

4-13 A fair coin is tossed three times and the random variable $\mathbf{x}$ equals the total number of heads. Find and sketch $F_x(x)$ and $f_x(x)$.

4-14 A fair coin is tossed 900 times and the random variable $\mathbf{x}$ equals the total number of heads. (*a*) Find $f_x(x)$: 1; exactly 2; approximately using (4-34). (*b*) Find $P\{435 \le \mathbf{x} \le 460\}$.

4-15 Show that, if $a \le \mathbf{x}(\zeta) \le b$ for every $\zeta \in S$, then $F(x) = 1$ for $x > b$ and $F(x) = 0$ for $x < a$.

4-16 Show that if $\mathbf{x}(\zeta) \le \mathbf{y}(\zeta)$ for every $\zeta \in S$, then $F_x(w) \ge F_y(w)$ for every w.

4-17 Show that if $\beta(t) = f(t \mid \mathbf{x} > t)$ is the conditional failure rate of the random variable $\mathbf{x}$ and $\beta(t) = kt$, then $f(x)$ is a Rayleigh density (see also Sec. 6-6).

4-18 Show that $P(A) = P(A \mid \mathbf{x} \le x)F(x) + P(A \mid \mathbf{x} > x)[1 - F(x)]$.

4-19 Show that

$$F_x(x \mid A) = \frac{P(A \mid \mathbf{x} \le x)F_x(x)}{P(A)}$$

4-20 Show that if $P(A \mid \mathbf{x} = x) = P(B \mid \mathbf{x} = x)$ for every $x \le x_0$, then $P(A \mid \mathbf{x} \le x_0) = P(B \mid \mathbf{x} \le x_0)$. *Hint:* Replace in (4-80) $P(A)$ and $f(x)$ by $P(A \mid \mathbf{x} \le x_0)$ and $f(x \mid \mathbf{x} \le x_0)$.

4-21 The probability of *heads* of a random coin is a random variable **p** uniform in the interval (0, 1). (*a*) Find $P\{0.3 \le \mathbf{p} \le 0.7\}$. (*b*) The coin is tossed 10 times and *heads* shows 6 times. Find the a posteriori probability that **p** is between 0.3 and 0.7.

4-22 The probability of *heads* of a random coin is a random variable **p** uniform in the interval (0.4, 0.6). (*a*) Find the probability that at the next tossing of the coin *heads* will show. (*b*) The coin is tossed 100 times and *heads* shows 60 times. Find the probability that at the next tossing *heads* will show.

4-23 A fair coin is tossed 900 times. Find the probability that the number of heads is between 420 and 465.

Answer: $\mathrm{G}(2) + \mathrm{G}(1) - 1 \simeq 0.819$.

4-24 A fair coin is tossed n times. Find n such that the probability that the number of heads is between $0.49n$ and $0.52n$ is at least 0.9.

Answer: $\mathrm{G}(0.04\sqrt{n}) + \mathrm{G}(0.02\sqrt{n}) \ge 1.9$; hence $n > 4556$.

4-25 If $P(A) = 0.6$ and k is the number of successes of A in n trials (a) show that $P\{550 \le k \le 650\} = 0.999$, for $n = 1000$. (b) Find n such that $P\{0.59n \le k \le 0.61n\} = 0.95$.

4-26 A system has 100 components. The probability that a specific component will fail in the interval (a, b) equals $e^{-a/T} - e^{-b/T}$. Find the probability that in the interval $(0, T/4)$, no more than 100 components will fail.

4-27 A coin is tossed an infinite number of times. Show that the probability that k heads are observed at the nth toss but not earlier equals $\binom{n-1}{k-1} p^k q^{n-k}$. [See also (4-63).]

4-28 Show that

$$\frac{1}{x}\left(1 - \frac{1}{x^2}\right) g(x) < 1 - \mathbf{G}(x) < \frac{1}{x} g(x) \qquad g(x) = \frac{1}{\sqrt{2\pi}} e^{-x^2/2} \quad x > 0$$

Hint: Prove the following inequalities and integrate from x to ∞:

$$-\frac{d}{dx}\left(\frac{1}{x} e^{-x^2/2}\right) > e^{-x^2/2} \qquad -\frac{d}{dx}\left[\left(\frac{1}{x} - \frac{1}{x^3}\right) e^{-x^2/2}\right] > e^{-x^2/2}$$

4-29 Suppose that in n trials, the probability that an event A occurs at least once equals P_1. Show that, if $P(A) = p$ and $pn \ll 1$, then $P_1 \simeq np$.

4-30 The probability that a driver will have an accident in 1 month equals 0.02. Find the probability that in 100 months he will have three accidents.

Answer: About $4e^{-2}/3$.

4-31 A fair die is rolled five times. Find the probability that *one* shows twice, *three* shows twice, and *six* shows once.

4-32 Show that (4-90) is a special case of (4-103) obtained with $r = 2$, $k_1 = k$, $k_2 = n - k$, $p_1 = p$, $p_2 = 1 - p$.

4-33 Players X and Y roll dice alternately starting with X. The player that rolls *eleven* wins. Show that the probability p that X wins equals 18/35.

Hint: Show that

$$P(A) = P(A \mid M)P(M) + P(A \mid \overline{M})P(\overline{M})$$

Set $A = \{X \text{ wins}\}$, $M = \{$*eleven* shows at first try$\}$. Note that $P(A) = p$, $P(A \mid M) = 1$, $P(M) = 2/36$, $P(A \mid \overline{M}) = 1 - p$.

4-34 We place at random n particles in $m > n$ boxes. Find the probability p that the particles will be found in n preselected boxes (one in each box). Consider the following cases: (a) M–B (Maxwell–Boltzmann)—the particles are distinct; all alternatives are possible, (b) B–E (Bose–Einstein)—the particles cannot be distinguished; all alternatives are possible, (c) F–D (Fermi–Dirac)—the particles cannot be distinguished; at most one particle is allowed in a box.

Answer:

	M–B	B–E	F–D
$p =$	$\dfrac{n!}{m^n}$	$\dfrac{n!(m-1)!}{(m+n-1)!}$	$\dfrac{n!(m-n)!}{m!}$

Hint: (a) The number N of all alternatives equals m^n. The number N_A of favorable alternatives equals the $n!$ permutations of the particles in the preselected boxes. (b) Place the $m - 1$ walls separating the boxes in line ending with the n particles. This corresponds to one alternative where all particles are in the last box. All other possibilities are obtained by a permutation of the $n + m - 1$ objects consisting of the $m - 1$ walls and the n particles. All the $(m - 1)!$ permutations of the walls and the $n!$ permutations of the particles count as one alternative. Hence $N = (m+n-1)!/(m-1)!n!$ and $N_A = 1$. (*c*) Since the particles are not distinguishable, N equals the number of ways of selecting n out of m objects: $N = \binom{m}{n}$ and $N_A = 1$.

4-35 Reasoning as in (4-107), show that, if

$$k_1 + k_2 + k_3 = n \qquad p_1 + p_2 + p_3 = 1 \qquad k_1 p_1 \ll 1 \qquad k_2 p_2 \ll 1$$

then

$$\frac{n!}{k_1!k_2!k_3!} \simeq \frac{n^{k_1+k_2}}{k_1!k_2!} \qquad p_3^{k_3} \simeq e^{-n(p_1+p_2)}$$

Use this to justify (4-119).

4-36 We place at random 200 points in the interval (0, 100). Find the probability that in the interval (0, 2) there will be one and only one point (*a*) exactly and (*b*) using the Poisson approximation.

CHAPTER

5

FUNCTIONS OF ONE RANDOM VARIABLE

5-1 THE RANDOM VARIABLE $g(\mathbf{x})$

Suppose that $\mathbf{x}$ is a random variable and $g(x)$ is a function of the real variable x. The expression

$$\mathbf{y} = g(\mathbf{x})$$

is a new random variable defined as follows: For a given ζ, $\mathbf{x}(\zeta)$ is a number and $g[\mathbf{x}(\zeta)]$ is another number specified in terms of $\mathbf{x}(\zeta)$ and $g(x)$. This number is the value $\mathbf{y}(\zeta) = g[\mathbf{x}(\zeta)]$ assigned to the random variable $\mathbf{y}$. Thus a function of a random variable $\mathbf{x}$ is a composite function $\mathbf{y} = g(\mathbf{x}) = g[\mathbf{x}(\zeta)]$ with domain the set S of experimental outcomes.

The distribution function $F_y(y)$ of the random variable so formed is the probability of the event $\{\mathbf{y} \le y\}$ consisting of all outcomes ζ such that $\mathbf{y}(\zeta) = g[\mathbf{x}(\zeta)] \le y$. Thus

$$F_y(y) = P\{\mathbf{y} \le y\} = P\{g(\mathbf{x}) \le y\} \tag{5-1}$$

For a specific y, the values of x such that $g(x) \le y$ form a set on the x axis denoted by R_y. Clearly, $g[\mathbf{x}(\zeta)] \le y$ if $\mathbf{x}(\zeta)$ is a number in the set R_y. Hence

$$F_y(y) = P\{\mathbf{x} \in R_y\} \tag{5-2}$$

This discussion leads to the conclusion that for $g(\mathbf{x})$ to be a random variable, the function $g(x)$ must have these properties:

1. Its domain must include the range of the random variable $\mathbf{x}$.
2. It must be a *Borel* function, that is, for every y, the set R_y such that $g(x) \le y$ must consist of the union and intersection of a countable number of intervals. Only then $\{\mathbf{y} \le y\}$ is an event.
3. The events $\{g(\mathbf{x}) = \pm\infty\}$ must have zero probability.

5-2 THE DISTRIBUTION OF $g(\mathbf{x})$

We shall express the distribution function $F_y(y)$ of the random variable $\mathbf{y} = g(\mathbf{x})$ in terms of the distribution function $F_x(x)$ of the random variable $\mathbf{x}$ and the function $g(x)$. For this purpose, we must determine the set R_y of the x axis such that $g(x) \le y$, and the probability that $\mathbf{x}$ is in this set. The method will be illustrated with several examples. Unless otherwise stated, it will be assumed that $F_x(x)$ is continuous.

1. We start with the function $g(x)$ in Fig. 5-1. As we see from the figure, $g(x)$ is between a and b for any x. This leads to the conclusion that if $y \ge b$, then $g(x) \le y$ for every x, hence $P\{\mathbf{y} \le y\} = 1$; if $y < a$, then there is no x such that $g(x) \le y$, hence $P\{\mathbf{y} \le y\} = 0$. Thus

$$F_y(y) = \begin{cases} 1 & y \ge b \\ 0 & y < a \end{cases}$$

With x_1 and $y_1 = g(x_1)$ as shown, we observe that $g(x) \le y_1$ for $x \le x_1$. Hence

$$F_y(y_1) = P\{\mathbf{x} \le x_1\} = F_x(x_1)$$

We finally note that

$$g(x) \le y_2 \qquad \text{if} \quad x \le x_2' \text{ or if } x_2'' \le x \le x_2'''$$

Hence

$$F_y(y_2) = P\{\mathbf{x} \le x_2'\} + P\{x_2'' \le \mathbf{x} \le x_2'''\} = F_x(x_2') + F_x(x_2''') - F_x(x_2'')$$

because the events $\{\mathbf{x} \le x_2'\}$ and $\{x_2'' \le \mathbf{x} \le x_2'''\}$ are mutually exclusive.

EXAMPLE 5-1 ▶

$$\mathbf{y} = a\mathbf{x} + b \tag{5-3}$$

To find $F_y(y)$, we must find the values of x such that $ax + b \le y$.
(*a*) If $a > 0$, then $ax + b \le y$ for $x \le (y - b)/a$ (Fig. 5-2*a*). Hence

$$F_y(y) = P\left\{\mathbf{x} \le \frac{y-b}{a}\right\} = F_x\left(\frac{y-b}{a}\right) \qquad a > 0$$

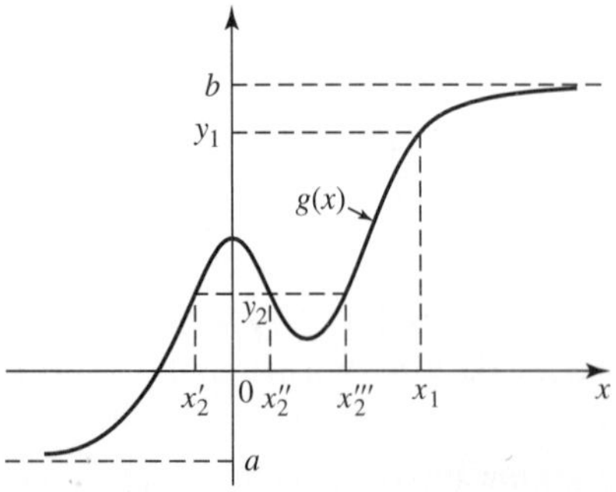

FIGURE 5-1

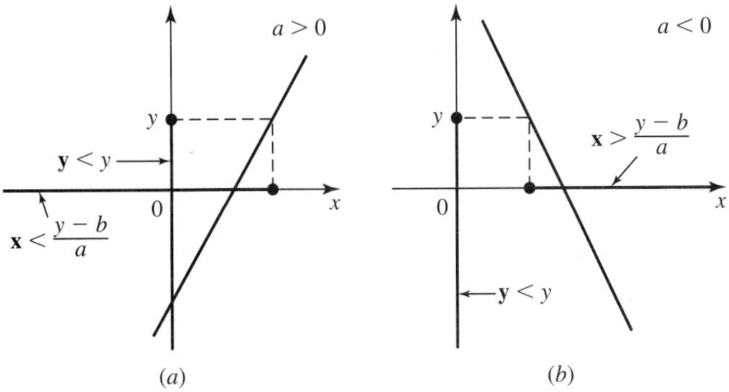

FIGURE 5-2

(*b*) If $a < 0$, then $ax + b \leq y$ for $x > (y - b)/a$ (Fig. 5-2*b*). Hence [see also (5-17)–(5-18)]

$$F_y(y) = P\left\{\mathbf{x} \geq \frac{y-b}{a}\right\} = 1 - F_x\left(\frac{y-b}{a}\right) \qquad a < 0$$

◀

EXAMPLE 5-2 ▶

$$\mathbf{y} = \mathbf{x}^2$$

If $y \geq 0$, then $x^2 \leq y$ for $-\sqrt{y} \leq x \leq \sqrt{y}$ (Fig. 5-3*a*). Hence

$$F_y(y) = P\{-\sqrt{y} \leq \mathbf{x} \leq \sqrt{y}\} = F_x(\sqrt{y}) - F_x(-\sqrt{y}) \qquad y > 0$$

If $y < 0$, then there are no values of x such that $x^2 < y$. Hence

$$F_y(y) = P\{\emptyset\} = 0 \qquad y < 0$$

By direct differentiation of $F_y(y)$, we get

$$f_y(y) = \begin{cases} \dfrac{1}{2\sqrt{y}}\left(f_x(\sqrt{y}) + f_x(-\sqrt{y})\right) & y > 0 \\ 0 & \text{otherwise} \end{cases} \tag{5-4}$$

If $f_x(x)$ represents an even function, then (5-4) reduces to

$$f_y(y) = \frac{1}{\sqrt{y}} f_x(\sqrt{y}) U(y) \tag{5-5}$$

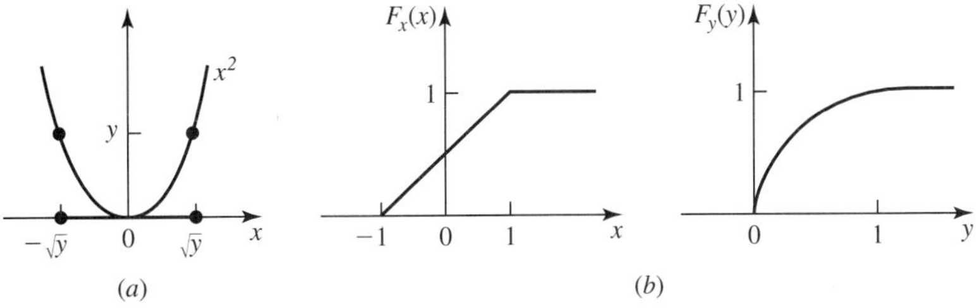

FIGURE 5-3

In particular if $\mathbf{x} \sim N(0, 1)$, so that

$$f_x(x) = \frac{1}{\sqrt{2\pi}} e^{-x^2/2} \tag{5-6}$$

and substituting this into (5-5), we obtain the p.d.f. of $\mathbf{y} = \mathbf{x}^2$ to be

$$f_y(y) = \frac{1}{\sqrt{2\pi y}} e^{-y/2} U(y) \tag{5-7}$$

On comparing this with (4-39), we notice that (5-7) represents a chi-square random variable with $n = 1$, since $\Gamma(1/2) = \sqrt{\pi}$. Thus, if $\mathbf{x}$ is a Gaussian random variable with $\mu = 0$, then $\mathbf{y} = \mathbf{x}^2$ represents a chi-square random variable with one degree of freedom. ◀

Special case If $\mathbf{x}$ is uniform in the interval $(-1, 1)$, then

$$F_x(x) = \frac{1}{2} + \frac{x}{2} \qquad |x| < 1$$

(Fig. 5-3*b*). Hence

$$F_y(y) = \sqrt{y} \qquad \text{for} \quad 0 \le y \le 1 \quad \text{and} \quad F_y(y) = \begin{cases} 1 & y > 1 \\ 0 & y < 0 \end{cases}$$

2. Suppose now that the function $g(x)$ is constant in an interval (x_0, x_1):

$$g(x) = y_1 \qquad x_0 < x \le x_1 \tag{5-8}$$

In this case

$$P\{\mathbf{y} = y_1\} = P\{x_0 < \mathbf{x} \le x_1\} = F_x(x_1) - F_x(x_0) \tag{5-9}$$

Hence $F_y(y)$ is discontinuous at $\mathbf{y} = y_1$ and its discontinuity equals $F_x(x_1) - F_x(x_0)$.

EXAMPLE 5-3

▶ Consider the function (Fig. 5-4)

$$g(x) = 0 \qquad \text{for} \quad -c \le x \le c \quad \text{and} \quad g(x) = \begin{cases} x - c & x > c \\ x + c & x < -c \end{cases} \tag{5-10}$$

In this case, $F_y(y)$ is discontinuous for $y = 0$ and its discontinuity equals $F_x(c) - F_x(-c)$. Furthermore,

$$\begin{aligned} &\text{If } y \ge 0 \qquad \text{then} \quad P\{\mathbf{y} \le y\} = P\{\mathbf{x} \le y + c\} = F_x(y + c) \\ &\text{If } y < 0 \qquad \text{then} \quad P\{\mathbf{y} \le y\} = P\{\mathbf{x} \le y - c\} = F_x(y - c) \end{aligned}$$ ◀

EXAMPLE 5-4

LIMITER

▶ The curve $g(x)$ of Fig. 5-5 is constant for $x \le -b$ and $x \ge b$ and in the interval $(-b, b)$ it is a straight line. With $\mathbf{y} = g(\mathbf{x})$, it follows that $F_y(y)$ is discontinuous for $y = g(-b) = -b$ and $y = g(b) = b$, respectively. Furthermore,

$$\begin{array}{lllll} \text{If } \quad y \ge b & \text{then} \quad g(x) \le y & \text{for every } x; & \text{hence} & F_y(y) = 1 \\ \text{If } \quad -b \le y < b & \text{then} \quad g(x) \le y & \text{for } x \le y; & \text{hence} & F_y(y) = F_x(y) \\ \text{If } \quad y < -b & \text{then} \quad g(x) \le y & \text{for no } x; & \text{hence} & F_y(y) = 0 \end{array}$$ ◀

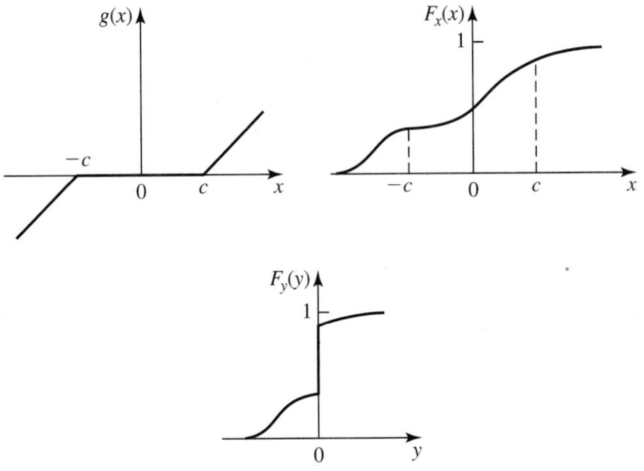

FIGURE 5-4

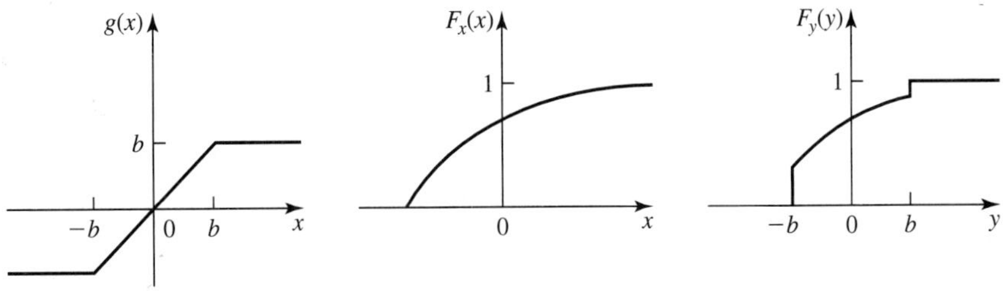

FIGURE 5-5

3. We assume next that $g(x)$ is a staircase function

$$g(x) = g(x_i) = y_i \qquad x_{i-1} < x \le x_i$$

In this case, the random variable $\mathbf{y} = g(\mathbf{x})$ is of discrete type taking the values y_i with

$$P\{\mathbf{y} = y_i\} = P\{x_{i-1} < \mathbf{x} \le x_i\} = F_x(x_i) - F_x(x_{i-1})$$

EXAMPLE 5-5

HARD LIMITER

▶ If

$$g(x) = \begin{cases} 1 & x > 0 \\ -1 & x \le 0 \end{cases} \tag{5-11}$$

then $\mathbf{y}$ takes the values ± 1 with

$$P\{\mathbf{y} = -1\} = P\{\mathbf{x} \le 0\} = F_x(0)$$

$$P\{\mathbf{y} = 1\} = P\{\mathbf{x} > 0\} = 1 - F_x(0)$$

Hence $F_y(y)$ is a staircase function as in Fig. 5-6. ◀

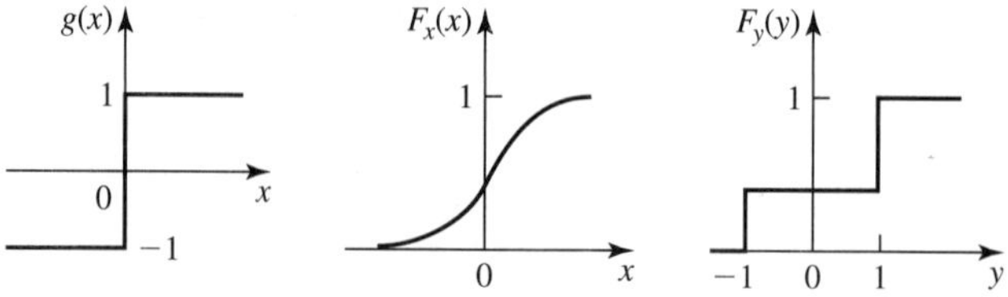

FIGURE 5-6

EXAMPLE 5-6

QUANTIZATION

If

$$g(x) = ns \qquad (n-1)s < x \le ns \tag{5-12}$$

then $\mathbf{y}$ takes the values $y_n = ns$ with

$$P\{\mathbf{y} = ns\} = P\{(n-1)s < \mathbf{x} \le ns\} = F_x(ns) - F_x(ns - s) \tag{5-13}$$

4. We assume, finally, that the function $g(x)$ is discontinuous at $x = x_0$ and such that

$$g(x) < g(x_0^-) \qquad \text{for} \quad x < x_0 \qquad g(x) > g(x_0^+) \qquad \text{for} \quad x > x_0$$

In this case, if y is between $g(x_0^-)$ and $g(x_0^+)$, then $g(x) < y$ for $x \le x_0$. Hence

$$F_y(y) = P\{\mathbf{x} \le x_0\} = F_x(x_0) \qquad g(x_0^-) \le y \le g(x_0^+)$$

EXAMPLE 5-7

Suppose that

$$g(x) = \begin{cases} x + c & x \ge 0 \\ x - c & x < 0 \end{cases} \tag{5-14}$$

is discontinuous (Fig. 5-7). Thus $g(x)$ is discontinuous for $x = 0$ with $g(0^-) = -c$ and

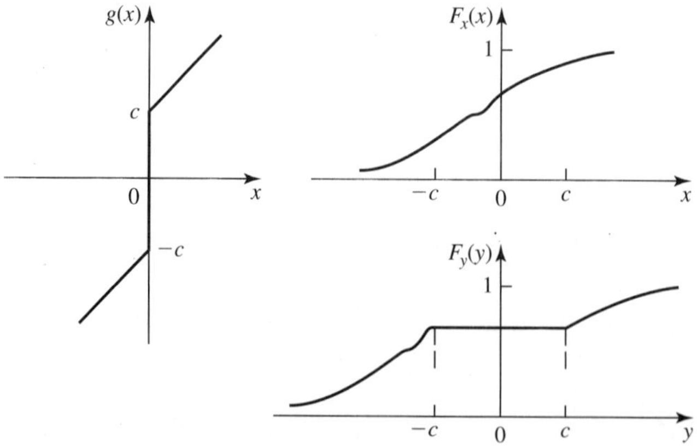

FIGURE 5-7

$g(0^+) = c$. Hence $F_y(y) = F_x(0)$ for $|y| \le c$. Furthermore,

If	$y \ge c$	then	$g(x) \le y$	for $x \le y - c$;	hence	$F_y(y) = F_x(y - c)$
If	$-c \le y \le c$	then	$g(x) \le y$	for $x \le 0$;	hence	$F_y(y) = F_x(0)$
If	$y \le -c$	then	$g(x) \le y$	for $x \le y + c$;	hence	$F_y(y) = F_x(y + c)$

◀

EXAMPLE 5-8 ▶ The function $g(x)$ in Fig. 5-8 equals 0 in the interval $(-c, c)$ and it is discontinuous for $x = \pm c$ with $g(c^+) = c$, $g(c^-) = 0$, $g(-c^-) = -c$, $g(-c^+) = 0$. Hence $F_y(y)$ is discontinuous for $y = 0$ and it is constant for $0 \le y \le c$ and $-c \le y \le 0$. Thus

If	$y \ge c$	then	$g(x) \le y$	for $x \le y$;	hence	$F_y(y) = F_x(y)$
If	$0 \le y < c$	then	$g(x) \le y$	for $x < c$;	hence	$F_y(y) = F_x(c)$
If	$-c \le y < c$	then	$g(x) \le y$	for $x \le -c$;	hence	$F_y(y) = F_x(-c)$
If	$y < -c$	then	$g(x) \le y$	for $x \le y$;	hence	$F_y(y) = F_x(y)$

◀

5. We now assume that the random variable $\mathbf{x}$ is of discrete type taking the values x_k with probability p_k. In this case, the random variable $\mathbf{y} = g(\mathbf{x})$ is also of discrete type taking the values $y_k = g(x_k)$.

If $y_k = g(x)$ for only one $x = x_k$, then

$$P\{\mathbf{y} = y_k\} = P\{\mathbf{x} = x_k\} = p_k$$

If, however, $y_k = g(x)$ for $x = x_k$ and $x = x_l$, then

$$P\{\mathbf{y} = y_k\} = P\{\mathbf{x} = x_k\} + P\{\mathbf{x} = x_l\} = p_k + p_l$$

EXAMPLE 5-9 ▶

$$\mathbf{y} = \mathbf{x}^2$$

(*a*) If $\mathbf{x}$ takes the values $1, 2, \ldots, 6$ with probability $1/6$, then $\mathbf{y}$ takes the values $1^2, 2^2, \ldots, 6^2$ with probability $1/6$.

(*b*) If, however, $\mathbf{x}$ takes the values $-2, -1, 0, 1, 2, 3$ with probability $1/6$, then $\mathbf{y}$ takes the values $0, 1, 4, 9$ with probabilities $1/6, 2/6, 2/6, 1/6$, respectively. ◀

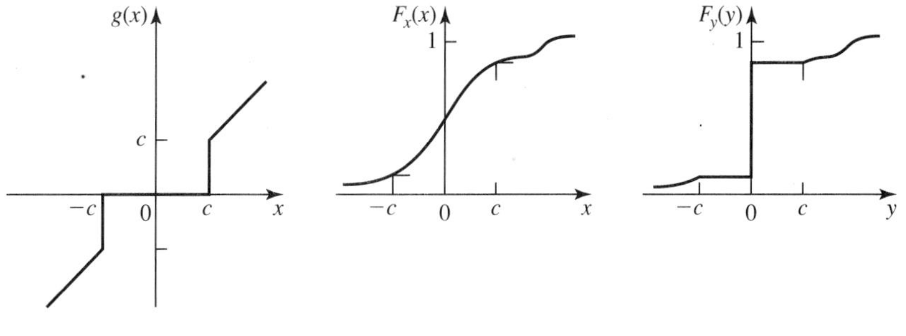

FIGURE 5-8

Determination of $f_y(y)$

We wish to determine the density of $\mathbf{y} = g(\mathbf{x})$ in terms of the density of $\mathbf{x}$. Suppose, first, that the set R of the y axis is not in the range of the function $g(x)$, that is, that $g(x)$ is not a point of R for any x. In this case, the probability that $g(\mathbf{x})$ is in R equals 0. Hence $f_y(y) = 0$ for $y \in R$. It suffices, therefore, to consider the values of y such that for some x, $g(x) = y$.

FUNDAMENTAL THEOREM. To find $f_y(y)$ for a specific y, we solve the equation $y = g(x)$. Denoting its real roots by x_n,

$$y = g(x_1) = \cdots = g(x_n) = \cdots \tag{5-15}$$

we shall show that

$$f_y(y) = \frac{f_x(x_1)}{|g'(x_1)|} + \cdots + \frac{f_x(x_n)}{|g'(x_n)|} + \cdots \tag{5-16}$$

where $g'(x)$ is the derivative of $g(x)$.

Proof. To avoid generalities, we assume that the equation $y = g(x)$ has three roots as in Fig. 5-9. As we know

$$f_y(y)\,dy = P\{y < \mathbf{y} \le y + dy\}$$

It suffices, therefore, to find the set of values x such that $y < g(x) \le y + dy$ and the probability that $\mathbf{x}$ is in this set. As we see from the figure, this set consists of the following three intervals

$$x_1 < x < x_1 + dx_1 \qquad x_2 + dx_2 < x < x_2 \qquad x_3 < x < x_3 + dx_3$$

where $dx_1 > 0$, $dx_3 > 0$ but $dx_2 < 0$. From this it follows that

$$\begin{aligned} P\{y < \mathbf{y} < y + dy\} = {} & P\{x_1 < \mathbf{x} < x_1 + dx_1\} \\ & + P\{x_2 + dx_2 < \mathbf{x} < x_2\} + P\{x_3 < \mathbf{x} < x_3 + dx_3\} \end{aligned}$$

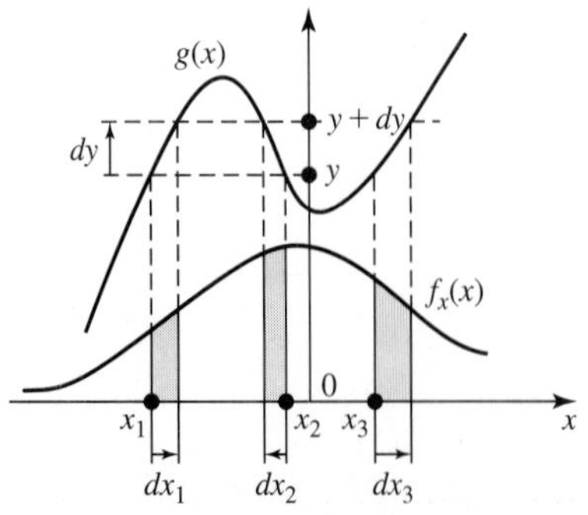

FIGURE 5-9

The right side equals the shaded area in Fig. 5-9. Since

$$P\{x_1 < \mathbf{x} < x_1 + dx_1\} = f_x(x_1)\,dx_1 \qquad dx_1 = dy/g'(x_1)$$

$$P\{x_2 + dx_2 < \mathbf{x} < x_2\} = f_x(x_2)\,|dx_2| \quad dx_2 = dy/g'(x_2)$$

$$P\{x_3 < \mathbf{x} < x_3 + dx_3\} = f_x(x_3)\,dx_3 \qquad dx_3 = dy/g'(x_3)$$

we conclude that

$$f_y(y)\,dy = \frac{f_x(x_1)}{g'(x_1)}\,dy + \frac{f_x(x_2)}{|g'(x_2)|}\,dy + \frac{f_x(x_3)}{g'(x_3)}\,dy$$

and (5-16) results.

We note, finally, that if $g(x) = y_1 = \text{constant}$ for every x in the interval (x_0, x_1), then [see (5-9)] $F_y(y)$ is discontinuous for $y = y_1$. Hence $f_y(y)$ contains an impulse $\delta(y - y_1)$ of area $F_x(x_1) - F_x(x_0)$.

Conditional density The conditional density $f_y(y \mid M)$ of the random variable $\mathbf{y} = g(\mathbf{x})$ assuming an event M is given by (5-5) if on the right side we replace the terms $f_x(x_i)$ by $f_x(x_i \mid M)$ (see, for example, Prob. 5-21).

Illustrations

We give next several applications of (5-2) and (5-16).

1.

$$\mathbf{y} = a\mathbf{x} + b \qquad g'(x) = a \tag{5-17}$$

The equation $y = ax + b$ has a single solution $x = (y - b)/a$ for every y. Hence

$$f_y(y) = \frac{1}{|a|} f_x\left(\frac{y - b}{a}\right) \tag{5-18}$$

Special case If $\mathbf{x}$ is uniform in the interval (x_1, x_2), then $\mathbf{y}$ is uniform in the interval $(ax_1 + b, ax_2 + b)$.

EXAMPLE 5-10 ▶ Suppose that the voltage $\mathbf{v}$ is a random variable given by

$$\mathbf{v} = i(\mathbf{r} + r_0)$$

where $i = 0.01$ A and $r_0 = 1000\ \Omega$. If the resistance $\mathbf{r}$ is a random variable uniform between 900 and 1100 Ω, then $\mathbf{v}$ is uniform between 19 and 21 V. ◀

2.

$$\mathbf{y} = \frac{1}{\mathbf{x}} \qquad g'(x) = -\frac{1}{x^2} \tag{5-19}$$

The equation $y = 1/x$ has a single solution $x = 1/y$. Hence

$$f_y(y) = \frac{1}{y^2} f_x\left(\frac{1}{y}\right) \tag{5-20}$$

Cauchy density: If $\mathbf{x}$ has a *Cauchy density* with parameter α,

$$f_x(x) = \frac{\alpha/\pi}{x^2 + \alpha^2} \qquad \text{then} \quad f_y(y) = \frac{1/\alpha\pi}{y^2 + 1/\alpha^2}$$

in (5-19) is also a Cauchy density with parameter $1/\alpha$.

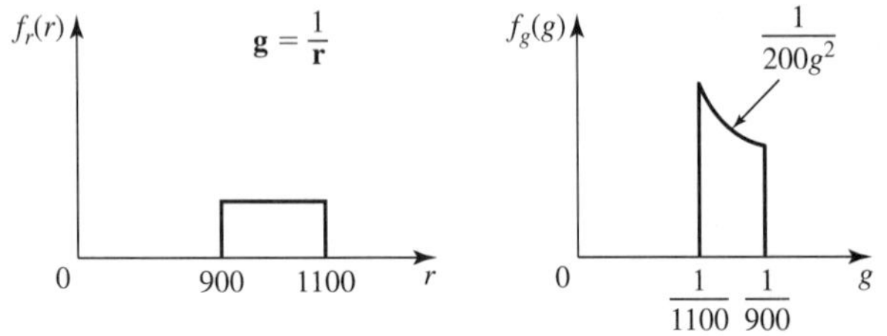

FIGURE 5-10

EXAMPLE 5-11 ▶ Suppose that the resistance $\mathbf{r}$ is uniform between 900 and 1100 Ω as in Fig. 5-10. We shall determine the density of the corresponding conductance

$$\mathbf{g} = 1/\mathbf{r}$$

Since $f_r(r) = 1/200$ S for r between 900 and 1100 it follows from (5-20) that

$$f_g(g) = \frac{1}{200g^2} \qquad \text{for} \quad \frac{1}{1100} < g < \frac{1}{900}$$

and 0 elsewhere. ◀

3.

$$\mathbf{y} = a\mathbf{x}^2 \quad a > 0 \quad g'(x) = 2ax \tag{5-21}$$

If $y \le 0$, then the equation $y = ax^2$ has no real solutions; hence $f_y(y) = 0$. If $y > 0$, then it has two solutions

$$x_1 = \sqrt{\frac{y}{a}} \qquad x_2 = -\sqrt{\frac{y}{a}}$$

and (5-16) yields [see also (5-4)]

$$f_y(y) = \frac{1}{2a\sqrt{y/a}}\left[f_x\left(\sqrt{\frac{y}{a}}\right) + f_x\left(-\sqrt{\frac{y}{a}}\right)\right] \qquad y > 0 \tag{5-22}$$

We note that $F_y(y) = 0$ for $y < 0$ and

$$F_y(y) = P\left\{-\sqrt{\frac{y}{a}} \le \mathbf{x} \le \sqrt{\frac{y}{a}}\right\} = F_x\left(\sqrt{\frac{y}{a}}\right) - F_x\left(-\sqrt{\frac{y}{a}}\right) \qquad y > 0$$

EXAMPLE 5-12 ▶ The voltage across a resistor is a random variable $\mathbf{e}$ uniform between 5 and 10 V. We shall determine the density of the power

$$\mathbf{w} = \frac{\mathbf{e}^2}{r} \qquad r = 1000\,\Omega$$

dissipated in r.

Since $f_e(e) = 1/5$ for e between 5 and 10 and 0 elsewhere, we conclude from (5-8) with $a = 1/r$ that

$$f_w(w) = \sqrt{\frac{10}{w}} \qquad \frac{1}{40} < w < \frac{1}{10}$$

and 0 elsewhere. ◀

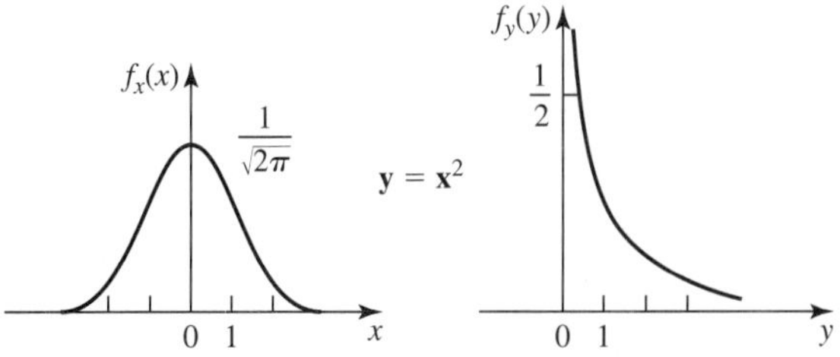

FIGURE 5-11

Special case Suppose that

$$f_x(x) = \frac{1}{\sqrt{2\pi}} e^{-x^2/2} \qquad \mathbf{y} = \mathbf{x}^2$$

With $a = 1$, it follows from (5-22) and the evenness of $f_x(x)$ that (Fig. 5-11)

$$f_y(y) = \frac{1}{\sqrt{y}} f_x(\sqrt{y}) = \frac{1}{\sqrt{2\pi y}} e^{-y/2} U(y)$$

We have thus shown that if $\mathbf{x}$ is an $N(0, 1)$ random variable, the random variable $\mathbf{y} = \mathbf{x}^2$ has a chi-square distribution with one degree of freedom [see (4-39) and also (5-7)].

4.

$$\mathbf{y} = \sqrt{\mathbf{x}} \quad g'(x) = \frac{1}{2\sqrt{x}} \tag{5-23}$$

The equation $y = \sqrt{x}$ has a single solution $x = y^2$ for $y > 0$ and no solution for $y < 0$. Hence

$$f_y(y) = 2y f_x(y^2) U(y) \tag{5-24}$$

The chi density Suppose that $\mathbf{x}$ has a chi-square density as in (4-39),

$$f_x(x) = \frac{1}{2^{n/2}\Gamma(n/2)} x^{n/2-1} e^{-x/2} U(y)$$

and $y = \sqrt{x}$. In this case, (5-24) yields

$$f_y(y) = \frac{2}{2^{n/2}\Gamma(n/2)} y^{n-1} e^{-y^2/2} U(y) \tag{5-25}$$

This function is called the **chi density** with n degrees of freedom. The following cases are of special interest.

Maxwell For $n = 3$, (5-25) yields the Maxwell density [see also (4-54)]

$$f_y(y) = \sqrt{2/\pi}\, y^2 e^{-y^2/2}$$

Rayleigh For $n = 2$, we obtain the Rayleigh density $f_y(y) = y e^{-y^2/2} U(y)$.

5.

$$\mathbf{y} = \mathbf{x} U(\mathbf{x}) \qquad g'(x) = U(x) \tag{5-26}$$

Clearly, $f_y(y) = 0$ and $F_y(y) = 0$ for $y < 0$ (Fig. 5-12). If $y > 0$, then the equation $y = xU(x)$ has a single solution $x_1 = y$. Hence

$$f_y(y) = f_x(y) \qquad F_y(y) = F_x(y) \qquad y > 0 \tag{5-27}$$

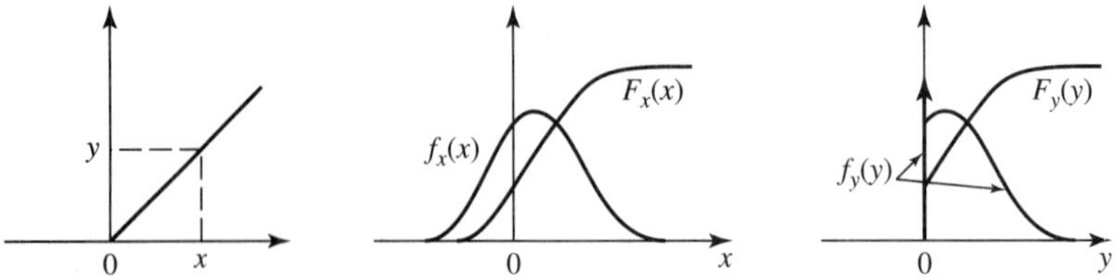

FIGURE 5-12
Half wave rectifier.

Thus $F_y(y)$ is discontinuous at $y = 0$ with discontinuity $F_y(0^+) - F_y(0^-) = F_x(0)$. Hence

$$f_y(y) = f_x(y)U(y) + F_x(0)\delta(y)$$

6. $$\mathbf{y} = e^{\mathbf{x}} \qquad g'(x) = e^x \tag{5-28}$$

If $y > 0$, then the equation $y = e^x$ has the single solution $x = \ln y$. Hence

$$f_y(y) = \frac{1}{y} f_x(\ln y) \qquad y > 0 \tag{5-29}$$

If $y < 0$, then $f_y(y) = 0$.

lognormal: If $\mathbf{x}$ is $N(\eta; \sigma)$, then

$$f_y(y) = \frac{1}{\sigma y \sqrt{2\pi}} e^{-(\ln y - \eta)^2/2\sigma^2} \tag{5-30}$$

This density is called *lognormal*.

7. $$\mathbf{y} = a \sin(\mathbf{x} + \theta) \qquad a > 0 \tag{5-31}$$

If $|y| > a$, then the equation $y = a \sin(x + \theta)$ has no solutions; hence $f_y(y) = 0$. If $|y| < a$, then it has infinitely many solutions (Fig. 5-13*a*)

$$x_n = \arcsin \frac{y}{a} - \theta \qquad n = - \cdots -1, 0, 1, \ldots$$

Since $g'(x_n) = a \cos(x_n + \theta) = \sqrt{a^2 - y^2}$, (5-5) yields

$$f_y(y) = \frac{1}{\sqrt{a^2 - y^2}} \sum_{n=-\infty}^{\infty} f_x(x_n) \qquad |y| < a \tag{5-32}$$

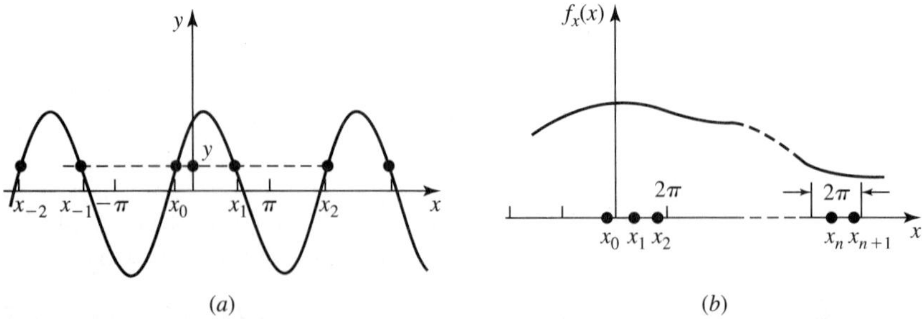

FIGURE 5-13

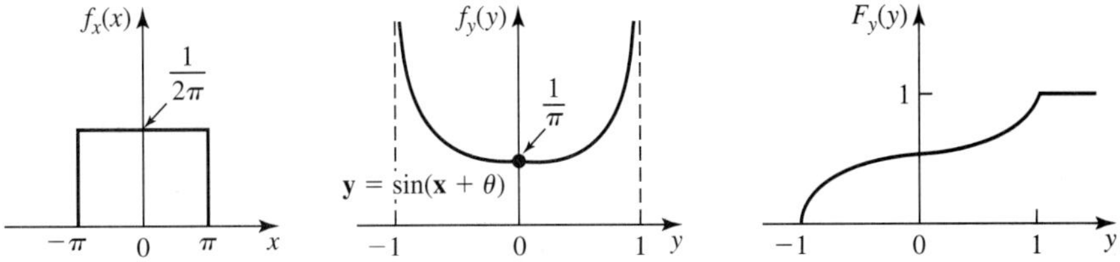

FIGURE 5-14

Special case: Suppose that $\mathbf{x}$ is uniform in the interval $(-\pi, \pi)$. In this case, the equation $y = a\sin(x + \theta)$ has exactly two solutions in the interval $(-\pi, \pi)$ for any θ (Fig. 5-14). The function $f_x(x)$ equals $1/2\pi$ for these two values and it equals 0 for any x_n outside the interval $(-\pi, \pi)$. Retaining the two nonzero terms in (5-32), we obtain

$$f_y(y) = \frac{2}{2\pi\sqrt{a^2 - y^2}} = \frac{1}{\pi\sqrt{a^2 - y^2}} \qquad |y| < a \tag{5-33}$$

To find $F_y(y)$, we observe that $\mathbf{y} \le y$ if $\mathbf{x}$ is either between $-\pi$ and x_0 or between x_1 and π (Fig. 5-13a). Since the total length of the two intervals equals $\pi + 2x_0 + 2\theta$, we conclude, dividing by 2π, that

$$F_y(y) = \frac{1}{2} + \frac{1}{\pi}\arcsin\frac{y}{a} \qquad |y| < a \tag{5-34}$$

We note that although $f_y(\pm a) = \infty$, the probability that $\mathbf{y} = \pm a$ is 0.

Smooth phase If the density $f_x(x)$ of $\mathbf{x}$ is sufficiently smooth so that it can be approximated by a constant in any interval of length 2π (see Fig. 5-13b), then

$$\pi \sum_{n=-\infty}^{\infty} f_x(x_n) \simeq \int_{-\infty}^{\infty} f_x(x)\,dx = 1$$

because in each interval of length 2π this sum has two terms. Inserting into (5-32), we conclude that the density of $\mathbf{y}$ is given approximately by (5-33).

EXAMPLE 5-13

▶ A particle leaves the origin under the influence of the force of gravity and its initial velocity v forms an angle φ with the horizontal axis. The path of the particle reaches the ground at a distance

$$\mathbf{d} = \frac{v^2}{g}\sin 2\varphi$$

from the origin (Fig. 5-15). Assuming that φ is a random variable uniform between 0 and $\pi/2$, we shall determine: (a) the density of $\mathbf{d}$ and (b) the probability that $\mathbf{d} \le d_0$.

SOLUTION

(a) Clearly,

$$\mathbf{d} = a\sin\mathbf{x} \qquad a = v^2/g$$

where the random variable $\mathbf{x} = 2\varphi$ is uniform between 0 and π. If $0 < d < a$, then the equation $d = a\sin x$ has exactly two solutions in the interval $(0, \pi)$. Reasoning as in

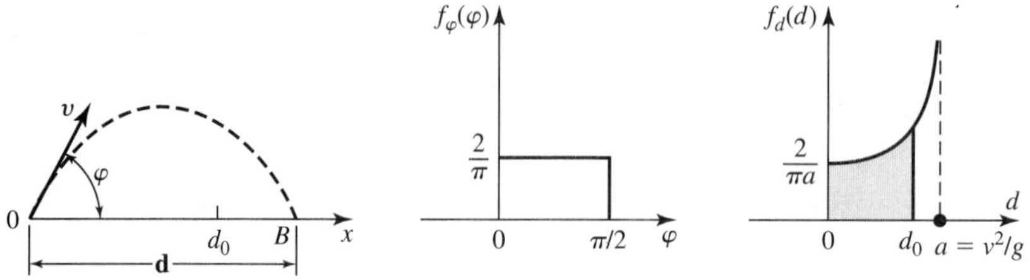

FIGURE 5-15

(5-33), we obtain

$$f_d(d) = \frac{2}{\pi\sqrt{a^2 - d^2}} \qquad 0 < d < a$$

and 0 otherwise.

(*b*) The probability that $\mathbf{d} \le d_0$ equals the shaded area in Fig. 5-15:

$$P\{\mathbf{d} \le d_0\} = F_d(d_0) = \frac{2}{\pi} \arcsin \frac{d_0}{a}$$

8.

$$\mathbf{y} = \tan \mathbf{x} \qquad \text{(5-35)}$$

The equation $y = \tan x$ has infinitely many solutions for any y (Fig. 5-16*a*)

$$x_n = \arctan y \qquad n = \ldots, -1, 0, 1, \ldots$$

Since $g'(x) = 1/\cos^2 x = 1 + y^2$, Eq. (5-16) yields

$$f_y(y) = \frac{1}{1+y^2} \sum_{n=-\infty}^{\infty} f_x(x_n) \qquad \text{(5-36)}$$

Special case If $\mathbf{x}$ is uniform in the interval $(-\pi/2, \pi/2)$, then the term $f_x(x_1)$ in (5-36) equals $1/\pi$ and all others are 0 (Fig. 5-16*b*). Hence $\mathbf{y}$ has a *Cauchy* density given by

$$f_y(y) = \frac{1/\pi}{1+y^2} \qquad \text{(5-37)}$$

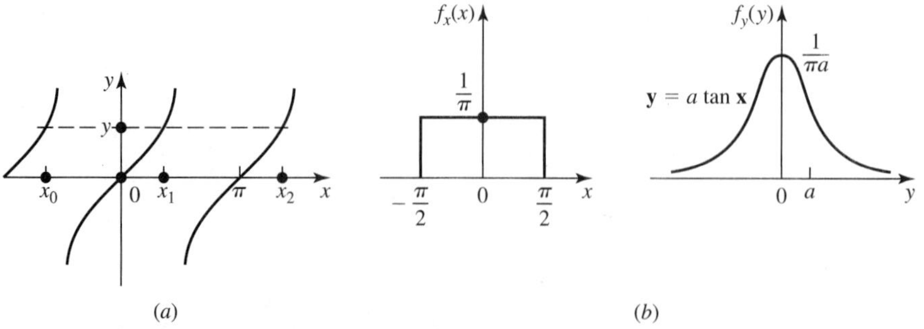

FIGURE 5-16

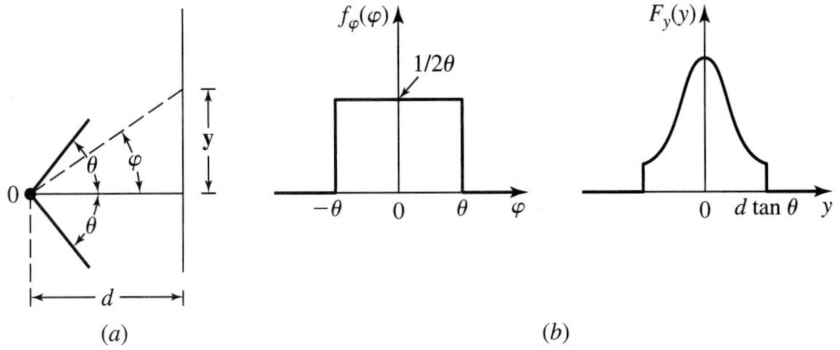

FIGURE 5-17

As we see from the figure, $\mathbf{y} \le y$ if $\mathbf{x}$ is between $-\pi/2$ and x_1. Since the length of this interval equals $x_1 + \pi/2$, we conclude, dividing by π, that

$$F_y(y) = \frac{1}{\pi}\left(x_1 + \frac{\pi}{2}\right) = \frac{1}{2} + \frac{1}{\pi}\text{arc tan } y \tag{5-38}$$

EXAMPLE 5-14

▶ A particle leaves the origin in a free motion as in Fig. 5-17 crossing the vertical line $x = d$ at

$$\mathbf{y} = d \tan \boldsymbol{\varphi}$$

Assuming that the angle $\boldsymbol{\varphi}$ is uniform in the interval $(-\theta, \theta)$, we conclude as in (5-37) that

$$f_y(y) = \frac{d/2\theta}{d^2 + y^2} \qquad \text{for} \quad |y| < d \tan\theta$$

and 0 otherwise. ◀

EXAMPLE 5-15

▶ Suppose $f_x(x) = 2x/\pi^2$, $0 < x < \pi$, and $\mathbf{y} = \sin \mathbf{x}$. Determine $f_y(y)$.

SOLUTION

Since $\mathbf{x}$ has zero probability of falling outside the interval $(0, \pi)$, $y = \sin x$ has zero probability of falling outside the interval $(0, 1)$ and $f_y(y) = 0$ outside this interval. For any $0 < y < 1$, from Fig. 5.18*b*, the equation $y = \sin x$ has an infinite number of solutions $\ldots, x_1, x_2, x_3, \ldots$, where $x_1 = \sin^{-1} y$ is the principal solution. Moreover, using the symmetry we also get $x_2 = \pi - x_1$ and so on. Further,

$$\frac{dy}{dx} = \cos x = \sqrt{1 - \sin^2 x} = \sqrt{1 - y^2}$$

so that

$$\left|\frac{dy}{dx}\right|_{x=x_i} = \sqrt{1 - y^2}$$

Using this in (5-16), we obtain

$$f_y(y) = \sum_{i=-\infty}^{+\infty} \frac{1}{\sqrt{1 - y^2}} f_x(x_i) \qquad 0 < y < 1 \tag{5-39}$$

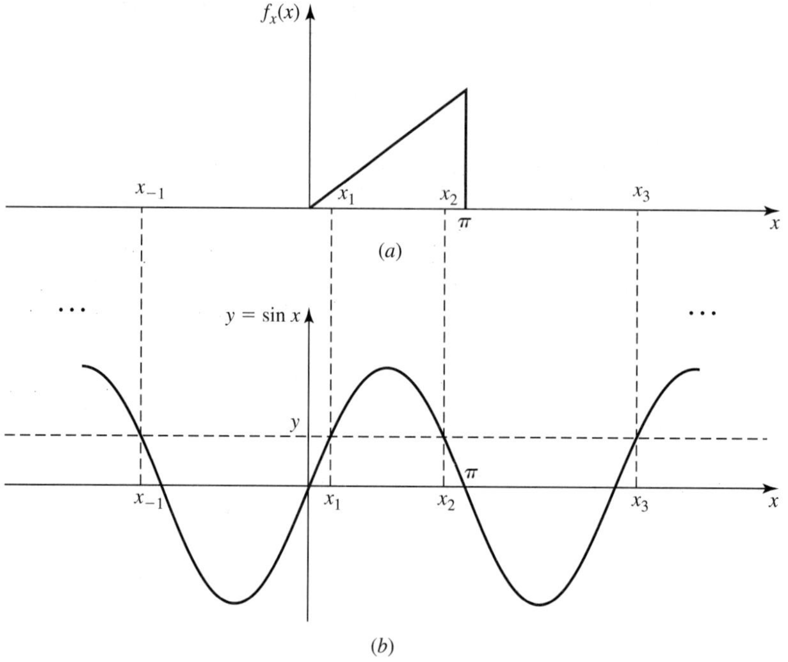

FIGURE 5-18

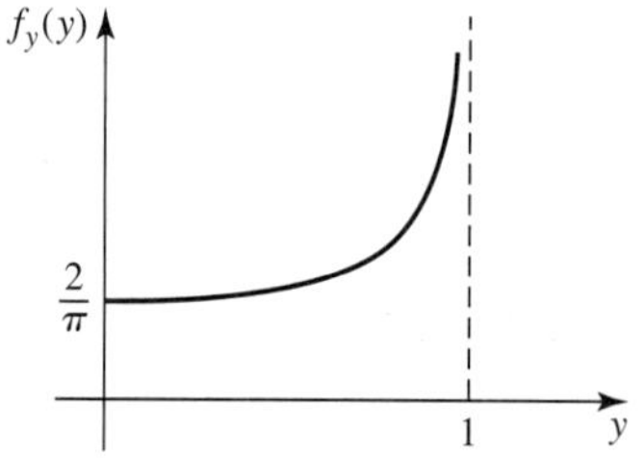

FIGURE 5-19

But from Fig. 5.18a, here $f_x(x_{-1}) = f_x(x_3) = f_x(x_4) = \cdots = 0$ (except for $f_x(x_1)$ and $f_x(x_2)$ the rest are all zeros). Thus (Fig. 5-19)

$$
\begin{aligned}
f_y(y) &= \frac{1}{\sqrt{1-y^2}}(f_x(x_1) + f_x(x_2)) = \frac{1}{\sqrt{1-y^2}}\left(\frac{2x_1}{\pi^2} + \frac{2x_2}{\pi^2}\right) \\
&= \frac{2(x_1 + \pi - x_1)}{\pi^2\sqrt{1-y^2}} = \begin{cases} \dfrac{2}{\pi\sqrt{1-y^2}} & 0 < y < 1 \\ 0 & \text{otherwise} \end{cases}
\end{aligned}
\tag{5-40}
$$

◀

THE INVERSE PROBLEM. In the preceding discussion, we were given a random variable **x** with known distribution $F_x(x)$ and a function $g(x)$ and we determined the distribution $F_y(y)$ of the random variable $\mathbf{y} = g(\mathbf{x})$. We consider now the inverse problem: We are given the distribution of **x** and we wish to find a function $g(x)$ such that the distribution of the random variable $\mathbf{y} = g(\mathbf{x})$ equals a specified function $F_y(y)$. This topic is developed further in Sec. 7-5. We start with two special cases.

From $F_x(x)$ to a uniform distribution. Given a random variable $\mathbf{x}$ with distribution $F_x(x)$, we wish to find a function $g(x)$ such that the random variable $\mathbf{u} = g(\mathbf{x})$ is uniformly distributed in the interval (0, 1). We maintain that $g(x) = F_x(x)$, that is, if

$$\mathbf{u} = F_x(\mathbf{x}) \quad \text{then } F_u(u) = u \text{ for } 0 \le u \le 1 \tag{5-41}$$

Proof. Suppose that x is an arbitrary number and $u = F_x(x)$. From the monotonicity of $F_x(x)$ it follows that $\mathbf{u} \le u$ iff $\mathbf{x} \le x$. Hence

$$F_u(u) = P\{\mathbf{u} \le u\} = P\{\mathbf{x} \le x\} = F_x(x) = u$$

and (5-41) results.

The random variable $\mathbf{u}$ can be considered as the output of a nonlinear memoryless system (Fig. 5-20) with input $\mathbf{x}$ and transfer characteristic $F_x(x)$. Therefore if we use $\mathbf{u}$ as the input to another system with transfer characteristic the inverse $F_x^{(-1)}(u)$ of the function $u = F_x(x)$, the resulting output will equal $\mathbf{x}$:

$$\text{If} \quad \mathbf{x} = F_x^{(-1)}(\mathbf{u}) \qquad \text{then} \quad P\{\mathbf{x} \le x\} = F_x(x)$$

From uniform to $F_y(y)$. Given a random variable $\mathbf{u}$ with uniform distribution in the interval (0, 1), we wish to find a function $g(u)$ such that the distribution of the random variable $\mathbf{y} = g(\mathbf{u})$ is a specified function $F_y(y)$. We maintain that $g(u)$ is the inverse of the function $u = F_y(y)$:

$$\text{If} \quad y = F_y^{(-1)}(\mathbf{u}) \qquad \text{then} \quad P\{\mathbf{y} \le y\} = F_y(y) \tag{5-42}$$

Proof. The random variable $\mathbf{u}$ in (5-41) is uniform and the function $F_x(x)$ is arbitrary. Replacing $F_x(x)$ by $F_y(y)$, we obtain (5-42) (see also Fig. 5-20).

From $F_x(x)$ to $F_y(y)$. We consider, finally, the general case: Given $F_x(x)$ and $F_y(y)$, find $g(x)$ such that the distribution of $\mathbf{y} = g(\mathbf{x})$ equals $F_y(y)$. To solve this problem, we form the random variable $\mathbf{u} = F_x(\mathbf{x})$ as in (5-41) and the random variable $\mathbf{y} = F^{(-1)}(\mathbf{u})$

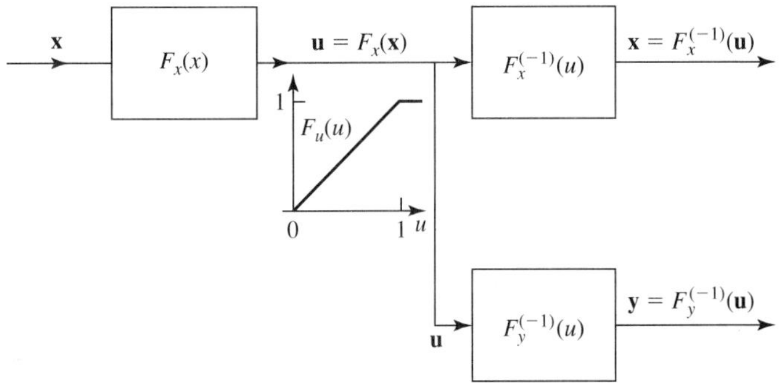

FIGURE 5-20

as in (5-42). Combining the two, we conclude:

$$\text{If} \quad \mathbf{y} = F_y^{(-1)}(F_x(\mathbf{x})) \qquad \text{then} \quad P\{\mathbf{y} \le y\} = F_y(y) \tag{5-43}$$

5-3 MEAN AND VARIANCE

The *expected value* or *mean* of a random variable $\mathbf{x}$ is by definition the integral

$$E\{\mathbf{x}\} = \int_{-\infty}^{\infty} x f(x)\, dx \tag{5-44}$$

This number will also be denoted by η_x or η.

EXAMPLE 5-16

▶ If $\mathbf{x}$ is uniform in the interval (x_1, x_2), then $f(x) = 1/(x_2 - x_1)$ in this interval. Hence

$$E\{\mathbf{x}\} = \frac{1}{x_2 - x_1} \int_{x_1}^{x_2} x\, dx = \frac{x_1 + x_2}{2}$$ ◀

We note that, if the vertical line $x = a$ is an axis of symmetry of $f(x)$ then $E\{\mathbf{x}\} = a$; in particular, if $f(-x) = f(x)$, then $E\{\mathbf{x}\} = 0$. In Example 5-16, $f(x)$ is symmetrical about the line $x = (x_1 + x_2)/2$.

Discrete type For discrete type random variables the integral in (5-44) can be written as a sum. Indeed, suppose that $\mathbf{x}$ takes the values x_i with probability p_i. In this case [see (4-15)]

$$f(x) = \sum_i p_i \delta(x - x_i) \tag{5-45}$$

Inserting into (5-44) and using the identity

$$\int_{-\infty}^{\infty} x\delta(x - x_i)\, dx = x_i$$

we obtain

$$E\{\mathbf{x}\} = \sum_i p_i x_i \qquad p_i = P\{\mathbf{x} = x_i\} \tag{5-46}$$

EXAMPLE 5-17

▶ If $\mathbf{x}$ takes the values $1, 2, \ldots, 6$ with probability $1/6$, then

$$E\{\mathbf{x}\} = \tfrac{1}{6}(1 + 2 + \cdots + 6) = 3.5$$ ◀

Conditional mean The conditional mean of a random variable $\mathbf{x}$ assuming an event M is given by the integral in (5-44) if $f(x)$ is replaced by the conditional density $f(x \mid M)$:

$$E\{\mathbf{x} \mid M\} = \int_{-\infty}^{\infty} x f(x \mid M)\, dx \tag{5-47}$$

For discrete-type random variables (5-47) yields

$$E\{\mathbf{x} \mid M\} = \sum_i x_i P\{\mathbf{x} = x_i \mid M\} \tag{5-48}$$

EXAMPLE 5-18 ▶ With $M = \{x \geq a\}$, it follows from (5-47) that

$$E\{\mathbf{x} \mid \mathbf{x} \geq a\} = \int_{-\infty}^{\infty} x f(x \mid \mathbf{x} \geq a)\, dx = \frac{\int_a^{\infty} x f(x)\, dx}{\int_a^{\infty} f(x)\, dx}$$ ◀

Lebesgue integral. The mean of a random variable can be interpreted as a Lebesgue integral. This interpretation is important in mathematics but it will not be used in our development. We make, therefore, only a passing reference.

We divide the x axis into intervals (x_k, x_{k+1}) of length Δx as in Fig. 5-21a. If Δx is small, then the Riemann integral in (5-44) can be approximated by a sum

$$\int_{-\infty}^{\infty} x f(x)\, dx \simeq \sum_{k=-\infty}^{\infty} x_k f(x_k)\, \Delta x \tag{5-49}$$

And since $f(x_k)\, \Delta x \simeq P\{x_k < \mathbf{x} < x_k + \Delta x\}$, we conclude that

$$E\{\mathbf{x}\} \simeq \sum_{k=-\infty}^{\infty} x_k P\{x_k < \mathbf{x} < x_k + \Delta x\}$$

Here, the sets $\{x_k < \mathbf{x} < x_k + \Delta x\}$ are differential events specified in terms of the random variable $\mathbf{x}$, and their union is the space S (Fig. 5-21b). Hence, to find $E\{\mathbf{x}\}$, we multiply the probability of each differential event by the corresponding value of $\mathbf{x}$ and sum over all k. The resulting limit as $\Delta x \to 0$ is written in the form

$$E\{\mathbf{x}\} = \int_S \mathbf{x}\, dP \tag{5-50}$$

and is called the *Lebesgue integral* of $\mathbf{x}$.

Frequency interpretation We maintain that the arithmetic average $\bar{x}$ of the observed values x_i of $\mathbf{x}$ tends to the integral in (5-44) as $n \to \infty$:

$$\bar{x} = \frac{x_1 + \cdots + x_n}{n} \to E\{\bar{\mathbf{x}}\} \tag{5-51}$$

Proof. We denote by Δn_k the number of x_i's that are between z_k and $z_k + \Delta x = z_{k+1}$. From this it follows that

$$x_1 + \cdots + x_n \simeq \sum z_k\, \Delta n_k$$

and since $f(z_k)\, \Delta x \simeq \Delta n_k / n$ [see (4-21)] we conclude that

$$\bar{x} \simeq \frac{1}{n} \sum z_k\, \Delta n_k = \sum z_k f(z_k)\, \Delta x \simeq \int_{-\infty}^{\infty} x f(x)\, dx$$

and (5-51) results.

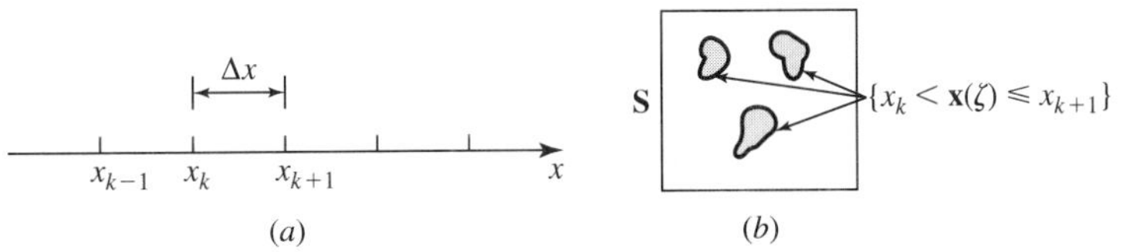

FIGURE 5-21

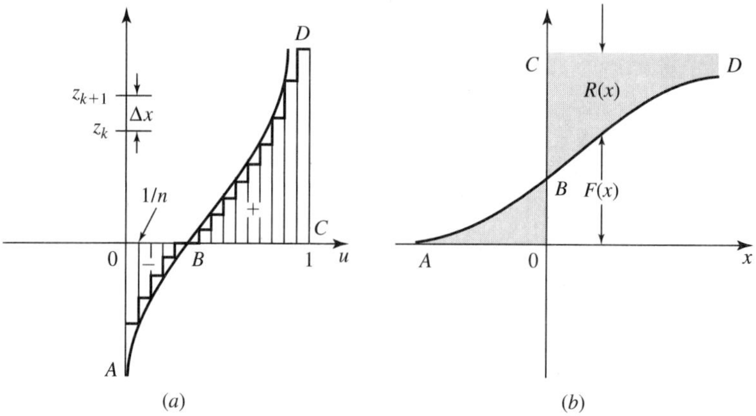

FIGURE 5-22

We shall use the above frequency interpretation to express the mean of $\mathbf{x}$ in terms of its distribution. From the construction of Fig. 5-22a it readily follows that $\bar{x}$ equals the area under the empirical percentile curve of $\mathbf{x}$. Thus

$$\bar{x} = (BCD) - (OAB)$$

where (BCD) and (OAB) are the shaded areas above and below the u axis, respectively. These areas equal the corresponding areas of Fig. 5-22b; hence

$$\bar{x} = \int_0^{\infty} [1 - F_n(x)]\,dx - \int_{-\infty}^{0} F_n(x)\,dx$$

where $F_n(x)$ is the empirical distribution of $\mathbf{x}$. With $n \to \infty$ this yields

$$E\{\mathbf{x}\} = \int_0^{\infty} R(x)\,dx - \int_{-\infty}^{0} F(x)\,dx, \qquad R(x) = 1 - F(x) = P\{\mathbf{x} > x\} \tag{5-52}$$

In particular, for a random variable that takes only nonnegative values, we also obtain

$$E\{\mathbf{x}\} = \int_0^{\infty} R(x)\,dx \tag{5-53}$$

Mean of $g(\mathbf{x})$. Given a random variable $\mathbf{x}$ and a function $g(x)$, we form the random variable $\mathbf{y} = g(\mathbf{x})$. As we see from (5-44), the mean of this random variable is given by

$$E\{\mathbf{y}\} = \int_{-\infty}^{\infty} y f_y(y)\,dy \tag{5-54}$$

It appears, therefore, that to determine the mean of $\mathbf{y}$, we must find its density $f_y(y)$. This, however, is not necessary. As the next basic theorem shows, $E\{\mathbf{y}\}$ can be expressed directly in terms of the function $g(x)$ and the density $f_x(x)$ of $\mathbf{x}$.

THEOREM 5-1

$$E\{g(\mathbf{x})\} = \int_{-\infty}^{\infty} g(x) f_x(x)\,dx \tag{5-55}$$

Proof. We shall sketch a proof using the curve $g(x)$ of Fig. 5-23. With $y = g(x_1) = g(x_2) = g(x_3)$ as in the figure, we see that

$$f_y(y)\,dy = f_x(x_1)\,dx_1 + f_x(x_2)\,dx_2 + f_x(x_3)\,dx_3$$

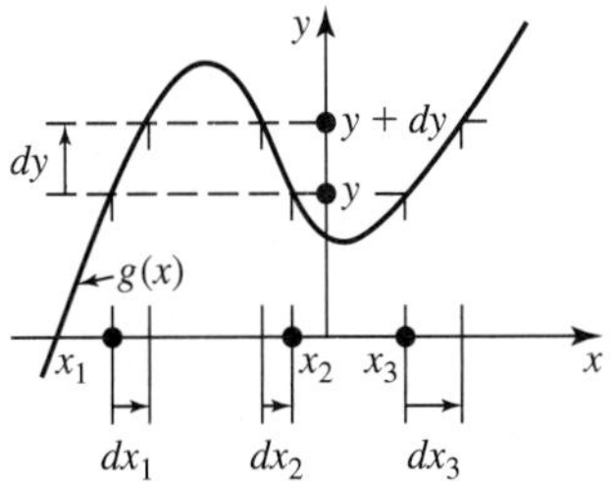

FIGURE 5-23

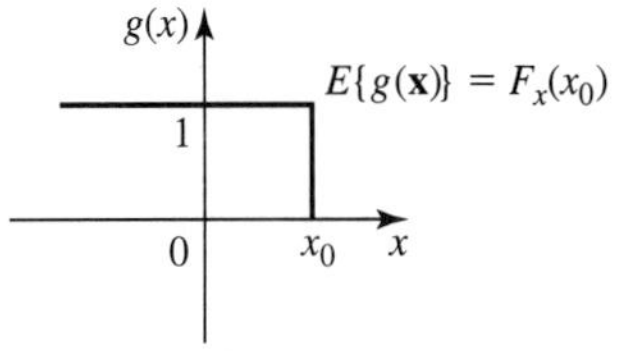

FIGURE 5-24

Multiplying by y, we obtain

$$yf_y(y)\,dy = g(x_1)f_x(x_1)\,dx_1 + g(x_2)f_x(x_2)\,dx_2 + g(x_3)f_x(x_3)\,dx_3$$

Thus to each differential in (5-54) there correspond one or more differentials in (5-55). As dy covers the y axis, the corresponding dx's are nonoverlapping and they cover the entire x axis. Hence the integrals in (5-54) and (5-55) are equal.

If $\mathbf{x}$ is of discrete type as in (5-45), then (5-55) yields

$$E\{g(\mathbf{x})\} = \sum_i g(x_i)P\{\mathbf{x} = x_i\} \tag{5-56}$$

◀

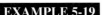

EXAMPLE 5-19 ▶ With x_0 an arbitrary number and $g(x)$ as in Fig. 5-24, (5-55) yields

$$E\{g(\mathbf{x})\} = \int_{-\infty}^{x_0} f_x(x)\,dx = F_x(x_0)$$

This shows that the distribution function of a random variable can be expressed as expected value. ◀

EXAMPLE 5-20 ▶ In this example, we show that the probability of any event A can be expressed as expected value. For this purpose we form the zero–one random variable $\mathbf{x}_A$ associated with the event A:

$$\mathbf{x}_A(\zeta) = \begin{cases} 1 & \zeta \in A \\ 0 & \zeta \notin A \end{cases}$$

Since this random variable takes the values 1 and 0 with respective probabilities $P(A)$ and $P(\overline{A})$, yields

$$E\{\mathbf{x}_A\} = 1 \times P(A) + 0 \times P(\overline{A}) = P(A)$$

◀

Linearity: From (5-55) it follows that

$$E\{a_1g_1(\mathbf{x}) + \cdots + a_ng_n(\mathbf{x})\} = a_1E\{g_1(\mathbf{x})\} + \cdots + a_nE\{g_n(\mathbf{x})\} \tag{5-57}$$

In particular, $E\{a\mathbf{x} + b\} = aE\{\mathbf{x}\} + b$

Complex random variables: If $\mathbf{z} = \mathbf{x} + j\mathbf{y}$ is a complex random variable, then its expected value is by definition

$$E\{\mathbf{z}\} = E\{\mathbf{x}\} + jE\{\mathbf{y}\}$$

From this and (5-55) it follows that if

$$g(\mathbf{x}) = g_1(\mathbf{x}) + jg_2(\mathbf{x})$$

is a complex function of the real random variable $\mathbf{x}$ then

$$E\{g(\mathbf{x})\} = \int_{-\infty}^{\infty} g_1(x)f(x)\,dx + j\int_{-\infty}^{\infty} g_2(x)f(x)\,dx = \int_{-\infty}^{\infty} g(x)f(x)\,dx \tag{5-58}$$

In other words, (5-55) holds even if $g(x)$ is complex.

Variance

Mean alone will not be able to truly represent the p.d.f. of any random variable. To illustrate this, consider two Gaussian random variables $\mathbf{x}_1 \sim N(0,1)$ and $\mathbf{x}_2 \sim N(0,3)$. Both of them have the same mean $\mu = 0$. However, as Fig. 5-25 shows, their p.d.fs are quite different. Here $\mathbf{x}_1$ is more concentrated around the mean, whereas $\mathbf{x}_2$ has a wider spread. Clearly, we need at least an additional parameter to measure this spread around the mean!

For a random variable $\mathbf{x}$ with mean μ, $\mathbf{x} - \mu$ represents the deviation of the random variable from its mean. Since this deviation can be either positive or negative, consider the quantity $(\mathbf{x} - \mu)^2$, and its average value $E[(\mathbf{x} - \mu)^2]$ represents the average square deviation of $\mathbf{x}$ around its mean. Define

$$\sigma_x^2 \stackrel{\Delta}{=} E[(\mathbf{x} - \mu)^2] > 0 \tag{5-59}$$

With $g(\mathbf{x}) = (\mathbf{x} - \mu)^2$ and using (5-55) we get

$$\sigma_x^2 = \int_{-\infty}^{+\infty} (x - \mu)^2 f_x(x)\,dx > 0 \tag{5-60}$$

The positive constant σ_x^2 is known as the *variance* of the random variable $\mathbf{x}$, and its positive square root $\sigma_x = \sqrt{E(\mathbf{x} - \mu)^2}$ is known as the standard deviation of $\mathbf{x}$. Note that the standard deviation represents the root mean square value of the random variable $\mathbf{x}$ around its mean μ.

From the definition it follows that σ^2 is the mean of the random variable $(\mathbf{x} - \eta)^2$. Thus

$$\text{Var}\{\mathbf{x}\} = \sigma^2 = E\{(\mathbf{x} - \eta)^2\} = E\{\mathbf{x}^2 - 2\mathbf{x}\eta + \eta^2\} = E\{\mathbf{x}^2\} - 2\eta E\{\mathbf{x}\} + \eta^2$$

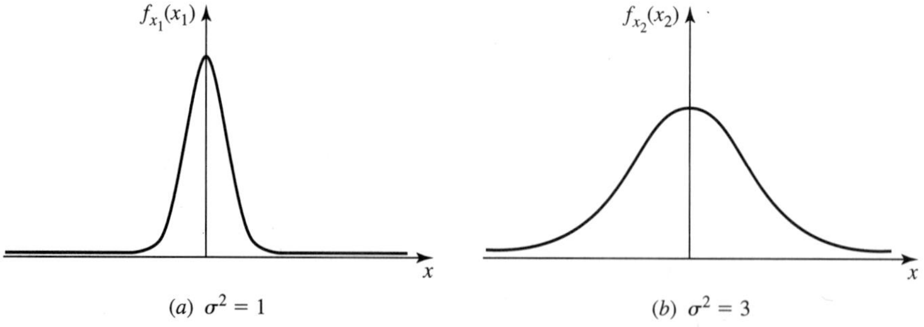

FIGURE 5-25

Hence

$$\sigma^2 = E\{\mathbf{x}^2\} - \eta^2 = E\{\mathbf{x}^2\} - (E\{\mathbf{x}\})^2 \tag{5-61}$$

or, for any random variable

$$E\{\mathbf{x}^2\} \geq (E\{\mathbf{x}\})^2$$

EXAMPLE 5-21 ▶ If $\mathbf{x}$ is uniform in the interval $(-c, c)$, then $\eta = 0$ and

$$\sigma^2 = E\{\mathbf{x}^2\} = \frac{1}{2c}\int_{-c}^{c} x^2\,dx = \frac{c^2}{3}$$ ◀

EXAMPLE 5-22 ▶ We have written the density of a normal random variable in the form

$$f(x) = \frac{1}{\sigma\sqrt{2\pi}} e^{-(x-\eta)^2/2\sigma^2}$$

where up to now η and σ^2 were two arbitrary constants. We show next that η is indeed the mean of $\mathbf{x}$ and σ^2 its variance.

Proof. Clearly, $f(x)$ is symmetrical about the line $x = \eta$; hence $E\{\mathbf{x}\} = \eta$. Furthermore,

$$\int_{-\infty}^{\infty} e^{-(x-\eta)^2/2\sigma^2}\,dx = \sigma\sqrt{2\pi}$$

because the area of $f(x)$ equals 1. Differentiating with respect to σ, we obtain

$$\int_{-\infty}^{\infty} \frac{(x-\eta)^2}{\sigma^3} e^{-(x-\eta)^2/2\sigma^2}\,dx = \sqrt{2\pi}$$

Multiplying both sides by $\sigma^2/\sqrt{2\pi}$, we conclude that $E(\mathbf{x}-\eta)^2 = \sigma^2$ and the proof is complete. ◀

Discrete type. If the random variable $\mathbf{x}$ is of discrete type as in (5-45), then

$$\sigma^2 = \sum_i p_i(x_i - \eta)^2 \qquad p_i = P\{\mathbf{x} = x_i\} \tag{5-62}$$

EXAMPLE 5-23 ▶ The random variable $\mathbf{x}$ takes the values 1 and 0 with probabilities p and $q = 1 - p$ respectively. In this case

$$E\{\mathbf{x}\} = 1 \times p + 0 \times q = p$$

$$E\{\mathbf{x}^2\} = 1^2 \times p + 0^2 \times q = p$$

Hence

$$\sigma^2 = E\{\mathbf{x}^2\} - E^2\{\mathbf{x}\} = p - p^2 = pq$$ ◀

EXAMPLE 5-24 ▶ A Poisson distributed random variable with parameter λ takes the values $0, 1, \ldots$ with probabilities

$$P\{\mathbf{x} = k\} = e^{-\lambda}\frac{\lambda^k}{k!}$$

We shall show that its mean and variance both equal λ:

$$E\{\mathbf{x}\} = \lambda \qquad E\{\mathbf{x}^2\} = \lambda^2 + \lambda \qquad \sigma^2 = \lambda \tag{5-63}$$

Proof. We differentiate twice the Taylor expansion of e^λ:

$$e^\lambda = \sum_{k=0}^{\infty} \frac{\lambda^k}{k!}$$

$$e^\lambda = \sum_{k=0}^{\infty} k\frac{\lambda^{k-1}}{k!} = \frac{1}{\lambda}\sum_{k=1}^{\infty} k\frac{\lambda^k}{k!}$$

$$e^\lambda = \sum_{k=1}^{\infty} k(k-1)\frac{\lambda^{k-2}}{k!} = \frac{1}{\lambda^2}\sum_{k=1}^{\infty} k^2\frac{\lambda^k}{k!} - \frac{1}{\lambda^2}\sum_{k=1}^{\infty} k\frac{\lambda^k}{k!}$$

Hence

$$E\{\mathbf{x}\} = e^{-\lambda}\sum_{k=1}^{\infty} k\frac{\lambda^k}{k!} = \lambda \qquad E\{\mathbf{x}^2\} = e^{-\lambda}\sum_{k=1}^{\infty} k^2\frac{\lambda^k}{k!} = \lambda^2 + \lambda$$

and (5-63) results.

Poisson points. As we have shown in (4-117), the number $\mathbf{n}$ of Poisson points in an interval of length t_0 is a Poisson distributed random variable with parameter $a = \lambda t_0$. From this it follows that

$$E\{\mathbf{n}\} = \lambda t_0 \qquad \sigma_n^2 = \lambda t_0 \tag{5-64}$$

This shows that the density λ of Poisson points equals the expected number of points per unit time. ◀

Notes 1. The variance σ^2 of a random variable $\mathbf{x}$ is a measure of the concentration of $\mathbf{x}$ near its mean η. Its relative frequency interpretation (empirical estimate) is the average of $(x_i - \eta)^2$:

$$\sigma^2 \simeq \frac{1}{n}\sum (x_i - \eta)^2 \tag{5-65}$$

where x_i are the observed values of $\mathbf{x}$. This average can be used as the estimate of σ^2 only if η is known. If it is unknown, we replace it by its estimate $\bar{x}$ and we change n to $n-1$. This yields the estimate

$$\sigma^2 \simeq \frac{1}{n-1}\sum (x_i - \bar{x})^2 \qquad \bar{x} = \frac{1}{n}\sum x_i \tag{5-66}$$

known as the *sample variance* of $\mathbf{x}$ [see (7-65)]. The reason for changing n to $n-1$ is explained later.

2. A simpler measure of the concentration of $\mathbf{x}$ near η is the first absolute central moment $M = E\{|\mathbf{x} - \eta|\}$. Its empirical estimate is the average of $|x_i - \eta|$:

$$M \simeq \frac{1}{n}\sum |x_i - \eta|$$

If η is unknown, it is replaced by $\bar{x}$. This estimate avoids the computation of squares.

5-4 MOMENTS

The following quantities are of interest in the study of random variables:

Moments

$$m_n = E\{\mathbf{x}^n\} = \int_{-\infty}^{\infty} x^n f(x)\,dx \tag{5-67}$$

Central moments

$$\mu_n = E\{(\mathbf{x} - \eta)^n\} = \int_{-\infty}^{\infty} (x - \eta)^n f(x)\, dx \tag{5-68}$$

Absolute moments

$$E\{|\mathbf{x}|^n\} \qquad E\{|\mathbf{x} - \eta|^n\} \tag{5-69}$$

Generalized moments

$$E\{(\mathbf{x} - a)^n\} \qquad E\{|\mathbf{x} - a|^n\} \tag{5-70}$$

We note that

$$\mu_n = E\{(\mathbf{x} - \eta)^n\} = E\left\{\sum_{k=0}^{n} \binom{n}{k} \mathbf{x}^k (-\eta)^{n-k}\right\}$$

Hence

$$\mu_n = \sum_{k=0}^{n} \binom{n}{k} m_k (-\eta)^{n-k} \tag{5-71}$$

Similarly,

$$m_n = E\{[(\mathbf{x} - \eta) + \eta]^n\} = E\left\{\sum_{k=0}^{n} \binom{n}{k} (\mathbf{x} - \eta)^k \eta^{n-k}\right\}$$

Hence

$$m_n = \sum_{k=0}^{n} \binom{n}{k} \mu_k \eta^{n-k} \tag{5-72}$$

In particular,

$$\mu_0 = m_0 = 1 \qquad m_1 = \eta \qquad \mu_1 = 0 \qquad \mu_2 = \sigma^2$$

and

$$\mu_3 = m_3 - 3\eta m_2 + 2\eta^3 \qquad m_3 = \mu_3 + 3\eta\sigma^2 + \eta^3$$

Notes 1. If the function $f(x)$ is interpreted as mass density on the x axis, then $E\{\mathbf{x}\}$ equals its center of gravity, $E\{\mathbf{x}^2\}$ equals the moment of inertia with respect to the origin, and σ^2 equals the central moment of inertia. The standard deviation σ is the radius of gyration.

2. The constants η and σ give only a limited characterization of $f(x)$. Knowledge of other moments provides additional information that can be used, for example, to distinguish between two densities with the same η and σ. In fact, if m_n is known for every n, then, under certain conditions, $f(x)$ is determined uniquely [see also (5-105)]. The underlying theory is known in mathematics as the *moment problem.*

3. The moments of a random variable are not arbitrary numbers but must satisfy various inequalities [see (5-92)]. For example [see (5-61)]

$$\sigma^2 = m_2 - m_1^2 \geq 0$$

Similarly, since the quadratic

$$E\{(\mathbf{x}^n - a)^2\} = m_{2n} - 2am_n + a^2$$

is nonnegative for any a, its discriminant cannot be positive. Hence

$$m_{2n} \geq m_n^2$$

Normal random variables. We shall show that if

$$f(x) = \frac{1}{\sigma\sqrt{2\pi}} e^{-x^2/2\sigma^2}$$

then

$$E\{\mathbf{x}^n\} = \begin{cases} 0 & n = 2k+1 \\ 1 \cdot 3 \cdots (n-1)\sigma^n & n = 2k \end{cases} \tag{5-73}$$

$$E\{|\mathbf{x}|^n\} = \begin{cases} 1 \cdot 3 \cdots (n-1)\sigma^n & n = 2k \\ 2^k k! \sigma^{2k+1}\sqrt{2/\pi} & n = 2k+1 \end{cases} \tag{5-74}$$

The odd moments of $\mathbf{x}$ are 0 because $f(-x) = f(x)$. To prove the lower part of (5-73), we differentiate k times the identity

$$\int_{-\infty}^{\infty} e^{-\alpha x^2}\, dx = \sqrt{\frac{\pi}{\alpha}}$$

This yields

$$\int_{-\infty}^{\infty} x^{2k} e^{-\alpha x^2}\, dx = \frac{1 \cdot 3 \cdots (2k-1)}{2^k} \sqrt{\frac{\pi}{\alpha^{2k+1}}}$$

and with $\alpha = 1/2\sigma^2$, (5-73) results.

Since $f(-x) = f(x)$, we have

$$E\{|\mathbf{x}|^{2k+1}\} = 2\int_0^{\infty} x^{2k+1} f(x)\, dx = \frac{2}{\sigma\sqrt{2\pi}} \int_0^{\infty} x^{2k+1} e^{-x^2/2\sigma^2}\, dx$$

With $y = x^2/2\sigma^2$, the above yields

$$\sqrt{\frac{2}{\pi}} \frac{(2\sigma^2)^{k+1}}{2\sigma} \int_0^{\infty} y^k e^{-y}\, dy$$

and (5-74) results because the last integral equals $k!$

We note in particular that

$$E\{\mathbf{x}^4\} = 3\sigma^4 = 3E^2\{\mathbf{x}^2\} \tag{5-75}$$

EXAMPLE 5-25

▶ If $\mathbf{x}$ has a *Rayleigh density*

$$f(x) = \frac{x}{\sigma^2} e^{-x^2/2\sigma^2} U(x)$$

then

$$E\{\mathbf{x}^n\} = \frac{1}{\sigma^2} \int_0^{\infty} x^{n+1} e^{-x^2/2\sigma^2}\, dx = \frac{1}{2\sigma^2} \int_{-\infty}^{\infty} |x|^{n+1} e^{-x^2/2\sigma^2}\, dx$$

From this and (5-74) it follows that

$$E\{\mathbf{x}^n\} = \begin{cases} 1 \cdot 3 \cdots n\sigma^n \sqrt{\pi/2} & n = 2k+1 \\ 2^k k! \sigma^{2k} & n = 2k \end{cases} \tag{5-76}$$

In particular,

$$E\{\mathbf{x}\} = \sigma\sqrt{\pi/2} \qquad \text{Var}\{\mathbf{x}\} = (2 - \pi/2)\sigma^2 \tag{5-77}$$

◀

EXAMPLE 5-26 ▶ If **x** has a *Maxwell density*

$$f(x) = \frac{\sqrt{2}}{\alpha^3\sqrt{\pi}} x^2 e^{-x^2/2\alpha^2} U(x)$$

then

$$E\{\mathbf{x}^n\} = \frac{1}{\alpha^3\sqrt{2\pi}} \int_{-\infty}^{\infty} |x|^{n+2} e^{-x^2/2\alpha^2}\, dx$$

and (5-74) yields

$$E\{\mathbf{x}^n\} = \begin{cases} 1 \cdot 3 \cdots (n+1)\alpha^n & n = 2k \\ 2^k k! \alpha^{2k-1}\sqrt{2/\pi} & n = 2k-1 \end{cases} \tag{5-78}$$

In particular,

$$E\{\mathbf{x}\} = 2\alpha\sqrt{2/\pi} \qquad E\{\mathbf{x}^2\} = 3\alpha^2 \tag{5-79}$$

◀

Poisson random variables. The moments of a Poisson distributed random variable are functions of the parameter λ:

$$m_n(\lambda) = E\{\mathbf{x}^n\} = e^{-\lambda}\sum_{k=0}^{\infty} k^n \frac{\lambda^k}{k!} \tag{5-80}$$

$$\mu_n(\lambda) = E\{(\mathbf{x}-\lambda)^n\} = e^{-\lambda}\sum_{k=0}^{\infty} (k-\lambda)^n \frac{\lambda^k}{k!} \tag{5-81}$$

We shall show that they satisfy the recursion equations

$$m_{n+1}(\lambda) = \lambda[m_n(\lambda) + m_n'(\lambda)] \tag{5-82}$$

$$\mu_{n+1}(\lambda) = \lambda[n\mu_{n-1}(\lambda) + \mu_n'(\lambda)] \tag{5-83}$$

Proof. Differentiating (5-80) with respect to λ, we obtain

$$m_n'(\lambda) = -e^{-\lambda}\sum_{k=0}^{\infty} k^n \frac{\lambda^k}{k!} + e^{-\lambda}\sum_{k=0}^{\infty} k^{n+1}\frac{\lambda^{k-1}}{k!} = -m_n(\lambda) + \frac{1}{\lambda} m_{n+1}(\lambda)$$

and (5-82) results. Similarly, from (5-81) it follows that

$$\mu_n'(\lambda) = -e^{-\lambda}\sum_{k=0}^{\infty}(k-\lambda)^n\frac{\lambda^k}{k!} - ne^{-\lambda}\sum_{k=0}^{\infty}(k-\lambda)^{n-1}\frac{\lambda^k}{k!}$$
$$+ e^{-\lambda}\sum_{k=0}^{\infty}(k-\lambda)^n k \frac{\lambda^{k-1}}{k!}$$

Setting $k = (k-\lambda) + \lambda$ in the last sum, we obtain $\mu_n' = -\mu_n - n\mu_{n-1} + (1/\lambda)$ $(\mu_{n+1} + \lambda\mu_n)$ and (5-83) results.

The preceding equations lead to the recursive determination of the moments m_n and μ_n. Starting with the known moments $m_1 = \lambda$, $\mu_1 = 0$, and $\mu_2 = \lambda$ [see (5-63)], we obtain $m_2 = \lambda(\lambda+1)$ and

$$m_3 = \lambda(\lambda^2+\lambda+2\lambda+1) = \lambda^3+3\lambda^2+\lambda \qquad \mu_3 = \lambda(\mu_2'+2\mu_1) = \lambda$$

ESTIMATE OF THE MEAN OF g(x). The mean of the random variable $\mathbf{y} = g(\mathbf{x})$ is given by

$$E\{g(\mathbf{x})\} = \int_{-\infty}^{\infty} g(x)f(x)\,dx \tag{5-84}$$

Hence, for its determination, knowledge of $f(x)$ is required. However, if $\mathbf{x}$ is concentrated near its mean, then $E\{g(\mathbf{x})\}$ can be expressed in terms of the moments μ_n of $\mathbf{x}$.

Suppose, first, that $f(x)$ is negligible outside an interval $(\eta-\varepsilon, \eta+\varepsilon)$ and in this interval, $g(x) \simeq g(\eta)$. In this case, (5-84) yields

$$E\{g(\mathbf{x})\} \simeq g(\eta)\int_{\eta-\varepsilon}^{\eta+\varepsilon} f(x)\,dx \simeq g(\eta)$$

This estimate can be improved if $g(x)$ is approximated by a polynomial

$$g(x) \simeq g(\eta) + g'(\eta)(x-\eta) + \cdots + g^{(n)}(\eta)\frac{(x-\eta)^n}{n!}$$

Inserting into (5-84), we obtain

$$E\{g(\mathbf{x})\} \simeq g(\eta) + g''(\eta)\frac{\sigma^2}{2} + \cdots + g^{(n)}(\eta)\frac{\mu_n}{n!} \tag{5-85}$$

In particular, if $g(x)$ is approximated by a parabola, then

$$\eta_y = E\{g(\mathbf{x})\} \simeq g(\eta) + g''(\eta)\frac{\sigma^2}{2} \tag{5-86}$$

And if it is approximated by a straight line, then $\eta_y \simeq g(\eta)$. This shows that the slope of $g(x)$ has no effect on η_y; however, as we show next, it affects the variance σ_y^2 of $\mathbf{y}$.

Variance. We maintain that the first-order estimate of σ_y^2 is given by

$$\sigma_y^2 \simeq |g'(\eta)|^2\sigma^2 \tag{5-87}$$

Proof. We apply (5-86) to the function $g^2(x)$. Since its second derivative equals $2(g')^2 + 2gg''$, we conclude that

$$\sigma_y^2 + \eta_y^2 = E\{g^2(\mathbf{x})\} \simeq g^2 + [(g')^2 + gg'']\sigma^2$$

Inserting the approximation (5-86) for η_y into the above and neglecting the σ^4 term, we obtain (5-87).

EXAMPLE 5-27

▶ A voltage $E = 120$ V is connected across a resistor whose resistance is a random variable $\mathbf{r}$ uniform between 900 and 1100 Ω. Using (5-85) and (5-86), we shall estimate the mean and variance of the resulting current

$$\mathbf{i} = \frac{E}{\mathbf{r}}$$

Clearly, $E\{\mathbf{r}\} = \eta = 10^3, \sigma^2 = 100^2/3$. With $g(r) = E/r$, we have

$$g(\eta) = 0.12 \qquad g'(\eta) = -12 \times 10^{-5} \qquad g''(\eta) = 24 \times 10^{-8}$$

Hence

$$E\{\mathbf{i}\} \simeq 0.12 + 0.0004\ A \qquad \sigma_i^2 \simeq 48 \times 10^{-6}\ A^2$$ ◀

A measure of the concentration of a random variable near its mean η is its variance σ^2. In fact, as the following theorem shows, the probability that **x** is outside an arbitrary interval $(\eta - \varepsilon, \eta + \varepsilon)$ is negligible if the ratio σ/ε is sufficiently small. This result, known as the *Chebyshev inequality,* is fundamental.

CHEBYSHEV (TCHEBYCHEFF) INEQUALITY

▶ For any $\varepsilon > 0$,

$$P\{|\mathbf{x} - \eta| \geq \varepsilon\} \leq \frac{\sigma^2}{\varepsilon^2} \tag{5-88}$$

Proof. The proof is based on the fact that

$$P\{|\mathbf{x} - \eta| \geq \varepsilon\} = \int_{-\infty}^{-\eta-\varepsilon} f(x)\,dx + \int_{\eta+\varepsilon}^{\infty} f(x)\,dx = \int_{|x-\eta|\geq\varepsilon} f(x)\,dx$$

Indeed

$$\sigma^2 = \int_{-\infty}^{\infty} (x-\eta)^2 f(x)\,dx \geq \int_{|x-\eta|\geq\varepsilon} (x-\eta)^2 f(x)\,dx \geq \varepsilon^2 \int_{|x-\eta|\geq\varepsilon} f(x)\,dx$$

and (5-88) results because the last integral equals $P\{|\mathbf{x} - \eta| \geq \varepsilon\}$. ◀

Notes 1. From (5-88) it follows that, if $\sigma = 0$, then the probability that **x** is outside the interval $(\eta - \varepsilon, \eta + \varepsilon)$ equals 0 for any ε; hence $\mathbf{x} = \eta$ with probability 1. Similarly, if

$$E\{\mathbf{x}^2\} = \eta^2 + \sigma^2 = 0 \qquad \text{then} \quad \eta = 0 \quad \sigma = 0$$

hence $\mathbf{x} = 0$ with probability 1.

2. For specific densities, the bound in (5-88) is too high. Suppose, for example, that **x** is normal. In this case, $P\{|x - \eta| \geq 3\sigma\} = 2 - 2G(3) = 0.0027$. Inequality (5-88), however, yields $P\{|\mathbf{x} - \eta| \geq 3\sigma\} \leq 1/9$.

The significance of Chebyshev's inequality is the fact that it holds for *any* $f(x)$ and can, therefore be used even if $f(x)$ is not known.

3. The bound in (5-88) can be reduced if various assumptions are made about $f(x)$ [see *Chernoff bound* (Prob. 5-35)].

MARKOV INEQUALITY

▶ If $f(x) = 0$ for $x < 0$, then, for any $\alpha > 0$,

$$P\{\mathbf{x} \geq \alpha\} \leq \frac{\eta}{\alpha} \tag{5-89}$$

Proof.

$$E\{\mathbf{x}\} = \int_0^{\infty} x f(x)\,dx \geq \int_{\alpha}^{\infty} x f(x)\,dx \geq \alpha \int_{\alpha}^{\infty} f(x)\,dx$$

and (5-89) results because the last integral equals $P\{\mathbf{x} \geq \alpha\}$. ◀

BIENAYMÉ INEQUALITY

▶ Suppose that $\mathbf{x}$ is an arbitrary random variable and a and n are two arbitrary numbers. Clearly, the random variable $|\mathbf{x} - a|^n$ takes only positive values. Applying (5-89), with $\alpha = \varepsilon^n$, we conclude that

$$P\{|\mathbf{x} - a|^n \geq \varepsilon^n\} \leq \frac{E\{|\mathbf{x} - a|^n\}}{\varepsilon^n} \tag{5-90}$$

Hence

$$P\{|\mathbf{x} - a| \geq \varepsilon\} \leq \frac{E\{|\mathbf{x} - a|^n\}}{\varepsilon^n} \tag{5-91}$$

This result is known as the *inequality of Bienaymé*. Chebyshev's inequality is a special case obtained with $a = \eta$ and $n = 2$. ◀

LYAPUNOV INEQUALITY

▶ Let $\beta_k = E\{|\mathbf{x}|^k\} < \infty$ represent the absolute moments of the random variable $\mathbf{x}$. Then for any k

$$\beta_{k-1}^{1/(k-1)} \leq \beta_k^{1/k} \qquad k \geq 1 \tag{5-92}$$

Proof. Consider the random variable

$$\mathbf{y} = a|\mathbf{x}|^{(k-1)/2} + |\mathbf{x}|^{(k+1)/2}$$

Then

$$E\{\mathbf{y}^2\} = a^2\beta_{k-1} + 2a\beta_k + \beta_{k+1} \geq 0$$

implying that the discriminant of the preceding quadratic must be nonpositive. Thus

$$\beta_k^2 \leq \beta_{k-1}\beta_{k+1} \quad \text{or} \quad \beta_k^{2k} \leq \beta_{k-1}^k \beta_{k+1}^k$$

This gives

$$\beta_1^2 \leq \beta_0\beta_2, \qquad \beta_2^4 \leq \beta_1^2\beta_3^2, \ldots, \beta_{n-1}^{2(n-1)} \leq \beta_{n-2}^{n-1}\beta_n^{n-1}$$

where $\beta_0 = 1$. Multiplying successively we get

$$\beta_1^2 \leq \beta_2, \qquad \beta_2^3 \leq \beta_3^2, \qquad \beta_3^4 \leq \beta_4^3, \ldots, \beta_{k-1}^k \leq \beta_k^{k-1}, \quad \text{or} \quad \beta_{k-1}^{1/(k-1)} \leq \beta_k^{1/k}$$

Thus, we also obtain

$$\beta_1 \leq \beta_2^{1/2} \leq \beta_3^{1/3} \leq \cdots \leq \beta_n^{1/n} \tag{5-93}$$

◀

5-5 CHARACTERISTIC FUNCTIONS

The *characteristic function* of a random variable is by definition the integral

$$\Phi_x(\omega) = \int_{-\infty}^{\infty} f(x)e^{j\omega x}\,dx \tag{5-94}$$

This function is maximum at the origin because $f(x) \geq 0$:

$$|\Phi_x(\omega)| \leq \Phi_x(0) = 1 \tag{5-95}$$

If $j\omega$ is changed to s, the resulting integral

$$\mathbf{\Phi}(s) = \int_{-\infty}^{\infty} f(x)e^{sx}\,dx \qquad \mathbf{\Phi}(j\omega) = \Phi_x(\omega) \tag{5-96}$$

is the *moment (generating) function* of $\mathbf{x}$.

The function

$$\Psi(\omega) = \ln \Phi_x(\omega) = \mathbf{\Psi}(j\omega) \tag{5-97}$$

is the *second characteristic function* of $\mathbf{x}$.

Clearly [see (5-58)]

$$\Phi_x(\omega) = E\{e^{j\omega\mathbf{x}}\} \qquad \mathbf{\Phi}(s) = E\{e^{s\mathbf{x}}\} \tag{5-98}$$

This leads to the fact that

$$\text{If } \mathbf{y} = a\mathbf{x} + b \quad \text{then} \quad \Phi_y(\omega) = e^{jb\omega}\Phi_x(a\omega) \tag{5-99}$$

because

$$E\{e^{j\omega\mathbf{y}}\} = E\left\{e^{j\omega(a\mathbf{x}+b)}\right\} = e^{jb\omega}E\{e^{ja\omega\mathbf{x}}\}$$

EXAMPLE 5-28

▶ We shall show that the characteristic function of an $N(\eta, \sigma)$ random variable $\mathbf{x}$ equals (see Table 5-2)

$$\Phi_x(\omega) = \exp\left\{j\eta\omega - \tfrac{1}{2}\sigma^2\omega^2\right\} \tag{5-100}$$

Proof. The random variable $\mathbf{z} = (\mathbf{x} - \eta)/\sigma$ is $N(0, 1)$ and its moment function equals

$$\mathbf{\Phi}_z(s) = \frac{1}{\sqrt{2\pi}} \int_{-\infty}^{\infty} e^{sz}e^{-z^2/2}\,dz$$

with

$$sz - \frac{z^2}{2} = -\frac{1}{2}(z-s)^2 + \frac{s^2}{2}$$

we conclude that

$$\mathbf{\Phi}_z(s) = e^{s^2/2} \int_{-\infty}^{\infty} \frac{1}{\sqrt{2\pi}} e^{-(z-s)^2/2}\,dz = e^{s^2/2} \tag{5-101}$$

And since $\mathbf{x} = \sigma\mathbf{z} + \eta$, (5-100) follows from (5-99) and (5-101) with $s = j\omega$. ◀

Inversion formula As we see from (5-94), $\Phi_x(\omega)$ is the Fourier transform of $f(x)$. Hence the properties of characteristic functions are essentially the same as the properties of Fourier transforms. We note, in particular, that $f(x)$ can be expressed in terms of $\Phi(\omega)$

$$f(x) = \frac{1}{2\pi} \int_{-\infty}^{\infty} \Phi_x(\omega)e^{-j\omega x}\,d\omega \tag{5-102}$$

Moment theorem. Differentiating (5-96) n times, we obtain

$$\mathbf{\Phi}^{(n)}(s) = E\{\mathbf{x}^n e^{s\mathbf{x}}\}$$

Hence

$$\mathbf{\Phi}^{(n)}(0) = E\{\mathbf{x}^n\} = m_n \tag{5-103}$$

Thus the derivatives of $\mathbf{\Phi}(s)$ at the origin equal the moments of $\mathbf{x}$. This justifies the name "moment function" given to $\mathbf{\Phi}(s)$.

In particular,

$$\mathbf{\Phi}'(0) = m_1 = \eta \qquad \mathbf{\Phi}''(0) = m_2 = \eta^2 + \sigma^2 \tag{5-104}$$

Note Expanding $\mathbf{\Phi}(s)$ into a series near the origin and using (5-103), we obtain

$$\mathbf{\Phi}(s) = \sum_{n=0}^{\infty} \frac{m_n}{n!} s^n \tag{5-105}$$

This is valid only if all moments are finite and the series converges absolutely near $s = 0$. Since $f(x)$ can be determined in terms of $\mathbf{\Phi}(s)$, (5-105) shows that, under the stated conditions, the density of a random variable is uniquely determined if all its moments are known.

EXAMPLE 5-29

▶ We shall determine the moment function and the moments of a random variable $\mathbf{x}$ with *gamma distribution:* (see also Table 5-2)

$$f(x) = \gamma x^{b-1} e^{-cx} U(x) \qquad \gamma = \frac{c^{b+1}}{\Gamma(b+1)}$$

From (4-35) it follows that

$$\mathbf{\Phi}(s) = \gamma \int_0^\infty x^{b-1} e^{-(c-s)x}\, dx = \frac{\gamma \Gamma(b)}{(c-s)^b} = \frac{c^b}{(c-s)^b} \tag{5-106}$$

Differentiating with respect to s and setting $s = 0$, we obtain

$$\mathbf{\Phi}^{(n)}(0) = \frac{b(b+1)\cdots(b+n-1)}{c^n} = E\{\mathbf{x}^n\}$$

With $n = 1$ and $n = 2$, this yields

$$E\{\mathbf{x}\} = \frac{b}{c} \qquad E\{\mathbf{x}^2\} = \frac{b(b+1)}{c^2} \qquad \sigma^2 = \frac{b}{c^2} \tag{5-107}$$

The ***exponential density*** is a special case obtained with $b = 1$, $c = \lambda$:

$$f(x) = \lambda e^{-\lambda x} U(x) \quad \mathbf{\Phi}(s) = \frac{\lambda}{\lambda - s} \quad E\{\mathbf{x}\} = \frac{1}{\lambda} \quad \sigma^2 = \frac{1}{\lambda^2} \tag{5-108}$$

Chi square: Setting $b = m/2$ and $c = 1/2$ in (5-106), we obtain the moment function of the chi-square density $\chi^2(m)$:

$$\mathbf{\Phi}(s) = \frac{1}{\sqrt{(1-2s)^m}} \qquad E\{\mathbf{x}\} = m \qquad \sigma^2 = 2m \tag{5-109}$$

◀

Cumulants. The cumulants λ_n of random variable $\mathbf{x}$ are by definition the derivatives

$$\frac{d^n \mathbf{\Psi}(0)}{ds^n} = \lambda_n \tag{5-110}$$

of its second moment function $\mathbf{\Psi}(s)$. Clearly [see (5-97)] $\mathbf{\Psi}(0) = \lambda_0 = 0$; hence

$$\mathbf{\Psi}(s) = \lambda_1 s + \frac{1}{2}\lambda_2 s^2 + \cdots + \frac{1}{n!}\lambda_n s^n + \cdots$$

We maintain that

$$\lambda_1 = \eta \qquad \lambda_2 = \sigma^2 \tag{5-111}$$

Proof. Since $\mathbf{\Phi} = e^{\mathbf{\Psi}}$, we conclude that

$$\mathbf{\Phi}' = \mathbf{\Psi}' e^{\mathbf{\Psi}} \qquad \mathbf{\Phi}' = [\mathbf{\Psi}'' + (\mathbf{\Psi}')^2]e^{\mathbf{\Psi}}$$

With $s = 0$, this yields

$$\mathbf{\Phi}'(0) = \mathbf{\Psi}'(0) = m_1 \qquad \mathbf{\Phi}''(0) = \mathbf{\Psi}''(0) + [\mathbf{\Psi}'(0)]^2 = m_2$$

and (5-111) results.

Discrete Type

Suppose that $\mathbf{x}$ is a discrete-type random variable taking the values x_i with probability p_i. In this case, (5-94) yields

$$\Phi_x(\omega) = \sum_i p_i e^{j\omega x_i} \tag{5-112}$$

Thus $\Phi_x(\omega)$ is a sum of exponentials. The moment function of $\mathbf{x}$ can be defined as in (5-96). However, if $\mathbf{x}$ takes only integer values, then a definition in terms of z transforms is preferable.

MOMENT GENERATING FUNCTIONS. If $\mathbf{x}$ is a lattice type random variable taking integer values, then its moment generating function is by definition the sum

$$\mathbf{\Gamma}(z) = E\{z^{\mathbf{x}}\} = \sum_{n=-\infty}^{+\infty} P\{\mathbf{x} = n\}z^n = \sum_{n=-\infty}^{\infty} p_n z^n \tag{5-113}$$

Thus $\mathbf{\Gamma}(1/z)$ is the ordinary z transform of the sequence $p_n = P\{\mathbf{x} = n\}$. With $\Phi_x(\omega)$ as in (5-112), this yields

$$\Phi_x(\omega) = \mathbf{\Gamma}(e^{j\omega}) = \sum_{n=-\infty}^{\infty} p_n e^{jn\omega}$$

Thus $\Phi_x(\omega)$ is the *discrete Fourier transform* (DFT) of the sequence $\{p_n\}$, and

$$\mathbf{\Psi}(s) = \ln \mathbf{\Gamma}(e^s) \tag{5-114}$$

Moment theorem. Differentiating (5-113) k times, we obtain

$$\mathbf{\Gamma}^{(k)}(z) = E\{\mathbf{x}(\mathbf{x}-1)\cdots(\mathbf{x}-k+1)z^{\mathbf{x}-k}\}$$

With $z = 1$, this yields

$$\mathbf{\Gamma}^{(k)}(1) = E\{\mathbf{x}(\mathbf{x}-1)\cdots(\mathbf{x}-k+1)\} \tag{5-115}$$

We note, in particular, that $\mathbf{\Gamma}(1) = 1$ and

$$\mathbf{\Gamma}'(1) = E\{\mathbf{x}\} \qquad \mathbf{\Gamma}''(1) = E\{\mathbf{x}^2\} - E\{\mathbf{x}\} \tag{5-116}$$

EXAMPLE 5-30

▶ (*a*) If $\mathbf{x}$ takes the values 0 and 1 with $P\{\mathbf{x} = 1\} = p$ and $P\{\mathbf{x} = 0\} = q$, then

$$\mathbf{\Gamma}(z) = pz + q$$

$$\mathbf{\Gamma}'(1) = E\{\mathbf{x}\} = p \qquad \mathbf{\Gamma}''(1) = E\{\mathbf{x}^2\} - E\{\mathbf{x}\} = 0$$

(*b*) If $\mathbf{x}$ has the binomial distribution $B(m, p)$ given by

$$p_n = P\{\mathbf{x} = n\} = \binom{m}{n} p^n q^{m-n} \qquad 0 \le n \le m$$

then

$$\mathbf{\Gamma}(z) = \sum_{n=0}^{m} \binom{m}{n} p^n q^{m-n} z^n = (pz + q)^m \tag{5-117}$$

and

$$\mathbf{\Gamma}'(1) = mp \qquad \mathbf{\Gamma}''(1) = m(m-1)p^2$$

Hence

$$E\{\mathbf{x}\} = mp \qquad \sigma^2 = mpq \tag{5-118}$$

◀

EXAMPLE 5-31

▶ If $\mathbf{x}$ is Poisson distributed with parameter λ,

$$P\{\mathbf{x} = n\} = e^{-\lambda} \frac{\lambda^n}{n!} \qquad n = 0, 1, \ldots$$

then

$$\mathbf{\Gamma}(z) = e^{-\lambda} \sum_{n=0}^{\infty} \lambda^n \frac{z^n}{n!} = e^{\lambda(z-1)} \tag{5-119}$$

In this case [see (5-114)]

$$\mathbf{\Psi}(s) = \lambda(e^s - 1) \qquad \mathbf{\Psi}'(0) = \lambda \qquad \mathbf{\Psi}''(0) = \lambda$$

and (5-111) yields $E\{\mathbf{x}\} = \lambda$, $\sigma^2 = \lambda$ in agreement with (5-63). ◀

We can use the characteristic function method to establish the DeMoivre–Laplace theorem in (4-90).

THEOREM 5-2

DEMOIVRE–LAPLACE THEOREM

▶ Let $\mathbf{x} \sim B(n, p)$. Then from (5-117), we obtain the characteristic function of the binomial random variable to be

$$\Phi_x(\omega) = (pe^{j\omega} + q)^n$$

and define

$$\mathbf{y} = \frac{\mathbf{x} - np}{\sqrt{npq}} \tag{5-120}$$

This gives

$$
\begin{aligned}
\Phi_y(\omega) &= E\{e^{j\mathbf{y}\omega}\} = e^{-np\omega/\sqrt{npq}}\,\Phi_x\left(\frac{\omega}{\sqrt{npq}}\right) \\
&= e^{-np\omega/\sqrt{npq}}(pe^{j\omega/\sqrt{npq}} + q)^n \\
&= (pe^{jq\omega/\sqrt{npq}} + qe^{-jp\omega/\sqrt{npq}})^n \\
&= \left\{p\left(1 + \frac{jq\omega}{\sqrt{npq}} - \frac{q^2\omega^2}{2npq} + \sum_{k=3}^{\infty}\frac{1}{k!}\left(\frac{jq\omega}{\sqrt{npq}}\right)^k\right)\right. \\
&\qquad \left. + q\left(1 - \frac{jp\omega}{\sqrt{npq}} - \frac{p^2\omega^2}{2npq} + \sum_{k=3}^{\infty}\frac{1}{k!}\left(\frac{-jp\omega}{\sqrt{npq}}\right)^k\right)\right\}^n \\
&= \left(1 - \frac{\omega^2}{2n}\{1 + \phi(n)\}\right)^n \to e^{-\omega^2/2}, \quad \text{as} \quad n \to \infty
\end{aligned}
\tag{5-121}
$$

since

$$
\phi(n) \triangleq 2\sum_{k=3}^{\infty}\frac{1}{k!}\left(\frac{j\omega}{\sqrt{n}}\right)^{k-2}\frac{pq^k + q(-p)^k}{(\sqrt{pq})^k} \to 0, \quad \text{as} \quad n \to \infty
$$

On comparing (5-121) with (5-100), we conclude that as $n \to \infty$, the random variable $\mathbf{y}$ tends to the standard normal distribution, or from (5-120), $\mathbf{x}$ tends to $N(np, npq)$. ◀

In Examples 5-32 and 5-33 we shall exhibit the usefulness of the moment generating function in solving problems. The next example is of historical interest, as it was first proposed and solved by DeMoivre.

EXAMPLE 5-32

▶ An event A occurs in a series of independent trials with constant probability p. If A occurs at least r times in succession, we refer to it as a run of length r. Find the probability of obtaining a run of length r for A in n trials.

SOLUTION

Let p_n denote the probability of the event X_n that represents a run of length r for A in n trials. A run of length r in $n+1$ trials can happen in only two mutually exclusive ways: either there is a run of length r in the first n trials, or a run of length r is obtained only in the last r trials of the $n+1$ trials and not before that. Thus

$$
X_{n+1} = X_n \cup B_{n+1} \tag{5-122}
$$

where

$$
\begin{aligned}
B_{n+1} &= \{\text{No run of length } r \text{ for } A \text{ in the first } n-r \text{ trials}\} \\
&\qquad \cap \{A \text{ does not occur in the } (n-r+1)\text{th trial}\} \\
&\qquad \cap \{\text{Run of length } r \text{ for } A \text{ in the last } r \text{ trials}\} \\
&= \overline{X}_{n-r} \cap \overline{A} \cap \underbrace{A \cap A \cap \cdots \cap A}_{r}
\end{aligned}
$$

Hence by the independence of these events

$$P\{B_{n+1}\} = (1 - p_{n-r})qp^r$$

so that from (5-122)

$$p_{n+1} = P\{X_{n+1}\} = P\{X_n\} + P\{B_{n+1}\} = p_n + (1 - p_{n-r})qp^r \tag{5-123}$$

The equation represents an ordinary difference equation with the obvious initial conditions

$$p_0 = p_1 = \cdots = p_{r-1} = 0 \quad \text{and} \quad p_r = p^r \tag{5-124}$$

From (5-123), although it is possible to obtain $p_{r+1} = p^r(1+q), \ldots, p_{r+m} = p^r(1+mq)$ for $m \le r-1$, the expression gets quite complicated for large values of n. The method of moment generating functions in (5-113) can be used to obtain a general expression for p_n. Toward this, let

$$q_n \triangleq 1 - p_n \tag{5-125}$$

so that (5-123) translates into (with n replaced by $n + r$)

$$q_{n+r+1} = q_{n+r} - qp^r q_n \qquad n \ge 0 \tag{5-126}$$

with the new initial conditions

$$q_0 = q_1 = \cdots = q_{r-1} = 1 \qquad q_r = 1 - p^r \tag{5-127}$$

Following (5-113), define the moment generating function

$$\phi(z) = \sum_{n=0}^{\infty} q_n z^n \tag{5-128}$$

and using (5-126) we obtain

$$\begin{aligned} qp^r\phi(z) &= \left(\sum_{n=0}^{\infty} q_{n+r}z^n - \sum_{n=0}^{\infty} q_{n+r+1}z^n\right) \\ &= \frac{\phi(z) - \sum_{k=0}^{r-1} q_k z^k}{z^r} - \frac{\phi(z) - \sum_{k=0}^{r} q_k z^k}{z^{r+1}} \\ &= \frac{(z-1)\phi(z) - \sum_{k=1}^{r} z^k + \left(\sum_{k=0}^{r-1} z^k + (1-p^r)z^r\right)}{z^{r+1}} \\ &= \frac{(z-1)\phi(z) + 1 - p^r z^r}{z^{r+1}} \end{aligned} \tag{5-129}$$

where we have made use of the initial conditions in (5-127). From (5-129) we get the desired moment generating function to be

$$\phi(z) = \frac{1 - p^r z^r}{1 - z + qp^r z^{r+1}} \tag{5-130}$$

$\phi(z)$ is a rational function in z, and the coefficient of z^n in its power series expansion gives q_n. More explicitly

$$\begin{aligned} \phi(z) &= (1 - p^r z^r)[1 - z(1 - qp^r z^r)]^{-1} \\ &= (1 - p^r z^r)[1 + \alpha_{1,r}z + \cdots + \alpha_{n,r}z^n + \cdots] \end{aligned} \tag{5-131}$$

so that the desired probability equals

$$q_n = \alpha_{n,r} - p^r \alpha_{n-r,r} \tag{5-132}$$

where $\alpha_{n,r}$ is the coefficient of z^n in the expansion of $[1 - z(1 - qp^r z^r)]^{-1}$. But

$$[1 - z(1 - qp^r z^r)]^{-1} = \sum_{m=0}^{\infty} z^m (1 - qp^r z^r)^m = \sum_{m=0}^{\infty}\sum_{k=0}^{m} \binom{m}{k} (-1)^k (qp^r)^k z^{m+kr}$$

Let $m + kr = n$ so that $m = n - kr$, and this expression simplifies to

$$[1 - z(1 - qp^r z^r)]^{-1} = \sum_{n=0}^{\infty} \sum_{k=0}^{\lfloor n/(r+1)\rfloor} \binom{n-kr}{k} (-1)^k (qp^r)^k z^n = \sum_{n=0}^{\infty} \alpha_{n,r} z^n$$

and the upper limit on k corresponds to the condition $n - kr \geq k$ so that $\binom{n-kr}{k}$ is well defined. Thus

$$\alpha_{n,r} = \sum_{k=0}^{\lfloor n/(r+1)\rfloor} \binom{n-kr}{k} (-1)^k (qp^r)^k \tag{5-133}$$

With $\alpha_{n,r}$ so obtained, finally the probability of r runs for A in n trials is given by

$$p_n = 1 - q_n = 1 - \alpha_{n,r} + p^r \alpha_{n-r,r} \tag{5-134}$$

For example, if $n = 25$, $r = 6$, $p = q = 1/2$, we get the probability of six successive heads in 25 trials to be 0.15775.

On a more interesting note, suppose for a regular commuter the morning commute takes 45 minutes under the best of conditions, the probability of which is assumed to be 1/5. Then there is a 67% chance for doing the trip within the best time at least once a week. However there is only about 13% chance of repeating it twice in a row in a week. This shows that especially the day after the "lucky day," one should allow extra travel time. Finally if the conditions for the return trip also are assumed to be the same, for a one week period the probability of doing two consecutive trips within the best time is 0.2733. ◀

The following problem has many varients and its solution goes back to Montmort (1708). It has been further generalized by Laplace and many others.

TABLE 5-1
Probability p_n in (5-134)

	$n = 5$		$n = 10$	
r	$p = 1/5$	$p = 1/3$	$p = 1/5$	$p = 1/3$
1	0.6723	0.8683	0.8926	0.9827
2	0.1347	0.3251	0.2733	0.5773
3	0.0208	0.0864	0.0523	0.2026
4	0.0029	0.0206	0.0093	0.0615
5	0.0003	0.0041	0.0016	0.0178
6	—	—	0.0003	0.0050

EXAMPLE 5-33

THE PAIRING PROBLEM

▶ A person writes n letters and addresses n envelopes. Then one letter is randomly placed into each envelope. What is the probability that at least one letter will reach its correct destination? What if $n \to \infty$?

SOLUTION

When a letter is placed into the envelope addressed to the intended person, let us refer to it as a coincidence. Let X_k represent the event that there are exactly k coincidences among the n envelopes. The events $X_0, X_1, \ldots, X_n$ form a partition since they are mutually exclusive and one of these events is bound to happen. Hence by the theorem of total probability

$$p_n(0) + p_n(1) + p_n(2) + \cdots + p_n(n) = 1 \tag{5-135}$$

where

$$p_n(k) \triangleq P\{X_k\} \tag{5-136}$$

To determine $p_n(k)$ let us examine the event X_k. There are $\binom{n}{k}$ number of ways of drawing k letters from a group of n, and to generate k coincidences, each such sequence should go into their intended envelopes with probability

$$\frac{1}{n} \cdot \frac{1}{n-1} \cdots \frac{1}{n-k+1}$$

while the remaining $n-k$ letters present no coincidences at all with probability $p_{n-k}(0)$. By the independence of these events, we get the probability of k coincidences for each sequence of k letters in a group of n to be

$$\frac{1}{n(n-1)\cdots(n-k+1)} p_{n-k}(0)$$

But there are $\binom{n}{k}$ such mutually exclusive sequences, and using (2-20) we get

$$p_n(k) = P\{X_k\} = \binom{n}{k} \frac{1}{n(n-1)\cdots(n-k+1)} p_{n-k}(0) = \frac{p_{n-k}(0)}{k!} \tag{5-137}$$

Since $p_n(n) = 1/n!$, equation (5-137) gives $p_0(0) = 1$. Substituting (5-137) into (5-135) term by term, we get

$$p_n(0) + \frac{p_{n-1}(0)}{1!} + \frac{p_{n-2}(0)}{2!} + \cdots + \frac{p_1(0)}{(n-1)!} + \frac{1}{n!} = 1 \tag{5-138}$$

which gives successively

$$p_1(0) = 0 \qquad p_2(0) = \tfrac{1}{2} \qquad p_3(0) = \tfrac{1}{6}$$

and to obtain an explicit expression for $p_n(0)$, define the moment generating function

$$\phi(z) = \sum_{n=0}^{\infty} p_n(0) z^n \tag{5-139}$$

Then

$$e^z \phi(z) = \left(\sum_{k=0}^{\infty} \frac{z^k}{k!}\right)\left(\sum_{n=0}^{\infty} p_n(0) z^n\right)$$

$$= 1 + z + z^2 + \cdots + z^n + \cdots = \frac{1}{1-z} \tag{5-140}$$

where we have made use of (5-138). Thus

$$\phi(z) = \frac{e^{-z}}{1-z} = \sum_{n=0}^{\infty}\left(\sum_{k=0}^{n}\frac{(-1)^k}{k!}\right)z^n$$

and on comparing with (5-139), we get

$$p_n(0) = \sum_{k=0}^{n}\frac{(-1)^k}{k!} \to \frac{1}{e} = 0.377879 \tag{5-141}$$

and using (5-137)

$$p_n(k) = \frac{1}{k!}\sum_{m=0}^{n-k}\frac{(-1)^m}{m!} \tag{5-142}$$

Thus

$$P\{\text{At least one letter reaches the correct destination}\}$$
$$= 1 - p_n(0) = 1 - \sum_{k=0}^{n}\frac{(-1)^k}{k!} \to 0.63212056 \tag{5-143}$$

Even for moderate n, this probability is close to 0.6321. Thus even if a mail delivery distributes letters in the most causal manner without undertaking any kind of sorting at all, there is still a 63% chance that at least one family will receive some mail addressed to them.

On a more serious note, by the same token, a couple trying to conceive has about 63% chance of succeeding in their efforts under normal conditions. The abundance of living organisms in Nature is a good testimony to the fact that odds are indeed tilted in favor of this process. ◀

Determination of the density of $g(\mathbf{x})$. We show next that characteristic functions can be used to determine the density $f_y(y)$ of the random variable $\mathbf{y} = g(\mathbf{x})$ in terms of the density $f_x(x)$ of $\mathbf{x}$.

From (5-58) it follows that the characteristic function

$$\Phi_y(\omega) = \int_{-\infty}^{\infty} e^{j\omega y} f_y(y)\,dy$$

of the random variable $\mathbf{y} = g(\mathbf{x})$ equals

$$\Phi_y(\omega) = E\{e^{j\omega g(\mathbf{x})}\} = \int_{-\infty}^{\infty} e^{j\omega g(x)} f_x(x)\,dx \tag{5-144}$$

If, therefore, the integral in (5-144) can be written in the form

$$\int_{-\infty}^{\infty} e^{j\omega y} h(y)\,dy$$

it will follow that (uniqueness theorem)

$$f_y(y) = h(y)$$

This method leads to simple results if the transformation $y = g(x)$ is one-to-one.

TABLE 5-2

Random variable	Probability density function $f_x(x)$	Mean	Variance	Characteristic function $\Phi_x(\omega)$
Normal or Gaussian $N(\mu, \sigma^2)$	$\frac{1}{\sqrt{2\pi\sigma^2}} e^{-(x-\mu)^2/2\sigma^2}$, $-\infty < x < \infty$	μ	σ^2	$e^{j\mu\omega-\sigma^2\omega^2/2}$
Log-normal	$\frac{1}{x\sqrt{2\pi\sigma^2}} e^{-(\ln x-\mu)^2/2\sigma^2}$, $x \geq 0$,			
Exponential $E(\lambda)$	$\lambda e^{-\lambda x}, x \geq 0, \lambda > 0$	$\frac{1}{\lambda}$	$\frac{1}{\lambda^2}$	$(1 - j\omega/\lambda)^{-1}$
Gamma $G(\alpha, \beta)$	$\frac{x^{\alpha-1}}{\Gamma(\alpha)\beta^\alpha} e^{-x/\beta}$, $x \geq 0, \alpha > 0, \beta > 0$	$\alpha\beta$	$\alpha\beta^2$	$(1 - j\omega\beta)^{-\alpha}$
Erlang-k	$\frac{(k\lambda)^k}{(k-1)!} x^{k-1} e^{-k\lambda x}$	$\frac{1}{\lambda}$	$\frac{1}{k\lambda^2}$	$(1 - j\omega/k\lambda)^{-k}$
Chi-square $\chi^2(n)$	$\frac{x^{n/2-1}}{2^{n/2}\Gamma(n/2)} e^{-x/2}, x \geq 0$	n	$2n$	$(1 - j2\omega)^{-n/2}$
Weibull	$\alpha x^{\beta-1} e^{-\alpha x^\beta/\beta}$, $x \geq 0, \alpha > 0, \beta > 0$	$\left(\frac{\beta}{\alpha}\right)^{1/\beta} \Gamma\left(1 + \frac{1}{\beta}\right)$	$\left(\frac{\beta}{\alpha}\right)^{2/\beta} \left[\Gamma\left(1 + \frac{2}{\beta}\right) - \left(\Gamma\left(1 + \frac{1}{\beta}\right)\right)^2\right]$	—
Rayleigh	$\frac{x}{\sigma^2} e^{-x^2/2\sigma^2}, x \geq 0$	$\sqrt{\frac{\pi}{2}}\sigma$	$(2 - \pi/2)\sigma^2$	$\left(1 + j\sqrt{\frac{\pi}{2}}\sigma\omega\right) e^{-\sigma^2\omega^2/2}$
Uniform $U(a, b)$	$\frac{1}{b-a}, a < x < b$	$\frac{a+b}{2}$	$\frac{(b-a)^2}{12}$	$\frac{e^{jb\omega} - e^{-ja\omega}}{j\omega(b-a)}$
Beta $\beta(\alpha, \beta)$	$\frac{\Gamma(\alpha+\beta)}{\Gamma(\alpha)\Gamma(\beta)} x^{\alpha-1}(1-x)^{\beta-1}$, $0 < x < 1, \alpha > 0, \beta > 0$	$\frac{\alpha}{\alpha+\beta}$	$\frac{\alpha\beta}{(\alpha+\beta)^2(\alpha+\beta+1)}$	—
Cauchy	$\frac{\alpha/\pi}{(x-\mu)^2 + \alpha^2}$, $-\infty < x < \infty, \alpha > 0$	—	∞	$e^{j\mu\omega} e^{-\alpha\|\omega\|}$
Rician	$\frac{x}{\sigma^2} e^{-\frac{x^2+a^2}{2\sigma^2}} I_0\left(\frac{ax}{\sigma^2}\right)$, $-\infty < x < \infty, a > 0$	$\sigma\frac{\sqrt{\pi}}{2}[(1+r)I_0(r/2) + rI_1(r/2)]e^{-r/2}$, $r = a^2/2\sigma^2$	—	—
Nakagami	$\frac{2}{\Gamma(m)}\left(\frac{m}{\Omega}\right)^m x^{2m-1} e^{-\frac{m}{\Omega}x^2}$, $x > 0$	$\frac{\Gamma(m+1/2)}{\Gamma(m)}\sqrt{\frac{\Omega}{m}}$	$\Omega\left(1 - \frac{1}{m}\left(\frac{\Gamma(m+1/2)}{\Gamma(m)}\right)^2\right)$	—

TABLE 5-2
(Continued)

Random variable	Probability density function $f_x(x)$	Mean	Variance	Characteristic function $\Phi_x(\omega)$
Students' $t(n)$	$\dfrac{\Gamma((n+1)/2)}{\sqrt{\pi n}\,\Gamma(n/2)}(1+x^2/n)^{-(n+1)/2}$, $-\infty < x < \infty$	0	$\dfrac{n}{n-2}, n > 2$	—
F-distribution	$\dfrac{\Gamma((m+n)/2)}{\Gamma(m/2)\Gamma(n/2)}\left(\dfrac{m}{n}\right)^{m/2} x^{m/2-1} \times \left(1+\dfrac{mx}{n}\right)^{-(m+n)/2}, x > 0$	$\dfrac{n}{n-2}, n > 2$	$\dfrac{n^2(2m+2n-4)}{m(n-2)^2(n-4)}, n > 4$	—
Bernoulli	$P(X=1)=p$, $P(X=0)=1-p=q$	p	$p(1-p)$	$pe^{j\omega}+q$
Binomial $B(n, p)$	$\dbinom{n}{k}p^k q^{n-k}$, $k=0,1,2,\ldots,n,\ p+q=1$	np	npq	$(pe^{j\omega}+q)^n$
Poisson $P(\lambda)$	$e^{-\lambda}\dfrac{\lambda^k}{k!}, k=0,1,2,\ldots,\infty$	λ	λ	$e^{-\lambda(1-e^{j\omega})}$
Hypergeometric	$\dfrac{\dbinom{M}{k}\dbinom{N-M}{n-k}}{\dbinom{N}{n}}$, $\max(0, M+n-N) \le k \le \min(M, n)$	$\dfrac{nM}{N}$	$n\dfrac{M}{N}\left(1-\dfrac{M}{N}\right)\left(1-\dfrac{n-1}{N-1}\right)$	—
Geometric	pq^k, $k=0,1,2,\ldots,\infty$	$\dfrac{q}{p}$	$\dfrac{q}{p^2}$	$\dfrac{p}{1-qe^{j\omega}}$
	or pq^{k-1}, $k=1,2,\ldots,\infty,\ p+q=1$	$\dfrac{1}{p}$	$\dfrac{q}{p^2}$	$\dfrac{p}{e^{-j\omega}-q}$
Pascal or negative binomial $NB(r, p)$	$\dbinom{r+k-1}{k}p^r q^k$, $k=0,1,2,\ldots,\infty$	$\dfrac{rq}{p}$	$\dfrac{rq}{p^2}$	$\left(\dfrac{p}{1-qe^{-j\omega}}\right)^r$
	or $\dbinom{k-1}{r-1}p^r q^{k-r}$, $k=r, r+1,\ldots,\infty,\ p+q=1$	$\dfrac{r}{p}$	$\dfrac{rq}{p^2}$	$\left(\dfrac{p}{e^{-j\omega}-q}\right)^r$
Discrete uniform	$1/N$, $k=1,2,\ldots,N$	$\dfrac{N+1}{2}$	$\dfrac{N^2-1}{12}$	$e^{j(N+1)\omega/2}\dfrac{\sin(N\omega/2)}{\sin(\omega/2)}$

EXAMPLE 5-34

▶ Suppose that $\mathbf{x}$ is $N(0;\sigma)$ and $\mathbf{y} = a\mathbf{x}^2$. Inserting into (5-144) and using the evenness of the integrand, we obtain

$$\Phi_y(\omega) = \int_{-\infty}^{\infty} e^{j\omega a x^2} f(x)\,dx = \frac{2}{\sigma\sqrt{2\pi}} \int_0^{\infty} e^{ja\omega x^2} e^{-x^2/2\sigma^2}\,dx$$

As x increases from 0 to ∞, the transformation $y = ax^2$ is one-to-one. Since

$$dy = 2ax\,dx = 2\sqrt{ay}\,dx$$

the last equation yields

$$\Phi_y(\omega) = \frac{2}{\sigma\sqrt{2\pi}} \int_0^{\infty} e^{j\omega y} e^{-y/2a\sigma^2} \frac{dy}{2\sqrt{ay}}$$

Hence

$$f_y(y) = \frac{e^{-y/2a\sigma^2}}{\sigma\sqrt{2\pi a y}} U(y) \tag{5-145}$$

in agreement with (5-7) and (5-22). ◀

EXAMPLE 5-35

▶ We assume finally that $\mathbf{x}$ is uniform in the interval $(-\pi/2, \pi/2)$ and $\mathbf{y} = \sin\mathbf{x}$. In this case

$$\Phi_y(\omega) = \int_{-\infty}^{\infty} e^{j\omega \sin x} f(x)\,dx = \frac{1}{\pi} \int_{-\pi/2}^{\pi/2} e^{j\omega \sin x}\,dx$$

As x increases from $-\pi/2$ to $\pi/2$, the function $y = \sin x$ increases from -1 to 1 and

$$dy = \cos x\,dx = \sqrt{1-y^2}\,dx$$

Hence

$$\Phi_y(\omega) = \frac{1}{\pi} \int_{-1}^{1} e^{j\omega y} \frac{dy}{\sqrt{1-y^2}}$$

This leads to the conclusion that

$$f_y(y) = \frac{1}{\pi\sqrt{1-y^2}} \qquad \text{for} \quad |y| < 1$$

and 0 otherwise, in agreement with (5-33). ◀

PROBLEMS

5-1 The random variable $\mathbf{x}$ is $N(5, 2)$ and $\mathbf{y} = 2\mathbf{x} + 4$. Find η_y, σ_y, and $f_y(y)$.

5-2 Find $F_y(y)$ and $f_y(y)$ if $\mathbf{y} = -4\mathbf{x} + 3$ and $f_x(x) = 2e^{-2x}U(x)$.

5-3 If the random variable $\mathbf{x}$ is $N(0, c^2)$ and $g(x)$ is the function in Fig. 5-4, find and sketch the distribution and the density of the random variable $\mathbf{y} = g(\mathbf{x})$.

5-4 The random variable $\mathbf{x}$ is uniform in the interval $(-2c, 2c)$. Find and sketch $f_y(y)$ and $F_y(y)$ if $\mathbf{y} = g(\mathbf{x})$ and $g(x)$ is the function in Fig. 5-3.

5-5 The random variable $\mathbf{x}$ is $N(0, b^2)$ and $g(x)$ is the function in Fig. 5-5. Find and sketch $f_y(y)$ and $F_y(y)$.

5-6 The random variable **x** is uniform in the interval (0, 1). Find the density of the random variable $\mathbf{y} = -\ln \mathbf{x}$.

5-7 We place at random 200 points in the interval (0, 100). The distance from 0 to the first random point is a random variable **z**. Find $F_z(z)$ (*a*) exactly and (*b*) using the Poisson approximation.

5-8 If $\mathbf{y} = \sqrt{\mathbf{x}}$, and **x** is an exponential random variable, show that **y** represents a Rayleigh random variable.

5-9 Express the density $f_y(y)$ of the random variable $\mathbf{y} = g(\mathbf{x})$ in terms of $f_x(x)$ if (*a*) $g(x) = |x|$; (*b*) $g(x) = e^{-x}U(x)$.

5-10 Find $F_y(y)$ and $f_y(y)$ if $F_x(x) = (1 - e^{-2x})U(x)$ and (*a*) $\mathbf{y} = (\mathbf{x} - 1)U(\mathbf{x} - 1)$; (*b*) $\mathbf{y} = \mathbf{x}^2$.

5-11 Show that, if the random variable **x** has a Cauchy density with $\alpha = 1$ and $\mathbf{y} = \arctan \mathbf{x}$, then **y** is uniform in the interval $(-\pi/2, \pi/2)$.

5-12 The random variable **x** is uniform in the interval $(-2\pi, 2\pi)$. Find $f_y(y)$ if (*a*) $\mathbf{y} = \mathbf{x}^3$, (*b*) $\mathbf{y} = \mathbf{x}^4$, and (*c*) $\mathbf{y} = 2\sin(3\mathbf{x} + 40°)$.

5-13 The random variable **x** is uniform in the interval $(-1, 1)$. Find $g(x)$ such that if $\mathbf{y} = g(\mathbf{x})$ then $f_y(y) = 2e^{-2y}U(y)$.

5-14 Given that random variable **x** is of continuous type, we form the random variable $\mathbf{y} = g(\mathbf{x})$. (*a*) Find $f_y(y)$ if $g(x) = 2F_x(x) + 4$. (*b*) Find $g(x)$ such that **y** is uniform in the interval (8, 10).

5-15 A fair coin is tossed 10 times and **x** equals the number of heads. (*a*) Find $F_x(x)$. (*b*) Find $F_y(y)$ if $\mathbf{y} = (\mathbf{x} - 3)^2$.

5-16 If **x** represents a beta random variable with parameters α and β, show that $1 - \mathbf{x}$ also represents a beta random variable with parameters β and α.

5-17 Let **x** represent a chi-square random variable with n degrees of freedom. Then $\mathbf{y} = \mathbf{x}^2$ is known as the chi-distribution with n degrees of freedom. Determine the p.d.f of **y**.

5-18 Let $\mathbf{x} \sim U(0, 1)$. Show that $\mathbf{y} = -2\log \mathbf{x}$ is $\chi^2(2)$.

5-19 If **x** is an exponential random variable with parameter λ, show that $\mathbf{y} = \mathbf{x}^{1/\beta}$ has a Weibull distribution.

5-20 If **t** is a random variable of continuous type and $\mathbf{y} = a \sin \omega\mathbf{t}$, show that

$$f_y(y) \xrightarrow[\omega\to\infty]{} \begin{cases} 1/\pi\sqrt{a^2 - y^2} & |y| < a \\ 0 & |y| > a \end{cases}$$

5-21 Show that if $\mathbf{y} = \mathbf{x}^2$, then

$$f_y(y \mid \mathbf{x} \geq 0) = \frac{U(y)}{1 - F_x(0)} \frac{f_x(\sqrt{y})}{2\sqrt{y}}$$

5-22 (*a*) Show that if $\mathbf{y} = a\mathbf{x} + b$, then $\sigma_y = |a|\sigma_x$. (*b*) Find η_y and σ_y if $\mathbf{y} = (\mathbf{x} - \eta_x)/\sigma_x$.

5-23 Show that if **x** has a Rayleigh density with parameter α and $\mathbf{y} = b + c\mathbf{x}^2$, then $\sigma_y^2 = 4c^2\sigma^4$.

5-24 If **x** is $N(0, 4)$ and $\mathbf{y} = 3\mathbf{x}^2$, find η_y, σ_y, and $f_y(y)$.

5-25 Let **x** represent a binomial random variable with parameters n and p. Show that (*a*) $E(\mathbf{x}) = np$; (*b*) $E[\mathbf{x}(\mathbf{x} - 1)] = n(n - 1)p^2$; (*c*) $E[\mathbf{x}(\mathbf{x} - 1)(\mathbf{x} - 2)] = n(n - 1)(n - 2)p^3$; (*d*) Compute $E(\mathbf{x}^2)$ and $E(\mathbf{x}^3)$.

5-26 For a Poisson random variable **x** with parameter λ show that (*a*) $P(0 < \mathbf{x} < 2\lambda) > (\lambda - 1)/\lambda$; (*b*) $E[\mathbf{x}(\mathbf{x} - 1)] = \lambda^2$, $E[\mathbf{x}(\mathbf{x} - 1)(\mathbf{x} - 2)] = \lambda^3$.

5-27 Show that if $\boldsymbol{U} = [A_1, \ldots, A_n]$ is a partition of S, then

$$E\{\mathbf{x}\} = E\{\mathbf{x} \mid A_1\}P(A_1) + \cdots + E\{\mathbf{x} \mid A_n\}P(A_n).$$

5-28 Show that if $\mathbf{x} \geq 0$ and $E\{\mathbf{x}\} = \eta$, then $P\{\mathbf{x} \geq \sqrt{\eta}\} \leq \sqrt{\eta}$.

5-29 Using (5-86), find $E\{\mathbf{x}^3\}$ if $\eta_x = 10$ and $\sigma_x = 2$.

5-30 If $\mathbf{x}$ is uniform in the interval (10,12) and $\mathbf{y} = \mathbf{x}^3$, (*a*) find $f_y(y)$; (*b*) find $E\{\mathbf{y}\}$: (i) exactly; (ii) using (5-86).

5-31 The random variable $\mathbf{x}$ is $N(100, 9)$. Find approximately the mean of the random variable $\mathbf{y} = 1/\mathbf{x}$ using (5-86).

5-32 (*a*) Show that if m is the median of $\mathbf{x}$, then

$$E\{|\mathbf{x} - a|\} = E\{|\mathbf{x} - m|\} + 2\int_a^m (x - a) f(x)\,dx$$

for any a. (*b*) Find c such that $E\{|\mathbf{x} - c|\}$ is minimum.

5-33 Show that if the random variable $\mathbf{x}$ is $N(\eta; \sigma^2)$, then

$$E\{|\mathbf{x}|\} = \sigma\sqrt{\frac{2}{\pi}}e^{-n^2/2\sigma^2} + 2\eta \mathrm{G}\left(\frac{\eta}{\sigma}\right) - \eta$$

5-34 Show that if $\mathbf{x}$ and $\mathbf{y}$ are two random variables with densities $f_x(x)$ and $f_y(y)$, respectively, then

$$E\{\log f_x(\mathbf{x})\} \geq E\{\log f_y(\mathbf{x})\}$$

5-35 (*Chernoff bound*) (*a*) Show that for any $\alpha > 0$ and for any real s,

$$P\{e^{s\mathbf{x}} \geq \alpha\} \leq \frac{\mathbf{\Phi}(s)}{\alpha} \qquad \text{where } \mathbf{\Phi}(s) = E\{e^{s\mathbf{x}}\} \tag{i}$$

Hint: Apply (5-89) to the random variable $\mathbf{y} = e^{s\mathbf{x}}$. (*b*) For any A,

$$P\{\mathbf{x} \geq A\} \leq e^{-sA}\mathbf{\Phi}(s) \qquad s > 0$$

$$P\{\mathbf{x} \leq A\} \leq e^{-sA}\mathbf{\Phi}(s) \qquad s < 0$$

(*Hint:* Set $\alpha = e^{sA}$ in (i).)

5-36 Show that for any random variable $\mathbf{x}$

$$[E(|\mathbf{x}|^m)]^{1/m} \leq [E(|\mathbf{x}|^n)]^{1/n} \qquad 1 < m < n < \infty$$

5-37 Show that (*a*) if $f(x)$ is a Cauchy density, then $\Phi(\omega) = e^{-\alpha|\omega|}$; (*b*) if $f(x)$ is a Laplace density, then $\Phi(\omega) = \alpha^2/(\alpha^2 + \omega^2)$.

5-38 (*a*) Let $\mathbf{x} \sim G(\alpha, \beta)$. Show that $E\{\mathbf{x}\} = \alpha\beta$, $\text{Var}\{\mathbf{x}\} = \alpha\beta^2$ and $\Phi_X(\omega) = (1 - \beta e^{j\omega})^{-\alpha}$.
(*b*) Let $\mathbf{x} \sim \chi^2(n)$. Show that $E\{\mathbf{x}\} = n$, $\text{Var}\{\mathbf{x}\} = 2n$ and $\Phi_x(\omega) = (1 - 2e^{j\omega})^{-n/2}$.
(*c*) Let $\mathbf{x} \sim B(n, p)$. Show that $E\{\mathbf{x}\} = np$, $\text{Var}\{\mathbf{x}\} = npq$ and $\Phi_x(\omega) = (pe^{j\omega} + q)^n$.
(*d*) Let $\mathbf{x} \sim NB(r, p)$. Show that $\Phi_x(\omega) = p^r(1 - qe^{j\omega})^{-r}$.

5-39 A random variable $\mathbf{x}$ has a *geometric* distribution if

$$P\{\mathbf{x} = k\} = pq^k \quad k = 0, 1, \ldots \quad p + q = 1$$

Find $\mathbf{\Gamma}(z)$ and show that $\eta_x = q/p, \sigma_x^2 = q/p^2$

5-40 Let $\mathbf{x}$ denote the event "*the number of failures that precede the nth success*" so that $\mathbf{x} + n$ represents the total number of trials needed to generate n successes. In that case, the event $\{\mathbf{x} = k\}$ occurs if and only if the last trial results in a success and among the previous $(\mathbf{x} + n - 1)$ trials there are $n - 1$ successes (or $\mathbf{x}$ failures). This gives an alternate formulation for the *Pascal* (or *negative binomial*) distribution as follows: (see Table 5-2)

$$P\{\mathbf{x} = k\} = \binom{n + k - 1}{k} p^n q^k = \binom{-n}{k} p^n (-q)^k \qquad k = 0, 1, 2, \ldots$$

find $\mathbf{\Gamma}(z)$ and show that $\eta_x = nq/p, \sigma_x^2 = nq/p^2$.

5-41 Let **x** be a negative binomial random variable with parameters r and p. Show that as $p \to 1$ and $r \to \infty$ such that $r(1-p) \to \lambda$, a constant, then

$$P(\mathbf{x} = n + r) \to e^{-\lambda}\frac{\lambda^n}{n!} \qquad n = 0, 1, 2, \ldots$$

5-42 Show that if $E\{\mathbf{x}\} = \eta$, then

$$E\{e^{s\mathbf{x}}\} = e^{s\eta}\sum_{n=0}^{\infty}\mu_n\frac{s^n}{n!} \qquad \mu_n = E\{(\mathbf{x}-\eta)^n\}$$

5-43 Show that if $\Phi_x(\omega_1) = 1$ for some $\omega_1 \neq 0$, then the random variable **x** is of lattice type taking the values $x_n = 2\pi n/\omega_1$.

Hint:

$$0 = 1 - \Phi_x(\omega_1) = \int_{-\infty}^{\infty}(1 - e^{j\omega_1 x})f_x(x)\,dx$$

5-44 The random variable **x** has zero mean, central moments μ_n, and cumulants λ_n. Show that $\lambda_3 = \mu_3$, $\lambda_4 = \mu_4 - 3\mu_2^2$; if **y** is $N(0; \sigma_y^2)$ and $\sigma_y = \sigma_x$, then $E\{\mathbf{x}^4\} = E\{\mathbf{y}^4\} + \lambda_4$.

5-45 The random variable **x** takes the values 0, 1, ... with $P\{\mathbf{x} = k\} = p_k$. Show that if

$$\mathbf{y} = (\mathbf{x} - 1)U(\mathbf{x} - 1) \qquad \text{then } \mathbf{\Gamma}_y(z) = p_0 + z^{-1}[\mathbf{\Gamma}_x(z) - p_0]$$

$$\eta_y = \eta_x - 1 + p_0 \qquad E\{\mathbf{y}^2\} = E\{\mathbf{x}^2\} - 2\eta_x + 1 - p_0$$

5-46 Show that, if $\Phi(\omega) = E\{e^{j\omega\mathbf{x}}\}$, then for any a_i,

$$\sum_{i=1}^{n}\sum_{j=1}^{n}\Phi(\omega_i - \omega_j)a_i a_j^* \ge 0$$

Hint:

$$E\left\{\left|\sum_{i=1}^{n} a_j e^{j\omega i\mathbf{x}}\right|^2\right\} \ge 0$$

5-47 We are given an even convex function $g(x)$ and a random variable **x** whose density $f(x)$ is symmetrical as in Fig. P5-47 with a single maximum at $x = \eta$. Show that the mean $E\{g(\mathbf{x} - a)\}$ of the random variable $g(\mathbf{x} - a)$ is minimum if $a = \eta$.

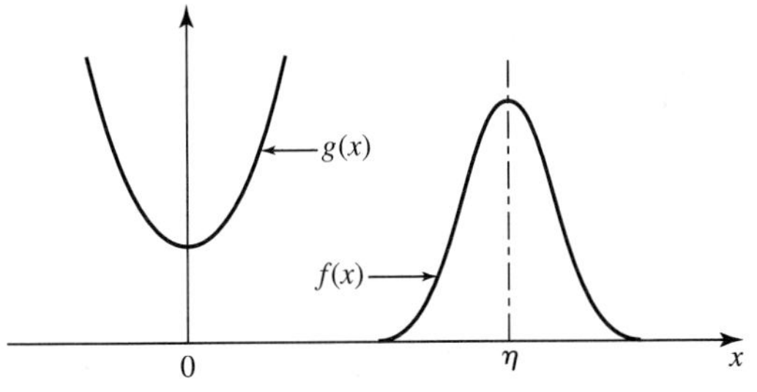

FIGURE P5-47

5-48 The random variable **x** is $N(0; \sigma^2)$. (*a*) Using characteristic functions, show that if $g(x)$ is a function such that $g(x)e^{-x^2/2\sigma^2} \to 0$ as $|x| \to \infty$, then (Price's theorem)

$$\frac{dE\{g(\mathbf{x})\}}{dv} = \frac{1}{2}E\left\{\frac{d^2g(\mathbf{x})}{d\mathbf{x}^2}\right\} \qquad v = \sigma^2 \qquad \text{(i)}$$

(b) The moments μ_n of $\mathbf{x}$ are functions of v. Using (i), show that

$$\mu_n(v) = \frac{n(n-1)}{2} \int_0^v \mu_{n-2}(\beta)\, d\beta$$

5-49 Show that, if $\mathbf{x}$ is an integer-valued random variable with moment function $\boldsymbol{\Gamma}(z)$ as in (5-113), then

$$P\{\mathbf{x} = k\} = \frac{1}{2\pi} \int_{-\pi}^{\pi} \boldsymbol{\Gamma}(e^{j\omega}) e^{-jk\omega}\, d\omega$$

5-50 A biased coin is tossed and the first outcome is noted. The tossing is continued until the outcome is the complement of the first outcome, thus completing the first run. Let $\mathbf{x}$ denote the length of the first run. Find the p.m.f of $\mathbf{x}$, and show that

$$E\{\mathbf{x}\} = \frac{p}{q} + \frac{q}{p}$$

5-51 A box contains N identical items of which $M < N$ are defective ones. A sample of size n is taken from the box, and let $\mathbf{x}$ represent the number of defective items in this sample.
(a) Find the distribution function of $\mathbf{x}$ if the n samples are drawn with replacement.
(b) If the n samples are drawn *without* replacement, then show that

$$P\{\mathbf{x} = k\} = \frac{\binom{M}{k}\binom{N-M}{n-k}}{\binom{N}{n}} \qquad \max(0, n + M - N) \le k \le \min(M, N)$$

Find the mean and variance of $\mathbf{x}$. The distribution in (b) is known as the *hypergeometric* distribution (see also Problem 3-5). The lottery distribution in (3-39) is an example of this distribution.
(c) In (b), let $N \to \infty$, $M \to \infty$, such that $M/N \to p$, $0 < p < 1$. Then show that the hypergeometric random variable can be approximated by a Binomial random variable with parameters n and p, provided $n \ll N$.

5-52 A box contains n white and m black marbles. Let $\mathbf{x}$ represent the number of draws needed for the rth white marble.
(a) If sampling is done with replacement, show that $\mathbf{x}$ has a negative binomial distribution with parameters r and $p = n/(m+n)$. (b) If sampling is done without replacement, then show that

$$P\{\mathbf{x} = k\} = \binom{k-1}{r-1} \frac{\binom{m+n-k}{n-r}}{\binom{m+n}{n}} \qquad k = r, r+1, \ldots, m+n$$

(c) For a given k and r, show that the probability distribution in (b) tends to a negative binomial distribution as $n + m \to \infty$. Thus, for large population size, sampling with or without replacement is the same.

CHAPTER

6

TWO RANDOM VARIABLES

6-1 BIVARIATE DISTRIBUTIONS

We are given two random variables $\mathbf{x}$ and $\mathbf{y}$, defined as in Sec. 4-1, and we wish to determine their joint statistics, that is, the probability that the point $(\mathbf{x}, \mathbf{y})$ is in a specified region[1] D in the xy plane. The distribution functions $F_x(x)$ and $F_y(y)$ of the given random variables determine their separate (marginal) statistics but not their joint statistics. In particular, the probability of the event

$$\{\mathbf{x} \le x\} \cap \{\mathbf{y} \le y\} = \{\mathbf{x} \le x, \mathbf{y} \le y\}$$

cannot be expressed in terms of $F_x(x)$ and $F_y(y)$. Here, we show that the joint statistics of the random variables $\mathbf{x}$ and $\mathbf{y}$ are completely determined if the probability of this event is known for every x and y.

Joint Distribution and Density

The joint (bivariate) distribution $F_{xy}(x, y)$ or, simply, $F(x, y)$ of two random variables $\mathbf{x}$ and $\mathbf{y}$ is the probability of the event

$$\{\mathbf{x} \le x, \mathbf{y} \le y\} = \{(\mathbf{x}, \mathbf{y}) \in D_1\}$$

where x and y are two arbitrary real numbers and D_1 is the quadrant shown in Fig. 6-1a:

$$F(x, y) = P\{\mathbf{x} \le x, \mathbf{y} \le y\} \tag{6-1}$$

[1] The region D is arbitrary subject only to the mild condition that it can be expressed as a countable union or intersection of rectangles.

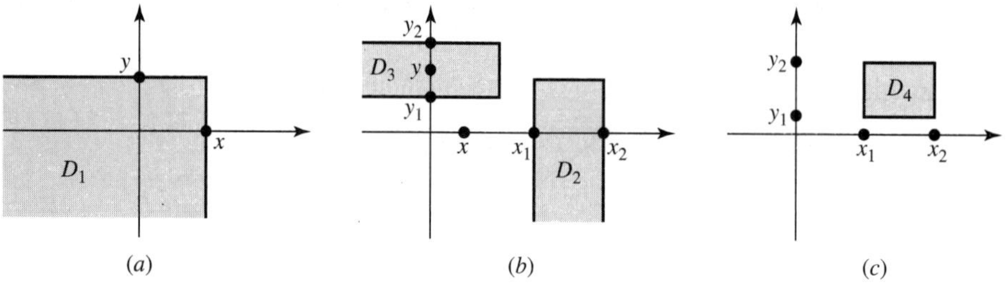

FIGURE 6-1

PROPERTIES

1. The function $F(x, y)$ is such that

$$F(-\infty, y) = 0, \quad F(x, -\infty) = 0, \quad F(\infty, \infty) = 1$$

Proof. As we know, $P\{\mathbf{x} = -\infty\} = P\{\mathbf{y} = -\infty\} = 0$. And since

$$\{\mathbf{x} = -\infty, \mathbf{y} \le y\} \subset \{\mathbf{x} = -\infty\} \qquad \{\mathbf{x} \le x, \mathbf{y} = -\infty\} \subset \{\mathbf{y} = -\infty\}$$

the first two equations follow. The last is a consequence of the identities

$$\{\mathbf{x} \le -\infty, \mathbf{y} \le -\infty\} = S \qquad P(S) = 1$$

2. The event $\{x_1 < \mathbf{x} \le x_2, \mathbf{y} \le y\}$ consists of all points $(\mathbf{x}, \mathbf{y})$ in the vertical half-strip D_2 and the event $\{\mathbf{x} \le x, y_1 < \mathbf{y} \le y_2\}$ consists of all points $(\mathbf{x}, \mathbf{y})$ in the horizontal half-strip D_3 of Fig. 6-1*b*. We maintain that

$$\{x_1 < \mathbf{x} \le x_2, \mathbf{y} \le y\} = F(x_2, y) - F(x_1, y) \tag{6-2}$$

$$\{\mathbf{x} \le x, y_1 < \mathbf{y} \le y_2\} = F(x, y_2) - F(x, y_1) \tag{6-3}$$

Proof. Clearly, for $x_2 > x_1$

$$\{\mathbf{x} \le x_2, \mathbf{y} \le y\} = \{\mathbf{x} \le x_1, \mathbf{y} \le y\} \cup \{x_1 < \mathbf{x} \le x_2, \mathbf{y} \le y\}$$

The last two events are mutually exclusive; hence [see (2-10)]

$$P\{\mathbf{x} \le x_2, \mathbf{y} \le y\} = P\{\mathbf{x} \le x_1, \mathbf{y} \le y\} + P\{x_1 < \mathbf{x} \le x_2, \mathbf{y} \le y\}$$

and (6-2) results. The proof of (6-3) is similar.

3.

$$P\{x_1 < \mathbf{x} \le x_2, y_1 < \mathbf{y} \le y_2\} = F(x_2, y_2) - F(x_1, y_2) - F(x_2, y_1) + F(x_1, y_1) \tag{6-4}$$

This is the probability that $(\mathbf{x}, \mathbf{y})$ is in the rectangle D_4 of Fig. 6-1*c*.

Proof. It follows from (6-2) and (6-3) because

$$\{x_1 < \mathbf{x} \le x_2, \mathbf{y} \le y_2\} = \{x_1 < \mathbf{x} \le x_2, \mathbf{y} \le y_1\} \cup \{x_1 < \mathbf{x} \le x_2, y_1 < \mathbf{y} \le y_2\}$$

and the last two events are mutually exclusive.

JOINT DENSITY. The joint density of **x** and **y** is by definition the function

$$f(x, y) = \frac{\partial^2 F(x, y)}{\partial x \partial y} \tag{6-5}$$

From this and property 1 it follows that

$$F(x, y) = \int_{-\infty}^{x} \int_{-\infty}^{y} f(\alpha, \beta)\, d\alpha\, d\beta \tag{6-6}$$

JOINT STATISTICS. We shall now show that the probability that the point $(\mathbf{x}, \mathbf{y})$ is in a region D of the xy plane equals the integral of $f(x, y)$ in D. In other words,

$$P\{(\mathbf{x}, \mathbf{y}) \in D\} = \int_D \int f(x, y)\, dx\, dy \tag{6-7}$$

where $\{(\mathbf{x}, \mathbf{y}) \in D\}$ is the event consisting of all outcomes ζ such that the point $[\mathbf{x}(\zeta), \mathbf{y}(\zeta)]$ is in D.

Proof. As we know, the ratio

$$\frac{F(x + \Delta x, y + \Delta y) - F(x, y + \Delta y) - F(x + \Delta x, y) + F(x, y)}{\Delta x\, \Delta y}$$

tends to $\partial F(x, y)/\partial x \partial y$ as $\Delta x \to 0$ and $\Delta y \to 0$. Hence [see (6-4) and (6-5)]

$$P\{x < \mathbf{x} \le x + \Delta x, y < \mathbf{y} \le y + \Delta y\} \simeq f(x, y)\, \Delta x\, \Delta y \tag{6-8}$$

We have thus shown that the probability that $(\mathbf{x}, \mathbf{y})$ is in a differential rectangle equals $f(x, y)$ times the area $\Delta x\, \Delta y$ of the rectangle. This proves (6-7) because the region D can be written as the limit of the union of such rectangles.

MARGINAL STATISTICS. In the study of several random variables, the statistics of each are called marginal. Thus $F_x(x)$ is the *marginal distribution* and $f_x(x)$ the *marginal density* of **x**. Here, we express the marginal statistics of **x** and **y** in terms of their joint statistics $F(x, y)$ and $f(x, y)$.

We maintain that

$$F_x(x) = F(x, \infty) \qquad F_y(y) = F(\infty, y) \tag{6-9}$$

$$f_x(x) = \int_{-\infty}^{\infty} f(x, y)\, dy \qquad f_y(y) = \int_{-\infty}^{\infty} f(x, y)\, dx \tag{6-10}$$

Proof. Clearly, $\{\mathbf{x} \le \infty\} = \{\mathbf{y} \le \infty\} = S$; hence

$$\{\mathbf{x} \le x\} = \{\mathbf{x} \le x, \mathbf{y} \le \infty\} \qquad \{\mathbf{y} \le y\} = \{\mathbf{x} \le \infty, \mathbf{y} \le y\}$$

The probabilistics of these two sides yield (6-9).

Differentiating (6-6), we obtain

$$\frac{\partial F(x, y)}{\partial x} = \int_{-\infty}^{y} f(x, \beta)\, d\beta \qquad \frac{\partial F(x, y)}{\partial y} = \int_{-\infty}^{y} f(\alpha, y)\, d\alpha \tag{6-11}$$

Setting $y = \infty$ in the first and $x = \infty$ in the second equation, we obtain (6-10) because [see (6-9)]

$$f_x(x) = \frac{\partial F(x, \infty)}{\partial x} \qquad f_y(x) = \frac{\partial F(\infty, y)}{\partial y}$$

EXISTENCE THEOREM. From properties 1 and 3 it follows that

$$F(-\infty, y) = 0 \qquad F(x, -\infty) = 0 \qquad F(\infty, \infty) = 1 \tag{6-12}$$

and

$$F(x_2, y_2) - F(x_1, y_2) - F(x_2, y_1) + F(x_1, y_1) \geq 0 \tag{6-13}$$

for every $x_1 < x_2$, $y_1 < y_2$. Hence [see (6-6) and (6-8)]

$$\int_{-\infty}^{\infty} \int_{-\infty}^{\infty} f(x, y)\, dx\, dy = 1 \qquad f(x, y) \geq 0 \tag{6-14}$$

Conversely, given $F(x, y)$ or $f(x, y)$ as before, we can find two random variables **x** and **y**, defined in some space S, with distribution $F(x, y)$ or density $f(x, y)$. This can be done by extending the existence theorem of Sec. 4-3 to joint statistics.

Probability Masses

The probability that the point (x, y) is in a region D of the plane can be interpreted as the probability mass in this region. Thus the mass in the entire plane equals 1. The mass in the half-plane $\mathbf{x} \leq x$ to the left of the line L_x of Fig. 6-2 equals $F_x(x)$. The mass in the half-plane $\mathbf{y} \leq y$ below the line L_y equals $F_y(y)$. The mass in the doubly-shaded quadrant $\{\mathbf{x} \leq x, \mathbf{y} \leq y\}$ equals $F(x, y)$.

Finally, the mass in the clear quadrant $(\mathbf{x} > x, \mathbf{y} > y)$ equals

$$P\{\mathbf{x} > x, \mathbf{y} > y\} = 1 - F_x(x) - F_y(y) + F(x, y) \tag{6-15}$$

The probability mass in a region D equals the integral [see (6-7)]

$$\int_D \int f(x, y)\, dx\, dy$$

If, therefore, $f(x, y)$ is a bounded function, it can be interpreted as surface mass density.

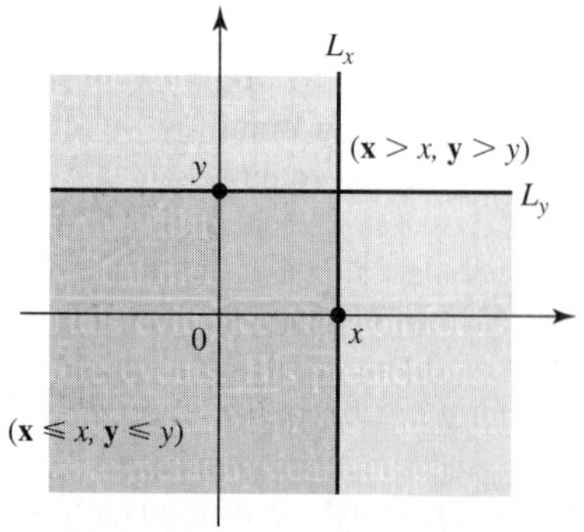

FIGURE 6-2

EXAMPLE 6-1

▶ Suppose that

$$f(x, y) = \frac{1}{2\pi\sigma^2} e^{-(x^2+y^2)/2\sigma^2} \tag{6-16}$$

We shall find the mass m in the circle $x^2 + y^2 \leq a^2$. Inserting (6-16) into (6-7) and using the transformation

$$x = r\cos\theta \qquad y = r\sin\theta$$

we obtain

$$m = \frac{1}{2\pi\sigma^2} \int_0^a \int_{-\pi}^{\pi} e^{-r^2/2\sigma^2} r\, dr\, d\theta = 1 - e^{-a^2/2\sigma^2} \tag{6-17}$$

◀

INDEPENDENCE

▶ Two random variables $\mathbf{x}$ and $\mathbf{y}$ are called (*statistically*) *independent* if the events $\{\mathbf{x} \in A\}$ and $\{\mathbf{y} \in B\}$ are independent [see (2-40)], that is, if

$$P\{\mathbf{x} \in A, \mathbf{y} \in B\} = P\{\mathbf{x} \in A\}P\{\mathbf{y} \in B\} \tag{6-18}$$

where A and B are two arbitrary sets on the x and y axes, respectively.

Applying this to the events $\{\mathbf{x} \leq x\}$ and $\{\mathbf{y} \leq y\}$, we conclude that, if the random variables $\mathbf{x}$ and $\mathbf{y}$ are independent, then

$$F(x, y) = F_x(x)F_y(y) \tag{6-19}$$

Hence

$$f(x, y) = f_x(x)f_y(y) \tag{6-20}$$

It can be shown that, if (6-19) or (6-20) is true, then (6-18) is also true; that is, the random variables $\mathbf{x}$ and $\mathbf{y}$ are independent [see (6-7)]. ◀

EXAMPLE 6-2

BUFFON'S NEEDLE

▶ A fine needle of length $2a$ is dropped at random on a board covered with parallel lines distance $2b$ apart where $b > a$ as in Fig. 6-3a. We shall show that the probability p that the needle intersects one of the lines equals $2a/\pi b$.

In terms of random variables the experiment just discussed can be phrased as: We denote by $\mathbf{x}$ the distance from the center of the needle to the nearest line and by $\boldsymbol{\theta}$ the angle between the needle and the direction perpendicular to the lines. We assume that the random variables $\mathbf{x}$ and $\boldsymbol{\theta}$ are independent, $\mathbf{x}$ is uniform in the interval $(0, b)$, and $\boldsymbol{\theta}$ is uniform in the interval $(0, \pi/2)$. From this it follows that

$$f(x, \theta) = f_x(x)f_\theta(\theta) = \frac{1}{b}\frac{2}{\pi} \qquad 0 \leq x \leq b \qquad 0 \leq \theta \leq \frac{\pi}{2}$$

and 0 elsewhere. Hence the probability that the point $(\mathbf{x}, \boldsymbol{\theta})$ is in a region D included in the rectangle R of Fig. 6-3b equals the area of D times $2/\pi b$.

The needle intersects the lines if $x < a\cos\theta$. Hence p equals the shaded area of Fig. 6-3b times $2/\pi b$:

$$p = P\{\mathbf{x} < a\cos\boldsymbol{\theta}\} = \frac{2}{\pi b}\int_0^{2/\pi} a\cos\theta\, d\theta = \frac{2a}{\pi b}$$

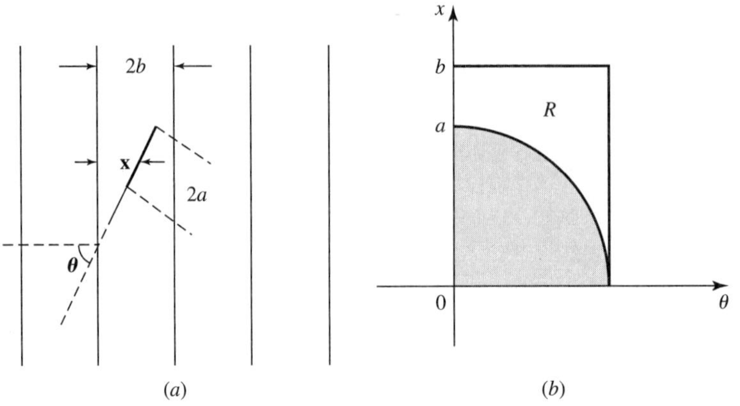

FIGURE 6-3

This can be used to determine experimentally the number π using the relative frequency interpretation of p: If the needle is dropped n times and it intersects the lines n_i times, then

$$\frac{n_i}{n} \simeq p = \frac{2a}{\pi b} \qquad \text{hence} \qquad \pi \simeq \frac{2an}{bn_i}$$

THEOREM 6-1

If the random variables **x** and **y** are independent, then the random variables

$$\mathbf{z} = g(\mathbf{x}) \qquad \mathbf{w} = h(\mathbf{y})$$

are also independent.

Proof. We denote by A_z the set of points on the x axis such that $g(x) \le z$ and by B_w the set of points on the y axis such that $h(y) \le w$. Clearly,

$$\{\mathbf{z} \le z\} = \{\mathbf{x} \in A_z\} \qquad \{\mathbf{w} \le w\} = \{\mathbf{y} \in B_w\} \tag{6-21}$$

Therefore the events $\{\mathbf{z} \le z\}$ and $\{\mathbf{w} \le w\}$ are independent because the events $\{\mathbf{x} \in A_z\}$ and $\{\mathbf{y} \in B_w\}$ are independent.

INDEPENDENT EXPERIMENTS. As in the case of events (Sec. 3-1), the concept of independence is important in the study of random variables defined on product spaces. Suppose that the random variable **x** is defined on a space S_1 consisting of the outcomes $\{\xi_1\}$ and the random variable **y** is defined on a space S_2 consisting of the outcomes $\{\xi_2\}$. In the combined experiment $S_1 \times S_2$ the random variables **x** and **y** are such that

$$\mathbf{x}(\xi_1\xi_2) = \mathbf{x}(\xi_1) \qquad \mathbf{y}(\xi_1\xi_2) = \mathbf{y}(\xi_2) \tag{6-22}$$

In other words, **x** depends on the outcomes of S_1 only, and **y** depends on the outcomes of S_2 only.

THEOREM 6-2

If the experiments S_1 and S_2 are independent, then the random variables **x** and **y** are independent.

Proof. We denote by A_x the set $\{\mathbf{x} \le x\}$ in S_1 and by B_y the set $\{\mathbf{y} \le y\}$ in S_2. In the space $S_1 \times S_2$,

$$\{\mathbf{x} \le x\} = A_x \times S_2 \qquad \{\mathbf{y} \le y\} = S_1 \times B_y$$

From the independence of the two experiments, it follows that [see (3-4)] the events $A_x \times S_2$ and $S_1 \times B_y$ are independent. Hence the events $\{\mathbf{x} \le x\}$ and $\{\mathbf{y} \le y\}$ are also independent. ◀

JOINT NORMALITY

▶ We shall say that the random variables **x** and **y** are *jointly normal* if their joint density is given by

$$f(x, y) = A \exp\left\{-\frac{1}{2(1-r^2)}\left(\frac{(x-\eta_1)^2}{\sigma_1^2} - 2r\frac{(x-\eta_1)(y-\eta_2)}{\sigma_1\sigma_2} + \frac{(y-\eta_2)^2}{\sigma_2^2}\right)\right\} \tag{6-23}$$

This function is positive and its integral equals 1 if

$$A = \frac{1}{2\pi\sigma_1\sigma_2\sqrt{1-r^2}} \qquad |r| < 1 \tag{6-24}$$

Thus $f(x, y)$ is an exponential and its exponent is a negative quadratic because $|r| < 1$. The function $f(x, y)$ will be denoted by

$$N(\eta_1, \eta_2, \sigma_1^2, \sigma_2^2, r) \tag{6-25}$$

As we shall presently see, η_1 and η_2 are the expected values of **x** and **y**, and σ_1^2 and σ_2^2 their variances. The significance of r will be given later in Example 6-30 (correlation coefficient).

We maintain that the marginal densities of **x** and **y** are given by

$$f_x(x) = \frac{1}{\sigma_1\sqrt{2\pi}}e^{-(x-\eta_1)^2/2\sigma_1^2} \qquad f_y(y) = \frac{1}{\sigma_2\sqrt{2\pi}}e^{-(y-\eta_2)^2/2\sigma_2^2} \tag{6-26}$$

Proof. To prove this, we must show that if (6-23) is inserted into (6-10), the result is (6-26). The bracket in (6-23) can be written in the form

$$(\cdots) = \left(\frac{x-\eta_1}{\sigma_1} - r\frac{y-\eta_2}{\sigma_2}\right)^2 + (1-r^2)\frac{(y-\eta_2)^2}{\sigma_2^2}$$

Hence

$$f_y(y) = \int_{-\infty}^{\infty} f(x, y)\, dx = Ae^{-(y-\eta_2)^2/2\sigma_2^2}\int_{-\infty}^{\infty} e^{-(x-\eta)^2/2(1-r^2)\sigma_1^2}$$

where

$$\eta = \eta_1 + r\frac{(y-\eta_2)\sigma_1}{\sigma_2}$$

The last integral represents a normal random variable with mean μ and variance $(1-r^2)\sigma_1^2$. Therefore the last integral is a constant (independent of x and y) $B = \sqrt{2\pi(1-r^2)\sigma_1^2}$. Therefore

$$f_y(y) = ABe^{(y-\eta_2)^2/2\sigma_2^2}$$

And since $f_y(y)$ is a density, its area must equal 1. This yields $AB = 1/\sigma_2\sqrt{2\pi}$, from which we obtain $A = 1/2\pi\sigma_1\sigma_2\sqrt{1-r^2}$ proving (6-24), and the second equation in (6-26). The proof of the first equation is similar. ◀

Notes 1. From (6-26) it follows that if two random variables are jointly normal, they are also marginally normal. However, as the next example shows, *the converse is not true*.

2. Joint normality can be defined also as follows: Two random variables **x** and **y** are jointly normal if the sum $a\mathbf{x} + b\mathbf{y}$ is normal for every a and b [see (7-56)].

EXAMPLE 6-3

▶ We shall construct two random variables **x** and **y** that are marginally but not jointly normal. Toward this, consider the function

$$f(x, y) = f_x(x) f_y(y)[1 + \rho\{2F_x(x) - 1\}\{2F_y(y) - 1\}] \qquad |\rho| < 1 \qquad (6\text{-}27)$$

where $f_x(x)$ and $f_y(y)$ are two p.d.fs with respective distribution functions $F_x(x)$ and $F_y(y)$. It is easy to show that $f(x, y) \geq 0$ for all x, y, and

$$\int_{-\infty}^{+\infty}\int_{-\infty}^{+\infty} f(x, y)\, dx\, dy = 1$$

which shows that (6-27) indeed represents a joint p.d.f. of two random variables **x** and **y**. Moreover, by direct integration

$$\int_{-\infty}^{+\infty} f(x, y)\, dy = f_x(x) + \rho(2F_x(x) - 1) f_x(x) \int_{-1}^{1} \frac{u\, du}{2} = f_x(x)$$

where we have made use of the substitution $u = 2F_y(y) - 1$. Similarly,

$$\int_{-\infty}^{+\infty} f(x, y)\, dx = f_y(y)$$

implying that $f_x(x)$ and $f_y(y)$ in (6-27) also represent the respective marginal p.d.f.s of **x** and **y**, respectively.

In particular, let $f_x(x)$ and $f_y(y)$ be normally distributed as in (6-26). In that case (6-27) represents a joint p.d.f. with normal marginals that is not jointly normal. ◀

Circular Symmetry

We say that the joint density of two random variables **x** and **y** is circularly symmetrical if it depends only on the distance from the origin, that is, if

$$f(x, y) = g(r) \qquad r = \sqrt{x^2 + y^2} \qquad (6\text{-}28)$$

THEOREM 6-3

▶ If the random variables **x** and **y** are circularly symmetrical and independent, then they are normal with zero mean and equal variance.

Proof. From (6-28) and (6-20) it follows that

$$g(\sqrt{x^2 + y^2}) = f_x(x) f_y(y) \qquad (6\text{-}29)$$

Since

$$\frac{\partial g(r)}{\partial x} = \frac{dg(r)}{dr}\frac{\partial r}{\partial x} \quad \text{and} \quad \frac{\partial r}{\partial x} = \frac{x}{r}$$

we conclude, differentiating (6-29) with respect to x, that

$$\frac{x}{r}g'(r) = f_x'(x)f_y(y)$$

Dividing both sides by $xg(r) = xf_x(x)f_y(y)$, we obtain

$$\frac{1}{r}\frac{g'(r)}{g(r)} = \frac{1}{x}\frac{f_x'(x)}{f_x(x)} \tag{6-30}$$

The right side of (6-30) is independent of y and the left side is a function of $r = \sqrt{x^2 + y^2}$. This shows that both sides are independent of x and y. Hence

$$\frac{1}{r}\frac{g'(r)}{g(r)} = \alpha = \text{constant}$$

and (6-28) yields

$$f(x, y) = g(\sqrt{x^2 + y^2}) = Ae^{\alpha(x^2+y^2)/2} \tag{6-31}$$

Thus the random variables **x** and **y** are normal with zero mean and variance $\sigma^2 = -1/\alpha$. ◀

DISCRETE TYPE RANDOM VARIABLES. Suppose the random variables **x** and **y** are of discrete type taking the values of x_i and y_k with respective probabilities

$$P\{\mathbf{x} = x_i\} = p_i \qquad P\{\mathbf{y} = y_k\} = q_k \tag{6-32}$$

Their joint statistics are determined in terms of the *joint probabilities*

$$P\{\mathbf{x} = x_i, \mathbf{y} = y_k\} = p_{ik} \tag{6-33}$$

Clearly,

$$\sum_{i,k} p_{ik} = 1$$

because, as i and k take all possible values, the events $\{\mathbf{x} = x_i, \mathbf{y} = y_k\}$ are mutually exclusive, and their union equals the certain event.

We maintain that the *marginal probabilities* p_i and q_k can be expressed in terms of the joint probabilities p_{ik}:

$$p_i = \sum_k p_{ik} \qquad q_k = \sum_i p_{ik} \tag{6-34}$$

This is the discrete version of (6-10).

Proof. The events $\{\mathbf{y} = y_k\}$ form a partition of S. Hence as k ranges over all possible values, the events $\{\mathbf{x} = x_i, \mathbf{y} = y_k\}$ are mutually exclusive and their union equals $\{\mathbf{x} = x_i\}$. This yields the first equation in (6-34) [see (2-41)]. The proof of the second is similar.

POINT MASSES. If the random variables **x** and **y** are of discrete type taking the values x_i and y_k, then the probability masses are 0 everywhere except at the point (x_i, y_k). We have, thus, only point masses and the mass at each point equals p_{ik} [see (6-33)]. The probability $p_i = P\{\mathbf{x} = x_i\}$ equals the sum of all masses p_{ik} on the line $x = x_i$ in agreement with (6-34).

If $i = 1, \ldots, M$ and $k = 1, \ldots, N$, then the number of possible point masses on the plane equals MN. However, as Example 6-4 shows, some of these masses might be 0.

EXAMPLE 6-4

▶ (*a*) In the fair-die experiment, **x** equals the number of dots shown and **y** equals twice this number:

$$\mathbf{x}(f_i) = i \qquad \mathbf{y}(f_i) = 2i \quad i = 1, \ldots, 6$$

In other words, $x_i = i$, $y_k = 2k$, and

$$p_{ik} = P\{\mathbf{x} = i, \mathbf{y} = 2k\} = \begin{cases} \dfrac{1}{6} & i = k \\ 0 & i \neq k \end{cases}$$

Thus there are masses only on the six points $(i, 2i)$ and the mass of each point equals $1/6$ (Fig. 6-4*a*).

(*b*) We toss the die twice obtaining the 36 outcomes $f_i f_k$ and we define **x** and **y** such that **x** equals the first number that shows, and **y** the second

$$\mathbf{x}(f_i, f_k) = i \qquad \mathbf{y}(f_i f_k) = k \qquad i, k = 1, \ldots, 6$$

Thus $x_i = i$, $y_k = k$, and $p_{ik} = 1/36$. We have, therefore, 36 point masses (Fig. 6-4*b*) and the mass of each point equals $1/36$. On the line $x = i$ there are six points with total mass $1/6$.

(*c*) Again the die is tossed twice but now

$$\mathbf{x}(f_i f_k) = |i - k| \qquad \mathbf{y}(f_i f_k) = i + k$$

In this case, **x** takes the values $0, 1, \ldots, 5$ and **y** the values $2, 3, \ldots, 12$. The number of possible points $6 \times 11 = 66$; however, only 21 have positive masses (Fig. 6-4*c*). Specifically, if $\mathbf{x} = 0$, then $\mathbf{y} = 2$, or $4, \ldots$, or 12 because if $\mathbf{x} = 0$, then $i = k$ and $\mathbf{y} = 2i$. There are, therefore, six mass points in this line and the mass of each point equals $1/36$. If $\mathbf{x} = 1$, then $\mathbf{y} = 3$, or $5, \ldots$, or 11. Thus, there are, five mass points on the line $x = 1$ and the mass of each point equals $2/36$. For example, if $\mathbf{x} = 1$ and $\mathbf{y} = 7$, then $i = 3$, $k = 4$, or $i = 4$, $k = 3$; hence $P\{\mathbf{x} = 1, \mathbf{y} = 7\} = 2/36$. ◀

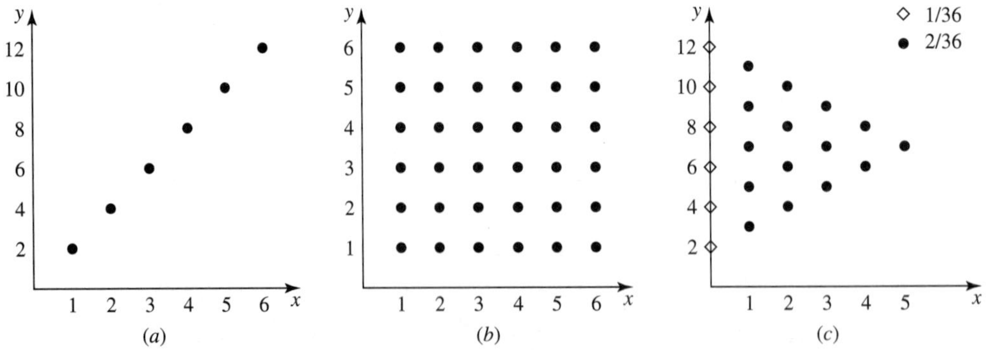

FIGURE 6-4

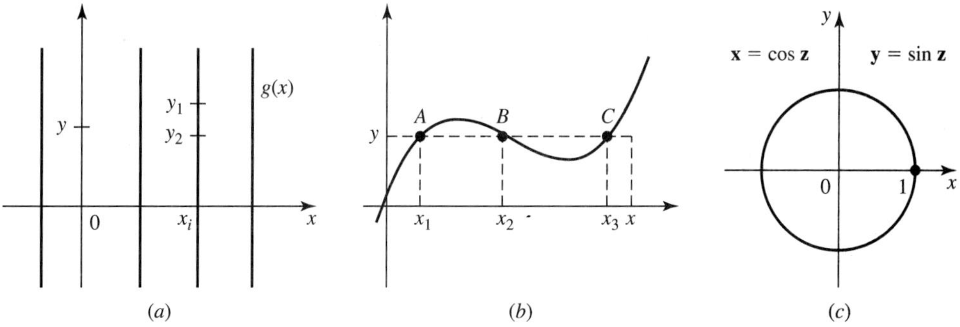

FIGURE 6-5

LINE MASSES. These cases lead to line masses:

1. If $\mathbf{x}$ is of discrete type taking the values x_i and $\mathbf{y}$ is of continuous type, then all probability masses are on the vertical lines $x = x_i$ (Fig. 6-5*a*). In particular, the mass between the point y_1 and y_2 on the line $x = x_i$ equals the probability of the event

$$\{\mathbf{x} = x_i, y_1 \le \mathbf{y} \le y_2\}$$

2. If $\mathbf{y} = g(\mathbf{x})$, then all the masses are on the curve $y = g(x)$. In this case, $F(x, y)$ can be expressed in terms of $F_x(x)$. For example, with x and y as in Fig. 6-5*b*, $F(x, y)$ equals the masses on the curve $y = g(x)$ to the left of the point A and between B and C equal $F_x(x_3) - F_x(x_2)$. Hence

$$F(x, y) = F_x(x_1) + F_x(x_3) - F_x(x_2) \qquad y = g(x_1) = g(x_2) = g(x_3)$$

3. If $\mathbf{x} = g(\mathbf{z})$ and $\mathbf{y} = h(\mathbf{z})$, then all probability masses are on the curve $x = g(z)$, $y = h(z)$ specified parametrically. For example, if $g(z) = \cos z$, $h(z) = \sin z$, then the curve is a circle (Fig. 6-5*c*). In this case, the joint statistics of $\mathbf{x}$ and $\mathbf{y}$ can be expressed in terms of $F_z(z)$.

If the random variables $\mathbf{x}$ and $\mathbf{y}$ are of discrete type as in (6-33) and *independent,* then

$$p_{ik} = p_i p_k \qquad (6\text{-}35)$$

This follows if we apply (6-19) to the events $\{\mathbf{x} = x_i\}$ and $\{\mathbf{y} = y_k\}$. This is the discrete version of (6-20).

EXAMPLE 6-5

▶ A die with $P\{f_i\} = p_i$ is tossed twice and the random variables $\mathbf{x}$ and $\mathbf{y}$ are such that

$$\mathbf{x}(f_i f_k) = i \qquad \mathbf{y}(f_i f_k) = k$$

Thus $\mathbf{x}$ equals the first number that shows and $\mathbf{y}$ equals the second; hence the random variables $\mathbf{x}$ and $\mathbf{y}$ are independent. This leads to the conclusion that

$$p_{ik} = P\{\mathbf{x} = i, \mathbf{y} = k\} = p_i p_k$$ ◀

6-2 ONE FUNCTION OF TWO RANDOM VARIABLES

Given two random variables $\mathbf{x}$ and $\mathbf{y}$ and a function $g(x, y)$, we form a new random variable $\mathbf{z}$ as

$$\mathbf{z} = g(\mathbf{x}, \mathbf{y}) \tag{6-36}$$

Given the joint p.d.f. $f_{xy}(x, y)$, how does one obtain $f_z(z)$, the p.d.f. of $\mathbf{z}$? Problems of this type are of interest from a practical standpoint. For example, a received signal in a communication scene usually consists of the desired signal buried in noise, and this formulation in that case reduces to $\mathbf{z} = \mathbf{x} + \mathbf{y}$. It is important to know the statistics of the incoming signal for proper receiver design. In this context, we shall analyze problems of the type shown in Fig. 6-6. Referring to (6-36), to start with,

$$\begin{aligned} F_z(z) &= P\{\mathbf{z}(\xi) \le z\} = P\{g(\mathbf{x}, \mathbf{y}) \le z\} = P\{(\mathbf{x}, \mathbf{y}) \in D_z\} \\ &= \iint_{x,y \in D_z} f_{xy}(x, y)\, dx\, dy \end{aligned} \tag{6-37}$$

where D_z in the xy plane represents the region where the inequality $g(x, y) \le z$ is satisfied (Fig. 6-7).

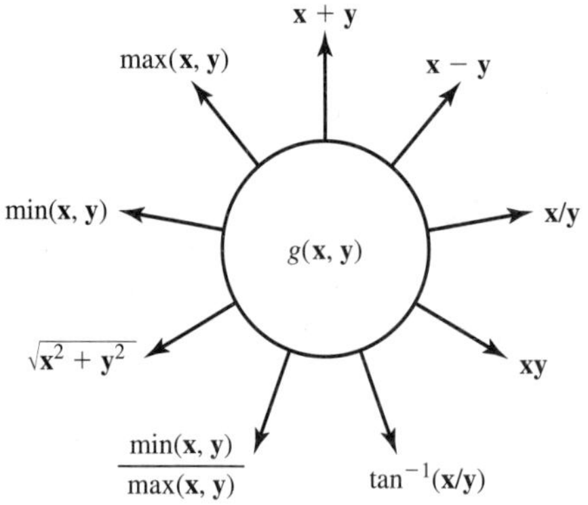

FIGURE 6-6

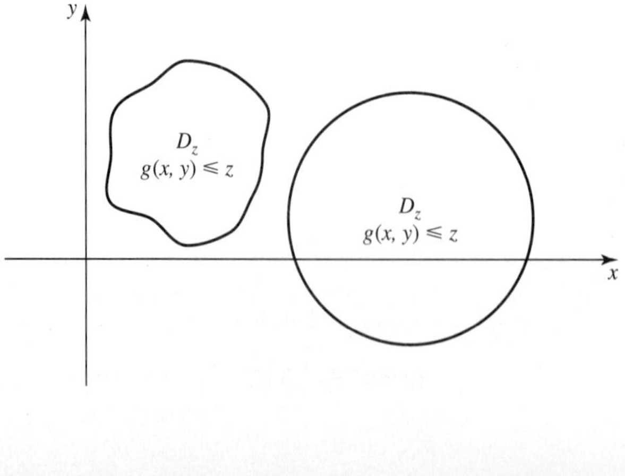

FIGURE 6-7

Note that D_z need not be simply connected. From (6-37), to determine $F_z(z)$ it is enough to find the region D_z for every z, and then evaluate the integral there.

We shall illustrate this method to determine the statistics of various functions of $\mathbf{x}$ and $\mathbf{y}$.

EXAMPLE 6-6

$\mathbf{z} = \mathbf{x} + \mathbf{y}$

▶ Let $\mathbf{z} = \mathbf{x} + \mathbf{y}$. Determine the p.d.f. $f_z(z)$.

From (6-37),

$$F_z(z) = P\{\mathbf{x} + \mathbf{y} \le z\} = \int_{y=-\infty}^{\infty} \int_{x=-\infty}^{z-y} f_{xy}(x, y)\, dx\, dy \tag{6-38}$$

since the region D_z of the xy plane where $x + y \le z$ is the shaded area in Fig. 6-8 to the left of the line $x + y \le z$. Integrating over the horizontal strip along the x axis first (inner integral) followed by sliding that strip along the y axis from $-\infty$ to $+\infty$ (outer integral) we cover the entire shaded area.

We can find $f_z(z)$ by differentiating $F_z(z)$ directly. In this context it is useful to recall the differentiation rule due to Leibnitz. Suppose

$$F_z(z) = \int_{a(z)}^{b(z)} f(x, z)\, dx \tag{6-39}$$

Then

$$f_z(z) = \frac{dF_z(z)}{dz} = \frac{db(z)}{dz} f(b(z), z) - \frac{da(z)}{dz} f(a(z), z) + \int_{a(z)}^{b(z)} \frac{\partial f(x, z)}{\partial z}\, dx \tag{6-40}$$

Using (6-40) in (6-38) we get

$$\begin{aligned} f_z(z) &= \int_{-\infty}^{\infty} \left(\frac{\partial}{\partial z} \int_{-\infty}^{z-y} f_{xy}(x, y)\, dx \right) dy \\ &= \int_{-\infty}^{\infty} \left(1 \cdot f_{xy}(z - y, y) - 0 + \int_{-\infty}^{z-y} \frac{\partial f_{xy}(x, y)}{\partial z} \right) dy \\ &= \int_{-\infty}^{\infty} f_{xy}(z - y, y)\, dy \end{aligned} \tag{6-41}$$

Alternatively, the integration in (6-38) can be carried out first along the y axis followed by the x axis as in Fig. 6-9 as well (see problem set).

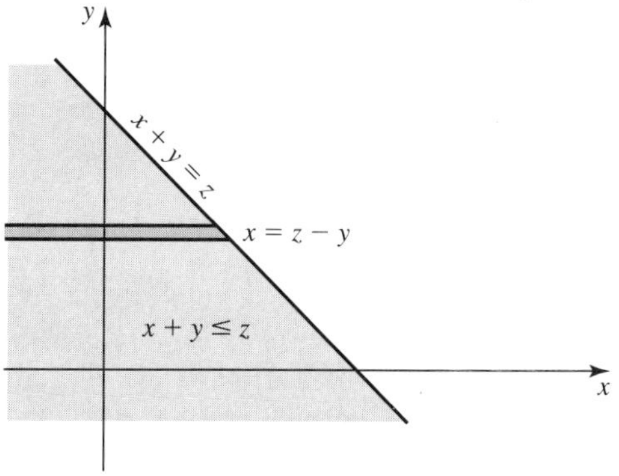

FIGURE 6-8

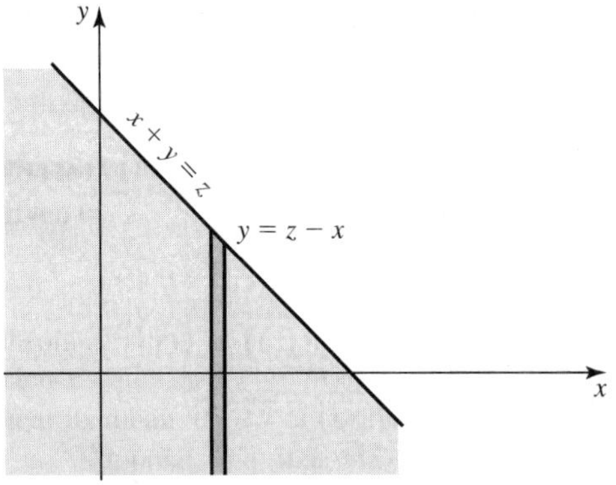

FIGURE 6-9

If **x** and **y** are independent, then

$$f_{xy}(x, y) = f_x(x) f_y(y) \tag{6-42}$$

and inserting (6-42) into (6-41) we get

$$f_z(z) = \int_{x=-\infty}^{\infty} f_x(z-y) f_y(y)\, dy = \int_{x=-\infty}^{\infty} f_x(x) f_y(z-x)\, dx \tag{6-43}$$

This integral is the convolution of the functions $f_x(z)$ and $f_y(z)$ expressed two different ways. We thus reach the following conclusion: If two random variables are *independent,* then the density of their sum equals the convolution of their densities.

As a special case, suppose that $f_x(x) = 0$ for $x < 0$ and $f_y(y) = 0$ for $y < 0$, then we can make use of Fig. 6-10 to determine the new limits for D_z.

In that case

$$F_z(z) = \int_{y=0}^{z} \int_{x=0}^{z-y} f_{xy}(x, y)\, dx\, dy$$

or

$$f_z(z) = \int_{y=0}^{z} \left(\frac{\partial}{\partial z} \int_{x=0}^{z-y} f_{xy}(x, y)\, dx \right) dy$$

$$= \begin{cases} \displaystyle\int_0^z f_{xy}(z-y, y)\, dy & z > 0 \\ 0 & z \le 0 \end{cases} \tag{6-44}$$

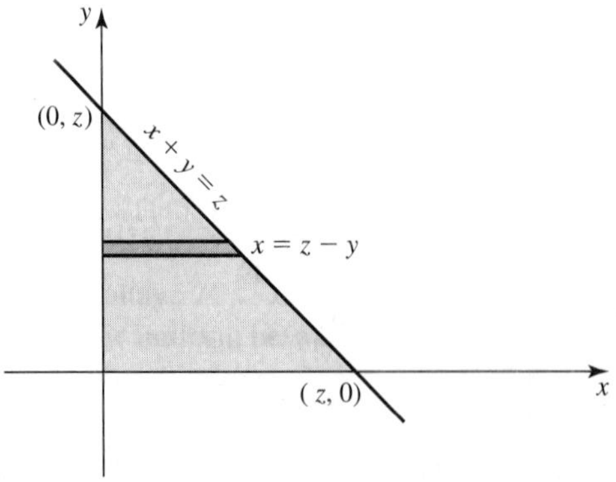

FIGURE 6-10

On the other hand, by considering vertical strips first in Fig. 6-10, we get

$$F_z(z) = \int_{x=0}^{z} \int_{y=0}^{z-x} f_{xy}(x, y)\, dy\, dx$$

or

$$f_z(z) = \int_{x=0}^{z} f_{xy}(x, z-x)\, dx$$

$$= \begin{cases} \int_0^z f_x(x) f_y(z-x)\, dx & z > 0 \\ 0 & z \le 0 \end{cases} \tag{6-45}$$

if **x** and **y** are independent random variables. ◀

EXAMPLE 6-7

▶ Suppose **x** and **y** are independent exponential random variables with common parameter λ. Then

$$f_x(x) = \lambda e^{-\lambda x} U(x) \qquad f_y(y) = \lambda e^{-\lambda y} U(y) \tag{6-46}$$

and we can make use of (6-45) to obtain the p.d.f. of $\mathbf{z} = \mathbf{x} + \mathbf{y}$.

$$f_z(z) = \int_0^z \lambda^2 e^{-\lambda x} e^{-\lambda(z-x)}\, dx = \lambda^2 e^{-\lambda z} \int_0^z dx$$

$$= z\lambda^2 e^{-\lambda z} U(z) \tag{6-47}$$

As Example 6-8 shows, care should be taken while using the convolution formula for random variables with finite range. ◀

EXAMPLE 6-8

▶ **x** and **y** are independent uniform random variables in the common interval (0, 1). Determine $f_z(z)$, where $\mathbf{z} = \mathbf{x} + \mathbf{y}$. Clearly,

$$\mathbf{z} = \mathbf{x} + \mathbf{y} \Rightarrow 0 < z < 2$$

and as Fig. 6-11 shows there are two cases for which the shaded areas are quite different in shape, and they should be considered separately.

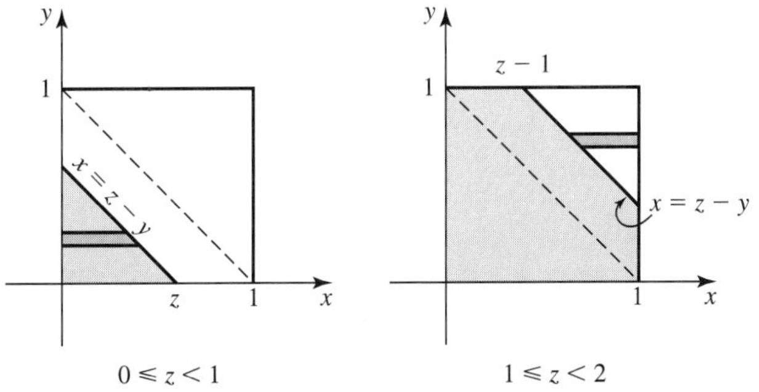

FIGURE 6-11

For $0 \le z < 1$,

$$F_z(z) = \int_{y=0}^{z} \int_{x=0}^{z-y} 1\, dx\, dy = \int_{y=0}^{z} (z-y)\, dy = \frac{z^2}{2} \qquad 0 < z < 1 \qquad (6\text{-}48)$$

For $1 \le z < 2$, notice that it is easy to deal with the unshaded region. In that case,

$$F_z(z) = 1 - P\{\mathbf{z} > z\} = 1 - \int_{y=z-1}^{1} \int_{x=z-y}^{1} 1\, dx\, dy$$

$$= 1 - \int_{y=z-1}^{1} (1 - z + y)\, dy = 1 - \frac{(2-z)^2}{2} \qquad 1 \le z < 2 \qquad (6\text{-}49)$$

Thus

$$f_z(z) = \frac{dF_z(z)}{dz} = \begin{cases} z & 0 \le z < 1 \\ 2 - z & 1 \le z < 2 \end{cases} \qquad (6\text{-}50)$$

By direct convolution of $f_x(x)$ and $f_y(y)$, we obtain the same result as above. In fact, for $0 \le z < 1$ (Fig. 6-12*a*)

$$f_z(z) = \int f_x(z-x) f_y(x)\, dx = \int_0^z 1\, dx = z \qquad (6\text{-}51)$$

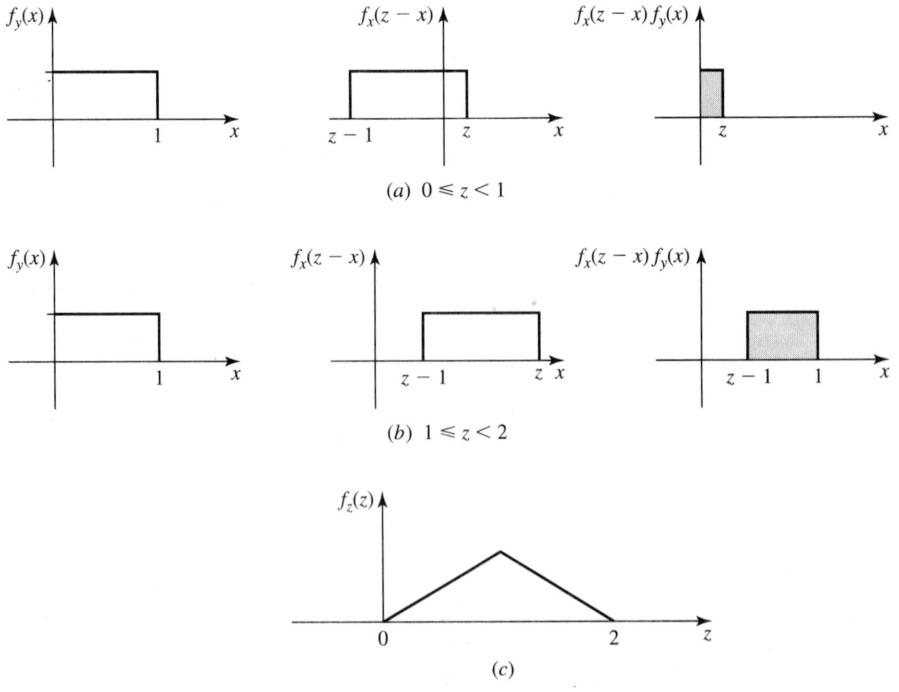

FIGURE 6-12

and for $1 \le z < 2$ (Fig. 6-12*b*)

$$f_z(z) = \int_{z-1}^{1} 1\, dx = 2 - z \tag{6-52}$$

Fig. 6-12*c* shows $f_z(z)$, which agrees with the convolution of two rectangular waveforms as well. ◀

EXAMPLE 6-9

$\mathbf{z} = \mathbf{x} - \mathbf{y}$

▶ Let $\mathbf{z} = \mathbf{x} - \mathbf{y}$. Determine $f_z(z)$.

From (6-37) and Fig. 6-13

$$F_z(z) = P\{\mathbf{x} - \mathbf{y} \le z\} = \int_{y=-\infty}^{\infty} \int_{x=-\infty}^{z+y} f_{xy}(x, y)\, dx\, dy$$

and hence

$$f_z(z) = \frac{dF_z(z)}{dz} = \int_{-\infty}^{\infty} f_{xy}(z + y, y)\, dy \tag{6-53}$$

If $\mathbf{x}$ and $\mathbf{y}$ are independent, then this formula reduces to

$$f_z(z) = \int_{-\infty}^{\infty} f_x(z + y) f_y(y)\, dy = f_x(-z) \otimes f_y(y) \tag{6-54}$$

which represents the convolution of $f_x(-z)$ with $f_y(z)$.

As a special case, suppose

$$f_x(x) = 0 \qquad x < 0, \qquad f_y(y) = 0 \qquad y < 0$$

In this case, $\mathbf{z}$ can be negative as well as positive, and that gives rise to two situations that should be analyzed separately, since the regions of integration for $z \ge 0$ and $z < 0$ are quite different.

For $z \ge 0$, from Fig. 6-14*a*

$$F_z(z) = \int_{y=0}^{\infty} \int_{x=0}^{z+y} f_{xy}(x, y)\, dx\, dy$$

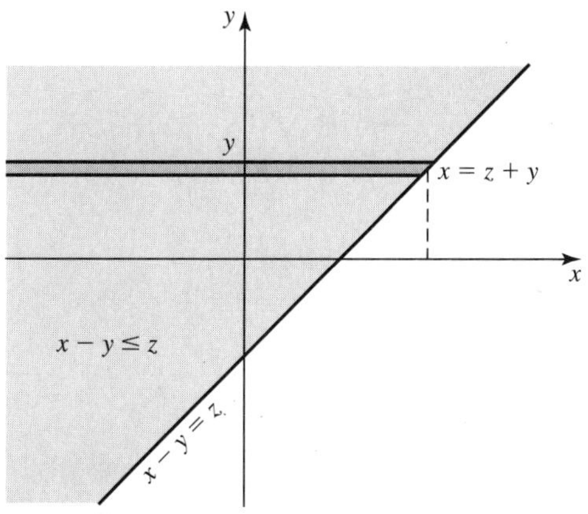

FIGURE 6-13

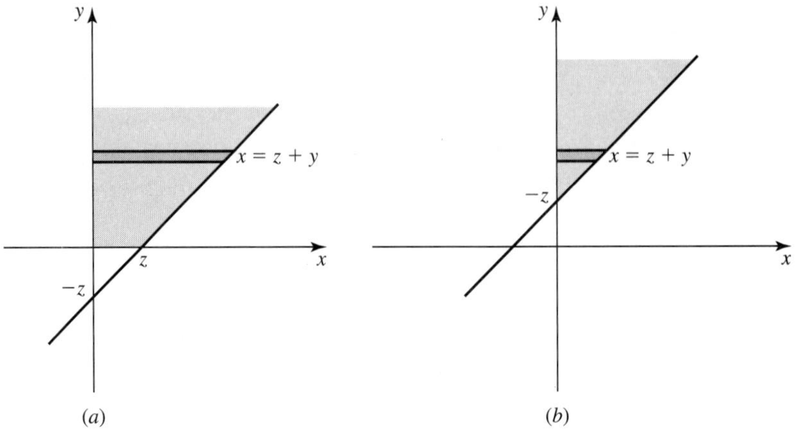

FIGURE 6-14

and for $z < 0$, from Fig. 6-14*b*

$$F_z(z) = \int_{y=-z}^{\infty} \int_{x=0}^{z+y} f_{xy}(x, y)\, dx\, dy$$

After differentiation, this gives

$$f_z(z) = \begin{cases} \displaystyle\int_0^{\infty} f_{xy}(z + y, y)\, dy & z \geq 0 \\ \displaystyle\int_{-z}^{\infty} f_{xy}(z + y, y)\, dy & z < 0 \end{cases} \tag{6-55}$$

◀

EXAMPLE 6-10

$\mathbf{z} = \mathbf{x}/\mathbf{y}$

▶ Let $\mathbf{z} = \mathbf{x}/\mathbf{y}$. Determine $f_z(z)$.

We have

$$F_z(z) = P\{\mathbf{x}/\mathbf{y} \leq z\} \tag{6-56}$$

The inequality $x/y \leq z$ can be rewritten as $\mathbf{x} \leq \mathbf{y}z$ if $\mathbf{y} > 0$, and $\mathbf{x} \geq \mathbf{y}z$ if $\mathbf{y} < 0$. Hence the event $\{\mathbf{x}/\mathbf{y} \leq z\}$ in (6-56) needs to be conditioned by the event $A = \{\mathbf{y} > 0\}$ and its compliment $\overline{A}$. Since $A \cup \overline{A} = S$, by the partition theorem, we have

$$\begin{aligned} P\{\mathbf{x}/\mathbf{y} \leq z\} &= P\{\mathbf{x}/\mathbf{y} \leq z \cap (A \cup \overline{A})\} \\ &= P\{\mathbf{x}/\mathbf{y} \leq z, \mathbf{y} > 0\} + P\{\mathbf{x}/\mathbf{y} \leq z, \mathbf{y} < 0\} \\ &= P\{\mathbf{x} \leq \mathbf{y}z, \mathbf{y} > 0\} + P\{\mathbf{x} \geq \mathbf{y}z, \mathbf{y} < 0\} \end{aligned} \tag{6-57}$$

Fig. 6-15*a* shows the area corresponding to the first term, and Fig. 6-15*b* shows that corresponding to the second term in (6-57).

Integrating over these two regions, we get

$$F_z(z) = \int_{y=0}^{\infty} \int_{x=-\infty}^{yz} f_{xy}(x, y)\, dx\, dy + \int_{y=-\infty}^{0} \int_{x=yz}^{\infty} f_{xy}(x, y)\, dx\, dy \tag{6-58}$$

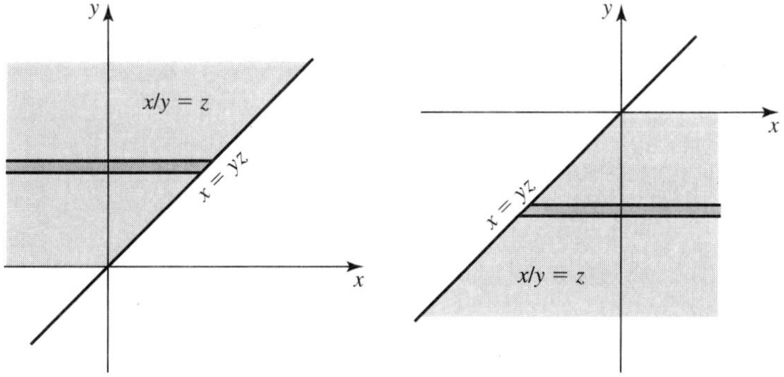

(*a*) $P(\mathbf{x} \le \mathbf{y}z, \mathbf{y} > 0)$ (*b*) $P(\mathbf{x} \ge \mathbf{y}z, \mathbf{y} < 0)$

FIGURE 6-15

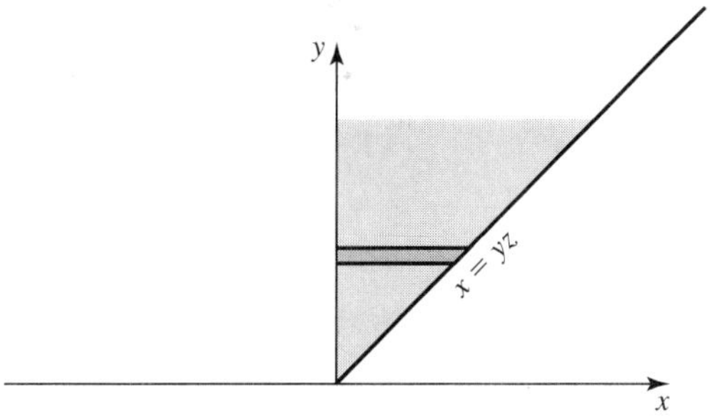

FIGURE 6-16

Differentiation gives

$$f_z(z) = \int_0^\infty y f_{xy}(yz, y)\,dy + \int_{-\infty}^0 -y f_{xy}(yz, y)\,dy$$

$$= \int_{-\infty}^\infty |y| f_{xy}(yz, y)\,dy \qquad (6\text{-}59)$$

Note that if **x** and **y** are non-negative random variables, then the area of integration reduces to that shown in Fig. 6-16.

This gives

$$F_z(z) = \int_{y=0}^\infty \int_{x=0}^{yz} f_{xy}(x, y)\,dx\,dy$$

or

$$f_z(z) = \int_{y=0}^\infty y f_{xy}(yz, y)\,dy \qquad (6\text{-}60)$$

◀

EXAMPLE 6-11 ▶ **x** and **y** are jointly normal random variables with zero mean and

$$f_{xy}(x, y) = \frac{1}{2\pi\sigma_1\sigma_2\sqrt{1-r^2}} e^{-\left[\frac{1}{2(1-r^2)}\left(\frac{x^2}{\sigma_1^2} - \frac{2rxy}{\sigma_1\sigma_2} + \frac{y^2}{\sigma_2^2}\right)\right]} \qquad (6\text{-}61)$$

Show that the ratio $\mathbf{z} = \mathbf{x}/\mathbf{y}$ has a Cauchy density centered at $r\sigma_1/\sigma_2$.

SOLUTION

Inserting (6-61) into (6-59) and using the fact that $f_{xy}(-x, -y) = f_{xy}(x, y)$, we obtain

$$f_z(z) = \frac{2}{2\pi\sigma_1\sigma_2\sqrt{1-r^2}} \int_0^\infty y e^{-y^2/2\sigma_0^2}\, dy = \frac{\sigma_0^2}{\pi\sigma_1\sigma_2\sqrt{1-r^2}}$$

where

$$\sigma_0^2 = \frac{1-r^2}{\left(z^2/\sigma_1^2\right) - (2rz/\sigma_1\sigma_2) + \left(1/\sigma_2^2\right)}$$

Thus

$$f_z(z) = \frac{\sigma_1\sigma_2\sqrt{1-r^2}/\pi}{\sigma_2^2(z - r\sigma_1/\sigma_2)^2 + \sigma_1^2(1-r^2)} \tag{6-62}$$

which represents a Cauchy random variable centered at $r\sigma_1/\sigma_2$. Integrating (6-62) from $-\infty$ to z, we obtain the corresponding distribution function to be

$$F_z(z) = \frac{1}{2} + \frac{1}{\pi} \arctan \frac{\sigma_2 z - r\sigma_1}{\sigma_1\sqrt{1-r^2}} \tag{6-63}$$

◀

As an application, we can use (6-63) to determine the probability masses m_1, m_2, m_3, and m_4 in the four quadrants of the xy plane for (6-61). From the spherical symmetry of (6-61), we have

$$m_1 = m_3 \qquad m_2 = m_4$$

But the second and fourth quadrants represent the region of the plane where $x/y < 0$. The probability that the point (x, y) is in that region equals, therefore, the probability that the random variable $\mathbf{z} = \mathbf{x}/\mathbf{y}$ is negative. Thus

$$m_2 + m_4 = P(\mathbf{z} \le 0) = F_z(0) = \frac{1}{2} - \frac{1}{\pi} \arctan \frac{r}{\sqrt{1-r^2}}$$

and

$$m_1 + m_3 = 1 - (m_2 + m_4) = \frac{1}{2} + \frac{1}{\pi} \arctan \frac{r}{\sqrt{1-r^2}}$$

If we define $\alpha = \text{arc tan}\, r/\sqrt{1-r^2}$, this gives

$$m_1 = m_3 = \frac{1}{4} + \frac{\alpha}{2\pi} \qquad m_2 = m_4 = \frac{1}{4} - \frac{\alpha}{2\pi} \tag{6-64}$$

Of course, we could have obtained this result by direct integration of (6-61) in each quadrant. However, this is simpler.

EXAMPLE 6-12

▶ Let $\mathbf{x}$ and $\mathbf{y}$ be independent gamma random variables with $\mathbf{x} \sim G(m, \alpha)$ and $\mathbf{y} \sim G(n, \alpha)$. Show that $\mathbf{z} = \mathbf{x}/(\mathbf{x} + \mathbf{y})$ has a beta distribution.

Proof.

$$f_{xy}(x, y) = f_x(x) f_y(y)$$

$$= \frac{1}{\alpha^{m+n}\Gamma(m)\Gamma(n)} x^{m-1} y^{n-1} e^{-(x+y)/\alpha} \qquad x > 0 \qquad y > 0 \tag{6-65}$$

Note that $0 < z < 1$, since **x** and **y** are non-negative random variables

$$F_z(z) = P\{\mathbf{z} \le z\} = P\left(\frac{\mathbf{x}}{\mathbf{x}+\mathbf{y}} \le z\right) = P\left(\mathbf{x} \le \mathbf{y}\frac{z}{1-z}\right)$$

$$= \int_0^\infty \int_0^{yz/(1-z)} f_{xy}(x, y)\,dx\,dy$$

where we have made use of Fig. 6-16. Differentiation with respect to z gives

$$\begin{aligned} f_z(z) &= \int_0^\infty \frac{y}{(1-z)^2} f_{xy}(yz/(1-z), y)\,dy \\ &= \int_0^\infty \frac{y}{(1-z)^2} \frac{1}{\alpha^{m+n}\Gamma(m)\Gamma(n)} \left(\frac{yz}{1-z}\right)^{m-1} y^{n-1} e^{-y/(1-z)\alpha}\,dy \\ &= \frac{1}{\alpha^{m+n}\Gamma(m)\Gamma(n)} \frac{z^{m-1}}{(1-z)^{m+1}} \int_0^\infty y^{m+n-1} e^{-y/\alpha(1-z)}\,dy \\ &= \frac{z^{m-1}(1-z)^{n-1}}{\Gamma(m)\Gamma(n)} \int_0^\infty u^{m+n-1} e^{-u}\,du = \frac{\Gamma(m+n)}{\Gamma(m)\Gamma(n)} z^{m-1}(1-z)^{n-1} \\ &= \begin{cases} \dfrac{1}{\beta(m,n)} z^{m-1}(1-z)^{n-1} & 0 < z < 1 \\ 0 & \text{otherwise} \end{cases} \end{aligned} \tag{6-66}$$

which represents a beta distribution. ◀

EXAMPLE 6-13

$\mathbf{z} = \mathbf{x}^2 + \mathbf{y}^2$

▶ Let $\mathbf{z} = \mathbf{x}^2 + \mathbf{y}^2$. Determine $f_z(z)$.

We have

$$F_z(z) = P\{\mathbf{x}^2 + \mathbf{y}^2 \le z\} = \iint_{x^2+y^2 \le z} f_{xy}(x, y)\,dx\,dy$$

But, $\mathbf{x}^2 + \mathbf{y}^2 \le z$ represents the area of a circle with radius $\sqrt{z}$, and hence (see Fig. 6-17)

$$F_z(z) = \int_{y=-\sqrt{z}}^{\sqrt{z}} \int_{x=-\sqrt{z-y^2}}^{\sqrt{z-y^2}} f_{xy}(x, y)\,dx\,dy$$

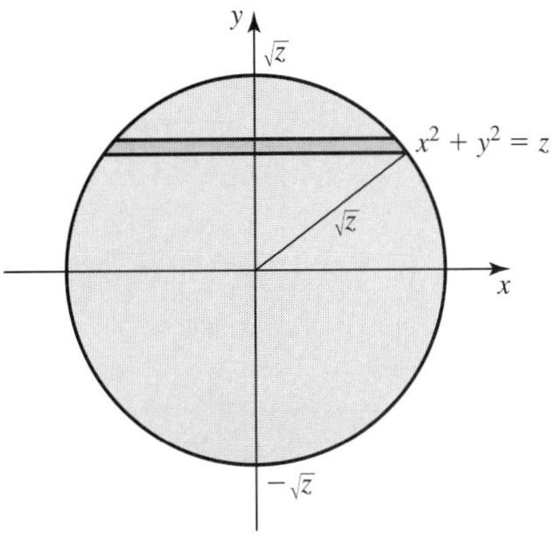

FIGURE 6-17

This gives

$$f_z(z) = \int_{-\sqrt{z}}^{\sqrt{z}} \frac{1}{2\sqrt{z-y^2}}\{f_{xy}(\sqrt{z-y^2}, y) + f_{xy}(-\sqrt{z-y^2}, y)\}\,dy \tag{6-67}$$

◀

As an illustration, consider Example 6-14.

EXAMPLE 6-14

▶ **x** and **y** are independent normal random variables with zero mean and common variance σ^2. Determine $f_z(z)$ for $\mathbf{z} = \mathbf{x}^2 + \mathbf{y}^2$.

SOLUTION

Using (6-67), we get

$$\begin{aligned} f_z(z) &= \int_{-\sqrt{z}}^{\sqrt{z}} \frac{1}{2\sqrt{z-y^2}}\left(2 \cdot \frac{1}{2\pi\sigma^2} e^{(z-y^2+y^2)/2\sigma^2}\right) dy \\ &= \frac{e^{-z/2\sigma^2}}{\pi\sigma^2} \int_0^{\sqrt{z}} \frac{1}{\sqrt{z-y^2}}\,dy = \frac{e^{-z/2\sigma^2}}{\pi\sigma^2} \int_0^{\pi/2} \frac{\sqrt{z}\cos\theta}{\sqrt{z}\cos\theta}\,d\theta \\ &= \frac{1}{2\sigma^2} e^{-z/2\sigma^2} U(z) \end{aligned} \tag{6-68}$$

where we have used the substitution $y = \sqrt{z}\sin\theta$. From (6-68), we have the following: If **x** and **y** are independent zero mean Gaussian random variables with common variance σ^2, then $\mathbf{x}^2 + \mathbf{y}^2$ is an exponential random variable with parameter $2\sigma^2$. ◀

EXAMPLE 6-15

$\mathbf{z} = \sqrt{\mathbf{x}^2 + \mathbf{y}^2}$

▶ Let $\mathbf{z} = \sqrt{\mathbf{x}^2 + \mathbf{y}^2}$. Find $f_z(z)$.

SOLUTION

From Fig. 6-17, the present case corresponds to a circle with radius z^2. Thus

$$F_z(z) = \int_{y=-z}^{z} \int_{x=-\sqrt{z^2-y^2}}^{\sqrt{z^2-y^2}} f_{xy}(x, y)\,dx\,dy$$

and by differentiation,

$$f_z(z) = \int_{-z}^{z} \frac{z}{\sqrt{z^2-y^2}}\{f_{xy}(\sqrt{z^2-y^2}, y) + f_{xy}(-\sqrt{z^2-y^2}, y)\}\,dy \tag{6-69}$$

In particular, if **x** and **y** are zero mean independent Gaussian random variables as in the previous example, then

$$\begin{aligned} f_z(z) &= 2\int_0^{z} \frac{z}{\sqrt{z^2-y^2}} \frac{2}{2\pi\sigma^2} e^{-(z^2-y^2+y^2)/2\sigma^2}\,dy \\ &= \frac{2z}{\pi\sigma^2} e^{-z^2/2\sigma^2} \int_0^{z} \frac{1}{\sqrt{z^2-y^2}}\,dy = \frac{2z}{\pi\sigma^2} e^{-z^2/2\sigma^2} \int_0^{\pi/2} \frac{z\cos\theta}{z\cos\theta}\,d\theta \\ &= \frac{z}{\sigma^2} e^{-z^2/2\sigma^2} U(z) \end{aligned} \tag{6-70}$$

which represents a Rayleigh distribution. Thus, if $\mathbf{w} = \mathbf{x} + i\mathbf{y}$, where **x** and **y** are real independent normal random variables with zero mean and equal variance, then the random variable $|\mathbf{w}| = \sqrt{\mathbf{x}^2 + \mathbf{y}^2}$ has a Rayleigh density. $\mathbf{w}$ is said to be a complex Gaussian random variable with zero mean, if its real and imaginary parts are independent. So far we have seen that the magnitude of a complex Gaussian random variable has Rayleigh distribution. What about its phase

$$\boldsymbol{\theta} = \tan^{-1}\left(\frac{\mathbf{y}}{\mathbf{x}}\right) \tag{6-71}$$

Clearly, the principal value of $\boldsymbol{\theta}$ lies in the interval $(-\pi/2, \pi/2)$. If we let $\mathbf{u} = \tan\boldsymbol{\theta} = \mathbf{y}/\mathbf{x}$, then from Example 6-11, $\mathbf{u}$ has a Cauchy distribution (see (6-62) with $\sigma_1 = \sigma_2$, $r = 0$)

$$f_u(u) = \frac{1/\pi}{u^2+1} \qquad -\infty < u < \infty$$

As a result, the principal value of $\boldsymbol{\theta}$ has the density function

$$f_\theta(\theta) = \frac{1}{|d\theta/du|} f_u(\tan\theta) = \frac{1}{(1/\sec^2\theta)} \frac{1/\pi}{\tan^2\theta + 1}$$

$$= \begin{cases} 1/\pi & -\pi/2 < \theta < \pi/2 \\ 0 & \text{otherwise} \end{cases} \tag{6-72}$$

However, in the representation $\mathbf{x} + j\mathbf{y} = \mathbf{r}e^{j\boldsymbol{\theta}}$, the variable $\boldsymbol{\theta}$ lies in the interval $(-\pi, \pi)$, and taking into account this scaling by a factor of two, we obtain

$$f_\theta(\theta) = \begin{cases} 1/2\pi & -\pi < \theta < \pi \\ 0 & \text{otherwise} \end{cases} \tag{6-73}$$

To summarize, the magnitude and phase of a zero mean complex Gaussian random variable have Rayleigh and uniform distributions respectively. Interestingly, as we will show later (Example 6-22), these two derived random variables are also statistically *independent* of each other! ◀

Let us reconsider Example 6-15 where $\mathbf{x}$ and $\mathbf{y}$ are independent Gaussian random variables with nonzero means μ_x and μ_y respectively. Then $\mathbf{z} = \sqrt{\mathbf{x}^2 + \mathbf{y}^2}$ is said to be a Rician random variable. Such a scene arises in fading multipath situations where there is a dominant constant component (mean) in addition to a zero mean Gaussian random variable. The constant component may be the line of sight signal and the zero mean Gaussian random variable part could be due to random multipath components adding up incoherently. The envelope of such a signal is said to be Rician instead of Rayleigh.

EXAMPLE 6-16

▶ Redo Example 6-15, where $\mathbf{x}$ and $\mathbf{y}$ are independent Gaussian random variables with nonzero means μ_x and μ_y respectively.

SOLUTION

Since

$$f_{xy}(x, y) = \frac{1}{2\pi\sigma^2} e^{-[(x-\mu_x)^2 + (y-\mu_y)^2]/2\sigma^2}$$

substituting this into (6-69) and letting $y = z\sin\theta$, $\mu = \sqrt{\mu_x^2 + \mu_y^2}$, $\mu_x = \mu\cos\phi$,

$\mu_y = \mu \sin\phi$, we get the Rician distribution to be

$$
\begin{aligned}
f_z(z) &= \frac{ze^{-(z^2+\mu^2)/2\sigma^2}}{2\pi\sigma^2} \int_{-\pi/2}^{\pi/2} \left(e^{z\mu\cos(\theta-\phi)/\sigma^2} + e^{-z\mu\cos(\theta+\phi)/\sigma^2}\right) d\theta \\
&= \frac{ze^{-(z^2+\mu^2)/2\sigma^2}}{2\pi\sigma^2} \left(\int_{-\pi/2}^{\pi/2} e^{z\mu\cos(\theta-\phi)/\sigma^2}\, d\theta + \int_{\pi/2}^{3\pi/2} e^{z\mu\cos(\theta-\phi)/\sigma^2}\, d\theta \right) \\
&= \frac{ze^{-(z^2+\mu^2)/2\sigma^2}}{\sigma^2} I_0\left(\frac{z\mu}{\sigma^2}\right)
\end{aligned}
\tag{6-74}
$$

where

$$
I_0(\eta) \triangleq \frac{1}{2\pi}\int_0^{2\pi} e^{\eta\cos(\theta-\phi)}\, d\theta = \frac{1}{\pi}\int_0^{\pi} e^{\eta\cos\theta}\, d\theta
$$

is the modified Bessel function of the first kind and zeroth order. ◀

Order Statistics

In general, given any n-tuple $\mathbf{x}_1, \mathbf{x}_2, \ldots, \mathbf{x}_n$, we can rearrange them in an increasing order of magnitude such that

$$
\mathbf{x}_{(1)} \le \mathbf{x}_{(2)} \le \cdots \le \mathbf{x}_{(n)}
$$

where $\mathbf{x}_{(1)} = \min(\mathbf{x}_1, \mathbf{x}_2, \ldots, \mathbf{x}_n)$, and $\mathbf{x}_{(2)}$ is the second smallest value among $\mathbf{x}_1, \mathbf{x}_2, \ldots,$ $\mathbf{x}_n$, and finally $\mathbf{x}_{(n)} = \max(\mathbf{x}_1, \mathbf{x}_2, \ldots, \mathbf{x}_n)$. The functions *min* and *max* are nonlinear operators, and represent special cases of the more general order statistics. If $\mathbf{x}_1, \mathbf{x}_2, \ldots, \mathbf{x}_n$ represent random variables, the function $\mathbf{x}_{(k)}$ that takes on the value $x_{(k)}$ in each possible sequence $(x_1, x_2, \ldots, x_n)$ is known as the kth-order statistic. $\{\mathbf{x}_{(1)}, \mathbf{x}_{(2)}, \ldots, \mathbf{x}_{(n)}\}$ represent the set of order statistics among n random variables. In this context

$$
\mathbf{R} = \mathbf{x}_{(n)} - \mathbf{x}_{(1)} \tag{6-75}
$$

represents the range, and when $n = 2$, we have the *max* and *min* statistics.

Order statistics is useful when relative magnitude of observations is of importance. When worst case scenarios have to be accounted for, then the function $max(\cdot)$ is quite useful. For example, let $\mathbf{x}_1, \mathbf{x}_2, \ldots, \mathbf{x}_n$ represent the recorded flood levels over the past n years at some location. If the objective is to construct a dam to prevent any more flooding, then the height H of the proposed dam should satisfy the inequality

$$
H > \max(\mathbf{x}_1, \mathbf{x}_2, \ldots, \mathbf{x}_n) \tag{6-76}
$$

with some finite probability. In that case, the p.d.f. of the random variable on the right side of (6-76) can be used to compute the desired height. In another case, if a bulb manufacturer wants to determine the average time to failure (μ) of its bulbs based on a sample of size n, the sample mean $(\mathbf{x}_1 + \mathbf{x}_2 + \cdots + \mathbf{x}_n)/n$ can be used as an estimate for μ. On the other hand, an estimate based on the least time to failure has other attractive features. This estimate $\min(\mathbf{x}_1, \mathbf{x}_2, \ldots, \mathbf{x}_n)$ may not be as good as the sample mean in terms of their respective variances, but the $min(\cdot)$ can be computed as soon as the first bulb fuses, whereas to compute the sample mean one needs to wait till the last of the lot extinguishes.

EXAMPLE 6-17

$\mathbf{z} = \max(\mathbf{x}, \mathbf{y})$
$\mathbf{w} = \min(\mathbf{x}, \mathbf{y})$

▶ Let $\mathbf{z} = \max(\mathbf{x}, \mathbf{y})$ and $\mathbf{w} = \min(\mathbf{x}, \mathbf{y})$. Determine $f_z(z)$ and $f_w(w)$.

$$\mathbf{z} = \max(\mathbf{x}, \mathbf{y}) = \begin{cases} \mathbf{x} & \mathbf{x} > \mathbf{y} \\ \mathbf{y} & \mathbf{x} \le \mathbf{y} \end{cases} \tag{6-77}$$

we have [see (6-57)]

$$\begin{aligned} F_z(z) &= P\{\max(\mathbf{x}, \mathbf{y}) \le z\} \\ &= P\{(\mathbf{x} \le z, \mathbf{x} > \mathbf{y}) \cup (\mathbf{y} \le z, \mathbf{x} \le \mathbf{y})\} \\ &= P\{\mathbf{x} \le z, \mathbf{x} > \mathbf{y}\} + P\{\mathbf{y} \le z, \mathbf{x} \le \mathbf{y}\} \end{aligned}$$

since $\{\mathbf{x} > \mathbf{y}\}$ and $\{\mathbf{x} \le \mathbf{y}\}$ are mutually exclusive sets that form a partition. Figure 6-18a and 6-18b show the regions satisfying the corresponding inequalities in each term seen here.

Figure 6-18c represents the total region, and from there

$$F_z(z) = P\{\mathbf{x} \le z, \mathbf{y} \le z\} = F_{xy}(z, z) \tag{6-78}$$

If $\mathbf{x}$ and $\mathbf{y}$ are independent, then

$$F_z(z) = F_x(z)F_y(z)$$

(a) $P\{\mathbf{x} \le z, \mathbf{x} > \mathbf{y}\}$

(b) $P\{\mathbf{y} \le z, \mathbf{x} \le \mathbf{y}\}$

(c)

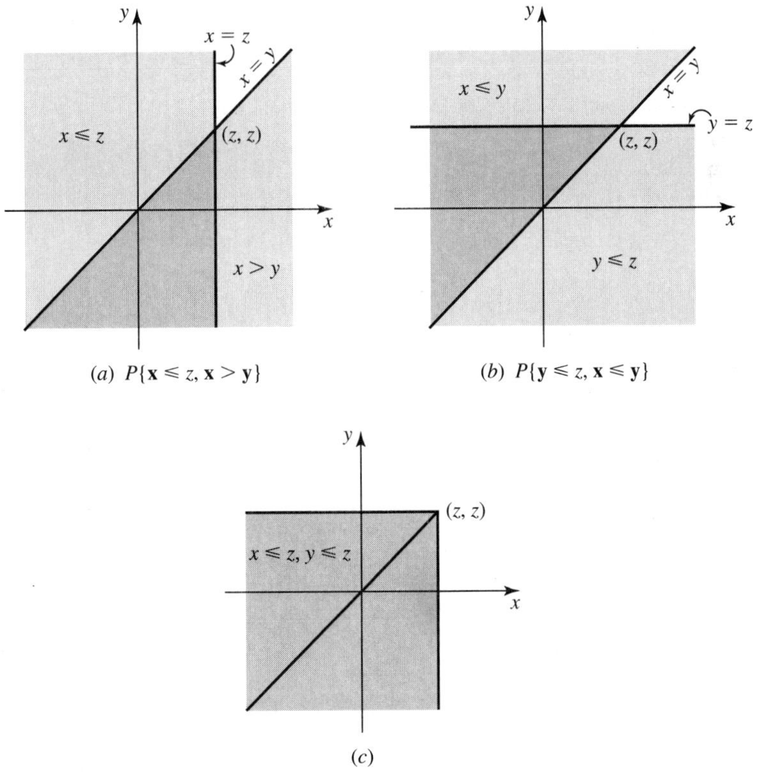

FIGURE 6-18

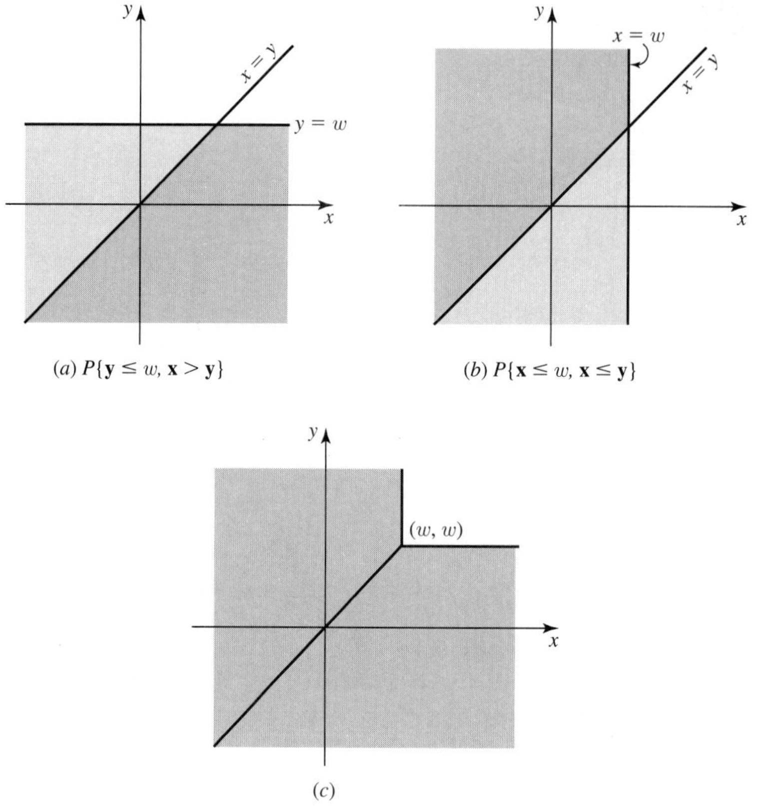

FIGURE 6-19

and hence

$$f_z(z) = F_x(z) f_y(z) + f_x(z) F_y(z) \tag{6-79}$$

Similarly,

$$\mathbf{w} = \min(\mathbf{x}, \mathbf{y}) = \begin{cases} \mathbf{y} & \mathbf{x} > \mathbf{y} \\ \mathbf{x} & \mathbf{x} \le \mathbf{y} \end{cases} \tag{6-80}$$

Thus,

$$\begin{aligned} F_w(w) &= P\{\min(\mathbf{x}, \mathbf{y}) \le w\} \\ &= P\{\mathbf{y} \le w, \mathbf{x} > \mathbf{y}\} + P\{\mathbf{x} \le w, \mathbf{x} \le \mathbf{y}\} \end{aligned}$$

Once again, the shaded areas in Fig. 6-19*a* and 6-19*b* show the regions satisfying these inequalities, and Fig. 6-19*c* shows them together.

From Fig. 6-19*c*,

$$\begin{aligned} F_w(w) &= 1 - P\{\mathbf{w} > w\} = 1 - P\{\mathbf{x} > w, \mathbf{y} > w\} \\ &= F_x(w) + F_y(w) - F_{xy}(w, w) \end{aligned} \tag{6-81}$$

where we have made use of (6-4) with $x_2 = y_2 = \infty$, and $x_1 = y_1 = w$. ◀

EXAMPLE 6-18

▶ Let **x** and **y** be independent exponential random variables with common parameter λ. Define $\mathbf{w} = \min(\mathbf{x}, \mathbf{y})$. Find $f_w(w)$.

SOLUTION

From (6-81)

$$F_w(w) = F_x(w) + F_y(w) - F_x(w)F_y(w)$$

and hence

$$f_w(w) = f_x(w) + f_y(w) - f_x(w)F_y(w) - F_x(w)f_y(w)$$

But $f_x(w) = f_y(w) = \lambda e^{-\lambda w}$, and $F_x(w) = F_y(w) = 1 - e^{-\lambda w}$, so that

$$f_w(w) = 2\lambda e^{\lambda w} - 2(1 - e^{-\lambda w})\lambda e^{-\lambda w} = 2\lambda e^{-2\lambda w}U(w) \tag{6-82}$$

Thus $\min(\mathbf{x}, \mathbf{y})$ is also exponential with parameter 2λ. ◀

EXAMPLE 6-19

▶ Suppose **x** and **y** are as given in Example 6-18. Define

$$z = \frac{\min(\mathbf{x}, \mathbf{y})}{\max(\mathbf{x}, \mathbf{y})}$$

Although $\min(\cdot)/\max(\cdot)$ represents a complicated function, by partitioning the whole space as before, it is possible to simplify this function. In fact

$$\mathbf{z} = \begin{cases} \mathbf{x}/\mathbf{y} & \mathbf{x} \le \mathbf{y} \\ \mathbf{y}/\mathbf{x} & \mathbf{x} > \mathbf{y} \end{cases} \tag{6-83}$$

As before, this gives

$$\begin{aligned} F_z(z) &= P\{\mathbf{x}/\mathbf{y} \le z, \mathbf{x} \le \mathbf{y}\} + P\{\mathbf{y}/\mathbf{x} \le z, \mathbf{x} > \mathbf{y}\} \\ &= P\{\mathbf{x} \le \mathbf{y}z, \mathbf{x} \le \mathbf{y}\} + P\{\mathbf{y} \le \mathbf{x}z, \mathbf{x} > \mathbf{y}\} \end{aligned}$$

Since **x** and **y** are both positive random variables in this case, we have $0 < z < 1$. The shaded regions in Fig. 6-20*a* and 6-20*b* represent the two terms in this sum.

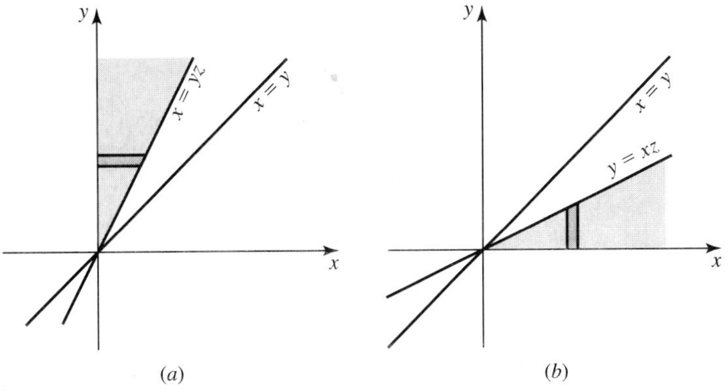

FIGURE 6-20

From Fig. 6-20,

$$F_z(z) = \int_0^{\infty}\int_{x=0}^{yz} f_{xy}(x, y)\,dx\,dy + \int_0^{\infty}\int_{y=0}^{xz} f_{xy}(x, y)\,dy\,dx$$

Hence

$$\begin{aligned} f_z(z) &= \int_0^{\infty} y f_{xy}(yz, y)\,dy + \int_0^{\infty} x f_{xy}(x, xz)\,dx \\ &= \int_0^{\infty} y\big(f_{xy}(yz, y) + f_{xy}(y, yz)\big)\,dy \\ &= \int_0^{\infty} y\lambda^2\big(e^{-\lambda(yz+y)} + e^{-\lambda(y+yz)}\big)\,dy \\ &= 2\lambda^2 \int_0^{\infty} y e^{-\lambda(1+z)y}\,dy = \frac{2}{(1+z)^2}\int_0^{\infty} u e^{-u}\,du \\ &= \begin{cases} \dfrac{2}{(1+z)^2} & 0 < z < 1 \\ 0 & \text{otherwise} \end{cases} \end{aligned} \tag{6-84}$$

◀

EXAMPLE 6-20

DISCRETE CASE

▶ Let **x** and **y** be independent Poisson random variables with parameters λ_1 and λ_2, respectively. Let $\mathbf{z} = \mathbf{x} + \mathbf{y}$. Determine the p.m.f. of **z**.

Since **x** and **y** both take values $\{0, 1, 2, \ldots\}$, the same is true for **z**. For any $n = 0, 1, 2, \ldots$, $\{\mathbf{x}+\mathbf{y} = n\}$ gives only a finite number of options for **x** and **y**. In fact, if $\mathbf{x} = 0$, then **y** must be n; if $\mathbf{x} = 1$, then **y** must be $n-1$, and so on. Thus the event $\{\mathbf{x}+\mathbf{y} = n\}$ is the union of mutually exclusive events $A_k = \{\mathbf{x} = k, \mathbf{y} = n-k\}$, $k = 0 \to n$.

$$\begin{aligned} P\{\mathbf{z} = n\} &= P\{\mathbf{x}+\mathbf{y} = n\} = P\left(\bigcup_{k=0}^{n}\{\mathbf{x} = k, \mathbf{y} = n-k\}\right) \\ &= \sum_{k=0}^{n} P\{\mathbf{x} = k, \mathbf{y} = n-k\} \end{aligned} \tag{6-85}$$

If **x** and **y** are also independent, then

$$P\{\mathbf{x} = k, \mathbf{y} = n-k\} = P\{\mathbf{x} = k\}P\{\mathbf{y} = n-k\}$$

and hence

$$\begin{aligned} P\{\mathbf{z} = n\} &= \sum_{k=0}^{n} P\{\mathbf{x} = k\}P\{\mathbf{y} = n-k\} \\ &= \sum_{k=0}^{n} e^{-\lambda_1}\frac{\lambda_1^k}{k!}\,e^{-\lambda_2}\frac{\lambda_2^{n-k}}{(n-k)!} = \frac{e^{-(\lambda_1+\lambda_2)}}{n!}\sum_{k=0}^{n}\frac{n!}{k!(n-k)!}\lambda_1^k\lambda_2^{n-k} \\ &= e^{-(\lambda_1+\lambda_2)}\frac{(\lambda_1+\lambda_2)^n}{n!}, \qquad n = 0, 1, 2, \ldots, \infty \end{aligned} \tag{6-86}$$

Thus **z** represents a Poisson random variable with parameter $\lambda_1 + \lambda_2$, indicating that sum of independent Poisson random variables is a Poisson random variable whose parameter is the sum of the parameters of the original random variables.

As Example 6-20 indicates, this procedure is too tedious in the discrete case. As we shall see in Sec. 6-5, the joint characteristic function or the moment generating function can be used to solve problems of this type in a much easier manner. ◀

6-3 TWO FUNCTIONS OF TWO RANDOM VARIABLES

In the spirit of the previous section, let us look at an immediate generalization. Suppose $\mathbf{x}$ and $\mathbf{y}$ are two random variables with joint p.d.f. $f_{xy}(x, y)$. Given two functions $g(x, y)$ and $h(x, y)$, define two new random variables

$$\mathbf{z} = g(\mathbf{x}, \mathbf{y}) \tag{6-87}$$

$$\mathbf{w} = h(\mathbf{x}, \mathbf{y}) \tag{6-88}$$

How does one determine their joint p.d.f. $f_{zw}(z, w)$? Obviously with $f_{zw}(z, w)$ in hand, the marginal p.d.f.s $f_z(z)$ and $f_w(w)$ can be easily determined.

The procedure for determining $f_{zw}(z, w)$ is the same as that in (6-36). In fact for given numbers z and w,

$$\begin{aligned} F_{zw}(z, w) &= P\{\mathbf{z}(\xi) \le z, \mathbf{w}(\xi) \le w\} = P\{g(\mathbf{x}, \mathbf{y}) \le z, h(\mathbf{x}, \mathbf{y}) \le w\} \\ &= P\{(\mathbf{x}, \mathbf{y}) \in D_{z,w}\} = \iint_{(x,y)\in D_{z,w}} f_{xy}(x, y)\, dx\, dy \end{aligned} \tag{6-89}$$

where $D_{z,w}$ is the region in the xy plane such that the inequalities $g(x, y) \le z$ and $h(x, y) \le w$ are *simultaneously* satisfied in Fig. 6-21.

We illustrate this technique in Example 6-21.

EXAMPLE 6-21

▶ Suppose $\mathbf{x}$ and $\mathbf{y}$ are independent uniformly distributed random variables in the interval $(0, \theta)$. Define $\mathbf{z} = \min(\mathbf{x}, \mathbf{y})$, $\mathbf{w} = \max(\mathbf{x}, \mathbf{y})$. Determine $f_{zw}(z, w)$.

SOLUTION

Obviously both z and w vary in the interval $(0, \theta)$. Thus

$$F_{zw}(z, w) = 0 \qquad \text{if} \quad z < 0 \quad \text{or} \quad w < 0 \tag{6-90}$$

$$F_{zw}(z, w) = P\{\mathbf{z} \le z, \mathbf{w} \le w\} = P\{\min(\mathbf{x}, \mathbf{y}) \le z, \max(\mathbf{x}, \mathbf{y}) \le w\} \tag{6-91}$$

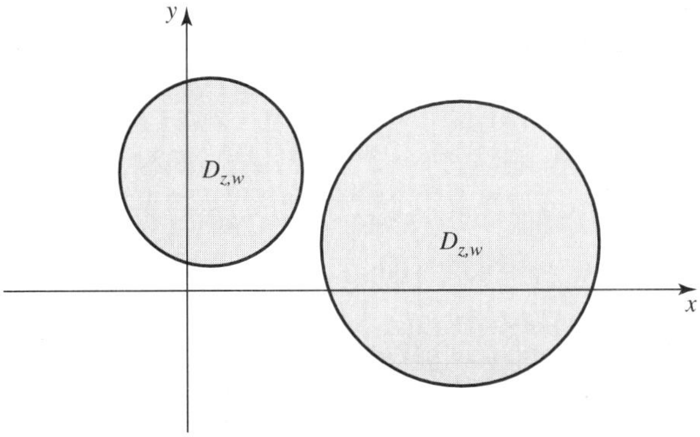

FIGURE 6-21

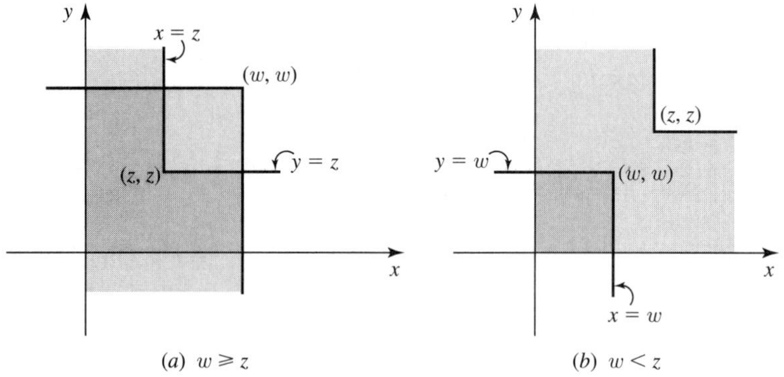

FIGURE 6-22

We must consider two cases: $w \geq z$ and $w < z$, since they give rise to different regions for $D_{z,w}$ (see Fig. 6-22*a* and 6-22*b*).

For $w \geq z$, from Fig. 6.22*a*, the region $D_{z,w}$ is represented by the doubly shaded area (see also Fig. 6-18*c* and Fig. 6-19*c*). Thus

$$F_{zw}(z, w) = F_{xy}(z, w) + F_{xy}(w, z) - F_{xy}(z, z) \qquad w \geq z \tag{6-92}$$

and for $w < z$, from Fig. 6.22*b*, we obtain

$$F_{zw}(z, w) = F_{xy}(w, w) \qquad w < z \tag{6-93}$$

with

$$F_{xy}(x, y) = F_x(x)F_y(y) = \frac{x}{\theta} \cdot \frac{y}{\theta} = \frac{xy}{\theta^2} \tag{6-94}$$

we obtain

$$F_{zw}(z, w) = \begin{cases} (2wz - z^2)/\theta^2 & 0 < z < w < \theta \\ w^2/\theta^2 & 0 < w < z < \theta \end{cases} \tag{6-95}$$

Thus

$$f_{zw}(z, w) = \begin{cases} 2/\theta^2 & 0 < z < w < \theta \\ 0 & \text{otherwise} \end{cases} \tag{6-96}$$

From (6-96), we also obtain

$$f_z(z) = \int_z^\theta f_{zw}(z, w)\, dw = \frac{2}{\theta}\left(1 - \frac{z}{\theta}\right) \qquad 0 < z < \theta \tag{6-97}$$

and

$$f_w(w) = \int_0^w f_{zw}(z, w)\, dz = \frac{2w}{\theta^2} \qquad 0 < w < \theta \tag{6-98}$$

◀

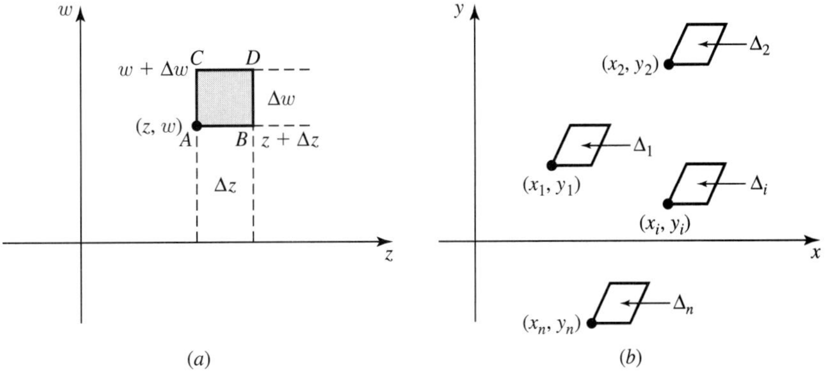

FIGURE 6-23

Joint Density

If $g(x, y)$ and $h(x, y)$ are continuous and differentiable functions, then, as in the case of one random variable [see (5-16)], it is possible to develop a formula to obtain the joint p.d.f. $f_{zw}(z, w)$ directly. Toward this, consider the equations

$$g(x, y) = z \qquad h(x, y) = w \tag{6-99}$$

For a given point (z, w), equation (6-99) can have many solutions. Let us say (x_1, y_1), $(x_2, y_2), \ldots, (x_n, y_n)$ represent these multiple solutions such that (see Fig. 6-23)

$$g(x_i, y_i) = z \qquad h(x_i, y_i) = w \tag{6-100}$$

Consider the problem of evaluating the probability

$$\begin{aligned} P\{z < \mathbf{z} \le z + \Delta z, w < \mathbf{w} \le w + \Delta w\} \\ = P\{z < g(\mathbf{x}, \mathbf{y}) \le z + \Delta z, w < h(\mathbf{x}, \mathbf{y}) \le w + \Delta w\} \end{aligned} \tag{6-101}$$

Using (6-8) we can rewrite (6-101) as

$$P\{z < \mathbf{z} \le z + \Delta z, w < \mathbf{w} \le w + \Delta w\} = f_{zw}(z, w)\, \Delta z\, \Delta w \tag{6-102}$$

But to translate this probability in terms of $f_{xy}(x, y)$, we need to evaluate the equivalent region for $\Delta z\, \Delta w$ in the xy plane. Toward this, referring to Fig. 6-24, we observe that the point A with coordinates (z, w) gets mapped onto the point A' with coordinates (x_i, y_i) (as well as to other points as in Fig 6.23b). As z changes to $z + \Delta z$ to point B in Fig. 6.24a, let B' represent its image in the xy plane. Similarly, as w changes to $w + \Delta w$ to C, let C' represent its image in the xy plane.

Finally D goes to D', and $A'B'C'D'$ represents the equivalent parallelogram in the xy plane with area Δ_i. Referring to Fig. 6-23, because of the nonoverlapping nature of these regions the probability in (6-102) can be alternatively expressed as

$$\sum_i P\{(\mathbf{x}, \mathbf{y}) \in \Delta_i\} = \sum_i f_{xy}(x_i, y_i)\, \Delta_i \tag{6-103}$$

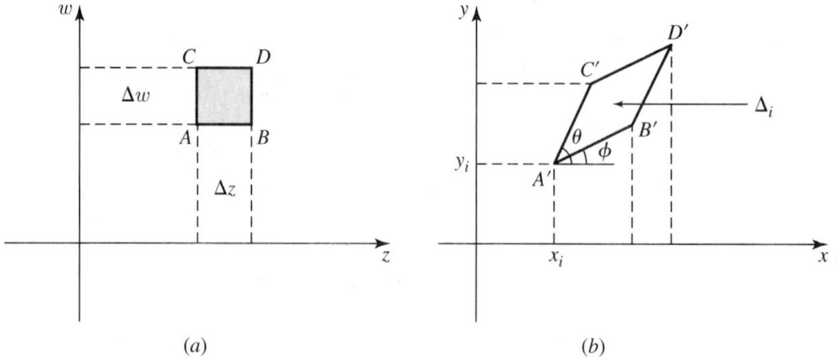

FIGURE 6-24

Equating (6-102) and (6-103) we obtain

$$f_{zw}(z, w) = \sum_i f_{xy}(x_i, y_i) \frac{\Delta_i}{\Delta z\, \Delta w} \tag{6-104}$$

To simplify (6-104), we need to evaluate the area Δ_i of the parallelograms in Fig. 6.24*b* in terms of $\Delta z\, \Delta w$. Toward this, let g_1 and h_1 denote the inverse transformation in (6-99), so that

$$x_i = g_1(z, w) \qquad y_i = h_1(z, w) \tag{6-105}$$

As the point (z, w) goes to $(x_i, y_i) \equiv A'$, the point $(z + \Delta z, w)$ goes to B', the point $(z, w + \Delta w)$ goes to C', and the point $(z + \Delta z, w + \Delta w)$ goes to D'. Hence the respective x and y coordinates of B' are given by

$$g_1(z + \Delta z, w) = g_1(z, w) + \frac{\partial g_1}{\partial z} \Delta z = x_i + \frac{\partial g_1}{\partial z} \Delta z \tag{6-106}$$

and

$$h_1(z + \Delta z, w) = h_1(z, w) + \frac{\partial h_1}{\partial z} \Delta z = y_i + \frac{\partial h_1}{\partial z} \Delta z \tag{6-107}$$

Similarly those of C' are given by

$$x_i + \frac{\partial g_1}{\partial w} \Delta w \qquad y_i + \frac{\partial h_1}{\partial w} \Delta w \tag{6-108}$$

The area of the parallelogram $A'B'C'D'$ in Fig. 6-24*b* is given by

$$\begin{aligned} \Delta_i &= (A'B')(A'C') \sin(\theta - \phi) \\ &= (A'B' \cos\phi)(A'C' \sin\theta) - (A'B' \sin\phi)(A'C' \cos\theta) \end{aligned} \tag{6-109}$$

But from Fig. 6-24*b*, and (6-106)–(6-108)

$$A'B' \cos\phi = \frac{\partial g_1}{\partial z} \Delta z \qquad A'C' \sin\theta = \frac{\partial h_1}{\partial w} \Delta w \tag{6-110}$$

$$A'B' \sin\phi = \frac{\partial h_1}{\partial z} \Delta z \qquad A'C' \cos\theta = \frac{\partial g_1}{\partial w} \Delta w \tag{6-111}$$

so that

$$\Delta_i = \left(\frac{\partial g_1}{\partial z}\frac{\partial h_1}{\partial w} - \frac{\partial g_1}{\partial w}\frac{\partial h_1}{\partial z}\right)\Delta z\,\Delta w \tag{6-112}$$

and

$$\frac{\Delta_i}{\Delta z\,\Delta w} = \left(\frac{\partial g_1}{\partial z}\frac{\partial h_1}{\partial w} - \frac{\partial g_1}{\partial w}\frac{\partial h_1}{\partial z}\right) = \begin{vmatrix} \dfrac{\partial g_1}{\partial z} & \dfrac{\partial g_1}{\partial w} \\ \dfrac{\partial h_1}{\partial z} & \dfrac{\partial h_1}{\partial w} \end{vmatrix} \tag{6-113}$$

The determinant on the right side of (6-113) represents the absolute value of the Jacobian $J(z, w)$ of the inverse transformation in (6-105). Thus

$$J(z, w) = \begin{vmatrix} \dfrac{\partial g_1}{\partial z} & \dfrac{\partial g_1}{\partial w} \\ \dfrac{\partial h_1}{\partial z} & \dfrac{\partial h_1}{\partial w} \end{vmatrix} \tag{6-114}$$

Substituting the absolute value of (6-114) into (6-104), we get

$$f_{zw}(z, w) = \sum_i |J(z, w)| f_{xy}(x_i, y_i) = \sum_i \frac{1}{|J(x_i, y_i)|} f_{xy}(x_i, y_i) \tag{6-115}$$

since

$$|J(z, w)| = \frac{1}{|J(x_i, y_i)|} \tag{6-116}$$

where the determinant $J(x_i, y_i)$ represents the Jacobian of the original transformation in (6-99) given by

$$J(x_i, y_i) = \begin{vmatrix} \dfrac{\partial g}{\partial x} & \dfrac{\partial g}{\partial y} \\ \dfrac{\partial h}{\partial x} & \dfrac{\partial h}{\partial y} \end{vmatrix}_{x=x_i, y=y_i} \tag{6-117}$$

We shall illustrate the usefulness of the formulas in (6-115) through various examples.

Linear Transformation

$$\mathbf{z} = a\mathbf{x} + b\mathbf{y} \qquad \mathbf{w} = c\mathbf{x} + d\mathbf{y} \tag{6-118}$$

If $ad - bc \neq 0$, then the system $ax + by = z$, $cx + dy = w$ has one and only one solution

$$x = Az + Bw \qquad y = Cz + Dw$$

Since $J(x, y) = ad - bc$, (6-115) yields

$$f_{zw}(z, w) = \frac{1}{|ad - bc|} f_{xy}(Az + Bw, Cz + Dw) \tag{6-119}$$

JOINT NORMALITY. From (6-119) it follows that if the random variables **x** and **y** are jointly normal as $N(\mu_x, \mu_y, \sigma_x^2, \sigma_y^2, \rho)$ and

$$\mathbf{z} = a\mathbf{x} + b\mathbf{y} \qquad \mathbf{w} = c\mathbf{x} + d\mathbf{y} \tag{6-120}$$

then **z** and **w** are also jointly normal since $f_{zw}(z, w)$ will be an exponential (similar to $f_{xy}(x, y)$) with a quadratic exponent in z and w. Using the notation in (6-25), **z** and **w** in (6-120) are jointly normal as $N(\mu_z, \mu_w, \sigma_z^2, \sigma_w^2, \rho_{zw})$, where by direct computation

$$\begin{aligned}
\mu_z &= a\mu_x + b\mu_y \\
\mu_w &= c\mu_x + d\mu_y \\
\sigma_z^2 &= a^2\sigma_x^2 + 2ab\rho\sigma_x\sigma_y + b^2\sigma_y^2 \\
\sigma_w^2 &= c^2\sigma_x^2 + 2cd\rho\sigma_x\sigma_y + d^2\sigma_y^2
\end{aligned} \tag{6-121}$$

and

$$\rho_{zw} = \frac{ac\sigma_x^2 + (ad + bc)\rho\sigma_x\sigma_y + bd\sigma_y^2}{\sigma_z\sigma_w}$$

In particular, *any* linear combination of two jointly normal random variables is normal.

EXAMPLE 6-22 ▶ Suppose **x** and **y** are zero mean independent Gaussian random variables with common variance σ^2. Define $\mathbf{r} = \sqrt{\mathbf{x}^2 + \mathbf{y}^2}$, $\boldsymbol{\theta} = \tan^{-1}(\mathbf{y}/\mathbf{x})$, where $|\theta| < \pi$. Obtain their joint density function.

SOLUTION
Here

$$f_{xy}(x, y) = \frac{1}{2\pi\sigma^2} e^{-(x^2+y^2)/2\sigma^2} \tag{6-122}$$

Since

$$r = g(x, y) = \sqrt{x^2 + y^2} \qquad \theta = h(x, y) = \tan^{-1}(y/x) \tag{6-123}$$

and θ is known to vary in the interval $(-\pi, \pi)$, we have one solution pair given by

$$x_1 = r\cos\theta \qquad y_1 = r\sin\theta \tag{6-124}$$

We can use (6-124) to obtain $J(r, \theta)$. From (6-114)

$$J(r, \theta) = \begin{vmatrix} \dfrac{\partial x_1}{\partial r} & \dfrac{\partial x_1}{\partial \theta} \\ \dfrac{\partial y_1}{\partial r} & \dfrac{\partial y_1}{\partial \theta} \end{vmatrix} = \begin{vmatrix} \cos\theta & -r\sin\theta \\ \sin\theta & r\cos\theta \end{vmatrix} = r \tag{6-125}$$

so that

$$|J(r, \theta)| = r \tag{6-126}$$

We can also compute $J(x, y)$ using (6-117). From (6-123),

$$J(x, y) = \begin{vmatrix} \frac{x}{\sqrt{x^2+y^2}} & \frac{y}{\sqrt{x^2+y^2}} \\ \frac{-y}{x^2+y^2} & \frac{x}{x^2+y^2} \end{vmatrix} = \frac{1}{\sqrt{x^2 + y^2}} = \frac{1}{r} \tag{6-127}$$

Notice that $|J(r,\theta)| = 1/|J(x,y)|$, agreeing with (6-116). Substituting (6-122), (6-124) and (6-126) or (6-127) into (6-115), we get

$$f_{r,\theta}(r,\theta) = r f_{xy}(x_1, y_1) = \frac{r}{2\pi\sigma^2} e^{-r^2/2\sigma^2} \qquad 0 < r < \infty \qquad |\theta| < \pi \tag{6-128}$$

Thus

$$f_r(r) = \int_{-\pi}^{\pi} f_{r,\theta}(r,\theta)\, d\theta = \frac{r}{\sigma^2} e^{-r^2/2\sigma^2} \qquad 0 < r < \infty \tag{6-129}$$

which represents a Rayleigh random variable with parameter σ^2, and

$$f_\theta(\theta) = \int_0^\infty f_{r,\theta}(r,\theta)\, dr = \frac{1}{2\pi} \qquad |\theta| < \pi \tag{6-130}$$

which represents a uniform random variable in the interval $(-\pi, \pi)$. Moreover by direct computation

$$f_{r,\theta}(r,\theta) = f_r(r) \cdot f_\theta(\theta) \tag{6-131}$$

implying that $\mathbf{r}$ and $\boldsymbol{\theta}$ are independent. We summarize these results in the following statement: If $\mathbf{x}$ and $\mathbf{y}$ are zero mean independent Gaussian random variables with common variance, then $\sqrt{\mathbf{x}^2 + \mathbf{y}^2}$ has a Rayleigh distribution, and $\tan^{-1}(\mathbf{y}/\mathbf{x})$ has a uniform distribution in $(-\pi, \pi)$ (see also Example 6-15). Moreover these two derived random variables are statistically independent. Alternatively, with $\mathbf{x}$ and $\mathbf{y}$ as independent zero mean random variables as in (6-122), $\mathbf{x} + j\mathbf{y}$ represents a complex Gaussian random variable. But

$$\mathbf{x} + j\mathbf{y} = \mathbf{r}e^{j\boldsymbol{\theta}} \tag{6-132}$$

with $\mathbf{r}$ and $\boldsymbol{\theta}$ as in (6-123), and hence we conclude that the magnitude and phase of a complex Gaussian random variable are independent with Rayleigh and uniform distributions respectively. The statistical independence of these derived random variables is an interesting observation. ◀

EXAMPLE 6-23

▶ Let $\mathbf{x}$ and $\mathbf{y}$ be independent exponential random variables with common parameter λ. Define $\mathbf{u} = \mathbf{x} + \mathbf{y}$, $\mathbf{v} = \mathbf{x} - \mathbf{y}$. Find the joint and marginal p.d.f. of $\mathbf{u}$ and $\mathbf{v}$.

SOLUTION

It is given that

$$f_{xy}(x,y) = \frac{1}{\lambda^2} e^{-(x+y)/\lambda} \qquad x > 0 \qquad y > 0 \tag{6-133}$$

Now since $u = x + y$, $v = x - y$, always $|v| < u$, and there is only one solution given by

$$x = \frac{u+v}{2} \qquad y = \frac{u-v}{2} \tag{6-134}$$

Moreover the Jacobian of the transformation is given by

$$J(x,y) = \begin{vmatrix} 1 & 1 \\ 1 & -1 \end{vmatrix} = -2$$

and hence

$$f_{uv}(u, v) = \frac{1}{2\lambda^2} e^{-u/\lambda} \qquad 0 < |v| < u < \infty \tag{6-135}$$

represents the joint p.d.f. of **u** and **v**. This gives

$$f_u(u) = \int_{-u}^{u} f_{uv}(u, v)\, dv = \frac{1}{2\lambda^2} \int_{-u}^{u} e^{-u/\lambda}\, dv = \frac{u}{\lambda^2} e^{-u/\lambda} \qquad 0 < u < \infty \tag{6-136}$$

and

$$f_v(v) = \int_{|v|}^{\infty} f_{uv}(u, v)\, du = \frac{1}{2\lambda^2} \int_{|v|}^{\infty} e^{-u/\lambda}\, du = \frac{1}{2\lambda} e^{-|v|/\lambda} \qquad -\infty < v < \infty \tag{6-137}$$

Notice that in this case $f_{uv}(u, v) \neq f_u(u) \cdot f_v(v)$, and the random variables **u** and **v** are not independent. ◀

As we show below, the general transformation formula in (6-115) making use of two functions can be made useful even when only one function is specified.

AUXILIARY VARIABLES. Suppose

$$\mathbf{z} = g(\mathbf{x}, \mathbf{y}) \tag{6-138}$$

where **x** and **y** are two random variables. To determine $f_z(z)$ by making use of the formulation in (6-115), we can define an auxiliary variable

$$\mathbf{w} = \mathbf{x} \quad \text{or} \quad \mathbf{w} = \mathbf{y} \tag{6-139}$$

and the p.d.f. of **z** can be obtained from $f_{zw}(z, w)$ by proper integration.

EXAMPLE 6-24

▶ Suppose $\mathbf{z} = \mathbf{x} + \mathbf{y}$ and let $\mathbf{w} = \mathbf{y}$ so that the transformation is one-to-one and the solution is given by $y_1 = w$, $x_1 = z - w$. The Jacobian of the transformation is given by

$$J(x, y) = \begin{vmatrix} 1 & 1 \\ 0 & 1 \end{vmatrix} = 1$$

and hence

$$f_{zw}(x, y) = f_{xy}(x_1, y_1) = f_{xy}(z - w, w)$$

or

$$f_z(z) = \int f_{zw}(z, w)\, dw = \int_{-\infty}^{+\infty} f_{xy}(z - w, w)\, dw \tag{6-140}$$

which agrees with (6-41). Note that (6-140) reduces to the convolution of $f_x(z)$ and $f_y(z)$ if **x** and **y** are independent random variables. ◀

Next, we consider a less trivial example along these lines.

EXAMPLE 6-25

▶ Let $\mathbf{x} \sim U(0, 1)$ and $\mathbf{y} \sim U(0, 1)$ be independent random variables. Define

$$\mathbf{z} = (-2 \ln \mathbf{x})^{1/2} \cos(2\pi \mathbf{y}) \tag{6-141}$$

Find the density function of **z**.

SOLUTION

We can make use of the auxiliary variable $\mathbf{w} = \mathbf{y}$ in this case. This gives the only solution to be

$$x_1 = e^{-[z\sec(2\pi w)]^2/2} \tag{6-142}$$

$$y_1 = w \tag{6-143}$$

and using (6-114)

$$J(z,w) = \begin{vmatrix} \dfrac{\partial x_1}{\partial z} & \dfrac{\partial x_1}{\partial w} \\ \dfrac{\partial y_1}{\partial z} & \dfrac{\partial y_1}{\partial w} \end{vmatrix} = \begin{vmatrix} -z\sec^2(2\pi w)e^{-[z\sec(2\pi w)]^2/2} & \dfrac{\partial x_1}{\partial w} \\ 0 & 1 \end{vmatrix}$$

$$= -z\sec^2(2\pi w)e^{-[z\sec(2\pi w)]^2/2} \tag{6-144}$$

Substituting (6-142) and (6-144) into (6-115), we obtain

$$f_{zw}(z,w) = z\sec^2(2\pi w)e^{-[z\sec(2\pi w)]^2/2} \qquad -\infty < z < +\infty \quad 0 < w < 1 \tag{6-145}$$

and

$$f_z(z) = \int_0^1 f_{zw}(z,w)\,dw = e^{-z^2/2}\int_0^1 z\sec^2(2\pi w)e^{-[z\tan(2\pi w)]^2/2}\,dw \tag{6-146}$$

Let $u = z\tan(2\pi w)$ so that $du = 2\pi z\sec^2(2\pi w)\,dw$. Notice that as w varies from 0 to 1, u varies from $-\infty$ to $+\infty$. Using this in (6-146), we get

$$f_z(z) = \frac{1}{\sqrt{2\pi}}e^{-z^2/2}\underbrace{\int_{-\infty}^{\infty} e^{-u^2/2}\frac{du}{\sqrt{2\pi}}}_{1} = \frac{1}{\sqrt{2\pi}}e^{-z^2/2} \qquad -\infty < z < \infty \tag{6-147}$$

which represents a zero mean Gaussian random variable with unit variance. Thus $\mathbf{z} \sim N(0,1)$. Equation (6-141) can be used as a practical procedure to generate Gaussian random variables from two independent uniformly distributed random sequences. ◀

EXAMPLE 6-26

▶ Let $\mathbf{z} = \mathbf{xy}$. Then with $\mathbf{w} = \mathbf{x}$ the system $xy = z$, $x = w$ has a single solution: $x_1 = w$, $y_1 = z/w$. In this case, $J(x,y) = -w$ and (6-115) yields

$$f_{zw}(z,w) = \frac{1}{|w|}f_{xy}\left(w, \frac{z}{w}\right)$$

Hence the density of the random variable $\mathbf{z} = \mathbf{xy}$ is given by

$$f_z(z) = \int_{-\infty}^{\infty} \frac{1}{|w|}f_{xy}\left(w, \frac{z}{w}\right)dw \tag{6-148}$$

Special case: We now assume that the random variables **x** and **y** are independent and each is uniform in the interval (0, 1). In this case, $z < w$ and

$$f_{xy}\left(w, \frac{z}{w}\right) = f_x(w)f_y\left(\frac{z}{w}\right) = 1$$

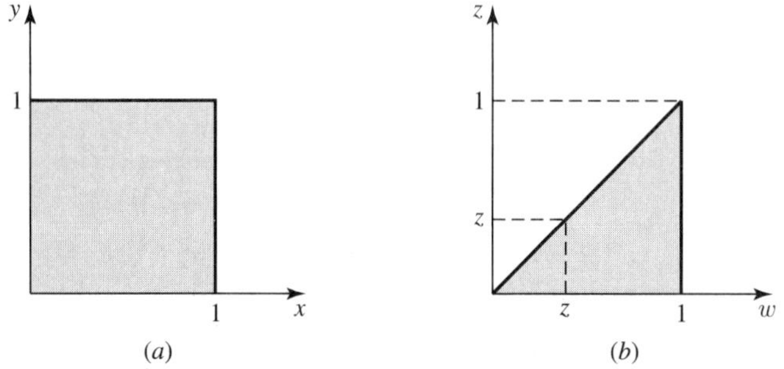

FIGURE 6-25

so that (see Fig. 6-25)

$$f_{zw}(z,w) = \begin{cases} 1/w & 0 < z < w < 1 \\ 0 & \text{otherwise} \end{cases} \tag{6-149}$$

Thus

$$f_z(z) = \int_z^1 \frac{1}{w}\,dw = \begin{cases} -\ln z & 0 < z < 1 \\ 0 & \text{elsewhere} \end{cases} \tag{6-150}$$

◀

EXAMPLE 6-27

▶ Let **x** and **y** be independent gamma random variables as in Example 6-12. Define $\mathbf{z} = \mathbf{x} + \mathbf{y}$ and $\mathbf{w} = \mathbf{x}/\mathbf{y}$. Show that **z** and **w** are independent random variables.

SOLUTION

Equations $\mathbf{z} = \mathbf{x} + \mathbf{y}$ and $\mathbf{w} = \mathbf{x}/\mathbf{y}$ generate one pair of solutions

$$x_1 = \frac{zw}{1+w} \qquad y_1 = \frac{z}{1+w}$$

Moreover

$$J(x,y) = \begin{vmatrix} 1 & 1 \\ 1/y & -x/y^2 \end{vmatrix} = -\frac{x+y}{y^2} = -\frac{(1+w)^2}{z}$$

Substituting these into (6-65) and (6-115) we get

$$\begin{aligned} f_{zw}(z,w) &= \frac{1}{\alpha^{m+n}\,\Gamma(m)\Gamma(n)}\frac{z}{(1+w)^2}\left(\frac{zw}{1+w}\right)^{m-1}\left(\frac{z}{1+w}\right)^{n-1} e^{-z/\alpha} \\ &= \frac{1}{\alpha^{m+n}}\frac{z^{m+n-1}}{\Gamma(m)\Gamma(n)}e^{-z/\alpha}\cdot\frac{w^{m-1}}{(1+w)^{m+n}} \\ &= \left(\frac{z^{m+n-1}}{\alpha^{m+n}\Gamma(m+n)}e^{-z/\alpha}\right)\cdot\left(\frac{\Gamma(m+n)}{\Gamma(m)\Gamma(n)}\frac{w^{m-1}}{(1+w)^{m+n}}\right) \\ &= f_z(z)f_w(w) \qquad z > 0 \quad w > 0 \end{aligned} \tag{6-151}$$

showing that **z** and **w** are independent random variables. Notice that $\mathbf{z} \sim G(m+n,\alpha)$ and **w** represents the ratio of two independent gamma random variables. ◀

EXAMPLE 6-28

THE STUDENT t DISTRIBUTION

▶ A random variable **z** has a Student t distribution[2] $t(n)$ with n degrees of freedom if for $-\infty < z < \infty$

$$f_z(z) = \frac{\gamma_1}{\sqrt{(1+z^2/n)^{n+1}}} \qquad \gamma_1 = \frac{\Gamma((n+1)/2)}{\sqrt{\pi n}\,\Gamma(n/2)} \tag{6-152}$$

We shall show that if **x** and **y** are two independent random variables, **x** is $N(0, 1)$, and **y** is $\chi^2(n)$:

$$f_x(x) = \frac{1}{\sqrt{2\pi}} e^{-x^2/2} \qquad f_y(y) = \frac{1}{2^{n/2}\Gamma(n/2)} y^{n/2-1} e^{-y/2} U(y) \tag{6-153}$$

then the random variable

$$\mathbf{z} = \frac{\mathbf{x}}{\sqrt{\mathbf{y}/n}}$$

has a $t(n)$ distribution. Note that the Student t distribution represents the ratio of a normal random variable to the square root of an independent χ^2 random variable divided by its degrees of freedom.

SOLUTION

We introduce the random variable $\mathbf{w} = \mathbf{y}$ and use (6-115) with

$$x = z\sqrt{\frac{w}{n}} \qquad y = w \qquad J(z, w) = \sqrt{\frac{w}{n}} \quad \text{or} \quad J(x, y) = \sqrt{\frac{n}{w}}$$

This yields

$$\begin{aligned} f_{zw}(z, w) &= \sqrt{\frac{w}{n}} \frac{1}{\sqrt{2\pi}} e^{-z^2 w/2n} \frac{w^{n/2-1}}{2^{n/2}\Gamma(n/2)} e^{-w/2} U(w) \\ &= \frac{w^{(n-1)/2}}{\sqrt{2\pi n}\, 2^{n/2}\Gamma(n/2)} e^{-w(1+z^2/n)/2} U(w) \end{aligned}$$

Integrating with respect to w after replacing $w(1+z^2/n)/2 = u$, we obtain

$$\begin{aligned} f_z(z) &= \frac{1}{\sqrt{\pi n}\,\Gamma(n/2)} \frac{1}{(1+z^2/n)^{(n+1)/2}} \int_0^\infty u^{(n-1)/2} e^{-u}\, du \\ &= \frac{\Gamma((n+1)/2)}{\sqrt{\pi n}\,\Gamma(n/2)} \frac{1}{(1+z^2/n)^{(n+1)/2}} \\ &= \frac{1}{\sqrt{n}\beta(1/2, n/2)} \frac{1}{(1+z^2/n)^{(n+1)/2}} \qquad -\infty < z < \infty \end{aligned} \tag{6-154}$$

For $n = 1$, (6-154) represents a Cauchy random variable. Notice that for each n, (6-154) generates a different p.d.f. As n gets larger, the t distribution tends towards the normal distribution. In fact from (6-154)

$$(1+z^2/n)^{-(n+1)/2} \to e^{-z^2/2} \quad \text{as} \quad n \to \infty$$

For small n, the t distributions have "fatter tails" compared to the normal distribution because of its polynomial form. Like the normal distribution, Student t distribution is important in statistics and is often available in tabular form. ◀

---

[2] *Student* was the pseudonym of the English statistician W. S. Gosset, who first introduced this law in empirical form (*The probable error of a mean,* Biometrica, 1908.) The first rigorous proof of this result was published by R. A. Fisher.

EXAMPLE 6-29

THE *F* DISTRIBUTION

▶ Let $\mathbf{x}$ and $\mathbf{y}$ be independent random variables such that $\mathbf{x}$ has a chi-square distribution with m degrees of freedom and $\mathbf{y}$ has a chi-square distribution with n degrees of freedom. Then the random variable

$$\mathbf{F} = \frac{\mathbf{x}/m}{\mathbf{y}/n} \tag{6-155}$$

is said to have an *F distribution with* (m, n) degrees of freedom. Show that the p.d.f. of $\mathbf{z} = \mathbf{F}$ is given by

$$f_z(z) = \begin{cases} \dfrac{\Gamma((m+n)/2)m^{m/2}n^{n/2}}{\Gamma(m/2)\Gamma(n/2)} \dfrac{z^{m/2-1}}{(n+mz)^{(m+n)/2}} & z > 0 \\ 0 & \text{otherwise} \end{cases} \tag{6-156}$$

SOLUTION

To compute the density of $\mathbf{F}$, using (6-153) we note that the density of $\mathbf{x}/m$ is given by

$$f_1(x) = \begin{cases} \dfrac{m(mx)^{m/2-1}e^{-mx/2}}{\Gamma(m/2)2^{m/2}} & x > 0 \\ 0 & \text{otherwise} \end{cases}$$

and that of $\mathbf{y}/n$ by

$$f_2(y) = \begin{cases} \dfrac{n(ny)^{n/2-1}e^{-ny/2}}{\Gamma(n/2)2^{n/2}} & y > 0 \\ 0 & \text{otherwise} \end{cases}$$

Using (6-60) from Example 6-10, the density of $\mathbf{z} = \mathbf{F}$ in (6-155) is given by

$$\begin{aligned} f_z(z) &= \int_0^\infty y \left(\frac{m(mzy)^{m/2-1}e^{-mzy/2}}{\Gamma(m/2)2^{m/2}} \right) \left(\frac{n(ny)^{n/2-1}e^{-ny/2}}{\Gamma(n/2)2^{n/2}} \right) dy \\ &= \frac{(m/2)^{m/2}(n/2)^{n/2}}{\Gamma(m/2)\Gamma(n/2)2^{(m+n)/2}} z^{m/2-1} \int_0^\infty y^{(m+n)/2-1} e^{y(n+mz)/2}\, dy \\ &= \frac{(m/2)^{m/2}(n/2)^{n/2}}{\Gamma(m/2)\Gamma(n/2)2^{(m+n)/2}} z^{m/2-1} \Gamma\left(\frac{m+n}{2}\right) \left(\frac{2}{n+mz}\right)^{(m+n)/2} \\ &= \frac{\Gamma((m+n)/2)m^{m/2}n^{n/2}}{\Gamma(m/2)\Gamma(n/2)} \frac{z^{m/2-1}}{(n+mz)^{(m+n)/2}} \\ &= \frac{(m/n)^{m/2}}{\beta(m/2, n/2)} z^{m/2-1}(1 + mz/n)^{-(m+n)/2} \qquad z > 0 \end{aligned} \tag{6-157}$$

and $f_z(z) = 0$ for $z \le 0$. The distribution in (6-157) is called Fisher's variance ration distribution. If $m = 1$ in (6-155), then from (6-154) and (6-157) we get $\mathbf{F} = [t(n)]^2$. Thus $F(1, n)$ and $t^2(n)$ have the same distribution. Moreover $F(1, 1) = t^2(1)$ represents the square of a Cauchy random variable. Both Student's t distribution and Fisher's F distribution play key roles in statistical tests of significance. ◀

6-4 JOINT MOMENTS

Given two random variables **x** and **y** and a function $g(x, y)$, we form the random variable $\mathbf{z} = g(\mathbf{x}, \mathbf{y})$. The expected value of this random variable is given by

$$E\{\mathbf{z}\} = \int_{-\infty}^{\infty} z f_z(z)\, dz \tag{6-158}$$

However, as the next theorem shows, $E\{\mathbf{z}\}$ can be expressed directly in terms of the function $g(x, y)$ and the joint density $f(x, y)$ of **x** and **y**.

THEOREM 6-4

$$E\{g(\mathbf{x}, \mathbf{y})\} = \int_{-\infty}^{\infty}\int_{-\infty}^{\infty} g(x, y) f(x, y)\, dx\, dy \tag{6-159}$$

Proof. The proof is similar to the proof of (5-55). We denote by ΔD_z the region of the xy plane such that $z < g(x, y) < z + dz$. Thus to each differential in (6-158) there corresponds a region ΔD_z in the xy plane. As dz covers the z axis, the regions ΔD_z are not overlapping and they cover the entire xy plane. Hence the integrals in (6-158) and (6-159) are equal.

We note that the expected value of $g(\mathbf{x})$ can be determined either from (6-159) or from (5-55) as a single integral

$$E\{g(\mathbf{x})\} = \int_{-\infty}^{\infty}\int_{-\infty}^{\infty} g(x) f(x, y)\, dx\, dy = \int_{-\infty}^{\infty} g(x) f_x(x)\, dx$$

This is consistent with the relationship (6-10) between marginal and joint densities.

If the random variables **x** and **y** are of discrete type taking the values x_i and y_k with probability p_{ik} as in (6-33), then

$$E\{g(\mathbf{x}, \mathbf{y})\} = \sum_i \sum_k g(x_i, y_k) p_{ik} \tag{6-160}$$

Linearity From (6-159) it follows that

$$E\left\{\sum_{k=1}^{n} a_k g_k(\mathbf{x}, \mathbf{y})\right\} = \sum_{k=1}^{n} a_k E\{g_k(\mathbf{x}, \mathbf{y})\} \tag{6-161}$$

This fundamental result will be used extensively.

We note in particular that

$$E\{\mathbf{x} + \mathbf{y}\} = E\{\mathbf{x}\} + E\{\mathbf{y}\} \tag{6-162}$$

Thus the expected value of the sum of two random variables equals the sum of their expected values. We should stress, however, that, in general,

$$E\{\mathbf{xy}\} \neq E\{\mathbf{x}\}E\{\mathbf{y}\}$$

Frequency interpretation As in (5-51)

$$\begin{aligned} E\{\mathbf{x} + \mathbf{y}\} &\simeq \frac{\mathbf{x}(\xi_1) + \mathbf{y}(\xi_i) + \cdots + \mathbf{x}(\xi_n) + \mathbf{y}(\xi_n)}{n} \\ &= \frac{\mathbf{x}(\xi_1) + \cdots + \mathbf{x}(\xi_n)}{n} + \frac{\mathbf{y}(\xi_1) + \cdots + \mathbf{y}(\xi_n)}{n} \\ &\simeq E\{\mathbf{x}\} + E\{\mathbf{y}\} \end{aligned}$$

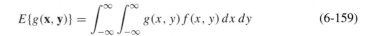

However, in general,

$$E\{\mathbf{xy}\} \simeq \frac{\mathbf{x}(\xi_1)\mathbf{y}(\xi_i) + \cdots + \mathbf{x}(\xi_n)\mathbf{y}(\xi_n)}{n}$$

$$\neq \frac{\mathbf{x}(\xi_1) + \cdots + \mathbf{x}(\xi_n)}{n} \times \frac{\mathbf{y}(\xi_1) + \cdots + \mathbf{y}(\xi_n)}{n} \simeq E\{\mathbf{x}\}E\{\mathbf{y}\}$$

In the case of one random variable, we defined the parameters mean and variance to represent its average behavior. How does one parametrically represent similar cross behavior between two random variables? Toward this, we can generalize the variance definition as shown next.

COVARIANCE. The covariance C or C_{xy} of two random variables $\mathbf{x}$ and $\mathbf{y}$ is by definition the number

$$C_{xy} = E\{(\mathbf{x} - \eta_x)(\mathbf{y} - \eta_y)\} \tag{6-163}$$

where $E\{\mathbf{x}\} = \eta_x$ and $E\{\mathbf{y}\} = \eta_y$. Expanding the product in (6-163) and using (6-161) we obtain

$$C_{xy} = E\{\mathbf{xy}\} - E\{\mathbf{x}\}E\{\mathbf{y}\} \tag{6-164}$$

Correlation coefficient The correlation coefficient ρ or ρ_{xy} of the random variables $\mathbf{x}$ and $\mathbf{y}$ is by definition the ratio

$$\rho_{xy} = \frac{C_{xy}}{\sigma_x \sigma_y} \tag{6-165}$$

We maintain that

$$|\rho_{xy}| \leq 1 \qquad |C_{xy}| \leq \sigma_x \sigma_y \tag{6-166}$$

Proof. Clearly,

$$E\{[a(\mathbf{x} - \eta_x) + (\mathbf{y} - \eta_y)]^2\} = a^2\sigma_x^2 + 2aC_{xy} + \sigma_y^2 \tag{6-167}$$

Equation (6-167) is a positive quadratic for any a; hence its discriminant is negative. In other words,

$$C_{xy}^2 - \sigma_x^2\sigma_y^2 \leq 0 \tag{6-168}$$

and (6-166) results.

We note that the random variables $\mathbf{x}$, $\mathbf{y}$ and $\mathbf{x} - \eta_x$, $\mathbf{y} - \eta_y$ have the same covariance and correlation coefficient.

EXAMPLE 6-30

▶ We shall show that the correlation coefficient of two jointly normal random variables is the parameter r in (6-23). It suffices to assume that $\eta_x = \eta_y = 0$ and to show that $E(\mathbf{xy}) = r\sigma_1\sigma_2$.

Since

$$\frac{x^2}{\sigma_1^2} - 2r\frac{xy}{\sigma_1\sigma_2} + \frac{y^2}{\sigma_2^2} = \left(\frac{x}{\sigma_1} - r\frac{y}{\sigma_2}\right)^2 + (1 - r^2)\frac{y^2}{\sigma_2^2}$$

we conclude with (6-23) that

$$E\{\mathbf{xy}\} = \frac{1}{\sigma_2\sqrt{2\pi}} \int_{-\infty}^{\infty} y e^{-y^2/2\sigma_2^2} \int_{-\infty}^{\infty} \frac{x}{\sigma_1\sqrt{2\pi(1-r^2)}} \exp\left(-\frac{(x - ry\sigma_1/\sigma_2)^2}{2\sigma_1^2(1-r^2)}\right) dx\, dy$$

The inner integral is a normal density with mean $ry\sigma_1/\sigma_2$ multiplied by x; hence it equals $ry\sigma_1/\sigma_2$. This yields

$$E\{\mathbf{xy}\} = r\sigma_1/\sigma_2 \int_{-\infty}^{\infty} \frac{1}{\sigma_2\sqrt{2\pi}} y^2 e^{-y^2/2\sigma_2^2}\, dy = r\sigma_1\sigma_2$$ ◀

Uncorrelatedness Two random variables are called uncorrelated if their covariance is 0. This can be phrased in the following equivalent forms

$$C_{xy} = 0 \qquad \rho_{xy} = 0 \qquad E\{\mathbf{xy}\} = E\{\mathbf{x}\}E\{\mathbf{y}\}$$

Orthogonality Two random variables are called orthogonal if

$$E\{\mathbf{xy}\} = 0$$

We shall use the notation

$$\mathbf{x} \perp \mathbf{y}$$

to indicate the random variables **x** and **y** are orthogonal.

Note (*a*) If **x** and **y** are uncorrelated, then $\mathbf{x} - \eta_x \perp \mathbf{y} - \eta_y$. (*b*) If **x** and **y** are uncorrelated and $\eta_x = 0$ or $\eta_y = 0$ then $\mathbf{x} \perp \mathbf{y}$.

Vector space of random variables. We shall find it convenient to interpret random variables as vectors in an abstract space. In this space, the second moment

$$E\{\mathbf{xy}\}$$

of the random variables **x** and **y** is by definition their *inner product* and $E\{\mathbf{x}^2\}$ and $E\{\mathbf{y}^2\}$ are the squares of their lengths. The ratio

$$\frac{E\{\mathbf{xy}\}}{\sqrt{E\{\mathbf{x}^2\}E\{\mathbf{y}^2\}}}$$

is the cosine of their angle.

We maintain that

$$E^2\{\mathbf{xy}\} \le E\{\mathbf{x}^2\}E\{\mathbf{y}^2\} \tag{6-169}$$

This is the *cosine inequality* and its proof is similar to the proof of (6-168): The quadratic

$$E\{(a\mathbf{x} - \mathbf{y})^2\} = a^2E\{\mathbf{x}^2\} - 2aE\{\mathbf{xy}\} + E\{\mathbf{y}^2\}$$

is positive for every a; hence its discriminant is negative and (6-169) results. If (6-169) is an equality, then the quadratic is 0 for some $a = a_0$, hence $\mathbf{y} = a_0\mathbf{x}$. This agrees with the geometric interpretation of random variables because, if (6-169) is an equality, then the vectors **x** and **y** are on the same line.

The following illustration is an example of the correspondence between vectors and random variables: Consider two random variables **x** and **y** such that $E\{\mathbf{x}^2\} = E\{\mathbf{y}^2\}$.

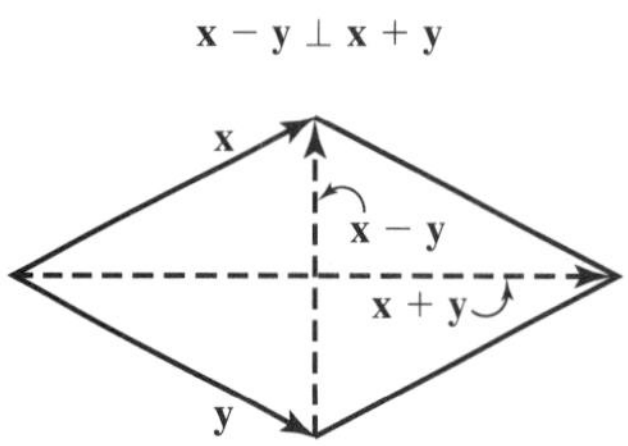

FIGURE 6-26

Geometrically, this means that the vectors **x** and **y** have the same length. If, therefore, we construct a parallelogram with sides **x** and **y**, it will be a rhombus with diagonals $\mathbf{x}+\mathbf{y}$ and $\mathbf{x}-\mathbf{y}$ (Fig. 6-26). These diagonals are perpendicular because

$$E\{(\mathbf{x}+\mathbf{y})(\mathbf{x}-\mathbf{y})\} = E\{\mathbf{x}^2 - \mathbf{y}^2\} = 0$$

THEOREM 6-5

▶ If two random variables are independent, that is, if [see also (6-20)]

$$f(x, y) = f_x(x) f_y(y) \tag{6-170}$$

then they are uncorrelated.

Proof. It suffices to show that

$$E\{\mathbf{xy}\} = E\{\mathbf{x}\}E\{\mathbf{y}\} \tag{6-171}$$

From (6-159) and (6-170) it follows that

$$E\{\mathbf{xy}\} = \int_{-\infty}^{\infty}\int_{-\infty}^{\infty} xy f_x(x) f_y(y)\, dx\, dy = \int_{-\infty}^{\infty} x f_x(x)\, dx \int_{-\infty}^{\infty} y f_y(y)\, dy$$

and (6-171) results.

If the random variables **x** and **y** are independent, then the random variables $g(\mathbf{x})$ and $h(\mathbf{y})$ are also independent [see (6-21)]. Hence

$$E\{g(\mathbf{x})h(\mathbf{y})\} = E\{g(\mathbf{x})\}E\{h(\mathbf{y})\} \tag{6-172}$$

This is not, in general, true if **x** and **y** are merely uncorrelated. ◀

As Example 6-31 shows if two random variables are uncorrelated they are not necessarily independent. However, for normal random variables uncorrelatedness is equivalent to independence. Indeed, if the random variables **x** and **y** are jointly normal and their correlation coefficient $r = 0$, then [see (6-23)] $f_{xy}(x, y) = f_x(x) f_y(y)$.

EXAMPLE 6-31

▶ Let $\mathbf{x} \sim U(0, 1)$, $\mathbf{y} \sim U(0, 1)$. Suppose **x** and **y** are independent. Define $\mathbf{z} = \mathbf{x}+\mathbf{y}$, $\mathbf{w} = \mathbf{x}-\mathbf{y}$. Show that **x** and **w** are not independent, but uncorrelated random variables.

SOLUTION

$z = x + y$, $w = x - y$ gives the only solution set to be

$$x = \frac{z+w}{2} \qquad y = \frac{z-w}{2}$$

Moreover $0 < z < 2$, $-1 < w < 1$, $z+w \le 2$, $z-w \le 2$, $z > |w|$ and $|J(z, w)| = 1/2$.

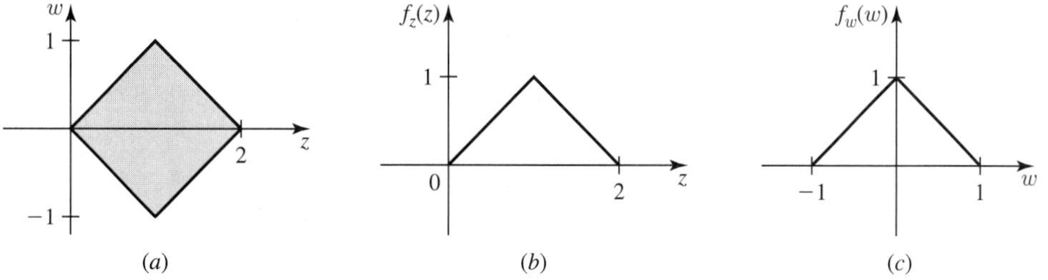

FIGURE 6-27

Thus (see the shaded region in Fig. 6-27)

$$f_{zw}(z, w) = \begin{cases} 1/2 & 0 < z < 2, -1 < w < 1, z + w \le 2, z - w \le 2, |w| < z \\ 0 & \text{otherwise} \end{cases} \tag{6-173}$$

and hence

$$f_z(z) = \int f_{zw}(z, w)\, dw = \begin{cases} \displaystyle\int_{-z}^{z} \frac{1}{2}\, dw = z & 0 < z < 1 \\ \displaystyle\int_{z-2}^{2-z} \frac{1}{2}\, dw = 2 - z & 1 < z < 2 \\ 0 & \text{otherwise} \end{cases} \tag{6-174}$$

and

$$f_w(w) = \int f_{zw}(z, w)\, dz = \int_{|w|}^{2-|w|} \frac{1}{2}\, dz = \begin{cases} 1 - |w| & -1 < w < 1 \\ 0 & \text{otherwise} \end{cases} \tag{6-175}$$

Clearly $f_{zw}(z, w) \ne f_z(z) f_w(w)$. Thus **z** and **w** are not independent. However,

$$E\{\mathbf{zw}\} = E\{(\mathbf{x} + \mathbf{y})(\mathbf{x} - \mathbf{y})\} = E\{\mathbf{x}^2\} - E\{\mathbf{y}^2\} = 0 \tag{6-176}$$

and

$$E\{\mathbf{w}\} = E\{\mathbf{x} - \mathbf{y}\} = 0 \tag{6-177}$$

and hence

$$\text{Cov}\{\mathbf{z}, \mathbf{w}\} = E\{\mathbf{zw}\} - E\{\mathbf{z}\}E\{\mathbf{w}\} = 0 \tag{6-178}$$

implying that **z** and **w** are uncorrelated random variables.

Variance of the sum of two random variables: If $\mathbf{z} = \mathbf{x} + \mathbf{y}$, then $\eta_z = \eta_x + \eta_y$; hence

$$\sigma_z^2 = E\{(\mathbf{z} - \eta_z)^2\} = E\{[(\mathbf{x} - \eta_x) + (\mathbf{y} - \eta_y)]^2\}$$

From this and (6-167) it follows that

$$\sigma_z^2 = \sigma_x^2 + 2\rho_{xy}\sigma_x\sigma_y + \sigma_y^2 \tag{6-179}$$

This leads to the conclusion that if $\rho_{xy} = 0$ then

$$\sigma_z^2 = \sigma_x^2 + \sigma_y^2 \tag{6-180}$$

Thus, if two random variables are uncorrelated, then the variance of their sum equals the sum of their variances.

It follows from (6-171) that this is also true if **x** and **y** are independent. ◀

Moments

The mean

$$m_{kr} = E\{\mathbf{x}^k\mathbf{y}^r\} = \int_{-\infty}^{\infty} x^k y^r f_{xy}(x, y)\, dx\, dy \tag{6-181}$$

of the product $\mathbf{x}^k\mathbf{y}^r$ is by definition a joint moment of the random variables **x** and **y** of order $k + r = n$.

Thus $m_{10} = \eta_x$, $m_{01} = \eta_y$ are the first-order moments and

$$m_{20} = E\{\mathbf{x}^2\} \qquad m_{11} = E\{\mathbf{xy}\} \qquad m_{02} = E\{\mathbf{y}^2\}$$

are the second-order moments.

The joint central moments of **x** and **y** are the moments of $\mathbf{x} - \eta_x$ and $\mathbf{y} - \eta_y$:

$$\mu_{kr} = E\{(\mathbf{x} - \eta_x)^k(\mathbf{y} - \eta_y)^r\} = \int_{-\infty}^{\infty} (x - \eta_x)^k (y - \eta_y)^r f_{xy}(x, y)\, dx\, dy \tag{6-182}$$

Clearly, $\mu_{10} = \mu_{01} = 0$ and

$$\mu_{11} = C_{xy} \qquad \mu_{20} = \sigma_x^2 \qquad \mu_{02} = \sigma_y^2$$

Absolute and generalized moments are defined similarly [see (5-69) and (5-70)].

For the determination of the joint statistics of **x** and **y** knowledge of their joint density is required. However, in many applications, only the first- and second-moments are used. These moments are determined in terms of the five parameters

$$\eta_x \quad \eta_y \quad \sigma_x^2 \quad \sigma_y^2 \quad \rho_{xy}$$

If **x** and **y** are jointly normal, then [see (6-23)] these parameters determine uniquely $f_{xy}(x, y)$.

EXAMPLE 6-32 ▶ The random variables **x** and **y** are jointly normal with

$$\eta_x = 10 \qquad \eta_y = 0 \qquad \sigma_x^2 = 4 \qquad \sigma_y^2 = 1 \qquad \rho_{xy} = 0.5$$

We shall find the joint density of the random variables

$$\mathbf{z} = \mathbf{x} + \mathbf{y} \qquad \mathbf{w} = \mathbf{x} - \mathbf{y}$$

Clearly,

$$\eta_z = \eta_x + \eta_y = 10 \quad \eta_w = \eta_x - \eta_y = 10$$

$$\sigma_z^2 = \sigma_x^2 + \sigma_y^2 + 2r_{xy}\sigma_x\sigma_y = 7 \qquad \sigma_w^2 = \sigma_x^2 + \sigma_y^2 - 2r_{xy}\sigma_x\sigma_y = 3$$

$$E(\mathbf{zw}) = E(\mathbf{x}^2 - \mathbf{y}^2) = (100 + 4) - 1 = 103$$

$$\rho_{zw} = \frac{E(\mathbf{zw}) - E(\mathbf{z})E(\mathbf{w})}{\sigma_z\sigma_w} = \frac{3}{\sqrt{7 \times 3}}$$

As we know [see (6-119)], the random variables **z** and **w** are jointly normal because they are linearly dependent on **x** and **y**. Hence their joint density is

$$N(10, 10, 7, 3, \sqrt{3/7})$$ ◀

ESTIMATE OF THE MEAN OF g(x, y). If the function $g(x, y)$ is sufficiently smooth near the point (η_x, η_y), then the mean η_g and variance σ_g^2 of $g(\mathbf{x}, \mathbf{y})$ can be estimated in terms of the mean, variance, and covariance of **x** and **y**:

$$\eta_g \simeq g + \frac{1}{2}\left(\frac{\partial^2 g}{\partial x^2}\sigma_x^2 + 2\frac{\partial^2 g}{\partial x \partial y}\rho_{xy}\sigma_x\sigma_y + \frac{\partial^2 g}{\partial y^2}\sigma_y^2\right) \tag{6-183}$$

$$\sigma_g^2 \simeq \left(\frac{\partial g}{\partial x}\right)^2 \sigma_x^2 + 2\left(\frac{\partial g}{\partial x}\right)\left(\frac{\partial g}{\partial y}\right)\rho_{xy}\sigma_x\sigma_y + \left(\frac{\partial g}{\partial y}\right)^2 \sigma_y^2 \tag{6-184}$$

where the function $g(x, y)$ and its derivatives are evaluated at $x = \eta_x$ and $y = \eta_y$.

Proof. We expand $g(x, y)$ into a series about the point (η_x, η_y):

$$g(x, y) = g(\eta_x, \eta_y) + (x - \eta_x)\frac{\partial g}{\partial x} + (y - \eta_y)\frac{\partial g}{\partial x} + \cdots \tag{6-185}$$

Inserting (6-185) into (6-159), we obtain the moment expansion of $E\{g(\mathbf{x}, \mathbf{y})\}$ in terms of the derivatives of $g(x, y)$ at (η_x, η_y) and the joint moments μ_{kr} of **x** and **y**. Using only the first five terms in (6-185), we obtain (6-183). Equation (6-184) follows if we apply (6-183) to the function $[g(x, y) - \eta_g]^2$ and neglect moments of order higher than 2.

6-5 JOINT CHARACTERISTIC FUNCTIONS

The *joint characteristic function* of the random variables **x** and **y** is by definition the integral

$$\Phi(\omega_1, \omega_2) = \int_{-\infty}^{\infty}\int_{-\infty}^{\infty} f(x, y)e^{j(\omega_1 x + \omega_2 y)}\, dx\, dy \tag{6-186}$$

From this and the two-dimensional *inversion formula* for Fourier transforms, it follows that

$$f(x, y) = \frac{1}{4\pi^2}\int_{-\infty}^{\infty}\int_{-\infty}^{\infty} \Phi(\omega_1, \omega_2)e^{-j(\omega_1 x + \omega_2 y)}\, d\omega_1\, d\omega_2 \tag{6-187}$$

Clearly,

$$\Phi(\omega_1, \omega_2) = E\left\{e^{j(\omega_1 \mathbf{x} + \omega_2 \mathbf{y})}\right\} \tag{6-188}$$

The logarithm

$$\Psi(\omega_1, \omega_2) = \ln\, \Phi(\omega_1, \omega_2) \tag{6-189}$$

of $\Phi(\omega_1, \omega_2)$ is the joint logarithmic-characteristic function of **x** and **y**.

The *marginal* characteristic functions

$$\Phi_x(\omega) = E\{e^{j\omega\mathbf{x}}\} \qquad \Phi_y(\omega) = E\{e^{j\omega\mathbf{y}}\} \tag{6-190}$$

of $\mathbf{x}$ and $\mathbf{y}$ can be expressed in terms of their joint characteristic function $\Phi(\omega_1, \omega_2)$. From (6-188) and (6-190) it follows that

$$\Phi_x(\omega) = \Phi(\omega, 0) \qquad \Phi_y(\omega) = \Phi(0, \omega) \tag{6-191}$$

We note that, if $\mathbf{z} = a\mathbf{x} + b\mathbf{y}$ then

$$\Phi_z(\omega) = E\left\{e^{j(a\mathbf{x}+b\mathbf{y})\omega}\right\} = \Phi(a\omega, b\omega) \tag{6-192}$$

Hence $\Phi_z(1) = \Phi(a, b)$.

Cramér-Wold theorem The material just presented shows that if $\Phi_z(\omega)$ is known for every a and b, then $\Phi(\omega_1, \omega_2)$ is uniquely determined. In other words, if the density of $a\mathbf{x} + b\mathbf{y}$ is known for every a and b, then the joint density $f(x, y)$ of $\mathbf{x}$ and $\mathbf{y}$ is uniquely determined.

Independence and convolution

If the random variables $\mathbf{x}$ and $\mathbf{y}$ are independent, then [see (6-172)]

$$E\left\{e^{j(\omega_1\mathbf{x}+\omega_2\mathbf{y})}\right\} = E\{e^{j\omega_1\mathbf{x}}\}E\{e^{j\omega_2\mathbf{y}}\}$$

From this it follows that

$$\Phi(\omega_1, \omega_2) = \Phi_x(\omega_1)\Phi_y(\omega_2) \tag{6-193}$$

Conversely, if (6-193) is true, then the random variables $\mathbf{x}$ and $\mathbf{y}$ are independent. Indeed, inserting (6-193) into the inversion formula (6-187) and using (5-102), we conclude that $f_xy(x, y) = f_x(x)f_y(y)$.

Convolution theorem If the random variables $\mathbf{x}$ and $\mathbf{y}$ are independent and $\mathbf{z} = \mathbf{x} + \mathbf{y}$, then

$$E\{e^{j\omega\mathbf{z}}\} = E\left\{e^{j\omega(\mathbf{x}+\mathbf{y})}\right\} = E\{e^{j\omega\mathbf{x}}\}E\{e^{j\omega\mathbf{y}}\}$$

Hence

$$\Phi_z(\omega) = \Phi_x(\omega)\Phi_y(\omega) \qquad \Psi_z(\omega) = \Psi_x(\omega) + \Psi_y(\omega) \tag{6-194}$$

As we know [see (6-43)], the density of $\mathbf{z}$ equals the convolution of $f_x(x)$ and $f_y(y)$. From this and (6-194) it follows that the characteristic function of the convolution of two densities equals the product of their characteristic functions.

EXAMPLE 6-33

▶ We shall show that if the random variables $\mathbf{x}$ and $\mathbf{y}$ are *independent* and Poisson distributed with parameters a and b, respectively, then their sum $\mathbf{z} = \mathbf{x} + \mathbf{y}$ is also Poisson distributed with parameter $a + b$.

Proof. As we know (see Example 5-31),

$$\Psi_x(\omega) = a(e^{j\omega} - 1) \qquad \Psi_y(\omega) = b(e^{j\omega} - 1)$$

Hence

$$\Psi_z(\omega) = \Psi_x(\omega) + \Psi_y(\omega) = (a + b)(e^{j\omega} - 1)$$

It can be shown that the converse is also true: If the random variables **x** and **y** are *independent* and their sum is Poisson distributed, then **x** and **y** are also Poisson distributed. The proof of this theorem is due to Raikov.[3] ◀

EXAMPLE 6-34

NORMAL RANDOM VARIABLES

▶ It was shown in Sec. 6-3 that if the random variables **x** and **y** are jointly normal, then the sum $a\mathbf{x} + b\mathbf{y}$ is also normal. Next we reestablish a special case of this result using (6-193): If **x** and **y** are *independent* and normal, then their sum $\mathbf{z} = \mathbf{x} + \mathbf{y}$ is also normal.

SOLUTION

In this case [see (5-100)]

$$\Psi_x(\omega) = j\eta_x\omega - \frac{1}{2}\sigma_x^2\omega^2 \qquad \Psi_y(\omega) = j\eta_y\omega - \frac{1}{2}\sigma_y^2\omega^2$$

Hence

$$\Psi_z(\omega) = j(\eta_x + \eta_y)\omega - \frac{1}{2}(\sigma_x^2 + \sigma_y^2)\omega^2$$

It can be shown that the converse is also true (Cramér's theorem): If the random variables **x** and **y** are *independent* and their sum is *normal,* then they are also *normal*. The proof of this difficult theorem will not be given.[4]

In a similar manner, it is easy to show that if **x** and **y** are independent identically distributed normal random variables, then $\mathbf{x}+\mathbf{y}$ and $\mathbf{x}-\mathbf{y}$ are independent (and normal). Interestingly, in this case also, the converse is true (Bernstein's theorem): If **x** and **y** are independent and identically distributed and if $\mathbf{x}+\mathbf{y}$ and $\mathbf{x}-\mathbf{y}$ are also independent, then all random variables $(\mathbf{x}, \mathbf{y}, \mathbf{x}+\mathbf{y}, \mathbf{x}-\mathbf{y})$ are normally distributed.

Darmois (1951) and Skitovitch (1954) have generalized this result as: If $\mathbf{x}_1$ and $\mathbf{x}_2$ are independent random variables and if two linear combinations $a_1\mathbf{x}_1 + a_2\mathbf{x}_2$ and $b_1\mathbf{x}_1 + b_2\mathbf{x}_2$ are also independent, where a_1, a_2, b_1, and b_2 represent nonzero coefficients, then *all* random variables are normally distributed. Thus if two nontrivial linear combinations of two independent random variables are also independent, then all of them represent normal random variables. ◀

More on Normal Random Variables

Let **x** and **y** be jointly Gaussian as $N(\eta_1, \eta_2, \sigma_1^2, \sigma_2^2, r)$ with p.d.f. as in (6-23) and (6-24). We shall show that the joint characteristic function of two jointly normal random variables is given by

$$\Phi(\omega_1, \omega_2) = e^{j(\eta_1\omega_1+\eta_2\omega_2)}e^{-(\omega_1^2\sigma_1^2+2r\sigma_1\sigma_2\omega_1\omega_2+\omega_2^2\sigma_2^2)/2} \tag{6-195}$$

Proof. This can be derived by inserting $f(x, y)$ into (6-186). The simpler proof presented here is based on the fact that the random variable $\mathbf{z} = \omega_1\mathbf{x} + \omega_2\mathbf{y}$ is normal and

$$\Phi_z(\omega) = e^{j\eta_z\omega-\sigma_z^2\omega^2/2} \tag{6-196}$$

---

[3] D. A. Raikov, "On the decomposition of Gauss and Poisson laws," *Izv. Akad. Nauk. SSSR*, Ser. Mat. 2, 1938, pp. 91–124.

[4] E. Lukacs, *Characteristic Functions,* Hafner Publishing Co., New York, 1960.

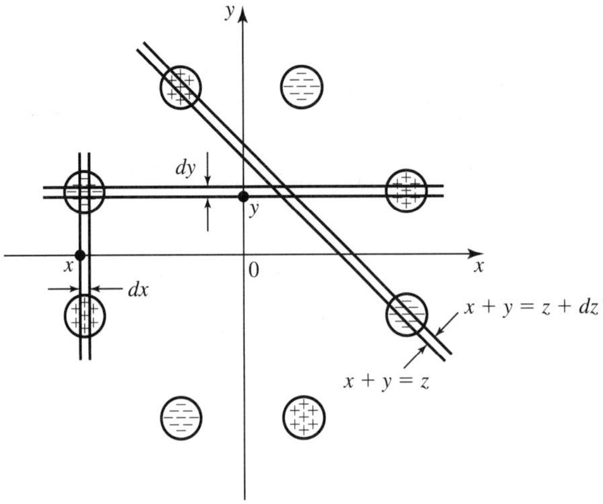

FIGURE 6-28

Since

$$\eta_z = \omega_1\eta_1 + \omega_2\eta_2 \qquad \sigma_z^2 = \omega_1^2\sigma_1^2 + 2r\omega_1\omega_2\sigma_1\sigma_2 + \omega_2^2\sigma_2^2$$

and $\Phi_z(\omega) = \Phi(\omega_1\omega, \omega_2\omega)$, (6-195) follows from (6-196) with $\omega = 1$.

This proof is based on the fact that the random variable $\mathbf{z} = \omega_1\mathbf{x} + \omega_2\mathbf{y}$ is normal for any ω_1 and ω_2; this leads to the conclusion: If it is known that the sum $a\mathbf{x} + b\mathbf{y}$ is normal for every a and b, then random variables **x** and **y** are jointly normal. We should stress, however, that this is not true if $a\mathbf{x} + b\mathbf{y}$ is normal for only a finite set of values of a and b. A counterexample can be formed by a simple extension of the construction in Fig. 6-28.

EXAMPLE 6-35

► We shall construct two random variables $\mathbf{x}_1$ and $\mathbf{x}_2$ with these properties: $\mathbf{x}_1$, $\mathbf{x}_2$, and $\mathbf{x}_1 + \mathbf{x}_2$ are normal but $\mathbf{x}_1$ and $\mathbf{x}_2$ are not jointly normal.

SOLUTION

Suppose that **x** and **y** are two jointly normal random variables with mass density $f(x, y)$. Adding and subtracting small masses in the region D of Fig. 6-28 consisting of eight circles as shown, we obtain a new function $f_1(x, y) = f(x, y) \pm \epsilon$ in D and $f_1(x, y) = f(x, y)$ everywhere else. The function $f_1(x, y)$ is a density; hence it defines two new random variables $\mathbf{x}_1$ and $\mathbf{y}_1$. These random variables are obviously not jointly normal. However, they are marginally normal because **x** and **y** are marginally normal and the masses in the vertical or horizontal strip have not changed. Furthermore, the random variable $\mathbf{z}_1 = \mathbf{x}_1 + \mathbf{y}_1$ is also normal because $\mathbf{z} = \mathbf{x} + \mathbf{y}$ is normal and the masses in any diagonal strip of the form $z \le x + y \le z + dz$ have not changed. ◄

THEOREM 6-6

MOMENT THEOREM

► The moment generating function of **x** and **y** is given by

$$\Phi(s_1, s_2) = E\{e^{s_1\mathbf{x}+s_2\mathbf{y}}\}.$$

Expanding the exponential and using the linearity of expected values, we obtain the series

$$\Phi(s_1, s_2) = \sum_{n=0}^{\infty} \frac{1}{n!} \sum_{n=0}^{n} \binom{n}{k} E\{\mathbf{x}^k \mathbf{y}^{n-k}\} s_1^k s_2^{n-k}$$

$$= 1 + m_{10}s_1 + m_{01}s_2 + \frac{1}{2}\left(m_{20}s_1^2 + 2m_{11}s_2 + m_{02}s_2^2\right) + \cdots \quad (6\text{-}197)$$

From this it follows that

$$\frac{\partial^k \partial^r}{\partial s_1^k \partial s_2^r} \Phi(0, 0) = m_{kr} \quad (6\text{-}198)$$

The derivatives of the function $\Psi(s_1, s_2) = \ln \Phi(s_1, s_2)$ are by definition the joint cumulants λ_{kr} of **x** and **y**. It can be shown that

$$\lambda_{10} = m_{10} \qquad \lambda_{01} = m_{01} \qquad \lambda_{20} = \mu_{20} \qquad \lambda_{02} = \mu_{02} \qquad \lambda_{11} = \mu_{11}$$

Hence

$$\Psi(s_1, s_2) = \eta_1 s_1 + \eta_2 s_2 + \frac{1}{2}\left(\sigma_1^2 s_1^2 + 2r\sigma_1\sigma_2 s_1 s_2 + \sigma_2^2 s_2^2\right) + \cdots$$

◀

EXAMPLE 6-36

▶ Using (6-197), we shall show that if the random variables **x** and **y** are jointly normal with zero mean, then

$$E\{\mathbf{x}^2\mathbf{y}^2\} = E\{\mathbf{x}^2\}E\{\mathbf{y}^2\} + 2E^2\{\mathbf{xy}\} \quad (6\text{-}199)$$

SOLUTION

As we see from (6-195)

$$\Phi(s_1, s_2) = e^{-A} \qquad A = \frac{1}{2}\left(\sigma_1^2 s_1^2 + 2Cs_1s_2 + \sigma_2^2 s_2^2\right)$$

where $C = E\{\mathbf{xy}\} = r\sigma_1\sigma_2$. To prove (6-199), we shall equate the coefficient

$$\frac{1}{4!}\binom{4}{2} E\{\mathbf{x}^2\mathbf{y}^2\}$$

of $s_1^2 s_2^2$ in (6-197) with the corresponding coefficient of the expansion of e^{-A}. In this expansion, the factor $s_1^2 s_2^2$ appear only in the terms

$$\frac{A^2}{2} = \frac{1}{8}\left(\sigma_1^2 s_1^2 + 2Cs_1s_2 + \sigma_2^2 s_2^2\right)^2$$

Hence

$$\frac{1}{4!}\binom{4}{2} E\{\mathbf{x}^2\mathbf{y}^2\} = \frac{1}{8}\left(2\sigma_1^2\sigma_2^2 + 4C^2\right)$$

and (6-199) results. ◀

THEOREM 6-7

PRICE'S THEOREM[5]

▶ Given two jointly normal random variables **x** and **y**, we form the mean

$$I = E\{g(\mathbf{x}, \mathbf{y})\} = \int_{-\infty}^{\infty}\int_{-\infty}^{\infty} g(x, y) f(x, y)\, dx\, dy \quad (6\text{-}200)$$

---

[5] R. Price, "A Useful Theorem for Nonlinear Devices Having Gaussian Inputs," *IRE, PGIT*, Vol. IT-4, 1958. See also A. Papoulis, "On an Extension of Price's Theorem," *IEEE Transactions on Information Theory*, Vol. IT-11, 1965.

of some function $g(\mathbf{x}, \mathbf{y})$ of $(\mathbf{x}, \mathbf{y})$. The above integral is a function $I(\mu)$ of the covariance μ of the random variables $\mathbf{x}$ and $\mathbf{y}$ and of four parameters specifying the joint density $f(x, y)$ of $\mathbf{x}$ and $\mathbf{y}$. We shall show that if $g(x, y)f(x, y) \to 0$ as $(x, y) \to \infty$, then

$$\frac{\partial^n I(\mu)}{\partial \mu^n} = \int_{-\infty}^{\infty}\int_{-\infty}^{\infty} \frac{\partial^{2n} g(x, y)}{\partial x^n \partial y^n} f(x, y)\,dx\,dy = E\left(\frac{\partial^{2n} g(\mathbf{x}, \mathbf{y})}{\partial \mathbf{x}^n \partial \mathbf{y}^n}\right) \qquad (6\text{-}201)$$

Proof. Inserting (6-187) into (6-200) and differentiating with respect to μ, we obtain

$$\frac{\partial^n I(\mu)}{\partial \mu^n} = \frac{(-1)^n}{4\pi^2} \int_{-\infty}^{\infty}\int_{-\infty}^{\infty} g(x, y)$$
$$\times \int_{-\infty}^{\infty}\int_{-\infty}^{\infty} \omega_1^n \omega_2^n \Phi(\omega_1, \omega_2) e^{-j(\omega_1 x + \omega_2 y)}\,d\omega_1\,d\omega_2\,dx\,dy$$

From this and the derivative theorem, it follows that

$$\frac{\partial^n I(\mu)}{\partial \mu^n} = \int_{-\infty}^{\infty}\int_{-\infty}^{\infty} g(x, y) \frac{\partial^{2n} f(x, y)}{\partial x^n \partial y^n}\,dx\,dy$$

After repeated integration by parts and using the condition at ∞, we obtain (6-201) (see also Prob. 5-48). ◀

EXAMPLE 6-37

▶ Using Price's theorem, we shall rederive (6-199). Setting $g(\mathbf{x}, \mathbf{y}) = \mathbf{x}^2\mathbf{y}^2$ into (6-201), we conclude with $n = 1$ that

$$\frac{\partial I(\mu)}{\partial \mu} = E\left(\frac{\partial^2 g(\mathbf{x}, \mathbf{y})}{\partial \mathbf{x}\,\partial \mathbf{y}}\right) = 4E\{\mathbf{xy}\} = 4\mu \qquad I(\mu) = \frac{4\mu^2}{2} + I(0)$$

If $\mu = 0$, the random variables $\mathbf{x}$ and $\mathbf{y}$ are independent; hence $I(0) = E(\mathbf{x}^2\mathbf{y}^2) = E(\mathbf{x}^2)E(\mathbf{y}^2)$ and (6-199) results. ◀

6-6 CONDITIONAL DISTRIBUTIONS

As we have noted, the conditional distributions can be expressed as conditional probabilities:

$$F_z(z \mid M) = P\{\mathbf{z} \le z \mid M\} = \frac{P\{\mathbf{z} \le z, M\}}{P(M)}$$

$$F_{zw}(z, w \mid M) = P\{\mathbf{z} \le z, \mathbf{w} \le w \mid M\} = \frac{P\{\mathbf{z} \le z, \mathbf{w} \le w, M\}}{P(M)} \qquad (6\text{-}202)$$

The corresponding densities are obtained by appropriate differentiations. In this section, we evaluate these functions for various special cases.

EXAMPLE 6-38

▶ We shall first determine the conditional distribution $F_y(y \mid \mathbf{x} \le x)$ and density $f_y(y \mid \mathbf{x} \le x)$.

With $M = \{\mathbf{x} \le x\}$, (6-202) yields

$$F_y(y \mid \mathbf{x} \le x) = \frac{P\{\mathbf{x} \le x, \mathbf{y} \le y\}}{P\{\mathbf{x} \le x\}} = \frac{F(x, y)}{F_x(x)}$$

$$f_y(y \mid \mathbf{x} \le x) = \frac{\partial F(x, y)/\partial y}{F_x(x)}$$

◀

EXAMPLE 6-39 ▶ We shall next determine the conditional distribution $F(x, y \mid M)$ for $M = \{x_1 < \mathbf{x} \le x_2\}$. In this case, $F(x, y \mid M)$ is given by

$$F(x, y \mid x_1 < \mathbf{x} \le x_2) = \frac{P\{\mathbf{x} \le x, \mathbf{y} \le y, x_1 < \mathbf{x} \le x_2\}}{P\{x_1 < \mathbf{x} \le x_2\}}$$

$$= \begin{cases} \dfrac{F(x_2, y) - F(x_1, y)}{F_x(x_2) - F_x(x_1)} & x > x_2 \\[2ex] \dfrac{F(x, y) - F(x_1, y)}{F_x(x_2) - F_x(x_1)} & x_1 < x \le x_2 \end{cases}$$

and it equals 0 for $x \le x_1$. Since $f = \partial^2 F/\partial x \partial y$, this yields

$$f(x, y \mid x_1 < \mathbf{x} \le x_2) = \frac{f(x, y)}{F_x(x_2) - F_x(x_1)} \qquad x_1 < x \le x_2 \tag{6-203}$$

and 0 otherwise.

The determination of the conditional density of $\mathbf{y}$ assuming $\mathbf{x} = x$ is of particular interest. This density cannot be derived directly from (6-202) because, in general, the event $\{\mathbf{x} = x\}$ has zero probability. It can, however, be defined as a limit. Suppose first that

$$M = \{x_1 < x \le x_2\}$$

In this case, (6-202) yields

$$F_y(y \mid x_1 < \mathbf{x} \le x_2) = \frac{P\{x_1 < \mathbf{x} \le x_2, \mathbf{y} \le y\}}{P\{x_1 < x \le x_2\}} = \frac{F(x_2, y) - F(x_1, y)}{F_x(x_2) - F_x(x_1)}$$

Differentiating with respect to y, we obtain

$$f_y(y \mid x_1 < \mathbf{x} \le x_2) = \frac{\int_{x_1}^{x_2} f(x, y)\, dx}{F_x(x_2) - F_x(x_1)} \tag{6-204}$$

because [see (6-6)]

$$\frac{\partial F(x, y)}{\partial y} = \int_{-\infty}^{x} f(\alpha, y)\, d\alpha$$

To determine $f_y(y \mid \mathbf{x} = x)$, we set $x_1 = x$ and $x_2 = x + \Delta x$ in (6-204). This yields

$$f_y(y \mid x < \mathbf{x} \le x + \Delta x) = \frac{\int_{x}^{x+\Delta x} f(\alpha, y)\, d\alpha}{F_x(x + \Delta x) - F_x(x)} \simeq \frac{f(x, y)\Delta x}{f_x(x)\Delta x}$$

Hence

$$f_y(y \mid \mathbf{x} = x) = \lim_{\Delta x \to 0} f_y(y \mid x < \mathbf{x} \le x + \Delta x) = \frac{f(x, y)}{f_x(x)}$$

If there is no fear of ambiguity, the function $f_y(y \mid \mathbf{x} = x) = f_{y|x}(y \mid x)$ will be written in the form $f(y \mid x)$. Defining $f(x \mid y)$ similarly, we obtain

$$f(y \mid x) = \frac{f(x, y)}{f(x)} \qquad f(x \mid y) = \frac{f(x, y)}{f(y)} \tag{6-205}$$

If the random variables $\mathbf{x}$ and $\mathbf{y}$ are independent, then

$$f(x, y) = f(x)f(y) \qquad f(y \mid x) = f(y) \qquad f(x \mid y) = f(x)$$

◀

Next we shall illustrate the method of obtaining conditional p.d.f.s through an example.

EXAMPLE 6-40 ▶ Given

$$f_{xy}(x, y) = \begin{cases} k & 0 < x < y < 1 \\ 0 & \text{otherwise} \end{cases} \tag{6-206}$$

determine $f_{x \mid y}(x \mid y)$ and $f_{y \mid x}(y \mid x)$.

SOLUTION

The joint p.d.f. is given to be a constant in the shaded region in Fig. 6-29. This gives

$$\iint f_{xy}(x, y)\, dx\, dy = \int_0^1 \int_0^y k\, dx\, dy = \int_0^1 ky\, dy = \frac{k}{2} = 1 \Rightarrow k = 2$$

Similarly

$$f_x(x) = \int f_{xy}(x, y)\, dy = \int_x^1 k\, dy = k(1 - x) \qquad 0 < x < 1 \tag{6-207}$$

and

$$f_y(y) = \int f_{xy}(x, y)\, dx = \int_0^y k\, dx = ky \qquad 0 < y < 1 \tag{6-208}$$

From (6-206)–(6-208), we get

$$f_{x \mid y}(x \mid y) = \frac{f_{xy}(x, y)}{f_y(y)} = \frac{1}{y} \qquad 0 < x < y < 1 \tag{6-209}$$

and

$$f_{y \mid x}(y \mid x) = \frac{f_{xy}(x, y)}{f_x(x)} = \frac{1}{1 - x} \qquad 0 < x < y < 1 \tag{6-210}$$

◀

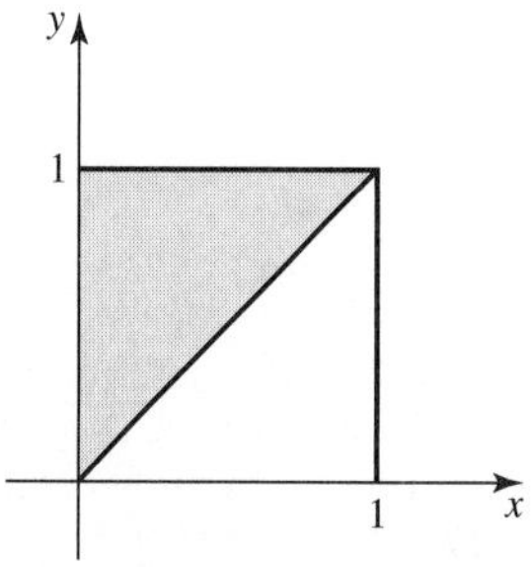

FIGURE 6-29

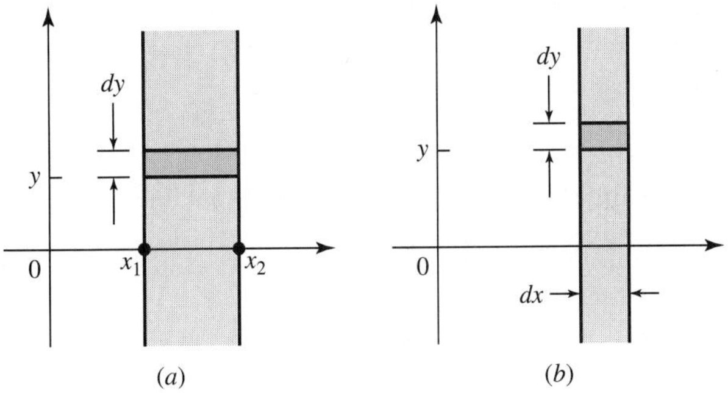

FIGURE 6-30

Notes 1. For a specific x, the function $f(x, y)$ is a *profile* of $f(x, y)$; that is, it equals the intersection of the surface $f(x, y)$ by the plane $x =$ constant. The conditional density $f(y \mid x)$ is the equation of this curve normalized by the factor $1/f(x)$ so as to make its area 1. The function $f(x \mid y)$ has a similar interpretation: It is the normalized equation of the intersection of the surface $f(x, y)$ by the plane $y =$ constant.

2. As we know, the product $f(y)dy$ equals the probability of the event $\{y < \mathbf{y} \le y + dy\}$. Extending this to conditional probabilities, we obtain

$$f_y(y \mid x_1 < \mathbf{x} \le x_2)\, dy = \frac{P\{x_1 < \mathbf{x} \le x_2,\, y < \mathbf{y} \le y + dy\}}{P\{x_1 < \mathbf{x} \le x_2\}}$$

This equals the mass in the rectangle of Fig. 6-30*a* divided by the mass in the vertical strip $x_1 < \mathbf{x} \le x_2$. Similarly, the product $f(y \mid x)dy$ equals the ratio of the mass in the differential rectangle $dx\, dy$ of Fig. 6-30*b* over the mass in the vertical strip $(x, x + dx)$.

3. The joint statistics of **x** and **y** are determined in terms of their joint density $f(x, y)$. Since

$$f(x, y) = f(y \mid x) f(x)$$

we conclude that they are also determined in terms of the marginal density $f(x)$ and the conditional density $f(y \mid x)$.

EXAMPLE 6-41

▶ We shall show that, if the random variables **x** and **y** are jointly normal with zero mean as in (6-61), then

$$f(y \mid x) = \frac{1}{\sigma_2 \sqrt{2\pi(1 - r^2)}} \exp\left(-\frac{(y - r\sigma_2 x/\sigma_1)^2}{2\sigma_2^2(1 - r^2)} \right) \tag{6-211}$$

Proof. The exponent in (6-61) equals

$$\frac{(y - r\sigma_2 x/\sigma_1)^2}{2\sigma_2^2(1 - r^2)} - \frac{x^2}{2\sigma_1^2}$$

Division by $f(x)$ removes the term $-x^2/2\sigma_1^2$ and (6-211) results.

The same reasoning leads to the conclusion that if **x** and **y** are jointly normal with $E\{\mathbf{x}\} = \eta_1$ and $E\{\mathbf{y}\} = \eta_2$, then $f(y \mid x)$ is given by (6-211) if y and x are replaced by $y - \eta_2$ and $x - \eta_1$, respectively. In other words, for a given x, $f(y \mid x)$ is a normal density with mean $\eta_2 + r\sigma_2(x - \eta_1)/\sigma_1$ and variance $\sigma_2^2(1 - r^2)$. ◀

BAYES' THEOREM AND TOTAL PROBABILITY. From (6-205) it follows that

$$f(x \mid y) = \frac{f(y \mid x) f(x)}{f(y)} \tag{6-212}$$

This is the density version of (2-43).

The denominator $f(y)$ can be expressed in terms of $f(y \mid x)$ and $f(x)$. Since

$$f(y) = \int_{-\infty}^{\infty} f(x, y)\, dx \quad \text{and} \quad f(x, y) = f(y \mid x) f(x)$$

we conclude that (total probability)

$$f(y) = \int_{-\infty}^{\infty} f(y \mid x) f(x)\, dx \tag{6-213}$$

Inserting into (6-212), we obtain Bayes' theorem for densities

$$f(x \mid y) = \frac{f(y \mid x) f(x)}{\int_{-\infty}^{\infty} f(y \mid x) f(x)\, dx} \tag{6-214}$$

Note As (6-213) shows, to remove the condition $\mathbf{x} = x$ from the conditional density $f(y \mid x)$, we multiply by the density $f(x)$ of $\mathbf{x}$ and integrate the product.

Equation (6-214) represents the p.d.f. version of Bayes' theorem. To appreciate the full significance of (6-214), we will look at a situation where observations are used to update our knowledge about unknown parameters. We shall illustrate this using the next example.

EXAMPLE 6-42 ▶ An unknown random phase $\boldsymbol{\theta}$ is uniformly distributed in the interval $(0, 2\pi)$, and $\mathbf{r} = \boldsymbol{\theta} + \mathbf{n}$, where $\mathbf{n} \sim N(0, \sigma^2)$. Determine $f(\theta \mid r)$.

SOLUTION

Initially almost nothing about the random phase $\boldsymbol{\theta}$ is known, so that we assume its a priori p.d.f. to be uniform in the interval $(0, 2\pi)$. In the equation $\mathbf{r} = \boldsymbol{\theta} + \mathbf{n}$, we can think of $\mathbf{n}$ as the noise contribution and $\mathbf{r}$ as the observation. In practical situations, it is reasonable to assume that $\boldsymbol{\theta}$ and $\mathbf{n}$ are independent. If we assume so, then

$$f(r \mid \boldsymbol{\theta} = \theta) \sim N(\theta, \sigma^2) \tag{6-215}$$

since it is given that $\boldsymbol{\theta}$ is a constant, and in that case $\mathbf{r} = \boldsymbol{\theta} + \mathbf{n}$ behaves like $\mathbf{n}$. Using (6-214), this gives the a posteriori p.d.f. of $\boldsymbol{\theta}$ given $\mathbf{r}$ to be (see Fig. 6-31*b*)

$$f(\theta \mid r) = \frac{f(r \mid \theta) f_\theta(\theta)}{\int_0^{2\pi} f(r \mid \theta) f_\theta(\theta)\, d\theta} = \frac{e^{-(r-\theta)^2/2\sigma^2}}{\int_0^{2\pi} e^{-(r-\theta)^2/2\sigma^2}\, d\theta}$$

$$= \varphi(r) e^{-(r-\theta)^2/2\sigma^2} \qquad 0 < \theta < 2\pi \tag{6-216}$$

where

$$\varphi(r) = \frac{1}{\int_0^{2\pi} e^{-(r-\theta)^2/2\sigma^2}\, d\theta}$$

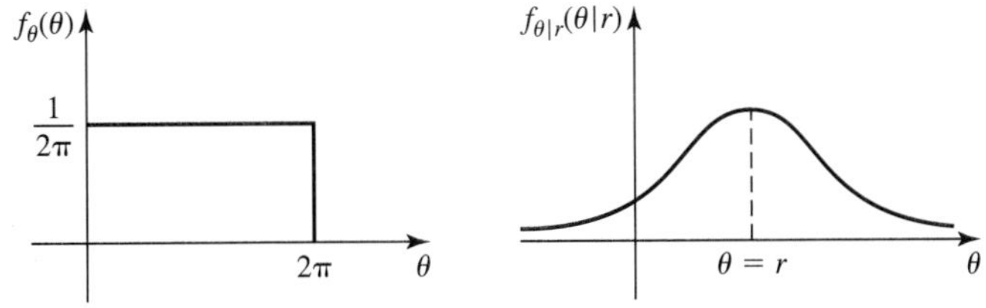

(*a*) *a priori* p.d.f of θ (*b*) *a posteriori* p.d.f of θ

FIGURE 6-31

Notice that knowledge about the observation **r** is reflected in the a posteriori p.d.f. of $\boldsymbol{\theta}$ in Fig. 6.31*b*. It is no longer flat as the a priori p.d.f. in Fig. 6.31*a*, and it shows higher probabilities in the neighborhood of $\boldsymbol{\theta} = r$.

Discrete Type Random Variables: Suppose that the random variables **x** and **y** are of discrete type

$$P\{\mathbf{x} = x_i\} = p_i \qquad P\{\mathbf{y} = y_k\} = q_k$$

$$P\{\mathbf{x} = x_i, \mathbf{y} = y_k\} = p_{ik} \qquad i = 1, \ldots, M \qquad k = 1, \ldots, N$$

where [see (6-34)]

$$p_i = \sum_k p_{ik} \qquad q_k = \sum_i p_{ik}$$

From the material just presented and (2-33) it follows that

$$P\{\mathbf{y} = y_k \,|\, \mathbf{x} = x_i\} = \frac{P\{\mathbf{x} = x_i, \mathbf{y} = y_k\}}{P\{\mathbf{x} = x_i\}} = \frac{p_{ik}}{p_i}$$

MARKOV MATRIX. We denote by π_{ik} the above conditional probabilities

$$P\{\mathbf{y} = y_k \,|\, \mathbf{x} = x_i\} = \pi_{ik}$$

and by P the $M \times N$ matrix whose elements are π_{ik}. Clearly,

$$\pi_{ik} = \frac{p_{ik}}{p_i} \tag{6-217}$$

Hence

$$\pi_{ik} \geq 0, \quad \sum_k \pi_{ik} = 1 \tag{6-218}$$

Thus the elements of the matrix P are nonnegative and the sum on each row equals 1. Such a matrix is called *Markov* (see Chap. 15 for further details). The conditional probabilities

$$P\{\mathbf{x} = x_i \,|\, \mathbf{y} = y_k\} = \pi^{ki} = \frac{p_{ik}}{q_k}$$

are the elements of an $N \times M$ Markov matrix.

If the random variables **x** and **y** are independent, then

$$p_{ik} = p_i q_k \qquad \pi_{ik} = q_k \qquad \pi^{ki} = p_i$$

We note that

$$\pi^{ki} = \pi_{ik}\frac{p_i}{q_k} \qquad q_k = \sum_i \pi_{ik} p_i \tag{6-219}$$

These equations are the discrete versions of Eqs. (6-212) and (6-213).

Next we examine some interesting results involving conditional distributions and independent binomial/Poisson random variables.

EXAMPLE 6-43

▶ Suppose **x** and **y** are independent binomial random variables with parameters (m, p) and (n, p) respectively. Then $\mathbf{x}+\mathbf{y}$ is also binomial with parameter $(m+n, p)$, so that

$$P\{\mathbf{x}=x \mid \mathbf{x}+\mathbf{y}=x+y\} = \frac{P\{\mathbf{x}=x\}P\{\mathbf{y}=y\}}{P\{\mathbf{x}+\mathbf{y}=x+y\}} = \frac{\binom{m}{x}\binom{n}{y}}{\binom{m+n}{x+y}} \tag{6-220}$$

Thus the conditional distribution of **x** given $\mathbf{x}+\mathbf{y}$ is hypergeometric. Show that the converse of this result which states that if **x** and **y** are nonnegative independent random variables such that $P\{\mathbf{x}=0\} > 0$, $P\{\mathbf{y}=0\} > 0$ and the conditional distribution of **x** given $\mathbf{x}+\mathbf{y}$ is hypergeometric as in (6-220), then **x** and **y** are binomial random variables.

SOLUTION

From (6-220)

$$\frac{P\{\mathbf{x}=x\}}{\binom{m}{x}}\,\frac{P\{\mathbf{y}=y\}}{\binom{n}{y}} = \frac{P\{\mathbf{x}+\mathbf{y}=x+y\}}{\binom{m+n}{x+y}}$$

Let

$$\frac{P\{\mathbf{x}=x\}}{\binom{m}{x}} = f(x) \qquad \frac{P\{\mathbf{y}=y\}}{\binom{n}{y}} = g(y) \qquad \frac{P\{\mathbf{x}+\mathbf{y}=x+y\}}{\binom{m+n}{x+y}} = h(x+y)$$

Then

$$h(x+y) = f(x)\,g(y)$$

and hence

$$h(1) = f(1)g(0) = f(0)g(1)$$

$$h(2) = f(2)g(0) = f(1)g(1) = f(0)g(2)$$

$$\vdots$$

$$h(k) = f(k)g(0) = f(k-1)g(1) = \cdots$$

Thus

$$f(k) = f(k-1)\frac{g(1)}{g(0)} = f(0)\left(\frac{g(1)}{g(0)}\right)^k$$

or

$$P\{\mathbf{x}=k\} = \binom{m}{k} P\{\mathbf{x}=0\}a^k \qquad k = 0, 1, \ldots \tag{6-221}$$

where $a = g(1)/g(0) > 0$. But $\sum_{k=0}^{m} P\{\mathbf{x}=k\} = 1$ gives $P\{\mathbf{x}=0\}(1+a)^m = 1$, or $P\{\mathbf{x}=0\} = q^m$, where $q = 1/(1+a) < 1$. Hence $a = p/q$, where $p = 1-q > 0$, and from (6-221) we obtain

$$P\{\mathbf{x}=k\} = \binom{m}{k} p^k q^{m-k} \qquad k = 0, 1, 2, \ldots m$$

Similarly, it follows that

$$P\{\mathbf{y}=r\} = \binom{n}{r} p^r q^{n-r} \qquad r = 0, 1, 2, \ldots n$$

and the proof is complete.

Similarly if **x** and **y** are independent Poisson random variables with parameters λ and μ respectively, then their sum is also Poisson with parameter $\lambda + \mu$ [see (6-86)], and, moreover,

$$P\{\mathbf{x}=k \mid \mathbf{x}+\mathbf{y}=n\} = \frac{P\{\mathbf{x}=k\}P\{\mathbf{y}=n-k\}}{P\{\mathbf{x}+\mathbf{y}=n\}} = \frac{e^{-\lambda}\frac{\lambda^k}{k!}e^{-\mu}\frac{\mu^{n-k}}{(n-k)!}}{e^{-(\lambda+\mu)}\frac{(\lambda+\mu)^n}{n!}}$$

$$= \binom{n}{k}\left(\frac{\lambda}{\lambda+\mu}\right)^k\left(\frac{\mu}{\lambda+\mu}\right)^{n-k} \qquad k = 0, 1, 2, \ldots n \qquad (6\text{-}222)$$

Thus if **x** and **y** are independent Poisson random variables, then the conditional density of **x** given **x** + **y** is Binomial. Interestingly, the converse is also true, when **x** and **y** are independent random variables. The proof is left as an exercise.

Equivalently, this shows that if $\mathbf{y} = \sum_{i=1}^{\mathbf{n}} \mathbf{x}_i$ where $\mathbf{x}_i$ are independent Bernoulli random variables as in (4-55) and **n** is a Poisson random variable with parameter λ as in (4-57), then $\mathbf{y} \sim P(p\lambda)$ and $\mathbf{z} = \mathbf{n} - \mathbf{y} \sim P((1-p)\lambda)$. Further, **y** and **z** are independent random variables. Thus, for example, if the total number of eggs that a bird lays follows a Poisson random variable with parameter λ, and if each egg survives with probability p, then the number of baby birds that survive is also Poisson with parameter $p\lambda$. ◀

System Reliability

We shall use the term *system* to identify a physical device used to perform a certain function. The device might be a simple element, a lightbulb, for example, or a more complicated structure. We shall call the time interval from the moment the system is put into operation until it fails the *time to failure*. This interval is, in general, random. It specifies, therefore, a random variable $\mathbf{x} \geq 0$. The distribution $F(t) = P\{\mathbf{x} \leq t\}$ of this random variable is the probability that the system fails prior to time t where we assume that $t = 0$ is the moment the system is put into operation. The difference

$$R(t) = 1 - F(t) = P\{\mathbf{x} > t\}$$

is the *system reliability*. It equals the probability that the system functions at time t.

The *mean time to failure* of a system is the mean of **x**. Since $F(x) = 0$ for $x < 0$, we conclude from (5-52) that

$$E\{\mathbf{x}\} = \int_0^\infty x f(x)\,dx = \int_0^\infty R(t)\,dt$$

The probability that a system functioning at time t fails prior to time $x > t$ equals

$$F(x \mid \mathbf{x} > t) = \frac{P\{\mathbf{x} \leq x, \mathbf{x} > t\}}{P\{\mathbf{x} > t\}} = \frac{F(x) - F(t)}{1 - F(t)}$$

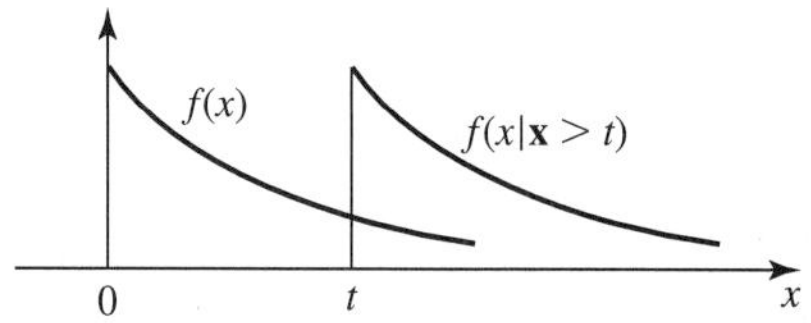

FIGURE 6-32

Differentiating with respect to x, we obtain

$$f(x \mid \mathbf{x} > t) = \frac{f(x)}{1 - F(t)} \qquad x > t \tag{6-223}$$

The product $f(x \mid \mathbf{x} > t)\,dx$ equals the probability that the system fails in the interval $(x, x + dx)$, assuming that it functions at time t.

EXAMPLE 6-44 ▶ If $f(x) = ce^{-cx}$, then $F(t) = 1 - e^{-ct}$ and (6-223) yields

$$f(x \mid \mathbf{x} > t) = \frac{ce^{-cx}}{e^{-ct}} = f(x - t)$$

This shows that the probability that a system functioning at time t fails in the interval $(x, x + dx)$ depends only on the difference $x - t$ (Fig. 6-32). We show later that this is true only if $f(x)$ is an exponential density. ◀

CONDITIONAL FAILURE RATE. The conditional density $f(x \mid \mathbf{x} > t)$ is a function of x and t. Its value at $x = t$ is a function only of t. This function is denoted by $\beta(t)$ and is called the *conditional failure rate* or, the *hazard rate* of the system. From (6-223) and the definition of hazard rate it follows that

$$\beta(t) = f(t \mid \mathbf{x} > t) = \frac{f(t)}{1 - F(t)} \tag{6-224}$$

The product $\beta(t)\,dt$ is the probability that a system functioning at time t fails in the interval $(t, t + dt)$. In Sec. 7-1 (Example 7-3) we interpret the function $\beta(t)$ as the expected failure rate.

EXAMPLE 6-45 ▶ (*a*) If $f(x) = ce^{-cx}$, then $F(t) = 1 - e^{-ct}$ and

$$\beta(t) = \frac{ce^{-ct}}{1 - (1 - e^{-ct})} = c$$

(*b*) If $f(x) = c^2 xe^{-cx}$, then $F(x) = 1 - cxe^{-cx} - e^{-cx}$ and

$$\beta(t) = \frac{c^2 te^{-ct}}{cte^{-ct} + e^{-ct}} = \frac{c^2 t}{1 + ct}$$ ◀

From (6-224) it follows that

$$\beta(t) = \frac{F'(t)}{1 - F(t)} = -\frac{R'(t)}{R(t)}$$

We shall use this relationship to express the distribution of $\mathbf{x}$ in terms of the function

$\beta(t)$. Integrating from 0 to x and using the fact that $\ln R(0) = 0$, we obtain

$$-\int_0^x \beta(t)\,dt = \ln R(x)$$

Hence

$$R(x) = 1 - F(x) = \exp\left\{-\int_0^x \beta(t)\,dt\right\}$$

And since $f(x) = F'(x)$, this yields

$$f(x) = \beta(x)\exp\left\{-\int_0^x \beta(t)\,dt\right\} \tag{6-225}$$

EXAMPLE 6-46

MEMORYLESS SYSTEMS

▶ A system is called *memoryless* if the probability that it fails in an interval (t, x), assuming that it functions at time t, depends only on the length of this interval. In other words, if the system works a week, a month, or a year after it was put into operation, it is as good as new. This is equivalent to the assumption that $f(x \mid \mathbf{x} > t) = f(x - t)$ as in Fig. 6-32. From this and (6-224) it follows that with $x = t$:

$$\beta(t) = f(t \mid \mathbf{x} > t) = f(t - t) = f(0) = c$$

and (6-225) yields $f(x) = ce^{-cx}$. Thus a system is memoryless iff $\mathbf{x}$ has an exponential density. ◀

EXAMPLE 6-47

▶ A special form of $\beta(t)$ of particular interest in reliability theory is the function

$$\beta(t) = ct^{b-1}$$

This is a satisfactory approximation of a variety of failure rates, at least near the origin. The corresponding $f(x)$ is obtained from (6-225):

$$f(x) = cx^{b-1}\exp\left\{-\frac{cx^b}{b}\right\} \tag{6-226}$$

This function is called the *Weibull density*. (See (4-43) and Fig. 4-16.) ◀

We conclude with the observation that the function $\beta(t)$ equals the value of the conditional density $f(x \mid \mathbf{x} > t)$ for $x = t$; however, $\beta(t)$ is not a density because its area is not one. In fact its area is infinite. This follows from (6-224) because $R(\infty) = 1 - F(\infty) = 0$.

INTERCONNECTION OF SYSTEMS. We are given two systems S_1 and S_2 with times to failure $\mathbf{x}$ and $\mathbf{y}$, respectively, and we connect them in parallel or in series or in standby as in Fig. 6-33, forming a new system S. We shall express the properties of S in terms of the joint distribution of the random variables $\mathbf{x}$ and $\mathbf{y}$.

Parallel: We say that the two systems are connected in parallel if S fails when both systems fail. Denoting by $\mathbf{z}$ the time to failure of S, we conclude that $\mathbf{z} = t$ when the

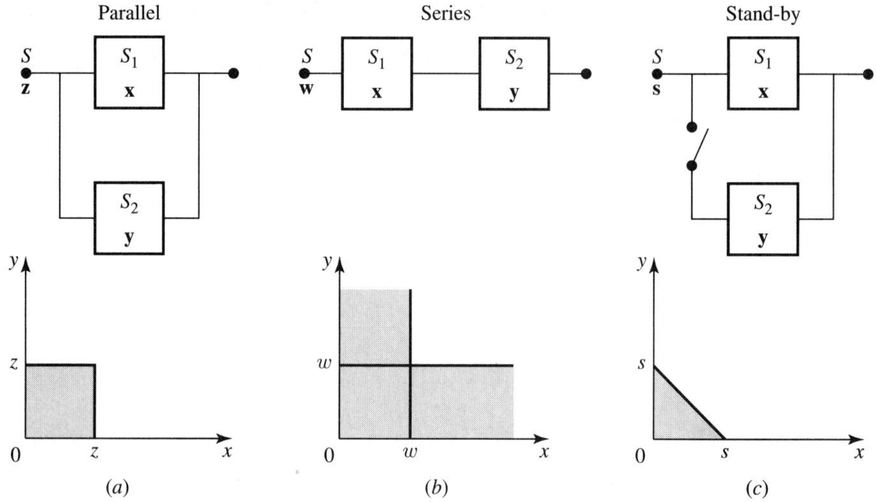

FIGURE 6-33

larger of the numbers **x** and **y** equals t. Hence [see (6-77)–(6-78)]

$$\mathbf{z} = \max(\mathbf{x}, \mathbf{y}) \qquad F_z(z) = F_{xy}(z, z)$$

If the random variables **x** and **y** are independent, $F_z(z) = F_x(z)F_y(z)$.

Series: We say that the two systems are connected in series if S fails when at least one of the two systems fails. Denoting by **w** the time to failure of S, we conclude that $\mathbf{w} = t$ when the smaller of the numbers **x** and **y** equals t. Hence [see (6-80)–(6-81)]

$$\mathbf{w} = \min(\mathbf{x}, \mathbf{y}) \qquad F_w(w) = F_x(w) + F_y(w) - F_{xy}(w, w)$$

If the random variables **x** and **y** are independent,

$$R_w(w) = R_x(w)R_y(w) \qquad \beta_w(t) = \beta_x(t) + \beta_y(t)$$

where $\beta_x(t)$, $\beta_y(t)$, and $\beta_w(t)$ are the conditional failure rates of systems S_1, S_2, and S, respectively.

Standby: We put system S_1 into operation, keeping S_2 in reserve. When S_1 fails, we put S_2 into operation. The system S so formed fails when S_2 fails. If t_1 and t_2 are the times of operation of S_1 and S_2, $t_1 + t_2$ is the time of operation of S. Denoting by **s** the time to failure of system S, we conclude that

$$\mathbf{s} = \mathbf{x} + \mathbf{y}$$

The distribution of **s** equals the probability that the point (**x**, **y**) is in the triangular shaded region of Fig. 6-33c. If the random variables **x** and **y** are independent, the density of **s** equals

$$f_s(s) = \int_0^s f_x(t)f_y(s-t)\,dt$$

as in (6-45).

6-7 CONDITIONAL EXPECTED VALUES

Applying theorem (5-55) to conditional densities, we obtain the conditional mean of $g(\mathbf{y})$:

$$E\{g(\mathbf{y}) \mid M\} = \int_{-\infty}^{\infty} g(y) f(y \mid M)\, dy \tag{6-227}$$

This can be used to define the conditional moments of $\mathbf{y}$.

Using a limit argument as in (6-205), we can also define the conditional mean $E\{g(\mathbf{y}) \mid x\}$. In particular,

$$\eta_{y|x} = E\{\mathbf{y} \mid x\} = \int_{-\infty}^{\infty} y f(y \mid x)\, dy \tag{6-228}$$

is the *conditional mean* of $\mathbf{y}$ assuming $\mathbf{x} = x$, and

$$\sigma_{y|x}^2 = E\{(\mathbf{y} - \eta_{y|x})^2 \mid x\} = \int_{-\infty}^{\infty} (y - \eta_{y|x})^2 f(y \mid x)\, dy \tag{6-229}$$

is its *conditional variance.*

We shall illustrate these calculations through an example.

EXAMPLE 6-48 ▶ Let

$$f_{xy}(x, y) = \begin{cases} 1 & 0 < |y| < x < 1 \\ 0 & \text{otherwise} \end{cases} \tag{6-230}$$

Determine $E\{\mathbf{x} \mid \mathbf{y}\}$ and $E\{\mathbf{y} \mid \mathbf{x}\}$.

SOLUTION

As Fig. 6-34 shows, $f_{xy}(x, y) = 1$ in the shaded area, and zero elsewhere. Hence

$$f_x(x) = \int_{-x}^{x} f_{xy}(x, y)\, dy = 2x \qquad 0 < x < 1$$

and

$$f_y(y) = \int_{-|y|}^{1} 1\, dx = 1 - |y| \qquad |y| < 1$$

This gives

$$f_{x|y}(x, y) = \frac{f_{xy}(x, y)}{f_y(y)} = \frac{1}{1 - |y|} \qquad 0 < |y| < x < 1 \tag{6-231}$$

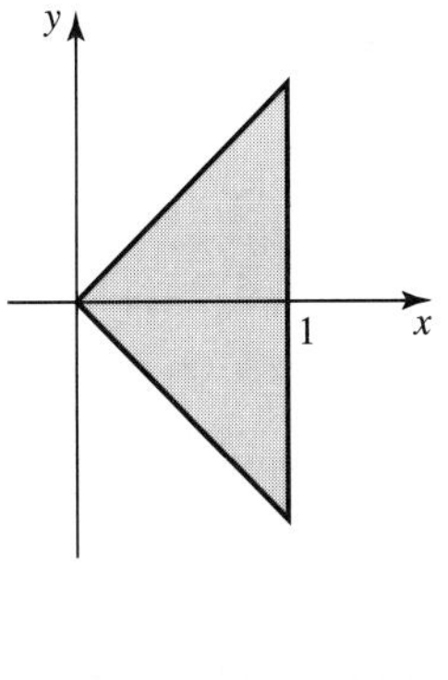

FIGURE 6-34

and

$$f_{y|x}(y\,|\,x) = \frac{f_{xy}(x,y)}{f_x(x)} = \frac{1}{2x} \qquad 0 < |y| < x < 1 \tag{6-232}$$

Hence

$$E\{\mathbf{x}\,|\,\mathbf{y}\} = \int x f_{x|y}(x\,|\,y)\,dx = \int_{|y|}^{1} \frac{x}{(1-|y|)}\,dx = \frac{1}{(1-|y|)}\frac{x^2}{2}\bigg|_{|y|}^{1}$$

$$= \frac{1-|y|^2}{2(1-|y|)} = \frac{1+|y|}{2} \qquad |y| < 1 \tag{6-233}$$

$$E\{\mathbf{y}\,|\,\mathbf{x}\} = \int y f_{y|x}(y\,|\,x)\,dy = \int_{-x}^{x} \frac{y}{2x}\,dy = \frac{1}{2x}\frac{y^2}{2}\bigg|_{-x}^{x} = 0 \qquad 0 < x < 1 \tag{6-234}$$

◀

For a given x, the integral in (6-228) is the center of gravity of the masses in the vertical strip $(x, x+dx)$. The locus of these points, as x varies from $-\infty$ to ∞, is the function

$$\varphi(x) = \int_{-\infty}^{\infty} y f(y\,|\,x)\,dy \tag{6-235}$$

known as the *regression line* (Fig. 6-35).

Note If the random variables $\mathbf{x}$ and $\mathbf{y}$ are functionally related, that is, if $\mathbf{y} = g(\mathbf{x})$, then the probability masses on the xy plane are on the line $y = g(x)$ (see Fig. 6-5b); hence $E\{\mathbf{y}\,|\,x\} = g(x)$.

Galton's law. The term *regression* has its origin in the following observation attributed to the geneticist Sir Francis Galton (1822–1911): "Population extremes *regress* toward their mean." This observation applied to parents and their adult children implies that children of tall (or short) parents are on the average shorter (or taller) than their parents. In statistical terms be phrased in terms of conditional expected values:

Suppose that the random variables $\mathbf{x}$ and $\mathbf{y}$ model the height of parents and their children respectively. These random variables have the same mean and variance, and they are positively correlated:

$$\eta_x = \eta_x = \eta \qquad \sigma_x = \sigma_y = \sigma \qquad r > 0$$

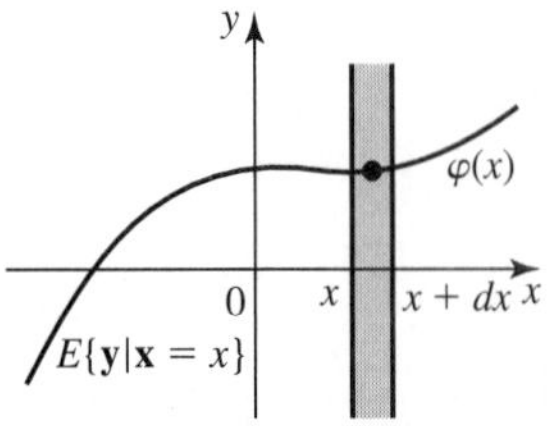

FIGURE 6-35

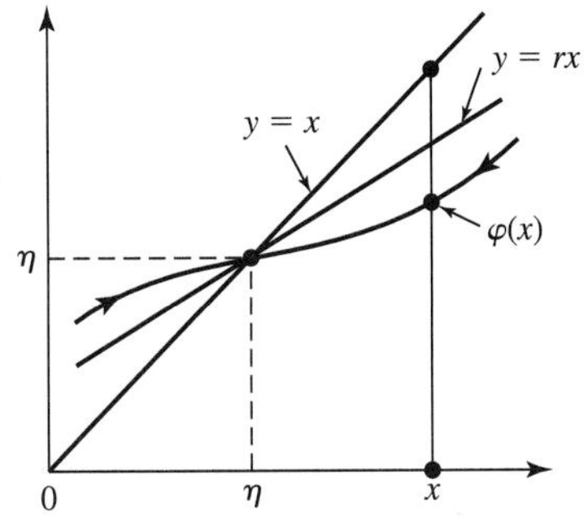

FIGURE 6-36

According to Galton's law, the conditional mean $E\{\mathbf{y} \mid x\}$ of the height of children whose parents height is x, is smaller (or larger) than x if $x > \eta$ (or $x < \eta$):

$$E\{\mathbf{y} \mid x\} = \varphi(x) \begin{cases} <x & \text{if} \quad x > \eta \\ >x & \text{if} \quad x < \eta \end{cases}$$

This shows that the regression line $\varphi(x)$ is below the line $y = x$ for $x > \eta$ and above this line if $x < \eta$ as in Fig. 6-36. If the random variables $\mathbf{x}$ and $\mathbf{y}$ are jointly normal, then [see (6-236) below] the regression line is the straight line $\varphi(x) = rx$. For arbitrary random variables, the function $\varphi(x)$ *does not* obey Galton's law. The term regression is used, however, to identify any conditional mean.

EXAMPLE 6-49

▶ If the random variables $\mathbf{x}$ and $\mathbf{y}$ are normal as in Example 6-41, then the function

$$E\{\mathbf{y} \mid x\} = \eta_2 + r\sigma_2 \frac{x - \eta_1}{\sigma_1} \tag{6-236}$$

is a straight line with slope $r\sigma_2/\sigma_1$ passing through the point (η_1, η_2). Since for normal random variables the conditional mean $E\{\mathbf{y} \mid x\}$ coincides with the maximum of $f(y \mid x)$, we conclude that the locus of the maxima of all profiles of $f(x, y)$ is the straight line (6-236).

From theorems (6-159) and (6-227) it follows that

$$E\{g(\mathbf{x}, \mathbf{y}) \mid M\} = \int_{-\infty}^{\infty} \int_{-\infty}^{\infty} g(x, y) f(x, y \mid M)\, dx\, dy \tag{6-237}$$

◀

This expression can be used to determine $E\{g(\mathbf{x}, \mathbf{y}) \mid x\}$; however, the conditional density $f(x, y \mid x)$ consists of line masses on the line x-constant. To avoid dealing with line masses, we shall define $E\{g(\mathbf{x}, \mathbf{y}) \mid x\}$ as a limit:

As we have shown in Example 6-39, the conditional density $f(x, y \mid x < \mathbf{x} < x + \Delta x)$ is 0 outside the strip $(x, x + \Delta x)$ and in this strip it is given by (6-203) where $x_1 = x$ and $x_2 = x + \Delta x$. It follows, therefore, from (6-237) with $M = \{x < \mathbf{x} \le x + \Delta x\}$ that

$$E\{g(\mathbf{x}, \mathbf{y}) \mid x < \mathbf{x} \le x + \Delta x\} = \int_{-\infty}^{\infty} \int_{x}^{x+\Delta x} g(\alpha, y) \frac{f(\alpha, y)\, d\alpha}{F_x(x + \Delta x) - F_x(x)}\, dy$$

As $\Delta x \to 0$, the inner integral tends to $g(x, y) f(x, y)/f(x)$. Defining $E\{g(\mathbf{x}, \mathbf{y}) \mid x\}$ as the limit of the above integral, we obtain

$$E\{g(\mathbf{x},\mathbf{y})\,|\,x\} = \int_{-\infty}^{\infty} g(x,y) f(y\,|\,x)\,dy \tag{6-238}$$

We also note that

$$E\{g(x,\mathbf{y})\,|\,x\} = \int_{-\infty}^{\infty} g(x,y) f(y\,|\,x)\,dy \tag{6-239}$$

because $g(x,\mathbf{y})$ is a function of the random variable $\mathbf{y}$, with x a parameter; hence its conditional expected value is given by (6-227). Thus

$$E\{g(\mathbf{x},\mathbf{y})\,|\,x\} = E\{g(x,\mathbf{y})\,|\,x\} \tag{6-240}$$

One might be tempted from the above to conclude that (6-240) follows directly from (6-227); however, this is not so. The functions $g(\mathbf{x},\mathbf{y})$ and $g(x,\mathbf{y})$ have the same expected value, assuming $\mathbf{x} = x$, but they are not equal. The first is a function $g(\mathbf{x},\mathbf{y})$ of the random variables $\mathbf{x}$ and $\mathbf{y}$, and for a specific ζ it takes the value $g[\mathbf{x}(\zeta), \mathbf{y}(\zeta)]$. The second is a function $g(x,\mathbf{y})$ of the real variable x and the random variable $\mathbf{y}$, and for a specific ζ it takes the value $g[x, \mathbf{y}(\zeta)]$ where x is an arbitrary number.

Conditional Expected Values as Random Variables

The conditional mean of $\mathbf{y}$, assuming $\mathbf{x} = x$, is a function $\varphi(x) = E\{\mathbf{y}\,|\,x\}$ of x given by (6-235). Using this function, we can construct the random variable $\varphi(\mathbf{x}) = E\{\mathbf{y}\,|\,\mathbf{x}\}$ as in Sec. 5-1. As we see from (5-55), the mean of this random variable equals

$$E\{\varphi(\mathbf{x})\} = \int_{-\infty}^{\infty} \varphi(x) f(x)\,dx = \int_{-\infty}^{\infty} f(x) \int_{-\infty}^{\infty} y f(y\,|\,x)\,dy\,dx$$

Since $f(x,y) = f(x) f(y\,|\,x)$, the last equation yields

$$E\{E\{\mathbf{y}\,|\,\mathbf{x}\}\} = \int_{-\infty}^{\infty}\int_{-\infty}^{\infty} y f(x,y)\,dx\,dy = E\{\mathbf{y}\} \tag{6-241}$$

This basic result can be generalized: The conditional mean $E\{g(\mathbf{x},\mathbf{y})\,|\,x\}$ of $g(\mathbf{x},\mathbf{y})$, assuming $\mathbf{x} = x$, is a function of the real variable x. It defines, therefore, the function $E\{g(\mathbf{x},\mathbf{y})\,|\,\mathbf{x}\}$ of the random variable $\mathbf{x}$. As we see from (6-159) and (6-237), the mean of $E\{g(\mathbf{x},\mathbf{y})\,|\,\mathbf{x}\}$ equals

$$\int_{-\infty}^{\infty} f(x) \int_{-\infty}^{\infty} g(x,y) f(y\,|\,x)\,dy\,dx = \int_{-\infty}^{\infty}\int_{-\infty}^{\infty} g(x,y) f(x,y)\,dx\,dy$$

But the last integral equals $E\{g(\mathbf{x},\mathbf{y})\}$; hence

$$E\{E\{g(\mathbf{x},\mathbf{y})\,|\,\mathbf{x}\}\} = E\{g(\mathbf{x},\mathbf{y})\} \tag{6-242}$$

We note, finally, that

$$E\{g_1(\mathbf{x})g_2(\mathbf{y})\,|\,x\} = E\{g_1(x)g_2(\mathbf{y})\,|\,x\} = g_1(x)E\{g_2(\mathbf{y})\,|\,x\} \tag{6-243}$$

$$E\{g_1(\mathbf{x})g_2(\mathbf{y})\} = E\{E\{g_1(\mathbf{x})g_2(\mathbf{y})\,|\,\mathbf{x}\}\} = E\{g_1(\mathbf{x})E\{g_2(\mathbf{y})\,|\,\mathbf{x}\}\} \tag{6-244}$$

EXAMPLE 6-50 ▶ Suppose that the random variables **x** and **y** are $N(0, 0, \sigma_1^2, \sigma_2^2, r)$. As we know

$$E\{\mathbf{x}^2\} = \sigma_1^2 \qquad E\{\mathbf{x}^4\} = 3\sigma_1^4$$

Furthermore, $f(y \mid x)$ is a normal density with mean $r\sigma_2 x/\sigma_1$ and variance $\sigma_2\sqrt{1-r^2}$. Hence

$$E\{\mathbf{y}^2 \mid x\} = \eta_{y|x}^2 + \sigma_{y|x}^2 = \left(\frac{r\sigma_2 x}{\sigma_1}\right)^2 + \sigma_2^2(1-r^2) \tag{6-245}$$

Using (6-244), we shall show that

$$E\{\mathbf{xy}\} = r\sigma_1\sigma_2 \qquad E\{\mathbf{x}^2\mathbf{y}^2\} = E\{\mathbf{x}^2\}E\{\mathbf{y}^2\} + 2E^2\{\mathbf{xy}\} \tag{6-246}$$

Proof.

$$E\{\mathbf{xy}\} = E\{\mathbf{x}E\{\mathbf{y} \mid \mathbf{x}\}\} = E\left\{r\sigma_2\frac{\mathbf{x}^2}{\sigma_1}\right\} = r\sigma_2\frac{\sigma_1^2}{\sigma_1}$$

$$E\{\mathbf{x}^2\mathbf{y}^2\} = E\{\mathbf{x}^2 E\{\mathbf{y}^2 \mid \mathbf{x}\}\} = E\left\{\mathbf{x}^2\left[r^2\sigma_2^2\frac{\mathbf{x}^2}{\sigma_1^2} + \sigma_2^2(1-r^2)\right]\right\}$$

$$= 3\sigma_1^4 r^2\frac{\sigma_2^2}{\sigma_1^2} + \sigma_1^2\sigma_2^2(1-r^2) = \sigma_1^2\sigma_2^2 + 2r^2\sigma_1^2\sigma_2^2$$

and the proof is complete [see also (6-199)]. ◀

PROBLEMS

6-1 **x** and **y** are independent, identically distributed (i.i.d.) random variables with common p.d.f.

$$f_x(x) = e^{-x}U(x) \qquad f_y(y) = e^{-y}U(y)$$

Find the p.d.f. of the following random variables (*a*) $\mathbf{x}+\mathbf{y}$, (*b*) $\mathbf{x}-\mathbf{y}$, (*c*) $\mathbf{xy}$, (*d*) $\mathbf{x}/\mathbf{y}$, (*e*) $\min(\mathbf{x},\mathbf{y})$, (*f*) $\max(\mathbf{x},\mathbf{y})$, (*g*) $\min(\mathbf{x},\mathbf{y})/\max(\mathbf{x},\mathbf{y})$.

6-2 **x** and **y** are independent and uniform in the interval $(0, a)$. Find the p.d.f. of (*a*) $\mathbf{x}/\mathbf{y}$, (*b*) $\mathbf{y}/(\mathbf{x}+\mathbf{y})$, (*c*) $|\mathbf{x}-\mathbf{y}|$.

6-3 The joint p.d.f. of the random variables **x** and **y** is given by

$$f_{xy}(x,y) = \begin{cases} 1 & \text{in the shaded area} \\ 0 & \text{otherwise} \end{cases}$$

Let $\mathbf{z} = \mathbf{x}+\mathbf{y}$. Find $F_z(z)$ and $f_z(z)$.

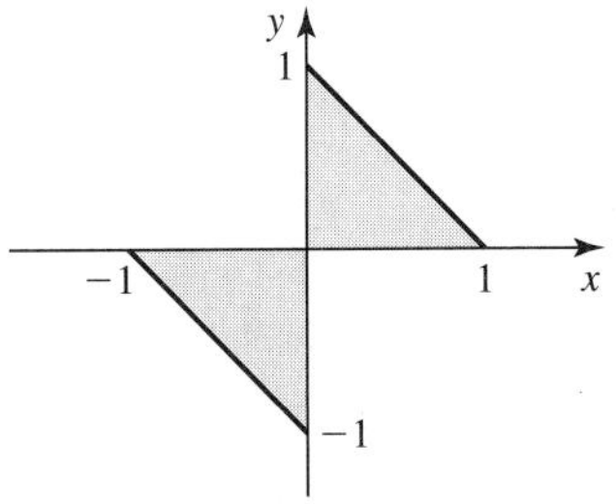

FIGURE P6-3

6-4 The joint p.d.f. of **x** and **y** is defined as

$$f_{xy}(x, y) = \begin{cases} 6x & x \geq 0, y \geq 0, x + y \leq 1 \\ 0 & \text{otherwise} \end{cases}$$

Define $\mathbf{z} = \mathbf{x} - \mathbf{y}$. Find the p.d.f. of **z**.

6-5 **x** and **y** are independent identically distributed normal random variables with zero mean and common variance σ^2, that is, $\mathbf{x} \sim N(0, \sigma^2)$, $\mathbf{y} \sim N(0, \sigma^2)$ and $f_{xy}(x, y) = f_x(x)f_y(y)$. Find the p.d.f. of (*a*) $\mathbf{z} = \sqrt{\mathbf{x}^2 + \mathbf{y}^2}$, (*b*) $\mathbf{w} = \mathbf{x}^2 + \mathbf{y}^2$, (*c*) $\mathbf{u} = \mathbf{x} - \mathbf{y}$.

6-6 The joint p.d.f. of **x** and **y** is given by

$$f_{xy}(x, y) = \begin{cases} 2(1 - x) & 0 < x \leq 1, 0 \leq y \leq 1 \\ 0 & \text{otherwise} \end{cases}$$

Determine the probability density function of $\mathbf{z} = \mathbf{xy}$.

6-7 Given

$$f_{xy}(x, y) = \begin{cases} x + y & 0 \leq x \leq 1, 0 \leq y \leq 1 \\ 0 & \text{otherwise} \end{cases}$$

Show that (*a*) $\mathbf{x} + \mathbf{y}$ has density $f_1(z) = z^2$, $0 < z < 1$, $f_1(z) = z(2 - z)$, $1 < z < 2$, and 0 elsewhere. (*b*) **xy** has density $f_2(z) = 2(1 - z)$, $0 < z < 1$, and 0 elsewhere. (*c*) $\mathbf{y}/\mathbf{x}$ has density $f_3(z) = (1 + z)/3$, $0 < z < 1$, $f_3(z) = (1 + z)/3z^3$, $z > 1$, and 0 elsewhere. (*d*) $\mathbf{y} - \mathbf{x}$ has density $f_4(z) = 1 - |z|$, $|z| < 1$, and 0 elsewhere.

6-8 Suppose **x** and **y** have joint density

$$f_{xy}(x, y) = \begin{cases} 1 & 0 \leq x \leq 2, 0 \leq y \leq 1, 2y \leq x \\ 0 & \text{otherwise} \end{cases}$$

Show that $\mathbf{z} = \mathbf{x} + \mathbf{y}$ has density

$$f_{xy}(x, y) = \begin{cases} (1/3)z & 0 < z < 2 \\ 2 - (2/3)z & 2 < z < 3 \\ 0 & \text{elsewhere} \end{cases}$$

6-9 **x** and **y** are uniformly distributed on the triangular region $0 \leq y \leq x \leq 1$. Show that (*a*) $\mathbf{z} = \mathbf{x}/\mathbf{y}$ has density $f_z(z) = 1/z^2$, $z \geq 1$, and $f_z(z) = 0$, otherwise. (*b*) Determine the density of **xy**.

6-10 **x** and **y** are uniformly distributed on the triangular region $0 < x \leq y \leq x + y \leq 2$. Find the p.d.f. of $\mathbf{x} + \mathbf{y}$ and $\mathbf{x} - \mathbf{y}$.

6-11 **x** and **y** are independent Gamma random variables with common parameters α and β. Find the p.d.f. of (*a*) $\mathbf{x} + \mathbf{y}$, (*b*) $\mathbf{x}/\mathbf{y}$, (*c*) $\mathbf{x}/(\mathbf{x} + \mathbf{y})$.

6-12 **x** and **y** are independent uniformly distributed random variables on $(0, 1)$. Find the joint p.d.f. of $\mathbf{x} + \mathbf{y}$ and $\mathbf{x} - \mathbf{y}$.

6-13 **x** and **y** are independent Rayleigh random variables with common parameter σ^2. Determine the density of $\mathbf{x}/\mathbf{y}$.

6-14 The random variables **x** and **y** are independent and $\mathbf{z} = \mathbf{x} + \mathbf{y}$. Find $f_y(y)$ if

$$f_x(x) = ce^{-cx}U(x) \qquad f_z(z) = c^2ze^{-cz}U(z)$$

6-15 The random variables **x** and **y** are independent and **y** is uniform in the interval $(0, 1)$. Show that, if $\mathbf{z} = \mathbf{x} + \mathbf{y}$, then

$$f_z(z) = F_x(z) - F_x(z - 1)$$

6-16 (*a*) The function $g(x)$ is monotone increasing and $\mathbf{y} = g(\mathbf{x})$. Show that

$$F_{xy}(x, y) = \begin{cases} F_x(x) & \text{if} \quad y > g(x) \\ F_y(y) & \text{if} \quad y < g(x) \end{cases}$$

(*b*) Find $F_{xy}(x, y)$ if $g(x)$ is monotone decreasing.

6-17 The random variables $\mathbf{x}$ and $\mathbf{y}$ are $N(0, 4)$ and independent. Find $f_z(z)$ and $F_z(z)$ if (*a*) $\mathbf{z} = 2\mathbf{x} + 3\mathbf{y}$, and (*b*) $\mathbf{z} = \mathbf{x}/\mathbf{y}$.

6-18 The random variables $\mathbf{x}$ and $\mathbf{y}$ are independent with

$$f_x(x) = \frac{x}{\alpha^2} e^{-x^2/2\alpha^2} U(x) \qquad f_y(y) = \begin{cases} 1/\pi\sqrt{1-y^2} & |y| < 1 \\ 0 & |y| > 1 \end{cases}$$

Show that the random variable $\mathbf{z} = \mathbf{xy}$ is $N(0, \alpha^2)$.

6-19 The random variables $\mathbf{x}$ and $\mathbf{y}$ are independent with Rayleigh densities

$$f_x(x) = \frac{x}{\alpha^2} e^{-x^2/2\alpha^2} U(x) \qquad f_y(y) = \frac{y}{\beta^2} e^{-y^2/2\beta^2} U(y)$$

(*a*) Show that if $\mathbf{z} = \mathbf{x}/\mathbf{y}$, then

$$f_z(z) = \frac{2\alpha^2}{\beta^2} \frac{z}{(z^2 + \alpha^2/\beta^2)^2} U(z) \tag{i}$$

(*b*) Using (i), show that for any $k > 0$,

$$P\{\mathbf{x} \le k\mathbf{y}\} = \frac{k^2}{k^2 + \alpha^2/\beta^2}$$

6-20 The random variables $\mathbf{x}$ and $\mathbf{y}$ are independent with exponential densities

$$f_x(x) = \alpha e^{-\alpha x} U(x) \qquad f_y(y) = \beta e^{-\beta y} U(y)$$

Find the densities of the following random variables:

$$(a)\ 2\mathbf{x} + \mathbf{y} \qquad (b)\ \mathbf{x} - \mathbf{y} \qquad (c)\ \frac{\mathbf{x}}{\mathbf{y}} \qquad (d)\ \max(\mathbf{x}, \mathbf{y}) \qquad (e)\ \min(\mathbf{x}, \mathbf{y})$$

6-21 The random variables $\mathbf{x}$ and $\mathbf{y}$ are independent and each is uniform in the interval $(0, a)$. Find the density of the random variable $\mathbf{z} = |\mathbf{x} - \mathbf{y}|$.

6-22 Show that (*a*) the convolution of two normal densities is a normal density, and (*b*) the convolution of two Cauchy densities is a Cauchy density.

6-23 The random variables $\mathbf{x}$ and $\mathbf{y}$ are independent with respective densities $\chi^2(m)$ and $\chi^2(n)$. Show that if (Example 6-29)

$$\mathbf{z} = \frac{\mathbf{x}/m}{\mathbf{y}/n} \qquad \text{then} \quad f_z(z) = \gamma \frac{z^{m/2-1}}{\sqrt{(1 + mz/n)^{m+n}}} U(z)$$

This distribution is denoted by $F(m, n)$ and is called the *Snedecor F* distribution. It is used in hypothesis testing (see Prob. 8-34).

6-24 Express $F_{zw}(z, w)$ in terms of $F_{xy}(x, y)$ if $\mathbf{z} = \max(\mathbf{x}, \mathbf{y})$, $\mathbf{w} = \min(\mathbf{x}, \mathbf{y})$.

6-25 Let $\mathbf{x}$ be the lifetime of a certain electric bulb, and $\mathbf{y}$ that of its replacement after the failure of the first bulb. Suppose $\mathbf{x}$ and $\mathbf{y}$ are independent with common exponential density function with parameter λ. Find the probability that the combined lifetime exceeds 2λ. What is the probability that the replacement outlasts the original component by λ?

6-26 $\mathbf{x}$ and $\mathbf{y}$ are independent uniformly distributed random variables in $(0, 1)$. Let

$$\mathbf{w} = \max(\mathbf{x}, \mathbf{y}) \qquad \mathbf{z} = \min(\mathbf{x}, \mathbf{y})$$

Find the p.d.f. of (*a*) $\mathbf{r} = \mathbf{w} - \mathbf{z}$, (*b*) $\mathbf{s} = \mathbf{w} + \mathbf{z}$.

6-27 Let $\mathbf{x}$ and $\mathbf{y}$ be independent identically distributed exponential random variables with common parameter λ. Find the p.d.f.s of (*a*) $\mathbf{z} = \mathbf{y}/\max(\mathbf{x}, \mathbf{y})$, (*b*) $\mathbf{w} = \mathbf{x}/\min(\mathbf{x}, 2\mathbf{y})$.

6-28 If $\mathbf{x}$ and $\mathbf{y}$ are independent exponential random variables with common parameter λ, show that $\mathbf{x}/(\mathbf{x}+\mathbf{y})$ is a uniformly distributed random variable in $(0, 1)$.

6-29 $\mathbf{x}$ and $\mathbf{y}$ are independent exponential random variables with common parameter λ. Show that

$$\mathbf{z} = \min(\mathbf{x}, \mathbf{y}) \quad \text{and} \quad \mathbf{w} = \max(\mathbf{x}, \mathbf{y}) - \min(\mathbf{x}, \mathbf{y})$$

are independent random variables.

6-30 Let $\mathbf{x}$ and $\mathbf{y}$ be independent random variables with common p.d.f. $f_x(x) = \beta^{-\alpha}\alpha x^{\alpha-1}$, $0 < x < \beta$, and zero otherwise ($\alpha \geq 1$). Let $\mathbf{z} = \min(\mathbf{x}, \mathbf{y})$ and $\mathbf{w} = \max(\mathbf{x}, \mathbf{y})$. (*a*) Find the p.d.f. of $\mathbf{x}+\mathbf{y}$. (*b*) Find the joint p.d.f. of $\mathbf{z}$ and $\mathbf{w}$. (*c*) Show that $\mathbf{z}/\mathbf{w}$ and $\mathbf{w}$ are independent random variables.

6-31 Let $\mathbf{x}$ and $\mathbf{y}$ be independent gamma random variables with parameters (α_1, β) and (α_2, β), respectively. (*a*) Determine the p.d.f.s of the random variables $\mathbf{x}+\mathbf{y}$, $\mathbf{x}/\mathbf{y}$, and $\mathbf{x}/(\mathbf{x}+\mathbf{y})$. (*b*) Show that $\mathbf{x}+\mathbf{y}$ and $\mathbf{x}/\mathbf{y}$ are independent random variables. (*c*) Show that $\mathbf{x}+\mathbf{y}$ and $\mathbf{x}/(\mathbf{x}+\mathbf{y})$ are independent gamma and beta random variables, respectively. The converse to (*b*) due to Lukacs is also true. It states that with $\mathbf{x}$ and $\mathbf{y}$ representing nonnegative random variables, if $\mathbf{x}+\mathbf{y}$ and $\mathbf{x}/\mathbf{y}$ are independent, then $\mathbf{x}$ and $\mathbf{y}$ are gamma random variables with common (second) parameter β.

6-32 Let $\mathbf{x}$ and $\mathbf{y}$ be independent normal random variables with zero mean and unit variances. (*a*) Find the p.d.f. of $\mathbf{x}/|\mathbf{y}|$ as well as that of $|\mathbf{x}|/|\mathbf{y}|$. (*b*) Let $\mathbf{u} = \mathbf{x}+\mathbf{y}$ and $\mathbf{v} = \mathbf{x}^2+\mathbf{y}^2$. Are $\mathbf{u}$ and $\mathbf{v}$ independent?

6-33 Let $\mathbf{x}$ and $\mathbf{y}$ be jointly normal random variables with parameters $\mu_x, \mu_y, \sigma_x^2, \sigma_y^2$, and r. Find a necessary and sufficient condition for $\mathbf{x}+\mathbf{y}$ and $\mathbf{x}-\mathbf{y}$ to be independent.

6-34 $\mathbf{x}$ and $\mathbf{y}$ are independent and identically distributed normal random variables with zero mean and variance σ^2. Define

$$\mathbf{u} = \frac{\mathbf{x}^2 - \mathbf{y}^2}{\sqrt{\mathbf{x}^2+\mathbf{y}^2}} \qquad \mathbf{v} = \frac{\mathbf{x}\mathbf{y}}{\sqrt{\mathbf{x}^2+\mathbf{y}^2}}$$

(*a*) Find the joint p.d.f. $f_{uv}(u, v)$ of the random variables $\mathbf{u}$ and $\mathbf{v}$. (*b*) Show that $\mathbf{u}$ and $\mathbf{v}$ are independent normal random variables. (*c*) Show that $[(\mathbf{x}-\mathbf{y})^2 - 2\mathbf{y}^2]/\sqrt{\mathbf{x}^2+\mathbf{y}^2}$ is also a normal random variable. Thus nonlinear functions of normal random variables *can* lead to normal random variables! (This result is due to Shepp.)

6-35 Suppose $\mathbf{z}$ has an F distribution with (m, n) degrees of freedom. (*a*) Show that $1/\mathbf{z}$ also has an F distribution with (n, m) degrees of freedom. (*b*) Show that $m\mathbf{z}/(m\mathbf{z}+n)$ has a beta distribution.

6-36 Let the joint p.d.f. of $\mathbf{x}$ and $\mathbf{y}$ be given by

$$f_{xy}(x, y) = \begin{cases} e^{-x} & 0 < y \leq x \leq \infty \\ 0 & \text{otherwise} \end{cases}$$

Define $\mathbf{z} = \mathbf{x}+\mathbf{y}$, $\mathbf{w} = \mathbf{x}-\mathbf{y}$. Find the joint p.d.f. of $\mathbf{z}$ and $\mathbf{w}$. Show that $\mathbf{z}$ is an exponential random variable.

6-37 Let

$$f_{xy}(x, y) = \begin{cases} 2e^{-(x+y)} & 0 < x < y < \infty \\ 0 & \text{otherwise} \end{cases}$$

Define $\mathbf{z} = \mathbf{x}+\mathbf{y}$, $\mathbf{w} = \mathbf{y}/\mathbf{x}$. Determine the joint p.d.f. of $\mathbf{z}$ and $\mathbf{w}$. Are $\mathbf{z}$ and $\mathbf{w}$ independent random variables?

6-38 The random variables $\mathbf{x}$ and $\boldsymbol{\theta}$ are independent and $\boldsymbol{\theta}$ is uniform in the interval $(-\pi, \pi)$. Show that if $\mathbf{z} = \mathbf{x}\cos(wt + \boldsymbol{\theta})$, then

$$f_z(z) = \frac{1}{\pi}\int_{-\infty}^{-|z|} \frac{f_x(y)}{\sqrt{y^2 - z^2}}\,dy + \frac{1}{\pi}\int_{|z|}^{\infty} \frac{f_x(y)}{\sqrt{y^2 - z^2}}\,dy$$

6-39 The random variables $\mathbf{x}$ and $\mathbf{y}$ are independent, $\mathbf{x}$ is $N(0, \sigma^2)$, and $\mathbf{y}$ is uniform in the interval $(0, \pi)$. Show that if $\mathbf{z} = \mathbf{x} + a\cos\mathbf{y}$, then

$$f_z(z) = \frac{1}{\pi\sigma\sqrt{2\pi}}\int_0^{\pi} e^{-(z - a\cos y)^2/2\sigma^2}\,dy$$

6-40 The random variables $\mathbf{x}$ and $\mathbf{y}$ are of discrete type, independent, with $P\{\mathbf{x} = n\} = a_n$, $P\{\mathbf{y} = n\} = b_n$, $n = 0, 1, \ldots$. Show that, if $\mathbf{z} = \mathbf{x} + \mathbf{y}$, then

$$P\{\mathbf{z} = n\} = \sum_{k=0}^{n} a_k b_{n-k}, \qquad n = 0, 1, 2, \ldots$$

6-41 The random variable $\mathbf{x}$ is of discrete type taking the values x_n with $P\{\mathbf{x} = x_n\} = p_n$ and the random variable $\mathbf{y}$ is of continuous type and independent of $\mathbf{x}$. Show that if $\mathbf{z} = \mathbf{x} + \mathbf{y}$ and $\mathbf{w} = \mathbf{xy}$, then

$$f_z(z) = \sum_n f_y(z - x_n)p_n \qquad f_w(w) = \sum_n \frac{1}{|x_n|} f_y\left(\frac{w}{x_n}\right) p_n$$

6-42 $\mathbf{x}$ and $\mathbf{y}$ are independent random variables with geometric p.m.f.

$$P\{\mathbf{x} = k\} = pq^k \quad k = 0, 1, 2, \ldots \qquad P\{\mathbf{y} = m\} = pq^m \quad m = 0, 1, 2, \ldots$$

Find the p.m.f. of (*a*) $\mathbf{x} + \mathbf{y}$ and (*b*) $\mathbf{x} - \mathbf{y}$.

6-43 Let $\mathbf{x}$ and $\mathbf{y}$ be independent identically distributed nonnegative discrete random variables with

$$P\{\mathbf{x} = k\} = P\{\mathbf{y} = k\} = p_k \qquad k = 0, 1, 2, \ldots$$

Suppose

$$P\{\mathbf{x} = k \mid \mathbf{x} + \mathbf{y} = k\} = P\{\mathbf{x} = k - 1 \mid \mathbf{x} + \mathbf{y} = k\} = \frac{1}{k+1} \qquad k \geq 0$$

Show that $\mathbf{x}$ and $\mathbf{y}$ are geometric random variables. (This result is due to Chatterji.)

6-44 $\mathbf{x}$ and $\mathbf{y}$ are independent, identically distributed binomial random variables with parameters n and p. Show that $\mathbf{z} = \mathbf{x} + \mathbf{y}$ is also a binomial random variable. Find its parameters.

6-45 Let $\mathbf{x}$ and $\mathbf{y}$ be independent random variables with common p.m.f.

$$P(\mathbf{x} = k) = pq^k \qquad k = 0, 1, 2, \ldots \qquad q = p - 1$$

(*a*) Show that $\min(\mathbf{x}, \mathbf{y})$ and $\mathbf{x} - \mathbf{y}$ are independent random variables. (*b*) Show that $\mathbf{z} = \min(\mathbf{x}, \mathbf{y})$ and $\mathbf{w} = \max(\mathbf{x}, \mathbf{y}) - \min(\mathbf{x}, \mathbf{y})$ are independent random variables.

6-46 Let $\mathbf{x}$ and $\mathbf{y}$ be independent Poisson random variables with parameters λ_1 and λ_2, respectively. Show that the conditional density function of $\mathbf{x}$ given $\mathbf{x} + \mathbf{y}$ is binomial.

6-47 The random variables $\mathbf{x}_1$ and $\mathbf{x}_2$ are jointly normal with zero mean. Show that their density can be written in the form

$$f(x_1, x_2) = \frac{1}{2\pi\sqrt{\Delta}}\exp\left\{-\frac{1}{2}XC^{-1}X^t\right\} \qquad C = \begin{bmatrix}\mu_{11} & \mu_{12}\\ \mu_{21} & \mu_{22}\end{bmatrix}$$

where $X\colon [x_1, x_2]$, $\mu_{ij} = E\{\mathbf{x}_i\mathbf{x}_j\}$, and $\Delta = \mu_{11}\mu_{22} - \mu_{12}^2$.

6-48 Show that if the random variables **x** and **y** are normal and independent, then

$$P\{\mathbf{xy} < 0\} = G\left(\frac{\eta_x}{\sigma_x}\right) + G\left(\frac{\eta_y}{\sigma_y}\right) - 2G\left(\frac{\eta_x}{\sigma_x}\right)G\left(\frac{\eta_y}{\sigma_y}\right)$$

6-49 The random variables **x** and **y** are $N(0;\sigma^2)$ and independent. Show that if $\mathbf{z} = |\mathbf{x} - \mathbf{y}|$, then $E\{\mathbf{z}\} = 2\sigma/\sqrt{\pi}$, $E\{\mathbf{z}^2\} = 2\sigma^2$.

6-50 Show that if **x** and **y** are two independent exponential random variables with $f_x(x) = e^{-x}U(x)$, $f_y(y) = e^{-y}U(y)$, and $\mathbf{z} = (\mathbf{x} - \mathbf{y})U(\mathbf{x} - \mathbf{y})$, then $E\{\mathbf{z}\} = 1/2$.

6-51 Show that for any **x**, **y** real or complex (*a*) $|E\{\mathbf{xy}\}|^2 \le E\{|\mathbf{x}|^2\}E\{|\mathbf{y}|^2\}$; (*b*) (*triangle inequality*) $\sqrt{E\{|\mathbf{x}+\mathbf{y}|^2\}} \le \sqrt{E\{|\mathbf{x}|^2\}} + \sqrt{E\{|\mathbf{y}|^2\}}$.

6-52 Show that, if the correlation coefficient $r_{xy} = 1$, then $\mathbf{y} = a\mathbf{x} + b$.

6-53 Show that, if $E\{\mathbf{x}^2\} = E\{\mathbf{y}^2\} = E\{\mathbf{xy}\}$, then $\mathbf{x} = \mathbf{y}$.

6-54 The random variable **n** is Poisson with parameter λ and the random variable **x** is independent of **n**. Show that, if $\mathbf{z} = \mathbf{nx}$ and

$$f_x(x) = \frac{\alpha}{\pi(\alpha^2 + x^2)} \qquad \text{then} \quad \Phi_z(\omega) = \exp\{\lambda e^{-\alpha|\omega|} - \lambda\}$$

6-55 Let **x** represent the number of successes and **y** the number of failures of n independent Bernoulli trials with p representing the probability of success in any one trial. Find the distribution of $\mathbf{z} = \mathbf{x} - \mathbf{y}$. Show that $E\{\mathbf{z}\} = n(2p - 1)$, $\text{Var}\{\mathbf{z}\} = 4np(1-p)$.

6-56 **x** and **y** are zero mean independent random variables with variances σ_1^2 and σ_2^2, respectively, that is, $\mathbf{x} \sim N(0, \sigma_1^2)$, $\mathbf{y} \sim N(0, \sigma_2^2)$. Let

$$\mathbf{z} = a\mathbf{x} + b\mathbf{y} + c \qquad c \neq 0$$

(*a*) Find the characteristic function $\Phi_z(u)$ of **z**. (*b*) Using $\Phi_z(u)$ conclude that **z** is also a normal random variable. (*c*) Find the mean and variance of **z**.

6-57 Suppose the conditional distribution of **x** given $\mathbf{y} = n$ is binomial with parameters n and p_1. Further, **y** is a binomial random variable with parameters M and p_2. Show that the distribution of **x** is also binomial. Find its parameters.

6-58 The random variables **x** and **y** are jointly distributed over the region $0 < x < y < 1$ as

$$f_{xy}(x, y) = \begin{cases} kx & 0 < x < y < 1 \\ 0 & \text{otherwise} \end{cases}$$

for some k. Determine k. Find the variances of **x** and **y**. What is the covariance between **x** and **y**?

6-59 **x** is a Poisson random variable with parameter λ and **y** is a normal random variable with mean μ and variance σ^2. Further **x** and **y** are given to be independent. (*a*) Find the joint characteristic function of **x** and **y**. (*b*) Define $\mathbf{z} = \mathbf{x} + \mathbf{y}$. Find the characteristic function of **z**.

6-60 **x** and **y** are independent exponential random variables with common parameter λ. Find (*a*) $E[\min(\mathbf{x}, \mathbf{y})]$, (*b*) $E[\max(2\mathbf{x}, \mathbf{y})]$.

6-61 The joint p.d.f. of **x** and **y** is given by

$$f_{xy}(x, y) = \begin{cases} 6x & x > 0,\ y > 0,\ 0 < x + y \le 1 \\ 0 & \text{otherwise} \end{cases}$$

Define $\mathbf{z} = \mathbf{x} - \mathbf{y}$. (*a*) Find the p.d.f. of **z**. (*b*) Find the conditional p.d.f. of **y** given **x**. (*c*) Determine $\text{Var}\{\mathbf{x} + \mathbf{y}\}$.

6-62 Suppose **x** represents the inverse of a chi-square random variable with one degree of freedom, and the conditional p.d.f. of **y** given **x** is $N(0, x)$. Show that **y** has a Cauchy distribution.

6-63 For any two random variables $\mathbf{x}$ and $\mathbf{y}$, let $\sigma_x^2 = \text{Var}\{\mathbf{x}\}$, $\sigma_y^2 = \text{Var}\{\mathbf{y}\}$ and $\sigma_{x+y}^2 = \text{Var}\{\mathbf{x}+\mathbf{y}\}$. (*a*) Show that

$$\frac{\sigma_{x+y}}{\sigma_x + \sigma_y} \leq 1$$

(*b*) More generally, show that for $p \geq 1$

$$\frac{\{E(|\mathbf{x}+\mathbf{y}|^p)\}^{1/p}}{\{E(|\mathbf{x}|^p)\}^{1/p} + \{E(|\mathbf{y}|^p)\}^{1/p}} \leq 1$$

6-64 $\mathbf{x}$ and $\mathbf{y}$ are jointly normal with parameters $N(\mu_x, \mu_y, \sigma_x^2, \sigma_y^2, \rho_{xy})$. Find (*a*) $E\{\mathbf{y} \mid \mathbf{x} = x\}$, and (*b*) $E\{\mathbf{x}^2 \mid \mathbf{y} = y\}$.

6-65 For any two random variables $\mathbf{x}$ and $\mathbf{y}$ with $E\{\mathbf{x}^2\} < \infty$, show that (*a*) $\text{Var}\{\mathbf{x}\} \geq E[\text{Var}\{\mathbf{x} \mid \mathbf{y}\}]$. (*b*) $\text{Var}\{\mathbf{x}\} = \text{Var}[E\{\mathbf{x} \mid \mathbf{y}\}] + E[\text{Var}\{\mathbf{x} \mid \mathbf{y}\}]$.

6-66 Let $\mathbf{x}$ and $\mathbf{y}$ be independent random variables with variances σ_1^2 and σ_2^2, respectively. Consider the sum

$$\mathbf{z} = a\mathbf{x} + (1-a)\mathbf{y} \qquad 0 \leq a \leq 1$$

Find a that minimizes the variance of $\mathbf{z}$.

6-67 Show that, if the random variable $\mathbf{x}$ is of discrete type taking the values x_n with $P\{\mathbf{x} = x_n\} = p_n$ and $\mathbf{z} = g(\mathbf{x}, \mathbf{y})$, then

$$E\{\mathbf{z}\} = \sum_n E\{g(x_n, \mathbf{y})\} p_n \qquad f_z(z) = \sum_n f_z(z \mid x_n) p_n$$

6-68 Show that, if the random variables $\mathbf{x}$ and $\mathbf{y}$ are $N(0, 0, \sigma^2, \sigma^2, r)$, then

(*a*)
$$E\{f_y(\mathbf{y} \mid x)\} = \frac{1}{\sigma\sqrt{2\pi(2-r^2)}} \exp\left\{-\frac{r^2x^2}{2\sigma^2(2-r^2)}\right\}$$

(*b*)
$$E\{f_x(\mathbf{x}) f_y(\mathbf{y})\} = \frac{1}{2\pi\sigma^2\sqrt{4-r^2}}$$

6-69 Show that if the random variables $\mathbf{x}$ and $\mathbf{y}$ are $N(0, 0, \sigma_1^2, \sigma_2^2, r)$ then

$$E\{|\mathbf{xy}|\} = \frac{2}{\pi}\int_0^C \arcsin\frac{\mu}{\sigma_1\sigma_2}\, d\mu + \frac{2\sigma_1\sigma_2}{\pi} = \frac{2\sigma_1\sigma_2}{\pi}(\cos\alpha + \alpha\sin\alpha)$$

where $r = \sin\alpha$ and $C = r\sigma_1\sigma_2$.
(*Hint:* Use (6-200) with $g(x, y) = |xy|$.)

6-70 The random variables $\mathbf{x}$ and $\mathbf{y}$ are $N(3, 4, 1, 4, 0.5)$. Find $f(y \mid x)$ and $f(x \mid y)$.

6-71 The random variables $\mathbf{x}$ and $\mathbf{y}$ are uniform in the interval $(-1, 1)$ and independent. Find the conditional density $f_r(r \mid M)$ of the random variable $\mathbf{r} = \sqrt{\mathbf{x}^2 + \mathbf{y}^2}$, where $M = \{\mathbf{r} \leq 1\}$.

6-72 Show that, if the random variables $\mathbf{x}$ and $\mathbf{y}$ are independent and $\mathbf{z} = \mathbf{x} + \mathbf{y}$, then $f_z(z \mid x) = f_y(z - x)$.

6-73 Show that, for any $\mathbf{x}$ and $\mathbf{y}$, the random variables $\mathbf{z} = F_x(\mathbf{x})$ and $\mathbf{w} = F_y(\mathbf{y} \mid \mathbf{x})$ are independent and each is uniform in the interval $(0, 1)$.

6-74 We have a pile of m coins. The probability of heads of the ith coin equals p_i. We select at random one of the coins, we toss it n times and heads shows k times. Show that the probability that we selected the rth coin equals

$$\frac{p_r^k(1-p_r)^{n-k}}{p_1^k(1-p_1)^{n-k} + \cdots + p_m^k(1-p_m)^{n-k}}$$

6-75 The random variable $\mathbf{x}$ has a Student t distribution $t(n)$. Show that $E\{\mathbf{x}^2\} = n/(n-2)$.

6-76 Show that if $\beta_x(t) = f_x(t \mid \mathbf{x} > t)$, $\beta_y(t \mid \mathbf{y} > t)$ and $\beta_x(t) = k\beta_y(t)$, then $1 - F_x(x) = [1 - F_y(x)]^k$.

6-77 Show that, for any $\mathbf{x}$, $\mathbf{y}$, and $\varepsilon > 0$,

$$P\{|\mathbf{x} - \mathbf{y}| > \varepsilon\} \le \frac{1}{\varepsilon^2} E\{|\mathbf{x} - \mathbf{y}|^2\}$$

6-78 Show that the random variables $\mathbf{x}$ and $\mathbf{y}$ are independent iff for any a and b:

$$E\{U(a - \mathbf{x})U(b - \mathbf{y})\} = E\{U(a - \mathbf{x})\}E\{U(b - \mathbf{y})\}$$

6-79 Show that

$$E\{\mathbf{y} \mid \mathbf{x} \le 0\} = \frac{1}{F_x(0)} \int_{-\infty}^{0} E\{\mathbf{y} \mid \mathbf{x}\} f_x(x)\, dx$$

CHAPTER

7

SEQUENCES OF RANDOM VARIABLES

7-1 GENERAL CONCEPTS

A *random vector* is a vector

$$\mathbf{X} = [\mathbf{x}_1, \ldots, \mathbf{x}_n] \tag{7-1}$$

whose components $\mathbf{x}_i$ are random variables.

The probability that $\mathbf{X}$ is in a region D of the n-dimensional space equals the probability masses in D:

$$P\{\mathbf{X} \in D\} = \int_D f(X)\,dX \qquad X = [x_1, \ldots, x_n] \tag{7-2}$$

where

$$f(X) = f(x_1, \ldots, x_n) = \frac{\partial^n F(x_1, \ldots, x_n)}{\partial x_1, \ldots, \partial x_n} \tag{7-3}$$

is the *joint* (or, *multivariate*) *density* of the random variables $\mathbf{x}_i$ and

$$F(X) = F(x_1, \ldots, x_n) = P\{\mathbf{x}_i \le x_1, \ldots, \mathbf{x}_n \le x_n\} \tag{7-4}$$

is their *joint distribution.*

If we substitute in $F(x_1, \ldots, x_n)$ certain variables by ∞, we obtain the joint distribution of the remaining variables. If we integrate $f(x_1, \ldots, x_n)$ with respect to certain variables, we obtain the joint density of the remaining variables. For example

$$F(x_1, x_3) = F(x_1, \infty, x_3, \infty)$$
$$f(x_1, x_3) = \int_{-\infty}^{\infty}\int_{-\infty}^{\infty} f(x_1, x_2, x_3, x_4)\,dx_2\,dx_4 \tag{7-5}$$

Note We have just identified various functions in terms of their independent variables. Thus $f(x_1, x_3)$ is the joint density of the random variables $\mathbf{x}_1$ and $\mathbf{x}_3$ and it is in general *different* from the joint density $f(x_2, x_4)$ of the random variables $\mathbf{x}_2$ and $\mathbf{x}_4$. Similarly, the density $f_i(x_i)$ of the random variable $\mathbf{x}_i$ will often be denoted by $f(x_i)$.

TRANSFORMATIONS. Given k functions

$$g_1(X), \ldots, g_k(X) \qquad X = [x_1, \ldots, x_n]$$

we form the random variables

$$\mathbf{y}_1 = g_1(\mathbf{X}), \ldots, \mathbf{y}_k = g_k(\mathbf{X}) \tag{7-6}$$

The statistics of these random variables can be determined in terms of the statistics of $\mathbf{X}$ as in Sec. 6-3. If $k < n$, then we could determine first the joint density of the n random variables $\mathbf{y}_1, \ldots, \mathbf{y}_k, \mathbf{x}_{k+1}, \ldots, \mathbf{x}_n$ and then use the generalization of (7-5) to eliminate the $\mathbf{x}$'s. If $k > n$, then the random variables $\mathbf{y}_{n+1}, \ldots, \mathbf{y}_k$ can be expressed in terms of $\mathbf{y}_1, \ldots, \mathbf{y}_n$. In this case, the masses in the k space are singular and can be determined in terms of the joint density of $\mathbf{y}_1, \ldots, \mathbf{y}_n$. It suffices, therefore, to assume that $k = n$.

To find the density $f_y(y_1, \ldots, y_n)$ of the random vector $\mathbf{Y} = [\mathbf{y}_1, \ldots, \mathbf{y}_n]$ for a specific set of numbers $y_1, \ldots, y_n$, we solve the system

$$g_1(X) = y_1, \ldots, g_n(X) = y_n \tag{7-7}$$

If this system has no solutions, then $f_y(y_1, \ldots, y_n) = 0$. If it has a single solution $X = [x_1, \ldots, x_n]$, then

$$f_y(y_1, \ldots, y_n) = \frac{f_x(x_1, \ldots, x_n)}{|J(x_1, \ldots, x_n)|} \tag{7-8}$$

where

$$J(x_1, \ldots, x_n) = \begin{vmatrix} \dfrac{\partial g_1}{\partial x_1} & \cdots & \dfrac{\partial g_1}{\partial x_n} \\ \cdots & \cdots & \cdots \\ \dfrac{\partial g_n}{\partial x_1} & \cdots & \dfrac{\partial g_n}{\partial x_n} \end{vmatrix} \tag{7-9}$$

is the jacobian of the transformation (7-7). If it has several solutions, then we add the corresponding terms as in (6-115).

Independence

The random variables $\mathbf{x}_1, \ldots, \mathbf{x}_n$ are called (mutually) independent if the events $\{\mathbf{x}_1 \le x_1\}, \ldots, \{\mathbf{x}_n \le x_n\}$ are independent. From this it follows that

$$\begin{aligned} F(x_1, \ldots, x_n) &= F(x_1) \cdots F(x_n) \\ f(x_1, \ldots, x_n) &= f(x_1) \cdots f(x_n) \end{aligned} \tag{7-10}$$

EXAMPLE 7-1

▶ Given n independent random variables $\mathbf{x}_i$ with respective densities $f_i(x_i)$, we form the random variables

$$\mathbf{y}_k = \mathbf{x}_1 + \cdots + \mathbf{x}_k \qquad k = 1, \ldots, n$$

We shall determine the joint density of $\mathbf{y}_k$. The system

$$x_1 = y_1, x_1 + x_2 = y_2, \ldots, x_1 + \cdots + x_n = y_n$$

has a unique solution

$$x_k = y_k - y_{k-1} \qquad 1 \le k \le n$$

and its jacobian equals 1. Hence [see (7-8) and (7-10)]

$$f_y(y_1, \ldots, y_n) = f_1(y_1) f_2(y_2 - y_1) \cdots f_n(y_n - y_{n-1}) \tag{7-11}$$

◀

From (7-10) it follows that any subset of the set $\mathbf{x}_i$ is a set of independent random variables. Suppose, for example, that

$$f(x_1, x_2, x_3) = f(x_1) f(x_2) f(x_3)$$

Integrating with respect to x_3, we obtain $f(x_1, x_2) = f(x_1) f(x_2)$. This shows that the random variables $\mathbf{x}_1$ and $\mathbf{x}_2$ are independent. Note, however, that if the random variables $\mathbf{x}_i$ are independent in pairs, they are not necessarily independent. For example, it is possible that

$$f(x_1, x_2) = f(x_1) f(x_2) \quad f(x_1, x_3) = f(x_1) f(x_3) \quad f(x_2, x_3) = f(x_2) f(x_3)$$

but $f(x_1, x_2, x_3) \neq f(x_1) f(x_2) f(x_3)$ (see Prob. 7-2).

Reasoning as in (6-21), we can show that if the random variables $\mathbf{x}_i$ are independent, then the random variables

$$\mathbf{y}_1 = g_1(\mathbf{x}_1), \ldots, \mathbf{y}_n = g_n(\mathbf{x}_n)$$

are also independent.

INDEPENDENT EXPERIMENTS AND REPEATED TRIALS. Suppose that

$$S^n = S_1 \times \cdots \times S_n$$

is a combined experiment and the random variables $\mathbf{x}_i$ depend only on the outcomes ζ_i of S_i:

$$\mathbf{x}_i(\zeta_1 \cdots \zeta_i \cdots \zeta_n) = \mathbf{x}_i(\zeta_i) \qquad i = 1, \ldots, n$$

If the experiments S_i are independent, then the random variables $\mathbf{x}_i$ are independent [see also (6-22)]. The following special case is of particular interest.

Suppose that $\mathbf{x}$ is a random variable defined on an experiment S and the experiment is performed n times generating the experiment $S^n = S \times \cdots \times S$. In this experiment, we define the random variables $\mathbf{x}_i$ such that

$$\mathbf{x}_i(\zeta_1 \cdots \zeta_i \cdots \zeta_n) = \mathbf{x}(\zeta_i) \qquad i = 1, \ldots, n \tag{7-12}$$

From this it follows that the distribution $F_i(x_i)$ of $\mathbf{x}_i$ equals the distribution $F_x(x)$ of the random variable $\mathbf{x}$. Thus, if an experiment is performed n times, the random variables $\mathbf{x}_i$ defined as in (7-12) are independent and they have the same distribution $F_x(x)$. These random variables are called i.i.d. (independent, identically distributed).

EXAMPLE 7-2

ORDER STATISTICS

▶ The *order statistics* of the random variables $\mathbf{x}_i$ are n random variables $\mathbf{y}_k$ defined as follows: For a specific outcome ζ, the random variables $\mathbf{x}_i$ take the values $\mathbf{x}_i(\zeta)$. Ordering these numbers, we obtain the sequence

$$\mathbf{x}_{r_1}(\zeta) \le \cdots \le \mathbf{x}_{r_k}(\zeta) \le \cdots \le \mathbf{x}_{r_n}(\zeta)$$

and we define the random variable $\mathbf{y}_k$ such that

$$\mathbf{y}_1(\zeta) = \mathbf{x}_{r_1}(\zeta) \le \cdots \le \mathbf{y}_k(\zeta) = \mathbf{x}_{r_k}(\zeta) \le \cdots \le \mathbf{y}_n(\zeta) = \mathbf{x}_{r_n}(\zeta) \tag{7-13}$$

We note that for a specific i, the values $\mathbf{x}_i(\zeta)$ of $\mathbf{x}_i$ occupy different locations in the above ordering as ζ changes.

We maintain that the density $f_k(y)$ of the kth statistic $\mathbf{y}_k$ is given by

$$f_k(y) = \frac{n!}{(k-1)!(n-k)!} F_x^{k-1}(y)[1 - F_x(y)]^{n-k} f_x(y) \tag{7-14}$$

where $F_x(x)$ is the distribution of the i.i.d. random variables $\mathbf{x}_i$ and $f_x(x)$ is their density.

Proof. As we know

$$f_k(y)\,dy = P\{y < \mathbf{y}_k \le y + dy\}$$

The event $B = \{y < \mathbf{y}_k \le y + dy\}$ occurs iff exactly $k-1$ of the random variables $\mathbf{x}_i$ are less than y and one is in the interval $(y, y + dy)$ (Fig. 7-1). In the original experiment S, the events

$$A_1 = \{\mathbf{x} \le y\} \qquad A_2 = \{y < \mathbf{x} \le y + dy\} \qquad A_3 = \{\mathbf{x} > y + dy\}$$

form a partition and

$$P(A_1) = F_x(y) \qquad P(A_2) = f_x(y)\,dy \qquad P(A_3) = 1 - F_x(y)$$

In the experiment S^n, the event B occurs iff A_1 occurs $k-1$ times, A_2 occurs once, and A_3 occurs $n-k$ times. With $k_1 = k-1$, $k_2 = 1$, $k_3 = n-k$, it follows from (4-102) that

$$P\{B\} = \frac{n!}{(k-1)!1!(n-k)!} P^{k-1}(A_1)P(A_2)P^{n-k}(A_3)$$

and (7-14) results.

Note that

$$f_1(y) = n[1 - F_x(y)]^{n-1} f_x(y) \qquad f_n(y) = nF_x^{n-1}(y) f_x(y)$$

These are the densities of the minimum $\mathbf{y}_1$ and the maximum $\mathbf{y}_n$ of the random variables $\mathbf{x}_i$.

Special Case. If the random variables $\mathbf{x}_i$ are exponential with parameter λ:

$$f_x(x) = \alpha e^{-\lambda x} U(x) \qquad F_x(x) = (1 - e^{-\lambda x}) U(x)$$

then

$$f_1(y) = n\lambda e^{-\lambda n y} U(y)$$

that is, their minimum $\mathbf{y}_1$ is also exponential with parameter $n\lambda$.

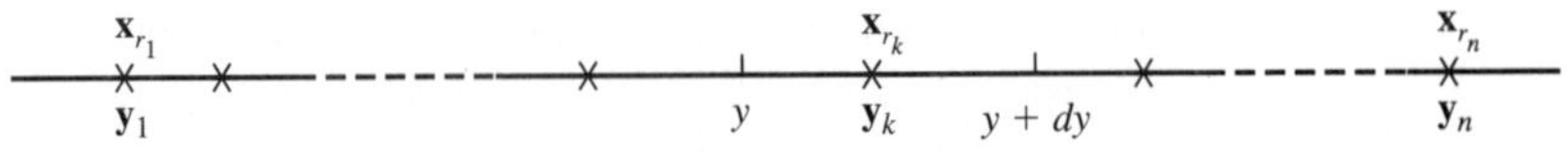

FIGURE 7-1

◀

EXAMPLE 7-3

▶ A system consists of m components and the time to failure of the ith component is a random variable $\mathbf{x}_i$ with distribution $F_i(x)$. Thus

$$1 - F_i(t) = P\{\mathbf{x}_i > t\}$$

is the probability that the ith component is good at time t. We denote by $\mathbf{n}(t)$ the number of components that are good at time t. Clearly,

$$\mathbf{n}(t) = \mathbf{n}_1 + \cdots + \mathbf{n}_m$$

where

$$\mathbf{n}_i = \begin{cases} 1 & \mathbf{x}_i > t \\ 0 & \mathbf{x}_i < t \end{cases} \qquad E\{\mathbf{n}_i\} = 1 - F_i(t)$$

Hence the mean $E\{\mathbf{n}(t)\} = \eta(t)$ of $\mathbf{n}(t)$ is given by

$$\eta(t) = 1 - F_1(t) + \cdots + 1 - F_m(t)$$

We shall assume that the random variables $\mathbf{x}_i$ have the same distribution $F(t)$. In this case,

$$\eta(t) = m[1 - F(t)]$$

Failure rate The difference $\eta(t) - \eta(t + dt)$ is the expected number of failures in the interval $(t, t + dt)$. The derivative $-\eta'(t) = mf(t)$ of $-\eta(t)$ is the rate of failure. The ratio

$$\beta(t) = -\frac{\eta'(t)}{\eta(t)} = \frac{f(t)}{1 - F(t)} \tag{7-15}$$

is called the *relative expected failure rate.* As we see from (6-221), the function $\beta(t)$ can also be interpreted as the conditional failure rate of each component in the system. Assuming that the system is put into operation at $t = 0$, we have $\mathbf{n}(0) = m$; hence $\eta(0) = E\{\mathbf{n}(0)\} = m$. Solving (7-15) for $\eta(t)$, we obtain

$$\eta(t) = m \exp\left\{-\int_0^t \beta(\tau)\, d\tau\right\}$$ ◀

EXAMPLE 7-4

MEASUREMENT ERRORS

▶ We measure an object of length η with n instruments of varying accuracies. The results of the measurements are n random variables

$$\mathbf{x}_i = \eta + \boldsymbol{\nu}_i \qquad E\{\boldsymbol{\nu}_i\} = 0 \qquad E\{\boldsymbol{\nu}_i^2\} = \sigma_i^2$$

where $\boldsymbol{\nu}_i$ are the measurement errors which we assume independent with zero mean. We shall determine the unbiased, minimum variance, linear estimation of η. This means the following: We wish to find n constants α_i such that the sum

$$\hat{\boldsymbol{\eta}} = \alpha_1 \mathbf{x}_1 + \cdots + \alpha_n \mathbf{x}_n$$

is a random variable with mean $E\{\hat{\boldsymbol{\eta}}\} = \alpha_1 E\{\mathbf{x}_1\} + \cdots + \alpha_n E\{\mathbf{x}_n\} = \eta$ and its variance

$$V = \alpha_1^2\sigma_1^2 + \cdots + \alpha_n^2\sigma_n^2$$

is minimum. Thus our problem is to minimize the above sum subject to the constraint

$$\alpha_1 + \cdots + \alpha_n = 1 \tag{7-16}$$

To solve this problem, we note that

$$V = \alpha_1^2\sigma_1^2 + \cdots + \alpha_n^2\sigma_n^2 - \lambda(\alpha_1 + \cdots + \alpha_n - 1)$$

for any λ (Lagrange multiplier). Hence V is minimum if

$$\frac{\partial V}{\partial \alpha_i} = 2\alpha_i\sigma_i^2 - \lambda = 0 \qquad \alpha_i = \frac{\lambda}{2\sigma_i^2}$$

Inserting into (7-16) and solving for λ, we obtain

$$\frac{\lambda}{2} = V = \frac{1}{1/\sigma_1^2 + \cdots + 1/\sigma_n^2}$$

Hence

$$\hat{\eta} = \frac{x_1/\sigma_1^2 + \cdots + x_n/\sigma_n^2}{1/\sigma_1^2 + \cdots + 1/\sigma_n^2} \tag{7-17}$$

Illustration. The voltage E of a generator is measured three times. We list here the results x_i of the measurements, the standard deviations σ_i of the measurement errors, and the estimate $\hat{E}$ of E obtained from (7-17):

$$x_i = 98.6,\ 98.8,\ 98.9 \qquad \sigma_i = 0.20,\ 0.25,\ 0.28$$

$$\hat{E} = \frac{x_1/0.04 + x_2/0.0625 + x_3/0.0784}{1/0.04 + 1/0.0625 + 1/0.0784} = 98.73$$ ◀

Group independence. We say that the group G_x of the random variables $\mathbf{x}_1, \ldots, \mathbf{x}_n$ is independent of the group G_y of the random variables $\mathbf{y}_1, \ldots, \mathbf{y}_k$ if

$$f(x_1, \ldots, x_n, y_1, \ldots, y_k) = f(x_1, \ldots, x_n) f(y_1, \ldots, y_k) \tag{7-18}$$

By suitable integration as in (7-5) we conclude from (7-18) that any subgroup of G_x is independent of any subgroup of G_y. In particular, the random variables $\mathbf{x}_i$ and $\mathbf{y}_i$ are independent for any i and j.

Suppose that S is a combined experiment $S_1 \times S_2$, the random variables $\mathbf{x}_i$ depend only on the outcomes of S_1, and the random variables $\mathbf{y}_j$ depend only on the outcomes of S_2. If the experiments S_1 and S_2 are independent, then the groups G_x and G_y are independent.

We note finally that if the random variables $\mathbf{z}_m$ depend only on the random variables $\mathbf{x}_i$ of G_x and the random variables $\mathbf{w}_r$ depend only on the random variables $\mathbf{y}_j$ of G_y, then the groups G_z and G_w are independent.

Complex random variables The statistics of the random variables

$$\mathbf{z}_1 = \mathbf{x}_1 + j\mathbf{y}_1, \ldots, \mathbf{z}_n = \mathbf{x}_n + j\mathbf{y}_n$$

are determined in terms of the joint density $f(x_1, \ldots, x_n, y_1, \ldots, y_n)$ of the $2n$ random variables $\mathbf{x}_i$ and $\mathbf{y}_j$. We say that the complex random variables $\mathbf{z}_i$ are independent if

$$f(x_1, \ldots, x_n, y_1, \ldots, y_n) = f(x_1, y_1) \cdots f(x_n, y_n) \tag{7-19}$$

Mean and Covariance

Extending (6-159) to n random variables, we conclude that the mean of $g(\mathbf{x}_1, \ldots, \mathbf{x}_n)$ equals

$$\int_{-\infty}^{\infty} \cdots \int_{-\infty}^{\infty} g(x_1, \ldots, x_n) f(x_1, \ldots, x_n)\, dx_1 \cdots dx_n \tag{7-20}$$

If the random variables $\mathbf{z}_i = \mathbf{x}_i + j\mathbf{y}_i$ are complex, then the mean of $g(\mathbf{z}_1, \ldots, \mathbf{z}_n)$ equals

$$\int_{-\infty}^{\infty} \cdots \int_{-\infty}^{\infty} g(z_1, \ldots, z_n) f(x_1, \ldots, x_n, y_1, \ldots, y_n)\, dx_1 \cdots dy_n$$

From this it follows that (linearity)

$$E\{a_1 g_1(\mathbf{X}) + \cdots + a_m g_m(\mathbf{X})\} = a_1 E\{g_1(\mathbf{X})\} + \cdots + a_m E\{g_m(\mathbf{X})\}$$

for any random vector $\mathbf{X}$ real or complex.

CORRELATION AND COVARIANCE MATRICES. The covariance C_{ij} of two real random variables $\mathbf{x}_i$ and $\mathbf{x}_j$ is defined as in (6-163). For complex random variables

$$C_{ij} = E\{(\mathbf{x}_i - \eta_i)(\mathbf{x}_j^* - \eta_j^*)\} = E\{\mathbf{x}_i \mathbf{x}_j^*\} - E\{\mathbf{x}_i\}E\{\mathbf{x}_j^*\}$$

by definition. The variance of $\mathbf{x}_i$ is given by

$$\sigma_i^2 = C_{ii} = E\{|\mathbf{x}_i - \eta_i|^2\} = E\{|\mathbf{x}_i|^2\} - |E\{\mathbf{x}_i\}|^2$$

The random variables $\mathbf{x}_i$ are called (mutually) *uncorrelated* if $C_{ij} = 0$ for every $i \neq j$. In this case, if

$$\mathbf{x} = \mathbf{x}_1 + \cdots + \mathbf{x}_n \qquad \text{then} \quad \sigma_x^2 = \sigma_1^2 + \cdots + \sigma_n^2 \tag{7-21}$$

EXAMPLE 7-5 ▶ The random variables

$$\overline{\mathbf{x}} = \frac{1}{n}\sum_{i=1}^{n} \mathbf{x}_i \qquad \overline{\mathbf{v}} = \frac{1}{n-1}\sum_{i=1}^{n} (\mathbf{x}_i - \overline{\mathbf{x}})^2$$

are by definition the *sample mean* and the *sample variance,* respectively, of $\mathbf{x}_i$. We shall show that, if the random variables $\mathbf{x}_i$ are uncorrelated with the same mean $E\{\mathbf{x}_i\} = \eta$ and variance $\sigma_i^2 = \sigma^2$, then

$$E\{\overline{\mathbf{x}}\} = \eta \qquad \sigma_{\bar{x}}^2 = \sigma^2/n \tag{7-22}$$

and

$$E\{\overline{\mathbf{v}}\} = \sigma^2 \tag{7-23}$$

Proof. The first equation in (7-22) follows from the linearity of expected values and the second from (7-21):

$$E\{\overline{\mathbf{x}}\} = \frac{1}{n}\sum_{i=1}^{n} E\{\mathbf{x}_i\} = \eta \qquad \sigma_{\bar{x}}^2 = \frac{1}{n^2}\sum_{i=1}^{n} \sigma_i^2 = \frac{\sigma^2}{n}$$

To prove (7-23), we observe that

$$E\{(\mathbf{x}_i - \eta)(\bar{\mathbf{x}} - \eta)\} = \frac{1}{n} E\{(\mathbf{x}_i - \eta)[(\mathbf{x}_1 - \eta) + \cdots + (\mathbf{x}_n - \eta)]\}$$
$$= \frac{1}{n} E\{(\mathbf{x}_i - \eta)(\mathbf{x}_i - \eta)\} = \frac{\sigma^2}{n}$$

because the random variables $\mathbf{x}_i$ and $\mathbf{x}_j$ are uncorrelated by assumption. Hence

$$E\{(\mathbf{x}_i - \bar{\mathbf{x}})^2\} = E\{[(\mathbf{x}_i - \eta) - (\bar{\mathbf{x}} - \eta)]^2\} = \sigma^2 + \frac{\sigma^2}{n} - 2\frac{\sigma^2}{n} = \frac{n-1}{n}\sigma^2$$

This yields

$$E\{\bar{\mathbf{v}}\} = \frac{1}{n-1}\sum_{i=1}^{n} E\{(\mathbf{x}_i - \bar{\mathbf{x}})^2\} = \frac{n}{n-1}\frac{n-1}{n}\sigma^2$$

and (7-23) results.

Note that if the random variables $\mathbf{x}_i$ are independent identically distributed (i.i.d.) with $E\{|\mathbf{x}_i - \eta|^4\} = \mu_4$, then (see Prob. 7-21)

$$\sigma_{\bar{v}}^2 = \frac{1}{n}\left(\mu_4 - \frac{n-3}{n-1}\sigma^4\right)$$ ◀

If the random variables $\mathbf{x}_1, \ldots, \mathbf{x}_n$, are independent, they are also uncorrelated. This follows as in (6-171) for real random variables. For complex random variables the proof is similar: If the random variables $\mathbf{z}_1 = \mathbf{x}_1 + j\mathbf{y}_1$ and $\mathbf{z}_2 = \mathbf{x}_2 + j\mathbf{y}_2$ are independent, then $f(x_1, x_2, y_1, y_2) = f(x_1, y_1) f(x_2, y_2)$. Hence

$$\int_{-\infty}^{\infty} \cdots \int_{-\infty}^{\infty} z_1 z_2^* f(x_1, x_2, y_1, y_2)\, dx_1\, dy_1\, dx_2\, dy_2$$
$$= \int_{-\infty}^{\infty}\int_{-\infty}^{\infty} z_1 f(x_1, y_1)\, dx_1\, dy_1 \int_{-\infty}^{\infty}\int_{-\infty}^{\infty} z_2^* f(x_2, y_2)\, dx_2\, dy_2$$

This yields $E\{\mathbf{z}_1\mathbf{z}_2^*\} = E\{\mathbf{z}_1\}E\{\mathbf{z}_2^*\}$ therefore, $\mathbf{z}_1$ and $\mathbf{z}_2$ are uncorrelated.

Note, finally, that if the random variables $\mathbf{x}_i$ are independent, then

$$E\{g_1(\mathbf{x}_1)\cdots g_n(\mathbf{x}_n)\} = E\{g_1(\mathbf{x}_1)\}\cdots E\{g_n(\mathbf{x}_n)\} \tag{7-24}$$

Similarly, if the groups $\mathbf{x}_1, \ldots, \mathbf{x}_n$ and $\mathbf{y}_1, \ldots, \mathbf{y}_k$ are independent, then

$$E\{g(\mathbf{x}_1, \ldots, \mathbf{x}_n)h(\mathbf{y}_1, \ldots, \mathbf{y}_k)\} = E\{g(\mathbf{x}_1, \ldots, \mathbf{x}_n)\}E\{h(\mathbf{y}_1, \ldots, \mathbf{y}_k)\}$$

The correlation matrix. We introduce the matrices

$$R_n = \begin{bmatrix} R_{11} & \cdots & R_{1n} \\ \cdots & \cdots & \cdots \\ R_{n1} & \cdots & R_{nn} \end{bmatrix} \qquad C_n = \begin{bmatrix} C_{11} & \cdots & C_{1n} \\ \cdots & \cdots & \cdots \\ C_{n1} & \cdots & C_{nn} \end{bmatrix}$$

where

$$R_{ij} = E\{\mathbf{x}_i\mathbf{x}_j^*\} = R_{ji}^* \qquad C_{ij} = R_{ij} - \eta_i\eta_j^* = C_{ji}^*$$

The first is the *correlation matrix* of the random vector $\mathbf{X} = [\mathbf{x}_1, \ldots, \mathbf{x}_n]$ and the second its *covariance matrix.* Clearly,

$$R_n = E\{\mathbf{X}^t\mathbf{X}^*\}$$

where $\mathbf{X}^t$ is the transpose of $\mathbf{X}$ (column vector). We shall discuss the properties of the matrix R_n and its determinant Δ_n. The properties of C_n are similar because C_n is the correlation matrix of the "centered" random variables $\mathbf{x}_i - \eta_i$.

THEOREM 7-1

The matrix R_n is *nonnegative definite.* This means that

$$Q = \sum_{i,j} a_i a_j^* R_{ij} = A R_n A^\dagger \geq 0 \tag{7-25}$$

where $A^\dagger$ is the complex conjugate transpose of the vector $A = [a_1, \ldots, a_n]$.

Proof. It follows readily from the linearity of expected values

$$E\{|a_1\mathbf{x}_1 + \cdots + a_n\mathbf{x}_n|^2\} = \sum_{i,j} a_i a_j^* E\{\mathbf{x}_i \mathbf{x}_j^*\} \tag{7-26}$$

If (7-25) is strictly positive, that is, if $Q > 0$ for any $A \neq 0$, then R_n is called *positive definite.*[1] The difference between $Q \geq 0$ and $Q > 0$ is related to the notion of linear dependence.

DEFINITION

The random variables $\mathbf{x}_i$ are called *linearly independent* if

$$E\{|a_1\mathbf{x}_1 + \cdots + a_n\mathbf{x}_n|^2\} > 0 \tag{7-27}$$

for any $A \neq 0$. In this case [see (7-26)], their correlation matrix R_n is positive definite.

The random variables $\mathbf{x}_i$ are called *linearly dependent* if

$$a_1\mathbf{x}_1 + \cdots + a_n\mathbf{x}_n = 0 \tag{7-28}$$

for some $A \neq 0$. In this case, the corresponding Q equals 0 and the matrix R_n is singular [see also (7-29)].

From the definition it follows that, if the random variables $\mathbf{x}_i$ are linearly independent, then any subset is also linearly independent.

The correlation determinant. The determinant Δ_n is real because $R_{ij} = R_{ji}^*$. We shall show that it is also nonnegative

$$\Delta_n \geq 0 \tag{7-29}$$

with equality iff the random variables $\mathbf{x}_i$ are linearly dependent. The familiar inequality $\Delta_2 = R_{11}R_{22} - R_{12}^2 \geq 0$ is a special case [see (6-169)].

Suppose, first, that the random variables $\mathbf{x}_i$ are linearly independent. We maintain that, in this case, the determinant Δ_n and all its principal minors are positive

$$\Delta_k > 0 \qquad k \leq n \tag{7-30}$$

[1] We shall use the abbreviation p.d. to indicate that R_n satisfies (7-25). The distinction between $Q \geq 0$ and $Q > 0$ will be understood from the context.

Proof. This is true for $n = 1$ because $\Delta_1 = R_{11} > 0$. Since the random variables of any subset of the set $\{\mathbf{x}_i\}$ are linearly independent, we can assume that (7-30) is true for $k \le n - 1$ and we shall show that $\Delta_n > 0$. For this purpose, we form the system

$$\begin{aligned} R_{11}a_1 + \cdots + R_{1n}a_n &= 1 \\ R_{21}a_1 + \cdots + R_{2n}a_n &= 0 \\ \cdots\cdots\cdots&\cdots\cdots\cdots \\ R_{n1}a_1 + \cdots + R_{nn}a_n &= 0 \end{aligned} \tag{7-31}$$

Solving for a_1, we obtain $a_1 = \Delta_{n-1}/\Delta_n$, where Δ_{n-1} is the correlation determinant of the random variables $\mathbf{x}_2, \ldots, \mathbf{x}_n$. Thus a_1 is a real number. Multiplying the jth equation by a_j^* and adding, we obtain

$$Q = \sum_{i,j} a_i a_j^* R_{ij} = a_1 = \frac{\Delta_{n-1}}{\Delta_n} \tag{7-32}$$

In this, $Q > 0$ because the random variables $\mathbf{x}_i$ are linearly independent and the left side of (7-27) equals Q. Furthermore, $\Delta_{n-1} > 0$ by the induction hypothesis; hence $\Delta_n > 0$.

We shall now show that, if the random variables $\mathbf{x}_i$ are linearly dependent, then

$$\Delta_n = 0 \tag{7-33}$$

Proof. In this case, there exists a vector $A \ne 0$ such that $a_1\mathbf{x}_1 + \cdots + a_n\mathbf{x}_n = 0$. Multiplying by $\mathbf{x}_i^*$ and taking expected values, we obtain

$$a_1R_{i1} + \cdots + a_nR_{in} = 0 \qquad i = 1, \ldots, n$$

This is a homogeneous system satisfied by the nonzero vector A; hence $\Delta_n = 0$.

Note, finally, that

$$\Delta_n \le R_{11}R_{22} \cdots R_{nn} \tag{7-34}$$

with equality iff the random variables $\mathbf{x}_i$ are (mutually) *orthogonal,* that is, if the matrix R_n is diagonal.

7-2 CONDITIONAL DENSITIES, CHARACTERISTIC FUNCTIONS, AND NORMALITY

Conditional densities can be defined as in Sec. 6-6. We shall discuss various extensions of the equation $f(y \mid x) = f(x, y)/f(x)$. Reasoning as in (6-205), we conclude that the conditional density of the random variables $\mathbf{x}_n, \ldots, \mathbf{x}_{k+1}$ assuming $\mathbf{x}_k, \ldots, \mathbf{x}_1$ is given by

$$f(x_n, \ldots, x_{k+1} \mid x_k, \ldots, x_1) = \frac{f(x_1, \ldots, x_k, \ldots, x_n)}{f(x_1, \ldots, x_k)} \tag{7-35}$$

The corresponding distribution function is obtained by integration:

$$F(x_n, \ldots, x_{k+1} \mid x_k, \ldots, x_1) = \int_{-\infty}^{x_n} \cdots \int_{-\infty}^{x_{k+1}} f(\alpha_n, \ldots, \alpha_{k+1} \mid x_k, \ldots, x_1)\, d\alpha_{k+1} \cdots d\alpha_n \tag{7-36}$$

For example,

$$f(x_1 \mid x_2, x_3) = \frac{f(x_1, x_2, x_3)}{f(x_2, x_3)} = \frac{dF(x_1 \mid x_2, x_3)}{dx_1}$$

Chain rule From (7-35) it follows that

$$f(x_1, \ldots, x_n) = f(x_n \mid x_{n-1}, \ldots, x_1) \cdots f(x_2 \mid x_1) f(x_1) \tag{7-37}$$

EXAMPLE 7-6

▶ We have shown that [see (5-41)] if $\mathbf{x}$ is a random variable with distribution $F(x)$, then the random variable $\mathbf{y} = F(\mathbf{x})$ is uniform in the interval (0, 1). The following is a generalization.

Given n arbitrary random variables $\mathbf{x}_i$ we form the random variables

$$\mathbf{y}_1 = F(\mathbf{x}_1) \qquad \mathbf{y}_2 = F(\mathbf{x}_2 \mid \mathbf{x}_1), \ldots, \mathbf{y}_n = F(\mathbf{x}_n \mid \mathbf{x}_{n-1}, \ldots, \mathbf{x}_1) \tag{7-38}$$

We shall show that these random variables are independent and each is uniform in the interval (0, 1).

SOLUTION

The random variables $\mathbf{y}_i$ are functions of the random variables $\mathbf{x}_i$ obtained with the transformation (7-38). For $0 \le \mathbf{y}_i \le 1$, the system

$$y_1 = F(x_1) \qquad y_2 = F(x_2 \mid x_1), \ldots, y_n = F(x_n \mid x_{n-1}, \ldots, x_1)$$

has a unique solution $x_1, \ldots, x_n$ and its jacobian equals

$$J = \begin{vmatrix} \dfrac{\partial y_1}{\partial x_1} & 0 & 0 & & 0 \\ \dfrac{\partial y_2}{\partial x_1} & \dfrac{\partial y_2}{\partial x_2} & 0 & \cdots & 0 \\ \cdots & \cdots & \cdots & \cdots & \cdots \\ \dfrac{\partial y_n}{\partial x_1} & \cdots & \cdots & \cdots & \dfrac{\partial y_n}{\partial x_n} \end{vmatrix}$$

This determinant is triangular; hence it equals the product of its diagonal elements

$$\frac{\partial y_k}{\partial x_k} = f(x_k \mid x_{k-1}, \ldots, x_1)$$

Inserting into (7-8) and using (7-37), we obtain

$$f(y_1, \ldots, y_n) = \frac{f(x_1, \ldots, x_n)}{f(x_1) f(x_2 \mid x_1) \cdots f(x_n \mid x_{n-1}, \ldots, x_1)} = 1$$

in the n-dimensional cube $0 \le y_i \le 1$, and 0 otherwise. ◀

From (7-5) and (7-35) it follows that

$$f(x_1 \mid x_3) = \int_{-\infty}^{\infty} f(x_1, x_2 \mid x_3)\, dx_2$$

$$f(x_1 \mid x_4) = \int_{-\infty}^{\infty} \int_{-\infty}^{\infty} f(x_1 \mid x_2, x_3, x_4) f(x_2, x_3 \mid x_4)\, dx_2\, dx_3$$

Generalizing, we obtain the following rule for removing variables on the left or on the right of the conditional line: To remove any number of variables on the left of the conditional line, we integrate with respect to them. To remove any number of variables to the right of the line, we multiply by their conditional density with respect to the remaining variables on the right, and we integrate the product. The following special case is used extensively (Chapman–Kolmogoroff equation, see also Chapter 16):

$$f(x_1 \mid x_3) = \int_{-\infty}^{\infty} f(x_1 \mid x_2, x_3) f(x_2 \mid x_3)\, dx_2 \tag{7-39}$$

Discrete type The above rule holds also for discrete type random variables provided that all densities are replaced by probabilities and all integrals by sums. We mention as an example the discrete form of (7-39): If the random variables $\mathbf{x}_1, \mathbf{x}_2, \mathbf{x}_3$ take the values a_i, b_k, c_r, respectively, then

$$P\{\mathbf{x}_1 = a_i \mid \mathbf{x}_3 = c_r\} = \sum_k P\{\mathbf{x}_1 = a_i \mid b_k, c_r\} P\{\mathbf{x}_2 = b_k \mid c_r\} \tag{7-40}$$

CONDITIONAL EXPECTED VALUES. The conditional mean of the random variables $g(\mathbf{x}_1, \ldots, \mathbf{x}_n)$ assuming M is given by the integral in (7-20) provided that the density $f(x_1, \ldots, x_n)$ is replaced by the conditional density $f(x_1, \ldots, x_n \mid M)$. Note, in particular, that [see also (6-226)]

$$E\{\mathbf{x}_1 \mid x_2, \ldots, x_n\} = \int_{-\infty}^{\infty} x_1 f(x_1 \mid x_2, \ldots, x_n)\, dx_1 \tag{7-41}$$

This is a function of $x_2, \ldots, x_n$; it defines, therefore, the random variable $E\{\mathbf{x}_1 \mid \mathbf{x}_2, \ldots, \mathbf{x}_n\}$. Multiplying (7-41) by $f(x_2, \ldots, x_n)$ and integrating, we conclude that

$$E\{E\{\mathbf{x}_1 \mid \mathbf{x}_2, \ldots, \mathbf{x}_n\}\} = E\{\mathbf{x}_1\} \tag{7-42}$$

Reasoning similarly, we obtain

$$\begin{aligned} E\{\mathbf{x}_1 \mid x_2, x_3\} &= E\{E\{\mathbf{x}_1 \mid x_2, x_3, \mathbf{x}_4\}\} \\ &= \int_{-\infty}^{\infty} E\{\mathbf{x}_1 \mid x_2, x_3, x_4\} f(x_4 \mid x_2, x_3)\, dx_4 \end{aligned} \tag{7-43}$$

This leads to the following generalization: To remove any number of variables on the right of the conditional expected value line, we multiply by their conditional density with respect to the remaining variables on the right and we integrate the product. For example,

$$E\{\mathbf{x}_1 \mid x_3\} = \int_{-\infty}^{\infty} E\{\mathbf{x}_1 \mid x_2, x_3\} f(x_2 \mid x_3)\, dx_2 \tag{7-44}$$

and for the discrete case [see (7-40)]

$$E\{\mathbf{x}_1 \mid c_r\} = \sum_k E\{\mathbf{x}_1 \mid b_k, c_r\} P\{\mathbf{x}_2 = b_k \mid c_r\} \tag{7-45}$$

EXAMPLE 7-7

▶ Given a discrete type random variable $\mathbf{n}$ taking the values $1, 2, \ldots$ and a sequence of random variables $\mathbf{x}_k$ independent of $\mathbf{n}$, we form the sum

$$\mathbf{s} = \sum_{k=1}^{\mathbf{n}} \mathbf{x}_k \tag{7-46}$$

This sum is a random variable specified as follows: For a specific ζ, $\mathbf{n}(\zeta)$ is an integer and $\mathbf{s}(\zeta)$ equals the sum of the numbers $\mathbf{x}_k(\zeta)$ for k from 1 to $\mathbf{n}(\zeta)$. We maintain that if the random variables $\mathbf{x}_k$ have the same mean, then

$$E\{\mathbf{s}\} = \eta E\{\mathbf{n}\} \qquad \text{where} \quad E\{\mathbf{x}_k\} = \eta \tag{7-47}$$

Clearly, $E\{\mathbf{x}_k \mid \mathbf{n} = n\} = E\{\mathbf{x}_k\}$ because $\mathbf{x}_k$ is independent of $\mathbf{n}$. Hence

$$E\{\mathbf{s} \mid \mathbf{n} = n\} = E\left\{ \sum_{k=1}^{n} \mathbf{x}_k \,\middle|\, \mathbf{n} = n \right\} = \sum_{k=1}^{n} E\{\mathbf{x}_k\} = \eta n$$

From this and (6-239) it follows that

$$E\{\mathbf{s}\} = E\{E\{\mathbf{s} \mid \mathbf{n}\}\} = E\{\eta \mathbf{n}\}$$

and (7-47) results.

We show next that if the random variables $\mathbf{x}_k$ are uncorrelated with the same variance σ^2, then

$$E\{\mathbf{s}^2\} = \eta^2 E\{\mathbf{n}^2\} + \sigma^2 E\{\mathbf{n}\} \tag{7-48}$$

Reasoning thus, we have

$$E\{\mathbf{s}^2 \mid \mathbf{n} = n\} = \sum_{i=1}^{n} \sum_{k=1}^{n} E\{\mathbf{x}_i \mathbf{x}_k\} \tag{7-49}$$

where

$$E\{\mathbf{x}_i \mathbf{x}_k\} = \begin{cases} \sigma^2 + \eta^2 & i = k \\ \eta^2 & i \neq k \end{cases}$$

The double sum in (7-49) contains n terms with $i = k$ and $n^2 - n$ terms with $i \neq k$; hence it equals

$$(\sigma^2 + \eta^2)n + \eta^2(n^2 - n) = \eta^2 n^2 + \sigma^2 n$$

This yields (7-48) because

$$E\{\mathbf{s}^2\} = E\{E\{\mathbf{s}^2 \mid \mathbf{n}\}\} = E\{\eta^2 \mathbf{n}^2 + \sigma^2 \mathbf{n}\}$$

Special Case. The number $\mathbf{n}$ of particles emitted from a substance in t seconds is a Poisson random variable with parameter λt. The energy $\mathbf{x}_k$ of the kth particle has a Maxwell distribution with mean $3kT/2$ and variance $3k^2T^2/2$ (see Prob. 7-5). The sum $\mathbf{s}$ in (7-46) is the total emitted energy in t seconds. As we know $E\{\mathbf{n}\} = \lambda t$, $E\{\mathbf{n}^2\} = \lambda^2 t^2 + \lambda t$ [see (5-64)]. Inserting into (7-47) and (7-48), we obtain

$$E\{\mathbf{s}\} = \frac{3kT\lambda t}{2} \qquad \sigma_s^2 = \frac{15k^2T^2\lambda t}{4}$$ ◀

Characteristic Functions and Normality

The characteristic function of a random vector is by definition the function

$$\Phi(\Omega) = E\{e^{j\Omega \mathbf{X}^t}\} = E\{e^{j(\omega_1 \mathbf{x}_1 + \cdots + \omega_n \mathbf{x}_n)}\} = \mathbf{\Phi}(j\Omega) \tag{7-50}$$

where

$$\mathbf{X} = [\mathbf{x}_1, \ldots, \mathbf{x}_n] \qquad \Omega = [\omega_1, \ldots, \omega_n]$$

As an application, we shall show that if the random variables $\mathbf{x}_i$ are independent with respective densities $f_i(x_i)$, then the density $f_z(z)$ of their sum $\mathbf{z} = \mathbf{x}_1 + \cdots + \mathbf{x}_n$ equals the convolution of their densities

$$f_z(z) = f_1(z) * \cdots * f_n(z) \tag{7-51}$$

Proof. Since the random variables $\mathbf{x}_i$ are independent and $e^{j\omega_i \mathbf{x}_i}$ depends only on $\mathbf{x}_i$, we conclude that from (7-24) that

$$E\left\{e^{j(\omega_1\mathbf{x}_1 + \cdots + \omega_n \mathbf{x}_n)}\right\} = E\{e^{j\omega_1\mathbf{x}_1}\} \cdots E\{e^{j\omega_n\mathbf{x}_n}\}$$

Hence

$$\Phi_z(\omega) = E\left\{e^{j\omega(\mathbf{x}_1 + \cdots + \mathbf{x}_n)}\right\} = \Phi_1(\omega) \cdots \Phi_n(\omega) \tag{7-52}$$

where $\Phi_i(\omega)$ is the characteristic function of $\mathbf{x}_i$. Applying the convolution theorem for Fourier transforms, we obtain (7-51).

EXAMPLE 7-8

▶ (*a*) **Bernoulli trials:** Using (7-52) we shall rederive the fundamental equation (3-13). We define the random variables $\mathbf{x}_i$ as follows: $\mathbf{x}_i = 1$ if heads shows at the ith trial and $\mathbf{x}_i = 0$ otherwise. Thus

$$P\{\mathbf{x}_i = 1\} = P\{h\} = p \qquad P\{\mathbf{x}_i = 0\} = P\{t\} = q \qquad \Phi_i(\omega) = pe^{j\omega} + q \tag{7-53}$$

The random variable $\mathbf{z} = \mathbf{x}_1 + \cdots + \mathbf{x}_n$ takes the values $0, 1, \ldots, n$ and $\{\mathbf{z} = k\}$ is the event $\{k$ heads in n tossings $\}$. Furthermore,

$$\Phi_z(\omega) = E\{e^{j\omega\mathbf{z}}\} = \sum_{k=0}^{n} P\{\mathbf{z} = k\}e^{jk\omega} \tag{7-54}$$

The random variables $\mathbf{x}_i$ are independent because $\mathbf{x}_i$ depends only on the outcomes of the ith trial and the trials are independent. Hence [see (7-52) and (7-53)]

$$\Phi_z(\omega) = (pe^{j\omega} + q)^n = \sum_{k=0}^{n} \binom{n}{k} p^k e^{jk\omega} q^{n-k}$$

Comparing with (7-54), we conclude that

$$P\{\mathbf{z} = k\} = P\{k \text{ heads}\} = \binom{n}{k} p^k q^{n-k} \tag{7-55}$$

(*b*) **Poisson theorem:** We shall show that if $p \ll 1$, then

$$P\{\mathbf{z} = k\} \simeq \frac{e^{-np}(np)^k}{k!}$$

as in (4-106). In fact, we shall establish a more general result. Suppose that the random variables $\mathbf{x}_i$ are independent and each takes the value 1 and 0 with respective probabilities p_i and $q_i = 1 - p_i$. If $p_i \ll 1$, then

$$e^{p_i(e^{j\omega}-1)} \simeq 1 + p_i(e^{j\omega} - 1) = p_i e^{j\omega} + q_i = \Phi_i(\omega)$$

With $\mathbf{z} = \mathbf{x}_1 + \cdots + \mathbf{x}_n$, it follows from (7-52) that

$$\Phi_z(\omega) \simeq e^{p_i(e^{j\omega}-1)} \cdots e^{p_n(e^{j\omega}-1)} = e^{a(e^{j\omega}-1)}$$

where $a = p_1 + \cdots + p_n$. This leads to the conclusion that [see (5-119)] the random variable $\mathbf{z}$ is approximately Poisson distributed with parameter a. It can be shown that the result is exact in the limit if

$$p_i \to 0 \quad \text{and} \quad p_1 + \cdots + p_n \to a \qquad \text{as} \quad n \to \infty$$ ◀

NORMAL VECTORS. Joint normality of n random variables $\mathbf{x}_i$ can be defined as in (6-23): Their joint density is an exponential whose exponent is a negative quadratic. We give next an equivalent definition that expresses the normality of n random variables in terms of the normality of a single random variable.

DEFINITION

▶ The random variables $\mathbf{x}_i$ are jointly normal iff the sum

$$a_1\mathbf{x}_1 + \cdots + a_n\mathbf{x}_n = A\mathbf{X}^t \tag{7-56}$$

is a normal random variable for any A.

We shall show that this definition leads to the following conclusions: If the random variables $\mathbf{x}_i$ have zero mean and covariance matrix C, then their joint characteristic function equals

$$\Phi(\Omega) = \exp\{-\tfrac{1}{2}\Omega C \Omega^t\} \tag{7-57}$$

Furthermore, their joint density equals

$$f(X) = \frac{1}{\sqrt{(2\pi)^n \Delta}} \exp\{-\tfrac{1}{2} X C^{-1} X^t\} \tag{7-58}$$

where Δ is the determinant of C.

Proof. From the definition of joint normality it follows that the random variable

$$\mathbf{w} = \omega_1\mathbf{x}_1 + \cdots + \omega_n\mathbf{x}_n = \Omega\mathbf{X}^t \tag{7-59}$$

is normal. Since $E\{\mathbf{x}_i\} = 0$ by assumption, this yields [see (7-26)]

$$E\{\mathbf{w}\} = 0 \qquad E\{\mathbf{w}^2\} = \sum_{i,j} \omega_i\omega_j C_{ij} = \sigma_w^2$$

Setting $\eta = 0$ and $\omega = 1$ in (5-100), we obtain

$$E\{e^{j\mathbf{w}}\} = \exp\left[-\frac{\sigma_w^2}{2}\right]$$

This yields

$$E\left\{e^{j\Omega\mathbf{X}^t}\right\} = \exp\left\{-\frac{1}{2}\sum_{i,j}\omega_i\omega_j C_{ij}\right\} \tag{7-60}$$

as in (7-57). The proof of (7-58) follows from (7-57) and the Fourier inversion theorem.

Note that if the random variables $\mathbf{x}_i$ are jointly normal and uncorrelated, they are independent. Indeed, in this case, their covariance matrix is diagonal and its diagonal elements equal σ_i^2.

Hence C^{-1} is also diagonal with diagonal elements $1/\sigma_i^2$. Inserting into (7-58), we obtain

$$f(x_1, \ldots, x_n) = \frac{1}{\sigma_1 \cdots \sigma_n \sqrt{(2\pi)^n}} \exp\left\{-\frac{1}{2}\left(\frac{x_1^2}{\sigma_1^2} + \cdots + \frac{x_n^2}{\sigma_n^2}\right)\right\}$$

◀

EXAMPLE 7-9

▶ Using characteristic functions, we shall show that if the random variables $\mathbf{x}_i$ are jointly normal with zero mean, and $E\{\mathbf{x}_i\mathbf{x}_j\} = C_{ij}$, then

$$E\{\mathbf{x}_1\mathbf{x}_2\mathbf{x}_3\mathbf{x}_4\} = C_{12}C_{34} + C_{13}C_{24} + C_{14}C_{23} \qquad (7\text{-}61)$$

Proof. We expand the exponentials on the left and right side of (7-60) and we show explicitly only the terms containing the factor $\omega_1\omega_2\omega_3\omega_4$:

$$E\left\{e^{j(\omega_1\mathbf{x}_1 + \cdots + \omega_4\mathbf{x}_4)}\right\} = \cdots + \frac{1}{4!}E\{(\omega_1\mathbf{x}_1 + \cdots + \omega_4\mathbf{x}_4)^4\} + \cdots$$

$$= \cdots + \frac{24}{4!}E\{\mathbf{x}_1\mathbf{x}_2\mathbf{x}_3\mathbf{x}_4\}\omega_1\omega_2\omega_3\omega_4$$

$$\exp\left\{-\frac{1}{2}\sum_{i,j}\omega_i\omega_j C_{ij}\right\} = +\frac{1}{2}\left(\frac{1}{2}\sum_{i,j}\omega_i\omega_j C_{ij}\right)^2 + \cdots$$

$$= \cdots + \frac{8}{8}(C_{12}C_{34} + C_{13}C_{24} + C_{14}C_{23})\omega_1\omega_2\omega_3\omega_4$$

Equating coefficients, we obtain (7-61). ◀

Complex normal vectors. A complex normal random vector is a vector $\mathbf{Z} = \mathbf{X} + j\mathbf{Y} = [\mathbf{z}_1, \ldots, \mathbf{z}_n]$ the components of which are n jointly normal random variables $\mathbf{z}_i = \mathbf{x}_i + j\mathbf{y}_i$. We shall assume that $E\{\mathbf{z}_i\} = 0$. The statistical properties of the vector $\mathbf{Z}$ are specified in terms of the joint density

$$f_Z(Z) = f(x_1, \ldots, x_n, y_1, \ldots, y_n)$$

of the $2n$ random variables $\mathbf{x}_i$ and $\mathbf{y}_j$. This function is an exponential as in (7-58) determined in terms of the $2n$ by $2n$ matrix

$$D = \begin{bmatrix} C_{XX} & C_{XY} \\ C_{YX} & C_{YY} \end{bmatrix}$$

consisting of the $2n^2 + n$ real parameters $E\{\mathbf{x}_i\mathbf{x}_j\}$, $E\{\mathbf{y}_i\mathbf{y}_j\}$, and $E\{\mathbf{x}_i\mathbf{y}_j\}$. The corresponding characteristic function

$$\Phi_Z(\Omega) = E\{\exp(j(u_1\mathbf{x}_1 + \cdots + u_n\mathbf{x}_n + v_1\mathbf{y}_1 + \cdots + v_n\mathbf{y}_n))\}$$

is an exponential as in (7-60):

$$\Phi_Z(\Omega) = \exp\left\{-\tfrac{1}{2}Q\right\} \qquad Q = [U \quad V]\begin{bmatrix} C_{XX} & C_{XY} \\ C_{YX} & C_{YY} \end{bmatrix}\begin{bmatrix} U^t \\ V^t \end{bmatrix}$$

where $U = [u_1, \ldots, u_n]$, $V = [v_1, \ldots, v_n]$, and $\Omega = U + jV$.

The covariance matrix of the complex vector $\mathbf{Z}$ is an n by n hermitian matrix

$$C_{ZZ} = E\{\mathbf{Z}^t\mathbf{Z}^*\} = C_{XX} + C_{YY} - j(C_{XY} - C_{YX})$$

with elements $E\{\mathbf{z}_i\mathbf{z}_j^*\}$. Thus, C_{ZZ} is specified in terms of n^2 real parameters. From this it follows that, unlike the real case, the density $f_Z(Z)$ of $\mathbf{Z}$ cannot in general be determined in terms of C_{ZZ} because $f_Z(Z)$ is a normal density consisting of $2n^2+n$ parameters. Suppose, for example, that $n=1$. In this case, $\mathbf{Z}=\mathbf{z}=\mathbf{x}+j\mathbf{y}$ is a scalar and $C_{ZZ}=E\{|\mathbf{z}|^2\}$. Thus, C_{ZZ} is specified in terms of the single parameter $\sigma_z^2=E\{\mathbf{x}^2+\mathbf{y}^2\}$. However, $f_z(z)=f(x,y)$ is a bivariate normal density consisting of the three parameters σ_x, σ_y, and $E\{\mathbf{xy}\}$. In the following, we present a special class of normal vectors that are statistically determined in terms of their covariance matrix. This class is important in modulation theory (see Sec. 10-3).

GOODMAN'S THEOREM[2]

▶ If the vectors $\mathbf{X}$ and $\mathbf{Y}$ are such that

$$C_{XX}=C_{YY} \qquad C_{XY}=-C_{YX}$$

and $\mathbf{Z}=\mathbf{X}+j\mathbf{Y}$, then

$$C_{ZZ}=2(C_{XX}-jC_{XY})$$

$$f_Z(Z)=\frac{1}{\pi^n|C_{ZZ}|}\exp\{-ZC_{ZZ}^{-1}Z^\dagger\} \tag{7-62}$$

$$\Phi_Z(\Omega)=\exp\left\{-\frac{1}{4}\Omega C_{ZZ}\Omega^\dagger\right\} \tag{7-63}$$

Proof. It suffices to prove (7-63); the proof of (7-62) follows from (7-63) and the Fourier inversion formula. Under the stated assumptions,

$$Q=[U \quad V]\begin{bmatrix} C_{XX} & C_{XY} \\ -C_{XY} & C_{XX}\end{bmatrix}\begin{bmatrix} U^t \\ V^t\end{bmatrix}$$
$$=UC_{XY}U^t+VC_{XY}U^t-UC_{XY}V^t+VC_{XX}V^t$$

Furthermore $C_{XX}^t=C_{XX}$ and $C_{XY}^t=-C_{XY}$. This leads to the conclusion that

$$VC_{XX}U^t=UC_{XX}V^t \qquad UC_{XY}U^t=VC_{XY}V^t=0$$

Hence

$$\tfrac{1}{2}\Omega C_{ZZ}\Omega^+=(U+jV)(C_{XX}-jC_{XY})(U^t-jV^t)=Q$$

and (7-63) results. ◀

Normal quadratic forms. Given n independent $N(0,1)$ random variables $\mathbf{z}_i$, we form the sum of their squares

$$\mathbf{x}=\mathbf{z}_1^2+\cdots+\mathbf{z}_n^2$$

Using characteristic functions, we shall show that the random variable $\mathbf{x}$ so formed has a chi-square distribution with n degrees of freedom:

$$f_x(x)=\gamma x^{n/2-1}e^{-x/2}U(x)$$

---

[2]N. R. Goodman, "Statistical Analysis Based on Certain Multivariate Complex Distribution," *Annals of Math. Statistics,* 1963, pp. 152–177.

Proof. The random variables $\mathbf{z}_i^2$ have a $\chi^2(1)$ distribution (see page 133); hence their characteristic functions are obtained from (5-109) with $m = 1$. This yields

$$\mathbf{\Phi}_i(s) = E\{e^{s\mathbf{z}_i^2}\} = \frac{1}{\sqrt{1-2s}}$$

From (7-52) and the independence of the random variables $\mathbf{z}_i^2$, it follows therefore that

$$\mathbf{\Phi}_x(s) = \mathbf{\Phi}_1(s) \cdots \mathbf{\Phi}_n(s) = \frac{1}{\sqrt{(1-2s)^n}}$$

Hence [see (5-109)] the random variable $\mathbf{x}$ is $\chi^2(n)$.

Note that

$$\frac{1}{\sqrt{(1-2s)^m}} \times \frac{1}{\sqrt{(1-2s)^n}} = \frac{1}{\sqrt{(1-2s)^{m+n}}}$$

This leads to the conclusion that if the random variables $\mathbf{x}$ and $\mathbf{y}$ are independent, $\mathbf{x}$ is $\chi^2(m)$ and $\mathbf{y}$ is $\chi^2(n)$, then the random variable

$$\mathbf{z} = \mathbf{x} + \mathbf{y} \quad \text{is} \quad \chi^2(m+n) \tag{7-64}$$

Conversely, if $\mathbf{z}$ is $\chi^2(m+n)$, $\mathbf{x}$ and $\mathbf{y}$ are independent, and $\mathbf{x}$ is $\chi^2(m)$, then $\mathbf{y}$ is $\chi^2(n)$. The following is an important application.

Sample variance. Given n i.i.d. $N(\eta, \sigma)$ random variables $\mathbf{x}_i$, we form their sample variance

$$\mathbf{s}^2 = \frac{1}{n-1}\sum_{i=1}^{n}(\mathbf{x}_i - \overline{\mathbf{x}})^2 \qquad \overline{\mathbf{x}} = \frac{1}{n}\sum_{i=1}^{n}\mathbf{x}_i \tag{7-65}$$

as in Example 7-4. We shall show that the random variable (see also Example 8-20)

$$\frac{(n-1)\mathbf{s}^2}{\sigma^2} = \sum_{i=1}^{n}\left(\frac{\mathbf{x}_i - \overline{\mathbf{x}}}{\sigma}\right)^2 \quad \text{is} \quad \chi^2(n-1) \tag{7-66}$$

Proof. We sum the identity

$$(\mathbf{x}_i - \eta)^2 = (\mathbf{x}_i - \overline{\mathbf{x}} + \overline{\mathbf{x}} - \eta)^2 = (\mathbf{x}_i - \overline{\mathbf{x}})^2 + (\overline{\mathbf{x}} - \eta)^2 + 2(\mathbf{x}_i - \overline{\mathbf{x}})(\overline{\mathbf{x}} - \eta)$$

from 1 to n. Since $\Sigma(x_i - \overline{x}) = 0$, this yields

$$\sum_{i=1}^{n}\left(\frac{\mathbf{x}_i - \eta}{\sigma}\right)^2 = \sum_{i=1}^{n}\left(\frac{\mathbf{x}_i - \overline{\mathbf{x}}}{\sigma}\right)^2 + n\left(\frac{\overline{\mathbf{x}} - \eta}{\sigma}\right)^2 \tag{7-67}$$

It can be shown that the random variables $\overline{\mathbf{x}}$ and $\mathbf{s}^2$ are independent (see Prob. 7-17 or Example 8-20). From this it follows that the two terms on the right of (7-67) are independent. Furthermore, the term

$$n\left(\frac{\overline{\mathbf{x}} - \eta}{\sigma}\right)^2 = \left(\frac{\overline{\mathbf{x}} - \eta}{\sigma/\sqrt{n}}\right)^2$$

is $\chi^2(1)$ because the random variable $\overline{\mathbf{x}}$ is $N(\eta, \sigma/\sqrt{n})$. Finally, the term on the left side is $\chi^2(n)$ and the proof is complete.

From (7-66) and (5-109) it follows that the mean of the random variable $(n-1)\mathbf{s}^2/\sigma^2$ equals $n-1$ and its variance equals $2(n-1)$. This leads to the conclusion that

$$E\{\mathbf{s}^2\} = (n-1)\frac{\sigma^2}{n-1} = \sigma^2 \qquad \text{Var}\,\{\mathbf{s}^2\} = 2(n-1)\frac{\sigma^4}{(n-1)^2} = \frac{2\sigma^4}{n-1}$$

EXAMPLE 7-10 ▶ We shall verify the above for $n = 2$. In this case,

$$\overline{\mathbf{x}} = \frac{\mathbf{x}_1 + \mathbf{x}_2}{2} \qquad \mathbf{s}^2 = (\mathbf{x}_1 - \overline{\mathbf{x}})^2 + (\mathbf{x}_2 - \overline{\mathbf{x}})^2 = \frac{1}{2}(\mathbf{x}_1 - \mathbf{x}_2)^2$$

The random variables $\mathbf{x}_1 + \mathbf{x}_2$ and $\mathbf{x}_1 - \mathbf{x}_2$ are independent because they are jointly normal and $E\{\mathbf{x}_1 - \mathbf{x}_2\} = 0$, $E\{(\mathbf{x}_1 - \mathbf{x}_2)(\mathbf{x}_1 + \mathbf{x}_2)\} = 0$. From this it follows that the random variables $\overline{\mathbf{x}}$ and $\mathbf{s}^2$ are independent. But the random variable $(\mathbf{x}_1 - \mathbf{x}_2)/\sigma\sqrt{2} = \mathbf{s}/\sigma$ is $N(0, 1)$; hence its square $\mathbf{s}^2/\sigma^2$ is $\chi^2(1)$ in agreement with (7-66). ◀

7-3 MEAN SQUARE ESTIMATION

The estimation problem is fundamental in the applications of probability and it will be discussed in detail later (Chap. 13). In this section first we introduce the main ideas using as illustration the estimation of a random variable $\mathbf{y}$ in terms of another random variable $\mathbf{x}$. Throughout this analysis, the optimality criterion will be the minimization of the mean square value (abbreviation: MS) of the estimation error.

We start with a brief explanation of the underlying concepts in the context of repeated trials, considering first the problem of estimating the random variable $\mathbf{y}$ by a constant.

Frequency interpretation As we know, the distribution function $F(y)$ of the random variable $\mathbf{y}$ determines completely its statistics. This does not, of course, mean that if we know $F(y)$ we can predict the value $\mathbf{y}(\zeta)$ of $\mathbf{y}$ at some future trial. Suppose, however, that we wish to estimate the unknown $\mathbf{y}(\zeta)$ by some number c. As we shall presently see, knowledge of $F(y)$ can guide us in the selection of c.

If $\mathbf{y}$ is estimated by a constant c, then, at a particular trial, the error $\mathbf{y}(\zeta) - c$ results and our problem is to select c so as to minimize this error in some sense. A reasonable criterion for selecting c might be the condition that, in a long series of trials, the average error is close to 0:

$$\frac{\mathbf{y}(\zeta_1) - c + \cdots + \mathbf{y}(\zeta_n) - c}{n} \simeq 0$$

As we see from (5-51), this would lead to the conclusion that c should equal the *mean* of $\mathbf{y}$ (Fig. 7-2*a*).

Another criterion for selecting c might be the minimization of the average of $|\mathbf{y}(\zeta) - c|$. In this case, the optimum c is the *median* of $\mathbf{y}$ [see (4-2)].

In our analysis, we consider only MS estimates. This means that c should be such as to minimize the average of $|\mathbf{y}(\zeta) - c|^2$. This criterion is in general useful but it is selected mainly because it leads to simple results. As we shall soon see, the best c is again the *mean* of $\mathbf{y}$.

Suppose now that at each trial we observe the value $\mathbf{x}(\zeta)$ of the random variable $\mathbf{x}$. On the basis of this observation it might be best to use as the estimate of $\mathbf{y}$ not the same number c at each trial, but a number that depends on the observed $\mathbf{x}(\zeta)$. In other words,

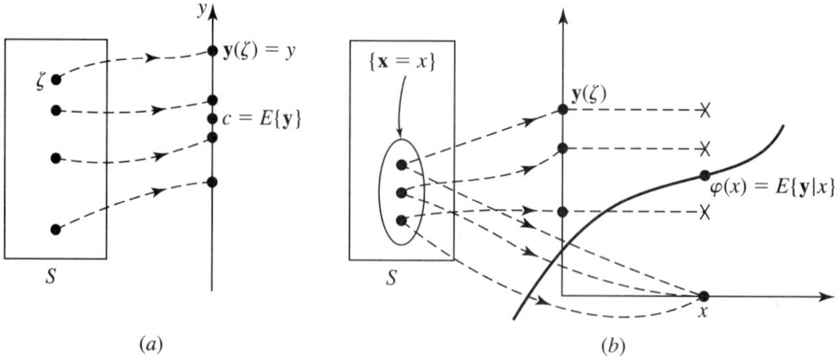

FIGURE 7-2

we might use as the estimate of $\mathbf{y}$ a function $c(\mathbf{x})$ of the random variable $\mathbf{x}$. The resulting problem is the optimum determination of this function.

It might be argued that, if at a certain trial we observe $\mathbf{x}(\zeta)$, then we can determine the outcome ζ of this trial, and hence also the corresponding value $\mathbf{y}(\zeta)$ of $\mathbf{y}$. This, however, is not so. The same number $\mathbf{x}(\zeta) = x$ is observed for every ζ in the set $\{\mathbf{x} = x\}$ (Fig. 7-2*b*). If, therefore, this set has many elements and the values of $\mathbf{y}$ are different for the various elements of this set, then the observed $\mathbf{x}(\zeta)$ does not determine uniquely $\mathbf{y}(\zeta)$. However, we know now that ζ is an element of the subset $\{\mathbf{x} = x\}$. This information reduces the uncertainty about the value of $\mathbf{y}$. In the subset $\{\mathbf{x} = x\}$, the random variable $\mathbf{x}$ equals x and the problem of determining $c(x)$ is reduced to the problem of determining the constant $c(x)$. As we noted, if the optimality criterion is the minimization of the MS error, then $c(x)$ must be the average of $\mathbf{y}$ in this set. In other words, $c(x)$ must equal the conditional mean of $\mathbf{y}$ assuming that $\mathbf{x} = x$.

We shall illustrate with an example. Suppose that the space S is the set of all children in a community and the random variable $\mathbf{y}$ is the height of each child. A particular outcome ζ is a specific child and $\mathbf{y}(\zeta)$ is the height of this child. From the preceding discussion it follows that if we wish to estimate $\mathbf{y}$ by a number, this number must equal the mean of $\mathbf{y}$. We now assume that each selected child is weighed. On the basis of this observation, the estimate of the height of the child can be improved. The weight is a random variable $\mathbf{x}$; hence the optimum estimate of $\mathbf{y}$ is now the conditional mean $E\{\mathbf{y} \mid x\}$ of $\mathbf{y}$ assuming $\mathbf{x} = x$ where x is the observed weight.

In the context of probability theory, the MS estimation of the random variable $\mathbf{y}$ by a constant c can be phrased as follows: Find c such that the second moment (MS error)

$$e = E\{(\mathbf{y} - c)^2\} = \int_{-\infty}^{\infty} (y - c)^2 f(y)\, dy \tag{7-68}$$

of the difference (error) $\mathbf{y} - c$ is minimum. Clearly, e depends on c and it is minimum if

$$\frac{de}{dc} = \int_{-\infty}^{\infty} 2(y - c) f(y)\, dy = 0$$

that is, if

$$c = \int_{-\infty}^{\infty} y f(y)\, dy$$

Thus

$$c = E\{\mathbf{y}\} = \int_{-\infty}^{\infty} y f(y)\, dy \tag{7-69}$$

This result is well known from mechanics: The moment of inertia with respect to a point c is minimum if c is the center of gravity of the masses.

NONLINEAR MS ESTIMATION. We wish to estimate **y** not by a constant but by a function $c(\mathbf{x})$ of the random variable **x**. Our problem now is to find the function $c(\mathbf{x})$ such that the MS error

$$e = E\{[\mathbf{y} - c(\mathbf{x})]^2\} = \int_{-\infty}^{\infty}\int_{-\infty}^{\infty} [y - c(x)]^2 f(x, y)\, dx\, dy \tag{7-70}$$

is minimum.

We maintain that

$$c(x) = E\{\mathbf{y} \mid x\} = \int_{-\infty}^{\infty} y f(y \mid x)\, dy \tag{7-71}$$

Proof. Since $f(x, y) = f(y \mid x) f(x)$, (7-70) yields

$$e = \int_{-\infty}^{\infty} f(x) \int_{-\infty}^{\infty} [y - c(x)]^2 f(y \mid x)\, dy\, dx$$

These integrands are positive. Hence e is minimum if the inner integral is minimum for every x. This integral is of the form (7-68) if c is changed to $c(x)$, and $f(y)$ is changed to $f(y \mid x)$. Hence it is minimum if $c(x)$ equals the integral in (7-69), provided that $f(y)$ is changed to $f(y \mid x)$. The result is (7-71).

Thus the optimum $c(x)$ is the regression line $\varphi(x)$ of Fig. 6-33.

As we noted in the beginning of the section, if $\mathbf{y} = g(\mathbf{x})$, then $E\{\mathbf{y} \mid x\} = g(x)$; hence $c(x) = g(x)$ and the resulting MS error is 0. This is not surprising because, if **x** is observed and $\mathbf{y} = g(\mathbf{x})$, then **y** is determined uniquely.

If the random variables **x** and **y** are independent, then $E\{\mathbf{y} \mid x\} = E\{\mathbf{y}\} = \text{constant}$. In this case, knowledge of **x** has no effect on the estimate of **y**.

Linear MS Estimation

The solution of the nonlinear MS estimation problem is based on knowledge of the function $\varphi(x)$. An easier problem, using only second-order moments, is the linear MS estimation of **y** in terms of **x**. The resulting estimate is not as good as the nonlinear estimate; however, it is used in many applications because of the simplicity of the solution.

The linear estimation problem is the estimation of the random variable **y** in terms of a linear function $A\mathbf{x} + B$ of **x**. The problem now is to find the constants A and B so as to minimize the MS error

$$e = E\{[\mathbf{y} - (A\mathbf{x} + B)]^2\} \tag{7-72}$$

We maintain that $e = e_m$ is minimum if

$$A = \frac{\mu_{11}}{\mu_{20}} = \frac{r\sigma_y}{\sigma_x} \qquad B = \eta_y - A\eta_x \tag{7-73}$$

and

$$e_m = \mu_{02} - \frac{\mu_{11}^2}{\mu_{20}} = \sigma_y^2(1 - r^2) \tag{7-74}$$

Proof. For a given A, e is the MS error of the estimation of $\mathbf{y} - A\mathbf{x}$ by the constant B. Hence e is minimum if $B = E\{\mathbf{y} - A\mathbf{x}\}$ as in (7-69). With B so determined, (7-72) yields

$$e = E\{[(\mathbf{y} - \eta_y) - A(\mathbf{x} - \eta_x)]^2\} = \sigma_y^2 - 2Ar\sigma_x\sigma_y + A^2\sigma_x^2$$

This is minimum if $A = r\sigma_y/\sigma_x$ and (7-73) results. Inserting into the preceding quadratic, we obtain (7-74).

TERMINOLOGY. In the above, the sum $A\mathbf{x} + B$ is the *nonhomogeneous* linear estimate of $\mathbf{y}$ in terms of $\mathbf{x}$. If $\mathbf{y}$ is estimated by a straight line $a\mathbf{x}$ passing through the origin, the estimate is called *homogeneous*.

The random variable $\mathbf{x}$ is the data of the estimation, the random variable $\boldsymbol{\varepsilon} = \mathbf{y} - (A\mathbf{x} + B)$ is the *error* of the estimation, and the number $e = E\{\boldsymbol{\varepsilon}^2\}$ is the MS error.

FUNDAMENTAL NOTE. In general, the nonlinear estimate $\varphi(x) = E[\mathbf{y} \mid x\}$ of $\mathbf{y}$ in terms of $\mathbf{x}$ is not a straight line and the resulting MS error $E\{[\mathbf{y} - \varphi(\mathbf{x})]^2\}$ is smaller than the MS error e_m of the linear estimate $A\mathbf{x} + B$. However, if the random variables $\mathbf{x}$ and $\mathbf{y}$ are jointly normal, then [see (7-60)]

$$\varphi(x) = \frac{r\sigma_y x}{\sigma_x} + \eta_y - \frac{r\sigma_y\eta_x}{\sigma_x}$$

is a straight line as in (7-73). In other words:

For normal random variables, nonlinear and linear MS estimates are identical.

The Orthogonality Principle

From (7-73) it follows that

$$E\{[\mathbf{y} - (A\mathbf{x} + B)]\mathbf{x}\} = 0 \tag{7-75}$$

This result can be derived directly from (7-72). Indeed, the MS error e is a function of A and B and it is minimum if $\partial e/\partial A = 0$ and $\partial e/\partial B = 0$. The first equation yields

$$\frac{\partial e}{\partial A} = E\{2[\mathbf{y} - (A\mathbf{x} + B)](-\mathbf{x})\} = 0$$

and (7-75) results. The interchange between expected value and differentiation is equivalent to the interchange of integration and differentiation.

Equation (7-75) states that the optimum linear MS estimate $A\mathbf{x} + B$ of $\mathbf{y}$ is such that the estimation error $\mathbf{y} - (A\mathbf{x} + B)$ is orthogonal to the data $\mathbf{x}$. This is known as the *orthogonality principle*. It is fundamental in MS estimation and will be used extensively. In the following, we reestablish it for the homogeneous case.

HOMOGENEOUS LINEAR MS ESTIMATION. We wish to find a constant a such that, if $\mathbf{y}$ is estimated by $a\mathbf{x}$, the resulting MS error

$$e = E\{(\mathbf{y} - a\mathbf{x})^2\} \tag{7-76}$$

is minimum. We maintain that a must be such that

$$E\{(\mathbf{y} - a\mathbf{x})\mathbf{x}\} = 0 \tag{7-77}$$

Proof. Clearly, e is minimum if $e'(a) = 0$; this yields (7-77). We shall give a second proof: We assume that a satisfies (7-77) and we shall show that e is minimum. With $\overline{a}$ an arbitrary constant,

$$\begin{aligned} E\{(\mathbf{y} - \overline{a}\mathbf{x})^2\} &= E\{[(\mathbf{y} - a\mathbf{x}) + (a - \overline{a})\mathbf{x}]^2\} \\ &= E\{(\mathbf{y} - a\mathbf{x})^2\} + (a - \overline{a})^2 E\{\mathbf{x}^2\} + 2(a - \overline{a})E\{(\mathbf{y} - a\mathbf{x})\mathbf{x}\} \end{aligned}$$

Here, the last term is 0 by assumption and the second term is positive. From this it follows that

$$E\{(\mathbf{y} - \overline{a}\mathbf{x})^2\} \geq E\{(\mathbf{y} - a\mathbf{x})^2\}$$

for any $\overline{a}$; hence e is minimum.

The linear MS estimate of $\mathbf{y}$ in terms of $\mathbf{x}$ will be denoted by $\hat{E}\{\mathbf{y} \mid \mathbf{x}\}$. Solving (7-77), we conclude that

$$\hat{E}\{\mathbf{y} \mid \mathbf{x}\} = a\mathbf{x} \qquad a = \frac{E\{\mathbf{xy}\}}{E\{\mathbf{x}^2\}} \tag{7-78}$$

MS error Since

$$e = E\{(\mathbf{y} - a\mathbf{x})\mathbf{y}\} - E\{(\mathbf{y} - a\mathbf{x})a\mathbf{x}\} = E\{\mathbf{y}^2\} - E\{(a\mathbf{x})^2\} - 2aE\{(\mathbf{y} - a\mathbf{x})\mathbf{x}\}$$

we conclude with (7-77) that

$$e = E\{(\mathbf{y} - a\mathbf{x})\mathbf{y}\} = E\{\mathbf{y}^2\} - E\{(a\mathbf{x})^2\} \tag{7-79}$$

We note finally that (7-77) is consistent with the orthogonality principle: The error $\mathbf{y} - a\mathbf{x}$ is orthogonal to the data $\mathbf{x}$.

Geometric interpretation of the orthogonality principle. In the vector representation of random variables (see Fig. 7-3), the difference $\mathbf{y} - a\mathbf{x}$ is the vector from the point $a\mathbf{x}$ on the $\mathbf{x}$ line to the point $\mathbf{y}$, and the length of that vector equals $\sqrt{e}$. Clearly, this length is minimum if $\mathbf{y} - a\mathbf{x}$ is perpendicular to $\mathbf{x}$ in agreement with (7-77). The right side of (7-79) follows from the pythagorean theorem and the middle term states that the square of the length of $\mathbf{y} - a\mathbf{x}$ equals the inner product of $\mathbf{y}$ with the error $\mathbf{y} - a\mathbf{x}$.

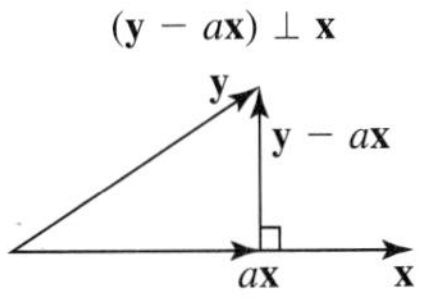

FIGURE 7-3

Risk and loss functions. We conclude with a brief comment on other optimality criteria limiting the discussion to the estimation of a random variable $\mathbf{y}$ by a constant c. We select a function $L(x)$ and we choose c so as to minimize the mean

$$R = E\{L(\mathbf{y} - c)\} = \int_{-\infty}^{\infty} L(y - c) f(y)\, dy$$

of the random variable $L(\mathbf{y} - c)$. The function $L(x)$ is called the *loss function* and the constant R is called the *average risk*. The choice of $L(x)$ depends on the applications. If $L(x) = x^2$, then $R = E\{(\mathbf{y} - c)^2\}$ is the MS error and as we have shown, it is minimum if $c = E\{\mathbf{y}\}$.

If $L(x) = |x|$, then $R = E\{|\mathbf{y} - c|\}$. We maintain that in this case, c equals the *median* $y_{0.5}$ of $\mathbf{y}$ (see also Prob. 5-32).

Proof. The average risk equals

$$R = \int_{-\infty}^{\infty} |y - c| f(y)\, dy = \int_{-\infty}^{c} (c - y) f(y)\, dy + \int_{c}^{\infty} (y - c) f(y)\, dy$$

Differentiating with respect to c, we obtain

$$\frac{dR}{dc} = \int_{-\infty}^{c} f(y)\, dy - \int_{c}^{\infty} f(y)\, dy = 2F(c) - 1$$

Thus R is minimum if $F(c) = 1/2$, that is, if $c = y_{0.5}$.

Next, we consider the general problem of estimating an unknown $\mathbf{s}$ in terms of n random variables $\mathbf{x}_1, \mathbf{x}_2, \ldots, \mathbf{x}_n$.

LINEAR ESTIMATION (GENERAL CASE). The linear MS estimate of $\mathbf{s}$ in terms of the random variables $\mathbf{x}_i$ is the sum

$$\hat{\mathbf{s}} = a_1\mathbf{x}_1 + \cdots + a_n\mathbf{x}_n \tag{7-80}$$

where $a_1, \ldots, a_n$ are n constants such that the MS value

$$P = E\{(\mathbf{s} - \hat{\mathbf{s}})^2\} = E\{[\mathbf{s} - (a_1\mathbf{x}_1 + \cdots + a_n\mathbf{x}_n)]^2\} \tag{7-81}$$

of the estimation error $\mathbf{s} - \hat{\mathbf{s}}$ is minimum.

Orthogonality principle. P is minimum if the error $\mathbf{s} - \hat{\mathbf{s}}$ is orthogonal to the data $\mathbf{x}_i$:

$$E\{[\mathbf{s} - (a_1\mathbf{x}_1 + \cdots + a_n\mathbf{x}_n)]\mathbf{x}_i\} = 0 \qquad i = 1, \ldots, n \tag{7-82}$$

Proof. P is a function of the constants a_i and it is minimum if

$$\frac{\partial P}{\partial a_i} = E\{-2[\mathbf{s} - (a_1\mathbf{x}_1 + \cdots + a_n\mathbf{x}_n)]\mathbf{x}_i\} = 0$$

and (7-82) results. This important result is known also as the *projection theorem.* Setting $i = 1, \dots, n$ in (7-82), we obtain the system

$$\begin{aligned} R_{11}a_1 + R_{21}a_2 + \cdots + R_{n1}a_n &= R_{01} \\ R_{12}a_1 + R_{22}a_2 + \cdots + R_{n2}a_n &= R_{02} \\ \cdots\cdots\cdots\cdots\cdots\cdots\cdots\cdots \\ R_{1n}a_1 + R_{2n}a_2 + \cdots + R_{nn}a_n &= R_{0n} \end{aligned} \tag{7-83}$$

where $R_{ij} = E\{\mathbf{x}_i\mathbf{x}_j\}$ and $R_{0j} = E\{\mathbf{s}\mathbf{x}_j\}$.

To solve this system, we introduce the row vectors

$$\mathbf{X} = [\mathbf{x}_1, \dots, \mathbf{x}_n] \qquad A = [a_1, \dots, a_n] \qquad R_0 = [R_{01}, \dots, R_{0n}]$$

and the data correlation matrix $R = E\{\mathbf{X}^t\mathbf{X}\}$, where $\mathbf{X}^t$ is the transpose of $\mathbf{X}$. This yields

$$AR = R_0 \qquad A = R_0 R^{-1} \tag{7-84}$$

Inserting the constants a_i so determined into (7-81), we obtain the least mean square (LMS) error. The resulting expression can be simplified. Since $\mathbf{s} - \hat{\mathbf{s}} \perp \mathbf{x}_i$ for every i, we conclude that $\mathbf{s} - \hat{\mathbf{s}} \perp \hat{\mathbf{s}}$; hence

$$P = E\{(\mathbf{s} - \hat{\mathbf{s}})\mathbf{s}\} = E\{\mathbf{s}^2\} - AR_0^t \tag{7-85}$$

Note that if the rank of R is $m < n$, then the data are linearly dependent. In this case, the estimate $\hat{\mathbf{s}}$ can be written as a linear sum involving a subset of m linearly independent components of the data vector $\mathbf{X}$.

Geometric interpretation. In the representation of random variables as vectors in an abstract space, the sum $\hat{\mathbf{s}} = a_1\mathbf{x}_1 + \cdots + a_n\mathbf{x}_n$ is a vector in the subspace S_n of the data $\mathbf{x}_i$ and the error $\boldsymbol{\varepsilon} = \mathbf{s} - \hat{\mathbf{s}}$ is the vector from $\mathbf{s}$ to $\hat{\mathbf{s}}$ as in Fig. 7-4*a*. The projection theorem states that the length of $\boldsymbol{\varepsilon}$ is minimum if $\boldsymbol{\varepsilon}$ is orthogonal to $\mathbf{x}_i$, that is, if it is perpendicular to the data subspace S_n. The estimate $\hat{\mathbf{s}}$ is thus the "projection" of $\mathbf{s}$ on S_n.

If $\mathbf{s}$ is a vector in S_n, then $\hat{\mathbf{s}} = \mathbf{s}$ and $P = 0$. In this case, the $n + 1$ random variables $\mathbf{s}, \mathbf{x}_1, \dots, \mathbf{x}_n$ are linearly dependent and the determinant Δ_{n+1} of their correlation matrix is 0. If $\mathbf{s}$ is perpendicular to S_n, then $\hat{\mathbf{s}} = 0$ and $P = E\{|\mathbf{s}|^2\}$. This is the case if $\mathbf{s}$ is orthogonal to all the data $\mathbf{x}_i$, that is, if $R_{0j} = 0$ for $j \neq 0$.

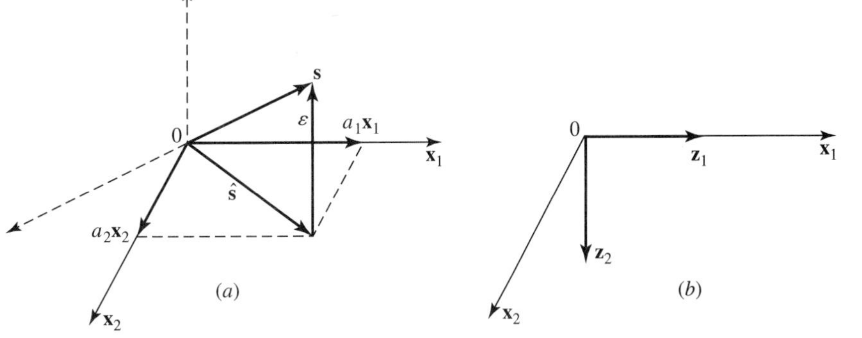

FIGURE 7-4

Nonhomogeneous estimation. The estimate (7-80) can be improved if a constant is added to the sum. The problem now is to determine $n+1$ parameters α_k such that if

$$\hat{\mathbf{s}} = \alpha_0 + \alpha_1 \mathbf{x}_1 + \cdots + \alpha_n \mathbf{x}_n \tag{7-86}$$

then the resulting MS error is minimum. This problem can be reduced to the homogeneous case if we replace the term α_0 by the product $\alpha_0 \mathbf{x}_0$, where $\mathbf{x}_0 \equiv 1$. Applying (7-82) to the enlarged data set

$$\mathbf{x}_0, \mathbf{x}_1, \ldots, \mathbf{x}_n \qquad \text{where} \quad E\{\mathbf{x}_0 \mathbf{x}_i\} = \begin{cases} E\{\mathbf{x}_i\} = \eta_i & i \neq 0 \\ 1 & i = 0 \end{cases}$$

we obtain

$$\begin{aligned} \alpha_0 + \eta_1 \alpha_1 + \cdots + \eta_n \alpha_n &= \eta_s \\ \eta_1 \alpha_0 + R_{11} \alpha_1 + \cdots + R_{1n} \alpha_n &= R_{01} \\ \cdots\cdots\cdots\cdots\cdots\cdots\cdots\cdots & \\ \eta_n \alpha_0 + R_{n1} \alpha_1 + \cdots + R_{nn} \alpha_n &= R_{0n} \end{aligned} \tag{7-87}$$

Note that, if $\eta_s = \eta_i = 0$, then (7-87) reduces to (7-83). This yields $\alpha_0 = 0$ and $\alpha_n = a_n$.

Nonlinear estimation. The nonlinear MS estimation problem involves the determination of a function $g(\mathbf{x}_1, \ldots, \mathbf{x}_n) = g(\mathbf{X})$ of the data $\mathbf{x}_i$ such as to minimize the MS error

$$P = E\{[\mathbf{s} - g(\mathbf{X})]^2\} \tag{7-88}$$

We maintain that P is minimum if

$$g(X) = E\{\mathbf{s} \mid X\} = \int_{-\infty}^{\infty} s f_s(s \mid X)\, ds \tag{7-89}$$

The function $f_s(s \mid X)$ is the conditional mean (regression surface) of the random variable $\mathbf{s}$ assuming $\mathbf{X} = X$.

Proof. The proof is based on the identity [see (7-42)]

$$P = E\{[\mathbf{s} - g(\mathbf{X})]^2\} = E\{E\{[\mathbf{s} - g(\mathbf{X})]^2 \mid \mathbf{X}\}\} \tag{7-90}$$

Since all quantities are positive, it follows that P is minimum if the conditional MS error

$$E\{[\mathbf{s} - g(\mathbf{X})]^2 \mid X\} = \int_{-\infty}^{\infty} [s - g(\mathbf{X})]^2 f_s(s \mid X)\, ds \tag{7-91}$$

is minimum. In the above integral, $g(X)$ is constant. Hence the integral is minimum if $g(X)$ is given by (7-89) [see also (7-71)].

The general orthogonality principle. From the projection theorem (7-82) it follows that

$$E\{[\mathbf{s} - \hat{\mathbf{s}}](c_1 \mathbf{x}_1 + \cdots + c_n \mathbf{x}_n)\} = 0 \tag{7-92}$$

for any $c_1, \ldots, c_n$. This shows that if $\hat{\mathbf{s}}$ is the linear MS estimator of $\mathbf{s}$, the estimation error $\mathbf{s} - \hat{\mathbf{s}}$ is orthogonal to any linear function $\mathbf{y} = c_1 \mathbf{x}_1 + \cdots + c_n \mathbf{x}_n$ of the data $\mathbf{x}_i$.

We shall now show that if $g(\mathbf{X})$ is the nonlinear MS estimator of $\mathbf{s}$, the estimation error $\mathbf{s} - g(\mathbf{X})$ is orthogonal to any function $w(\mathbf{X})$, linear or nonlinear, of the data $\mathbf{x}_i$:

$$E\{[\mathbf{s} - g(\mathbf{X})]w(\mathbf{X})\} = 0 \tag{7-93}$$

Proof. We shall use the following generalization of (7-60):

$$E\{[\mathbf{s} - g(\mathbf{X})]w(\mathbf{X})\} = E\{w(\mathbf{X})E\{\mathbf{s} - g(\mathbf{X}) \mid \mathbf{X}\}\} \tag{7-94}$$

From the linearity of expected values and (7-89) it follows that

$$E\{\mathbf{s} - g(\mathbf{X}) \mid X\} = E\{\mathbf{s} \mid X\} - E\{g(\mathbf{X}) \mid X\} = 0$$

and (7-93) results.

Normality. Using the material just developed, we shall show that if the random variables $\mathbf{s}, \mathbf{x}_1, \ldots, \mathbf{x}_n$ are jointly normal with zero mean, the linear and nonlinear estimators of $\mathbf{s}$ are equal:

$$\hat{\mathbf{s}} = a_1\mathbf{x}_1 + \cdots + a_n\mathbf{x}_n = g(\mathbf{X}) = E\{s \mid \mathbf{X}\} \tag{7-95}$$

Proof. To prove (7-93), it suffices to show that $\hat{\mathbf{s}} = E\{\mathbf{s} \mid \mathbf{X}\}$. The random variables $\mathbf{s} - \hat{\mathbf{s}}$ and $\mathbf{x}_i$ are jointly normal with zero mean and orthogonal; hence they are independent. From this it follows that

$$E\{\mathbf{s} - \hat{\mathbf{s}} \mid X\} = E\{\mathbf{s} - \hat{\mathbf{s}}\} = 0 = E\{\mathbf{s} \mid X\} - E\{\hat{\mathbf{s}} \mid X\}$$

and (7-95) results because $E\{\hat{\mathbf{s}} \mid \mathbf{X}\} = \hat{\mathbf{s}}$.

Conditional densities of normal random variables. We shall use the preceding result to simplify the determination of conditional densities involving normal random variables. The conditional density $f_s(s \mid X)$ of $\mathbf{s}$ assuming $\mathbf{X}$ is the ratio of two exponentials the exponents of which are quadratics, hence it is normal. To determine it, it suffices, therefore, to find the conditional mean and variance of $\mathbf{s}$. We maintain that

$$E\{\mathbf{s} \mid X\} = \hat{s} \qquad E\{(\mathbf{s} - \hat{\mathbf{s}})^2 \mid X\} = E\{(\mathbf{s} - \hat{\mathbf{s}})^2\} = P \tag{7-96}$$

The first follows from (7-95). The second follows from the fact that $\mathbf{s} - \hat{\mathbf{s}}$ is orthogonal and, therefore, independent of $\mathbf{X}$. We thus conclude that

$$f(s \mid x_1, \ldots, x_n) = \frac{1}{\sqrt{2\pi P}} e^{-[s-(a_1x_1+\cdots+a_nx_n)]^2/2P} \tag{7-97}$$

EXAMPLE 7-11

▶ The random variables $\mathbf{x}_1$ and $\mathbf{x}_2$ are jointly normal with zero mean. We shall determine their conditional density $f(x_2 \mid x_1)$. As we know [see (7-78)]

$$E\{\mathbf{x}_2 \mid x_1\} = ax_1 \qquad a = \frac{R_{12}}{R_{11}}$$

$$\sigma^2_{x_2 \mid x_1} = P = E\{(\mathbf{x}_2 - a\mathbf{x}_1)\mathbf{x}_2\} = R_{22} - aR_{12}$$

Inserting into (7-97), we obtain

$$f(x_2 \mid x_1) = \frac{1}{\sqrt{2\pi P}} e^{-(x_2 - ax_1)^2/2P}$$ ◀

EXAMPLE 7-12 ▶ We now wish to find the conditional density $f(x_3 \mid x_1, x_2)$. In this case,

$$E\{\mathbf{x}_3 \mid \mathbf{x}_1, \mathbf{x}_2\} = a_1\mathbf{x}_1 + a_2\mathbf{x}_2$$

where the constants a_1 and a_2 are determined from the system

$$R_{11}a_1 + R_{12}a_2 = R_{13} \qquad R_{12}a_1 + R_{22}a_2 = R_{23}$$

Furthermore [see (7-96) and (7-85)]

$$\sigma^2_{x_3 \mid x_1, x_2} = P = R_{33} - (R_{13}a_1 + R_{23}a_2)$$

and (7-97) yields

$$f(x_3 \mid x_1, x_2) = \frac{1}{\sqrt{2\pi P}} e^{-(x_3 - a_1x_1 - a_2x_2)^2/2P}$$ ◀

EXAMPLE 7-13 ▶ In this example, we shall find the two-dimensional density $f(x_2, x_3 \mid x_1)$. This involves the evaluation of five parameters [see (6-23)]; two conditional means, two conditional variances, and the conditional covariance of the random variables $\mathbf{x}_2$ and $\mathbf{x}_3$ assuming $\mathbf{x}_1$.

The first four parameters are determined as in Example 7-11:

$$E\{\mathbf{x}_2 \mid x_1\} = \frac{R_{12}}{R_{11}}x_1 \qquad E\{\mathbf{x}_3 \mid x_1\} = \frac{R_{13}}{R_{11}}x_1$$

$$\sigma^2_{x_2 \mid x_1} = R_{22} - \frac{R_{12}^2}{R_{11}} \qquad \sigma^2_{x_3 \mid x_1} = R_{33} - \frac{R_{13}^2}{R_{11}}$$

The conditional covariance

$$C_{x_2x_3 \mid x_1} = E\left\{\left(\mathbf{x}_2 - \frac{R_{12}}{R_{11}}\mathbf{x}_1\right)\left(\mathbf{x}_3 - \frac{R_{13}}{R_{11}}\mathbf{x}_1\right)\middle|\, \mathbf{x}_1 = x_1\right\} \tag{7-98}$$

is found as follows: We know that the errors $\mathbf{x}_2 - R_{12}\mathbf{x}_1/R_{11}$ and $\mathbf{x}_3 - R_{13}\mathbf{x}_1/R_{11}$ are independent of $\mathbf{x}_1$. Hence the condition $\mathbf{x}_1 = x_1$ in (7-98) can be removed. Expanding the product, we obtain

$$C_{x_2x_3 \mid x_1} = R_{23} - \frac{R_{12}R_{13}}{R_{11}}$$

This completes the specification of $f(x_2, x_3 \mid x_1)$. ◀

Orthonormal Data Transformation

If the data $\mathbf{x}_i$ are orthogonal, that is, if $R_{ij} = 0$ for $i \neq j$, then R is a diagonal matrix and (7-83) yields

$$a_i = \frac{R_{0i}}{R_{ii}} = \frac{E\{\mathbf{s}\mathbf{x}_i\}}{E\{\mathbf{x}_i^2\}} \tag{7-99}$$

Thus the determination of the projection $\hat{\mathbf{s}}$ of $\mathbf{s}$ is simplified if the data $\mathbf{x}_i$ are expressed in terms of an orthonormal set of vectors. This is done as follows. We wish to find a set $\{\mathbf{z}_k\}$ of n orthonormal random variables $\mathbf{z}_k$ *linearly equivalent* to the data set $\{\mathbf{x}_k\}$. By this we mean that each $\mathbf{z}_k$ is a linear function of the elements of the set $\{\mathbf{x}_k\}$ and each $\mathbf{x}_k$ is a linear function of the elements of the set $\{\mathbf{z}_k\}$. The set $\{\mathbf{z}_k\}$ is not unique. We shall determine it using the Gram–Schmidt method (Fig. 7-4*b*). In this method, each $\mathbf{z}_k$ depends only on the first k data $\mathbf{x}_1, \ldots, \mathbf{x}_k$. Thus

$$\begin{aligned}
\mathbf{z}_1 &= \gamma_1^1 \mathbf{x}_1 \\
\mathbf{z}_2 &= \gamma_1^2 \mathbf{x}_1 + \gamma_2^2 \mathbf{x}_2 \\
&\cdots\cdots\cdots\cdots\cdots\cdots \\
\mathbf{z}_n &= \gamma_1^n \mathbf{x}_1 + \gamma_2^n \mathbf{x}_2 + \cdots + \gamma_n^n \mathbf{x}_n
\end{aligned} \tag{7-100}$$

In the notation γ_r^k, k is a superscript identifying the kth equation and r is a subscript taking the values 1 to k. The coefficient γ_1^1 is obtained from the normalization condition

$$E\{\mathbf{z}_1^2\} = (\gamma_1^1)^2 R_{11} = 1$$

To find the coefficients γ_1^2 and γ_2^2, we observe that $\mathbf{z}_2 \perp \mathbf{x}_1$ because $\mathbf{z}_2 \perp \mathbf{z}_1$ by assumption. From this it follows that

$$E\{\mathbf{z}_2 \mathbf{x}_1\} = 0 = \gamma_1^2 R_{11} + \gamma_2^2 R_{21}$$

The condition $E\{\mathbf{z}_2^2\} = 1$ yields a second equation. Similarly, since $\mathbf{z}_k \perp \mathbf{z}_r$ for $r < k$, we conclude from (7-100) that $\mathbf{z}_k \perp \mathbf{x}_r$ if $r < k$. Multiplying the kth equation in (7-100) by $\mathbf{x}_r$ and using the preceding development, we obtain

$$E\{\mathbf{z}_k \mathbf{x}_r\} = 0 = \gamma_1^k R_{1r} + \cdots + \gamma_k^k R_{kr} \qquad 1 \le r \le k-1 \tag{7-101}$$

This is a system of $k - 1$ equations for the k unknowns $\gamma_1^k, \ldots, \gamma_k^k$. The condition $E\{\mathbf{z}_k^2\} = 1$ yields one more equation.

The system (7-100) can be written in a vector form

$$\mathbf{Z} = \mathbf{X}\Gamma \tag{7-102}$$

where $\mathbf{Z}$ is a row vector with elements $\mathbf{z}_k$. Solving for $\mathbf{X}$, we obtain

$$\begin{aligned}
\mathbf{x}_1 &= l_1^1 \mathbf{z}_1 \qquad\qquad \mathbf{X} = \mathbf{Z}\Gamma^{-1} = \mathbf{Z}L \\
\mathbf{x}_2 &= l_1^2 \mathbf{z}_1 + l_2^2 \mathbf{z}_2 \\
&\cdots\cdots\cdots\cdots\cdots\cdots \\
\mathbf{x}_n &= l_1^n \mathbf{z}_1 + l_2^n \mathbf{z}_2 + \cdots + l_n^n \mathbf{z}_n
\end{aligned} \tag{7-103}$$

In the above, the matrix Γ and its inverse are upper triangular

$$\Gamma = \begin{bmatrix} \gamma_1^1 & \gamma_1^2 & \cdots & \gamma_1^n \\ & \gamma_2^2 & \cdots & \gamma_2^n \\ & & \cdots & \cdots \\ & 0 & & \\ & & & \gamma_n^n \end{bmatrix} \qquad L = \begin{bmatrix} l_1^1 & l_1^2 & \cdots & l_1^n \\ & l_2^2 & \cdots & l_2^n \\ & & \cdots & \cdots \\ & 0 & & \\ & & & l_n^n \end{bmatrix}$$

Since $E\{\mathbf{z}_i \mathbf{z}_j\} = \delta[i-j]$ by construction, we conclude that

$$E\{\mathbf{Z}^t \mathbf{Z}\} = 1_n = E\{\Gamma^t \mathbf{X}^t \mathbf{X} \Gamma\} = \Gamma^t E\{\mathbf{X}^t \mathbf{X}\} \Gamma \tag{7-104}$$

where 1_n is the identity matrix. Hence

$$\Gamma^t R \Gamma = 1_n \qquad R = L^t L \qquad R^{-1} = \Gamma\Gamma^t \tag{7-105}$$

We have thus expressed the matrix R and its inverse R^{-1} as products of an upper triangular and a lower triangular matrix [see also Cholesky factorization (13-79)].

The orthonormal base $\{\mathbf{z}_n\}$ in (7-100) is the finite version of the innovations process $\mathbf{i}[n]$ introduced in Sec. (11-1). The matrices Γ and L correspond to the whitening filter and to the innovations filter respectively and the factorization (7-105) corresponds to the spectral factorization (11-6).

From the linear equivalence of the sets $\{\mathbf{z}_k\}$ and $\{\mathbf{x}_k\}$, it follows that the estimate (7-80) of the random variable $\mathbf{s}$ can be expressed in terms of the set $\{\mathbf{z}_k\}$:

$$\hat{\mathbf{s}} = b_1\mathbf{z}_1 + \cdots + b_n\mathbf{z}_n = B\mathbf{Z}^t$$

where again the coefficients b_k are such that

$$\mathbf{s} - \hat{\mathbf{s}} \perp \mathbf{z}_k \qquad 1 \le k \le n$$

This yields [see (7-104)]

$$E\{(\mathbf{s} - B\mathbf{Z}^t)\mathbf{Z}\} = 0 = E\{\mathbf{s}\mathbf{Z}\} - B$$

from which it follows that

$$B = E\{\mathbf{s}\mathbf{Z}\} = E\{\mathbf{s}\mathbf{X}\Gamma\} = R_0\Gamma \tag{7-106}$$

Returning to the estimate (7-80) of $\mathbf{s}$, we conclude that

$$\hat{\mathbf{s}} = B\mathbf{Z}^t = B\Gamma^t\mathbf{X}^t = A\mathbf{X}^t \qquad A = B\Gamma^t \tag{7-107}$$

This simplifies the determination of the vector A if the matrix Γ is known.

7-4 STOCHASTIC CONVERGENCE AND LIMIT THEOREMS

A fundamental problem in the theory of probability is the determination of the asymptotic properties of random sequences. In this section, we introduce the subject, concentrating on the clarification of the underlying concepts. We start with a simple problem.

Suppose that we wish to measure the length a of an object. Due to measurement inaccuracies, the instrument reading is a sum

$$\mathbf{x} = a + \boldsymbol{\nu}$$

where $\boldsymbol{\nu}$ is the error term. If there are no systematic errors, then $\boldsymbol{\nu}$ is a random variable with zero mean. In this case, if the standard deviation σ of $\boldsymbol{\nu}$ is small compared to a, then the observed value $\mathbf{x}(\zeta)$ of $\mathbf{x}$ at a single measurement is a satisfactory estimate of the unknown length a. In the context of probability, this conclusion can be phrased as follows: The mean of the random variable $\mathbf{x}$ equals a and its variance equals σ^2. Applying Tchebycheff's inequality, we conclude that

$$P\{|\mathbf{x} - a| < \varepsilon\} > 1 - \frac{\sigma^2}{\varepsilon^2} \tag{7-108}$$

If, therefore, $\sigma \ll \varepsilon$, then the probability that $|\mathbf{x} - a|$ is less than that ε is close to 1. From this it follows that "almost certainly" the observed $\mathbf{x}(\zeta)$ is between $a - \varepsilon$ and

$a + \varepsilon$, or equivalently, that the unknown a is between $\mathbf{x}(\zeta) - \varepsilon$ and $\mathbf{x}(\zeta) + \varepsilon$. In other words, the reading $\mathbf{x}(\zeta)$ of a single measurement is "almost certainly" a satisfactory estimate of the length a as long as $\sigma \ll a$. If σ is not small compared to a, then a single measurement does not provide an adequate estimate of a. To improve the accuracy, we perform the measurement a large number of times and we average the resulting readings. The underlying probabilistic model is now a product space

$$S^n = S \times \cdots \times S$$

formed by repeating n times the experiment S of a single measurement. If the measurements are independent, then the ith reading is a sum

$$\mathbf{x}_i = a + \boldsymbol{\nu}_i$$

where the noise components $\boldsymbol{\nu}_i$ are independent random variables with zero mean and variance σ^2. This leads to the conclusion that the *sample mean*

$$\overline{\mathbf{x}} = \frac{\mathbf{x}_1 + \cdots + \mathbf{x}_n}{n} \tag{7-109}$$

of the measurements is a random variable with mean a and variance σ^2/n. If, therefore, n is so large that $\sigma^2 \ll na^2$, then the value $\overline{\mathbf{x}}(\zeta)$ of the sample mean $\overline{\mathbf{x}}$ in a single performance of the experiment S^n (consisting of n independent measurements) is a satisfactory estimate of the unknown a.

To find a bound of the error in the estimate of a by $\overline{\mathbf{x}}$, we apply (7-108). To be concrete, we assume that n is so large that $\sigma^2/na^2 = 10^{-4}$, and we ask for the probability that $\mathbf{x}$ is between $0.9a$ and $1.1a$. The answer is given by (7-108) with $\varepsilon = 0.1a$.

$$P\{0.9a < \overline{\mathbf{x}} < 1.1a\} \geq 1 - \frac{100\sigma^2}{n} = 0.99$$

Thus, if the experiment is performed $n = 10^4\sigma^2/a^2$ times, then "almost certainly" in 99 percent of the cases, the estimate $\overline{\mathbf{x}}$ of a will be between $0.9a$ and $1.1a$. Motivated by the above, we introduce next various convergence modes involving sequences of random variables.

DEFINITION

▶ A *random sequence* or a *discrete-time random process* is a sequence of random variables

$$\mathbf{x}_1, \ldots, \mathbf{x}_n, \ldots \tag{7-110}$$

◀

For a specific ζ, $\mathbf{x}_n(\zeta)$ is a sequence of numbers that might or might not converge. This suggests that the notion of convergence of a random sequence might be given several interpretations:

Convergence everywhere (e) As we recall, a sequence of numbers x_n tends to a limit x if, given $\varepsilon > 0$, we can find a number n_0 such that

$$|x_n - x| < \varepsilon \qquad \text{for every} \quad n > n_0 \tag{7-111}$$

We say that a random sequence $\mathbf{x}_n$ converges everywhere if the sequence of numbers $\mathbf{x}_n(\zeta)$ converges as in (7-111) for every ζ. The limit is a number that depends, in general, on ζ. In other words, the limit of the random sequence $\mathbf{x}_n$ is a random variable $\mathbf{x}$:

$$\mathbf{x}_n \to \mathbf{x} \qquad \text{as} \quad n \to \infty$$

Convergence almost everywhere (a.e.) If the set of outcomes ζ such that

$$\lim \mathbf{x}_n(\zeta) = \mathbf{x}(\zeta) \qquad \text{as} \quad n \to \infty \tag{7-112}$$

exists and its probability equals 1, then we say that the sequence $\mathbf{x}_n$ converges almost everywhere (or with probability 1). This is written in the form

$$P\{\mathbf{x}_n \to \mathbf{x}\} = 1 \qquad \text{as} \quad n \to \infty \tag{7-113}$$

In (7-113), $\{\mathbf{x}_n \to \mathbf{x}\}$ is an event consisting of all outcomes ζ such that $\mathbf{x}_n(\zeta) \to \mathbf{x}(\zeta)$.

Convergence in the MS sense (MS) The sequence $\mathbf{x}_n$ tends to the random variable $\mathbf{x}$ in the MS sense if

$$E\{|\mathbf{x}_n - \mathbf{x}|^2\} \to 0 \qquad \text{as} \quad n \to \infty \tag{7-114}$$

This is called
it limit in the mean and it is often written in the form

$$\text{l.i.m. } \mathbf{x}_n = \mathbf{x} \qquad n \to \infty$$

Convergence in probability (p) The probability $P\{|\mathbf{x} - \mathbf{x}_n| > \varepsilon\}$ of the event $\{|\mathbf{x} - \mathbf{x}_n| > \varepsilon\}$ is a sequence of numbers depending on ε. If this sequence tends to 0:

$$P\{|\mathbf{x} - \mathbf{x}_n| > \varepsilon\} \to 0 \qquad n \to \infty \tag{7-115}$$

for any $\varepsilon > 0$, then we say that the sequence $\mathbf{x}_n$ tends to the random variable $\mathbf{x}$ in probability (or in measure). This is also called stochastic convergence.

Convergence in distribution (d) We denote by $F_n(x)$ and $F(x)$, respectively, the distribution of the random variables $\mathbf{x}_n$ and $\mathbf{x}$. If

$$F_n(x) \to F(x) \qquad n \to \infty \tag{7-116}$$

for every point x of continuity of $F(x)$, then we say that the sequence $\mathbf{x}_n$ tends to the random variable $\mathbf{x}$ in distribution. We note that, in this case, the sequence $\mathbf{x}_n(\zeta)$ need not converge for any ζ.

Cauchy criterion As we noted, a deterministic sequence x_n converges if it satisfies (7-111). This definition involves the limit x of x_n. The following theorem, known as the Cauchy criterion, establishes conditions for the convergence of x_n that avoid the use of x: If

$$|x_{n+m} - x_n| \to 0 \qquad \text{as} \quad n \to \infty \tag{7-117}$$

for any $m > 0$, then the sequence x_n converges.

The above theorem holds also for random sequence. In this case, the limit must be interpreted accordingly. For example, if

$$E\{|\mathbf{x}_{n+m} - \mathbf{x}_n|^2\} \to 0 \qquad \text{as} \quad n \to \infty$$

for every $m > 0$, then the random sequence $\mathbf{x}_n$ converges in the MS sense.

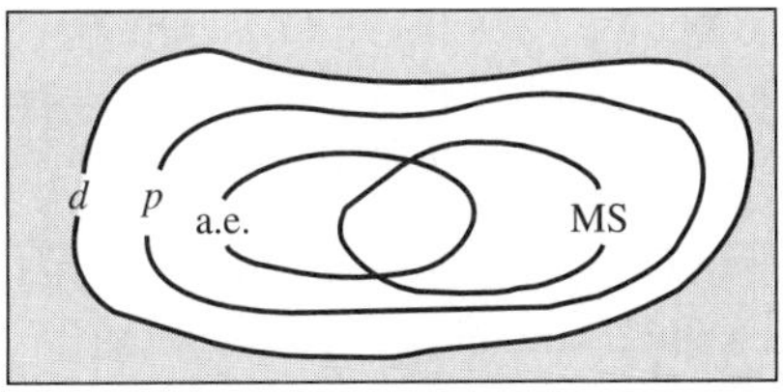

FIGURE 7-5

Comparison of convergence modes. In Fig. 7-5, we show the relationship between various convergence modes. Each point in the rectangle represents a random sequence. The letter on each curve indicates that all sequences in the interior of the curve converge in the stated mode. The shaded region consists of all sequences that do not converge in any sense. The letter *d* on the outer curve shows that if a sequence converges at all, then it converges also in distribution. We comment next on the less obvious comparisons:

If a sequence converges in the MS sense, then it also converges in probability. Indeed, Tchebycheff's inequality yields

$$P\{|\mathbf{x}_n - \mathbf{x}| > \varepsilon\} \le \frac{E\{|\mathbf{x}_n - \mathbf{x}|^2\}}{\varepsilon^2}$$

If $\mathbf{x}_n \to \mathbf{x}$ in the MS sense, then for a fixed $\varepsilon > 0$ the right side tends to 0; hence the left side also tends to 0 as $n \to \infty$ and (7-115) follows. The converse, however, is not necessarily true. If $\mathbf{x}_n$ is not bounded, then $P\{|\mathbf{x}_n - x| > \varepsilon\}$ might tend to 0 but not $E\{|\mathbf{x}_n - \mathbf{x}|^2\}$. If, however, $\mathbf{x}_n$ vanishes outside some interval $(-c, c)$ for every $n > n_0$, then p convergence and MS convergence are equivalent.

It is self-evident that a.e. convergence in (7-112) implies p convergence. We shall show by a heuristic argument that the converse is not true. In Fig. 7-6, we plot the difference $|\mathbf{x}_n - \mathbf{x}|$ as a function of n, where, for simplicity, sequences are drawn as curves. Each curve represents, thus, a particular sequence $|\mathbf{x}_n(\zeta) - \mathbf{x}(\zeta)|$. Convergence in probability means that for a *specific* $n > n_0$, only a *small* percentage of these curves will have ordinates that exceed ε (Fig. 7-6*a*). It is, of course, possible that not even one of these curves will remain less than ε for *every* $n > n_0$. Convergence a.e., on the other hand, demands that most curves will be below ε for every $n > n_0$ (Fig. 7-6*b*).

The law of large numbers (Bernoulli). In Sec. 3.3 we showed that if the probability of an event A in a given experiment equals p and the number of successes of A in n trials

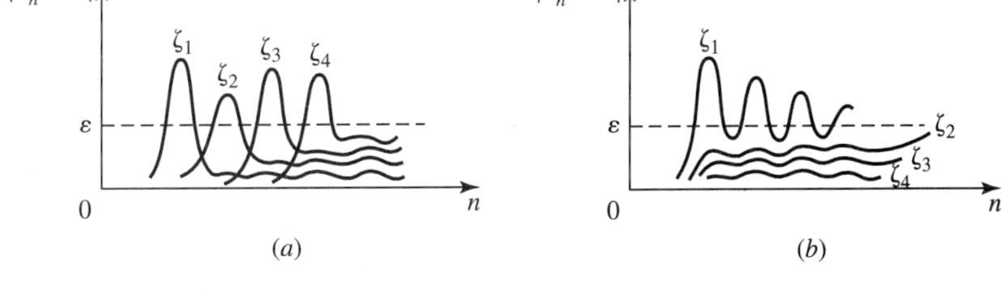

FIGURE 7-6

equals k, then

$$P\left\{\left|\frac{k}{n} - p\right| < \varepsilon\right\} \to 1 \qquad \text{as} \quad n \to \infty \tag{7-118}$$

We shall reestablish this result as a limit of a sequence of random variables. For this purpose, we introduce the random variables

$$\mathbf{x}_i = \begin{cases} 1 & \text{if } A \text{ occurs at the } i\text{th trial} \\ 0 & \text{otherwise} \end{cases}$$

We shall show that the sample mean

$$\overline{\mathbf{x}}_n = \frac{\mathbf{x}_1 + \cdots + \mathbf{x}_n}{n}$$

of these random variables tends to p in probability as $n \to \infty$.

Proof. As we know

$$E\{\mathbf{x}_i\} = E\{\overline{\mathbf{x}}_n\} = p \qquad \sigma_{x_i}^2 = pq \qquad \sigma_{\overline{x}_n}^2 = \frac{pq}{n}$$

Furthermore, $pq = p(1-p) \le 1/4$. Hence [see (5-88)]

$$P\{|\overline{\mathbf{x}}_n - p| < \varepsilon\} \ge 1 - \frac{pq}{n\varepsilon^2} \ge 1 - \frac{1}{4n\varepsilon^2} \xrightarrow[n\to\infty]{} 1$$

This reestablishes (7-118) because $\overline{\mathbf{x}}_n(\zeta) = k/n$ if A occurs k times.

The strong law of large numbers (Borel). It can be shown that $\overline{\mathbf{x}}_n$ tends to p not only in probability, but also with probability 1 (a.e.). This result, due to Borel, is known as the strong law of large numbers. The proof will not be given. We give below only a heuristic explanation of the difference between (7-118) and the strong law of large numbers in terms of relative frequencies.

Frequency interpretation We wish to estimate p within an error $\varepsilon = 0.1$, using as its estimate the sample mean $\overline{\mathbf{x}}_n$. If $n \ge 1000$, then

$$P\{|\overline{\mathbf{x}}_n - p| < 0.1\} \ge 1 - \frac{1}{4n\varepsilon^2} \ge \frac{39}{40}$$

Thus, if we repeat the experiment at least 1000 times, then in 39 out of 40 such runs, our error $|\overline{\mathbf{x}}_n - p|$ will be less than 0.1.

Suppose, now, that we perform the experiment 2000 times and we determine the sample mean $\overline{\mathbf{x}}_n$ not for one n but for every n between 1000 and 2000. The Bernoulli version of the law of large numbers leads to the following conclusion: If our experiment (the toss of the coin 2000 times) is repeated a large number of times, then, for a *specific* n larger than 1000, the error $|\overline{\mathbf{x}}_n - p|$ will exceed 0.1 only in one run out of 40. In other words, 97.5% of the runs will be "good." We cannot draw the conclusion that in the good runs the error will be less than 0.1 for *every* n between 1000 and 2000. This conclusion, however, is correct, but it can be deduced only from the strong law of large numbers.

Ergodicity. Ergodicity is a topic dealing with the relationship between statistical averages and sample averages. This topic is treated in Sec. 11-1. In the following, we discuss certain results phrased in the form of limits of random sequences.

Markov 's theorem. We are given a sequence $\mathbf{x}_i$ of random variables and we form their sample mean

$$\overline{\mathbf{x}}_n = \frac{\mathbf{x}_1 + \cdots + \mathbf{x}_n}{n}$$

Clearly, $\overline{\mathbf{x}}_n$ is a random variable whose values $\overline{\mathbf{x}}_n(\zeta)$ depend on the experimental outcome ζ. We maintain that, if the random variables $\mathbf{x}_i$ are such that the mean $\overline{\eta}_n$ of $\overline{\mathbf{x}}_n$ tends to a limit η and its variance $\overline{\sigma}_n$ tends to 0 as $n \to \infty$:

$$E\{\overline{\mathbf{x}}_n\} = \overline{\eta}_n \underset{n\to\infty}{\longrightarrow} \eta \qquad \overline{\sigma}_n^2 = E\{(\overline{\mathbf{x}}_n - \overline{\eta}_n)^2\} \underset{n\to\infty}{\longrightarrow} 0 \tag{7-119}$$

then the random variable $\overline{\mathbf{x}}_n$ tends to η in the MS sense

$$E\{(\overline{\mathbf{x}}_n - \eta)^2\} \underset{n\to\infty}{\longrightarrow} 0 \tag{7-120}$$

Proof. The proof is based on the simple inequality

$$|\overline{\mathbf{x}}_n - \eta|^2 \le 2|\overline{\mathbf{x}}_n - \overline{\eta}_n|^2 + 2|\overline{\eta}_n - \eta|^2$$

Indeed, taking expected values of both sides, we obtain

$$E\{(\overline{\mathbf{x}}_n - \eta)^2\} \le 2E\{(\overline{\mathbf{x}}_n - \overline{\eta}_n)^2\} + 2(\overline{\eta}_n - \eta)^2$$

and (7-120) follows from (7-119).

COROLLARY

▶ (Tchebycheff's condition.) If the random variables $\mathbf{x}_i$ are uncorrelated and

$$\frac{\sigma_1^2 + \cdots + \sigma_n^2}{n^2} \underset{n\to\infty}{\longrightarrow} 0 \tag{7-121}$$

then

$$\overline{\mathbf{x}}_n \underset{n\to\infty}{\longrightarrow} \eta = \lim_{n\to\infty} \frac{1}{n} \sum_{i=1}^{n} E\{\mathbf{x}_i\}$$

in the MS sense.

Proof. It follows from the theorem because, for uncorrelated random variables, the left side of (7-121) equals $\overline{\sigma}_n^2$.

We note that Tchebycheff's condition (7-121) is satisfied if $\sigma_i < K < \infty$ for every i. This is the case if the random variables $\mathbf{x}_i$ are i.i.d. with finite variance. ◀

Khinchin We mention without proof that if the random variables $\mathbf{x}_i$ are i.i.d., then their sample mean $\overline{\mathbf{x}}_n$ tends to η even if nothing is known about their variance. In this case, however, $\overline{\mathbf{x}}_n$ tends to η in probability only. The following is an application:

EXAMPLE 7-14

▶ We wish to determine the distribution $F(x)$ of a random variable $\mathbf{x}$ defined in a certain experiment. For this purpose we repeat the experiment n times and form the random variables $\mathbf{x}_i$ as in (7-12). As we know, these random variables are i.i.d. and their common distribution equals $F(x)$. We next form the random variables

$$\mathbf{y}_i(x) = \begin{cases} 1 & \text{if} \quad \mathbf{x}_i \le x \\ 0 & \text{if} \quad \mathbf{x}_i > x \end{cases}$$

where x is a fixed number. The random variables $\mathbf{y}_i(x)$ so formed are also i.i.d. and their mean equals

$$E\{\mathbf{y}_i(x)\} = 1 \times P\{\mathbf{y}_i = 1\} = P\{\mathbf{x}_i \le x\} = F(x)$$

Applying Khinchin's theorem to $\mathbf{y}_i(x)$, we conclude that

$$\frac{\mathbf{y}_1(x) + \cdots + \mathbf{y}_n(x)}{n} \xrightarrow[n\to\infty]{} F(x)$$

in probability. Thus, to determine $F(x)$, we repeat the original experiment n times and count the number of times the random variable $\mathbf{x}$ is less than x. If this number equals k and n is sufficiently large, then $F(x) \simeq k/n$. The above is thus a restatement of the relative frequency interpretation (4-3) of $F(x)$ in the form of a limit theorem. ◀

The Central Limit Theorem

Given n independent random variables $\mathbf{x}_i$, we form their sum

$$\mathbf{x} = \mathbf{x}_1 + \cdots + \mathbf{x}_n$$

This is a random variable with mean $\eta = \eta_1 + \cdots + \eta_n$ and variance $\sigma^2 = \sigma_1^2 + \cdots + \sigma_n^2$. The central limit theorem (CLT) states that under certain general conditions, the distribution $F(x)$ of $\mathbf{x}$ approaches a normal distribution with the same mean and variance:

$$F(x) \simeq G\left(\frac{x-\eta}{\sigma}\right) \tag{7-122}$$

as n increases. Furthermore, if the random variables $\mathbf{x}_i$ are of continuous type, the density $f(x)$ of $\mathbf{x}$ approaches a normal density (Fig. 7-7a):

$$f(x) \simeq \frac{1}{\sigma\sqrt{2\pi}} e^{-(x-\eta)^2/2\sigma^2} \tag{7-123}$$

This important theorem can be stated as a limit: If $\mathbf{z} = (\mathbf{x} - \eta)/\sigma$ then

$$F_z(Z) \xrightarrow[n\to\infty]{} G(z) \qquad f_z(z) \xrightarrow[n\to\infty]{} \frac{1}{\sqrt{2\pi}} e^{-z^2/2}$$

for the general and for the continuous case, respectively. The proof is outlined later.

The CLT can be expressed as a property of convolutions: The convolution of a large number of positive functions is approximately a normal function [see (7-51)].

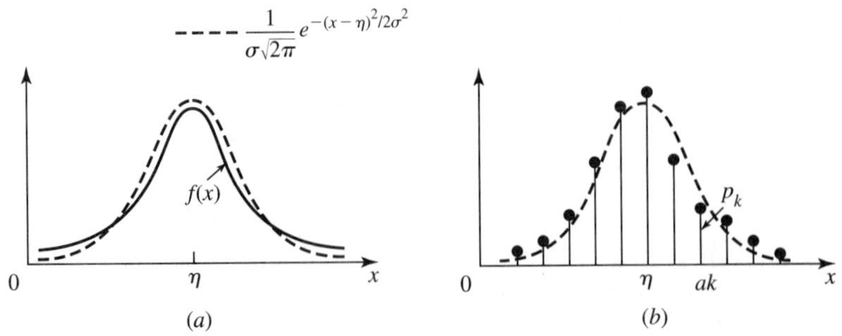

FIGURE 7-7

The nature of the CLT approximation and the required value of n for a specified error bound depend on the form of the densities $f_i(x)$. If the random variables $\mathbf{x}_i$ are i.i.d., the value $n = 30$ is adequate for most applications. In fact, if the functions $f_i(x)$ are smooth, values of n as low as 5 can be used. The next example is an illustration.

EXAMPLE 7-15

▶ The random variables $\mathbf{x}_i$ are i.i.d. and uniformly distributed in the interval (0, 1). We shall compare the density $f_x(x)$ of their sum $\mathbf{x}$ with the normal approximation (7-111) for $n = 2$ and $n = 3$. In this problem,

$$\eta_i = \frac{T}{2} \qquad \sigma_i^2 = \frac{T^2}{12} \qquad \eta = n\frac{T}{2} \qquad \sigma^2 = n\frac{T^2}{12}$$

$n = 2$ $f(x)$ is a triangle obtained by convolving a pulse with itself (Fig. 7-8)

$$\eta = T \qquad \sigma^2 = \frac{T^2}{6} \qquad f(x) \simeq \frac{1}{T}\sqrt{\frac{3}{\pi}}e^{-3(x-T)^2/T^2}$$

$n = 3$ $f(x)$ consists of three parabolic pieces obtained by convolving a triangle with a pulse

$$\eta = \frac{3T}{2} \qquad \sigma^2 = \frac{T^2}{4} \qquad f(x) \simeq \frac{1}{T}\sqrt{\frac{2}{\pi}}e^{-2(x-1.5T)^2/T^2}$$

As we can see from the figure, the approximation error is small even for such small values of n. ◀

For a discrete-type random variable, $F(x)$ is a staircase function approaching a normal distribution. The probabilities p_k, however, that $\mathbf{x}$ equals specific values x_k are, in general, unrelated to the normal density. Lattice-type random variables are an exception: If the random variables $\mathbf{x}_i$ take equidistant values ak_i, then $\mathbf{x}$ takes the values ak and for large n, the discontinuities $p_k = P\{\mathbf{x} = ak\}$ of $F(x)$ at the points $x_k = ak$ equal the samples of the normal density (Fig. 7-7b):

$$P\{\mathbf{x} = ak\} \simeq \frac{1}{\sigma\sqrt{2\pi}}e^{-(ak-\eta)^2/2\sigma^2} \tag{7-124}$$

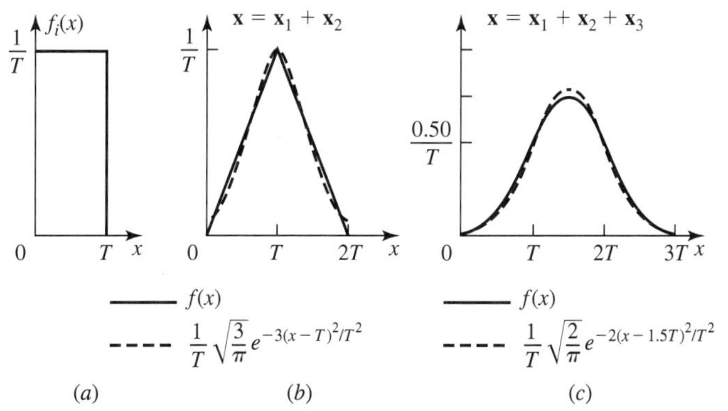

FIGURE 7-8

We give next an illustration in the context of Bernoulli trials. The random variables $\mathbf{x}_i$ of Example 7-7 are i.i.d. taking the values 1 and 0 with probabilities p and q respectively; hence their sum $\mathbf{x}$ is of lattice type taking the values $k = 0, \ldots, n$. In this case,

$$E\{\mathbf{x}\} = nE\{\mathbf{x}_i\} = np \qquad \sigma_x^2 = n\sigma_1^2 = npq$$

Inserting into (7-124), we obtain the approximation

$$P\{\mathbf{x} = k\} = \binom{n}{k} p^k q^{n-k} \simeq \frac{1}{\sqrt{2\pi npq}} e^{-(k-np)^2/2npq} \tag{7-125}$$

This shows that the *DeMoivre–Laplace theorem* (4-90) is a special case of the lattice-type form (7-124) of the central limit theorem.

EXAMPLE 7-16

▶ A fair coin is tossed six times and $\mathbf{x}_i$ is the zero–one random variable associated with the event {heads at the ith toss}. The probability of k heads in six tosses equals

$$P\{\mathbf{x} = k\} = \binom{6}{k} \frac{1}{2^6} = p_k \qquad \mathbf{x} = \mathbf{x}_1 + \cdots + \mathbf{x}_6$$

In the following table we show the above probabilities and the samples of the normal curve $N(\eta, \sigma^2)$ (Fig. 7-9) where

$$\eta = np = 3 \qquad \sigma^2 = npq = 1.5$$

k	0	1	2	3	4	5	6
p_k	0.016	0.094	0.234	0.312	0.234	0.094	0.016
$N(\eta, \sigma)$	0.016	0.086	0.233	0.326	0.233	0.086	0.016

◀

ERROR CORRECTION. In the approximation of $f(x)$ by the normal curve $N(\eta, \sigma^2)$, the error

$$\varepsilon(x) = f(x) - \frac{1}{\sigma\sqrt{2\pi}} e^{-x^2/2\sigma^2}$$

results where we assumed, shifting the origin, that $\eta = 0$. We shall express this error in terms of the moments

$$m_n = E\{\mathbf{x}^n\}$$

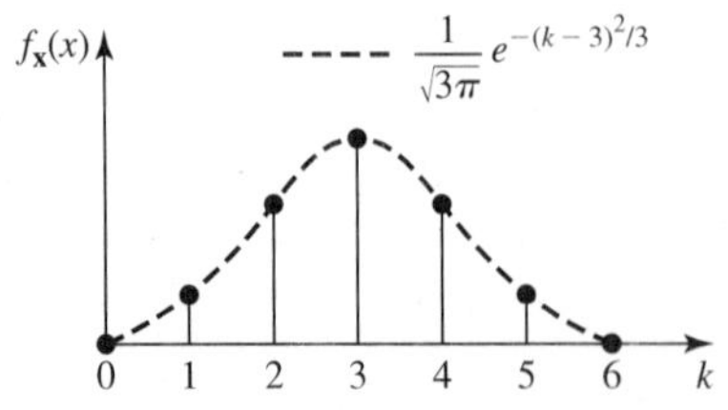

FIGURE 7-9

of $\mathbf{x}$ and the *Hermite polynomials*

$$H_k(x) = (-1)^k e^{x^2/2} \frac{d^k}{dx^k} e^{-x^2/2}$$

$$= x^k - \binom{k}{2} x^{k-2} + 1 \cdot 3 \binom{k}{4} x^{k-4} + \cdots \qquad (7\text{-}126)$$

These polynomials form a complete orthogonal set on the real line:

$$\int_{-\infty}^{\infty} e^{-x^2/2} H_n(x) H_m(x)\, dx = \begin{cases} n!\sqrt{2\pi} & n = m \\ 0 & n \neq m \end{cases}$$

Hence $\varepsilon(x)$ can be written as a series

$$\varepsilon(x) = \frac{1}{\sigma\sqrt{2\pi}} e^{-x^2/2\sigma^2} \sum_{k=3}^{\infty} C_k H_k\left(\frac{x}{\sigma}\right) \qquad (7\text{-}127)$$

The series starts with $k = 3$ because the moments of $\varepsilon(x)$ of order up to 2 are 0. The coefficients C_n can be expressed in terms of the moments m_n of $\mathbf{x}$. Equating moments of order $n = 3$ and $n = 4$, we obtain [see (5-73)]

$$3!\sigma^3 C_3 = m_3 \qquad 4!\sigma^4 C_4 = m_4 - 3\sigma^4$$

First-order correction. From (7-126) it follows that

$$H_3(x) = x^3 - 3x \qquad H_4(x) = x^4 - 6x^2 + 3$$

Retaining the first nonzero term of the sum in (7-127), we obtain

$$f(x) \simeq \frac{1}{\sigma\sqrt{2\pi}} e^{-x^2/2\sigma^2} \left[1 + \frac{m_3}{6\sigma^3}\left(\frac{x^3}{\sigma^3} - \frac{3x}{\sigma}\right)\right] \qquad (7\text{-}128)$$

If $f(x)$ is even, then $m_3 = 0$ and (7-127) yields

$$f(x) \simeq \frac{1}{\sigma\sqrt{2\pi}} e^{-x^2/2\sigma^2} \left[1 + \frac{1}{24}\left(\frac{m_4}{\sigma^4} - 3\right)\left(\frac{x^4}{\sigma^4} - \frac{6x^2}{\sigma^2} + 3\right)\right] \qquad (7\text{-}129)$$

EXAMPLE 7-17

▶ If the random variables $\mathbf{x}_i$ are i.i.d. with density $f_i(x)$ as in Fig. 7-10*a*, then $f(x)$ consists of three parabolic pieces (see also Example 7-12) and $N(0, 1/4)$ is its normal approximation. Since $f(x)$ is even and $m_4 = 13/80$ (see Prob. 7-4), (7-129) yields

$$f(x) \simeq \sqrt{\frac{2}{\pi}} e^{-2x^2} \left(1 - \frac{4x^4}{15} + \frac{2x^2}{5} - \frac{1}{20}\right) \equiv \bar{f}(x)$$

In Fig. 7-10*b*, we show the error $\varepsilon(x)$ of the normal approximation and the first-order correction error $f(x) - \bar{f}(x)$. ◀

ON THE PROOF OF THE CENTRAL LIMIT THEOREM. We shall justify the approximation (7-123) using characteristic functions. We assume for simplicity that $\eta_i = 0$. Denoting by $\Phi_i(\omega)$ and $\Phi(\omega)$, respectively, the characteristic functions of the random variables $\mathbf{x}_i$ and $\mathbf{x} = \mathbf{x}_1 + \cdots + \mathbf{x}_n$, we conclude from the independence of $\mathbf{x}_i$ that

$$\Phi(\omega) = \Phi_1(\omega) \cdots \Phi_n(\omega)$$

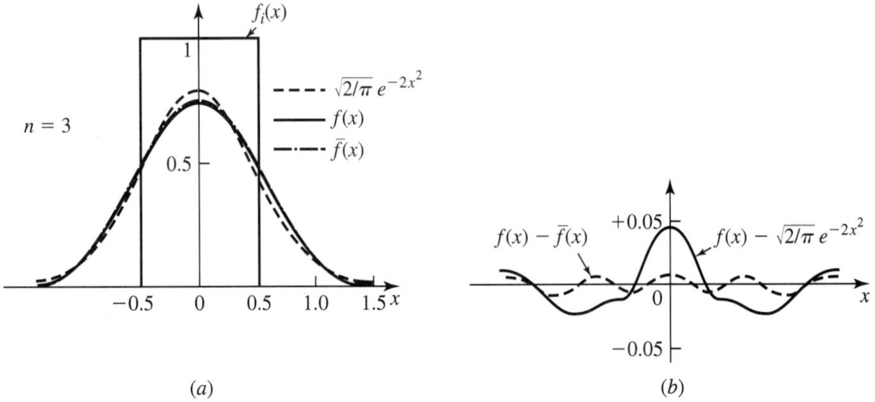

FIGURE 7-10

Near the origin, the functions $\Psi_i(\omega) = \ln \Phi_i(\omega)$ can be approximated by a parabola:

$$\Psi_i(\omega) \simeq -\tfrac{1}{2}\sigma_i^2\omega^2 \qquad \Phi_i(\omega) = e^{-\sigma_i^2\omega^2/2} \quad \text{for } |\omega| < \varepsilon \tag{7-130}$$

If the random variables $\mathbf{x}_i$ are of continuous type, then [see (5-95) and Prob. 5-29]

$$\Phi_i(0) = 1 \qquad |\Phi_i(\omega)| < 1 \quad \text{for } |\omega| \neq 0 \tag{7-131}$$

Equation (7-131) suggests that for small ε and large n, the function $\Phi(\omega)$ is negligible for $|\omega| > \varepsilon$, (Fig. 7-11a). This holds also for the exponential $e^{-\sigma^2\omega^2/2}$ if $\sigma \to \infty$ as in (7-135). From our discussion it follows that

$$\Phi(\omega) \simeq e^{-\sigma_1^2\omega^2/2} \cdots e^{-\sigma_n^2\omega^2/2} = e^{-\sigma^2\omega^2/2} \quad \text{for all } \omega \tag{7-132}$$

in agreement with (7-123).

The exact form of the theorem states that the normalized random variable

$$\mathbf{z} = \frac{\mathbf{x}_1 + \cdots + \mathbf{x}_n}{\sigma} \qquad \sigma^2 = \sigma_1^2 + \cdots + \sigma_n^2$$

tends to an $N(0, 1)$ random variable as $n \to \infty$:

$$f_z(z) \xrightarrow[n\to\infty]{} \frac{1}{\sqrt{2\pi}} e^{-z^2/2} \tag{7-133}$$

A general proof of the theorem is given later. In the following, we sketch a proof under the assumption that the random variables $\mathbf{x}_i$ are i.i.d. In this case

$$\Phi_1(\omega) = \cdots = \Phi_n(\omega) \qquad \sigma = \sigma_i\sqrt{n}$$

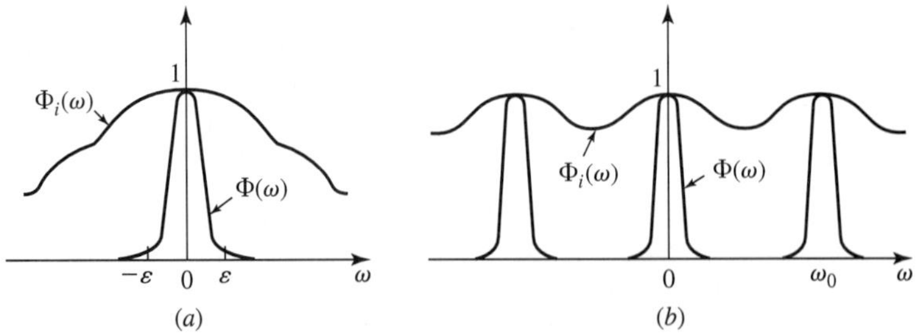

FIGURE 7-11

Hence,

$$\Phi_z(\omega) = \Phi_i^n = \left(\frac{\omega}{\sigma_i\sqrt{n}}\right)$$

Expanding the functions $\Psi_i(\omega) = \ln \Phi_i(\omega)$ near the origin, we obtain

$$\Psi_i(\omega) = -\frac{\sigma_i^2\omega^2}{2} + O(\omega^3)$$

Hence,

$$\Psi_z(\omega) = n\Psi_i\left(\frac{\omega}{\sigma_i\sqrt{n}}\right) = -\frac{\omega^2}{2} + O\left(\frac{1}{\sqrt{n}}\right) \underset{n\to\infty}{\longrightarrow} -\frac{\omega^2}{2} \tag{7-134}$$

This shows that $\Phi_z(\omega) \to e^{-\omega^2/2}$ as $n \to \infty$ and (7-133) results.

As we noted, the theorem is not always true. The following is a set of sufficient conditions:

(*a*)
$$\sigma_1^2 + \cdots + \sigma_n^2 \underset{n\to\infty}{\longrightarrow} \infty \tag{7-135}$$

(*b*) There exists a number $\alpha > 2$ and a finite constant K such that

$$\int_{-\infty}^{\infty} x^\alpha f_i(x)\,dx < K < \infty \qquad \text{for all } i \tag{7-136}$$

These conditions are not the most general. However, they cover a wide range of applications. For example, (7-135) is satisfied if there exists a constant $\varepsilon > 0$ such that $\sigma_i > \varepsilon$ for all i. Condition (7-136) is satisfied if all densities $f_i(x)$ are 0 outside a finite interval $(-c, c)$ no matter how large.

Lattice type The preceding reasoning can also be applied to discrete-type random variables. However, in this case the functions $\Phi_i(\omega)$ are periodic (Fig. 7-11*b*) and their product takes significant values only in a small region near the points $\omega = 2\pi n/a$. Using the approximation (7-124) in each of these regions, we obtain

$$\Phi(\omega) \simeq \sum_n e^{-\sigma^2(\omega - n\omega_0)^2/2} \qquad \omega_0 = \frac{2\pi}{a} \tag{7-137}$$

As we can see from (11A-1), the inverse of the above yields (7-124).

The Berry–Esseén theorem[3] This theorem states that if

$$E\{\mathbf{x}_i^3\} \le c\sigma_i^2 \quad \text{all } i \tag{7-138}$$

where c is some constant, then the distribution $\overline{F}(x)$ of the normalized sum

$$\overline{\mathbf{x}} = \frac{\mathbf{x}_1 + \cdots + \mathbf{x}_n}{\sigma} \qquad \sigma_1^2 + \cdots + \sigma_n^2 = \sigma^2$$

is close to the normal distribution $G(x)$ in the following sense

$$|\overline{F}(x) - G(x)| < \frac{4c}{\sigma} \tag{7-139}$$

---

[3] A. Papoulis: "Narrow-Band Systems and Gaussianity," *IEEE Transactions on Information Theory,* January 1972.

The central limit theorem is a corollary of (7-139) because (7-139) leads to the conclusion that

$$\overline{F}(x) \to G(x) \qquad \text{as} \quad \sigma \to \infty \tag{7-140}$$

This proof is based on condition (7-138). This condition, however, is not too restrictive. It holds, for example, if the random variables $\mathbf{x}_i$ are i.i.d. and their third moment is finite.

We note, finally, that whereas (7-140) establishes merely the convergence in distribution of $\bar{\mathbf{x}}$ to a normal random variable, (7-139) gives also a *bound* of the deviation of $\overline{F}(x)$ from normality.

The central limit theorem for products. Given n independent positive random variables $\mathbf{x}_i$, we form their product:

$$\mathbf{y} = \mathbf{x}_1\mathbf{x}_2 \cdots \mathbf{x}_n \qquad \mathbf{x}_i > 0$$

THEOREM 7-2

For large n, the density of $\mathbf{y}$ is approximately *lognormal:*

$$f_y(y) \simeq \frac{1}{y\sigma\sqrt{2\pi}} \exp\left\{-\frac{1}{2\sigma^2}(\ln y - \eta)^2\right\} U(y) \tag{7-141}$$

where

$$\eta = \sum_{i=1}^{n} E\{\ln \mathbf{x}_i\} \qquad \sigma^2 = \sum_{i=1}^{n} \text{Var}(\ln \mathbf{x}_i)$$

Proof. The random variable

$$\mathbf{z} = \ln \mathbf{y} = \ln \mathbf{x}_1 + \cdots + \ln \mathbf{x}_n$$

is the sum of the random variables $\ln \mathbf{x}_i$. From the CLT it follows, therefore, that for large n, this random variable is nearly normal with mean η and variance σ^2. And since $\mathbf{y} = e^{\mathbf{z}}$, we conclude from (5-30) that $\mathbf{y}$ has a lognormal density. The theorem holds if the random variables $\ln \mathbf{x}_i$ satisfy the conditions for the validity of the CLT.

EXAMPLE 7-18

Suppose that the random variables $\mathbf{x}_i$ are uniform in the interval (0, 1). In this case,

$$E\{\ln \mathbf{x}_i\} = \int_0^1 \ln x\, dx = -1 \qquad E\{(\ln \mathbf{x}_i)^2\} = \int_0^1 (\ln \mathbf{x})^2\, dx = 2$$

Hence $\eta = -n$ and $\sigma^2 = n$. Inserting into (7-141), we conclude that the density of the product $\mathbf{y} = \mathbf{x}_1 \cdots \mathbf{x}_n$ equals

$$f_y(y) \simeq \frac{1}{y\sqrt{2\pi n}} \exp\left\{-\frac{1}{2n}(\ln y + n)^2\right\} U(y)$$

7-5 RANDOM NUMBERS: MEANING AND GENERATION

Random numbers (RNs) are used in a variety of applications involving computer generation of statistical data. In this section, we explain the underlying ideas concentrating on the meaning and generation of random numbers. We start with a simple illustration of the role of statistics in the numerical solution of deterministic problems.

MONTE CARLO INTEGRATION. We wish to evaluate the integral

$$I = \int_0^1 g(x)\,dx \tag{7-142}$$

For this purpose, we introduce a random variable $\mathbf{x}$ with uniform distribution in the interval (0, 1) and we form the random variable $\mathbf{y} = g(\mathbf{x})$. As we know,

$$E\{g(\mathbf{x})\} = \int_0^1 g(x) f_x(x)\,dx = \int_0^1 g(x)\,dx \tag{7-143}$$

hence $\eta_y = I$. We have thus expressed the unknown I as the expected value of the random variable $\mathbf{y}$. This result involves only concepts; it does not yield a numerical method for evaluating I. Suppose, however, that the random variable $\mathbf{x}$ models a physical quantity in a real experiment. We can then estimate I using the relative frequency interpretation of expected values: We repeat the experiment a large number of times and observe the values x_i of $\mathbf{x}$; we compute the corresponding values $y_i = g(x_i)$ of $\mathbf{y}$ and form their average as in (5-51). This yields

$$I = E\{g(\mathbf{x})\} \simeq \frac{1}{n}\sum g(x_i) \tag{7-144}$$

This suggests the method described next for determining I:

The data x_i, no matter how they are obtained, are random numbers; that is, they are numbers having certain properties. If, therefore, we can numerically generate such numbers, we have a method for determining I. To carry out this method, we must reexamine the meaning of random numbers and develop computer programs for generating them.

THE DUAL INTERPRETATION OF RANDOM NUMBERS. "What are random numbers? Can they be generated by a computer? Is it possible to generate truly random number sequences?" Such questions do not have a generally accepted answer. The reason is simple. As in the case of probability (see Chap. 1), the term *random numbers* has two distinctly different meanings. The first is theoretical: Random numbers are mental constructs defined in terms of an abstract model. The second is empirical: Random numbers are sequences of real numbers generated either as physical data obtained from a random experiment or as computer output obtained from a deterministic program. The duality of interpretation of random numbers is apparent in the following extensively quoted definitions[4]:

> A sequence of numbers is random if it has every property that is shared by all infinite sequences of independent samples of random variables from the uniform distribution. (J. M. Franklin)
>
> A random sequence is a vague notion embodying the ideas of a sequence in which each term is unpredictable to the uninitiated and whose digits pass a certain number of tests, traditional with statisticians and depending somewhat on the uses to which the sequence is to be put. (D. H. Lehmer)

[4]D. E. Knuth: *The Art of Computer Programming,* Addison-Wesley, Reading, MA, 1969.

It is obvious that these definitions cannot have the same meaning. Nevertheless, both are used to define random number sequences. To avoid this confusing ambiguity, we shall give two definitions: one theoretical, the other empirical. For these definitions we shall rely solely on the uses for which random numbers are intended: Random numbers are used to apply statistical techniques to other fields. It is natural, therefore, that they are defined in terms of the corresponding probabilistic concepts and their properties as physically generated numbers are expressed directly in terms of the properties of real data generated by random experiments.

CONCEPTUAL DEFINITION. A sequence of numbers x_i is called random if it equals the samples $x_i = \mathbf{x}_i(\zeta)$ of a sequence $\mathbf{x}_i$ of i.i.d. random variables $\mathbf{x}_i$ defined in the space of repeated trials.

It appears that this definition is the same as Franklin's. There is, however, a subtle but important difference. Franklin says that the sequence x_i has every property shared by i.i.d. random variables; we say that x_i equals the samples of the i.i.d. random variables $\mathbf{x}_i$. In this definition, all theoretical properties of random numbers are the same as the corresponding properties of random variables. There is, therefore, no need for a new theory.

EMPIRICAL DEFINITION. A sequence of numbers x_i is called random if its statistical properties are the same as the properties of random data obtained from a random experiment.

Not all experimental data lead to conclusions consistent with the theory of probability. For this to be the case, the experiments must be so designed that data obtained by repeated trials satisfy the i.i.d. condition. This condition is accepted only after the data have been subjected to a variety of tests and in any case, it can be claimed only as an approximation. The same applies to computer-generated random numbers. Such uncertainties, however, cannot be avoided no matter how we define physically generated sequences. The advantage of the above definition is that it shifts the problem of establishing the randomness of a sequence of numbers to an area with which we are already familiar. We can, therefore, draw directly on our experience with random experiments and apply the well-established tests of randomness to computer-generated random numbers.

Generation of Random Number Sequences

Random numbers used in Monte Carlo calculations are generated mainly by computer programs; however, they can also be generated as observations of random data obtained from real experiments: The tosses of a fair coin generate a random sequence of 0's (heads) and 1's (tails); the distance between radioactive emissions generates a random sequence of exponentially distributed samples. We accept number sequences so generated as random because of our long experience with such experiments. Random number sequences experimentally generated are not, however, suitable for computer use, for obvious reasons. An efficient source of random numbers is a computer program with small memory, involving simple arithmetic operations. We outline next the most commonly used programs.

Our objective is to generate random number sequences with arbitrary distributions. In the present state of the art, however, this cannot be done directly. The available algorithms only generate sequences consisting of integers z_i uniformly distributed in an interval $(0, m)$. As we show later, the generation of a sequence x_i with an arbitrary distribution is obtained indirectly by a variety of methods involving the uniform sequence z_i.

The most general algorithm for generating a random number sequence z_i is an equation of the form

$$z_n = f(z_{n-1}, \ldots, z_{n-r}) \mod m \tag{7-145}$$

where $f(z_{n-1}, \ldots, z_{n-r})$ is a function depending on the r most recent past values of z_n. In this notation, z_n is the remainder of the division of the number $f(z_{n-1}, \ldots, z_{n-r})$ by m. This is a nonlinear recursion expressing z_n in terms of the constant m, the function f, and the initial conditions $z_1, \ldots, z_{r-1}$. The quality of the generator depends on the form of the function f. It might appear that good random number sequences result if this function is complicated. Experience has shown, however, that this is not the case. Most algorithms in use are linear recursions of order 1. We shall discuss the homogeneous case.

LEHMER'S ALGORITHM. The simplest and one of the oldest random number generators is the recursion

$$z_n = az_{n-1} \mod m \qquad z_0 = 1 \quad n \geq 1 \tag{7-146}$$

where m is a large prime number and a is an integer. Solving, we obtain

$$z_n = a^n \mod m \tag{7-147}$$

The sequence z_n takes values between 1 and $m - 1$; hence at least two of its first m values are equal. From this it follows that z_n is a periodic sequence for $n > m$ with period $m_o \leq m - 1$. A periodic sequence is not, of course, random. However, if for the applications for which it is intended the required number of sample does not exceed m_o, periodicity is irrelevant. For most applications, it is enough to choose for m a number of the order of 10^9 and to search for a constant a such that $m_o = m - 1$. A value for m suggested by Lehmer in 1951 is the prime number $2^{31} - 1$.

To complete the specification of (7-146), we must assign a value to the multiplier a. Our first condition is that the period m_o of the resulting sequence z_0 equal $m - 1$.

DEFINITION ▶ An integer a is called the *primitive root* of m if the smallest n such that

$$a^n = 1 \mod m \text{ is } n = m - 1 \tag{7-148}$$

From the definition it follows that the sequence a^n is periodic with period $m_o = m - 1$ iff a is a primitive root of m. Most primitive roots do not generate good random number sequences. For a final selection, we subject specific choices to a variety of tests based on tests of randomness involving real experiments. Most tests are carried out not in terms of the integers z_i but in terms of the properties of the numbers

$$u_i = \frac{z_i}{m} \tag{7-149}$$

These numbers take essentially all values in the interval (0, 1) and the purpose of testing is to establish whether they are the values of a sequence $\mathbf{u}_i$ of continuous-type i.i.d. random variables uniformly distributed in the interval (0, 1). The i.i.d. condition leads to the following equations:

For every u_i in the interval (0, 1) and for every n,

$$P\{\mathbf{u}_i \le u_i\} = u_i \tag{7-150}$$

$$P\{\mathbf{u}_1 \le u_1, \ldots, \mathbf{u}_n \le u_n\} = P\{\mathbf{u}_1 \le u_1\} \cdots P\{\mathbf{u}_n \le u_n\} \tag{7-151}$$

To establish the validity of these equations, we need an infinite number of tests. In real life, however, we can perform only a finite number of tests. Furthermore, all tests involve approximations based on the empirical interpretation of probability. We cannot, therefore, claim with certainty that a sequence of real numbers is truly random. We can claim only that a particular sequence is reasonably random for certain applications or that one sequence is more random than another. In practice, a sequence u_n is accepted as random not only because it passes the standard tests but also because it has been used with satisfactory results in many problems.

Over the years, several algorithms have been proposed for generating "good" random number sequences. Not all, however, have withstood the test of time. An example of a sequence z_n that seems to meet most requirements is obtained from (7-146) with $a = 7^5$ and $m = 2^{31} - 1$:

$$z_n = 16{,}807 z_{n-1} \quad \text{mod}\, 2{,}147{,}483{,}647 \tag{7-152}$$

This sequence meets most standard tests of randomness and has been used effectively in a variety of applications.[5]

We conclude with the observation that most tests of randomness are applications, direct or indirect, of well-known tests of various statistical hypotheses. For example, to establish the validity of (7-150), we apply the Kolmogoroff–Smirnov test, page 361, or the chi-square test, page 361–362. These tests are used to determine whether given experimental data fit a particular distribution. To establish the validity of (7-151), we apply the chi-square test, page 363. This test is used to determine the independence of various events.

In addition to direct testing, a variety of special methods have been proposed for testing indirectly the validity of equations (7-150)–(7-151). These methods are based on well-known properties of random variables and they are designed for particular applications. The generation of random vector sequences is an application requiring special tests. ◀

Random vectors. We shall construct a multidimensional sequence of random numbers using the following properties of subsequences. Suppose that $\mathbf{x}$ is a random variable with distribution $F(x)$ and x_i is the corresponding random number sequence. It follows from (7-150) and (7-151) that every subsequence of x_i is a random number sequence with

[5] S. K. Park and K. W. Miller "Random Number Generations: Good Ones Are Hard to Find," *Communications of the ACM,* vol. 31, no. 10, October 1988.

distribution $F(x)$. Furthermore, if two subsequences have no common elements, they are the samples of two independent random variables. From this we conclude that the odd-subscript and even-subscript sequences

$$x_i^o = x_{2i-1} \qquad x_i^e = x_{2i} \qquad i = 1, 2, \ldots$$

are the samples of two i.i.d. random variables $\mathbf{x}^o$ and $\mathbf{x}^e$ with distribution $F(x)$. Thus, starting from a scalar random number sequence, x_i, we constructed a vector random number sequence (x_i^o, x_i^e). Proceeding similarly, we can construct random number sequences of any dimensionality. Using superscripts to identify various random variables and their samples, we conclude that the random number sequences

$$x_i^k = x_{mi-m+k} \qquad k = 1, \ldots, m \qquad i = 1, 2, \ldots \tag{7-153}$$

are the samples of m i.i.d. random variables $\mathbf{x}^1, \ldots, \mathbf{x}^m$ with distribution $F(x)$.

Note that a sequence of numbers might be sufficiently random for scalar but not for vector applications. If, therefore, a random number sequence x_i is to be used for multidimensional applications, it is desirable to subject it to special tests involving its subsequences.

Random Number Sequences with Arbitrary Distributions

In the following, the letter $\mathbf{u}$ will identify a random variable with uniform distribution in the interval (0, 1); the corresponding random number sequence will be identified by u_i. Using the sequence u_i, we shall present a variety of methods for generating sequences with arbitrary distributions. In this analysis, we shall make frequent use of the following:

If x_i are the samples of the random variable $\mathbf{x}$, then $y_i = g(x_i)$ are the samples of the random variable $\mathbf{y} = g(\mathbf{x})$. For example, if x_i is a random number sequence with distribution $F_x(x)$, then $y_i = a + bx_i$ is a random number sequence with distribution $F_x[(y - a)/b]$ if $b > 0$, and $1 - F_x[(y - a)/b]$ if $b < 0$. From this it follows, for example, that $v_i = 1 - u_i$ is a random number sequence uniform in the interval (0, 1).

PERCENTILE TRANSFORMATION METHOD. Consider a random variable $\mathbf{x}$ with distribution $F_x(x)$. We have shown in Sec. 5-2 that the random variable $\mathbf{u} = F_x(\mathbf{x})$ is uniform in the interval (0, 1) no matter what the form of $F_x(x)$ is. Denoting by $F_x^{(-1)}(u)$ the inverse of $F_x(x)$, we conclude that $\mathbf{x} = F_x^{(-1)}(\mathbf{u})$ (see Fig. 7-12). From this it follows that

$$x_i = F_x^{(-1)}(u_i) \tag{7-154}$$

is a random number sequence with distribution $F_x(x)$, [see also (5-42)]. Thus, to find a random number sequence x_i with distribution a given function $F_x(x)$, it suffices to determine the inverse of $F_x(x)$ and to compute $F_x^{(-1)}(u_i)$. Note that the numbers x_i are the u_i percentiles of $F_x(x)$.

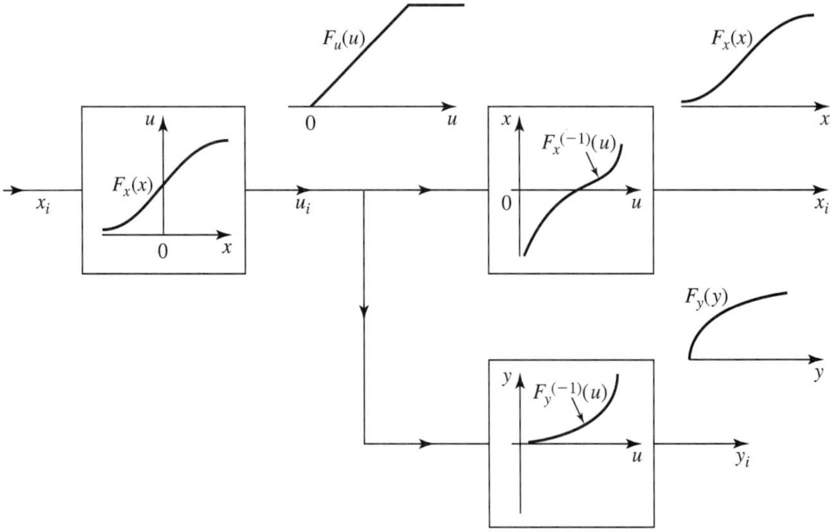

FIGURE 7-12

EXAMPLE 7-19 ▶ We wish to generate a random number sequence x_i with exponential distribution. In this case,

$$F_x(x) = 1 - e^{-x/\lambda} \qquad \mathbf{x} = -\lambda \ln(1 - \mathbf{u})$$

Since $1 - \mathbf{u}$ is a random variable with uniform distribution, we conclude that the sequence

$$x_i = -\lambda \ln u_i \tag{7-155}$$

has an exponential distribution. ◀

EXAMPLE 7-20 ▶ We wish to generate a random number sequence x_i with Rayleigh distribution. In this case,

$$F_x(x) = 1 - e^{-x^2/2} \qquad F_x^{(-1)}(\mathbf{u}) = \sqrt{-2\ln(1 - \mathbf{u})}$$

Replacing $1 - \mathbf{u}$ by $\mathbf{u}$, we conclude that the sequence

$$x_i = \sqrt{-2 \ln u_i}$$

has a Rayleigh distribution. ◀

Suppose now that we wish to generate the samples x_i of a discrete-type random variable $\mathbf{x}$ taking the values a_k with probability

$$p_k = P\{\mathbf{x} = a_k\} \qquad k = 1, \ldots, m$$

In this case, $F_x(x)$ is a staircase function (Fig. 7-13) with discontinuities at the points a_k, and its inverse is a staircase function with discontinuities at the points

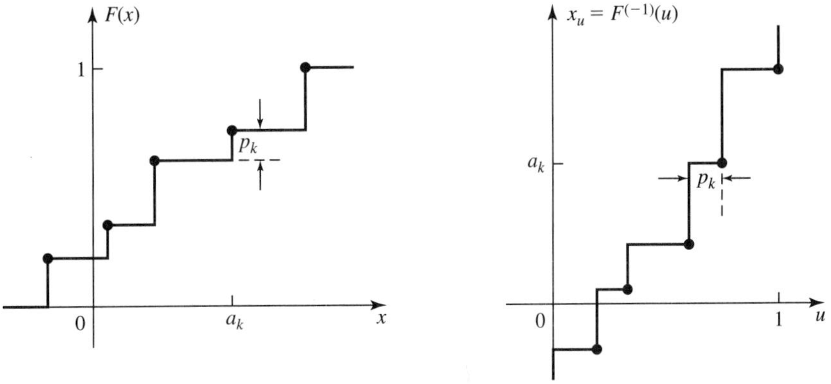

FIGURE 7-13

$F_x(a_k) = p_1 + \cdots + p_k$. Applying (7-154), we obtain the following rule for generating the random number sequence x_i:

$$\text{Set} \quad x_i = a_k \qquad \text{iff} \quad p_1 + \cdots + p_{k-1} \le u_i < p_1 + \cdots + p_k \tag{7-156}$$

EXAMPLE 7-21 ▶ The sequence

$$x_i = \begin{cases} 0 & \text{if} \quad 0 < u_i < p \\ 1 & \text{if} \quad p < u_i < 1 \end{cases}$$

takes the values 0 and 1 with probability p and $1 - p$ respectively. It specifies, therefore, a *binary* random number sequence.

The sequence

$$x_i = k \qquad \text{iff} \quad 0.1k < u_i < 0.1(k+1) \qquad k = 0, 1, \ldots, 9$$

takes the values $0, 1, \ldots, 9$ with equal probability. It specifies, therefore, a *decimal* random number sequence with uniform distribution.

Setting

$$a_k = k \qquad p_k = \binom{n}{k} p^k q^{n-k} \qquad k = 0, 1, \ldots, m$$

into (7-15), we obtain a random number sequence with *binomial* distribution.

Setting

$$a_k = k \qquad p_k = e^{-\lambda}\frac{\lambda^k}{k!} \qquad k = 0, 1, \ldots$$

into (7-15) we obtain a random number sequence with *Poisson* distribution. ◀

Suppose now that we are given not a uniform sequence, but a sequence x_i with distribution $F_x(x)$. We wish to find a sequence y_i with distribution $F_y(y)$. As we know, $y_i = F_y^{(-1)}(u_i)$ is a random number sequence with distribution $F_y(y)$. Hence (see Fig. 7-12) the composite function

$$y_i = F_y^{(-1)}(F_x(x_i)) \tag{7-157}$$

generates a random number sequence with distribution $F_y(y)$ [see also (5-43)].

EXAMPLE 7-22

▶ We are given a random number sequence $x_i > 0$ with distribution $F_x(x) = 1 - e^{-x} - xe^{-x}$ and we wish to generate a random number sequence $y_i > 0$ with distribution $F_y(y) = 1 - e^{-y}$. In this example $F_y^{(-1)}(u) = -\ln(1-u)$; hence

$$F_y^{(-1)}(F_x(x)) = -\ln[1 - F_x(x)] = -\ln(e^{-x} + xe^{-x})$$

Inserting into (7-145), we obtain

$$y_i = -\ln(e^{-x_i} + x_i e^{-x_i})$$ ◀

REJECTION METHOD. In the percentile transformation method, we used the inverse of the function $F_x(x)$. However, inverting a function is not a simple task. To overcome this difficulty, we develop next a method that avoids inversion. The problem under consideration is the generation of a random number sequence y_i with distribution $F_y(y)$ in terms of the random number sequence x_i as in (7-157).

The proposed method is based on the relative frequency interpretation of the conditional density

$$f_x(x \mid M)\,dx = \frac{P\{x < \mathbf{x} \le x + dx, M\}}{P(M)} \tag{7-158}$$

of a random variable $\mathbf{x}$ assuming M (see page 98). In the following method, the event M is expressed in terms of the random variable $\mathbf{x}$ and another random variable $\mathbf{u}$, and it is so chosen that the resulting function $f_x(x \mid M)$ equals $f_y(y)$. The sequence y_i is generated by setting $y_i = x_i$ if M occurs, rejecting x_i otherwise. The problem has a solution only if $f_y(x) = 0$ in every interval in which $f_x(x) = 0$. We can assume, therefore, without essential loss of generality, that the ratio $f_x(x)/f_y(x)$ is bounded from below by some positive constant a:

$$\frac{f_x(x)}{f_y(x)} \ge a > 0 \qquad \text{for every } x$$

Rejection theorem. If the random variables $\mathbf{x}$ and $\mathbf{u}$ are independent and

$$M = \{\mathbf{u} \le r(\mathbf{x})\} \qquad \text{where} \quad r(x) = a\frac{f_y(x)}{f_x(x)} \le 1 \tag{7-159}$$

then

$$f_x(x \mid M) = f_y(x) \tag{7-160}$$

Proof. The joint density of the random variables $\mathbf{x}$ and $\mathbf{u}$ equals $f_x(x)$ in the strip $0 < u < 1$ of the xu plane, and 0 elsewhere. The event M consists of all outcomes such that the point $(\mathbf{x}, \mathbf{u})$ is in the shaded area of Fig. 7-14 below the curve $u = r(x)$. Hence

$$P(M) = \int_{-\infty}^{\infty} r(x) f_x(x)\,dx = a \int_{-\infty}^{\infty} f_y(x)\,dx = a$$

The event $\{x < \mathbf{x} \le x + dx, M\}$ consists of all outcomes such that the point $(\mathbf{x}, \mathbf{u})$ is in the strip $x < \mathbf{x} \le x + dx$ below the curve $u = r(x)$. The probability masses in this strip

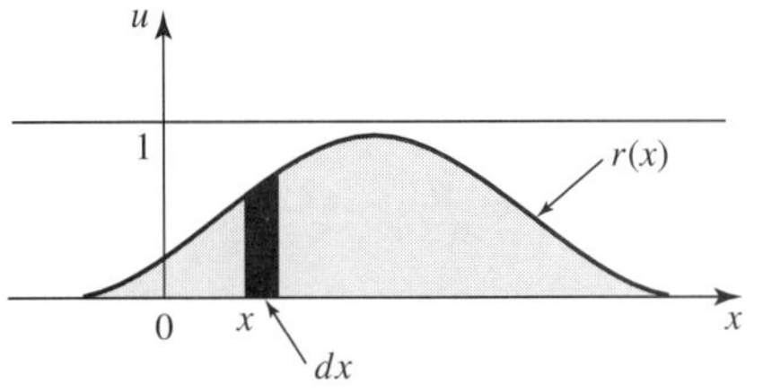

FIGURE 7-14

equal $f_x(x)r(x)\,dx$. Hence

$$P\{x < \mathbf{x} \le x + dx, M\} = f_x(x)r(x)\,dx$$

Inserting into (7-158), we obtain (7-159).

From the rejection theorem it follows that the subsequence of x_i such that $u_i \le r(x_i)$ forms a sequence of random numbers that are the samples of a random variable $\mathbf{y}$ with density $f_x(y \mid M) = f_y(y)$. This leads to the following rule for generating the sequence y_i: Form the two-dimensional random number sequence (x_i, u_i).

$$\text{Set} \quad y_i = x_i \qquad \text{if} \quad u_i \le a\frac{f_y(x_i)}{f_x(x_i)}; \qquad \text{reject } x_i \text{ otherwise} \tag{7-161}$$

EXAMPLE 7-23

▶ We are given a random number sequence x_i with exponential distribution and we wish to construct a random number sequence y_i with truncated normal distribution:

$$f_x(x) = e^{-x}U(x) \qquad f_y(y) = \frac{2}{\sqrt{2\pi}}e^{-y^2/2}U(y)$$

For $x > 0$,

$$\frac{f_y(x)}{f_x(x)} = \sqrt{\frac{2e}{\pi}}e^{-(x-1)^2/2} \le \sqrt{\frac{2e}{\pi}}$$

Setting $a = \sqrt{\pi/2e}$, we obtain the following rule for generating the sequence y_i:

$$\text{Set} \quad y_i = x_i \qquad \text{if} \quad u_i < e^{-(x_i-1)^2/2}; \quad \text{reject } x_i \text{ otherwise}$$ ◀

MIXING METHOD. We develop next a method generating a random number sequence x_i with density $f(x)$ under the following assumptions: The function $f(x)$ can be expressed as the weighted sum of m densities $f_k(m)$:

$$f(x) = p_1 f_1(x) + \cdots + p_m f_m(x) \qquad p_k > 0 \tag{7-162}$$

Each component $f_k(x)$ is the density of a known random number sequence x_i^k.

In the mixing method, we generate the sequence x_i by a mixing process involving certain subsequences of the m sequences x_i^k selected according to the following rule:

$$\text{Set} \quad x_i = x_i^k \qquad \text{if} \quad p_1 + \cdots + p_{k-1} \le u_i < p_1 + \cdots + p_k \tag{7-163}$$

Mixing theorem. If the sequences u_i and $x_i^1, \ldots, x_i^m$ are mutually independent, then the density $f_x(x)$ of the sequence x_i specified by (7-163) equals

$$f_x(x) = p_1 f_1(x) + \cdots + p_m f_m(x) \tag{7-164}$$

Proof. The sequence x_i is a mixture of m subsequences. The density of the subsequence of the kth sequence x_i^k equals $f_k(x)$. This subsequence is also a subsequence of x_i conditioned on the event

$$A_k = \{p_1 + \cdots + p_{k-1} \le \mathbf{u} < p_1 + \cdots + p_k\}$$

Hence its density also equals $f_x(x \mid A_k)$. This leads to the conclusion that

$$f_x(x \mid A_k) = f_k(x)$$

From the total probability theorem (4-74), it follows that

$$f_x(x) = f_x(x \mid A_1)P(A_1) + \cdots + f_x(x \mid A_m)P(A_m)$$

And since $P(A_k) = p_k$, (7-164) results. Comparing with (7-162), we conclude that the density $f_x(x)$ generated by (7-164) equals the given function $f(x)$.

EXAMPLE 7-24

▶ The Laplace density $0.5e^{-|x|}$ can be written as a sum

$$f(x) = 0.5e^{-x}U(x) + 0.5e^{x}U(-x)$$

This is a special case of (7-162) with

$$f_1(x) = e^{-x}U(x) \qquad f_2(x) = e^{x}U(-x) \qquad p_1 = p_2 = 0.5$$

A sequence x_i with density $f(x)$ can, therefore, be realized in terms of the samples of two random variables $\mathbf{x}^1$ and $\mathbf{x}^2$ with the above densities. As we have shown in Example 7-19, if the random variable $\mathbf{v}$ is uniform in the interval (0, 1), then the density of the random variable $\mathbf{x}^1 = -\ln \mathbf{v}$ equals $f_1(x)$; similarly, the density of the random variable $\mathbf{x}^2 = \ln \mathbf{v}$ equals $f_2(x)$. This yields the following rule for generating a random number sequence x_i with Laplace distribution: Form two independent uniform sequences u_i and v_i:

$$\text{Set} \quad x_i = -\ln v_i \quad \text{if} \quad 0 \le u_i < 0.5$$
$$\text{Set} \quad x_i = \ln v_i \quad \text{if} \quad 0.5 \le u_i < 1$$

◀

GENERAL TRANSFORMATIONS. We now give various examples for generating a random number sequence w_i with specified distribution $F_w(w)$ using the transformation

$$\mathbf{w} = g(\mathbf{x}^1, \ldots, \mathbf{x}^m)$$

where $\mathbf{x}^k$ are m random variables with known distributions. To do so, we determine g such that the distribution of $\mathbf{w}$ equals $F_w(w)$. The desired sequence is given by

$$w_i = g\left(x_i^1, \ldots, x_i^m\right)$$

Binomial Random Numbers. If $\mathbf{x}^k$ are m i.i.d. random variables taking the values 0 and 1 with probabilities p and q, respectively, their sum has a binomial distribution. From this it follows that if x_i^k are m binary sequences, their sum

$$w_i = x_i^1 + \cdots + x_i^m$$

is a random number sequence with binomial distribution. The m sequences x_i^k can be realized as subsequences of a single binary sequence $\mathbf{x}_i$ as in (7-153).

Erlang Random Numbers. The sum $\mathbf{w} = \mathbf{x}^1 + \cdots + \mathbf{x}^m$ of m i.i.d. random variables $\mathbf{x}^k$ with density $e^{-cx}U(x)$ has an Erlang density [see (4-37)–(4-38)]:

$$f_w(w) \sim w^{m-1}e^{-w}U(w) \tag{7-165}$$

From this it follows that the sum $w_i = w_i^1 + \cdots + w_i^m$ of m exponentially distributed random number sequences w_i^k is a random number sequence with Erlang distribution.

The sequences x_i^k can be generated in terms of m subsequences of a single sequence u_i (see Example 7-19):

$$w_i = -\frac{1}{c}\left(\ln u_i^1 + \cdots + \ln u_i^m\right) \tag{7-166}$$

Chi-square Random Numbers. We wish to generate a random number sequence w_i with density

$$f_w(w) \sim w^{n/2-1}e^{-w/2}U(w)$$

For $n = 2m$, this is a special case of (7-153) with $c = 1/2$. Hence w_i is given by (7-166).

To find w_i for $n = 2m + 1$, we observe that if $\mathbf{y}$ is $\chi^2(2m)$ and $\mathbf{z}$ is $N(0, 1)$ and independent of $\mathbf{y}$, the sum $\mathbf{w} = \mathbf{y} + \mathbf{z}^2$ is $\chi^2(2m+1)$ [see (7-64)]; hence the sequence

$$w_i = -2\left(\ln u_i^1 + \cdots + \ln u_i^m\right) + (z_i)^2$$

has a $\chi^2(2m+1)$ distribution.

Student *t* Random Numbers. Given two independent random variables $\mathbf{x}$ and $\mathbf{y}$ with distributions $N(0, 1)$ and $\chi^2(n)$ respectively, we form the random variable $\mathbf{w} = \mathbf{x}/\sqrt{\mathbf{y}/n}$. As we know, $\mathbf{w}$ has a $t(n)$ distribution (see Example 6-28). From this it follows that, if x_i and y_i are samples of $\mathbf{x}$ and $\mathbf{y}$, the sequence

$$w_i = \frac{x_i}{\sqrt{y_i/n}}$$

has a $t(n)$ distribution.

Lognormal Random Numbers. If $\mathbf{z}$ is $N(0, 1)$ and $\mathbf{w} = e^{a+b\mathbf{z}}$, then $\mathbf{w}$ has a lognormal distribution [see (5-25)]:

$$f_w(w) = \frac{1}{bw\sqrt{2\pi}}\exp\left\{-\frac{(\ln w - a)^2}{2b^2}\right\}$$

Hence, if z_i is an $N(0, 1)$ sequence, the sequence

$$w_i = e^{a+bz_i}$$

has a lognormal distribution.

Random Number sequences with normal distributions. Several methods are available for generating normal random variables. We next give various illustrations. The percentile transformation method is not used because of the difficulty of inverting the normal distribution. The mixing method is used extensively because the normal density is a smooth curve; it can, therefore, be approximated by a sum as in (7-162). The major

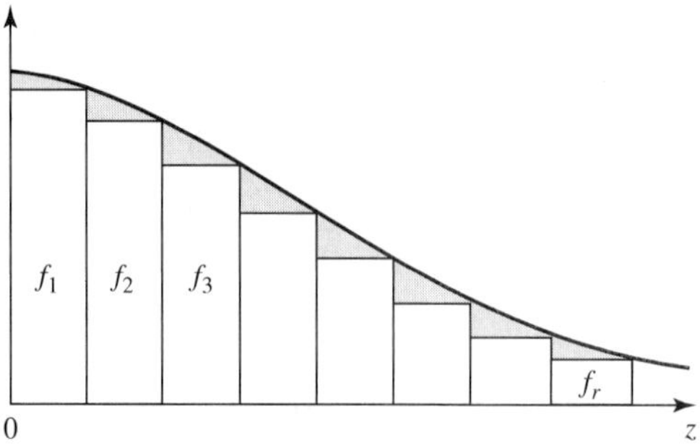

FIGURE 7-15

components (unshaded) of this sum are rectangles (Fig. 7-15) that can be realized by sequences of the form $au_i + b$. The remaining components (shaded) are more complicated; however, since their areas are small, they need not be realized exactly. Other methods involve known properties of normal random variables. For example, the central limit theorem leads to the following method.

Given m independent random variables $\mathbf{u}^k$, we form the sum

$$\mathbf{z} = \mathbf{u}^1 + \cdots + \mathbf{u}^m$$

If m is large, the random variable $\mathbf{z}$ is approximately normal [see (7-123)]. From this it follows that if u_1^k are m independent random number sequences their sum

$$z_i = u_i^1 + \cdots + u_i^m$$

is approximately a normal random number sequence. This method is not very efficient. The following three methods are more efficient and are used extensively.

Rejection and mixing (G. Marsaglia). In Example 7-23, we used the rejection method to generate a random variable sequence y_i with a truncated normal density

$$f_y(y) = \frac{2}{\sqrt{2\pi}} e^{-y^2/2} U(y)$$

The normal density can be written as a sum

$$f_z(z) = \frac{1}{\sqrt{2\pi}} e^{-z^2/2} = \frac{1}{2} f_y(z) + \frac{1}{2} f_y(-z) \tag{7-167}$$

The density $f_y(y)$ is realized by the sequence y_i as in Example 7-23 and the density $f_y(-y)$ by the sequence $-y_i$. Applying (7-163), we conclude that the following rule generates an $N(0, 1)$ sequence z_i:

$$\begin{aligned} &\text{Set} \quad z_i = y_i \qquad \text{if} \quad 0 \le u_i < 0.5 \\ &\text{Set} \quad z_i = -y_i \qquad \text{if} \quad 0.5 \le u_i < 1 \end{aligned} \tag{7-168}$$

Polar coordinates. We have shown that, if the random variables $\mathbf{r}$ and $\boldsymbol{\varphi}$ are independent, $\mathbf{r}$ has the Rayleigh density $f_r(r) = re^{-r^2/2}$ and $\boldsymbol{\varphi}$ is uniform in the interval $(-\pi, \pi)$, then (see Example 6-15) the random variables

$$\mathbf{z} = \mathbf{r}\cos\boldsymbol{\varphi} \qquad \mathbf{w} = \mathbf{r}\sin\boldsymbol{\varphi} \tag{7-169}$$

are $N(0, 1)$ and independent. Using this, we shall construct two independent normal random number sequences z_i and w_i as follows: Clearly, $\boldsymbol{\varphi} = \pi(2\mathbf{u} - 1)$; hence

$\varphi_i = \pi(2u_i - 1)$. As we know, $\mathbf{r} = \sqrt{2\mathbf{x}} = \sqrt{-2\ln \mathbf{v}}$, where $\mathbf{x}$ is a random variable with exponential distribution and $\mathbf{v}$ is uniform in the interval (0, 1). Denoting by x_i and v_i the samples of the random variables $\mathbf{x}$ and $\mathbf{v}$, we conclude that $r_i = \sqrt{2x_i} = \sqrt{-2\ln v_i}$ is a random number sequence with Rayleigh distribution. From this and (7-169) it follows that if u_i and v_i are two independent random number sequences uniform in the interval (0, 1), then the sequences

$$z_i = \sqrt{-2\ln v_i}\cos\pi(2u_i - 1) \qquad w_i = \sqrt{-2\ln v_i}\sin\pi(2u_i - 1) \tag{7-170}$$

are $N(0, 1)$ and independent.

The Box–Muller method. The rejection method was based on the following: If x_i is a random number sequence with distribution $F(x)$, its subsequence y_i conditioned on an event M is a random number sequence with distribution $F(x \mid M)$. Using this, we shall generate two independent $N(0, 1)$ sequences z_i and w_i in terms of the samples x_i and y_i of two independent random variables $\mathbf{x}$ and $\mathbf{y}$ uniformly distributed in the interval $(-1, 1)$. We shall use for M the event

$$M = \{\mathbf{q} \le 1\} \qquad \mathbf{q} = \sqrt{\mathbf{x}^2 + \mathbf{y}^2}$$

The joint density of $\mathbf{x}$ and $\mathbf{y}$ equals 1/4 in the square $|x| < 1$, $|y| < 1$ of Fig. 7-16 and 0 elsewhere. Hence

$$P(M) = \frac{\pi}{4} \qquad P\{\mathbf{q} \le q\} = \frac{\pi q^2}{4} \qquad \text{for} \quad q < 1$$

But $\{\mathbf{q} \le q, M\} = \{\mathbf{q} \le q\}$, for $q < 1$ because $\{\mathbf{q} < q\}$ is a subset of M. Hence

$$F_q(q \mid M) = \frac{P\{\mathbf{q} \le q, M\}}{P(M)} = q^2 \qquad f_q(q \mid M) = 2q \qquad 0 \le q < 1 \tag{7-171}$$

Writing the random variables $\mathbf{x}$ and $\mathbf{y}$ in polar form:

$$\mathbf{x} = \mathbf{q}\cos\boldsymbol{\varphi} \qquad \mathbf{y} = \mathbf{q}\sin\boldsymbol{\varphi} \qquad \tan\boldsymbol{\varphi} = \mathbf{y}/\mathbf{x} \tag{7-172}$$

we conclude as in (7-171) that the joint density of the random variables $\mathbf{q}$ and $\boldsymbol{\varphi}$ is such that

$$f_{q\varphi}(q, \varphi \mid M)\, dq\, d\varphi = \frac{P\{q \le \mathbf{q} < q + dq, \varphi \le \boldsymbol{\varphi} < \varphi + d\varphi\}}{P(M)} = \frac{q\, dq\, d\varphi/4}{\pi/4}$$

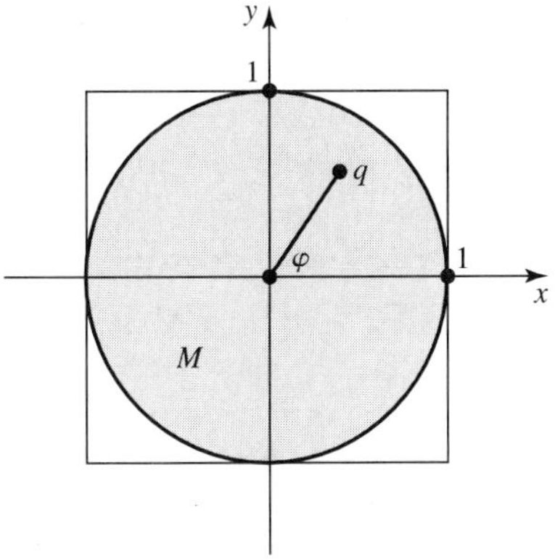

FIGURE 7-16

for $0 \le q < 1$ and $|\varphi| < \pi$. From this it follows that the random variables $\mathbf{q}$ and φ are conditionally independent and

$$f_q(q \mid M) = 2q \qquad f_\varphi(\varphi) = 1/2\pi \qquad 0 \le q \le 1 \qquad -\pi < \varphi < \pi$$

THEOREM 7-3 If $\mathbf{x}$ and $\mathbf{y}$ are two independent random variables uniformly distributed in the interval $(-1, 1)$ and $\mathbf{q} = \sqrt{\mathbf{x}^2 + \mathbf{y}^2}$, then the random variables

$$\mathbf{z} = \frac{\mathbf{x}}{\mathbf{q}}\sqrt{-4\ln\mathbf{q}} \qquad \mathbf{w} = \frac{\mathbf{y}}{\mathbf{q}}\sqrt{-4\ln\mathbf{q}} \tag{7-173}$$

are conditionally $N(0, 1)$ and independent:

$$f_{zw}(z, w \mid M) = f_z(z \mid M) f_w(w \mid M) = \frac{1}{2\pi} e^{-(z^2+w^2)/2}$$

Proof. From (7-172) it follows that

$$\mathbf{z} = \sqrt{-4\ln\mathbf{q}}\cos\varphi \qquad \mathbf{w} = \sqrt{-4\ln\mathbf{q}}\sin\varphi$$

This system is similar to the system (7-169). To prove the theorem, it suffices, therefore, to show that the conditional density of the random variable $\mathbf{r} = \sqrt{-4\ln\mathbf{q}}$ assuming M equals $re^{-r^2/2}$. To show this, we apply (5-16). In our case,

$$q(r) = e^{-r^2/4} \qquad q'(r) = \frac{-r}{2} e^{-r^2/4} = \frac{1}{r'(q)} \qquad f_q(q \mid M) = 2q$$

Hence

$$f_r(r \mid M) = f_q(q \mid M)|q'(r)| = 2e^{-r^2/4}\frac{r}{2}e^{-r^2/4} = re^{-r^2/2}$$

This shows that the conditional density of the random variable $\mathbf{r}$ is Rayleigh as in (7-169).

The preceding theorem leads to the following rule for generating the sequences z_i and w_i: Form two independent sequences $x_i = 2u_i - 1$, $y_i = 2v_i - 1$.

$$\text{If} \quad q_i = \sqrt{x_i^2 + y_i^2} < 1, \qquad \text{set} \quad z_i = \frac{x_i}{q_i}\sqrt{-4\ln q_i} \qquad w_i = \frac{y_i}{q_i}\sqrt{-4\ln q_i}$$

Reject (x_i, y_i) otherwise.

COMPUTERS AND STATISTICS. In this section, so far we have analyzed the dual meaning of random numbers and their computer generation. We conclude with a brief outline of the general areas of interaction between computers and statistics:

1. Statistical methods are used to solve numerically a variety of deterministic problems.

   Examples include: evaluation of integrals, solution of differential equations; determination of various mathematical constants. The solutions are based on the availability of random number sequences. Such sequences can be obtained from random experiments; in most cases, however, they are computer generated. We shall give a simple illustration of the two approaches in the context of Buffon's needle. The objective in this problem is the statistical estimation of the number π. The method proposed in Example 6-2 involves the performance of a physical experiment. We introduce the event $A = \{\mathbf{x} < a\cos\boldsymbol{\theta}\}$, where $\mathbf{x}$ (distance from the nearest line) and $\boldsymbol{\theta}$ (angle of the needle) are two independent random variables uniform in the intervals $(0, a)$ and $(0, \pi/2)$, respectively. This event occurs if the needle intersects one of the lines and its probability equals $\pi b/2a$. From this it follows that

$$P(A) \simeq \frac{n_A}{n} \qquad \pi \simeq \frac{2a}{nb} n_A \tag{7-174}$$

where n_A is the number of intersections in n trials. This estimate can be obtained without experimentation. We form two independent random number sequences x_i and θ_i with distributions $F_x(x)$ and $F_\theta(\theta)$, respectively, and we denote by n_A the number of times $x_i < a \cos\theta_i$. With n_A so determined the computer generated estimate of π is obtained from (7-174).

2. Computers are used to solve a variety of deterministic problems originating in statistics.

Examples include: evaluation of the mean, the variance, or other averages used in parameter estimation and hypothesis testing; classification and storage of experimental data; use of computers as instructional tools. For example, graphical demonstration of the law of large numbers or the central limit theorem. Such applications involve mostly routine computer programs unrelated to statistics. There is, however, another class of deterministic problems the solution of which is based on statistical concepts and random number sequences. A simple illustration is:

We are given m random variables $\mathbf{x}_1, \ldots, \mathbf{x}_n$ with known distributions and we wish to estimate the distribution of the random variable $\mathbf{y} = g(\mathbf{x}_1, \ldots, \mathbf{x}_n)$. This problem can, in principle, be solved analytically; however, its solution is, in general, complex. See, for example, the problem of determining the exact distribution of the random variable $\mathbf{q}$ used in the chi-square test (8-325). As we can show next, the determination of $F_y(y)$ is simplified if we use Monte Carlo techniques. Assuming for simplicity that $m = 1$, we generate a random number sequence x_i of length n with distribution the known function $F_x(x)$ and we form the random number sequence $y_i = g(x_i)$. To determine $F_y(y)$ for a specific y, we count the number n_y of samples y_i such that $y_i < y$. Inserting into (4-3), we obtain the estimate

$$F_y(y) \simeq \frac{n_y}{n} \tag{7-175}$$

A similar approach can be used to determine the u percentile x_u of $\mathbf{x}$ or to decide whether x_u is larger or smaller than a given number (see hypothesis testing, Sec. 8-4).

3. Computers are used to simulate random experiments or to verify a scientific theory.

This involves the familiar methods of simulating physical systems where now all inputs and responses are replaced by appropriate random number sequences.

PROBLEMS

7-1 Show that if $F(x, y, z)$ is a joint distribution, then for any $x_1 \le x_2$, $y_1 \le y_2$, $z_1 \le z_2$:

$$F(x_2, y_2, z_2) + F(x_1, y_1, z_1) + F(x_1, y_2, z_1) + F(x_2, y_1, z_1)$$
$$- F(x_1, y_2, z_2) - F(x_2, y_1, z_2) - F(x_2, y_2, z_1) - F(x_1, y_1, z_1) \ge 0$$

7-2 The events A, B, and C are such that

$$P(A) = P(B) = P(C) = 0.5$$

$$P(AB) = P(AC) = P(BC) = P(ABC) = 0.25$$

Show that the zero–one random variables associated with these events are not independent; they are, however, independent in pairs.

7-3 Show that if the random variables $\mathbf{x}$, $\mathbf{y}$, and $\mathbf{z}$ are jointly normal and independent in pairs, they are independent.

7-4 The random variables $\mathbf{x}_i$ are i.i.d. and uniform in the interval $(-0.5, 0.5)$. Show that

$$E\{(\mathbf{x}_1 + \mathbf{x}_2 + \mathbf{x}_3)^4\} = \tfrac{13}{80}$$

7-5 (*a*) Reasoning as in (6-31), show that if the random variables $\mathbf{x}$, $\mathbf{y}$, and $\mathbf{z}$ are independent and their joint density has spherical symmetry:

$$f(x, y, z) = f\left(\sqrt{x^2 + y^2 + z^2}\right)$$

then they are normal with zero mean and equal variance.
(*b*) The components $\mathbf{v}_x$, $\mathbf{v}_y$, and $\mathbf{v}_z$ of the velocity $\mathbf{v} = \sqrt{\mathbf{v}_x^2 + \mathbf{v}_y^2 + \mathbf{v}_z^2}$ of a particle are independent random variables with zero mean and variance kT/m. Furthermore, their joint density has spherical symmetry. Show that $\mathbf{v}$ has a Maxwell density and

$$E\{\mathbf{v}\} = 2\sqrt{\frac{2kT}{\pi m}} \qquad E\{\mathbf{v}^2\} = \frac{3kT}{m} \qquad E\{\mathbf{v}^4\} = \frac{15k^2T^2}{m^2}$$

7-6 Show that if the random variables $\mathbf{x}$, $\mathbf{y}$, and $\mathbf{z}$ are such that $r_{xy} = r_{yz} = 1$, then $r_{xz} = 1$.

7-7 Show that

$$E\{\mathbf{x}_1\mathbf{x}_2 \mid \mathbf{x}_3\} = E\{E\{\mathbf{x}_1\mathbf{x}_2 \mid \mathbf{x}_2, \mathbf{x}_3\} \mid \mathbf{x}_3\} = E\{\mathbf{x}_2 E\{\mathbf{x}_1 \mid \mathbf{x}_2, \mathbf{x}_3\} \mid \mathbf{x}_3\}$$

7-8 Show that $\hat{E}\{\mathbf{y} \mid \mathbf{x}_1\} = \hat{E}\{\hat{E}\{\mathbf{y} \mid \mathbf{x}_1, \mathbf{x}_2\} \mid \mathbf{x}_1\}$ where $\hat{E}\{\mathbf{y} \mid \mathbf{x}_1, \mathbf{x}_2\} = a_1\mathbf{x}_1 + a_2\mathbf{x}_2$ is the linear MS estimate of $\mathbf{y}$ terms of $\mathbf{x}_1$ and $\mathbf{x}_2$.

7-9 Show that if

$$\mathbf{x}_i \ge 0, \qquad E\{\mathbf{x}_i^2\} = M \qquad \text{and} \quad \mathbf{s} = \sum_{i=1}^{n} \mathbf{x}_i$$

then

$$E\{\mathbf{s}^2\} \le ME\{\mathbf{n}^2\}$$

7-10 We denote by $\mathbf{x}_m$ a random variable equal to the number of tosses of a coin until heads shows for the mth time. Show that if $P\{h\} = p$, then $E\{\mathbf{x}_m\} = m/p$.
Hint: $E\{\mathbf{x}_m - \mathbf{x}_{m-1}\} = E\{\mathbf{x}_1\} = p + 2pq + \cdots + npq^{n-1} + \cdots = 1/p$.

7-11 The number of daily accidents is a Poisson random variable $\mathbf{n}$ with parameter a. The probability that a single accident is fatal equals p. Show that the number $\mathbf{m}$ of fatal accidents in one day is a Poisson random variable with parameter ap.
Hint:
$$E\{e^{j\omega\mathbf{m}} \mid \mathbf{n} = n\} = \sum_{k=0}^{n} e^{j\omega k}\binom{n}{k} p^k q^{n-k} = (pe^{j\omega} + q)^n$$

7-12 The random variables $\mathbf{x}_k$ are independent with densities $f_k(x)$ and the random variable $\mathbf{n}$ is independent of $\mathbf{x}_k$ with $P\{\mathbf{n} = k\} = p_k$. Show that if

$$\mathbf{s} = \sum_{k=1}^{\mathbf{n}} \mathbf{x}_k \qquad \text{then} \quad f_s(s) = \sum_{k=1}^{\infty} p_k[f_1(s) * \cdots * f_k(s)]$$

7-13 The random variables $\mathbf{x}_i$ are i.i.d. with moment function $\mathbf{\Phi}_x(s) = E\{e^{s\mathbf{x}_i}\}$. The random variable $\mathbf{n}$ takes the values 0, 1, ... and its moment function equals $\mathbf{\Gamma}_n(z) = E\{\mathbf{z}^n\}$. Show that if

$$\mathbf{y} = \sum_{i=1}^{\mathbf{n}} \mathbf{x}_i \qquad \text{then} \quad \mathbf{\Phi}_y(s) = E\{e^{s\mathbf{y}}\} = \mathbf{\Gamma}_n[\mathbf{\Phi}_x(s)]$$

Hint: $E\{e^{s\mathbf{y}} \mid \mathbf{n} = k\} = E\{e^{s(\mathbf{x}_1 + \cdots + \mathbf{x}_k)}\} = \mathbf{\Phi}_x^k(s)$.
Special case: If $\mathbf{n}$ is Poisson with parameter a, then $\mathbf{\Phi}_y(s) = e^{a\mathbf{\Phi}_x(s) - a}$.

7-14 The random variables $\mathbf{x}_i$ are i.i.d. and uniform in the interval (0, 1). Show that if $\mathbf{y} = \max \mathbf{x}_i$, then $F(y) = y^n$ for $0 \le y \le 1$.

7-15 Given a random variable $\mathbf{x}$ with distribution $F_x(x)$, we form its order statistics $\mathbf{y}_k$ as in Example 7-2, and their extremes

$$\mathbf{z} = \mathbf{y}_n = \mathbf{x}_{\max} \qquad \mathbf{w} = \mathbf{y}_1 = \mathbf{x}_{\min}$$

Show that

$$f_{zw}(z, w) = \begin{cases} n(n-1) f_x(z) f_x(w)[F_x(z) - F_x(w)]^{n-2} & z > w \\ 0 & z < w \end{cases}$$

7-16 Given n independent $N(\eta_i, 1)$ random variables $\mathbf{z}_i$, we form the random variable $\mathbf{w} = \mathbf{z}_1^2 + \cdots + \mathbf{z}_n^2$. This random variable is called *noncentral chi-square* with n degrees of freedom and eccentricity $e = \eta_1^2 + \cdots + \eta_n^2$. Show that its moment generating function equals

$$\mathbf{\Phi}_w(s) = \frac{1}{\sqrt{(1-2s)^n}} \exp\left\{\frac{es}{1-2s}\right\}$$

7-17 Show that if the random variables $\mathbf{x}_i$ are i.i.d. and normal, then their sample mean $\overline{\mathbf{x}}$ and sample variances $\mathbf{s}^2$ are two independent random variables.

7-18 Show that, if $\alpha_0 + \alpha_1 \mathbf{x}_1 + \alpha_2 \mathbf{x}_2$ is the nonhomogeneous linear MS estimate of $\mathbf{s}$ in terms of $\mathbf{x}_1$, and $\mathbf{x}_2$, then

$$\hat{E}\{\mathbf{s} - \eta_s \mid \mathbf{x}_1 - \eta_1, \mathbf{x}_2 - \eta_2\} = \alpha_1(\mathbf{x}_1 - \eta_1) + \alpha_2(\mathbf{x}_2 - \eta_2)$$

7-19 Show that

$$\hat{E}\{\mathbf{y} \mid \mathbf{x}_1\} = \hat{E}\{\hat{E}\{\mathbf{y} \mid \mathbf{x}_1, \mathbf{x}_2\} \mid \mathbf{x}_1\}$$

7-20 We place at random n points in the interval (0, 1) and we denote by $\mathbf{x}$ and $\mathbf{y}$ the distance from the origin to the first and last point respectively. Find $F(x)$, $F(y)$, and $F(x, y)$.

7-21 Show that if the random variables $\mathbf{x}_i$ are i.i.d. with zero mean, variance σ^2, and sample variance $\overline{\mathbf{v}}$ (see Example 7-5), then

$$\sigma_{\bar{v}}^2 = \frac{1}{n}\left(E\{\mathbf{x}_i^4\} - \frac{n-3}{n-1}\sigma^4\right)$$

7-22 The random variables $\mathbf{x}_i$ are $N(0; \sigma)$ and independent. Show that if

$$\mathbf{z} = \frac{\sqrt{\pi}}{2n} \sum_{i=1}^{n} |\mathbf{x}_{2i} - \mathbf{x}_{2i-1}| \qquad \text{then} \quad E\{\mathbf{z}\} = \sigma \qquad \sigma_z^2 = \frac{\pi - 2}{2n}\sigma^2$$

7-23 Show that if R is the correlation matrix of the random vector $\mathbf{X}$: $[\mathbf{x}_1, \ldots, \mathbf{x}_n]$ and R^{-1} is its inverse, then

$$E\{\mathbf{X}R^{-1}\mathbf{X}^t\} = n$$

7-24 Show that if the random variables $\mathbf{x}_i$ are of continuous type and independent, then, for sufficiently large n, the density of $\sin(\mathbf{x}_1 + \cdots + \mathbf{x}_n)$ is nearly equal to the density of $\sin \mathbf{x}$, where $\mathbf{x}$ is a random variable uniform in the interval $(-\pi, \pi)$.

7-25 Show that if $a_n \to a$ and $E\{|\mathbf{x}_n - a_n|^2\} \to 0$, then $\mathbf{x}_n \to a$ in the MS sense as $n \to \infty$.

7-26 Using the Cauchy criterion, show that a sequence $\mathbf{x}_n$ tends to a limit in the MS sense iff the limit of $E\{\mathbf{x}_n\mathbf{x}_m\}$ as $n, m \to \infty$ exists.

7-27 An infinite sum is by definition a limit:

$$\sum_{k=1}^{\infty} \mathbf{x}_k = \lim_{n\to\infty} \mathbf{y}_n \qquad \mathbf{y}_n = \sum_{k=1}^{n} \mathbf{x}_k$$

Show that if the random variables $\mathbf{x}_k$ are independent with zero mean and variance σ_k^2, then the sum exists in the MS sense iff

$$\sum_{k=1}^{\infty} \sigma_k^2 < \infty$$

Hint:

$$E\{(\mathbf{y}_{n+m} - \mathbf{y}_n)^2\} = \sum_{k=n+1}^{n+m} \sigma_k^2$$

7-28 The random variables $\mathbf{x}_i$ are i.i.d. with density $ce^{-cx}U(x)$. Show that, if $\mathbf{x} = \mathbf{x}_1 + \cdots + \mathbf{x}_n$, then $f_x(x)$ is an Erlang density.

7-29 Using the central limit theorem, show that for large n:

$$\frac{c^n}{(n-1)!}x^{n-1}e^{-cx} \simeq \frac{c}{\sqrt{2\pi n}}e^{-(cx-n)^2/2n} \qquad x > 0$$

7-30 The resistors $\mathbf{r}_1, \mathbf{r}_2, \mathbf{r}_3$, and $\mathbf{r}_4$ are independent random variables and each is uniform in the interval (450, 550). Using the central limit theorem, find $P\{1900 \le \mathbf{r}_1 + \mathbf{r}_2 + \mathbf{r}_3 + \mathbf{r}_4 \le 2100\}$.

7-31 Show that the central limit theorem does not hold if the random variables $\mathbf{x}_i$ have a Cauchy density.

7-32 The random variables $\mathbf{x}$ and $\mathbf{y}$ are uncorrelated with zero mean and $\sigma_x = \sigma_y = \sigma$. Show that if $\mathbf{z} = \mathbf{x} + j\mathbf{y}$, then

$$f_z(z) = f(x, y) = \frac{1}{2\pi\sigma^2}e^{-(x^2+y^2)/2\sigma^2} = \frac{1}{\pi\sigma_z^2}e^{-|z|^2/\sigma_z^2}$$

$$\Phi_z(\Omega) = \exp\left\{-\frac{1}{2}(\sigma^2u^2 + \sigma^2v_z^2)\right\} = \exp\left\{-\frac{1}{4}\sigma_z^2|\Omega|^2\right\}$$

where $\Omega = u + jv$. This is the scalar form of (7-62)–(7-63).

CHAPTER

8

STATISTICS

8-1 INTRODUCTION

Probability is a mathematical discipline developed as an abstract model and its conclusions are *deductions* based on the axioms. Statistics deals with the applications of the theory to real problems and its conclusions are *inferences* based on observations. Statistics consists of two parts: analysis and design.

Analysis, or mathematical statistics, is part of probability involving mainly repeated trials and events the probability of which is close to 0 or to 1. This leads to inferences that can be accepted as near certainties (see pages 11-12). *Design,* or applied statistics, deals with data collection and construction of experiments that can be adequately described by probabilistic models. In this chapter, we introduce the basic elements of mathematical statistics.

We start with the observation that the connection between probabilistic concepts and reality is based on the approximation

$$p \simeq \frac{n_A}{n} \tag{8-1}$$

relating the probability $p = P(A)$ of an event A to the number n_A of successes of A in n trials of the underlying physical experiment. We used this empirical formula to give the relative frequency interpretation of all probabilistic concepts. For example, we showed that the mean η of a random variable $\mathbf{x}$ can be approximated by the average

$$\hat{\eta} = \frac{1}{n} \sum x_i = \bar{x} \tag{8-2}$$

of the observed values x_i of $\mathbf{x}$, and its distribution $F(x)$ by the empirical distribution

$$\hat{F}(x) = \frac{n_x}{n} \tag{8-3}$$

where n_x is the number of x_i's that do not exceed x. These relationships are empirical

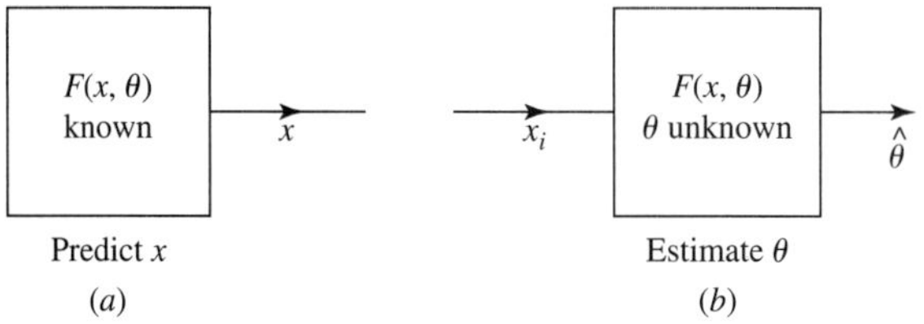

FIGURE 8-1

point estimates of the parameters η and $F(x)$ and a major objective of statistics is to give them an exact interpretation.

In a statistical investigation, we deal with two general classes of problems. In the first class, we assume that the probabilistic model is known and we wish to make predictions concerning future observations. For example, we know the distribution $F(x)$ of a random variable $\mathbf{x}$ and we wish to predict the average $\bar{x}$ of its n

future samples or we know the probability p of an event A and we wish to predict the number n_A of successes of A in n future trials. In both cases, we proceed from the model to the observations (Fig. 8-1a). In the second class, one or more parameters θ_i of the model are unknown and our objective is either to *estimate* their values (parameter estimation) or to *decide* whether θ_i is a set of known constants θ_{0i} (hypothesis testing). For example, we observe the values x_i of a random variable $\mathbf{x}$ and we wish to estimate its mean η or to decide whether to accept the hypothesis that $\eta = 5.3$. We toss a coin 1000 times and heads shows 465 times. Using this information, we wish to estimate the probability p of heads or to decide whether the coin is fair. In both cases, we proceed from the observations to the model (Fig. 8-1b). In this chapter, we concentrate on parameter estimation and hypothesis testing. As a preparation, we comment briefly on the prediction problem.

Prediction. We are given a random variable $\mathbf{x}$ with known distribution and we wish to predict its value x at a future trial. A *point prediction* of $\mathbf{x}$ is the determination of a constant c chosen so as to minimize in some sense the error $\mathbf{x} - c$. At a specific trial, the random variable $\mathbf{x}$ can take one of many values. Hence the value that it actually takes cannot be predicted; it can only be estimated. Thus prediction of a random variable $\mathbf{x}$ is the estimation of its next value x by a constant c. If we use as the criterion for selecting c the minimization of the MS error $E\{(\mathbf{x} - c)^2\}$, then $c = E\{\mathbf{x}\}$. This problem was considered in Sec. 7-3.

An *interval prediction* of $\mathbf{x}$ is the determination of two constants c_1 and c_2 such that

$$P\{c_1 < \mathbf{x} < c_2\} = \gamma = 1 - \delta \tag{8-4}$$

where γ is a given constant called the *confidence coefficient.* Equation (8-4) states that if we predict that the value x of $\mathbf{x}$ at the next trial will be in the interval (c_1, c_2), our prediction will be correct in $100\gamma\%$ of the cases. The problem in interval prediction is to find c_1 and c_2 so as to minimize the difference $c_2 - c_1$ subject to the constraint (8-4). The selection of γ is dictated by two conflicting requirements. If γ is close to 1, the prediction that x will be in the interval (c_1, c_2) is reliable but the difference $c_2 - c_1$ is large; if γ is reduced, $c_2 - c_1$ is reduced but the estimate is less reliable. Typical values of γ are 0.9, 0.95, and 0.99. For optimum prediction, we assign a value to γ and we determine c_1 and c_2 so as to minimize the difference $c_2 - c_1$ subject to the constraint (8-4). We can show that (see Prob. 8-6) if the density $f(x)$ of $\mathbf{x}$ has a single maximum,

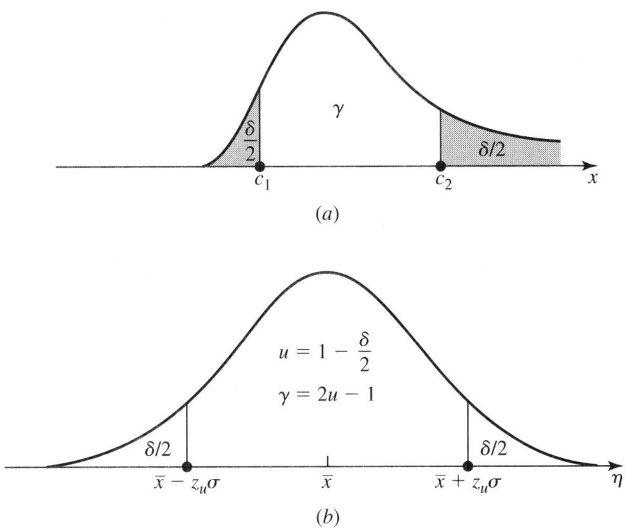

FIGURE 8-2

$c_2 - c_1$ is minimum if $f(c_1) = f(c_2)$. This yields c_1 and c_2 by trial and error. A simpler suboptimal solution is easily found if we determine c_1 and c_2 such that

$$P\{\mathbf{x} < c_1\} = \frac{\delta}{2} \qquad P\{\mathbf{x} > c_2\} = \frac{\delta}{2} \tag{8-5}$$

This yields $c_1 = x_{\delta/2}$ and $c_2 = x_{1-\delta/2}$ where x_u is the u percentile of $\mathbf{x}$ (Fig. 8-2*a*). This solution is optimum if $f(x)$ is symmetrical about its mean η because then $f(c_1) = f(c_2)$. If $\mathbf{x}$ is also normal, then $x_u = \eta + z_u\sigma$, where z_u is the standard normal percentile (Fig. 8-2*b*).

EXAMPLE 8-1

▶ The life expectancy of batteries of a certain brand is modeled by a normal random variable with $\eta = 4$ years and $\sigma = 6$ months. Our car has such a battery. Find the prediction interval of its life expectancy with $\gamma = 0.95$.

In this example, $\delta = 0.05$, $z_{1-\delta/2} = z_{0.975} = 2 = -z_{\delta/2}$. This yields the interval $4 \pm 2 \times 0.5$. We can thus expect with confidence coefficient 0.95 that the life expectancy of our battery will be between 3 and 5 years. ◀

As a second application, we shall estimate the number n_A of successes of an event A in n trials. The point estimate of n_A is the product np. The interval estimate (k_1, k_2) is determined so as to minimize the difference $k_2 - k_1$ subject to the constraint

$$P\{k_1 < n_A < k_2\} = \gamma$$

We shall assume that n is large and $\gamma = 0.997$. To find the constants k_1 and k_2, we set $\mathbf{x} = n_A$ into (4-91)–(4-92). This yields

$$P\{np - 3\sqrt{npq} < n_A < np + 3\sqrt{npq}\} = 0.997 \tag{8-6}$$

because $2G(3) - 1 \simeq 0.997$. Hence we predict with confidence coefficient 0.997 that n_A will be in the interval $np \pm 3\sqrt{npq}$.

EXAMPLE 8-2

▶ We toss a fair coin 100 times and we wish to estimate the number n_A of heads with $\gamma = 0.997$. In this problem $n = 100$ and $p = 0.5$. Hence

$$k_1 = np - 3\sqrt{npq} = 35 \qquad k_2 = np - 3\sqrt{npq} = 65$$

We predict, therefore, with confidence coefficient 0.997 that the number of heads will be between 35 and 65. ◀

Example 8-2 illustrates the role of statistics in the applications of probability to real problems: The event $A = \{\text{heads}\}$ is defined in the experiment S of the single toss of a coin. The given information that $P(A) = 0.5$ cannot be used to make a reliable prediction about the occurrence of A at a single performance of S. The event

$$B = \{35 < n_A < 65\}$$

is defined in the experiment S_n of repeated trials and its probability equals $P(B) = 0.997$. Since $P(B) \simeq 1$ we can claim with near certainty that B will occur at a single performance of the experiment S_n. We have thus changed the "subjective" knowledge about A based on the given information that $P(A) = 0.5$ to the "objective" conclusion that B will almost certainly occur, based on the derived probability that $P(B) \simeq 1$. Note, however, that both conclusions are inductive inferences; the difference between them is only quantitative.

8-2 ESTIMATION

Suppose that the distribution of a random variable $\mathbf{x}$ is a function $F(x, \theta)$ of known form depending on a parameter θ, scalar or vector. We wish to estimate θ. To do so, we repeat the underlying physical experiment n times and we denote by x_i the observed values of $\mathbf{x}$. Using these observations, we shall find a point estimate and an interval estimate of θ.

A *point estimate* is a function $\hat{\theta} = g(X)$ of the observation vector $X = [x_1, \ldots, x_n]$. The corresponding random variable $\hat{\boldsymbol{\theta}} = g(\mathbf{X})$ is the *point estimator* of θ (see Sec. 8-3). Any function of the sample vector $\mathbf{X} = [\mathbf{x}_1, \ldots, \mathbf{x}_n]$ is called a *statistic*.[1] Thus a point estimator is a statistic.

We shall say that $\hat{\boldsymbol{\theta}}$ is an unbiased estimator of the parameter θ if $E\{\hat{\boldsymbol{\theta}}\} = \theta$. Otherwise, it is called biased with bias $b = E\{\hat{\boldsymbol{\theta}}\} - \theta$. If the function $g(X)$ is properly selected, the estimation error $\hat{\boldsymbol{\theta}} - \theta$ decreases as n increases. If it tends to 0 in probability as $n \to \infty$, then $\hat{\boldsymbol{\theta}}$ is called a *consistent* estimator. The sample mean $\bar{\mathbf{x}}$ of $\mathbf{x}$ is an unbiased estimator of its mean η. Furthermore, its variance σ^2/n tends to 0 as $n \to \infty$. From this it follows that $\bar{\mathbf{x}}$ tends to η in the MS sense, therefore, also in probability. In other words, $\bar{\mathbf{x}}$ is a consistent estimator of η. Consistency is a desirable property; however, it is a theoretical concept. In reality, the number n of trials might be large but it is finite. The objective of estimation is thus the selection of a function $g(X)$ minimizing in some sense the estimation error $g(\mathbf{X}) - \theta$. If $g(X)$ is chosen so as to minimize the MS error

$$e = E\{[g(\mathbf{X}) - \theta]^2\} = \int_R [g(X) - \theta]^2 f(X, \theta)\, dX \tag{8-7}$$

---

[1] This interpretation of the term *statistic* applies only for Chap. 8. In all other chapters, *statistics* means *statistical properties*.

then the estimator $\hat{\boldsymbol{\theta}} = g(\mathbf{X})$ is called the *best estimator*. The determination of best estimators is not, in general, simple because the integrand in (8-7) depends not only on the function $g(X)$ but also on the unknown parameter θ. The corresponding prediction problem involves the same integral but it has a simple solution because in this case, θ is known (see Sec. 7-3).

In the following, we shall select the function $g(X)$ empirically. In this choice we are guided by the following: Suppose that θ is the mean $\theta = E\{q(\mathbf{x})\}$ of some function $q(\mathbf{x})$ of $\mathbf{x}$. As we have noted, the sample mean

$$\hat{\boldsymbol{\theta}} = \frac{1}{n}\sum q(\mathbf{x}_i) \tag{8-8}$$

of $q(\mathbf{x})$ is a consistent estimator of θ. If, therefore, we use the sample mean $\hat{\boldsymbol{\theta}}$ of $q(\mathbf{x})$ as the point estimator of θ, our estimate will be satisfactory at least for large n. In fact, it turns out that in a number of cases it is the best estimate.

INTERVAL ESTIMATES. We measure the length θ of an object and the results are the samples $x_i = \theta + \nu_i$ of the random variable $\mathbf{x} = \theta + \boldsymbol{\nu}$, where $\boldsymbol{\nu}$ is the measurement error. Can we draw with near certainty a conclusion about the true value of θ? We cannot do so if we claim that θ equals its point estimate $\hat{\theta}$ or any other constant. We can, however, conclude with near certainty that θ equals $\hat{\theta}$ within specified tolerance limits. This leads to the following concept.

An *interval estimate* of a parameter θ is an interval (θ_1, θ_2), the endpoints of which are functions $\theta_1 = g_1(X)$ and $\theta_2 = g_2(X)$ of the observation vector X. The corresponding random interval $(\boldsymbol{\theta}_1, \boldsymbol{\theta}_2)$ is the *interval estimator* of θ. We shall say that (θ_1, θ_2) is a γ *confidence interval* of θ if

$$P\{\boldsymbol{\theta}_1 < \theta < \boldsymbol{\theta}_2\} = \gamma \tag{8-9}$$

The constant γ is the *confidence coefficient* of the estimate and the difference $\delta = 1 - \gamma$ is the *confidence level*. Thus γ is a subjective measure of our confidence that the unknown θ is in the interval (θ_1, θ_2). If γ is close to 1 we can expect with near certainty that this is true. Our estimate is correct in 100γ percent of the cases. The objective of interval estimation is the determination of the functions $g_1(X)$ and $g_2(X)$ so as to minimize the length $\theta_2 - \theta_1$ of the interval (θ_1, θ_2) subject to the constraint (8-9). If $\hat{\boldsymbol{\theta}}$ is an unbiased estimator of the mean η of $\mathbf{x}$ and the density of $\mathbf{x}$ is symmetrical about η, then the optimum interval is of the form $\eta \pm a$ as in (8-10). In this section, we develop estimates of the commonly used parameters. In the selection of $\hat{\theta}$ we are guided by (8-8) and in all cases we assume that n is large. This assumption is necessary for good estimates and, as we shall see, it simplifies the analysis.

Mean

We wish to estimate the mean η of a random variable $\mathbf{x}$. We use as the point estimate of η the value

$$\bar{x} = \frac{1}{n}\sum x_i$$

of the sample mean $\bar{\mathbf{x}}$ of $\mathbf{x}$. To find an interval estimate, we must determine the distribution of $\bar{\mathbf{x}}$. In general, this is a difficult problem involving multiple convolutions. To simplify it we shall assume that $\bar{\mathbf{x}}$ is normal. This is true if $\mathbf{x}$ is normal and it is approximately true for any $\mathbf{x}$ if n is large (CLT).

KNOWN VARIANCE. Suppose first that the variance σ^2 of $\mathbf{x}$ is known. The normality assumption leads to the conclusion that the point estimator $\bar{\mathbf{x}}$ of η is $N(\eta, \sigma/\sqrt{n})$. Denoting by z_u the u percentile of the standard normal density, we conclude that

$$P\left\{\eta - z_{1-\delta/2}\frac{\sigma}{\sqrt{n}} < \bar{\mathbf{x}} < \eta + z_{1-\delta/2}\frac{\sigma}{\sqrt{n}}\right\} = \mathbb{G}(z_{1-\delta/2}) - \mathbb{G}(-z_{1-\delta/2})$$
$$= 1 - \frac{\delta}{2} - \frac{\delta}{2} \tag{8-10}$$

because $z_u = -z_{1-u}$ and $\mathbb{G}(-z_{1-u}) = \mathbb{G}(z_u) = u$. This yields

$$P\left\{\bar{\mathbf{x}} - z_{1-\delta/2}\frac{\sigma}{\sqrt{n}} < \eta < \bar{\mathbf{x}} + z_{1-\delta/2}\frac{\sigma}{\sqrt{n}}\right\} = 1 - \delta = \gamma \tag{8-11}$$

We can thus state with confidence coefficient γ that η is in the interval $\bar{x} \pm z_{1-\delta/2}\sigma/\sqrt{n}$. The determination of a confidence interval for η thus proceeds as discussed next.

Observe the samples x_i of $\mathbf{x}$ and form their average $\bar{x}$. Select a number $\gamma = 1 - \delta$ and find the standard normal percentile z_u for $u = 1-\delta/2$. Form the interval $\bar{x} \pm z_u\sigma/\sqrt{n}$.

This also holds for discrete-type random variables provided that n is large [see (7-122)]. The choice of the confidence coefficient γ is dictated by two conflicting requirements: If γ is close to 1, the estimate is reliable but the size $2z_u\sigma/\sqrt{n}$ of the confidence interval is large; if γ is reduced, z_u is reduced but the estimate is less reliable. The final choice is a compromise based on the applications. In Table 8-1 we list z_u for the commonly used values of u. The listed values are determined from Table 4-1 by interpolation.

TCHEBYCHEFF INEQUALITY. Suppose now that the distribution of $\bar{\mathbf{x}}$ is not known. To find the confidence interval of η, we shall use (5-88): We replace x by $\bar{x}$ and σ by $\sigma/\sqrt{n}$, and we set $\varepsilon = \sigma/n\delta$. This yields (5-88)

$$P\left\{\bar{\mathbf{x}} - \frac{\sigma}{\sqrt{n\delta}} < \eta < \bar{\mathbf{x}} + \frac{\sigma}{\sqrt{n\delta}}\right\} > 1 - \delta = \gamma \tag{8-12}$$

This shows that the exact γ confidence interval of η is contained in the interval $\bar{x} \pm \sigma/\sqrt{n\delta}$. If, therefore, we claim that η is in this interval, the probability that we are correct is larger than γ. This result holds regardless of the form of $F(x)$ and, surprisingly, it is not very different from the estimate (8-11). Indeed, suppose that $\gamma = 0.95$; in this case, $1/\sqrt{\delta} = 4.47$. Inserting into (8-12), we obtain the interval $\bar{x} \pm 4.47\sigma/\sqrt{n}$. The

TABLE 8-1

$z_{1-u} = -z_u \qquad u = \dfrac{1}{\sqrt{2\pi}}\displaystyle\int_{-\infty}^{z_u} e^{-z^2/2}\,dz$

u	0.90	0.925	0.95	0.975	0.99	0.995	0.999	0.9995
z_u	1.282	1.440	1.645	1.967	2.326	2.576	3.090	3.291

corresponding interval (8-11), obtained under the normality assumption, is $\bar{x} \pm 2\sigma/\sqrt{n}$ because $z_{0.975} \simeq 2$.

UNKNOWN VARIANCE. If σ is unknown, we cannot use (8-11). To estimate η, we form the sample variance

$$s^2 = \frac{1}{n-1}\sum_{i=1}^{n}(x_i - \bar{x})^2 \tag{8-13}$$

This is an unbiased estimate of σ^2 [see (7-23)] and it tends to σ^2 as $n \to \infty$. Hence, for large n, we can use the approximation $s \simeq \sigma$ in (8-11). This yields the approximate confidence interval

$$\bar{x} - z_{1-\delta/2}\frac{s}{\sqrt{n}} < \eta < \bar{x} + z_{1-\delta/2}\frac{s}{\sqrt{n}} \tag{8-14}$$

We shall find an exact confidence interval under the assumption that $\mathbf{x}$ is normal. In this case [see (7-66)] the ratio

$$\frac{\bar{\mathbf{x}} - \eta}{\mathbf{s}/\sqrt{n}} \tag{8-15}$$

has a Student t distribution with $n-1$ degrees of freedom. Denoting by t_u its u percentiles, we conclude that

$$P\left\{-t_u < \frac{\bar{\mathbf{x}} - \eta}{\mathbf{s}/\sqrt{n}} < t_u\right\} = 2u - 1 = \gamma \tag{8-16}$$

This yields the interval

$$\bar{x} - t_{1-\delta/2}\frac{s}{\sqrt{n}} < \eta < \bar{x} + t_{1-\delta/2}\frac{s}{\sqrt{n}} \tag{8-17}$$

In Table 8-2 we list $t_u(n)$ for n from 1 to 20. For $n > 20$, the $t(n)$ distribution is nearly normal with zero mean and variance $n/(n-2)$ (see Prob. 6-75).

EXAMPLE 8-3

▶ The voltage V of a voltage source is measured 25 times. The results of the measurement[2] are the samples $x_i = V + \nu_i$ of the random variable $\mathbf{x} = V + \boldsymbol{\nu}$ and their average equals $\bar{x} = 112$ V. Find the 0.95 confidence interval of V.

(*a*) Suppose that the standard deviation of $\mathbf{x}$ due to the error $\boldsymbol{\nu}$ is $\sigma = 0.4$ V. With $\delta = 0.05$, Table 8-1 yields $z_{0.975} \simeq 2$. Inserting into (8-11), we obtain the interval

$$\bar{x} + z_{0.975}\sigma/\sqrt{n} = 112 \pm 2 \times 0.4/\sqrt{25} = 112 \pm 0.16\text{ V}$$

(*b*) Suppose now that σ is unknown. To estimate it, we compute the sample variance and we find $s^2 = 0.36$. Inserting into (8-14), we obtain the approximate estimate

$$\bar{x} \pm z_{0.975}s/\sqrt{n} = 112 \pm 2 \times 0.6/\sqrt{25} = 112 \pm 0.24\text{ V}$$

Since $t_{0.975}(25) = 2.06$, the exact estimate (8-17) yields 112 ± 0.247 V. ◀

---

[2]In most examples of this chapter, we shall not list all experimental data. To avoid lengthly tables, we shall list only the relevant averages.

TABLE 8-2
Student t Percentiles $t_u(n)$

n \ u	.9	.95	.975	.99	.995
1	3.08	6.31	12.7	31.8	63.7
2	1.89	2.92	4.30	6.97	9.93
3	1.64	2.35	3.18	4.54	5.84
4	1.53	2.13	2.78	3.75	4.60
5	1.48	2.02	2.57	3.37	4.03
6	1.44	1.94	2.45	3.14	3.71
7	1.42	1.90	2.37	3.00	3.50
8	1.40	1.86	2.31	2.90	3.36
9	1.38	1.83	2.26	2.82	3.25
10	1.37	1.81	2.23	2.76	3.17
11	1.36	1.80	2.20	2.72	3.11
12	1.36	1.78	2.18	2.68	3.06
13	1.35	1.77	2.16	2.65	3.01
14	1.35	1.76	2.15	2.62	2.98
15	1.34	1.75	2.13	2.60	2.95
16	1.34	1.75	2.12	2.58	2.92
17	1.33	1.74	2.11	2.57	2.90
18	1.33	1.73	2.10	2.55	2.88
19	1.33	1.73	2.09	2.54	2.86
20	1.33	1.73	2.09	2.53	2.85
22	1.32	1.72	2.07	2.51	2.82
24	1.32	1.71	2.06	2.49	2.80
26	1.32	1.71	2.06	2.48	2.78
28	1.31	1.70	2.05	2.47	2.76
30	1.31	1.70	2.05	2.46	2.75

For $n \geq 30$: $t_u(n) \simeq z_u \sqrt{\dfrac{n}{n-2}}$

In the following three estimates the distribution of $\mathbf{x}$ is specified in terms of a single parameter. We cannot, therefore, use (8-11) directly because the constants η and σ are related.

EXPONENTIAL DISTRIBUTION. We are given a random variable $\mathbf{x}$ with density

$$f(x, \lambda) = \frac{1}{\lambda} e^{-x/\lambda} U(x)$$

and we wish to find the γ confidence interval of the parameter λ. As we know, $\eta = \lambda$ and $\sigma = \lambda$; hence, for large n, the sample mean $\bar{\mathbf{x}}$ of $\mathbf{x}$ is $N(\lambda, \lambda/\sqrt{n})$. Inserting into (8-11), we obtain

$$P\left\{\lambda - z_u \frac{\lambda}{\sqrt{n}} < \bar{\mathbf{x}} < \lambda + z_u \frac{\lambda}{\sqrt{n}}\right\} = \gamma = 2u - 1$$

This yields

$$P\left\{\frac{\overline{\mathbf{x}}}{1+z_u/\sqrt{n}} < \lambda < \frac{\overline{\mathbf{x}}}{1-z_u/\sqrt{n}}\right\} = \gamma \tag{8-18}$$

and the interval $\overline{x}/(1 \pm z_u/\sqrt{n})$ results.

EXAMPLE 8-4 ▶ The time to failure of a light bulb is a random variable $\mathbf{x}$ with exponential distribution. We wish to find the 0.95 confidence interval of λ. To do so, we observe the time to failure of 64 bulbs and we find that their average $\overline{x}$ equals 210 hours. Setting $z_u/\sqrt{n} \simeq 2/\sqrt{64} = 0.25$ into (8-18), we obtain the interval

$$168 < \lambda < 280$$

We thus expect with confidence coefficient 0.95 that the mean time to failure $E\{\mathbf{x}\} = \lambda$ of the bulb is between 168 and 280 hours. ◀

POISSON DISTRIBUTION. Suppose that the random variable $\mathbf{x}$ is Poisson distribution with parameter λ:

$$P\{\mathbf{x} = k\} = e^{-\lambda}\frac{\lambda^k}{k!} \qquad k = 0, 1, \ldots$$

In this case, $\eta = \lambda$ and $\sigma^2 = \lambda$; hence, for large n, the distribution of $\overline{\mathbf{x}}$ is approximately $N(\lambda, \sqrt{\lambda/n})$ [see (7-122)]. This yields

$$P\left\{|\overline{\mathbf{x}} - \lambda| < z_u\sqrt{\frac{\lambda}{n}}\right\} = \gamma$$

The points of the $\overline{x}\lambda$ plane that satisfy the inequality $|\overline{x} - \lambda| < z_u\sqrt{\lambda/n}$ are in the interior of the parabola

$$(\lambda - \overline{x})^2 = \frac{z_u^2}{n}\lambda \tag{8-19}$$

From this it follows that the γ confidence interval of λ is the vertical segment (λ_1, λ_2) of Fig. 8-3, where λ_1 and λ_2 are the roots of the quadratic (8-19).

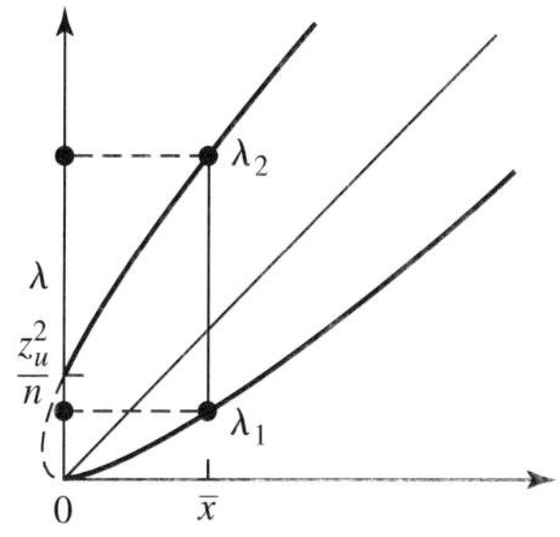

FIGURE 8-3

EXAMPLE 8-5 ▶ The number of particles emitted from a radioactive substance per second is a Poisson random variable $\mathbf{x}$ with parameter λ. We observe the emitted particles x_i in 64 consecutive seconds and we find that $\bar{x} = 6$. Find the 0.95 confidence interval of λ. With $z_u^2/n = 0.0625$, (8-19) yields the quadratic

$$(\lambda - 6)^2 = 0.0625\lambda$$

Solving, we obtain $\lambda_1 = 5.42$, $\lambda_2 = 6.64$. We can thus claim with confidence coefficient 0.95 that $5.42 < \lambda < 6.64$. ◀

PROBABILITY. We wish to estimate the probability $p = P(A)$ of an event A. To do so, we form the zero–one random variable $\mathbf{x}$ associated with this event. As we know, $E\{\mathbf{x}\} = p$ and $\sigma_x^2 = pq$. Thus the estimation of p is equivalent to the estimation of the mean of the random variable $\mathbf{x}$.

We repeat the experiment n times and we denote by k the number of successes of A. The ratio $\bar{x} = k/n$ is the point estimate of p. To find its interval estimate, we form the sample mean $\bar{\mathbf{x}}$ of $\mathbf{x}$. For large n, the distribution of $\bar{\mathbf{x}}$ is approximately $N(p, \sqrt{pq/n})$. Hence

$$P\left\{ |\bar{\mathbf{x}} - p| < z_u \sqrt{\frac{pq}{n}} \right\} = \gamma = 2u - 1$$

The points of the $\bar{x}p$ plane that satisfy the inequality $|\bar{x} - p| < z_u\sqrt{pq/n}$ are in the interior of the ellipse

$$(p - \bar{x})^2 = z_u^2 \frac{p(1-p)}{n} \qquad \bar{x} = \frac{k}{n} \tag{8-20}$$

From this it follows that the γ confidence interval of p is the vertical segment (p_1, p_2) of Fig. 8-4. The endpoints p_1 and p_2 of this segment are the roots of (8-20). For $n > 100$ the following approximation can be used:

$$\begin{matrix} p_1 \\ p_2 \end{matrix} \simeq \bar{x} \pm z_u \sqrt{\frac{\bar{x}(1-\bar{x})}{n}} \qquad p_1 < p < p_2 \tag{8-21}$$

This follows from (8-20) if we replace on the right side the unknown p by its point estimate $\bar{x}$.

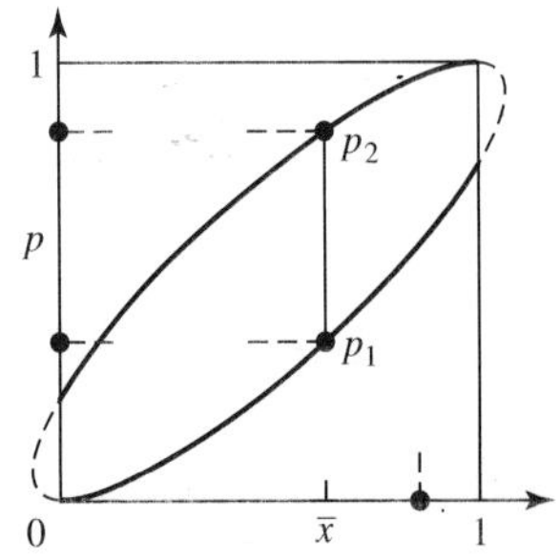

FIGURE 8-4

EXAMPLE 8-6 ▶ In a preelection poll, 500 persons were questioned and 240 responded Republican. Find the 0.95 confidence interval of the probability $p = \{\text{Republican}\}$. In this example, $z_u \simeq 2$, $n = 500$, $\bar{x} = 240/500 = 0.48$, and (8-21) yields the interval 0.48 ± 0.045.

In the usual reporting of the results, the following wording is used: We estimate that 48% of the voters are Republican. The margin of error is $\pm 4.5\%$. This only specifies the point estimate and the confidence interval of the poll. The confidence coefficient (0.95 in this case) is rarely mentioned. ◀

Variance

We wish to estimate the variance $v = \sigma^2$ of a *normal* random variable $\mathbf{x}$ in terms of the n samples x_i of $\mathbf{x}$.

KNOWN MEAN. We assume first that the mean η of $\mathbf{x}$ is known and we use as the point estimator of v the average

$$\hat{\mathbf{v}} = \frac{1}{n}\sum_{i=1}^{n}(\mathbf{x}_i - \eta)^2 \tag{8-22}$$

As we know,

$$E\{\hat{\mathbf{v}}\} = v \qquad \sigma_{\hat{v}}^2 = \frac{2\sigma^4}{n} \xrightarrow[n\to\infty]{} 0$$

Thus $\hat{\mathbf{y}}$ is a consistent estimator of σ^2. We shall find an interval estimate. The random variable $n\hat{\mathbf{v}}/\sigma^2$ has a $\chi^2(n)$ density [see (4-39)]. This density is not symmetrical; hence the interval estimate of σ^2 is not centered at σ^2. To determine it, we introduce two constants c_1 and c_2 such that (Fig. 8-5*a*)

$$P\left\{\frac{n\hat{\mathbf{v}}}{\sigma^2} < c_1\right\} = \frac{\delta}{2} \qquad P\left\{\frac{n\hat{\mathbf{v}}}{\sigma^2} > c_2\right\} = \frac{\delta}{2}$$

This yields $c_1 = \chi^2_{\delta/2}(n)$, $c_2 = \chi^2_{1-\delta/2}(n)$, and the interval

$$\frac{n\hat{v}}{\chi_{1-\delta/2}(n)} < \sigma^2 < \frac{n\hat{v}}{\chi_{\delta/2}(n)} \tag{8-23}$$

results. This interval does not have minimum length. The minimum interval is such that

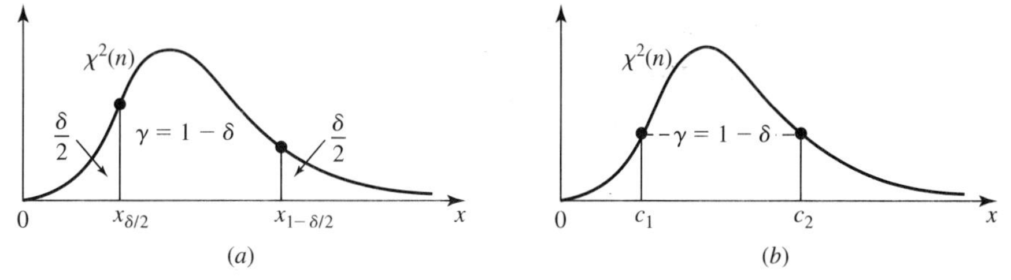

FIGURE 8-5

TABLE 8-3
Chi-square percentiles $\chi_u^2(n)$

n \ u	.005	.01	.025	.05	.1	.9	.95	.975	.99	.995
1	0.00	0.00	0.00	0.00	0.02	2.71	3.84	5.02	6.63	7.88
2	0.01	0.02	0.05	0.10	0.21	4.61	5.99	7.38	9.21	10.60
3	0.07	0.11	0.22	0.35	0.58	6.25	7.81	9.35	11.34	12.84
4	0.21	0.30	0.48	0.71	1.06	7.78	9.49	11.14	13.28	14.86
5	0.41	0.55	0.83	1.15	1.61	9.24	11.07	12.83	15.09	16.75
6	0.68	0.87	1.24	1.64	2.20	10.64	12.59	14.45	16.81	18.55
7	0.99	1.24	1.69	2.17	2.83	12.02	14.07	16.01	18.48	20.28
8	1.34	1.65	2.18	2.73	3.49	13.36	15.51	17.53	20.09	21.96
9	1.73	2.09	2.70	3.33	4.17	14.68	16.92	19.02	21.67	23.59
10	2.16	2.56	3.25	3.94	4.87	15.99	18.31	20.48	23.21	25.19
11	2.60	3.05	3.82	4.57	5.58	17.28	19.68	21.92	24.73	26.76
12	3.07	3.57	4.40	5.23	6.30	18.55	21.03	23.34	26.22	28.30
13	3.57	4.11	5.01	5.89	7.04	19.81	22.36	24.74	27.69	29.82
14	4.07	4.66	5.63	6.57	7.79	21.06	23.68	26.12	29.14	31.32
15	4.60	5.23	6.26	7.26	8.55	22.31	25.00	27.49	30.58	32.80
16	5.14	5.81	6.91	7.96	9.31	23.54	26.30	28.85	32.00	34.27
17	5.70	6.41	7.56	8.67	10.09	24.77	27.59	30.19	33.41	35.72
18	6.26	7.01	8.23	9.39	10.86	25.99	28.87	31.53	34.81	37.16
19	6.84	7.63	8.91	10.12	11.65	27.20	30.14	32.85	36.19	38.58
20	7.43	8.26	9.59	10.85	12.44	28.41	31.41	34.17	37.57	40.00
22	8.6	9.5	11.0	12.3	14.0	30.8	33.9	36.8	40.3	42.8
24	9.9	10.9	12.4	13.8	15.7	33.2	36.4	39.4	43.0	45.6
26	11.2	12.2	13.8	15.4	17.3	35.6	38.9	41.9	45.6	48.3
28	12.5	13.6	15.3	16.9	18.9	37.9	41.3	44.5	48.3	51.0
30	13.8	15.0	16.8	18.5	20.6	40.3	43.8	47.0	50.9	53.7
40	20.7	22.2	24.4	26.5	29.1	51.8	55.8	59.3	63.7	66.8
50	28.0	29.7	32.4	34.8	37.7	63.2	67.5	71.4	76.2	79.5

For $n \geq 50$: $\chi_u^2(n) \simeq \frac{1}{2}(z_u + \sqrt{2n-1})^2$

$f_\chi(c_1) = f_\chi(c_2)$ (Fig. 8-5*b*); however, its determination is not simple. In Table 8-3, we list the percentiles $\chi_u^2(n)$ of the $\chi^2(n)$ distribution.

UNKNOWN MEAN. If η is unknown, we use as the point estimate of σ^2 the sample variance s^2 [see (8-13)]. The random variable $(n-1)\mathbf{s}^2/\sigma^2$ has a $\chi^2(n-1)$ distribution. Hence

$$P\left\{\chi_{\delta/2}^2(n-1) < \frac{(n-1)\mathbf{s}^2}{\sigma^2} < \chi_{1-\delta/2}^2(n-1)\right\} = \gamma$$

This yields the interval

$$\frac{(n-1)s^2}{\chi_{1-\delta/2}^2(n-1)} < \sigma^2 < \frac{(n-1)s^2}{\chi_{\delta/2}(n-1)} \tag{8-24}$$

EXAMPLE 8-7

▶ A voltage source V is measured six times. The measurements are modeled by the random variable $\mathbf{x} = V + \boldsymbol{\nu}$. We assume that the error $\boldsymbol{\nu}$ is $N(0, \sigma)$. We wish to find the 0.95 interval estimate of σ^2.

(a) Suppose first that the source is a known standard with $V = 110$ V. We insert the measured values $x_i = 110 + \nu_i$ of V into (8-22) and we find $\hat{v} = 0.25$. From Table 8-3 we obtain

$$\chi^2_{0.025}(6) = 1.24 \qquad \chi^2_{0.975}(6) = 14.45$$

and (8-23) yields $0.104 < \sigma^2 < 1.2$. The corresponding interval for σ is $0.332 < \sigma < 1.096$ V.

(b) Suppose now that V is unknown. Using the same data, we compute s^2 from (8-13) and we find $s^2 = 0.30$. From Table 8-3 we obtain

$$\chi^2_{0.025}(5) = 0.83 \qquad \chi^2_{0.975}(5) = 12.83$$

and (8-24) yields $0.117 < \sigma^2 < 1.8$. The corresponding interval for σ is $0.342 < \sigma < 1.344$ V. ◀

PERCENTILES. The u percentile of a random variable $\mathbf{x}$ is by definition a number x_u such that $F(x_u) = u$. Thus x_u is the inverse function $F^{(-1)}(u)$ of the distribution $F(x)$ of x. We shall estimate x_u in terms of the samples x_i of $\mathbf{x}$. To do so, we write the n observations x_i in ascending order and we denote by y_k the kth number so obtained. The corresponding random variables $\mathbf{y}_k$ are the order statistics of $\mathbf{x}$ [see (7-13)].

From the definition it follows that $y_k < x_u$ iff at least k of the samples x_i are less than x_u; similarly, $y_{k+r} > x_u$ iff at least $k+r$ of the samples x_i are greater than x_u. Finally, $y_k < x_u < y_{k+r}$ iff at least k and at most $k + r - 1$ of the samples x_i are less than x_u. This leads to the conclusion that the event $\{\mathbf{y}_k < x_u < \mathbf{y}_{k+r}\}$ occurs iff the number of successes of the event $\{\mathbf{x} \le x_u\}$ in n repetitions of the experiment S is at least k and at most $k + r - 1$. And since $P\{\mathbf{x} \le x_u\} = u$, it follows from (3-24) with $p = u$ that

$$P\{\mathbf{y}_k < x_u < \mathbf{y}_{k+r}\} = \sum_{m=k}^{k+r-1} \binom{n}{m} u^m (1-u)^{n-m} \tag{8-25}$$

Using this basic relationship, we shall find the γ confidence interval of x_u for a specific u. To do so, we must find an integer k such that the sum in (8-25) equals γ for the smallest possible r. This is a complicated task involving trial and error. A simple solution can be obtained if n is large. Using the normal approximation (4-99) with $p = nu$, we obtain

$$P\{\mathbf{y}_k < x_u < \mathbf{y}_{k+r}\} \simeq \mathrm{G}\left(\frac{k + r - 0.5 - nu}{\sqrt{nu(1-u)}}\right) - \mathrm{G}\left(\frac{k - 0.5 - nu}{\sqrt{nu(n-u)}}\right) = \gamma$$

This follows from (4.99) with $p = nu$. For a specific γ, r is minimum if nu is near the center of the interval $(k, k+r)$. This yields

$$k \simeq nu - z_{1-\delta/2}\sqrt{nu(1-u)} \qquad k + r \simeq nu + z_{1-\delta/2}\sqrt{nu(1-u)} \tag{8-26}$$

to the nearest integer.

EXAMPLE 8-8

▶ We observe 100 samples of $\mathbf{x}$ and we wish to find the 0.95 confidence interval of the median $x_{0.5}$ of $\mathbf{x}$. With $u = 0.5$, $nu = 50$, $z_{0.975} \simeq 2$, (8-26) yields $k = 40$, $k + r = 60$. Thus we can claim with confidence coefficient 0.95 that the median of $\mathbf{x}$ is between y_{40} and y_{60}. ◀

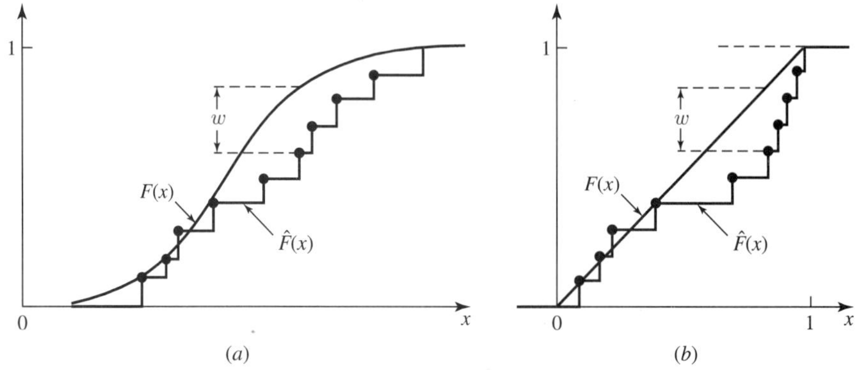

FIGURE 8-6

DISTRIBUTIONS. We wish to estimate the distribution $F(x)$ of a random variable $\mathbf{x}$ in terms of the samples x_i of $\mathbf{x}$. For a specific x, $F(x)$ equals the probability of the event $\{\mathbf{x} \leq x\}$; hence its point estimate is the ratio n_x/n, where n_x is the number of x_i's that do not exceed x. Repeating this for every x, we obtain the empirical estimate

$$\hat{F}(x) = \frac{n_x}{n}$$

of the distribution $F(x)$ [see also (4-3)]. This estimate is a staircase function (Fig. 8-6*a*) with discontinuities at the points x_i.

INTERVAL ESTIMATES. For a *specific* x, the interval estimate of $F(x)$ is obtained from (8-20) with $p = F(x)$ and $\overline{x} = \hat{F}(x)$. Inserting into (8-21), we obtain the interval

$$\hat{F}(x) \pm \frac{z_u}{\sqrt{n}}\sqrt{\hat{F}(x)[1 - \hat{F}(x)]}$$

We can thus claim with confidence coefficient $\gamma = 2u - 1$ that the unknown $F(x)$ is in the above interval. Note that the length of this interval depends on x.

We shall now find an interval estimate $\hat{F}(x) \pm c$ of $F(x)$, where c is a constant. The empirical estimate $\hat{F}(x)$ depends on the samples x_i of $\mathbf{x}$. It specifies, therefore, a family of staircase functions $\hat{\mathbf{F}}(x)$, one for each set of samples x_i. The constant c is such that

$$P\{|\hat{\mathbf{F}}(x) - F(x)| \leq c\} = \gamma \tag{8-27}$$

for *every* x and the γ confidence region of $F(x)$ is the strip $\hat{F}(x) \pm c$. To find c, we form the maximum

$$\mathbf{w} = \max|\hat{\mathbf{F}}(x) - F(x)| \tag{8-28}$$

(least upper bound) of the distance between $\hat{\mathbf{F}}(x)$ and $F(x)$. Suppose that $w = \mathbf{w}(\xi)$ is a specific value of $\mathbf{w}$. From (8-28) it follows that $w < c$ iff $\hat{F}(x) - F(x) < c$ for every x. Hence

$$\gamma = P\{\mathbf{w} \leq c\} = F_w(c)$$

It suffices, therefore, to find the distribution of $\mathbf{w}$. We shall show first that the function $F_w(w)$ does not depend on $F(x)$. As we know [see (5-41)], the random variable $\mathbf{y} = F(\mathbf{x})$

is uniform in the interval (0, 1) for any $F(x)$. The function $y = F(x)$ transforms the points x_i to the points $y_i = F(x_i)$ and the random variable $\mathbf{w}$ to itself (see Fig. 8-6*b*). This shows that $F_w(w)$ does not depend on the form of $F(x)$. For its determination it suffices, therefore, to assume that $\mathbf{x}$ is uniform. However, even with this simplification, it is not simple to find $F_w(w)$. We give next an approximate solution due to *Kolmogorov:*

For large n:

$$F_w(w) \simeq 1 - 2e^{-2nw^2} \tag{8-29}$$

From this it follows that $\gamma = F_w(c) \simeq 1 - 2e^{-2nc^2}$. We can thus claim with confidence coefficient γ that the unknown $F(x)$ is between the curves $\hat{F}(x)+c$ and $\hat{F}(x)-c$, where

$$c = \sqrt{-\frac{1}{2n}\log\frac{1-\gamma}{2}} \tag{8-30}$$

This approximation is satisfactory if $w > 1/\sqrt{n}$.

Bayesian Estimation

We return to the problem of estimating the parameter θ of a distribution $F(x, \theta)$. In our earlier approach, we viewed θ as an unknown constant and the estimate was based solely on the observed values x_i of the random variable $\mathbf{x}$. This approach to estimation is called *classical*. In certain applications, θ is not totally unknown. If, for example, θ is the probability of six in the die experiment, we expect that its possible values are close to $1/6$ because most dice are reasonably fair. In *bayesian* statistics, the available prior information about θ is used in the estimation problem. In this approach, the unknown parameter θ is viewed as the value of a random variable $\boldsymbol{\theta}$ and the distribution of $\mathbf{x}$ is interpreted as the conditional distribution $F_x(x \mid \theta)$ of $\mathbf{x}$ assuming $\boldsymbol{\theta} = \theta$. The prior information is used to assign somehow a density $f_\theta(\theta)$ to the random variable $\boldsymbol{\theta}$, and the problem is to estimate the value θ of $\boldsymbol{\theta}$ in terms of the observed values x_i of $\mathbf{x}$ and the density of $\boldsymbol{\theta}$. The problem of estimating the unknown parameter θ is thus changed to the problem of estimating the value θ of the random variable $\boldsymbol{\theta}$. Thus, in bayesian statistics, estimation is changed to prediction.

We shall introduce the method in the context of the following problem. We wish to estimate the inductance θ of a coil. We measure θ n times and the results are the samples $x_i = \theta + \nu_i$ of the random variable $\mathbf{x} = \theta + \boldsymbol{\nu}$. If we interpret θ as an unknown number, we have a classical estimation problem. Suppose, however, that the coil is selected from a production line. In this case, its inductance θ can be interpreted as the value of a random variable $\boldsymbol{\theta}$ modeling the inductances of all coils. This is a problem in bayesian estimation. To solve it, we assume first that no observations are available, that is, that the specific coil has not been measured. The available information is now the *prior* density $f_\theta(\theta)$ of $\boldsymbol{\theta}$, which we assume known and our problem is to find a constant $\hat{\theta}$ close in some sense to the unknown θ, that is, to the true value of the inductance of the particular coil. If we use the least mean squares (LMS) criterion for selecting $\hat{\theta}$, then [see (6-236)]

$$\hat{\theta} = E\{\boldsymbol{\theta}\} = \int_{-\infty}^{\infty} \theta f_\theta(\theta)\, d\theta$$

To improve the estimate, we measure the coil n times. The problem now is to estimate θ in terms of the n samples x_i of $\mathbf{x}$. In the general case, this involves the estimation of the value θ of a random variable $\boldsymbol{\theta}$ in terms of the n samples x_i of $\mathbf{x}$. Using again the MS criterion, we obtain

$$\hat{\theta} = E\{\boldsymbol{\theta} \mid X\} = \int_{-\infty}^{\infty} \theta f_\theta(\theta \mid X)\, d\theta \tag{8-31}$$

[see (7-89)] where $X = [x_1, \ldots, x_n]$ and

$$f_\theta(\theta \mid X) = \frac{f(X \mid \theta)}{f(X)} f_\theta(\theta) \tag{8-32}$$

In (8-32), $f(X \mid \theta)$ is the conditional density of the n random variables $\mathbf{x}_i$ assuming $\boldsymbol{\theta} = \theta$. If these random variables are conditionally independent, then

$$f(X \mid \theta) = f(x_1 \mid \theta) \cdots f(x_n \mid \theta) \tag{8-33}$$

where $f(x \mid \theta)$ is the conditional density of the random variable $\mathbf{x}$ assuming $\boldsymbol{\theta} = \theta$. These results hold in general. In the measurement problem, $f(x \mid \theta) = f_\nu(x - \theta)$.

We conclude with the clarification of the meaning of the various densities used in bayesian estimation, and of the underlying model, in the context of the measurement problem. The density $f_\theta(\theta)$, called *prior* (prior to the measurements), models the inductances of all coils. The density $f_\theta(\theta \mid X)$, called *posterior* (after the measurements), models the inductances of all coils of measured inductance x. The conditional density $f_x(x \mid \theta) = f_\nu(x - \theta)$ models all measurements of a particular coil of true inductance θ. This density, considered as a function of θ, is called the *likelihood* function. The unconditional density $f_x(x)$ models all measurements of all coils. Equation (8-33) is based on the reasonable assumption that the measurements of a given coil are independent.

The bayesian model is a product space $S = S_\theta \times S_x$, where S_θ is the space of the random variable $\boldsymbol{\theta}$ and S_x is the space of the random variable $\mathbf{x}$. The space S_θ is the space of all coils and S_x is the space of all measurements of a particular coil. Finally, S is the space of all measurements of all coils. The number θ has two meanings: It is the value of the random variable $\boldsymbol{\theta}$ in the space S_θ; it is also a parameter specifying the density $f(x \mid \theta) = f_\nu(x - \theta)$ of the random variable $\mathbf{x}$ in the space S_x.

EXAMPLE 8-9

▶ Suppose that $\mathbf{x} = \theta + \boldsymbol{\nu}$ where $\boldsymbol{\nu}$ is an $N(0, \sigma)$ random variable and θ is the value of an $N(\theta_0, \sigma_0)$ random variable $\boldsymbol{\theta}$ (Fig. 8-7). Find the bayesian estimate $\hat{\theta}$ of θ.

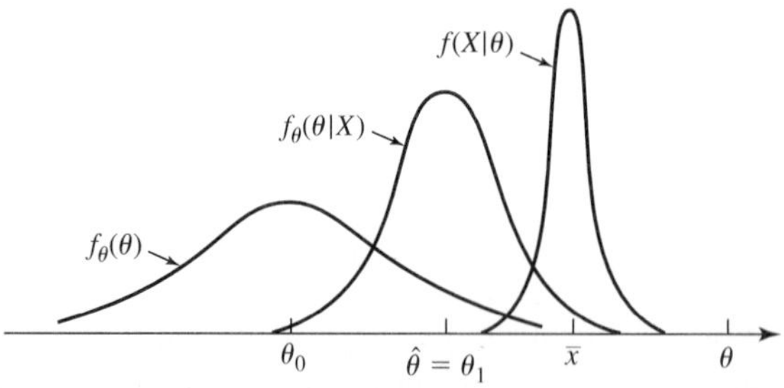

FIGURE 8-7

The density $f(x \mid \theta)$ of $\mathbf{x}$ is $N(\theta, \sigma)$. Inserting into (8-32), we conclude that (see Prob. 8-37) the function $f_\theta(\theta \mid X)$ is $N(\theta_1, \sigma_1)$ where

$$\sigma_1^2 = \frac{\sigma^2}{n} \times \frac{\sigma_0^2}{\sigma_0^2 + \sigma^2/n} \qquad \theta_1 = \frac{\sigma_1^2}{\sigma_0^2}\theta_0 + \frac{n\sigma_1^2}{\sigma^2}\overline{x}$$

From this it follows that $E\{\boldsymbol{\theta} \mid X\} = \theta_1$; in other words, $\hat{\theta} = \theta_1$.

Note that the classical estimate of θ is the average $\overline{x}$ of x_i. Furthermore, its prior estimate is the constant θ_0. Hence $\hat{\theta}$ is the weighted average of the prior estimate θ_0 and the classical estimate $\overline{x}$. Note further that as n tends to ∞, $\sigma_1 \to 0$ and $n\sigma_1^2/\sigma^2 \to 1$; hence $\hat{\theta}$ tends to $\overline{x}$. Thus, as the number of measurements increases, the bayesian estimate $\hat{\theta}$ approaches the classical estimate $\overline{x}$; the effect of the prior becomes negligible. ◀

We present next the estimation of the probability $p = P(A)$ of an event A. To be concrete, we assume that A is the event "heads" in the coin experiment. The result is based on Bayes' formula [see (4-81)]

$$f(x \mid M) = \frac{P(M \mid x) f(x)}{\int_{-\infty}^{\infty} P(M \mid x) f(x)\,dx} \tag{8-34}$$

In bayesian statistics, p is the value of a random variable $\mathbf{p}$ with prior density $f(p)$. In the absence of any observations, the LMS estimate $\hat{p}$ is given by

$$\hat{p} = \int_0^1 p f(p)\,dp \tag{8-35}$$

To improve the estimate, we toss the coin at hand n times and we observe that "heads" shows k times. As we know,

$$P\{M \mid \mathbf{p} = p\} = p^k q^{n-k} \qquad M = \{k \text{ heads}\}$$

Inserting into (8-34), we obtain the posterior density

$$f(p \mid M) = \frac{p^k q^{n-k} f(p)}{\int_0^1 p^k q^{n-k} f(p)\,dp} \tag{8-36}$$

Using this function, we can estimate the probability of heads at the next toss of the coin. Replacing $f(p)$ by $f(p \mid M)$ in (8-35), we conclude that the updated estimate $\hat{p}$ of p is the conditional estimate of $\mathbf{p}$ assuming M:

$$\int_0^1 p f(p \mid M)\,dp \tag{8-37}$$

Note that for large n, the factor $\varphi(p) = p^k(1-p)^{n-k}$ in (8-36) has a sharp maximum at $p = k/n$. Therefore, if $f(p)$ is smooth, the product $f(p)\varphi(p)$ is concentrated near k/n (Fig. 8-8a). However, if $f(p)$ has a sharp peak at $p = 0.5$ (this is the case for reasonably fair coins), then for moderate values of n, the product $f(p)\varphi(p)$ has two maxima: one near k/n and the other near 0.5 (Fig. 8-8b). As n increases, the sharpness of $\varphi(p)$ prevails and $f(p \mid M)$ is maximum near k/n (Fig. 8-8c). (See also Example 4-19.)

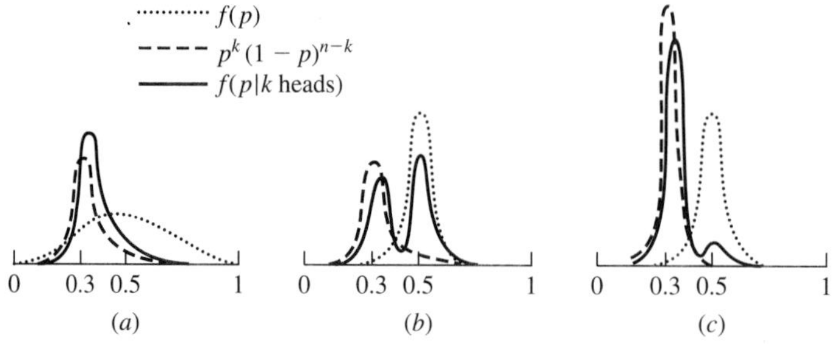

FIGURE 8-8

Note Bayesian estimation is a controversial subject. The controversy has its origin on the dual interpretation of the physical meaning of probability. In the first interpretation, the probability $P(A)$ of an event A is an "objective" measure of the relative frequency of the occurrence of A in a large number of trials. In the second interpretation, $P(A)$ is a "subjective" measure of our state of knowledge concerning the occurrence of A in a single trial. This dualism leads to two different interpretations of the meaning of parameter estimation. In the coin experiment, these interpretations take the following form:

In the classical (objective) approach, p is an unknown number. To estimate its value, we toss the coin n times and use as an estimate of p the ratio $\hat{p} = k/n$. In the bayesian (subjective) approach, p is also an unknown number, however, we interpret it as the value of a random variable $\boldsymbol{\theta}$, the density of which we determine using whatever knowledge we might have about the coin. The resulting estimate of p is determined from (8-37). If we know nothing about p, we set $f(p) = 1$ and we obtain the estimate $\hat{p} = (k+1)/(n+2)$ [see (4-88)]. Conceptually, the two approaches are different. However, practically, they lead in most estimates of interest to similar results if the size n of the available sample is large. In the coin problem, for example, if n is large, k is also large with high probability; hence $(k+1)/(n+2) \simeq k/n$. If n is not large, the results are different but unreliable for either method. The mathematics of bayesian estimation are also used in classical estimation problems if θ is the value of a random variable the density of which can be determined objectively in terms of averages. This is the case in the problem considered in Example 8-9.

Next, we examine the classical parameter estimation problem in some greater detail.

8-3 PARAMETER ESTIMATION

Let $\mathbf{x} = (\mathbf{x}_1, \mathbf{x}_2, \ldots, \mathbf{x}_n)$ denote n random variables representing observations $\mathbf{x}_1 = x_1$, $\mathbf{x}_2 = x_2, \ldots, \mathbf{x}_n = x_n$, and suppose we are interested in estimating an unknown nonrandom parameter θ based on this data. For example, assume that these random variables have a common normal distribution $N(\mu, \sigma^2)$, where μ is unknown. We make n observations $x_1, x_2, \ldots, x_n$. What can we say about μ given these observations? Obviously the underlying assumption is that the measured data has something to do with the unknown parameter θ. More precisely, we assume that the joint probability density function (p.d.f.) of $\mathbf{x}_1, \mathbf{x}_2, \ldots, \mathbf{x}_n$ given by $f_x(x_1, x_2, \ldots, x_n; \theta)$ depends on θ. We form an estimate for the unknown parameter θ as a function of these data samples. The problem of point estimation is to pick a one-dimensional statistic $T(\mathbf{x}_1, \mathbf{x}_2, \ldots, \mathbf{x}_n)$ that best estimates θ in some optimum sense. The method of maximum likelihood (ML) is often used in this context.

Maximum Likelihood Method

The principle of maximum likelihood assumes that the sample data set is representative of the population $f_x(x_1, x_2, \ldots, x_n; \theta)$, and chooses that value for θ that most

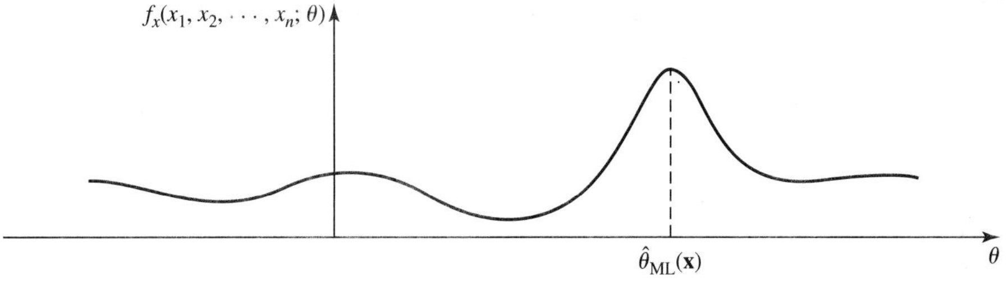

FIGURE 8-9

likely caused the observed data to occur, i.e., once observations $x_1, x_2, \ldots, x_n$ are given, $f_x(x_1, x_2, \ldots, x_n; \theta)$ is a function of θ alone, and the value of θ that maximizes the above p.d.f. is the most likely value for θ, and it is chosen as its ML estimate $\hat{\theta}_{ML}(\mathbf{x})$ (see Fig. 8-9).

Given $\mathbf{x}_1 = x_1, \mathbf{x}_2 = x_2, \ldots, \mathbf{x}_n = x_n$, the joint p.d.f. $f_x(x_1, x_2, \ldots, x_n; \theta)$ is defined to be the likelihood function, and the ML estimate can be determined either from the likelihood equation as

$$\hat{\theta}_{ML} = \sup_{\theta} f_x(x_1, x_2, \ldots, x_n; \theta) \tag{8-38}$$

or by using the log-likelihood function

$$L(x_1, x_2, \ldots, x_n; \theta) \triangleq \log f_x(x_1, x_2, \ldots, x_n; \theta) \tag{8-39}$$

If $L(x_1, x_2, \ldots, x_n; \theta)$ is differentiable and a supremum $\hat{\theta}_{ML}$ exists, then that must satisfy the equation

$$\left.\frac{\partial \log f_x(x_1, x_2, \ldots, x_n; \theta)}{\partial \theta}\right|_{\theta=\hat{\theta}_{ML}} = 0 \tag{8-40}$$

EXAMPLE 8-10

▶ Let $\mathbf{x}_1, \mathbf{x}_2, \ldots, \mathbf{x}_n$ be i.i.d. uniformly distributed random variables in the interval $(0, \theta)$ with common p.d.f.

$$f_{x_i}(x_i; \theta) = \frac{1}{\theta} \qquad 0 < x_i < \theta \tag{8-41}$$

where θ is unknown. The likelihood function in this case is given by

$$\begin{aligned} f(\mathbf{x}_1 = x_1, \mathbf{x}_2 = x_2, \ldots, \mathbf{x}_n = x_n; \theta) &= \frac{1}{\theta^n}, \qquad 0 < x_i \le \theta \qquad i = 1 \to n \\ &= \frac{1}{\theta^n} \qquad 0 \le \max(x_1, x_2, \ldots, x_n) \le \theta \end{aligned} \tag{8-42}$$

The likelihood function is maximized by the minimum value of θ, and since $\theta \ge \max(x_1, x_2, \ldots, x_n)$, we get

$$\hat{\theta}_{ML}(\mathbf{x}) = \max(\mathbf{x}_1, \mathbf{x}_2, \ldots, \mathbf{x}_n) \tag{8-43}$$

to be the ML estimate for θ. ◀

EXAMPLE 8-11 ▶ Let $\mathbf{x}_1, \mathbf{x}_2, \ldots, \mathbf{x}_n$ be i.i.d. gamma random variables with unknown parameters α and β. Thus $x_i \geq 0$, and

$$f_x(x_1, x_2, \ldots, x_n ; \alpha, \beta) = \frac{\beta^{n\alpha}}{(\Gamma(\alpha))^n} \prod_{i=1}^{n} x_i^{\alpha-1} e^{-\beta \sum_{i=1}^{n} x_i} \tag{8-44}$$

This gives the log-likelihood function to be

$$\begin{aligned} L(x_1, x_2, \ldots, x_n; \alpha, \beta) &= \log f_x(x_1, x_2, \ldots, x_n ; \alpha, \beta) \\ &= n\alpha \log\beta - n \log\Gamma(\alpha) + (\alpha - 1)\left(\sum_{i=1}^{n} \log x_i\right) - \beta \sum_{i=1}^{n} x_i \end{aligned} \tag{8-45}$$

Differentiating L with respect to α and β we get

$$\frac{\partial L}{\partial \alpha} = n \log\beta - \frac{n}{\Gamma(\alpha)}\Gamma'(\alpha) + \sum_{i=1}^{n} \log x_i \bigg|_{\alpha,\beta=\hat{\alpha},\hat{\beta}} = 0 \tag{8-46}$$

$$\frac{\partial L}{\partial \beta} = \frac{n\alpha}{\beta} - \sum_{i=1}^{n} x_i \bigg|_{\alpha,\beta=\hat{\alpha},\hat{\beta}} = 0 \tag{8-47}$$

Thus

$$\hat{\beta}_{\text{ML}} x_i = \frac{\hat{\alpha}_{\text{ML}}}{\frac{1}{n}\sum_{i=1}^{n} x_i} \tag{8-48}$$

and substituting (8-48) into (8-46), it gives

$$\log\hat{\alpha}_{\text{ML}} - \frac{\Gamma'(\hat{\alpha}_{\text{ML}})}{\Gamma(\hat{\alpha}_{\text{ML}})} = \log\left(\frac{1}{n}\sum_{i=1}^{n} x_i\right) - \frac{1}{n}\sum_{i=1}^{n} \log x_i \tag{8-49}$$

Notice that (8-49) is highly nonlinear in $\hat{\alpha}_{\text{ML}}$. ◀

In general the (log)-likelihood function can have more than one solution, or no solutions at all. Further, the (log)-likelihood function may not be even differentiable, or it can be extremely complicated to solve explicitly [see (8-49)].

From (8-38), the maximum likelihood method tries to find the mode of the p.d.f. $f_x(x_1, x_2, \ldots, x_n ; \theta)$. Since mode is in general a poorer estimate than the mean or the median, small-sample properties of this method are usually poor. For large sample size, the mode approaches the mean and median, and as we shall see later the ML estimator has many desirable large-sample properties [see (8-284)–(8-289)].

To study the problem of nonrandom parameter estimation in some systematic manner, it is useful to introduce the notion of sufficient statistic and other related concepts.

Sufficient Statistic

A function $T(\mathbf{x}) = T(\mathbf{x}_1, \mathbf{x}_2, \ldots, \mathbf{x}_n)$ is said to be a sufficient statistic for θ, if $T(\mathbf{x})$ contains all information about θ that is contained in the data set, i.e., given the probability density function

$$P\{\mathbf{x}_1 = x_1, \mathbf{x}_2 = x_2, \ldots, \mathbf{x}_n = x_n; \theta\} \tag{8-50}$$

if

$$P\{\mathbf{x}_1 = x_1, \mathbf{x}_2 = x_2, \ldots, \mathbf{x}_n = x_n; \theta \mid T(\mathbf{x}) = t\} \tag{8-51}$$

does not depend on θ for all possible values of x_i, then $T(\mathbf{x})$ must contain all information about θ that is contained in the sample set $\mathbf{x}$. Once $T(\mathbf{x})$ is known, the sample set $\mathbf{x}$ contains no further information about θ, and hence it is called the sufficient statistic for θ.

Thus a statistic $T(\mathbf{x})$ is said to be sufficient for the parameter θ, if the conditional p.d.f.

$$f_{x \mid T}(x_1, x_2, \ldots, x_n; \theta \mid T(\mathbf{x}) = t) \tag{8-52}$$

does not depend on θ.

Notice that the outcome $\mathbf{x}_1, \mathbf{x}_2, \ldots, \mathbf{x}_n$ is always trivially sufficient, but more interestingly if a function $T(\mathbf{x})$ is sufficient, we need to concentrate only on T since it exhausts all the information that the sample set $\mathbf{x}$ has about θ. The following example shows a statistics that is sufficient.

EXAMPLE 8-12 ▶ Let $\mathbf{x}_1, \mathbf{x}_2, \ldots, \mathbf{x}_n$ be i.i.d. $\sim P(\lambda)$, and consider the function

$$T(\mathbf{x}_1, \mathbf{x}_2, \ldots, \mathbf{x}_n) = \sum_{i=1}^{n} \mathbf{x}_i \tag{8-53}$$

Then

$$P\{\mathbf{x}_1 = x_1, \mathbf{x}_2 = x_2, \ldots, \mathbf{x}_n = x_n \mid T(\mathbf{x}) = t\}$$

$$= \begin{cases} \dfrac{P\{\mathbf{x}_1 = x_1, \mathbf{x}_2 = x_2, \ldots, \mathbf{x}_n = t - (x_1 + x_2 + \cdots + x_{n-1})\}}{P\{T = \sum_{i=1}^{n} x_i\}} & \text{if } \ t = \sum_{i=1}^{n} x_i \\ 0 & t \neq \sum_{i=1}^{n} x_i \end{cases} \tag{8-54}$$

But as we know $T(\mathbf{x}) \sim P(n\lambda)$. Thus

$$P\left(T = \sum_{i=1}^{n} x_i\right) = e^{-n\lambda} \frac{(n\lambda)^{\sum_{i=1}^{n} x_i}}{\left(\sum_{i=1}^{n} x_i\right)!} = e^{-n\lambda} \frac{(n\lambda)^t}{t!} \tag{8-55}$$

Thus

$$\frac{P\{\mathbf{x}_1 = x_1, \mathbf{x}_2 = x_2, \ldots, \mathbf{x}_n = t - (x_1 + x_2 + \cdots + x_{n-1})\}}{P\{T = \sum_{i=1}^{n} x_i\}}$$

$$= \frac{e^{-\lambda} \frac{\lambda^{x_1}}{(x_1)!} e^{-\lambda} \frac{\lambda^{x_2}}{(x_2)!} \cdots e^{-\lambda} \frac{\lambda^{t-x_1-x_2-\cdots-x_{n-1}}}{(t-x_1-x_2-\cdots-x_{n-1})!}}{e^{-n\lambda} \frac{(n\lambda)^t}{t!}} = \frac{\left(\sum_{i=1}^{n} x_i\right)!}{x_1! x_2! \cdots x_n! n^t} \tag{8-56}$$

is independent of $\theta = \lambda$. Thus, $T(\mathbf{x}) = \sum_{i=1}^{n} \mathbf{x}_i$ is sufficient for λ. ◀

EXAMPLE 8-13

A STATISTICS THAT IS NOT SUFFICIENT

▶ Let $\mathbf{x}_1, \mathbf{x}_2$ be i.i.d. $\sim N(\mu, \sigma^2)$. Suppose μ is the only unknown parameter, and consider some arbitrary function of $\mathbf{x}_1$ and $\mathbf{x}_2$, say

$$T_1 = \mathbf{x}_1 + 2\mathbf{x}_2 \tag{8-57}$$

Clearly, $T_1 \sim N(3\mu, 5\sigma^2)$, and

$$\begin{aligned}
f(\mathbf{x}_1, \mathbf{x}_2 \mid T_1) &= \frac{f(\mathbf{x}_1, \mathbf{x}_2, T = \mathbf{x}_1 + 2\mathbf{x}_2)}{f(T_1 = \mathbf{x}_1 + 2\mathbf{x}_2)} \\
&= \begin{cases} \dfrac{f_{x_1,x_2}(x_1, x_2)}{f_{x_1+2x_2}(x_1 + 2x_2)} & \text{if} \quad T_1 = x_1 + 2x_2 \\ 0 & T_1 \neq x_1 + 2x_2 \end{cases} \\
&= \frac{\frac{1}{2\pi\sigma^2} e^{-[(x_1-\mu)^2+(x_2-\mu)^2]/2\sigma^2}}{\frac{1}{\sqrt{10\pi\sigma^2}} e^{-(x_1+2x_2-3\mu)^2/10\sigma^2}} \\
&= \sqrt{\frac{5}{2\pi\sigma^2}} \frac{e^{-\left[x_1^2+x_2^2-2\mu(x_1+x_2)+2\mu^2\right]/2\sigma^2}}{e^{-[(x_1+2x_2)^2-6\mu(x_1+2x_2)+9\mu^2)]/10\sigma^2}} \\
&= \sqrt{\frac{5}{2\pi\sigma^2}} e^{-[(2x_1-x_2)^2-2\mu(2x_1-x_2)+\mu^2]/10\sigma^2}
\end{aligned} \tag{8-58}$$

is *not* independent of μ. Hence $T_1 = \mathbf{x}_1 + 2\mathbf{x}_2$ is *not sufficient* for μ.

However $T(\mathbf{x}) = \mathbf{x}_1 + \mathbf{x}_2$ is sufficient for μ, since $T(\mathbf{x}_1, \mathbf{x}_2) = \mathbf{x}_1 + \mathbf{x}_2 \sim N(2\mu, 2\sigma^2)$ and

$$\begin{aligned}
f(\mathbf{x}_1, \mathbf{x}_2 \mid T) &= \frac{f(\mathbf{x}_1, \mathbf{x}_2, T = \mathbf{x}_1 + \mathbf{x}_2)}{f(T_1 = \mathbf{x}_1 + \mathbf{x}_2)} \\
&= \begin{cases} \dfrac{f_{x_1,x_2}(x_1, x_2)}{f_{x_1+x_2}(x_1 + x_2)}, & \text{if} \quad T = x_1 + x_2 \\ 0 & \text{otherwise} \end{cases} \\
&= \frac{f_{x_1,x_2}(x_1, x_2)}{f_{x_1+x_2}(x_1 + x_2)} = \frac{\frac{1}{2\pi\sigma^2} e^{-[(x_1-\mu)^2+(x_2-\mu)^2]/2\sigma^2}}{\frac{1}{\sqrt{4\pi\sigma^2}} e^{-(x_1+x_2-2\mu)^2/4\sigma^2}} \\
&= \frac{1}{\sqrt{\pi\sigma^2}} \frac{e^{-\left(x_1^2+x_2^2-2\mu(x_1+x_2)+2\mu^2\right)/2\sigma^2}}{e^{-[(x_1+x_2)^2/2-2\mu(x_1+x_2)+2\mu^2]/2\sigma^2}} \\
&= \frac{1}{\sqrt{\pi\sigma^2}} e^{-(x_1-x_2)^2/4\sigma^2}
\end{aligned} \tag{8-59}$$

is independent of μ. Hence $T(\mathbf{x}_1, \mathbf{x}_2) = \mathbf{x}_1 + \mathbf{x}_2$ is sufficient for μ. ◀

How does one find the sufficient statistics in general? Fortunately, rather than go through an exhaustive trial and error procedure as in Example 8-13, sufficient statistic can be determined from the following theorem.

THE FACTORIZATION THEOREM

Let $\mathbf{x}_1, \mathbf{x}_2, \ldots, \mathbf{x}_n$ be a set of discrete random variables with p.m.f. $P(\mathbf{x}_1 = x_1, \mathbf{x}_2 = x_2, \ldots, \mathbf{x}_n = x_n; \theta)$. Then $T(\mathbf{x}_1, \mathbf{x}_2, \ldots, \mathbf{x}_n)$ is sufficient if and only if

$$P\{\mathbf{x}_1 = x_1, \mathbf{x}_2 = x_2, \ldots, \mathbf{x}_n = x_n; \theta\} = h(x_1, x_2, \ldots, x_n) g_\theta(T(\mathbf{x}) = t) \quad (8\text{-}60)$$

where $h(\cdot)$ is a nonnegative function of $\mathbf{x}_1, \mathbf{x}_2, \ldots, \mathbf{x}_n$ that does not depend on θ, and $g(\cdot)$ is a nonnegative function of θ and T only (and *not* of $\mathbf{x}_1, \mathbf{x}_2, \ldots, \mathbf{x}_n$ in any other manner).

Proof. Sufficiency. Suppose

$$P\{\mathbf{x}; \theta\} = h(\mathbf{x}) g_\theta(T) \quad (8\text{-}61)$$

Then

$$P\{\mathbf{x} = x\,; \theta \mid T(\mathbf{x}) = t\} = \begin{cases} \dfrac{P\{\mathbf{x} = x; \theta,\ T = t\}}{P\{T = t\}} & T = t \\ 0 & T \neq t \end{cases}$$

$$= \begin{cases} \dfrac{P\{\mathbf{x} = x; \theta\}}{P\{T = t\}} & T = t \\ 0 & T \neq t \end{cases} \quad (8\text{-}62)$$

If $T(\mathbf{x}) = t$, then

$$P\{T(\mathbf{x}) = t\} = \sum_{T\{\mathbf{x}\}=t} P\{\mathbf{x} = x; \theta\} = \sum_{T(\mathbf{x})=t} h(\mathbf{x}) g_\theta(T = t)$$

$$= g_\theta(T = t) \sum_{T(\mathbf{x})=t} h(\mathbf{x}) \quad (8\text{-}63)$$

Hence

$$P\{\mathbf{x} = x; \theta \mid T(\mathbf{x}) = t\} = \begin{cases} \dfrac{h(\mathbf{x})}{\sum_{T(\mathbf{x})=t} h(\mathbf{x})} & T(\mathbf{x}) = t \\ 0 & T(x) \neq t \end{cases} \quad (8\text{-}64)$$

which does not depend on θ. Hence T is sufficient.

Conversely, suppose $T(\mathbf{x}) = t$ is sufficient. Then

$$P\{\mathbf{x} = x; \theta\} = P\{\mathbf{x} = x; \theta, T(\mathbf{x}) = t\} \qquad \text{if} \quad T(\mathbf{x}) = t$$

$$= P\{\mathbf{x} = x; \theta \mid T(\mathbf{x}) = t\}\, P\{T(\mathbf{x}) = t\} \quad (8\text{-}65)$$

Since T is sufficient, $P\{\mathbf{x} = x; \theta \mid T = t\}$ is independent of θ, say $h(\mathbf{x})$. Thus

$$P\{\mathbf{x} = x; \theta\} = h(\mathbf{x}) P(T = t) \quad (8\text{-}66)$$

But

$$P\{T(\mathbf{x}) = t\} = \sum_{T(\mathbf{x})=t} P\{\mathbf{x} = x; \theta\} = g_\theta(T = t) \quad (8\text{-}67)$$

Thus if T is sufficient, we obtain

$$P\{\mathbf{x} = x; \theta\} = h(\mathbf{x}) g_\theta(T = t) \quad (8\text{-}68)$$

that agrees with (8-60).

Next, we consider two examples to illustrate the usefulness of this theorem in finding the sufficient statistic.

EXAMPLE 8-14

▶ Let $\mathbf{x}_1, \mathbf{x}_2, \ldots, \mathbf{x}_n$ be i.i.d. $\sim N(\mu, \sigma^2)$. Thus

$$f(\mathbf{x}; \mu, \sigma^2) = \left(\frac{1}{2\pi\sigma^2}\right)^{n/2} e^{-\sum_{i=1}^n (x_i-\mu)^2/2\sigma^2}$$

$$= \left(\frac{1}{2\pi\sigma^2}\right)^{n/2} e^{-\left(\sum_{i=1}^n x_i^2/2\sigma^2 - \mu\sum_{i=1}^n x_i/\sigma^2 + n\mu^2/2\sigma^2\right)} \tag{8-69}$$

Hence the set $(\sum_{i=1}^n \mathbf{x}_i, \sum_{i=1}^n \mathbf{x}_i^2)$ is sufficient for the parameter pair (μ, σ^2). Note that if σ^2 is known, then $\sum_{i=1}^n \mathbf{x}_i$ is sufficient for μ; and similarily if μ is known then $\sum_{i=1}^n \mathbf{x}_i^2$ is sufficient for the unknown parameter σ^2. ◀

EXAMPLE 8-15

▶ Let $\mathbf{x}_1, \mathbf{x}_2, \ldots, \mathbf{x}_n$ be i.i.d. $\sim U(0, \theta)$. Then

$$f_x(x_1, x_2, \ldots, x_n; \theta) = \begin{cases} \dfrac{1}{\theta^n} & 0 < x_1, x_2, \ldots, x_n < \theta \\ 0 & \text{otherwise} \end{cases} \tag{8-70}$$

Define

$$I_{(a,b)} = \begin{cases} 1 & b > a \\ 0 & \text{otherwise} \end{cases} \tag{8-71}$$

and the random variables

$$\max \mathbf{x} \triangleq \max(\mathbf{x}_1, \mathbf{x}_2, \ldots, \mathbf{x}_n)$$

$$\min \mathbf{x} \triangleq \min(\mathbf{x}_1, \mathbf{x}_2, \ldots, \mathbf{x}_n)$$

Then

$$f_x(x_1, x_2, \ldots, x_n; \theta) = \frac{1}{\theta^n} I_{(0, \min \mathbf{x})} I_{(\max \mathbf{x}, \theta)} = h(\mathbf{x}) g_\theta(T) \tag{8-72}$$

where

$$h(\mathbf{x}) \triangleq I_{(0, \min \mathbf{x})} \tag{8-73}$$

$$g_\theta(T) \triangleq \frac{1}{\theta^n} I_{(\max \mathbf{x}, \theta)} \tag{8-74}$$

Thus

$$T(\mathbf{x}_1, \mathbf{x}_2, \ldots, \mathbf{x}_n) = \max(\mathbf{x}_1, \mathbf{x}_2, \ldots, \mathbf{x}_n) \tag{8-75}$$

is sufficient for θ for uniformly distributed random variables.

In a similar manner, if $\mathbf{x}_i$ represents a uniformly distributed discrete random variable with

$$P\{\mathbf{x}_i = k\} = 1/N \qquad k = 1, 2, \ldots, N \tag{8-76}$$

and $\mathbf{x}_1, \mathbf{x}_2, \ldots, \mathbf{x}_n$ are i.i.d., then using the same argument,

$$T(\mathbf{x}) = \max(\mathbf{x}_1, \mathbf{x}_2, \ldots, \mathbf{x}_n) \tag{8-77}$$

also represents the sufficient statistic for the unknown discrete parameter N. ◀

Next, we consider important properties that estimators of nonrandom parameters should posses ideally. To start with, it is desirable that estimates for unknown nonrandom parameters be unbiased and possess low variances.

Unbiased Estimators

Recall that an estimator $T(\mathbf{x})$ is said to be an unbiased estimator for θ, if

$$E[T(\mathbf{x})] = \theta \tag{8-78}$$

If $T_1(\mathbf{x})$ and $T_2(\mathbf{x})$ are both unbiased estimators for θ, clearly the one with lower variance is preferable. In this context, it is reasonable to ask: (i) How does one find an unbiased estimator for θ with the lowest possible variance? (ii) Is it possible to obtain an expression for the lower bound for the variance of all unbiased estimators for θ?

The answers to both these questions are positive, and they are given by Rao in his 1945 paper.[3] Cramer also obtained the desired bound quite independently in 1946. We begin with the Cramer–Rao bound that gives the lower bound for the variances of *all* unbiased estimators for θ.

CRAMER-RAO (CR) LOWER BOUND

▶ Let $T(\mathbf{x})$ represent any unbiased estimate for the unknown parameter θ based on observations of the random variables $\mathbf{x}_1, \mathbf{x}_2, \ldots, \mathbf{x}_n$ under the joint probability density function $f(\mathbf{x}_1 = x_1, \mathbf{x}_2 = x_2, \ldots, \mathbf{x}_n = x_n; \theta)$ denoted by $f(\mathbf{x}; \theta)$. Then

$$\operatorname{Var}\{T(\mathbf{x})\} \geq \frac{1}{E\left\{\left(\frac{\partial}{\partial\theta}\log f(\mathbf{x};\theta)\right)^2\right\}} = \frac{-1}{E\left\{\frac{\partial^2}{\partial\theta^2}\log f(\mathbf{x};\theta)\right\}} \tag{8-79}$$

provided the following regularity conditions are satisfied:

$$\frac{\partial}{\partial\theta}\int f(\mathbf{x};\theta)\,dx = \int \frac{\partial f(\mathbf{x};\theta)}{\partial\theta}\,dx = 0 \tag{8-80}$$

$$\frac{\partial}{\partial\theta}\int T(\mathbf{x}) f(\mathbf{x};\theta)\,dx = \int T(\mathbf{x})\frac{\partial f(\mathbf{x};\theta)}{\partial\theta}\,dx \tag{8-81}$$

Here the integrals represent n-fold integration.

Proof. Using the unbiased property, we have

$$E\{T(\mathbf{x}) - \theta\} = \int_{-\infty}^{\infty} \{T(\mathbf{x}) - \theta\} f(\mathbf{x};\theta)\,dx = 0 \tag{8-82}$$

Differentiate with respect to θ on both sides to obtain

$$\int_{-\infty}^{\infty} (T(\mathbf{x}) - \theta)\frac{\partial f(\mathbf{x};\theta)}{\partial\theta}\,dx - \int_{-\infty}^{\infty} f(\mathbf{x};\theta)\,dx = 0 \tag{8-83}$$

---

[3] C. R. Rao, *Information and the Accuracy Attainable in the Estimation of Statistical Parameters,* Bulletin of the Calcutta Mathematical Society, Vol. 37, pp. 81–89, 1945.

where we have made use of the regularity conditions given in (8-80) and (8-81). Thus

$$\int_{-\infty}^{\infty} (T(\mathbf{x}) - \theta)\frac{\partial f(\mathbf{x};\theta)}{\partial\theta}\,dx = 1 \tag{8-84}$$

But

$$\frac{\partial \log f(\mathbf{x};\theta)}{\partial\theta} = \frac{1}{f(\mathbf{x};\theta)}\frac{\partial f(\mathbf{x};\theta)}{\partial\theta} \tag{8-85}$$

so that

$$\frac{\partial f(\mathbf{x};\theta)}{\partial\theta} = f(\mathbf{x};\theta)\frac{\partial \log f(\mathbf{x};\theta)}{\partial\theta} \tag{8-86}$$

and (8-84) becomes

$$\int_{-\infty}^{\infty} (T(\mathbf{x}) - \theta) f(\mathbf{x};\theta)\frac{\partial \log f(\mathbf{x};\theta)}{\partial\theta}\,dx = 1 \tag{8-87}$$

Rewrite this expression as

$$\int_{-\infty}^{\infty} \left\{(T(\mathbf{x}) - \theta)\sqrt{f(\mathbf{x};\theta)}\right\}\left\{\sqrt{f(\mathbf{x};\theta)}\frac{\partial \log f(\mathbf{x};\theta)}{\partial\theta}\right\}dx = 1 \tag{8-88}$$

By Cauchy–Schwarz inequality, we obtain

$$1 \le \int_{-\infty}^{\infty} \left\{(T(\mathbf{x}) - \theta)\sqrt{f(\mathbf{x};\theta)}\right\}^2 dx \int_{-\infty}^{\infty} \left\{\sqrt{f(\mathbf{x};\theta)}\frac{\partial \log f(\mathbf{x};\theta)}{\partial\theta}\right\}^2 dx \tag{8-89}$$

or

$$\int_{-\infty}^{\infty} (T(\mathbf{x}) - \theta)^2 f(\mathbf{x};\theta)\,dx \int_{-\infty}^{\infty} \left(\frac{\partial \log f(\mathbf{x};\theta)}{\partial\theta}\right)^2 f(\mathbf{x}\,|\,\theta)\,dx \ge 1 \tag{8-90}$$

But

$$\int_{-\infty}^{\infty} (T(\mathbf{x}) - \theta)^2 f(\mathbf{x};\theta)\,dx = \mathrm{Var}\{T(\mathbf{x})\} \tag{8-91}$$

so that we obtain

$$\mathrm{Var}\{T(\mathbf{x})\}\cdot E\left\{\left(\frac{\partial \log f(\mathbf{x};\theta)}{\partial\theta}\right)^2\right\} \ge 1 \tag{8-92}$$

which gives the desired bound

$$\mathrm{Var}\{T(\mathbf{x})\} \ge \frac{1}{E\left\{\left(\frac{\partial \log f(\mathbf{x};\theta)}{\partial\theta}\right)^2\right\}} \tag{8-93}$$

Also integrating the identity in (8-86) we get

$$\int_{-\infty}^{\infty} f(\mathbf{x};\theta)\frac{\partial \log f(\mathbf{x};\theta)}{\partial\theta}\,dx = \int_{-\infty}^{\infty} \frac{\partial f(\mathbf{x};\theta)}{\partial\theta}\,dx = \frac{\partial}{\partial\theta}\int_{-\infty}^{\infty} f(\mathbf{x};\theta)\,dx = 0 \tag{8-94}$$

where we have once again made use of the regularity condition in (8-80). Differentiating this expression again with respect to θ, we get

$$\int_{-\infty}^{\infty} \frac{\partial f(\mathbf{x};\theta)}{\partial\theta}\frac{\partial \log f(\mathbf{x};\theta)}{\partial\theta}\,dx + \int_{-\infty}^{\infty} f(\mathbf{x};\theta)\frac{\partial^2 \log f(\mathbf{x};\theta)}{\partial^2\theta}\,dx = 0 \qquad \text{(8-95)}$$

where we have assumed that $\partial f(\mathbf{x};\theta)/\partial\theta$ is once again differentiable. Substituting (8-86) into (8-95), we obtain

$$\int_{-\infty}^{\infty} f(\mathbf{x};\theta)\left(\frac{\partial \log f(\mathbf{x};\theta)}{\partial\theta}\right)^2 dx + \int_{-\infty}^{\infty} f(\mathbf{x};\theta)\frac{\partial^2 \log f(\mathbf{x};\theta)}{\partial\theta^2}\,dx = 0 \qquad \text{(8-96)}$$

which is the same as

$$E\left\{\left(\frac{\partial \log f(\mathbf{x};\theta)}{\partial\theta}\right)^2\right\} = -E\left\{\frac{\partial^2 \log f(\mathbf{x};\theta)}{\partial\theta^2}\right\} \triangleq J_{11}(\theta) \qquad \text{(8-97)}$$

Note $J_{11}(\theta)$ is often referred to as the Fisher information contained in the data set about the parameter θ. Thus we obtain

$$\text{Var}\{T(\mathbf{x})\} \geq \frac{1}{E\left\{\left(\dfrac{\partial \log f(\mathbf{x};\theta)}{\partial\theta}\right)^2\right\}} = \frac{-1}{E\left(\dfrac{\partial^2 \log f(\mathbf{x};\theta)}{\partial\theta^2}\right)} \qquad \text{(8-98)}$$

If $\mathbf{x}_1, \mathbf{x}_2, \ldots, \mathbf{x}_n$ are i.i.d. random variables, then

$$\log f(x_1, x_2, \ldots, x_n;\theta) = \log f(x_1;\theta)f(x_2;\theta)\cdots f(x_n;\theta) = \sum_{i=1}^{n} \log f(x_i;\theta) \qquad \text{(8-99)}$$

represents a sum of n independent random variables, and hence

$$E\left\{\left(\frac{\partial \log f(\mathbf{x}_1, \mathbf{x}_2, \ldots, \mathbf{x}_n;\theta)}{\partial\theta}\right)^2\right\} = \sum_{i=1}^{n} E\left\{\left(\frac{\partial \log f(\mathbf{x}_i;\theta)}{\partial\theta}\right)^2\right\}$$

$$= nE\left\{\left(\frac{\partial \log f(\mathbf{x}_i;\theta)}{\partial\theta}\right)^2\right\} \qquad \text{(8-100)}$$

so that the CR bound simplifies into

$$\text{Var}\{T(\mathbf{x})\} \geq \frac{1}{nE\left\{\left(\dfrac{\partial \log f(\mathbf{x}_i;\theta)}{\partial\theta}\right)^2\right\}} = \frac{-1}{nE\left(\dfrac{\partial^2 \log f(\mathbf{x}_i;\theta)}{\partial^2\theta}\right)} \qquad \text{(8-101)}$$

◀

Returning to the general situation in (8-98), from the Cauchy–Schwarz inequality, equality holds good in (8-89) and (8-98) if and only if

$$\frac{\partial \log f(\mathbf{x};\theta)}{\partial\theta} = \{T(\mathbf{x}) - \theta\}a(\theta) \qquad \text{(8-102)}$$

In that case the unbiased estimator $T(\mathbf{x})$ has the lowest possible variance and it is said to be an *efficient* estimator for θ.

Clearly if an efficient estimator exists, then that is the best estimator from the minimum variance point of view. There is no reason to assume that such an estimator would exist in all cases. In general, such an estimator may or may not exist and it depends on the situation at hand.

However, if an efficient estimator exists, then it is easy to show that it is in fact the ML estimator. To see this, notice that the maximum likelihood estimator in (8-40) satisfies the equation

$$\left.\frac{\partial \log f(\mathbf{x};\theta)}{\partial \theta}\right|_{\theta=\hat{\theta}_{\mathrm{ML}}} = 0 \tag{8-103}$$

Substituting this into the efficiency equation, we get

$$\left.\frac{\partial \log f(\mathbf{x};\theta)}{\partial \theta}\right|_{\theta=\hat{\theta}_{\mathrm{ML}}} = \{T(\mathbf{x}) - \theta\}a(\theta)|_{\theta=\hat{\theta}_{\mathrm{ML}}} = 0 \tag{8-104}$$

It is easy to show that $a(\theta) > 0$, so that

$$\hat{\theta}_{\mathrm{ML}} = T(\mathbf{x}) \tag{8-105}$$

is the only solution to (8-104), and this shows that if an efficient estimator $T(\mathbf{x})$ exists, then it is necessarily the ML estimator. To complete the proof, we need to show that $a(\theta) > 0$, and towards this differentiate (8-102) once again with respect to θ to get

$$\frac{\partial^2 \log f(\mathbf{x};\theta)}{\partial \theta^2} = -a(\theta) + \{T(\mathbf{x}) - \theta\}a'(\theta) \tag{8-106}$$

Thus

$$a(\theta) = -E\left\{\frac{\partial^2 \log f(\mathbf{x};\theta)}{\partial \theta^2}\right\} = E\left\{\left(\frac{\partial \log f(\mathbf{x};\theta)}{\partial \theta}\right)^2\right\} = J_{11} > 0 \tag{8-107}$$

where we have made use of (8-82) and (8-97), thus completing the proof. Using (8-107) in (8-102), the p.d.f. of an efficient estimator has the form

$$f(\mathbf{x};\theta) = h(\mathbf{x})e^{\int J_{11}\{T(\mathbf{x})-\theta\}\,d\theta} \tag{8-108}$$

and it represents the exponential family of p.d.f.s. In particular if J_{11} is independent of θ, then (8-108) reduces to the Gaussian p.d.f. itself.

An efficient estimator need not exist in all cases; and in those cases the ML estimator may be not even unbiased. Nevertheless, the parameter still can have several unbiased estimators, and among them one such estimator has the lowest variance. How does one go ahead to obtain such an estimator?

Before we proceed to answer this important question, let us consider a few examples to illustrate the CR bound.

EXAMPLE 8-16

▶ Let $\mathbf{x}_i \sim P(\lambda)$, i.i.d. $i = 1 \to n$; and λ is unknown. Since $E(\mathbf{x}_i) = \lambda$, we have

$$\bar{\mathbf{x}} = \frac{\mathbf{x}_1 + \mathbf{x}_2 + \cdots + \mathbf{x}_n}{n} \tag{8-109}$$

is an unbiased estimator for λ. Further

$$\mathrm{Var}\{\bar{\mathbf{x}}\} = \frac{1}{n^2}\sum_{i=1}^{n}\mathrm{Var}\{\mathbf{x}_i\} = \frac{n\lambda}{n^2} = \frac{\lambda}{n} \tag{8-110}$$

To examine the efficiency of this estimator, let us compute the CR bound in this case.

$$P\{\mathbf{x}_1, \mathbf{x}_2, \ldots, \mathbf{x}_n\,;\lambda\} = \prod_{i=1}^{n} P\{\mathbf{x}_i\,;\lambda\} = \prod_{i=1}^{n} e^{-\lambda}\frac{\lambda^{\mathbf{x}_i}}{\mathbf{x}_i!} = e^{-n\lambda}\frac{\lambda^{\left(\sum_{i=1}^{n}\mathbf{x}_i\right)}}{\prod_{i=1}^{n}\mathbf{x}_i!} \tag{8-111}$$

$$\log P\{\mathbf{x}_1, \mathbf{x}_2, \ldots, \mathbf{x}_n\,;\lambda\} = -n\lambda + \left(\sum_{i=1}^{n}\mathbf{x}_i\right)\log\lambda - \log\left(\prod_{i=1}^{n}\mathbf{x}_i!\right) \tag{8-112}$$

and

$$\frac{\partial \log P\{\mathbf{x}_1, \mathbf{x}_2, \ldots, \mathbf{x}_n\,;\lambda\}}{\partial\lambda} = -n + \frac{\sum_{i=1}^{n}\mathbf{x}_i}{\lambda} \tag{8-113}$$

$$\frac{\partial^2 \log P\{\mathbf{x}_1, \mathbf{x}_2, \ldots, \mathbf{x}_n\,;\lambda\}}{\partial^2\lambda} = -\frac{\sum_{i=1}^{n}\mathbf{x}_i}{\lambda^2} \tag{8-114}$$

so that

$$-E\left[\frac{\partial^2 \log P\{\mathbf{x}_1, \mathbf{x}_2, \ldots, \mathbf{x}_n\,;\lambda\}}{\partial^2\lambda}\right] = \frac{\sum_{i=1}^{n}E\{\mathbf{x}_i\}}{\lambda^2} = \frac{n\lambda}{\lambda^2} = \frac{n}{\lambda} \tag{8-115}$$

and hence the CR bound is given by

$$\sigma_{\mathrm{CR}}^2 = \frac{\lambda}{n} \tag{8-116}$$

Thus the variance of $\overline{\mathbf{x}}$ equals the CR bound, and hence $\overline{\mathbf{x}}$ is an efficient estimator in this case. From (8-105), it must also be the maximum likelihood estimator for λ. In fact,

$$\left.\frac{\partial \log P(\mathbf{x}_1, \mathbf{x}_2, \ldots, \mathbf{x}_n\,;\lambda)}{\partial\lambda}\right|_{\lambda=\hat{\lambda}_{\mathrm{ML}}} = -n + \left.\frac{\sum_{i=1}^{n}\mathbf{x}_i}{\lambda}\right|_{\lambda=\hat{\lambda}_{\mathrm{ML}}} = 0 \tag{8-117}$$

so that

$$\hat{\lambda}_{\mathrm{ML}}(\mathbf{x}) = \frac{\sum_{i=1}^{n}\mathbf{x}_i}{\lambda} = \overline{\mathbf{x}} \tag{8-118}$$

It is important to verify the validity of the regularity conditions, otherwise the CR bound may not make much sense. To illustrate this point, consider the case where $\mathbf{x}_i$s are i.i.d. $\sim U(0,\theta)$, $i = 1 \to n$. Thus

$$f_{x_i}(x_i;\theta) = \frac{1}{\theta} \qquad 0 < x_i < \theta \tag{8-119}$$

From Examples 8-10 and 8-15, we know that

$$T(\mathbf{x}) = \max(\mathbf{x}_1, \mathbf{x}_2, \ldots, \mathbf{x}_n) \tag{8-120}$$

represents the maximum likelihood estimator as well as the sufficient statistic for θ in this case. The probability distribution function of $T(\mathbf{x})$ is given by

$$\begin{aligned} F_T(t) &= P\{T \le t\} = P\{\max(\mathbf{x}_1, \mathbf{x}_2, \ldots, \mathbf{x}_n) \le t\} \\ &= P\{\mathbf{x}_1 \le t, \mathbf{x}_2 \le t, \ldots, \mathbf{x}_n \le t\} \\ &= [P\{\mathbf{x}_i \le t\}]^n = \left(\frac{t}{\theta}\right)^n \qquad 0 < t < \theta \end{aligned} \tag{8-121}$$

so that the p.d.f. of T is given by

$$f_T(t) = \frac{dF_T(t)}{dt} = \frac{nt^{n-1}}{\theta^n} \qquad 0 < t < \theta \tag{8-122}$$

This gives

$$E[T(\mathbf{x})] = \int_0^\theta t f_T(t)\,dt = \frac{n}{\theta^n}\int_0^\theta t^n\,dt = \frac{n}{\theta^n}\frac{\theta^{n+1}}{n+1} = \frac{n}{n+1}\theta \tag{8-123}$$

and hence and ML estimator is not unbiased for θ. However,

$$\hat{\theta}(\mathbf{x}) = \left(\frac{n+1}{n}\right)\max(\mathbf{x}_1, \mathbf{x}_2, \ldots, \mathbf{x}_n) \tag{8-124}$$

is an unbiased estimator for θ. Moreover

$$\text{Var}\{\hat{\theta}(\mathbf{x})\} = \left(\frac{n+1}{n}\right)^2 \text{Var}\{T(\mathbf{x})\} = \left(\frac{n+1}{n}\right)^2\left[\frac{n\theta^2}{n+2} - \left(\frac{n\theta}{n+1}\right)^2\right] = \frac{\theta^2}{n(n+2)} \tag{8-125}$$

However, using (8-119) the Cramer–Rao bound in this case is given by

$$\frac{1}{nE\left\{\left(\frac{\partial \log f_{x_i}(\mathbf{x}_i\,;\,\theta)}{\partial\theta}\right)^2\right\}} = \frac{1}{nE\left\{\left(\frac{\partial(-\log\theta)}{\partial\theta}\right)^2\right\}} = \frac{1}{n(-1/\theta)^2} = \frac{\theta^2}{n} \tag{8-126}$$

Note that

$$\text{Var}\{\hat{\theta}(\mathbf{x})\} < \frac{\theta^2}{n} \tag{8-127}$$

the CR bound in this case. This apparent contradiction occurs because the regularity conditions are not satisfied here. Notice that

$$\frac{\partial}{\partial\theta}\int_0^\theta f(x;\theta)\,dx = 0 \tag{8-128}$$

as required in (8-80), but

$$\int_0^\theta \frac{\partial f(x;\theta)}{\partial\theta}\,dx = \int_0^\theta \frac{\partial}{\partial\theta}\left(\frac{1}{\theta}\right)dx = \int_0^\theta -\frac{1}{\theta^2}\,dx = \frac{1}{\theta} \neq 0 \tag{8-129}$$

The second regularity condition in (8-80) does not hold here, and hence the CR bound is meaningless in this case. ◀

As we mentioned earlier, all unbiased estimators need not be efficient. To construct an example for an unbiased estimator that is *not* efficient, consider the situation in Example 8-17.

EXAMPLE 8-17

▶ Let $\mathbf{x}_i \sim N(\mu, \sigma^2)$, where μ is unknown and σ^2 is known. Further, $\mathbf{x}_i$s are i.i.d. for $i = 1 \rightarrow n$.

It is easy to verify that

$$T(\mathbf{x}) = \frac{\mathbf{x}_1 + \mathbf{x}_2 + \cdots + \mathbf{x}_n}{n} = \bar{\mathbf{x}} \tag{8-130}$$

is an unbiased estimator for μ that is also efficient. In fact $\bar{\mathbf{x}} \sim N(\mu, \sigma^2/n)$. But suppose our parameter of interest is not μ, but say μ^2. Thus $\theta = \mu^2$ is unknown, and we need an unbiased estimator for θ. To start with

$$f(x_1, x_2, \ldots, x_n; \theta) = \left(\frac{1}{2\pi\sigma^2}\right)^{n/2} e^{-\sum_{i=1}^{n}(x_i - \sqrt{\theta})^2/2\sigma^2} \tag{8-131}$$

so that

$$\frac{\partial \log f(x_1, x_2, \ldots, x_n; \theta)}{\partial \theta} = \frac{\sum_{i=1}^{n}(x_i - \sqrt{\theta})}{2\sigma^2\sqrt{\theta}} \tag{8-132}$$

which gives

$$E\left\{\left(\frac{\partial \log f(\mathbf{x}_1, \mathbf{x}_2, \ldots, \mathbf{x}_n; \theta)}{\partial \theta}\right)^2\right\} = \frac{\sum_{i=1}^{n} E\{\mathbf{x}_i - \mu\}^2}{4\sigma^4\mu^2} = \frac{n\sigma^2}{4\sigma^4\mu^2} = \frac{n}{4\sigma^2\mu^2} \tag{8-133}$$

Thus the CR bound for $\theta = \mu^2$ is given by

$$\sigma_{\mathrm{CR}}^2 \triangleq \frac{4\sigma^2\mu^2}{n} \le \mathrm{Var}\{\hat{\theta}\} \tag{8-134}$$

where $\hat{\theta}$ represents any unbiased estimator for $\theta = \mu^2$. To obtain an unbiased estimator for θ, let us examine the estimator $\bar{\mathbf{x}}^2$. We obtain

$$E\{\bar{\mathbf{x}}^2\} = \mathrm{Var}\{\bar{\mathbf{x}}\} + [E\{\bar{\mathbf{x}}\}]^2 = \frac{\sigma^2}{n} + \mu^2 \tag{8-135}$$

so that

$$\hat{\theta}(\mathbf{x}) = \bar{\mathbf{x}}^2 - \frac{\sigma^2}{n} \tag{8-136}$$

is an unbiased estimator for $\theta = \mu^2$. Its variance is given by

$$\begin{aligned} E\{[\hat{\theta} - \mu^2]^2\} &= E\left\{\left[\bar{\mathbf{x}}^2 - \left(\frac{\sigma^2}{n} + \mu^2\right)\right]^2\right\} = E\left\{\left[\bar{\mathbf{x}}^2 - E\{\bar{\mathbf{x}}^2\}\right]^2\right\} \\ &= E\{\bar{\mathbf{x}}^4\} - [E\{\bar{\mathbf{x}}^2\}]^2 = E\{\bar{\mathbf{x}}^4\} - \left(\frac{\sigma^2}{n} + \mu^2\right)^2 \end{aligned} \tag{8-137}$$

But if $\mathbf{y} \sim N(\mu, \sigma^2)$, then by direct computation, we get

$$\begin{aligned} E\{\mathbf{y}\} &= \mu \\ E\{\mathbf{y}^2\} &= \sigma^2 + \mu^2 \\ E\{\mathbf{y}^3\} &= \mu^3 + 3\mu\sigma^2 \\ E\{\mathbf{y}^4\} &= \mu^4 + 6\mu^2\sigma^2 + 3\sigma^4 \end{aligned} \tag{8-138}$$

Since $\bar{\mathbf{x}} \sim N(\mu, \sigma^2/n)$, we obtain

$$E\{\bar{\mathbf{x}}^4\} = \mu^4 + 6\mu^2\frac{\sigma^2}{n} + \frac{3\sigma^4}{n^2} \tag{8-139}$$

so that using (8-137) and (8-139),

$$\begin{aligned}\operatorname{Var}\{\overline{\theta}(\mathbf{x})\} &= E\{(\hat{\theta}-\mu^2)^2\} = E\{\overline{\mathbf{x}}^4\} - \left(\frac{\sigma^2}{n}+\mu^2\right)^2 \\ &= \mu^4 + \frac{6\mu^2\sigma^2}{n} + \frac{3\sigma^4}{n^2} - \left(\frac{\sigma^4}{n^2}+\mu^4+\frac{2\mu^2\sigma^2}{n}\right) \\ &= \frac{4\mu^2\sigma^2}{n} + \frac{2\sigma^4}{n^2} \end{aligned} \tag{8-140}$$

Thus

$$\operatorname{Var}\{\hat{\theta}\} = \frac{4\mu^2\sigma^2}{n} + \frac{2\sigma^4}{n^2} > \frac{4\mu^2\sigma^2}{n} = \sigma_{\text{CR}}^2 \tag{8-141}$$

Clearly the variance of the unbiased estimator for $\theta = \mu^2$ exceeds the corresponding CR bound, and hence it is not an efficient estimator. How good is the above unbiased estimator? Is it possible to obtain another unbiased estimator for $\theta = \mu^2$ with even lower variance than (8-140)? To answer these questions it is important to introduce the concept of the uniformly minimum variance unbiased estimator (UMVUE). ◀

UMVUE

Let $\{T(\mathbf{x})\}$ represent the set of all unbiased estimators for θ. Suppose $T_0(\mathbf{x}) \in \{T(\mathbf{x})\}$, and if

$$E\{(T_0(\mathbf{x})-\theta)^2\} \le E\{(T(\mathbf{x})-\theta)^2\} \tag{8-142}$$

for all $T(\mathbf{x})$ in $\{T(\mathbf{x})\}$, then $T_0(\mathbf{x})$ is said to be an UMVUE for θ. Thus UMVUE for θ represents an unbiased estimator with the *lowest* possible variance. It is easy to show that UMVUEs are unique.

If not, suppose T_1 and T_2 are both UMVUE for θ, and let $T_1 \neq T_2$. Then

$$E\{T_1\} = E\{T_2\} = \theta \tag{8-143}$$

and

$$\operatorname{Var}\{T_1\} = \operatorname{Var}\{T_2\} = \sigma^2 \tag{8-144}$$

Let $T = (T_1 + T_2)/2$. Then T is unbiased for θ, and since T_1 and T_2 are both UMVUE, we must have

$$\operatorname{Var}\{T\} \ge \operatorname{Var}\{T_1\} = \sigma^2 \tag{8-145}$$

But

$$\begin{aligned}\operatorname{Var}\{T\} &= E\left(\frac{T_1+T_2}{2}-\theta\right)^2 = \frac{\operatorname{Var}\{T_1\}+\operatorname{Var}\{T_2\}+2\operatorname{Cov}\{T_1,T_2\}}{4} \\ &= \frac{\sigma^2+\operatorname{Cov}\{T_1,T_2\}}{2} \ge \sigma^2 \end{aligned} \tag{8-146}$$

or

$$\mathrm{Cov}\{T_1, T_2\} \geq \sigma^2 \tag{8-147}$$

But $\mathrm{Cov}\{T_1, T_2\} \leq \sigma^2$ always, and hence we have

$$\mathrm{Cov}\{T_1, T_2\} = \sigma^2 \Rightarrow \rho_{T_1,T_2} = 1 \tag{8-148}$$

Thus T_1 and T_2 are completely correlated. Thus for some a

$$P\{T_1 = aT_2\} = 1 \qquad \text{for all } \theta \tag{8-149}$$

Since T_1 and T_2 are both unbiased for θ, we have $a = 1$, and

$$P\{T_1 = T_2\} = 1 \qquad \text{for all } \theta \tag{8-150}$$

Q.E.D.

How does one find the UMVUE for θ? The Rao–Blackwell theorem gives a complete answer to this problem.

RAO–BLACKWELL THEOREM

▶ Let $h(\mathbf{x})$ represent any unbiased estimator for θ, and $T(\mathbf{x})$ be a sufficient statistic for θ under $f(\mathbf{x}, \theta)$. Then the conditional expectation $E[h(\mathbf{x}) \mid T(\mathbf{x})]$ is independent of θ, and it is the UMVUE for θ.

Proof. Let

$$T_0(\mathbf{x}) = E[h(\mathbf{x}) \mid T(\mathbf{x})] \tag{8-151}$$

Since

$$E\{T_0(\mathbf{x})\} = E\{E[h(\mathbf{x}) \mid T(\mathbf{x})]\} = E\{h(\mathbf{x})\} = \theta, \tag{8-152}$$

$T_0(\mathbf{x})$ is an unbiased estimator for θ, and it is enough to show that

$$\mathrm{Var}\{T_0(\mathbf{x})\} \leq \mathrm{Var}\{h(\mathbf{x})\}, \tag{8-153}$$

which is the same as

$$E\left\{T_0^2(\mathbf{x})\right\} \leq E\{h^2(\mathbf{x})\} \tag{8-154}$$

But

$$E\{h^2(\mathbf{x})\} = E\{E[h^2(\mathbf{x}) \mid T(\mathbf{x})]\} \tag{8-155}$$

and hence it is sufficient to show that

$$E\left\{T_0^2(\mathbf{x})\right\} = E\{[E(h \mid T)]^2\} \leq E[E(h^2 \mid T)] \tag{8-156}$$

where we have made use of (8-151). Clearly, for (8-156) to hold, it is sufficient to show that

$$[E(h \mid T)]^2 \leq E(h^2 \mid T) \tag{8-157}$$

Infact, by Cauchy–Schwarz

$$[E(h \mid T)]^2 \leq E(h^2 \mid T)E(1 \mid T) = E(h^2 \mid T) \tag{8-158}$$

and hence the desired inequality in (8-153) follows. Equality holds in (8-153)–(8-154) if and only if

$$E\{T_0^2(\mathbf{x})\} = E\{[E(h \mid T)]^2\} = E\{h^2\} \tag{8-159}$$

or

$$E\{[E(h \mid T)]^2 - h^2\} = E\{[(E(h \mid T)]^2 - E(h^2 \mid T)\} = -E[\mathrm{Var}\{h \mid T\}] = 0 \tag{8-160}$$

Since $\mathrm{Var}\{h \mid T\} \geq 0$, this happens if and only if

$$\mathrm{Var}\{h \mid T\} = 0$$

i.e., if and only if

$$[E\{h \mid T\}]^2 = E[h^2 \mid T] \tag{8-161}$$

which will be the case if and only if

$$h = E[h(\mathbf{x}) \mid T] \tag{8-162}$$

i.e., if h is a function of the sufficient statistic $T = t$ alone, then it is UMVUE. ◀

Thus according to Rao–Blackwell theorem an unbiased estimator for θ that is a function of its sufficient statistic is the UMVUE for θ. Otherwise the conditional expectation of any unbiased estimator based on its sufficient statistic gives the UMVUE. The significance of the sufficient statistic in obtaining the UMVUE must be clear from the Rao–Blackwell theorem.

If an efficient estimator does not exist, then the next best estimator is its UMVUE, since it has the minimum possible realizable variance among all unbiased estimators for θ.

Returning to the estimator for $\theta = \mu^2$ in Example 8-17, where $\mathbf{x}_i \sim N(\mu, \sigma^2)$, $i = 1 \to n$ are i.i.d. random variables, we have seen that

$$\hat{\theta}(\overline{\mathbf{x}}) = \overline{\mathbf{x}}^2 - \frac{\sigma^2}{n} \tag{8-163}$$

is an unbiased estimator for $\theta = \mu^2$. Clearly $T(\mathbf{x}) = \overline{\mathbf{x}}$ is a sufficient statistic for μ as well as $\mu^2 = \theta$. Thus the above unbiased estimator is a function of the sufficient statistic alone, since

$$\hat{\theta}(\overline{\mathbf{x}}) = T^2(\mathbf{x}) - \frac{\sigma^2}{n} \tag{8-164}$$

From Rao–Blackwell theorem it is the UMVUE for $\theta = \mu^2$.

Notice that Rao–Blackwell theorem states that all UMVUEs depend only on their sufficient statistic, and on *no other* functional form of the data. This is consistent with the notion of sufficiency, since in that case $T(\mathbf{x})$ contains all information about θ contained in the data set $\mathbf{x} = (\mathbf{x}_1, \mathbf{x}_2, \ldots, \mathbf{x}_n)$, and hence the best possible estimator should depend only upon the sufficient statistic. Thus if $h(\mathbf{x})$ is an unbiased estimator for θ, from Rao–Blackwell theorem it is possible to improve upon its variance by conditioning it on $T(\mathbf{x})$. Thus compared to $h(\mathbf{x})$,

$$\begin{aligned} T_1 &= E\{h(\mathbf{x}) \mid T(\mathbf{x}) = t\} \\ &= \int h(\mathbf{x}) f(x_1, x_2, \ldots, x_n; \theta \mid T(\mathbf{x}) = t)\, dx_1\, dx_2 \cdots dx_n \end{aligned} \tag{8-165}$$

is a better estimator for θ. In fact, notice that $f(x_1, x_2, \ldots, x_n; \theta \mid T(\mathbf{x}) = t)$ does not depend on θ, since $T(\mathbf{x})$ is sufficient. Thus $T_1 = E[h(\mathbf{x}) \mid T(\mathbf{x}) = t]$ is independent of θ and is a function of $T(\mathbf{x}) = t$ alone. Thus T_1 itself is an estimate for θ. Moreover,

$$E\{T_1\} = E\{E[h(\mathbf{x}) \mid T(\mathbf{x}) = t]\} = E[h(\mathbf{x})] = \theta \tag{8-166}$$

Further, using the identity[4]

$$\text{Var}\{\mathbf{x}\} = E[\text{Var}\{\mathbf{x} \mid \mathbf{y}\}] + \text{Var}[E\{\mathbf{x} \mid \mathbf{y}\}] \tag{8-167}$$

we obtain

$$\text{Var}\{h(\mathbf{x})\} = E[\text{Var}\{h \mid T\}] + \text{Var}[E\{h \mid T\}] = \text{Var}\{T_1\} + E[\text{Var}\{h \mid T\}] \tag{8-168}$$

Thus

$$\text{Var}\{T_1\} < \text{Var}\{h(\mathbf{x})\} \tag{8-169}$$

once again proving the UMVUE nature of $T_1 = E[h(\mathbf{x}) \mid T(\mathbf{x}) = t]$. Next we examine the so called "taxi problem," to illustrate the usefulness of the concepts discussed above.

The Taxi Problem

Consider the problem of estimating the total number of taxis in a city by observing the "serial numbers" of a few of them by a stationary observer.[5] Let $x_1, x_2, \ldots, x_n$ represent n such random serial numbers, and our problem is to estimate the total number N of taxis based on these observations.

It is reasonable to assume that the probability of observing a particular serial number is $1/N$, so that these observed random variables $\mathbf{x}_1, \mathbf{x}_2, \ldots, \mathbf{x}_n$ can be assumed to be discrete valued random variables that are uniformly distributed in the interval $(1, N)$, where N is an unknown constant to be determined. Thus

$$P\{\mathbf{x}_i = k\} = \frac{1}{N} \qquad k = 1, 2, \ldots, N \qquad i = 1, 2, \ldots, n \tag{8-170}$$

In that case, we have seen that the statistic in (8-77) (refer to Example 8-15)

$$T(\mathbf{x}_1, \mathbf{x}_2, \ldots, \mathbf{x}_n) = \max(\mathbf{x}_1, \mathbf{x}_2, \ldots, \mathbf{x}_n) \tag{8-171}$$

is sufficient for N. In search of its UMVUE, let us first compute the mean of this sufficient

---

[4]

$$\text{Var}\{\mathbf{x} \mid \mathbf{y}\} \triangleq E\{\mathbf{x}^2 \mid \mathbf{y}\} - [E\{\mathbf{x} \mid \mathbf{y}\}]^2$$

and

$$\text{Var}[E\{\mathbf{x} \mid \mathbf{y}\}] = E[E\{\mathbf{x} \mid \mathbf{y}\}]^2 - (E[E\{\mathbf{x} \mid \mathbf{y}\}])^2$$

so that

$$E[\text{Var}\{\mathbf{x} \mid \mathbf{y}\}] + \text{Var}[E\{\mathbf{x} \mid \mathbf{y}\}] = E[E\{\mathbf{x}^2 \mid \mathbf{y}\}] - (E[E\{\mathbf{x} \mid \mathbf{y}\}])^2 = E\{\mathbf{x}^2\} - [E\{\mathbf{x}\}]^2 = \text{Var}\{\mathbf{x}\}$$

[5]The stationary observer is assumed to be standing at a street corner, so that the observations can be assumed to be done with replacement.

statistic. To compute the p.m.f. of the sufficient statistic in (8-171), we proceed as:

$$\begin{aligned}
P\{T=k\} &= P[\max(\mathbf{x}_1, \mathbf{x}_2, \ldots, \mathbf{x}_n) = k] \\
&= P[\max(\mathbf{x}_1, \mathbf{x}_2, \ldots, \mathbf{x}_n) \le k] - P[\max(\mathbf{x}_1, \mathbf{x}_2, \ldots, \mathbf{x}_n) < k] \\
&= P[\max(\mathbf{x}_1, \mathbf{x}_2, \ldots, \mathbf{x}_n) \le k] - P[\max(\mathbf{x}_1, \mathbf{x}_2, \ldots, \mathbf{x}_n) \le k-1] \\
&= P[\mathbf{x}_1 \le k, \mathbf{x}_2 \le k, \ldots, \mathbf{x}_n \le k] \\
&\quad - P[\mathbf{x}_1 \le k-1, \mathbf{x}_2 \le k-1, \ldots, \mathbf{x}_n \le k-1] \\
&= [P\{\mathbf{x}_i \le k\}]^n - [P\{\mathbf{x}_i \le k-1\}]^n \\
&= \left(\frac{k}{N}\right)^n - \left(\frac{k-1}{N}\right)^n \qquad k = 1, 2, \ldots, N
\end{aligned} \tag{8-172}$$

so that

$$\begin{aligned}
E\{T(\mathbf{x})\} &= N^{-n} \sum_{k=1}^{N} k\{k^n - (k-1)^n\} \\
&= N^{-n} \sum_{k=1}^{N} \{k^{n+1} - (k-1+1)(k-1)^n\} \\
&= N^{-n} \sum_{k=1}^{N} \{(k^{n+1} - (k-1)^{n+1}) - (k-1)^n\} \\
&= N^{-n} \left\{ N^{n+1} - \sum_{k=1}^{N} (k-1)^n \right\} = N - N^{-n} \sum_{k=1}^{N-1} k^n
\end{aligned} \tag{8-173}$$

For large N, we can approximate this sum as:

$$\sum_{k=1}^{N} (k-1)^n = 1^n + 2^n + \cdots + (N-1)^n \simeq \int_0^N y^n \, dy = \frac{N^{n+1}}{n+1} \qquad \text{for large } N \tag{8-174}$$

Thus

$$E\{\max(\mathbf{x})\} \simeq N^{-n} \left(N^{n+1} - \frac{N^{n+1}}{n+1} \right) = \frac{nN}{n+1} \tag{8-175}$$

so that

$$T_1(\mathbf{x}) = \left(\frac{n+1}{n}\right) T(\mathbf{x}) = \left(\frac{n+1}{n}\right) \max(\mathbf{x}_1, \mathbf{x}_2, \ldots, \mathbf{x}_n) \tag{8-176}$$

is nearly an unbiased estimator for N, especially when N is large. Since it depends only on the sufficient statistic, from Rao-Blackwell theorem it is also a nearly UMVUE for N. It is easy to compute the variance of this nearly UMVUE estimator in (8-176).

$$\text{Var}\{T_1(\mathbf{x})\} = \left(\frac{n+1}{n}\right)^2 \text{Var}\{T(\mathbf{x})\} \tag{8-177}$$

But

$$\operatorname{Var}\{T(\mathbf{x})\} = E\{T^2(\mathbf{x})\} - [E\{T(\mathbf{x})\}]^2 = E\{T^2(\mathbf{x})\} - \frac{n^2N^2}{(n+1)^2} \tag{8-178}$$

so that

$$\operatorname{Var}\{T_1(\mathbf{x})\} = \left(\frac{n+1}{n}\right)^2 E\{T^2(\mathbf{x})\} - N^2 \tag{8-179}$$

Now

$$\begin{aligned}
E\{T^2(\mathbf{x})\} &= \sum_{k=1}^{N} k^2 P\{T = k\} = N^{-n} \sum_{k=1}^{N} k^2\{k^n - (k-1)^n\} \\
&= N^{-n} \sum_{k=1}^{N} \{k^{n+2} - (k-1+1)^2(k-1)^n\} \\
&= N^{-n} \left\{ \sum_{k=1}^{N} [k^{n+2} - (k-1)^{n+2}] - 2\sum_{k=1}^{N}(k-1)^{n+1} - \sum_{k=1}^{N}(k-1)^n \right\} \\
&\simeq N^{-n} \left(N^{n+2} - 2\int_0^N y^{n+1}\,dy - \int_0^N y^n\,dy \right) \\
&= N^{-n} \left(N^{n+2} - \frac{2N^{n+2}}{n+2} - \frac{N^{n+1}}{n+1} \right) = \frac{nN^2}{n+2} - \frac{N}{n+1}
\end{aligned} \tag{8-180}$$

Substituting (8-180) into (8-179) we get the variance to be

$$\operatorname{Var}\{T_1(\mathbf{x})\} \simeq \frac{(n+1)^2}{n(n+2)}N^2 - \frac{(n+1)N}{n^2} - N^2 = \frac{N^2}{n(n+2)} - \frac{(n+1)N}{n^2} \tag{8-181}$$

Returning to the estimator in (8-176), however, to be precise it is not an unbiased estimator for N, so that the Rao–Blackwell theorem is not quite applicable here. To remedy this situation, all we need is an unbiased estimator for N, however trivial it is.

At this point, we can examine the other "extreme" statistics given by

$$N(\mathbf{x}) = \min(\mathbf{x}_1, \mathbf{x}_2, \ldots, \mathbf{x}_n) \tag{8-182}$$

which represents the smallest observed number among the n observations. Notice that this new statistics surely represents relevant information, because it indicates that there are $N(\mathbf{x}) - 1$ taxis with numbers smaller than the smallest observed number $N(\mathbf{x})$. Assuming that there are as many taxis with numbers larger than the largest observed number, we can choose

$$T_0(\mathbf{x}) = \max(\mathbf{x}_1, \mathbf{x}_2, \ldots, \mathbf{x}_n) + \min(\mathbf{x}_1, \mathbf{x}_2, \ldots, \mathbf{x}_n) - 1 \tag{8-183}$$

to be a reasonable estimate for N. To examine whether this is an unbiased estimator for

N, we need to evaluate the p.m.f. for $\min(\mathbf{x}_1, \mathbf{x}_2, \ldots, \mathbf{x}_n)$. we have[6]

$$P[\min(\mathbf{x}_1, \mathbf{x}_2, \ldots, \mathbf{x}_n) = j]$$
$$= P[\min(\mathbf{x}_1, \mathbf{x}_2, \ldots, \mathbf{x}_n) \le j] - P[\min(\mathbf{x}_1, \mathbf{x}_2, \ldots, \mathbf{x}_n) \le j-1]. \quad (8\text{-}184)$$

But

$$P[\min(\mathbf{x}_1, \mathbf{x}_2, \ldots, \mathbf{x}_n) \le j] = 1 - P\{\mathbf{x}_1 > j, \mathbf{x}_2 > j, \ldots, \mathbf{x}_n > j\}$$
$$= 1 - \prod_{i=1}^{n} P\{\mathbf{x}_i > j\} = 1 - \prod_{i=1}^{n} [1 - P\{\mathbf{x}_i \le j\}]$$
$$= 1 - \prod_{i=1}^{n} \left(1 - \frac{j}{N}\right) = 1 - \frac{(N-j)^n}{N^n} \quad (8\text{-}185)$$

Thus

$$P[\min(\mathbf{x}_1, \mathbf{x}_2, \ldots, \mathbf{x}_n) = j] = 1 - \frac{(N-j)^n}{N^n} - \left(1 - \frac{(N-j+1)^n}{N^n}\right)$$
$$= \frac{(N-j+1)^n - (N-j)^n}{N^n} \quad (8\text{-}186)$$

Thus

$$E\{\min(\mathbf{x}_1, \mathbf{x}_2, \ldots, \mathbf{x}_n)\} = \sum_{j=1}^{N} j P[\min(\mathbf{x}_1, \mathbf{x}_2, \ldots, \mathbf{x}_n) = j]$$
$$= \sum_{j=1}^{N} j \frac{(N-j+1)^n - (N-j)^n}{N^n} \quad (8\text{-}187)$$

Let $N - j + 1 = k$ so that $j = N - k + 1$ and the summation in (8-187) becomes

$$N^{-n} \sum_{k=1}^{N} (N-k+1)[k^n - (k-1)^n]$$
$$= N^{-n} \left\{ N \sum_{k=1}^{N} [k^n - (k-1)^n] - \sum_{k=1}^{N} [(k-1)k^n - (k-1)^{n+1}] \right\}$$
$$= N^{-n} \left\{ N^{n+1} - \sum_{k=1}^{N} \{k^{n+1} - (k-1)^{n+1}\} + \sum_{k=1}^{n} k^n \right\}$$
$$= N^{-n} \left\{ N^{n+1} - N^{n+1} + \sum_{k=1}^{N} k^n \right\}$$
$$= N^{-n} \left\{ \sum_{k=1}^{N-1} k^n + N^n \right\} = 1 + N^{-n} \sum_{k=1}^{N-1} k^n \quad (8\text{-}188)$$

---

[6]The p.m.f. for $\max(\mathbf{x}_1, \mathbf{x}_2, \ldots, \mathbf{x}_n)$ and its expected value are given in (8-172)–(8-173).

Thus

$$E[\min(\mathbf{x}_1, \mathbf{x}_2, \ldots, \mathbf{x}_n)] = 1 + N^{-n} \sum_{k=1}^{N-1} k^n \tag{8-189}$$

From (8-171)–(8-173), we have

$$E[\max(\mathbf{x}_1, \mathbf{x}_2, \ldots, \mathbf{x}_n)] = N - N^{-n} \sum_{k=1}^{N-1} k^n \tag{8-190}$$

so that together with (8-183) we get

$$\begin{aligned} E[T_0(\mathbf{x})] &= E[\max(\mathbf{x}_1, \mathbf{x}_2, \ldots, \mathbf{x}_n)] + E[\min(\mathbf{x}_1, \mathbf{x}_2, \ldots, \mathbf{x}_n)] - 1 \\ &= \left(N - N^{-n} \sum_{k=1}^{N-1} k^n\right) + \left(1 + N^{-n} \sum_{k=1}^{N-1} k^n\right) - 1 \\ &= N \end{aligned} \tag{8-191}$$

Thus $T_0(\mathbf{x})$ is an unbiased estimator for N. However it is not a function of the sufficient statistic $\max(\mathbf{x}_1, \mathbf{x}_2, \ldots, \mathbf{x}_n)$ alone, and hence it is *not* the UMVUE for N. To obtain the UMVUE, the procedure is straight forward. In fact, from Rao–Blackwell theorem,

$$E\{T_0(\mathbf{x}) \mid T(\mathbf{x}) = \max(\mathbf{x}_1, \mathbf{x}_2, \ldots, \mathbf{x}_n)\} \tag{8-192}$$

is the unique UMVUE for N. However, because of the $\min(\mathbf{x}_1, \mathbf{x}_2, \ldots, \mathbf{x}_n)$ statistics present in $T_0(\mathbf{x})$, computing this conditional expectation could turn out to be a difficult task. To minimize the computations involved in the Rao–Blackwell conditional expectation, it is clear that one should begin with an unbiased estimator that is a simpler function of the data. In that context, consider the simpler estimator

$$T_2(\mathbf{x}) = \mathbf{x}_1 \tag{8-193}$$

$$E\{T_2\} = E\{\mathbf{x}_1\} = \sum_{k=1}^{N} kP\{\mathbf{x}_1 = k\} = \sum_{k=1}^{N} k\frac{1}{N} = \frac{N(N+1)}{2N} = \frac{N+1}{2} \tag{8-194}$$

so that

$$h(\mathbf{x}) = 2\mathbf{x}_1 - 1 \tag{8-195}$$

is an unbiased estimator for N. We are in a position to apply the Rao–Blackwell theorem to obtain the UMVUE here. From the theorem

$$\begin{aligned} E[h(\mathbf{x}) \mid T(\mathbf{x})] &= E\{2\mathbf{x}_1 - 1 \mid \max(\mathbf{x}_1, \mathbf{x}_2, \ldots, \mathbf{x}_n) = t\} \\ &= \sum_{k=1}^{N} (2k-1) P\{\mathbf{x}_1 = k \mid \max(\mathbf{x}_1, \mathbf{x}_2, \ldots, \mathbf{x}_n) = t\} \end{aligned} \tag{8-196}$$

is the "best" unbiased estimator from the minimum variance point of view. Using (8-172)

$$P\{\mathbf{x}_1 = k \mid \max(\mathbf{x}_1, \mathbf{x}_2, \ldots, \mathbf{x}_n) = t\} = \frac{P\{\mathbf{x}_1 = k, T(\mathbf{x}) = t\}}{P\{T(\mathbf{x}) = t\}}$$

$$= \frac{P\{\mathbf{x}_1 = k,\ T(\mathbf{x}) \le t\} - P\{\mathbf{x}_1 = k,\ T(\mathbf{x}) \le t-1\}}{P\{\max T(\mathbf{x}) \le t\} - P\{T(\mathbf{x}) \le t-1\}}$$

$$= \frac{P\{\mathbf{x}_1 = k, \mathbf{x}_2 \le t, \ldots, \mathbf{x}_n \le t\} - P\{\mathbf{x}_1 = k, \mathbf{x}_2 \le t-1, \ldots, \mathbf{x}_n \le t-1\}}{P\{\mathbf{x}_1 \le t, \mathbf{x}_2 \le t, \ldots, \mathbf{x}_n \le t\} - P\{\mathbf{x}_1 \le t-1, \mathbf{x}_2 \le t-1, \ldots, \mathbf{x}_n \le t-1)\}}$$

$$= \begin{cases} \dfrac{t^{n-1} - (t-1)^{n-1}}{t^n - (t-1)^n} & \text{if } \ k = 1, 2, \ldots, t-1 \\[2ex] \dfrac{t^{n-1}}{t^n - (t-1)^n} & \text{if } \ k = t \end{cases} \tag{8-197}$$

Thus

$$E\{h(\mathbf{x}) \mid T(\mathbf{x})\} = E\{(2\mathbf{x}_1 - 1) \mid \max(\mathbf{x}_1, \mathbf{x}_2, \ldots, \mathbf{x}_n)\}$$

$$= \sum_{k=1}^{t} (2k-1) P\{\mathbf{x}_1 = k \mid \max(\mathbf{x}_1, \mathbf{x}_2, \ldots, \mathbf{x}_n) = t\}$$

$$= \left(\frac{t^{n-1} - (t-1)^{n-1}}{t^n - (t-1)^n} \right) \sum_{k=1}^{t-1} (2k-1) + \frac{t^{n-1}}{t^n - (t-1)^n} (2t-1)$$

$$= \frac{t^{n-1} - (t-1)^{n-1}}{t^n - (t-1)^n} (t-1)^2 + \frac{t^{n-1}(2t-1)}{t^n - (t-1)^n}$$

$$= \frac{t^{n+1} - 2t^n + t^{n-1} - (t-1)^{n+1} + 2t^n - t^{n-1}}{t^n - (t-1)^n}$$

$$= \frac{t^{n+1} - (t-1)^{n+1}}{t^n - (t-1)^n} \tag{8-198}$$

is the UMVUE for N where $t = \max(\mathbf{x}_1, \mathbf{x}_2, \ldots, \mathbf{x}_n)$. Thus if $\mathbf{x}_1, \mathbf{x}_2, \ldots, \mathbf{x}_n$ represent n random observations from a numbered population, then[7]

$$\hat{N}_n = \frac{[\max(\mathbf{x}_1, \mathbf{x}_2, \ldots, \mathbf{x}_n)]^{n+1} - [\max(\mathbf{x}_1, \mathbf{x}_2, \ldots, \mathbf{x}_n) - 1]^{n+1}}{[\max(\mathbf{x}_1, \mathbf{x}_2, \ldots, \mathbf{x}_n)]^n - [\max(\mathbf{x}_1, \mathbf{x}_2, \ldots, \mathbf{x}_n) - 1]^n} \tag{8-199}$$

is the minimum variance unbiased estimator for the total size of the population. The calculation of the variance of this UMVUE is messy and somewhat difficult, nevertheless we are assured that it is the best estimator with minimum possible variance.

As an example, let

$$82, 124, 312, 45, 218, 151 \tag{8-200}$$

represent six independent observations from a population. Then

$$\max(\mathbf{x}_1, \mathbf{x}_2, \ldots, \mathbf{x}_6) = 312 \tag{8-201}$$

[7]S.M. Stigler, *Completeness and Unbiased Estimation, AM. Stat.* 26, pp. 28-29, 1972.

and hence

$$\hat{N}_6 = \frac{312^7 - 311^7}{312^6 - 311^6} = 363 \tag{8-202}$$

is the "best" estimate for the total size of the population. We can note that the nearly unbiased estimator, and hence the nearly UMVUE $T_1(\mathbf{x})$ in (8-176) gives the answer

$$T_1(\mathbf{x}) = \frac{6+1}{6} 312 = 364 \tag{8-203}$$

to be the best estimate for N, thus justifying its name!

Cramer–Rao Bound for the Multiparameter Case

It is easy to extend the Cramer–Rao bound to the multiparameter situation. Let $\underline{\theta} \stackrel{\Delta}{=} (\theta_1, \theta_2, \ldots, \theta_m)^t$ represent the unknown set of parameters under the p.d.f. $f(\mathbf{x}; \underline{\theta})$ and

$$\underline{T}(\mathbf{x}) = [T_1(\mathbf{x}),\ T_2(\mathbf{x}), \ldots,\ T_m(\mathbf{x})]^t \tag{8-204}$$

an unbiased estimator vector for $\underline{\theta}$. Then the covariance matrix for $\underline{T}(\mathbf{x})$ is given by

$$\text{Cov}\{\underline{T}(\mathbf{x})\} = E[\{\underline{T}(\mathbf{x}) - \underline{\theta}\}\{\underline{T}(\mathbf{x}) - \underline{\theta}\}^t] \tag{8-205}$$

Note that $\text{Cov}\{\underline{T}(\mathbf{x})\}$ is of size $m \times m$ and represents a positive definite matrix. The Cramer–Rao bound in this case has the form[8]

$$\text{Cov}\{\underline{T}(\mathbf{x})\} \geq J^{-1}(\underline{\theta}) \tag{8-206}$$

where $J(\underline{\theta})$ represents the $m \times m$ Fisher information matrix associated with the parameter set $\underline{\theta}$ under the p.d.f. $f(\mathbf{x}; \underline{\theta})$. The entries of the Fisher information matrix are given by

$$J_{ij} \stackrel{\Delta}{=} E\left(\frac{\partial \log f(\mathbf{x}; \underline{\theta})}{\partial \theta_i} \frac{\partial \log f(\mathbf{x}; \underline{\theta})}{\partial \theta_j}\right) \qquad i, j = 1 \to m \tag{8-207}$$

To prove (8-206), we can make use of (8-87) that is valid for every parameter θ_k, $k = 1 \to m$. From (8-87) we obtain

$$E\left(\{T_k(\mathbf{x}) - \theta_k\}\frac{\partial \log f(\mathbf{x}; \underline{\theta})}{\partial \theta_k}\right) = 1 \qquad k = 1 \to m \tag{8-208}$$

Also

$$\begin{aligned} E\left(\{T_k(\mathbf{x}) - \theta_k\}\frac{\partial \log f(\mathbf{x}; \underline{\theta})}{\partial \theta_j}\right) &= \int_{-\infty}^{\infty} \{T_k(\mathbf{x}) - \theta_k\}\frac{\partial f(\mathbf{x}; \underline{\theta})}{\partial \theta_j}\, dx \\ &= \frac{\partial}{\partial \theta_j}\int_{-\infty}^{\infty} T_k(\mathbf{x}) f(\mathbf{x}; \underline{\theta})\, dx - \theta_k \int_{-\infty}^{\infty} \frac{\partial f(\mathbf{x}; \underline{\theta})}{\partial \theta_j}\, dx \\ &= \frac{\partial \theta_k}{\partial \theta_j} = 0 \qquad k \neq j = 1 \to m \end{aligned} \tag{8-209}$$

[8]In (8-206), the notation $A \geq B$ is used to indicate that the matrix $A - B$ is a nonnegative-definite matrix. Strict inequality as in (8-218) would imply positive-definiteness.

To exploit (8-208) and (8-209), define the $2m \times 1$ vector

$$\mathbf{Z} = \begin{bmatrix} T_1(\mathbf{x}) - \theta_1 \\ T_2(\mathbf{x}) - \theta_2 \\ \vdots \\ T_m(\mathbf{x}) - \theta_m \\ \hline \dfrac{\partial \log f(\mathbf{x}; \underline{\theta})}{\partial \theta_1} \\ \dfrac{\partial \log f(\mathbf{x}; \underline{\theta})}{\partial \theta_2} \\ \vdots \\ \dfrac{\partial \log f(\mathbf{x}; \underline{\theta})}{\partial \theta_m} \end{bmatrix} \triangleq \begin{bmatrix} \mathbf{Y}_1 \\ \hline \mathbf{Y}_2 \end{bmatrix} \tag{8-210}$$

Then using the regularity conditions

$$E\{\mathbf{Z}\} = 0 \tag{8-211}$$

and hence

$$\text{Cov}\{\mathbf{Z}\} = E\{\mathbf{Z}\mathbf{Z}^t\} = E\begin{pmatrix} \mathbf{Y}_1\mathbf{Y}_1^t & \mathbf{Y}_1\mathbf{Y}_2^t \\ \mathbf{Y}_2\mathbf{Y}_1^t & \mathbf{Y}_2\mathbf{Y}_2^t \end{pmatrix} \tag{8-212}$$

But

$$E\{\mathbf{Y}_1\mathbf{Y}_1^t\} = \text{Cov}\{T(\mathbf{x})\} \tag{8-213}$$

and from (8-208)–(8-209)

$$E\{\mathbf{Y}_1\mathbf{Y}_2^t\} = I_m \tag{8-214}$$

the identity matrix, and from (8-207)

$$E\{\mathbf{Y}_2\mathbf{Y}_2^t\} = J \tag{8-215}$$

the Fisher information matrix with entries as in (8-207). Thus

$$\text{Cov}\{\mathbf{Z}\} = \begin{pmatrix} \text{Cov}\{T(\mathbf{x})\} & I \\ I & J \end{pmatrix} \geq 0 \tag{8-216}$$

Using the matrix identity

$$\begin{pmatrix} A & B \\ C & D \end{pmatrix} \begin{pmatrix} I & 0 \\ -D^{-1}C & I \end{pmatrix} = \begin{pmatrix} A - BD^{-1}C & B \\ 0 & D \end{pmatrix} \tag{8-217}$$

with $A = \text{Cov}\{T(\mathbf{x})\}$, $B = C = I$ and $D = J$, we have

$$\begin{pmatrix} I & 0 \\ -D^{-1}C & I \end{pmatrix} = \begin{pmatrix} I & 0 \\ -J^{-1} & I \end{pmatrix} > 0 \tag{8-218}$$

and hence

$$\begin{pmatrix} A - BD^{-1}C & B \\ 0 & D \end{pmatrix} = \begin{pmatrix} \text{Cov}\{T(\mathbf{x})\} - J^{-1} & I \\ 0 & J \end{pmatrix} \geq 0 \tag{8-219}$$

which gives

$$\text{Cov}\{T(\mathbf{x})\} - J^{-1} \geq 0$$

or

$$\text{Cov}\{T(\mathbf{x})\} \geq J^{-1}(\underline{\theta}) \tag{8-220}$$

the desired result. In particular,

$$\text{Var}\{T_k(\mathbf{x})\} \geq J^{kk} \stackrel{\Delta}{=} (J^{-1})_{kk} \qquad k = 1 \to m \tag{8-221}$$

Interestingly (8-220)–(8-221) can be used to evaluate the degradation factor for the CR bound caused by the presence of other unknown parameters in the scene. For example, when θ_1 is the only unknown, the CR bound is given by J_{11}, whereas it is given by J^{11} in presence of other parameters $\theta_2, \theta_3, \ldots, \theta_m$. With

$$J = \left(\begin{array}{c|c} J_{11} & \underline{b}^T \\ \hline \underline{b} & G \end{array}\right) \tag{8-222}$$

in (8-215), from (8-217) we have

$$J^{11} = \frac{1}{J_{11} - \underline{b}^T G^{-1} \underline{b}} = \frac{1}{J_{11}} \left(\frac{1}{1 - \underline{b}^T G^{-1} \underline{b}/J_{11}} \right) \tag{8-223}$$

Thus $1/[1 - \underline{b}^T G^{-1} \underline{b}/J_{11}] > 1$ represents the effect of the remaining unknown parameters on the bound for θ_1. As a result, with one additional parameter, we obtain the increment in the bound for θ_1 to be

$$J^{11} - \frac{1}{J_{11}} = \frac{J_{12}^2}{J_{11} J_{22} - J_{12}^2} \geq 0$$

Improvements on the Cramer-Rao Bound: The Bhattacharya Bound

Referring to (8-102)–(8-108) and the related discussion there, it follows that if an efficient estimator does not exist, then the Cramer–Rao bound will not be tight. Thus the variance of the UMVUE in such situations will be larger than the Cramer–Rao bound. An interesting question in these situations is to consider techniques for possibly tightening the Cramer–Rao lower bound. Recalling that the CR bound only makes use of the first order derivative of the joint p.d.f. $f(\mathbf{x};\theta)$, it is natural to analyze the situation using higher order derivatives of $f(\mathbf{x};\theta)$, assuming that they exist.

To begin with, we can rewrite (16-52) as

$$\begin{aligned}
&\int_{-\infty}^{\infty} \{T(\mathbf{x}) - \theta\} \frac{\partial f(\mathbf{x};\theta)}{\partial \theta}\, dx \\
&\quad = \int_{-\infty}^{\infty} \left\{ \{T(\mathbf{x}) - \theta\} \frac{1}{f(\mathbf{x};\theta)} \frac{\partial f(\mathbf{x};\theta)}{\partial \theta} \right\} f(\mathbf{x};\theta)\, dx \\
&\quad = E\left(\{T(\mathbf{x}) - \theta\} \frac{1}{f(\mathbf{x};\theta)} \frac{\partial f(\mathbf{x};\theta)}{\partial \theta} \right) = 1
\end{aligned} \tag{8-224}$$

Similarly

$$E\left(\{T(\mathbf{x})-\theta\}\frac{1}{f(\mathbf{x};\theta)}\frac{\partial^k f(\mathbf{x};\theta)}{\partial\theta^k}\right)$$
$$=\frac{\partial^k}{\partial\theta^k}\int_{-\infty}^{\infty}T(\mathbf{x})f(\mathbf{x};\theta)\,dx-\theta\int_{-\infty}^{\infty}\frac{\partial^k f(\mathbf{x};\theta)}{\partial\theta^k}\,dx$$
$$=\frac{\partial^k}{\partial\theta^k}\int_{-\infty}^{\infty}T(\mathbf{x})f(\mathbf{x};\theta)\,dx=\frac{\partial^k\theta}{\partial\theta^k}=0 \qquad k\geq 2 \tag{8-225}$$

where we have repeatedly made use of the regularity conditions and higher order differentiability of $f(\mathbf{x};\theta)$. Motivated by (8-224) and (8-225), define the Bhattacharya random vector to be

$$\mathbf{Y}_m=\left[T(\mathbf{x})-\theta,\ \frac{1}{f}\frac{\partial f}{\partial\theta},\ \frac{1}{f}\frac{\partial^2 f}{\partial\theta^2},\ldots,\ \frac{1}{f}\frac{\partial^m f}{\partial\theta^m}\right]^t \tag{8-226}$$

Notice that $E\{\mathbf{Y}_m\}=0$, and using (8-224) and (8-225) we get the covariance matrix of $\mathbf{Y}_m$ to be

$$E\{\mathbf{Y}_m\mathbf{Y}_m^T\}=\begin{pmatrix} \mathrm{Var}\{T\} & 1 & 0 & 0 & \ldots & 0\\ 1 & J_{11} & B_{12} & B_{13} & \ldots & B_{1m}\\ 0 & B_{12} & B_{22} & B_{23} & \ldots & B_{2m}\\ 0 & B_{13} & B_{23} & B_{33} & \ldots & B_{3m}\\ \vdots & \vdots & \vdots & \vdots & \ldots & \vdots\\ 0 & B_{1m} & B_{2m} & B_{3m} & \ldots & B_{mm}\end{pmatrix} \tag{8-227}$$

where

$$B_{ij}\triangleq E\left\{\left(\frac{1}{f(\mathbf{x};\theta)}\frac{\partial^i f(\mathbf{x};\theta)}{\partial\theta^i}\right)\left(\frac{1}{f(\mathbf{x};\theta)}\frac{\partial^j f(\mathbf{x};\theta)}{\partial\theta^j}\right)\right\} \tag{8-228}$$

In particular,

$$B_{11}=J_{11}=E\left\{\left(\frac{\partial\log f(\mathbf{x};\theta)}{\partial\theta}\right)^2\right\} \tag{8-229}$$

represents the Fisher information about θ contained in the p.d.f. $f(\mathbf{x}\,|\,\theta)$. We will proceed under the assumption that $(1/f)(\partial^k f/\partial\theta^k)$, $k=1\rightarrow m$, are linearly independent (not completely correlated with each other) so that the matrix

$$B(m)\triangleq\begin{pmatrix} J_{11} & B_{12} & B_{13} & \ldots & B_{1m}\\ B_{12} & B_{22} & B_{23} & \ldots & B_{2m}\\ B_{13} & B_{23} & B_{33} & \ldots & B_{3m}\\ \vdots & \vdots & \vdots & \ldots & \vdots\\ B_{1m} & B_{2m} & B_{3m} & \ldots & B_{mm}\end{pmatrix} \tag{8-230}$$

is a full rank symmetric (positive definite) matrix for $m\geq 1$. Using the determinantal identity [refer to (8-217)]

$$\det\begin{pmatrix} A & B\\ C & D\end{pmatrix}=\det D\det(A-BD^{-1}C) \tag{8-231}$$

on (8-227) and (8-230) with $A = \text{Var}\{T\}$, $B = [1, 0, 0, \ldots, 0] = C^t$ and $D = B(m)$, we get

$$\det E\{\mathbf{Y}_m \mathbf{Y}_m^t\} = \det B(m)[\text{Var}\{T\} - BB^{-1}(m)C]$$
$$= \det B(m)[\text{Var}\{T\} - B^{11}(m)] \geq 0 \tag{8-232}$$

or

$$\text{Var}\{T(\mathbf{x})\} \geq B^{11}(m) \triangleq [B^{-1}(m)]_{11} \tag{8-233}$$

Since $B^{11}(m) \geq 1/[B_{11}(m)] = 1/J_{11}$, clearly (8-233) always represents an improvement over the CR bound. Using another well known matrix identity,[9] we get

$$B^{-1}(m+1) = \left(\begin{array}{c|c} B^{-1}(m) + q_m \mathbf{c}\mathbf{c}^t & -q_m \mathbf{c} \\ \hline -q_m \mathbf{c}^t & q_m \end{array}\right) \tag{8-234}$$

where

$$\mathbf{c} = \begin{pmatrix} c_1 \\ c_2 \\ \vdots \\ c_m \end{pmatrix} = B^{-1}(m)\mathbf{b}$$

with

$$\mathbf{b} = [B_{1,m+1}, B_{2,m+1}, \ldots, B_{m,m+1}]^t$$

and

$$q_m = \frac{1}{B_{m+1,m+1} - \mathbf{b}^t B^{-1}(m)\mathbf{b}} > 0$$

From (8-234)

$$B^{11}(m+1) = B^{11}(m) + c_1^2 q_m \tag{8-235}$$

Thus the new bound in (8-233) represents a monotone nondecreasing sequence, i.e.,

$$B^{11}(m+1) \geq B^{11}(m) \qquad m = 1, 2, \ldots \tag{8-236}$$

Higher Order Improvement Factors

To obtain $B^{11}(m)$ explicitly in (8-233), we resort to (8-230). Let M_{ij} denote the minor of B_{ij} in (8-230). Then

$$B^{11}(m) = \frac{M_{11}}{\det B(m)} = \frac{M_{11}}{J_{11}M_{11} - \sum_{k=2}^{m}(-1)^k B_{1k} M_{1k}} = \frac{1}{J_{11}}\left(\frac{1}{1-\varepsilon}\right) \tag{8-237}$$

9

$$\begin{pmatrix} A & B \\ C & D \end{pmatrix}^{-1} = \begin{pmatrix} A^{-1} + F_1 E F_2 & -F_1 E \\ -E F_2 & E \end{pmatrix}$$

where $E = (D - CA^{-1}B)^{-1}$, $F_1 = A^{-1}B$, and $F_2 = CA^{-1}$.

where

$$\varepsilon = \sum_{k=2}^{m} \frac{(-1)^k B_{1k} M_{1k}}{J_{11} M_{11}} \tag{8-238}$$

Clearly, since $B^{11}(m) > 0$ and represents an improvement over $1/J_{11}$, ε is always strictly less than unity and greater than or equal to zero; $0 \le \varepsilon < 1$. Moreover, it is also a function of n and m. Thus

$$B^{11}(m) = \frac{1}{J_{11}} + \frac{\varepsilon(n,m)}{J_{11}} + \frac{\varepsilon^2(n,m)}{J_{11}} + \cdots \tag{8-239}$$

Note that the first term is independent of m and the remaining terms represent the improvement over the CR bound using this approach. Clearly, if $T(\mathbf{x})$ is first-order efficient then $\varepsilon = 0$, and hence the higher order terms are present only when an efficient estimator does not exist. These terms can be easily computed for small values of m. In fact for $m = 2$, (8-230)–(8-233) yields,

$$\text{Var}\{T(\mathbf{x})\} \ge \frac{1}{J_{11}} + \frac{B_{12}^2}{J_{11}\left(J_{11}B_{22} - B_{12}^2\right)} \tag{8-240}$$

When the observations are independent and identically distributed, we have[10]

$$\begin{aligned} J_{11} &= nj_{11} \\ B_{12} &= nb_{12} \\ B_{22} &= nb_{22} + 2n(n-1)j_{11}^2 \end{aligned} \tag{8-241}$$

and hence

$$\begin{aligned} \text{Var}\{T(\mathbf{x})\} &\ge \frac{1}{nj_{11}} + \frac{b_{12}^2}{2n^2 j_{11}^4 \left[1 + \left(j_{11}b_{22} - b_{12}^2 - 2j_{11}^3\right)/2nj_{11}^3\right]} \\ &= \frac{1}{nj_{11}} + \frac{b_{12}^2}{2n^2 j_{11}^4} + o(1/n^2) \end{aligned} \tag{8-242}$$

Similarly for $m = 3$, we start with

$$\begin{vmatrix} \text{Var}\{T\} & 1 & 0 & 0 \\ 1 & J_{11} & B_{12} & B_{13} \\ 0 & B_{12} & B_{22} & B_{23} \\ 0 & B_{13} & B_{23} & B_{33} \end{vmatrix} \ge 0 \tag{8-243}$$

and for independent observations, (8-241) together with

$$\begin{aligned} B_{13} &= nb_{13} \\ B_{23} &= nb_{23} + 6n(n-1)b_{12}j_{11} \\ B_{33} &= nb_{33} + 9n(n-1)\left(b_{22}j_{11} + b_{12}^2\right) + 6n(n-1)(n-2)j_{11}^3 \end{aligned} \tag{8-244}$$

we have

$$\text{Var}\{T\} \ge \frac{1}{nj_{11}} + \frac{b_{12}^2}{2n^2 j_{11}^4} + \frac{c(3)}{n^3} + o(1/n^3) \tag{8-245}$$

---

[10]Here $j_{11}, b_{12}, b_{22}, b_{13}, \ldots$ correspond to their counterparts $J_{11}, B_{12}, B_{22}, B_{13}, \ldots$ evaluated for $n = 1$.

where $c(3)$ turns out to be

$$c(3) = \frac{2j_{11}^2 b_{13}^2 + b_{12}^2(6j_{11}^3 + 21b_{12}^2 - 3j_{11}b_{22} - 12j_{11}b_{13})}{12j_{11}^7} \tag{8-246}$$

In general, if we consider $\mathbf{Y}_m$, then from (16-110) we have

$$\mathrm{Var}\{T(\mathbf{x})\} \geq \frac{a(m)}{n} + \frac{b(m)}{n^2} + \frac{c(m)}{n^3} + \frac{d(m)}{n^4} + \cdots + o(1/n^m) \tag{8-247}$$

From (8-239) and (8-241), it is easily seen that

$$a(m) = \frac{1}{j_{11}} \stackrel{\Delta}{=} a \tag{8-248}$$

i.e., the $1/n$ term is independent of m and equals the CR bound. From (8-242) and (8-245) we obtain

$$b(2) = b(3) = \frac{b_{12}^2}{2j_{11}^4} \tag{8-249}$$

Infact, it can be shown that

$$b(m) = b(2) = \frac{b_{12}^2}{2j_{11}^4} \stackrel{\Delta}{=} b \tag{8-250}$$

Thus, the $1/n^2$ term is also independent of m. However, this is no longer true for $c(k), d(k), \ldots, (k \geq 3)$. Their exact values depend on m because of the contributions from higher indexed terms on the right. To sum up, if an estimator $T(\mathbf{x})$ is no longer efficient, but the $1/n$ term as well as the $1/n^2$ term in its variance agree with the corresponding terms in (8-247)–(8-250), then $T(\mathbf{x})$ is said to be second order efficient. Next, we will illustrate the role of these higher order terms through various examples.

EXAMPLE 8-18

▶ Let $\mathbf{x}_i, i = 1 \to n$, represent independent and identically distributed (i.i.d.) samples from $N(\mu, 1)$, and let μ be unknown. Further, let $\theta = \mu^2$ be the parameter of interest. From Example 8-17, $\mathbf{z} = (1/n)\sum_{i=1}^n \mathbf{x}_i = \bar{\mathbf{x}}$ represents the sufficient statistic in this case, and hence the UMVUE for θ must be a function of $\mathbf{z}$ alone. From (8-135)–(8-136), since

$$T = \mathbf{z}^2 - 1/n \tag{8-251}$$

is an unbiased estimator for μ^2, from the Rao–Blackwell theorem it is also the UMVUE for θ. Moreover, from (8-140)

$$\mathrm{Var}\{T\} = E\left(\mathbf{z}^4 - \frac{2\mathbf{z}^2}{n} + \frac{1}{n^2}\right) - \mu^4 = \frac{4\mu^2}{n} + \frac{2}{n^2} \tag{8-252}$$

Clearly, no unbiased estimator for θ can have lower variance than (8-252). In this case using Prob. 8-41 the CR bound is [see also (8-134)]

$$\frac{1}{J_{11}} = \frac{(\partial\theta/\partial\mu)^2}{nE\{(\partial \log f/\partial\mu)^2\}} = \frac{(2\mu)^2}{nE\{(\mathbf{x}-\mu)^2\}} = \frac{4\mu^2}{n} \tag{8-253}$$

Though, $1/J_{11} < \text{Var}\{T\}$ here, it represents the first-order term that agrees with the corresponding $1/n$ term in (8-252). To evaluate the second-order term, from (8-132) with $n = 1$ we have

$$\frac{1}{f}\frac{\partial^2 f}{\partial\theta^2} = \frac{\partial^2 \log f}{\partial\theta^2} + \left(\frac{\partial \log f}{\partial\theta}\right)^2 = -\frac{\mathbf{x}}{4\mu^3} + \left(\frac{\mathbf{x}-\mu}{2\mu}\right)^2 \tag{8-254}$$

so that

$$b_{12} = E\left[\left(\frac{\mathbf{x}-\mu}{2\mu}\right)\left(\frac{-\mathbf{x}}{4\mu^3} + \left(\frac{\mathbf{x}-\mu}{2\mu}\right)^2\right)\right] = \frac{-1}{8\mu^4} \tag{8-255}$$

and from (8-248)–(8-250), we have

$$\frac{b}{n^2} = \frac{b_{12}^2}{2n^2 j_{11}^4} = \frac{2}{n^2} \tag{8-256}$$

Comparing (8-252) with (8-253) and (8-256), we observe that the UMVUE in (8-251) is in fact second-order efficient. ◀

EXAMPLE 8-19 ▶ Let $\mathbf{x}_i$, $i = 1 \to n$, be independent and identically distributed Poisson random variables with common parameter λ. Also let $\theta = \lambda^2$ be the unknown parameter. Once again $\mathbf{z} = \overline{\mathbf{x}}$ is the sufficient statistic for λ and hence θ. Since

$$E\{\mathbf{z}\} = \lambda \quad \text{and} \quad E\{\mathbf{z}^2\} = \lambda^2 + \lambda/n \tag{8-257}$$

we have

$$T = \mathbf{z}^2 - \frac{\mathbf{z}}{n} \tag{8-258}$$

to be an unbiased estimator and the UMVUE for $\theta = \lambda^2$. By direct calculation,

$$\text{Var}\{T\} = E\left(\mathbf{z}^4 - \frac{2}{n}\mathbf{z}^3 + \frac{\mathbf{z}^2}{n^2}\right) - \lambda^4 = \frac{4\lambda^3}{n} + \frac{2\lambda^2}{n^2} \tag{8-259}$$

Next we will compute the CR bound. Since for $n = 1$

$$\log f = -\lambda + \mathbf{x}\log\lambda - \log(\mathbf{x}!) = -\sqrt{\theta} + \mathbf{x}\log(\sqrt{\theta}) - \log(\mathbf{x}!) \tag{8-260}$$

we have

$$\frac{\partial \log f}{\partial\theta} = \frac{-1}{2\lambda} + \frac{\mathbf{x}}{2\lambda^2} \tag{8-261}$$

$$\frac{\partial^2 \log f}{\partial\theta^2} = \frac{1}{4\lambda^3} - \frac{\mathbf{x}}{2\lambda^4} \tag{8-262}$$

or

$$j_{11} = -E\left(\frac{\partial^2 \log f}{\partial\theta^2}\right) = -\frac{1}{4\lambda^3} + \frac{\mathbf{x}}{2\lambda^4} = \frac{1}{4\lambda^3} \tag{8-263}$$

Thus, the CR bound in this case is $1/nj_{11} = 4\lambda^3/n$ and it agrees with the first term in (8-259). To determine whether $T(\mathbf{x})$ is second-order efficient, using (8-261) together with (8-262) gives [see (8-254)]

$$b_{12} = E\left\{\frac{\partial \log f}{\partial \theta}\left(\frac{\partial^2 \log f}{\partial \theta^2} + \left(\frac{\partial \log f}{\partial \theta}\right)^2\right)\right\}$$

$$= E\left\{\left(\frac{-1}{2\lambda} + \frac{\mathbf{x}}{2\lambda^2}\right)\left(\left(\frac{1}{4\lambda^3} - \frac{\mathbf{x}}{2\lambda^4}\right) + \left(\frac{1}{2\lambda} - \frac{\mathbf{x}}{2\lambda^2}\right)^2\right)\right\} = \frac{-1}{8\lambda^5} \tag{8-264}$$

and hence the second-order term

$$\frac{b}{n^2} = \frac{b_{12}^2}{2n^2 j_{11}^4} = \frac{2\lambda^2}{n^2} \tag{8-265}$$

which also agrees with second term in (8-259). Again, T given by (8-258) is a second-order efficient estimator. Next, we will consider a multiparameter example that is efficient only in the second-order Bhattacharya sense. ◀

EXAMPLE 8-20

▶ Let $\mathbf{x}_i \sim N(\mu, \sigma^2)$, $i = 1 \to n$, represent independent and identically distributed Gaussian random variables, where both μ and σ^2 are unknown. From Example 8-14, $\mathbf{z}_1 = \overline{\mathbf{x}}$ and $\mathbf{z}_2 = \sum_{i=1}^n \mathbf{x}_i^2$ form the sufficient statistic in this case, and since the estimates

$$\hat{\mu} = \frac{1}{n}\sum_{i=1}^n \mathbf{x}_i = \overline{\mathbf{x}} = \mathbf{z}_1 \tag{8-266}$$

and

$$\hat{\sigma}^2 = \sum_{i=1}^n \frac{(\mathbf{x}_i - \overline{\mathbf{x}})^2}{n-1} = \frac{\mathbf{z}_2 - n\mathbf{z}_1^2}{n-1} \tag{8-267}$$

are two independent unbiased estimates that are also functions of $\mathbf{z}_1$ and $\mathbf{z}_2$ only, they are UMVUEs for μ and σ^2, respectively. To verify that $\hat{\boldsymbol{\mu}}$ and $\hat{\boldsymbol{\sigma}}^2$ are in fact independent random variables, let

$$A = \begin{bmatrix} 1 & -1 & & & & & \\ 1 & 1 & -2 & & & \mathbf{0} & \\ 1 & 1 & 1 & -3 & & & \\ \vdots & \vdots & \vdots & \vdots & \ddots & & \\ 1 & 1 & 1 & 1 & \cdots & 1 & -(n-1) \\ 1 & 1 & 1 & 1 & \cdots & 1 & 1 \end{bmatrix} \tag{8-268}$$

so that

$$AA^T = \begin{bmatrix} 1 & -1 & & & & & \\ 1 & 1 & -2 & & & \mathbf{0} & \\ 1 & 1 & 1 & -3 & & & \\ \vdots & \vdots & \vdots & \vdots & \ddots & & \\ 1 & 1 & 1 & 1 & \cdots & 1 & -(n-1) \\ 1 & 1 & 1 & 1 & \cdots & 1 & 1 \end{bmatrix}\begin{bmatrix} 1 & 1 & 1 & \cdots & 1 & 1 \\ -1 & 1 & 1 & \cdots & 1 & 1 \\ & -2 & 1 & \cdots & 1 & 1 \\ & & -3 & \cdots & \vdots & \vdots \\ & \mathbf{0} & & \ddots & 1 & 1 \\ & & & & -(n-1) & 1 \end{bmatrix}$$

$$= \begin{bmatrix} 2 & & & & \\ & 6 & & \mathbf{0} & \\ & & \ddots & & \\ & \mathbf{0} & & n(n-1) & \\ & & & & n \end{bmatrix} \tag{8-269}$$

Thus

$$U = \begin{bmatrix} \frac{1}{\sqrt{2}} & & & & \\ & \frac{1}{\sqrt{6}} & & \mathbf{0} & \\ & & \ddots & & \\ & \mathbf{0} & & \frac{1}{\sqrt{n(n-1)}} & \\ & & & & \frac{1}{\sqrt{n}} \end{bmatrix} A = \begin{bmatrix} \frac{1}{\sqrt{2}} & -\frac{1}{\sqrt{2}} & & \mathbf{0} & \\ \frac{1}{\sqrt{6}} & \frac{1}{\sqrt{6}} & -\frac{1}{\sqrt{6}} & & \\ \vdots & \vdots & \vdots & \ddots & \\ \frac{1}{\sqrt{n}} & \frac{1}{\sqrt{n}} & \frac{1}{\sqrt{n}} & \cdots & \frac{1}{\sqrt{n}} \end{bmatrix} \tag{8-270}$$

is an orthogonal matrix ($UU^t = I$). Let $\mathbf{x} = (\mathbf{x}_1, \mathbf{x}_2, \ldots, \mathbf{x}_n)^t$, and define

$$\mathbf{y} = \begin{pmatrix} \mathbf{y}_1 \\ \mathbf{y}_2 \\ \vdots \\ \mathbf{y}_n \end{pmatrix} = U\mathbf{x} = \begin{bmatrix} \frac{1}{\sqrt{2}} & -\frac{1}{\sqrt{2}} & & \mathbf{0} & \\ \frac{1}{\sqrt{6}} & \frac{1}{\sqrt{6}} & -\frac{1}{\sqrt{6}} & & \\ \vdots & \vdots & \vdots & \ddots & \\ \frac{1}{\sqrt{n}} & \frac{1}{\sqrt{n}} & \frac{1}{\sqrt{n}} & \cdots & \frac{1}{\sqrt{n}} \end{bmatrix} \begin{bmatrix} \mathbf{x}_1 \\ \mathbf{x}_2 \\ \vdots \\ \mathbf{x}_n \end{bmatrix} = \begin{bmatrix} \frac{\mathbf{x}_1 - \mathbf{x}_2}{\sqrt{2}} \\ \frac{\mathbf{x}_1 + \mathbf{x}_2 - 2\mathbf{x}_3}{\sqrt{6}} \\ \vdots \\ \frac{\mathbf{x}_1 + \mathbf{x}_2 + \cdots + \mathbf{x}_n}{\sqrt{n}} \end{bmatrix} \tag{8-271}$$

which gives

$$\mathbf{y}_n = \sqrt{n}\,\bar{\mathbf{x}} \sim N(\sqrt{n}\mu, \sigma^2) \tag{8-272}$$

From (8-271)

$$E[\mathbf{y}\mathbf{y}^t] = U\,E[\mathbf{x}\mathbf{x}^t]\,U^t = U(\sigma^2 I)U^t = \sigma^2 I \tag{8-273}$$

and hence $(\mathbf{y}_1, \mathbf{y}_2, \ldots, \mathbf{y}_n)$ are independent random variables since $\mathbf{x}_1, \mathbf{x}_2, \ldots, \mathbf{x}_n$ are independent random variables. Moreover,

$$\sum_{i=1}^{n} \mathbf{y}_i^2 = \mathbf{y}^t\mathbf{y} = \mathbf{x}^t U^t U \mathbf{x} = \mathbf{x}^t\mathbf{x} = \sum_{i=1}^{n} \mathbf{x}_i^2 \tag{8-274}$$

and hence

$$(n-1)\hat{\sigma}^2 = \sum_{i=1}^{n}(\mathbf{x}_i - \bar{\mathbf{x}})^2 = \sum_{i=1}^{n}\mathbf{x}_i^2 - n\bar{\mathbf{x}}^2 = \sum_{i=1}^{n}\mathbf{y}_i^2 - \mathbf{y}_n^2 = \sum_{i=1}^{n-1}\mathbf{y}_i^2 \tag{8-275}$$

where we have used (8-272) and (8-274). From (8-273) and (8-275)

$$\hat{\mu} = \bar{\mathbf{x}} = \frac{1}{\sqrt{n}}\mathbf{y}_n \quad \text{and} \quad \hat{\sigma}^2 = \frac{1}{n-1}\sum_{i=1}^{n-1}\mathbf{y}_i^2 \tag{8-276}$$

are independent random variables since $\mathbf{y}_1, \mathbf{y}_2, \ldots, \mathbf{y}_n$ are all independent random variables. Moreover, since $\mathbf{y}_i \sim N(0, \sigma^2), i = 1 \to n-1, \mathbf{y}_n \sim N(\sqrt{n}\mu, \sigma^2)$, using (8-272) and (8-275) we obtain

$$\text{Var}\{\hat{\mu}\} = \sigma^2/n \quad \text{and} \quad \text{Var}\{\hat{\sigma}^2\} = 2\sigma^4/(n-1) \tag{8-277}$$

To determine whether these estimators are efficient, let $\underline{\theta} = (\mu, \sigma^2)^t$ represent the multiparameter vector. Then with

$$\mathbf{Z}_k \triangleq \left(\frac{1}{f}\frac{\partial^k f}{\partial \mu^k}, \frac{1}{f}\frac{\partial^k f}{\partial (\sigma^2)^k}\right)^t \tag{8-278}$$

a direct calculation shows

$$\mathbf{Z}_1 = \begin{pmatrix} \dfrac{\partial \log f(\mathbf{x}\,;\underline{\theta})}{\partial \mu} \\ \dfrac{\partial \log f(\mathbf{x}\,;\underline{\theta})}{\partial \sigma^2} \end{pmatrix} = \begin{pmatrix} \dfrac{\sum_{i=1}^n (\mathbf{x}_i - \mu)}{\sigma^2} \\ -\dfrac{n}{2\sigma^2} + \dfrac{\sum_{i=1}^n (\mathbf{x}_i - \mu)^2}{2\sigma^4} \end{pmatrix}$$

and the 2×2 Fisher information matrix J to be [see (8-206)–(8-207)]

$$J = E\left(Z_1 Z_1^T\right) = \begin{bmatrix} n/\sigma^2 & 0 \\ 0 & n/2\sigma^4 \end{bmatrix} \tag{8-279}$$

Note that in (8-279)

$$\begin{aligned} J_{22} &= E\left[\left(-\frac{n}{2\sigma^2} + \frac{\sum_{i=1}^n (\mathbf{x}_i - \mu)^2}{2\sigma^4}\right)^2\right] \\ &= \frac{1}{\sigma^4}\left(\frac{n^2}{4} - \frac{n\sum_{i=1}^n E[(\mathbf{x}_i - \mu)^2]}{2\sigma^2} + \frac{\sum_{i=1}^n E[(\mathbf{x}_i - \mu)^4]}{4\sigma^4}\right) \\ &\quad + \frac{1}{\sigma^4}\left(\frac{\sum_{i=1}^n \sum_{j=1, i\neq j}^n E[(\mathbf{x}_i - \mu)^2(\mathbf{x}_j - \mu)^2]}{4\sigma^4}\right) \\ &= \frac{1}{\sigma^4}\left(\frac{n^2}{4} - \frac{n^2\sigma^2}{2\sigma^2} + \frac{3n\sigma^4}{4\sigma^4} + \frac{n(n-1)\sigma^4}{4\sigma^4}\right) \\ &= \frac{1}{\sigma^4}\left(\frac{n^2}{4} - \frac{n^2}{2} + \frac{3n + n^2 - n}{4}\right) = \frac{n}{2\sigma^4} \end{aligned}$$

Thus $\text{Var}\{\hat{\mu}\} = \sigma^2/n = J^{11}$ whereas $\text{Var}\{\hat{\sigma}^2\} = 2\sigma^4/(n-1) > 2\sigma^4/n = J^{22}$, implying that $\hat{\mu}$ is an efficient estimator whereas $\hat{\sigma}^2$ is not efficient. However, since $\hat{\sigma}^2$ is a function of the sufficient statistic $\mathbf{z}_1$ and $\mathbf{z}_2$ only, to check whether it is second-order efficient, after some computations, the block structured Bhattacharya matrix in (8-230) with $m = 2$ turns out to be

$$B(2) = E\left(\begin{bmatrix} \mathbf{Z}_1 \\ \mathbf{Z}_2 \end{bmatrix} \begin{bmatrix} \mathbf{Z}_1^t \mathbf{Z}_2^t \end{bmatrix}\right) = \begin{bmatrix} n/\sigma^2 & 0 & 0 & 0 \\ 0 & n/2\sigma^4 & n/\sigma^4 & 0 \\ 0 & n/\sigma^4 & 2n^2/\sigma^4 & 0 \\ 0 & 0 & 0 & n(n+2)/2\sigma^8 \end{bmatrix} \tag{8-280}$$

and this gives the "extended inverse" Fisher information matrix at the (1,1) block location of $B^{-1}(2)$ to be

$$[B(2)]_{11}^{-1} \triangleq \begin{bmatrix} \sigma^2/n & 0 \\ 0 & 2\sigma^4/(n-1) \end{bmatrix} \tag{8-281}$$

Comparing (8-277) and (8-281) we have

$$E[(\hat{\underline{\theta}} - \underline{\theta})(\hat{\underline{\theta}} - \underline{\theta})^t] = \begin{bmatrix} \text{Var}\{\hat{\mu}\} & 0 \\ 0 & \text{Var}\{\hat{\sigma}^2\} \end{bmatrix} = [B(2)]_{11}^{-1}. \tag{8-282}$$

Thus (8-233) is satisfied with equality in this case for $m = 2$ and hence $\hat{\sigma}^2$ is a second-order efficient estimator. ◀

More on Maximum Likelihood Estimators

Maximum likelihood estimators (MLEs) have many remarkable properties, especially for large sample size. To start with, it is easy to show that if an ML estimator exists, then it is only a function of the sufficient statistic T for that family of p.d.f. $f(\mathbf{x};\theta)$. This follows from (8-60), for in that case

$$f(\mathbf{x};\theta) = h(\mathbf{x})g_\theta(T(\mathbf{x})) \tag{8-283}$$

and (8-40) yields

$$\frac{\partial \log f(\mathbf{x};\theta)}{\partial\theta} = \left.\frac{\partial \log g_\theta(T)}{\partial\theta}\right|_{\theta=\hat{\theta}_{\text{ML}}} = 0 \tag{8-284}$$

which shows $\hat{\theta}_{\text{ML}}$ to be a function of T alone. However, this does not imply that the MLE is itself a sufficient statistic all the time, although this is usually true.

We have also seen that if an efficient estimator exists for the Cramer–Rao bound, then it is the ML estimator.

Suppose $\psi(\theta)$ is an unknown parameter that depends on θ. Then the ML estimator for ψ is given by $\psi(\hat{\boldsymbol{\theta}}_{\text{ML}})$, i.e.,

$$\hat{\psi}_{\text{ML}} = \psi(\hat{\boldsymbol{\theta}}_{\text{ML}}) \tag{8-285}$$

Note that this important property is not characteristic of unbiased estimators.

If the ML estimator is a unique solution to the likelihood equation, then under some additional restrictions and the regularity conditions, for large sample size we also have

(i)
$$E\{\hat{\boldsymbol{\theta}}_{\text{ML}}\} \to \theta \tag{8-286}$$

(ii)
$$\text{Var}\{\hat{\boldsymbol{\theta}}_{\text{ML}}\} \to \sigma^2_{\text{CR}} \triangleq E\left[\left(\frac{\partial \log f(\mathbf{x};\theta)}{\partial\theta}\right)^2\right]^{-1} \tag{8-287}$$

and

(iii)
$$\hat{\theta}_{\text{ML}}(\mathbf{x}) \text{ is also asymptotically normal.} \tag{8-288}$$

Thus, as $n \to \infty$

$$\frac{\hat{\theta}_{\text{ML}}(\mathbf{x}) - \theta}{\sigma_{\text{CR}}} \to N(0, 1) \tag{8-289}$$

i.e., asymptotically ML estimators are consistent and possess normal distributions.

8-4 HYPOTHESIS TESTING

A statistical hypothesis is an assumption about the value of one or more parameters of a statistical model. Hypothesis testing is a process of establishing the validity of a hypothesis. This topic is fundamental in a variety of applications: Is Mendel's theory of heredity valid? Is the number of particles emitted from a radioactive substance Poisson distributed? Does the value of a parameter in a scientific investigation equal a specific constant? Are two events independent? Does the mean of a random variable change if certain factors of the experiment are modified? Does smoking decrease life expectancy? Do voting patterns depend on sex? Do IQ scores depend on parental education? The list is endless.

We shall introduce the main concepts of hypothesis testing in the context of the following problem: The distribution of a random variable $\mathbf{x}$ is a known function $F(x, \theta)$ depending on a parameter θ. We wish to test the assumption $\theta = \theta_0$ against the assumption $\theta \neq \theta_0$. The assumption that $\theta = \theta_0$ is denoted by H_0 and is called the *null hypothesis.* The assumption that $\theta \neq \theta_0$ is denoted by H_1 and is called the *alternative hypothesis.* The values that θ might take under the alternative hypothesis form a set Θ_1 in the parameter space. If Θ_1 consists of a single point $\theta = \theta_1$, the hypothesis H_1 is called *simple;* otherwise, it is called *composite.* The null hypothesis is in most cases simple.

The purpose of hypothesis testing is to establish whether experimental evidence supports the rejection of the null hypothesis. The decision is based on the location of the observed sample X of $\mathbf{x}$. Suppose that under hypothesis H_0 the density $f(X, \theta_0)$ of the sample vector $\mathbf{X}$ is negligible in a certain region D_c of the sample space, taking significant values only in the complement $\overline{D}_c$ of D_c. It is reasonable then to reject H_0 if X is in D_c and to accept H_0 if X is in $\overline{D}_c$. The set D_c is called the *critical region* of the test and the set $\overline{D}_c$ is called the *region of acceptance* of H_0. The test is thus specified in terms of the set D_c.

We should stress that the purpose of hypothesis testing is not to determine whether H_0 or H_1 is true. It is to establish whether the evidence supports the rejection of H_0. The terms "accept" and "reject" must, therefore, be interpreted accordingly. Suppose, for example, that we wish to establish whether the hypothesis H_0 that a coin is fair is true. To do so, we toss the coin 100 times and observe that heads show k times. If $k = 15$, we reject H_0, that is, we decide on the basis of the evidence that the fair-coin hypothesis should be rejected. If $k = 49$, we accept H_0, that is, we decide that the evidence does not support the rejection of the fair-coin hypothesis. The evidence alone, however, does not lead to the conclusion that the coin is fair. We could have as well concluded that $p = 0.49$.

In hypothesis testing two kinds of errors might occur depending on the location of X:

1. Suppose first that H_0 is true. If $X \in D_c$, we reject H_0 even though it is true. We then say that a *Type I error* is committed. The probability for such an error is denoted by α and is called the *significance level* of the test. Thus

$$\alpha = P\{\mathbf{X} \in D_c \mid H_0\} \tag{8-290}$$

The difference $1 - \alpha = P\{X \notin D_c \mid H_0\}$ equals the probability that we accept H_0 when true. In this notation, $P\{\cdots \mid H_0\}$ is not a conditional probability. The symbol H_0 merely indicates that H_0 is true.

2. Suppose next that H_0 is false. If $X \notin D_c$, we accept H_0 even though it is false. We then say that a *Type II error* is committed. The probability for such an error is a function $\beta(\theta)$ of θ called the *operating characteristic* (OC) of the test. Thus

$$\beta(\theta) = P\{\mathbf{X} \notin D_c \mid H_1\} \tag{8-291}$$

The difference $1 - \beta(\theta)$ is the probability that we reject H_0 when false. This is denoted by $P(\theta)$ and is called the *power of the test.* Thus

$$P(\theta) = 1 - \beta(\theta) = P\{\mathbf{X} \in D_c \mid H_1\} \tag{8-292}$$

Fundamental note Hypothesis testing is not a part of statistics. It is part of *decision theory* based on statistics. Statistical consideration alone cannot lead to a decision. They merely lead to the following probabilistic statements:

$$\begin{aligned} &\text{If } H_0 \text{ is true, then } P\{\mathbf{X} \in D_c\} = \alpha \\ &\text{If } H_0 \text{ is false, then } P\{\mathbf{X} \notin D_c\} = \beta(\theta) \end{aligned} \tag{8-293}$$

Guided by these statements, we "reject" H_0 if $\mathbf{X} \in D_c$ and we "accept" H_0 if $\mathbf{X} \notin D_c$. These decisions are not based on (8-293) alone. They take into consideration other, often subjective, factors, for example, our prior knowledge concerning the truth of H_0, or the consequences of a wrong decision.

The test of a hypothesis is specified in terms of its critical region. The region D_c is chosen so as to keep the probabilities of both types of errors small. However, both probabilities cannot be arbitrarily small because a decrease in α results in an increase in β. In most applications, it is more important to control α. The selection of the region D_c proceeds thus as follows:

Assign a value to the Type I error probability α and search for a region D_c of the sample space so as to minimize the Type II error probability for a specific θ. If the resulting $\beta(\theta)$ is too large, increase α to its largest tolerable value; if $\beta(\theta)$ is still too large, increase the number n of samples.

A test is called *most powerful* if $\beta(\theta)$ is minimum. In general, the critical region of a most powerful test depends on θ. If it is the same for every $\theta \in \Theta_1$, the test is *uniformly most powerful.* Such a test does not always exist. The determination of the critical region of a most powerful test involves a search in the n-dimensional sample space. In the following, we introduce a simpler approach.

TEST STATISTIC. Prior to any experimentation, we select a function

$$\mathbf{q} = g(\mathbf{X})$$

of the sample vector $\mathbf{X}$. We then find a set R_c of the real line where under hypothesis H_0 the density of $\mathbf{q}$ is negligible, and we reject H_0 if the value $q = g(X)$ of $\mathbf{q}$ is in R_c. The set R_c is the *critical region* of the test; the random variable $\mathbf{q}$ is the *test statistic.* In the selection of the function $g(X)$ we are guided by the point estimate of θ.

In a hypothesis test based on a test statistic, the two types of errors are expressed in terms of the region R_c of the real line and the density $f_q(q, \theta)$ of the test statistic $\mathbf{q}$:

$$\alpha = P\{\mathbf{q} \in R_c \mid H_0\} = \int_{R_c} f_q(q, \theta_0)\, dq \tag{8-294}$$

$$\beta(\theta) = P\{\mathbf{q} \notin R_c \mid H_1\} = \int_{\bar{R}_c} f_q(q, \theta)\, dq \tag{8-295}$$

To carry out the test, we determine first the function $f_q(q, \theta)$. We then assign a value to α and we search for a region R_c minimizing $\beta(\theta)$. The search is now limited to the real line. We shall assume that the function $f_q(q, \theta)$ has a single maximum. This is the case for most practical tests.

Our objective is to test the hypothesis $\theta = \theta_0$ against each of the hypotheses $\theta \neq \theta_0$, $\theta > \theta_0$, and $\theta < \theta_0$. To be concrete, we shall assume that the function $f_q(q, \theta)$ is concentrated on the right of $f_q(q, \theta_0)$ for $\theta > \theta_0$ and on its left for $\theta < \theta_0$ as in Fig. 8-10.

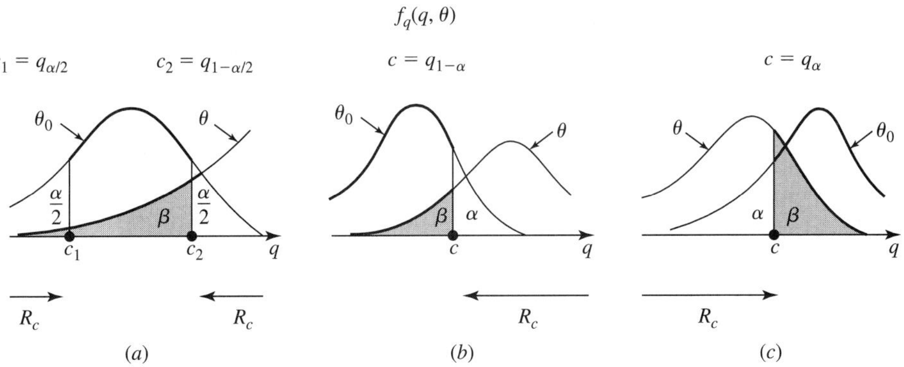

FIGURE 8-10

$H_1: \theta \neq \theta_0$

Under the stated assumptions, the most likely values of $\mathbf{q}$ are on the right of $f_q(q, \theta_0)$ if $\theta > \theta_0$ and on its left if $\theta < \theta_0$. It is, therefore, desirable to reject H_0 if $\mathbf{q} < c_1$ or if $\mathbf{q} > c_2$. The resulting critical region consists of the half-lines $q < c_1$ and $q > c_2$. For convenience, we shall select the constants c_1 and c_2 such that

$$P\{\mathbf{q} < c_1 \mid H_0\} = \frac{\alpha}{2} \qquad P\{\mathbf{q} > c_2 \mid H_0\} = \frac{\alpha}{2}$$

Denoting by q_u the u percentile of $\mathbf{q}$ under hypothesis H_0, we conclude that $c_1 = q_{\alpha/2}$, $c_2 = q_{1-\alpha/2}$. This yields the following test:

$$\text{Accept } H_0 \text{ iff } q_{\alpha/2} < q < q_{1-\alpha/2} \tag{8-296}$$

The resulting OC function equals

$$\beta(\theta) = \int_{q_{\alpha/2}}^{q_{1-\alpha/2}} f_q(q, \theta)\, dq \tag{8-297}$$

$H_1: \theta > \theta_0$

Under hypothesis H_1, the most likely values of q are on the right of $f_q(q, \theta)$. It is, therefore, desirable to reject H_0 if $q > c$. The resulting critical region is now the half-line $q > c$, where c is such that

$$P\{\mathbf{q} > c \mid H_0\} = \alpha \qquad c = q_{1-\alpha}$$

and the following test results:

$$\text{Accept } H_0 \text{ iff } q < q_{1-\alpha} \tag{8-298}$$

The resulting OC function equals

$$\beta(\theta) = \int_{-\infty}^{c} f_q(q, \theta)\, dq \tag{8-299}$$

$H_1: \theta < \theta_0$

Proceeding similarly, we obtain the critical region $q < c$ where c is such that

$$P\{\mathbf{q} < c \mid H_0\} = \alpha \qquad c = q_\alpha$$

This yields the following test:

$$\text{Accept } H_0 \text{ iff } q > q_\alpha \tag{8-300}$$

The resulting OC function equals

$$\beta(\theta) = \int_c^\infty f_q(q, \theta)\, dq \tag{8-301}$$

The test of a hypothesis thus involves the following steps: Select a test statistic $\mathbf{q} = g(\mathbf{X})$ and determine its density. Observe the sample X and compute the function $q = g(X)$. Assign a value to α and determine the critical region R_c. Reject H_0 iff $q \in R_c$.

In the following, we give several illustrations of hypothesis testing. The results are based on (8-296)–(8-301). In certain cases, the density of $\mathbf{q}$ is known for $\theta = \theta_0$ only. This suffices to determine the critical region. The OC function $\beta(\theta)$, however, cannot be determined.

MEAN. We shall test the hypothesis $H_0\colon \eta = \eta_0$ that the mean η of a random variable $\mathbf{x}$ equals a given constant η_0.

Known variance. We use as the test statistic the random variable

$$\mathbf{q} = \frac{\overline{\mathbf{x}} - \eta_0}{\sigma/\sqrt{n}} \tag{8-302}$$

Under the familiar assumptions, $\overline{x}$ is $N(\eta, \sigma/\sqrt{n})$; hence $\mathbf{q}$ is $N(\eta_q, 1)$ where

$$\eta_q = \frac{\eta - \eta_0}{\sigma/\sqrt{n}} \tag{8-303}$$

Under hypothesis H_0, $\mathbf{q}$ is $N(0, 1)$. Replacing in (8-296)–(8-301) the q_u percentile by the standard normal percentile z_u, we obtain the following test:

$$H_1\colon \eta \neq \eta_0 \qquad \text{Accept } H_0 \text{ iff } z_{\alpha/2} < q < z_{1-\alpha/2} \tag{8-304}$$

$$\beta(\eta) = P\{|\mathbf{q}| < z_{1-\alpha/2} \mid H_1\} = \mathrm{G}(z_{1-\alpha/2} - \eta_q) - \mathrm{G}(z_{\alpha/2} - \eta_q) \tag{8-305}$$

$$H_1\colon \eta > \eta_0 \qquad \text{Accept } H_0 \text{ iff } q < z_{1-\alpha} \tag{8-306}$$

$$\beta(\eta) = P\{\mathbf{q} < z_{1-\alpha} \mid H_1\} = \mathrm{G}(z_{1-\alpha} - \eta_q) \tag{8-307}$$

$$H_1\colon \eta < \eta_0 \qquad \text{Accept } H_0 \text{ iff } q > z_\alpha \tag{8-308}$$

$$\beta(\eta) = P\{\mathbf{q} > z_\alpha \mid H_1\} = 1 - \mathrm{G}(z_\alpha - \eta_q) \tag{8-309}$$

Unknown variance. We assume that $\mathbf{x}$ is normal and use as the test statistic the random variable

$$\mathbf{q} = \frac{\overline{\mathbf{x}} - \eta_0}{\mathbf{s}/\sqrt{n}} \tag{8-310}$$

where $\mathbf{s}^2$ is the sample variance of $\mathbf{x}$. Under hypothesis H_0, the random variable $\mathbf{q}$ has a Student t distribution with $n - 1$ degrees of freedom. We can, therefore, use (8-296),

(8-298) and (8-300) where we replace q_u by the tabulated $t_u(n-1)$ percentile. To find $\beta(\eta)$, we must find the distribution of $\mathbf{q}$ for $\eta \neq \eta_0$.

EXAMPLE 8-21 ▶ We measure the voltage V of a voltage source 25 times and we find $\overline{x} = 110.12$ V (see also Example 8-3). Test the hypothesis $V = V_0 = 110$ V against $V \neq 110$ V with $\alpha = 0.05$. Assume that the measurement error $\boldsymbol{\nu}$ is $N(0, \sigma)$.

(*a*) Suppose that $\sigma = 0.4$ V. In this problem, $z_{1-\alpha/2} = z_{0.975} = 2$:

$$q = \frac{110.12 - 110}{0.4/\sqrt{25}} = 1.5$$

Since 1.5 is in the interval $(-2, 2)$, we accept H_0.

(*b*) Suppose that σ is unknown. From the measurements we find $s = 0.6$ V. Inserting into (8-310), we obtain

$$q = \frac{110.12 - 110}{0.6/\sqrt{25}} = 1$$

Table 8-3 yields $t_{1-\alpha/2}(n-1) = t_{0.975}(25) = 2.06 = -t_{0.025}$. Since 1 is in the interval $(-2.06, 2.06)$, we accept H_0. ◀

PROBABILITY. We shall test the hypothesis H_0: $p = p_0 = 1 - q_0$ that the probability $p = P(A)$ of an event A equals a given constant p_0, using as data the number k of successes of A in n trials. The random variable $\mathbf{k}$ has a binomial distribution and for large n it is $N(np, \sqrt{npq})$. We shall assume that n is large.

The test will be based on the test statistic

$$\mathbf{q} = \frac{\mathbf{k} - np_0}{\sqrt{np_0q_0}} \tag{8-311}$$

Under hypothesis H_0, $\mathbf{q}$ is $N(0, 1)$. The test thus proceeds as in (8-304)–(8-309).

To find the OC function $\beta(p)$, we must determine the distribution of $\mathbf{q}$ under the alternative hypothesis. Since $\mathbf{k}$ is normal, $\mathbf{q}$ is also normal with

$$\eta_q = \frac{np - np_0}{\sqrt{np_0q_0}} \qquad \sigma_q^2 = \frac{npq}{np_0q_0}$$

This yields the following test:

$$H_1\colon p \neq p_0 \qquad \text{Accept } H_0 \text{ iff } z_{\alpha/2} < q < z_{1-\alpha/2} \tag{8-312}$$

$$\beta(p) = P\{|\mathbf{q}| < z_{1-\alpha/2} \mid H_1\} = \mathrm{G}\left(\frac{z_{1-\alpha/2} - \eta_q}{\sqrt{pq/p_0q_0}}\right) - \mathrm{G}\left(\frac{z_{\alpha/2} - \eta_q}{\sqrt{pq/p_0q_0}}\right) \tag{8-313}$$

$$H_1\colon p > p_0 \qquad \text{Accept } H_0 \text{ iff } q < z_{1-\alpha} \tag{8-314}$$

$$\beta(p) = P\{\mathbf{q} < z_{1-\alpha} \mid H_1\} = \mathrm{G}\left(\frac{z_{1-\alpha} - \eta_q}{\sqrt{pq/p_0q_0}}\right) \tag{8-315}$$

$$H_1\colon p < p_0 \qquad \text{Accept } H_0 \text{ iff } q > z_{\alpha} \tag{8-316}$$

$$\beta(p) = P\{\mathbf{q} > z_{\alpha} \mid H_1\} = 1 - \mathrm{G}\left(\frac{z_{\alpha} - \eta_q}{\sqrt{pq/p_0q_0}}\right) \tag{8-317}$$

EXAMPLE 8-22 ▶ We wish to test the hypothesis that a coin is fair against the hypothesis that it is loaded in favor of "heads":

$$H_0: p = 0.5 \qquad \text{against} \qquad H_1: p > 0.5$$

We toss the coin 100 times and "heads" shows 62 times. Does the evidence support the rejection of the null hypothesis with significance level $\alpha = 0.05$?

In this example, $z_{1-\alpha} = z_{0.95} = 1.645$. Since

$$q = \frac{62 - 50}{\sqrt{25}} = 2.4 > 1.645$$

the fair-coin hypothesis is rejected. ◀

VARIANCE. The random variable $\mathbf{x}$ is $N(\eta, \sigma)$. We wish to test the hypothesis $H_0: \sigma = \sigma_0$.

Known mean. We use as test statistic the random variable

$$\mathbf{q} = \sum_i \left(\frac{\mathbf{x}_i - \eta}{\sigma_0} \right)^2 \tag{8-318}$$

Under hypothesis H_0, this random variable is $\chi^2(n)$. We can, therefore, use (8-296) where q_u equals the $\chi_u^2(n)$ percentile.

Unknown mean. We use as the test statistic the random variable

$$\mathbf{q} = \sum_i \left(\frac{\mathbf{x}_i - \overline{\mathbf{x}}}{\sigma_0} \right)^2 \tag{8-319}$$

Under hypothesis H_0, this random variable is $\chi^2(n-1)$. We can, therefore, use (8-296) with $q_u = \chi_u^2(n-1)$.

EXAMPLE 8-23 ▶ Suppose that in Example 8-21, the variance σ^2 of the measurement error is unknown. Test the hypothesis $H_0: \sigma = 0.4$ against $H_1: \sigma > 0.4$ with $\alpha = 0.05$ using 20 measurements $x_i = V + \nu_i$.

(*a*) Assume that $V = 110$ V. Inserting the measurements x_i into (8-318), we find

$$q = \sum_{i=1}^{20} \left(\frac{x_i - 110}{0.4} \right)^2 = 36.2$$

Since $\chi_{1-\alpha}^2(n) = \chi_{0.95}^2(20) = 31.41 < 36.2$, we reject H_0.

(*b*) If V is unknown, we use (8-319). This yields

$$q = \sum_{i=1}^{20} \left(\frac{x_i - \overline{x}}{0.4} \right)^2 = 22.5$$

Since $\chi_{1-\alpha}^2(n-1) = \chi_{0.95}^2(19) = 30.14 > 22.5$, we accept H_0. ◀

DISTRIBUTIONS. In this application, H_0 does not involve a parameter; it is the hypothesis that the distribution $F(x)$ of a random variable $\mathbf{x}$ equals a given function $F_0(x)$. Thus

$$H_0: F(x) \equiv F_0(x) \qquad \text{against} \qquad H_1: F(x) \neq F_0(x)$$

The Kolmogorov–Smirnov test. We form the random process $\hat{\mathbf{F}}(x)$ as in the estimation problem (see (8-27)–(8-30)) and use as the test statistic the random variable

$$\mathbf{q} = \max_x |\mathbf{F}(x) - F_0(x)| \tag{8-320}$$

This choice is based on the following observations: For a specific ζ, the function $\hat{F}(x)$ is the empirical estimate of $F(x)$ [see (4-3)]; it tends, therefore, to $F(x)$ as $n \to \infty$. From this it follows that

$$E\{\hat{\mathbf{F}}(x)\} = F(x) \qquad \hat{\mathbf{F}}(x) \xrightarrow[n\to\infty]{} F(x)$$

This shows that for large n, $\mathbf{q}$ is close to 0 if H_0 is true and it is close to $\max |F(x) - F_0(x)|$ if H_1 is true. It leads, therefore, to the conclusion that we must reject H_0 if q is larger than some constant c. This constant is determined in terms of the significance level $\alpha = P\{\mathbf{q} > c \mid H_0\}$ and the distribution of $\mathbf{q}$. Under hypothesis H_0, the test statistic $\mathbf{q}$ equals the random variable $\mathbf{w}$ in (8-28). Using the Kolmogorov approximation (8-29), we obtain

$$\alpha = P\{\mathbf{q} > c \mid H_0\} \simeq 2e^{-2nc^2} \tag{8-321}$$

The test thus proceeds as follows: Form the empirical estimate $\hat{F}(x)$ of $F(x)$ and determine q from (8-320).

$$\text{Accept } H_0 \text{ iff } q < \sqrt{-\frac{1}{2n}\log\frac{\alpha}{2}} \tag{8-322}$$

The resulting Type II error probability is reasonably small only if n is large.

Chi-Square Tests

We are given a partition $U = [A_1, \ldots, A_m]$ of the space S and we wish to test the hypothesis that the probabilities $p_i = P(A_i)$ of the events A_i equal m given constants p_{0i}:

$$H_0: p_i = p_{0i}, \text{ all } i \qquad \text{against} \qquad H_1: p_i \neq p_{0i}, \text{ some } i \tag{8-323}$$

using as data the number of successes k_i of each of the events A_i in n trials. For this purpose, we introduce the sum

$$\mathbf{q} = \sum_{i=1}^{m} \frac{(\mathbf{k}_i - np_{0i})^2}{np_{0i}} \tag{8-324}$$

known as *Pearson's test statistic*. As we know, the random variables $\mathbf{k}_i$ have a binomial distribution with mean np_i and variance np_iq_i. Hence the ratio $\mathbf{k}_i/n$ tends to p_i as $n \to \infty$. From this it follows that the difference $|\mathbf{k}_i - np_{0i}|$ is small if $p_i = p_{0i}$ and it increases as $|p_i - p_{0i}|$ increases. This justifies the use of the random variable $\mathbf{q}$ as a test statistic and the set $q > c$ as the critical region of the test.

To find c, we must determine the distribution of $\mathbf{q}$. We shall do so under the assumption that n is large. For moderate values of n, we use computer simulation [see (8-334)]. With this assumption, the random variables $\mathbf{k}_i$ are nearly normal with mean kp_i. Under hypothesis H_0, the random variable $\mathbf{q}$ has a $\chi^2(m-1)$ distribution. This follows from the fact that the constants p_{0i} satisfy the constraint $\sum p_{0i} = 1$. The proof, however, is rather involved.

This leads to this test: Observe the numbers k_i and compute the sum q in (8-324); find $\chi^2_{1-\alpha}(m-1)$ from Table 8-3.

$$\text{Accept } H_0 \text{ iff } q < \chi^2_{1-\alpha}(m-1) \tag{8-325}$$

We note that the chi-square test is reduced to the test (8-312)–(8-317) involving the probability p of an event A. In this case, the partition U equals $[A, \overline{A}]$ and the statistic $\mathbf{q}$ in (8-324) equals $(k - np_0)^2/np_0q_0$, where $p_0 = p_{01}$, $q_0 = p_{02}$, $k = k_1$, and $n - k = k_2$ (see Prob. 8-40).

EXAMPLE 8-24 ▶ We roll a die 300 times and we observe that f_i shows $k_i = 55\ 43\ 44\ 61\ 40\ 57$ times. Test the hypothesis that the die is fair with $\alpha = 0.05$. In this problem, $p_{0i} = 1/6$, $m = 6$, and $np_{0i} = 50$. Inserting into (8-324), we obtain

$$q = \sum_{i=1}^{6} \frac{(k_i - 50)^2}{50} = 7.6$$

Since $\chi^2_{0.95}(5) = 11.07 > 7.6$, we accept the fair-die hypothesis. ◀

The chi-square test is used in *goodness-of-fit* tests involving the agreement between experimental data and theoretical models. We next give two illustrations.

TESTS OF INDEPENDENCE. We shall test the hypothesis that two events B and C are independent:

$$H_0\colon P(B \cap C) = P(B)P(C) \qquad \text{against} \qquad H_1\colon P(B \cap C) \neq P(B)P(C) \tag{8-326}$$

under the assumption that the probabilities $b = P(B)$ and $c = P(C)$ of these events are known. To do so, we apply the chi-square test to the partition consisting of the four events

$$A_1 = B \cap C \qquad A_2 = B \cap \overline{C} \qquad A_3 = \overline{B} \cap C \qquad A_4 = \overline{B} \cap \overline{C}$$

Under hypothesis H_0, the components of each of the events A_i are independent. Hence

$$p_{01} = bc \qquad p_{02} = b(1-c) \qquad p_{03} = (1-b)c \qquad p_{04} = (1-b)(1-c)$$

This yields the test:

$$\text{Accept } H_0 \text{ iff } \sum_{k=1}^{4} \frac{(k_i - np_{0i})^2}{np_{0i}} < \chi^2_{1-\alpha}(3) \tag{8-327}$$

In (8-327), k_i is the number of occurrences of the event A_i; for example, k_2 is the number of times B occurs but C does not occur.

EXAMPLE 8-25 ▶ In a certain university, 60% of all first-year students are male and 75% of all entering students graduate. We select at random the records of 299 males and 101 females and we find that 168 males and 68 females graduated. Test the hypothesis that the events $B = \{\text{male}\}$ and $C = \{\text{graduate}\}$ are independent with $\alpha = 0.05$. In this problem, $m = 400$, $P(B) = 0.6$, $P(C) = 0.75$, $p_{0i} = 0.45\ 0.15\ 0.3\ 0.1$, $k_i = 168\ 68\ 131\ 33$, and (8-324) yields

$$q = \sum_{i=1}^{4} \frac{(k_i - 400p_{0i})^2}{400p_{0i}} = 4.1$$

Since $\chi^2_{0.95}(3) = 7.81 > 4.1$, we accept the independence hypothesis. ◀

TESTS OF DISTRIBUTIONS. We introduced earlier the problem of testing the hypothesis that the distribution $F(x)$ of a random variable $\mathbf{x}$ equals a given function $F_0(x)$. The resulting test is reliable only if the number of available samples x_j of $\mathbf{x}$ is very large. In the following, we test the hypothesis that $F(x) = F_0(x)$ not at every x but only at a set of $m - 1$ points a_i (Fig. 8-11):

$$H_0\colon F(a_i) = F_0(a_i),\ 1 \le i \le m-1 \qquad \text{against} \qquad H_1\colon F(a_i) \ne F_0(a_i),\ \text{some } i \tag{8-328}$$

We introduce the m events

$$A_i = \{a_{i-1} < \mathbf{x} < a_i\} \qquad i = 1, \dots, m$$

where $a_0 = -\infty$ and $a_m = \infty$. These events form a partition of S. The number k_i of successes of A_i equals the number of samples x_j in the interval (a_{i-1}, a_i). Under hypothesis H_0,

$$P(A_i) = F_0(a_i) - F_0(a_{i-1}) = p_{0i}$$

Thus, to test the hypothesis (8-328), we form the sum q in (8-324) and apply (8-325). If H_0 is rejected, then the hypothesis that $F(x) = F_0(x)$ is also rejected.

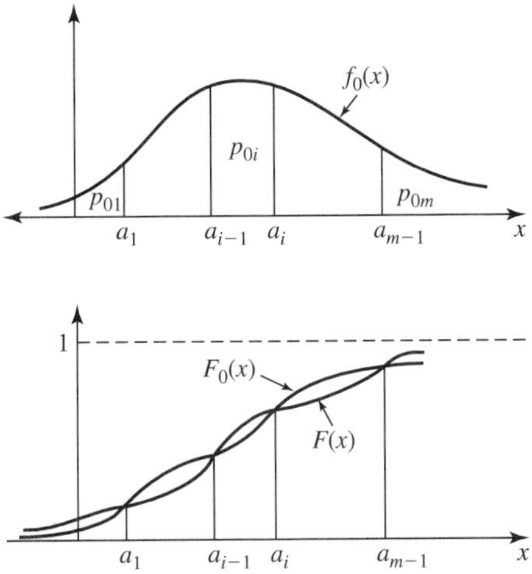

FIGURE 8-11

EXAMPLE 8-26 ▶ We have a list of 500 computer-generated decimal numbers x_j and we wish to test the hypothesis that they are the samples of a random variable $\mathbf{x}$ uniformly distributed in the interval (0,1). We divide this interval into 10 subintervals of length 0.1 and we count the number k_i of samples x_j that are in the ith subinterval. The results are

$$k_i = 43 \quad 56 \quad 42 \quad 38 \quad 59 \quad 61 \quad 41 \quad 57 \quad 46 \quad 57$$

In this problem, $m = 500$, $p_{0i} = 0.1$, and

$$q = \sum_{i=1}^{10} \frac{(k_i - 50)^2}{50} = 13.8$$

Since $\chi^2_{0.95}(9) = 16.9 > 13.8$ we accept the uniformity hypothesis. ◀

Likelihood Ratio Test

We conclude with a general method for testing any hypothesis, simple or composite. We are given a random variable $\mathbf{x}$ with density $f(x, \theta)$, where θ is an arbitrary parameter, scalar or vector, and we wish to test the hypothesis $H_0: \theta \in \Theta_0$ against $H_1: \theta \in \Theta_1$. The sets Θ_0 and Θ_1 are subsets of the parameter space $\Theta = \Theta_0 \cup \Theta_1$.

The density $f(X, \theta)$, considered as a function of θ, is the likelihood function of $\mathbf{X}$. We denote by θ_m the value of θ for which $f(\mathbf{X}, \theta)$ is maximum in the space Θ. Thus θ_m is the ML estimate of θ. The value of θ for which $f(X, \theta)$ is maximum in the set Θ_0 will be denoted by θ_{m0}. If H_0 is the simple hypothesis $\theta = \theta_0$, then $\theta_{m0} = \theta_0$. The ML test is a test based on the statistic

$$\boldsymbol{\lambda} = \frac{f(\mathbf{X}, \theta_{m0})}{f(\mathbf{X}, \theta_m)} \tag{8-329}$$

Note that

$$0 \le \boldsymbol{\lambda} \le 1$$

because $f(X, \theta_{m0}) \le f(X, \theta_m)$. We maintain that $\boldsymbol{\lambda}$ is concentrated near 1 if H_0 is true. As we know, the ML estimate θ_m of θ tends to its true value θ^* as $n \to \infty$. Furthermore, under the null hypothesis, θ^* is in the set Θ_0; hence $\lambda \to 1$ as $n \to \infty$. From this it follows that we must reject H_0 if $\lambda < c$. The constant c is determined in terms of the significance level α of the test.

Suppose, first, that H_0 is the simple hypothesis $\theta = \theta_0$. In this case,

$$\alpha = P\{\boldsymbol{\lambda} \le c \mid H_0\} = \int_0^c f_\lambda(\lambda, \theta_0)\, d\lambda \tag{8-330}$$

This leads to the test: Using the samples x_i of $\mathbf{x}$, form the likelihood function $f(X, \theta)$. Find θ_m and θ_{m0} and form the ratio $\lambda = f(X, \theta_{m0})/f(X, \theta_m)$:

$$\text{Reject } H_0 \text{ iff } \lambda < \lambda_\alpha \tag{8-331}$$

where λ_α is the α percentile of the test statistic $\boldsymbol{\lambda}$ under hypothesis H_0.

If H_0 is a composite hypothesis, c is the smallest constant such that $P\{\boldsymbol{\lambda} \le c\} < \lambda_\alpha$ for every $\theta \in \Theta_0$.

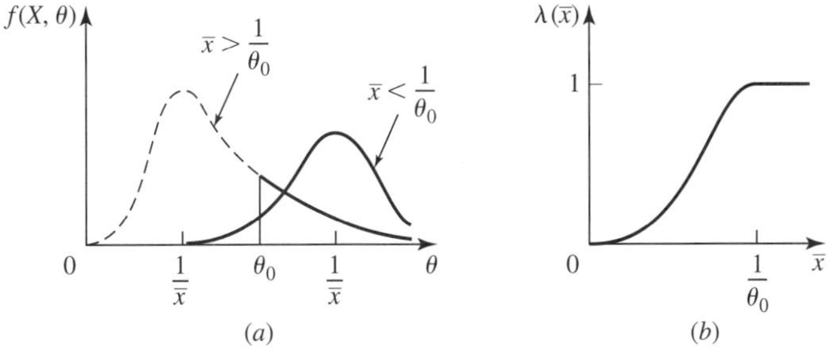

FIGURE 8-12

EXAMPLE 8-27 ▶ Suppose that $f(x, \theta) \sim \theta e^{-\theta x} U(x)$. We shall test the hypothesis

$$H_0: 0 < \theta \le \theta_0 \qquad \text{against} \qquad H_1: \theta > \theta_0$$

In this problem, Θ_0 is the segment $0 < \theta \le \theta_0$ of the real line and Θ is the half-line $\theta > 0$. Thus both hypotheses are composite. The likelihood function

$$f(\mathbf{X}, \theta) = \theta^n e^{-n\bar{x}\theta}$$

is shown in Fig. 8-12*a* for $\bar{x} > 1/\theta_0$ and $\bar{x} < 1/\theta_0$. In the half-line $\theta > 0$ this function is maximum for $\theta = 1/\bar{x}$. In the interval $0 < \theta \le \theta_0$ it is maximum for $\theta = 1/\bar{x}$ if $\bar{x} > 1/\theta_0$ and for $\theta = \theta_0$ if $\bar{x} < 1/\theta_0$. Hence

$$\theta_m = \frac{1}{x} \qquad \theta_{m0} = \begin{cases} 1/\bar{x} & \text{for} \quad \bar{x} > 1/\theta_0 \\ \theta_0 & \text{for} \quad \bar{x} < 1/\theta_0 \end{cases}$$

The likelihood ratio equals (Fig. 8-12*b*)

$$\lambda = \begin{cases} 1 & \text{for} \quad \bar{x} > 1/\theta_0 \\ (\bar{x}\theta_0)^n e^{-n\theta_0\bar{x}+n\theta_0} & \text{for} \quad \bar{x} < 1/\theta_0 \end{cases}$$

We reject H_0 if $\lambda < c$ or, equivalently, if $\bar{x} < c_1$, where c_1 equals the α percentile of the random variable $\bar{\mathbf{x}}$. ◀

To carry out a likelihood ratio test, we must determine the density of $\boldsymbol{\lambda}$. This is not always a simple task. The following theorem simplifies the problem for large n.

ASYMPTOTIC PROPERTIES. We denote by m and m_0 the number of free parameters in Θ and Θ_0, respectively, that is, the number of parameters that take noncountably many values. It can be shown that if $m > m_0$, then the distribution of the random variable $\mathbf{w} = -2\log\boldsymbol{\lambda}$ approaches a chi-square distribution with $m - m_0$ degrees of freedom as $n \to \infty$. The function $w = -2\log\lambda$ is monotone decreasing; hence $\lambda < c$ iff $w > c_1 = -2\log c$. From this it follows that

$$\alpha = P\{\boldsymbol{\lambda} < c\} = P\{\mathbf{w} > c_1\}$$

where $c_1 = \chi^2_{1-\alpha}(m - m_0)$, and (8-331) yields this test

$$\text{Reject } H_0 \text{ iff } -2\log\lambda > \chi^2_{1-\alpha}(m - m_0) \tag{8-332}$$

We give next an example illustrating the theorem.

EXAMPLE 8-28 ▶ We are given an $N(\eta, 1)$ random variable $\mathbf{x}$ and we wish to test the simple hypotheses $\eta = \eta_0$ against $\eta \neq \eta_0$. In this problem $\eta_{m0} = \eta_0$ and

$$f(X, \eta) = \frac{1}{\sqrt{(2\pi)^n}} \exp\left\{-\frac{1}{2}\sum(x_i - \eta)^2\right\}$$

This is maximum if the sum [see (7-67)]

$$\sum(x_i - \eta)^2 = \sum(x_i - \overline{x})^2 + n(\overline{x} - \eta)^2$$

is minimum, that is, if $\eta = \overline{x}$. Hence $\eta_m = \overline{x}$ and

$$\lambda = \frac{\exp\left\{-\frac{1}{2}\sum(x_i - \eta_0)^2\right\}}{\exp\left\{-\frac{1}{2}\sum(x_i - \overline{x})^2\right\}} = \exp\left\{-\frac{n}{2}(\overline{x} - \eta_0)^2\right\}$$

From this it follows that $\lambda > c$ iff $|\overline{x} - \eta_0| < c_1$. This shows that the likelihood ratio test of the mean of a normal random variable is equivalent to the test (8-304).

Note that in this problem, $m = 1$ and $m_0 = 0$. Furthermore,

$$\mathbf{w} = -2\log\boldsymbol{\lambda} = n(\overline{\mathbf{x}} - \eta_0)^2 = \left(\frac{\overline{\mathbf{x}} - \eta_0}{1/\sqrt{n}}\right)^2$$

But the right side is a random variable with $\chi^2(1)$ distribution. Hence the random variable $\mathbf{w}$ has a $\chi^2(m - m_0)$ distribution not only asymptotically, but for any n. ◀

COMPUTER SIMULATION IN HYPOTHESIS TESTING. As we have seen, the test of a hypothesis H_0 involves the following steps: We determine the value X of the random vector $\mathbf{X} = [\mathbf{x}_1, \ldots, \mathbf{x}_m]$ in terms of the observations x_k of the m random variables $\mathbf{x}_k$ and compute the corresponding value $q = g(X)$ of the test statistic $\mathbf{q} = g(\mathbf{X})$. We accept H_0 if q is not in the critical region of the test, for example, if q is in the interval (q_a, q_b), where q_a and q_b are appropriately chosen values of the u percentile q_u of $\mathbf{q}$ [see (8-296)]. This involves the determination of the distribution $F(q)$ of $\mathbf{q}$ and the inverse $q_u = F^{(-1)}(u)$ of $F(q)$. The inversion problem can be avoided if we use the following approach.

The function $F(q)$ is monotone increasing. Hence,

$$q_a < q < q_b \text{ iff } a = F(q_a) < F(q) < F(q_b) = b$$

This shows that the test $q_a < q < q_b$ is equivalent to the test

$$\text{Accept } H_0 \text{ iff } a < F(q) < b \tag{8-333}$$

involving the determination of the distribution $F(q)$ of $\mathbf{q}$. As we have shown in Sec. 7-3, the function $F(q)$ can be determined by computer simulation [see (7-175)]:

To estimate numerically $F(q)$ we construct the random variable vector sequence

$$X_i = [x_{1,i}, \ldots, x_{m,i}] \qquad i = 1, \ldots, n$$

where $x_{k,i}$ are the computer generated samples of the m random variables $\mathbf{x}_k$. Using the sequence X_i, we form the random number sequence $q_i = g(X_i)$ and we count the number n_q of q_i's that are smaller than the computed q. Inserting into (7-175), we obtain the estimate $F(q) \simeq n_q/n$. With $F(q)$ so determined, (8-333) yields the test

$$\text{Accept } H_0 \text{ iff } a < \frac{n_q}{n} < b \tag{8-334}$$

In this discussion, $q = g(X)$ is a number determined in terms of the experimental data x_k. The sequence q_i, however, is computer generated.

The approach we have described is used if it is difficult to determine analytically, the function $F(q)$. This is the case in the determination of Pearson's test statistic (8-324).

PROBLEMS

8-1 The diameter of cylindrical rods coming out of a production line is a normal random variable $\mathbf{x}$ with $\sigma = 0.1$ mm. We measure $n = 9$ units and find that the average of the measurements is $\overline{x} = 91$ mm. (*a*) Find c such that with a 0.95 confidence coefficient, the mean η of $\mathbf{x}$ is in the interval $\overline{x} \pm c$. (*b*) We claim that η is in the interval (90.95, 91.05). Find the confidence coefficient of our claim.

8-2 The length of a product is a random variable $\mathbf{x}$ with $\sigma = 1$ mm and unknown mean. We measure four units and find that $\overline{x} = 203$ mm. (*a*) Assuming that $\mathbf{x}$ is a normal random variable, find the 0.95 confidence interval of η. (*b*) The distribution of $\mathbf{x}$ is unknown. Using Tchebycheff's inequality, find c such that with confidence coefficient 0.95, η is in the interval $203 \pm c$.

8-3 We know from past records that the life length of type A tires is a random variable $\mathbf{x}$ with $\sigma = 5000$ miles. We test 64 samples and find that their average life length is $\overline{x} = 25{,}000$ miles. Find the 0.9 confidence interval of the mean of $\mathbf{x}$.

8-4 We wish to determine the length a of an object. We use as an estimate of a the average $\overline{x}$ of n measurements. The measurement error is approximately normal with zero mean and standard deviation 0.1 mm. Find n such that with 95 % confidence, $\overline{x}$ is within ± 0.2 mm of a.

8-5 The random variable $\mathbf{x}$ is uniformly distributed in the interval $\theta - 2 < x < \theta + 2$. We observe 100 samples x_i and find that their average equals $\overline{x} = 30$. Find the 0.95 confidence interval of θ.

8-6 Consider a random variable $\mathbf{x}$ with density $f(x) = xe^{-x}U(x)$. Predict with 95% confidence that the next value of x will be in the interval (a, b). Show that the length $b - a$ of this interval is minimum if a and b are such that

$$f(a) = f(b) \qquad P\{a < \mathbf{x} < b\} = 0.95$$

Find a and b.

8-7 (*Estimation–prediction*) The time to failure of electric bulbs of brand A is a normal random variable with $\sigma = 10$ hours and unknown mean. We have used 20 such bulbs and have observed that the average $\overline{x}$ of their time to failure is 80 hours. We buy a new bulb of the same brand and wish to predict with 95% confidence that its time to failure will be in the interval $80 \pm c$. Find c.

8-8 Suppose that the time between arrivals of patients in a dentist's office constitutes samples of a random variable $\mathbf{x}$ with density $\theta e^{-\theta x}U(x)$. The 40th patient arrived 4 hours after the first. Find the 0.95 confidence interval of the mean arrival time $\eta = 1/\theta$.

8-9 The number of particles emitted from a radioactive substance in 1 second is a Poisson distributed random variable with mean λ. It was observed that in 200 seconds, 2550 particles were emitted. Find the 0.95 confidence interval of λ.

8-10 Among 4000 newborns, 2080 are male. Find the 0.99 confidence interval of the probability $p = P\{\text{male}\}$.

8-11 In an exit poll of 900 voters questioned, 360 responded that they favor a particular proposition. On this basis, it was reported that 40% of the voters favor the proposition. (*a*) Find the margin of error if the confidence coefficient of the results is 0.95. (*b*) Find the confidence coefficient if the margin of error is $\pm 2\%$.

8-12 In a market survey, it was reported that 29% of respondents favor product A. The poll was conducted with confidence coefficient 0.95, and the margin of error was $\pm 4\%$. Find the number of respondents.

8-13 We plan a poll for the purpose of estimating the probability p of Republicans in a community. We wish our estimate to be within ± 0.02 of p. How large should our sample be if the confidence coefficient of the estimate is 0.95?

8-14 A coin is tossed once, and heads shows. Assuming that the probability p of heads is the value of a random variable $\mathbf{p}$ uniformly distributed in the interval (0.4, 0.6), find its bayesian estimate.

8-15 The time to failure of a system is a random variable $\mathbf{x}$ with density $f(x,\theta) = \theta e^{-\theta x} U(x)$. We wish to find the bayesian estimate $\hat{\theta}$ of θ in terms of the sample mean $\bar{\mathbf{x}}$ of the n samples x_i of $\mathbf{x}$. We assume that θ is the value of a random variable $\boldsymbol{\theta}$ with prior density $f_\theta(\theta) = ce^{-c\theta} U(\theta)$. Show that

$$\hat{\theta} = \frac{n+1}{c + n\bar{x}} \xrightarrow[n\to\infty]{} \frac{1}{\bar{x}}$$

8-16 The random variable $\mathbf{x}$ has a Poisson distribution with mean θ. We wish to find the bayesian estimate $\hat{\theta}$ of θ under the assumption that θ is the value of a random variable $\boldsymbol{\theta}$ with prior density $f_\theta(\theta) \sim \theta^b e^{-c\theta} U(\theta)$. Show that

$$\hat{\theta} = \frac{n\bar{x} + b + 1}{n + c}$$

8-17 Suppose that the IQ scores of children in a certain grade are the samples of an $N(\eta, \sigma)$ random variable $\mathbf{x}$. We test 10 children and obtain the following averages: $\bar{x} = 90$, $s = 5$. Find the 0.95 confidence interval of η and of σ.

8-18 The random variables $\mathbf{x}_i$ are i.i.d. and $N(0, \sigma)$. We observe that $x_1^2 + \cdots + x_{10}^2 = 4$. Find the 0.95 confidence interval of σ.

8-19 The readings of a voltmeter introduces an error $\boldsymbol{\nu}$ with mean 0. We wish to estimate its standard deviation σ. We measure a calibrated source $V = 3$ V four times and obtain the values 2.90, 3.15, 3.05, and 2.96. Assuming that $\boldsymbol{\nu}$ is normal, find the 0.95 confidence interval of σ.

8-20 The random variable $\mathbf{x}$ has the Erlang density $f(x) \sim c^4 x^3 e^{-cx} U(x)$. We observe the samples $x_i = 3.1, 3.4, 3.3$. Find the ML estimate $\hat{c}$ of c.

8-21 The random variable $\mathbf{x}$ has the truncated exponential density $f(x) = ce^{-c(x-x_0)} U(x - x_0)$. Find the ML estimate $\hat{c}$ of c in terms of the n samples x_i of $\mathbf{x}$.

8-22 The time to failure of a bulb is a random variable $\mathbf{x}$ with density $ce^{-cx} U(x)$. We test 80 bulbs and find that 200 hours later, 62 of them are still good. Find the ML estimate of c.

8-23 The random variable $\mathbf{x}$ has a Poisson distribution with mean θ. Show that the ML estimate of θ equals $\bar{x}$.

8-24 Show that if $L(x,\theta) = \log f(x,\theta)$ is the likelihood function of a random variable $\mathbf{x}$, then

$$E\left\{\left|\frac{\partial L(\mathbf{x},\theta)}{\partial \theta}\right|^2\right\} = -E\left\{\frac{\partial^2 L(\mathbf{x},\theta)}{\partial \theta^2}\right\}$$

8-25 We are given a random variable $\mathbf{x}$ with mean η and standard deviation $\sigma = 2$, and we wish to test the hypothesis $\eta = 8$ against $\eta = 8.7$ with $\alpha = 0.01$ using as the test statistic the sample mean $\bar{x}$ of n samples. (*a*) Find the critical region R_c of the test and the resulting β if $n = 64$. (*b*) Find n and R_c if $\beta = 0.05$.

8-26 A new car is introduced with the claim that its average mileage in highway driving is at least 28 miles per gallon. Seventeen cars are tested, and the following mileage is obtained:

19	20	24	25	26	26.8	27.2	27.5	
28	28.2	28.4	29	30	31	32	33.3	35

Can we conclude with significance level at most 0.05 that the claim is true?

8-27 The weights of cereal boxes are the values of a random variable $\mathbf{x}$ with mean η. We measure 64 boxes and find that $\overline{x} = 7.7$ oz. and $s = 1.5$ oz. Test the hypothesis H_0: $\eta = 8$ oz. against H_1: $\eta \neq 8$ oz. with $\alpha = 0.1$ and $\alpha = 0.01$.

8-28 Brand A batteries cost more than brand B batteries. Their life lengths are two normal and independent random variables $\mathbf{x}$ and $\mathbf{y}$. We test 16 batteries of brand A and 26 batteries of brand B and find these values, in hours:

$$\overline{x} = 4.6 \qquad s_x = 1.1 \qquad \overline{y} = 4.2 \qquad s_y = 0.9$$

Test the hypothesis $\eta_x = \eta_y$ against $\eta_x > \eta_y$ with $\alpha = 0.05$.

8-29 A coin is tossed 64 times, and heads shows 22 times. (*a*) Test the hypothesis that the coin is fair with significance level 0.05. (*b*) We toss a coin 16 times, and heads shows k times. If k is such that $k_1 \leq k \leq k_2$, we accept the hypothesis that the coin is fair with significance level $\alpha = 0.05$. Find k_1 and k_2 and the resulting β error.

8-30 In a production process, the number of defective units per hour is a Poisson distributed random variable $\mathbf{x}$ with parameter $\lambda = 5$. A new process is introduced, and it is observed that the hourly defectives in a 22-hour period are

$$x_i = 3 \quad 0 \quad 5 \quad 4 \quad 2 \quad 6 \quad 4 \quad 1 \quad 5 \quad 3 \quad 7 \quad 4 \quad 0 \quad 8 \quad 3 \quad 2 \quad 4 \quad 3 \quad 6 \quad 5 \quad 6 \quad 9$$

Test the hypothesis $\lambda = 5$ against $\lambda < 5$ with $\alpha = 0.05$.

8-31 A die is tossed 102 times, and the ith face shows $k_i = 18, 15, 19, 17, 13$, and 20 times. Test the hypothesis that the die is fair with $\alpha = 0.05$ using the chi-square test.

8-32 A computer prints out 1000 numbers consisting of the 10 integers $j = 0, 1, \ldots, 9$. The number n_j of times j appears equals

$$n_j = 85 \quad 110 \quad 118 \quad 91 \quad 78 \quad 105 \quad 122 \quad 94 \quad 101 \quad 96$$

Test the hypothesis that the numbers j are uniformly distributed between 0 and 9, with $\alpha = 0.05$.

8-33 The number $\mathbf{x}$ of particles emitted from a radioactive substance in 1 second is a Poisson random variable with mean θ. In 50 seconds, 1058 particles are emitted. Test the hypothesis $\theta_0 = 20$ against $\theta \neq 20$ with $\alpha = 0.05$ using the asymptotic approximation.

8-34 The random variables $\mathbf{x}$ and $\mathbf{y}$ are $N(\eta_x, \sigma_x)$ and $N(\eta_y, \sigma_y)$, respectively, and independent. Test the hypothesis $\sigma_x = \sigma_y$ against $\sigma_x \neq \sigma_y$ using as the test statistic the ratio (see Example 6-29)

$$\mathbf{q} = \frac{1}{m}\sum_{i=1}^{m}(\mathbf{x}_i - \eta_x)^2 \bigg/ \frac{1}{n}\sum_{i=1}^{n}(\mathbf{y}_i - \eta_y)^2$$

8-35 Show that the variance of a random variable with the Student t distribution $t(n)$ equals $n/(n-2)$.

8-36 Find the probability p_5 that in a men's tennis tournament the final match will last five sets. (*a*) Assume that the probability p that a player wins a set equals 0.5. (*b*) Use bayesian statistic with uniform prior (see *law of succession*).

8-37 Show that in the measurement problem of Example 8-9, the bayesian estimate $\hat{\theta}$ of the parameter θ equals

$$\hat{\theta} = \frac{\sigma_1^2}{\sigma_0^2}\theta_0 + \frac{n\sigma_1^2}{\sigma^2}\bar{x} \quad \text{where} \quad \sigma_1^2 = \frac{\sigma^2}{n} \times \frac{\sigma_0^2}{\sigma_0^2 + \sigma^2/n}$$

8-38 Using the ML method, find the γ confidence interval of the variance $v = \sigma^2$ of an $N(\eta, \sigma)$ random variable with known mean.

8-39 Show that if $\hat{\boldsymbol{\theta}}_1$ and $\hat{\boldsymbol{\theta}}_2$ are two unbiased minimum variance estimators of a parameter θ, then $\hat{\boldsymbol{\theta}}_1 = \hat{\boldsymbol{\theta}}_2$. *Hint*: Form the random variable $\hat{\boldsymbol{\theta}} = (\hat{\boldsymbol{\theta}}_1 + \hat{\boldsymbol{\theta}}_2)/2$. Show that $\sigma_{\hat{\theta}}^2 = \sigma^2(1+r)/2 \le \sigma^2$, where σ^2 is the common variance of $\hat{\boldsymbol{\theta}}_1$ and $\hat{\boldsymbol{\theta}}_2$ and r is their correlation coefficient.

8-40 The number of successes of an event A in n trials equals k_1. Show that

$$\frac{(k_1 - np_1)^2}{np_1} + \frac{(k_2 - np_2)^2}{np_2} = \frac{(k_1 - np_1)^2}{np_1 p_2}$$

where $k_2 = n - k_1$ and $P(A) = p_1 = 1 - p_2$.

8-41 Let $T(x)$ represent an unbiased estimator for the unknown parameter $\psi(\theta)$ based on the random variables $(\mathbf{x}_1, \mathbf{x}_2, \cdots \mathbf{x}_n) = \mathbf{x}$ under joint density function $f(\mathbf{x}; \theta)$. Show that the Cramer-Rao lower bound for the parameter $\psi(\theta)$ satisfies the inequality

$$\text{Var}\{T(\mathbf{x})\} \ge \frac{[\psi'(\theta)]^2}{E\left\{\left(\frac{\partial \log f(\mathbf{x};\theta)}{\partial\theta}\right)^2\right\}}$$

PART II

STOCHASTIC PROCESSES

CHAPTER
9
GENERAL CONCEPTS

9-1 DEFINITIONS

As we recall, a random variable **x** is a rule for assigning to every outcome ζ of an experiment S a *number* $\mathbf{x}(\zeta)$. A stochastic process $\mathbf{x}(t)$ is a rule for assigning to every ζ a *function* $\mathbf{x}(t, \zeta)$. Thus a stochastic process is a family of time functions depending on the parameter ζ or, equivalently, a function of t and ζ. The domain of ζ is the set of all experimental outcomes and the domain of t is a set R of real numbers.

If R is the real axis, then $\mathbf{x}(t)$ is a *continuous-time* process. If R is the set of integers, then $\mathbf{x}(t)$ is a *discrete-time* process. A discrete-time process is, thus, a sequence of random variables. Such a sequence will be denoted by $\mathbf{x}_n$ as in Sec. 7-4, or, to avoid double indices, by $\mathbf{x}[n]$.

We shall say that $\mathbf{x}(t)$ is a *discrete-state* process if its values are countable. Otherwise, it is a *continuous-state* process.

Most results in this investigation will be phrased in terms of continuous-time processes. Topics dealing with discrete-time processes will be introduced either as illustrations of the general theory, or when their discrete-time version is not self-evident.

We shall use the notation $\mathbf{x}(t)$ to represent a stochastic process omitting, as in the case of random variables, its dependence on ζ. Thus $\mathbf{x}(t)$ has the following interpretations:

1. It is a family (or an *ensemble*) of functions $\mathbf{x}(t, \zeta)$. In this interpretation, t and ζ are variables.
2. It is a single time function (or a *sample* of the given process). In this case, t is a variable and ζ is fixed.
3. If t is fixed and ζ is variable, then $\mathbf{x}(t)$ is a random variable equal to the *state* of the given process at time t.
4. If t and ζ are fixed, then $\mathbf{x}(t)$ is a *number*.

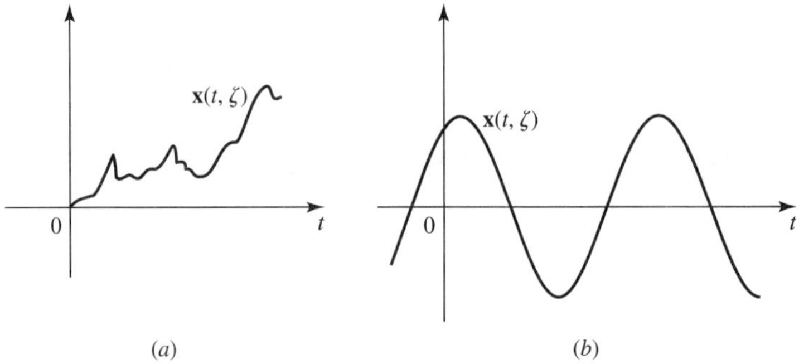

FIGURE 9-1

A physical example of a stochastic process is the motion of microscopic particles in collision with the molecules in a fluid (*brownian motion*). The resulting process $\mathbf{x}(t)$ consists of the motions of all particles (ensemble). A single realization $\mathbf{x}(t, \zeta_i)$ of this process (Fig. 9-1a) is the motion of a specific particle (sample). Another example is the voltage

$$\mathbf{x}(t) = \mathbf{r}\cos(\omega t + \boldsymbol{\varphi})$$

of an ac generator with random amplitude $\mathbf{r}$ and phase $\boldsymbol{\varphi}$. In this case, the process $\mathbf{x}(t)$ consists of a family of pure sine waves and a single sample is the function (Fig. 9-1b)

$$\mathbf{x}(t, \zeta_i) = \mathbf{r}(\zeta_i)\cos[\omega t + \boldsymbol{\varphi}(\zeta_i)]$$

According to our definition, both examples are stochastic processes. There is, however, a fundamental difference between them. The first example (regular) consists of a family of functions that cannot be described in terms of a finite number of parameters. Furthermore, the future of a sample $\mathbf{x}(t, \zeta)$ of $\mathbf{x}(t)$ cannot be determined in terms of its past. Finally, under certain conditions, the statistics[1] of a regular process $\mathbf{x}(t)$ can be determined in terms of a single sample (see Sec. 12-1). The second example (predictable) consists of a family of pure sine waves and it is completely specified in terms of the random variables $\mathbf{r}$ and $\boldsymbol{\varphi}$. Furthermore, if $\mathbf{x}(t, \zeta)$ is known for $t \le t_o$, then it is determined for $t > t_o$. Finally, a single sample $\mathbf{x}(t, \zeta)$ of $\mathbf{x}(t)$ does not specify the properties of the entire process because it depends only on the particular values $\mathbf{r}(\zeta)$ and $\boldsymbol{\varphi}(\zeta)$ of $\mathbf{r}$ and $\boldsymbol{\varphi}$. A formal definition of regular and predictable processes is given in Sec. 11-3.

Equality. We shall say that two stochastic processes $\mathbf{x}(t)$ and $\mathbf{y}(t)$ are equal (everywhere) if their respective samples $\mathbf{x}(t, \zeta)$ and $\mathbf{y}(t, \zeta)$ are identical for every ζ. Similarly, the equality $\mathbf{z}(t) = \mathbf{x}(t) + \mathbf{y}(t)$ means that $\mathbf{z}(t, \zeta) = \mathbf{x}(t, \zeta) + \mathbf{y}(t, \zeta)$ for every ζ. Derivatives, integrals, or any other operations involving stochastic processes are defined similarly in terms of the corresponding operations for each sample.

As in the case of limits, the above definitions can be relaxed. We give below the meaning of MS equality and in App. 9A we define MS derivatives and integrals. Two

[1]Keep in mind that *statistics* hereafter will mean statistical properties.

processes $\mathbf{x}(t)$ and $\mathbf{y}(t)$ are equal in the MS sense iff

$$E\{|\mathbf{x}(t) - \mathbf{y}(t)|^2\} = 0 \tag{9-1}$$

for every t. Equality in the MS sense leads to the following conclusions: We denote by A_t the set of outcomes ζ such that $\mathbf{x}(t, \zeta) = \mathbf{y}(t, \zeta)$ for a *specific* t, and by A_∞ the set of outcomes ζ such that $\mathbf{x}(t, \zeta) = \mathbf{y}(t, \zeta)$ for *every* t. From (9-1) it follows that $\mathbf{x}(t, \zeta) - \mathbf{y}(t, \zeta) = 0$ with probability 1; hence $P(A_t) = P(S) = 1$. It does not follow, however, that $P(A_\infty) = 1$. In fact, since A_∞ is the intersection of all sets A_t as t ranges over the entire axis, $P(A_\infty)$ might even equal 0.

Statistics of Stochastic Processes

A stochastic process is a noncountable infinity of random variables, one for each t. For a specific t, $\mathbf{x}(t)$ is a random variable with distribution

$$F(x, t) = P\{\mathbf{x}(t) \le x\} \tag{9-2}$$

This function depends on t, and it equals the probability of the event $\{\mathbf{x}(t) \le x\}$ consisting of all outcomes ζ such that, at the specific time t, the samples $\mathbf{x}(t, \zeta)$ of the given process do not exceed the number x. The function $F(x, t)$ will be called the *first-order distribution* of the process $\mathbf{x}(t)$. Its derivative with respect to x:

$$f(x, t) = \frac{\partial F(x, t)}{\partial x} \tag{9-3}$$

is the *first-order density* of $\mathbf{x}(t)$.

Frequency interpretation If the experiment is performed n times, then n functions $\mathbf{x}(t, \zeta_i)$ are observed, one for each trial (Fig. 9-2). Denoting by $n_t(x)$ the number of trials such that at time t the ordinates of the observed functions do not exceed x (solid lines), we conclude as in (4-3) that

$$F(x, t) \simeq \frac{n_t(x)}{n} \tag{9-4}$$

The *second-order distribution* of the process $\mathbf{x}(t)$ is the joint distribution

$$F(x_1, x_2; t_1, t_2) = P\{\mathbf{x}(t_1) \le x_1, \mathbf{x}(t_2) \le x_2\} \tag{9-5}$$

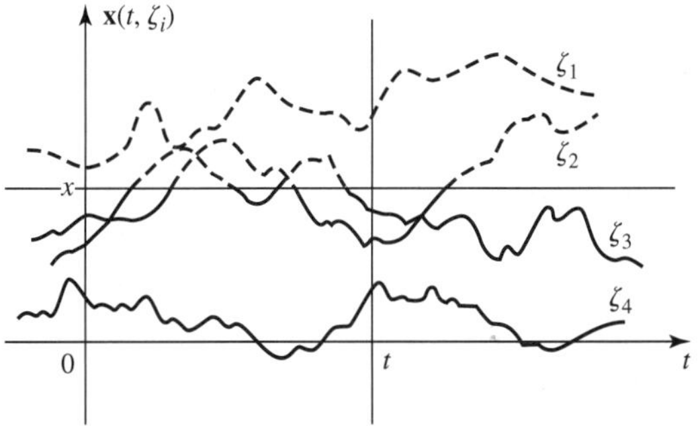

FIGURE 9-2

of the random variables $\mathbf{x}(t_1)$ and $\mathbf{x}(t_2)$. The corresponding density equals

$$f(x_1, x_2; t_1, t_2) = \frac{\partial^2 F(x_1, x_2; t_1, t_2)}{\partial x_1 \partial x_2} \tag{9-6}$$

We note that (consistency conditions)

$$F(x_1; t_1) = F(x_1, \infty; t_1, t_2) \qquad f(x_1, t_1) = \int_{-\infty}^{\infty} f(x_1, x_2; t_1, t_2)\, dx_2$$

as in (6-9) and (6-10).

The *nth-order distribution* of $\mathbf{x}(t)$ is the joint distribution $F(x_1, \ldots, x_n; t_1, \ldots, t_n)$ of the random variables $\mathbf{x}(t_1), \ldots, \mathbf{x}(t_n)$.

SECOND-ORDER PROPERTIES. For the determination of the statistical properties of a stochastic process, knowledge of the function $F(x_1, \ldots, x_n; t_1, \ldots, t_n)$ is required for every x_i, t_i, and n. However, for many applications, only certain averages are used, in particular, the expected value of $\mathbf{x}(t)$ and of $\mathbf{x}^2(t)$. These quantities can be expressed in terms of the second-order properties of $\mathbf{x}(t)$ defined as follows:

Mean The mean $\eta(t)$ of $\mathbf{x}(t)$ is the expected value of the random variable $\mathbf{x}(t)$:

$$\eta(t) = E\{\mathbf{x}(t)\} = \int_{-\infty}^{\infty} x f(x, t)\, dx \tag{9-7}$$

Autocorrelation The autocorrelation $R(t_1, t_2)$ of $\mathbf{x}(t)$ is the expected value of the product $\mathbf{x}(t_1)\mathbf{x}(t_2)$:

$$R(t_1, t_2) = E\{\mathbf{x}(t_1)\mathbf{x}(t_2)\} = \int_{-\infty}^{\infty}\int_{-\infty}^{\infty} x_1 x_2 f(x_1, x_2; t_1, t_2)\, dx_1\, dx_2 \tag{9-8}$$

The value of $R(t_1, t_2)$ on the diagonal $t_1 = t_2 = t$ is the *average power* of $\mathbf{x}(t)$:

$$E\{\mathbf{x}^2(t)\} = R(t, t)$$

The *autocovariance* $C(t_1, t_2)$ of $\mathbf{x}(t)$ is the covariance of the random variables $\mathbf{x}(t_1)$ and $\mathbf{x}(t_2)$:

$$C(t_1, t_2) = R(t_1, t_2) - \eta(t_1)\eta(t_2) \tag{9-9}$$

and its value $C(t, t)$ on the diagonal $t_1 = t_2 = t$ equals the variance of $\mathbf{x}(t)$.

Note The following is an explanation of the reason for introducing the function $R(t_1, t_2)$ even in problems dealing only with average power: Suppose that $\mathbf{x}(t)$ is the input to a linear system and $\mathbf{y}(t)$ is the resulting output. In Sec. 9-2 we show that the mean of $\mathbf{y}(t)$ can be expressed in terms of the mean of $\mathbf{x}(t)$. However, the average power of $\mathbf{y}(t)$ cannot be found if only $E\{\mathbf{x}^2(t)\}$ is given. For the determination of $E\{\mathbf{y}^2(t)\}$, knowledge of the function $R(t_1, t_2)$ is required, not just on the diagonal $t_1 = t_2$, but for every t_1 and t_2. The following identity is a simple illustration

$$E\{[\mathbf{x}(t_1) + \mathbf{x}(t_2)]^2\} = R(t_1, t_1) + 2R(t_1, t_2) + R(t_2, t_2)$$

This follows from (9-8) if we expand the square and use the linearity of expected values.

EXAMPLE 9-1 ▶ An extreme example of a stochastic process is a deterministic signal $\mathbf{x}(t) = f(t)$. In this case,

$$\eta(t) = E\{f(t)\} = f(t) \qquad R(t_1, t_2) = E\{f(t_1)f(t_2)\} = f(t_1)f(t_2)$$ ◀

EXAMPLE 9-2 ▶ Suppose that $\mathbf{x}(t)$ is a process with

$$\eta(t) = 3 \qquad R(t_1, t_2) = 9 + 4e^{-0.2|t_1 - t_2|}$$

We shall determine the mean, the variance, and the covariance of the random variables $\mathbf{z} = \mathbf{x}(5)$ and $\mathbf{w} = \mathbf{x}(8)$.

Clearly, $E\{\mathbf{z}\} = \eta(5) = 3$ and $E\{\mathbf{w}\} = \eta(8) = 3$. Furthermore,

$$E\{\mathbf{z}^2\} = R(5,5) = 13 \qquad E\{\mathbf{w}^2\} = R(8,8) = 13$$

$$E\{\mathbf{zw}\} = R(5,8) = 9 + 4e^{-0.6} = 11.195$$

Thus $\mathbf{z}$ and $\mathbf{w}$ have the same variance $\sigma^2 = 4$ and their covariance equals $C(5,8) = 4e^{-0.6} = 2.195$. ◀

EXAMPLE 9-3 ▶ The integral

$$\mathbf{s} = \int_a^b \mathbf{x}(t)\,dt$$

of a stochastic process $\mathbf{x}(t)$ is a random variable $\mathbf{s}$ and its value $\mathbf{s}(\zeta)$ for a specific outcome ζ is the area under the curve $\mathbf{x}(t,\zeta)$ in the interval (a, b) (see also App. 9A). Interpreting the above as a Riemann integral, we conclude from the linearity of expected values that

$$\eta_s = E\{\mathbf{s}\} = \int_a^b E\{\mathbf{x}(t)\}\,dt = \int_a^b \eta(t)\,dt \tag{9-10}$$

Similarly, since

$$\mathbf{s}^2 = \int_a^b\!\!\int_a^b \mathbf{x}(t_1)\mathbf{x}(t_2)\,dt_1\,dt_2$$

we conclude, using again the linearity of expected values, that

$$E\{\mathbf{s}^2\} = \int_a^b\!\!\int_a^b E\{\mathbf{x}(t_1)\mathbf{x}(t_2)\}\,dt_1\,dt_2 = \int_a^b\!\!\int_a^b R(t_1, t_2)\,dt_1\,dt_2 \tag{9-11}$$

◀

EXAMPLE 9-4 ▶ We shall determine the autocorrelation $R(t_1, t_2)$ of the process

$$\mathbf{x}(t) = \mathbf{r}\cos(\omega t + \boldsymbol{\varphi})$$

where we assume that the random variables $\mathbf{r}$ and $\boldsymbol{\varphi}$ are independent and $\boldsymbol{\varphi}$ is uniform in the interval $(-\pi, \pi)$.

Using simple trigonometric identities, we find

$$E\{\mathbf{x}(t_1)\mathbf{x}(t_2)\} = \tfrac{1}{2}E\{\mathbf{r}^2\}E\{\cos\omega(t_1 - t_2) + \cos(\omega t_1 + \omega t_2 + 2\boldsymbol{\varphi})\}$$

and since

$$E\{\cos(\omega t_1 + \omega t_2 + 2\boldsymbol{\varphi})\} = \frac{1}{2\pi}\int_{-\pi}^{\pi} \cos(\omega t_1 + \omega t_2 + 2\varphi)\, d\varphi = 0$$

we conclude that

$$R(t_1, t_2) = \tfrac{1}{2} E\{\mathbf{r}^2\} \cos\omega(t_1 - t_2) \tag{9-12}$$

◀

EXAMPLE 9-5

POISSON PROCESS

▶ In Sec. 4-5 we introduced the concept of Poisson points and we showed that these points are specified by the following properties:

P_1: The number $\mathbf{n}(t_1, t_2)$ of the points $\mathbf{t}_i$ in an interval (t_1, t_2) of length $t = t_2 - t_1$ is a Poisson random variable with parameter λt:

$$P\{\mathbf{n}(t_1, t_2) = k\} = \frac{e^{-\lambda t}(\lambda t)^k}{k!} \tag{9-13}$$

P_2: If the intervals (t_1, t_2) and (t_3, t_4) are nonoverlapping, then the random variables $\mathbf{n}(t_1, t_2)$ and $\mathbf{n}(t_3, t_4)$ are independent.

Using the points $\mathbf{t}_i$, we form the stochastic process

$$\mathbf{x}(t) = \mathbf{n}(0, t)$$

shown in Fig. 9-3*a*. This is a discrete-state process consisting of a family of increasing staircase functions with discontinuities at the points $\mathbf{t}_i$.

For a specific t, $\mathbf{x}(t)$ is a Poisson random variable with parameter λt; hence

$$E\{\mathbf{x}(t)\} = \eta(t) = \lambda t$$

We shall show that its autocorrelation equals

$$R(t_1, t_2) = \begin{cases} \lambda t_2 + \lambda^2 t_1 t_2 & t_1 \ge t_2 \\ \lambda t_1 + \lambda^2 t_1 t_2 & t_1 \le t_2 \end{cases} \tag{9-14}$$

or equivalently that

$$C(t_1, t_2) = \lambda \min(t_1, t_2) = \lambda t_1 U(t_2 - t_1) + \lambda t_2 U(t_1 - t_2)$$

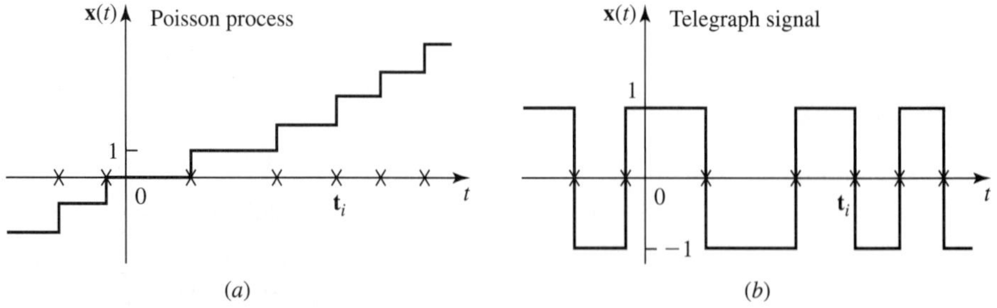

FIGURE 9-3

Proof. The preceding is true for $t_1 = t_2$ because [see (5-63)]

$$E\{\mathbf{x}^2(t)\} = \lambda t + \lambda^2 t^2 \tag{9-15}$$

Since $R(t_1, t_2) = R(t_2, t_1)$, it suffices to prove (9-14) for $t_1 < t_2$. The random variables $\mathbf{x}(t_1)$ and $\mathbf{x}(t_2) - \mathbf{x}(t_1)$ are independent because the intervals $(0, t_1)$ and (t_1, t_2) are nonoverlapping. Furthermore, they are Poisson distributed with parameters λt_1 and $\lambda(t_2 - t_1)$ respectively. Hence

$$E\{\mathbf{x}(t_1)[\mathbf{x}(t_2) - \mathbf{x}(t_1)]\} = E\{\mathbf{x}(t_1)\}E\{\mathbf{x}(t_2) - \mathbf{x}(t_1)\} = \lambda t_1 \lambda(t_2 - t_1)$$

Using the identity

$$\mathbf{x}(t_1)\mathbf{x}(t_2) = \mathbf{x}(t_1)[\mathbf{x}(t_1) + \mathbf{x}(t_2) - \mathbf{x}(t_1)]$$

we conclude from the above and (9-15) that

$$R(t_1, t_2) = \lambda t_1 + \lambda^2 t_1^2 + \lambda t_1 \lambda(t_2 - t_1)$$

and (9-14) results.

Nonuniform case If the points $\mathbf{t}_i$ have a nonuniform density $\lambda(t)$ as in (4-124), then the preceding results still hold provided that the product $\lambda(t_2 - t_1)$ is replaced by the integral of $\lambda(t)$ from t_1 to t_2.

Thus

$$E\{\mathbf{x}(t)\} = \int_0^t \lambda(\alpha)\, d\alpha \tag{9-16}$$

and

$$R(t_1, t_2) = \int_0^{t_1} \lambda(t)\, dt \left[1 + \int_0^{t_2} \lambda(t)\, dt\right] \qquad t_1 \le t_2 \tag{9-17}$$

◀

EXAMPLE 9-6

TELEGRAPH SIGNAL

▶ Using the Poisson points $\mathbf{t}_i$, we form a process $\mathbf{x}(t)$ such that $\mathbf{x}(t) = 1$ if the number of points in the interval $(0, t)$ is even, and $\mathbf{x}(t) = -1$ if this number is odd (Fig. 9-3b).

Denoting by $p(k)$ the probability that the number of points in the interval $(0, t)$ equals k, we conclude that [see (9-13)]

$$\begin{aligned} P\{\mathbf{x}(t) = 1\} &= p(0) + p(2) + \cdots \\ &= e^{-\lambda t}\left[1 + \frac{(\lambda t)^2}{2!} + \cdots\right] = e^{-\lambda t}\cosh \lambda t \\ P\{\mathbf{x}(t) = -1\} &= p(1) + p(3) + \cdots \\ &= e^{-\lambda t}\left[\lambda t + \frac{(\lambda t)^3}{3!} + \cdots\right] = e^{-\lambda t}\sinh \lambda t \end{aligned}$$

Hence

$$E\{\mathbf{x}(t)\} = e^{-\lambda t}(\cosh \lambda t - \sinh \lambda t) = e^{-2\lambda t} \tag{9-18}$$

To determine $R(t_1, t_2)$, we note that, if $t = t_1 - t_2 > 0$ and $\mathbf{x}(t_2) = 1$, then $\mathbf{x}(t_1) = 1$ if the number of points in the interval (t_1, t_2) is even. Hence

$$P\{\mathbf{x}(t_1) = 1 \mid \mathbf{x}(t_2) = 1\} = e^{-\lambda t}\cosh \lambda t \qquad t = t_1 - t_2$$

Multiplying by $P\{\mathbf{x}(t_2) = 1\}$, we obtain

$$P\{\mathbf{x}(t_1) = 1, \mathbf{x}(t_2) = 1\} = e^{-\lambda t} \cosh \lambda t e^{-\lambda t_2} \cosh \lambda t_2$$

Similarly,

$$P\{\mathbf{x}(t_1) = -1, \mathbf{x}(t_2) = -1\} = e^{-\lambda t} \cosh \lambda t e^{-\lambda t_2} \sinh \lambda t_2$$

$$P\{\mathbf{x}(t_1) = 1, \mathbf{x}(t_2) = -1\} = e^{-\lambda t} \sinh \lambda t e^{-\lambda t_2} \sinh \lambda t_2$$

$$P\{\mathbf{x}(t_1) = -1, \mathbf{x}(t_2) = 1\} = e^{-\lambda t} \sinh \lambda t e^{-\lambda t_2} \cosh \lambda t_2$$

Since the product $\mathbf{x}(t_1)\mathbf{x}(t_2)$ equals 1 or -1, we conclude omitting details that

$$R(t_1, t_2) = e^{-2\lambda|t_1 - t_2|} \tag{9-19}$$

This process is called *semirandom* telegraph signal because its value $\mathbf{x}(0) = 1$ at $t = 0$ is not random. To remove this certainty, we form the product

$$\mathbf{y}(t) = \mathbf{a}\mathbf{x}(t)$$

where $\mathbf{a}$ is a random variable taking the values $+1$ and -1 with equal probability and is independent of $\mathbf{x}(t)$. The process $\mathbf{y}(t)$ so formed is called *random* telegraph signal. Since $E\{\mathbf{a}\} = 0$ and $E\{\mathbf{a}^2\} = 1$, the mean of $\mathbf{y}(t)$ equals $E\{\mathbf{a}\}E\{\mathbf{x}(t)\} = 0$ and its autocorrelation is given by

$$E\{\mathbf{y}(t_1)\mathbf{y}(t_2)\} = E\{\mathbf{a}^2\}E\{\mathbf{x}(t_1)\mathbf{x}(t_2)\} = e^{-2\lambda|t_1 - t_2|}$$

We note that as $t \to \infty$ the processes $\mathbf{x}(t)$ and $\mathbf{y}(t)$ have asymptotically equal statistics. ◀

More on Poisson processes. If $\mathbf{x}_1(t)$ and $\mathbf{x}_2(t)$ represent two independent Poisson processes with parameters $\lambda_1 t$ and $\lambda_2 t$, respectively, it easily follows [as in (6-86)] that their sum $\mathbf{x}_1(t) + \mathbf{x}_2(t)$ is also a Poisson process with parameter $(\lambda_1 + \lambda_2)t$. What about the difference of two independent Poisson processes? What can we say about the distribution of such a process? Let

$$\mathbf{y}(t) = \mathbf{x}_1(t) - \mathbf{x}_2(t) \tag{9-20}$$

where $\mathbf{x}_1(t)$ and $\mathbf{x}_2(t)$ are two independent Poisson processes as just defined. Then

$$\begin{aligned} P\{\mathbf{y}(t) = n\} &= \sum_{k=0}^{\infty} P\{\mathbf{x}_1(t) = n + k\}P\{\mathbf{x}_2(t) = k\} \\ &= \sum_{k=0}^{\infty} e^{-\lambda_1 t} \frac{(\lambda_1 t)^{n+k}}{(n+k)!} e^{-\lambda_2 t} \frac{(\lambda_2 t)^k}{k!} \\ &= e^{-(\lambda_1+\lambda_2)t} \left(\frac{\lambda_1}{\lambda_2}\right)^{n/2} \sum_{k=0}^{\infty} \frac{(t\sqrt{\lambda_1\lambda_2})^{n+2k}}{k!(n+k)!} \\ &= e^{-(\lambda_1+\lambda_2)t} \left(\frac{\lambda_1}{\lambda_2}\right)^{n/2} I_{|n|}(2\sqrt{\lambda_1\lambda_2}t) \qquad n = 0, \pm 1, \pm 2, \ldots \end{aligned} \tag{9-21}$$

where

$$I_n(x) \triangleq \sum_{k=0}^{\infty} \frac{(x/2)^{n+2k}}{k!(n+k)!} \tag{9-22}$$

represents the modified Bessel function of order n. From (9-15) and (9-20), it follows that

$$E\{\mathbf{y}(t)\} = (\lambda_1 - \lambda_2)t \qquad \text{Var}\{\mathbf{y}(t)\} = (\lambda_1 + \lambda_2)t \tag{9-23}$$

Thus the difference of two independent Poisson processes is *not* Poisson. However, it is easy to show that a *random selection* from a Poisson process yields a Poisson process!

Random selection of Poisson points. Let $\mathbf{x}(t) \sim P(\lambda t)$ represent a Poisson process with parameter λt as before, and suppose each occurrence of $\mathbf{x}(t)$ gets tagged independently with probability p. Let $\mathbf{y}(t)$ represent the total number of tagged events in the interval $(0, t)$ and let $\mathbf{z}(t)$ be the total number of untagged events in $(0, t)$. Then

$$\mathbf{y}(t) \sim P(\lambda pt) \qquad \mathbf{z}(t) \sim P(\lambda qt) \tag{9-24}$$

where $q = 1 - p$.

Proof. Let A_n represent the event "n events occur in $(0, t)$ and k of them are tagged." Then

$$\begin{aligned} P(A_n) &= P\{k \text{ events are tagged} \mid \mathbf{x}(t) = n\}\, P\{\mathbf{x}(t) = n\} \\ &= \binom{n}{k} p^k q^{n-k} e^{-\lambda t} \frac{(\lambda t)^n}{n!} \end{aligned}$$

Also the event $\{\mathbf{y}(t) = k\}$ represents the mutually exclusive union of the events A_k, $A_{k+1}, \ldots$. Thus

$$\{\mathbf{y}(t) = k\} = \bigcup_{n=k}^{\infty} A_n$$

so that

$$\begin{aligned} P\{\mathbf{y}(t) = k\} &= \sum_{n=k}^{\infty} P(A_n) = e^{-\lambda t} \sum_{n=k}^{\infty} \frac{(\lambda t)^n}{k!(n-k)!} p^k q^{n-k} \\ &= e^{-\lambda t} \frac{(\lambda pt)^k}{k!} \sum_{r=0}^{\infty} \frac{(\lambda qt)^r}{r!} \\ &= e^{-\lambda(1-q)t} \frac{(\lambda pt)^k}{k!} = e^{-\lambda pt} \frac{(\lambda pt)^k}{k!} \qquad k = 0, 1, 2, \ldots \end{aligned} \tag{9-25}$$

represents a Poisson process with parameter λpt. Similarly the untagged events $\mathbf{z}(t)$ form an independent Poisson process with parameter λqt. (See also page 227.) For example, if customers arrive at a counter according to a Poisson process with parameter λt, and the probability of a customer being male is p, then the male customers form a Poisson process with parameter λpt, and the female customers form an independent Poisson process with parameter λqt. (See (10-90) for a deterministic selection of Poisson points.)

Next we will show that the conditional probability of a subset of a Poisson event is in fact binomial.

Poisson points and binomial distribution. For $t_1 < t_2$ consider the conditional probability

$$
\begin{aligned}
P\{\mathbf{x}(t_1) = k \mid \mathbf{x}(t_2) = n\} \\
&= \frac{P\{\mathbf{x}(t_1) = k, \mathbf{x}(t_2) = n\}}{P\{\mathbf{x}(t_2) = n\}} \\
&= \frac{P\{\mathbf{x}(t_1) = k, \mathbf{n}(t_1, t_2) = n - k\}}{P\{\mathbf{x}(t_2) = n\}} \\
&= \frac{e^{-\lambda t_1}(\lambda t_1)^k}{k!} \frac{e^{-\lambda(t_2 - t_1)}[\lambda(t_2 - t_1)]^{n-k}}{(n-k)!} \frac{n!}{e^{-\lambda t_2}(\lambda t_2)^n} \\
&= \binom{n}{k}\left(\frac{t_1}{t_2}\right)^k\left(1 - \frac{t_1}{t_2}\right)^{n-k} \sim B\left(n, \frac{t_1}{t_2}\right) \qquad k = 0, 1, 2, \ldots, n \qquad \text{(9-26)}
\end{aligned}
$$

which proves our claim. In particular, let $k = n = 1$, and let Δ be the subinterval in the beginning of an interval of length T. Then from (9-26) we obtain

$$P\{\mathbf{n}(\Delta) = 1 \mid \mathbf{n}(t, t + T) = 1\}\frac{\Delta}{T}$$

But the event $\{\mathbf{n}(\Delta) = 1\}$ is equivalent to $\{t < \mathbf{t}_i < t + \Delta\}$, where $\mathbf{t}_i$ denotes the random arrival instant. Hence the last expression represents

$$P\{t < \mathbf{t}_i < t + \Delta \mid \mathbf{n}(t, t + T\} = 1) = \frac{\Delta}{T} \qquad \text{(9-27)}$$

i.e., given that only one Poisson occurrence has taken place in an interval of length T, the conditional p.d.f. of the corresponding arrival instant is uniform in that interval. In other words, a Poisson arrival is equally likely to happen anywhere in an interval T, given that only one occurrence has taken place in that interval.

More generally if $\mathbf{t}_1 < \mathbf{t}_2 < \cdots < \mathbf{t}_n < T$ represents the n arrival instants of a Poisson process in the interval $(0, T)$, then the joint conditional distribution of $\mathbf{t}_1, \mathbf{t}_2, \ldots, \mathbf{t}_n$ given $\mathbf{x}(T) = n$ simplifies into

$$
\begin{aligned}
P\{\mathbf{t}_1 \le x_1, \mathbf{t}_2 \le x_2, \ldots, \mathbf{t}_n \le x_n \mid \mathbf{x}(T) = n\} \\
&= \frac{P\{\mathbf{t}_1 \le x_1, \mathbf{t}_2 \le x_2, \ldots, \mathbf{t}_n \le x_n, \mathbf{x}(t) = n\}}{P\{\mathbf{x}(T) = n\}} \\
&= \frac{1}{e^{-\lambda T}\dfrac{(\lambda T)^n}{n!}} \sum_{\{m_1, m_2, \ldots, m_n\}} \prod_{i=1}^{n} e^{-\lambda(x_i - x_{i-1})} \frac{[\lambda(x_i - x_{i-1})]^{m_i}}{m_i!} \\
&= \sum_{m_1, m_2, \ldots, m_n} \frac{n!}{m_1! m_2! \cdots m_n!}\left(\frac{x_1}{T}\right)^{m_1}\left(\frac{x_2 - x_1}{T}\right)^{m_2} \cdots \left(\frac{x_n - x_{n-1}}{T}\right)^{m_n} \qquad \text{(9-28)}
\end{aligned}
$$

with $x_0 = 0$. The summation is over all nonnegative integers $\{m_1, m_2, \ldots, m_n\}$ for which $m_1 + m_2 + \cdots + m_n = n$ and $m_1 + m_2 + \cdots + m_k \ge k = 1, 2, \ldots, n - 1$. On comparing with (4-102), the above formula in (9-28) represents the distribution of n independent points arranged in increasing order, each of which is uniformly distributed over the interval $(0, T)$. It follows that a Poisson process $\mathbf{x}(t)$ distributes points at random over the infinite interval $(0, \infty)$ the same way the uniform random variable distributes points in a finite interval.

General Properties

The statistical properties of a real stochastic process $\mathbf{x}(t)$ are completely determined[2] in terms of its nth-order distribution

$$F(x_1, \ldots, x_n; t_1, \ldots, t_n) = P\{\mathbf{x}(t_1) \le x_1, \ldots, \mathbf{x}(t_n) \le x_n\} \tag{9-29}$$

The joint statistics of two real processes $\mathbf{x}(t)$ and $\mathbf{y}(t)$ are determined in terms of the joint distribution of the random variables

$$\mathbf{x}(t_1), \ldots, \mathbf{x}(t_n), \mathbf{y}(t_1'), \ldots, \mathbf{y}(t_m')$$

The *complex process* $\mathbf{z}(t) = \mathbf{x}(t) + j\mathbf{y}(t)$ is specified in terms of the joint statistics of the real processes $\mathbf{x}(t)$ and $\mathbf{y}(t)$.

A *vector process* (n-dimensional process) is a family of n stochastic processes.

CORRELATION AND COVARIANCE. The autocorrelation of a process $\mathbf{x}(t)$, real or complex, is by definition the mean of the product $\mathbf{x}(t_1)\mathbf{x}^*(t_2)$. This function will be denoted by $R(t_1, t_2)$ or $R_x(t_1, t_2)$ or $R_{xx}(t_1, t_2)$. Thus

$$R_{xx}(t_1, t_2) = E\{\mathbf{x}(t_1)\mathbf{x}^*(t_2)\} \tag{9-30}$$

where the conjugate term is associated with the second variable in $R_{xx}(t_1, t_2)$. From this it follows that

$$R(t_2, t_1) = E\{\mathbf{x}(t_2)\mathbf{x}^*(t_1)\} = R^*(t_1, t_2) \tag{9-31}$$

We note, further, that

$$R(t, t) = E\{|\mathbf{x}(t)|^2\} \ge 0 \tag{9-32}$$

The last two equations are special cases of this property: The autocorrelation $R(t_1, t_2)$ of a stochastic process $\mathbf{x}(t)$ is a *positive definite* (p.d.) function, that is, for any a_i and a_j:

$$\sum_{i,j} a_i a_j^* R(t_i, t_j) \ge 0 \tag{9-33}$$

This is a consequence of the identity

$$0 \le E\left\{\left|\sum_i a_i \mathbf{x}(t_i)\right|^2\right\} = \sum_{i,j} a_i a_j^* E\{\mathbf{x}(t_i)\mathbf{x}^*(t_j)\}$$

We show later that the converse is also true: Given a p.d. function $R(t_1, t_2)$, we can find a process $\mathbf{x}(t)$ with autocorrelation $R(t_1, t_2)$.

[2]There are processes (nonseparable) for which this is not true. However, such processes are mainly of mathematical interest.

EXAMPLE 9-7

▶ (*a*) If $\mathbf{x}(t) = \mathbf{a}e^{j\omega t}$, then

$$R(t_1, t_2) = E\{\mathbf{a}e^{j\omega t_1}\mathbf{a}^*e^{-j\omega t_2}\} = E\{|\mathbf{a}|^2\}e^{j\omega(t_1-t_2)}$$

(*b*) Suppose that the random variables $\mathbf{a}_i$ are uncorrelated with zero mean and variance σ_i^2. If

$$\mathbf{x}(t) = \sum_i \mathbf{a}_i e^{j\omega_i t}$$

then (9-30) yields

$$R(t_1, t_2) = \sum_i \sigma_i^2 e^{j\omega_i(t_1-t_2)}$$

◀

The ***autocovariance*** $C(t_1, t_2)$ of a process $\mathbf{x}(t)$ is the covariance of the random variables $\mathbf{x}(t_1)$ and $\mathbf{x}(t_2)$:

$$C(t_1, t_2) = R(t_1, t_2) - \eta(t_1)\eta^*(t_2) \qquad (9\text{-}34)$$

In (9-34), $\eta(t) = E\{\mathbf{x}(t)\}$ is the *mean* of $\mathbf{x}(t)$.

The ratio

$$r(t_1, t_2) = \frac{C(t_1, t_2)}{\sqrt{C(t_1, t_1)C(t_2, t_2)}} \qquad (9\text{-}35)$$

is the ***correlation coefficient***[3] of the process $\mathbf{x}(t)$.

Note The autocovariance $C(t_1, t_2)$ of a process $\mathbf{x}(t)$ is the autocorrelation of the *centered process*

$$\tilde{\mathbf{x}}(t) = \mathbf{x}(t) - \eta(t)$$

Hence it is p.d.

The correlation coefficient $r(t_1, t_2)$ of $\mathbf{x}(t)$ is the autocovariance of the *normalized process* $\mathbf{x}(t)/\sqrt{C(t, t)}$; hence it is also p.d. Furthermore [see (6-166)]

$$|r(t_1, t_2)| \le 1 \qquad r(t, t) = 1 \qquad (9\text{-}36)$$

EXAMPLE 9-8

▶ If

$$\mathbf{s} = \int_a^b \mathbf{x}(t)\,dt \qquad \text{then} \qquad \mathbf{s} - \eta_s = \int_a^b \tilde{\mathbf{x}}(t)\,dt$$

where $\tilde{\mathbf{x}}(t) = \mathbf{x}(t) - \eta_x(t)$. Using (9-11), we conclude from the note that

$$\sigma_s^2 = E\{|\mathbf{s} - \eta_s|^2\} \int_a^b\int_a^b C_x(t_1, t_2)\,dt_1\,dt_2 \qquad (9\text{-}37)$$

The ***cross-correlation*** of two processes $\mathbf{x}(t)$ and $\mathbf{y}(t)$ is the function

$$R_{xy}(t_1, t_2) = E\{\mathbf{x}(t_1)\mathbf{y}^*(t_2)\} = R_{yx}^*(t_2, t_1) \qquad (9\text{-}38)$$

◀

[3] In optics, $C(t_1, t_2)$ is called the *coherence function* and $r(t_1, t_2)$ is called the *complex degree of coherence* (see Papoulis, 1968 [19]).

Similarly,

$$C_{xy}(t_1, t_2) = R_{xy}(t_1, t_2) - \eta_x(t_1)\eta_y^*(t_2) \tag{9-39}$$

is their ***cross-covariance***.

Two processes $\mathbf{x}(t)$ and $\mathbf{y}(t)$ are called (mutually) *orthogonal* if

$$R_{xy}(t_1, t_2) = 0 \qquad \text{for every} \quad t_1 \text{ and } t_2 \tag{9-40}$$

They are called ***uncorrelated*** if

$$C_{xy}(t_1, t_2) = 0 \qquad \text{for every} \quad t_1 \text{ and } t_2 \tag{9-41}$$

a-dependent processes In general, the values $\mathbf{x}(t_1)$ and $\mathbf{x}(t_2)$ of a stochastic process $\mathbf{x}(t)$ are statistically dependent for any t_1 and t_2. However, in most cases this dependence decreases as $|t_1 - t_2| \to \infty$. This leads to the following concept: A stochastic process $\mathbf{x}(t)$ is called *a-dependent* if all its values $\mathbf{x}(t)$ for $t < t_o$ and for $t > t_o + a$ are mutually *independent*. From this it follows that

$$C(t_1, t_2) = 0 \qquad \text{for} \quad |t_1 - t_2| > a \tag{9-42}$$

A process $\mathbf{x}(t)$ is called *correlation a-dependent* if its autocorrelation satisfies (9-42). Clearly, if $\mathbf{x}(t)$ is correlation a-dependent, then any linear combination of its values for $t < t_o$ is uncorrelated with any linear combination of its values for $t > t_o + a$.

White noise We shall say that a process $\mathbf{v}(t)$ is white noise if its values $\mathbf{v}(t_i)$ and $\mathbf{v}(t_j)$ are uncorrelated for every t_i and $t_j \neq t_i$:

$$C(t_i, t_j) = 0 \qquad t_i \neq t_j$$

As we explain later, the autocovariance of a nontrivial white-noise process must be of the form

$$C(t_1, t_2) = q(t_1)\delta(t_1 - t_2) \qquad q(t) \geq 0 \tag{9-43}$$

If the random variables $\mathbf{v}(t_i)$ and $\mathbf{v}(t_j)$ are not only uncorrelated but also independent, then $\mathbf{v}(t)$ will be called *strictly* white noise. Unless otherwise stated, it will be assumed that the mean of a white-noise process is identically 0.

EXAMPLE 9-9 ▶ Suppose that $\mathbf{v}(t)$ is white noise and

$$\mathbf{x}(t) = \int_0^t \mathbf{v}(\alpha)\, d\alpha \tag{9-44}$$

Inserting (9-43) into (9-44), we obtain

$$E\{\mathbf{x}^2(t)\} = \int_0^t \int_0^t q(t_1)\delta(t_1 - t_2)\, dt_2\, dt_1 = \int_0^t q(t_1)\, dt_1 \tag{9-45}$$

because

$$\int_0^t \delta(t_1 - t_2)\, dt_2 = 1 \qquad \text{for} \quad 0 < t_1 < t$$ ◀

Uncorrelated and independent increments If the increments $\mathbf{x}(t_2) - \mathbf{x}(t_1)$ and $\mathbf{x}(t_4) - \mathbf{x}(t_3)$ of a process $\mathbf{x}(t)$ are uncorrelated (independent) for any $t_1 < t_2 < t_3 < t_4$, then we say that $\mathbf{x}(t)$ is a process with uncorrelated (independent) increments. The Poisson process is a process with independent increments. The integral (9-44) of white noise is a process with uncorrelated increments.

Independent processes If two processes $\mathbf{x}(t)$ and $\mathbf{y}(t)$ are such that the random variables $\mathbf{x}(t_1), \ldots, \mathbf{x}(t_n)$ and $\mathbf{y}(t'_1), \ldots, \mathbf{y}(t'_n)$ are mutually independent, then these processes are called independent.

NORMAL PROCESSES

▶ A process $\mathbf{x}(t)$ is called normal, if the random variables $\mathbf{x}(t_1), \ldots, \mathbf{x}(t_n)$ are jointly normal for any n and $t_1, \ldots, t_n$.

The statistics of a normal process are completely determined in terms of its mean $\eta(t)$ and autocovariance $C(t_1, t_2)$. Indeed, since

$$E\{\mathbf{x}(t)\} = \eta(t) \qquad \sigma_x^2(t) = C(t, t)$$

we conclude that the first-order density $f(x, t)$ of $\mathbf{x}(t)$ is the normal density $N[\eta(t); \sqrt{C(t,t)}]$.

Similarly, since the function $r(t_1, t_2)$ in (9-35) is the correlation coefficient of the random variables $\mathbf{x}(t_1)$ and $\mathbf{x}(t_2)$, the second-order density $f(x_1, x_2; t_1, t_2)$ of $\mathbf{x}(t)$ is the jointly normal density

$$N[\eta(t_1), \eta(t_2); \sqrt{C(t_1, t_1)}, \sqrt{C(t_2, t_2)}; r(t_1, t_2)]$$

The nth-order characteristic function of the process $\mathbf{x}(t)$ is given by [see (7-60)]

$$\exp\left\{ j \sum_i \eta(t_i)\omega_i - \frac{1}{2} \sum_{i,k} C(t_i, t_k)\omega_i\omega_k \right\} \tag{9-46}$$

Its inverse $f(x_1, \ldots, x_n, t_1, \ldots, t_n)$ is the nth-order density of $\mathbf{x}(t)$. ◀

EXISTENCE THEOREM. Given an arbitrary function $\eta(t)$ and a p.d. function $C(t_1, t_2)$, we can construct a normal process with mean $\eta(t)$ and autocovariance $C(t_1, t_2)$. This follows if we use in (9-46) the given functions $\eta(t)$ and $C(t_1, t_2)$. The inverse of the resulting characteristic function is a density because the function $C(t_1, t_2)$ is p.d. by assumption.

EXAMPLE 9-10

▶ Suppose that $\mathbf{x}(t)$ is a normal process with

$$\eta(t) = 3 \qquad C(t_1, t_2) = 4e^{-0.2|t_1 - t_2|}$$

(*a*) Find the probability that $\mathbf{x}(5) \le 2$.

Clearly, $\mathbf{x}(5)$ is a normal random variable with mean $\eta(5) = 3$ and variance $C(5, 5) = 4$. Hence

$$P\{\mathbf{x}(5) \le 2\} = G(-1/2) = 0.309$$

(*b*) Find the probability that $|\mathbf{x}(8) - \mathbf{x}(5)| \le 1$.

The difference $\mathbf{s} = \mathbf{x}(8) - \mathbf{x}(5)$ is a normal random variable with mean $\eta(8) - \eta(5) = 0$ and variance

$$C(8, 8) + C(5, 5) - 2C(8, 5) = 8(1 - e^{-0.6}) = 3.608$$

Hence

$$P\{|\mathbf{x}(8) - \mathbf{x}(5)| \le 1\} = 2G(1/1.9) - 1 = 0.4$$

◀

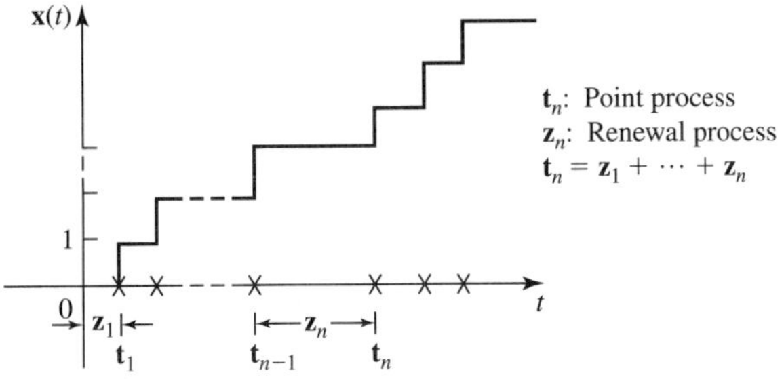

FIGURE 9-4

POINT AND RENEWAL PROCESSES. A *point process* is a set of random points $\mathbf{t}_i$ on the time axis. To every point process we can associate a stochastic process $\mathbf{x}(t)$ equal to the number of points $\mathbf{t}_i$ in the interval $(0, t)$. An example is the Poisson process. To every point process $\mathbf{t}_i$ we can associate a sequence of random variables $\mathbf{z}_n$ such that

$$\mathbf{z}_1 = \mathbf{t}_1 \qquad \mathbf{z}_2 = \mathbf{t}_2 - \mathbf{t}_1 \cdots \mathbf{z}_n = \mathbf{t}_n - \mathbf{t}_{n-1}$$

where $\mathbf{t}_1$ is the first random point to the right of the origin. This sequence is called a *renewal process*. An example is the life history of lightbulbs that are replaced as soon as they fail. In this case, $\mathbf{z}_i$ is the total time the ith bulb is in operation and $\mathbf{t}_i$ is the time of its failure.

We have thus established a correspondence between the following three concepts (Fig. 9-4): (a) a point process $\mathbf{t}_i$, (b) a discrete-state stochastic process $\mathbf{x}(t)$ increasing in unit steps at the points $\mathbf{t}_i$, (c) a renewal process consisting of the random variables $\mathbf{z}_i$ and such that $\mathbf{t}_n = \mathbf{z}_1 + \cdots + \mathbf{z}_n$.

Stationary Processes

STRICT SENSE STATIONARY. A stochastic process $\mathbf{x}(t)$ is called *strict-sense stationary* (abbreviated SSS) if its statistical properties are invariant to a shift of the origin. This means that the processes $\mathbf{x}(t)$ and $\mathbf{x}(t+c)$ have the same statistics for any c.

Two processes $\mathbf{x}(t)$ and $\mathbf{y}(t)$ are called *jointly stationary* if the joint statistics of $\mathbf{x}(t)$ and $\mathbf{y}(t)$ are the same as the joint statistics of $\mathbf{x}(t+c)$ and $\mathbf{y}(t+c)$ for any c.

A complex process $\mathbf{z}(t) = \mathbf{x}(t) + j\mathbf{y}(t)$ is stationary if the processes $\mathbf{x}(t)$ and $\mathbf{y}(t)$ are jointly stationary.

From the definition it follows that the nth-order density of an SSS process must be such that

$$f(x_1, \ldots, x_n; t_1, \ldots, t_n) = f(x_1, \ldots, x_n; t_1 + c, \ldots, t_n + c) \tag{9-47}$$

for any c.

From this it follows that $f(x; t) = f(x; t + c)$ for any c. Hence the first-order density of $\mathbf{x}(t)$ is independent of t:

$$f(x; t) = f(x) \tag{9-48}$$

Similarly, $f(x_1, x_2; t_1 + c, t_2 + c)$ is independent of c for any c, in particular for $c = -t_2$. This leads to the conclusion that

$$f(x_1, x_2, t_1, t_2) = f(x_1, x_2; \tau) \qquad \tau = t_1 - t_2 \tag{9-49}$$

Thus the joint density of the random variables $\mathbf{x}(t+\tau)$ and $\mathbf{x}(t)$ is independent of t and it equals $f(x_1, x_2; \tau)$.

WIDE SENSE STATIONARY. A stochastic process $\mathbf{x}(t)$ is called *wide-sense stationary* (abbreviated WSS) if its mean is constant

$$E\{\mathbf{x}(t)\} = \eta \tag{9-50}$$

and its autocorrelation depends only on $\tau = t_1 - t_2$:

$$E\{\mathbf{x}(t+\tau)\mathbf{x}^*(t)\} = R(\tau) \tag{9-51}$$

Since τ is the distance from t to $t+\tau$, the function $R(\tau)$ can be written in the symmetrical form

$$R(\tau) = E\left\{\mathbf{x}\left(t+\frac{\tau}{2}\right)\mathbf{x}^*\left(t-\frac{\tau}{2}\right)\right\} \tag{9-52}$$

Note in particular that

$$E\{|\mathbf{x}(t)|^2\} = R(0)$$

Thus the average power of a stationary process is independent of t and it equals $R(0)$.

EXAMPLE 9-11

▶ Suppose that $\mathbf{x}(t)$ is a WSS process with autocorrelation

$$R(\tau) = Ae^{-\alpha|\tau|}$$

We shall determine the second moment of the random variable $\mathbf{x}(8) - \mathbf{x}(5)$. Clearly,

$$\begin{aligned} E\{[\mathbf{x}(8) - \mathbf{x}(5)]^2\} &= E\{\mathbf{x}^2(8)\} + E\{\mathbf{x}^2(5)\} - 2E\{\mathbf{x}(8)\mathbf{x}(5)\} \\ &= R(0) + R(0) - 2R(3) = 2A - 2Ae^{-3\alpha} \end{aligned}$$ ◀

Note As Example 9-11 suggests, the autocorrelation of a stationary process $\mathbf{x}(t)$ can be defined as average power. Assuming for simplicity that $\mathbf{x}(t)$ is real, we conclude from (9-51) that

$$E\{[\mathbf{x}(t+\tau) - \mathbf{x}(t)]^2\} = 2[R(0) - R(\tau)] \tag{9-53}$$

From (9-51) it follows that the autocovariance of a WSS process depends only on $\tau = t_1 - t_2$:

$$C(\tau) = R(\tau) - |\eta|^2 \tag{9-54}$$

and its correlation coefficient [see (9-35)] equals

$$r(\tau) = C(\tau)/C(0) \tag{9-55}$$

Thus $C(\tau)$ is the covariance, and $r(\tau)$ the correlation coefficient of the random variables $\mathbf{x}(t+\tau)$ and $\mathbf{x}(t)$.

Two processes $\mathbf{x}(t)$ and $\mathbf{y}(t)$ are called jointly WSS if each is WSS and their cross-correlation depends only on $\tau = t_1 - t_2$:

$$R_{xy}(\tau) = E\{\mathbf{x}(t+\tau)\mathbf{y}^*(t)\} \qquad C_{xy}(\tau) = R_{x,y}(\tau) - \eta_x\eta_y^* \tag{9-56}$$

If $\mathbf{x}(t)$ is WSS white noise, then [see (9-43)]

$$C(\tau) = q\delta(\tau) \tag{9-57}$$

If $\mathbf{x}(t)$ is an a-dependent process, then $C(\tau) = 0$ for $|\tau| > a$. In this case, the constant a is called the *correlation time* of $\mathbf{x}(t)$. This term is also used for arbitrary processes and it is defined as the ratio

$$\tau_c = \frac{1}{C(0)} \int_0^\infty C(\tau)\, d\tau \tag{9-58}$$

In general $C(\tau) \neq 0$ for every τ. However, for most regular processes

$$C(\tau) \underset{|\tau|\to\infty}{\longrightarrow} 0 \qquad R(\tau) \underset{|\tau|\to\infty}{\longrightarrow} |\eta|^2$$

EXAMPLE 9-12

▶ If $\mathbf{x}(t)$ is WSS and

$$\mathbf{s} = \int_{-T}^{T} \mathbf{x}(t)\, dt$$

then [see (9-37)]

$$\sigma_s^2 = \int_{-T}^{T}\int_{-T}^{T} C(t_1 - t_2)\, dt_1\, dt_2 = \int_{-2T}^{2T} (2T - |\tau|) C(\tau)\, d\tau \tag{9-59}$$

The last equality follows with $\tau = t_1 - t_2$ (see Fig. 9-5); the details, however, are omitted [see also (9-156)].

Special cases. (a) If $C(\tau) = q\delta(\tau)$, then

$$\sigma_s^2 = q \int_{-2T}^{2T} (2T - |\tau|)\delta(\tau)\, d\tau = 2Tq$$

(b) If the process $\mathbf{x}(t)$ is a-dependent and $a \ll T$, then (9-59) yields

$$\sigma_s^2 = \int_{-2T}^{2T} (2T - |\tau|) C(\tau)\, d\tau \simeq 2T \int_{-a}^{a} C(\tau)\, d\tau$$

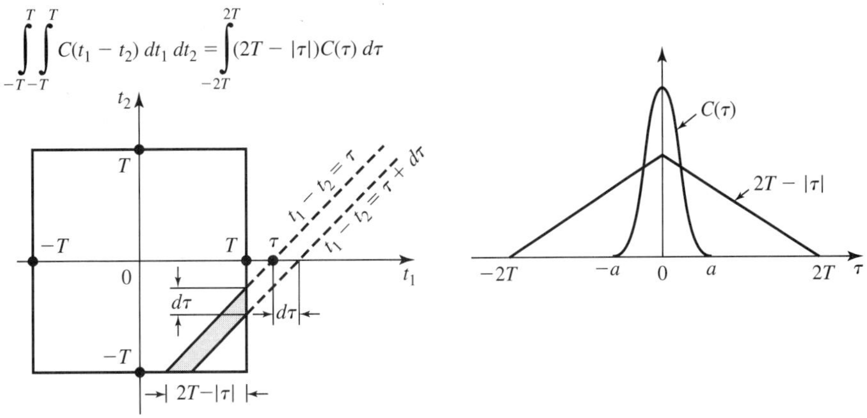

FIGURE 9-5

This shows that, in the evaluation of the variance of **s**, an a-dependent process with $a \ll T$ can be replaced by white noise as in (9-57) with

$$q = \int_{-a}^{a} C(\tau)\, d\tau$$ ◀

If a process is SSS, then it is also WSS. This follows readily from (9-48) and (9-49). The converse, however, is not in general true. As we show next, normal processes are an important exception.

Indeed, suppose that $\mathbf{x}(t)$ is a normal WSS process with mean η and autocovariance $C(\tau)$. As we see from (9-46), its nth-order characteristic function equals

$$\exp\left\{ j\eta \sum_i \omega_i - \frac{1}{2} \sum_{i,k} C(t_i - t_k)\omega_i\omega_k \right\} \tag{9-60}$$

This function is invariant to a shift of the origin. And since it determines completely the statistics of $\mathbf{x}(t)$, we conclude that $\mathbf{x}(t)$ is SSS.

EXAMPLE 9-13

▶ We shall establish necessary and sufficient conditions for the stationarity of the process

$$\mathbf{x}(t) = \mathbf{a}\cos\omega t + \mathbf{b}\sin\omega t \tag{9-61}$$

The mean of this process equals

$$E\{\mathbf{x}(t)\} = E\{\mathbf{a}\}\cos\omega t + E\{\mathbf{b}\}\sin\omega t$$

This function must be independent of t. Hence the condition

$$E\{\mathbf{a}\} = E\{\mathbf{b}\} = 0 \tag{9-62}$$

is necessary for both forms of stationarity. We shall assume that it holds.

Wide sense. The process $\mathbf{x}(t)$ is WSS iff the random variables **a** and **b** are uncorrelated with equal variance:

$$E\{\mathbf{ab}\} = 0 \qquad E\{\mathbf{a}^2\} = E\{\mathbf{b}^2\} = \sigma^2 \tag{9-63}$$

If this holds, then

$$R(\tau) = \sigma^2 \cos\omega\tau \tag{9-64}$$

Proof. If $\mathbf{x}(t)$ is WSS, then

$$E\{\mathbf{x}^2(0)\} = E\{\mathbf{x}^2(\pi/2\omega)\} = R(0)$$

But $\mathbf{x}(0) = \mathbf{a}$ and $\mathbf{x}(\pi/2\omega) = \mathbf{b}$; hence $E\{\mathbf{a}^2\} = E\{\mathbf{b}^2\}$. Using the above, we obtain

$$\begin{aligned} E\{\mathbf{x}(t+\tau)\mathbf{x}(t)\} &= E\{[\mathbf{a}\cos\omega(t+\tau) + \mathbf{b}\sin\omega(t+\tau)][\mathbf{a}\cos\omega t + \mathbf{b}\sin\omega t]\} \\ &= \sigma^2\cos\omega\tau + E\{\mathbf{ab}\}\sin\omega(2t+\tau) \end{aligned} \tag{9-65}$$

This is independent of t only if $E\{\mathbf{ab}\} = 0$ and (9-63) results.

Conversely, if (9-63) holds, then, as we see from (9-65), the autocorrelation of $\mathbf{x}(t)$ equals $\sigma^2\cos\omega\tau$; hence $\mathbf{x}(t)$ is WSS.

Strict sense. The process $\mathbf{x}(t)$ is SSS iff the joint density $f(a, b)$ of the random variables $\mathbf{a}$ and $\mathbf{b}$ has circular symmetry, that is, if

$$f(a, b) = f(\sqrt{a^2 + b^2}) \tag{9-66}$$

Proof. If $\mathbf{x}(t)$ is SSS, then the random variables

$$\mathbf{x}(0) = \mathbf{a} \qquad \mathbf{x}(\pi/2\omega) = \mathbf{b}$$

and

$$\mathbf{x}(t) = \mathbf{a}\cos\omega t + \mathbf{b}\sin\omega t \qquad \mathbf{x}(t + \pi/2\omega) = \mathbf{b}\cos\omega t - \mathbf{a}\sin\omega t$$

have the same joint density for every t. Hence $f(a, b)$ must have circular symmetry.

We shall now show that, if $f(a, b)$ has circular symmetry, then $\mathbf{x}(t)$ is SSS. With τ a given number and

$$\mathbf{a}_1 = \mathbf{a}\cos\omega\tau + \mathbf{b}\sin\omega\tau \qquad \mathbf{b}_1 = \mathbf{b}\cos\omega\tau - \mathbf{a}\sin\omega\tau$$

we form the process

$$\mathbf{x}_1(t) = \mathbf{a}_1\cos\omega t + \mathbf{b}_1\sin\omega t = \mathbf{x}(t + \tau)$$

Clearly, the statistics of $\mathbf{x}(t)$ and $\mathbf{x}_1(t)$ are determined in terms of the joint densities $f(a, b)$ and $f(a_1, b_1)$ of the random variables $\mathbf{a}, \mathbf{b}$ and $\mathbf{a}_1, \mathbf{b}_1$. But the random variables $\mathbf{a}, \mathbf{b}$ and $\mathbf{a}_1, \mathbf{b}_1$ have the same joint density. Hence the processes $\mathbf{x}(t)$ and $\mathbf{x}(t + \tau)$ have the same statistics for every τ. ◀

COROLLARY

▶ If the process $\mathbf{x}(t)$ is SSS and the random variables $\mathbf{a}$ and $\mathbf{b}$ are independent, then they are normal.

Proof. It follows from (9-66) and (6-31). ◀

EXAMPLE 9-14

▶ Given a random variable $\boldsymbol{\omega}$ with density $f(\omega)$ and a random variable $\boldsymbol{\varphi}$ uniform in the interval $(-\pi, \pi)$ and independent of $\boldsymbol{\omega}$, we form the process

$$\mathbf{x}(t) = a\cos(\boldsymbol{\omega} t + \boldsymbol{\varphi}) \tag{9-67}$$

We shall show that $\mathbf{x}(t)$ is WSS with zero mean and autocorrelation

$$R(\tau) = \frac{a^2}{2}E\{\cos\boldsymbol{\omega}\tau\} = \frac{a^2}{2}\operatorname{Re}\Phi_{\omega}(\tau) \tag{9-68}$$

where

$$\Phi_{\omega}(\tau) = E\{e^{j\boldsymbol{\omega}\tau}\} = E\{\cos\boldsymbol{\omega}\tau\} + jE\{\sin\boldsymbol{\omega}\tau\} \tag{9-69}$$

is the characteristic function of $\boldsymbol{\omega}$.

Proof. Clearly [see (6-235)]

$$E\{\cos(\boldsymbol{\omega} t + \boldsymbol{\varphi})\} = E\{E\{\cos(\boldsymbol{\omega} t + \boldsymbol{\varphi}) \mid \boldsymbol{\omega}\}\}$$

From the independence of $\boldsymbol{\omega}$ and $\boldsymbol{\varphi}$, it follows that

$$E\{\cos(\boldsymbol{\omega} t + \boldsymbol{\varphi}) \mid \omega\} = \cos\omega t\, E\{\cos\boldsymbol{\varphi}\} - \sin\omega t\, E\{\sin\boldsymbol{\varphi}\}$$

Hence $E\{\mathbf{x}(t)\} = 0$ because

$$E\{\cos\boldsymbol{\varphi}\} = \frac{1}{2\pi}\int_{-\pi}^{\pi}\cos\varphi\, d\varphi = 0 \qquad E\{\sin\boldsymbol{\varphi}\} = \frac{1}{2\pi}\int_{-\pi}^{\pi}\sin\varphi\, d\varphi = 0$$

Reasoning similarly, we obtain $E\{\cos(2\boldsymbol{\omega} t + \boldsymbol{\omega}\tau + 2\boldsymbol{\varphi})\} = 0$. And since

$$2\cos[\boldsymbol{\omega}(t+\tau) + \boldsymbol{\varphi}]\cos(\boldsymbol{\omega} t + \boldsymbol{\varphi}) = \cos\boldsymbol{\omega}\tau + \cos(2\boldsymbol{\omega} t + \boldsymbol{\omega}\tau + 2\boldsymbol{\varphi})$$

we conclude that

$$R(\tau) = a^2 E\{\cos[\boldsymbol{\omega}(t+\tau) + \boldsymbol{\varphi}]\cos(\boldsymbol{\omega} t + \boldsymbol{\varphi})\} = \frac{a^2}{2}E\{\cos\boldsymbol{\omega}\tau\}$$

Further, with $\boldsymbol{\omega}$ and $\boldsymbol{\varphi}$ as above, the process

$$\mathbf{z}(t) = ae^{j(\boldsymbol{\omega} t + \boldsymbol{\varphi})}$$

is WSS with zero mean and autocorrelation

$$E\{\mathbf{z}(t+\tau)\mathbf{z}^*(t)\} = a^2 E\{e^{j\boldsymbol{\omega} t}\} = a^2\Phi_\omega(\tau)$$ ◀

CENTERING. Given a process $\mathbf{x}(t)$ with mean $\eta(t)$ and autocovariance $C_x(t_1, t_2)$, we form difference

$$\tilde{\mathbf{x}}(t) = \mathbf{x}(t) - \eta(t) \tag{9-70}$$

This difference is called the *centered process* associated with the process $\mathbf{x}(t)$. Note that

$$E\{\tilde{\mathbf{x}}(t)\} = 0 \qquad R_{\tilde{x}}(t_1, t_2) = C_x(t_1, t_2)$$

From this it follows that if the process $\mathbf{x}(t)$ is covariance stationary, that is, if $C_x(t_1, t_2) = C_x(t_1 - t_2)$, then its centered process $\tilde{\mathbf{x}}(t)$ is WSS.

OTHER FORMS OF STATIONARITY. A process $\mathbf{x}(t)$ is *asymptotically stationary* if the statistics of the random variables $\mathbf{x}(t_1 + c), \ldots, \mathbf{x}(t_n + c)$ do not depend on c if c is large. More precisely, the function

$$f(x_1, \ldots, x_n, t_1 + c, \ldots, t_n + c)$$

tends to a limit (that does not depend on c) as $c \to \infty$. The semirandom telegraph signal is an example.

A process $\mathbf{x}(t)$ is *Nth-order stationary* if (9-47) holds not for every n, but only for $n \le N$.

A process $\mathbf{x}(t)$ is *stationary in an interval* if (9-47) holds for every t_i and $t_i + c$ in this interval.

We say that $\mathbf{x}(t)$ is a process with *stationary increments* if its increments $\mathbf{y}(t) = \mathbf{x}(t+h) - \mathbf{x}(t)$ form a stationary process for every h. The Poisson process is an example.

MEAN SQUARE PERIODICITY. A process $\mathbf{x}(t)$ is called MS periodic if

$$E\{|\mathbf{x}(t+T) - \mathbf{x}(t)|^2\} = 0 \tag{9-71}$$

for every t. From this it follows that, for a specific t,

$$\mathbf{x}(t+T) = \mathbf{x}(t) \tag{9-72}$$

with probability 1. It does not, however, follow that the set of outcomes ζ such that $\mathbf{x}(t+T, \zeta) = \mathbf{x}(t, \zeta)$ for all t has probability 1.

As we see from (9-72) the mean of an MS periodic process is periodic. We shall examine the properties of $R(t_1, t_2)$.

THEOREM 9-1

A process $\mathbf{x}(t)$ is MS periodic iff its autocorrelation is *doubly periodic,* that is, if

$$R(t_1 + mT, t_2 + nT) = R(t_1, t_2) \tag{9-73}$$

for every integer m and n.

Proof. As we know [see (6-167)]

$$E^2\{\mathbf{zw}\} \le E\{\mathbf{z}^2\}E\{\mathbf{w}^2\}$$

With $\mathbf{z} = \mathbf{x}(t_1)$ and $\mathbf{w} = \mathbf{x}(t_2 + T) - \mathbf{x}(t_2)$ this yields

$$E^2\{\mathbf{x}(t_1)[\mathbf{x}(t_2+T) - \mathbf{x}(t_2)]\} \le E\{\mathbf{x}^2(t_1)\}E\{[\mathbf{x}(t_2+T) - \mathbf{x}(t_2)]^2\}$$

If $\mathbf{x}(t)$ is MS periodic, then the last term in the last equation is 0. Equating the left side to 0, we obtain

$$R(t_1, t_2 + T) - R(t_1, t_2) = 0$$

Repeated application of this yields (9-73).

Conversely, if (9-73) is true, then

$$R(t+T, t+T) = R(t+T, t) = R(t, t)$$

Hence

$$E\{[\mathbf{x}(t+T) - \mathbf{x}(t)]^2\} = R(t+T, t+T) + R(t,t) - 2R(t+T, t) = 0$$

therefore $\mathbf{x}(t)$ is MS periodic.

9-2 SYSTEMS WITH STOCHASTIC INPUTS

Given a stochastic process $\mathbf{x}(t)$, we assign according to some rule to each of its samples $\mathbf{x}(t, \zeta_i)$ a function $\mathbf{y}(t, \zeta_i)$. We have thus created another process

$$\mathbf{y}(t) = T[\mathbf{x}(t)]$$

whose samples are the functions $\mathbf{y}(t, \zeta_i)$. The process $\mathbf{y}(t)$ so formed can be considered as the output of a *system* (transformation) with input the process $\mathbf{x}(t)$. The system is completely specified in terms of the operator T, that is, the rule of correspondence between the samples of the input $\mathbf{x}(t)$ and the output $\mathbf{y}(t)$.

The system is *deterministic* if it operates only on the variable t treating ζ as a parameter. This means that if two samples $\mathbf{x}(t, \zeta_1)$ and $\mathbf{x}(t, \zeta_2)$ of the input are identical in t, then the corresponding samples $\mathbf{y}(t, \zeta_1)$ and $\mathbf{y}(t, \zeta_2)$ of the output are also identical in t. The system is called *stochastic* if T operates on both variables t and ζ. This means that there exist two outcomes ζ_1 and ζ_2 such that $\mathbf{x}(t, \zeta_1) = \mathbf{x}(t, \zeta_2)$ identically in t but $\mathbf{y}(t, \zeta_1) \ne \mathbf{y}(t, \zeta_2)$. These classifications are based on the terminal properties of the system. If the system is specified in terms of physical elements or by an equation, then it is deterministic (stochastic) if the elements or the coefficients of the defining equations are deterministic (stochastic). Throughout this book we shall consider only deterministic systems.

In principle, the statistics of the output of a system can be expressed in terms of the statistics of the input. However, in general this is a complicated problem. We consider next two important special cases.

Memoryless Systems

A system is called memoryless if its output is given by

$$\mathbf{y}(t) = g[\mathbf{x}(t)]$$

where $g(x)$ is a function of x. Thus, at a given time $t = t_1$, the output $\mathbf{y}(t_1)$ depends only on $\mathbf{x}(t_1)$ and not on any other past or future values of $\mathbf{x}(t)$.

From this it follows that the first-order density $f_y(y; t)$ of $\mathbf{y}(t)$ can be expressed in terms of the corresponding density $f_x(x; t)$ of $\mathbf{x}(t)$ as in Sec. 5-2. Furthermore,

$$E\{\mathbf{y}(t)\} = \int_{-\infty}^{\infty} g(x) f_x(x; t)\, dx$$

Similarly, since $\mathbf{y}(t_1) = g[\mathbf{x}(t_1)]$ and $\mathbf{y}(t_2) = g[\mathbf{x}(t_2)]$, the second-order density $f_y(y_1, y_2; t_1, t_2)$ of $\mathbf{y}(t)$ can be determined in terms of the corresponding density $f_x(x_1, x_2; t_1, t_2)$ of $\mathbf{x}(t)$ as in Sec. 6-3. Furthermore,

$$E\{\mathbf{y}(t_1)\mathbf{y}(t_2)\} = \int_{-\infty}^{\infty}\int_{-\infty}^{\infty} g(x_1)g(x_2) f_x(x_1, x_2; t_1, t_2)\, dx_1\, dx_2$$

The nth-order density $f_y(y_1, \ldots, y_n; t_1, \ldots, t_n)$ of $\mathbf{y}(t)$ can be determined from the corresponding density of $\mathbf{x}(t)$ as in (7-8), where the underlying transformation is the system

$$\mathbf{y}(t_1) = g[\mathbf{x}(t_1)], \ldots, \mathbf{y}(t_n) = g[\mathbf{x}(t_n)] \tag{9-74}$$

STATIONARITY. Suppose that the input to a memoryless system is an SSS process $\mathbf{x}(t)$. We shall show that the resulting output $\mathbf{y}(t)$ is also SSS.

Proof. To determine the nth-order density of $\mathbf{y}(t)$, we solve the system

$$g(x_1) = y_1, \ldots, g(x_n) = y_n \tag{9-75}$$

If this system has a unique solution, then [see (7-8)]

$$f_y(y_1, \ldots, y_n; t_1, \ldots, t_n) = \frac{f_x(x_1, \ldots, x_n; t_1, \ldots, t_n)}{|g'(x_1) \cdots g'(x_n)|} \tag{9-76}$$

From the stationarity of $\mathbf{x}(t)$ it follows that the numerator in (9-76) is invariant to a shift of the time origin. And since the denominator does not depend on t, we conclude that the left side does not change if t_i is replaced by $t_i + c$. Hence $\mathbf{y}(t)$ is SSS. We can similarly show that this is true even if (9-75) has more than one solution.

Notes 1. If $\mathbf{x}(t)$ is stationary of order N, then $\mathbf{y}(t)$ is stationary of order N.
2. If $\mathbf{x}(t)$ is stationary in an interval, then $\mathbf{y}(t)$ is stationary in the same interval.
3. If $\mathbf{x}(t)$ is WSS stationary, then $\mathbf{y}(t)$ might not be stationary in any sense.

Square-law detector. A square-law detector is a memoryless syste[
equals

$$\mathbf{y}(t) = \mathbf{x}^2(t)$$

We shall determine its first- and second-order densities. If $y > 0$, then the system $y = x^2$ has the two solutions $\pm\sqrt{y}$. Furthermore, $y'(x) = \pm 2\sqrt{y}$; hence

$$f_y(y;t) = \frac{1}{2\sqrt{y}}[f_x(\sqrt{y};t) + f_x(-\sqrt{y};t)]$$

If $y_1 > 0$ and $y_2 > 0$, then the system

$$y_1 = x_1^2 \qquad y_2 = x_2^2$$

has the four solutions $(\pm\sqrt{y_1}, \pm\sqrt{y_2})$. Furthermore, its jacobian equals $\pm 4\sqrt{y_1 y_2}$; hence

$$f_y(y_1, y_2; t_1, t_2) = \frac{1}{4\sqrt{y_1 y_2}} \sum f_x(\pm\sqrt{y_1}, \pm\sqrt{y_2}; t_1, t_2)$$

where the summation has four terms.

Note that, if $\mathbf{x}(t)$ is SSS, then $f_x(x;t) = f_x(x)$ is independent of t and $f_x(x_1, x_2; t_1, t_2) = f_x(x_1, x_2; \tau)$ depends only on $\tau = t_1 - t_2$. Hence $f_y(y)$ is independent of t and $f_y(y_1, y_2; \tau)$ depends only on $\tau = t_1 - t_2$.

EXAMPLE 9-15

▶ Suppose that $\mathbf{x}(t)$ is a normal stationary process with zero mean and autocorrelation $R_x(\tau)$. In this case, $f_x(x)$ is normal with variance $R_x(0)$.

If $\mathbf{y}(t) = \mathbf{x}^2(t)$ (Fig. 9-6), then $E\{\mathbf{y}(t)\} = R_x(0)$ and [see (5-22)]

$$f_y(y) = \frac{1}{\sqrt{2\pi R_x(0) y}} e^{-y/2R_x(0)} U(y)$$

We shall show that

$$R_y(\tau) = R_x^2(0) + 2R_x^2(\tau) \tag{9-77}$$

Proof. The random variables $\mathbf{x}(t+\tau)$ and $\mathbf{x}(t)$ are jointly normal with zero mean. Hence [see (6-199)]

$$E\{\mathbf{x}^2(t+\tau)\mathbf{x}^2(t)\} = E\{\mathbf{x}^2(t+\tau)\}E\{\mathbf{x}^2(t)\} + 2E^2\{\mathbf{x}(t+\tau)\mathbf{x}(t)\}$$

and (9-77) results.

Note in particular that

$$E\{\mathbf{y}^2(t)\} = R_y(0) = 3R_x^2(0) \qquad \sigma_y^2 = 2R_x^2(0)$$ ◀

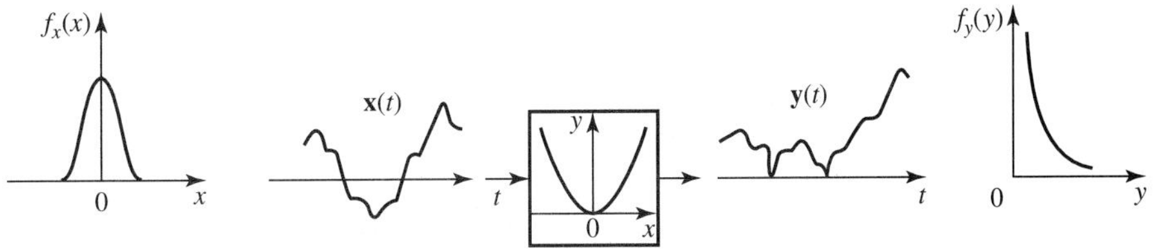

FIGURE 9-6

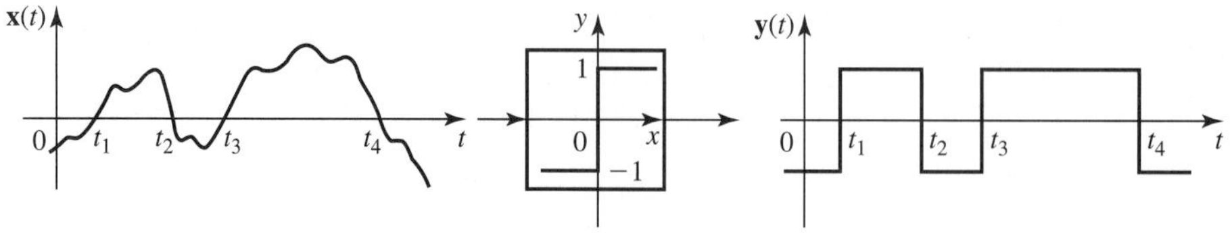

FIGURE 9-7

Hard limiter. Consider a memoryless system with

$$g(x) = \begin{cases} 1 & x > 0 \\ -1 & x < 0 \end{cases} \tag{9-78}$$

(Fig. 9-7). Its output $\mathbf{y}(t)$ takes the values ± 1 and

$$P\{\mathbf{y}(t) = 1\} = P\{\mathbf{x}(t) > 0\} = 1 - F_x(0)$$

$$P\{\mathbf{y}(t) = -1\} = P\{\mathbf{x}(t) < 0\} = F_x(0)$$

Hence

$$E\{\mathbf{y}(t)\} = 1 \times P\{\mathbf{y}(t) = 1\} - 1 \times P\{\mathbf{y}(t) = -1\} = 1 - 2F_x(0)$$

The product $\mathbf{y}(t+\tau)\mathbf{y}(t)$ equals 1 if $\mathbf{x}(t+\tau)\mathbf{x}(t) > 0$ and it equals -1 otherwise. Hence

$$R_y(\tau) = P\{\mathbf{x}(t+\tau)\mathbf{x}(t) > 0\} - P\{\mathbf{x}(t+\tau)\mathbf{x}(t) < 0\} \tag{9-79}$$

Thus, in the probability plane of the random variables $\mathbf{x}(t+\tau)$ and $\mathbf{x}(t)$, $R_y(\tau)$ equals the masses in the first and third quadrants minus the masses in the second and fourth quadrants.

EXAMPLE 9-16

▶ We shall show that if $\mathbf{x}(t)$ is a normal stationary process with zero mean, then the autocorrelation of the output of a hard limiter equals

$$R_y(\tau) = \frac{2}{\pi} \arcsin \frac{R_x(\tau)}{R_x(0)} \tag{9-80}$$

This result is known as the *arcsine law*.[4]

Proof. The random variables $\mathbf{x}(t+\tau)$ and $\mathbf{x}(t)$ are jointly normal with zero mean, variance $R_x(0)$, and correlation coefficient $R_x(\tau)/R_x(0)$. Hence [see (6-64)].

$$P\{\mathbf{x}(t+\tau)\mathbf{x}(t) > 0\} = \frac{1}{2} + \frac{\alpha}{\pi} \qquad \sin\alpha = \frac{R_x(\tau)}{R_x(0)}$$

$$P\{\mathbf{x}(t+\tau)\mathbf{x}(t) < 0\} = \frac{1}{2} - \frac{\alpha}{\pi}$$

Inserting in (9-79), we obtain

$$R_y(\tau) = \frac{1}{2} + \frac{\alpha}{\pi} - \left(\frac{1}{2} - \frac{\alpha}{\pi}\right) = \frac{2\alpha}{\pi}$$

and (9-80) follows. ◀

[4]J. L. Lawson and G. E. Uhlenbeck: *Threshold Signals,* McGraw-Hill Book Company, New York, 1950.

EXAMPLE 9-17

BUSSGANG'S THEOREM

▶ Using Price's theorem, we shall show that if the input to a memoryless system $y = g(x)$ is a zero-mean normal process $\mathbf{x}(t)$, the cross-correlation of $\mathbf{x}(t)$ with the resulting output $\mathbf{y}(t) = g[\mathbf{x}(t)]$ is proportional to $R_{xx}(\tau)$:

$$R_{xy}(\tau) = KR_{xx}(\tau) \qquad \text{where} \quad K = E\{g'[\mathbf{x}(t)]\} \tag{9-81}$$

Proof. For a specific τ, the random variables $\mathbf{x} = \mathbf{x}(t)$ and $\mathbf{z} = \mathbf{x}(t+\tau)$ are jointly normal with zero mean and covariance $\mu = E\{\mathbf{xz}\} = R_{xx}(\tau)$. With

$$I = E\{\mathbf{z}g(\mathbf{x})\} = E\{\mathbf{x}(t+\tau)\mathbf{y}(t)\} = R_{xy}(\tau)$$

it follows from (6-201) that

$$\frac{\partial I}{\partial \mu} = E\left\{\frac{\partial^2[\mathbf{z}g(\mathbf{x})]}{\partial x\,\partial z}\right\} = E\{g'[\mathbf{x}(t)]\} = K \tag{9-82}$$

If $\mu = 0$, the random variables $\mathbf{x}(t+\tau)$ and $\mathbf{x}(t)$ are independent; hence $I = 0$. Integrating (9-82) with respect to μ, we obtain $I = K\mu$ and (9-81) results.

Special cases.[5] (*a*) (Hard limiter) Suppose that $g(x) = \operatorname{sgn} x$ as in (9-78). In this case, $g'(x) = 2\delta(x)$; hence

$$K = E\{2\delta(\mathbf{x})\} = 2\int_{-\infty}^{\infty} \delta(x) f(x)\,dx = 2f(0)$$

where

$$f(x) = \frac{1}{\sqrt{2\pi R_{xx}(0)}} \exp\left\{-\frac{x^2}{2R_{xx}(0)}\right\}$$

is the first-order density of $\mathbf{x}(t)$. Inserting into (9-81), we obtain

$$R_{xy}(\tau) = R_{xx}(\tau)\sqrt{\frac{2}{\pi R_{xx}(0)}} \qquad \mathbf{y}(t) = \operatorname{sgn}\mathbf{x}(t) \tag{9-83}$$

(*b*) (Limiter) Suppose next that $\mathbf{y}(t)$ is the output of a limiter

$$g(x) = \begin{cases} x & |x| < c \\ c & |x| > c \end{cases} \qquad g'(x) = \begin{cases} 1 & |x| < c \\ 0 & |x| > c \end{cases}$$

In this case,

$$K = \int_{-c}^{c} f(x)\,dx = 2G\left(\frac{c}{\sqrt{R_{xx}(0)}}\right) - 1 \tag{9-84}$$

◀

Linear Systems

The notation

$$\mathbf{y}(t) = L[\mathbf{x}(t)] \tag{9-85}$$

will indicate that $\mathbf{y}(t)$ is the output of a *linear* system with input $\mathbf{x}(t)$. This means that

$$L[\mathbf{a}_1\mathbf{x}_1(t) + \mathbf{a}_2\mathbf{x}_2(t)] = \mathbf{a}_1 L[\mathbf{x}_1(t)] + \mathbf{a}_2 L[\mathbf{x}_2(t)] \tag{9-86}$$

for any $\mathbf{a}_1, \mathbf{a}_2, \mathbf{x}_1(t), \mathbf{x}_2(t)$.

---

[5] H. E. Rowe, "Memoryless Nonlinearities with Gaussian Inputs," *BSTJ,* vol. 67, no. 7, September 1982.

This is the familiar definition of linearity and it also holds if the coefficients $\mathbf{a}_1$ and $\mathbf{a}_2$ are random variables because, as we have assumed, the system is deterministic, that is, it operates only on the variable t.

Note If a system is specified by its internal structure or by a differential equation, then (9-86) holds only if $\mathbf{y}(t)$ is the *zero-state* response. The response due to the initial conditions (zero-input response) will not be considered.

A system is called *time-invariant* if its response to $\mathbf{x}(t+c)$ equals $\mathbf{y}(t+c)$. We shall assume throughout that all linear systems under consideration are time-invariant.

It is well known that the output of a linear system is a convolution

$$\mathbf{y}(t) = \mathbf{x}(t) * h(t) = \int_{-\infty}^{\infty} \mathbf{x}(t-\alpha)h(\alpha)\,d\alpha \tag{9-87}$$

where

$$h(t) = L[\delta(t)]$$

is its impulse response. In the following, most systems will be specified by (9-87). However, we start our investigation using the operational notation (9-85) to stress the fact that various results based on the next theorem also hold for arbitrary linear operators involving one or more variables.

The following observations are immediate consequences of the linearity and time invariance of the system.

If $\mathbf{x}(t)$ is a normal process, then $\mathbf{y}(t)$ is also a normal process. This is an extension of the familiar property of linear transformations of normal random variables and can be justified if we approximate the integral in (9-87) by a sum:

$$\mathbf{y}(t_i) \simeq \sum_k \mathbf{x}(t_i - \alpha_k)h(\alpha_k)\Delta(\alpha)$$

If $\mathbf{x}(t)$ is SSS, then $\mathbf{y}(t)$ is also SSS. Indeed, since $\mathbf{y}(t+c) = L[\mathbf{x}(t+c)]$ for every c, we conclude that if the processes $\mathbf{x}(t)$ and $\mathbf{x}(t+c)$ have the same statistical properties, so do the processes $\mathbf{y}(t)$ and $\mathbf{y}(t+c)$. We show later [see (9-142)] that if $\mathbf{x}(t)$ is WSS, the processes $\mathbf{x}(t)$ and $\mathbf{y}(t)$ are jointly WSS.

Fundamental theorem. For any linear system

$$E\{L[\mathbf{x}(t)]\} = L[E\{\mathbf{x}(t)\}] \tag{9-88}$$

In other words, the mean $\eta_y(t)$ of the output $\mathbf{y}(t)$ equals the response of the system to the mean $\eta_x(t)$ of the input (Fig. 9-8*a*)

$$\eta_y(t) = L[\eta_x(t)] \tag{9-89}$$

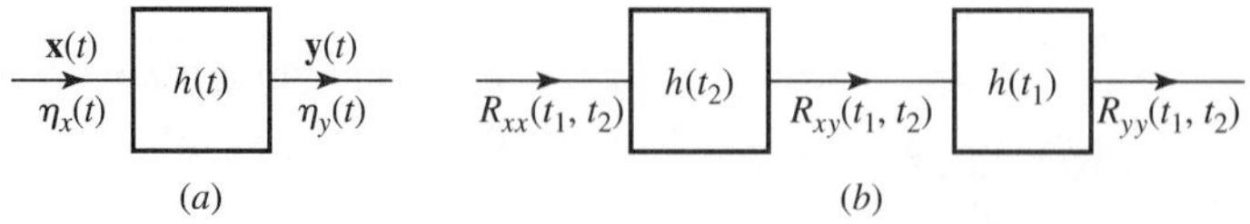

FIGURE 9-8

This is a simple extension of the linearity of expected values to arbitrary linear operators. In the context of (9-87) it can be deduced if we write the integral as a limit of a sum. This yields

$$E\{\mathbf{y}(t)\} = \int_{-\infty}^{\infty} E\{\mathbf{x}(t-\alpha)\}h(\alpha)\,d\alpha = \eta_x(t) * h(t) \tag{9-90}$$

Frequency interpretation At the ith trial the input to our system is a function $\mathbf{x}(t, \zeta_i)$ yielding as output the function $\mathbf{y}(t, \zeta_i) = L[\mathbf{x}(t, \zeta_i)]$. For large n,

$$E\{\mathbf{y}(t)\} \simeq \frac{\mathbf{y}(t,\zeta_1) + \cdots + \mathbf{y}(t,\zeta_n)}{n} = \frac{L[\mathbf{x}(t,\zeta_1)] + \cdots + L[\mathbf{x}(t,\zeta_n)]}{n}$$

From the linearity of the system it follows that the last term above equals

$$L\left[\frac{\mathbf{x}(t,\zeta_1) + \cdots + \mathbf{x}(t,\zeta_n)}{n}\right]$$

This agrees with (9-88) because the fraction is nearly equal to $E\{\mathbf{x}(t)\}$.

Notes 1. From (9-89) it follows that if

$$\tilde{\mathbf{x}}(t) = \mathbf{x}(t) - \eta_x(t) \qquad \tilde{\mathbf{y}}(t) = \mathbf{y}(t) - \eta_y(t)$$

then

$$L[\tilde{\mathbf{x}}(t)] = L[\mathbf{x}(t)] - L[\eta_x(t)] = \tilde{\mathbf{y}}(t) \tag{9-91}$$

Thus the response of a linear system to the centered input $\tilde{\mathbf{x}}(t)$ equals the centered output $\tilde{\mathbf{y}}(t)$.

2. Suppose that

$$\mathbf{x}(t) = f(t) + \boldsymbol{\nu}(t) \qquad E\{\boldsymbol{\nu}(t)\} = 0$$

In this case, $E\{\mathbf{x}(t)\} = f(t)$; hence

$$\eta_y(t) = f(t) * h(t)$$

Thus, if $\mathbf{x}(t)$ is the sum of a deterministic signal $f(t)$ and a random component $\boldsymbol{\nu}(t)$, then for the determination of the mean of the output we can ignore $\boldsymbol{\nu}(t)$ provided that the system is linear and $E\{\boldsymbol{\nu}(t)\} = 0$.

Theorem (9-88) can be used to express the joint moments of any order of the output $\mathbf{y}(t)$ of a linear system in terms of the corresponding moments of the input. The following special cases are of fundamental importance in the study of linear systems with stochastic inputs.

OUTPUT AUTOCORRELATION. We wish to express the autocorrelation $R_{yy}(t_1, t_2)$ of the output $\mathbf{y}(t)$ of a linear system in terms of the autocorrelation $R_{xx}(t_1, t_2)$ of the input $\mathbf{x}(t)$. As we shall presently see, it is easier to find first the cross-correlation $R_{xy}(t_1, t_2)$ between $\mathbf{x}(t)$ and $\mathbf{y}(t)$.

THEOREM 9-2

▶ (*a*)

$$R_{xy}(t_1, t_2) = L_2[R_{xx}(t_1, t_2)] \tag{9-92}$$

In the notation just established, L_2 means that the system operates on the variable t_2, treating t_1 as a parameter. In the context of (9-87) this means that

$$R_{xy}(t_1, t_2) = \int_{-\infty}^{\infty} R_{xx}(t_1, t_2 - \alpha)h(\alpha)\,d\alpha \tag{9-93}$$

(*b*) $$R_{yy}(t_1, t_2) = L_1[R_{xy}(t_1, t_2)] \tag{9-94}$$

In this case, the system operates on t_1.

$$R_{yy}(t_1, t_2) = \int_{-\infty}^{\infty} R_{xy}(t_1 - \alpha, t_2)h(\alpha)\,d\alpha \tag{9-95}$$

Proof. Multiplying (9-85) by $\mathbf{x}(t_1)$ and using (9-86), we obtain

$$\mathbf{x}(t_1)\mathbf{y}(t) = L_t[\mathbf{x}(t_1)\mathbf{x}(t)]$$

where L_t means that the system operates on t. Hence [see (9-88)]

$$E\{\mathbf{x}(t_1)\mathbf{y}(t)\} = L_t[E\{\mathbf{x}(t_1)\mathbf{x}(t)\}]$$

and (9-92) follows with $t = t_2$. The proof of (9-94) is similar: We multiply (9-85) by $\mathbf{y}(t_2)$ and use (9-88). This yields

$$E\{\mathbf{y}(t)\mathbf{y}(t_2)\} = L_t[E\{\mathbf{x}(t)\mathbf{y}(t_2)\}]$$

and (9-94) follows with $t = t_1$. ◀

The preceding theorem is illustrated in Fig. 9-8*b*: If $R_{xx}(t_1, t_2)$ is the input to the given system and the system operates on t_2, the output equals $R_{xy}(t_1, t_2)$. If $R_{xy}(t_1, t_2)$ is the input and the system operates on t_1, the output equals $R_{yy}(t_1, t_2)$.

Inserting (9-93) into (9-95), we obtain

$$R_{yy}(t_1, t_2) = \int_{-\infty}^{\infty}\int_{-\infty}^{\infty} R_{xx}(t_1 - \alpha, t_2 - \beta)h(\alpha)h(\beta)\,d\alpha\,d\beta$$

This expresses $R_{yy}(t_1, t_2)$ directly in terms of $R_{xx}(t_1, t_2)$. However, conceptually and operationally, it is preferable to find first $R_{xy}(t_1, t_2)$.

EXAMPLE 9-18

▶ A stationary process $\mathbf{v}(t)$ with autocorrelation $R_{vv}(\tau) = q\delta(\tau)$ (white noise) is applied at $t = 0$ to a linear system with

$$h(t) = e^{-ct}U(t)$$

We shall show that the autocorrelation of the resulting output $\mathbf{y}(t)$ equals

$$R_{yy}(t_1, t_2) = \frac{q}{2c}(1 - e^{-2ct_1})e^{-c|t_2 - t_1|} \tag{9-96}$$

for $0 < t_1 < t_2$.

Proof. We can use the preceding results if we assume that the input to the system is the process

$$\mathbf{x}(t) = \mathbf{v}(t)U(t)$$

With this assumption, all correlations are 0 if $t_1 < 0$ or $t_2 < 0$. For $t_1 > 0$ and $t_2 > 0$,

$$R_{xx}(t_1, t_2) = E\{\mathbf{v}(t_1)\mathbf{v}(t_2)\} = q\delta(t_1 - t_2)$$

As we see from (9-92), $R_{xy}(t_1, t_2)$ equals the response of the system to $q\delta(t_1 - t_2)$ considered as a function of t_2. Since $\delta(t_1 - t_2) = \delta(t_2 - t_1)$ and $L[\delta(t_2 - t_1)] = h(t_2 - t_1)$ (time invariance), we conclude that

$$R_{xy}(t_1, t_2) = qh(t_2 - t_1) = qe^{-c(t_2 - t_1)}U(t_2 - t_1)U(t_1)$$

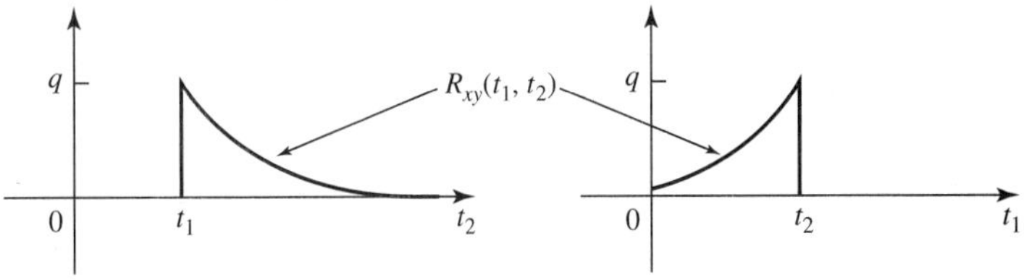

FIGURE 9-9

In Fig. 9-9, we show $R_{xy}(t_1, t_2)$ as a function of t_1 and t_2. Inserting into (9-95), we obtain

$$R_{yy}(t_1, t_2) = q \int_0^{t_1} e^{c(t_1-\alpha-t_2)} e^{-c\alpha}\, d\alpha \quad t_1 < t_2$$

and (9-96) results.

Note that

$$E\{\mathbf{y}^2(t)\} = R_{yy}(t, t) = \frac{q}{2c}(1 - e^{-2ct}) = q \int_0^t h^2(\alpha)\, d\alpha$$

◀

COROLLARY

▶ The autocovariance $C_{yy}(t_1, t_2)$ of $\mathbf{y}(t)$ is the autocorrelation of the process $\tilde{\mathbf{y}}(t) = \mathbf{y}(t) - \eta_y(t)$ and, as we see from (9-91), $\tilde{\mathbf{y}}(t)$ equals $L[\tilde{\mathbf{x}}(t)]$. Applying (9-93) and (9-95) to the centered processes $\tilde{\mathbf{x}}(t)$ and $\tilde{\mathbf{y}}(t)$, we obtain

$$\begin{aligned} C_{xy}(t_1, t_2) &= C_{xx}(t_1, t_2) * h(t_2) \\ C_{yy}(t_1, t_2) &= C_{xy}(t_1, t_2) * h(t_1) \end{aligned} \tag{9-97}$$

where the convolutions are in t_1 and t_2, respectively. ◀

Complex processes The preceding results can be readily extended to complex processes and to systems with complex-valued $h(t)$. Reasoning as in the real case, we obtain

$$\begin{aligned} R_{xy}(t_1, t_2) &= R_{xx}(t_1, t_2) * h^*(t_2) \\ R_{yy}(t_1, t_2) &= R_{xy}(t_1, t_2) * h(t_1) \end{aligned} \tag{9-98}$$

Response to white noise. We shall determine the average power $E\{|\mathbf{y}(t)|^2\}$ of the output of a system driven by white noise. This is a special case of (9-98), however, because of its importance it is stated as a theorem.

THEOREM 9-3

▶ If the input to a linear system is white noise with autocorrelation

$$R_{xx}(t_1, t_2) = q(t_1)\delta(t_1 - t_2)$$

then

$$E\{|\mathbf{y}(t)|^2\} = q(t) * |h(t)|^2 = \int_{-\infty}^{\infty} q(t-\alpha)|h(\alpha)|^2\, d\alpha \tag{9-99}$$

Proof. From (9-98) it follows that

$$R_{xy}(t_1, t_2) = q(t_1)\delta(t_2 - t_1) * h^*(t_2) = q(t_1)h^*(t_2 - t_1)$$

$$R_{yy}(t_1, t_2) = \int_{-\infty}^{\infty} q(t_1 - \alpha)h^*[t_2 - (t_1 - \alpha)]h(\alpha)\,d\alpha$$

and with $t_1 = t_2 = t$, (9-99) results.

Special cases. (*a*) If $\mathbf{x}(t)$ is stationary white noise, then $q(t) = q$ and (9-99) yields

$$E\{\mathbf{y}^2(t)\} = qE \qquad \text{where} \quad E = \int_{-\infty}^{\infty} |h(t)|^2\,dt$$

is the energy of $h(t)$.

(*b*) If $h(t)$ is of short duration relative to the variations of $q(t)$, then

$$E\{\mathbf{y}^2(t)\} \simeq q(t) \int_{-\infty}^{\infty} |h(\alpha)|^2\,d\alpha = Eq(t) \tag{9-100}$$

This relationship justifies the term *average intensity* used to describe the function $q(t)$.

(*c*) If $R_{\nu\nu}(\tau) = q\delta(\tau)$ and $\boldsymbol{\nu}(t)$ is applied to the system at $t = 0$, then $q(t) = qU(t)$ and (9-99) yields

$$E\{\mathbf{y}^2(t)\} = q \int_{-\infty}^{t} |h(\alpha)|^2\,d\alpha$$

◀

EXAMPLE 9-19

▶ The integral

$$\mathbf{y} = \int_{o}^{t} \boldsymbol{\nu}(\alpha)\,d\alpha$$

can be considered as the output of a linear system with input $\mathbf{x}(t) = \boldsymbol{\nu}(t)U(t)$ and impulse response $h(t) = U(t)$. If, therefore, $\boldsymbol{\nu}(t)$ is white noise with average intensity $q(t)$, then $\mathbf{x}(t)$ is white noise with average intensity $q(t)U(t)$ and (9-99) yields

$$E\{\mathbf{y}^2(t)\} = q(t)U(t) * U(t) = \int_0^t q(\alpha)\,d\alpha$$

◀

Differentiators. A differentiator is a linear system whose output is the derivative of the input

$$L[\mathbf{x}(t)] = \mathbf{x}'(t)$$

We can, therefore, use the preceding results to find the mean and the autocorrelation of $\mathbf{x}'(t)$.

From (9-89) it follows that

$$\eta_{x'}(t) = L[\eta_x(t)] = \eta'_x(t) \tag{9-101}$$

Similarly [see (9-92)]

$$R_{xx'}(t_1, t_2) = L_2[R_{xx}(t_1, t_2)] = \frac{\partial R_{xx}(t_1, t_2)}{\partial t_2} \tag{9-102}$$

because, in this case, L_2 means differentiation with respect to t_2. Finally,

$$R_{x'x'}(t_1, t_2) = L_1[R_{xx'}(t_1, t_2)] = \frac{\partial R_{xx'}(t_1, t_2)}{\partial t_1} \tag{9-103}$$

Combining, we obtain

$$R_{x'x'}(t_1, t_2) = \frac{\partial^2 R_{xx}(t_1, t_2)}{\partial t_1 \partial t_2} \tag{9-104}$$

Stationary processes If $\mathbf{x}(t)$ is WSS, then $\eta_x(t)$ is constant; hence

$$E\{\mathbf{x}'(t)\} = 0 \tag{9-105}$$

Furthermore, since $R_{xx}(t_1, t_2) = R_{xx}(\tau)$, we conclude with $\tau = t_1 - t_2$ that

$$\frac{\partial R_{xx}(t_1 - t_2)}{\partial t_2} = -\frac{dR_{xx}(\tau)}{d\tau} \qquad \frac{\partial^2 R_{xx}(t_1 - t_2)}{\partial t_1 \partial t_2} = -\frac{d^2 R_{xx}(\tau)}{d\tau^2}$$

Hence

$$R_{xx'}(\tau) = -R'_{xx}(\tau) \qquad R_{x'x'}(\tau) = -R''_{xx}(\tau) \tag{9-106}$$

Poisson impulses. If the input $\mathbf{x}(t)$ to a differentiator is a Poisson process, the resulting output $\mathbf{z}(t)$ is a train of impulses (Fig. 9-10)

$$\mathbf{z}(t) = \sum_i \delta(t - \mathbf{t}_i) \tag{9-107}$$

We maintain that $\mathbf{z}(t)$ is a stationary process with mean

$$\eta_z = \lambda \tag{9-108}$$

and autocorrelation

$$R_{zz}(\tau) = \lambda^2 + \lambda\delta(\tau) \tag{9-109}$$

Proof. The first equation follows from (9-101) because $\eta_x(t) = \lambda t$. To prove the second, we observe that [see (9-14)]

$$R_{xx}(t_1, t_2) = \lambda^2 t_1 t_2 + \lambda \min(t_1, t_2) \tag{9-110}$$

And since $\mathbf{z}(t) = \mathbf{x}'(t)$, (9-102) yields

$$R_{xz}(t_1, t_2) = \frac{\partial R_{xx}(t_1, t_2)}{\partial t_2} = \lambda^2 t_1 + \lambda U(t_1 - t_2)$$

This function is plotted in Fig. 9-10*b*, where the independent variable is t_1. As we see, it is discontinuous for $t_1 = t_2$ and its derivative with respect to t_1 contains the impulse

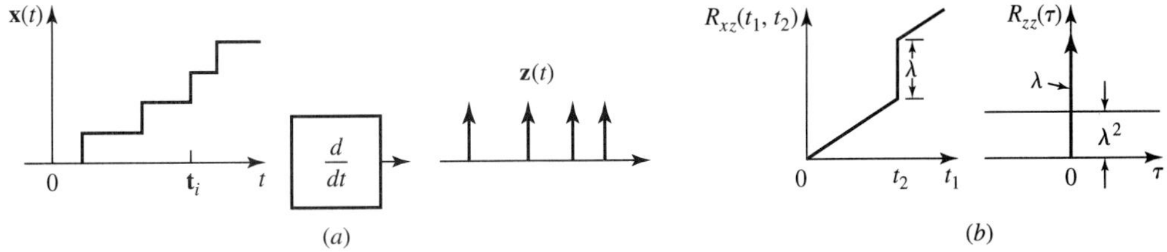

FIGURE 9-10

$\lambda\delta(t_1 - t_2)$. This yields [see (9-103)]

$$R_{zz}(t_1, t_2) = \frac{\partial R_{xz}(t_1, t_2)}{\partial t_1} = \lambda^2 + \lambda\delta(t_1 - t_2)$$

DIFFERENTIAL EQUATIONS. A deterministic differential equation with random excitation is an equation of the form

$$a_n \mathbf{y}^{(m)}(t) + \cdots + a_0 \mathbf{y}(t) = \mathbf{x}(t) \tag{9-111}$$

where the coefficients a_k are given numbers and the driver $\mathbf{x}(t)$ is a stochastic process. We shall consider its solution $\mathbf{y}(t)$ under the assumption that the initial conditions are 0. With this assumption, $\mathbf{y}(t)$ is unique (zero-state response) and it satisfies the linearity condition (9-86). We can, therefore, interpret $\mathbf{y}(t)$ as the output of a linear system specified by (9-111).

In general, the determination of the complete statistics of $\mathbf{y}(t)$ is complicated. In the following, we evaluate only its second-order moments using the preceding results. The above system is an operator L specified as follows: Its output $\mathbf{y}(t)$ is a process with zero initial conditions satisfying (9-111).

Mean. As we know [see (9-89)] the mean $\eta_y(t)$ of $\mathbf{y}(t)$ is the output of L with input $\eta_x(t)$. Hence it satisfies the equation

$$a_n \eta_y^{(n)}(t) + \cdots + a_0 \eta_y(t) = \eta_x(t) \tag{9-112}$$

and the initial conditions

$$\eta_y(0) = \cdots = \eta_y^{(n-1)}(0) = 0 \tag{9-113}$$

This result can be established directly: Clearly,

$$E\{\mathbf{y}^{(k)}(t)\} = \eta_y^{(k)}(t) \tag{9-114}$$

Taking expected values of both sides of (9-111) and using (9-114), we obtain (9-112). Equation (9-113) follows from (9-114) because $\mathbf{y}^{(k)}(0) = 0$ by assumption.

Correlation. To determine $R_{xy}(t_1, t_2)$, we use (9-92)

$$R_{xy}(t_1, t_2) = L_2[R_{xx}(t_1, t_2)]$$

In this case, L_2 means that $R_{xy}(t_1, t_2)$ satisfies the differential equation

$$a_n \frac{\partial^n R_{xy}(t_1, t_2)}{\partial t_2^n} + \cdots + a_0 R_{xy}(t_1, t_2) = R_{xx}(t_1, t_2) \tag{9-115}$$

with the initial conditions

$$R_{xy}(t_1, 0) = \cdots = \frac{\partial^{n-1} R_{xy}(t_1, 0)}{\partial t_2^{n-1}} = 0 \tag{9-116}$$

Similarly, since [see (9-94)]

$$R_{yy}(t_1, t_2) = L_1[R_{xy}(t_1, t_2)]$$

we conclude as earlier that

$$a_n \frac{\partial^n R_{yy}(t_1, t_2)}{\partial t_1^n} + \cdots + a_0 R_{yy}(t_1, t_2) = R_{xy}(t_1, t_2) \tag{9-117}$$

$$R_{yy}(0, t_2) = \cdots = \frac{\partial^{n-1} R_{yy}(0, t_2)}{\partial t_1^{n-1}} = 0 \tag{9-118}$$

These results can be established directly: From (9-111) it follows that

$$\mathbf{x}(t_1)\left[a_n \mathbf{y}^{(n)}(t_2) + \cdots + a_0 \mathbf{y}(t_2)\right] = \mathbf{x}(t_1)\mathbf{x}(t_2)$$

This yields (9-115) because [see (9-132)]

$$E\left\{\mathbf{x}(t_1)\mathbf{y}^{(k)}(t_2)\right\} = \partial^k R_{xy}(t_1, t_2)/\partial t_2^k$$

Similarly, (9-117) is a consequence of the identity

$$\left[a_n \mathbf{y}^{(n)}(t_1) + \cdots + a_0 \mathbf{y}(t_1)\right]\mathbf{y}(t_2) = \mathbf{x}(t_1)\mathbf{y}(t_2)$$

because

$$E\left\{\mathbf{y}^{(k)}(t_1)\mathbf{y}(t_2)\right\} = \partial^k R_{yy}(t_1, t_2)/\partial t_1^k$$

Finally, the expected values of

$$\mathbf{x}(t_1)\mathbf{y}^{(k)}(0) = 0 \qquad \mathbf{y}^{(k)}(0)\mathbf{y}(t_2) = 0$$

yield (9-116) and (9-118).

General moments. The moments of any order of the output $\mathbf{y}(t)$ of a linear system can be expressed in terms of the corresponding moments of the input $\mathbf{x}(t)$. As an illustration, we shall determine the third-order moment

$$R_{yyy}(t_1, t_2, t_3) = E\{\mathbf{y}_1(t)\mathbf{y}_2(t)\mathbf{y}_3(t)\}$$

of $\mathbf{y}(t)$ in terms of the third-order moment $R_{xxx}(t_1, t_2, t_3)$ of $\mathbf{x}(t)$. Proceeding as in (9-92), we obtain

$$\begin{aligned} E\{\mathbf{x}(t_1)\mathbf{x}(t_2)\mathbf{y}(t_3)\} &= L_3[E\{\mathbf{x}(t_1)\mathbf{x}(t_2)\mathbf{x}(t_3)\}] \\ &= \int_{-\infty}^{\infty} R_{xxx}(t_1, t_2, t_3 - \gamma)h(\gamma)\,d\gamma \end{aligned} \tag{9-119}$$

$$\begin{aligned} E\{\mathbf{x}(t_1)\mathbf{y}(t_2)\mathbf{y}(t_3)\} &= L_2[E\{\mathbf{x}(t_1)\mathbf{x}(t_2)\mathbf{y}(t_3)\}] \\ &= \int_{-\infty}^{\infty} R_{xxy}(t_1, t_2 - \beta, t_3)h(\beta)\,d\beta \end{aligned} \tag{9-120}$$

$$\begin{aligned} E\{\mathbf{y}(t_1)\mathbf{y}(t_2)\mathbf{y}(t_3)\} &= L_1[E\{\mathbf{x}(t_1)\mathbf{y}(t_2)\mathbf{y}(t_3)\}] \\ &= \int_{-\infty}^{\infty} R_{xyy}(t_1 - \alpha, t_2, t_3)h(\alpha)\,d\alpha \end{aligned} \tag{9-121}$$

Note that for the evaluation of $R_{yyy}(t_1, t_2, t_3)$ for specific times t_1, t_2, and t_3, the function $R_{xxx}(t_1, t_2, t_3)$ must be known for every t_1, t_2, and t_3.

Vector Processes and Multiterminal Systems

We consider now systems with n inputs $\mathbf{x}_i(t)$ and r outputs $\mathbf{y}_j(t)$. As a preparation, we introduce the notion of autocorrelation and cross-correlation for vector processes starting with a review of the standard matrix notation.

The expression $A = [a_{ij}]$ will mean a matrix with elements a_{ij}. The notation

$$A^t = [a_{ji}] \qquad A^* = [a^*_{ij}] \qquad A^\dagger = [a^*_{ji}]$$

will mean the transpose, the conjugate, and the conjugate transpose of A.

A column vector will be identified by $A = [a_i]$. Whether A is a vector or a general matrix will be understood from the context. If $A = [a_i]$ and $B = [b_j]$ are two vectors with m elements each, the product $A^t B = a_1 b_1 + \cdots + a_m b_m$ is a number, and the product $AB^t = [a_i b_j]$ is an $m \times m$ matrix with elements $a_i b_j$.

A vector process $\mathbf{X}(t) = [\mathbf{x}_i(t)]$ is a vector, the components of which are stochastic processes. The mean $\eta(t) = E\{\mathbf{X}(t)\} = [\eta_i(t)]$ of $\mathbf{X}(t)$ is a vector with components $\eta_i(t) = E\{\mathbf{x}_i(t)\}$. The autocorrelation $R(t_1, t_2)$ or $R_{xx}(t_1, t_2)$ of a vector process $\mathbf{X}(t)$ is an $m \times m$ matrix

$$R(t_1, t_2) = E\{\mathbf{X}(t_1)\mathbf{X}^\dagger(t_2)\} \tag{9-122}$$

with elements $E\{\mathbf{x}_i(t_1)\mathbf{x}^*_j(t_2)\}$. We define similarly the cross-correlation matrix

$$R_{xy}(t_1, t_2) = E\{\mathbf{X}(t_1)\mathbf{Y}^\dagger(t_2)\} \tag{9-123}$$

of the vector processes

$$\mathbf{X}(t) = [\mathbf{x}_i(t)] \quad i = 1, \ldots, m \qquad \mathbf{Y}(t) = [\mathbf{y}_j(t)] \quad j = 1, \ldots, r \tag{9-124}$$

A multiterminal system with m inputs $\mathbf{x}_i(t)$ and r outputs $\mathbf{y}_j(t)$ is a rule for assigning to an m vector $\mathbf{X}(t)$ an r vector $\mathbf{Y}(t)$. If the system is linear and time-invariant, it is specified in terms of its impulse response matrix. This is an $r \times m$ matrix

$$H(t) = [h_{ji}(t)] \qquad i = 1, \ldots, m \quad j = 1, \ldots, r \tag{9-125}$$

defined as: Its component $h_{ji}(t)$ is the response of the jth output when the ith input equals $\delta(t)$ and all other inputs equal 0. From this and the linearity of the system, it follows that the response $\mathbf{y}_j(t)$ of the jth output to an arbitrary input $\mathbf{X}(t) = [\mathbf{x}_i(t)]$ equals

$$\mathbf{y}_j(t) = \int_{-\infty}^{\infty} h_{j1}(\alpha)\mathbf{x}_1(t-\alpha)\,d\alpha + \cdots + \int_{-\infty}^{\infty} h_{jm}(\alpha)\mathbf{x}_m(t-\alpha)\,d\alpha$$

Hence

$$\mathbf{Y}(t) = \int_{-\infty}^{\infty} H(\alpha)\mathbf{X}(t-\alpha)\,d\alpha \tag{9-126}$$

In this material, $\mathbf{X}(t)$ and $\mathbf{Y}(t)$ are column vectors and $H(t)$ is an $r \times m$ matrix. We shall use this relationship to determine the autocorrelation $R_{yy}(t_1, t_2)$ of $\mathbf{Y}(t)$. Premultiplying the conjugate transpose of (9-126) by $\mathbf{X}(t_1)$ and setting $t = t_2$, we obtain

$$\mathbf{X}(t_1)\mathbf{Y}^\dagger(t_2) = \int_{-\infty}^{\infty} \mathbf{X}(t_1)\mathbf{X}^\dagger(t_2-\alpha)H^\dagger(\alpha)\,d\alpha$$

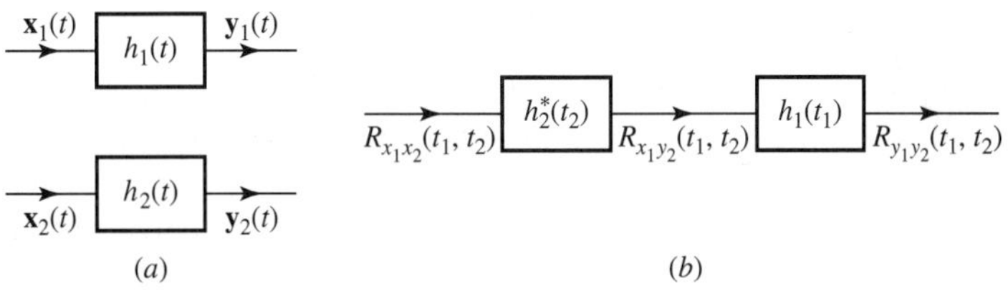

FIGURE 9-11

Hence

$$R_{xy}(t_1, t_2) = \int_{-\infty}^{\infty} R_{xx}(t_1, t_2 - \alpha) H^{\dagger}(\alpha)\, d\alpha \tag{9-127}$$

Postmultiplying (9-126) by $\mathbf{Y}^{\dagger}(t_2)$ and setting $t = t_1$, we obtain

$$R_{yy}(t_1, t_2) = \int_{-\infty}^{\infty} H(\alpha) R_{xy}(t_1 - \alpha, t_2)\, d\alpha \tag{9-128}$$

as in (9-98). These results can be used to express the cross-correlation of the outputs of several scalar systems in terms of the cross-correlation of their inputs. The next example is an illustration.

EXAMPLE 9-20

▶ In Fig. 9-11 we show two systems with inputs $\mathbf{x}_1(t)$ and $\mathbf{x}_2(t)$ and outputs

$$\mathbf{y}_1(t) = \int_{-\infty}^{\infty} h_1(\alpha)\mathbf{x}_1(t - \alpha)\, d\alpha \qquad \mathbf{y}_2(t) = \int_{-\infty}^{\infty} h_2(\alpha)\mathbf{x}_2(t - \alpha)\, d\alpha \tag{9-129}$$

These signals can be considered as the components of the output vector $\mathbf{Y}^t(t) = [\mathbf{y}_1(t), \mathbf{y}_2(t)]$ of a 2×2 system with input vector $\mathbf{X}^t(t) = [\mathbf{x}_1(t), \mathbf{x}_2(t)]$ and impulse response matrix

$$H(t) = \begin{bmatrix} h_1(t) & 0 \\ 0 & h_2(t) \end{bmatrix}$$

Inserting into (9-127)–(9-128), we obtain

$$R_{x_1y_2}(t_1, t_2) = \int_{-\infty}^{\infty} R_{x_1x_2}(t_1, t_2 - \alpha) h_2^*(\alpha)\, d\alpha \tag{9-130}$$

$$R_{y_1y_2}(t_1, t_2) = \int_{-\infty}^{\infty} h_1(\alpha) R_{x_1y_2}(t_1 - \alpha, t_2)\, d\alpha \tag{9-131}$$

Thus, to find $R_{x_1y_2}(t_1, t_2)$, we use $R_{x_1x_2}(t_1, t_2)$ as the input to the conjugate $h_2^*(t)$ of $h_2(t)$, operating on the variable t_2. To find $R_{y_1y_2}(t_1, t_2)$, we use $R_{x_1y_2}(t_1, t_2)$ as the input to $h_1(t)$ operating on the variable t_1 (Fig. 9-11). ◀

EXAMPLE 9-21

▶ The derivatives $\mathbf{y}_1(t) = \mathbf{z}^{(m)}(t)$ and $\mathbf{y}_2(t) = \mathbf{w}^{(n)}(t)$ of two processes $\mathbf{z}(t)$ and $\mathbf{w}(t)$ can be considered as the responses of two differentiators with inputs $\mathbf{x}_1(t) = \mathbf{z}(t)$ and $\mathbf{x}_2(t) = \mathbf{w}(t)$. Applying (9-130) suitably interpreted, we conclude that

$$E\{\mathbf{z}^{(m)}(t_1)\mathbf{w}^{(n)}(t_2)\} = \frac{\partial^{m+n} R_{zw}(t_1, t_2)}{\partial t_1^m \partial t_2^n} \tag{9-132}$$

◀

9-3 THE POWER SPECTRUM

In signal theory, spectra are associated with Fourier transforms. For deterministic signals, they are used to represent a function as a superposition of exponentials. For random signals, the notion of a spectrum has two interpretations. The first involves transforms of averages; it is thus essentially deterministic. The second leads to the representation of the process under consideration as superposition of exponentials with random coefficients. In this section, we introduce the first interpretation. The second is treated in Sec. 11-4. We shall consider only stationary processes. For nonstationary processes the notion of a spectrum is of limited interest.

DEFINITIONS. The *power spectrum* (or *spectral density)* of a WSS process $\mathbf{x}(t)$, real or complex, is the Fourier transform $S(\omega)$ of its autocorrelation $R(\tau) = E\{\mathbf{x}(t+\tau)\mathbf{x}^*(t)\}$:

$$S(\omega) = \int_{-\infty}^{\infty} R(\tau)e^{-j\omega\tau}\,d\tau \tag{9-133}$$

Since $R(-\tau) = R^*(\tau)$ it follows that $S(\omega)$ is a real function of ω.

From the Fourier inversion formula, it follows that

$$R(\tau) = \frac{1}{2\pi}\int_{-\infty}^{\infty} S(\omega)e^{j\omega\tau}\,d\omega \tag{9-134}$$

If $\mathbf{x}(t)$ is a real process, then $R(\tau)$ is real and even; hence $S(\omega)$ is also real and even. In this case,

$$\begin{aligned} S(\omega) &= \int_{-\infty}^{\infty} R(\tau)\cos\omega\tau\,d\tau = 2\int_{0}^{\infty} R(\tau)\cos\omega\tau\,d\tau \\ R(\tau) &= \frac{1}{2\pi}\int_{-\infty}^{\infty} S(\omega)\cos\omega\tau\,d\omega = \frac{1}{\pi}\int_{0}^{\infty} S(\omega)\cos\omega\tau\,d\omega \end{aligned} \tag{9-135}$$

The *cross-power spectrum* of two processes $\mathbf{x}(t)$ and $\mathbf{y}(t)$ is the Fourier transform $S_{xy}(\omega)$ of their cross-correlation $R_{xy}(\tau) = E\{\mathbf{x}(t+\tau)\mathbf{y}^*(t)\}$:

$$S_{xy}(\omega) = \int_{-\infty}^{\infty} R_{xy}(\tau)e^{-j\omega\tau}\,d\tau \qquad R_{xy}(\tau) = \frac{1}{2\pi}\int_{-\infty}^{\infty} S_{xy}(\omega)e^{j\omega\tau}\,d\omega \tag{9-136}$$

The function $S_{xy}(\omega)$ is, in general, complex even when both processes $\mathbf{x}(t)$ and $\mathbf{y}(t)$ are real. In all cases,

$$S_{xy}(\omega) = S_{yx}^*(\omega) \tag{9-137}$$

because $R_{xy}(-\tau) = E\{\mathbf{x}(t-\tau)\mathbf{y}^*(t)\} = \mathbf{R}_{yx}^*(\tau)$.

In Table 9-1 we list a number of frequently used autocorrelations and the corresponding spectra. Note that in all cases, $S(\omega)$ is positive. As we shall soon show, this is true for every spectrum.

TABLE 9-1

$$R(\tau) = \frac{1}{2\pi}\int_{-\infty}^{\infty} S(\omega)e^{j\omega\tau}\,d\omega \leftrightarrow S(\omega) = \int_{-\infty}^{\infty} R(\tau)e^{-j\omega\tau}\,d\tau$$

$\delta(\tau) \leftrightarrow 1$	$1 \leftrightarrow 2\pi\delta(\omega)$
$e^{j\beta\tau} \leftrightarrow 2\pi\delta(\omega-\beta)$	$\cos\beta\tau \leftrightarrow \pi\delta(\omega-\beta)+\pi\delta(\omega+\beta)$
$e^{-\alpha\|\tau\|} \leftrightarrow \dfrac{2\alpha}{\alpha^2+\omega^2}$	$e^{-\alpha\tau^2} \leftrightarrow \sqrt{\dfrac{\pi}{\alpha}}e^{-\omega^2/4\alpha}$

$$e^{-\alpha|\tau|}\cos\beta\tau \leftrightarrow \frac{\alpha}{\alpha^2+(\omega-\beta)^2}+\frac{\alpha}{\alpha^2+(\omega+\beta)^2}$$

$$2e^{-\alpha\tau^2}\cos\beta\tau \leftrightarrow \sqrt{\frac{\pi}{\alpha}}\left[e^{-(\omega-\beta)^2/4\alpha}+e^{-(\omega+\beta)^2/4\alpha}\right]$$

$$\begin{cases} 1-\dfrac{|\tau|}{T} & |\tau|<T \\ 0 & |\tau|>T \end{cases} \leftrightarrow \frac{4\sin^2(\omega T/2)}{T\omega^2}$$

$$\frac{\sin\sigma\tau}{\pi\tau} \leftrightarrow \begin{cases} 1 & |\omega|<\sigma \\ 0 & |\omega|>\sigma \end{cases}$$

EXAMPLE 9-22 ▶ A random telegraph signal is a process $\mathbf{x}(t)$ taking the values $+1$ and -1 as in Example 9-6:

$$\mathbf{x}(t) = \begin{cases} 1 & \mathbf{t}_{2i} < t < \mathbf{t}_{2i+1} \\ -1 & \mathbf{t}_{2i-1} < t < \mathbf{t}_{2i} \end{cases}$$

where $\mathbf{t}_i$ is a set of Poisson points with average density λ. As we have shown in (9-19), its autocorrelation equals $e^{-2\lambda|\tau|}$. Hence

$$S(\omega) = \frac{4\lambda}{4\lambda^2+\omega^2}$$ ◀

For most processes $R(\tau) \to \eta^2$, where $\eta = E\{\mathbf{x}(t)\}$ (see Sec. 11-4). If, therefore, $\eta \neq 0$, then $S(\omega)$ contains an impulse at $\omega = 0$. To avoid this, it is often convenient to express the spectral properties of $\mathbf{x}(t)$ in terms of the Fourier transform $S^c(\omega)$ of its autocovariance $C(\tau)$. Since $R(\tau) = C(\tau) + \eta^2$, it follows that

$$S(\omega) = S^c(\omega) + 2\pi\eta^2\delta(\omega) \tag{9-138}$$

The function $S^c(\omega)$ is called the *covariance spectrum* of $\mathbf{x}(t)$.

EXAMPLE 9-23 ▶ We have shown in (9-109) that the autocorrelation of the Poisson impulses

$$\mathbf{z}(t) = \frac{d}{dt}\sum_i U(t-\mathbf{t}_i) = \sum_i \delta(t-\mathbf{t}_i)$$

equals $R_z(\tau) = \lambda^2 + \lambda\delta(\tau)$. From this it follows that

$$S_z(\omega) = \lambda + 2\pi\lambda^2\delta(\omega) \qquad S_z^c(\omega) = \lambda$$ ◀

We shall show that given an arbitrary positive function $S(\omega)$, we can find a process $\mathbf{x}(t)$ with power spectrum $S(\omega)$.

(*a*) Consider the process

$$\mathbf{x}(t) = ae^{j(\boldsymbol{\omega} t - \boldsymbol{\varphi})} \tag{9-139}$$

where a is a real constant, $\boldsymbol{\omega}$ is a random variable with density $f_\omega(\omega)$, and $\boldsymbol{\varphi}$ is a random variable independent of $\boldsymbol{\omega}$ and uniform in the interval $(0, 2\pi)$. As we know, this process is WSS with zero mean and autocorrelation

$$R_x(\tau) = a^2 E\{e^{j\boldsymbol{\omega}\tau}\} = a^2 \int_{-\infty}^{\infty} f_\omega(\omega) e^{j\omega\tau}\, d\omega$$

From this and the uniqueness property of Fourier transforms, it follows that [see (9-134)] the power spectrum of $\mathbf{x}(t)$ equals

$$S_x(\omega) = 2\pi a^2 f_\omega(\omega) \tag{9-140}$$

If, therefore,

$$f_\omega(\omega) = \frac{S(\omega)}{2\pi a^2} \qquad a^2 = \frac{1}{2\pi}\int_{-\infty}^{\infty} S(\omega)\, d\omega = R(0)$$

then $f_\omega(\omega)$ is a density and $S_x(\omega) = S(\omega)$. To complete the specification of $\mathbf{x}(t)$, it suffices to construct a random variable $\boldsymbol{\omega}$ with density $S(\omega)/2\pi a^2$ and insert it into (9-139).

(*b*) We show next that if $S(-\omega) = S(\omega)$, we can find a real process with power spectrum $S(\omega)$. To do so, we form the process

$$\mathbf{y}(t) = a\cos(\boldsymbol{\omega} t + \boldsymbol{\varphi}) \tag{9-141}$$

In this case (see Example 9-14)

$$R_y(\tau) = \frac{a^2}{2} E\{\cos \boldsymbol{\omega}\tau\} = \frac{a^2}{2}\int_{-\infty}^{\infty} f(\omega)\cos\omega\tau\, d\omega$$

From this it follows that if $f_\omega(\omega) = S(\omega)/\pi a^2$, then $S_y(\omega) = S(\omega)$.

EXAMPLE 9-24

DOPPLER EFFECT

▶ A harmonic oscillator located at point P of the x axis (Fig. 9-12) moves in the x direction with velocity $\mathbf{v}$. The emitted signal equals $e^{j\omega_0 t}$ and the signal received by an observer located at point O equals

$$\mathbf{s}(t) = ae^{j\omega_0(t - \mathbf{r}/c)}$$

where c is the velocity of propagation and $\mathbf{r} = r_0 + \mathbf{v}t$. We assume that $\mathbf{v}$ is a random variable with density $f_v(v)$. Clearly,

$$\mathbf{s}(t) = ae^{j(\boldsymbol{\omega} t - \varphi)} \qquad \boldsymbol{\omega} = \omega_0\left(1 - \frac{\mathbf{v}}{c}\right) \qquad \varphi = \frac{r_0\omega_0}{c}$$

hence the spectrum of the received signal is given by (9-140)

$$S(\omega) = 2\pi a^2 f_\omega(\omega) = \frac{2\pi a^2 c}{\omega_0} f_v\left[\left(1 - \frac{\omega}{\omega_0}\right)c\right] \tag{9-142}$$

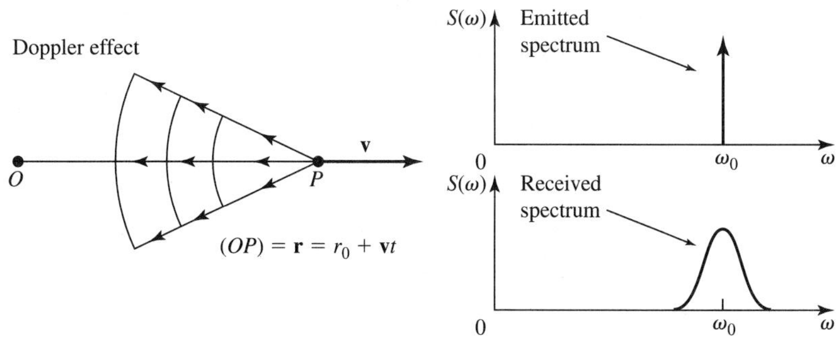

FIGURE 9-12

Note that if $\mathbf{v} = 0$, then

$$\mathbf{s}(t) = ae^{j(\omega_0 t - \varphi)} \qquad R(\tau) = a^2 e^{j\omega_0\tau} \qquad S(\omega) = 2\pi a^2 \delta(\omega - \omega_0)$$

This is the spectrum of the emitted signal. Thus the motion causes broadening of the spectrum.

This development holds also if the motion forms an angle with the x axis provided that $\mathbf{v}$ is replaced by its projection $\mathbf{v}_x$ on OP. The next case is of special interest. Suppose that the emitter is a particle in a gas of temperature T. In this case, the x component of its velocity is a normal random variable with zero mean and variance kT/m (see Prob. 9-5). Inserting into (9-142), we conclude that

$$S(\omega) = \frac{2\pi a^2 c}{\omega_0\sqrt{2\pi kT/m}} \exp\left\{-\frac{mc^2}{2kT}\left(1 - \frac{\omega}{\omega_0}\right)^2\right\}$$

$$R(\tau) = a^2 \exp\left\{-\frac{kT\omega_0^2\tau^2}{2mc^2}\right\} e^{j\omega_0\tau}$$ ◀

Line spectra.

(*a*) We have shown in Example 9-7 that the process

$$\mathbf{x}(t) = \sum_i \mathbf{c}_i e^{j\omega_i t}$$

is WSS if the random variables $\mathbf{c}_i$ are uncorrelated with zero mean. From this and Table 9-1 it follows that

$$R(\tau) = \sum_i \sigma_i^2 e^{j\omega_i\tau} \qquad S(\omega) = 2\pi \sum_i \sigma_i^2 \delta(\omega - \omega_i) \tag{9-143}$$

where $\sigma_i^2 = E\{\mathbf{c}_i^2\}$. Thus $S(\omega)$ consists of lines. In Sec. 13-2 we show that such a process is predictable, that is, its present value is uniquely determined in terms of its past.

(*b*) Similarly, the process

$$\mathbf{y}(t) = \sum_i (\mathbf{a}_i \cos\omega_i t + \mathbf{b}_i \sin\omega_i t)$$

is WSS iff the random variables $\mathbf{a}_i$ and $\mathbf{b}_i$ are uncorrelated with zero mean and $E\{\mathbf{a}_i^2\} = E\{\mathbf{b}_i^2\} = \sigma_i^2$. In this case,

$$R(\tau) = \sum_i \sigma_i^2 \cos \omega_i \tau \qquad S(\omega) = \pi \sum_i \sigma_i^2[\delta(\omega - \omega_i) + \delta(\omega + \omega_i)] \tag{9-144}$$

Linear systems. We shall express the autocorrelation $R_{yy}(\tau)$ and power spectrum $S_{yy}(\omega)$ of the response

$$\mathbf{y}(t) = \int_{-\infty}^{\infty} \mathbf{x}(t - \alpha)h(\alpha)\, d\alpha \tag{9-145}$$

of a linear system in terms of the autocorrelation $R_{xx}(\tau)$ and power spectrum $S_{xx}(\omega)$ of the input $\mathbf{x}(t)$.

THEOREM 9-4

$$R_{xy}(\tau) = R_{xx}(\tau) * h^*(-\tau) \qquad R_{yy}(\tau) = R_{xy}(\tau) * h(\tau) \tag{9-146}$$

$$S_{xy} = S_{xx}(\omega)H^*(\omega) \qquad S_{yy}(\omega) = S_{xy}(\omega)H(\omega) \tag{9-147}$$

Proof. The two equations in (9-146) are special cases of (9-211) and (9-212). However, because of their importance they will be proved directly. Multiplying the conjugate of (9-145) by $\mathbf{x}(t + \tau)$ and taking expected values, we obtain

$$E\{\mathbf{x}(t + \tau)\mathbf{y}^*(t)\} = \int_{-\infty}^{\infty} E\{\mathbf{x}(t + \tau)\mathbf{x}^*(t - \alpha)\}h^*(\alpha)\, d\alpha$$

Since $E\{\mathbf{x}(t + \tau)\mathbf{x}^*(t - \alpha)\} = R_{xx}(\tau + \alpha)$, this yields

$$R_{xy}(\tau) = \int_{-\infty}^{\infty} R_{xx}(\tau + \alpha)h^*(\alpha)\, d\alpha = \int_{-\infty}^{\infty} R_{xx}(\tau - \beta)h^*(-\beta)\, d\beta$$

Proceeding similarly, we obtain

$$E\{\mathbf{y}(t)\mathbf{y}^*(t - \tau)\} = \int_{-\infty}^{\infty} E\{\mathbf{x}(t - \alpha)\mathbf{y}^*(t - \tau)\}h(\alpha)\, d\alpha$$

$$= \int_{-\infty}^{\infty} R_{xy}(\tau - \alpha)h(\alpha)\, d\alpha$$

Equation (9-147) follows from (9-146) and the convolution theorem.

COROLLARY

Combining the two equations in (9-146) and (9-147), we obtain

$$R_{yy}(\tau) = R_{xx}(\tau) * h(\tau) * h^*(-\tau) = R_{xx}(\tau) * \rho(\tau) \tag{9-148}$$

$$S_{yy}(\omega) = S_{xx}(\omega)H(\omega)H^*(\omega) = S_{xx}(\omega)|H(\omega)|^2 \tag{9-149}$$

where

$$\rho(\tau) = h(\tau) * h^*(-\tau) = \int_{-\infty}^{\infty} h(t + \tau)h^*(t)\, dt \leftrightarrow |H(\omega)|^2 \tag{9-150}$$

Note, in particular, that if $\mathbf{x}(t)$ is white noise with average power q, then

$$R_{xx}(\tau) = q\delta(\tau) \qquad S_{xx}(\omega) = q$$
$$S_{yy}(\omega) = q|H(\omega)|^2 \qquad R_{yy}(\tau) = q\rho(\tau) \tag{9-151}$$

From (9-149) and the inversion formula (9-134), it follows that

$$E\{|\mathbf{y}(t)|^2\} = R_{yy}(0) = \frac{1}{2\pi}\int_{-\infty}^{\infty} S_{xx}(\omega)|H(\omega)|^2\,d\omega \ge 0 \tag{9-152}$$

This equation describes the filtering properties of a system when the input is a random process. It shows, for example, that if $H(\omega) = 0$ for $|\omega| > \omega_0$ and $S_{xx}(\omega) = 0$ for $|\omega| < \omega_0$, then $E\{\mathbf{y}^2(t)\} = 0$. ◀

Note The preceding results hold if all correlations are replaced by the corresponding covariances and all spectra by the corresponding covariance spectra. This follows from the fact that the response to $\mathbf{x}(t) - \eta_x$ equals $\mathbf{y}(t) - \eta_y$. For example, (9-149) and (9-155) yield

$$S^c_{yy}(\omega) = S^c_{xx}(\omega)|H(\omega)|^2 \tag{9-153}$$

$$\operatorname{Var}\mathbf{y}(t) = \frac{1}{2\pi}\int_{-\infty}^{\infty} S^c_{xx}(\omega)|H(\omega)|^2\,d\omega \tag{9-154}$$

EXAMPLE 9-25

▶ (*a*) (*Moving average*) The integral

$$\mathbf{y}(t) = \frac{1}{2T}\int_{t-T}^{t+T} \mathbf{x}(\alpha)\,d\alpha$$

is the average of the process $\mathbf{x}(t)$ in the interval $(t-T, t+T)$. Clearly, $\mathbf{y}(t)$ is the output of a system with input $\mathbf{x}(t)$ and impulse response a rectangular pulse as in Fig. 9-13. The corresponding $\rho(\tau)$ is a triangle. In this case,

$$H(\omega) = \frac{1}{2T}\int_{-T}^{T} e^{-j\omega\tau}\,d\tau = \frac{\sin T\omega}{T\omega} \qquad S_{yy}(\omega) = S_{xx}(\omega)\frac{\sin^2 T\omega}{T^2\omega^2}$$

Thus $H(\omega)$ takes significant values only in an interval of the order of $1/T$ centered at the origin. Hence the moving average suppresses the high-frequency components of the input. It is thus a simple low-pass filter.

Since $\rho(\tau)$ is a triangle, it follows from (9-148) that

$$R_{yy}(\tau) = \frac{1}{2T}\int_{-2T}^{2T}\left(1 - \frac{|\alpha|}{2T}\right) R_{xx}(\tau - \alpha)\,d\alpha \tag{9-155}$$

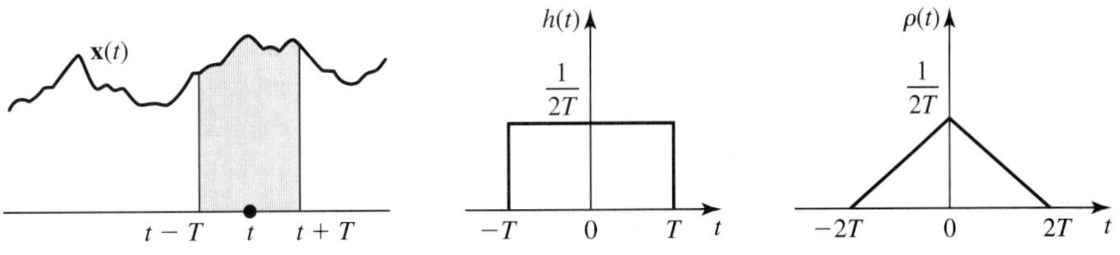

FIGURE 9-13

We shall use this result to determine the variance of the integral

$$\eta_T = \frac{1}{2T}\int_{-T}^{T} \mathbf{x}(t)\,dt$$

Clearly, $\boldsymbol{\eta}_T = \mathbf{y}(0)$; hence

$$\text{Var}\{\boldsymbol{\eta}_T\} = C_{yy}(0) = \frac{1}{2T}\int_{-2T}^{2T}\left(1 - \frac{|\alpha|}{2T}\right) C_{xx}(\alpha)\,d\alpha \tag{9-156}$$

(*b*) (*High-pass filter*) The process $\mathbf{z}(t) = \mathbf{x}(t) - \mathbf{y}(t)$ is the output of a system with input $\mathbf{x}(t)$ and system function

$$H(\omega) = 1 - \frac{\sin T\omega}{T\omega}$$

This function is nearly 0 in an interval of the order of $1/T$ centered at the origin, and it approaches 1 for large ω. It acts, therefore, as a high-pass filter suppressing the low frequencies of the input. ◀

EXAMPLE 9-26

DERIVATIVES

▶ The derivative $\mathbf{x}'(t)$ of a process $\mathbf{x}(t)$ can be considered as the output of a linear system with input $\mathbf{x}(t)$ and system function $j\omega$. From this and (9-147), it follows that

$$S_{xx'}(\omega) = -j\omega S_{xx}(\omega) \qquad S_{x'x'}(\omega) = \omega^2 S_{xx}(\omega)$$

Hence

$$R_{xx'}(\tau) = -\frac{dR_{xx}(\tau)}{d\tau} \qquad R_{x'x'}(\tau) = -\frac{d^2R_{xx}(\tau)}{d\tau^2}$$

The nth derivative $\mathbf{y}(t) = \mathbf{x}^{(n)}(t)$ of $\mathbf{x}(t)$ is the output of a system with input $\mathbf{x}(t)$ and system function $(j\omega)^n$. Hence

$$S_{yy}(\omega) = |j\omega|^{2n} \qquad R_{yy}(\tau) = (-1)^n R^{(2n)}(\tau) \tag{9-157}$$

◀

EXAMPLE 9-27

▶ (*a*) The differential equation

$$\mathbf{y}'(t) + c\mathbf{y}(t) = \mathbf{x}(t) \qquad \text{all } t$$

specifies a linear system with input $\mathbf{x}(t)$, output $\mathbf{y}(t)$, and system function $1/(j\omega + c)$. We assume that $\mathbf{x}(t)$ is white noise with $R_{xx}(\tau) = q\delta(\tau)$. Applying (9-149), we obtain

$$S_{yy}(\omega) = \frac{S_{xx}(\omega)}{\omega^2 + c^2} = \frac{q}{\omega^2 + c^2} \qquad R_{yy}(\tau) = \frac{q}{2c}e^{-c|\tau|}$$

Note that $E\{\mathbf{y}^2(t)\} = R_{yy}(0) = q/2c$.

(*b*) Similarly, if

$$\mathbf{y}''(t) + b\mathbf{y}'(t) + c\mathbf{y}(t) = \mathbf{x}(t) \qquad S_{xx}(\omega) = q$$

then

$$H(\omega) = \frac{1}{-\omega^2 + jb\omega + c} \qquad S_{yy}(\omega) = \frac{q}{(c - \omega^2)^2 + b^2\omega^2}$$

To find $R_{yy}(\tau)$, we shall consider three cases:

$\underline{b^2 < 4c}$

$$R_{yy}(\tau) = \frac{q}{2bc} e^{-\alpha|\tau|} \left(\cos \beta\tau + \frac{\alpha}{\beta} \sin \beta|\tau| \right) \qquad \alpha = \frac{b}{2} \qquad \alpha^2 + \beta^2 = c$$

$\underline{b^2 = 4c}$

$$R_{yy}(\tau) = \frac{q}{2bc} e^{-\alpha|\tau|}(1 + \alpha|\tau|) \qquad \alpha = \frac{b}{2}$$

$\underline{b^2 > 4c}$

$$R_{yy}(\tau) = \frac{q}{4\gamma bc} \left[(\alpha + \gamma) e^{-(\alpha-\gamma)|\tau|} - (\alpha - \gamma) e^{-(\alpha+\gamma)|\tau|} \right]$$

$$\alpha = \frac{b}{2} \qquad \alpha^2 - \gamma^2 = c$$

In all cases, $E\{\mathbf{y}^2(t)\} = q/2bc$. ◀

EXAMPLE 9-28

HILBERT TRANSFORMS

▶ A system with system function (Fig. 9-14)

$$H(\omega) = -j \operatorname{sgn} \omega = \begin{cases} -j & \omega > 0 \\ j & \omega < 0 \end{cases} \tag{9-158}$$

is called a *quadrature filter*. The corresponding impulse response equals $1/\pi t$ (Papoulis, 1977 [20]). Thus $H(\omega)$ is all-pass with -90° phase shift; hence its response to $\cos \omega t$ equals $\cos(\omega t - 90^\circ) = \sin \omega t$ and its response to $\sin \omega t$ equals $\sin(\omega t - 90^\circ) = -\cos \omega t$.

The response of a quadrature filter to a real process $\mathbf{x}(t)$ is denoted by $\hat{\mathbf{x}}(t)$ and it is called the *Hilbert transform* of $\mathbf{x}(t)$. Thus

$$\hat{\mathbf{x}}(t) = \mathbf{x}(t) * \frac{1}{\pi t} = \frac{1}{\pi} \int_{-\infty}^{\infty} \frac{\mathbf{x}(\alpha)}{t - \alpha} d\alpha \tag{9-159}$$

From (9-147) and (9-137) it follows that (Fig. 9-14)

$$\begin{aligned} S_{x\hat{x}}(\omega) &= j S_{xx}(\omega) \operatorname{sgn} \omega = -S_{\hat{x}x}(\omega) \\ S_{\hat{x}\hat{x}}(\omega) &= S_{xx}(\omega) \end{aligned} \tag{9-160}$$

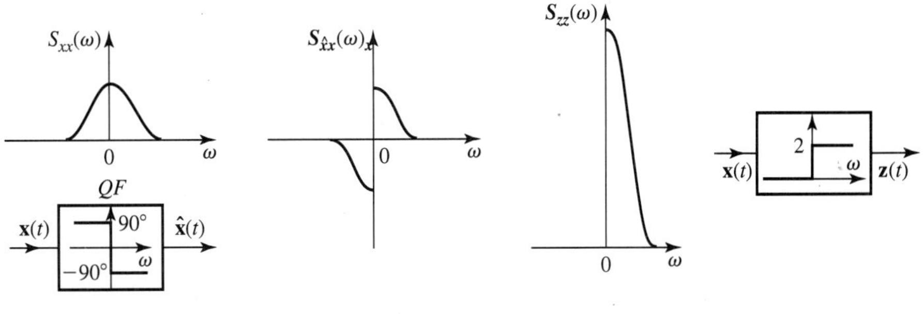

FIGURE 9-14

The complex process

$$\mathbf{z}(t) = \mathbf{x}(t) + j\hat{\mathbf{x}}(t)$$

is called the *analytic signal* associated with $\mathbf{x}(t)$. Clearly, $\mathbf{z}(t)$ is the response of the system

$$1 + j(-j \operatorname{sgn} \omega) = 2U(\omega)$$

with input $\mathbf{x}(t)$. Hence [see (9-149)]

$$S_{zz}(\omega) = 4S_{xx}(\omega)U(\omega) = 2S_{xx}(\omega) + 2jS_{\hat{x}x}(\omega) \tag{9-161}$$

$$R_{zz}(\tau) = 2R_{xx}(\tau) + 2jR_{\hat{x}x}(\tau) \tag{9-162}$$

◀

THE WIENER–KHINCHIN THEOREM. From (9-134) it follows that

$$E\{\mathbf{x}^2(t)\} = R(0) = \frac{1}{2\pi}\int_{-\infty}^{\infty} S(\omega)\,d\omega \ge 0 \tag{9-163}$$

This shows that the area of the power spectrum of any process is positive. We shall show that

$$S(\omega) \ge 0 \tag{9-164}$$

for every ω.

Proof. We form an ideal bandpass system with system function

$$H(\omega) = \begin{cases} 1 & \omega_1 < \omega < \omega_2 \\ 0 & \text{otherwise} \end{cases}$$

and apply $\mathbf{x}(t)$ to its input. From (9-152) it follows that the power spectrum $S_{yy}(\omega)$ of the resulting output $\mathbf{y}(t)$ equals

$$S_{yy}(\omega) = \begin{cases} S(\omega) & \omega_1 < \omega < \omega_2 \\ 0 & \text{otherwise} \end{cases}$$

Hence

$$0 \le E\{\mathbf{y}^2(t)\} = \frac{1}{2\pi}\int_{-\infty}^{\infty} S_{yy}(\omega)\,d\omega = \frac{1}{2\pi}\int_{\omega_1}^{\omega_2} S(\omega)\,d\omega \tag{9-165}$$

Thus the area of $S(\omega)$ in any interval is positive. This is possible only if $S(\omega) \ge 0$ everywhere.

We have shown on page 410 that if $S(\omega)$ is a positive function, then we can find a process $\mathbf{x}(t)$ such that $S_{xx}(\omega) = S(\omega)$. From this it follows that a function $S(\omega)$ is a power spectrum iff it is positive. In fact, we can find an exponential with random frequency $\boldsymbol{\omega}$ as in (9-140) with power spectrum an arbitrary positive function $S(\omega)$.

We shall use (9-165) to express the power spectrum $S(\omega)$ of a process $\mathbf{x}(t)$ as the average power of another process $\mathbf{y}(t)$ obtained by filtering $\mathbf{x}(t)$. Setting $\omega_1 = \omega_0 + \delta$ and $\omega_2 = \omega_0 - \delta$, we conclude that if δ is sufficiently small,

$$E\{\mathbf{y}^2(t)\} \simeq \frac{\delta}{\pi} S(\omega_0) \tag{9-166}$$

This shows the *localization* of the average power of $\mathbf{x}(t)$ on the frequency axis.

Integrated spectrum. In mathematics, the spectral properties of a process $\mathbf{x}(t)$ are expressed in terms of the integrated spectrum $F(\omega)$ defined as the integral of $S(\omega)$:

$$F(\omega) = \int_{-\infty}^{\omega} S(\alpha)\,d\alpha \tag{9-167}$$

From the positivity of $S(\omega)$, it follows that $F(\omega)$ is a nondecreasing function of ω. Integrating the inversion formula (9-134) by parts, we can express the autocorrelation $R(\tau)$ of $\mathbf{x}(t)$ as a Riemann–Stieltjes integral:

$$R(\tau) = \frac{1}{2\pi}\int_{-\infty}^{\infty} e^{j\omega\tau}\,dF(\omega) \tag{9-168}$$

This approach avoids the use of singularity functions in the spectral representation of $R(\tau)$ even when $S(\omega)$ contains impulses. If $S(\omega)$ contains the terms $\beta_i\delta(\omega-\omega_i)$, then $F(\omega)$ is discontinuous at ω_i and the discontinuity jump equals β_i.

The integrated covariance spectrum $F^c(\omega)$ is the integral of the covariance spectrum. From (9-138) it follows that $F(\omega) = F^c(\omega) + 2\pi\eta^2 U(\omega)$.

Vector spectra. The vector process $\mathbf{X}(t) = [\mathbf{x}_i(t)]$ is WSS if its components $\mathbf{x}_i(t)$ are jointly WSS. In this case, its autocorrelation matrix depends only on $\tau = t_1 - t_2$. From this it follows that [see (9-127)–(9-128)]

$$R_{xy}(\tau) = \int_{-\infty}^{\infty} R_{xx}(\tau+\alpha)H^{\dagger}(\alpha)\,d\alpha \qquad R_{yy}(\tau) = \int_{-\infty}^{\infty} H(\alpha)R_{xy}(\tau-\alpha)\,d\alpha \tag{9-169}$$

The power spectrum of a WSS vector process $\mathbf{X}(t)$ is a square matrix $S_{xx}(\omega) = [S_{ij}(\omega)]$, the elements of which are the Fourier transforms $S_{ij}(\omega)$ of the elements $R_{ij}(\tau)$ of its autocorrelation matrix $R_{xx}(\tau)$. Defining similarly the matrices $S_{xy}(\omega)$ and $S_{yy}(\omega)$, we conclude from (9-169) that

$$S_{xy}(\omega) = S_{xx}(\omega)\overline{H}^{\dagger}(\omega) \qquad S_{yy}(\omega) = \overline{H}(\omega)S_{xy}(\omega) \tag{9-170}$$

where $\overline{H}(\omega) = [H_{ji}(\omega)]$ is an $m \times r$ matrix with elements the Fourier transforms $H_{ji}(\omega)$ of the elements $h_{ji}(t)$ of the impulse response matrix $H(t)$. Thus

$$S_{yy}(\omega) = \overline{H}(\omega)S_{xx}(\omega)\overline{H}^{\dagger}(\omega) \tag{9-171}$$

This is the extension of (9-149) to a multiterminal system.

EXAMPLE 9-29

▶ The derivatives

$$\mathbf{y}_1(t) = \mathbf{z}^{(m)}(t) \qquad \mathbf{y}_2(t) = \mathbf{w}^{(n)}(t)$$

of two WSS processes $\mathbf{z}(t)$ and $\mathbf{w}(t)$ can be considered as the responses of two differentiators with inputs $\mathbf{z}(t)$ and $\mathbf{w}(t)$ and system functions $H_1(\omega) = (j\omega)^m$ and $H_2(\omega) = (j\omega)^n$. Proceeding as in (9-132), we conclude that the cross-power spectrum of $\mathbf{z}^{(m)}(t)$ and $\mathbf{w}^{(n)}(t)$ equals $(j\omega)^m(-j\omega)^n S_{zw}(\omega)$. Hence

$$E\{\mathbf{z}^{(m)}(t+\tau)\mathbf{z}^{(n)}(t)\} = (-1)^n\frac{d^{m+n}R_{zw}(\tau)}{d\tau^{m+n}} \tag{9-172}$$

◀

PROPERTIES OF CORRELATIONS. If a function $R(\tau)$ is the autocorrelation of a WSS process $\mathbf{x}(t)$, then [see (9-164)] its Fourier transform $S(\omega)$ is positive. Furthermore, if $R(\tau)$ is a function with positive Fourier transform, we can find a process $\mathbf{x}(t)$ as in (9-139) with autocorrelation $R(\tau)$. Thus a necessary and sufficient condition for a function $R(\tau)$ to be an autocorrelation is the positivity of its Fourier transform. The conditions for a function $R(\tau)$ to be an autocorrelation can be expressed directly in terms of $R(\tau)$. We have shown in (9-93) that the autocorrelation $R(\tau)$ of a process $\mathbf{x}(t)$ is p.d., that is,

$$\sum_{i,j} a_i a_j^* R(\tau_i - \tau_j) \geq 0 \tag{9-173}$$

for every a_i, a_j, τ_i, and τ_j. It can be shown that the converse is also true[6]: If $R(\tau)$ is a p.d. function, then its Fourier transform is positive (see (9-197)). Thus a function $R(\tau)$ has a positive Fourier transform iff it is positive definite.

A sufficient condition. To establish whether $R(\tau)$ is p.d., we must show either that it satisfies (9-173) or that its transform is positive. This is not, in general, a simple task. The following is a simple sufficient condition.

Polya's criterion. It can be shown that a function $R(\tau)$ is p.d. if it is concave for $\tau > 0$ and it tends to a finite limit as $\tau \to \infty$ (see Yaglom, 1987 [30]).

Consider, for example, the function $w(\tau) = e^{-\alpha|\tau|^c}$. If $0 < c < 1$, then $w(\tau) \to 0$ as $\tau \to \infty$ and $w''(\tau) > 0$ for $\tau > 0$; hence $w(\tau)$ is p.d. because it satisfies Polya's criterion. Note, however, that it is p.d. also for $1 \leq c \leq 2$ even though it does not satisfy this criterion.

Necessary conditions. The autocorrelation $R(\tau)$ of any process $\mathbf{x}(t)$ is maximum at the origin because [see (9-134)]

$$|R(\tau)| \leq \frac{1}{2\pi}\int_{-\infty}^{\infty} S(\omega)\,d\omega = R(0) \tag{9-174}$$

We show next that if $R(\tau)$ is not periodic, it reaches its maximum only at the origin.

THEOREM 9-5

If $R(\tau_1) = R(0)$ for some $\tau_1 \neq 0$, then $R(\tau)$ is periodic with period τ_1:

$$R(\tau + \tau_1) = R(\tau) \qquad \text{for all } \tau \tag{9-175}$$

Proof. From Schwarz's inequality

$$E^2\{\mathbf{zw}\} \leq E\{\mathbf{z}^2\}E\{\mathbf{w}^2\} \tag{9-176}$$

it follows that

$$E^2\{[\mathbf{x}(t+\tau+\tau_1) - \mathbf{x}(t+\tau)]\mathbf{x}(t)\} \leq E\{[\mathbf{x}(t+\tau+\tau_1) - \mathbf{x}(t+\tau)]^2\}E\{\mathbf{x}^2(t)\}$$

Hence

$$[R(\tau+\tau_1) - R(\tau)]^2 \leq 2[R(0) - R(\tau_1)]R(0) \tag{9-177}$$

If $R(\tau_1) = R(0)$, then the right side is 0; hence the left side is also 0 for every τ. This yields (9-175).

[6] S. Bocher: *Lectures on Fourier Integrals,* Princeton Univ. Press, Princeton, NJ, 1959.

COROLLARY ▶ If $R(\tau_1) = R(\tau_2) = R(0)$ and the numbers τ_1 and τ_2 are noncommensurate, that is, their ratio is irrational, then $R(\tau)$ is constant.

Proof. From the theorem it follows that $R(\tau)$ is periodic with periods τ_1 and τ_2. This is possible only if $R(\tau)$ is constant. ◀

Continuity. If $R(\tau)$ is continuous at the origin, it is continuous for every τ.

Proof. From the continuity of $R(\tau)$ at $\tau = 0$ it follows that $R(\tau_1) \to R(0)$; hence the left side of (9-177) also tends to 0 for every τ as $\tau_1 \to 0$.

EXAMPLE 9-30 ▶ Using the theorem, we shall show that the truncated parabola

$$w(\tau) = \begin{cases} a^2 - \tau^2 & |\tau| < a \\ 0 & |\tau| > a \end{cases}$$

is not an autocorrelation.

If $w(\tau)$ is the autocorrelation of some process $\mathbf{x}(t)$, then [see (9-157)] the function

$$-w''(\tau) = \begin{cases} 2 & |\tau| < a \\ 0 & |\tau| > a \end{cases}$$

is the autocorrelation of $\mathbf{x}'(t)$. This is impossible because $-w''(\tau)$ is continuous for $\tau = 0$ but not for $\tau = a$. ◀

MS continuity and periodicity. We shall say that the process $\mathbf{x}(t)$ is MS continuous if

$$E\{[\mathbf{x}(t+\varepsilon) - \mathbf{x}(t)]^2\} \to 0 \qquad \text{as} \quad \varepsilon \to 0 \tag{9-178}$$

Since $E\{[\mathbf{x}(t+\varepsilon) - \mathbf{x}(t)]^2\} = 2[R(0) - R(\varepsilon)]$, we conclude that if $\mathbf{x}(t)$ is MS continuous, $R(0) - R(\varepsilon) \to 0$ as $\varepsilon \to 0$. Thus a WSS process $\mathbf{x}(t)$ is MS continuous iff its autocorrelation $R(\tau)$ is continuous for all τ.

We shall say that the process $\mathbf{x}(t)$ is MS periodic with period τ_1 if

$$E\{[\mathbf{x}(t+\tau_1) - \mathbf{x}(t)]^2\} = 0 \tag{9-179}$$

Since the left side equals $2[R(0) - R(\tau_1)]$, we conclude that $R(\tau_1) = R(0)$; hence [see (9-175)] $R(\tau)$ is periodic. This leads to the conclusion that a WSS process $\mathbf{x}(t)$ is MS periodic iff its autocorrelation is periodic.

Cross-correlation. Using (9-176), we shall show that the cross-correlation $R_{xy}(\tau)$ of two WSS processes $\mathbf{x}(t)$ and $\mathbf{y}(t)$ satisfies the inequality

$$R_{xy}^2(\tau) \le R_{xx}(0)R_{yy}(0) \tag{9-180}$$

Proof. From (9-176) it follows that

$$E^2\{\mathbf{x}(t+\tau)\mathbf{y}^*(t)\} \le E\{|\mathbf{x}(t+\tau)|^2\}E\{|\mathbf{y}(t)|^2\} = R_{xx}(0)R_{yy}(0)$$

and (9-180) results.

COROLLARY For any a and b,

$$\left|\int_a^b S_{xy}(\omega)\,d\omega\right|^2 \le \int_a^b S_{xx}(\omega)\,d\omega \int_a^b S_{yy}(\omega)\,d\omega \tag{9-181}$$

Proof. Suppose that $\mathbf{x}(t)$ and $\mathbf{y}(t)$ are the inputs to the ideal filters

$$H_1(\omega) = H_2(\omega) = \begin{cases} 1 & a < \omega < b \\ 0 & \text{otherwise} \end{cases}$$

Denoting by $\mathbf{z}(t)$ and $\mathbf{w}(t)$, respectively, the resulting outputs, we conclude that

$$R_{zz}(0) = \frac{1}{2\pi}\int_a^b S_{xx}(\omega)\,d\omega \qquad R_{ww}(0) = \frac{1}{2\pi}\int_a^b S_{yy}(\omega)\,d\omega$$

$$R_{zw}(0) = \frac{1}{2\pi}\int_a^b S_{zw}(\omega)\,d\omega$$

and (9-181) follows because $R_{zw}^2(0) \le R_{zz}(0)R_{ww}(0)$.

9-4 DISCRETE-TIME PROCESSES

A digital (or discrete-time) process is a sequence $\mathbf{x}_n$ of random variables. To avoid double subscripts, we shall use also the notation $\mathbf{x}[n]$ where the brackets will indicate that n is an integer. Most results involving analog (or continuous-time) processes can be readily extended to digital processes. We outline the main concepts.

The autocorrelation and autocovariance of $\mathbf{x}[n]$ are given by

$$R[n_1, n_2] = E\{\mathbf{x}[n_1]\mathbf{x}^*[n_2]\} \qquad C[n_1, n_2] = R[n_1, n_2] - \eta[n_1]\eta^*[n_2] \tag{9-182}$$

respectively where $\eta[n] = E\{\mathbf{x}[n]\}$ is the mean of $\mathbf{x}[n]$.

A process $\mathbf{x}[n]$ is SSS if its statistical properties are invariant to a shift of the origin. It is WSS if $\eta[n] = \eta = \text{constant}$ and

$$R[n+m, n] = E\{\mathbf{x}[n+m]\mathbf{x}^*[n]\} = R[m] \tag{9-183}$$

A process $\mathbf{x}[n]$ is strictly white noise if the random variables $\mathbf{x}[n_i]$ are independent. It is white noise if the random variables $\mathbf{x}[n_i]$ are uncorrelated. The autocorrelation of a white-noise process with zero mean is thus given by

$$R[n_1, n_2] = q[n_1]\delta[n_1 - n_2] \qquad \text{where} \quad \delta[n] = \begin{cases} 1 & n = 0 \\ 0 & n \ne 0 \end{cases} \tag{9-184}$$

and $q[n] = E\{\mathbf{x}^2[n]\}$. If $\mathbf{x}[n]$ is also stationary, then $R[m] = q\delta[m]$. Thus a WSS white noise is a sequence of i.i.d. random variables with variance q.

The delta response $h[n]$ of a linear system is its response to the delta sequence $\delta[n]$. Its system function is the z transform of $h[n]$:

$$\mathbf{H}(z) = \sum_{n=-\infty}^{\infty} h[n]z^{-n} \tag{9-185}$$

If $\mathbf{x}[n]$ is the input to a digital system, the resulting output is the digital convolution of $\mathbf{x}[n]$ with $h[n]$:

$$\mathbf{y}[n] = \sum_{k=-\infty}^{\infty} \mathbf{x}[n-k]h[k] = \mathbf{x}[n] * h[n] \tag{9-186}$$

From this it follows that $\eta_y[n] = \eta_x[n] * h[n]$. Furthermore,

$$R_{xy}[n_1, n_2] = \sum_{k=-\infty}^{\infty} R_{xx}[n_1, n_2 - k]h^*[k] \tag{9-187}$$

$$R_{yy}[n_1, n_2] = \sum_{r=-\infty}^{\infty} R_{xy}[n_1 - r, n_2]h[r] \tag{9-188}$$

If $\mathbf{x}[n]$ is white noise with average intensity $q[n]$ as in (9-184), then, [see (9-99)],

$$E\{\mathbf{y}^2[n]\} = q[n] * |h[n]|^2 \tag{9-189}$$

If $\mathbf{x}[n]$ is WSS, then $\mathbf{y}[n]$ is also WSS with $\eta_y = \eta_x \mathbf{H}(1)$. Furthermore,

$$\begin{aligned} R_{xy}[m] &= R_{xx}[m] * h^*[-m] \qquad & R_{yy}[m] &= R_{xy}[m] * h[m] \\ R_{yy}[m] &= R_{xx}[m] * \rho[m] \qquad & \rho[m] &= \sum_{k=-\infty}^{\infty} h[m+k]h^*[k] \end{aligned} \tag{9-190}$$

as in (9-146) and (9-148).

THE POWER SPECTRUM. Given a WSS process $\mathbf{x}[n]$, we form the z transform $\mathbf{S}(z)$ of its autocorrelation $R[m]$:

$$\mathbf{S}(z) = \sum_{m=-\infty}^{\infty} R[m]z^{-m} \tag{9-191}$$

The power spectrum of $\mathbf{x}[n]$ is the function

$$S(\omega) = \mathbf{S}(e^{j\omega}) = \sum_{m=-\infty}^{\infty} R[m]e^{-jm\omega} \geq 0 \tag{9-192}$$

Thus $\mathbf{S}(e^{j\omega})$ is the discrete Fourier transform (DFT) of $R[m]$. The function $\mathbf{S}(e^{j\omega})$ is periodic with period 2π and Fourier series coefficients $R[m]$. Hence

$$R[m] = \frac{1}{2\pi}\int_{-\pi}^{\pi} \mathbf{S}(e^{j\omega})e^{jm\omega}\,d\omega \tag{9-193}$$

It suffices, therefore, to specify $\mathbf{S}(e^{j\omega})$ for $|\omega| < \pi$ only (see Fig. 9-15).

If $\mathbf{x}[n]$ is a real process, then $R[-m] = R[m]$ and (9-192) yields

$$\mathbf{S}(e^{j\omega}) = R[0] + 2\sum_{m=0}^{\infty} R[m]\cos m\omega \tag{9-194}$$

This shows that the power spectrum of a real process is a function of $\cos\omega$ because $\cos m\omega$ is a function of $\cos\omega$.

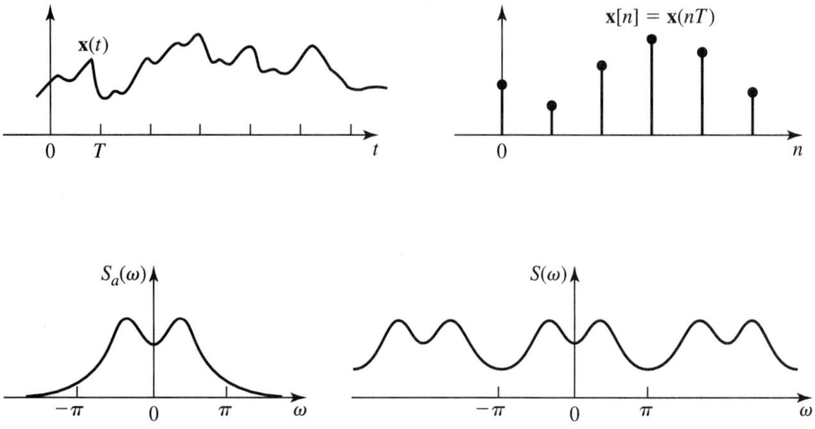

FIGURE 9-15

The nonnegativity condition in (9-173) can be expressed in terms of certain Hermitian Toeplitz matrices. Let

$$r_k \triangleq R[k] \tag{9-195}$$

and define

$$\mathbf{T}_n = \begin{pmatrix} r_0 & r_1 & r_2 & \cdot & \cdots & r_n \\ r_1^* & r_0 & r_1 & r_2 & \cdots & r_{n-1} \\ r_2^* & r_1^* & r_0 & r_1 & \cdots & r_{n-2} \\ \cdot & & & & & \\ \cdot & & & & & \\ \cdot & & & & & \\ r_n^* & r_{n-1}^* & \cdot & \cdots & r_1^* & r_0 \end{pmatrix} \tag{9-196}$$

In that case

$$S(\omega) \geq 0 \Leftrightarrow \mathbf{T}_n \geq 0 \qquad n = 0 \to \infty \tag{9-197}$$

i.e., the nonnegative nature of the spectrum is equivalent to the nonnegativity of every Hemitian Toeplitz matrix $\mathbf{T}_n$, $n = 0 \to \infty$ in (16-10). To prove this result due to Schur, first assume that $S(\omega) \geq 0$ in (9-192). Then letting

$$\mathbf{a} = [a_0, a_1, a_2, \ldots, a_n]^t \tag{9-198}$$

we have

$$\begin{aligned} \mathbf{a}^\dagger \mathbf{T}_n \mathbf{a} &= \sum_{i=0}^{n} \sum_{m=0}^{n} a_i^* a_m r_{i-m} \\ &= \sum_{i=0}^{n} \sum_{m=0}^{n} a_i^* a_m \frac{1}{2\pi} \int_{-\pi}^{\pi} S(\omega) e^{j(i-m)\omega}\, d\omega \\ &= \frac{1}{2\pi} \int_{-\pi}^{\pi} S(\omega) \left| \sum_{m=0}^{n} a_m e^{-jm\omega} \right|^2 d\omega \geq 0 \end{aligned} \tag{9-199}$$

Since **a** is arbitrary, this gives

$$S(\omega) \geq 0 \Rightarrow \mathbf{T}_n \geq 0 \qquad n = 0 \rightarrow \infty \tag{9-200}$$

Conversely, assume that every $\mathbf{T}_n$, $n = 0 \rightarrow \infty$ are nonnegative definite matrices. Further, for any ρ, $0 < \rho < 1$, and ω_0, $0 < \omega_0 < 2\pi$, define the vector **a** in (9-198) with

$$a_m = \sqrt{1-\rho^2}\,\rho^m e^{jm\omega_0}$$

Then $\mathbf{T}_n$ nonnegative implies that

$$0 \leq a^\dagger \mathbf{T}_n a = \frac{1}{2\pi}\int_{-\pi}^{\pi}(1-\rho^2)\left|\sum_{m=0}^{n}\rho^m e^{jm(\omega-\omega_0)}\right|^2 S(\omega)\,d\omega$$

and letting $n \rightarrow \infty$, the above intergrad tends to

$$\frac{1}{2\pi}\int_{-\pi}^{\pi}\frac{1-\rho^2}{1-2\rho\cos(\omega-\omega_0)+\rho^2}S(\omega)\,d\omega \geq 0 \tag{9-201}$$

The left-hand side of (9-201) represents the Poisson integral, and its interior ray limit as $\rho \rightarrow 1-0$ equals $S(\omega_0)$ for almost all ω_0. Thus

$$\mathbf{T}_n \geq 0 \qquad n \rightarrow 0 \rightarrow \infty \Rightarrow S(\omega) \geq 0 \qquad \textit{almost everywhere (a.e.)} \tag{9-202}$$

More interestingly, subject to the additional constraint, known as the Paley–Wiener criterion

$$\frac{1}{2\pi}\int_{-\pi}^{\pi}\ln S(\omega)\,d\omega > -\infty \tag{9-203}$$

every $\mathbf{T}_k$, $k = 0 \rightarrow \infty$, must be positive definite. This follows from (9-199). In fact, if some $\mathbf{T}_k$ is singular, then there exists a nontrivial vector **a** such $\mathbf{T}_k\mathbf{a} = 0$ and, from (9-199),

$$\mathbf{a}^\dagger \mathbf{T}_k \mathbf{a} = \frac{1}{2\pi}\int_{-\pi}^{\pi} S(\omega)\left|\sum_{m=0}^{k} a_m e^{jm\omega}\right|^2 d\omega = 0$$

Since $S(\omega) \geq 0$, *a.e.*, this expression gives

$$S(\omega)\left|\sum_{m=0}^{k} a_m e^{jm\omega}\right|^2 = 0 \quad a.e.$$

and $\sum_{m=0}^{k} a_m e^{-jm\omega} \neq 0$, *a.e.*, implies

$$S(\omega) = 0 \quad a.e.$$

and

$$\int_{-\pi}^{\pi}\ln S(\omega)\,d\omega = -\infty$$

contradicting (9-203). Hence subject to (9-203), every

$$\mathbf{T}_k > 0 \qquad k = 0 \rightarrow \infty \tag{9-204}$$

The integrability condition together with (9-203) enables the factorization of the power spectral density in terms of a specific function with certain interesting properties.

More precisely, there exists a unique function

$$H(z) = \sum_{k=0}^{\infty} h_k z^{-k} \qquad h_0 > 0 \quad |z| > 1 \tag{9-205}$$

that is analytic together with its inverse in $|z| > 1$ (minimum-phase function, see Appendix 12A) satisfying

$$\sum_{k=0}^{\infty} |h_k|^2 < \infty \tag{9-206}$$

and

$$S(\omega) = |H(e^{-j\omega})|^2 \quad a.e. \tag{9-207}$$

if and only if $S(\omega)$ as well as $\ln S(\omega)$ are integrable functions. Here $H(e^{-j\omega})$ is defined as the exterior radial limit of $H(z)$ on the unit circle, i.e.,

$$H(e^{-j\omega}) = \lim_{r\to 1+0} H(re^{-j\omega}) \tag{9-208}$$

EXAMPLE 9-31 ▶ If $R[m] = a^{|m|}$, then

$$\mathbf{S}(z) = \sum_{m=-\infty}^{-1} a^{-m} z^{-m} + \sum_{m=0}^{\infty} a^m z^{-m} = \frac{az}{1-az} + \frac{z}{z-a}$$

$$= \frac{a^{-1} - a}{(a^{-1} + a) - (z^{-1} + z)}$$

Hence

$$S(\omega) = \mathbf{S}(e^{j\omega}) = \frac{a^{-1} - a}{a^{-1} + a - 2\cos\omega}$$ ◀

EXAMPLE 9-32 ▶ Proceeding as in the analog case, we can show that the process

$$\mathbf{x}[n] = \sum_i \mathbf{c}_i e^{j\omega_i n}$$

is WSS iff the coefficients $\mathbf{c}_i$ are uncorrelated with zero mean. In this case,

$$R[m] = \sum_i \sigma_i^2 e^{j\beta_i |m|} \qquad S(\omega) = 2\pi \sum_i \sigma_i^2 \delta(\omega - \beta_i) \qquad |\omega| < \pi \tag{9-209}$$

where $\sigma_i^2 = E\{\mathbf{c}_i^2\}$, $\omega_i = 2\pi k_i + \beta_i$, and $|\beta_i| < \pi$. ◀

From (9-190) and the convolution theorem, it follows that if $\mathbf{y}[n]$ is the output of a linear system with input $\mathbf{x}[n]$, then

$$\begin{aligned} \mathbf{S}_{xy}(e^{j\omega}) &= \mathbf{S}_{xx}(e^{j\omega})\mathbf{H}^*(e^{j\omega}) \\ \mathbf{S}_{yy}(e^{j\omega}) &= \mathbf{S}_{xy}(e^{j\omega})\mathbf{H}(e^{j\omega}) \\ \mathbf{S}_{yy}(e^{j\omega}) &= \mathbf{S}_{xx}(e^{j\omega})|\mathbf{H}(e^{j\omega})|^2 \end{aligned} \tag{9-210}$$

If $h[n]$ is real, $\mathbf{H}^*(e^{j\omega}) = \mathbf{H}(e^{-j\omega})$. In this case,

$$\mathbf{S}_{yy}(z) = \mathbf{S}_{xx}(z)\mathbf{H}(z)\mathbf{H}(1/z) \tag{9-211}$$

EXAMPLE 9-33 ▶ The first difference

$$\mathbf{y}[n] = \mathbf{x}[n] - \mathbf{x}[n-1]$$

of a process $\mathbf{x}[n]$ can be considered as the output of a linear system with input $\mathbf{x}[n]$ and system function $\mathbf{H}(z) = 1 - z^{-1}$. Applying (9-211), we obtain

$$\mathbf{S}_{yy}(z) = \mathbf{S}_{xx}(z)(1 - z^{-1})(1 - z) = \mathbf{S}_{xx}(z)(2 - z - z^{-1})$$

$$R_{yy}[m] = -R_{xx}[m+1] + 2R_{xx}[m] - R_{xx}[m-1]$$

If $\mathbf{x}[n]$ is white noise with $\mathbf{S}_{xx}(z) = q$, then

$$\mathbf{S}_{yy}(e^{j\omega}) = q(2 - e^{j\omega} - e^{-j\omega}) = 2q(1 - \cos\omega)$$ ◀

EXAMPLE 9-34 ▶ The recursion equation

$$\mathbf{y}[n] - a\mathbf{y}[n-1] = \mathbf{x}[n]$$

specifies a linear system with input $\mathbf{x}[n]$ and system function $\mathbf{H}(z) = 1/(1 - az^{-1})$. If $\mathbf{S}_{xx}(z) = q$, then (see Example 9-31)

$$\mathbf{S}_{yy}(z) = \frac{q}{(1 - az^{-1})(1 - az)} \qquad R_{yy}[m] = \frac{q}{a^{-1} - a} a^{|m|}$$ ◀

From (9-210) it follows that

$$E\{|\mathbf{y}[n]|^2\} = R_{yy}[0] = \frac{1}{2\pi}\int_{-\pi}^{\pi} \mathbf{S}_{xx}(e^{j\omega})|\mathbf{H}(e^{j\omega})|^2\, d\omega \tag{9-212}$$

Using this identity, we shall show that the power spectrum of a process $\mathbf{x}[n]$ real or complex is a positive function:

$$\mathbf{S}_{xx}(e^{j\omega}) \geq 0 \tag{9-213}$$

Proof. We form an ideal bandpass filter with center frequency ω_0 and bandwidth 2Δ and apply (9-212). For small Δ,

$$E\{|\mathbf{y}[n]|^2\} = \frac{1}{2\pi}\int_{\omega_0-\Delta}^{\omega_0+\Delta} \mathbf{S}_{xx}(e^{j\omega})\, d\omega \simeq \frac{\Delta}{\pi}\mathbf{S}_{xx}(e^{j\omega_0})$$

and (9-213) results because $E\{\mathbf{y}^2[n]\} \geq 0$ and ω_0 is arbitrary.

SAMPLING. In many applications, the digital processes under consideration are obtained by sampling various analog processes. We relate next the corresponding correlations and spectra.

Given an analog process $\mathbf{x}(t)$, we form the digital process

$$\mathbf{x}[n] = \mathbf{x}(nT)$$

where T is a given constant. From this it follows that

$$\eta[n] = \eta_a(nT) \qquad R[n_1, n_2] = R_a(n_1 T, n_2 T) \tag{9-214}$$

where $\eta_a(t)$ is the mean and $R_a(t_1, t_2)$ the autocorrelation of $\mathbf{x}(t)$. If $\mathbf{x}(t)$ is a stationary process, then $\mathbf{x}[n]$ is also stationary with mean $\eta = \eta_a$ and autocorrelation

$$R[m] = R_a(mT)$$

From this it follows that the power spectrum of $\mathbf{x}[n]$ equals (Fig. 9-15)

$$\mathbf{S}(e^{j\omega}) = \sum_{m=-\infty}^{\infty} R_a(mT)e^{-jm\omega} = \frac{1}{T}\sum_{n=-\infty}^{\infty} S_a\left(\frac{\omega + 2\pi n}{T}\right) \tag{9-215}$$

where $S_a(\omega)$ is the power spectrum of $\mathbf{x}(t)$. The above is a consequence of Poisson's sum formula [see (10A-1)].

EXAMPLE 9-35 ▶ Suppose that $\mathbf{x}(t)$ is a WSS process consisting of M exponentials as in (9-143):

$$\mathbf{x}(t) = \sum_{i=1}^{M} \mathbf{c}_i e^{j\omega_i t} \qquad S_a(\omega) = 2\pi \sum_{i=1}^{M} \sigma_i^2 \delta(\omega - \omega_i)$$

where $\sigma_i^2 = E\{\mathbf{c}_i^2\}$. We shall determine the power spectrum $\mathbf{S}(e^{j\omega})$ of the process $\mathbf{x}[n] = \mathbf{x}(nT)$. Since $\delta(\omega/T) = T\delta(\omega)$, it follows from (9-215) that

$$\mathbf{S}(e^{j\omega}) = 2\pi \sum_{n=-\infty}^{\infty} \sum_{i=1}^{M} \sigma_i^2 \delta(\omega - T\omega_i + 2\pi n)$$

In the interval $(-\pi, \pi)$, this consists of M lines:

$$\mathbf{S}(e^{j\omega}) = 2\pi \sum_{i=1}^{M} \sigma_i^2 \delta(\omega - \beta_i) \qquad |\omega| < \pi$$

where $\beta_i = T\omega_i - 2\pi n_i$ and such that $|\beta_i| < \pi$. ◀

APPENDIX 9A
CONTINUITY, DIFFERENTIATION, INTEGRATION

In the earlier discussion, we routinely used various limiting operations involving stochastic processes, with the tacit assumption that these operations hold for every sample involved. This assumption is, in many cases, unnecessarily restrictive. To give some idea of the notion of limits in a more general case, we discuss next conditions for the existence of MS limits and we show that these conditions can be phrased in terms of second-order moments (see also Sec. 7-4).

STOCHASTIC CONTINUITY. A process $\mathbf{x}(t)$ is called MS continuous if

$$E\{[\mathbf{x}(t+\varepsilon) - \mathbf{x}(t)]^2\} \xrightarrow[\varepsilon \to 0]{} 0 \tag{9A-1}$$

THEOREM 9A-1 We maintain that $\mathbf{x}(t)$ is MS continuous if its autocorrelation is continuous.

Proof. Clearly,

$$E\{[\mathbf{x}(t+\varepsilon)-\mathbf{x}(t)]^2\} = R(t+\varepsilon, t+\varepsilon) - 2R(t+\varepsilon, t) + R(t,t)$$

If, therefore, $R(t_1, t_2)$ is continuous, then the right side tends to 0 as $\varepsilon \to 0$ and (9A-1) results.

Note Suppose that (9A-1) holds for *every* t in an interval I. From this it follows that [see (9-1)] almost all samples of $\mathbf{x}(t)$ will be continuous at a *particular* point of I. It does not follow, however, that these samples will be continuous for *every* point in I. We mention as illustrations the Poisson process and the Wiener process. As we see from (9-14) and (10-5), both processes are MS continuous. However, the samples of the Poisson process are discontinuous at the points $\mathbf{t}_i$, whereas almost all samples of the Wiener process are continuous.

COROLLARY If $\mathbf{x}(t)$ is MS continuous, then its mean is continuous

$$\eta(t+\varepsilon) \to \eta(t) \qquad \varepsilon \to 0 \tag{9A-2}$$

Proof. As we know

$$E\{[\mathbf{x}(t+\varepsilon)-\mathbf{x}(t)]^2\} \geq E^2\{[\mathbf{x}(t+\varepsilon)-\mathbf{x}(t)]\}$$

Hence (9A-2) follows from (9A-1). Thus

$$\lim_{\varepsilon\to 0} E\{\mathbf{x}(t+\varepsilon)\} = E\left\{\lim_{\varepsilon\to 0} \mathbf{x}(t+\varepsilon)\right\} \tag{9A-3}$$

STOCHASTIC DIFFERENTIATION. A process $\mathbf{x}(t)$ is MS differentiable if

$$\frac{\mathbf{x}(t+\varepsilon)-\mathbf{x}(t)}{\varepsilon} \xrightarrow[\varepsilon\to 0]{} \mathbf{x}'(t) \tag{9A-4}$$

in the MS sense, that is, if

$$E\left\{\left[\frac{\mathbf{x}(t+\varepsilon)-\mathbf{x}(t)}{\varepsilon} - \mathbf{x}'(t)\right]^2\right\} \xrightarrow[\varepsilon\to 0]{} 0 \tag{9A-5}$$

THEOREM 9A-2 The process $\mathbf{x}(t)$ is MS differentiable if $\partial^2 R(t_1,t_2)/\partial t_1 \partial t_2$ exists.

Proof. It suffices to show that (Cauchy criterion)

$$E\left\{\left[\frac{\mathbf{x}(t+\varepsilon_1)-\mathbf{x}(t)}{\varepsilon_1} - \frac{\mathbf{x}(t+\varepsilon_2)-\mathbf{x}(t)}{\varepsilon_2}\right]^2\right\} \xrightarrow[\varepsilon_1,\varepsilon_2\to 0]{} 0 \tag{9A-6}$$

We use this criterion because, unlike (9A-5), it does not involve the unknown $\mathbf{x}'(t)$. Clearly,

$$E\{[\mathbf{x}(t+\varepsilon_1)-\mathbf{x}(t)][\mathbf{x}(t+\varepsilon_2)-\mathbf{x}(t)]\}$$
$$= R(t+\varepsilon_1, t+\varepsilon_2) - R(t+\varepsilon_1, t) - R(t, t+\varepsilon_2) + R(t,t)$$

The right side divided by $\varepsilon_1\varepsilon_2$ tends to $\partial^2 R(t,t)/\partial t\,\partial t$ which, by assumption, exists. Expanding the square in (9A-6), we conclude that its left side tends to

$$\frac{\partial^2 R(t,t)}{\partial t\,\partial t} - \frac{\partial^2 R(t,t)}{\partial t\,\partial t} + \frac{\partial^2 R(t,t)}{\partial t\,\partial t} = 0$$

COROLLARY The proof of the above theorem yields

$$E\{\mathbf{x}'(t)\} = E\left\{\lim_{\varepsilon\to 0}\frac{\mathbf{x}(t+\varepsilon)-\mathbf{x}(t)}{\varepsilon}\right\} = \lim_{\varepsilon\to 0} E\left\{\frac{\mathbf{x}(t+\varepsilon)-\mathbf{x}(t)}{\varepsilon}\right\}$$

Note The autocorrelation of a Poisson process $\mathbf{x}(t)$ is discontinuous at the points t_i; hence $\mathbf{x}'(\mathbf{t})$ does not exist at these points. However, as in the case of deterministic signals, it is convenient to introduce random impulses and to interpret $\mathbf{x}'(t)$ as in (9-107).

STOCHASTIC INTEGRALS. A process $\mathbf{x}(t)$ is MS integrable if the limit

$$\int_a^b \mathbf{x}(t)\,dt = \lim_{\Delta t_i\to 0}\sum_i \mathbf{x}(t_i)\,\Delta t_i \tag{9A-7}$$

exists in the MS sense.

THEOREM 9A-3 The process $\mathbf{x}(t)$ is MS integrable if

$$\int_a^b\int_a^b |R(t_1,t_2)|\,dt_1\,dt_2 < \infty \tag{9A-8}$$

Proof. Using again the Cauchy criterion, we must show that

$$E\left\{\left|\sum_i \mathbf{x}(t_i)\,\Delta t_i - \sum_k \mathbf{x}(t_k)\,\Delta t_k\right|^2\right\} \xrightarrow[\Delta t_i,\Delta t_k\to 0]{} 0$$

This follows if we expand the square and use the identity

$$E\left\{\sum_i \mathbf{x}(t_i)\,\Delta t_i \sum_k \mathbf{x}(t_k)\,\Delta t_k\right\} = \sum_{i,k} R(t_i,t_k)\,\Delta t_i\,\Delta t_k$$

because the right side tends to the integral of $R(t_1,t_2)$ as Δt_i and Δt_k tend to 0.

COROLLARY From the proof of the theorem it follows that

$$E\left\{\left|\int_a^b x(t)\,dt\right|^2\right\} = \int_a^b\int_a^b R(t_1,t_2)\,dt_1\,dt_2 \tag{9A-9}$$

as in (9-11).

APPENDIX 9B
SHIFT OPERATORS AND STATIONARY PROCESSES

An SSS process can be generated by a succession of shifts $T\mathbf{x}$ of a single random variable $\mathbf{x}$ where T is a one-to-one measure preserving transformation (mapping) of the probability space S into itself. This difficult topic is of fundamental importance in mathematics. Here, we give a brief explanation of the underlying concept, limiting the discussion to the discrete-time case.

A *transformation* T of S into itself is a rule for assigning to each element ζ_i of S another element of S:

$$\tilde{\zeta}_i = T\zeta_i \tag{9B-1}$$

called the *image* of ζ_i. The images $\tilde{\zeta}_i$ of all elements ζ_i of a subset A of S form another subset

$$\tilde{A} = TA$$

of S called the image of A.

We shall assume that the transformation T has these properties.

P_1: It is one-to-one. This means that

$$\text{if} \quad \zeta_i \neq \zeta_j \qquad \text{then} \quad \tilde{\zeta}_i \neq \tilde{\zeta}_j$$

P_2: It is measure preserving. This means that if A is an event, then its image $\tilde{A}$ is also an event and

$$P(\tilde{A}) = P(A) \tag{9B-2}$$

Suppose that $\mathbf{x}$ is a random variable and that T is a transformation as before. The expression $T\mathbf{x}$ will mean another random variable

$$\mathbf{y} = T\mathbf{x} \qquad \text{such that} \quad \mathbf{y}(\tilde{\zeta}_i) = \mathbf{x}(\zeta_i) \tag{9B-3}$$

where ζ_i is the unique inverse of $\tilde{\zeta}_i$. This specifies $\mathbf{y}$ for every element of S because (see P_1) the set of elements $\tilde{\zeta}_i$ equals S.

The expression $\mathbf{z} = T^{-1}\mathbf{x}$ will mean that $\mathbf{x} = T\mathbf{z}$. Thus

$$\mathbf{z} = T^{-1}\mathbf{x} \qquad \text{iff} \quad \mathbf{z}(\zeta_i) = \mathbf{x}(\tilde{\zeta}_i)$$

We can define similarly $T^2\mathbf{x} = T(T\mathbf{x}) = T\mathbf{y}$ and

$$T^n\mathbf{x} = T(T^{n-1}\mathbf{x}) = T^{-1}(T^{n+1}\mathbf{x})$$

for any n positive or negative.

From (9B-3) it follows that if, for some ζ_i, $\mathbf{x}(\zeta_i) \le w$, then $\mathbf{y}(\tilde{\zeta}_i) = \mathbf{x}(\zeta_i) \le w$. Hence the event $\{\mathbf{y} \le w\}$ is the image of the event $\{\mathbf{x} \le w\}$. This yields [see (9B-2)]

$$P\{\mathbf{x} \le w\} = P\{\mathbf{y} \le w\} \qquad \mathbf{y} = T\mathbf{x} \tag{9B-4}$$

for any w. We thus conclude that the random variables $\mathbf{x}$ and $T\mathbf{x}$ have the same distribution $F_x(x)$.

Given a random variable $\mathbf{x}$ and a transformation T as before, we form the random process

$$\mathbf{x}_0 = \mathbf{x} \qquad \mathbf{x}_n = T^n\mathbf{x} \qquad n = -\infty, \ldots, \infty \tag{9B-5}$$

It follows from (9B-4) that the random variables $\mathbf{x}_n$ so formed have the same distribution. We can similarly show that their joint distributions of any order are invariant to a shift of the origin. Hence the process $\mathbf{x}_n$ so formed is SSS.

It can be shown that the converse is also true: Given an SSS process $\mathbf{x}_n$, we can find a random variable $\mathbf{x}$ and a one-to-one measuring preserving transformation of the space S into itself such that for all essential purposes, $\mathbf{x}_n = T^n\mathbf{x}$. The proof of this difficult result will not be given.

PROBLEMS

9-1 In the fair-coin experiment, we define the process $\mathbf{x}(t)$ as follows: $\mathbf{x}(t) = \sin \pi t$ if heads show, $\mathbf{x}(t) = 2t$ if tails show. (*a*) Find $E\{\mathbf{x}(t)\}$. (*b*) Find $F(x, t)$ for $t = 0.25$, $t = 0.5$, and $t = 1$.

9-2 The process $\mathbf{x}(t) = e^{\mathbf{a}t}$ is a family of exponentials depending on the random variable $\mathbf{a}$. Express the mean $\eta(t)$, the autocorrelation $R(t_1, t_2)$, and the first-order density $f(x, t)$ of $\mathbf{x}(t)$ in terms of the density $f_a(a)$ of $\mathbf{a}$.

9-3 Suppose that $\mathbf{x}(t)$ is a Poisson process as in Fig. 9-3 such that $E\{\mathbf{x}(9)\} = 6$. (*a*) Find the mean and the variance of $\mathbf{x}(8)$. (*b*) Find $P\{\mathbf{x}(2) \le 3\}$. (*c*) Find $P\{\mathbf{x}(4) \le 5 \,|\, \mathbf{x}(2) \le 3\}$.

9-4 The random variable $\mathbf{c}$ is uniform in the interval $(0, T)$. Find $R_x(t_1, t_2)$ if (*a*) $\mathbf{x}(t) = U(t - \mathbf{c})$, (*b*) $\mathbf{x}(t) = \delta(t - \mathbf{c})$.

9-5 The random variables $\mathbf{a}$ and $\mathbf{b}$ are independent $N(0; \sigma)$ and p is the probability that the process $\mathbf{x}(t) = \mathbf{a} - \mathbf{b}t$ crosses the t axis in the interval $(0, T)$. Show that $\pi p = \arctan T$.

Hint: $p = P\{0 \le \mathbf{a}/\mathbf{b} \le T\}$.

9-6 Show that if

$$R_v(t_1, t_2) = q(t_1)\delta(t_1 - t_2)$$

$\mathbf{w}''(t) = \mathbf{v}(t)U(t)$ and $\mathbf{w}(0) = \mathbf{w}'(0) = 0$, then

$$E\{\mathbf{w}^2(t)\} = \int_0^t (t - \tau)q(\tau)\,d\tau$$

9-7 The process $\mathbf{x}(t)$ is real with autocorrelation $R(\tau)$. (*a*) Show that

$$P\{|\mathbf{x}(t + \tau) - \mathbf{x}(t)| \ge a\} \le 2[R(0) - R(\tau)]/a^2$$

(*b*) Express $P\{|\mathbf{x}(t + \tau) - \mathbf{x}(t)| \ge a\}$ in terms of the second-order density $f(x_1, x_2; \tau)$ of $\mathbf{x}(t)$.

9-8 The process $\mathbf{x}(t)$ is WSS and normal with $E\{\mathbf{x}(t)\} = 0$ and $R(\tau) = 4e^{-2|\tau|}$. (*a*) Find $P\{\mathbf{x}(t) \le 3\}$. (*b*) Find $E\{[\mathbf{x}(t + 1) - \mathbf{x}(t - 1)]^2\}$.

9-9 Show that the process $\mathbf{x}(t) = \mathbf{c}w(t)$ is WSS iff $E\{\mathbf{c}\} = 0$ and $w(t) = e^{j(\omega t + \theta)}$.

9-10 The process $\mathbf{x}(t)$ is normal WSS and $E\{\mathbf{x}(t)\} = 0$. Show that if $\mathbf{z}(t) = \mathbf{x}^2(t)$, then $C_{zz}(\tau) = 2C_{xx}^2(\tau)$.

9-11 Find $E\{\mathbf{y}(t)\}$, $E\{\mathbf{y}^2(t)\}$, and $R_{yy}(\tau)$ if

$$\mathbf{y}''(t) + 4\mathbf{y}'(t) + 13\mathbf{y}(t) = 26 + \boldsymbol{\nu}(t) \qquad R_{\nu\nu}(\tau) = 10\delta(\tau)$$

Find $P\{\mathbf{y}(t) \le 3\}$ if $\boldsymbol{\nu}(t)$ is normal.

9-12 Show that: If $\mathbf{x}(t)$ is a process with zero mean and autocorrelation $f(t_1)f(t_2)w(t_1 - t_2)$, then the process $\mathbf{y}(t) = \mathbf{x}(t)/f(t)$ is WSS with autocorrelation $w(\tau)$. If $\mathbf{x}(t)$ is white noise with autocorrelation $q(t_1)\delta(t_1 - t_2)$, then the process $\mathbf{z}(t) = \mathbf{x}(t)/\sqrt{q(t)}$ is WSS white noise with autocorrelation $\delta(\tau)$.

9-13 Show that $|R_{xy}(\tau)| \le \frac{1}{2}[R_{xx}(0) + R_{yy}(0)]$.

9-14 Show that if the processes $\mathbf{x}(t)$ and $\mathbf{y}(t)$ are WSS and $E\{|\mathbf{x}(0) - \mathbf{y}(0)|^2\} = 0$, then $R_{xx}(\tau) \equiv R_{xy}(\tau) \equiv R_{yy}(\tau)$.

Hint: Set $\mathbf{z} = \mathbf{x}(t + \tau)$, $\mathbf{w} = \mathbf{x}^*(t) - \mathbf{y}^*(t)$ in (9-176).

9-15 Show that if $\mathbf{x}(t)$ is a complex WSS process, then

$$E\{|\mathbf{x}(t+\tau)-\mathbf{x}(t)|^2\} = 2\,\mathrm{Re}\,[R(0)-R(\tau)]$$

9-16 Show that if φ is a random variable with $\Phi(\lambda) = E\{e^{j\lambda\varphi}\}$ and $\Phi(1) = \Phi(2) = 0$, then the process $\mathbf{x}(t) = \cos(\omega t + \varphi)$ is WSS. Find $E\{\mathbf{x}(t)\}$ and $R_x(\tau)$ if φ is uniform in the interval $(-\pi, \pi)$.

9-17 Given a process $\mathbf{x}(t)$ with orthogonal increments and such that $\mathbf{x}(0) = 0$, show that (a) $R(t_1, t_2) = R(t_1, t_1)$ for $t_1 \le t_2$, and (b) if $E\{[\mathbf{x}(t_1) - \mathbf{x}(t_2)]^2\} = q|t_1 - t_2|$ then the process $\mathbf{y}(t) = [\mathbf{x}(t+\varepsilon) - \mathbf{x}(t)]/\varepsilon$ is WSS and its autocorrelation is a triangle with area q and base 2ε.

9-18 Show that if $R_{xx}(t_1, t_2) = q(t_1)\delta(t_1 - t_2)$ and $\mathbf{y}(t) = \mathbf{x}(t) * h(t)$ then

$$E\{\mathbf{x}(t)\mathbf{y}(t)\} = h(0)q(t)$$

9-19 The process $\mathbf{x}(t)$ is normal with $\eta_x = 0$ and $R_x(\tau) = 4e^{-3|\tau|}$. Find a memoryless system $g(x)$ such that the first-order density $f_y(y)$ of the resulting output $\mathbf{y}(t) = g[\mathbf{x}(t)]$ is uniform in the interval (6, 9).

9-20 Show that if $\mathbf{x}(t)$ is an SSS process and $\boldsymbol{\varepsilon}$ is a random variable independent of $\mathbf{x}(t)$, then the process $\mathbf{y}(t) = \mathbf{x}(t - \boldsymbol{\varepsilon})$ is SSS.

9-21 Show that if $\mathbf{x}(t)$ is a stationary process with derivative $\mathbf{x}'(t)$, then for a given t the random variables $\mathbf{x}(t)$ and $\mathbf{x}'(t)$ are orthogonal and uncorrelated.

9-22 Given a normal process $\mathbf{x}(t)$ with $\eta_x = 0$ and $R_x(\tau) = 4e^{-2|\tau|}$, we form the random variables $\mathbf{z} = \mathbf{x}(t+1)$, $\mathbf{w} = \mathbf{x}(t-1)$, (a) find $E\{\mathbf{zw}\}$ and $E\{(\mathbf{z}+\mathbf{w})^2\}$, (b) find

$$f_z(z) \qquad P\{\mathbf{z} < 1\} \qquad f_{zw}(z, w)$$

9-23 Show that if $\mathbf{x}(t)$ is normal with autocorrelation $R(\tau)$, then

$$P\{\mathbf{x}'(t) \le a\} = G\left[\frac{a}{\sqrt{-R''(0)}}\right]$$

9-24 Show that if $\mathbf{x}(t)$ is a normal process with zero mean and $\mathbf{y}(t) = \operatorname{sgn} \mathbf{x}(t)$, then

$$R_y(\tau) = \frac{2}{\pi}\sum_{n=1}^{\infty}\frac{1}{n}[J_0(n\pi) - (-1)^n]\sin\left[n\pi\frac{R_x(\tau)}{R_x(0)}\right]$$

where $J_0(x)$ is the Bessel function.

Hint: Expand the arcsine in (9-80) into a Fourier series.

9-25 Show that if $\mathbf{x}(t)$ is a normal process with zero mean and $\mathbf{y}(t) = Ie^{a\mathbf{x}(t)}$, then

$$\eta_y == I\exp\left\{\frac{a^2}{2}R_x(0)\right\} \qquad R_y(\tau) = I^2\exp\{a^2[R_x(0) + R_x(\tau)]\}$$

9-26 Show that (a) if

$$\mathbf{y}(t) = a\mathbf{x}(ct) \qquad \text{then} \quad R_y(\tau) = a^2R_x(c\tau)$$

(b) if $R_x(\tau) \to 0$ as $\tau \to \infty$ and

$$\mathbf{z}(t) = \lim_{\varepsilon\to\infty}\sqrt{\varepsilon}\,\mathbf{x}(\varepsilon t) \qquad \text{then} \quad R_z(\tau) = q\delta(\tau) \qquad q = \int_{-\infty}^{\infty}R_x(\tau)\,d\tau$$

9-27 Show that if $\mathbf{x}(t)$ is white noise, $h(t) = 0$ outside the interval $(0, T)$, and $\mathbf{y}(t) = \mathbf{x}(t) * h(t)$ then $R_{yy}(t_1, t_2) = 0$ for $|t_1 - t_2| > T$.

9-28 Show that if

$$R_{xx}(t_1, t_2) = q(t_1)\delta(t_1 - t_2) \qquad E\{\mathbf{y}^2(t)\} = I(t)$$

and

(*a*) $$\mathbf{y}(t) = \int_0^t h(t, \alpha)\mathbf{x}(\alpha)\,d\alpha \quad \text{then} \quad I(t) = \int_0^t h^2(t, \alpha)q(\alpha)\,d\alpha$$

(*b*) $$\mathbf{y}'(t) + c(t)\mathbf{y}(t) = \mathbf{x}(t) \quad \text{then} \quad I'(t) + 2c(t)I(t) = q(t)$$

9-29 Find $E\{\mathbf{y}^2(t)\}$ (*a*) if $R_{xx}(\tau) = 5\delta(\tau)$ and

$$\mathbf{y}'(t) + 2\mathbf{y}(t) = \mathbf{x}(t) \qquad \text{all } t \tag{i}$$

(*b*) if (i) holds for $t > 0$ only and $\mathbf{y}(t) = 0$ for $t \le 0$.

Hint: Use (9-99).

9-30 The input to a linear system with $h(t) = Ae^{-\alpha t}U(t)$ is a process $\mathbf{x}(t)$ with $R_x(\tau) = N\delta(\tau)$ applied at $t = 0$ and disconnected at $t = T$. Find and sketch $E\{\mathbf{y}^2(t)\}$.

Hint: Use (9-99) with $q(t) = N$ for $0 < t < T$ and 0 otherwise.

9-31 Show that if

$$\mathbf{s} = \int_0^{10} \mathbf{x}(t)\,dt \quad \text{then} \quad E\{\mathbf{s}^2\} = \int_{-10}^{10} (10 - |\tau|)R_x(\tau)\,d\tau$$

Find the mean and variance of $\mathbf{s}$ if $E\{\mathbf{x}(t)\} = 8$, $R_x(\tau) = 64 + 10e^{-2|\tau|}$.

9-32 The process $\mathbf{x}(t)$ is WSS with $R_{xx}(\tau) = 5\delta(\tau)$ and

$$\mathbf{y}'(t) + 2\mathbf{y}(t) = \mathbf{x}(t) \tag{i}$$

Find $E\{\mathbf{y}^2(t)\}$, $R_{xy}(t_1, t_2)$, $R_{yy}(t_1, t_2)$ (*a*) if (i) holds for all t, (*b*) if $\mathbf{y}(0) = 0$ and (i) holds for $t \ge 0$.

9-33 Find $S(\omega)$ if (*a*) $R(\tau) = e^{-\alpha\tau^2}$, (*b*) $R(\tau) = e^{-\alpha\tau^2}\cos\omega_0\tau$.

9-34 Show that the power spectrum of an SSS process $\mathbf{x}(t)$ equals

$$S(\omega) = \int_{-\infty}^{\infty}\int_{-\infty}^{\infty} x_1 x_2 G(x_1, x_2; \omega)\,dx_1\,dx_2$$

where $G(x_1, x_2; \omega)$ is the Fourier transform in the variable τ of the second-order density $f(x_1, x_2; \tau)$ of $\mathbf{x}(t)$.

9-35 Show that if $\mathbf{y}(t) = \mathbf{x}(t + a) - \mathbf{x}(t - a)$, then

$$R_y(\tau) = 2R_x(\tau) - R_x(\tau + 2a) - R_x(\tau - 2a) \qquad S_y(\omega) = 4S_x(\omega)\sin^2 a\omega$$

9-36 Using (9-135), show that

$$R(0) - R(\tau) \ge \frac{1}{4^n}[R(0) - R(2^n\tau)]$$

Hint:

$$1 - \cos\theta = 2\sin^2\frac{\theta}{2} \ge 2\sin^2\frac{\theta}{2}\cos^2\frac{\theta}{2} = \frac{1}{4}(1 - \cos 2\theta)$$

9-37 The process $\mathbf{x}(t)$ is normal with zero mean and $R_x(\tau) = Ie^{-\alpha|\tau|}\cos\beta\tau$. Show that if $\mathbf{y}(t) = \mathbf{x}^2(t)$, then $C_y(\tau) = I^2 e^{-2\alpha|\tau|}(1 + \cos 2\beta\tau)$. Find $S_y(\omega)$.

9-38 Show that if $R(\tau)$ is the inverse Fourier transform of a function $S(\omega)$ and $S(\omega) \ge 0$, then, for any a_i,

$$\sum_{i,k} a_i a_k^* R(\tau_i - \tau_k) \ge 0$$

Hint:

$$\int_{-\infty}^{\infty} S(\omega)\left|\sum_i a_i e^{j\omega\tau_i}\right|^2 d\omega \geq 0$$

9-39 Find $R(\tau)$ if (*a*) $S(\omega) = 1/(1+\omega^4)$, (*b*) $S(\omega) = 1/(4+\omega^2)^2$.

9-40 Show that, for complex systems, (9-149) and (9-194) yield

$$\mathbf{S}_{yy}(s) = \mathbf{S}_{xx}(s)\mathbf{H}(s)\mathbf{H}^*(-s^*) \qquad \mathbf{S}_{yy}(z) = \mathbf{S}_{xx}(z)\mathbf{H}(z)\mathbf{H}^*(1/z^*)$$

9-41 The process $\mathbf{x}(t)$ is normal with zero mean. Show that if $\mathbf{y}(t) = \mathbf{x}^2(t)$, then

$$S_y(\omega) = 2\pi R_x^2(0)\delta(\omega) + 2S_x(\omega) * S_x(\omega)$$

Plot $S_y(\omega)$ if $S_x(\omega)$ is (*a*) ideal LP, (*b*) ideal BP.

9-42 The process $\mathbf{x}(t)$ is WSS with $E\{\mathbf{x}(t)\} = 5$ and $R_{xx}(\tau) = 25 + 4e^{-2|\tau|}$. If $\mathbf{y}(t) = 2\mathbf{x}(t) + 3\mathbf{x}'(t)$, find η_y, $R_{yy}(\tau)$, and $S_{yy}(\omega)$.

9-43 The process $\mathbf{x}(t)$ is WSS and $R_{xx}(\tau) = 5\delta(\tau)$. (*a*) Find $E\{\mathbf{y}^2(t)\}$ and $S_{yy}(\omega)$ if $\mathbf{y}'(t)+3\mathbf{y}(t) = \mathbf{x}(t)$. (*b*) Find $E\{\mathbf{y}^2(t)\}$ and $R_{xy}(t_1, t_2)$ if $\mathbf{y}'(t) + 3\mathbf{y}(t) = \mathbf{x}(t)U(t)$. Sketch the functions $R_{xy}(2, t_2)$ and $R_{xy}(t_1, 3)$.

9-44 Given a complex process $\mathbf{x}(t)$ with autocorrelation $R(\tau)$, show that if $|R(\tau_1)| = |R(0)|$, then

$$R(\tau) = e^{j\omega_0\tau}\omega(\tau) \qquad \mathbf{x}(t) = e^{j\omega_0 t}\mathbf{y}(t)$$

where $\omega(\tau)$ is a periodic function with period τ_1 and $\mathbf{y}(t)$ is an MS periodic process with the same period.

9-45 Show that (*a*) $E\{\mathbf{x}(t)\check{\mathbf{x}}(t)\} = 0$, (*b*) $\check{\check{\mathbf{x}}}(t) = -\mathbf{x}(t)$.

9-46 (*Stochastic resonance*) The input to the system

$$\mathbf{H}(s) = \frac{1}{s^2 + 2s + 5}$$

is a WSS process $\mathbf{x}(t)$ with $E\{\mathbf{x}^2(t)\} = 10$. Find $S_x(\omega)$ such that the average power $E\{\mathbf{y}^2(t)\}$ of the resulting output $\mathbf{y}(t)$ is maximum.

Hint: $|\mathbf{H}(j\omega)|$ is maximum for $\omega = \sqrt{3}$.

9-47 Show that if $R_x(\tau) = Ae^{j\omega_0\tau}$, then $R_{xy}(\tau) = Be^{j\omega_0\tau}$ for any $\mathbf{y}(t)$.

Hint: Use (9-180).

9-48 Given a system $H(\omega)$ with input $\mathbf{x}(t)$ and output $\mathbf{y}(t)$, show that (*a*) if $\mathbf{x}(t)$ is WSS and $R_{xx}(\tau) = e^{j\alpha\tau}$, then

$$R_{yx}(\tau) = e^{j\alpha\tau}H(\alpha) \qquad R_{yy}(\tau) = e^{j\alpha\tau}|H(\alpha)|^2$$

(*b*) If $R_{xx}(t_1, t_2) = e^{j(\alpha t_1 - \beta t_2)}$, then

$$R_{yx}(t_1, t_2) = e^{j(\alpha t_1 - \beta t_2)}H(\alpha) \qquad R_{yy}(t_1, t_2) = e^{j(\alpha t_1 - \beta t_2)}H(\alpha)H^*(\beta)$$

9-49 Show that if $S_{xx}(\omega)S_{yy}(\omega) \equiv 0$, then $S_{xy}(\omega) \equiv 0$.

9-50 Show that if $\mathbf{x}[n]$ is WSS and $R_x[1] = R_x[0]$, then $R_x[m] = R_x[0]$ for every m.

9-51 Show that if $R[m] = E\{\mathbf{x}[n+m]\mathbf{x}[n]\}$, then

$$R[0]R[2] > 2R^2[1] - R^2[0]$$

9-52 Given a random variable $\boldsymbol{\omega}$ with density $f(\omega)$ such that $f(\omega) = 0$ for $|\omega| > \pi$, we form the process $\mathbf{x}[n] = Ae^{jn\boldsymbol{\omega}}\pi$. Show that $S_x(\omega) = 2\pi A^2 f(\omega)$ for $|\omega| < \pi$.

9-53 (*a*) Find $E\{\mathbf{y}^2(t)\}$ if $\mathbf{y}(0) = \mathbf{y}'(0) = 0$ and

$$\mathbf{y}''(t) + 7\mathbf{y}'(t) + 10\mathbf{y}(t) = \mathbf{x}(t) \qquad R_x(\tau) = 5\delta(\tau)$$

(*b*) Find $E\{\mathbf{y}^2[n]\}$ if $\mathbf{y}[-1] = \mathbf{y}[-2] = 0$ and

$$8\mathbf{y}[n] - 6\mathbf{y}[n-1] + \mathbf{y}[n-2] = \mathbf{x}[n] \qquad R_x[m] = 5\delta[m]$$

9-54 The process $\mathbf{x}[n]$ is WSS with $R_{xx}[m] = 5\delta[m]$ and

$$\mathbf{y}[n] - 0.5\mathbf{y}[n-1] = \mathbf{x}[n] \qquad \text{(i)}$$

Find $E\{\mathbf{y}^2[n]\}$, $R_{xy}[m_1, m_2]$, $R_{yy}[m_1, m_2]$ (*a*) if (i) holds for all n, (*b*) if $\mathbf{y}[-1] = 0$ and (i) holds for $n \geq 0$.

9-55 Show that (*a*) if $R_x[m_1, m_2] = q[m_1]\delta[m_1 - m_2]$ and

$$\mathbf{s} = \sum_{n=0}^{N} a_n \mathbf{x}[n] \qquad \text{then} \quad E\{\mathbf{s}^2\} = \sum_{n=0}^{N} a_n^2 q[n]$$

(*b*) If $R_{xx}(t_1, t_2) = q(t_1)\delta(t_1 - t_2)$ and

$$\mathbf{s} = \int_0^T a(t)\mathbf{x}(t)\,dt \qquad \text{then} \quad E\{\mathbf{s}^2\} = \int_0^T a^2(t)q(t)\,dt$$

CHAPTER 10

RANDOM WALKS AND OTHER APPLICATIONS

10-1 RANDOM WALKS[1]

Consider a sequence of independent random variables that assume values $+1$ and -1 with probabilities p and $q = 1 - p$, respectively. A natural example is the sequence of Bernoulli trials $\mathbf{x}_1, \mathbf{x}_2, \ldots, \mathbf{x}_n, \ldots$ with probability of success equal to p in each trial, where $\mathbf{x}_k = +1$ if the kth trial results in a success and $\mathbf{x}_k = -1$ otherwise. Let $\mathbf{s}_n$ denote the partial sum

$$\mathbf{s}_n = \mathbf{x}_1 + \mathbf{x}_2 + \cdots + \mathbf{x}_n \qquad \mathbf{s}_0 = 0 \tag{10-1}$$

that represents the accumulated positive or negative excess at the nth trial. In a random walk model, the particle takes a unit step up or down at regular intervals, and $\mathbf{s}_n$ represents the position of the particle at the nth step (see Fig. 10-1). The random walk is said to be symmetric if $p = q = 1/2$ and unsymmetric if $p \neq q$. In the gambler's ruin problem discussed in Example 3-15, $\mathbf{s}_n$ represents the accumulated wealth of the player A at the nth stage. Many "real life" phenomena can be modeled quite faithfully using a random walk. The motion of gas molecules in a diffusion process, thermal noise phenomena, and the stock value variations of a particular stock are supposed to vary in consequence of successive collisions/occurrences of some sort of random impulses. In particular, this model will enable us to study the long-time behavior of a prolonged series of individual observations.

In this context, the following events and their probabilities are of special interest. In n successive steps, "return to the origin (or zero)" that represents the return of the

[1]The phrase *random walk* was first used by George Polya in his 1921 paper on that subject. (See *Random Walks of George Polya* by G. L. Alexanderson, published by *The Mathematical Association of America* (2000) for references.)

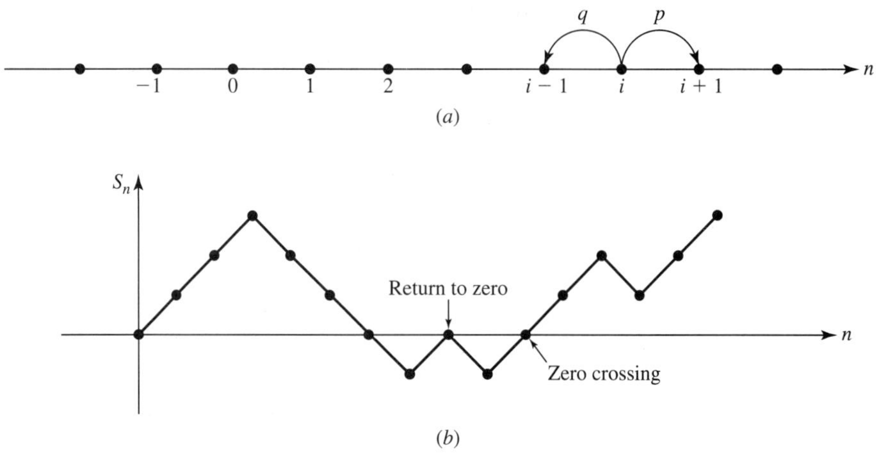

FIGURE 10-1
Random walk.

random walk to the starting point is a noteworthy event since the process starts all over again from that point onward. In particular, the events "the first return (or visit) to the origin," and more generally "the rth return to the origin," "waiting time for the first gain (first visit to $+1$)," "first passage through $r > 0$ (waiting time for rth gain)" are also of interest. In addition, the number of sign changes (zero crossings), the level of maxima and minima and their corresponding probabilities are also of great interest.

To compute the probabilities of these events, let $\{\mathbf{s}_n = r\}$ represent the event "at stage n, the particle is at the point r," and $p_{n,r}$ its probability. Thus

$$p_{n,r} \triangleq P\{\mathbf{s}_n = r\} = \binom{n}{k} p^k q^{n-k} \tag{10-2}$$

where k represents the number of successes in n trials and $n - k$ the number of failures. But the net gain

$$r = k - (n - k) = 2k - n \tag{10-3}$$

or $k = (n + r)/2$, so that

$$p_{n,r} = \binom{n}{(n+r)/2} p^{(n+r)/2} q^{(n-r)/2} \tag{10-4}$$

where the binomial coefficient is understood to be zero unless $(n + r)/2$ is an integer between 0 and n, both inclusive. Note that n and r must be therefore odd or even together.

Return to the origin. If the accumulated number of successes and failures are equal at stage n, then $\mathbf{s}_n = 0$, and the random walk has returned to the origin. In that case $r = 0$ in (10-3) or $n = 2k$ so that n is necessarily even, and the probability of return at the $2n$th trial is given by

$$P\{\mathbf{s}_{2n} = 0\} = \binom{2n}{n} (pq)^n \triangleq u_{2n} \tag{10-5}$$

with $u_0 = 1$. Alternatively

$$\begin{aligned}
u_{2n} &= \frac{(2n)!}{n!n!}(pq)^n \\
&= \frac{2n(2n-2)\cdots 4\cdot 2}{n!}\cdot\frac{(2n-1)(2n-3)\cdots 3\cdot 1}{n!}(pq)^n \\
&= \frac{2^n n!}{n!}\cdot\frac{2^n(-1)^n(-1/2)(-3/2)\cdots(-1/2-(n-1))}{n!}(pq)^n \\
&= (-1)^n\binom{-1/2}{n}(4pq)^n
\end{aligned} \qquad (10\text{-}6)$$

so that the moment generating function of the sequence $\{u_{2n}\}$ is given by

$$U(z) = \sum_{n=0}^{\infty} u_{2n}\, z^{2n} = \sum_{n=0}^{\infty}\binom{-1/2}{n}(-4pqz)^n = \frac{1}{\sqrt{1-4pqz^2}} \qquad (10\text{-}7)$$

Since $U(1) = \sum_{n=0}^{\infty} u_{2n} \neq 1$, the sequence $\{u_{2n}\}$ in (10-6) does not represent a probability distribution. In fact, for $p = q = 1/2$, we obtain $U(1) = \sum_{n=0}^{\infty} u_{2n} = \infty$, and from the second part of Borel-Cantelli lemma in (2-70) (see page 43), returns to equilibrium occur repeatedly or infinitely often.

The first return to origin. Among the returns to origin or equilibrium point, *the first return to the origin* commands special attention. A first return to zero occurs at stage $2n$ if the event

$$B_n = \{\mathbf{s}_1 \neq 0, \mathbf{s}_2 \neq 0, \ldots, \mathbf{s}_{2n-1} \neq 0, \mathbf{s}_{2n} = 0\} \qquad (10\text{-}8)$$

occurs. Let v_{2n} denote the probability of this event. Thus

$$v_{2n} = P(B_n) = P\{\mathbf{s}_1 \neq 0, \mathbf{s}_2 \neq 0, \ldots, \mathbf{s}_{2n-1} \neq 0, \mathbf{s}_{2n} = 0\} \qquad (10\text{-}9)$$

with $v_0 = 0$. The probabilities u_{2n} and v_{2n} can be related in a noteworthy manner. A visit to the origin at stage $2n$ is either the first return with probability v_{2n}, or the first return occurs at an earlier stage $2k < 2n$ with probability v_{2k}, and it is followed by an independent new return to zero after $2n - 2k$ stages with probability u_{2n-2k}, for $k = 1, 2, \ldots, n$. Since these events are mutually exclusive and exhaustive, we get the fundamental identity

$$u_{2n} = v_{2n} + v_{2n-2}u_2 + \cdots + v_2 u_{2n-2} = \sum_{k=1}^{n} v_{2k}\, u_{2n-2k} \qquad n \geq 1 \qquad (10\text{-}10)$$

We can use (10-10) to compute the moment generating function of the sequence $\{v_{2n}\}$. Since $u_0 = 1$, we get

$$\begin{aligned}
U(z) &= 1 + \sum_{n=1}^{\infty} u_{2n}\, z^{2n} = 1 + \sum_{n=1}^{\infty}\left(\sum_{k=1}^{n} v_{2k}\, u_{2n-2k}\right) z^{2n} \\
&= 1 + \sum_{m=0}^{\infty} u_{2m}\, z^{2m} \cdot \sum_{k=0}^{\infty} v_{2k}\, z^{2k} = 1 + U(z)V(z)
\end{aligned} \qquad (10\text{-}11)$$

or

$$U(z) = \frac{1}{1 - V(z)} \tag{10-12}$$

and hence

$$V(z) = \sum_{n=0}^{\infty} v_{2n}\, z^{2n} = 1 - \frac{1}{U(z)} = 1 - \sqrt{1 - 4pqz^2} \tag{10-13}$$

$$\begin{aligned} v_{2n} &= (-1)^{n-1}\binom{1/2}{n}(4pq)^n \\ &= \frac{(-1)^{n-1}(1/2)(-1/2)(-3/2)\cdots(3/2-n)}{n!}(4pq)^n \\ &= \frac{(2n-3)(2n-5)\cdots 3\cdot 1}{2^n\, n!}(4pq)^n \\ &= \frac{(2n-2)!}{2^{n-1}2^n \cdot n!\,(n-1)!}(4pq)^n \\ &= \frac{1}{2n-1}\binom{2n-1}{n}2(pq)^n \qquad n \geq 1 \end{aligned} \tag{10-14}$$

More importantly, we can use (10-13) to compute the probability that the particle will sooner or later return to the origin. Clearly in that case one of the mutually exclusive events B_2 or $B_4, \ldots$ must happen. Hence

$$\begin{aligned} P\left\{\begin{matrix}\textit{particle will ever} \\ \textit{return to the origin}\end{matrix}\right\} &= \sum_{n=0}^{\infty} P(B_n) = \sum_{n=0}^{\infty} v_{2n} \\ &= V(1) = 1 - \sqrt{1 - 4pq} \\ &= 1 - |p - q| = \begin{cases} 1 - |p-q| < 1 & p \neq q \\ 1 & p = q \end{cases} \end{aligned} \tag{10-15}$$

Thus if $p \neq q$, the probability that the particle never will return to the origin ($P(\mathbf{s}_{2k} \neq 0)$, $k \geq 1$) is $|p - q| \neq 0$ (finite), and *if* $p = q = 1/2$, then with probability 1 the particle will return to the origin. In the latter case, the return to origin is a certain event and $\{v_{2n}\}$ represents the probability distribution for the waiting time for the first return to origin. The expected value of this random variable is given by

$$\mu = V'(1) = \begin{cases} \dfrac{4pq}{|p-q|} & p \neq q \\ \infty & p = q \end{cases} \tag{10-16}$$

In gambling terminology this means that in a fair game with infinite resources on both sides, sooner or later one should be able to recover all losses, since return to the break even point is bound to happen. How long would it take to realize that break even point? From (10-16), the expected number of trials to the break even point in a fair game is infinite. Thus even in a fair game, a player with finite resources may never get to this point, let alone realize a positive net gain. For example, the probability that no break even occurs in 100 trials in a fair game is around 0.08. The mean value for the first return to the origin in a fair game is infinite, implying that the chance fluctuations in an individual prolonged coin tossing game are governed entirely differently from the familiar pattern of the normal distribution. From (10-7) and (10-13) we also obtain

$$1 - V(z) = (1 - 4pqz^2)U(z)$$

or

$$-v_{2n} = u_{2n} - 4pq\, u_{2n-2}$$

which gives the interesting identity

$$v_{2n} = 4pq\, u_{2n-2} - u_{2n} \tag{10-17}$$

In a fair game (symmetric random walk), $p = q = 1/2$, and (10-17) reduces to

$$v_{2n} = u_{2n-2} - u_{2n} \tag{10-18}$$

from which we also obtain the identity

$$\begin{aligned} u_{2n} &= v_{2n+2} + u_{2n+2} \\ &= v_{2n+2} + v_{2n+4} + v_{2n+6} + \cdots \end{aligned} \tag{10-19}$$

The right side of (10-19) represents the probability of the event *"the first return to origin occurs after* $2n$ *steps,"* which is the same as the event $\{\mathbf{s}_1 \neq 0, \mathbf{s}_2 \neq 0, \ldots, \mathbf{s}_{2n} \neq 0\}$. Thus (10-19) states that in a fair game the probability of $\{\mathbf{s}_{2n} = 0\}$ equals the probability that $\{\mathbf{s}_1, \mathbf{s}_2, \ldots, \mathbf{s}_{2n}\}$ are all different from zero. Thus in a symmetric random walk, we get the curious identity

$$P\{\mathbf{s}_1 \neq 0, \mathbf{s}_2 \neq 0, \ldots, \mathbf{s}_{2n} \neq 0\} = P\{\mathbf{s}_{2n} = 0\} = u_{2n} \tag{10-20}$$

Later returns to the origin. The event *"the first return to the origin"* naturally leads to the more general event *"rth return to the origin at 2nth trial."* Let $v_{2n}^{(r)}$ represent the probability of this cumulative event. Since the trials following the first return to zero form a probabilistic replica of the whole sequence, repeating the same arguments used in deriving (10-10), we get

$$v_{2n}^{(r)} = \sum_{k=0}^{n} v_{2k}\, v_{2n-2k}^{(r-1)} \tag{10-21}$$

This gives the generating function of $\{v_{2n}^{(r)}\}$ to be

$$\begin{aligned} V^{(r)}(z) &\triangleq \sum_{n=0}^{\infty} v_{2n}^{(r)} z^{2n} \\ &= \sum_{k=0}^{\infty} v_{2k} z^{2k} \sum_{m=0}^{\infty} v_{2m}^{(r-1)} z^{2m} \\ &= V(z)V^{(r-1)}(z) = V^r(z) \end{aligned} \tag{10-22}$$

Thus $V^{(r)}(z)$ is given by the rth power of $V(z)$. An easy calculation using (10-13) shows that $V^r(z)$ satisfies the identity

$$V^r(z) = 2V^{r-1}(z) - 4pqz^2\, V^{r-2}(z) \tag{10-23}$$

from which we get the recursive formula

$$v_{2n}^{(r)} = 2v_{2n}^{(r-1)} - 4pq\, v_{2n-2}^{(r-2)} \tag{10-24}$$

that starts with $v_{2n}^{(1)}$ given by (10-14). By induction, it follows that the probability for the desired event "rth return to zero at the $2n$th step" is given by

$$v_{2n}^{(r)} = \frac{r}{2n-r}\binom{2n-r}{n} 2^r (pq)^n \tag{10-25}$$

In the special case when $p = q = 1/2$, the probability of the rth return to origin at the $2n$th step is given by

$$v_{2n}^{(r)} = \frac{r}{2n-r}\binom{2n-r}{n} 2^{-(2n-r)} \tag{10-26}$$

Clearly $\xi_{r,N} = \sum_{2n \le N} v_{2n}^{(r)}$ gives the probability that the rth return to origin happens by the Nth step. When r is also large, we can use the De Moivre–Laplace theorem in (4-90) to approximate the above sum. Let $2n - r = m$ in (10-26) so that $n = (m+r)/2$ and using (4-90) we get

$$\begin{aligned}\binom{2n-r}{n} 2^{-(2n-r)} &= \binom{m}{\frac{m+r}{2}} (1/2)^{(m+r)/2} (1/2)^{(m-r)/2} \\ &\simeq \frac{1}{\sqrt{2\pi m/4}} e^{-[(m+r)/2-(m/2)]^2/(2m/4)} \\ &= \sqrt{\frac{2}{\pi m}} e^{-r^2/2m}\end{aligned} \tag{10-27}$$

so that

$$v_m^{(r)} = \sqrt{\frac{2}{\pi}} \frac{r}{m^{3/2}} e^{-r^2/2m} \tag{10-28}$$

It should be borne in mind that $\binom{m}{(m+r)/2}$ and $v_m(r)$ are nonzero only if m and r are of the same parity, and hence for a given r the above approximation is valid only if m together with r is either even or odd. For any $m = cr^2$, counting all such nonzero (alternate) terms, we get

$$\eta_c \triangleq \sum_{2n-r \le cr^2} v_{2n}^{(r)} = \sum_{m=0}^{cr^2} v_m^{(r)} = \frac{1}{2}\int_0^{cr^2} \sqrt{\frac{2}{\pi}} \frac{r}{m^{3/2}} e^{-r^2/2m}\, dm \tag{10-29}$$

For a given r, the integration in (10-29) should be performed only over those alternate terms involving m (of same parity as r) for which the approximation is valid, and the factor $1/2$ outside the integral accounts for the total number of such terms. From the exponent in (10-29), only those terms for which r^2/m is neither too small nor too large will play a significant role in the summation there. Substituting $r^2/m = x^2$ in (10-29), it simplifies into

$$\eta_c = \sum_{m=0}^{cr^2} v_m^{(r)} = 2\int_{1/\sqrt{c}}^{\infty} \frac{1}{\sqrt{2\pi}} e^{-x^2/2}\, dx \tag{10-30}$$

For fixed c, (10-30) gives the probability that r returns to origin occur before the instant $t = cr^2$. Thus for $c = 10$ we get the preceding probability to be 0.7566 so that in a fair game, to observe r returns to zero with 75% confidence, one must be prepared to play $10r^2$ games. In other words, the waiting time to the rth return to zero in a fair game increases as the square of r (or r increases only as $\sqrt{n}$). Since the random walk starts from scratch every time the particle returns to zero, the time to the rth return to zero can be interpreted as the sum of r independent waiting times, all of which add up to cr^2 in this case, and hence their average is proportional to r. To ensure the probability of occurrence for the rth return to zero close to one, from (10-30) c must be roughly of the order of r (large), so that the earlier average of r independent waiting times increases as $cr = r^2$, implying that at least one among the r waiting times is very likely to be of the same magnitude as the whole sum, namely, r^2. This result shows that some of the waiting times between successive returns to zero in long runs can be quite large. In other words, returns to zero are rare events in a long run. Since the number of zero-crossings does not exceed the number of returns to zero, it follows that the zero crossings are also rare events in long runs, a rather unexpected result that reveals the peculiar nature of chance fluctuations in random walks. For example, in a prolonged fair game "common sense" might tell us that the number of sign changes (or returns to zero) should increase in proportion to the duration of the game. In a game that lasts twice as long, one should expect twice as many returns to zeros and sign changes. However, the above analysis indicates that this is not true. The number of returns to zero in fact increases only as fast as the square root of the number of trials, and from (10-26)–(10-30) with $cr^2 = N$, the probability that fewer than $a\sqrt{N}$ returns to zero occur prior to N trials equals

$$p_a = 2\int_a^\infty \frac{1}{\sqrt{2\pi}} e^{-x^2/2}\,dx \qquad \text{as} \quad n \to \infty \tag{10-31}$$

Thus the median for the total number of returns to zero is about $\frac{2}{3}\sqrt{N}$, and hence in a run of length 10,000 it is as likely that there occur fewer than 66 returns to zero than more of them. There is only about a 30% probability that the total number of returns to zero will exceed 100 in 10,000 trials. More generally the probability p_a in (10-31) decreases as a increases. Thus regardless of the number of tosses, low values for the number of returns to zero are more probable in a long sequence of trials.

Figure 10-2 shows the results of simulation for 10,000 tosses of a fair coin in four separate trial runs (upper half of Fig. 10-2). Reversing the order of occurrences, we get the lower figures, which also represent legitimate random walks. Thus Fig. 10-2(a) and 10-2(b) refer to one set of forward and reversed pair of random walks. Starting from their respective origins the reversed random walks have their own returns to zero and changes of sign. The actual number of returns to zero and sign changes for each random walk are listed in Table 10-1. The low values for the actual number of returns to zero in these cases may appear as somewhat counterintuitive and surprisingly low only because of our intuition having been exposed to rather "commonsense" interpretations!

After studying the first and subsequent returns to the origin, the next significant event is where the first visit to +1 takes place at the nth step. In gambling terminology this refers to a net gain for the first time for player A.

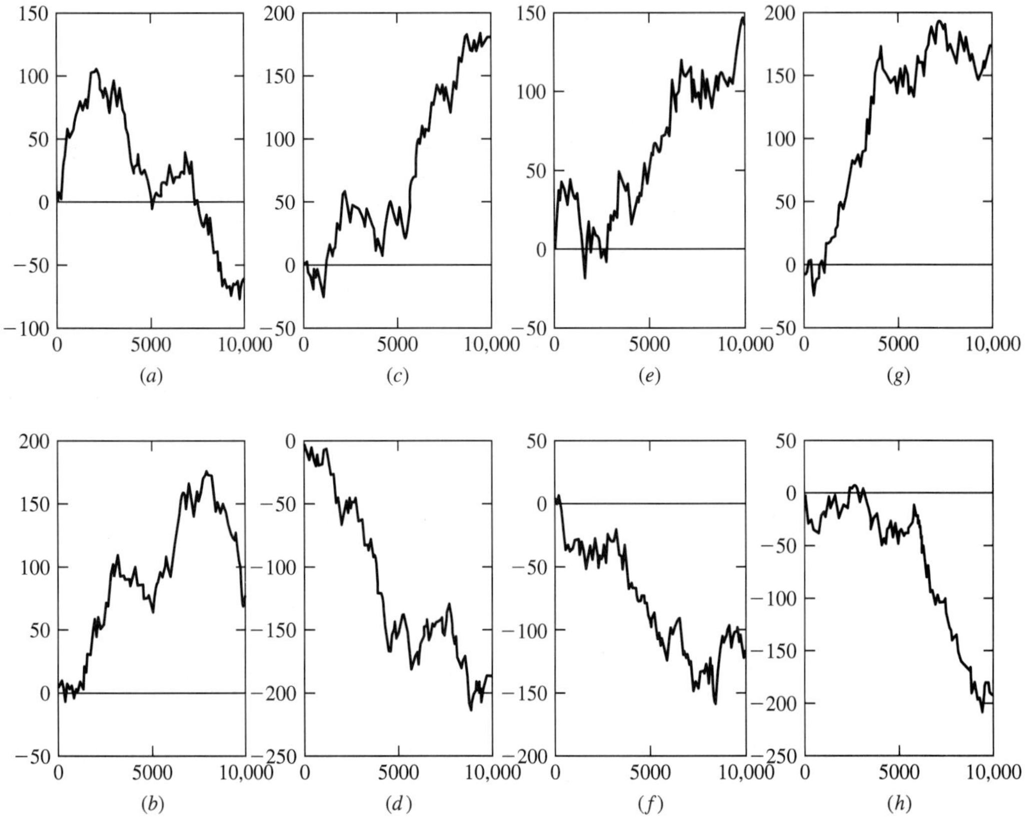

FIGURE 10-2
Random walks: The upper half represents four different random walks each generated from 10,000 tosses of a fair coin. The lower half refers to reversed random walks generated from the respective forward walk above.

TABLE 10-1
Total number of trials equal 10,000 in each run

Random walk		Number of returns to zero	Number of sign changes (zero-crossings)
I	*a*	60	28
	b	49	19
II	*c*	46	19
	d	40	25
III	*e*	47	29
	f	30	15
IV	*g*	35	21
	h	54	30

a, c, e, g: forward walks; b, d, f, h: reversed walks.

Waiting time for a gain. The event

$$\{\mathbf{s}_1 \le 0, \mathbf{s}_2 \le 0, \ldots, \mathbf{s}_{n-1} \le 0, \mathbf{s}_n = +1\} \tag{10-32}$$

signifies *the first visit to* $+1$, or *the first passage through* $+1$. Let ϕ_n represent the probability of the event in (10-32), with $\phi_0 = 0$. To start with $\phi_1 = p$, and if (10-32) holds for some $n > 1$, then $\mathbf{s}_1 = -1$ with probability q, and there must exist a smallest integer $k < n$ such that $\mathbf{s}_k = 0$, and it is followed by a new first passage though $+1$ in the remaining $(n-k)$ steps, $k = 1, 2, \ldots, n-2$. Now $P(\mathbf{s}_1 = -1) = q$,

$$\begin{aligned} P\{\mathbf{s}_1 = -1, \mathbf{s}_2 < 0, \ldots, \mathbf{s}_{k-1} < 0, \mathbf{s}_k = 0\} \\ = P\{\mathbf{s}_1 = 0, \mathbf{s}_2 \le 0, \ldots, \mathbf{s}_{k-1} \le 0, \mathbf{s}_k = +1\} = \phi_{k-1} \end{aligned} \tag{10-33}$$

and the probability of the third event mentioned above is ϕ_{n-k} so that from their independence, and summing over all possible k (mutually exclusive) events, we get

$$\phi_n = q(\phi_1\phi_{n-2} + \phi_2\phi_{n-3} + \cdots + \phi_{n-2}\phi_1) \qquad n > 1 \tag{10-34}$$

The corresponding generating function is given by

$$\begin{aligned} \Phi(z) = \sum_{n=1}^{\infty} \phi_n z^n &= pz + \sum_{n=2}^{\infty} q\left\{\sum_{k=1}^{n-1} \phi_k\phi_{n-k-1}\right\} z^n \\ &= pz + qz\sum_{m=1}^{\infty}\phi_m z^m \sum_{k=1}^{\infty}\phi_k z^k = pz + qz\Phi^2(z) \end{aligned} \tag{10-35}$$

Of the two roots of this quadratic equation, one of them is unbounded near $z = 0$, and the unique bounded solution of $\phi(z)$ is given by the second solution

$$\Phi(z) = \frac{1 - \sqrt{1 - 4pqz^2}}{2qz} \tag{10-36}$$

Using (10-13) we get

$$V(z) = 2qz\,\Phi(z) \tag{10-37}$$

so that from (10-14)

$$\phi_{2n-1} = \frac{v_{2n}}{2q} = \frac{(-1)^{n-1}}{2q}\binom{1/2}{n}(4pq)^n \qquad n \ge 1 \tag{10-38}$$

and $\phi_{2n} = 0$. From (10-36), we also get

$$\Phi(1) = \sum_{n=1}^{\infty} \phi_n = \frac{1 - \sqrt{1 - 4pq}}{2q} = \frac{1 - |p - q|}{2q} \tag{10-39}$$

so that

$$\sum_{n=1}^{\infty} \phi_n = \begin{cases} p/q & p < q \\ 1 & p \ge q \end{cases} \tag{10-40}$$

Thus if $p < q$, the probability that the accumulated gain $\mathbf{s}_n$ remains negative forever equals $(q - p)/q$. However, if $p \ge q$, this probability is zero, implying that sooner or later $\mathbf{s}_n$ will become positive with probability 1. In that case, ϕ_n represents the probability distribution of the waiting time for the first passage though $+1$, and from (10-37) its expected value is given by

$$\Phi'(1) = \frac{V'(1)}{2q} - \phi(1) = \left\{ \frac{1}{|p-q|} - 1 \right\} \frac{1}{2q} = \begin{cases} 1/(p-q) & p > q \\ \infty & p = q \\ p/q(q-p) & q > p \end{cases} \tag{10-41}$$

Once again in fair game, although a positive net gain is bound to occur, the number of trials preceding the first positive sum has infinite expectation. It could take an exceedingly long time for that event to happen. This leads us to the following question: What about the probability of obtaining a considerably large gain? How long would it take for that event to happen?

First passage through maxima. More generally, we can consider the event "*the first passage though $r > 0$ occurs at the nth step*," and denote its probability by $\phi_n^{(r)}$. Here r stands for the desired maximum gain. Since the trials following the first passage through $+1$ form a replica of the whole sequence, the waiting times between successive incremental first passages are independent, so that the waiting time for the first passage through a positive gain r is the sum of r independent random variables each with common distribution $\{\phi_n\}$. This gives the generating function of $\phi_n^{(r)}$ to be (see also (10-22))

$$\Phi^{(r)}(z) \triangleq \sum_{n=1}^{\infty} \phi_n^{(r)} z^r = \Phi^r(z) \tag{10-42}$$

Using (10-22) and (10-37), we get

$$V^{(r)}(z) = (2q)^r z^r \Phi^{(r)}(z) \tag{10-43}$$

and hence

$$v_{2n}^{(r)} = (2q)^r \phi_{2n-r}^{(r)} \tag{10-44}$$

or[2]

$$\phi_m^{(r)} = \frac{r}{m} \binom{m}{(m+r)/2} p^{(m+r)/2} q^{(m-r)/2} \tag{10-45}$$

where we have made use of (10-25). In the special case when $p = q = 1/2$, the probability for the first passage through r at the nth step is given by

$$\phi_n^{(r)} = \frac{r}{n} \binom{n}{(n+r)/2} 2^{-n} \tag{10-46}$$

Clearly, $\sum_{n=0}^{N} \phi_n(r)$ given the probability that the first passage through r occurs before the Nth step. To compute this probability for large values of N, once again we can proceed as in (10-27)–(10-29). Using (10-27), when n and r are of same parity

$$\phi_n^{(r)} \simeq \sqrt{\frac{2}{\pi}} \frac{r}{n^{3/2}} e^{-r^2/2n} \tag{10-47}$$

[2]The binomial coefficient in (10-45) is to be interpreted as zero if $(m+r)/2$ is not an integer.

and hence for $N = cr^2$, as before

$$\sum_{n=0}^{N} \phi_n^{(r)} \simeq \frac{1}{2}\int_0^{cr^2} \sqrt{\frac{2}{\pi}}\frac{r}{n^{3/2}}e^{-r^2/2n}\,dn = 2\int_{1/\sqrt{c}}^{\infty} \frac{1}{\sqrt{2\pi}}e^{-x^2/2}\,dx \qquad (10\text{-}48)$$

For fixed c, (10-48) gives the probability that the first passage through gain r occurs before the instant $t = cr^2$. Notice that (10-48) is identical to (10-30) and it enables us to conclude that in a fair game the two events "*r returns to origin occur before the instant t*" and "*the first passage through r occurs before the instant t*" have the same probability of occurrence as $t \to \infty$.

For $c = 10$, we get the above probability to be 0.7566 so that in a fair game of \$1 stakes, to secure a gain of \$100 with 75% probability of success, one must be prepared to play through $m = cr^2 = 100{,}000$ trials! If the stakes are increased to \$10, the same goal can be realized in 1000 trials with the same probability of success. Similarly to achieve a modest \$3 gain with 75% probability of success in a fair game, one need to play through 90 trials. On the other hand, referring back to the game of craps (Examples 3-16 and 3-17) from Table 3-3, a slightly better gain of \$4 can be realized there with 75% probability of success in about 67 trials ($a = \$16,\ b = \$4$ play). Even though the game of craps is slightly disadvantageous to the player compared to a fair game ($p = 0.492929$ versus $p = 0.5$), nevertheless it appears that for the same success rate a better return (\$4 versus \$3) is possible with the game of craps. How does one explain this apparent anomaly of an unfair game being more favorable? The somewhat higher return for the game of craps is due to the willingness of the player to lose a higher capital (lose \$16 with 25% probability) while aiming for a modest \$4 gain. In a fair game, the probability of losing \$16 in 90 trials is only 9.4%. The risk levels are different in these two games and the payoff is better at higher risk levels in a carefully chosen game.

The Wiener Process

To study the limiting behavior of the random walk as $n \to \infty$, let T represent the duration of a step. Then

$$\mathbf{x}(nT) = \mathbf{s}_n = \mathbf{x}_1 + \mathbf{x}_2 + \cdots + \mathbf{x}_n \qquad (10\text{-}49)$$

represents the random walk in (10-1). Let the modified step size be s, so that the independent random variables $\mathbf{x}_i$ can take values $\pm s$, with $E\{\mathbf{x}_i\} = 0$, $E\{\mathbf{x}_i^2\} = s^2$. From this it follows that

$$E\{\mathbf{x}(nT)\} = 0 \qquad E\{\mathbf{x}^2(nT)\} = ns^2 \qquad (10\text{-}50)$$

As we know, if n is large and k is in the $\sqrt{npq}$ vicinity of np, then

$$\binom{n}{k}p^k q^{n-k} \simeq \frac{1}{\sqrt{2\pi npq}}e^{-(k-np)^2/2npq}$$

From this and (10-2), with $p = q = 0.5$ and $m = 2k - n$ it follows that

$$P\{\mathbf{x}(nT) = ms\} \simeq \frac{1}{\sqrt{n\pi/2}}e^{-m^2/2n}$$

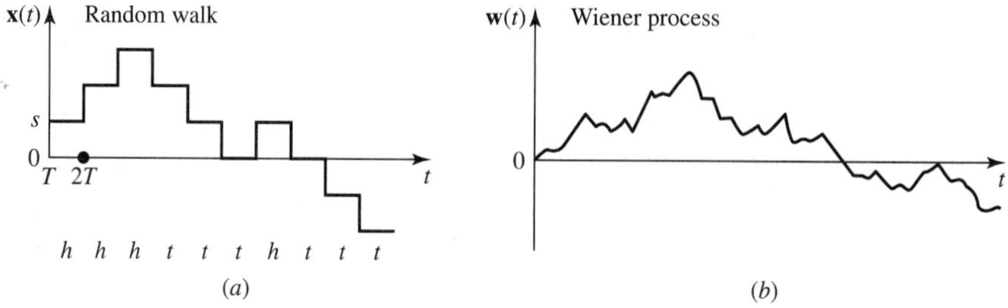

FIGURE 10-3

for m of the order of $\sqrt{n}$. Hence

$$P\{\mathbf{x}(t) \le ms\} \simeq \mathrm{G}(m/\sqrt{n}) \qquad nT - T < t \le nT \tag{10-51}$$

where $\mathrm{G}(x)$ is the $N(0, 1)$ distribution defined in (4-92).

Note that if $n_1 < n_2 \le n_3 < n_4$, then the increments $\mathbf{x}(n_4T) - \mathbf{x}(n_3T)$ and $\mathbf{x}(n_2T) - \mathbf{x}(n_1T)$ of $\mathbf{x}(t)$ are independent.

To examine the limiting form of the random walk as $n \to \infty$ or, equivalently, as $T \to 0$, note that

$$E\{\mathbf{x}^2(t)\} = ns^2 = \frac{ts^2}{T} \qquad t = nT$$

Hence, to obtain meaningful results, we shall assume that s tends to 0 as $\sqrt{T}$:

$$s^2 = \alpha T$$

The limit of $\mathbf{x}(t)$ as $T \to 0$ is then a continuous-state process (Fig. 10-3b)

$$\mathbf{w}(t) = \lim \mathbf{x}(t) \qquad T \to 0$$

known as the *Wiener process*.

We shall show that the first-order density $f(w, t)$ of $\mathbf{w}(t)$ is normal with zero mean and variance αt:

$$f(w, t) = \frac{1}{\sqrt{2\pi\alpha t}} e^{-w^2/2\alpha t} \tag{10-52}$$

Proof. If $w = ms$ and $t = nT$, then

$$\frac{m}{\sqrt{n}} = \frac{w/s}{\sqrt{t/T}} = \frac{w}{\sqrt{\alpha t}}$$

Inserting into (10-51), we conclude that

$$P\{\mathbf{w}(t) \le w\} = \mathrm{G}\left(\frac{w}{\sqrt{\alpha t}}\right)$$

and (10-52) results.

We show next that the autocorrelation of $\mathbf{w}(t)$ equals

$$R(t_1, t_2) = \alpha \min(t_1, t_2) \tag{10-53}$$

Indeed, if $t_1 < t_2$, then the difference $\mathbf{w}(t_2) - \mathbf{w}(t_1)$ is independent of $\mathbf{w}(t_1)$. Hence

$$E\{[\mathbf{w}(t_2) - \mathbf{w}(t_1)]\mathbf{w}(t_1)\} = E\{[\mathbf{w}(t_2) - \mathbf{w}(t_1)]\}E\{\mathbf{w}(t_1)\} = 0$$

This yields

$$E\{\mathbf{w}(t_1)\mathbf{w}(t_2)\} = E\{\mathbf{w}^2(t_1)\} = \frac{t_1 s^2}{T} = \alpha t_1$$

as in (10-53). The proof is similar if $t_1 > t_2$.

Note finally that if $t_1 < t_2 < t_3 < t_4$ then the increments $\mathbf{w}(t_4) - \mathbf{w}(t_3)$ and $\mathbf{w}(t_2) - \mathbf{w}(t_1)$ of $\mathbf{w}(t)$ are independent.

Generalized random walk. The random walk can be written as a sum

$$\mathbf{x}(t) = \sum_{k=1}^{n} \mathbf{c}_k U(t - kT) \qquad (n-1)T < t \le nT \tag{10-54}$$

where $\mathbf{c}_k$ is a sequence of i.i.d. random variables taking the values s and $-s$ with equal probability. In the generalized random walk, the random variables $\mathbf{c}_k$ take the values s and $-s$ with probability p and q, respectively. In this case,

$$E\{\mathbf{c}_k\} = (p-q)s \qquad E\{c_k^2\} = s^2 \qquad \sigma_{c_k}^2 = 4pqs^2$$

From this it follows that

$$E\{\mathbf{x}(t)\} = n(p-q)s \qquad \text{Var}\,\mathbf{x}(t) = 4npqs^2 \tag{10-55}$$

For large n, the process $\mathbf{x}(t)$ is nearly normal with

$$E\{\mathbf{x}(t)\} \simeq \frac{t}{T}(p-q)s \qquad \text{Var}\,\mathbf{x}(t) \simeq \frac{4t}{T}pqs^2 \tag{10-56}$$

Brownian Motion

The term *brownian motion* is used to describe the movement of a particle in a liquid, subjected to collisions and other forces. Macroscopically, the position $\mathbf{x}(t)$ of the particle can be modeled as a stochastic process satisfying a second-order differential equation:

$$m\mathbf{x}''(t) + f\mathbf{x}'(t) + c\mathbf{x}(t) = \mathbf{F}(t) \tag{10-57}$$

where $\mathbf{F}(t)$ is the collision force, m is the mass of the particle, f is the coefficient of friction, and $c\mathbf{x}(t)$ is an external force which we assume proportional to $\mathbf{x}(t)$. On a macroscopic scale, the process $\mathbf{F}(t)$ can be viewed as *normal* white noise with zero mean and power spectrum

$$S_F(\omega) = 2kTf \tag{10-58}$$

where T is the absolute temperature of the medium and $k = 1.37 \times 10^{-23}$ Joule-degrees is the Boltzmann constant. We shall determine the statistical properties of $\mathbf{x}(t)$ for various cases.

BOUND MOTION. We assume first that the restoring force $c\mathbf{x}(t)$ is different from 0. For sufficiently large t, the position $\mathbf{x}(t)$ of the particle approaches a stationary state with zero mean and power spectrum (see Example 9-27)

$$S_x(\omega) = \frac{2kTf}{(c - m\omega^2)^2 + f^2\omega^2} \tag{10-59}$$

To determine the statistical properties of $\mathbf{x}(t)$, it suffices to find its autocorrelation. We shall do so under the assumption that the roots of the equation $ms^2 + fs + c = 0$ are complex

$$s_{1,2} = -\alpha \pm j\beta \qquad \alpha = \frac{f}{2m} \qquad \alpha^2 + \beta^2 = \frac{c}{m}$$

Replacing b, c, and q in Example 9-27b by f/m, c/m, and $2kTf/m^2$, respectively, we obtain

$$R_x(\tau) = \frac{kT}{c} e^{-\alpha|\tau|} \left(\cos\beta\tau + \frac{\alpha}{\beta} \sin\beta|\tau| \right) \tag{10-60}$$

Thus, for a specific t, $\mathbf{x}(t)$ is a normal random variable with mean 0 and variance $R_x(0) = kT/c$. Hence its density equals

$$f_x(x) = \sqrt{\frac{c}{2\pi kT}} e^{-cx^2/2kT} \tag{10-61}$$

The conditional density of $\mathbf{x}(t)$ assuming $\mathbf{x}(t_0) = x_0$ is a normal curve with mean ax_0 and variance P, where (see Example 7-11)

$$a = \frac{R_x(\tau)}{R_x(0)} \qquad P = R_x(0)(1 - a^2) \qquad \tau = t - t_0$$

FREE MOTION. We say that a particle is in free motion if the restoring force is 0. In this case, (10-57) yields

$$m\mathbf{x}''(t) + f\mathbf{x}'(t) = \mathbf{F}(t) \tag{10-62}$$

The solution of this equation is not a stationary process. We shall express its properties in terms of the properties of the velocity $\mathbf{v}(t)$ of the particle. Since $\mathbf{v}(t) = \mathbf{x}'(t)$, (10-65) yields

$$m\mathbf{v}'(t) + f\mathbf{v}(t) = \mathbf{F}(t) \tag{10-63}$$

The steady state solution of this equation is a stationary process with

$$S_v(\omega) = \frac{2kTf}{m^2\omega^2 + f^2} \qquad R_v(\tau) = \frac{kT}{m} e^{-f|\tau|/m} \tag{10-64}$$

From the preceding, it follows that $\mathbf{v}(t)$ is a normal process with zero mean, variance kT/m, and density

$$f_v(v) = \sqrt{\frac{m}{2\pi kT}} e^{-mv^2/2kT} \tag{10-65}$$

The conditional density of $\mathbf{v}(t)$ assuming $\mathbf{v}(0) = v_0$ is normal with mean av_0 and variance P (see Example 7-11) where

$$a = \frac{R_v(t)}{R_v(0)} = e^{-ft/m} \qquad P = \frac{kT}{m}(1 - a^2) = \frac{kT}{m}(1 - e^{-2ft/m})$$

In physics, (10-63) is called the *Langevin equation,* its solution the *Ornstein–Uhlenbeck* process, and its spectrum *Lorenzian.*

The position $\mathbf{x}(t)$ of the particle is the integral of its velocity:

$$\mathbf{x}(t) = \int_0^t \mathbf{v}(\alpha)\,d\alpha \tag{10-66}$$

From this and (9-11) it follows that

$$E\{\mathbf{x}^2(t)\} = \int_0^t\int_0^t R_v(\alpha-\beta)\,d\alpha\,d\beta = \frac{kT}{m}\int_0^t\int_0^t e^{-f|\alpha-\beta|/m}\,d\alpha\,d\beta$$

Hence

$$E\{\mathbf{x}^2(t)\} = \frac{2kT}{f}\left(t - \frac{m}{f} + \frac{m}{f}e^{-ft/m}\right) \tag{10-67}$$

Thus, the position of a particle in free motion is a nonstationary normal process with zero mean and variance the right side of (10-67).

For $t \gg m/f$, (10-67) yields

$$E\{\mathbf{x}^2(t)\} \simeq \frac{2kT}{f}t = 2D^2 t \qquad D^2 \equiv \frac{kT}{f} \tag{10-68}$$

The parameter D is the *diffusion constant.* This result will be presently rederived.

THE WIENER PROCESS. We now assume that the acceleration term $m\mathbf{x}''(t)$ of a particle in free motion is small compared to the friction term $f\mathbf{x}'(t)$; this is the case if $f \gg m/t$. Neglecting the term $m\mathbf{x}''(t)$ in (10-62), we conclude that

$$f\mathbf{x}'(t) = \mathbf{F}(t) \qquad \mathbf{x}(t) = \frac{1}{f}\int_0^t \mathbf{F}(\alpha)\,d\alpha$$

Because $\mathbf{F}(t)$ is white noise with spectrum $2kTf$, it follows from (9-45) with $\mathbf{v}(t) = \mathbf{F}(t)/f$ and $q(t) = 2kT/f$ that

$$E\{\mathbf{x}^2(t)\} = \frac{2kT}{f}t = \alpha t \qquad \alpha \equiv \frac{2kT}{f} = 2D^2$$

Thus, $\mathbf{x}(t)$ is a nonstationary normal process with density

$$f_{x(t)}(x) = \frac{1}{\sqrt{2\pi\alpha t}}e^{-x^2/2\alpha t}$$

We maintain that it is also a process with independent increments. Because it is normal, it suffices to show that it is a process with orthogonal increments, that is

$$E\{[\mathbf{x}(t_2) - \mathbf{x}(t_1)][\mathbf{x}(t_4) - \mathbf{x}(t_3)]\} = 0 \tag{10-69}$$

for $t_1 < t_2 < t_3 < t_4$. This follows from the fact that $\mathbf{x}(t_i) - \mathbf{x}(t_j)$ depends only on the values of $\mathbf{F}(t)$ in the interval (t_i, t_j) and $\mathbf{F}(t)$ is white noise. Using this, we shall show that

$$R_x(t_1, t_2) = \alpha \min(t_1, t_2) \tag{10-70}$$

To do so, we observe from (10-69) that if $t_1 < t_2$, then

$$E\{\mathbf{x}(t_1)\mathbf{x}(t_2)\} = E\{\mathbf{x}(t_1)[\mathbf{x}(t_2) - \mathbf{x}(t_1) + \mathbf{x}(t_1)]\} = E\{\mathbf{x}^2(t_1)\} = \alpha t_1$$

and (10-70) results. Thus the position of a particle in free motion with negligible acceleration has the following properties:

It is normal with zero mean, variance αt and autocorrelation $\alpha \min(t_1, t_2)$. It is a process with independent increments.

A process with these properties is called the *Wiener process*. As we have seen, it is the limiting form of the position of a particle in free motion as $t \to \infty$; it is also the limiting form of the random walk process as $n \to \infty$.

We note finally that the conditional density of $\mathbf{x}(t)$ assuming $\mathbf{x}(t_0) = x_0$ is normal with mean ax_0 and variance P, where (see Example 7-11)

$$a = \frac{R_x(t, t_0)}{R_x(t_0, t_0)} = 1 \qquad P = R(t, t) - aR(t, t_0) = \alpha t - \alpha t_0$$

Hence

$$f_{x(t)}(x \mid \mathbf{x}(t_0) = x_0) = \frac{1}{\sqrt{2\pi\alpha(t - t_0)}} e^{-(x-x_0)^2/2\alpha(t-t_0)} \tag{10-71}$$

Diffusion equations. The right side of (10-71) is a function depending on the four parameters x, x_0, t, and t_0. Denoting this function by $\pi(x, x_0; t, t_0)$, we conclude by repeated differentiation that

$$\frac{\partial \pi}{\partial t} = D^2 \frac{\partial^2 \pi}{\partial x^2} \qquad \frac{\partial \pi}{\partial t_0} = -D^2 \frac{\partial^2 \pi}{\partial x_0^2} \tag{10-72}$$

where $D^2 = \alpha/2$. These equations are called *diffusion equations*.

Thermal Noise

Thermal noise is the distribution of voltages and currents in a network due to the thermal electron agitation. In the following, we discuss the statistical properties of thermal noise ignoring the underlying physics. The analysis is based on a model consisting of noiseless reactive elements and noisy resistors.

A noisy resistor is modeled by a noiseless resistor R in series with a voltage source $\mathbf{n}_e(t)$ or in parallel with a current source $\mathbf{n}_i(t) = \mathbf{n}_e(t)/R$ as in Fig. 10-4. It is assumed that $\mathbf{n}_e(t)$ is a normal process with zero mean and flat spectrum

$$S_{n_e}(\omega) = 2kTR \qquad S_{n_i}(\omega) = \frac{S_{n_e}(\omega)}{R^2} = 2kTG \tag{10-73}$$

where k is the Boltzmann constant, T is the absolute temperature of the resistor, and $G = 1/R$ is its conductance. Furthermore, the noise sources of the various network resistors are mutually independent processes. Note the similarity between the spectrum (10-73) of thermal noise and the spectrum (10-58) of the collision forces in brownian motion.

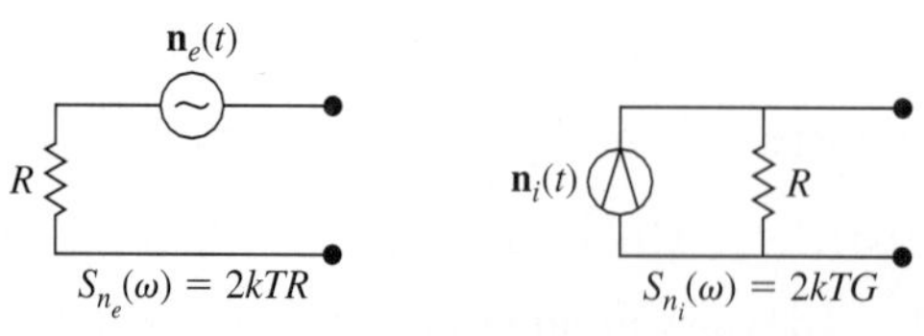

FIGURE 10-4

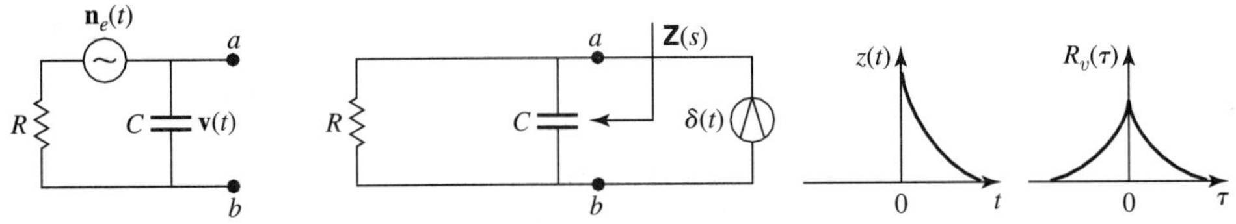

FIGURE 10-5

Using Fig. 10-5 and the properties of linear systems, we shall derive the spectral properties of general network responses starting with an example.

EXAMPLE 10-1

▶ The circuit of Fig. 10-5 consists of a resistor R and a capacitor C. We shall determine the spectrum of the voltage $\mathbf{v}(t)$ across the capacitor due to thermal noise.

The voltage $\mathbf{v}(t)$ can be considered as the output of a system with input the noise voltage $\mathbf{n}_e(t)$ and system function

$$\mathbf{H}(s) = \frac{1}{1 + RCs}$$

Applying (9-149), we obtain

$$S_v(\omega) = S_{n_e}(\omega)|H(\omega)|^2 = \frac{2kTR}{1 + \omega^2 R^2 C^2}$$
$$R_v(\tau) = \frac{kT}{C} e^{-|\tau|/RC} \tag{10-74}$$

The following consequences are illustrations of Nyquist's theorem to be discussed presently: We denote by $\mathbf{Z}(s)$ the impedance across the terminals a and b and by $z(t)$ its inverse transform

$$\mathbf{Z}(s) = \frac{R}{1 + RCs} \qquad z(t) = \frac{1}{C} e^{-t/RC} U(t)$$

The function $z(t)$ is the voltage across C due to an impulse current $\delta(t)$ (Fig. 10-5). Comparing with (10-74), we obtain

$$S_v(\omega) = 2kT \operatorname{Re} \mathbf{Z}(j\omega) \qquad \operatorname{Re} \mathbf{Z}(j\omega) = \frac{R}{1 + \omega^2 R^2 C^2}$$
$$R_v(\tau) = kTz(\tau) \qquad \tau > 0 \qquad R_v(0) = kTz(0^+)$$
$$E\{\mathbf{v}^2(t)\} = R_v(0) = \frac{kT}{C} \qquad \frac{1}{C} = \lim_{\omega \to \infty} j\omega \mathbf{Z}(j\omega)$$ ◀

Given a passive, reciprocal network, we denote by $\mathbf{v}(t)$ the voltage across two arbitrary terminals a and b and by $\mathbf{Z}(s)$ the impedance from a to b (Fig. 10-6).

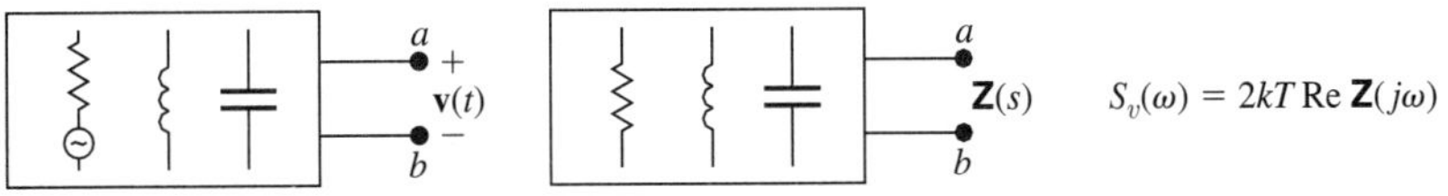

FIGURE 10-6

$$S_{n_i}(\omega) = \frac{2kT}{R} \qquad H(\omega) = \frac{V(\omega)}{I(\omega)}, \qquad \text{Re } \mathbf{Z}(j\omega) = \frac{|H(\omega)|^2}{R}$$

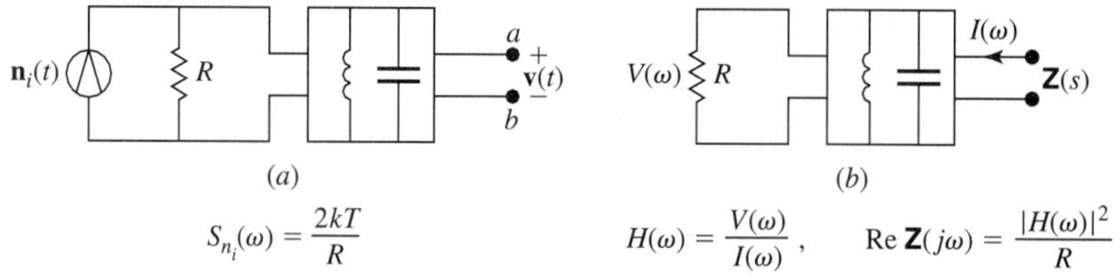

FIGURE 10-7

THEOREM 10-1

NYQUIST THEOREM

▶ The power spectrum of $\mathbf{v}(t)$ equals

$$S_v(\omega) = 2kT \text{ Re } \mathbf{Z}(j\omega) \qquad (10\text{-}75)$$

Proof. We shall assume that there is only one resistor in the network. The general case can be established similarly if we use the independence of the noise sources. The resistor is represented by a noiseless resistor in parallel with a current source $\mathbf{n}_i(t)$ and the remaining network contains only reactive elements (Fig. 10-7a). Thus $\mathbf{v}(t)$ is the output of a system with input $\mathbf{n}_i(t)$ and system function $H(\omega)$. From the reciprocity theorem it follows that $H(\omega) = V(\omega)/I(\omega)$ where $I(\omega)$ is the amplitude of a sine wave from a to b (Fig. 10-7b) and $V(\omega)$ is the amplitude of the voltage across R. The input power equals $|I(\omega)|^2 \text{ Re } \mathbf{Z}(j\omega)$ and the power delivered to the resistance equals $|V(\omega)|^2/R$. Since the connecting network is lossless by assumption, we conclude that

$$|I(\omega)|^2 \text{ Re } \mathbf{Z}(j\omega) = \frac{|V(\omega)|^2}{R}$$

Hence

$$|H(\omega)|^2 = \frac{|V(\omega)|^2}{|I(\omega)|^2} = R \text{ Re } \mathbf{Z}(j\omega)$$

and (10-75) results because

$$S_v(\omega) = S_{n_i}(\omega)|H(\omega)|^2 \qquad S_{n_i}(\omega) = \frac{2kT}{R} \qquad \blacktriangleleft$$

COROLLARY 1

▶ The autocorrelation of $\mathbf{v}(t)$ equals

$$R_v(\tau) = kT z(\tau) \qquad \tau > 0 \qquad (10\text{-}76)$$

where $z(t)$ is the inverse transform of $\mathbf{Z}(s)$.

Proof. Since $\mathbf{Z}(-j\omega) = \mathbf{Z}^*(j\omega)$, it follows from (10-75) that

$$S_v(\omega) = kT[\mathbf{Z}(j\omega) + \mathbf{Z}(-j\omega)]$$

and (10-76) results because the inverse of $\mathbf{Z}(-j\omega)$ equals $z(-t)$ and $z(-t) = 0$ for $t > 0$. ◀

COROLLARY 2

▶ The average power of $\mathbf{v}(t)$ equals

$$E\{\mathbf{v}^2(t)\} = \frac{kT}{C} \qquad \text{where} \qquad \frac{1}{C} = \lim_{\omega \to \infty} j\omega \mathbf{Z}(j\omega) \qquad (10\text{-}77)$$

where C is the input capacity.

Proof. As we know (initial value theorem)

$$z(0^+) = \lim s\mathbf{Z}(s) \qquad s \to \infty$$

and (10-77) follows from (10-76) because

$$E\{\mathbf{v}^2(t)\} = R_v(0) = kTz(0^+)$$

◀

Currents. From Thévenin's theorem it follows that, terminally, a noisy network is equivalent to a noiseless network with impedance $\mathbf{Z}(s)$ in series with a voltage source $\mathbf{v}(t)$. The power spectrum $S_v(\omega)$ of $\mathbf{v}(t)$ is the right side of (10-75), and it leads to this version of Nyquist's theorem:

The power spectrum of the short-circuit current $\mathbf{i}(t)$ from a to b due to thermal noise equals

$$S_i(\omega) = 2kT \text{ Re } \mathbf{Y}(j\omega) \qquad \mathbf{Y}(s)\frac{1}{\mathbf{Z}(s)} \tag{10-78}$$

Proof. From Thévenin's theorem it follows that

$$S_i(\omega) = S_v(\omega)|\mathbf{Y}(j\omega)|^2 = \frac{2kT \text{ Re } \mathbf{Z}(j\omega)}{|\mathbf{Z}(j\omega)|^2}$$

and (10-78) results.

The current version of the corollaries is left as an exercise.

10-2 POISSON POINTS AND SHOT NOISE

Given a set of Poisson points $\mathbf{t}_i$ and a fixed point t_0, we form the random variable $\mathbf{z} = \mathbf{t}_1 - t_0$, where $\mathbf{t}_1$ is the first random point to the right of t_0 (Fig. 10-8). We shall show that $\mathbf{z}$ has an exponential distribution:

$$f_z(z) = \lambda e^{-\lambda z} \qquad F_z(z) = 1 - e^{-\lambda z} \qquad z > 0 \tag{10-79}$$

Proof. For a given $z > 0$, the function $F_z(z)$ equals the probability of the event $\{\mathbf{z} \le z\}$. This event occurs if $\mathbf{t}_1 < t_0 + z$, that is, if there is at least one random point in the interval $(t_0, t_0 + z)$. Hence

$$F_z(z) = P\{\mathbf{z} \le z\} = P\{\mathbf{n}(t_0, t_0 + z) > 0\} = 1 - P\{\mathbf{n}(t_0, t_0 + z) = 0\}$$

and (10-79) results because the probability that there are no points in the interval $(t_0, t_0 + z)$ equals $e^{-\lambda z}$.

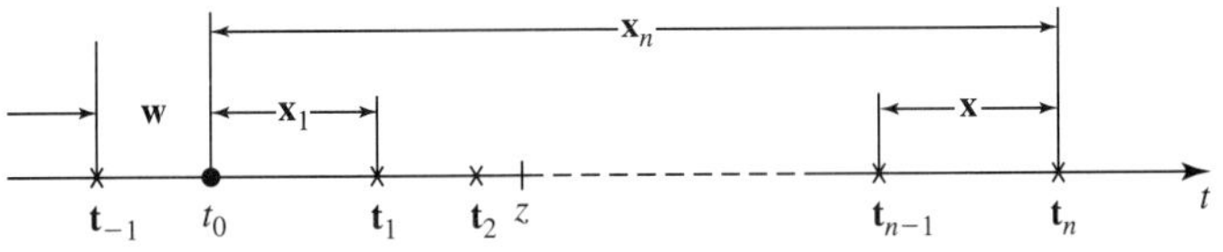

FIGURE 10-8
Poisson points.

We can show similarly that if $\mathbf{w} = t_0 - \mathbf{t}_{-1}$ is the distance from t_0 to the first point $\mathbf{t}_{-1}$, to the left of t_0 then

$$f_w(w) = \lambda e^{-\lambda w} \qquad F_w(w) = 1 - e^{-\lambda w} \qquad w > 0 \tag{10-80}$$

We shall now show that the distance $\mathbf{x}_n = \mathbf{t}_n - t_0$ from t_0 the nth random point $\mathbf{t}_n$ to the right of t_0 (Fig. 10-8) has a gamma distribution:

$$f_n(x) = \frac{\lambda^n}{(n-1)!} x^{n-1} e^{-\lambda x} \qquad x > 0 \tag{10-81}$$

Proof. The event $\{\mathbf{x}_n \le x\}$ occurs if there are at least n points in the interval $(t_0, t_0 + x)$. Hence

$$F_n(x) = P\{\mathbf{x}_n \le x\} = 1 - P\{\mathbf{n}(t_0, t_0 + x) < n\} = 1 - \sum_{k=0}^{n-1} \frac{(\lambda x)^k}{k!} e^{-\lambda x}$$

Differentiating, we obtain (10-81).

Distance between random points. We show next that the distance

$$\mathbf{x} = \mathbf{x}_n - \mathbf{x}_{n-1} = \mathbf{t}_n - \mathbf{t}_{n-1}$$

between two consecutive points $\mathbf{t}_{n-1}$ and $\mathbf{t}_n$ has an exponential distribution:

$$f_x(x) = \lambda e^{-\lambda x} \tag{10-82}$$

Proof. From (10-81) and (5-106) it follows that the moment function of $\mathbf{x}_n$ equals

$$\mathbf{\Phi}_n(s) = \frac{\lambda^n}{(\lambda - s)^n} \tag{10-83}$$

Furthermore, the random variables $\mathbf{x}$ and $\mathbf{x}_{n-1}$ are independent and $\mathbf{x}_n = \mathbf{x} + \mathbf{x}_{n-1}$. Hence, if $\mathbf{\Phi}_x(s)$ is the moment function of $\mathbf{x}$, then

$$\mathbf{\Phi}_n(s) = \mathbf{\Phi}_x(s)\mathbf{\Phi}_{n-1}(s)$$

Comparing with (10-83), we obtain $\mathbf{\Phi}_x(s) = \lambda/(\lambda - s)$ and (10-82) results.

An apparent paradox. We should stress that the notion of the "distance $\mathbf{x}$ between two consecutive points of a point process" is ambiguous. In Fig. 10-8, we interpreted $\mathbf{x}$ as the distance between $\mathbf{t}_{n-1}$ and $\mathbf{t}_n$, where $\mathbf{t}_n$ was the nth random point to the right of some *fixed* point t_0. This interpretation led to the conclusion that the density of $\mathbf{x}$ is an exponential as in (10-82). The same density is obtained if we interpret $\mathbf{x}$ as the distance between consecutive points to the left of t_0. Suppose, however, that $\mathbf{x}$ is interpreted as follows:

Given a fixed point t_a, we denote by $\mathbf{t}_l$ and $\mathbf{t}_r$ the random points nearest to t_a on its left and right, respectively (Fig. 10-9a). We maintain that the density of the distance $\mathbf{x} = \mathbf{t}_r - \mathbf{t}_l$ between these two points equals

$$f(x) = \lambda^2 x e^{-\lambda x} \tag{10-84}$$

Indeed the random variables

$$\mathbf{x}_l = \mathbf{t}_a - \mathbf{t}_l \qquad \text{and} \qquad \mathbf{x}_r = \mathbf{t}_r - \mathbf{t}_a$$

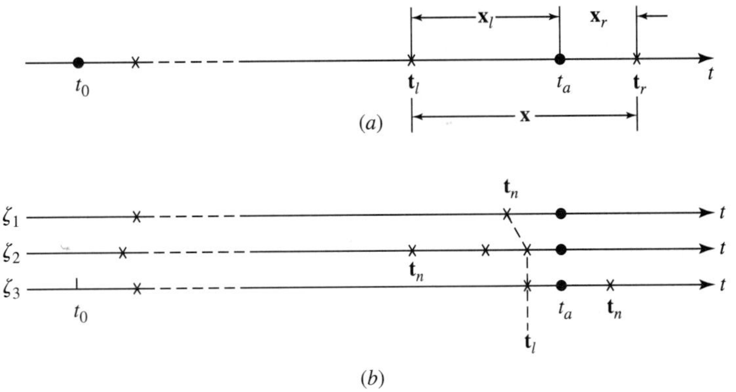

FIGURE 10-9

are independent with exponential density as in (10-79); furthermore, $\mathbf{x} = \mathbf{x}_r + \mathbf{x}_l$. This yields (10-84) because the convolution of two exponentials is the density in (10-84).

Thus, although $\mathbf{x}$ is again the "distance between two consecutive points," its density is not an exponential. This apparent paradox is a consequence of the ambiguity in the specification of the identity of random points. Suppose, for example, that we identify the points $\mathbf{t}_i$ by their order i, where the count starts from some fixed point t_0, and we observe that in one particular realization of the point process, the point $\mathbf{t}_l$, defined as above, equals $\mathbf{t}_n$. In other realizations of the process, the random variables $\mathbf{t}_l$ might equal some other point in this identification (Fig. 10-9*b*). The same argument shows that the point $\mathbf{t}_r$ does not coincide with the ordered point $\mathbf{t}_{n+1}$ for all realizations. Hence we should not expect that the random variable $\mathbf{x} = \mathbf{t}_r - \mathbf{t}_l$ has the same density as the random variable $\mathbf{t}_{n+1} - \mathbf{t}_n$.

CONSTRUCTIVE DEFINITION. Given a sequence $\mathbf{w}_n$ of positive i.i.d. (independent, identically distributed) random variables with density

$$f(w) = \lambda e^{-\lambda w} \qquad w > 0 \tag{10-85}$$

we form a set of points $\mathbf{t}_n$ as in Fig. 10-10*a*, where $t = 0$ is an arbitrary origin and

$$\mathbf{t}_n = \mathbf{w}_1 + \mathbf{w}_2 + \cdots + \mathbf{w}_n \tag{10-86}$$

We maintain that the points so formed are Poisson distributed with parameter λ.

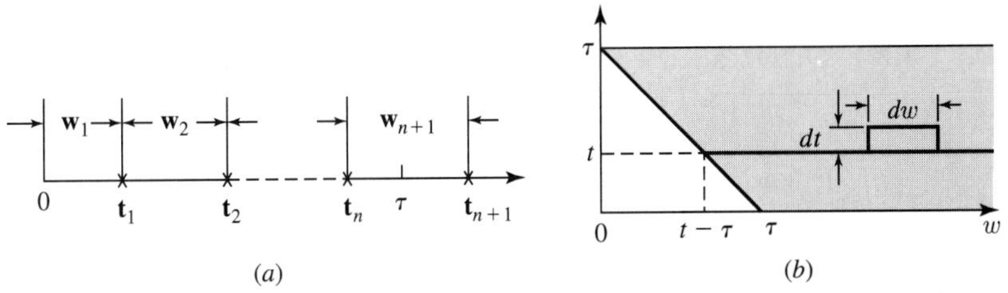

FIGURE 10-10

Proof. From the independence of the random variables $\mathbf{w}_n$, it follows that the random variables $\mathbf{t}_n$ and $\mathbf{w}_{n+1}$ are independent, the density $f_n(t)$ of $\mathbf{t}_n$ is given by (10-81)

$$f_n(t) = \frac{\lambda^n}{(n-1)!} t^{n-1} e^{-\lambda t} \tag{10-87}$$

and the joint density of $\mathbf{t}_n$ and $\mathbf{w}_{n+1}$ equals the product $f_n(t) f(w)$. If $\mathbf{t}_n < \tau$ and $\mathbf{t}_{n+1} = \mathbf{t}_n + \mathbf{w}_{n+1} > \tau$ then there are exactly n points in the interval $(0, \tau)$. As we see from Fig. 10-10b, the probability of this event equals

$$\int_0^\tau \int_{\tau-t}^\infty \lambda e^{-\lambda w} \frac{\lambda^n}{(n-1)!} t^{n-1} e^{-\lambda t}\, dw\, dt$$

$$= \int_0^\tau e^{-\lambda(t-\tau)} \frac{\lambda^n}{(n-1)!} t^{n-1} e^{-\lambda t}\, dt = e^{-\lambda t} \frac{(\lambda\tau)^n}{n!}$$

Thus the points $\mathbf{t}_n$ so constructed have property P_1. We can show similarly that they have also property P_2.

POISSON PROCESSES AND GEOMETRIC DISTRIBUTION

▶ Let $\mathbf{x}(t) \sim P(\lambda t)$ and $\mathbf{y}(t) \sim P(\mu t)$ represent two independent Poisson processes, and $\mathbf{n}$ the number of occurrences of $\mathbf{x}(t)$ between *any* two successive occurrences of $\mathbf{y}(t)$. Then with $\mathbf{z}$ representing the random interval between two successive occurrences of $\mathbf{y}(t)$, we have

$$P\{\mathbf{n} = k \mid \mathbf{z} = t\} = e^{-\lambda t} \frac{(\lambda t)^k}{k!}$$

so that

$$P\{\mathbf{n} = k\} = \int_0^\infty P\{\mathbf{n} = k \mid \mathbf{z} = t\}\, f_z(t)\, dt \tag{10-88}$$

But from, (10-82) the interarrival duration $\mathbf{z}$ is exponentially distributed with parameter μ so that $f_z(t) = \mu e^{-\mu t}$, $t \geq 0$, and substituting this into (10-88), we get

$$P\{\mathbf{n} = k\} = \int_0^\infty e^{-\lambda t} \frac{(\lambda t)^k}{k!} \mu\, e^{-\mu t}\, dt$$

$$= \mu\lambda^k \int_0^\infty \frac{t^k}{k!} e^{-(\lambda+\mu)t}\, dt = \frac{\mu}{\lambda+\mu} \left(\frac{\lambda}{\lambda+\mu}\right)^k \int_0^\infty \frac{x^k}{k!} e^{-x}\, dx$$

$$= \frac{\mu}{\lambda+\mu} \left(\frac{\lambda}{\lambda+\mu}\right)^k \qquad k = 0, 1, 2, \ldots \tag{10-89}$$

Thus the number of occurrences (count) of a Poisson process between *any* two successive occurrences of another independent Poisson process has a geometric distribution. It can be shown that counts corresponding to different interarrival times are independent geometric random variables. For example, if $\mathbf{x}(t)$ and $\mathbf{y}(t)$ represent the arrival and departure Poisson processes at a counter, then from (10-89) the number of arrivals between any two successive departures has a geometric distribution. Similarly the number of departures between any two arrivals is also a geometric random variable. ◀

POISSON PROCESSES AND ERLANG PROCESSES

▶ Suppose every kth outcome of a Poisson process $\mathbf{x}(\mathbf{t})$ is systematically tagged to generate a new process $\mathbf{y}(t)$. Then

$$P\{\mathbf{y}(t) = n\} = P\{nk \le \mathbf{x}(t) \le (n+1)k - 1\}$$
$$= \sum_{r=nk}^{(n+1)k-1} e^{-\lambda t} \frac{(\lambda t)^r}{r!} \qquad n \ge 0 \tag{10-90}$$

Using (10-86)–(10-87) and the definition of $\mathbf{y}(t)$, the interarrival time between any two successive occurrences of $\mathbf{y}(t)$ is a gamma random variable. If $\lambda = k\mu$, then the interarrival time represents an Erlang-k random variable and $\mathbf{y}(t)$ is an Erlang-k process. (See also (4-38).)

Interestingly from (9-25), a random selection of a Poisson process yields another Poisson process, while a systematic selection from a Poisson process as above results in an Erlang-k process. For example, suppose Poisson arrivals at a main counter are immediately redirected sequentially to k service counters such that each counter gets every kth customer. The interarrival times at the service counters in that case will follow independent Erlang-k distributions, whereas a random assignment at the main counter would have preserved the exponential nature of the interarrival times at the service counters (why?). ◀

POISSON POINTS REDEFINED. Poisson points are realistic models for a large class of point processes: photon counts, electron emissions, telephone calls, data communications, visits to a doctor, arrivals at a park. The reason is that in these and other applications, the properties of the underlying points can be derived from certain general conditions that lead to Poisson distributions. As we show next, these conditions can be stated in a variety of forms that are equivalent to the two conditions used in Sec. 4-5 to specify random Poisson points (see page 118).

I. If we place at random N points in an interval of length T where $N \gg 1$, then the resulting point process is nearly Poisson with parameter N/T. This is exact in the limit as N and T tend to ∞ [see (4-117)].

II. If the distances $\mathbf{w}_n$ between two consecutive points $\mathbf{t}_{n-1}$ and $\mathbf{t}_n$ of a point process are independent and exponentially distributed, as in (10-85), then this process is Poisson. (See also (9-28).)

This can be phrased in an equivalent form: If the distance $\mathbf{w}$ from an arbitrary point t_0 to the next point of a point process is a random variable whose density does not depend on the choice of t_0, then the process is Poisson. The reason for this equivalence is that this assumption leads to the conclusion that

$$f(w \mid \mathbf{w} \ge t_0) = f(w - t_0) \tag{10-91}$$

and the only function satisfying (10-91) is an exponential (see Example 6-43). In queueing theory, the above is called the *Markov* or *memoryless property.*

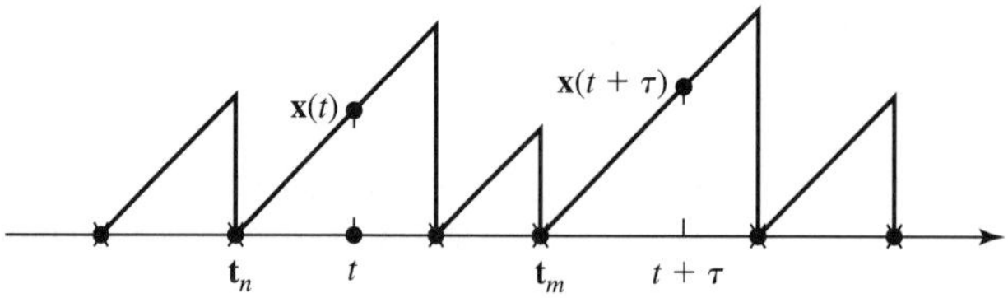

FIGURE 10-11

III. If the number of points $\mathbf{n}(t, t+dt)$ in an interval $(t, t+dt)$ is such that:

(a) $P\{\mathbf{n}(t, t+dt) = 1\}$ is of the order of dt;
(b) $P\{\mathbf{n}(t, t+dt) > 1\}$ is of order higher than dt;
(c) these probabilities do not depend on the state of the point process outside the interval $(t, t+dt)$;

then the process is Poisson (see Sec. 16-1).

IV. Suppose, finally, that:

(a) $P\{\mathbf{n}(a, b) = k\}$ depends only on k and on the length of the interval (a, b);
(b) if the intervals (a_i, b_i) are nonoverlapping, then the random variables $\mathbf{n}(a_i, b_i)$ are independent;
(c) $P\{\mathbf{n}(a, b) = \infty\} = 0$.

These conditions lead again to the conclusion that the probability $p_k(\tau)$ of having k points in any interval of length τ equals

$$p_k(\tau) = e^{-\lambda\tau}(\lambda\tau)^k/k! \tag{10-92}$$

The proof is omitted.

Linear interpolation. The process

$$\mathbf{x}(t) = t - \mathbf{t}_n \qquad \mathbf{t}_n \le t < \mathbf{t}_{n+1} \tag{10-93}$$

of Fig. 10-11 consists of straight line segments of slope 1 between two consecutive random points $\mathbf{t}_n$ and $\mathbf{t}_{n+1}$. For a specific t, $\mathbf{x}(t)$ equals the distance $\mathbf{w} = t - \mathbf{t}_n$ from t to the nearest point $\mathbf{t}_n$ to the left of t; hence the first-order distribution of $\mathbf{x}(t)$ is exponential as in (10-80). From this it follows that

$$E\{\mathbf{x}(t)\} = \frac{1}{\lambda} \qquad E\{\mathbf{x}^2(t)\} = \frac{2}{\lambda^2} \tag{10-94}$$

THEOREM 10-2 ▶ The autocovariance of $\mathbf{x}(t)$ equals

$$C(\tau) = \frac{1}{\lambda^2}(1 + \lambda|\tau|)e^{-\lambda|\tau|} \tag{10-95}$$

Proof. We denote by $\mathbf{t}_m$ and $\mathbf{t}_n$ the random points to the left of the points $t+\tau$ and t, respectively. Suppose first, that $\mathbf{t}_m = \mathbf{t}_n$; in this case $\mathbf{x}(t+\tau) = t + \tau - \mathbf{t}_n$ and $\mathbf{x}(t) = t - \mathbf{t}_n$. Hence [see (10-93)]

$$C(\tau) = E\{(t+\tau-\mathbf{t}_n)(t-\mathbf{t}_n)\} = E\{(t-\mathbf{t}_n)^2\} + \tau E\{t - \mathbf{t}_n\} = \frac{2}{\lambda^2} + \frac{\tau}{\lambda}$$

Suppose, next, that $\mathbf{t}_m \neq \mathbf{t}_n$; in this case

$$C(\tau) = E\{(t+\tau-\mathbf{t}_m)(t-\mathbf{t}_n)\} = E\{t+\tau-\mathbf{t}_m\}E\{t-\mathbf{t}_n\} = \frac{1}{\lambda^2}$$

Clearly, $\mathbf{t}_m = \mathbf{t}_n$ if there are no random points in the interval $(t+\tau, t)$; hence $P\{\mathbf{t}_m = \mathbf{t}_n\} = e^{-\lambda\tau}$. Similarly, $\mathbf{t}_m \neq \mathbf{t}_n$ if there is at least one random point in the interval $(t+\tau, t)$; hence $P\{\mathbf{t}_n \neq \mathbf{t}_m\} = 1 - e^{-\lambda\tau}$. And since [see (4-74)]

$$R(\tau) = E\{\mathbf{x}(t+\tau)\mathbf{x}(t) \mid \mathbf{t}_m = \mathbf{t}_n\}P\{\mathbf{t}_m = \mathbf{t}_n\} + E\{\mathbf{x}(t+\tau)\mathbf{x}(t) \mid \mathbf{t}_n \neq \mathbf{t}_m\}P\{\mathbf{t}_n \neq \mathbf{t}_m\}$$

we conclude that

$$R(\tau) = \left(\frac{2}{\lambda^2} + \frac{\tau}{\lambda}\right)e^{-\lambda\tau} + \frac{1}{\lambda^2}(1 - e^{-\lambda\tau})$$

for $\tau > 0$. Subtracting $1/\lambda^2$, we obtain (10-95). ◀

Shot Noise

Given a set of Poisson points $\mathbf{t}_i$ with average density λ and a real function $h(t)$, we form the sum

$$\mathbf{s}(t) = \sum_i h(t - \mathbf{t}_i) \tag{10-96}$$

This sum is an SSS process known as *shot noise*. Here, we discuss its second-order properties.

From the definition it follows that $\mathbf{s}(t)$ can be represented as the output of a linear system (Fig. 10-12) with impulse response $h(t)$ and input the Poisson impulses

$$\mathbf{z}(t) = \sum_i \delta(t - \mathbf{t}_i) \tag{10-97}$$

This representation agrees with the generation of shot noise in physical problems: The process $\mathbf{s}(t)$ is the output of a dynamic system activated by a sequence of impulses (particle emissions, for example) occurring at the random times $\mathbf{t}_i$.

As we know, $\eta_z = \lambda$; hence

$$E\{\mathbf{s}(t)\} = \lambda \int_{-\infty}^{\infty} h(t)\,dt = \lambda H(0) \tag{10-98}$$

Furthermore, since (see Example 9-22)

$$S_{zz}(\omega) = 2\pi\lambda^2\delta(\omega) + \lambda \tag{10-99}$$

it follows from (9-149) that

$$S_{ss}(\omega) = 2\pi\lambda^2 H^2(0)\delta(\omega) + \lambda|H(\omega)|^2 \tag{10-100}$$

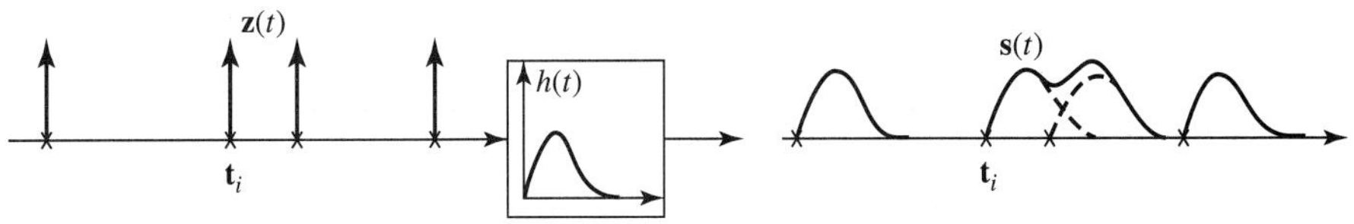

FIGURE 10-12

because $|H(\omega)|^2\delta(\omega) = H^2(0)\delta(\omega)$. The inverse of this yields

$$R_{ss}(\tau) = \lambda^2 H^2(0) + \lambda\rho(\tau) \qquad C_{ss}(\tau) = \lambda\rho(\tau) \tag{10-101}$$

Campbell's theorem. The mean η_s and variance σ_s^2 of the shot-noise process $\mathbf{s}(t)$ equal

$$\eta_s = \lambda \int_{-\infty}^{\infty} h(t)\,dt \qquad \sigma_s^2 = \lambda\rho(0) = \lambda \int_{-\infty}^{\infty} h^2(t)\,dt \tag{10-102}$$

Proof. It follows from (10-101) because $\sigma_s^2 = C_{ss}(0)$.

EXAMPLE 10-2

▶ If

$$h(t) = e^{-\alpha t}U(t) \qquad H(\omega) = \frac{1}{\alpha + j\omega}$$

then

$$\eta_s = \frac{\lambda}{\alpha} \qquad \sigma_s^2 = \frac{\lambda}{2\alpha}$$

$$S_{ss}(\omega) = \frac{2\pi\lambda^2}{\alpha^2}\delta(\omega) + \frac{\lambda}{\alpha^2 + \omega^2} \qquad C_{ss}(\tau) = \frac{\lambda}{2\alpha}e^{-\alpha|\tau|}$$ ◀

EXAMPLE 10-3

ELECTRON TRANSIT

▶ Suppose that $h(t)$ is a triangle as in Fig. 10-13*a*. Since

$$\int_0^T kt\,dt = \frac{kT^2}{2} \qquad \int_0^T k^2t^2\,dt = \frac{k^2T^3}{3}$$

it follows from (10-102) that

$$\eta_s = \frac{\lambda k T^2}{2} \qquad \sigma_s^2 = \frac{\lambda k^2 T^3}{3}$$

In this case

$$H(\omega) = \int_0^T kte^{-j\omega t}\,dt = e^{-j\omega T/2}\frac{2k\sin\omega T/2}{j\omega^2} - e^{-j\omega T}\frac{kT}{j\omega}$$

Inserting into (10-100), we obtain (Fig. 10-13*b*).

$$S_{ss}(\omega) = 2\pi\eta_s^2\delta(\omega) + \frac{\lambda k^2}{\omega^4}(2 - 2\cos\omega T + \omega^2T^2 - 2\omega T\sin\omega T)$$ ◀

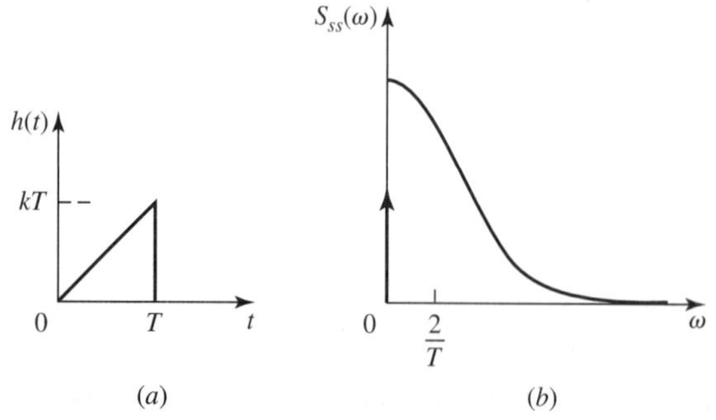

FIGURE 10-13

Generalized Poisson processes and shot noise. Given a set of Poisson points $\mathbf{t}_i$ with average density λ, we form the process

$$\mathbf{x}(t) = \sum_i \mathbf{c}_i U(t - \mathbf{t}_i) = \sum_{i=1}^{\mathbf{n}(t)} \mathbf{c}_i \tag{10-103}$$

where $\mathbf{c}_i$ is a sequence of i.i.d. random variables independent of the points $\mathbf{t}_i$ with mean η_c and variance σ_c^2. Thus $\mathbf{x}(t)$ is a staircase function as in Fig. 9-3 with jumps at the points $\mathbf{t}_i$ equal to $\mathbf{c}_i$. The process $\mathbf{n}(t)$ is the number of Poisson points in the interval $(0, t)$; hence $E\{\mathbf{n}(t)\} = \lambda t$ and $E\{\mathbf{n}^2(t)\} = \lambda^2 + \lambda t$. For a specific t, $\mathbf{x}(t)$ is a sum as in (7-46). From this it follows that

$$\begin{aligned} E\{\mathbf{x}(t)\} &= \eta_c E\{\mathbf{n}(t)\} = \eta_c \lambda t \\ E\{\mathbf{x}^2(t)\} &= \eta_c^2 E\{\mathbf{n}^2(t)\} + \sigma_c^2 E\{\mathbf{n}(t)\} = \eta_c^2(\lambda t + \lambda^2 t^2) + \sigma_c^2 \lambda t \end{aligned} \tag{10-104}$$

Proceeding as in Example 9-5, we obtain

$$C_{xx}(t_1, t_2) = \left(\eta_c^2 + \sigma_c^2\right)\lambda \min(t_1, t_2) \tag{10-105}$$

We next form the impulse train

$$\mathbf{z}(t) = \mathbf{x}'(t) = \sum_i \mathbf{c}_i \delta(t - \mathbf{t}_i) \tag{10-106}$$

From (10-105) it follows as in (9-109) that

$$E\{\mathbf{z}(t)\} = \frac{d}{dt} E\{\mathbf{x}(t)\} = \eta_c \lambda \tag{10-107}$$

$$C_{zz}(t_1, t_2) = \frac{\partial^2 C_{xx}(t_1, t_2)}{\partial t_1 \, \partial t_2} = \left(\eta_c^2 + \sigma_c^2\right)\lambda \delta(\tau) \tag{10-108}$$

where $\tau = t_2 - t_1$. Convolving $\mathbf{z}(t)$ with a function $h(t)$, we obtain the generalized shot noise

$$\mathbf{s}(t) = \sum_i \mathbf{c}_i h(t - \mathbf{t}_i) = \mathbf{z}(t) * h(t) \tag{10-109}$$

This yields

$$E\{\mathbf{s}(t)\} = E\{\mathbf{z}(t)\} * h(\mathbf{t}) = \eta_c \lambda \int_{-\infty}^{\infty} h(t)\, dt \tag{10-110}$$

$$C_{ss}(\tau) = C_{zz}(\tau) * h(\tau) * h(-\tau) = \left(\eta_c^2 + \sigma^2\right)\lambda \rho(\tau) \tag{10-111}$$

$$\operatorname{Var}\{\mathbf{s}(t)\} = C_{ss}(0) = \left(\eta_c^2 + \sigma_c^2\right)\lambda \int_{-\infty}^{\infty} h^2(t)\, dt \tag{10-112}$$

This is the extension of Campbell's theorem to a shot-noise process with random coefficients.

Equation (10-103) can be given yet another interesting interpretation.

COMPOUND POISSON PROCESSES

▶ In an ordinary Poisson process only one event occurs at any arrival instant. Instead, consider a random number of events $\mathbf{c}_i$ occurring simultaneously as a cluster at an instant $\mathbf{t}_i$ as in (10-103) such that the total number of clusters in time t constitute an ordinary Poisson process $\mathbf{n}(t)$. Each cluster has a random number of occurrences that is independent of other clusters. Let $\mathbf{c}_i$ denote the number of occurrences in the ith cluster, and $\mathbf{x}(t)$ the total number of occurrences in the interval $(0, t)$. Then $\mathbf{x}(t)$ represents a *compound Poisson process*. For example, $\mathbf{c}_i$ may represent the number of cars involved in the ith automobile accident (or number of houses involved in the ith fire incident) in some interval, and if the *number* of accidents (fires) in that interval is assumed to be a Poisson process, then the total number of claims during that interval has a compound Poisson distribution. Let

$$p_k = P\{\mathbf{c}_i = k\} \qquad k = 0, 1, 2, \ldots \tag{10-113}$$

represent the common probability mass function of occurrences in *any* cluster, and

$$P(z) = E\{z^{\mathbf{c}_i}\} = \sum_{k=0}^{\infty} p_k z^k \tag{10-114}$$

its moment generation function. For any t, the moment generating function of the compound Poisson process $\mathbf{x}(t)$ is given by

$$\begin{aligned}
\Phi_x(z) &= E\{z^{\mathbf{x}(t)}\} = E\{E(z^{\mathbf{x}(t)} \mid \mathbf{n}(t) = k)\} \\
&= E\left\{E\left(z^{\sum_{i=1}^{k} \mathbf{c}_i} \mid \mathbf{n}(t) = k\right)\right\} \\
&= \sum_{k=0}^{\infty} [E(z^{\mathbf{c}_i})]^k P\{\mathbf{n}(t) = k\} \\
&= \sum_{k=0}^{\infty} [P(z)]^k e^{-\lambda t} \frac{(\lambda t)^k}{k!} = e^{-\lambda t[1-P(z)]}
\end{aligned} \tag{10-115}$$

We can use this identity to determine the probability of n occurrences in the interval $(0, t)$. Toward this, let

$$P^k(z) \triangleq \sum_{n=0}^{\infty} p_n^{(k)} z^n \tag{10-116}$$

where $\{p_n^{(k)}\}$ represents the k-fold convolution of the sequence $\{p_n\}$ with itself. Substituting (10-116) into the first expression in (10-115) we get the interesting identity

$$P\{\mathbf{x}(t) = n\} = \sum_{k=0}^{\infty} e^{-\lambda t} \frac{(\lambda t)^k}{k!} p_n^{(k)} \tag{10-117}$$

Equations (10-114) and (10-115) can be used also to show that every compound Poisson distribution is a linear combination (with integer coefficients) of independent Poisson processes. In fact substituting (10-114) into (10-115) we get

$$\Phi_x(z) = e^{-\lambda_1 t(1-z)} e^{-\lambda_2 t(1-z^2)} \cdots e^{-\lambda_k t(1-z^k)} \cdots \tag{10-118}$$

where $\lambda_k = \lambda p_k$ and hence it follows that

$$\mathbf{x}(t) = \mathbf{m}_1(t) + 2\mathbf{m}_2(t) + \cdots + k\mathbf{m}_k(t) + \cdots \tag{10-119}$$

where $\mathbf{m}_k(t)$ are independent Poisson processes with parameter λp_k, $k = 1, 2, \ldots$.

More generally, we can use these observations to show that *any* linear combination of independent Poisson processes is a compound Poisson process. Thus

$$\mathbf{y}(t) = \sum_{k=1}^{n} a_k \mathbf{x}_k(t) \tag{10-120}$$

is a compound Poisson process, where a_k are *arbitrary* constants and $\mathbf{x}_k(t)$ are independent Poisson processes with parameter $\lambda_k t$, since

$$\Phi_y(z) = E\{z^{\mathbf{y}(t)}\} = \prod_{k=1}^{n} E\{z^{a_k \mathbf{x}_k(t)}\}$$

$$= \prod_{k=1}^{n} e^{-\lambda_k t(1-z^{a_k})} = e^{-\lambda t(1-P_0(z))} \tag{10-121}$$

where

$$P_0(z) = \sum_{k=1}^{n} \frac{\lambda_k}{\lambda} z^{a_k} \qquad \lambda \triangleq \sum_{k=1}^{n} \lambda_k \tag{10-122}$$

On comparing (10-121) and (10-122) with (10-114) and (10-115) we conclude that

$$\mathbf{y}(t) = \sum_{i=1}^{\mathbf{z}(t)} \mathbf{c}_i \tag{10-123}$$

as in (10-103), where $\mathbf{z}(t) \sim P\{\lambda t\}$ and

$$P\{\mathbf{c}_i = a_k\} = \frac{\lambda_k}{\lambda} \qquad k = 1, 2, \ldots, n \tag{10-124}$$

◀

10-3 MODULATION[3]

Given two real jointly WSS processes $\mathbf{a}(t)$ and $\mathbf{b}(t)$ with zero mean and a constant ω_0, we form the process

$$\mathbf{x}(t) = \mathbf{a}(t)\cos\omega_0 t - \mathbf{b}(t)\sin\omega_0 t$$
$$= \mathbf{r}(t)\cos[\omega_0 t + \boldsymbol{\varphi}(t)] \tag{10-125}$$

where

$$\mathbf{r}(t) = \sqrt{\mathbf{a}^2(t) + \mathbf{b}^2(t)} \qquad \tan\boldsymbol{\varphi}(t) = \frac{\mathbf{b}(t)}{\mathbf{a}(t)}$$

This process is called modulated with *amplitude modulation* $\mathbf{r}(t)$ and *phase modulation* $\boldsymbol{\varphi}(t)$.

We shall show that $\mathbf{x}(t)$ is WSS iff the processes $\mathbf{a}(t)$ and $\mathbf{b}(t)$ are such that

$$R_{aa}(\tau) = R_{bb}(\tau) \qquad R_{ab}(\tau) = -R_{ba}(\tau) \tag{10-126}$$

---

[3] A. Papoulis: "Random Modulation: A Review," *IEEE Transactions on Acoustics, Speech, and Signal Processing,* vol. ASSP-31, 1983.

Proof. Clearly,

$$E\{\mathbf{x}(t)\} = E\{\mathbf{a}(t)\}\cos\omega_0 t - E\{\mathbf{b}(t)\}\sin\omega_0 t = 0$$

Furthermore,

$$\mathbf{x}(t+\tau)\mathbf{x}(t) = [\mathbf{a}(t+\tau)\cos\omega_0(t+\tau) - \mathbf{b}(t+\tau)\sin\omega_0(\mathbf{t}+\tau)] \times [\mathbf{a}(t)\cos\omega_0 t - \mathbf{b}(t)\sin\omega_0 t]$$

Multiplying, taking expected values, and using appropriate trigonometric identities, we obtain

$$\begin{aligned} 2E\{\mathbf{x}(t+\tau)\mathbf{x}(t)\} = {} & [R_{aa}(\tau) + R_{bb}(\tau)]\cos\omega_0\tau + [R_{ab}(\tau) - R_{ba}(\tau)]\sin\omega_0\tau \\ & + [R_{aa}(\tau) - R_{bb}(\tau)]\cos\omega_0(2t+\tau) \\ & - [R_{ab}(\tau) + R_{ba}(\tau)]\sin\omega_0(2t+\tau) \end{aligned} \tag{10-127}$$

If (10-126) is true, then (10-127) yields

$$R_{xx}(\tau) = R_{aa}(\tau)\cos\omega_0\tau + R_{ab}(\tau)\sin\omega_0\tau \tag{10-128}$$

Conversely, if $\mathbf{x}(t)$ is WSS, then the second and third lines in (10-127) must be independent of t. This is possible only if (10-126) is true.

We introduce the "dual" process

$$\mathbf{y}(t) = \mathbf{b}(t)\cos\omega_0 t + \mathbf{a}(t)\sin\omega_0 t \tag{10-129}$$

This process is also WSS and

$$R_{yy}(\tau) = R_{xx}(\tau) \qquad R_{xy}(\tau) = -R_{yx}(\tau) \tag{10-130}$$

$$R_{xy}(\tau) = R_{ab}(\tau)\cos\omega_0\tau - R_{aa}(\tau)\sin\omega_0\tau \tag{10-131}$$

This follows from (10-127) if we change one or both factors of the product $\mathbf{x}(t+\tau)\mathbf{x}(t)$ with $\mathbf{y}(t+\tau)$ or $\mathbf{y}(t)$.

Complex representation. We introduce the processes

$$\begin{aligned} \mathbf{w}(t) &= \mathbf{a}(t) + j\mathbf{b}(t) = \mathbf{r}(t)e^{j\varphi(t)} \\ \mathbf{z}(t) &= \mathbf{x}(t) + j\mathbf{y}(t) = \mathbf{w}(t)e^{j\omega_0 t} \end{aligned} \tag{10-132}$$

Thus

$$\mathbf{x}(t) = \operatorname{Re}\mathbf{z}(t) = \operatorname{Re}[\mathbf{w}(t)e^{j\omega_0 t}] \tag{10-133}$$

and

$$\mathbf{a}(t) + j\mathbf{b}(t) = \mathbf{w}(t) = \mathbf{z}(t)e^{-j\omega_0 t}$$

This yields

$$\begin{aligned} \mathbf{a}(t) &= \mathbf{x}(t)\cos\omega_0 t + \mathbf{y}(t)\sin\omega_0 t \\ \mathbf{b}(t) &= \mathbf{y}(t)\cos\omega_0 t - \mathbf{x}(t)\sin\omega_0 t \end{aligned} \tag{10-134}$$

Correlations and spectra. The autocorrelation of the complex process $\mathbf{w}(t)$ equals

$$R_{ww}(\tau) = E\{[\mathbf{a}(t+\tau) + j\mathbf{b}(t+\tau)][\mathbf{a}(t) - j\mathbf{b}(t)]\}$$

Expanding and using (10-126), we obtain

$$R_{ww}(\tau) = 2R_{aa}(\tau) - 2jR_{ab}(\tau) \tag{10-135}$$

Similarly,

$$R_{zz}(\tau) = 2R_{xx}(\tau) - 2jR_{xy}(\tau) \tag{10-136}$$

We note, further, that

$$R_{zz}(\tau) = e^{j\omega_0\tau} R_{ww}(\tau) \tag{10-137}$$

From this it follows that

$$\begin{aligned} S_{ww}(\omega) &= 2S_{aa}(\omega) - 2jS_{ab}(\omega) \\ S_{zz}(\omega) &= 2S_{xx}(\omega) - 2jS_{xy}(\omega) \end{aligned} \tag{10-138}$$

$$S_{zz}(\omega) = S_{ww}(\omega - \omega_0) \tag{10-139}$$

The functions $S_{xx}(\omega)$ and $S_{zz}(\omega)$ are real and positive. Furthermore [see (10-130)]

$$R_{xy}(-\tau) = -R_{yx}(-\tau) = -R_{xy}(\tau)$$

This leads to the conclusion that the function $-jS_{xy}(\omega) = B_{xy}(\omega)$ is real and (Fig. 10-14a)

$$|B_{xy}(\omega)| \le S_{xx}(\omega) \qquad B_{xy}(-\omega) = -B_{xy}(\omega) \tag{10-140}$$

And since $S_{xx}(-\omega) = S_{xx}(\omega)$, we conclude from the second equation in (10-138) that

$$\begin{aligned} 4S_{xx}(\omega) &= S_{zz}(\omega) + S_{zz}(-\omega) \\ 4jS_{xy}(\omega) &= S_{zz}(-\omega) - S_{zz}(\omega) \end{aligned} \tag{10-141}$$

Single sideband If $\mathbf{b}(t) = \hat{\mathbf{a}}(t)$ is the Hilbert transform of $\mathbf{a}(t)$, then [see (9-160)] the constraint (10-126) is satisfied and the first equation in (10-138) yields

$$S_{ww}(\omega) = 4S_{aa}(\omega)U(\omega)$$

(Fig. 10-14b) because

$$S_{a\hat{a}}(\omega) = jS_{aa}(\omega)\operatorname{sgn}\omega$$

The resulting spectra are shown in Fig. 10-14b. We note, in particular, that $S_{xx}(\omega) = 0$ for $|\omega| < \omega_0$.

RICE'S REPRESENTATION. In (10-125) we assumed that the carrier frequency ω_0 and the processes $\mathbf{a}(t)$ and $\mathbf{b}(t)$ were given. We now consider the converse problem: Given a WSS process $\mathbf{x}(t)$ with zero mean, find a constant ω_0 and two processes $\mathbf{a}(t)$ and $\mathbf{b}(t)$ such that $\mathbf{x}(t)$ can be written in the form (10-125). To do so, it suffices to find the constant ω_0 and the dual process $\mathbf{y}(t)$ [see (10-134)]. This shows that the representation of $\mathbf{x}(t)$ in the form (10-125) is not unique because, not only ω_0 is arbitrary, but also the process $\mathbf{y}(t)$ can be chosen arbitrarily subject only to the constraint (10-130). The question then arises whether, among all possible representations of $\mathbf{x}(t)$, there is one that is optimum. The answer depends, of course, on the optimality criterion. As we shall presently explain, if $\mathbf{y}(t)$ equals the Hilbert transform $\hat{\mathbf{x}}(t)$ of $\mathbf{x}(t)$, then (10-125)

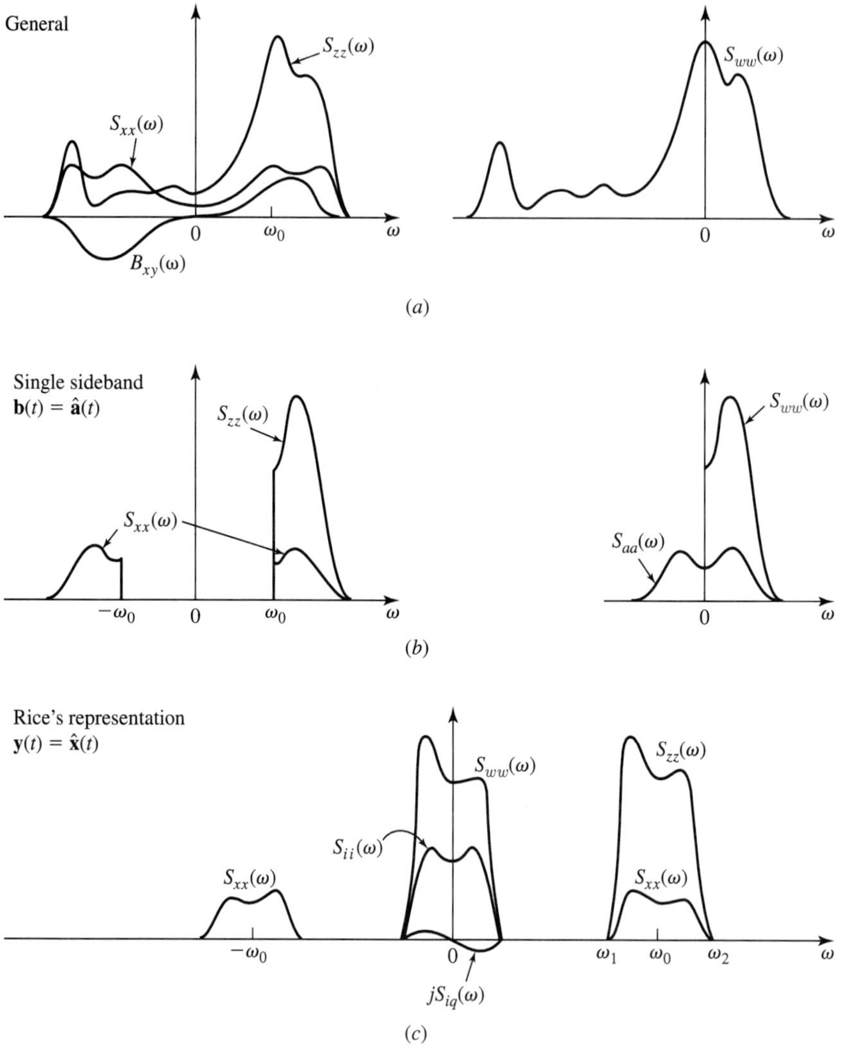

FIGURE 10-14

is optimum in the sense of minimizing the average rate of variation of the envelope of $\mathbf{x}(t)$.

Hilbert transforms. As we know [see (9-160)]

$$R_{\hat{x}\hat{x}}(\tau) = R_{xx}(\tau) \qquad R_{x\hat{x}}(\tau) = -R_{x\hat{x}}(\tau) \tag{10-142}$$

We can, therefore, use $\hat{\mathbf{x}}(t)$ to form the processes

$$\begin{aligned}\mathbf{z}(t) &= \mathbf{x}(t) + j\hat{\mathbf{x}}(t) = \mathbf{w}(t)e^{j\omega_0 t}\\ \mathbf{w}(t) &= \mathbf{i}(t) + j\mathbf{q}(t) = \mathbf{z}(t)e^{-j\omega_0 t}\end{aligned} \tag{10-143}$$

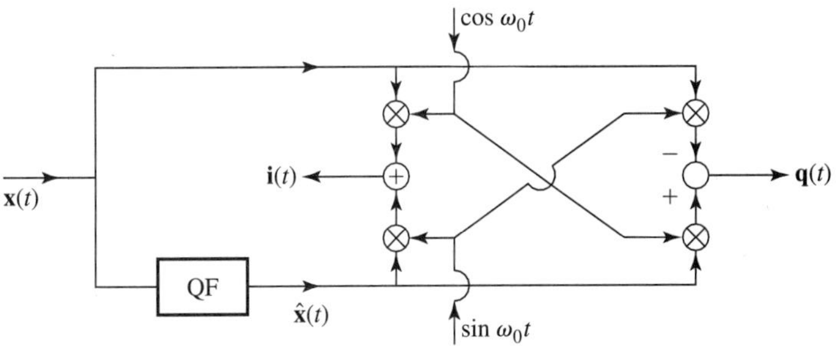

FIGURE 10-15

as in (10-132) where now (Fig. 10-14*c*)

$$\mathbf{y}(t) = \hat{\mathbf{x}}(t) \qquad \mathbf{a}(t) = \mathbf{i}(t) \qquad \mathbf{b}(t) = \mathbf{q}(t)$$

Inserting into (10-125), we obtain

$$\mathbf{x}(t) = \mathbf{i}(t)\cos\omega_0 t - \mathbf{q}(t)\sin\omega_0 t \tag{10-144}$$

This is known as *Rice's representation*. The process $\mathbf{i}(t)$ is called the *inphase* component and the process $\mathbf{q}(t)$ the *quadrature* component of $\mathbf{x}(t)$. Their realization is shown in Fig. 10-15 [see (10-134)]. These processes depend, not only on $\mathbf{x}(t)$, but also on the choice of the carrier frequency ω_0.

From (9-149) and (10-138) it follows that

$$S_{zz}(\omega) = 4S_{xx}(\omega)U(\omega) \tag{10-145}$$

Bandpass processes. A process $\mathbf{x}(t)$ is called bandpass (Fig. 10-14*c*) if its spectrum $S_{xx}(\omega)$ is 0 outside an interval (ω_1, ω_2). It is called narrowband or *quasimonochromatic* if its bandwidth $\omega_2 - \omega_1$ is small compared with the center frequency. It is called *monochromatic* if $S_{xx}(\omega)$ is an impulse function. The process $\mathbf{a}\cos\omega_0 t + \mathbf{b}\sin\omega_0 t$ is monochromatic.

The representations (10-125) or (10-144) hold for an arbitrary $\mathbf{x}(t)$. However, they are useful mainly if $\mathbf{x}(t)$ is bandpass. In this case, the complex envelope $\mathbf{w}(t)$ and the processes $\mathbf{i}(t)$ and $\mathbf{q}(t)$ are low-pass (LP) because

$$\begin{aligned} S_{ww}(\omega) &= S_{zz}(\omega + \omega_0) \\ S_{ii}(\omega) &= S_{qq}(\omega) = \tfrac{1}{4}[S_{ww}(\omega) + S_{ww}(-\omega)] \end{aligned} \tag{10-146}$$

We shall show that if the process $\mathbf{x}(t)$ is bandpass and $\omega_1 + \omega_c \le 2\omega_0$, then the inphase component $\mathbf{i}(t)$ and the quadrature component $\mathbf{q}(t)$ can be obtained as responses of the system of Fig. 10-16*a* where the LP filters are ideal with cutoff frequency ω_c such that

$$\omega_2 - \omega_0 < \omega_c \qquad \omega_1 - \omega_0 > -\omega_c \tag{10-147}$$

Proof. It suffices to show that (linearity) the response of the system of Fig. 10-16*b* equals $\mathbf{w}(t)$. Clearly,

$$2\mathbf{x}(t) = \mathbf{z}(t) + \mathbf{z}^*(t) \qquad \mathbf{w}^*(t) = \mathbf{z}^*(t)e^{j\omega_0 t}$$

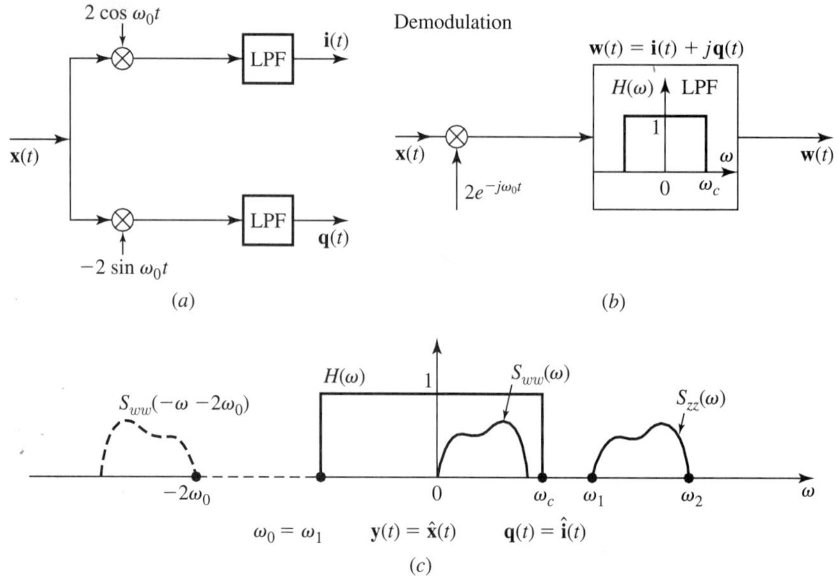

FIGURE 10-16

Hence

$$2\mathbf{x}(t)e^{-j\omega_0 t} = \mathbf{w}(t) + \mathbf{w}^*(t)e^{-j2\omega_0 t}$$

The spectra of the processes $\mathbf{w}(t)$ and $\mathbf{w}^*(t)e^{-j2\omega_0 t}$ equal $S_{ww}(\omega)$ and $S_{ww}(-\omega - 2\omega_0)$, respectively. Under the stated assumptions, the first is in the band of the LP filter $H(\omega)$ and the second outside the band. Therefore, the response of the filter equals $\mathbf{w}(t)$.

We note, finally, that if $\omega_0 \leq \omega_1$, then $S_{ww}(\omega) = 0$ for $\omega < 0$. In this case, $\mathbf{q}(t)$ is the Hilbert transform of $\mathbf{i}(t)$. Since $\omega_2 - \omega_1 \leq 2\omega_0$, this is possible only if $\omega_2 \leq 3\omega_1$. In Fig. 10-16*c*, we show the corresponding spectra for $\omega_0 = \omega_1$.

Optimum envelope. We are given an arbitrary process $\mathbf{x}(t)$ and we wish to determine a constant ω_0 and a process $\mathbf{y}(t)$ so that, in the resulting representation (10-125), the complex envelope $\mathbf{w}(t)$ of $\mathbf{x}(t)$ is smooth in the sense of minimizing $E\{|\mathbf{w}'(t)|^2\}$. As we know, the power spectrum of $\mathbf{w}'(t)$ equals

$$\omega^2 S_{ww}(\omega) = \omega^2 S_{zz}(\omega + \omega_0)$$

Our problem, therefore, is to minimize the integral[4]

$$M = 2\pi E\{|\mathbf{w}'(t)|^2\} = \int_{-\infty}^{\infty} (\omega - \omega_0)^2 S_{zz}(\omega)\, d\omega \tag{10-148}$$

subject to the constraint that $S_{xx}(\omega)$ is specified.

[4]L. Mandel: "Complex Representation of Optical Fields in Coherence Theory," *Journal of the Optical Society of America,* vol. 57, 1967. See also N. M. Blachman: *Noise and Its Effect on Communication,* Krieger Publishing Company, Malabar, FL, 1982.

RICE'S REPRESENTATION

▶ Rice's representation (10-144) is optimum and the optimum carrier frequency ω_0 is the center of gravity $\overline{\omega}_0$ of $S_{xx}(\omega)U(\omega)$.

Proof. Suppose, first, that $S_{xx}(\omega)$ is specified. In this case, M depends only on ω_0. Differentiating the right side of (10-148) with respect to ω_0, we conclude that M is minimum if ω_0 equals the center of gravity

$$\overline{\omega}_0 = \frac{\int_{-\infty}^{\infty} \omega S_{zz}(\omega)\,d\omega}{\int_{-\infty}^{\infty} S_{zz}(\omega)\,d\omega} = \frac{\int_0^{\infty} \omega B_{xy}(\omega)\,d\omega}{\int_0^{\infty} S_{xx}(\omega)\,d\omega} \tag{10-149}$$

of $S_{zz}(\omega)$. The second equality in (10-149) follows from (10-138) and (10-140). Inserting (10-149) into (10-148), we obtain

$$M = \int_{-\infty}^{\infty} \left(\omega^2 - \overline{\omega}_0^2\right) S_{zz}(\omega)\,d\omega = 2\int_{-\infty}^{\infty} \left(\omega^2 - \overline{\omega}_0^2\right) S_{xx}(\omega)\,d\omega \tag{10-150}$$

We wish now to choose $S_{zz}(\omega)$ so as to minimize M. Since $S_{xx}(\omega)$ is given, M is minimum if $\overline{\omega}_0$ is maximum. As we see from (10-149), this is the case if $|B_{xy}(\omega)| = S_{xx}(\omega)$ because $|B_{xy}(\omega)| \le S_{xx}(\omega)$. We thus conclude that $-jS_{xy}(\omega) = S_{xx}(\omega)\mathrm{sgn}\,\omega$ and (10-138) yields

$$S_{zz}(\omega) = 4S_{xx}(\omega)U(\omega)$$ ◀

Instantaneous frequency. With $\varphi(t)$ as in (10-125), the process

$$\boldsymbol{\omega}_i(t) = \omega_0 + \boldsymbol{\varphi}'(t) \tag{10-151}$$

is called the instantaneous frequency of $\mathbf{x}(t)$. Since

$$\mathbf{z} = \mathbf{r}e^{j(\omega_0 t + \boldsymbol{\varphi})} = \mathbf{x} + j\mathbf{y}$$

we have

$$\mathbf{z}'\mathbf{z}^* = \mathbf{r}\mathbf{r}' + j\mathbf{r}^2\boldsymbol{\omega}_i = (\mathbf{x}' + j\mathbf{y}')(\mathbf{x} - j\mathbf{y}) \tag{10-152}$$

This yields $E\{\mathbf{r}\mathbf{r}'\} = 0$ and

$$E\{\mathbf{r}^2\boldsymbol{\omega}_i\} = \frac{1}{2\pi}\int_{-\infty}^{\infty} \omega S_{zz}(\omega)\,d\omega \tag{10-153}$$

because the cross-power spectrum of $\mathbf{z}'$ and $\mathbf{z}$ equals $j\omega S_{zz}(\omega)$.

The instantaneous frequency of a process $\mathbf{x}(t)$ is not a uniquely defined process because the dual process $\mathbf{y}(t)$ is not unique. In Rice's representation $\mathbf{y} = \hat{\mathbf{x}}$, hence

$$\boldsymbol{\omega}_i = \frac{\mathbf{x}\hat{\mathbf{x}}' - \mathbf{x}'\hat{\mathbf{x}}}{\mathbf{r}^2} \qquad \mathbf{r}^2 = \mathbf{x}^2 + \hat{\mathbf{x}}^2 \tag{10-154}$$

In this case [see (10-145) and (10-149)] the optimum carrier frequency $\overline{\omega}_0$ equals the weighted average of $\boldsymbol{\omega}_i$:

$$\overline{\omega}_0 = \frac{E\{\mathbf{r}^2\boldsymbol{\omega}_i\}}{E\{\mathbf{r}^2\}}$$

Frequency Modulation

The process

$$\mathbf{x}(t) = \cos[\omega_0 t + \lambda\boldsymbol{\varphi}(t) + \boldsymbol{\varphi}_0] \qquad \boldsymbol{\varphi}(t) = \int_0^t \mathbf{c}(\alpha)\,d\alpha \tag{10-155}$$

is FM with instantaneous frequency $\omega_0 + \lambda\mathbf{c}(t)$ and modulation index λ. The corresponding complex processes equal

$$\mathbf{w}(t) = e^{j\lambda\boldsymbol{\varphi}(t)} \qquad \mathbf{z}(t) = \mathbf{w}(t)e^{j(\omega_0 t + \boldsymbol{\varphi}_0)} \tag{10-156}$$

We shall study their spectral properties.

THEOREM 10-3

▶ If the process $\mathbf{c}(t)$ is SSS and the random variable $\boldsymbol{\varphi}_0$ is independent of $\mathbf{c}(t)$ and such that

$$E\{e^{j\boldsymbol{\varphi}_0}\} = E\{e^{j2\boldsymbol{\varphi}_0}\} = 0 \tag{10-157}$$

then the process $\mathbf{x}(t)$ is WSS with zero mean. Furthermore,

$$\begin{aligned} R_{xx}(\tau) &= \tfrac{1}{2}\mathrm{Re}\,R_{zz}(\tau) \\ R_{zz}(\tau) &= R_{ww}(\tau)e^{j\omega_0\tau} \qquad R_{ww}(\tau) = E\{\mathbf{w}(\tau)\} \end{aligned} \tag{10-158}$$

Proof. From (10-157) it follows that $E\{\mathbf{x}(t)\} = 0$ because

$$E\{\mathbf{z}(t)\} = E\left\{e^{j[\omega_0 t + \lambda\boldsymbol{\varphi}(t)]}\right\}E\{e^{j\boldsymbol{\varphi}_0}\} = 0$$

Furthermore,

$$E\{\mathbf{z}(t+\tau)\mathbf{z}(t)\} = E\left\{e^{j[\omega_0(2t+\tau)+\lambda\boldsymbol{\varphi}(t+\tau)+\lambda\boldsymbol{\varphi}(t)]}\right\}E\{e^{j2\boldsymbol{\varphi}_0}\} = 0$$

$$E\{\mathbf{z}(t+\tau)\mathbf{z}^*(t)\} = e^{j\omega_0\tau}E\left\{\exp\left[j\lambda\int_t^{t+\tau}\mathbf{c}(\alpha)\,d\alpha\right]\right\} = e^{j\omega_0\tau}E\{\mathbf{w}(\tau)\}$$

The last equality is a consequence of the stationarity of the process $\mathbf{c}(t)$. Since $2\mathbf{x}(t) = \mathbf{z}(t) + \mathbf{z}^*(t)$, we conclude from the above that

$$4E\{\mathbf{x}(t+\tau)\mathbf{x}(t)\} = R_{zz}(\tau) + R_{zz}(-\tau)$$

and (10-158) results because $R_{zz}(-\tau) = R^*_{zz}(\tau)$. ◀

Definitions A process $\mathbf{x}(t)$ is *phase modulated* if the statistics of $\boldsymbol{\varphi}(t)$ are known. In this case, its autocorrelation can simply be found because

$$E\{\mathbf{w}(t)\} = E\left\{e^{j\lambda\boldsymbol{\varphi}(t)}\right\} = \Phi_\varphi(\lambda, t) \tag{10-159}$$

where $\Phi_\varphi(\lambda, t)$ is the characteristic function of $\boldsymbol{\varphi}(t)$.

A process $\mathbf{x}(t)$ is *frequency modulated* if the statistics of $\mathbf{c}(t)$ are known. To determine $\Phi_\varphi(\lambda, t)$, we must now find the statistics of the integral of $\mathbf{c}(t)$. However, in general this is not simple. The normal case is an exception because then $\Phi_\varphi(\lambda, t)$ can be expressed in terms of the mean and variance of $\boldsymbol{\varphi}(t)$ and, as we know [see (9-156)]

$$\begin{aligned} E\{\boldsymbol{\varphi}(t)\} &= \int_0^t E\{\mathbf{c}(\alpha)\}\,d\alpha = \eta_c t \\ E\{\boldsymbol{\varphi}^2(t)\} &= 2\int_0^t R_c(\alpha)(t-\alpha)\,d\alpha \end{aligned} \tag{10-160}$$

For the determination of the power spectrum $S_{xx}(\omega)$ of $\mathbf{x}(t)$, we must find the function $\Phi_\varphi(\lambda, t)$ and its Fourier transform. In general, this is difficult. However, as the next theorem shows, if λ is large, then $S_{xx}(\omega)$ can be expressed directly in terms of the density $f_c(c)$ of $\mathbf{c}(t)$.

THEOREM 10-4

WOODWARD'S THEOREM[5]

If the process $\mathbf{c}(t)$ is continuous and its density $f_c(c)$ is bounded, then for large λ:

$$S_{xx}(\omega) \simeq \frac{\pi}{2\lambda}\left[f_c\left(\frac{\omega-\omega_0}{\lambda}\right) + f_c\left(\frac{-\omega-\omega_0}{\lambda}\right)\right] \qquad (10\text{-}161)$$

Proof. If τ_0 is sufficiently small, then $\mathbf{c}(t) \simeq \mathbf{c}(0)$, and

$$\varphi(t) = \int_0^t \mathbf{c}(\alpha)\,d\alpha \simeq \mathbf{c}(0)t \qquad |t| < \tau_0 \qquad (10\text{-}162)$$

Inserting into (10-159), we obtain

$$E\{\mathbf{w}(\tau)\} \simeq E\left\{e^{j\lambda\tau\mathbf{c}(0)}\right\} = \Phi_c(\lambda\tau) \qquad |\tau| < \tau_0 \qquad (10\text{-}163)$$

where

$$\Phi_c(\mu) = E\left\{e^{j\mu\mathbf{c}(t)}\right\}$$

is the characteristic function of $\mathbf{c}(t)$. From this and (10-158) it follows that

$$R_{zz}(\tau) \simeq \Phi_c(\lambda\tau)e^{j\omega_0 t} \qquad |\tau| < \tau_0 \qquad (10\text{-}164)$$

If λ is sufficiently large, then $\Phi_c(\lambda\tau) \simeq 0$ for $|\tau| > \tau_0$ because $\Phi_c(\mu) \to 0$ as $\mu \to \infty$. Hence (10-164) is a satisfactory approximation for every τ in the region where $\Phi_c(\lambda\tau)$ takes significant values. Transforming both sides of (10-164) and using the inversion formula

$$f_c(c) = \frac{1}{2\pi}\int_{-\infty}^{\infty} \Phi_c(\mu)e^{-j\mu c}\,d\mu$$

we obtain

$$S_{zz}(\omega) = \int_{-\infty}^{\infty} \Phi_c(\lambda\tau)e^{j\omega_0\tau}e^{-j\omega\tau}\,d\tau = \frac{2\pi}{\lambda}f_c\left(\frac{\omega-\omega_0}{\lambda}\right)$$

and (10-161) follows from (10-141). ◀

NORMAL PROCESSES. Suppose now that $\mathbf{c}(t)$ is normal with zero mean. In this case $\varphi(t)$ is also normal with zero mean. Hence [see (10-160)]

$$\Phi_\varphi(\lambda, \tau) = \exp\left\{-\tfrac{1}{2}\lambda^2\sigma_\varphi^2(\tau)\right\}$$
$$\sigma_\varphi^2(\tau) = 2\int_0^\tau R_c(\alpha)(\tau-\alpha)\,d\alpha \qquad (10\text{-}165)$$

In general, the Fourier transform of $\Phi_\varphi(\lambda, \tau)$ is found only numerically. However, as we show next, explicit formulas can be obtained if λ is large or small. We introduce

[5] P. M. Woodward: "The Spectrum of Random Frequency Modulation," *Telecommunications Research,* Great Malvern, Worcs., England, Memo 666, 1952.

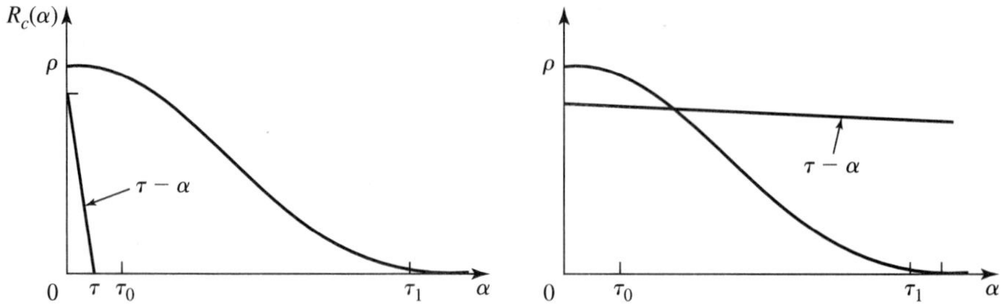

FIGURE 10-17

the "correlation time" τ_c of $\mathbf{c}(t)$:

$$\tau_c = \frac{1}{\rho}\int_0^\infty R_c(\alpha)\,d\alpha \qquad \rho = R_c(0) \tag{10-166}$$

and we select two constants τ_0 and τ_1 such that

$$R_c(\tau) \simeq \begin{cases} 0 & |\tau| > \tau_1 \\ \rho & |\tau| < \tau_0 \end{cases}$$

Inserting into (10-165), we obtain (Fig. 10-17)

$$\sigma_\varphi^2(\tau) \simeq \begin{cases} \rho\tau^2 & |\tau| < \tau_0 \\ 2\rho\tau\tau_c & \tau > \tau_1 \end{cases} \qquad \left.\begin{matrix} e^{-\rho\lambda^2\tau^2/2} \\ e^{-\rho\lambda^2\tau\tau_c} \end{matrix}\right\} \simeq R_{ww}(\tau) \tag{10-167}$$

It is known from the asymptotic properties of Fourier transforms that the behavior of $R_{ww}(\tau)$ for small (large) τ determines the behaviors of $S_{ww}(\omega)$ for large (small) ω. Since

$$\begin{aligned} e^{-\rho\lambda^2\tau^2/2} &\leftrightarrow \frac{1}{\lambda}\sqrt{\frac{2\pi}{\rho}}e^{-\omega^2/2\rho\lambda^2} \\ e^{-\rho\lambda^2\tau_c|\tau|} &\leftrightarrow \frac{2\rho\tau_c\lambda^2}{\omega^2+\rho^2\tau_c^2\lambda^4} \end{aligned} \tag{10-168}$$

we conclude that $S_{ww}(\omega)$ is *lorenzian* near the origin and it is asymptotically *normal* as $\omega \to \infty$. As we show next, these limiting cases give an adequate description of $S_{ww}(\omega)$ for large or small λ.

Wideband FM. If λ is such that

$$\rho\lambda^2\tau_0^2 \gg 1$$

then $R_{ww}(\tau) \simeq 0$ for $|\tau| > \tau_0$. This shows that we can use the upper approximation in (10-167) for every significant value of τ. The resulting spectrum equals

$$S_{ww}(\omega) \simeq \frac{1}{\lambda}\sqrt{\frac{2\pi}{\rho}}e^{-\omega^2/2\rho\lambda^2} = \frac{2\pi}{\lambda}f_c\left(\frac{\omega}{\lambda}\right) \tag{10-169}$$

in agreement with Woodward's theorem. The last equality in (10-169) follows because $\mathbf{c}(t)$ is normal with variance $E\{\mathbf{c}^2(t)\} = \rho$.

Narrowband FM. If λ is such that

$$\rho\lambda^2\tau_1\tau_c \ll 1$$

then $R_{ww}(\tau) \simeq 1$ for $|\tau| < \tau_1$. This shows that we can use the lower approximation in (10-167) for every significant value of τ. Hence

$$S_{ww}(\omega) \simeq \frac{2\rho\tau_c\lambda^2}{\omega^2 + \rho^2\tau_c^2\lambda^4} \qquad (10\text{-}170)$$

10-4 CYCLOSTATIONARY PROCESSES[6]

A process $\mathbf{x}(t)$ is called strict-sense cyclostationary (SSCS) with period T if its statistical properties are invariant to a shift of the origin by integer multiples of T, or, equivalently, if

$$F(x_1, \ldots, x_n; t_1 + mT, \ldots, t_n + mT) = F(x_1, \ldots, x_n; t_1, \ldots, t_n) \qquad (10\text{-}171)$$

for every integer m.

A process $\mathbf{x}(t)$ is called wide-sense cyclostationary (WSCS) if

$$\eta(t + mT) = \eta(t) \qquad R(t_1 + mT, t_2 + mT) = R(t_1, t_2) \qquad (10\text{-}172)$$

for every integer m.

It follows from the definition that if $\mathbf{x}(t)$ is SSCS, it is also WSCS. The following theorems show the close connection between stationary and cyclostationary processes.

THEOREM 10-5

SSCS AND SSS

▶ If $\mathbf{x}(t)$ is an SSCS process and $\boldsymbol{\theta}$ is a random variable uniform in the interval $(0, T)$ and independent of $\mathbf{x}(t)$, then the *shifted* process

$$\bar{\mathbf{x}}(t) = \mathbf{x}(t - \boldsymbol{\theta}) \qquad (10\text{-}173)$$

obtained by a random shift of the origin is SSS and its nth-order distribution equals

$$\bar{F}(x_1, \ldots, x_n; t_1, \ldots, t_n) = \frac{1}{T}\int_0^T F(x_1, \ldots, x_n; t_1 - \alpha, \ldots, t_n - \alpha)\, d\alpha \qquad (10\text{-}174)$$

Proof. To prove the theorem, it suffices to show that the probability of the event

$$A = \{\bar{\mathbf{x}}(t_1 + c) \le x_1, \ldots, \bar{\mathbf{x}}(t_n + c) \le x_n\}$$

is independent of c and it equals the right side of (10-174). As we know [see (4-80)]

$$P(A) = \frac{1}{T}\int_0^T P(A \mid \boldsymbol{\theta} = \theta)\, d\theta \qquad (10\text{-}175)$$

Furthermore,

$$P(A \mid \boldsymbol{\theta} = \theta) = P\{\mathbf{x}(t_1 + c - \theta) \le x_1, \ldots, \mathbf{x}(t_n + c - \theta) \le x_n \mid \theta\}$$

And since $\boldsymbol{\theta}$ is independent of $\mathbf{x}(t)$, we conclude that

$$P\{A \mid \boldsymbol{\theta} = \theta\} = F(x_1, \ldots, x_n; t_1 + c - \theta, \ldots, t_n + c - \theta)$$

Inserting into (10-175) and using (10-171), we obtain (10-174). ◀

---

[6]N. A. Gardner and L. E. Franks: Characteristics of Cyclostationary Random Signal Processes, *IEEE Transactions in Information Theory,* vol. IT-21, 1975.

THEOREM 10-6

WSCS AND WSS

If $\mathbf{x}(t)$ is a WSCS process, then the shifted process $\bar{\mathbf{x}}(t)$ is WSS with mean

$$\bar{\eta} = \frac{1}{T}\int_0^T \eta(t)\,dt \tag{10-176}$$

and autocorrelation

$$\bar{R}(\tau) = \frac{1}{T}\int_0^T R(t+\tau, t)\,dt \tag{10-177}$$

Proof. From (6-240) and the independence of $\boldsymbol{\theta}$ from $\mathbf{x}(t)$, it follows that

$$E\{\mathbf{x}(t-\boldsymbol{\theta})\} = E\{\eta(t-\boldsymbol{\theta})\} = \frac{1}{T}\int_0^T \eta(t-\theta)\,d\theta$$

and (10-176) results because $\eta(t)$ is periodic. Similarly,

$$E\{\mathbf{x}(t+\tau-\boldsymbol{\theta})\mathbf{x}(t-\boldsymbol{\theta})\} = E\{R(t+\tau-\boldsymbol{\theta}, t-\boldsymbol{\theta})\}$$
$$= \frac{1}{T}\int_0^T R(t+\tau-\theta, t-\theta)\,d\theta$$

This yields (10-177) because $R(t+\tau, t)$ is a periodic function of t. ◀

Pulse-Amplitude Modulation (PAM)

An important example of a cyclostationary process is the random signal

$$\mathbf{x}(t) = \sum_{n=-\infty}^{\infty} \mathbf{c}_n h(t-nT) \tag{10-178}$$

where $h(t)$ is a given function with Fourier transform $H(\omega)$ and $\mathbf{c}_n$ is a stationary sequence of random variables with autocorrelation $R_c[m] = E\{\mathbf{c}_{n+m}\mathbf{c}_n\}$ and power spectrum

$$\mathbf{S}_c(e^{j\omega}) = \sum_{m=-\infty}^{\infty} R_c[m]e^{-jm\omega} \tag{10-179}$$

THEOREM 10-7

PAM SPECTRUM

The power spectrum $\bar{S}_x(\omega)$ of the shifted process $\bar{\mathbf{x}}(t)$ equals

$$\bar{S}_x(\omega) = \frac{1}{T}\mathbf{S}_c(e^{j\omega})|H(\omega)|^2 \tag{10-180}$$

Proof. We form the impulse train

$$\mathbf{z}(t) = \sum_{n=-\infty}^{\infty} \mathbf{c}_n \delta(t-nT) \tag{10-181}$$

Clearly, $\mathbf{z}(t)$ is the derivative of the process $\mathbf{w}(t)$ of Fig. 10-18:

$$\mathbf{w}(t) = \sum_{n=-\infty}^{\infty} \mathbf{c}_n U(t-nT) \qquad \mathbf{z}(t) = \mathbf{w}'(t) \tag{10-182}$$

The process $\mathbf{w}(t)$ is cyclostationary with autocorrelation

$$R_w(t_1, t_2) = \sum_n \sum_r R_c(n-r)U(t_1-nT)U(t_2-rT)$$

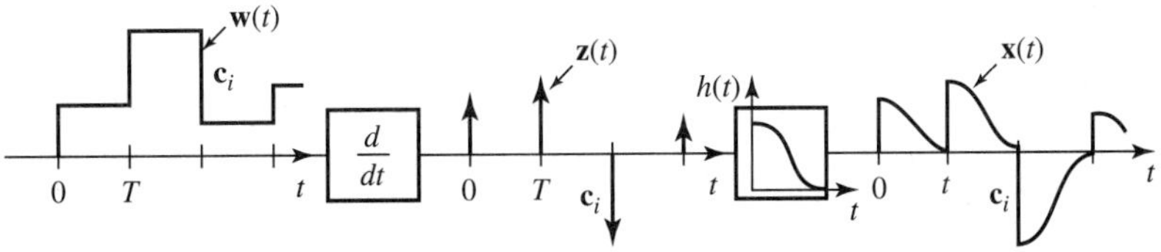

FIGURE 10-18

From (9-103) it follows that

$$R_z(t_1, t_2) = \frac{\partial^2 R_w(t_1, t_2)}{\partial t_1 \, \partial t_2} = \sum_n \sum_r R_c[n-r]\delta(t_1 - nT)\delta(t_2 - rT)$$

This yields

$$R_z(t+\tau, t) = \sum_{m=-\infty}^{\infty} R_c[m] \sum_{r=-\infty}^{\infty} \delta[t+\tau-(m+r)T)\delta(t-rT] \tag{10-183}$$

We shall find first the autocorrelation $\overline{R}_z(\tau)$ and the power spectrum $\overline{S}_z(\omega)$ of the shifted process $\overline{\mathbf{z}}(t) = \mathbf{z}(t - \boldsymbol{\theta})$. Inserting (10-183) into (10-177) and using the identity

$$\int_0^T \delta[t+\tau-(m+r)T]\delta(t-rT)\,dt = \delta(\tau - mT)$$

we obtain

$$\overline{R}_z(\tau) = \frac{1}{T}\sum_{m=-\infty}^{\infty} R_c[m]\delta(\tau - mT) \tag{10-184}$$

From this it follows that

$$\overline{S}_z(\omega) = \frac{1}{T}\sum_{n=-\infty}^{\infty} R_c[m]e^{-jmT\omega} = \frac{1}{T}S_c(\omega) \tag{10-185}$$

The process $\mathbf{x}(t)$ is the output of a linear system with input $\mathbf{z}(t)$. Thus

$$\mathbf{x}(t) = \mathbf{z}(t) * h(t) \qquad \overline{\mathbf{x}}(t) = \overline{\mathbf{z}}(t) * h(t)$$

Hence [see (10-185) and (9-149)] the power spectrum of the shifted PAM process $\overline{\mathbf{x}}(t)$ is given by (10-180). ◀

COROLLARY ▶ If the process $\mathbf{c}_n$ is white noise with $S_c(\omega) = q$, then

$$\overline{S}_x(\omega) = \frac{q}{T}|H(\omega)|^2 \qquad \overline{R}_x(\tau) = \frac{q}{T}h(t) * h(-t) \tag{10-186}$$

◀

EXAMPLE 10-4 ▶ Suppose that $h(t)$ is a pulse and $\mathbf{c}_n$ is a white-noise process taking the values ± 1 with equal probability:

$$h(t) = \begin{cases} 1 & 0 \le t < T \\ 0 & \text{otherwise} \end{cases} \qquad \mathbf{c}_n = \mathbf{x}(nT) \qquad R_c[m] = \delta[m]$$

The resulting process $\mathbf{x}(t)$ is called *binary transmission*. It is SSCS taking the values ± 1

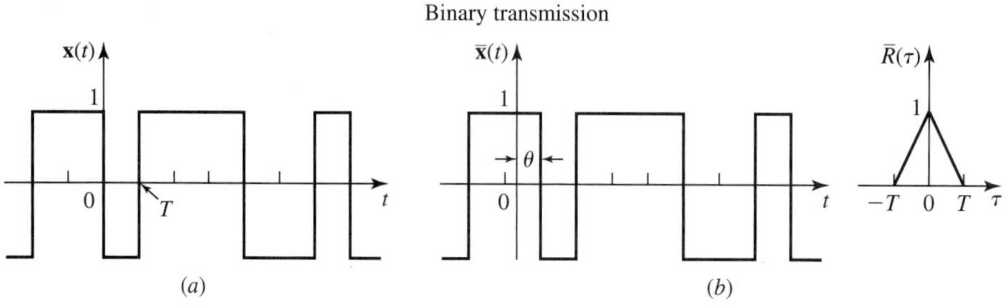

FIGURE 10-19

in every interval $(nT - T, nT)$, the shifted process $\bar{\mathbf{x}}(t) = \mathbf{x}(t - \boldsymbol{\theta})$ is stationary. From (10-180) it follows that

$$\overline{S}_x(\omega) = \frac{4\sin^2(\omega T/2)}{T\omega^2}$$

because $\mathbf{S}_c(z) = 1$. Hence $\overline{R}_x(\tau)$ is a triangle as in Fig. 10-19. ◀

10-5 BANDLIMITED PROCESSES AND SAMPLING THEORY

A process $\mathbf{x}(t)$ is called bandlimited (abbreviated BL) if its spectrum vanishes for $|\omega| > \sigma$ and it has finite power:

$$S(\omega) = 0 \qquad |\omega| > \sigma, \qquad R(0) < \infty \tag{10-187}$$

In this section we establish various identities involving linear functionals of BL processes. To do so, we express the two sides of each identity as responses of linear systems. The underlying reasoning is based on the following:

THEOREM 10-8

▶ Suppose that $\mathbf{w}_1(t)$ and $\mathbf{w}_2(t)$ are the responses of the systems $T_1(\omega)$ and $T_2(\omega)$ to a BL process $\mathbf{x}(t)$ (Fig. 10-20). We shall show that if

$$T_1(\omega) = T_2(\omega) \qquad \text{for} \quad |\omega| \le \sigma \tag{10-188}$$

then

$$\mathbf{w}_1(t) = \mathbf{w}_2(t) \tag{10-189}$$

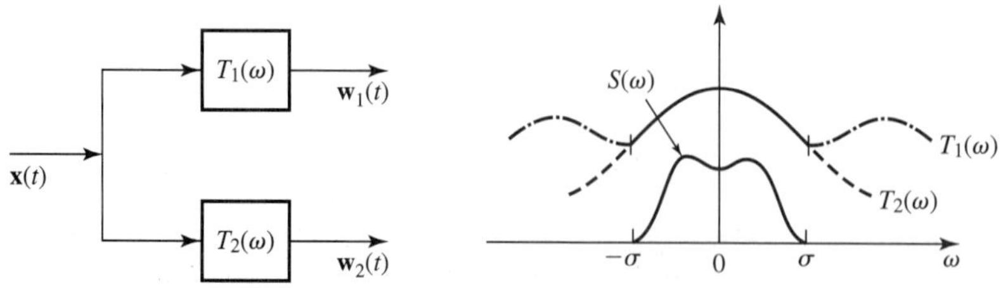

FIGURE 10-20

Proof. The difference $\mathbf{w}_1(t) - \mathbf{w}_2(t)$ is the response of the system $T_1(\omega) - T_2(\omega)$ to the input $\mathbf{x}(t)$. Since $S(\omega) = 0$ for $|\omega| > \sigma$, we conclude from (9-152) and (10-188) that

$$E\{|\mathbf{w}_1(t) - \mathbf{w}_2(t)|^2\} = \frac{1}{2\pi}\int_{-\sigma}^{\sigma} S(\omega)|T_1(\omega) - T_2(\omega)|^2\, d\omega = 0$$

Hence[7] $\mathbf{w}_1(t) = \mathbf{w}_2(t)$. ◀

Taylor series. If $\mathbf{x}(t)$ is BL, then [see (9-134)]

$$R(\tau) = \frac{1}{2\pi}\int_{-\sigma}^{\sigma} S(\omega)e^{j\omega\tau}\, d\omega \tag{10-190}$$

In (10-190), the limits of integration are finite and the area $2\pi R(0)$ of $S(\omega)$ is also finite. We can therefore differentiate under the integral sign

$$R^{(n)}(\tau) = \frac{1}{2\pi}\int_{-\sigma}^{\sigma} (j\omega)^n S(\omega)e^{j\omega\tau}\, d\omega \tag{10-191}$$

This shows that the autocorrelation of a BL process is an entire function; that is, it has derivatives of any order for every τ. From this it follows that $\mathbf{x}^{(n)}(t)$ exists for any n (see App. 9A).

We maintain that

$$\mathbf{x}(t+\tau) = \sum_{n=0}^{\infty} \mathbf{x}^{(n)}(t)\frac{\tau^n}{n!} \tag{10-192}$$

Proof. We shall prove (10-192) using (10-189). As we know

$$e^{j\omega\tau} = \sum_{n=0}^{\infty} (j\omega)^n \frac{\tau^n}{n!} \qquad \text{all} \quad \omega \tag{10-193}$$

The processes $\mathbf{x}(t+\tau)$ and $\mathbf{x}^{(n)}(t)$ are the responses of the systems $e^{j\omega\tau}$ and $(j\omega)^n$, respectively, to the input $\mathbf{x}(t)$. If, therefore, we use as systems $T_1(\omega)$ and $T_2(\omega)$ in (10-188) the two sides of (10-193), the resulting responses will equal the two sides of (10-192). And since (10-193) is true for all ω, (10-192) follows from (10-189).

Bounds. Bandlimitedness is often associated with slow variation. The following is an analytical formulation of this association.

If $\mathbf{x}(t)$ is BL, then

$$E\{[\mathbf{x}(t+\tau) - \mathbf{x}(t)]^2\} \le \sigma^2\tau^2 R(0) \tag{10-194}$$

or, equivalently,

$$2[R(0) - R(\tau)] \le \sigma^2\tau^2 R(0) \tag{10-195}$$

Proof. The familiar inequality $|\sin\varphi| \le |\varphi|$ yields

$$1 - \cos\omega\tau = 2\sin^2\frac{\omega\tau}{2} \le \frac{\omega^2\tau^2}{2}$$

---

[7]All identities in this section are interpreted in the MS sense.

Since $S(\omega) \geq 0$, it follows from the above expression and (9-135) that

$$R(0) - R(\tau) = \frac{1}{2\pi}\int_{-\sigma}^{\sigma} S(\omega)(1 - \cos\omega\tau)\,d\omega$$

$$\leq \frac{1}{2\pi}\int_{-\sigma}^{\sigma} S(\omega)\frac{\omega^2\tau^2}{2}\,d\omega \leq \frac{\sigma^2\tau^2}{4\pi}\int_{-\sigma}^{\sigma} S(\omega)\,d\omega = \frac{\sigma^2\tau^2}{2}R(0)$$

as in (10-195).

Sampling Expansions

The sampling theorem for deterministic signals states that if $f(t) \leftrightarrow F(\omega)$ and $F(\omega) = 0$ for $|\omega| > \sigma$, then the function $f(t)$ can be expressed in terms of its samples $f(nT)$, where $T = \pi/\sigma$ is the *Nyquist interval.* The resulting expansion applied to the autocorrelation $R(\tau)$ of a BL process $\mathbf{x}(t)$ takes the following form:

$$R(\tau) = \sum_{n=-\infty}^{\infty} R(nT)\frac{\sin\sigma(\tau - nT)}{\sigma(\tau - nT)} \tag{10-196}$$

We shall establish a similar expansion for the process $\mathbf{x}(t)$.

THEOREM 10-9

STOCHASTIC SAMPLING THEOREM

If $\mathbf{x}(t)$ is a BL process, then

$$\mathbf{x}(t + \tau) = \sum_{n=-\infty}^{\infty} \mathbf{x}(t + nT)\frac{\sin\sigma(\tau - nT)}{\sigma(\tau - nT)} \qquad T = \frac{\pi}{\sigma} \tag{10-197}$$

for every t and τ. This is a slight extension of (10-196). This extension will permit us to base the proof of the theorem on (10-189).

Proof. We consider the exponential $e^{j\omega\tau}$ as a function of ω, viewing τ as a parameter, and we expand it into a Fourier series in the interval $(-\sigma \leq \omega \leq \sigma)$. The coefficients of this expansion equal

$$a_n = \frac{1}{2\sigma}\int_{-\sigma}^{\sigma} e^{j\omega\tau}e^{-jnT\omega}\,d\omega = \frac{\sin\sigma(\tau - nT)}{\sigma(\tau - nT)}$$

Hence

$$e^{j\omega\tau} = \sum_{n=-\infty}^{\infty} e^{jnT\omega}\frac{\sin\sigma(\tau - nT)}{\sigma(\tau - nT)} \qquad |\omega| \leq \sigma \tag{10-198}$$

We denote by $T_1(\omega)$ and $T_2(\omega)$ the left and right side respectively of (10-197). Clearly, $T_1(\omega)$ is a delay line and its response $\mathbf{w}_1(t)$ to $\mathbf{x}(t)$ equals $\mathbf{x}(t + \tau)$. Similarly, the response $\mathbf{w}_2(t)$ of $T_2(\omega)$ to $\mathbf{x}(t)$ equals the right side of (10-197). Since $T_1(\omega) = T_2(\omega)$ for $|\omega| < \sigma$, (10-197) follows from (10-189). ◀

Past samples. A deterministic BL signal is determined only if all its samples, past and future, are known. This is not necessary for random signals. We show next that a BL process $\mathbf{x}(t)$ can be approximated arbitrarily closely by a sum involving only its past samples $\mathbf{x}(nT_0)$ provided that $T_0 < T$. We illustrate first with an example.[8]

---

[8] L. A. Wainstein and V. Zubakov: *Extraction of Signals in Noise,* Prentice-Hall, Englewood Cliffs, NJ, 1962.

EXAMPLE 10-5

SAMPLING BASED ON PAST SAMPLES

▶ Consider the process

$$\hat{\mathbf{x}}(t) = n\mathbf{x}(t - T_0) - \binom{n}{2}\mathbf{x}(t - 2T_0) + \cdots - (-1)^n\mathbf{x}(t - nT_0) \qquad (10\text{-}199)$$

The difference

$$\mathbf{y}(t) = \mathbf{x}(t) - \hat{\mathbf{x}}(t) = \sum_{k=0}^{n}(-1)^k\binom{n}{k}\mathbf{x}(t - kT_0)$$

is the response of the system

$$H(\omega) = \sum_{k=0}^{n}(-1)^k\binom{n}{k}e^{-jkT_0\omega} = \left(1 - e^{-j\omega T_0}\right)^n$$

with input $\mathbf{x}(t)$. Since $|H(\omega)| = |2\sin(\omega T_0/2)|^n$, we conclude from (9-45) that

$$E\{\mathbf{y}^2(t)\} = \frac{1}{2\pi}\int_{-\sigma}^{\sigma} S(\omega)\left(2\sin\frac{\omega T_0}{2}\right)^{2n} d\omega \qquad (10\text{-}200)$$

If $T_0 < \pi/3\sigma$, then $2\sin|\omega T_0/2| < 2\sin(\pi/6) = 1$ for $|\omega| < \sigma$. From this it follows that the integrand in (10-200) tends to 0 as $n \to \infty$. Therefore, $E\{\mathbf{y}^2(t)\} \to 0$ and

$$\hat{\mathbf{x}}(t) \to \mathbf{x}(t) \qquad \text{as} \quad n \to \infty$$

Note that this holds only if $T_0 < T/3$; furthermore, the coefficients $\binom{n}{k}$ of $\hat{\mathbf{x}}(t)$ tend to ∞ as $n \to \infty$. ◀

We show next that $\mathbf{x}(t)$ can be approximated arbitrarily closely by a sum involving only its past samples $\mathbf{x}(t - kT_0)$, where T_0 is a number smaller than T but otherwise arbitrary.

THEOREM 10-10

▶ Given a number $T_0 < T$ and a constant $\varepsilon > 0$, we can find a set of coefficients a_k such that

$$E\{|\mathbf{x}(t) - \hat{\mathbf{x}}(t)|^2\} < \varepsilon \qquad \hat{\mathbf{x}}(t) = \sum_{k=1}^{n} a_k\mathbf{x}(t - kT_0) \qquad (10\text{-}201)$$

where n is a sufficiently large constant.

Proof. The process $\hat{\mathbf{x}}(t)$ is the response of the system

$$P(\omega) = \sum_{k=1}^{n} a_k e^{-jkT_0\omega} \qquad (10\text{-}202)$$

with input $\mathbf{x}(t)$. Hence

$$E\{|\mathbf{x}(t) - \hat{\mathbf{x}}(t)|^2\} = \frac{1}{2\pi}\int_{-\sigma}^{\sigma} S(\omega)|1 - P(\omega)|^2\,d\omega$$

It suffices, therefore, to find a sum of exponentials with positive exponents only, approximating 1 arbitrarily closely. This cannot be done for every $|\omega| < \sigma_0 = \pi/T_0$ because $P(\omega)$ is periodic with period $2\sigma_0$. We can show, however, that if $\sigma_0 > \sigma$, we can find $P(\omega)$ such that the differences $|1 - P(\omega)|$ can be made arbitrarily small for $|\omega| < \sigma$ as in Fig. 10-21. The proof follows from

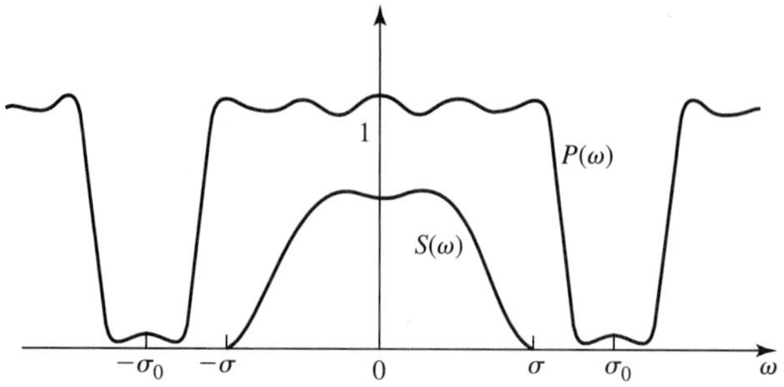

FIGURE 10-21

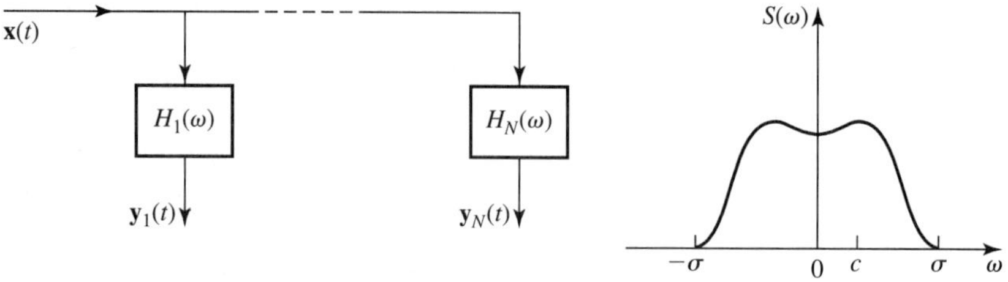

FIGURE 10-22

the Weierstrass approximation theorem and the Fejer–Riesz factorization theorem; the details, however, are not simple.[9]

Note that, as in Example 10-5, the coefficients a_k tend to ∞ as $\varepsilon \to 0$. This is based on the fact that we cannot find a sum $P(\omega)$ of exponentials as in (10-202) such that $|1 - P(\omega)| = 0$ for every ω in an interval. This would violate the Paley–Wiener condition (9-203) or more generally (11-9). ◀

THE PAPOULIS SAMPLING EXPANSION.[10] The sampling expansion holds only if $T \le \pi/\sigma$. The following theorem states that if we have access to the samples of the outputs $\mathbf{y}_1(t), \ldots, \mathbf{y}_N(t)$ of N linear systems $H_1(\omega), \ldots, H_N(\omega)$ driven by $\mathbf{x}(t)$ (Fig. 10-22), then we can increase the sampling interval from π/σ to $N\pi/\sigma$.

We introduce the constants

$$c = \frac{2\sigma}{N} = \frac{2\pi}{\overline{T}} \qquad \overline{T} = NT \tag{10-203}$$

and the N functions

$$P_1(\omega, t), \ldots, P_N(\omega, t)$$

---

[9] A. Papoulis: "A Note on the Predictability of Band-Limited Processes," *Proceedings of the IEEE,* vol. 13, no. 8, 1985.

[10] A. Papoulis: "New Results in Sampling Theory," *Hawaii Intern Conf. System Sciences,* January, 1968. (See also Papoulis, 1968 [19], pp. 132–137.)

defined as the solutions of the system

$$
\begin{aligned}
H_1(\omega)P_1(\omega,\tau) + \cdots + H_N(\omega)P_N(\omega,\tau) &= 1\\
H_1(\omega+c)P_1(\omega,\tau) + \cdots + H_N(\omega+c)P_N(\omega,\tau) &= e^{jc\tau}\\
\cdots\cdots\cdots\cdots\cdots\cdots\cdots\cdots\cdots\cdots\cdots\cdots &\\
H_1(\omega+Nc-c)P_1(\omega,\tau) + \cdots + H_N(\omega+Nc-c)P_N(\omega,\tau) &= e^{j(N-1)c\tau}
\end{aligned}
\qquad (10\text{-}204)
$$

In (10-204), ω takes all values in the interval $(-\sigma, -\sigma+c)$ and τ is arbitrary.

We next form the N functions

$$p_k(\tau) = \frac{1}{c}\int_{-\sigma}^{-\sigma+c} P_k(\omega,\tau)e^{j\omega\tau}\,d\omega \qquad 1 \le k \le N \qquad (10\text{-}205)$$

THEOREM 10-11

$$\mathbf{x}(t+\tau) = \sum_{n=-\infty}^{\infty} [\mathbf{y}_1(t+n\overline{T})p_1(\tau-n\overline{T}) + \cdots + \mathbf{y}_N(t+n\overline{T})p_N(\tau-n\overline{T})] \qquad (10\text{-}206)$$

Proof. The process $\mathbf{y}_i(t+n\overline{T})$ is the response of the system $H_i(\omega)e^{jn\overline{T}\omega}$ to the input $\mathbf{x}(t)$. Therefore, if we use as systems $T_1(\omega)$ and $T_2(\omega)$ in Fig. 10-20 the two sides of the identity

$$e^{j\omega\tau} = H_1(\omega)\sum_{n=-\infty}^{\infty} p_1(\tau-n\overline{T})e^{jn\omega\overline{T}} + \cdots + H_N(\omega)\sum_{n=-\infty}^{\infty} p_N(\tau-n\overline{T})e^{jn\omega\overline{T}} \qquad (10\text{-}207)$$

the resulting responses will equal the two sides of (10-206). To prove (10-206), it suffices, therefore, to show that (10-207) is true for every $|\omega| \le \sigma$.

The coefficients $H_k(\omega+kc)$ of the system (10-207) are independent of τ and the right side consists of periodic functions of τ with period $\overline{T} = 2\pi/c$ because $e^{jkc\overline{T}} = 1$. Hence the solutions $P_k(\omega,\tau)$ are periodic

$$P_k(\omega, \tau - n\overline{T}) = P_k(\omega,\tau)$$

From this and (10-205) it follows that

$$p_k(\tau-n\overline{T}) = \frac{1}{c}\int_{-\sigma}^{-\sigma+c} P_k(\omega,\tau)e^{j\omega(\tau-n\overline{T})}\,d\omega$$

This shows that if we expand the function $P_k(\omega,\tau)e^{j\omega\tau}$ into a Fourier series in the interval $(-\sigma, -\sigma+c)$, the coefficient of the expansion will equal $p_k(\tau-n\overline{T})$. Hence

$$P_k(\omega,\tau)e^{j\omega\tau} = \sum_{n=-\infty}^{\infty} p_k(\tau-n\overline{T})e^{jn\omega\overline{T}} \qquad -\sigma < \omega < -\sigma+c \qquad (10\text{-}208)$$

Multiplying each of the equations in (10-204) by $e^{j\omega\tau}$ and using (10-208) and the identity

$$e^{jn(\omega+kc)\overline{T}} = e^{jn\omega\overline{T}}$$

we conclude that (10-207) is true for every ω in the interval $(-\sigma, \sigma)$. ◀

Random Sampling

We wish to estimate the Fourier transform $F(\omega)$ of a deterministic signal $f(t)$ in terms of a sum involving the samples of $f(t)$. If we approximate the integral of $f(t)e^{-j\omega t}$ by its Riemann sum, we obtain the estimate

$$F(\omega) \simeq F_*(\omega) \equiv \sum_{n=-\infty}^{\infty} T f(nT) e^{-jn\omega T} \tag{10-209}$$

From the Poisson sum formula (10A-1), it follows that $F_*(\omega)$ equals the sum of $F(\omega)$ and its displacements

$$F_*(\omega) = \sum_{n=-\infty}^{\infty} F(\omega + 2n\sigma) \qquad \sigma = \frac{\pi}{T}$$

Hence $F_*(\omega)$ can be used as the estimate of $F(\omega)$ in the interval $(-\sigma, \sigma)$ only if $F(\omega)$ is negligible outside this interval. The difference $F(\omega) - F_*(\omega)$ is called *aliasing error*. In the following, we replace in (10-209) the equidistant samples $f(nT)$ of $f(t)$ by its samples $f(\mathbf{t}_i)$ at a random set of points $\mathbf{t}_i$ and we examine the nature of the resulting error.[11]

We maintain that if $\mathbf{t}_i$ is a Poisson point process with average density λ, then the sum

$$\mathbf{P}(\omega) = \frac{1}{\lambda} \sum_i f(\mathbf{t}_i) e^{-j\omega \mathbf{t}_i} \tag{10-210}$$

is an unbiased estimate of $F(\omega)$. Furthermore, if the energy

$$E = \int_{-\infty}^{\infty} f^2(t)\, dt$$

of $f(t)$ is finite, then $\mathbf{P}(\omega) \to F(\omega)$ as $\lambda \to \infty$. To prove this, it suffices to show that

$$E\{\mathbf{P}(\omega)\} = F(\omega) \qquad \sigma^2_{P(\omega)} = \frac{E}{\lambda} \tag{10-211}$$

Proof. Clearly,

$$\int_{-\infty}^{\infty} f(t) e^{-j\omega t} \sum_i \delta(t - \mathbf{t}_i)\, dt = \sum_i f(\mathbf{t}_i) e^{-j\omega \mathbf{t}_i} \tag{10-212}$$

Comparing with (10-210), we obtain

$$\mathbf{P}(\omega) = \frac{1}{\lambda} \int_{-\infty}^{\infty} f(t)\mathbf{z}(t) e^{-j\omega t}\, dt \qquad \text{where} \quad \mathbf{z}(t) = \sum_i \delta(t - \mathbf{t}_i) \tag{10-213}$$

is a Poisson impulse train as in (9-107) with

$$E\{\mathbf{z}(t)\} = \lambda \qquad C_z(t_1, t_2) = \lambda \delta(t_1 - t_2) \tag{10-214}$$

Hence

$$E\{\mathbf{P}(\omega)\} = \frac{1}{\lambda} \int_{-\infty}^{\infty} f(t) E\{\mathbf{z}(t)\} e^{-j\omega t}\, dt = F(\omega)$$

$$\sigma^2_{P(\omega)} = \frac{1}{\lambda^2} \int_{-\infty}^{\infty}\int_{-\infty}^{\infty} f(t_1) f(t_2) \lambda \delta(t_1 - t_2)\, dt_1\, dt_2 = \frac{1}{\lambda} \int_{-\infty}^{\infty} f^2(t_2)\, dt_2$$

and (10-211) results.

[11] E. Masry: "Poisson Sampling and Spectral Estimation of Continuous-Time Processes," *IEEE Transactions on Information Theory,* vol. IT-24, 1978. See also F. J. Beutler: "Alias Free Randomly Timed Sampling of Stochastic Processes," *IEEE Transactions on Information Theory,* vol. IT-16, 1970.

From (10-211) it follows that, for a satisfactory estimate of $F(\omega)$, λ must be such that

$$|F(\omega)| \gg \sqrt{\frac{E}{\lambda}} \tag{10-215}$$

EXAMPLE 10-6

▶ Suppose that $f(t)$ is a sum of sine waves in the interval $(-a, a)$:

$$f(t) = \sum_k c_k e^{j\omega_k t} \qquad |t| < a$$

and it equals 0 for $|t| > a$. In this case,

$$F(\omega) = \sum_k 2c_k \frac{\sin a(\omega - \omega_k)}{\omega - \omega_k} \qquad E \simeq 2a \sum_k |c_k|^2 \tag{10-216}$$

where we neglected cross-products in the evaluation of E. If a is sufficiently large, then

$$F(\omega_k) \simeq 2ac_k$$

This shows that if

$$\sum_i |c_i|^2 \ll 2a\lambda |c_k|^2 \qquad \text{then} \quad \mathbf{P}(\omega_k) \simeq F(\omega_k)$$

Thus with random sampling we can detect line spectra of any frequency even if the average rate λ is small, provided that the observation interval $2a$ is large. ◀

10-6 DETERMINISTIC SIGNALS IN NOISE

A central problem in the applications of stochastic processes is the estimation of a signal in the presence of noise. This problem has many aspects (see Chap. 13). In the following, we discuss two cases that lead to simple solutions. In both cases the signal is a deterministic function $f(t)$ and the noise is a random process $\mathbf{v}(t)$ with zero mean.

The Matched Filter Principle

The following problem is typical in radar: A signal of known form is reflected from a distant target. The received signal is a sum

$$\mathbf{x}(t) = f(t) + \mathbf{v}(t) \qquad E\{\mathbf{v}(t)\} = 0$$

where $f(t)$ is a shifted and scaled version of the transmitted signal and $\mathbf{v}(t)$ is a WSS process with known power spectrum $S(\omega)$. We assume that $f(t)$ is known and we wish to establish its presence and location. To do so, we apply the process $\mathbf{x}(t)$ to a linear filter with impulse response $h(t)$ and system function $H(\omega)$. The resulting output $\mathbf{y}(t) = \mathbf{x}(t) * h(t)$ is a sum

$$\mathbf{y}(t) = \int_{-\infty}^{\infty} \mathbf{x}(t-\alpha)h(\alpha)d(\alpha) = y_f(t) + \mathbf{y}_\nu(t) \tag{10-217}$$

where

$$y_f(t) = \int_{-\infty}^{\infty} f(t-\alpha)h(\alpha)\,d\alpha = \frac{1}{2\pi}\int_{-\infty}^{\infty} F(\omega)H(\omega)e^{j\omega t}\,d\omega \qquad (10\text{-}218)$$

is the response due to the signal $f(t)$, and $\mathbf{y}_\nu(t)$ is a random component with average power

$$E\{\mathbf{y}_\nu^2(t)\} = \frac{1}{2\pi}\int_{-\infty}^{\infty} S(\omega)|H(\omega)|^2\,d\omega \qquad (10\text{-}219)$$

Since $\mathbf{y}_\nu(t)$ is due to $\boldsymbol{\nu}(t)$ and $E\{\boldsymbol{\nu}(t)\} = 0$, we conclude that $E\{\mathbf{y}_\nu(t)\} = 0$ and $E\{\mathbf{y}(t)\} = y_f(t)$. Our objective is to find $H(\omega)$ so as to maximize the *signal-to-noise* ratio

$$\rho = \frac{|y_f(t_0)|}{\sqrt{E\{\mathbf{y}_\nu^2(t_0)\}}} \qquad (10\text{-}220)$$

at a specific time t_0.

WHITE NOISE. Suppose, first, that $S(\omega) = \sigma_0^2$. Applying Schwarz's inequality (10B-1) to the second integral in (10-218), we conclude that

$$\rho^2 \le \frac{\int |F(\omega)e^{j\omega t_0}|^2\,d\omega \int |H(\omega)|^2\,d\omega}{2\pi\sigma_0^2 \int |H(\omega)|^2\,d\omega} = \frac{E_f}{\sigma_0^2} \qquad (10\text{-}221)$$

where $E_f = (1/2\pi)\int |F(\omega)|^2\,d\omega$ is the energy of $f(t)$. The above is an equality if [see (10B-2)]

$$H(\omega) = kF^*(\omega)e^{-j\omega t_0} \qquad h(t) = kf(t_0 - t) \qquad (10\text{-}222)$$

This determines the optimum $H(\omega)$ within a constant factor k. The system so obtained is called the *matched filter*. The resulting signal-to-noise ratio is maximum and it equals $\sqrt{E_f/\sigma_0^2}$.

COLORED NOISE. The solution is not so simple if $S(\omega)$ is not a constant. In this case, first multiply and divide the integrand of (10-218) by $\sqrt{S(\omega)}$ and then apply Schwarz's inequality. This yields

$$\begin{aligned}|2\pi y_f(t_0)|^2 &= \left|\int \frac{F(\omega)}{\sqrt{S(\omega)}}\sqrt{S(\omega)}\,H(\omega)e^{j\omega t_0}\,d\omega\right|^2 \\ &\le \int \frac{|F(\omega)|^2}{S(\omega)}\,d\omega \int S(\omega)|H(\omega)|^2\,d\omega\end{aligned}$$

Inserting into (10-220), we obtain

$$\rho^2 \le \frac{\int \frac{|F(\omega)|^2}{S(\omega)}\,d\omega \int S(\omega)|H(\omega)|^2\,d\omega}{2\pi\int S(\omega)|H(\omega)|^2\,d\omega} = \frac{1}{2\pi}\int \frac{|F(\omega)|^2}{S(\omega)}\,d\omega \qquad (10\text{-}223)$$

Equality holds if

$$\sqrt{S(\omega)}\,H(\omega) = k\frac{F^*(\omega)e^{-j\omega t_0}}{\sqrt{S(\omega)}}$$

Thus the signal-to-noise ratio is maximum if

$$H(\omega) = k\frac{F^*(\omega)}{S(\omega)}e^{-j\omega t_0} \tag{10-224}$$

TAPPED DELAY LINE. The matched filter is in general noncausal and difficult to realize.[12] A suboptimal but simpler solution results if $H(\omega)$ is a tapped delay line:

$$H(\omega) = a_0 + a_1 e^{-j\omega T} + \cdots + a_m e^{-jm\omega T} \tag{10-225}$$

In this case,

$$y_f(t_0) = \sum_{i=0}^{m} a_i f(t_0 - iT) \qquad \mathbf{y}_\nu(t) = \sum_{i=0}^{m} a_i \mathbf{v}(t - iT) \tag{10-226}$$

and our problem is to find the $m+1$ constants a_i so as to maximize the resulting signal-to-noise ratio. It can be shown that (see Prob. 10-26) the unknown constants are the solutions of the system

$$\sum_{i=0}^{m} a_i R(nT - iT) = kf(t_0 - nT) \qquad n = 0, \ldots, m \tag{10-227}$$

where $R(\tau)$ is the autocorrelation of $\mathbf{v}(t)$ and k is an arbitrary constant.

Smoothing

We wish to estimate an unknown signal $f(t)$ in terms of the observed value of the sum $\mathbf{x}(t) = f(t) + \mathbf{v}(t)$. We assume that the noise $\mathbf{v}(t)$ is white with known autocorrelation $R(t_1, t_2) = q(t_1)\delta(t_1 - t_2)$. Our estimator is again the response $\mathbf{y}(t)$ of the filter $h(t)$:

$$\mathbf{y}(t) = \int_{-\infty}^{\infty} \mathbf{x}(t - \tau)h(\tau)\,d\tau \tag{10-228}$$

The estimator is biased with bias

$$b = y_f(t) - f(t) = \int_{-\infty}^{\infty} f(t - \tau)h(\tau)\,d\tau - f(t) \tag{10-229}$$

and variance [see (9-99)]

$$\sigma^2 = E\{\mathbf{y}_\nu^2(t)\} = \int_{-\infty}^{\infty} q(t - \tau)h^2(\tau)\,d\tau \tag{10-230}$$

Our objective is to find $h(t)$ so as to minimize the MS error

$$e = E\{[\mathbf{y}(t) - f(t)]^2\} = b^2 + \sigma^2$$

[12] More generally, for the optimum causal transmitter-receiver pair given the target response as well as the interference and noise spectral characteristics, see S. U. Pillai, H. S. Oh, D. C. Youla and J. R. Guerci, "Optimum Transmit-Receiver Design in the Presence of Signal-Dependent Interference and Channel Noise." *IEEE Trans. on Information Theory,* vol. 46, no. 2, pp. 577–584, March 2000.

We shall assume that $h(t)$ is an even positive function of unit area and finite duration:

$$h(-t) = h(t) \qquad \int_{-T}^{T} h(t)\,dt = 1 \qquad h(t) > 0 \tag{10-231}$$

where T is a constant to be determined. If T is small, $y_f(t) \simeq f(t)$, hence the bias is small; however, the variance is large. As T increases, the variance decreases but the bias increases. The determination of the optimum shape and duration of $h(t)$ is in general complicated. We shall develop a simple solution under the assumption that the functions $f(t)$ and $q(t)$ are smooth in the sense that $f(t)$ can be approximated by a parabola and $q(t)$ by a constant in any interval of length $2T$. From this assumption it follows that (Taylor expansion)

$$f(t-\tau) \simeq f(t) - \tau f'(t) + \frac{\tau^2}{2} f''(t) \qquad q(t-\tau) \simeq q(t) \tag{10-232}$$

for $|\tau| < T$. And since the interval of integration in (10-229) and (10-230) is $(-T, T)$, we conclude that

$$b \simeq \frac{f''(t)}{2} \int_{-T}^{T} \tau^2 h(\tau)\,d\tau \qquad \sigma^2 \simeq q(t) \int_{-T}^{T} h^2(\tau)\,d\tau \tag{10-233}$$

because the function $h(t)$ is even and its area equals 1. The resulting MS error equals

$$e \simeq \tfrac{1}{4} M^2 [f''(t)]^2 + E q(t) \tag{10-234}$$

where $M = \int_{-T}^{T} t^2 h(t)\,dt$ and $E = \int_{-T}^{T} h^2(t)\,d\tau$.

To separate the effects of the shape and the size of $h(t)$ on the MS error, we introduce the normalized filter

$$w(t) = T h(Tt) \tag{10-235}$$

The function $w(t)$ is of unit area and $w(t) = 0$ for $|t| > 1$. With

$$M_w = \int_{-1}^{1} t^2 w(t)\,dt = \frac{M}{T^2} \qquad E_w = \int_{-1}^{1} w^2(t)\,dt = TE$$

it follows from (10-231) and (10-234) that

$$b \simeq \frac{T^2}{2} M_w f''(t) \qquad \sigma^2 = \frac{E_w}{T} q(t) \tag{10-236}$$

$$e = \frac{1}{4} T^2 M_w^2 [f''(t)]^2 + \frac{E_w}{T} q(t) \tag{10-237}$$

Thus e depends on the shape of $w(t)$ and on the constant T.

THE TWO-TO-ONE RULE.[13] We assume first that $w(t)$ is specified. In Fig. 10-23 we plot the bias b, the variance σ^2, and the MS error e as functions of T. As T increases, b increases, and σ^2 decreases. Their sum e is minimum for

$$T = T_m = \left(\frac{E_w q(t)}{M_w^2 [f''(t)]^2} \right)^{1/5} \tag{10-238}$$

[13] A. Papoulis, Two-to-One Rule in Data Smoothing, *IEEE Trans. Inf. Theory*, September, 1977.

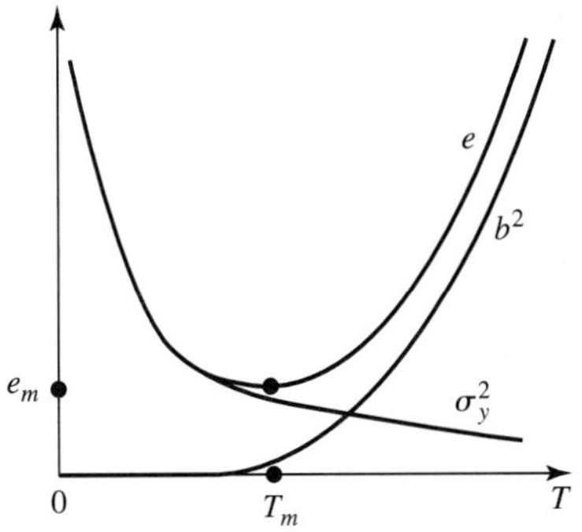

FIGURE 10-23

Inserting into (10-236), we conclude, omitting the simple algebra, that

$$\sigma = 2b \tag{10-239}$$

Thus if $w(t)$ is of specified shape and T is chosen so as to minimize the MS error e, then the standard deviation of the estimation error equals twice its bias.

MOVING AVERAGE. A simple estimator of $f(t)$ is the moving average

$$\mathbf{y}(t) = \frac{1}{2T}\int_{t-T}^{t+T} \mathbf{x}(\tau)\,d\tau$$

of $\mathbf{x}(t)$. This is a special case of (10-228), where the normalized filter $w(t)$ equals a pulse of width 2. In this case

$$M_w = \frac{1}{2}\int_{-1}^{1} t^2\,dt = \frac{1}{3} \qquad E_w = \frac{1}{4}\int_{-1}^{1} dt = \frac{1}{2}$$

Inserting into (10-238), we obtain

$$T_m = \sqrt[5]{\frac{9q(t)}{2[f''(t)]^2}} \qquad e = 5b^2 = \frac{5q(t)}{8T_m} \tag{10-240}$$

THE PARABOLIC WINDOW. We wish now to determine the shape of $w(t)$ so as to minimize the sum in (10-237). Since $h(t)$ needs to be determined within a scale factor, it suffices to assume that E_w has a constant value. Thus our problem is to find a positive even function $w(t)$ vanishing for $|t| > 1$ and such that its second moment M_w is minimum. It can be shown that (see footnote 13, page 486)

$$w(t) = \begin{cases} 0.75(1-t^2) & |t| < 1 \\ 0 & |t| > 1 \end{cases} \quad E_w = \tfrac{3}{5} \qquad M_w = \tfrac{1}{5} \tag{10-241}$$

Thus the optimum $w(t)$ is a truncated parabola. With $w(t)$ so determined, the optimum filter is

$$h(t) = \frac{1}{T_m} w\left(\frac{t}{T_m}\right)$$

where T_m is the constant in (10-238). This filter is, of course, time varying because the scaling factor T_m depends on t.

10-7 BISPECTRA AND SYSTEM IDENTIFICATION[14]

Correlations and spectra are the most extensively used concepts in the applications of stochastic processes. These concepts involve only second-order moments. In certain applications, moments of higher order are also used. In the following, we introduce the transform of the third-order moment

$$R_{xxx}(t_1, t_2, t_3) = E\{\mathbf{x}(t_1)\mathbf{x}(t_2)\mathbf{x}(t_3)\} \tag{10-242}$$

of a process $\mathbf{x}(t)$ and we apply it to the phase problem in system identification. We assume that $\mathbf{x}(t)$ is a real SSS process with zero mean. From the stationarity of $\mathbf{x}(t)$ it follows that the function $R_{xxx}(t_1, t_2, t_3)$ depends only on the differences

$$t_1 - t_3 = \mu \qquad t_2 - t_3 = \nu$$

Setting $t_3 = t$ in (10-242) and omitting subscripts, we obtain

$$R(t_1, t_2, t_3) = R(\mu, \nu) = E\{\mathbf{x}(t+\mu)\mathbf{x}(t+\nu)\mathbf{x}(t)\} \tag{10-243}$$

DEFINITION

▶ The bispectrum $S(u, v)$ of the process $\mathbf{x}(t)$ is the two-dimensional Fourier transform of its third-order moment $R(\mu, \nu)$:

$$S(u, v) = \iint_{-\infty}^{\infty} R(\mu, \nu) e^{-j(u\mu + v\nu)}\, d\mu\, d\nu \tag{10-244}$$

The function $R(\mu, \nu)$ is real; hence

$$S(-u, -v) = S^*(u, v) \tag{10-245}$$

If $\mathbf{x}(t)$ is white noise then

$$R(\mu, \nu) = Q\delta(\mu)\delta(\nu) \qquad S(u, v) = Q \tag{10-246}$$

◀

Notes 1. The third-order moment of a normal process with zero mean is identically zero. This is a consequence of the fact that the joint density of three jointly normal random variables with zero mean is symmetrical with respect to the origin.

2. The autocorrelation of a white noise process with third-order moment as in (10-246) is an impulse $q\delta(\tau)$; in general, however, $q \neq Q$. For example if $\mathbf{x}(t)$ is normal white noise, then $Q = 0$ but $q \neq 0$. Furthermore, whereas $q > 0$ for all nontrivial processes, Q might be negative.

SYMMETRIES. The function $R(t_1, t_2, t_3)$ is invariant to the six permutations of the numbers t_1, t_2, and t_3. For stationary processes,

$$t_1 - t_3 = \mu \qquad t_2 - t_3 = \nu \qquad t_1 - t_2 = \mu - \nu$$

1	t_1, t_2, t_3	μ, ν	u, v	4	t_3, t_2, t_1	$-\mu, -\mu+\nu$	$-u-v, v$
2	t_2, t_1, t_3	ν, μ	v, u	5	t_2, t_3, t_1	$-\mu+\nu, -\mu$	$v, -u-v$
3	t_3, t_1, t_2	$-\nu, \mu-\nu$	$-u-v, u$	6	t_1, t_3, t_2	$\mu-\nu, -\nu$	$u, -u-v$

[14]D. R. Brillinger: "An Introduction to Polyspectra," *Annals of Math Statistics,* vol. 36. Also C. L. Nikias and M. R. Raghuveer (1987): "Bispectrum Estimation; Digital Processing Framework," *IEEE Proceedings,* vol. 75, 1965.

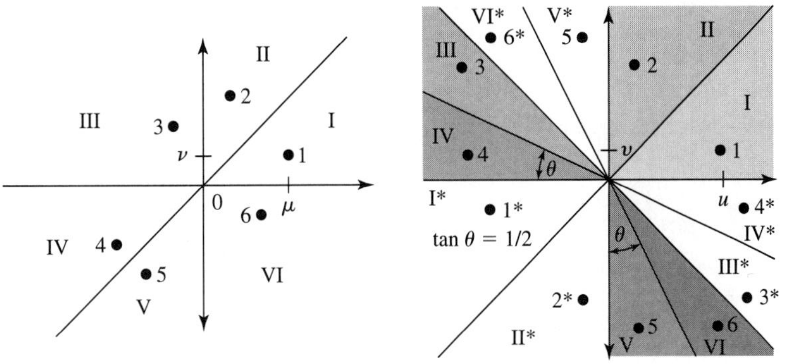

FIGURE 10-24

This yields the identities

$$\begin{aligned} R(\mu, \nu) &= R(\nu, \mu) = R(-\nu, \mu - \nu) = R(-\mu, -\mu + \nu) \\ &= R(-\mu + \nu, -\mu) = R(\mu - \nu, -\nu) \end{aligned} \qquad (10\text{-}247)$$

Hence if we know the function $R(\mu, \nu)$ in any one of the six regions of Fig. 10-24, we can determine it everywhere.

From (10-244) and (10-247) it follows that

$$\begin{aligned} S(u, v) &= S(v, u) = S(-u - v, u) = S(-u - v, v) \\ &= S(v, -u - v) = S(u, -u - v) \end{aligned} \qquad (10\text{-}248)$$

Combining with (10-245), we conclude that if we know $S(u, v)$ in any one of the 12 regions of Fig. 10-24, we can determine it everywhere.

Linear Systems

We have shown in (9-119)–(9-121) that if $\mathbf{x}(t)$ is the input to a linear system, the third-order moment of the resulting output $\mathbf{y}(t)$ equals

$$R_{yyy}(t_1, t_2, t_3) = \iiint_{-\infty}^{\infty} R_{xxx}(t_1-\alpha, t_2-\beta, t_3-\gamma)h(\alpha)h(\beta)h(\gamma)\,d\alpha\,d\beta\,d\gamma \qquad (10\text{-}249)$$

For stationary processes, $R_{xxx}(t_1-\alpha, t_2-\beta, t_3-\gamma) = R_{xxx}(\mu+\gamma-\alpha, \nu+\gamma-\beta)$; hence

$$R_{yyy}(\mu, \nu) = \iiint_{-\infty}^{\infty} R_{xxx}(\mu+\gamma-\alpha, \nu+\gamma-\beta)h(\alpha)h(\beta)h(\gamma)\,d\alpha\,d\beta\,d\gamma \qquad (10\text{-}250)$$

Using this relationship, we shall express the bispectrum $S_{yyy}(u, v)$ of $\mathbf{y}(t)$ in terms of the bispectrum $S_{xxx}(u, v)$ of $\mathbf{x}(t)$.

THEOREM 10-12

$$S_{yyy}(u, v) = S_{xxx}(u, v)H(u)H(v)H^*(u+v) \qquad (10\text{-}251)$$

Proof. Taking transformations of both sides of (10-249) and using the identity

$$\iint_{-\infty}^{\infty} R_{xxx}(\mu+\gamma-\alpha, \nu+\gamma-\beta)e^{-j(u\mu+v\nu)}\,d\mu\,d\nu$$

$$= S_{xxx}(u, v)e^{j[u(\gamma-\alpha)+v(\gamma-\beta)]}$$

we obtain

$$S_{yyy}(u, v) = S_{xxx}(u, v) \iiint_{-\infty}^{\infty} e^{j[u(\gamma-\alpha)+v(\gamma-\beta)]} h(\alpha)h(\beta)h(\gamma)\, d\alpha\, d\beta\, d\gamma$$

Expressing this integral as a product of three one-dimensional integrals, we obtain (10-251). ◀

EXAMPLE 10-7 ▶ Using (10-249), we shall determine the bispectrum of the shot noise

$$\mathbf{s}(t) = \sum_i h(t - \mathbf{t}_i) = \mathbf{z}(t) * h(t) \qquad \mathbf{z}(t) = \sum_i \delta(t - \mathbf{t}_i)$$

where $\mathbf{t}_i$ is a Poisson point process with average density λ.

To do so, we form the centered impulse train $\tilde{\mathbf{z}}(t) = \mathbf{z}(t) - \lambda$ and the centered shot noise $\tilde{\mathbf{s}}(t) = \tilde{\mathbf{z}}(t) * h(t)$. As we know (see Prob. 10-28)

$$R_{\tilde{z}\tilde{z}\tilde{z}}(\mu, \nu) = \lambda\delta(\mu)\delta(\nu) \quad \text{hence} \quad S_{\tilde{z}\tilde{z}\tilde{z}}(u, v) = \lambda$$

From this it follows that

$$S_{\tilde{s}\tilde{s}\tilde{s}}(u, v) = \lambda H(u)H(v)H^*(u + v)$$

and since $S_{\tilde{s}\tilde{s}}(\omega) = \lambda|H(\omega)|^2$, we conclude from Prob. 10-27 with $c = E\{\mathbf{s}(t)\} = \lambda H(0)$ that

$$\begin{aligned} S_{sss}(u, v) = {} & \lambda H(u)H(v)H^*(u + v) \\ & + 2\pi\lambda^2 H(0)[|H(u)|^2\delta(v) + |H(v)|^2\delta(u) + |H(u)|^2\delta(u + v)] \\ & + 4\pi^2\lambda^4 H^3(0)\delta(u)\delta(v) \end{aligned}$$ ◀

System Identification

A linear system is specified terminally in terms of its system function

$$H(\omega) = A(\omega)e^{j\varphi(\omega)}$$

System identification is the problem of determining $H(\omega)$. This problem is central in system theory and it has been investigated extensively. In this discussion, we apply the notion of spectra and polyspectra in the determination of $A(\omega)$ and $\varphi(\omega)$.

SPECTRA. Suppose that the input to the system $H(\omega)$ is a WSS process $\mathbf{x}(t)$ with power spectrum $S_{xx}(\omega)$. As we know,

$$S_{xy}(\omega) = S_{xx}(\omega)H^*(\omega) \tag{10-252}$$

This relationship expresses $H(\omega)$ in terms of the spectra $S_{xx}(\omega)$ and $S_{xy}(\omega)$ or, equivalently, in terms of the second-order moments $R_{xx}(\tau)$ and $R_{xy}(\tau)$. The problem of estimating these functions is considered in Chap. 12. In a number of applications, we cannot estimate $R_{xy}(\tau)$ either because we do not have access to the input $\mathbf{x}(t)$ of the system or because we cannot form the product $\mathbf{x}(t + \tau)\mathbf{y}(t)$ in real time. In such cases, an alternative method is used based on the assumption that $\mathbf{x}(t)$ is white noise. With this assumption (9-149) yields

$$S_{yy}(\omega) = S_{xx}(\omega)|H(\omega)|^2 = qA^2(\omega) \tag{10-253}$$

This relationship determines the amplitude $A(\omega)$ of $H(\omega)$ in terms of $S_{yy}(\omega)$ within a constant factor. It involves, however, only the estimation of the power spectrum $S_{yy}(\omega)$ of the output of the system. If the system is minimum phase (see page 499, and also Appendix 12A, page 574), then $H(\omega)$ is completely determined from (10-253) because, then, $\varphi(\omega)$ can be expressed in terms of $A(\omega)$. In general, however, this is not the case. The phase of an arbitrary system cannot be determined in terms of second-order moment of its output. It can, however, be determined if the third-order moment of $\mathbf{y}(t)$ is known.

PHASE DETERMINATION. We assume that $\mathbf{x}(t)$ is an SSS white-noise process with $S_{xxx}(u, v) = Q$. Inserting into (10-251), we obtain

$$S_{yyy}(u, v) = QH(u)H(v)H^*(u+v) \tag{10-254}$$

The function $S_{yyy}(u, v)$ is, in general, complex:

$$S_{yyy}(u, v) = B(u, v)e^{j\theta(u,v)} \tag{10-255}$$

Inserting (10-255) into (10-254) and equating amplitudes and phases, we obtain

$$B(u, v) = QA(u)A(v)A(u+v) \tag{10-256}$$

$$\theta(u, v) = \varphi(u) + \varphi(v) - \varphi(u+v) \tag{10-257}$$

We shall use these equations to express $A(\omega)$ in terms of $B(u, v)$ and $\varphi(\omega)$ in terms of $\theta(u, v)$. Setting $v = 0$ in (10-256), we obtain

$$QA^2(\omega) = \frac{1}{A(0)}B(\omega, 0) \qquad QA^3(0) = B(0, 0) \tag{10-258}$$

Since Q is in general unknown, $A(\omega)$ can be determined only within a constant factor. The phase $\varphi(\omega)$ can be determined only within a linear term because if it satisfies (10-257), so does the sum $\varphi(\omega) + c\omega$ for any c. We can assume therefore that $\varphi'(0) = 0$. To find $\varphi(\omega)$, we differentiate (10-257) with respect to v and we set $v = 0$. This yields

$$\theta_v(u, 0) = -\varphi'(u) \qquad \varphi(\omega) = -\int_0^\omega \theta_v(u, 0)\, du \tag{10-259}$$

where $\theta_v(u, v) = \partial\theta(u, v)/\partial v$. The above is the solution of (10-257).

In a numerical evaluation of $\varphi(\omega)$, we proceed as follows: Clearly, $\theta(u, 0) = \varphi(u) + \varphi(0) - \varphi(u) = \varphi(0) = 0$ for every u. From this it follows that

$$\theta_v(u, 0) = \lim \frac{1}{\Delta}\theta(u, \Delta) \qquad \text{as} \quad \Delta \to 0$$

Hence $\theta_v(u, 0) \simeq \theta(u, \Delta)/\Delta$ for sufficiently small Δ. Inserting into (10-259), we obtain the approximations

$$\varphi(\omega) \simeq -\frac{1}{\Delta}\int_0^\omega \theta(u, \Delta)\, du \qquad \varphi(n\Delta) \simeq -\sum_{k=1}^{n} \theta(k\Delta, \Delta) \tag{10-260}$$

This is the solution of the digital version

$$\theta(k\Delta, r\Delta) = \varphi(k\Delta) + \varphi(r\Delta) - \varphi(k\Delta + r\Delta) \tag{10-261}$$

of (10-257) where $(k\Delta, r\Delta)$ are points in the sector I of Fig. 10-24. As we see from (10-261) $\varphi(n\Delta)$ is determined in terms of the values of $\theta(k\Delta, \Delta)$ of $\theta(u, \Delta)$ on the

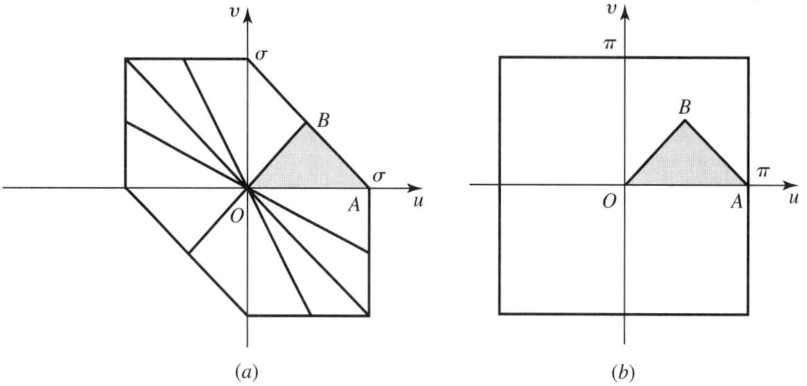

FIGURE 10-25

horizontal line $v = \Delta$. Hence the system (10-261) is overdetermined. This is used to improve the estimate of $\varphi(\omega)$ if $\theta(u, v)$ is not known exactly but it is estimated in terms of a single sample of $\mathbf{y}(t)$.[15] The corresponding problem of spectral estimation is considered in Chap. 12.

Note If the bispectrum $S(u, v)$ of a process $\mathbf{x}(t)$ equals the right side of (10-254) and $H(\omega) = 0$ for $|\omega| > \sigma$, then

$$S(u, v) = 0 \qquad \text{for} \quad |u| > \sigma \quad \text{or} \quad |v| > \sigma \quad \text{or} \quad |u + v| > \sigma$$

Thus, $S(u, v) = 0$ outside the hexagon of Fig. 10-25a. From this and the symmetries of Fig. 10-24 it follows that $S(u, v)$ is uniquely determined in terms of its values in the triangle OAB of Fig. 10-25a.

Digital processes. The preceding concepts can be readily extended to digital processes. We cite only the definition of bispectra.

Given an SSS digital process $\mathbf{x}[n]$, we form its third-order moment

$$R[k, r] = E\{\mathbf{x}[n + k]\mathbf{x}[n + r]\mathbf{x}[n]\} \tag{10-262}$$

The bispectrum of $\mathbf{x}[n]$ is the two-dimensional DFT of $R[k, r]$:

$$S(u, v) = \sum_{k=-\infty}^{\infty} \sum_{r=-\infty}^{\infty} R[k, r]e^{-j(uk+vr)} \tag{10-263}$$

This function is doubly periodic with period 2π:

$$S(u + 2\pi m, v + 2\pi n) = S(u, v) \tag{10-264}$$

It is therefore determined in terms of its values in the square $|u| \le \pi$, $|v| \le \pi$ of Fig. 10-25b. Furthermore, it has the 12 symmetries of Fig. 10-24.

Suppose now that $\mathbf{x}[n]$ is the output of a system $H(z)$ with input a white noise process. Proceeding as in (10-251) and (10-254) we conclude that its bispectrum

[15] T. Matsuoka and T. J. Ulrych: "Phase Estimation Using the Bispectrum," *IEEE Proceedings,* vol. 72, 1984.

equals

$$S(u, v) = QH(e^{ju})H(e^{jv})H\left(e^{-j(u+v)}\right)$$

From the above expression it follows that $S(u, v)$ is determined here in terms of its values in the triangle OAB of Fig. 10-25b (see Prob. 10-29).

APPENDIX 10A THE POISSON SUM FORMULA

If

$$F(u) = \int_{-\infty}^{\infty} f(x)e^{-jux}\,dx$$

is the Fourier transform of $f(x)$ then for any c

$$\sum_{n=-\infty}^{\infty} f(x+nc) = \frac{1}{c}\sum_{n=-\infty}^{\infty} F(nu_0)e^{jnu_0x} \qquad u_0 = \frac{2\pi}{c} \tag{10A-1}$$

Proof. Clearly

$$\sum_{n=-\infty}^{\infty} \delta(x+nc) = \frac{1}{c}\sum_{n=-\infty}^{\infty} e^{jnu_0x} \tag{10A-2}$$

because the left side is periodic and its Fourier series coefficients equal

$$\frac{1}{c}\int_{-c/2}^{c/2} \delta(x)e^{-jnu_0x}\,dx = \frac{1}{c}$$

Furthermore, $\delta(x+nc) * f(x) = f(x+nc)$ and

$$e^{jnu_0x} * f(x) = \int_{-\infty}^{\infty} e^{jnu_0(x-\alpha)}f(\alpha)\,d\alpha = e^{jnu_0x}F(nu_0)$$

Convolving both sides of (10A-2) with $f(x)$ and using the last equation, we obtain (10A-1).

APPENDIX 10B THE SCHWARZ INEQUALITY

We shall show that

$$\left|\int_a^b f(x)g(x)\,dx\right|^2 \le \int_a^b |f(x)|^2\,dx \int_a^b |g(x)|^2\,dx \tag{10B-1}$$

with equality iff

$$f(x) = kg^*(x) \tag{10B-2}$$

Proof. Clearly

$$\left|\int_a^b f(x)g(x)\,dx\right| \le \int_a^b |f(x)||g(x)|\,dx$$

Equality holds only if the product $f(x)g(x)$ is real. This is the case if the angles of $f(x)$ and $g(x)$ are opposite as in (10B-2). It suffices, therefore, to assume that the functions $f(x)$ and $g(x)$ are real. The quadratic

$$\begin{aligned} I(z) &= \int_a^b [f(x) - zg(x)]^2\,dx \\ &= z^2 \int_a^b g^2(x)\,dx - 2z \int_a^b f(x)g(x)\,dx + \int_a^b f^2(x)\,dx \end{aligned}$$

is nonnegative for every real z. Hence, its discriminant cannot be positive. This yields (10B-1). If the discriminant of $I(z)$ is zero, then $I(z)$ has a real (double) root $z = k$. This shows that $I(k) = 0$ and (10B-2) follows.

PROBLEMS

10-1 Find the first-order characteristic function (*a*) of a Poisson process, and (*b*) of a Wiener process.

Answer: (*a*) $e^{\lambda t(e^{j\omega}-1)}$; (*b*) $e^{-\alpha t\omega^2/2}$

10-2 (*Two-dimensional random walk*). The coordinates $\mathbf{x}(t)$ and $\mathbf{y}(t)$ of a moving object are two independent random-walk processes with the same s and T as in Fig. 10-3*a*. Show that if $\mathbf{z}(t) = \sqrt{\mathbf{x}^2(t) + \mathbf{y}^2(t)}$ is the distance of the object from the origin and $t \gg T$, then for z of the order of $\sqrt{\alpha t}$:

$$f_z(z, t) \simeq \frac{z}{\alpha t} e^{-z^2/2\alpha t} U(z) \qquad \alpha = \frac{s^2}{T}$$

10-3 In the circuit of Fig. P10-3, $\mathbf{n}_e(t)$ is the voltage due to thermal noise. Show that

$$S_v(\omega) = \frac{2kTR}{(1 - \omega^2 LC)^2 + \omega^2 R^2 C^2} \qquad S_i(\omega) = \frac{2kTR}{R^2 + \omega^2 L^2}$$

and verify Nyquist's theorems (10-75) and (10-78).

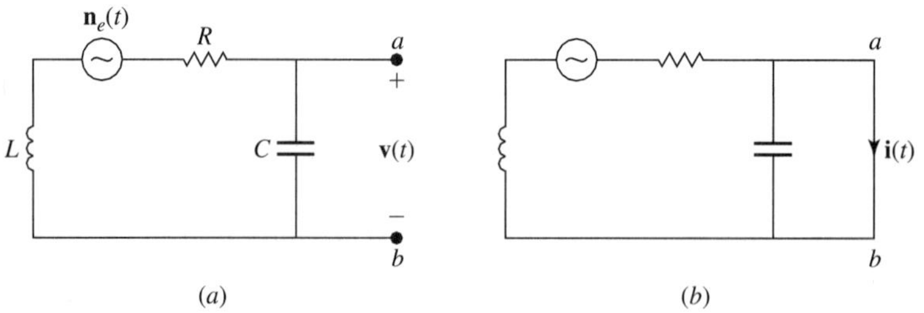

FIGURE P10-3

10-4 A particle in free motion satisfies the equation

$$m\mathbf{x}''(t) + f\mathbf{x}'(t) = \mathbf{F}(t) \qquad S_F(\omega) = 2kTf$$

Show that if $\mathbf{x}(0) = \mathbf{x}'(0) = 0$, then

$$E\{\mathbf{x}^2(t)\} = 2D^2\left(t - \frac{3}{4\alpha} + \frac{1}{\alpha}e^{-2\alpha t} - \frac{1}{4\alpha}e^{-4\alpha t}\right)$$

where $D^2 = kT/f$ and $\alpha = f/2m$.

Hint: Use (9-99) with

$$h(t) = \frac{1}{f}(1 - e^{-2\alpha t})U(t) \qquad q(t) = 2kTfU(t)$$

10-5 The position of a particle in underdamped harmonic motion is a normal process with autocorrelation as in (10-60). Show that its conditional density assuming $\mathbf{x}(0) = x_0$ and $\mathbf{x}'(0) = \mathbf{v}(0) = v_0$ equals

$$f_{x(t)}(x \mid x_0, v_0) = \frac{1}{\sqrt{2\pi P}}e^{-(x - ax_0 - bv_0)^2/2P}$$

Find the constants a, b, and P.

10-6 Given a Wiener process $\mathbf{w}(t)$ with parameter α, we form the processes

$$\mathbf{x}(t) = \mathbf{w}(t^2) \qquad \mathbf{y}(t) = \mathbf{w}^2(t) \qquad \mathbf{z}(t) = |\mathbf{w}(t)|$$

Show that $\mathbf{x}(t)$ is normal with zero mean. Furthermore, if $t_1 < t_2$, then

$$R_x(t_1, t_2) = \alpha t_1^2 \qquad R_y(t_1, t_2) = \alpha^2 t_1(2t_1 + t_2)$$

$$R_z(t_1, t_2) = \frac{2\alpha}{\pi}\sqrt{t_1 t_2}(\cos\theta + \theta\sin\theta) \qquad \sin\theta = \sqrt{\frac{t_1}{t_2}}$$

10-7 The process $\mathbf{s}(t)$ is shot noise with $\lambda = 3$ as in (10-96) where $h(t) = 2$ for $0 \le t \le 10$ and $h(t) = 0$ otherwise. Find $E\{\mathbf{s}(t)\}$, $E\{\mathbf{s}^2(t)\}$, and $P\{\mathbf{s}(7) = 0\}$.

10-8 The input to a real system $H(\omega)$ is a WSS process $\mathbf{x}(t)$ and the output equals $\mathbf{y}(t)$. Show that if

$$R_{xx}(\tau) = R_{yy}(\tau) \qquad R_{xy}(-\tau) = -R_{xy}(\tau)$$

as in (10-130), then $H(\omega) = jB(\omega)$ where $B(\omega)$ is a function taking only the values $+1$ and -1.

Special case: If $\mathbf{y}(t) = \hat{\mathbf{x}}(t)$, then $B(\omega) = -\text{sgn}\,\omega$.

10-9 Show that if $\hat{\mathbf{x}}(t)$ is the Hilbert transform of $\mathbf{x}(t)$ and

$$\mathbf{i}(t) = \mathbf{x}(t)\cos\omega_0 t + \hat{\mathbf{x}}(t)\sin\omega_0 t \qquad \mathbf{q}(t) = \hat{\mathbf{x}}(t)\cos\omega_0 t - \mathbf{x}(t)\sin\omega_0 t$$

then (Fig. P10-9)

$$S_i(\omega) = S_q(\omega) = \frac{S_w(\omega) + S_w(-\omega)}{4} \qquad S_{qi}(\omega) = \frac{S_w(\omega) + S_w(-\omega)}{4j}$$

where $S_w(\omega) = 4S_x(\omega + \omega_0)U(\omega + \omega_0)$.

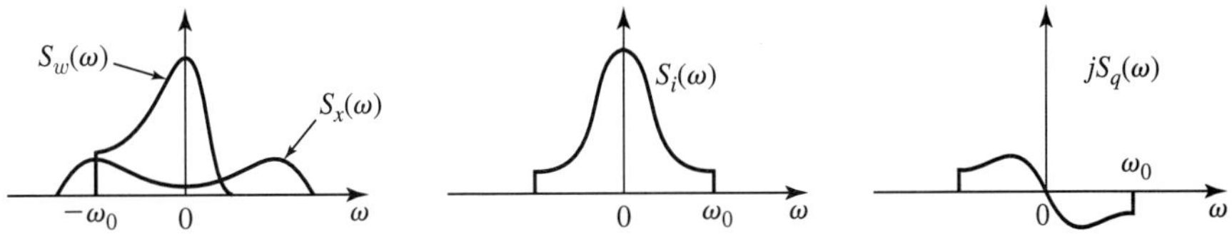

FIGURE P10-9

10-10 Show that if $\mathbf{w}(t)$ and $\mathbf{w}_\tau(t)$ are the complex envelopes of the processes $\mathbf{x}(t)$ and $\mathbf{x}(t-\tau)$ respectively, then $\mathbf{w}_\tau(t) = \mathbf{w}(t-\tau)e^{-j\omega_0\tau}$.

10-11 Show that if $\mathbf{w}(t)$ is the optimum complex envelope of $\mathbf{x}(t)$ [see (10-148)], then

$$E\{|\mathbf{w}'(t)|^2\} = -2\left[R_x''(0) + \omega_0^2 R_x(0)\right]$$

10-12 Show that if the process $\mathbf{x}(t)\cos\omega t + \mathbf{y}(t)\sin\omega t$ is normal and WSS, then its statistical properties are determined in terms of the variance of the process $\mathbf{z}(t) = \mathbf{x}(t) + j\mathbf{y}(t)$.

10-13 Show that if $\boldsymbol{\theta}$ is a random variable uniform in the interval $(0, T)$ and $f(t)$ is a periodic function with period T, then the process $\mathbf{x}(t) = f(t-\boldsymbol{\theta})$ is stationary and

$$S_x(\omega) = \frac{1}{T}\left|\int_0^T f(t)e^{-j\omega t}\,dt\right|^2 \sum_{m=-\infty}^{\infty} \delta\left(\omega - \frac{2\pi}{T}m\right)$$

10-14 Show that if

$$\boldsymbol{\varepsilon}_N(t) = \mathbf{x}(t) - \sum_{n=-N}^{N} \mathbf{x}(nT)\frac{\sin\sigma(t-nT)}{\sigma(t-nT)} \qquad \sigma = \frac{\pi}{T}$$

then

$$E\{\boldsymbol{\varepsilon}_N^2(t)\} = \frac{1}{2\pi}\int_{-\infty}^{\infty} S(\omega)\left|e^{j\omega t} - \sum_{n=-N}^{N}\frac{\sin\sigma(t-nT)}{\sigma(t-nT)}e^{jn\omega T}\right|^2 d\omega$$

and if $S(\omega) = 0$ for $|\omega| > \sigma$, then $E\{\boldsymbol{\varepsilon}_N^2(t)\} \to 0$ as $N \to \infty$.

10-15 Show that if $\mathbf{x}(t)$ is BL as in (10-187), then[16] for $|\tau| < \pi/\sigma$:

$$\frac{2\tau^2}{\pi^2}|R''(0)| \le R(0) - R(\tau) \le \frac{\tau^2}{2}|R''(0)|$$

$$E\{[\mathbf{x}(t+\tau) - \mathbf{x}(t)]^2\} \ge \frac{4\tau^2}{\pi^2}E\{[\mathbf{x}'(t)]^2\}$$

Hint: If $0 < \varphi < \pi/2$ then $2\varphi/\pi < \sin\varphi < \varphi$.

10-16 A WSS process $\mathbf{x}(t)$ is BL as in (10-187) and its samples $\mathbf{x}(n\pi/\sigma)$ are uncorrelated. Find $S_x(\omega)$ if $E\{\mathbf{x}(t)\} = \eta$ and $E\{\mathbf{x}^2(t)\} = I$.

10-17 Find the power spectrum $S(\omega)$ of a process $\mathbf{x}(t)$ if $S(\omega) = 0$ for $|\omega| > \pi$ and

$$E\{\mathbf{x}(n+m)\mathbf{x}(n)\} = N\delta[m]$$

10-18 Show that if $S(\omega) = 0$ for $|\omega| > \sigma$, then

$$R(\tau) \ge R(0)\cos\sigma\tau \qquad \text{for} \quad |\tau| < \pi/2\sigma$$

10-19 Show that if $\mathbf{x}(t)$ is BL as in (10-187) and $\Delta = 2\pi/\sigma$, then

$$\mathbf{x}(t) = 4\sin^2\frac{\sigma t}{2}\sum_{n=-\infty}^{\infty}\left[\frac{\mathbf{x}(n\Delta)}{(\sigma t - 2n\pi)^2} + \frac{\mathbf{x}'(n\Delta)}{\sigma(\sigma t - 2n\pi)}\right]$$

Hint: Use (10-206) with $N = 2$, $H_1(\omega) = 1$, $H_2(\omega) = j\omega$.

10-20 Find the mean and the variance of $\mathbf{P}(\omega_0)$ if $\mathbf{t}_i$ is a Poisson point process and

$$\mathbf{P}(\omega) = \frac{1}{\lambda}\sum_i \cos\omega_0\mathbf{t}_i\cos\omega\mathbf{t}_i \qquad |\mathbf{t}_i| < a$$

---

[16] A. Papoulis: "An Estimation of the Variation of a Bandlimited Process," *IEEE, PGIT,* 1984.

10-21 Given a WSS process $\mathbf{x}(t)$ and a set of Poisson points $\mathbf{t}_i$ independent of $\mathbf{x}(t)$ and with average density λ, we form the sum

$$\mathbf{X}_c(\omega) = \sum_{|\mathbf{t}_i|<c} \mathbf{x}(\mathbf{t}_i)e^{-j\omega \mathbf{t}_i}$$

Show that if $E\{\mathbf{x}(t)\} = 0$ and $\int_{-\infty}^{\infty} |R_x(\tau)|\,d\tau < \infty$, then for large c,

$$E\{|\mathbf{X}_c(\omega)|^2\} = 2cS_x(\omega) + \frac{2c}{\lambda}R_x(0)$$

10-22 We are given the data $\mathbf{x}(t) = f(t) + \mathbf{n}(t)$, where $R_n(\tau) = N\delta(\tau)$ and $E\{\mathbf{n}(t) = 0\}$. We wish to estimate the integral

$$g(t) = \int_0^t f(\alpha)\,d\alpha$$

knowing that $g(T) = 0$. Show that if we use as the estimate of $g(t)$ the process $\mathbf{w}(t) = \mathbf{z}(t) - \mathbf{z}(T)t/T$, where

$$\mathbf{z}(t) = \int_0^t \mathbf{x}(\alpha)\,d\alpha \qquad \text{then} \quad E\{\mathbf{w}(t)\} = g(t) \qquad \sigma_w^2 = Nt\left(1 - \frac{t}{T}\right)$$

10-23 (*Cauchy inequality*) Show that

$$\left|\sum_i a_i b_i\right|^2 \le \sum_i |a_i|^2 \sum_i |b_i|^2 \qquad \text{(i)}$$

with equality iff $a_i = kb_i^*$.

10-24 The input to a system $\mathbf{H}(z)$ is the sum $\mathbf{x}[n] = f[n] + \mathbf{v}[n]$, where $f[n]$ is a known sequence with z transform $\mathbf{F}(z)$. We wish to find $\mathbf{H}(z)$ such that the ratio $y_f^2[0]/E\{\mathbf{y}_v^2[n]\}$ of the output $\mathbf{y}[n] = y_f[n] + \mathbf{y}_v[n]$ is maximum. Show that (*a*) if $\mathbf{v}[n]$ is white noise, then $\mathbf{H}(z) = k\mathbf{F}(z^{-1})$, and (*b*) if $\mathbf{H}(z)$ is a finite impulse response (FIR) filter that is, if $\mathbf{H}(z) = a_0 + a_1 z^{-1} + \cdots + a_N z^{-N}$, then its weights a_m are the solutions of the system

$$\sum_{m=0}^{N} R_v[n-m]a_m = kf[-n] \qquad n = 0, \ldots, N$$

10-25 If $R_n(\tau) = N\delta(\tau)$ and

$$\mathbf{x}(t) = A\cos\omega_0 t + \mathbf{n}(t) \qquad H(\omega) = \frac{1}{\alpha + j\omega}$$

$$\mathbf{y}(t) = B\cos(\omega_0 + t + \varphi) + \mathbf{y}_n(t)$$

where $\mathbf{y}_n(t)$ is the component of the output $\mathbf{y}(t)$ due to $\mathbf{n}(t)$, find the value of α that maximizes the signal-to-noise ratio

$$\frac{|B|^2}{E\{\mathbf{y}_n^2(t)\}}$$

Answer: $\alpha = \omega_0$.

10-26 In the detection problem of pages 483–485, we apply the process $\mathbf{x}(t) = f(t) + \mathbf{v}(t)$ to the tapped delay line (10-225). Show that: (*a*) The signal-to-noise (SNR) ratio r is maximum if the coefficients a_i satisfy (10-227); (*b*) the maximum r equals $\sqrt{y_f(t_0)/k}$.

10-27 Given an SSS process $\mathbf{x}(t)$ with zero mean, power spectrum $S(\omega)$, and bispectrum $S(u, v)$, we form the process $\mathbf{y}(t) = \mathbf{x}(t) + c$. Show that

$$S_{yyy}(u, v) = S(u, v) + 2\pi c[S(u)\delta(v) + S(v)\delta(u) + S(u)\delta(u + v)] + 4\pi^2 c^3 \delta(u)\delta(v)$$

10-28 Given a Poisson process $\mathbf{x}(t)$, we form its centered process $\tilde{\mathbf{x}}(t) = \mathbf{x}(t) - \lambda t$ and the centered Poisson impulses

$$\tilde{\mathbf{z}}(t) = \frac{d\tilde{\mathbf{x}}(t)}{dt} = \sum_i \delta(t - \mathbf{t}_i) - \lambda$$

Show that

$$E\{\tilde{\mathbf{x}}(t_1)\tilde{\mathbf{x}}(t_2)\tilde{\mathbf{x}}(t_3)\} = \lambda \min(t_1, t_2, t_3)$$

$$E\{\tilde{\mathbf{z}}(t_1)\tilde{\mathbf{z}}(t_2)\tilde{\mathbf{z}}(t_3)\} = \lambda\delta(t_1 - t_2)\delta(t_1 - t_3)$$

Hint: Use (9-103) and the identity

$$\min(t_1, t_2, t_3) = t_1 U(t_2 - t_1) U(t_3 - t_1) + t_2 U(t_1 - t_2) U(t_3 - t_2) + t_3 U(t_1 - t_3) U(t_2 - t_3)$$

10-29 Show that the function

$$S(u, v) = H(e^{ju}) H(e^{jv}) H\left(e^{-j(u+v)}\right)$$

is determined in terms of its values in the triangle of Fig. 10-25*b*.

Outline: Form the function

$$S_a(u, v) = H_a(u) H_a(v) H_a(-ju - jv) \qquad \text{where} \quad H_a(\omega) = \begin{cases} H(e^{j\omega}) & |\omega| \le \pi \\ 0 & |\omega| > 0 \end{cases}$$

Clearly, $S_a(u, v) = S(u, v)$ for $|u|, |v|, |u + v| < \pi$ and 0 otherwise. The function $S_a(u, v)$ is a bispectrum of a bandlimited process, $\mathbf{x}(t)$ with $\sigma = \pi$; hence (see note page 492) it is determined from its values in the triangle of Fig. 10-25*a*. Inserting into (10-258) and (10-259) we obtain $H_a(\omega)$. This yields $H(e^{j\omega})$ and $S(u, v)$.

CHAPTER 11

SPECTRAL REPRESENTATION

11-1 FACTORIZATION AND INNOVATIONS

In this section, we consider the problem of representing a real WSS process $\mathbf{x}(t)$ as the response of a minimum-phase system $\mathbf{L}(s)$ with input a white-noise process $\mathbf{i}(t)$. The term *minimum-phase* has the following meaning: The system $\mathbf{L}(s)$ is causal and its impulse response $l(t)$ has finite energy; the system $\boldsymbol{\Gamma}(s) = 1/\mathbf{L}(s)$ is causal and its impulse response $\gamma(t)$ has finite energy. Thus a system $\mathbf{L}(s)$ is minimum-phase if the functions $\mathbf{L}(s)$ and $1/\mathbf{L}(s)$ are analytic in the right-hand plane Re $s > 0$. A process $\mathbf{x}(t)$ that can be so represented will be called *regular*. From the definition it follows that $\mathbf{x}(t)$ is a regular process if it is linearly equivalent with a white-noise process $\mathbf{i}(t)$ in the sense that (see Fig. 11-1)

$$\mathbf{i}(t) = \int_0^\infty \gamma(\alpha)\mathbf{x}(t-\alpha)\,d\alpha \qquad R_{ii}(\tau) = \delta(\tau) \tag{11-1}$$

$$\mathbf{x}(t) = \int_0^\infty l(\alpha)\mathbf{i}(t-\alpha)\,d\alpha \qquad E\{\mathbf{x}^2(t)\} = \int_0^\infty l^2(t)\,dt < \infty \tag{11-2}$$

The last equality follows from (9-100). This shows that the power spectrum $\mathbf{S}(s)$ of a regular process can be written as a product

$$\mathbf{S}(s) = \mathbf{L}(s)\mathbf{L}(-s) \qquad S(\omega) = |\mathbf{L}(j\omega)|^2 \tag{11-3}$$

where $\mathbf{L}(s)$ is a minimum-phase function uniquely determined in terms of $S(\omega)$. The function $\mathbf{L}(s)$ will be called the *innovations filter* of $\mathbf{x}(t)$ and its inverse $\boldsymbol{\Gamma}(s)$ the *whitening filter* of $\mathbf{x}(t)$. The process $\mathbf{i}(t)$ will be called the *innovations* of $\mathbf{x}(t)$. It is the output of the filter $\mathbf{L}(s)$ with input $\mathbf{x}(t)$.

The problem of determining the function $\mathbf{L}(s)$ can be phrased as follows: Given a positive even function $S(\omega)$ of finite area, find a minimum-phase function $\mathbf{L}(s)$ such that $|\mathbf{L}(j\omega)|^2 = S(\omega)$. It can be shown that this problem has a solution if $S(\omega)$ satisfies the

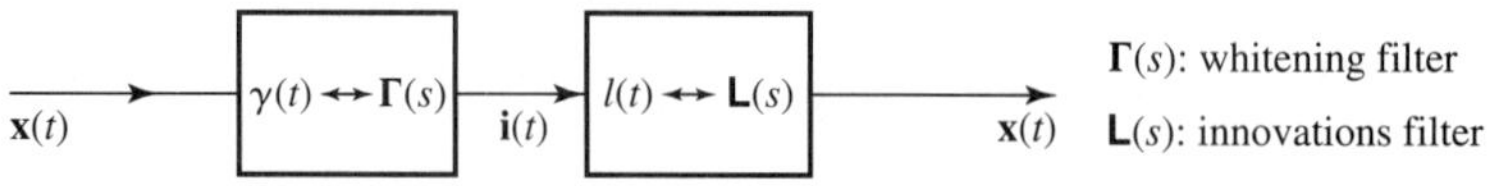

FIGURE 11-1

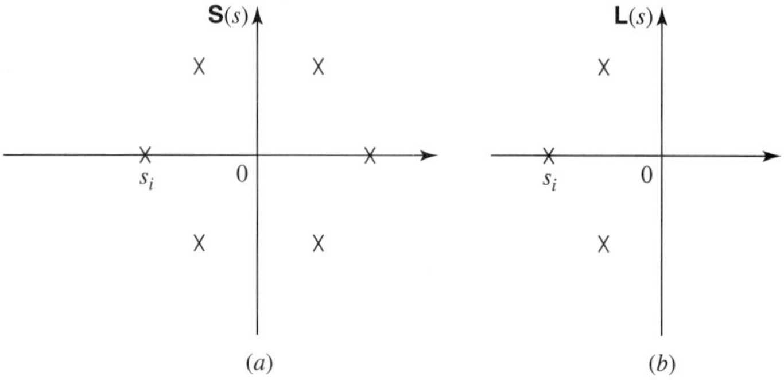

FIGURE 11-2

Paley–Wiener condition[1]

$$\int_{-\infty}^{\infty} \frac{|\log S(\omega)|}{1+\omega^2}\, d\omega < \infty \tag{11-4}$$

This condition is not satisfied if $S(\omega)$ consists of lines, or, more generally, if it is band-limited. As we show later, processes with such spectra are predictable. In general, the problem of factoring $S(\omega)$ as in (11-3) is not simple. In the following, we discuss an important special case.

Rational spectra. A rational spectrum is the ratio of two polynomials in ω^2 because $S(-\omega) = S(\omega)$:

$$S(\omega) = \frac{A(\omega^2)}{B(\omega^2)} \qquad \mathbf{S}(s) = \frac{A(-s^2)}{B(-s^2)} \tag{11-5}$$

This shows that if s_i is a root (zero or pole) of $\mathbf{S}(s)$, $-s_i$ is also a root. Furthermore, all roots are either real or complex conjugate. From this it follows that the roots of $\mathbf{S}(s)$ are symmetrical with respect to the $j\omega$ axis (Fig. 11-2*a*). Hence they can be separated into two groups: The "left" group consists of all roots s_i with Re $s_i < 0$, and the "right" group consists of all roots with Re $s_i > 0$. The minimum-phase factor $\mathbf{L}(s)$ of $\mathbf{S}(s)$ is a ratio of two polynomials formed with the left roots of $\mathbf{S}(s)$:

$$\mathbf{S}(s) = \frac{N(s)N(-s)}{D(s)D(-s)} \qquad \mathbf{L}(s) = \frac{N(s)}{D(s)} \qquad \mathbf{L}^2(0) = \mathbf{S}(0)$$

[1]N. Wiener, R. E. A. C. Paley: Fourier Transforms in the Complex Domain, *American Mathematical Society College,* 1934 (see also Papoulis, 1962 [20]).

EXAMPLE 11-1 ▶ If $S(\omega) = N/(\alpha^2 + \omega^2)$ then

$$\mathbf{S}(s) = \frac{N}{\alpha^2 - s^2} = \frac{N}{(\alpha + s)(\alpha - s)} \qquad \mathbf{L}(s) = \frac{\sqrt{N}}{\alpha + s}$$ ◀

EXAMPLE 11-2 ▶ If $S(\omega) = (49 + 25\omega^2)/(\omega^4 + 10\omega^2 + 9)$, then

$$\mathbf{S}(s) = \frac{49 - 25s^2}{(1 - s^2)(9 - s^2)} \qquad \mathbf{L}(s) = \frac{7 + 5s}{(1 + s)(3 + s)}$$ ◀

EXAMPLE 11-3 ▶ If $S(\omega) = 25/(\omega^4 + 1)$ then

$$\mathbf{S}(s) = \frac{25}{s^4 + 1} = \frac{25}{(s^2 + \sqrt{2}s + 1)(s^2 - \sqrt{2}s + 1)} \qquad \mathbf{L}(s) = \frac{5}{s^2 + \sqrt{2}s + 1}$$ ◀

Discrete-Time Processes

A discrete-time system is minimum-phase if its system function $\mathbf{L}(z)$ and its inverse $\boldsymbol{\Gamma}(z) = 1/\mathbf{L}(z)$ are analytic in the exterior $|z| > 1$ of the unit circle. A real WSS digital process $\mathbf{x}[n]$ is regular if its spectrum $\mathbf{S}(z)$ can be written as a product

$$\mathbf{S}(z) = \mathbf{L}(z)\mathbf{L}(1/z) \qquad \mathbf{S}(e^{j\omega}) = |\mathbf{L}(e^{j\omega})|^2 \tag{11-6}$$

Denoting by $l[n]$ and $\gamma[n]$, respectively, the delta responses of $\mathbf{L}(z)$ and $\boldsymbol{\Gamma}(z)$, we conclude that a regular process $\mathbf{x}[n]$ is linearly equivalent with a white-noise process $\mathbf{i}[n]$ (see Fig. 11-3):

$$\mathbf{i}[n] = \sum_{k=0}^{\infty} \gamma[k]\mathbf{x}[n-k] \qquad R_{ii}[m] = \delta[m] \tag{11-7}$$

$$\mathbf{x}[n] = \sum_{k=0}^{\infty} l[k]\mathbf{i}[n-k] \qquad E\{\mathbf{x}^2[n]\} = \sum_{k=0}^{\infty} l^2[k] < \infty \tag{11-8}$$

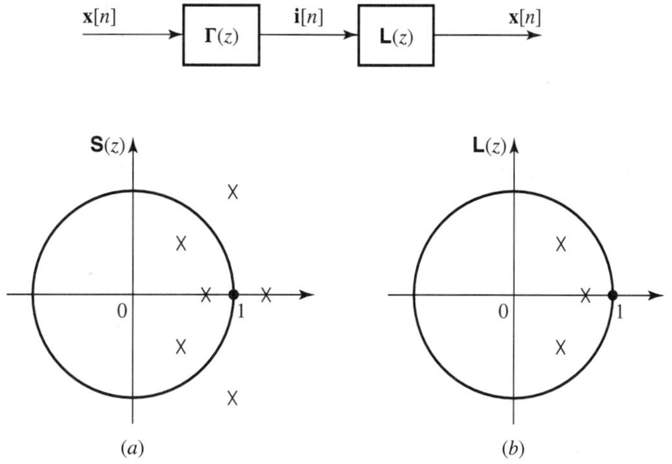

FIGURE 11-3

The process $\mathbf{i}[n]$ is the innovations of $\mathbf{x}[n]$ and the function $\mathbf{L}(z)$ its innovations filter. The whitening filter of $\mathbf{x}[n]$ is the function $\mathbf{\Gamma}(z) = 1/\mathbf{L}(z)$.

It can be shown that the power spectrum $\mathbf{S}(e^{j\omega})$ of a process $\mathbf{x}[n]$ can be factored as in (11-6) if it satisfies the Paley–Wiener condition

$$\int_{-\pi}^{\pi} |\log S(\omega)\, d\omega| < \infty \tag{11-9}$$

If the power spectrum $S(\omega)$ is an integrable function, then (11-9) reduces to (9-203). (See also (9-207).)

Rational spectra. The power spectrum $\mathbf{S}(e^{j\omega})$ of a real process is a function of $\cos\omega = (e^{j\omega} + e^{-j\omega})/2$ [see (9-193)]. From this it follows that $\mathbf{S}(z)$ is a function of $z + 1/z$. If therefore, z_i is a root of $\mathbf{S}(z)$, $1/z_i$ is also a root. We thus conclude that the roots of $\mathbf{S}(z)$ are symmetrical with respect to the unit circle (Fig. 11-3); hence they can be separated into two groups: The "inside" group consists of all roots z_i such that $|z_i| < 1$ and the "outside" group consists of all roots such that $|z_i| > 1$. The minimum-phase factor $\mathbf{L}(z)$ of $\mathbf{S}(z)$ is a ratio of two polynomials consisting of the inside roots of $\mathbf{S}(z)$:

$$\mathbf{S}(z) = \frac{N(z)N(1/z)}{D(z)D(1/z)} \qquad \mathbf{L}(z) = \frac{N(z)}{D(z)} \qquad \mathbf{L}^2(1) = \mathbf{S}(1)$$

EXAMPLE 11-4

▶ If $S(\omega) = (5 - 4\cos\omega)/(10 - 6\cos\omega)$ then

$$\mathbf{S}(z) = \frac{5 - 2(z + z^{-1})}{10 - 3(z + z^{-1})} = \frac{2(z - 1/2)(z - 2)}{3(z - 1/3)(z - 3)} \qquad \mathbf{L}(z) = \frac{2z - 1}{3z - 1}$$ ◀

11-2 FINITE-ORDER SYSTEMS AND STATE VARIABLES

In this section, we consider systems specified in terms of differential equations or recursion equations. As a preparation, we review briefly the meaning of finite-order systems and state variables starting with the analog case. The systems under consideration are multiterminal with m inputs $\mathbf{x}_i(t)$ and r outputs $\mathbf{y}_j(t)$ forming the column vectors $\mathbf{X}(t) = [\mathbf{x}_i(t)]$ and $\mathbf{Y}(t) = [\mathbf{y}_j(t)]$ as in (9-124).

At a particular time $t = t_1$, the output $\mathbf{Y}(t)$ of a system is in general specified only if the input $\mathbf{X}(t)$ is known for every t. Thus, to determine $\mathbf{Y}(t)$ for $t > t_0$, we must know $\mathbf{X}(t)$ for $t > t_0$ and for $t \le t_0$. For a certain class of systems, this is not necessary. The values of $\mathbf{Y}(t)$ for $t > t_0$ are completely specified if we know $\mathbf{X}(t)$ for $t > t_0$ and, in addition, the values of a finite number of parameters. These parameters specify the "state" of the system at time $t = t_0$ in the sense that their values determine the effect of the past $t < t_0$ of $\mathbf{X}(t)$ on the future $t > t_0$ of $\mathbf{Y}(t)$. The values of these parameters depend on t_0; they are, therefore, functions $\mathbf{z}_i(t)$ of t. These functions are called *state variables*. The number n of state variables is called the *order* of the system. The vector

$$\mathbf{Z}(t) = [\mathbf{z}_i(t)] \qquad i = 1, \ldots, n$$

is called the *state vector;* this vector is not unique. We shall say that the system is in *zero state* at $t = t_0$ if $\mathbf{Z}(t_0) = 0$.

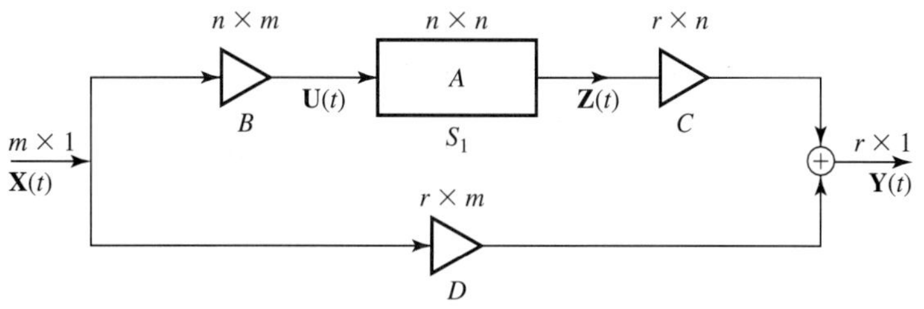

FIGURE 11-4

We shall consider here only linear, time-invariant, real, causal systems. Such systems are specified in terms of the following equations:

$$\frac{d\mathbf{Z}(t)}{dt} = A\mathbf{Z}(t) + B\mathbf{X}(t) \tag{11-10a}$$

$$\mathbf{Y}(t) = C\mathbf{Z}(t) + D\mathbf{X}(t) \tag{11-10b}$$

In (11-10*a*) and (11-10*b*), A, B, C, and D are matrices with real constant elements, of order $n \times n$, $n \times m$, $r \times n$, and $r \times m$, respectively. In Fig. 11-4 we show a block diagram of the system S specified terminally in terms of these equations. It consists of a dynamic system S_1 with input $\mathbf{U}(t) = B\mathbf{X}(t)$ and output $\mathbf{Z}(t)$, and of three memoryless systems (multipliers). If the input $\mathbf{X}(t)$ of the system S is specified for every t, or, if $\mathbf{X}(t) = 0$ for $t < 0$ and the system is in zero state at $t = 0$, then the response $\mathbf{Y}(t)$ of S for $t > 0$ equals

$$\mathbf{Y}(t) = \int_0^\infty H(\alpha)\mathbf{X}(t-\alpha)\,d\alpha \tag{11-11}$$

where $H(t)$ is the impulse response matrix of S. This follows from (9-87) and the fact that $H(t) = 0$ for $t < 0$ (causality assumption).

We shall determine the matrix $H(t)$ starting with the system S_1. As we see from (11-10*a*), the output $\mathbf{Z}(t)$ of this system satisfies the equation

$$\frac{d\mathbf{Z}(t)}{dt} - A\mathbf{Z}(t) = \mathbf{U}(t) \tag{11-12}$$

The impulse response of the system S_1 is an $n \times n$ matrix $\Phi(t) = [\varphi_{ji}(t)]$ called the *transition matrix* of S. The function $\varphi_{ji}(t)$ equals the value of the jth state variable $\mathbf{z}_j(t)$ when the ith element $\mathbf{u}_i(t)$ of the input $\mathbf{U}(t)$ of S_1 equals $\delta(t)$ and all other elements are 0. From this it follows that [see (9-126)]

$$\mathbf{Z}(t) = \int_0^\infty \Phi(\alpha)\mathbf{U}(t-\alpha)\,d\alpha = \int_0^\infty \Phi(\alpha)B\mathbf{X}(t-\alpha)\,d\alpha \tag{11-13}$$

Inserting into (11-10*b*), we obtain

$$\begin{aligned}\mathbf{Y}(t) &= \int_0^\infty C\Phi(\alpha)B\mathbf{X}(t-\alpha)\,d\alpha + D\mathbf{X}(t)\\ &= \int_0^\infty [C\Phi(\alpha)B\mathbf{X}(t-\alpha) + \delta(\alpha)D\mathbf{X}(t-\alpha)]\,d\alpha\end{aligned} \tag{11-14}$$

where $\delta(t)$ is the (scalar) impulse function. Comparing with (11-11), we conclude that the impulse response matrix of the system S equals

$$H(t) = C\Phi(t)B + \delta(t)D \tag{11-15}$$

From the definition of $\Phi(t)$ it follows that

$$\frac{d\Phi(t)}{dt} - A\Phi(t) = \delta(t)1_n \tag{11-16}$$

where 1_n is the identity matrix of order n. The Laplace transform $\mathbf{\Phi}(s)$ of $\Phi(t)$ is the system function of the system S_1. Taking transforms of both sides of (11-16), we obtain

$$s\mathbf{\Phi}(s) - A\mathbf{\Phi}(s) = 1_n \qquad \mathbf{\Phi}(s) = (s1_n - A)^{-1} \tag{11-17}$$

Hence

$$\Phi(t) = e^{At} \qquad t > 0 \tag{11-18}$$

This is a direct generalization of the scalar case; however, the determination of the elements $\varphi_{ji}(t)$ of $\Phi(t)$ is not trivial. Each element is a sum of exponentials of the form

$$\varphi_{ji}(t) = \sum_k p_{ji,k}(t)e^{s_k t} \qquad t > 0$$

where s_k are the eigenvalues of the matrix A and $p_{ji,k}(t)$ are polynomials in t of degree equal to the multiplicity of s_k. There are several methods for determining these polynomials. For small n, it is simplest to replace (11-16) by n systems of n scalar equations.

Inserting $\Phi(t)$ into (11-15), we obtain

$$\begin{aligned} H(t) &= Ce^{At}B + \delta(t)D \\ \mathbf{H}(s) &= C(s1_n - A)^{-1}B + D \end{aligned} \tag{11-19}$$

Suppose now that the input to the system S is a WSS process $\mathbf{X}(t)$. We shall comment briefly on the spectral properties of the resulting output, limiting the discussion to the state vector $\mathbf{Z}(t)$. The system S_1 is a special case of S obtained with $B = C = 1_n$ and $D = 0$. In this case, $\mathbf{Z}(t) = \mathbf{Y}(t)$ and

$$\frac{d\mathbf{Y}(t)}{dt} - A\mathbf{Y}(t) = \mathbf{X}(t) \qquad \mathbf{H}(s) = (s1_n - A)^{-1} \tag{11-20}$$

Inserting into (9-170), we conclude that

$$\begin{aligned} \mathbf{S}_{xy}(s) &= \mathbf{S}_{xx}(s)(-s1_n - A)^{-1} \\ \mathbf{S}_{yy}(s) &= (s1_n - A^t)^{-1}\mathbf{S}_{xy}(s) \\ \mathbf{S}_{yy}(s) &= (s1_n - A^t)^{-1}\mathbf{S}_{xx}(s)(-s1_n - A)^{-1} \end{aligned} \tag{11-21}$$

Differential equations. The equation

$$\mathbf{y}^{(n)}(t) + a_1\mathbf{y}^{(n-1)}(t) + \cdots + a_n\mathbf{y}(t) = \mathbf{x}(t) \tag{11-22}$$

specifies a system S with input $\mathbf{x}(t)$ and output $\mathbf{y}(t)$. This system is of finite order because $\mathbf{y}(t)$ is determined for $t > 0$ in terms of the values of $\mathbf{x}(t)$ for $t \geq 0$ and the initial conditions

$$\mathbf{y}(0), \mathbf{y}'(0), \ldots, \mathbf{y}^{(n-1)}(0)$$

It is, in fact, a special case of the system of Fig. 11-4 if we set $m = r = 1$,

$$\mathbf{z}_1(t) = \mathbf{y}(t) \qquad \mathbf{z}_2(t) = \mathbf{y}'(t) \cdots \mathbf{z}_n(t) = \mathbf{y}^{(n-1)}(t)$$

$$A = \begin{bmatrix} 0 & 1 & \cdots & 0 \\ 0 & 0 & \cdots & 0 \\ \cdot & \cdot & \cdots & \cdot \\ -a_n & -a_{n-1} & \cdots & -a_1 \end{bmatrix} \qquad B = \begin{bmatrix} 0 \\ 0 \\ \cdots \\ 1 \end{bmatrix} \qquad C^t = \begin{bmatrix} 1 \\ 0 \\ \cdots \\ 0 \end{bmatrix}$$

and $D = 0$. Inserting this into (11-19), we conclude after some effort that

$$\mathbf{H}(s) = \frac{1}{s^n + a_1 s^{n-1} + \cdots + a_n}$$

This result can be derived simply from (11-22).

Multiplying both sides of (11-22) by $\mathbf{x}(t - \tau)$ and $\mathbf{y}(t + \tau)$, we obtain

$$R_{yx}^{(n)}(\tau) + a_1 R_{yx}^{(n-1)}(\tau) + \cdots + a_n R_{yx}(\tau) = R_{xx}(\tau) \tag{11-23}$$

$$R_{yy}^{(n)}(\tau) + a_1 R_{yy}^{(n-1)}(\tau) + \cdots + a_n R_{yy}(\tau) = R_{xy}(\tau) \tag{11-24}$$

for all τ. This is a special case of (9-146).

Finite-order processes. We shall say that a process $\mathbf{x}(t)$ is of finite order if its innovations filter $\mathbf{L}(s)$ is a rational function of s:

$$\mathbf{S}(s) = \mathbf{L}(s)\mathbf{L}(-s) \qquad \mathbf{L}(s) = \frac{b_0 s^m + b_1 s^{m-1} + \cdots + b_m}{s^n + a_1 s^{n-1} + \cdots + a_n} = \frac{N(s)}{D(s)} \tag{11-25}$$

where $N(s)$ and $D(s)$ are two Hurwitz polynomials. The process $\mathbf{x}(t)$ is the response of the filter $\mathbf{L}(s)$ with input the white-noise process $\mathbf{i}(t)$:

$$\mathbf{x}^{(n)}(t) + a_1 \mathbf{x}^{(n-1)}(t) + \cdots + a_n \mathbf{x}(t) = b_0 \mathbf{i}^{(m)}(t) + \cdots + b_m \mathbf{i}(t) \tag{11-26}$$

The past $\mathbf{x}(t - \tau)$ of $\mathbf{x}(t)$ depends only on the past of $\mathbf{i}(t)$; hence it is orthogonal to the right side of (11-26) for every $\tau > 0$. From this it follows as in (11-24) that

$$R^{(n)}(\tau) + a_1 R^{(n-1)}(\tau) + \cdots + a_n R(\tau) = 0 \qquad \tau > 0 \tag{11-27}$$

Assuming that the roots s_i of $D(s)$ are simple, we conclude from (11-27) that

$$R(\tau) = \sum_{i=1}^{n} \alpha_i e^{s_i \tau} \qquad \tau > 0$$

The coefficients α_i can be determined from the initial value theorem. Alternatively, to find $R(\tau)$, we expand $\mathbf{S}(s)$ into partial fractions:

$$\mathbf{S}(s) = \sum_{i=1}^{n} \frac{\alpha_i}{s - s_i} + \sum_{i=1}^{n} \frac{\alpha_i}{-s - s_i} = \mathbf{S}^+(s) + \mathbf{S}^-(s) \tag{11-28}$$

The first sum is the transform of the causal part $R^+(\tau) = R(\tau)U(\tau)$ of $R(\tau)$ and the second of its anticausal part $R^-(\tau) = R(\tau)U(-\tau)$. Since $R(-\tau) = R(\tau)$, this yields

$$R(\tau) = R^+(|\tau|) = \sum_{i=1}^{n} \alpha_i e^{s_i|\tau|} \tag{11-29}$$

EXAMPLE 11-5 ▶ If $\mathsf{L}(s) = 1/(s+\alpha)$, then

$$\mathsf{S}(s) = \frac{1}{(s+\alpha)(-s+\alpha)} = \frac{1/2\alpha}{s+\alpha} + \frac{1/2\alpha}{-s+\alpha}$$

Hence $R(\tau) = (1/2\alpha)e^{-\alpha|\tau|}$. ◀

EXAMPLE 11-6 ▶ The differential equation

$$\mathbf{x}''(t) + 3\mathbf{x}'(t) + 2\mathbf{x}(t) = \mathbf{i}(t) \qquad R_{ii}(\tau) = \delta(\tau)$$

specifies a process $\mathbf{x}(t)$ with autocorrelation $R(\tau)$. From (11-27) it follows that

$$R''(\tau) + 3R'(\tau) + 2R(\tau) = 0 \qquad \text{hence} \quad R(\tau) = c_1 e^{-\tau} + c_2 e^{-2\tau}$$

for $\tau > 0$. To find the constants c_1 and c_2, we shall determine $R(0)$ and $R'(0)$. Clearly,

$$\mathsf{S}(s) = \frac{1}{(s^2+3s+2)(s^2-3s+2)} = \frac{s/12 + 1/4}{s^2+3s+2} + \frac{-s/12+1/4}{s^2-3s+2}$$

The first fraction on the right is the transform of $R^+(\tau)$; hence

$$R^+(0^+) = \lim_{s\to\infty} s\mathsf{S}^+(s) = \tfrac{1}{12} = c_1 + c_2 = R(0)$$

Similarly,

$$R'(0^+) = \lim_{s\to\infty} s\left(s\mathsf{S}^+(s) - \tfrac{1}{12}\right) = 0 = -c_1 - 2c_2$$

This yields $R(\tau) = \frac{1}{6}e^{-|\tau|} - \frac{1}{12}e^{-2|\tau|}$.

Note finally that $R(\tau)$ can be expressed in terms of the impulse response $l(t)$ of the innovations filter $\mathsf{L}(s)$:

$$R(\tau) = l(\tau) * l(-\tau) = \int_0^{\infty} l(|\tau| + \alpha) l(\alpha)\, d\alpha \tag{11-30}$$

◀

Discrete-Time Systems

The digital version of the system of Fig. 10-6 is a finite-order system S specified by the equations:

$$\mathbf{Z}[k+1] = A\mathbf{Z}[k] + B\mathbf{X}[k] \tag{11-31a}$$

$$\mathbf{Y}[k] = C\mathbf{Z}[k] + D\mathbf{X}[k] \tag{11-31b}$$

where k is the discrete time, $\mathbf{X}[k]$ the input vector, $\mathbf{Y}[k]$ the output vector, and $\mathbf{Z}[k]$ the state vector. The system is stable if the eigenvalues z_i of the $n \times n$ matrix A are such that $|z_i| < 1$. The preceding results can be readily extended to digital systems. Note, in particular, that the system function of S is the z transform

$$\mathbf{H}(z) = C(z1_n - A)^{-1}B + D \tag{11-32}$$

of the delta response matrix

$$H[k] = C\Phi[k]B + \delta[k]D \qquad k \geq 0 \tag{11-33}$$

We shall discuss in some detail scalar systems driven by white noise. This material is used in Sec. 12-3.

FINITE-ORDER PROCESSES. Consider a real digital process $\mathbf{x}[n]$ with innovations filter $\mathbf{L}(z)$ and power spectrum $\mathbf{S}(z)$:

$$\mathbf{S}(z) = \mathbf{L}(z)\mathbf{L}(1/z) \qquad \mathbf{L}(z) = \sum_{n=0}^{\infty} l[n]z^{-n} \tag{11-34}$$

where n is now the discrete time. If we know $\mathbf{L}(z)$, we can find the autocorrelation $R[m]$ of $\mathbf{x}[n]$ either from the inversion formula (9-192) or from the convolution theorem

$$R[m] = l[m] * l[-m] = \sum_{k=0}^{\infty} l[|m| + k]l[k] \tag{11-35}$$

We shall discuss the properties of $R[m]$ for the class of finite-order processes.

The power spectrum $S(\omega)$ of a finite-order process $\mathbf{x}[n]$ is a rational function of $\cos\omega$; hence its innovations filter is a rational function of z:

$$\mathbf{L}(z) = \frac{N(z)}{D(z)} = \frac{b_0 + b_1 z^{-1} + \cdots + b_M z^{-M}}{1 + a_1 z^{-1} + \cdots + a_N z^{-N}} \tag{11-36}$$

To find its autocorrelation, we determine $l[n]$ and insert the result into (11-35). Assuming that the roots z_i of $D(z)$ are simple and $M \leq N$, we obtain

$$\mathbf{L}(z) = \sum_i \frac{\gamma_i}{1 - z_i z^{-1}} \qquad l[n] = \sum_i \gamma_i z_i^n U[n]$$

Alternatively, we expand $\mathbf{S}(z)$:

$$\mathbf{S}(z) = \sum_i \frac{\alpha_i}{1 - z_i z^{-1}} + \sum_i \frac{\alpha_i}{1 - z_i z} \qquad R[m] = \sum_i \alpha_i z_i^{|m|} \tag{11-37}$$

Note that $\alpha_i = \gamma_i L(1/z_i)$.

The process $\mathbf{x}[n]$ satisfies the recursion equation

$$\mathbf{x}[n] + a_1\mathbf{x}[n-1] + \cdots + a_N\mathbf{x}[n-N] = b_0\mathbf{i}[n] + \cdots + b_M\mathbf{i}[n-M] \tag{11-38}$$

where $\mathbf{i}[n]$ is its innovations. We shall use this equation to relate the coefficients of $\mathbf{L}(z)$ to the sequence $R[m]$ starting with two special cases.

AUTOREGRESSIVE PROCESSES. The process $\mathbf{x}[n]$ is called autoregressive (AR) if

$$\mathbf{L}(z) = \frac{b_0}{1 + a_1 z^{-1} + \cdots + a_N z^{-N}} \tag{11-39}$$

In this case, (11-38) yields

$$\mathbf{x}[n] + a_1\mathbf{x}[n-1] + \cdots + a_N\mathbf{x}[n-N] = b_0\mathbf{i}[n] \tag{11-40}$$

The past $\mathbf{x}[n-m]$ of $\mathbf{x}[n]$ depends only on the past of $\mathbf{i}[n]$; furthermore, $E\{\mathbf{i}^2[n]\} = 1$. From this it follows that $E\{\mathbf{x}[n]\mathbf{i}[n]\} = b_0$ and $E\{\mathbf{x}[n-m]\mathbf{i}[n]\} = 0$ for $m > 0$. Multiplying (11-40) by $\mathbf{x}[n-m]$ and setting $m = 0, 1, \ldots,$ we obtain the equations

$$\begin{aligned} R[0] + a_1 R[1] + \cdots + a_N R[N] &= b_0^2 \\ R[1] + a_1 R[0] + \cdots + a_N R[N-1] &= 0 \\ \cdots\cdots\cdots\cdots\cdots&\cdots\cdots\cdots \\ R[N] + a_1 R[N-1] + \cdots + a_N R[0] &= 0 \end{aligned} \tag{11-41a}$$

and

$$R[m] + a_1 R[m-1] + \cdots + a_N R[m-N] = 0 \tag{11-41b}$$

for $m > N$. The first $N+1$ of these are called the *Yule–Walker* equations. They are used in Sec. 12-3 to express the $N+1$ parameters a_k and b_0 in terms of the first $N+1$ values of $R[m]$. Conversely, if $\mathbf{L}(z)$ is known, we find $R[m]$ for $|m| \le N$ solving the system (11-41a) and we determine $R[m]$ recursively from (11-41b) for $m > N$.

EXAMPLE 11-7

▶ Suppose that

$$\mathbf{x}[n] - a\mathbf{x}[n-1] = \boldsymbol{\nu}[n] \qquad R_{\nu\nu}[m] = b\delta[m]$$

This is a special case of (11-40) with $D(z) = 1 - az^{-1}$ and $z_1 = a$. Hence

$$R[0] - aR[1] = b \qquad R[m] = \alpha a^{|m|} \qquad \alpha = \frac{b}{1-a^2}$$ ◀

LINE SPECTRA. Suppose that $\mathbf{x}[n]$ satisfies the homogeneous equation

$$\mathbf{x}[n] + a_1\mathbf{x}[n-1] + \cdots + a_N\mathbf{x}[n-N] = 0 \tag{11-42}$$

This is a special case of (11-40) if we set $b_0 = 0$. Solving for $\mathbf{x}[n]$, we obtain

$$\mathbf{x}[n] = \mathbf{c}_1 z_1^n + \cdots + \mathbf{c}_N z_N^n \qquad D(z_i) = 0 \tag{11-43}$$

If $\mathbf{x}[n]$ is a stationary process, only the terms with $z_i = e^{j\omega_i}$ can appear. Furthermore, their coefficients $\mathbf{c}_k$ must be uncorrelated with zero mean. From this it follows that if $\mathbf{x}[n]$ is a WSS process satisfying (11-42), its autocorrelation must be a sum of exponentials as in Example 9-31:

$$R[m] = \sum \alpha_i e^{j\omega_i |m|} \qquad S(\omega) = 2\pi \sum \alpha_i \delta(\omega - \beta_i) \qquad |\omega| < \pi \tag{11-44}$$

where $\alpha_i = E\{\mathbf{c}_i^2\}$ and $\beta_i = \omega_i - 2\pi k_i$ as in (9-209).

MOVING AVERAGE PROCESSES. A process $\mathbf{x}[n]$ is a moving average (MA) if

$$\mathbf{x}[n] = b_0\mathbf{i}[n] + \cdots + b_M\mathbf{i}[n-M] \tag{11-45}$$

In this case, $\mathbf{L}(z)$ is a polynomial and its inverse $l[n]$ has a finite length (FIR filter):

$$\mathbf{L}(z) = b_0 + b_1 z^{-1} + \cdots + b_M Z^{-M} \qquad l[n] = b_0\delta[n] + \cdots + b_M\delta[n-M] \tag{11-46}$$

Since $l[n] = 0$ for $n > m$, (11-35) yields

$$R[m] = \sum_{k=0}^{M-m} l[m+k]l[k] = \sum_{k=0}^{M-m} b_{k+m}b_k \tag{11-47}$$

for $0 \le m \le M$ and 0 for $m > M$. Explicitly,

$$\begin{aligned} R[0] &= b_0^2 + b_1^2 + \cdots + b_M^2 \\ R[1] &= b_0 b_1 + b_1 b_2 + \cdots + b_{M-1} b_M \\ &\cdots\cdots\cdots\cdots\cdots\cdots \\ R[M] &= b_0 b_M \end{aligned}$$

EXAMPLE 11-8 ▶ Suppose that $\mathbf{x}[n]$ is the arithmetic average of the M values of $\mathbf{i}[n]$:

$$\mathbf{x}[n] = \frac{1}{M}(\mathbf{i}[n] + \mathbf{i}[n-1] + \cdots + \mathbf{i}[n-M+1])$$

In this case,

$$\mathbf{L}(z) = \frac{1}{M}(1 + z^{-1} + \cdots + z^{-M+1}) = \frac{1 - z^{-M}}{M(1 - z^{-1})}$$

$$R[m] = \frac{1}{M^2}\sum_{k=0}^{M-1-|m|} 1 = \frac{M - |m|}{M^2} = \frac{1}{M}\left(1 - \frac{|m|}{M}\right) \qquad |m| \le M$$

$$\mathbf{S}(z) = \mathbf{L}(z)\mathbf{L}(1/z) = \frac{2 - z^{-M} - z^M}{M^2(2 - z^{-1} - z)} \qquad \mathbf{S}(e^{j\omega}) = \frac{\sin^2 \frac{M\omega}{2}}{M^2 \sin^2 \frac{\omega}{2}}$$ ◀

AUTOREGRESSIVE MOVING AVERAGE. We shall say that $\mathbf{x}[n]$ is an autoregressive moving average (ARMA) process if it satisfies the equation

$$\mathbf{x}[n] + a_1\mathbf{x}[n-1] + \cdots + a_N\mathbf{x}[n-N] = b_0\mathbf{i}[n] + \cdots + b_M\mathbf{i}[n-M] \tag{11-48}$$

Its innovations filter $\mathbf{L}(z)$ is the fraction in (11-36). Again, $\mathbf{i}[n]$ is white noise; hence

$$E\{\mathbf{x}[n-m]\mathbf{i}[n-r]\} = 0 \qquad \text{for} \quad m < r$$

Multiplying (11-48) by $\mathbf{x}[n-m]$ and using the equation just stated, we conclude that

$$R[m] + a_1 R[m-1] + \cdots + a_N R[m-N] = 0 \qquad m > M \tag{11-49}$$

Note that, unlike the AR case, this is true only for $m > M$.

11-3 FOURIER SERIES AND KARHUNEN–LOÈVE EXPANSIONS

A process $\mathbf{x}(t)$ is MS periodic with period T if $E\{|\mathbf{x}(t+T) - \mathbf{x}(t)|^2\} = 0$ for all t. A WSS process is MS periodic if its autocorrelation $R(\tau)$ is periodic with period $T = 2\pi/\omega_0$ [see (9-178)]. Expanding $R(\tau)$ into Fourier series, we obtain

$$R(\tau) = \sum_{n=-\infty}^{\infty} \gamma_n e^{jn\omega_0\tau} \qquad \gamma_n = \frac{1}{T}\int_0^T R(\tau)e^{-jn\omega_0\tau}\,d\tau \tag{11-50}$$

Given a WSS periodic process $\mathbf{x}(t)$ with period T, we form the sum

$$\hat{\mathbf{x}}(t) = \sum_{n=-\infty}^{\infty} \mathbf{c}_n e^{jn\omega_0 t} \qquad \mathbf{c}_n = \frac{1}{T}\int_0^T \mathbf{x}(t)e^{-jn\omega_0 t}\,dt \tag{11-51}$$

THEOREM 11-1

The sum in (11-51) equals $\mathbf{x}(t)$ in the MS sense:

$$E\{|\mathbf{x}(t) - \hat{\mathbf{x}}(t)|^2\} = 0 \tag{11-52}$$

Furthermore, the random variables $\mathbf{c}_n$ are uncorrelated with zero mean for $n \neq 0$, and their variance equals γ_n:

$$E\{\mathbf{c}_n\} = \begin{cases} \eta_x & n = 0 \\ 0 & n \neq 0 \end{cases} \qquad E\{\mathbf{c}_n\mathbf{c}_m^*\} = \begin{cases} \gamma_n & n = m \\ 0 & n \neq m \end{cases} \tag{11-53}$$

Proof. We form the products

$$\mathbf{c}_n\mathbf{x}^*(\alpha) = \frac{1}{T}\int_0^T \mathbf{x}(t)\mathbf{x}^*(\alpha)e^{-jn\omega_0 t}\,dt$$

$$\mathbf{c}_n\mathbf{c}_n^* = \frac{1}{T}\int_0^T \mathbf{c}_n\mathbf{x}^*(t)e^{jm\omega_0 t}\,dt$$

and we take expected values. This yields

$$E\{\mathbf{c}_n\mathbf{x}^*(\alpha)\} = \frac{1}{T}\int_0^T R(t-\alpha)e^{-jn\omega_0 t}\,dt = \gamma_n e^{-jn\omega_0 \alpha}$$

$$E\{\mathbf{c}_n\mathbf{c}_m^*\} = \frac{1}{T}\int_0^T \gamma_n e^{-jn\omega_0 t}e^{jm\omega_0 t}\,dt = \begin{cases} \gamma_n & n = m \\ 0 & n \neq m \end{cases}$$

and (11-53) results.

To prove (11-52), we observe, using the above, that

$$E\{|\hat{\mathbf{x}}(t)|^2\} = \sum E\{|\mathbf{c}_n|^2\} = \sum \gamma_n = R(0) = E\{|\mathbf{x}(t)|^2\}$$

$$E\{\hat{\mathbf{x}}(t)\mathbf{x}^*(t)\} = \sum E\{\mathbf{c}_n\mathbf{x}^*(t)\}e^{jn\omega_0 t} = \sum \gamma_n = E\{\hat{\mathbf{x}}^*(t)\mathbf{x}(t)\}$$

and (11-51) follows readily.

Suppose now that the WSS process $\mathbf{x}(t)$ is not periodic. Selecting an arbitrary constant T, we form again the sum $\hat{\mathbf{x}}(t)$ as in (11-51). It can be shown that (see Prob. 11-12) $\hat{\mathbf{x}}(t)$ equals $\mathbf{x}(t)$ not for all t, but only in the interval $(0, T)$:

$$E\{|\hat{\mathbf{x}}(t) - \mathbf{x}(t)|^2\} = 0 \qquad 0 < t < T \tag{11-54}$$

Unlike the periodic case, however, the coefficients $\mathbf{c}_n$ of this expansion are not orthogonal (they are nearly orthogonal for large n). In the following, we show that an arbitrary process $\mathbf{x}(t)$, stationary or not, can be expanded into a series with orthogonal coefficients.

The Karhunen–Loève Expansion

The Fourier series is a special case of the expansion of a process $\mathbf{x}(t)$ into a series of the form

$$\hat{\mathbf{x}}(t) = \sum_{n=1}^{\infty} \mathbf{c}_n \varphi_n(t) \qquad 0 < t < T \tag{11-55}$$

where $\varphi_n(t)$ is a set of orthonormal functions in the interval $(0, T)$:

$$\int_0^T \varphi_n(t)\varphi_m^*(t)\,dt = \delta[n - m] \tag{11-56}$$

and the coefficients $\mathbf{c}_n$ are random variables given by

$$\mathbf{c}_n = \int_0^T \mathbf{x}(t)\varphi_n^*(t)\,dt \tag{11-57}$$

In this development, we consider the problem of determining a set of orthonormal functions $\varphi_n(t)$ such that: (*a*) the sum in (11-55) equals $\mathbf{x}(t)$; (*b*) the coefficients $\mathbf{c}_n$ are orthogonal.

To solve this problem, we form the *integral equation*

$$\int_0^T R(t_1, t_2)\varphi(t_2)\,dt_2 = \lambda\varphi(t_1) \qquad 0 < t_1 < T \tag{11-58}$$

where $R(t_1, t_2)$ is the autocorrelation of the process $\mathbf{x}(t)$. It is well known from the theory of integral equations that the eigenfunctions $\varphi_n(t)$ of (11-58) are orthonormal as in (11-56) and they satisfy the identity

$$R(t, t) = \sum_{n=1}^{\infty} \lambda_n |\varphi_n(t)|^2 \tag{11-59}$$

where λ_n are the corresponding eigenvalues. This is a consequence of the p.d. character of $R(t_1, t_2)$.

Using this, we shall show that if $\varphi_n(t)$ are the eigenfunctions of (11-58), then

$$E\{|\mathbf{x}(t) - \hat{\mathbf{x}}(t)|^2\} = 0 \qquad 0 < t < T \tag{11-60}$$

and

$$E\{\mathbf{c}_n \mathbf{c}_m^*\} = \lambda_n \delta[n - m] \tag{11-61}$$

Proof. From (11-57) and (11-58) it follows that

$$\begin{aligned} E\{\mathbf{c}_n \mathbf{x}^*(\alpha)\} &= \int_0^T R^*(\alpha, t)\varphi_n^*(t)\,dt = \lambda_n \varphi_n^*(\alpha) \\ E\{\mathbf{c}_n \mathbf{c}_m^*\} &= \lambda_m \int_0^T \varphi_n^*(t)\varphi_m(t)\,dt = \lambda_n \delta[n - m] \end{aligned} \tag{11-62}$$

Hence

$$E\{\mathbf{c}_n\hat{\mathbf{x}}^*(t)\} = \sum_{m=1}^{\infty} E\{\mathbf{c}_n\mathbf{c}_m^*\}\varphi_m^*(t) = \lambda_n\varphi_n^*(t)$$

$$E\{\hat{\mathbf{x}}(t)\mathbf{x}^*(t)\} = \sum_{n=1}^{\infty} \lambda_n\varphi_n(t)\varphi_n^*(t) = R(t,t)$$

$$= E\{\hat{\mathbf{x}}^*(t)\mathbf{x}(t)\} = E\{|\mathbf{x}(t)|^2\} = E\{|\hat{\mathbf{x}}(t)|^2\}$$

and (11-60) results.

It is of interest to note that the converse of the above is also true: If $\varphi_n(t)$ is an orthonormal set of functions and

$$\mathbf{x}(t) = \sum_{n=1}^{\infty} \mathbf{c}_n\varphi_n(t) \qquad E\{\mathbf{c}_n\mathbf{c}_m^*\} = \begin{cases} \sigma_n^2 & n = m \\ 0 & n \neq m \end{cases}$$

then the functions $\varphi_n(t)$ must satisfy (11-58) with $\lambda = \sigma_n^2$.

Proof. From the assumptions it follows that $\mathbf{c}_n$ is given by (11-57). Furthermore,

$$E\{\mathbf{x}(t)\mathbf{c}_m^*\} = \sum_{n=1}^{\infty} E\{\mathbf{c}_n\mathbf{c}_m^*\}\varphi_n(t) = \sigma_m^2\varphi_m(t)$$

$$E\{\mathbf{x}(t)\mathbf{c}_m^*\} = \int_0^T E\{\mathbf{x}(t)\mathbf{x}^*(\alpha)\}\varphi_m(\alpha)\,d\alpha = \int_0^T R(t,\alpha)\varphi_m(\alpha)\,d\alpha$$

This completes the proof.

The sum in (11-55) is called the Karhunen–Loève (K–L) expansion of the process $\mathbf{x}(t)$. In this expansion, $\mathbf{x}(t)$ need not be stationary. If it is stationary, then the origin can be chosen arbitrarily. We shall illustrate with two examples.

EXAMPLE 11-9

▶ Suppose that the process $\mathbf{x}(t)$ is ideal low-pass with autocorrelation

$$R(\tau) = \frac{\sin a\tau}{\pi\tau}$$

We shall find its K–L expansion. Shifting the origin appropriately, we conclude from (11-58) that the functions $\varphi_n(t)$ must satisfy the integral equation

$$\int_{-T/2}^{T/2} \frac{\sin a(t-\tau)}{\pi(t-\tau)}\varphi_n(\tau)\,d\tau = \lambda_n\varphi_n(t) \qquad \text{(11-63)}$$

The solutions of this equation are known as *prolate-spheroidal* functions.[2] ◀

EXAMPLE 11-10

▶ We shall determine the K–L expansion (11-55) of the Wiener process $\mathbf{w}(t)$ introduced in Sec. 10-1. In this case [see (10-53)]

$$R(t_1,t_2) = \alpha\min(t_1,t_2) = \begin{cases} \alpha t_2 & t_2 < t_1 \\ \alpha t_1 & t_2 > t_1 \end{cases}$$

[2]D. Slepian, H. J. Landau, and H. O. Pollack: "Prolate Spheroidal Wave Functions," *Bell System Technical Journal,* vol. 40, 1961.

Inserting into (11-58), we obtain

$$\alpha \int_0^{t_1} t_2 \varphi(t_2)\, dt_2 + \alpha t_1 \int_{t_1}^{T} \varphi(t_2)\, dt_2 = \lambda \varphi(t_1) \tag{11-64}$$

To solve this integral equation, we evaluate the appropriate endpoint conditions and differentiate twice. This yields

$$\varphi(0) = 0 \qquad \alpha \int_{t_1}^{T} \varphi(t_2)\, dt_2 = \lambda \varphi'(t_1)$$

$$\varphi'(T) = 0 \qquad \lambda \varphi''(t) + \alpha \varphi(t) = 0$$

Solving the last equation, we obtain

$$\varphi_n(t) = \sqrt{\frac{2}{T}} \sin \omega_n t \qquad \omega_n = \sqrt{\frac{\alpha}{\lambda_n}} = \frac{(2n+1)\pi}{2T}$$

Thus, in the interval $(0, T)$, the Wiener process can be written as a sum of sine waves

$$\mathbf{w}(t) = \sqrt{\frac{2}{T}} \sum_{n=1}^{\infty} \mathbf{c}_n \sin \omega_n t \qquad \mathbf{c}_n = \sqrt{\frac{2}{T}} \int_0^T \mathbf{w}(t) \sin \omega_n t\, dt$$

where the coefficients $\mathbf{c}_n$ are uncorrelated with variance $E\{\mathbf{c}_n^2\} = \lambda_n$. ◀

11-4 SPECTRAL REPRESENTATION OF RANDOM PROCESSES

The Fourier transform of a stochastic process $\mathbf{x}(t)$ is a stochastic process $\mathbf{X}(\omega)$ given by

$$\mathbf{X}(\omega) = \int_{-\infty}^{\infty} \mathbf{x}(t) e^{-j\omega t}\, dt \tag{11-65}$$

The integral is interpreted as an MS limit. Reasoning as in (11-52), we can show that (inversion formula)

$$\mathbf{x}(t) = \frac{1}{2\pi} \int_{-\infty}^{\infty} \mathbf{X}(\omega) e^{j\omega t}\, d\omega \tag{11-66}$$

in the MS sense. The properties of Fourier transforms also hold for random signals. For example, if $\mathbf{y}(t)$ is the output of a linear system with input $\mathbf{x}(t)$ and system function $H(\omega)$, then $\mathbf{Y}(\omega) = \mathbf{X}(\omega)H(\omega)$.

The mean of $\mathbf{X}(\omega)$ equals the Fourier transform of the mean of $\mathbf{x}(t)$. We shall express the autocorrelation of $\mathbf{X}(\omega)$ in terms of the two-dimensional Fourier transform:

$$\Gamma(u, v) = \int_{-\infty}^{\infty} \int_{-\infty}^{\infty} R(t_1, t_2) e^{-j(ut_1 + vt_2)}\, dt_1\, dt_2 \tag{11-67}$$

of the autocorrelation $R(t_1, t_2)$ of $\mathbf{x}(t)$. Multiplying (11-65) by its conjugate and taking expected values, we obtain

$$E\{\mathbf{X}(u)\mathbf{X}^*(v)\} = \int_{-\infty}^{\infty} \int_{-\infty}^{\infty} E\{\mathbf{x}(t_1)\mathbf{x}^*(t_2)\} e^{-j(ut_1 - vt_2)}\, dt_1\, dt_2$$

Hence

$$E\{\mathbf{X}(u)\mathbf{X}^*(v)\} = \Gamma(u, -v) \tag{11-68}$$

Using (11-68), we shall show that, if $\mathbf{x}(t)$ is nonstationary white noise with average power $q(t)$, then $\mathbf{X}(\omega)$ is a stationary process and its autocorrelation equals the Fourier transform $Q(\omega)$ of $q(t)$:

THEOREM 11-2 ▶ If $R(t_1, t_2) = q(t_1)\delta(t_1 - t_2)$, then

$$E\{\mathbf{X}(\omega + \alpha)\mathbf{X}^*(\alpha)\} = Q(\omega) = \int_{-\infty}^{\infty} q(t)e^{-j\omega t}\,dt \tag{11-69}$$

Proof. From the identity

$$\int_{-\infty}^{\infty}\int_{-\infty}^{\infty} q(t_1)\delta(t_1 - t_2)e^{-j(ut_1+vt_2)}\,dt_1\,dt_2 = \int_{-\infty}^{\infty} q(t_2)e^{-j(u+v)t_2}\,dt_2$$

it follows that $\Gamma(u, v) = Q(u + v)$. Hence [see (11-68)]

$$E\{\mathbf{X}(\omega + \alpha)\mathbf{X}^*(\alpha)\} = \Gamma(\omega + \alpha, -\alpha) = Q(\omega)$$

Note that if the process $\mathbf{x}(t)$ is *real,* then

$$E\{\mathbf{X}(u)\mathbf{X}(v)\} = \Gamma(u, v) \tag{11-70}$$

Furthermore,

$$\mathbf{X}(-\omega) = \mathbf{X}^*(\omega) \qquad \Gamma(-u, -v) = \Gamma^*(u, v) \tag{11-71}$$

◀

Covariance of energy spectrum. To find the autocovariance of $|\mathbf{X}(\omega)|^2$, we must know the fourth-order moments of $\mathbf{X}(\omega)$. However, if the process $\mathbf{x}(t)$ is normal, the results can be expressed in terms of the function $\Gamma(u, v)$. We shall assume that the process $\mathbf{x}(t)$ is *real* with

$$\mathbf{X}(\omega) = \mathbf{A}(\omega) + j\mathbf{B}(\omega) \qquad \Gamma(u, v) = \Gamma_r(u, v) + j\Gamma_i(u, v) \tag{11-72}$$

From (11-68) and (11-70) it follows that

$$\begin{aligned}
2E\{\mathbf{A}(u)\mathbf{A}(v)\} &= \Gamma_r(u, v) + \Gamma_r(u, -v)\\
2E\{\mathbf{A}(v)\mathbf{B}(u)\} &= \Gamma_i(u, v) + \Gamma_i(u, -v)\\
2E\{\mathbf{B}(u)\mathbf{B}(v)\} &= \Gamma_r(u, v) - \Gamma_r(u, -v)\\
2E\{\mathbf{A}(u)\mathbf{B}(v)\} &= \Gamma_i(u, v) - \Gamma_i(u, -v)
\end{aligned} \tag{11-73}$$

THEOREM 11-3 ▶ If $\mathbf{x}(t)$ is a real normal process with zero mean, then

$$\text{Cov}\{|\mathbf{X}(u)|^2, |\mathbf{X}(v)|^2\} = \Gamma^2(u, -v) + \Gamma^2(u, v) \tag{11-74}$$

Proof. From the normality of $\mathbf{x}(t)$ it follows that the processes $\mathbf{A}(\omega)$ and $\mathbf{B}(\omega)$ are jointly normal with zero mean. Hence [see (6-197)]

$$\begin{aligned}
&E\{|\mathbf{X}(u)|^2|\mathbf{X}(v)|^2\} - E\{|\mathbf{X}(u)|^2\}E\{|\mathbf{X}(v)|^2\}\\
&\quad = E\{[\mathbf{A}^2(u) + \mathbf{B}^2(u)][\mathbf{A}^2(v) + \mathbf{B}^2(v)]\} - E\{\mathbf{A}^2(u) + \mathbf{B}^2(u)\}E\{\mathbf{A}^2(v) + \mathbf{B}^2(v)\}\\
&\quad = 2E^2\{\mathbf{A}(u)\mathbf{A}(v)\} + 2E^2\{\mathbf{B}(u)\mathbf{B}(v)\} + 2E^2\{\mathbf{A}(u)\mathbf{B}(v)\} + 2E^2\{\mathbf{A}(v)\mathbf{B}(u)\}
\end{aligned}$$

Inserting (11-73) into the above, we obtain (11-74). ◀

STATIONARY PROCESSES. Suppose that $\mathbf{x}(t)$ is a stationary process with autocorrelation $R(t_1, t_2) = R(t_1 - t_2)$ and power spectrum $S(\omega)$. We shall show that

$$\Gamma(u, v) = 2\pi S(u)\delta(u + v) \tag{11-75}$$

Proof. With $t_1 = t_2 + \tau$, it follows from (11-67) that the two-dimensional transform of $R(t_1 - t_2)$ equals

$$\int_{-\infty}^{\infty}\int_{-\infty}^{\infty} R(t_1 - t_2)e^{-j(ut_1+vt_2)}\,dt_1\,dt_2 = \int_{-\infty}^{\infty} e^{-j(u+v)t_2}\int_{-\infty}^{\infty} R(\tau)e^{-ju\tau}\,d\tau\,dt_2$$

Hence

$$\Gamma(u, v) = S(u)\int_{-\infty}^{\infty} e^{-j(u+v)t_2}\,dt_2$$

This yields (11-75) because $\int e^{-j\omega t}\,dt = 2\pi\delta(\omega)$.

From (11-75) and (11-68) it follows that

$$E\{\mathbf{X}(u)\mathbf{X}^*(v)\} = 2\pi S(u)\delta(u - v) \tag{11-76}$$

This shows that the Fourier transform of a stationary process is nonstationary white noise with average power $2\pi S(u)$. It can be shown that the converse is also true (see Prob. 11-12): The process $\mathbf{x}(t)$ in (11-66) is WSS iff $E\{\mathbf{X}(\omega)\} = 0$ for $\omega \neq 0$, and

$$E\{\mathbf{X}(u)\mathbf{X}^*(v)\} = Q(u)\delta(u - v) \tag{11-77}$$

Real processes. If $\mathbf{x}(t)$ is real, then $\mathbf{A}(-\omega) = \mathbf{A}(\omega)$, $\mathbf{B}(-\omega) = \mathbf{B}(\omega)$, and

$$\mathbf{x}(t) = \frac{1}{\pi}\int_0^{\infty} \mathbf{A}(\omega)\cos\omega t\,d\omega - \frac{1}{\pi}\int_0^{\infty} \mathbf{B}(\omega)\sin\omega t\,d\omega \tag{11-78}$$

It suffices, therefore, to specify $\mathbf{A}(\omega)$ and $\mathbf{B}(\omega)$ for $\omega \geq 0$ only. From (11-68) and (11-70) it follows that

$$E\{[\mathbf{A}(u) + j\mathbf{B}(u)][\mathbf{A}(v) \pm j\mathbf{B}(v)]\} = 0 \qquad u \neq \pm v$$

Equating real and imaginary parts, we obtain

$$E\{\mathbf{A}(u)\mathbf{A}(v)\} = E\{\mathbf{A}(u)\mathbf{B}(v)\} = E\{\mathbf{B}(u)\mathbf{B}(v)\} = 0 \qquad \text{for} \quad u \neq v \tag{11-79a}$$

With $u = \omega$ and $v = -\omega$, (11-9) yields $E\{\mathbf{X}(\omega)\mathbf{X}(\omega)\} = 0$ for $\omega \neq 0$; hence

$$E\{\mathbf{A}^2(\omega)\} = E\{\mathbf{B}^2(\omega)\} \qquad E\{\mathbf{A}(\omega)\mathbf{B}(\omega)\} = 0 \tag{11-79b}$$

It can be shown that the converse is also true (see Prob. 11-13). Thus a real process $\mathbf{x}(t)$ is WSS if the coefficients $\mathbf{A}(\omega)$ and $\mathbf{B}(\omega)$ of its expansion (11-78) satisfy (11-79) and $E\{\mathbf{A}(\omega)\} = E\{\mathbf{B}(\omega)\} = 0$ for $\omega \neq 0$.

Windows. Given a WSS process $\mathbf{x}(t)$ and a function $w(t)$ with Fourier transform $W(\omega)$, we form the process $\mathbf{y}(t) = w(t)\mathbf{x}(t)$. This process is nonstationary with autocorrelation

$$R_{yy}(t_1, t_2) = w(t_1)w^*(t_2)R(t_1 - t_2)$$

The Fourier transform of $R_{yy}(t_1, t_2)$ equals

$$\Gamma_{yy}(u, v) = \int_{-\infty}^{\infty}\int_{-\infty}^{\infty} w(t_1)w^*(t_2)R(t_1 - t_2)e^{-j(ut_1+vt_2)}\,dt_1\,dt_2$$

Proceeding as in the proof of (11-75), we obtain

$$\Gamma_{yy}(u, v) = \frac{1}{2\pi}\int_{-\infty}^{\infty} W(u-\beta)W^*(-v-\beta)S(\beta)\,d\beta \qquad (11\text{-}80)$$

From (11-68) and the development resulting in (11-80) it follows that the autocorrelation of the Fourier transform

$$\mathbf{Y}(\omega) = \int_{-\infty}^{\infty} w(t)\mathbf{x}(t)e^{-j\omega t}\,dt \qquad (11\text{-}81)$$

of $\mathbf{y}(t)$ equals

$$E\{\mathbf{Y}(u)\mathbf{Y}^*(v)\} = \Gamma_{yy}(u, -v) = \frac{1}{2\pi}\int_{-\infty}^{\infty} W(u-\beta)W^*(v-\beta)S(\beta)\,d\beta$$

Hence

$$E\{|\mathbf{Y}(\omega)|^2\} = \frac{1}{2\pi}\int_{-\infty}^{\infty} |W(\omega-\beta)|^2 S(\beta)\,d\beta \qquad (11\text{-}82)$$

EXAMPLE 11-11 ▶ The integral

$$\mathbf{X}_T(\omega) = \int_{-T}^{T} \mathbf{x}(t)e^{-j\omega t}\,dt$$

is the transform of the segment $\mathbf{x}(t)p_T(t)$ of the process $\mathbf{x}(t)$. This is a special case of (11-81) with $w(t) = p_T(t)$ and $W(\omega) = 2\sin T\omega/\omega$. If, therefore, $\mathbf{x}(t)$ is a stationary process, then [see (11-82)]

$$E\{|\mathbf{X}_T(\omega)|^2\} = S(\omega) * \frac{2\sin^2 T\omega}{\pi\omega^2} \qquad (11\text{-}83)$$

◀

Fourier–Stieltjes Representation of WSS Processes[3]

We shall express the spectral representation of a WSS process $\mathbf{x}(t)$ in terms of the integral

$$\mathbf{Z}(\omega) = \int_0^{\omega} \mathbf{X}(\alpha)\,d\alpha \qquad (11\text{-}84)$$

We have shown that the Fourier transform $\mathbf{X}(\omega)$ of $\mathbf{x}(t)$ is nonstationary white noise with average power $2\pi S(\omega)$. From (11-76) it follows that, $\mathbf{Z}(\omega)$ is a process with orthogonal increments:

For any $\omega_1 < \omega_2 < \omega_3 < \omega_4$:

$$E\{[\mathbf{Z}(\omega_2) - \mathbf{Z}(\omega_1)][\mathbf{Z}^*(\omega_4) - \mathbf{Z}^*(\omega_3)]\} = 0 \qquad (11\text{-}85a)$$

$$E\{|\mathbf{Z}(\omega_2) - \mathbf{Z}(\omega_1)|^2\} = 2\pi\int_{\omega_1}^{\omega_2} S(\omega)\,d\omega \qquad (11\text{-}85b)$$

[3]H. Cramer: *Mathematical Methods of Statistics.* Princeton Univ. Press, Princeton, N.J., 1946.

Clearly,

$$d\mathbf{Z}(\omega) = \mathbf{X}(\omega)\,d\omega \tag{11-86}$$

hence the inversion formula (11-66) can be written as a Fourier–Stieltjes integral:

$$\mathbf{x}(t) = \frac{1}{2\pi}\int_{-\infty}^{\infty} e^{j\omega t}\,d\mathbf{Z}(\omega) \tag{11-87}$$

With $\omega_1 = u$, $\omega_2 = u + du$ and $\omega_3 = v$, $\omega_4 = v + dv$, (11-85) yields

$$\begin{aligned} E\{d\mathbf{Z}(u)\,d\mathbf{Z}^*(v)\} &= 0 \qquad u \neq v \\ E\{|d\mathbf{Z}(u)|^2\} &= 2\pi S(u)\,du \end{aligned} \tag{11-88}$$

The last equation can be used to define the spectrum $S(\omega)$ of WSS process $\mathbf{x}(t)$ in terms of the process $\mathbf{Z}(\omega)$.

WOLD'S DECOMPOSITION. Using (11-85), we shall show that an arbitrary WSS process $\mathbf{x}(t)$ can be written as a sum:

$$\mathbf{x}(t) = \mathbf{x}_r(t) + \mathbf{x}_p(t) \tag{11-89}$$

where $\mathbf{x}_r(t)$ is a *regular* process and $\mathbf{x}_p(t)$ is a *predictable* process consisting of exponentials:

$$\mathbf{x}_p(t) = \mathbf{c}_0 + \sum_i \mathbf{c}_i e^{j\omega_i t} \qquad E\{\mathbf{c}_i\} = 0 \tag{11-90}$$

Furthermore, the two processes are orthogonal:

$$E\{\mathbf{x}_r(t+\tau)\mathbf{x}_p^*(t)\} = 0 \tag{11-91}$$

This expansion is called *Wold's decomposition*. In Sec. 13-2, we determine the processes $\mathbf{x}_r(t)$ and $\mathbf{x}_p(t)$ as the responses of two linear systems with input $\mathbf{x}(t)$. We also show that $\mathbf{x}_p(t)$ is predictable in the sense that it is determined in terms of its past; the process $\mathbf{x}_r(t)$ is not predictable.

We shall prove (11-89) using the properties of the integrated transform $\mathbf{Z}(\omega)$ of $\mathbf{x}(t)$. The process $\mathbf{Z}(\omega)$ is a family of functions. In general, these functions are discontinuous at a set of points ω_i for almost every outcome. We expand $\mathbf{Z}(\omega)$ as a sum (Fig. 11-5)

$$\mathbf{Z}(\omega) = \mathbf{Z}_r(\omega) + \mathbf{Z}_p(\omega) \tag{11-92}$$

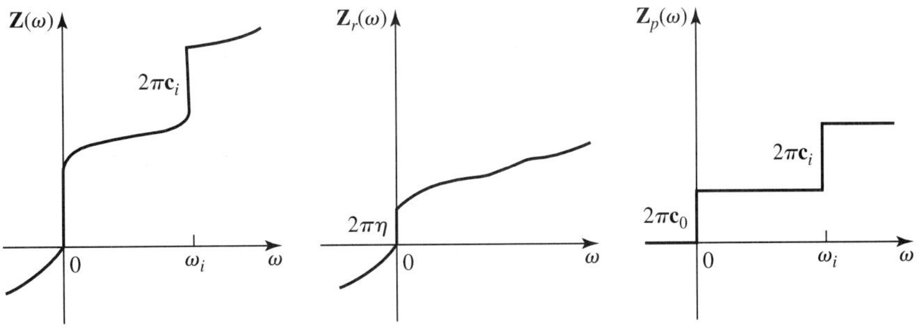

FIGURE 11-5

where $\mathbf{Z}_r(\omega)$ is a continuous process for $\omega \neq 0$ and $\mathbf{Z}_p(\omega)$ is a staircase function with discontinuities at ω_i. We denote by $2\pi\mathbf{c}_i$ the discontinuity jumps at $\omega_i \neq 0$. These jumps equal the jumps of $\mathbf{Z}_p(\omega)$. We write the jump at $\omega = 0$ as a sum $2\pi(\eta + \mathbf{c}_0)$, where $\eta = E\{\mathbf{x}(t)\}$, and we associate the term $2\pi\eta$ with $\mathbf{Z}_r(\omega)$. Thus at $\omega = 0$ the process $\mathbf{Z}_r(\omega)$ is discontinuous with jump equal to $2\pi\eta$. The jump of $\mathbf{Z}_p(\omega)$ at $\omega = 0$ equals $2\pi\mathbf{c}_0$. Inserting (11-92) into (11-87), we obtain the decomposition (11-89) of $\mathbf{x}(t)$, where $\mathbf{x}_r(t)$ and $\mathbf{x}_p(t)$ are the components due to $\mathbf{Z}_r(\omega)$ and $\mathbf{Z}_p(\omega)$, respectively.

From (11-85) it follows that $\mathbf{Z}_r(\omega)$ and $\mathbf{Z}_p(\omega)$ are two processes with orthogonal increments and such that

$$E\{\mathbf{Z}_r(u)\mathbf{Z}_p^*(v)\} = 0 \qquad E\{\mathbf{c}_i\mathbf{c}_j^*\} = \begin{cases} k_i & i = j \\ 0 & i \neq j \end{cases} \tag{11-93}$$

The first equation shows that the processes $\mathbf{x}_r(t)$ and $\mathbf{x}_p(t)$ are orthogonal as in (11-89); the second shows that the coefficients $\mathbf{c}_i$ of $\mathbf{x}_p(t)$ are orthogonal. This also follows from the stationarity of $\mathbf{x}_p(t)$.

We denote by $S_r(\omega)$ and $S_p(\omega)$ the spectra and by $F_r(\omega)$ and $F_p(\omega)$ the integrated spectra of $\mathbf{x}_r(t)$ and $\mathbf{x}_p(t)$ respectively. From (11-89) and (11-91) it follows that

$$S(\omega) = S_r(\omega) + S_p(\omega) \qquad F(\omega) = F_r(\omega) + F_p(\omega) \tag{11-94}$$

The term $F_r(\omega)$ is continuous for $\omega \neq 0$; for $\omega = 0$ it is discontinuous with a jump equal to $2\pi\eta^2$. The term $F_p(\omega)$ is a staircase function, discontinuous at the points ω_i with jumps equal to $2\pi k_i$. Hence

$$S_p(\omega) = 2\pi k_0 \delta(\omega) + 2\pi \sum_i k_i \delta(\omega - \omega_i) \tag{11-95}$$

The impulse at the origin of $S(\omega)$ equals $2\pi(k_0 + \eta^2)\delta(\omega)$.

EXAMPLE 11-12 ▶ Consider the process

$$\mathbf{y}(t) = \mathbf{a}\mathbf{x}(t) \qquad E\{\mathbf{a}\} = 0$$

where $\mathbf{x}(t)$ is a regular process independent of $\mathbf{a}$. We shall determine its Wold decomposition.

From the assumptions it follows that

$$E\{\mathbf{y}(t)\} = 0 \qquad R_{yy}(\tau) = E\{\mathbf{a}^2\mathbf{x}(t+\tau)\mathbf{x}(t)\} = \sigma_a^2 R_{xx}(\tau)$$

The spectrum of $\mathbf{x}(t)$ equals $S_{xx}^c(\omega) + 2\pi\eta_x^2\delta(\omega)$. Hence

$$S_{yy}(\omega) = \sigma_a^2 S_{xx}^c(\omega) + 2\pi\sigma_a^2\eta_x^2\delta(\omega)$$

From the regularity of $\mathbf{x}(t)$ it follows that its covariance spectrum $S_{xx}^c(\omega)$ has no impulses. Since $\eta_y = 0$, we conclude from (11-95) that $S_p(\omega) = 2\pi k_0\delta(\omega)$, where $k_0 = \sigma_a^2\eta_x^2$. This yields

$$\mathbf{y}_p(t) = \eta_x\mathbf{a} \qquad \mathbf{y}_r(t) = \mathbf{a}[\mathbf{x}(t) - \eta_x]$$ ◀

DISCRETE-TIME PROCESSES. Given a discrete-time process $\mathbf{x}[n]$, we form its discrete Fourier transform (DFT)

$$\mathbf{X}(\omega) = \sum_{n=-\infty}^{\infty} \mathbf{x}[n]e^{-jn\omega} \tag{11-96}$$

This yields

$$\mathbf{x}[n] = \frac{1}{2\pi}\int_{-\pi}^{\pi} \mathbf{X}(\omega)e^{jn\omega}\,d\omega \tag{11-97}$$

From the definition it follows that the process $\mathbf{X}(\omega)$ is periodic with period 2π. It suffices, therefore, to study its properties for $|\omega| < \pi$ only. The preceding results properly modified also hold for discrete-time processes. We shall discuss only the digital version of (11-76):

If $\mathbf{x}[n]$ is a WSS process with power spectrum $S(\omega)$, then its DFT $\mathbf{X}(\omega)$ is nonstationary white noise with autocovariance

$$E\{\mathbf{X}(u)\mathbf{X}^*(v)\} = 2\pi S(u)\delta(u-v) \qquad -\pi < u, v < \pi \tag{11-98}$$

Proof. The proof is based on the identity

$$\sum_{n=-\infty}^{\infty} e^{-jn\omega} = 2\pi\delta(\omega) \qquad |\omega| < \pi$$

Clearly,

$$E\{\mathbf{X}(u)\mathbf{X}^*(v)\} = \sum_{n=-\infty}^{\infty}\sum_{m=-\infty}^{\infty} E\{\mathbf{x}[n+m]\mathbf{x}^*[m]\}\exp\{-j[(m+n)u - nv]\}$$

$$= \sum_{n=-\infty}^{\infty} e^{-jn(u-v)} \sum_{m=-\infty}^{\infty} R[m]e^{-jmu}$$

and (11-98) results.

BISPECTRA AND THIRD ORDER MOMENTS. Consider a real SSS process $\mathbf{x}(t)$ with Fourier transform $\mathbf{X}(\omega)$ and third-order moment $R(\mu, \nu)$ [see (10-243)]. Generalizing (11-76), we shall express the third-order moment of $\mathbf{X}(\omega)$ in terms of the bispectrum $S(u, v)$ of $\mathbf{x}(t)$.

THEOREM 11-4

$$E\{\mathbf{X}(u)\mathbf{X}(v)\mathbf{X}^*(w)\} = 2\pi S(u, v)\delta(u+v-w) \tag{11-99}$$

Proof. From (11-65) it follows that the left side of (11-99) equals

$$\int_{-\infty}^{\infty}\int_{-\infty}^{\infty}\int_{-\infty}^{\infty} E\{\mathbf{x}(t_1)\mathbf{x}(t_2)\mathbf{x}(t_3)\}e^{-j(ut_1+vt_2-wt_3)}\,dt_1\,dt_2\,dt_3$$

With $t_1 = t_3 + \mu$ and $t_2 = t_3 + \nu$, the last equation yields

$$\int_{-\infty}^{\infty}\int_{-\infty}^{\infty} R\{\mu, \nu\}e^{-j(u\mu+v\nu)}\,d\mu\,d\nu \int_{-\infty}^{\infty} e^{-j(u+v-w)t_3}\,dt_3$$

and (11-99) results because the last integral equals $2\pi\delta(u+v-w)$. ◀

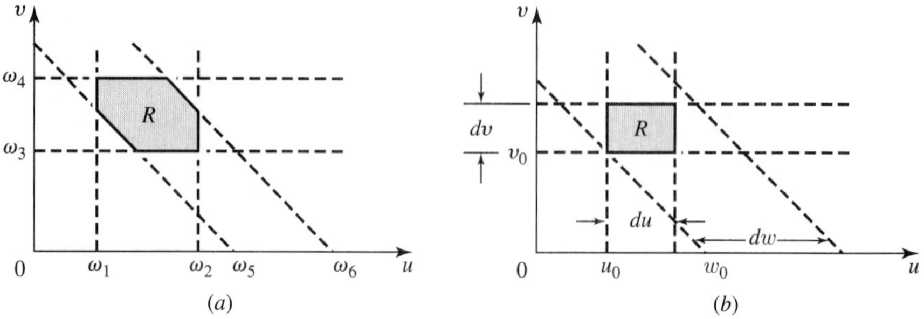

FIGURE 11-6

We have thus shown that the third-order moment of $\mathbf{X}(\omega)$ is 0 everywhere in the uvw space except on the plane $w = u + v$, where it equals a surface singularity with density $2\pi S(u, v)$. Using this result, we shall determine the third-order moment of the increments

$$\mathbf{Z}(\omega_i) - \mathbf{Z}(\omega_k) = \int_{\omega_i}^{\omega_k} \mathbf{X}(\omega)\, d\omega \tag{11-100}$$

of the integrated transforms $\mathbf{Z}(\omega)$ of $\mathbf{x}(t)$.

THEOREM 11-5

$$E\{[\mathbf{Z}(\omega_2) - \mathbf{Z}(\omega_1)]\{\mathbf{Z}(\omega_4) - \mathbf{Z}(\omega_3)][\mathbf{Z}^*(\omega_6) - \mathbf{Z}^*(\omega_5)]\}$$
$$= 2\pi \int_R\int S(u, v)\, du\, dv \tag{11-101}$$

where R is the set of points common to the three regions

$$\omega_1 < u < \omega_2 \qquad \omega_3 < v < \omega_4 \qquad \omega_5 < w < \omega_6$$

(shaded in Fig. 11-6a) of the uv plane.

Proof. From (11-99) and (11-100) it follows that the left side of (11-101) equals

$$\int_{\omega_1}^{\omega_2}\int_{\omega_3}^{\omega_4}\int_{\omega_5}^{\omega_6} 2\pi S(u, v)\delta(u + v - w)\, du\, dv\, dw$$
$$= 2\pi \int_{\omega_1}^{\omega_2}\int_{\omega_3}^{\omega_4} S(u, v)\, du\, dv \int_{\omega_5}^{\omega_6} \delta(u + v - w)\, dw$$

The last integral equals one for $\omega_5 < u + v < \omega_6$ and 0 otherwise. Hence the right side equals the integral of $2\pi S(u, v)$ in the set R as in (11-101).

COROLLARY

Consider the differentials

$$d\mathbf{Z}(u_0) = \mathbf{X}(u_0)\, du \qquad d\mathbf{Z}(v_0) = \mathbf{X}(v_0)\, dv \qquad d\mathbf{Z}(w_0) = \mathbf{X}(w_0)\, dw$$

We maintain that

$$E\{d\mathbf{Z}(u_0)\, d\mathbf{Z}(v_0)\, d\mathbf{Z}^*(w_0)\} = 2\pi S(u_0, v_0)\, du\, dv \tag{11-102}$$

if $w_0 = u_0 + v_0$ and $dw \geq du + dv$; it is zero if $w_0 \neq u_0 + v_0$.

Proof. Setting

$$\omega_1 = u_0 \qquad \omega_3 = v_0 \qquad \omega_5 = w_0 = u_0 + v_0$$
$$\omega_2 = u_0 + du \qquad \omega_4 = v_0 + dv \qquad \omega_6 \geq w_0 + du + dv$$

into (11-101), we obtain (11-102) because the set R is the shaded rectangle of Fig. 11-6*b*.

We conclude with the observation that equation (11-102) can be used to define the bispectrum of a SSS process $\mathbf{x}(t)$ in terms of $\mathbf{Z}(\omega)$. ◀

PROBLEMS

11-1 Find $R_x[m]$ and the whitening filter of $\mathbf{x}[n]$ if

$$\mathbf{S}_x(\omega) = \frac{\cos 2\omega + 1}{12\cos 2\omega - 70\cos\omega + 62}$$

11-2 Find the innovations filter of the process $\mathbf{x}(t)$ if

$$S_x(\omega) = \frac{\omega^4 + 64}{\omega^4 + 10\omega^2 + 9}$$

11-3 Show that if $l_s[n]$ is the delta response of the innovations filter of $\mathbf{s}[n]$, then

$$R_s[0] = \sum_{n=0}^{\infty} l_s^2[n]$$

11-4 The process $\mathbf{x}(t)$ is WSS and

$$\mathbf{y}''(t) + 3\mathbf{y}'(t) + 2\mathbf{y}(t) = \mathbf{x}(t)$$

Show that (*a*)

$$\begin{aligned} R''_{yx}(\tau) + 3R'_{yx}(\tau) + 2R_{yx}(\tau) &= R_{xx}(\tau) \\ R''_{yy}(\tau) + 3R'_{yy}(\tau) + 2R_{yy}(\tau) &= R_{xy}(\tau) \end{aligned} \qquad \text{all } \tau$$

(*b*) If $R_{xx}(\tau) = q\delta(\tau)$, then $R_{yx}(\tau) = 0$ for $\tau < 0$ and for $\tau > 0$:

$$R''_{yx}(\tau) + 3R'_{yx}(\tau) + 2R_{yx}(\tau) = 0 \qquad R_{yx}(0) = 0 \qquad R'_{yx}(0^+) = q$$

$$R''_{yy}(\tau) + 3R'_{yy}(\tau) + 2R_{yy}(\tau) = 0 \qquad R_{yy}(0) = \frac{q}{12} \qquad R'_{yy}(0) = 0$$

11-5 Show that if $\mathbf{s}[n]$ is AR and $\mathbf{v}[n]$ is white noise orthogonal to $\mathbf{s}[n]$, then the process $\mathbf{x}[n] = \mathbf{s}[n] + \mathbf{v}[n]$ is ARMA. Find $\mathbf{S}_x(z)$ if $R_s[m] = 2^{-|m|}$ and $\mathbf{S}_v(z) = 5$.

11-6 Show that if $\mathbf{x}(t)$ is a WSS process and

$$\mathbf{s} = \frac{1}{n}\sum_{k=1}^{n} \mathbf{x}(kT) \qquad \text{then} \quad E\{\mathbf{s}^2\} = \frac{1}{2\pi n^2}\int_{-\infty}^{\infty} S_x(\omega)\frac{\sin^2 n\omega T/2}{\sin^2 \omega T/2}\,d\omega$$

11-7 Show that if $R_x(\tau) = e^{-c|\tau|}$, then the Karhunen–Loève expansion of $\mathbf{x}(t)$ in the interval $(-a, a)$ is the sum

$$\hat{\mathbf{x}}(t) = \sum_{n=1}^{\infty} (\beta_n \mathbf{b}_n \cos\omega_n t + \beta'_n \mathbf{b}'_n \sin\omega'_n t)$$

where

$$\tan a\omega_n = \frac{c}{\omega_n} \qquad \cot a\omega'_n = \frac{-c}{\omega'_n} \qquad \beta_n = (a + c\lambda_n)^{-1/2} \qquad \beta'_n = (a - c\lambda'_n)^{-1/2}$$

$$E\{\mathbf{b}_n^2\} = \lambda_n = \frac{2c}{c^2 + \omega_n^2} \qquad E\{\mathbf{b}_n'^2\} = \lambda'_n = \frac{2c}{c^2 + \omega_n'^2}$$

11-8 Show that if $\mathbf{x}(t)$ is WSS and

$$\mathbf{X}_T(\omega) = \int_{-T/2}^{T/2} \mathbf{x}(t)e^{-j\omega t}\,dt \qquad \text{then} \quad E\left\{\frac{\partial}{\partial T}|\mathbf{X}_T(\omega)|^2\right\} = \int_{-T}^{T} R_x(\tau)e^{-j\omega\tau}\,d\tau$$

11-9 Find the mean and the variance of the integral

$$\mathbf{X}(\omega) = \int_{-a}^{a} [5\cos 3t + \mathbf{v}(t)]e^{-j\omega t}\,dt$$

if $E\{\mathbf{v}(t)\} = 0$ and $R_v(\tau) = 2\delta(\tau)$.

11-10 Show that if

$$E\{\mathbf{x}_n\mathbf{x}_k\} = \sigma_n^2\delta[n-k] \qquad \mathbf{X}(\omega) = \sum_{n=-\infty}^{\infty} \mathbf{x}_n e^{-jn\omega T}$$

and $E\{\mathbf{x}_n\} = 0$, then $E\{\mathbf{X}(\omega)\} = 0$ and

$$E\{\mathbf{X}(u)\mathbf{X}^*(v)\} = \sum_{n=-\infty}^{\infty} \sigma_n^2 e^{-jn(u-v)T}$$

11-11 Given a nonperiodic WSS process $\mathbf{x}(t)$, we form the sum $\hat{\mathbf{x}}(t) = \Sigma \mathbf{c}_n e^{-jn\omega_0 t}$ as in (11-51). Show that (*a*) $E\{|\mathbf{x}(t) - \hat{\mathbf{x}}(t)|^2\} = 0$ for $0 < t < T$. (*b*) $E\{\mathbf{c}_n\mathbf{c}_m^*\} = (1/T)\int_0^T \beta_n(\alpha)e^{jn\omega_0\alpha}\,d\alpha$, where $\beta_n(\alpha) = (1/T)\int_0^T R(\tau-\alpha)e^{-jn\omega_0\tau}\,d\tau$ are the coefficients of the Fourier expansion of $R(\tau-\alpha)$ in the interval $(0, T)$. (*c*) For large T, $E\{\mathbf{c}_n\mathbf{c}_m^*\} \simeq S(n\omega_0)\delta(n-m)/T$.

11-12 Show that, if the process $\mathbf{X}(\omega)$ is white noise with zero mean and autocovariance $Q(u)\delta(u-v)$, then its inverse Fourier transform $\mathbf{x}(t)$ is WSS with power spectrum $Q(\omega)/2\pi$.

11-13 Given a real process $\mathbf{x}(t)$ with Fourier transform $\mathbf{X}(\omega) = \mathbf{A}(\omega) + j\mathbf{B}(\omega)$, show that if the processes $\mathbf{A}(\omega)$ and $\mathbf{B}(\omega)$ satisfy (11-79) and $E\{\mathbf{A}(\omega)\} = E\{\mathbf{B}(\omega)\} = 0$, then $\mathbf{x}(t)$ is WSS.

11-14 We use as an estimate of the Fourier transform $F(\omega)$ of a signal $f(t)$ the integral

$$\mathbf{X}_T(\omega) = \int_{-T}^{T} [f(t) + \boldsymbol{\nu}(t)]e^{-j\omega t}\,dt$$

where $\boldsymbol{\nu}(t)$ is the measurement noise. Show that if $S_{\nu\nu}(\omega) = q$ and $E\{\boldsymbol{\nu}(t)\} = 0$, then

$$E\{\mathbf{X}_T(\omega)\} = \int_{-\infty}^{\infty} F(y)\frac{\sin T(\omega-y)}{\pi(\omega-y)}\,dy \qquad \text{Var}\,\mathbf{X}_T(\omega) = 2qT$$

CHAPTER

12

SPECTRUM ESTIMATION

12-1 ERGODICITY

A central problem in the applications of stochastic processes is the estimation of various statistical parameters in terms of real data. Most parameters can be expressed as expected values of some functional of a process $\mathbf{x}(t)$. The problem of estimating the mean of a given process $\mathbf{x}(t)$ is, therefore, central in this investigation. We start with this problem.

For a specific t, $\mathbf{x}(t)$ is a random variable; its mean $\eta(t) = E\{\mathbf{x}(t)\}$ can, therefore, be estimated as in Sec. 8-2: We observe n samples $\mathbf{x}(t, \zeta_i)$ of $\mathbf{x}(t)$ and use as the point estimate of $E\{\mathbf{x}(t)\}$ the average

$$\hat{\eta}(t) = \frac{1}{n}\sum_i \mathbf{x}(t, \zeta_i)$$

As we know, $\hat{\eta}(t)$ is a consistent estimate of $\eta(t)$; however, it can be used only if a large number of realizations $\mathbf{x}(t, \zeta_i)$ of $\mathbf{x}(t)$ are available. In many applications, we know only a *single sample* of $\mathbf{x}(t)$. Can we then estimate $\eta(t)$ in terms of the time average of the given sample? This is not possible if $E\{\mathbf{x}(t)\}$ depends on t. However, if $\mathbf{x}(t)$ is a regular stationary process, its time average tends to $E\{\mathbf{x}(t)\}$ as the length of the available sample tends to ∞. Ergodicity is a topic dealing with the underlying theory.

Mean-Ergodic Processes

We are given a real stationary process $\mathbf{x}(t)$ and we wish to estimate its mean $\eta = E\{\mathbf{x}(t)\}$. For this purpose, we form the *time average*

$$\eta_T = \frac{1}{2T}\int_{-T}^{T} \mathbf{x}(t)\,dt \tag{12-1}$$

Clearly, $\boldsymbol{\eta}_T$ is a random variable with mean

$$E\{\boldsymbol{\eta}_T\} = \frac{1}{2T}\int_{-T}^{T} E\{\mathbf{x}(t)\}\,dt = \eta$$

Thus $\boldsymbol{\eta}_T$ is an unbiased estimator of η. If its variance $\sigma_T^2 \to 0$ as $T \to \infty$, then $\boldsymbol{\eta}_T \to \eta$ in the MS sense. In this case, the time average $\boldsymbol{\eta}_T(\zeta)$ computed from a single realization of $\mathbf{x}(t)$ is close to η with probability close to 1. If this is true, we shall say that the process $\mathbf{x}(t)$ is *mean-ergodic*. Thus a process $\mathbf{x}(t)$ is mean-ergodic if its time average $\boldsymbol{\eta}_T$ tends to the ensemble average η as $T \to \infty$.

To establish the ergodicity of a process, it suffices to find σ_T and to examine the conditions under which $\sigma_T \to 0$ as $T \to \infty$. As Examples 12-1 and 12-2 show, not all processes are mean-ergodic.

EXAMPLE 12-1 ▶ Suppose that $\mathbf{c}$ is a random variable with mean η_c and

$$\mathbf{x}(t) = \mathbf{c} \qquad \eta = E\{\mathbf{x}(t)\} = E\{\mathbf{c}\} = \eta_c$$

In this case, $\mathbf{x}(t)$ is a family of straight lines and $\boldsymbol{\eta}_T = \mathbf{c}$. For a specific sample, $\boldsymbol{\eta}_T(\zeta) = \mathbf{c}(\zeta)$ is a constant different from η if $\mathbf{c}(\zeta) \neq \eta$. Hence $\mathbf{x}(t)$ is not mean-ergodic. ◀

EXAMPLE 12-2 ▶ Given two mean-ergodic processes $\mathbf{x}_1(t)$ and $\mathbf{x}_2(t)$ with means η_1 and η_2, we form the sum

$$\mathbf{x}(t) = \mathbf{x}_1(t) + \mathbf{c}\mathbf{x}_2(t)$$

where $\mathbf{c}$ is a random variable independent of $\mathbf{x}_2(t)$ taking the values 0 and 1 with probability 0.5. Clearly,

$$E\{\mathbf{x}(t)\} = E\{\mathbf{x}_1(t)\} + E\{\mathbf{c}\}E\{\mathbf{x}_2(t)\} = \eta_1 + 0.5\eta_2$$

If $\mathbf{c}(\zeta) = 0$ for a particular ζ, then $\mathbf{x}(t) = \mathbf{x}_1(t)$ and $\boldsymbol{\eta}_T \to \eta_1$ as $T \to \infty$. If $\mathbf{c}(\zeta) = 1$ for another ζ, then $\mathbf{x}(t) = \mathbf{x}_1(t) + \mathbf{x}_2(t)$ and $\boldsymbol{\eta}_T \to \eta_1 + \eta_2$ as $T \to \infty$. Hence $\mathbf{x}(t)$ is not mean-ergodic. ◀

VARIANCE. To determine the variance σ_T^2 of the time average $\boldsymbol{\eta}_T$ of $\mathbf{x}(t)$, we start with the observation that

$$\boldsymbol{\eta}_T = \mathbf{w}(0) \qquad \text{where} \quad \mathbf{w}(t) = \frac{1}{2T}\int_{t-T}^{t+T} \mathbf{x}(\alpha)\,d\alpha \tag{12-2}$$

is the moving average of $\mathbf{x}(t)$. As we know, $\mathbf{w}(t)$ is the output of a linear system with input $\mathbf{x}(t)$ and with impulse response a pulse centered at $t = 0$. Hence $\mathbf{w}(t)$ is stationary and its autocovariance equals

$$C_{ww}(\tau) = \frac{1}{2T}\int_{-2T}^{2T} C(\tau - \alpha)\left(1 - \frac{|\alpha|}{2T}\right)d\alpha \tag{12-3}$$

where $C(\tau)$ is the autocovariance of $\mathbf{x}(t)$ [see (9-155)]. Since $\sigma_T^2 = \text{Var}\,\mathbf{w}(0) = C_{ww}(0)$ and $C(-\alpha) = C(\alpha)$, this yields

$$\sigma_T^2 = \frac{1}{2T}\int_{-2T}^{2T} C(\alpha)\left(1 - \frac{|\alpha|}{2T}\right)d\alpha = \frac{1}{T}\int_0^{2T} C(\alpha)\left(1 - \frac{\alpha}{2T}\right)d\alpha \qquad (12\text{-}4)$$

This fundamental result leads to the following conclusion: A process $\mathbf{x}(t)$ with autocovariance $C(\tau)$ is mean-ergodic iff

$$\frac{1}{T}\int_0^{2T} C(\alpha)\left(1 - \frac{\alpha}{2T}\right)d\alpha \xrightarrow[T\to\infty]{} 0 \qquad (12\text{-}5)$$

The determination of the variance of $\boldsymbol{\eta}_T$ is useful not only in establishing the ergodicity of $\mathbf{x}(t)$ but also in determining a confidence interval for the estimate η_T of η. Indeed, from Tchebycheff's inequality it follows that the probability that the unknown η in the interval $\boldsymbol{\eta}_T \pm 10\,\sigma_T$ is larger than 0.99 [see (5-88)]. Hence η_T is a satisfactory estimate of η if T is such that $\sigma_T \ll \eta$.

EXAMPLE 12-3

▶ Suppose that $C(\tau) = qe^{-c|\tau|}$ as in (10-63). In this case,

$$\sigma_T^2 = \frac{q}{T}\int_0^{2T} e^{-c\tau}\left(1 - \frac{\tau}{2T}\right)d\tau = \frac{q}{cT}\left(1 - \frac{1 - e^{-2cT}}{2cT}\right)$$

Clearly, $\sigma_T \to 0$ as $T \to \infty$; hence $\mathbf{x}(t)$ is mean-ergodic. If $T \gg 1/c$, then $\sigma_T^2 \simeq q/cT$. ◀

EXAMPLE 12-4

▶ Suppose that $\mathbf{x}(t) = \eta + \boldsymbol{\nu}(t)$, where $\boldsymbol{\nu}(t)$ is white noise with $R_{\nu\nu}(\tau) = q\delta(\tau)$. In this case, $C(\tau) = R_{\nu\nu}(\tau)$ and (12-4) yields

$$\sigma_T^2 = \frac{1}{2T}\int_{-2T}^{2T} q\delta(\tau)\left(1 - \frac{|\tau|}{2T}\right)d\tau = \frac{q}{2T}$$

Hence $\mathbf{x}(t)$ is mean-ergodic. ◀

It is clear from (12-5) that the ergodicity of a process depends on the behavior of $C(\tau)$ for large τ. If $C(\tau) = 0$ for $\tau > a$, that is, if $\mathbf{x}(t)$ is a-dependent and $T \gg a$, then

$$\sigma_T^2 = \frac{1}{T}\int_0^a C(\tau)\left(1 - \frac{\tau}{2T}\right)d\tau \simeq \frac{1}{T}\int_0^a C(\tau)\,d\tau < \frac{a}{T}C(0) \xrightarrow[T\to\infty]{} 0$$

because $|C(\tau)| < C(0)$; hence $\mathbf{x}(t)$ is mean-ergodic.

In many applications, the random variables $\mathbf{x}(t+\tau)$ and $\mathbf{x}(t)$ are nearly uncorrelated for large τ, that is, $C(\tau) \to 0$ as $\tau \to \infty$. The above suggests that if this is the case, then $\mathbf{x}(t)$ is mean-ergodic and for large T the variance of $\boldsymbol{\eta}_T$ can be approximated by

$$\sigma_T^2 \simeq \frac{1}{T}\int_0^{2T} C(\tau)\,d\tau \simeq \frac{1}{T}\int_0^{\infty} C(\tau)\,d\tau = \frac{\tau_c}{T}C(0) \qquad (12\text{-}6)$$

where τ_c is the correlation time of $\mathbf{x}(t)$ defined in (9-58). This result will be justified presently.

THEOREM 12-1

SLUTSKY'S THEOREM

A process $\mathbf{x}(t)$ is mean-ergodic iff

$$\frac{1}{T}\int_0^T C(\tau)\,d\tau \xrightarrow[T\to\infty]{} 0 \tag{12-7}$$

Proof. (*a*) We show first that if $\sigma_T \to 0$ as $T \to \infty$, then (12-7) is true. The covariance of the random variables $\boldsymbol{\eta}_T$ and $\mathbf{x}(0)$ equals

$$\text{Cov}[\boldsymbol{\eta}_T, \mathbf{x}(0)] = E\left\{\frac{1}{2T}\int_{-T}^{T}[\mathbf{x}(t)-\eta][\mathbf{x}(0)-\eta]\,dt\right\} = \frac{1}{2T}\int_{-T}^{T} C(t)\,dt$$

But [see (6-164)]

$$\text{Cov}^2[\boldsymbol{\eta}_T, \mathbf{x}(0)] \le \text{Var}\,\boldsymbol{\eta}_T\,\text{Var}\,\mathbf{x}(0) = \sigma_T^2 C(0)$$

Hence (12-7) holds if $\sigma_T \to 0$.

(*b*) We show next that if (12-7) is true, then $\sigma_T \to 0$ as $T \to \infty$. From (12-7) it follows that given $\varepsilon > 0$, we can find a constant c_0 such that

$$\frac{1}{t}\int_c^t C(\tau)\,d\tau < \varepsilon \qquad \text{for every} \qquad c > c_0 \tag{12-8}$$

The variance of $\boldsymbol{\eta}_T$ equals [see (12-4)]

$$\sigma_T^2 = \frac{1}{T}\int_0^{2T_0} + \frac{1}{T}\int_{2T_0}^{2T} C(\tau)\left(1-\frac{\tau}{2T}\right)d\tau$$

The integral from 0 to $2T_0$ is less than $2T_0C(0)/T$ because $|C(\tau)| \le C(0)$. Hence

$$\sigma_T^2 < \frac{2T_0}{T}C(0) + \frac{1}{T}\int_{2T_0}^{2T} C(\tau)\left(1-\frac{\tau}{2T}\right)d\tau$$

But (see Fig. 12-1)

$$\int_{2T_0}^{2T} C(\tau)(2T-\tau)\,d\tau = \int_{2T_0}^{2T} C(\tau)\int_{\tau}^{2T} dt\,d\tau = \int_{2T_0}^{2T}\int_{2T_0}^{t} C(\tau)\,d\tau\,dt$$

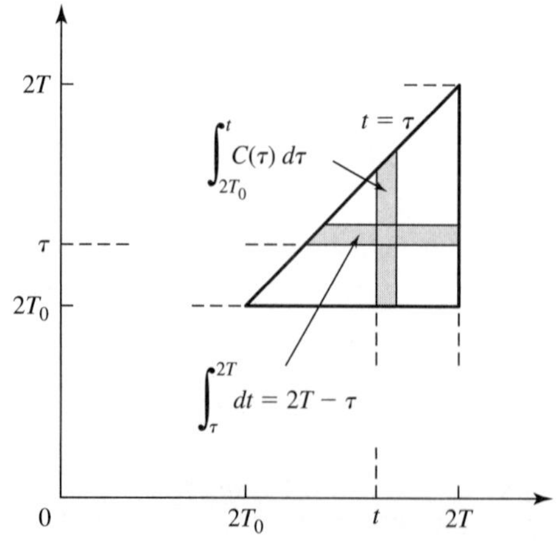

FIGURE 12-1

From (12-8) it follows that the inner integral on the right is less than εt; hence

$$\sigma_T^2 < \frac{2T_0}{T} C(0) + \frac{\varepsilon}{T^2} \int_{2T_0}^{2T} t\, dt \xrightarrow[T\to\infty]{} 2\varepsilon$$

and since ε is arbitrary, we conclude that $\sigma_T \to 0$ as $T \to \infty$. ◀

EXAMPLE 12-5 ▶ Consider the process

$$\mathbf{x}(t) = \mathbf{a}\cos\omega t + \mathbf{b}\sin\omega t + c$$

where **a** and **b** are two uncorrelated random variables with zero mean and equal variance. As we know [see (11-106)], the process $\mathbf{x}(t)$ is WSS with mean c and autocovariance $\sigma^2 \cos\omega\tau$. We shall show that it is mean-ergodic. This follows from (12-7) and the fact that

$$\frac{1}{T}\int_0^T C(\tau)\, d\tau = \frac{\sigma^2}{T}\int_0^T \cos\omega\tau\, d\tau = \frac{\sigma^2}{\omega T}\sin\omega T \xrightarrow[T\to\infty]{} 0$$ ◀

Sufficient conditions. (a) If

$$\int_0^\infty C(\tau)\, d\tau < \infty \tag{12-9}$$

then (12-7) holds; hence the process $\mathbf{x}(t)$ is mean-ergodic.

(b) If $R(\tau) \to \eta^2$ or, equivalently, if

$$C(\tau) \to 0 \qquad \text{as} \quad \tau \to \infty \tag{12-10}$$

then $\mathbf{x}(t)$ is mean-ergodic.

Proof. If (12-10) is true, then given $\varepsilon > 0$, we can find a constant T_0 such that $|C(\tau)| < \varepsilon$ for $\tau > T_0$; hence

$$\frac{1}{T}\int_0^T C(\tau)\, d\tau = \frac{1}{T}\int_0^{T_0} C(\tau)\, d\tau + \frac{1}{T}\int_{T_0}^T C(\tau)\, d\tau$$
$$< \frac{T_0}{T}C(0) + \varepsilon\frac{T - T_0}{T} \xrightarrow[T\to\infty]{} \varepsilon$$

and since ε is arbitrary, we conclude that (12-7) is true.

Condition (12-10) is satisfied if the random variables $\mathbf{x}(t+\tau)$ and $\mathbf{x}(t)$ are *uncorrelated* for large τ.

Note The time average η_T is an unbiased estimator of η; however, it is not best. An estimator with smaller variance results if we use the weighted average

$$\eta_w = \int_{-T}^{T} w(t)\mathbf{x}(t)\, dt$$

and select the function $w(t)$ appropriately (see also Example 7-4).

DISCRETE-TIME PROCESSES. We outline next, without elaboration, the discrete-time version of the preceding results. We are given a real stationary process $\mathbf{x}[n]$ with autocovariance $C[m]$ and we form the time average

$$\eta_M = \frac{1}{N} \sum_{n=-M}^{M} \mathbf{x}[n] \qquad N = 2M + 1 \tag{12-11}$$

This is an unbiased estimator of the mean of $\mathbf{x}[n]$ and its variance equals

$$\sigma_M^2 = \frac{1}{N} \sum_{m=-2M}^{2M} C[m]\left(1 - \frac{|m|}{N}\right) \tag{12-12}$$

The process $\mathbf{x}[n]$ is mean-ergodic if the right side of (12-12) tends to 0 as $M \to \infty$.

THEOREM 12-2

SLUTSKY'S THEOREM (DISCRETE CASE)

▶ The process $\mathbf{x}[n]$ is mean-ergodic iff

$$\frac{1}{M} \sum_{m=0}^{M} C[m] \xrightarrow[m\to\infty]{} 0 \tag{12-13}$$

We can show as in (12-10) that if $C[m] \to 0$ as $m \to \infty$, then $\mathbf{x}[n]$ is mean-ergodic.

For large M,

$$\sigma_M^2 \simeq \frac{1}{M} \sum_{m=0}^{M} C[m] \tag{12-14}$$

◀

EXAMPLE 12-6

▶ (*a*) Suppose that the centered process $\tilde{\mathbf{x}}[n] = \mathbf{x}[n] - \eta$ is white noise with autocovariance $P\delta[m]$. In this case,

$$C[m] = P\delta[m] \qquad \sigma_M^2 \simeq \frac{1}{N} \sum_{m=-M}^{M} P\delta[m] = \frac{P}{N}$$

Thus $\mathbf{x}[n]$ is mean-ergodic and the variance of η_M equals P/N. This agrees with (7-22): The random variables $\mathbf{x}[n]$ are i.i.d. with variance $C[0] = P$, and the time average η_M is their sample mean.

(*b*) Suppose now that $C[m] = Pa^{|m|}$ as in Example 9-31. In this case, (12-14) yields

$$\sigma_M^2 \simeq \frac{1}{N} \sum_{m=-\infty}^{\infty} Pa^{|m|} = \frac{P(1+a)}{N(1-a)}$$

Note that if we replace $\mathbf{x}[n]$ by white noise as in (*a*) with the same P and use as estimate of η the time average of N_1 terms, the variance P/N_1 of the resulting estimator will equal σ_M^2 if

$$N_1 = N\frac{1-a}{1+a}$$

◀

Sampling. In a numerical estimate of the mean of a continuous-time process $\mathbf{x}(t)$, the time-average η_T is replaced by the average

$$\eta_N = \frac{1}{N}\sum \mathbf{x}(t_n)$$

of the N samples $\mathbf{x}(t_n)$ of $\mathbf{x}(t)$. This is an unbiased estimate of η and its variance equals

$$\sigma_N^2 = \frac{1}{N^2}\sum_n \sum_k C(t_n - t_k)$$

where $C(\tau)$ is the autocovariance of $\mathbf{x}(t)$. If the samples are equidistant, then the random variables $\mathbf{x}(t_n) = \mathbf{x}(nT_0)$ form a discrete-time process with autovariance $C(mT_0)$. In this case, the variance σ_N^2 of η_N is given by (12-12) if we replace $C[m]$ by $C(mT_0)$.

SPECTRAL INTERPRETATION OF ERGODICITY. We shall express the ergodicity conditions in terms of the properties of the covariance spectrum

$$S^c(\omega) = S(\omega) - 2\pi\eta^2\delta(\omega)$$

of the process $\mathbf{x}(t)$. The variance σ_T^2 of η_T equals the variance of the moving average $\mathbf{w}(t)$ of $\mathbf{x}(t)$ [see (12-2)]. As we know,

$$S_{ww}^c(\omega) = S^c(\omega)\frac{\sin^2 T\omega}{T^2\omega^2} \tag{12-15}$$

hence

$$\sigma_T^2 = \frac{1}{2\pi}\int_{-\infty}^{\infty} S^c(\omega)\frac{\sin^2 T\omega}{T^2\omega^2}\,d\omega \tag{12-16}$$

The fraction in (12-16) takes significant values only in an interval of the order of $1/T$ centered at the origin. The ergodicity conditions of $\mathbf{x}(t)$ depend, therefore, only on the behavior of $S^c(\omega)$ near the origin.

Suppose first that the process $\mathbf{x}(t)$ is regular. In this case, $S^c(\omega)$ does not have an impulse at $\omega = 0$. If, therefore, T is sufficiently large, we can use the approximation $S^c(\omega) \simeq S^c(0)$ in (12-16). This yields

$$\sigma_T^2 \simeq \frac{S^c(0)}{2\pi}\int_{-\infty}^{\infty}\frac{\sin^2 T\omega}{T^2\omega^2}\,d\omega = \frac{S^c(0)}{2T} \xrightarrow[T\to\infty]{} 0 \tag{12-17}$$

Hence $\mathbf{x}(t)$ is mean-ergodic.

Suppose now that

$$S^c(\omega) = S_1^c(\omega) + 2\pi k_0\delta(\omega) \qquad S_1^c(0) < \infty \tag{12-18}$$

Inserting into (12-16), we conclude as in (12-17) that

$$\sigma_T^2 \simeq \frac{1}{2T}S_1(0) + k_0 \xrightarrow[T\to\infty]{} k_0$$

Hence $\mathbf{x}(t)$ is not mean-ergodic. This case arises if in Wold's decomposition (11-89) the constant term $\mathbf{c}_0$ is different from 0, or, equivalently, if the Fourier transform $\mathbf{X}(\omega)$ of $\mathbf{x}(t)$ contains the impulse $2\pi\mathbf{c}_0\delta(\omega)$.

EXAMPLE 12-7 ▶ Consider the process

$$\mathbf{y}(t) = \mathbf{a}\mathbf{x}(t) \qquad E\{\mathbf{a}\} = 0$$

where $\mathbf{x}(t)$ is a mean-ergodic process independent of the random variable $\mathbf{a}$. Clearly, $E\{\mathbf{y}(t)\} = 0$ and

$$S_{yy}^{c}(\omega) = \sigma_a^2 S_{xx}^2(\omega) + 2\pi\sigma_a^2\eta_x^2\delta(\omega)$$

as in Example 11-12. This shows that the process $\mathbf{y}(t)$ is not mean-ergodic. ◀

The preceding discussion leads to the following equivalent conditions for mean ergodicity:

1. σ_T must tend to 0 as $T \to \infty$.
2. In Wold's decomposition (11-89) the constant random term $\mathbf{c}_0$ must be 0.
3. The integrated power spectrum $F^c(\omega)$ must be continuous at the origin.
4. The integrated Fourier transform $\mathbf{Z}(\omega)$ must be continuous at the origin.

Analog estimators. The mean η of a process $\mathbf{x}(t)$ can be estimated by the response of a physical system with input $\mathbf{x}(t)$. A simple example is a normalized integrator of finite integration time. This is a linear device with impulse response the rectangular pulse $p(t)$ of Fig. 12-2. For $t > T_0$ the output of the integrator equals

$$\mathbf{y}(t) = \frac{1}{T_0}\int_{t-T_0}^{t} \mathbf{x}(\alpha)\,d\alpha$$

If T_0 is large compared to the correlation time τ_c of $\mathbf{x}(t)$, then the variance of $\mathbf{y}(t)$ equals $2\tau_c C(0)/T_0$. This follows from (12-6) with $T_0 = 2T$.

Suppose now that $\mathbf{x}(t)$ is the input to a system with impulse response $h(t)$ of unit area and energy E:

$$\mathbf{w}(t) = \int_0^t \mathbf{x}(\alpha)h(t-\alpha)\,d\alpha \qquad E = \int_0^\infty h^2(t)\,dt$$

We assume that $C(\tau) \simeq 0$ for $\tau > T_1$ and $h(t) \simeq 0$ for $t > T_0 > T_1$ as in Fig. 12-2. From these assumptions it follows that $E\{\mathbf{w}(t)\} = \eta$ and $\sigma_w^2 \simeq EC(0)\tau_c$ for $t > T_0$. If, therefore, $EC(0)\tau_c \ll \eta^2$ then $\mathbf{w}(t) \simeq \eta$ for $t > T_0$. These conditions are satisfied if the system is low-pass, that is, if $H(\omega) \simeq 0$ for $|\omega| < \omega_c$ and $\omega_c \ll \eta^2/C(0)\tau_c$.

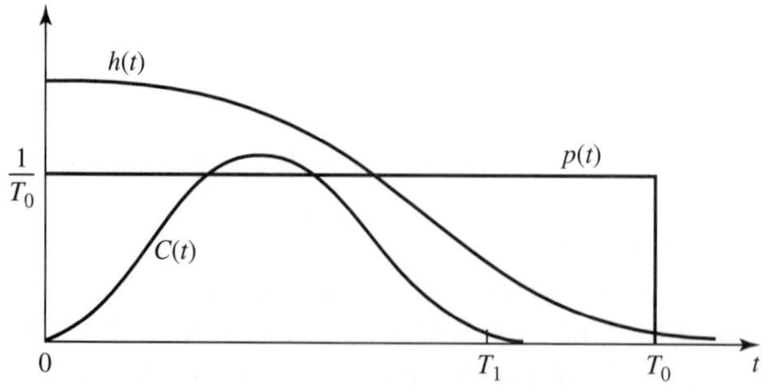

FIGURE 12-2

Covariance-Ergodic Processes

We shall now determine the conditions that an SSS process $\mathbf{x}(t)$ must satisfy such that its autocovariance $C(\lambda)$ can be estimated as a time average. The results are essentially the same for the estimates of the autocorrelation $R(\lambda)$ of $\mathbf{x}(t)$.

VARIANCE. We start with the estimate of the variance

$$V = C(0) = E\{|\mathbf{x}(t) - \eta|^2\} = E\{\mathbf{x}^2(t)\} - \eta^2 \tag{12-19}$$

of $\mathbf{x}(t)$.

Known mean. Suppose, first, that η is known. We can then assume, replacing the process $\mathbf{x}(t)$ by its centered process $\mathbf{x}(t) - \eta$, that

$$E\{\mathbf{x}(t)\} = 0 \qquad V = E\{\mathbf{x}^2(t)\}$$

Our problem is thus to estimate the mean V of the process $\mathbf{x}^2(t)$. Proceeding as in (12-1), we use as the estimate of V the time average

$$\mathbf{V}_T = \frac{1}{2T}\int_{-T}^{T} \mathbf{x}^2(t)\,dt \tag{12-20}$$

This estimate is unbiased and its variance is given by (12-4) where we replace the function $C(\tau)$ by the autocovariance

$$C_{x^2x^2}(\tau) = E\{\mathbf{x}^2(t+\tau)\mathbf{x}^2(t)\} - E^2\{\mathbf{x}^2(t)\} \tag{12-21}$$

of the process $\mathbf{x}^2(t)$. Applying (12-7) to this process, we conclude that $\mathbf{x}(t)$ is variance-ergodic iff

$$\frac{1}{T}\int_0^T E\{\mathbf{x}^2(t+\tau)\mathbf{x}^2(t)\}\,dt \xrightarrow[T\to\infty]{} C^2(0) \tag{12-22}$$

To test the validity of (12-22), we need the fourth-order moments of $\mathbf{x}(t)$. If, however, $\mathbf{x}(t)$ is a normal process, then [see (9-77)]

$$C_{x^2x^2}(\tau) = 2C^2(\tau) \tag{12-23}$$

From this and (12-22) it follows that a normal process is variance-ergodic iff

$$\frac{1}{T}\int_0^T C^2(\tau)\,d\tau \xrightarrow[T\to\infty]{} 0 \tag{12-24}$$

Using the simple inequality (see Prob. 12-10)

$$\left|\frac{1}{T}\int_0^T C(\tau)\,d\tau\right|^2 \le \frac{1}{T}\int_0^T C^2(\tau)\,d\tau$$

we conclude with (12-7) and (12-24) that if a normal process is variance-ergodic, it is also mean-ergodic. The converse, however, is not true. This theorem has the following spectral interpretation: The process $\mathbf{x}(t)$ is mean-ergodic iff $S^c(\omega)$ has no impulses at the origin; it is variance-ergodic iff $S^c(\omega)$ has no impulses anywhere.

EXAMPLE 12-8 ▶ Suppose that the process

$$\mathbf{x}(t) = \mathbf{a}\cos\omega t + \mathbf{b}\sin\omega t + \eta$$

is normal and stationary. Clearly, $\mathbf{x}(t)$ is mean-ergodic because it does not contain a random constant. However, it is not variance-ergodic because the square

$$|\mathbf{x}(t) - \eta|^2 = \tfrac{1}{2}(\mathbf{a}^2 + \mathbf{b}^2) + \tfrac{1}{2}(\mathbf{a}^2\cos 2\omega t - \mathbf{b}^2\cos 2\omega t) + \mathbf{ab}\sin 2\omega t$$

of $\mathbf{x}(t) - \eta$ contains the random constant $(\mathbf{a}^2 + \mathbf{b}^2)/2$. ◀

Unknown mean. If η is unknown, we evaluate its estimator $\boldsymbol{\eta}_T$ from (12-1) and form the average

$$\hat{\mathbf{V}}_T = \frac{1}{2T}\int_{-T}^{T}[\mathbf{x}(t) - \boldsymbol{\eta}_T]^2\,dt = \frac{1}{2T}\int_{-T}^{T}\mathbf{x}^2(t)\,dt - \boldsymbol{\eta}_T^2$$

The determination of the statistical properties of $\hat{\mathbf{V}}_T$ is difficult. The following observations, however, simplify the problem. In general, $\hat{\mathbf{V}}_T$ is a biased estimator of the variance V of $\mathbf{x}(t)$. However, if T is large, the bias can be neglected in the determination of the estimation error; furthermore, the variance of $\hat{\mathbf{V}}_T$ can be approximated by the variance of the known-mean estimator $\mathbf{V}_T$. In many cases, the MS error $E\{(\hat{\mathbf{V}}_T - V)^2\}$ is smaller than $E\{(\mathbf{V}_T - V)^2\}$ for moderate values of T. It might thus be preferable to use $\hat{\mathbf{V}}_T$ as the estimator of V even when η is known.

AUTOCOVARIANCE. We shall establish the ergodicity conditions for the autocovariance $C(\lambda)$ of the process $\mathbf{x}(t)$ under the assumption that $E\{\mathbf{x}(t)\} = 0$. We can do so, replacing $\mathbf{x}(t)$ by $\mathbf{x}(t) - \eta$ if η is known. If it is unknown, we replace $\mathbf{x}(t)$ by $\mathbf{x}(t) - \boldsymbol{\eta}_T$. In this case, the results are approximately correct if T is large.

For a specific λ, the product $\mathbf{x}(t+\lambda)\mathbf{x}(t)$ is an SSS process with mean $C(\lambda)$. We can, therefore, use as the estimate of $C(\lambda)$ the time average

$$\mathbf{C}_T(\lambda) = \frac{1}{2T}\int_{-T}^{T}\mathbf{z}(t)\,dt \qquad \mathbf{z}(t) = \mathbf{x}(t+\lambda)\mathbf{x}(t) \tag{12-25}$$

This is an unbiased estimator of $C(\lambda)$ and its variance is given by (12-4) if we replace the autocovariance of $\mathbf{x}(t)$ by the autocovariance

$$C_{zz}(\tau) = E\{\mathbf{x}(t+\lambda+\tau)\mathbf{x}(t+\tau)\mathbf{x}(t+\lambda)\mathbf{x}(t)\} - C^2(\lambda)$$

of the process $\mathbf{z}(t)$. Applying Slutsky's theorem, we conclude that the process $\mathbf{x}(t)$ is covariance-ergodic iff

$$\frac{1}{T}\int_0^T C_{zz}(\tau)\,d\tau \xrightarrow[T\to\infty]{} 0 \tag{12-26}$$

If $\mathbf{x}(t)$ is a normal process,

$$C_{zz}(\tau) = C(\lambda+\tau)C(\lambda-\tau) + C^2(\tau) \tag{12-27}$$

In this case, (12-6) yields

$$\operatorname{Var}\mathbf{C}_T(\lambda) \simeq \frac{1}{T}\int_0^{2T}[C(\lambda+\tau)C(\lambda-\tau) + C^2(\tau)]\,d\tau \tag{12-28}$$

From (12-27) it follows that if $C(\tau) \to 0$, then $C_{zz}(\tau) \to 0$ as $\tau \to \infty$; hence $\mathbf{x}(t)$ is covariance-ergodic.

Cross-covariance. We comment briefly on the estimate of the cross-covariance $C_{xy}(\tau)$ of two zero-mean processes $\mathbf{x}(t)$ and $\mathbf{y}(t)$. As in (12-25), the time average

$$\hat{\mathbf{C}}_{xy}(\tau) = \frac{1}{2T} \int_{-T}^{T} \mathbf{x}(t+\tau)\mathbf{y}(t)\,dt \tag{12-29}$$

is an unbiased estimate of $C_{xy}(\tau)$ and its variance is given by (12-4) if we replace $C(\tau)$ by $C_{xy}(\tau)$. If the functions $C_{xx}(\tau)$, $C_{yy}(\tau)$, and $C_{xy}(\tau)$ tend to 0 as $\tau \to \infty$ then the processes $\mathbf{x}(t)$ and $\mathbf{y}(t)$ are cross-covariance-ergodic (see Prob. 12-9).

NONLINEAR ESTIMATORS. The numerical evaluation of the estimate $C_t(\lambda)$ of $C(\lambda)$ involves the evaluation of the integral of the product $\mathbf{x}(t+\lambda)\mathbf{x}(t)$ for various values of λ. We show next that the computations can in certain cases be simplified if we replace one or both factors of this product by some function[1] of $\mathbf{x}(t)$. We shall assume that the process $\mathbf{x}(t)$ is normal with zero mean.

The arcsine law. We have shown in (9-80) that if $\mathbf{y}(t)$ is the output of a hard limiter with input $\mathbf{x}(t)$:

$$\mathbf{y}(t) = \operatorname{sgn} \mathbf{x}(t) = \begin{cases} 1 & \mathbf{x}(t) > 0 \\ -1 & \mathbf{x}(t) < 0 \end{cases}$$

then

$$C_{yy}(\tau) = \frac{2}{\pi} \arcsin \frac{C_{xx}(\tau)}{C_{xx}(0)} \tag{12-30}$$

The estimate of $C_{yy}(\tau)$ is given by

$$\hat{\mathbf{C}}_{yy}(\tau) = \frac{1}{2T} \int_{-T}^{T} \operatorname{sgn} \mathbf{x}(t+\tau) \operatorname{sgn} \mathbf{x}(t)\,dt \tag{12-31}$$

This integral is simple to determine because the integrand equals ± 1. Thus

$$\hat{\mathbf{C}}_{yy}(\tau) = \left(\frac{\mathbf{T}_\tau^+}{T} - 1 \right)$$

where T_τ^+ is the total time that $\mathbf{x}(t+\tau)\mathbf{x}(t) > 0$. This yields the estimate

$$\hat{\mathbf{C}}_{xx}(\tau) = \hat{\mathbf{C}}_{xx}(0) \sin \left[\frac{\pi}{2} \hat{\mathbf{C}}_{yy}(\tau) \right]$$

of $C_{xx}(\tau)$ within a factor.

[1] S. Cambanis and E. Masry: "On the Reconstruction of the Covariance of Stationary Gaussian Processes Through Zero-Memory Nonlinearities," *IEEE Transactions on Information Theory,* vol. IT-24, 1978.

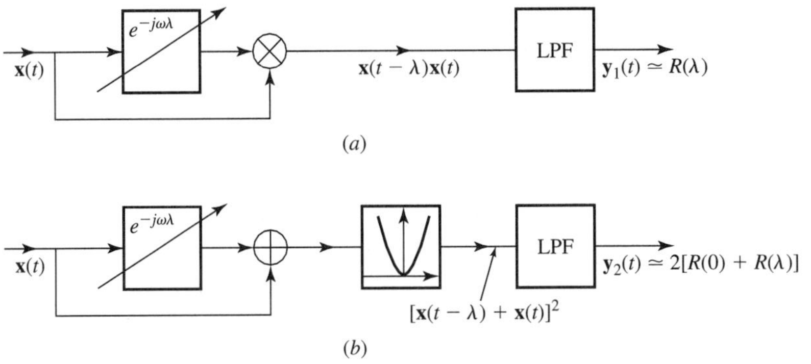

FIGURE 12-3

Bussgang's theorem. We have shown in (9-81) that the cross-covariance of the processes $\mathbf{x}(t)$ and $\mathbf{y}(t) = \operatorname{sgn} \mathbf{x}(t)$ is proportional to $C_{xx}(\tau)$:

$$C_{xy}(\tau) = K C_{xx}(\tau) \qquad K = \sqrt{\frac{2}{\pi C_{xx}(0)}} \tag{12-32}$$

To estimate $C_{xx}(\tau)$, it suffices, therefore, to estimate $C_{xy}(\tau)$. Using (12-29), we obtain

$$\hat{\mathbf{C}}_{xx}(\tau) = \frac{1}{K}\hat{\mathbf{C}}_{xy}(\tau) = \frac{1}{2KT}\int_{-T}^{T} \mathbf{x}(t+\tau)\operatorname{sgn}\mathbf{x}(t)\,dt \tag{12-33}$$

CORRELOMETERS AND SPECTROMETERS. A correlometer is a physical device measuring the autocorrelation $R(\lambda)$ of a process $\mathbf{x}(t)$. In Fig. 12-3 we show two correlometers. The first consists of a delay element, a multiplier, and a low-pass (LP) filter. The input to the LP filter is the process $\mathbf{x}(t-\lambda)\mathbf{x}(t)$; the output $\mathbf{y}_1(t)$ is the estimate of the mean $R(\lambda)$ of the input. The second consists of a delay element, an adder, a square-law detector, and an LP filter. The input to the LP filter is the process $[\mathbf{x}(t-\lambda)+\mathbf{x}(t)]^2$; the output $\mathbf{y}_2(t)$ is the estimate or the mean $2[R(0)+R(\lambda)]$ of the input.

A spectrometer is a physical device measuring the Fourier transform $S(\omega)$ of $R(\lambda)$. This device consists of a bandpass filter $B(\omega)$ with input $\mathbf{x}(t)$ and output $\mathbf{y}(t)$, in series with a square-law detector and an LP filter (Fig. 12-4). The input to the LP filter is the process $\mathbf{y}^2(t)$; its output $\mathbf{z}(t)$ is the estimate of the mean $E\{\mathbf{y}^2(t)\}$ of the input. Suppose that $B(\omega)$ is a narrow-band filter of unit energy with center frequency ω_0 and bandwidth $2c$. If the function $S(\omega)$ is continuous at ω_0 and c is sufficiently small, then $S(\omega) \simeq S(\omega_0)$ for $|\omega-\omega_0| < c$; hence [see (9-152)]

$$E\{\mathbf{y}^2(t)\} = \frac{1}{2\pi}\int_{-\infty}^{\infty} S(\omega)B^2(\omega)\,d\omega \simeq \frac{S(\omega_0)}{2\pi}\int_{\omega_0-c}^{\omega_0+c} B^2(\omega)\,d\omega = S(\omega_0)$$

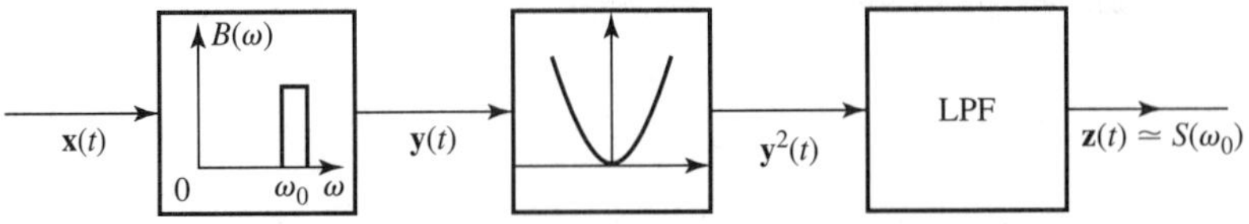

FIGURE 12-4

as in (9-166). This yields

$$\mathbf{z}(t) \simeq E\{\mathbf{y}^2(t)\} \simeq S(\omega_0)$$

We give next the optical realization of the correlometer of Fig. 12-3*b* and the spectrometer of Fig. 12-4.

The Michelson interferometer. The device of Fig. 12-5 is an optical correlometer. It consists of a light source S, a beam-splitting surface B, and two mirrors. Mirror M_1 is in a fixed position and mirror M_2 is movable. The light from the source S is a random signal $\mathbf{x}(t)$ traveling with velocity c and it reaches a square-law detector D along paths 1 and 2 as shown. The lengths of these paths equal l and $l + 2d$, respectively, where d is the displacement of mirror M_2 from its equilibrium position.

The signal reaching the detector is thus the sum

$$A\mathbf{x}(t - t_0) + A\mathbf{x}(t - t_0 - \lambda)$$

where A is the attenuation in each path, $t_0 = l/c$ is the delay along path 1, and $\lambda = 2d/c$ is the additional delay due to the displacement of mirror M_2. The detector output is the signal

$$\mathbf{z}(t) = A^2[\mathbf{x}(t - t_0 - \lambda) + \mathbf{x}(t - t_0)]^2$$

Clearly,

$$E\{\mathbf{z}(t)\} = 2A^2[R(0) + R(\lambda)]$$

If, therefore, we use $\mathbf{z}(t)$ as the input to a low-pass filter, its output $\mathbf{y}(t)$ will be proportional to $R(0) + R(\lambda)$ provided that the process $\mathbf{x}(t)$ is correlation-ergodic and the band of the filter is sufficiently narrow.

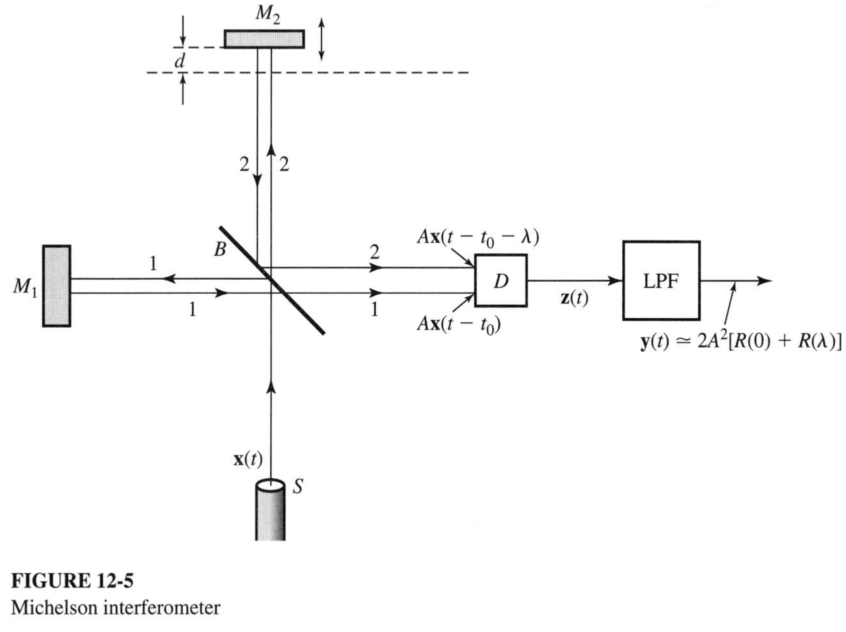

FIGURE 12-5
Michelson interferometer

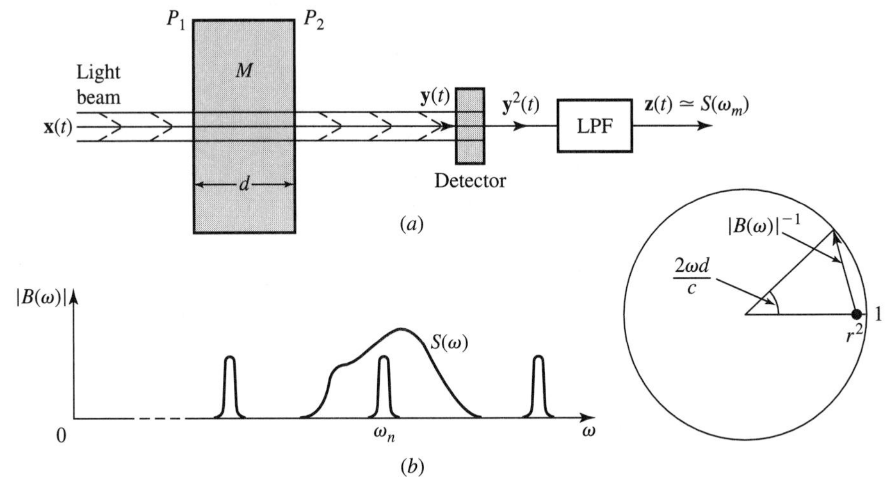

FIGURE 12-6
Fabry-Perot interferometer

The Fabry–Pérot interferometer. The device of Fig. 12-6 is an optical spectrometer. The bandpass filter consists of two highly reflective plates P_1 and P_2 distance d apart and the input is a light beam $\mathbf{x}(t)$ with power spectrum $S(\omega)$. The frequency response of the filter is proportional to

$$B(\omega) = \frac{1}{1 - r^2 e^{-j2\omega d/c}} \qquad r \simeq 1$$

where r is the reflection coefficient of each plate and c is the velocity of light in the medium M between the plates. The function $B(\omega)$ is shown in Fig. 12-6*b*. It consists of a sequence of bands centered at

$$\omega_n = \frac{\pi n d}{c}$$

whose bandwidth tends to 0 as $r \to 1$. If only the mth band of $B(\omega)$ overlaps with $S(\omega)$ and $r \simeq 1$, then the output $\mathbf{z}(t)$ of the LP filter is proportional to $S(\omega_m)$. To vary ω_m, we can either vary the distance d between the plates or the dielectric constant of the medium M.

Distribution-Ergodic Processes

Any parameter of a probabilistic model that can be expressed as the mean of some function of an SSS process $\mathbf{x}(t)$ can be estimated by a time average. For a specific x, the distribution of $\mathbf{x}(t)$ is the mean of the process $\mathbf{y}(t) = U[x - \mathbf{x}(t)]$:

$$\mathbf{y}(t) = \begin{cases} 1 & \mathbf{x}(t) \le x \\ 0 & \mathbf{x}(t) > x \end{cases} \qquad E\{\mathbf{y}(t)\} = P\{\mathbf{x}(t) \le x\} = F(x)$$

Hence $F(x)$ can be estimated by the time average of $\mathbf{y}(t)$. Inserting into (12-1), we obtain

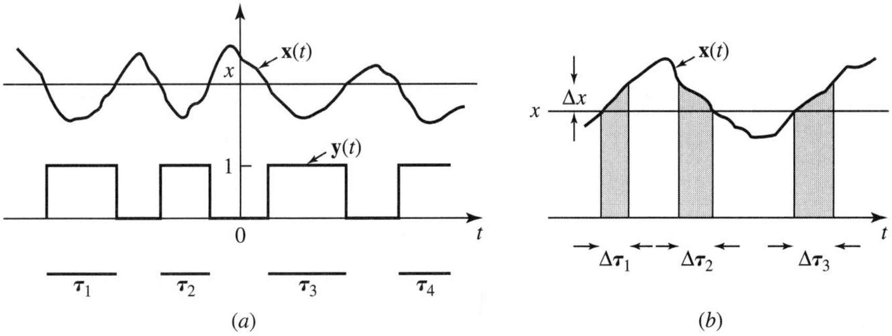

FIGURE 12-7

the estimator

$$\mathbf{F}_T(x) = \frac{1}{2T}\int_{-T}^{T} \mathbf{y}(t)\,dt = \frac{\boldsymbol{\tau}_1 + \cdots + \boldsymbol{\tau}_n}{2T} \tag{12-34}$$

where $\boldsymbol{\tau}_i$ are the lengths of the time intervals during which $\mathbf{x}(t)$ is less than x (Fig. 12-7*a*).

To find the variance of $\mathbf{F}_T(x)$, we must first find the autocovariance of $\mathbf{y}(t)$. The product $\mathbf{y}(t+\tau)\mathbf{y}(t)$ equals 1 if $\mathbf{x}(t+\tau) \le x$ and $\mathbf{x}(t) \le x$; otherwise, it equals 0. Hence

$$R_y(\tau) = P\{\mathbf{x}(t+\tau) \le x, \mathbf{x}(t) \le x\} = F(x, x; \tau)$$

where $F(x, x; \tau)$ is the second-order distribution of $\mathbf{x}(t)$. The variance of $\mathbf{F}_T(x)$ is obtained from (12-4) if we replace $C(\tau)$ by the autocovariance $F(x, x; \tau) - F^2(x)$ of $\mathbf{y}(t)$. From (12-7) it follows that a process $\mathbf{x}(t)$ is distribution-ergodic iff

$$\frac{1}{T}\int_0^T F(x, x; \tau)\,d\tau \underset{T\to\infty}{\longrightarrow} F^2(x) \tag{12-35}$$

A sufficient condition is obtained from (12-10): A process $\mathbf{x}(t)$ is distribution-ergodic if $F(x, x; \tau) \to F^2(x)$ as $\tau \to \infty$. This is the case if the random variables $\mathbf{x}(t)$ and $\mathbf{x}(t+\tau)$ are *independent* for large τ.

Density. To estimate the density of $\mathbf{x}(t)$, we form the time intervals $\Delta\tau_i$ during which $\mathbf{x}(t)$ is between x and $x + \Delta x$ (Fig. 12-7*b*). From (12-34) it follows that

$$f(x)\,\Delta x \simeq F(x + \Delta x) - F(x) \simeq \frac{1}{2T}\sum_i \Delta\tau_i$$

Thus $f(x)\,\Delta x$ equals the percentage of time that a single sample of $\mathbf{x}(t)$ is between x and $x + \Delta x$. This can be used to design an analog estimator of $f(x)$.

12-2 SPECTRUM ESTIMATION

We wish to estimate the power spectrum $S(\omega)$ of a real process $\mathbf{x}(t)$ in terms of a single realization of a finite segment

$$\mathbf{x}_T(t) = \mathbf{x}(t)p_T(t) \qquad p_T(t) = \begin{cases} 1 & |t| < T \\ 0 & |t| > T \end{cases} \tag{12-36}$$

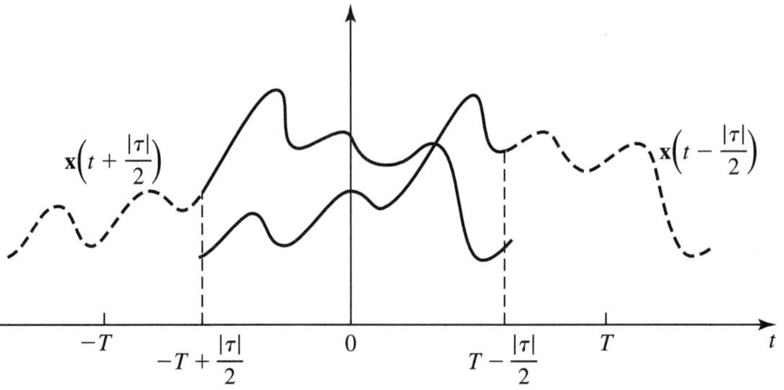

FIGURE 12-8

of $\mathbf{x}(t)$. The spectrum $S(\omega)$ is not the mean of some function of $\mathbf{x}(t)$. It cannot, therefore, be estimated directly as a time average. It is, however, the Fourier transform of the autocorrelation

$$R(\tau) = E\left\{\mathbf{x}\left(t+\frac{\tau}{2}\right)\mathbf{x}\left(t-\frac{\tau}{2}\right)\right\}$$

It will be determined in terms of the estimate of $R(\tau)$. This estimate cannot be computed from (12-25) because the product $\mathbf{x}(t+\tau/2)\mathbf{x}(t-\tau/2)$ is available only for t in the interval $(-T+|\tau|/2, T-|\tau|/2)$ (Fig. 12-8). Changing $2T$ to $2T-|\tau|$, we obtain the estimate

$$\mathbf{R}^{T}(\tau) = \frac{1}{2T-|\tau|}\int_{T+|\tau|/2}^{T-|\tau|/2} \mathbf{x}\left(t+\frac{\tau}{2}\right)\mathbf{x}\left(t-\frac{\tau}{2}\right)dt \tag{12-37}$$

This integral specifies $\mathbf{R}^T(\tau)$ for $|\tau| < 2T$; for $|\tau| > 2T$ we set $\mathbf{R}^T(\tau) = 0$. This estimate is unbiased; however, its variance increases as $|\tau|$ increases because the length $2T-|\tau|$ of the integration interval decreases. Instead of $\mathbf{R}^T(\tau)$, we shall use the product

$$\mathbf{R}_T(\tau) = \left(1-\frac{|\tau|}{2T}\right)\mathbf{R}^T(\tau) \tag{12-38}$$

This estimator is biased; however, its variance is smaller than the variance of $\mathbf{R}^T(\tau)$. The main reason we use it is that its transform is proportional to the energy spectrum of the segment $\mathbf{x}_T(t)$ of $\mathbf{x}(t)$ [see (12-39)].

The Periodogram

The periodogram of a process $\mathbf{x}(t)$ is by definition the process

$$\mathbf{S}_T(\omega) = \frac{1}{2T}\left|\int_{-T}^{T}\mathbf{x}(t)e^{-j\omega t}\,dt\right|^2 \tag{12-39}$$

This integral is the Fourier transform of the known segment $\mathbf{x}_T(t)$ of $\mathbf{x}(t)$:

$$\mathbf{S}_T(\omega) = \frac{1}{2T}|\mathbf{X}_T(\omega)|^2 \qquad \mathbf{X}_T(\omega) = \int_{-T}^{T}\mathbf{x}(t)e^{-j\omega t}\,dt$$

We shall express $\mathbf{S}_T(\omega)$ in terms of the estimator $\mathbf{R}_T(\tau)$ of $R(\tau)$.

THEOREM 12-3

$$\mathbf{S}_T(\omega) = \int_{-2T}^{2T} \mathbf{R}_T(\tau) e^{-j\omega\tau} \, d\tau \tag{12-40}$$

Proof. The integral in (12-37) is the convolution of $\mathbf{x}_T(t)$ with $\mathbf{x}_T(-t)$ because $\mathbf{x}_T(t) = 0$ for $|t| > T$. Hence

$$\mathbf{R}_T(\tau) = \frac{1}{2T} \mathbf{x}_T(\tau) * \mathbf{x}_T(-\tau) \tag{12-41}$$

Since $\mathbf{x}_T(t)$ is real, the transform of $\mathbf{x}_T(-t)$ equals $\mathbf{X}_T^*(\omega)$. This shows that (convolution theorem) the transform of $\mathbf{R}_T(\tau)$ equals the right side of (12-39).

In the early years of signal analysis, the spectral properties of random processes were expressed in terms of their periodogram. This approach yielded reliable results so long as the integrations were based on analog techniques of limited accuracy. With the introduction of digital processing, the accuracy was improved and, paradoxically, the computed spectra exhibited noisy behavior. This apparent paradox can be readily explained in terms of the properties of the periodogram: The integral in (12-40) depends on all values of $\mathbf{R}_T(\tau)$ for τ large and small. The variance of $\mathbf{R}_T(\tau)$ is small for small τ only, and it increases as $\tau \to 2T$. As a result, $\mathbf{S}_T(\omega)$ approaches a white-noise process with mean $S(\omega)$ as T increases [see (12-57)].

To overcome this behavior of $\mathbf{S}_T(\omega)$, we can do one of two things: (1) We replace in (12-40) the term $\mathbf{R}_T(\tau)$ by the product $w(\tau)\mathbf{R}_T(\tau)$, where $w(\tau)$ is a function (window) close to 1 near the origin, approaching 0 as $\tau \to 2T$. This deemphasizes the unreliable parts of $\mathbf{R}_T(\tau)$, thus reducing the variance of its transform. (2) We convolve $\mathbf{S}_T(\omega)$ with a suitable window as in (10-228).

We continue with the determination of the bias and the variance of $\mathbf{S}_T(\omega)$.

SPECTRAL BIAS

From (12-38) and (12-40) it follows that

$$E\{\mathbf{S}_T(\omega)\} = \int_{-2T}^{2T} \left(1 - \frac{|\tau|}{2T}\right) R(\tau) e^{-j\omega\tau} \, d\tau$$

Since

$$\left(1 - \frac{|\tau|}{2T}\right) p_T(\tau) \leftrightarrow \frac{2\sin^2 T\omega}{T\omega^2}$$

we conclude that [see also (11-83)]

$$E\{\mathbf{S}_T(\omega)\} = \int_{-\infty}^{\infty} \frac{\sin^2 T(\omega - y)}{\pi T(\omega - y)^2} S(y) \, dy \tag{12-42}$$

This shows that the mean of the periodogram is a smoothed version of $S(\omega)$; however, the smoothing kernel $\sin^2 T(\omega - y)/\pi T(\omega - y)^2$ takes significant values only in an interval of the order of $1/T$ centered at $y = \omega$. If, therefore, T is sufficiently large, we can set $S(y) \simeq S(\omega)$ in (12-42) for every point of continuity of $S(\omega)$. Hence for large T,

$$E\{\mathbf{S}_T(\omega)\} \simeq S(\omega) \int_{-\infty}^{\infty} \frac{\sin^2 T(\omega - y)}{\pi T(\omega - y)^2} \, dy = S(\omega) \tag{12-43}$$

From this it follows that $\mathbf{S}_T(\omega)$ is asymptotically an unbiased estimator of $S(\omega)$.

Data window. If $S(\omega)$ is not nearly constant in an interval of the order of $1/T$, the periodogram is a biased estimate of $S(\omega)$. To reduce the bias, we replace in (12-39) the process $\mathbf{x}(t)$ by the product $c(t)\mathbf{x}(t)$. This yields the *modified periodogram*

$$\mathbf{S}_c(\omega) = \frac{1}{2T}\left|\int_{-T}^{T} c(t)\mathbf{x}(t)e^{-j\omega t}\,dt\right|^2 \tag{12-44}$$

The factor $c(t)$ is called the *data window.* Denoting by $C(\omega)$ its Fourier transform, we conclude that [see (11-82)]

$$E\{\mathbf{S}_c(\omega)\} = \frac{1}{4\pi T} S(\omega) * C^2(\omega) \tag{12-45}$$

VARIANCE. For the determination of the variance of $\mathbf{S}_T(\omega)$, knowledge of the fourth-order moments of $\mathbf{x}(t)$ is required. For normal processes, all moments can be expressed in terms of $R(\tau)$. Furthermore, as $T \to \infty$, the fourth-order moments of most processes approach the corresponding moments of a normal process with the same autocorrelation (see Papoulis, 1977 [22]). We can assume, therefore, without essential loss of generality, that $\mathbf{x}(t)$ is normal with zero mean.

THEOREM 12-4

SPECTRAL VARIANCE

► For large T:

$$\operatorname{Var} \mathbf{S}_T(\omega) \simeq \begin{cases} 2S^2(0) & \omega = 0 \\ S^2(\omega) & |\omega| \gg 1/T \end{cases} \tag{12-46}$$

at every point of continuity of $S(\omega)$.

Proof. The Fourier transform of the autocorrelation $R(t_1 - t_2)p_T(t_1)p_T(t_2)$ of the process $\mathbf{x}_T(t)$ equals

$$\Gamma(u, v) = \int_{-\infty}^{\infty} \frac{2\sin T\alpha \sin T(u+v-\alpha)}{\pi\alpha(u+v-\alpha)} S(u-\alpha)\,d\alpha \tag{12-47}$$

This follows from (11-80) with $W(\omega) = 2\sin T\omega/\omega$. The fraction in (12-47) takes significant values only if the terms αT and $(u+v-\alpha)T$ are of the order of 1; hence, the entire fraction is negligible if $|u+v| \gg 1/T$. Setting $u = v = \omega$, we conclude that $\Gamma(\omega, \omega) \simeq 0$ and

$$\Gamma(\omega, -\omega) = \int_{-\infty}^{\infty} \frac{2\sin^2 T\alpha}{\pi\alpha^2} S(\omega - \alpha)\,d\alpha$$

$$\simeq S(\omega)\int_{-\infty}^{\infty} \frac{2\sin^2 T\alpha}{\pi\alpha^2}\,d\alpha = 2TS(\omega) \tag{12-48}$$

for $|\omega| \gg 1/T$ and since [see (11-74)]

$$\operatorname{Var} \mathbf{S}_T(\omega) = \frac{1}{4T^2}[\Gamma^2(\omega, -\omega) + \Gamma^2(\omega, \omega)]$$

and $\Gamma(0, 0) = S(0)$, (12-46) follows. ◄

Note For a specific τ, no matter how large, the estimate $\mathbf{R}_T(\tau) \to R(\tau)$ as $T \to \infty$. Its transform $\mathbf{S}_T(\omega)$, however, does not tend to $S(\omega)$ as $T \to \infty$. The reason is that the convergence of $\mathbf{R}_T(\tau)$ to $R(\tau)$ is not uniform in τ, that is, given $\varepsilon > 0$, we cannot find a constant T_0 independent of τ such that $|\mathbf{R}_T(\tau) - R(\tau)| < \varepsilon$ for every τ, and every $T > T_0$.

Proceeding similarly, we can show that the variance of the spectrum $\mathbf{S}_c(\omega)$ obtained with the data window $c(t)$ is essentially equal to the variance of $\mathbf{S}_T(\omega)$. This shows that use of data windows does not reduce the variance of the estimate. To improve the estimation, we must replace in (12-40) the sample autocorrelation $\mathbf{R}_T(\tau)$ by the product $w(\tau)\mathbf{R}_T(\tau)$, or, equivalently, we must smooth the periodogram $\mathbf{S}_T(\omega)$.

Note Data windows might be useful if we smooth $\mathbf{S}_T(\omega)$ by an ensemble average: Suppose that we have access to N independent samples $\mathbf{x}(t, \zeta_i)$ of $\mathbf{x}(t)$, or, we divide a single long sample into N essentially independent pieces, each of duration $2T$. We form the periodograms $\mathbf{S}_T(\omega, \zeta_i)$ of each sample and their average

$$\overline{\mathbf{S}}_T(\omega) = \frac{1}{N}\sum \mathbf{S}_T(\omega, \zeta_i) \tag{12-49}$$

As we know,

$$E\{\overline{\mathbf{S}}_T(\omega)\} = S(\omega) * \frac{\sin^2 \omega T}{\pi T \omega^2} \qquad \text{Var } \overline{\mathbf{S}}_T(\omega) \simeq \frac{1}{N} S^2(\omega) \qquad \omega \neq 0 \tag{12-50}$$

If N is large, the variance of $\overline{\mathbf{S}}_T(\omega)$ is small. However, its bias might be significant. Use of data windows is in this case desirable.

Smoothed Spectrum

We shall assume as before that T is large and $\mathbf{x}(t)$ is normal. To improve the estimate, we form the smoothed spectrum

$$\mathbf{S}_w(\omega) = \frac{1}{2\pi}\int_{-\infty}^{\infty} \mathbf{S}_T(\omega - y)W(y)\,dy = \int_{-2T}^{2T} w(\tau)\mathbf{R}_T(\tau)e^{-j\omega\tau}\,d\tau \tag{12-51}$$

where

$$w(t) = \frac{1}{2\pi}\int_{-\infty}^{\infty} W(\omega)e^{j\omega t}\,d\omega$$

The function $w(t)$ is called the *lag window* and its transform $W(\omega)$ the *spectral window.* We shall assume that $W(-\omega) = W(\omega)$ and

$$w(0) = 1 = \frac{1}{2\pi}\int_{-\infty y}^{\infty} W(\omega)\,d\omega \qquad W(\omega) \geq 0 \tag{12-52}$$

Bias. From (12-42) it follows that

$$E\{\mathbf{S}_w(\omega)\} = \frac{1}{2\pi}E\{\mathbf{S}_T(\omega)\} * W(\omega) = \frac{1}{2\pi}S(\omega) * \frac{\sin^2 T\omega}{\pi T\omega^2} * W(\omega)$$

Assuming that $W(\omega)$ is nearly constant in any interval of length $1/T$, we obtain the large T approximation

$$E\{\mathbf{S}_w(\omega)\} \simeq \frac{1}{2\pi}S(\omega) * W(\omega) \tag{12-53}$$

Variance. We shall determine the variance of $\mathbf{S}_w(\omega)$ using the identity [see (11-74)]

$$\text{Cov}[\mathbf{S}_T(u), \mathbf{S}_T(v)] = \frac{1}{4T^2}[\Gamma^2(u, -v) + \Gamma^2(u, v)] \tag{12-54}$$

This problem is in general complicated. We shall outline an approximate solution based on the following assumptions: The constant T is large in the sense that the functions

$S(\omega)$ and $W(\omega)$ are nearly constant in any interval of length $1/T$. The width of $W(\omega)$, that is, the constant σ such that $W(\omega) \simeq 0$ for $|\omega| > \sigma$, is small in the sense that $S(\omega)$ is nearly constant in any interval of length 2σ.

Reasoning as in the proof of (12-48), we conclude from (12-47) that $\Gamma(u, v) \simeq 0$ for $u + v \gg 1/T$ and

$$\Gamma(u, -v) \simeq S(u) \int_{-\infty}^{\infty} \frac{2 \sin T(u - v - \alpha) \sin T\alpha}{\pi(u - v - \alpha)\alpha} \, d\alpha = S(u) \frac{2 \sin T(u - v)}{u - v}$$

This is the generalization of (12-48). Inserting into (12-54), we obtain

$$\text{Cov}[\mathbf{S}_T(u), \mathbf{S}_T(v)] \simeq \frac{\sin^2 T(u - v)}{T^2(u - v)^2} S^2(u) \tag{12-55}$$

Equation (12-46) is a special case obtained with $u = v = \omega$.

THEOREM 12-5 ▶ For $|\omega| \gg 1/T$

$$\text{Var}\, \mathbf{S}_w(\omega) \simeq \frac{E_w}{2T} S^2(\omega) \tag{12-56}$$

where

$$E_w = \frac{1}{2\pi} \int_{-\infty}^{\infty} W^2(\omega) \, d\omega$$

Proof. The smoothed spectrum $\mathbf{S}_w(\omega)$ equals the convolution of $\mathbf{S}_T(\omega)$ with the spectral window $W(\omega)/2\pi$. From this and (9-96) it follows mutatis mutandis that the variance of $\mathbf{S}_w(\omega)$ is a double convolution involving the covariance of $\mathbf{S}_T(\omega)$ and the window $W(\omega)$. The fraction in (12-55) is negligible for $|u - v| \gg 1/T$. In any interval of length $1/T$, the function $W(\omega)$ is nearly constant by assumption. This leads to the conclusion that in the evaluation of the variance of $\mathbf{S}_w(\omega)$, the covariance of $\mathbf{S}_T(\omega)$ can be approximated by an impulse of area equal to the area

$$S^2(u) \int_{-\infty}^{\infty} \frac{\sin^2 T(u - v)}{T^2(u - v)^2} \, dv = \frac{\pi}{T} S^2(u)$$

of the right side of (12-55). This yields

$$\text{Cov}[\mathbf{S}_T(u), \mathbf{S}_T(v)] = q(u)\delta(u - v) \qquad q(u) = \frac{\pi}{T} S^2(u) \tag{12-57}$$

From the above and (9-100) it follows that

$$\text{Var}\, \mathbf{S}_w(\omega) \simeq \frac{\pi}{T} \int_{-\infty}^{\infty} \mathbf{S}^2(\omega - y) \frac{W^2(y)}{4\pi^2} \, dy = \frac{S^2(\omega)}{2T} \int_{-\infty}^{\infty} \frac{W^2(y)}{2\pi} \, dy$$

and (12-56) results. ◀

WINDOW SELECTION. The selection of the window pair $w(t) \leftrightarrow W(\omega)$ depends on two conflicting requirements: For the variance of $\mathbf{S}_w(\omega)$ to be small, the energy E_w of the lag window $w(t)$ must be small compared to T. From this it follows that $w(t)$ must approach 0 as $t \to 2T$. We can assume, therefore, without essential loss of generality that $w(t) = 0$ for $|t| > M$, where M is a fraction of $2T$. Thus

$$\mathbf{S}_w(\omega) = \int_{-M}^{M} w(t) \mathbf{R}_T(t) e^{-j\omega t} \, dt \qquad M < 2T$$

The mean of $\mathbf{S}_w(\omega)$ is a smoothed version of $S(\omega)$. To reduce the effect of the resulting bias, we must use a spectral window $W(\omega)$ of short duration. This is in conflict with the requirement that M be small (uncertainty principle). The final choice of M is a compromise between bias and variance. The quality of the estimate depends on M and on the shape of $w(t)$. To separate the shape factor from the size factor, we express $w(t)$ as a scaled version of a normalized window $w_0(t)$ of size 2:

$$w(t) = w_0\left(\frac{t}{M}\right) \leftrightarrow W(\omega) = MW_0(M\omega) \tag{12-58}$$

where

$$w_0(t) = 0 \qquad \text{for} \quad |t| > 1$$

The critical parameter in the selection of a window is the scaling factor M. In the absence of any prior information, we have no way of determining the optimum size of M. The following considerations, however, are useful: A reasonable measure of the reliability of the estimation is the ratio

$$\frac{\operatorname{Var} \mathbf{S}_w(\omega)}{S^2(\omega)} \simeq \frac{E_w}{2T} = \alpha \tag{12-59}$$

For most windows in use, E_w is between $0.5M$ and $0.8M$ (see Table 12-1). If we set $\alpha = 0.2$ as the largest acceptable α, we must set $M \le T/2$. If nothing is known about $S(\omega)$, we estimate it several times using windows of decreasing size. We start with $M = T/2$ and observe the form of the resulting estimate $\mathbf{S}_w(\omega)$. This estimate might not be very reliable; however, it gives us some idea of the form of $S(\omega)$. If we see that the estimate is nearly constant in any interval of the order of $1/M$, we conclude that the initial

TABLE 12-1

$w(t)$	$W(\omega)$
1. Bartlett	
$1 - \lvert t\rvert$	$\dfrac{4\sin^2 \omega/2}{\omega^2}$
$m_2 = \infty \quad E_w = \frac{2}{3} \quad n = 2$	
2. Tukey	
$\frac{1}{2}(1 + \cos \pi t)$	$\dfrac{\pi^2 \sin \omega}{\omega(\pi^2 - \omega^2)}$
$m_2 = \frac{\pi^2}{2} \quad E_w = \frac{3}{4} \quad n = 3$	
3. Parzen	
$[3(1 - 2\lvert t\rvert)p_1(t)] * [3(1 - 2\lvert t\rvert)p_1(t)]$	$\dfrac{3}{4}\left(\dfrac{\sin \omega/4}{\omega/4}\right)^4$
$m_2 = 12 \quad E_w = 0.539 \quad n = 4$	
4. Papoulis[1]	
$\dfrac{1}{\pi}\lvert \sin \pi t\rvert + (1 - \lvert t\rvert) \cos \pi t$	$8\pi^2 \dfrac{\cos^2(\omega/2)}{(\pi^2 - \omega^2)^2}$
$m_2 = \pi^2 \quad E_w = 0.587 \quad n = 4$	

[1]A. Papoulis: "Minimum Bias Windows for High Resolution Spectral Estimates," *IEEE Transactions on Information Theory,* vol. IT-19, 1973.

choice $M = T/2$ is too large. A reduction of M will not appreciably affect the bias but it will yield a smaller variance. We repeat this process until we obtain a balance between bias and variance. As we show later, for optimum balance, the standard deviation of the estimate must equal twice its bias. The quality of the estimate depends, of course, on the size of the available sample. If, for the given T, the resulting $\mathbf{S}_w(\omega)$ is not smooth for $M = T/2$, we conclude that T is not large enough for a satisfactory estimate.

To complete the specification of the window, we must select the form of $w_0(t)$. In this selection, we are guided by the following considerations:

1. The window $W(\omega)$ must be positive and its area must equal 2π as in (12-52). This ensures the positivity and consistency of the estimation.
2. For small bias, the "duration" of $W(\omega)$ must be small. A measure of duration is the second moment

$$m_2 = \frac{1}{2\pi}\int_{-\infty}^{\infty} \omega^2 W(\omega)\, d\omega \tag{12-60}$$

3. The function $W(\omega)$ must go to 0 rapidly as ω increases (small sidelobes). This reduces the effect of distant peaks in the estimate of $S(\omega)$. As we know, the asymptotic properties of $W(\omega)$ depend on the continuity properties of its inverse $w(t)$. Since $w(t) = 0$ for $|t| > M$, the condition that $W(\omega) \to 0$ as A/ω^n as $n \to \infty$ leads to the requirement that the derivatives of $w(t)$ of order up to $n - 1$ be zero at the end-points $\pm M$ of the lag window $w(t)$:

$$w(\pm M) = w'(\pm M) = \cdots = w^{(n-1)}(\pm M) = 0 \tag{12-61}$$

4. The energy E_w of $w(t)$ must be small. This reduces the variance of the estimate.

Over the years, a variety of windows have been proposed. They meet more or less the stated requirements but most of them are selected empirically. Optimality criteria leading to windows that do not depend on the form of the unknown $S(w)$ are difficult to generate. However, as we show next, for high-resolution estimates (large T) the last example of Table 12-1 minimizes the bias. In this table and in Fig. 12-9, we list the most

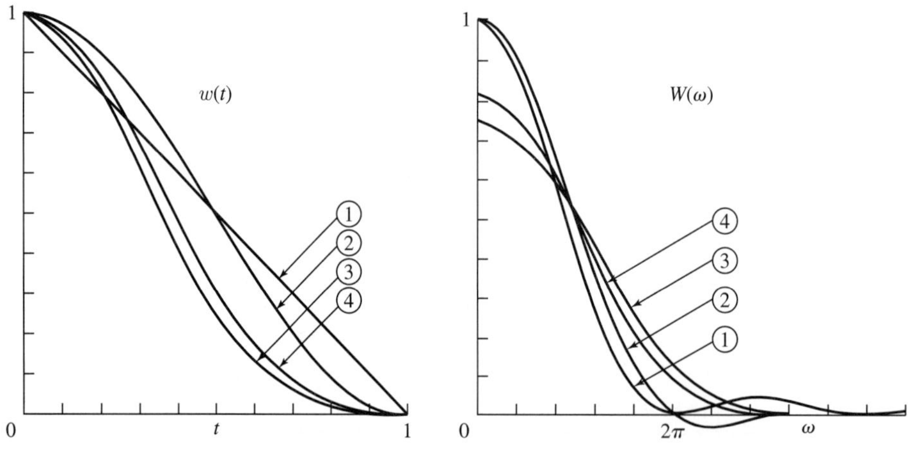

FIGURE 12-9

common window pairs $w(t) \leftrightarrow W(\omega)$. We also show the values of the second moment m_2, the energy E_w, and the exponent n of the asymptotic attenuation A/w^n of $W(\omega)$. In all cases, $w(t) = 0$ for $|t| > 1$.

OPTIMUM WINDOWS. We introduce next three classes of windows. In all cases, we assume that the data size T and the scaling factor M are large (high-resolution estimates) in the sense that we can use the parabolic approximation of $S(\omega - \alpha)$ in the evaluation of the bias. This yields [see (10-232)]

$$\frac{1}{2\pi}\int_{-\infty}^{\infty} S(\omega-\alpha)W(\alpha)\,d\alpha \simeq S(\omega) + \frac{S''(\omega)}{4\pi}\int_{-\infty}^{\infty} \alpha^2 W(\alpha)\,d\alpha \tag{12-62}$$

Note that since $W(w) > 0$, the above is an equality if we replace the term $S''(\omega)$ by $S''(\omega + \delta)$, where δ is a constant in the region where $W(\omega)$ takes significant values.

Minimum bias data window. The modified periodogram $\mathbf{S}_c(\omega)$ obtained with the data window $c(t)$ is a biased estimator of $S(\omega)$. Inserting (12-62) into (12-45), we conclude that the bias equals

$$\begin{aligned} B_c(\omega) &= \frac{1}{2\pi}\int_{-\infty}^{\infty} S(\omega-\alpha)C^2(\alpha)\,d\alpha - S(\omega) \\ &\simeq \frac{S''(\omega)}{4\pi}\int_{-\infty}^{\infty} \alpha^2 C^2(\alpha)\,d\alpha \end{aligned} \tag{12-63}$$

We have thus expressed the bias as a product where the first factor depends only on $S(\omega)$ and the second depends only on $C(\omega)$.This separation enables us to find $C(\omega)$ so as to minimize $B_c(\omega)$. To do so, it suffices to minimize the second moment

$$M_2 = \frac{1}{2\pi}\int_{-\infty}^{\infty} \omega^2 C^2(\omega)\,d\omega = \int_{-T}^{T} |c'(t)|^2\,dt \tag{12-64}$$

of $C^2(\omega)$ subject to the constraints

$$\frac{1}{2\pi}\int_{-\infty}^{\infty} C^2(\omega)\,d\omega = 1 \qquad C(-\omega) = C(\omega)$$

It can be shown that[2] the optimum data window is a truncated cosine (Fig. 12-10):

$$c(t) = \begin{cases} \dfrac{1}{\sqrt{T}}\cos\dfrac{\pi}{2T}t & |t| < T \\ 0 & |t| > T \end{cases} \leftrightarrow C(\omega) = 4\pi\sqrt{T}\,\frac{\cos T\omega}{\pi^2 - 4T^2\omega^2} \tag{12-65}$$

The resulting second moment M_2 equals 1. Note that if no data window is used, then $c(t) = 1$ and $M_2 = 2$. Thus the optimum data window yields a 50% reduction of the bias.

---

[2] A. Papoulis: "Apodization for Optimum Imaging of Smooth Objects," *J. Opt. Soc. Am.*, vol. 62, December, 1972.

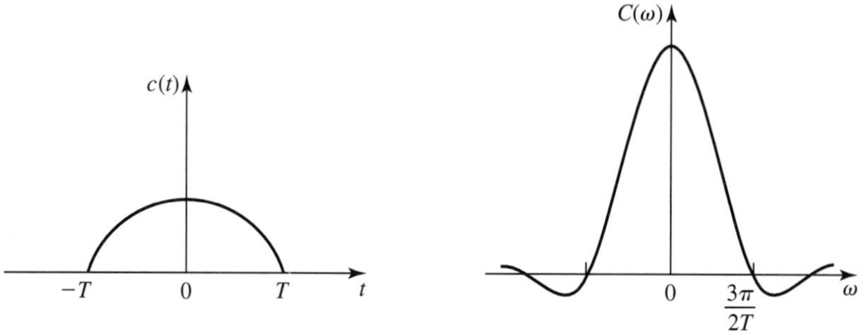

FIGURE 12-10

Minimum bias spectral window. From (12-53) and (12-62) it follows that the bias of $\mathbf{S}_w(\omega)$ equals

$$B(\omega) = \frac{1}{2\pi}\int_{-\infty}^{\infty} S(\omega-\alpha)W(\alpha)\,d\alpha - S(\omega) \simeq \frac{m_2}{2}S''(\omega) \qquad (12\text{-}66)$$

where m_2 is the second moment of $W(\omega)/2\pi$. To minimize $B(\omega)$, it suffices, therefore, to minimize m_2 subject to the constraints

$$W(\omega) \geq 0 \qquad W(-\omega) = W(\omega) \qquad \frac{1}{2\pi}\int_{-\infty}^{\infty} W(\omega)\,d\omega = 1 \qquad (12\text{-}67)$$

This is the same as the problem just considered if we replace $2T$ by M and we set

$$W(\omega) = C^2(\omega) \qquad w(t) = c(t) * c(-t)$$

This yields the pair (Fig. 12-11)

$$w(t) = \begin{cases} \dfrac{1}{\pi}\left|\sin\dfrac{\pi}{M}t\right| + \left(1 - \dfrac{|t|}{M}\right)\cos\dfrac{\pi}{M}t & |t| \leq M \\ 0 & |t| > M \end{cases} \qquad (12\text{-}68)$$

$$W(\omega) = 8M\pi^2 \frac{\cos^2(M\omega/2)}{(\pi^2 - M^2\omega^2)^2} \qquad (12\text{-}69)$$

Thus the last window in Table 12-1 minimizes the bias in high-resolution spectral estimates.

LMS spectral window. We shall finally select the spectral window $W(\omega)$ so as to minimize the MS estimation error

$$e = B^2(\omega) + \text{Var}\,\mathbf{S}_w(\omega) \qquad (12\text{-}70)$$

We have shown that for sufficiently large values of T, the periodogram $\mathbf{S}_T(\omega)$ can be written as a sum $S(\omega) + \boldsymbol{\nu}(\omega)$, where $\boldsymbol{\nu}(\omega)$ is a nonstationary white noise process with autocorrelation $\pi S^2(u)\delta(u-v)/T$ as in (12-57). Thus our problem is the estimation of a deterministic function $S(\omega)$ in the presence of additive noise $\boldsymbol{\nu}(\omega)$. This problem was considered in Sec. 10-6. We shall reestablish the results in the context of spectral estimation.

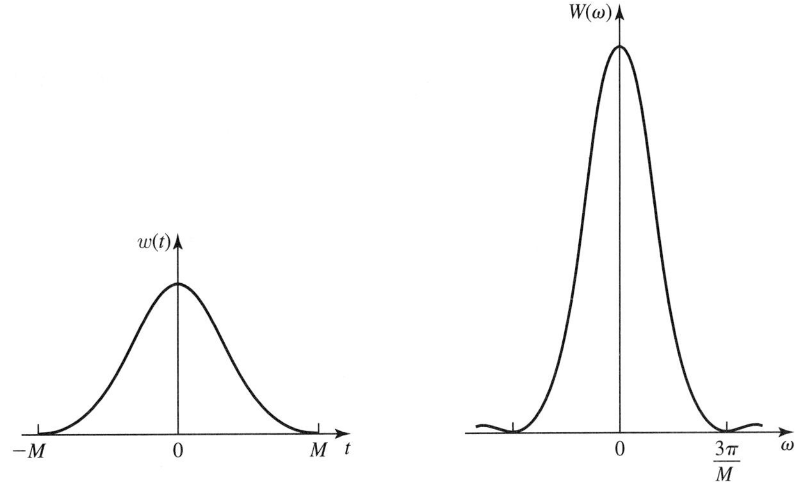

FIGURE 12-11

We start with a rectangular window of size 2Δ and area 1. The resulting estimate of $S(\omega)$ is the moving average

$$\mathbf{S}_\Delta(\omega) = \frac{1}{2\Delta}\int_{-\Delta}^{\Delta} \mathbf{S}_T(\omega - \alpha)\, d\alpha \tag{12-71}$$

of $\mathbf{S}_T(\omega)$. The rectangular window was used first by Daniell[3] in the early years of spectral estimation. It is a special case of the spectral window $W(\omega)/2\pi$. Note that the corresponding lag window $\sin \Delta t / 2\pi \Delta t$ is not time-limited.

With the familiar large-T assumption, the periodogram $\mathbf{S}_T(\omega)$ is an unbiased estimator of $S(\omega)$. Hence the bias of $\mathbf{S}_\Delta(\omega)$ equals

$$\frac{1}{2\Delta}\int_{-\Delta}^{\Delta} S(\omega - y)\, dy - S(\omega) \simeq \frac{S''(\omega)}{2\Delta}\int_{-\Delta}^{\Delta} y^2\, dy = S''(\omega)\frac{\Delta^2}{6}$$

and its variance equals

$$\frac{\pi S^2(\omega)}{4\Delta^2 T}\int_{-\Delta}^{\Delta} d\omega = \frac{\pi S^2(\omega)}{2\Delta T}$$

This follows from (12-71) because $\mathbf{S}_T(\omega)$ is white noise as in (12-57) with $q(u) = \pi S^2(u)/T$ [see also (10-236)]. This leads to the conclusion that

$$\operatorname{Var} \mathbf{S}_\Delta(\omega) = \frac{\pi S^2(\omega)}{2\Delta T} \qquad e = \frac{\Delta^4}{36}[S''(\omega)]^2 + \frac{\pi S^2(\omega)}{2\Delta T} \tag{12-72}$$

Proceeding as in (10-240), we conclude that e is minimum if

$$\Delta = \left(\frac{9\pi}{2T}\right)^{0.2}\left(\frac{S(\omega)}{S''(\omega)}\right)^{0.4}$$

The resulting bias equals twice the standard deviation of $\mathbf{S}_\Delta(\omega)$ (see two-to-one rule).

---

[3] P. J. Daniell: Discussion on "Symposium on Autocorrelation in Time Series," *J. Roy. Statist. Soc. Suppl.*, **8**, 1946.

Suppose finally that the spectral window is a function of unknown form. We wish to determine its shape so as to minimize the MS error e. Proceeding as in (10-241), we can show that e is minimum if the window is a truncated parabola:

$$\mathbf{S}_w(\omega) = \frac{3}{4\Delta}\int_{-\Delta}^{\Delta} \mathbf{S}_T(\omega - y)\left(1 - \frac{y^2}{\Delta^2}\right)dy \qquad \Delta = \left(\frac{15\pi}{T}\right)^{0.2}\left(\frac{s(\omega)}{S''(\omega)}\right)^{0.4} \tag{12-73}$$

This window was first suggested by Priestley.[4] Note that unlike the earlier windows, it is frequency dependent and its size is a function of the unknown spectrum $S(\omega)$ and its second derivative. To determine $\mathbf{S}_w(\omega + \delta)$ we must therefore estimate first not only $S(\omega)$ but also $S''(\omega)$. Using these estimates we determine Δ for the next step.

12-3 EXTRAPOLATION AND SYSTEM IDENTIFICATION

In the preceding discussion, we computed the estimate $\mathbf{R}_T(\tau)$ of $R(\tau)$ for $|\tau| < M$ and used as the estimate of $S(\omega)$ the Fourier transform $\mathbf{S}_w(\omega)$ of the product $w(t)\mathbf{R}_T(t)$. The portion of $\mathbf{R}_T(\tau)$ for $|\tau| > M$ was not used. In this section, we shall assume that $S(\omega)$ belongs to a class of functions that can be specified in terms of certain parameters, and we shall use the estimated part of $R(\tau)$ to determine these parameters. In our development, we shall not consider the variance problem. We shall assume that the portion of $R(\tau)$ for $|\tau| < M$ is known exactly. This is a realistic assumption if $T \gg M$ because $\mathbf{R}_T(\tau) \to R(\tau)$ for $|\tau| < M$ as $T \to \infty$. A physical problem leading to the assumption that $R(\tau)$ is known exactly but only for $|\tau| < M$ is the Michelson interferometer. In this example, the time of observation is arbitrarily large; however, $R(\tau)$ can be determined only for $|\tau| < M$, where M is a constant proportional to the maximum displacement of the moving mirror (Fig. 12-5).

Our problem can thus be phrased as follows: We are given a finite segment

$$R_M(\tau) = \begin{cases} R(\tau) & |\tau| < M \\ 0 & |\tau| > M \end{cases}$$

of the autocorrelation $R(\tau)$ of a process $\mathbf{x}(t)$ and we wish to estimate its power spectrum $S(\omega)$. This is essentially a deterministic problem: We wish to find the Fourier transform $S(\omega)$ of a function $R(\tau)$ knowing only the segment $R_M(\tau)$ of $R(\tau)$ and the fact that $S(\omega) \geq 0$. This problem does not have a unique solution. Our task then is to find a particular $S(\omega)$ that is close in some sense to the unknown spectrum. In the early years of spectral estimation, the function $S(\omega)$ was estimated with the method of windows (Blackman and Tukey[5]). In this method, the unknown $R(\tau)$ is replaced by 0 and the known or estimated part is tapered by a suitable factor $w(\tau)$. In recent years, a different approach has been used: It is assumed that $S(\omega)$ can be specified in terms of a finite number of parameters (parametric extrapolation) and the problem is reduced to the estimation of these parameters. In this section we concentrate on the extrapolation method starting with brief coverage of the method of windows.

---

[4] M. B. Priestley: "Basic Considerations in the Estimation of Power Spectra," *Technometrics*, **4**, 1962.

[5] R. B. Blackman and J. W. Tukey: *The Measurement of Power Spectra*, Dover, New York, 1959.

METHOD OF WINDOWS. The continuous-time version of this method is treated in the last section in the context of the bias reduction problem: We use as the estimate of $S(\omega)$ the integral

$$S_w(\omega) = \int_{-M}^{M} w(\tau)R(\tau)e^{-j\omega\tau}\,d\tau = \frac{1}{2\pi}\int_{-\infty}^{\infty} S(\omega-\alpha)W(\alpha)\,d\alpha \qquad \text{(12-74)}$$

and we select $w(t)$ so as to minimize in some sense the estimation error $S_w(\omega) - S(\omega)$. If M is large in the sense that $S(\omega-\alpha) \simeq S(\omega)$ for $|\alpha| \le 1/M$, we can use the approximation [see (12-62)]

$$S_w(\omega) - S(\omega) \simeq \frac{S''(\omega)}{4\pi}\int_{-\infty}^{\infty} \alpha^2 W(\alpha)\,d\alpha$$

This is minimum if

$$w(\tau) = \frac{1}{\pi}\left|\sin\frac{\pi}{M}\tau\right| + \left(1 - \frac{|\tau|}{M}\right)\cos\frac{\pi}{M}\tau \qquad |\tau| < M$$

The discrete-time version of this method is similar: We are given a finite segment

$$R_L[m] = \begin{cases} R[m] & |m| \le L \\ 0 & |m| > L \end{cases} \qquad \text{(12-75)}$$

of the autocorrelation $R[m] = E\{\mathbf{x}[n+m]\mathbf{x}[n]\}$ of a process $\mathbf{x}[n]$ and we wish to estimate its power spectrum

$$S(\omega) = \sum_{m=-\infty}^{\infty} R[m]e^{-jm\omega}$$

We use as the estimate of $S(\omega)$ the DFT

$$S_w(\omega) = \sum_{m=-L}^{L} w[m]R[m]e^{-jm\omega} = \frac{1}{2\pi}\int_{-\pi}^{\pi} S(\omega-\alpha)W(\alpha)\,d\alpha \qquad \text{(12-76)}$$

of the product $w[m]R[m]$, where $w[m] \leftrightarrow W(\omega)$ is a DFT pair. The criteria for selecting $w[m]$ are the same as in the continuous-time case. In fact, if M is large, we can choose for $w[m]$ the samples

$$w[m] = w(Mm/L) \qquad m = 0, \ldots, L \qquad \text{(12-77)}$$

of an analog window $w(t)$ where M is the size of $w(t)$.

In a real problem, the data $R_L[m]$ are not known exactly. They are estimated in terms of the J samples of $\mathbf{x}[n]$:

$$\mathbf{R}_L[m] = \frac{1}{J}\sum_{n} \mathbf{x}[n+m]\mathbf{x}[n] \qquad \text{(12-78)}$$

The mean and variance of $\mathbf{R}_L[m]$ can be determined as in the analog case. The details, however, will not be given. In the upcoming material, we assume that $R_L[m]$ is known exactly. This assumption is satisfactory if $J \gg L$.

Extrapolation Method

The spectral estimation problem is essentially numerical. This involves digital data even if the given process is analog. We shall, therefore, carry out the analysis in digital form. In the extrapolation method we assume that $S(z)$ is of known form. We shall assume that it is rational

$$\mathbf{S}(z) = \mathbf{L}(z)\mathbf{L}(1/z) \qquad \mathbf{L}(z) = \frac{b_0 + b_1 z^{-1} + \cdots + b_M z^{-M}}{1 + a_1 z^{-1} + \cdots + a_N z^{-N}} = \frac{N(z)}{D(z)} \tag{12-79}$$

We select the rational model for the following reasons: The numerical evaluation of its unknown parameters is relatively simple. An arbitrary spectrum can be closely approximated by a rational model of sufficiently large order. Spectra involving responses of dynamic systems are often rational.

SYSTEM IDENTIFICATION. The rational model leads directly to the solution of the identification problem (see also Sec. 10-7): We wish to determine the system function $\mathbf{H}(z)$ of a system driven by white noise in terms of the measurements of its output $\mathbf{x}[n]$. As we know, the power spectrum of the output is proportional to $\mathbf{H}(z)\mathbf{H}(1/z)$. If, therefore, the system is of finite order and minimum phase, then $\mathbf{H}(z)$ is proportional to $\mathbf{L}(z)$. To determine $\mathbf{H}(z)$, it suffices, therefore, to determine the $M + N + 1$ parameters of $\mathbf{L}(z)$. We shall do so under the assumption that $R_L[m]$ is known exactly for $|m| \le M + N + 1$.

We should stress that the proposed model is only an abstraction. In a real problem, $R[m]$ is not known exactly. Furthermore, $\mathbf{S}(z)$ might not be rational; even if it is, the constants M and N might not be known. However, the method leads to reasonable approximations if $R_L[m]$ is replaced by its time-average estimate $\mathbf{R}_L[m]$ and L is large.

AUTOREGRESSIVE PROCESSES. Our objective is to determine the $M + N + 1$ coefficients b_i and a_k specifying the spectrum $\mathbf{S}(z)$ in terms of the first $M + N + 1$ values $R_L[m]$ of $R[m]$. We start with the assumption that

$$\mathbf{L}(z) = \frac{\sqrt{P_N}}{1 + a_1 z^{-1} + \cdots + a_N z^{-N}} = \frac{\sqrt{P_N}}{D(z)} \tag{12-80}$$

This is a special case of (11-36) with $M = 0$ and $b_0 = \sqrt{P_N}$. As we know, the process $\mathbf{x}[n]$ satisfies the equation

$$\mathbf{x}[n] + a_1\mathbf{x}[n-1] + \cdots + a_N\mathbf{x}[n-N] = \boldsymbol{\varepsilon}[n] \tag{12-81}$$

where $\boldsymbol{\varepsilon}[n]$ is white noise with average power P_N. Our problem is to find the $N + 1$ coefficients a_k and P_N. To do so, we multiply (12-81) by $\mathbf{x}[n-m]$ and take expected values. With $m = 0, \ldots, N$, this yields the *Yule–Walker* equations

$$\begin{aligned} R[0] + a_1 R[1] + \cdots + a_N R[N] &= P_N \\ R[1] + a_1 R[0] + \cdots + a_N R[N-1] &= 0 \\ \cdots\cdots\cdots\cdots\cdots\cdots\cdots\cdots\cdots\cdots& \\ R[N] + a_1 R[N-1] + \cdots + a_N R[0] &= 0 \end{aligned} \tag{12-82}$$

This is a system of $N + 1$ equations involving the $N + 1$ unknowns a_k and P_N, and it has a unique solution if the determinant Δ_N of the correlation matrix D_N of $\mathbf{x}[n]$ is strictly positive. We note, in particular, that

$$P_N = \frac{\Delta_{N+1}}{\Delta_N} \qquad \Delta_N > 0 \tag{12-83}$$

If $\Delta_{N+1} = 0$, then $P_N = 0$ and $\boldsymbol{\varepsilon}_N[m] = 0$. In this case, the unknown $S(\omega)$ consists of lines [see (11-44)].

To find $\mathbf{L}(z)$, it suffices, therefore, to solve the system (12-82). This involves the inversion of the matrix D_N. The problem of inversion can be simplified because the matrix D_N is Toeplitz; that is, it is symmetrical with respect to its diagonal. We give later a simple method for determining a_k and P_N based on this property (Levinson's algorithm).

MOVING AVERAGE PROCESSES. If $\mathbf{x}[n]$ is an MA process, then

$$\mathbf{S}(z) = \mathbf{L}(z)\mathbf{L}(1/z) \qquad \mathbf{L}(z) = b_0 + b_1 z^{-1} + \cdots + b_M z^{-M} \tag{12-84}$$

In this case, $R[m] = 0$ for $|m| > M$ [see (11-47)]; hence $\mathbf{S}(z)$ can be expressed directly in terms of $R[m]$:

$$\mathbf{S}(z) = \sum_{m=-M}^{M} R[m] z^{-m} \qquad \mathbf{S}(e^{j\omega}) = \left| \sum_{m=0}^{M} b_m e^{-jm\omega} \right|^2 \tag{12-85}$$

In the identification problem, our objective is to find not the function $\mathbf{S}(z)$, but the $M+1$ coefficients b_m of $\mathbf{L}(z)$. One method for doing so is the factorization $\mathbf{S}(z) = \mathbf{L}(z)\mathbf{L}(1/z)$ of $\mathbf{S}(z)$ as in Sec. 11-1. This method involves the determination of the roots of $\mathbf{S}(z)$. We discuss later a method that avoids factorization (see pages 561–562).

ARMA PROCESSES. We assume now that $\mathbf{x}[n]$ is an ARMA process:

$$\mathbf{L}(z) = \frac{b_0 + b_1 z^{-1} + \cdots + b_M z^{-M}}{1 + a_1 z^{-1} + \cdots + a_N z^{-N}} = \frac{N(z)}{D(z)} \tag{12-86}$$

In this case, $\mathbf{x}[n]$ satisfies the equation[6]

$$\mathbf{x}[n] + a_1\mathbf{x}[n-1] + \cdots + a_N\mathbf{x}[n-N] = b_0\mathbf{i}[n] + \cdots + b_M\mathbf{i}[n-M] \tag{12-87}$$

where $\mathbf{i}[n]$ is its innovations. Multiplying both sides of (12-87) by $\mathbf{x}[n-m]$ and taking expected values, we conclude as in (11-49) that

$$R[m] + a_1 R[m-1] + \cdots + a_N R[m-N] = 0 \qquad m > M \tag{12-88}$$

Setting $m = M+1, M+2, \ldots, M+N$ into (12-88), we obtain a system of N equations. The solution of this system yields the N unknowns $a_1, \ldots, a_N$.

To complete the specification of $\mathbf{L}(z)$, it suffices to find the $M+1$ constants $b_0, \ldots, b_M$. To do so, we form a filter with input $\mathbf{x}[n]$, and system function (Fig. 12-12)

$$D(z) = 1 + a_1 z^{-1} + \cdots + a_N z^{-N}$$

---

[6]M. Kaveh: "High Resolution Spectral Estimation for Noisy Signals," *IEEE Transactions on Acoustics, Speech, and Signal Processing,* vol. **ASSP-27,** 1979. See also J. A. Cadzow: "Spectral Estimation: An Overdetermined Rational Model Equation Approach," *IEEE Proceedings,* vol. **70,** 1982.

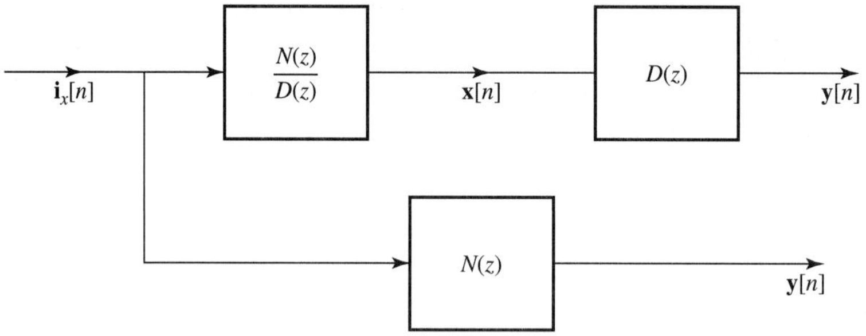

FIGURE 12-12

The resulting output $\mathbf{y}[n]$ is called the *residual sequence*. Inserting into (9-210), we obtain

$$\mathbf{S}_{yy}(z) = \mathbf{S}(z)D(z)D(1/z) = N(z)N(1/z)$$

From this it follows that $\mathbf{y}[n]$ is an MA process, and its innovations filter equals

$$\mathbf{L}_y(z) = N(z) = b_0 + b_1 z^{-1} + \cdots + b_M z^{-M} \tag{12-89}$$

To determine the constants b_i, it suffices, therefore, to find the autocorrelation $R_{yy}[m]$ for $|m| \le M$. Since $\mathbf{y}[n]$ is the output of the filter $D(z)$ with input $\mathbf{x}[n]$, it follows from (11-47) with $a_0 = 1$ that

$$R_{yy}[m] = R[m] * d[m] * d[-m] \qquad d[m] = \sum_{k=0}^{N} a_k \delta[m-k]$$

This yields

$$R_{yy}[m] = \sum_{i=-N}^{N} R[m-i]\rho[i] \qquad \rho[m] = \sum_{k=m}^{N} a_{k-m} a_k = \rho[-m] \tag{12-90}$$

for $0 \le m \le M$ and 0 for $m > M$. With $R_{yy}[m]$ so determined, we proceed as in the MA case.

The determination of the ARMA model involves thus the following steps:

Find the constants a_k from (12-88); this yields $D(z)$.
Find $R_{yy}[m]$ from (12-90).
Find the roots of the polynomial

$$\mathbf{S}_{yy}(z) = \sum_{m=-M}^{M} R_{yy}[m] z^{-m} = N(z)N(1/z)$$

Form the Hurwitz factor $N(z)$ of $\mathbf{S}_{yy}(z)$.

LATTICE FILTERS AND LEVINSON'S ALGORITHM. An MA filter is a polynomial in z^{-1}. Such a filter is usually realized by a ladder structure as in Fig. 12-14*a*. A lattice filter is an alternate realization of an MA filter in the form of Fig. 12-14*b*. In the context of spectral estimation, lattice filters are used to simplify the solution of the Yule–Walker equations and the factorization of polynomials. Furthermore, as we show later, they are also used to give a convenient description of the properties of extrapolating spectra. Related applications are developed in the next chapter in the solution of the prediction problem.

The polynomial

$$D(z) = 1 - a_1^N z^{-1} - \cdots - a_N^N z^{-N} = 1 - \sum_{k=1}^{N} a_k^N z^{-k}$$

specifies an MA filter with $\mathbf{H}(z) = D(z)$. The superscript in a_k^N identifies the order of the filter. If the input to this filter is an AR process $\mathbf{x}[n]$ with $\mathbf{L}(z)$ as in (12-80) and $a_k^N = -a_k$, then the resulting output

$$\boldsymbol{\varepsilon}[n] = \mathbf{x}[n] - a_1^N \mathbf{x}[n-1] - \cdots - a_N^N \mathbf{x}[n-N] \tag{12-91}$$

is white noise as in (12-81). The filter $D(z)$ is usually realized by the ladder structure of Fig. 12-14*a*. We shall show that the lattice filter of Fig. 12-14*b* is an equivalent realization. We start with $N = 1$.

In Fig. 12-13*a* we show the ladder realization of an MA filter of order 1 and its mirror image. The input to both systems is the process $\mathbf{x}[n]$; the outputs equal

$$\mathbf{y}[n] = \mathbf{x}[n] - a_1^1 \mathbf{x}[n-1] \qquad \mathbf{z}[n] = -a_1^1 \mathbf{x}[n] + \mathbf{x}[n-1]$$

The corresponding system functions equal

$$1 - a_1^1 z^{-1} \qquad -a_1^1 + z^{-1}$$

In Fig. 12-13*b* we show a lattice filter of order 1. It has a single input $\mathbf{x}[n]$ and two outputs

$$\hat{\boldsymbol{\varepsilon}}_1[n] = \mathbf{x}[n] - K_1 \mathbf{x}[n-1] \qquad \check{\boldsymbol{\varepsilon}}_1[n] = -K_1 \mathbf{x}[n] + \mathbf{x}[n-1]$$

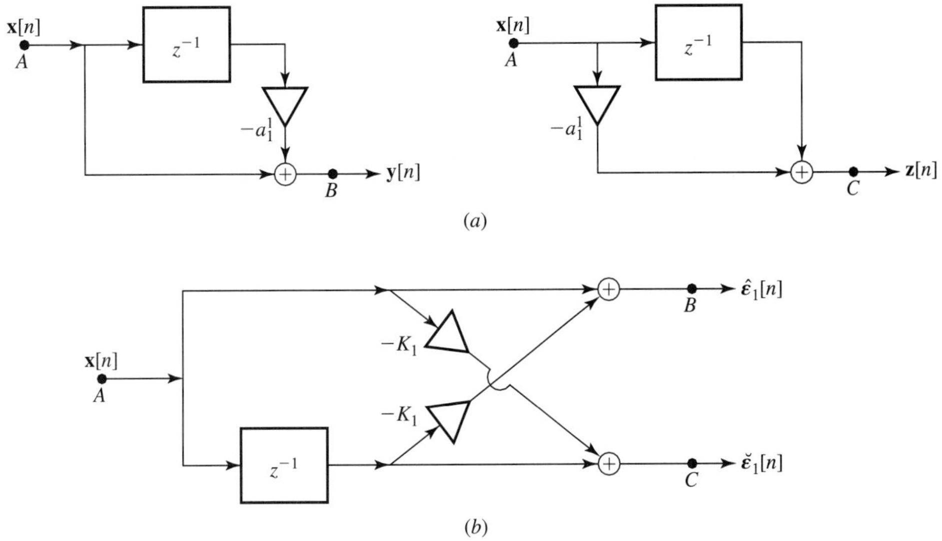

FIGURE 12-13
Ladder and lattice filters of order one.

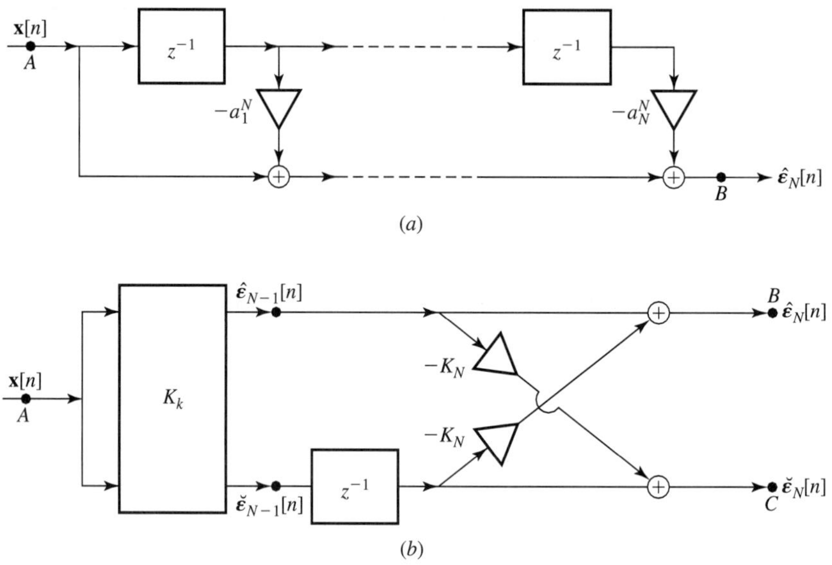

FIGURE 12-14
Ladder and lattice filters.

The corresponding system functions are

$$\hat{\mathbf{E}}_1(z) = 1 - K_1 z^{-1} \qquad \check{\mathbf{E}}_1(z) = -K_1 + z^{-1} = z^{-1}\hat{\mathbf{E}}_1(1/z)$$

If $K_1 = a_1^1$, then the lattice filter of Fig. 12-13b is equivalent to the two MA filters of Fig. 12-13a.

In Fig. 12-14b we show a lattice filter of order N formed by cascading N first-order filters. The input to this filter is the process $\mathbf{x}[n]$. The resulting outputs are denoted by $\hat{\varepsilon}_N[n]$ and $\check{\varepsilon}_N[n]$ and are called *forward* and *backward,* respectively. As we see from the diagram these signals satisfy the equations

$$\hat{\varepsilon}_N[n] = \hat{\varepsilon}_{N-1}[n] - K_N \check{\varepsilon}_{N-1}[n-1] \qquad (12\text{-}92a)$$

$$\check{\varepsilon}_N[n] = \check{\varepsilon}_{N-1}[n-1] - K_N \hat{\varepsilon}_{N-1}[n] \qquad (12\text{-}92b)$$

Denoting by $\hat{\mathbf{E}}_N(z)$ and $\check{\mathbf{E}}_N(z)$ the system functions from the input A to the upper output B and lower output C, respectively, we conclude that

$$\hat{\mathbf{E}}_N(z) = \hat{\mathbf{E}}_{N-1}(z) - K_N z^{-1}\check{\mathbf{E}}_{N-1}(z) \qquad (12\text{-}93a)$$

$$\check{\mathbf{E}}_N(z) = z^{-1}\check{\mathbf{E}}_{N-1}(z) - K_N \hat{\mathbf{E}}_{N-1}(z) \qquad (12\text{-}93b)$$

where $\hat{\mathbf{E}}_{N-1}(z)$ and $\check{\mathbf{E}}_{N-1}(z)$ are the forward and backward system functions of the lattice of the first $N-1$ sections. From (12-93) it follows by a simple induction that

$$\check{\mathbf{E}}_N(z) = z^{-N}\hat{\mathbf{E}}_N(1/z) \qquad (12\text{-}94)$$

The lattice filter is thus specified in terms of the N constants K_k. These constants are called *reflection coefficients.*

Since $\hat{\mathsf{E}}_1(z) = 1 - K_1 z^{-1}$ and $\check{\mathsf{E}}_1(z) = -K_1 + z^{-1}$, we conclude from (12-93) that the functions $\hat{\mathsf{E}}_1(z)$ and $\check{\mathsf{E}}_N(z)$ are polynomials in z^{-1} of the form

$$\hat{\mathsf{E}}_N(z) = 1 - a_1^N z^{-1} - \cdots - a_N^N z^{-N} \tag{12-95}$$

$$\check{\mathsf{E}}_N(z) = z^{-N} - a_1^N z^{-N+1} - \cdots - a_N^N \tag{12-96}$$

where a_k^N are N constants that are specified in terms of the reflection coefficients K_k. We shall refer to $\hat{\mathsf{E}}_n(z)$ as the normalized Levinson polynomial of degree N (constant term is unity) and $\check{\mathsf{E}}_n(z)$ represents its reciprocal polynomial.

LEVINSON'S ALGORITHM.[7] We denote by a_k^{N-1} the coefficients of the lattice filter of the first $N-1$ sections:

$$\hat{\mathsf{E}}_{N-1}(z) = 1 - a_1^{N-1} z^{-1} - \cdots - a_{N-1}^{N-1} z^{-(N-1)}$$

From (12-94) it follows that

$$z^{-1}\check{\mathsf{E}}_{N-1}(z) = z^{-N}\hat{\mathsf{E}}_{N-1}(1/z)$$

Inserting into (12-93a) and equating coefficients of equal powers of z, we obtain

$$\begin{aligned} a_k^N &= a_k^{N-1} - K_N a_{N-k}^{N-1} \qquad k = 1, \ldots, N-1 \\ a_N^N &= K_N \end{aligned} \tag{12-97}$$

We have thus expressed the coefficients a_k^N of a lattice of order N in terms of the coefficients a_k^{N-1} and the last reflection coefficient K_N. Starting with $a_1^1 = K_1$, we can express recursively the N parameters a_k^N in terms of the N reflection coefficients K_k.

Conversely, if we know a_k^N, we find K_k using inverse recursion: The coefficient K_N equals a_N^N. To find K_{N-1}, it suffices to find the polynomial $\hat{\mathsf{E}}_{N-1}(z)$. Multiplying (12-93$b$) by K_N and adding to (12-93a), we obtain

$$\left(1 - K_N^2\right)\hat{\mathsf{E}}_{N-1}(z) = \hat{\mathsf{E}}_N(z) + K_N z^{-N}\hat{\mathsf{E}}_N(1/z) \tag{12-98}$$

This expresses $\hat{\mathsf{E}}_{N-1}(z)$ in terms of $\hat{\mathsf{E}}_N(z)$ because $K_N = a_N^N$. With $\hat{\mathsf{E}}_{N-1}(z)$ so determined, we set $K_{N-1} = a_{N-1}^{N-1}$. Continuing this process, we find $\hat{\mathsf{E}}_{N-k}(z)$ and K_{N-k} for every $k < N$.

Minimum-phase properties of Levinson Polynomials. We shall relate the location of the roots z_i^N of the polynomial $\hat{\mathsf{E}}_N(z)$ to the magnitude of the reflection coefficients K_k.

THEOREM 12-6

LEVINSON POLYNOMIAL ROOTS

If

$$|K_k| < 1 \quad \text{for all } k \le N \qquad \text{then} \quad \left|z_i^N\right| < 1 \quad \text{for all } i \le N \tag{12-99}$$

Proof. By induction. The theorem is true for $N = 1$ because $\hat{\mathsf{E}}_1(z) = 1 - K_1 z^{-1}$; hence $|z_1^1| = |K_1| < 1$. Suppose that $|z_j^{N-1}| \le 1$ for all $j \le N-1$ where z_j^{N-1} are the roots of $\hat{\mathsf{E}}_{N-1}(z)$. From this it follows that the function

$$A_{N-1}(z) = \frac{z^{-N}\hat{\mathsf{E}}_{N-1}(1/z)}{\hat{\mathsf{E}}_{N-1}(z)} \tag{12-100}$$

[7] N. Levinson: "The Wiener RMS Error Criterion in Filter Design and Prediction," *Journal of Mathematics and Physics,* vol. **25**, 1947. See also J. Durbin: "The Fitting of Time Series Models," *Revue L'Institut Internationale de Statisque,* vol. **28**, 1960.

is all-pass. Since $\hat{\mathbf{E}}_N(z_i^N) = 0$ by assumption, we conclude from (12-93a) and (12-94) that

$$\hat{\mathbf{E}}_N(z_i) = \hat{\mathbf{E}}_{N-1}(z_i) - K_N z^{-N} \hat{\mathbf{E}}_{N-1}(1/z_i) = 0$$

Hence

$$\left|A_{N-1}\left(z_i^N\right)\right| = \frac{1}{|K_N|} > 1$$

This shows that $|z_i^N| < 1$ [see (12B-2)]. ◀

THEOREM 12-7

ROOTS AND REFLECTION COEFFICIENTS

▶ If

$$\left|z_i^N\right| < 1 \quad \text{for all} \quad i \le N \qquad \text{then} \quad |K_k| < 1 \quad \text{for all} \quad k \le N \qquad (12\text{-}101)$$

Proof. The product of the roots of the polynomial $\hat{\mathbf{E}}_N(z)$ equals the last coefficient a_N^N. Hence

$$K_N = a_N^N = z_1^N \cdots z_N^N \qquad |K_N| < 1$$

Thus (12-100) is true for $k = N$. To show that it is true for $k = N - 1$, it suffices to show that $|z_j^{N-1}| < 1$ for $j \le N - 1$. To do so, we form the all-pass function

$$A_N(z) = \frac{z^{-N}\hat{\mathbf{E}}_N(1/z)}{\hat{\mathbf{E}}_N(z)} \qquad (12\text{-}102)$$

Since $\hat{\mathbf{E}}_{N-1}(z_j^{N-1}) = 0$ it follows from (12-98) that

$$\left|A_N\left(z_j^{N-1}\right)\right| = \frac{1}{|K_N|} > 1$$

Hence $|z_j^{N-1}| < 1$ and $|K_{N-1}| = |a_{N-1}^{N-1}| = |z_1^{N-1} \cdots z_{N-1}^{N-1}| < 1$. Proceeding similarly, we conclude that $|K_k| < 1$ for all $k \le N$. ◀

ROOTS ON THE UNIT CIRCLE

▶ If $|K_k| < 1$ for $k \le N - 1$ and $|K_N| = 1$, then

$$\left|z_i^N\right| = 1 \qquad \text{for all} \quad i \le N \qquad (12\text{-}103)$$

Proof. From the theorem it follows that $|z_j^{N-1}| < 1$ because $|K_k| < 1$ for all $k \le N - 1$. Hence the function $A_{N-1}(z)$ in (12-100) is all-pass and $|A_{N-1}(z_i^N)| = 1/|K_N| = 1$. This leads to the conclusion that $|z_i^N| = 1$ [see (12B-2)]. ◀

We have thus established the equivalence between the Levinson polynomial $\hat{\mathbf{E}}_N(z)$ and a set of N constants K_k. We have shown further that the Levinson polynomial is strictly Hurwitz, iff $|K_k| < 1$ for all k.

Inverse lattice realization of AR systems. An *inverse lattice* is a modification of a lattice as in Fig. 12-15. In this modification, the input is at point B and the outputs are at points A and C. Furthermore, the multipliers from the lower to the upper line are changed from $-K_k$ to K_k. Denoting by $\hat{\boldsymbol{\varepsilon}}_N[n]$ the input at point B and by $\check{\boldsymbol{\varepsilon}}_{N-1}[n]$ the resulting output at C, we observe from the figure that

$$\hat{\boldsymbol{\varepsilon}}_{N-1}[n] = \hat{\boldsymbol{\varepsilon}}_N[n] + K_N \check{\boldsymbol{\varepsilon}}_{N-1}[n-1] \qquad (12\text{-}104a)$$

$$\check{\boldsymbol{\varepsilon}}_N[n] = \check{\boldsymbol{\varepsilon}}_{N-1}[n-1] - K_N \hat{\boldsymbol{\varepsilon}}_{N-1}[n] \qquad (12\text{-}104b)$$

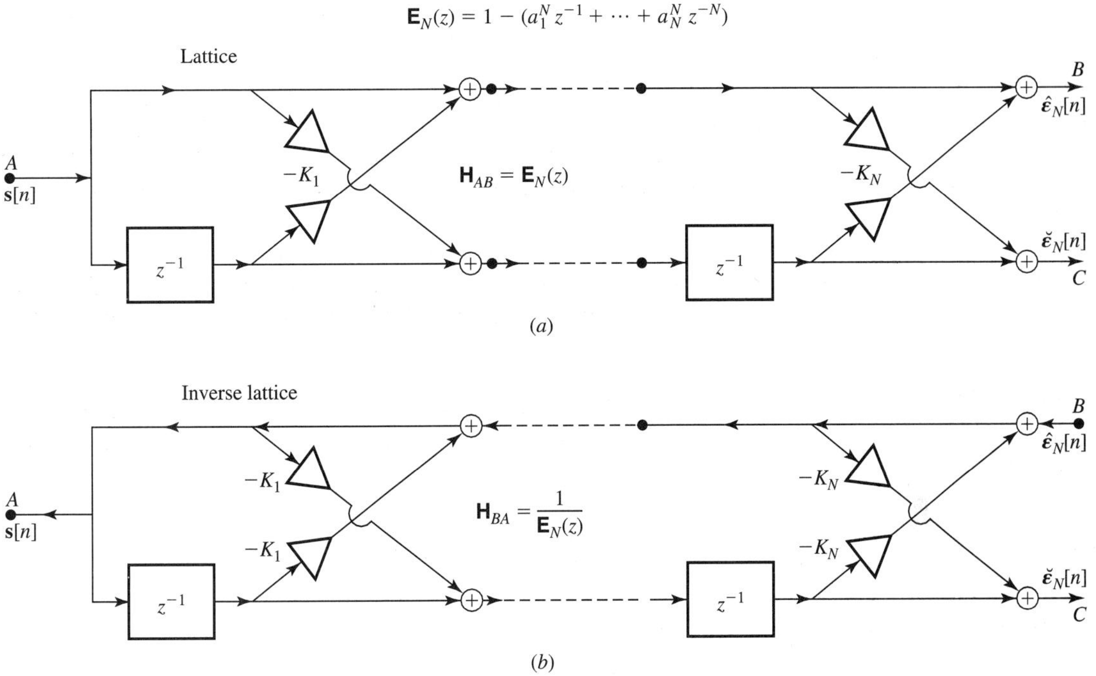

FIGURE 12-15
Lattice and inverse lattice filters.

These equations are identical with the two equations in (12-92). From this it follows that the system function from B to A equals

$$\frac{1}{\hat{\mathbf{E}}_N(z)} = \frac{1}{1 - a_1^N z^{-1} - \cdots - a_N^N z^{-N}}$$

We have thus shown that an AR system can be realized by an inverse lattice. The coefficients a_k^N and K_k satisfy Levinson's algorithm (12-97).

Iterative solution of the Yule–Walker equations. Consider an AR process $\mathbf{x}[n]$ with innovations filter $\mathbf{L}(z) = \sqrt{P_N}/D(z)$ as in (12-80). We form the lattice equivalent of the MA system $\mathbf{D}(z)$ with $a_k^N = -a_k$, and use $\mathbf{x}[n]$ as its input. As we know [see (12-95)] the forward and backward responses are given by

$$\begin{aligned} \hat{\boldsymbol{\varepsilon}}_N[n] &= \mathbf{x}[n] - a_1^N \mathbf{x}[n-1] - \cdots - a_N^N \mathbf{x}[n-N] \\ \check{\boldsymbol{\varepsilon}}_N[n] &= \mathbf{x}[n-N] - a_1^N \mathbf{x}[n-N+1] - \cdots - a_N^N \mathbf{x}[n] \end{aligned} \tag{12-105}$$

Denoting by $\hat{\mathbf{S}}_N(z)$ and $\check{\mathbf{S}}_N(z)$ the spectra of $\hat{\boldsymbol{\varepsilon}}_N[n]$ and $\check{\boldsymbol{\varepsilon}}_N[n]$, respectively, we conclude from (12-105) that

$$\hat{\mathbf{S}}_N(z) = \mathbf{S}(z)\hat{\mathbf{E}}_N(z)\hat{\mathbf{E}}_N(1/z) = P_N$$

$$\check{\mathbf{S}}_N(z) = \mathbf{S}(z)\check{\mathbf{E}}_N(z)\check{\mathbf{E}}_N(1/z) = P_N$$

From this it follows that $\hat{\varepsilon}_N[n]$ and $\check{\varepsilon}_N[n]$ are two white-noise processes and

$$E\{\hat{\varepsilon}_N^2[n]\} = E\{\check{\varepsilon}_N^2[n]\} = P_N \tag{12-106a}$$

$$E\{\mathbf{x}[n-m]\hat{\varepsilon}_N[n]\} = \begin{cases} P_N & m = 0 \\ 0 & 1 \le m \le N \end{cases} \tag{12-106b}$$

$$E\{\mathbf{x}[n-m]\check{\varepsilon}_N[n]\} = \begin{cases} 0 & 0 \le m \le N-1 \\ P_N & m = N \end{cases} \tag{12-106c}$$

These equations also hold for all filters of lower order. We shall use them to express recursively the parameters a_k^N, K_N, and P_N in terms of the $N+1$ constants $R[0], \ldots, R[N]$.

For $N = 1$ (12-82) yields

$$R[0] - a_1^1 R[1] = P_1 \qquad R[1] - a_1^1 R[0] = 0$$

Setting $P_0 = R[0]$, we obtain

$$a_1^1 = K_1 = \frac{R[1]}{R[0]} \qquad P_1 = \left(1 - K_1^2\right) P_0$$

Suppose now that we know the $N+1$ parameters a_k^{N-1}, K_{N-1}, and P_N. From Levinson's algorithm (12-97) it follows that we can determine a_k^N if K_N is known. To complete the iteration, it suffices, therefore, to find K_N and P_N.

We maintain that

$$P_{N-1}K_N = R[N] - \sum_{k=1}^{N-1} a_k^{N-1} R[N-k] \tag{12-107}$$

$$P_N = \left(1 - K_N^2\right) P_{N-1} \tag{12-108}$$

The first equation yields K_N in terms of the known parameters a_k^{N-1}, $R[m]$, and P_{N-1}. With K_N so determined, P_N is determined from the second equation.

Proof. Multiplying (12-92*a*) by $\mathbf{x}[n-N]$ and using the identities

$$E\{\hat{\varepsilon}_{N-1}[n]\mathbf{x}[n-N]\} = R[N] - \sum_{k=1}^{N-1} a_k^{N-1} R[k]$$

$$E\{\hat{\varepsilon}_N[n]\mathbf{x}[n]\} = P_N \qquad E\{\hat{\varepsilon}_{N-1}[n-1]\mathbf{x}[n]\} = P_{N-1}$$

we obtain (12-107). From (12-92*a*) and the identities

$$E\{\hat{\varepsilon}_N[n]\mathbf{x}[n]\} = P_N \qquad E\{\hat{\varepsilon}_{N-1}[n-1]\mathbf{x}[n]\} = P_{N-1}$$

$$E\{\check{\varepsilon}_{N-1}[n-1]\mathbf{x}[n]\} = R[N] - \sum_{k=1}^{N-1} a_k^{N-1} R[N-k] = P_{N-1}K_N$$

it follows similarly that $P_N = P_{N-1} - K_N^2 P_{N-1}$ and (12-108) results.

Since $P_k \ge 0$ for every k, it follows from (12-108) that

$$|K_k| \le 1 \qquad \text{and} \qquad P_0 \ge P_1 \ge \cdots \ge P_N \ge 0 \tag{12-109}$$

If $|K_N| = 1$ but $|K_k| < 1$ for all $k < N$, then

$$P_0 > P_1 > \cdots > P_N = 0 \qquad (12\text{-}110)$$

As we show next this is the case if $S(\omega)$ consists of lines.

Line spectra and hidden periodicities. If $P_N = 0$, then $\hat{\boldsymbol{\varepsilon}}_N[n] = 0$; hence the process $\mathbf{x}[n]$ satisfies the homogeneous recursion equation

$$\mathbf{x}[n] = a_1^N \mathbf{x}[n-1] + \cdots + a_N^N \mathbf{x}[n-N] \qquad (12\text{-}111)$$

This shows that $\mathbf{x}[n]$ is a predictable process, that is, it can be expressed in terms of its N past values. Furthermore,

$$R[m] - a_1^N R[m-1] - \cdots - a_N^N R[m-N] = 0 \qquad (12\text{-}112)$$

As we know [see (12-103)] the roots z_i^N of the characteristic polynomial $\hat{\mathbf{E}}_N(z)$ of this equation are on the unit circle: $z_i^N = e^{j\omega_i}$. From this it follows that

$$R[m] = \sum_{i=1}^{N} \alpha_i e^{j\omega_i m} \qquad S(\omega) = 2\pi \sum_{i=1}^{N} \alpha_i \delta(\omega - \omega_i) \qquad (12\text{-}113)$$

And since $S(\omega) \ge 0$, we conclude that $\alpha_i \ge 0$.

Solving (12-111), we obtain

$$\mathbf{x}[n] = \sum_{i=1}^{N} \mathbf{c}_i e^{j\omega_i n} \qquad E\{\mathbf{c}_i\} = 0 \qquad E\{\mathbf{c}_i \mathbf{c}_k\} = \begin{cases} \alpha_i & i = k \\ 0 & i \ne k \end{cases} \qquad (12\text{-}114)$$

THEOREM 12-8

CARATHEODORY'S THEOREM

▶ We show next that if $R[m]$ is a p.d. sequence and its correlation matrix is of rank N, that is, if

$$\Delta_N > 0 \qquad \Delta_{N+1} = 0 \qquad (12\text{-}115)$$

then $R[m]$ is a sum of exponentials with positive coefficients:

$$R[m] = \sum_{i=1}^{N} \alpha_i e^{j\omega_i m} \qquad \alpha_i > 0 \qquad (12\text{-}116)$$

Proof. Since $R[m]$ is a p.d. sequence, we can construct a process $\mathbf{x}[n]$ with autocorrelation $R[m]$. Applying Levinson's algorithm, we obtain a sequence of constants K_k and P_k. The iteration stops at the Nth step because $P_N = \Delta_{N+1}/\Delta_N = 0$. This shows that the process $\mathbf{x}[n]$ satisfies the recursion equation (12-111). ◀

Detection of hidden periodicities.[8] We shall use the development in (13-111)–(13-116) to solve the following problem: We wish to determine the frequencies ω_i of a process $\mathbf{x}[n]$ consisting of at most N exponentials as in (12-114). The available information is the sum

$$\mathbf{y}[n] = \mathbf{x}[n] + \boldsymbol{\nu}[n] \qquad E\{\boldsymbol{\nu}^2[n]\} = q \qquad (12\text{-}117)$$

where $\boldsymbol{\nu}[n]$ is white noise independent of $\mathbf{x}[n]$.

[8] V. F. Pisarenko: "The Retrieval of Harmonics," *Geophysical Journal of the Royal Astronomical Society,* 1973.

Using J samples of $\mathbf{y}[n]$, we estimate its autocorrelation

$$R_{yy}[m] = R_{xx}[m] + q\delta[m] \tag{12-118}$$

as in (12-78). The correlation matrix D_{N+1} of $\mathbf{x}[n]$ is thus given by

$$D_{N+1} = \begin{bmatrix} R_{yy}[0]-q & R_{yy}[1] & \cdots & R_{yy}[N] \\ R_{yy}[1] & R_{yy}[0]-q & \cdots & R_{yy}[N-1] \\ \cdots & \cdots & \cdots & \cdots \\ R_{yy}[N] & R_{yy}[N-1] & \cdots & R_{yy}[0]-q \end{bmatrix} \tag{12-119}$$

In this expression, $R_{yy}[m]$ is known but q is unknown. We know, however, that $\Delta_{N+1} = 0$ because $\mathbf{x}[n]$ consists of N lines. Hence q is an eigenvalue of D_{N+1}. It is, in fact, the smallest eigenvalue q_0 because $D_{N+1} > 0$ for $q < q_0$. With $R_{xx}[m]$ so determined, we proceed as before: Using Levinson's algorithm, we find the coefficients a_k^N and the roots $e^{j\omega_i}$ of the resulting polynomial $\hat{\mathbf{E}}_N(z)$. If q_0 is a simple eigenvalue, then all roots are distinct and $\mathbf{x}[n]$ is a sum of N exponentials. If, however, q_0 is a multiple root with multiplicity N_0 then $\mathbf{x}[n]$ consists of $N - N_0 + 1$ exponentials.

This analysis leads to the following extension of Caratheodory's theorem: The $N+1$ values $R[0], \ldots, R[N]$ of a strictly p.d. sequence $R[m]$ can be expressed in the form

$$R[m] = q_0\delta[m] + \sum_{i=1}^{N} \alpha_i e^{j\omega_i m} \tag{12-120}$$

where q_0 and α_i are positive constants and ω_i are real frequencies.

BURG'S ITERATION[9]

▶ Levinson's algorithm is used to determine recursively the coefficients a_k^N of the innovations filter $\mathbf{L}(z)$ of an AR process $\mathbf{x}[n]$ in terms of $R[m]$. In a real problem the data $R[m]$ are not known exactly. They are estimated from the J samples of $\mathbf{x}[n]$ and these estimates are inserted into (12-107) and (12-108) yielding the estimates of K_N and P_N. The results are then used to estimate a_k^N from (12-97). A more direct approach, suggested by Burg, avoids the estimation of $R[m]$. It is based on the observation that Levinson's algorithm expresses recursively the coefficients a_k^N in terms of K_N and P_N. The estimates of these coefficients can, therefore, be obtained directly in terms of the estimates of K_N and P_N. These estimates are based on the following identities [see (12-106)]:

$$\begin{aligned} P_{N-1}K_N &= E\{\hat{\boldsymbol{\varepsilon}}_{N-1}[n]\check{\boldsymbol{\varepsilon}}_{N-1}[n-1]\} \\ P_N &= \tfrac{1}{2}E\left\{\hat{\boldsymbol{\varepsilon}}_N^2[n] + \check{\boldsymbol{\varepsilon}}_N^2[n]\right\} \end{aligned} \tag{12-121}$$

Replacing expected values by time averages, we obtain the following iteration: Start with

$$\mathbf{P}_0 = \frac{1}{J}\sum_{n=1}^{J} \mathbf{x}^2[n] \qquad \hat{\boldsymbol{\varepsilon}}_0[n] = \check{\boldsymbol{\varepsilon}}_0[n] = \mathbf{x}[n]$$

---

[9] J. P. Burg: Maximum entropy spectral analysis, presented at the International Meeting of the Society for the Exploration of Geophysics, Orlando, FL, 1967. Also J. P. Burg, "*Maximum entropy spectral analysis*" Ph.D. Diss., Dept. Geophysics, Stanford Univ., Stanford, CA, May 1975.

Find $\mathbf{K}_{N-1}, \mathbf{P}_{N-1}, \mathbf{a}_K^{N-1}, \hat{\boldsymbol{\varepsilon}}_{N-1}[n], \check{\boldsymbol{\varepsilon}}_{N-1}$. Set

$$\mathbf{K}_N = \frac{\sum_{n=N+1}^{J} \hat{\boldsymbol{\varepsilon}}_{N-1}[n]\check{\boldsymbol{\varepsilon}}_{N-1}[n-1]}{\frac{1}{2}\sum_{n=N+1}^{J}\left(\hat{\boldsymbol{\varepsilon}}_{N-1}^2[n]+\check{\boldsymbol{\varepsilon}}_{N-1}^2[n-1]\right)} \tag{12-122}$$

$$\mathbf{P}_N = \left(1 - \mathbf{K}_N^2\right) P_{N-1} \tag{12-123}$$

$$\begin{aligned} \mathbf{a}_k^N &= \mathbf{a}_k^{N-1} - \mathbf{K}_N \mathbf{a}_{N-k}^{N-1} \qquad k = 1, \ldots, N-1 \\ \mathbf{a}_N^N &= K_N \end{aligned} \tag{12-124}$$

$$\begin{aligned} \hat{\boldsymbol{\varepsilon}}_N[n] &= \mathbf{x}[n] - \sum_{k=1}^{N-1} \mathbf{a}_k^N \mathbf{x}[n-k] \\ \check{\boldsymbol{\varepsilon}}_N[n] &= \mathbf{x}[n-N] - \sum_{k=1}^{N} \mathbf{a}_{N-k}^N \mathbf{x}[n-N+k] \end{aligned} \tag{12-125}$$

This completes the Nth iteration step. Note that

$$|\mathbf{K}_N| \le 1 \qquad \mathbf{P}_N \ge 0$$

This follows readily if we apply Cauchy's inequality (see Prob. 10-23) to the numerator of (12-122). ◀

Levinson's algorithm yields the correct spectrum $\mathbf{S}(z)$ only if $\mathbf{x}[n]$ is an AR process. If it is not, the result is only an approximation. If $R[m]$ is known exactly, the approximation improves as N increases. However, if $R[m]$ is estimated as above, the error might increase because the number of terms in (12-49) equals $J - N - 1$ and it decreases as N increases. The determination of an optimum N is in general difficult.

THEOREM 12-9

FEJÉR–RIESZ THEOREM AND LEVINSON'S ALGORITHM

▶ Given a polynomial spectrum

$$W(e^{j\omega}) = \sum_{n=-N}^{N} w_n e^{-jn\omega} \ge 0 \tag{12-126}$$

we can find a Hurwitz polynomial

$$Y(z) = \sum_{n=0}^{N} y_n z^{-n} \tag{12-127}$$

such that $W(e^{j\omega}) = |\mathbf{Y}(e^{j\omega})|^2$. This theorem has extensive applications. We used it in Sec. 11-1 (spectral factorization) and in the estimation of the spectrum of an MA and an ARMA process. The construction of the polynomial $Y(z)$ involves the determination of the roots of $W(z)$. This is not a simple problem particularly if $W(e^{j\omega})$ is known only as a function of ω. We discuss next a method for determining $Y(z)$ involving Levinson's algorithm and Fourier series.

We compute, first, the Fourier series coefficients

$$R[m] = \frac{1}{2\pi}\int_{-\pi}^{\pi} \frac{1}{W(e^{j\omega})} e^{-jm\omega}\, d\omega \qquad 0 \le m \le N \tag{12-128}$$

of the inverse spectrum $S(e^{j\omega}) = 1/W(e^{j\omega})$. The numbers $R[m]$ so obtained are the values of a p.d. sequence because $S(e^{j\omega}) \geq 0$. Applying Levinson's algorithm to the numbers $R[m]$ so computed, we obtain $N+1$ constants a_k^N and P_N. This yields

$$S(e^{j\omega}) = \frac{1}{W(e^{j\omega})} = \frac{P_N}{\left|1 - \sum_{n=1}^{N} a_n^N e^{-jn\omega}\right|^2}$$

Hence

$$Y(z) = \frac{1}{\sqrt{P_N}}\left(1 - \sum_{n=1}^{N} a_n^N z^{-n}\right)$$

as in (12-127). This method thus avoids the factorization problem. ◀

12-4 THE GENERAL CLASS OF EXTRAPOLATING SPECTRA AND YOULA'S PARAMETRIZATION

We consider now the following problem[10]: Given $n+1$ values (data)

$$r_0, r_1, \ldots, r_n$$

of the autocorrelation sequence $\{r_m\}$ of a process $\mathbf{x}[k]$ we wish to find all its positive definite extrapolations, that is, we wish to find the family C_n of spectra $\mathbf{S}(e^{j\omega}) \geq 0$ such that the first $n+1$ coefficients of their Fourier series expansion equal the given data. The sequences $\{r_m\}$ of the class C_n and their spectra will be called admissible.

Known as the trigonometric moment problem, it has been the subject of extensive study for a long time [14]. In view of the considerable mathematical interest as well as the practical significance of the moment problem in interpolation theory, system identification, power gain approximation theory and rational approximation of nonrational spectra, it is best to review this problem in some detail. Towards this, note that a member of the class C_n is the AR spectrum

$$\mathbf{S}(z) = \mathbf{L}(z)\mathbf{L}(1/z) \qquad \mathbf{L}(z) = \sqrt{P_n}/\mathbf{E}_n(z)$$

where $\mathbf{E}_n(z) = \hat{\mathbf{E}}_n(z)$ is the forward filter of order n obtained from an n-step Levinson algorithm. The continuation of the corresponding r_m is obtained from (11-41b):

$$r_m = \sum_{k=1}^{n} a_k^n r_{m-k} \qquad m > n$$

To find all members of the class C_n, we can continue the algorithm by assigning arbitrary values

$$|K_k| \leq 1 \qquad k = n+1, n+2, \ldots$$

to the reflection coefficients. The resulting values of r_m can be determined recursively [see (12-107)]

$$r_m = \sum_{k=1}^{m-1} a_k^{m-1} r_{m-k} + P_{m-1} K_m \tag{12-129}$$

[10] A. Papoulis: "Levinson's Algorithm, Wold's Decomposition, and Spectral Estimation," *SIAM Review*, vol. **27**, 1985.

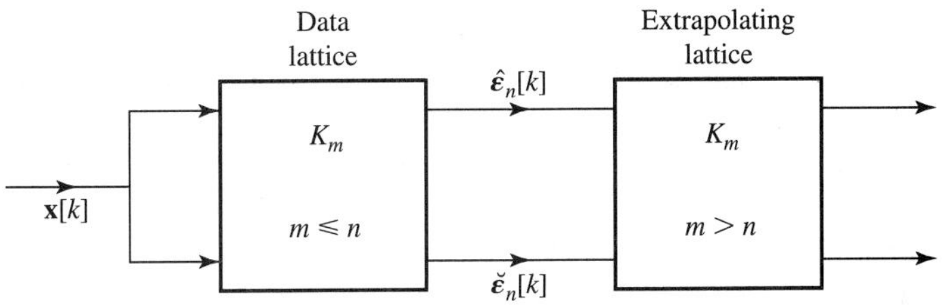

FIGURE 12-16

This shows that the admissible values of r_m at the mth iteration are in an interval of length $2P_{m-1}$:

$$\sum_{k=1}^{m-1} a_k^{m-1} r_{m-k} - P_{m-1} \le r_m \le \sum_{k=1}^{m-1} a_k^{m-1} r_{m-k} + P_{m-1} \tag{12-130}$$

because $|K_m| \le 1$. At the endpoints of the interval, $|K_m| = 1$; in this case, $P_m = 0$ and $\Delta_{m+1} = 0$. As we have shown, the corresponding spectrum $S(\omega)$ consists of lines. If $|K_{m_0}| < 1$ and $K_m = 0$ for $m > m_0$, then $\mathbf{S}(z)$ is an AR spectrum of order m_0. In Fig. 12-16, we show the iteration lattice. The first n sections are uniquely determined in terms of the data. The remaining sections form a four-terminal lattice specified in terms of the arbitrarily chosen reflection coefficients $K_{n+1}, K_{n+2}, \ldots$.

Interestingly, Youla has given a complete parametrization to the class of all such admissible extensions by making use of the positive-real and bounded-real function concepts from classical network theory [33].

Youla's Parametrization of Admissible Extensions[11]

Let $S(\omega)$ represent the spectral density of a zero mean, real, second-order stationary stochastic process $\mathbf{x}(nT)$ with finite power and covariance sequence $\{r_k\}_{k=-\infty}^{\infty}$. Then,

$$S(\omega) = \sum_{k=-\infty}^{\infty} r_k e^{jk\omega} \ge 0 \qquad \omega \text{ real} \tag{12-131}$$

is periodic with period 2π and

$$\int_{-\pi}^{\pi} S(\omega)\, d\omega < \infty \tag{12-132}$$

Under these conditions, $r_k = r_{-k}$ are real, well defined and $r_k \to 0$ as $|k| \to \infty$. As we have seen in Sec. 9-4 the nonnegativity property of the power spectral density can be characterized in terms of certain Toeplitz matrices generated from $r_0, r_1, \ldots, r_n$ in (9-196) and their determinants Δ_n, $n = 0 \to \infty$. Thus the nonnegativity property of

[11] D. C. Youla, "The FEE: A new tunable high-resolution spectral estimator," Part I, Technical note, no. 3, Department of Electrical Engineering, Polytechnic Institute of New York, Brooklyn, New York, 1980: also available as RADC Rep. RADC-TR-81-397, AD A114996, February 1982.

the power spectral density function is equivalent to the nonnegative definiteness of the Toeplitz matrices in (9-196) for every n. In addition, if the power spectral density also satisfies the Paley–Wiener criterion in (9-203), then the nonnegativity property of the power spectral density implies positive definiteness for all T_n, that is, $\Delta_n > 0, n = 0 \to \infty$. Moreover, under these conditions there exists a function $H(z)$, analytic together with its inverse in $|z| < 1$, such that [see (9-203)–(9-207)]

$$S(\omega) = |H(e^{j\omega})|^2 \qquad a.e. \tag{12-133}$$

This minimum phase factor $H(z)$ is also unique up to sign, admits a power series expansion[12]

$$H(z) = \sum_{k=0}^{\infty} h_k z^k \qquad |z| < 1 \tag{12-134}$$

such that $\sum_{k=0}^{\infty} h_k^2 < \infty$.

Given $(n+1)$ partial covariances $r_0, r_1, \ldots, r_n$, from a zero-mean, stationary stochastic process whose power spectral density function satisfies (12-132) and (9-203), the spectral estimation problem is to obtain all possible solutions for the power spectral density that interpolate the given data; that is, such a solution $K(\omega)$ should satisfy

$$K(\omega) \geq 0$$

and

$$\frac{1}{2\pi}\int_{-\pi}^{\pi} K(\omega)e^{-jk\omega}\,d\omega = r_k \qquad |k| = 0 \to n$$

in addition to satisfying (12-132) and (9-203).[13]

To see the general development, it is useful to introduce the notions of positive-real (p.r.) and bounded-real (b.r.) functions.

POSITIVE FUNCTIONS AND BOUNDED FUNCTIONS. A function $Z(z)$ is said to be positive if (*i*) $Z(z)$ is analytic in $|z| < 1$ and (*ii*) Re $Z(z) \geq 0$ in $|z| < 1$. If, in addition, $Z(z)$ is also real for real z, then it is said to be positive-real or (p.r.). Such functions can be shown to be free of poles and zeros in the interior of the unit circle and, moreover, their poles and zeros on the boundary of the unit circle, if any, must be simple with positive residues for the poles.

Similarly a function $\rho(z)$ is said to be bounded if (*i*) it is analytic in $|z| < 1$ and (*ii*) $|\rho(z)| \leq 1$ in $|z| < 1$. If in addition $\rho(z)$ is also real for real z, then it is said to be bounded-real. For example, $e^{-(1-z)}$, z^n, $(1+2z)/(2+z)$ are all bounded-real functions.

---

[12] In this subsection we shall use the variable z rather than z^{-1} so that the region of convergence is the compact region interior to the unit circle (see note in Appendix 12A).

[13] As (9-204) shows, equation (9-203) implies $T_k > 0, k = 0 \to \infty$. However, if the covariances form a singular sequence satisfying $\Delta_r = 0$ for some r, then there exists an $m \leq r$ such that $\Delta_{m-1} > 0$ and $\Delta_m = 0$ and the given covariances have the unique extension $r_k = \sum_{i=1}^{m} P_i e^{jk\omega_i}$, $k \geq 0$, and this corresponds to a line spectra with m discrete components. Here $P_i > 0, 0 \leq \omega_i < 2\pi$, for $i = 1 \to m$ are unique positive constants that can be obtained from the unique eigenvector associated with the zero eigenvalue of $\mathbf{T}_m$. This is Caratheodory's theorem in (12-116) [14], and p. 81 of [25].

Positive-real functions and bounded functions are intimately related. If $Z(z)$ is positive real, then

$$\rho(z) = \frac{Z(z) - R}{Z(z) + R} \tag{12-135}$$

is bounded-real for every $R > 0$, since $1 - |\rho(z)|^2 \geq 0$ in $|z| < 1$ and $\rho(z)$ is analytic in that region. Let

$$Z(z) = r_0 + 2\sum_{k=0}^{\infty} r_k z^k \qquad |z| < 1 \tag{12-136}$$

where $\{r_k\}_{k=0}^{\infty}$ represents the covariance sequence in (12-131). Then since

$$Z(z) = \frac{1}{2\pi}\int_{-\pi}^{\pi} \left(\frac{e^{j\omega} + z}{e^{j\omega} - z}\right) S(\omega)\, d\omega$$

it follows readily that

$$\text{Re}\, \frac{1}{2\pi}\int_{-\pi}^{\pi} \left(\frac{1 - |z|^2}{|e^{j\omega} - z|^2}\right) S(\omega)\, d\omega \geq 0 \tag{12-137}$$

i.e., $Z(z)$ defined by (12-136) is positive-real. Referring to (10-184) and (12-137), it follows that $Z(z)$ given by (12-136) represents a positive-real function iff every hermitian Toeplitz matrix $\mathbf{T}_n$ generated from $r_0, r_1, \ldots, r_n, n = 0 \to \infty$ as in (10-182) is nonnegative definite. It can be shown that, for such functions the interior radial limit

$$Z(e^{j\omega}) = \lim_{\rho \to 1-0} Z(\rho e^{j\omega}) \tag{12-138}$$

exists for almost all ω and hence its real part is nonnegative almost everywhere on the unit circle, that is,

$$K(\omega) \triangleq \text{Re}\, Z(e^{j\omega}) = \sum_{k=-\infty}^{\infty} r_k e^{jk\omega} \geq 0 \qquad 0 < \omega < 2\pi \tag{12-139}$$

where $r_{-k} \triangleq r_k^*$. Since $K(\omega)$ is also uniformly bounded for almost all ω, it is an integrable function, and associated with every positive-real function there exists a power spectrum defined as in (12-139) with finite power. Conversely associated with every power spectrum $S(\omega)$, the function $Z(z)$ defined in (12-136), with r_k's from (12-131), represents a positive-real function. Thus, there exists a one-to-one relationship between positive-real functions and power spectral density functions.

In the rational case, $Z(z)$ is rational, and since every power spectral density can be represented uniquely in terms of its minimum phase factor $H(z)$ as in (12-133), together with (12-139), we have[14]

$$K(\theta) = \text{Re}\, Z(e^{j\omega}) = |H(e^{j\omega})|^2 \tag{12-140}$$

Clearly, $H(z)$ is free of zeros on $|z| = 1$ if and only if $K(\omega)$ is free of zeros in $0 \leq \omega < 2\pi$.

[14]This is also true in the nonrational case under certain restrictions.

Equation (12-140) can be rewritten in a more convenient form by introducing the para-conjugate notation

$$H_*(z) \triangleq H^*(1/z^*) = H(1/z) \tag{12-141}$$

where H^* denotes the complex conjugate of H so that $H_*(e^{j\theta}) = H^*(e^{j\theta})$ and (12-140) translates into the identity

$$\frac{Z(z) + Z_*(z)}{2} = H(z)H_*(z) \tag{12-142}$$

Here $Z(z)$ is free of poles on $|z| = 1$, and for every rational minimum phase system $H(z)$ there exists a unique rational positive-real function $Z(z)$ that is free of poles on $|z| = 1$, and of degree[15] equal to that of $H(z)$. Thus, for every minimum phase rational transfer function, there exists a unique rational positive-real function that is free of poles on $|z| = 1$.

As Youla has shown, by making use of an algorithm due to Schur, every such positive-real function $Z(z)$ has a unique representation in terms of a set of ideal line elements and a *unique* positive-real function. Thus any such $Z(z)$ can be represented as the input impedance of a cascade of equidelay ideal (TEM) lines[16] closed on a unique positive-real function. This alternate representation of $Z(z)$ can be further exploited to identify the correct model order and system parameters in the rational case, and to obtain stable (minimum phase) rational approximations in the nonrational case[17] [25].

Referring back to (12-135), in particular, since $Z(0) = \text{Re } Z(0) > 0$, it follows that

$$\rho_0(z) = \frac{Z(z) - Z(0)}{Z(z) + Z(0)} \tag{12-143}$$

is another bounded-real function. In the rational case, $\rho(z)$ is rational, and for the rational positive-real function $Z(z)$, (12-143) defines a rational bounded-real function. Moreover from (12-143), $\rho_0(z)$ has at least a simple zero at $z = 0$, and hence the real rational function

$$\rho_1(z) = \frac{1}{z}\rho_0(z) = \frac{1}{z}\frac{Z(z) - Z(0)}{Z(z) + Z(0)} \tag{12-144}$$

is analytic in $|z| < 1$. Since $\rho_0(z)$ is also analytic on $z = e^{j\omega}$,

$$|\rho_1(e^{j\omega})| = |\rho_0(e^{j\omega})| \leq 1$$

and from maximum-modulus theorem, $|\rho_1(z)| \leq 1$ in $|z| \leq 1$, i.e., $\rho_1(z)$ is also bounded-real. Further, since $z = 0$ is *not* a pole of $\rho_1(z)$, we have degree of $\rho_1(z)$ given by $\delta[\rho_1(z)] \leq \delta[Z(z)]$ with inequality if and only if the factor $1/z$ cancels a pole of the bounded-real function $\rho_0(z)$ at $z = \infty$. To observe this degree reduction condition ex-

---

[15] The degree of a rational system $\delta[H(z)]$ equals the totality of its poles (or zeros), including those at infinity with multiplicities counted.

[16] In the present context, transverse electromagnetic (TEM) lines can be thought of as pure multiplier/delay elements and z represents the two-way round-trip delay operator common to all lines (see Fig. 12.18 later in the section).

[17] S. U. Pillai, T. I. Shim and D. C. Youla, "A New Technique for ARMA-System Identification and Rational Approximation," *IEEE Trans. on Signal Processing,* vol. **41**, no. 3, pp. 1281–1304, March 1993.

plicitly, write

$$Z(z) = \frac{b_0 + b_1 z + \cdots + b_p z^p}{a_0 + a_1 z + \cdots + a_p a^p}$$

then from (12-144)

$$\rho_1(z) = \frac{1}{z}\rho_0(z) = \frac{(a_0 b_1 - b_0 a_1) + (a_0 b_2 - b_0 a_2)z + \cdots + (a_0 b_p - b_0 a_p)z^{p-1}}{2a_0 b_0 + (a_0 b_1 + b_0 a_1)z + \cdots + (a_0 b_p + b_0 a_p)z^p} \tag{12-145}$$

and $\delta[\rho_1(z)] < \delta[Z(z)] = p$, iff the denominator term in (12-145) is of degree $p - 1$; that is,

$$\delta[\rho_1(z)] = p - 1 \Leftrightarrow a_0 b_p + b_0 a_p = 0$$

or, iff

$$b_0/a_0 = -b_p/a_p$$

But, $b_0/a_0 = Z(0)$ and $b_p/a_p = Z(1/z)|_{z=0} = 0$, and hence degree reduction occurs if and only if

$$Z(z) + Z_*(z) \to 0 \tag{12-146}$$

as $z \to 0$, a result known as Richards's theorem in classical network synthesis theory [33]. Making use of (12-142), this degree reduction condition is

$$\delta(\rho_1(z)) = p - 1 \Leftrightarrow Z(z) + Z_*(z) = 2H(z)H_*(z) \to 0 \quad \text{as} \quad z \to 0 \tag{12-147}$$

that is, for degree reduction to occur, the "even part" of $Z(z)$ must possess a zero at $z = 0$. This condition can be further exploited for rational system identification (see reference in footnote 17 and [25] for details).

Let $Z_1(z)$ represent the positive-real function associated with $\rho_1(z)$ normalized to $Z(0)$. Thus,

$$\rho_1(z) = \frac{Z_1(z) - Z(0)}{Z_1(z) + Z(0)} \tag{12-148}$$

In that case, under the identification $z = e^{-2s\tau}$ ($s = \sigma + j\omega$ represents the complex frequency variable), (12-144) yields the interesting configuration in Fig. 12-17, in which $Z(z)$ is realized as the input impedance of an ideal (TEM) line of "characteristic

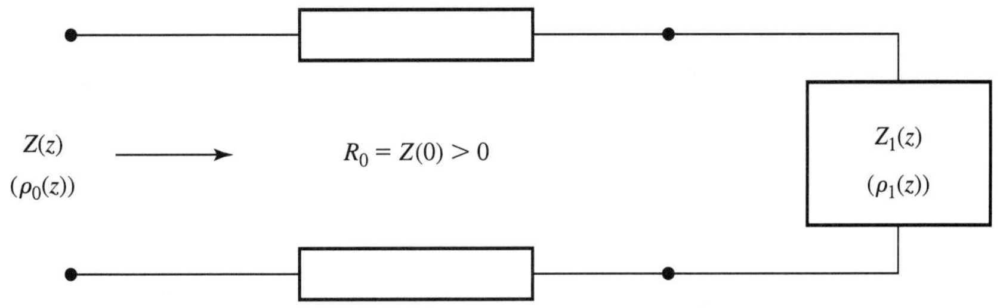

FIGURE 12-17
Positive-real/bounded-real functions and line extraction.

impedance" $R_0 = Z(0)$ and one-way delay $\tau(>0)$, closed on a new positive-real function $Z_1(z)$ given by

$$Z_1(z) = R_0 \frac{1+\rho_1(z)}{1-\rho_1(z)}$$

Within this setup, (12-144) can be rewritten as

$$\frac{Z_1(z) - Z(0)}{Z_1(z) + Z(0)} = \frac{1}{z}\frac{Z(z) - Z(0)}{Z(z) + Z(0)} \tag{12-149}$$

Repeating this procedure using the basic Richards transformation formula (12-149) together with (12-148), more generally at the rth stage, we have

$$\rho_{r+1}(z) \triangleq \frac{Z_{r+1}(z) - Z_r(0)}{Z_{r+1}(z) + Z_r(0)} = \frac{1}{z}\frac{Z_r(z) - Z_r(0)}{Z_r(z) + Z_r(0)} \qquad r \geq 0 \tag{12-150}$$

With $R_r = Z_r(0)$ and letting

$$\rho_{r+1}(0) = \frac{R_{r+1} - R_r}{R_{r+1} + R_r} \triangleq s_{r+1} \qquad r \geq 0 \tag{12-151}$$

represent the $(r+1)$st junction "mismatch" reflection coefficient that is bounded by unity, the single-step update rule (Schur algorithm)

$$\rho_{r+1}(z) = \frac{z\rho_{r+2}(z) + \rho_{r+1}(0)}{1 + z\rho_{r+1}(0)\rho_{r+2}(z)} = \frac{z\rho_{r+2}(z) + s_{r+1}}{1 + zs_{r+1}\rho_{r+2}(z)} \qquad r \geq 0 \tag{12-152}$$

that follows readily from (12-150), by expressing $Z_{r+1}(z)$ in terms of $\rho_{r+1}(z)$ as well as $\rho_{r+2}(z)$, also proves to be extremely useful for model order determination. With this setup, to make progress in the spectrum extension problem, it is best to express the input reflection coefficient $\rho_0(z)$ in (12-143) (normalized to R_0) associated with $Z(z)$ in (12-135), in terms of $\rho_{n+1}(z)$, the reflection coefficient of the termination $Z_{n+1}(z)$ after $n+1$ stages of the Schur algorithm described in (12-150)–(12-152), (see Fig. 12-18). (From (12-150), $\rho_{n+1}(z)$ is normalized with respect to the characteristic impedance R_n of the last line.) Toward this, from (12-152) with $r = 0$ we have

$$\rho_1(z) = \frac{s_1 + z\rho_2(z)}{1 + zs_1\rho_2(z)} \tag{12-153}$$

and (12-144) gives

$$\rho_0(z) = z\rho_1(z) \tag{12-154}$$

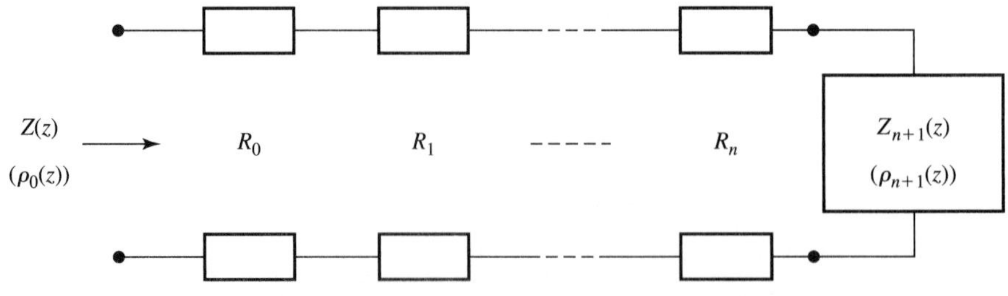

FIGURE 12-18
Cascade representation.

Since $\rho_0(0) = s_0 = 0$, notice that (12-154) in fact has the same form as in (12-152). Thus, for the sake of uniformity, we can rewrite (12-154) as

$$\rho_0(z) = \frac{s_0 + z\rho_1(z)}{1 + zs_0\rho_1(z)} \tag{12-155}$$

and with (12-153) in (12-155), we obtain

$$\rho_0(z) = \frac{(s_0 + zs_1) + z(s_0s_1 + z)\rho_2(z)}{(1 + zs_0s_1) + z(s_1 + zs_0)\rho_2(z)} = \frac{zs_1 + z^2\rho_2(z)}{1 + zs_1\rho_2(z)} \tag{12-156}$$

Continuing this iteration for $(n-2)$ more steps, we have

$$\rho_0(z) = \frac{h_{n-1}(z) + z\tilde{g}_{n-1}(z)\rho_n(z)}{g_{n-1}(z) + z\tilde{h}_{n-1}(z)\rho_n(z)} \tag{12-157}$$

where

$$\tilde{g}_n(z) \triangleq z^n g_{n*}(z) = z^n g_n^*(1/z^*) \tag{12-158}$$

represents the polynomial reciprocal to $g_n(z)$. Once again, updating $\rho_n(z)$ with the help of (12-152), Eq. (12-157) becomes

$$\begin{aligned}\rho_0(z) &= \frac{[h_{n-1}(z) + zs_n\tilde{g}_{n-1}(z)] + z[z\tilde{g}_{n-1}(z) + s_nh_{n-1}(z)]\rho_{n+1}(z)}{[g_{n-1}(z) + zs_n\tilde{h}_{n-1}(z)] + z[z\tilde{h}_{n-1}(z) + s_ng_{n-1}(z)]\rho_{n+1}(z)} \\ &= \frac{h_n(z) + z\tilde{g}_n(z)\,\rho_{n+1}(z)}{g_n(z) + z\tilde{h}_n(z)\,\rho_{n+1}(z)}\end{aligned} \tag{12-159}$$

where, for $n \geq 1$,

$$\begin{aligned}\alpha g_n(z) &\triangleq g_{n-1}(z) + zs_n\tilde{h}_{n-1}(z) \\ \alpha h_n(z) &\triangleq h_{n-1}(z) + zs_n\tilde{g}_{n-1}(z)\end{aligned} \tag{12-160}$$

and α is a suitable normalization constant yet to be determined. From (12-154)–(12-155), the above iterations start with

$$g_0(z) = 1 \qquad \text{and} \qquad h_0(z) = 0 \tag{12-161}$$

Notice that (12-160) gives $\alpha g_1(z) = 1$, $\alpha h_1(z) = zs_1$, $\alpha\tilde{g}_1(z) = z$, $\alpha\tilde{h}_1(z) = s_1$, all of which agree with (12-156). A direct computation also shows

$$\alpha^2[g_n(z)g_{n*}(z) - h_n(z)h_{n*}(z)] = \left(1 - s_n^2\right)[g_{n-1}(z)g_{n-1^*}(z) - h_{n-1}(z)h_{n-1^*}(z)] \tag{12-162}$$

and hence by setting

$$\alpha = \sqrt{1 - s_n^2} \tag{12-163}$$

we have

$$g_n(z)g_{n*}(z) - h_n(z)h_{n*}(z) = 1 \tag{12-164}$$

a relation known as the Feltketter identity[18] in network theory [33].

---

[18] It is easy to see that $h_n(0) = 0$ for every n, and hence we must have $h_n(z) = h_1z + h_2z^2 + \cdots + h_nz^n$. However, for (12-163) to hold, the highest term generated by $g_n(z)g_{n*}(z)$ must cancel with that of $h_n(z)h_{n*}(z)$ and hence $g_n(z)$ is at most of degree $(n-1)$.

As pointed out by Youla, (12-159) can be interpreted as the input reflection coefficient of an $(n+1)$-stage transmission line terminated upon a load $Z_{n+1}(z)$ with reflection coefficient $\rho_{n+1}(z)$ (see Fig. 12-18). Here the input reflection coefficient $\rho_0(z)$ is normalized with respect to the characteristic impedance R_0 of the first line and the termination $Z_{n+1}(z)$ is normalized with respect to the characteristic impedance R_n of the $(n+1)$th line, that is [see (12-150)],

$$\rho_{n+1}(z) = \frac{Z_{n+1}(z) - R_n}{Z_{n+1}(z) + R_n} \tag{12-165}$$

Naturally, $g_n(z)$ and $h_n(z)$ characterize the $(n+1)$-stage transmission line though the single step update rule,

$$\sqrt{1 - s_n^2}\, g_n(z) = g_{n-1}(z) + z s_r \tilde{h}_{n-1}(z) \tag{12-166}$$

and

$$\sqrt{1 - s_n^2}\, h_n(z) = h_{n-1}(z) + z s_n \tilde{g}_{n-1}(z) \tag{12-167}$$

Using Rouche's theorem together with an induction argument on (12-166), with the help of (12-161), it follows that $g_n(z), n = 1, 2, \ldots$ are free of zeros in the closed unit circle (strict Hurwitz Polynomials) provided $|s_k| < 1, k = 1 \to n$. But (12-191) shows $\mathbf{T}_n > 0$ implies $|s_k| < 1, k = 1 \to n$. As a result, (12-164) together with maximum modulus gives $h_n(z)/g_n(z)$ to be bounded-real functions. Moreover, from (12-159) and (12-164), we obtain the key identity

$$\rho_0(z) - \frac{h_n(z)}{g_n(z)} = \frac{z^{n+1}\rho_{n+1}(z)}{g_n^2(z)[1 + z\rho_{n+1}(z)\tilde{h}_n(z)/g_n(z)]} \tag{12-168}$$

Since $|g_n(z)| > 0$ and $|h_n(z)/g_n(z)| < 1$ in $|z| \le 1$, (12-168) implies

$$\rho_0(z) - \frac{h_n(z)}{g_n(z)} = O(z^{n+1})$$

that is, the power series expansions about $z = 0$ of the bounded-real functions $\rho_0(z)$ and $h_n(z)/g_n(z)$ *agree* for the first $(n+1)$ terms. Since $\rho_0(z)$ contains an arbitrary bounded-real function $\rho_{n+1}(z)$, clearly that function *does not affect* the first $(n+1)$ coefficients in the expansion of $\rho_0(z)$. As a result, the first $(n+1)$ coefficients $r_0, r_1, \ldots, r_n$ in the expansion of the corresponding driving-point positive-real function

$$Z(z) = R_0\left(\frac{1 + \rho_0(z)}{1 - \rho_0(z)}\right) = r_0 + 2\sum_{k=1}^{n} r_k z^k + O(z^{n+1}) \tag{12-169}$$

in (12-150) are also *not contaminated* by the choice of $\rho_{n+1}(z)$, since $r_k, k = 0 \to n$ depend only on the first $(n+1)$ coefficients of $\rho_0(z)$.

To understand this from the transmission line point of view, let the terminated structure in Fig. 12-18 be excited by an input current source $i(t) = \delta(t)$. This current impulse launches a voltage impulse $r_0\delta(t)$ into the first line, which reaches the junction separating the first and second line after τ seconds. At this point, part of the incident impulse is reflected back to the driving point and part of it is transmitted toward the next line. In the same way, after τ more seconds, the transmitted part reaches the next junction, where once again it is partly reflected and transmitted. This process continues until the transmitted part reaches the terminated passive load $Z_{n+1}(z)$. Since interaction with this passive load can begin only after the lapse of $(n+1)\tau$ seconds, the input voltage response $v(t)$ over the interval $0 \le t \le 2(n+1)\tau$ is determined entirely by the $(r+1)$ line structure. Moreover, since 2τ represents the two-way round-trip delay common to all lines, under the identification $z = e^{-2s\tau}$, it follows that the first $(n+1)$ coefficients of the driving-point impedance in the expansion (12-169) are determined by the characteristic impedances $R_0, R_1, \ldots, R_n$ of the first $(n+1)$ lines and *not* by the termination $\rho_{n+1}(z)$. Using (12-159) in (12-169), the input impedance is given by

$$Z(z) = R_0 \frac{1+\rho_0(z)}{1-\rho_0(z)} = 2\frac{Q_n(z) + z\rho_{n+1}(z)\tilde{Q}_n(z)}{P_n(z) - z\rho_{n+1}(z)\tilde{P}_n(z)} \tag{12-170}$$

where we define

$$P_n(z) \triangleq \frac{g_n(z) - h_n(z)}{\sqrt{R_0}} \tag{12-171}$$

and

$$Q_n(z) \triangleq \sqrt{R_0}\,\frac{g_n(z) + h_n(z)}{2} \tag{12-172}$$

$P_n(z)$ and $Q_n(z)$ are known as the Levinson polynomials of the first and second kind respectively.[19] Finally, using (12-164), a direct calculation shows

$$P_n(z)Q_{n*}(z) + P_{n*}(z)Q_n(z) \equiv 1 \tag{12-173}$$

and with the help of this (even part condition), "the even part" of $Z(z)$ in (12-170) takes the simple form[20]

$$\frac{Z(z) + Z_*(z)}{2} = \frac{1 - \rho_{n+1}(z)\rho_{n+1*}(z)}{[P_n(z) - z\rho_{n+1}(z)\tilde{P}_n(z)][P_n(z) - z\rho_{n+1}(z)\tilde{P}_n(z)]_*} \tag{12-174}$$

From (12-139), since the real part of every positive-real function on the unit circle corresponds to a power spectral density function, we have[21]

$$K(\omega) = \text{Re}\, Z(e^{j\omega}) = \frac{1 - |\rho_{n+1}(e^{j\omega})|^2}{|D_n(e^{j\omega})|^2} \ge 0 \tag{12-175}$$

where

$$D_n(z) \triangleq P_n(z) - z\rho_{n+1}(z)\tilde{P}_n(z) \qquad n = 0 \to \infty \tag{12-176}$$

Using (12-166) and (12-167) in (12-171), we also obtain the recursion

$$\sqrt{1 - s_n^2}\, P_n(z) = P_{n-1}(z) - z s_n \tilde{P}_{n-1}(z) \qquad n = 1 \to \infty \tag{12-177}$$

that begins under the initialization (use (12-161) and (12-171))

$$P_0(z) = 1/\sqrt{r_0} \tag{12-178}$$

Once again, since $|s_n| < 1$ for all n, arguing as before, it follows that $P_n(z), n = 0, 1, \ldots$ are free of zeros in $|z| \le 1$ and this together with $|\rho_{n+1}(z)| \le 1$ in $|z| \le 1$, enables

---

[19] $P_n(z)$ is the same as $\hat{\mathbf{E}}_n(z)$ in (12-93*a*) and (12-95) except for a normalization constant and the variable z (instead of z^{-1}). Notice that $\tilde{P}_n(z) = z^n P_n^*(1/z^*)$ represents the polynomial reciprocal to $P_n(z)$.

[20] Using (12-170) together with (12-173) we obtain more directly $Z(z) - \frac{2Q_n(z)}{P_n(z)} = O(z^{n+1})$ that confirms (12-169).

[21] Observe that on the unit circle $[Z(z) + Z_*(z)]/2 = [Z(z) + Z^*(z)]/2 = \text{Re}\, Z(z)$.

us to conclude from (12-176) that $D_n(z)$, $n = 0, 1, \ldots$ are analytic and free of zeros in $|z| < 1$. Let $\Gamma_{n+1}(z)$ represent the solution of the equation

$$1 - |\rho_{n+1}(e^{j\omega})|^2 = |\Gamma_{n+1}(e^{j\omega})|^2$$

that is analytic and free of zeros in $|z| < 1$. From (9-203)–(9-207), this factorization is possible iff

$$\int_{-\pi}^{\pi} \ln(1 - |\rho_{n+1}(e^{j\omega})|^2)\, d\omega > -\infty \qquad (12\text{-}179)$$

(Equation (12-179) is automatic if $\rho_{n+1}(z)$ is rational.) In that case, from (12-175)

$$K(\omega) = |H(e^{j\omega})|^2$$

where up to sign

$$H(z) = \frac{\Gamma_{n+1}(z)}{D_r(z)} = \frac{\Gamma_{n+1}(z)}{P_n(z) - z\rho_{n+1}(z)\tilde{P}_n(z)} \qquad (12\text{-}180)$$

represents the Wiener factor associated with $K(\omega)$ in (12-175), since $H(z)$ is analytic together with its inverse in $|z| < 1$.

Clearly $K(\omega)$ is parameterized by the bounded-real function $\rho_{n+1}(z)$ and for *every* b.r. function (rational or nonrational), from (12-169) and (12-175),

$$K(\omega) = \sum_{k=-n}^{n} r_k e^{jk\omega} + \textit{higher order terms} \qquad (12\text{-}181)$$

Thus, (12-175) represents the class of all spectral extensions that interpolate the given autocorrelation sequence $r_0, r_1, \ldots, r_n$. As remarked before, r_k, $|k| = 0 \rightarrow n$, are determined completely from the cascade structure description given by $g_n(z)$ and $h_n(z)$, or as (12-171), (12-175), and (12-176) suggests, entirely from the Levinson polynomials $P_n(z)$. We shall refer to (12-175) and (12-180) as Youla's parametrization formulas for the class of all spectra and the underlying Wiener factors respectively.

The inverse problem of generating $P_n(z)$, given $r_0, r_1, \ldots, r_n$, can be solved by considering the special case $\rho_{n+1}(z) \equiv 0$. From (12-165), this corresponds to the situation where $Z_{n+1}(z) \equiv R_n$, that is, the terminating load is matched to the characteristic impedance of the last line. In that case, from (12-175) and (12-176)

$$K(\omega) = K_0(\omega) = \frac{1}{|P_n(e^{j\omega})|^2} \qquad (12\text{-}182)$$

represents the maximum entropy extension, since the entropy associated with any arbitrary extension in (16-50)–(16-55) is given by

$$\begin{aligned} E_\rho &\triangleq \frac{1}{2\pi}\int_{-\pi}^{\pi} \ln K(\omega)\, d\omega = \ln|H(0)|^2 \\ &= -\ln|P_n(0)|^2 - \ln(1/|\Gamma_{n+1}(0)|^2) \qquad (12\text{-}183) \end{aligned}$$

and it is maximized for $\Gamma_{n+1}(0) = 1$, which corresponds to $\Gamma_{n+1}(z) \equiv 1$ and $\rho_{n+1}(z) \equiv 0$. From (12-170), the corresponding impedance has the form

$$Z_0(z) = \frac{2Q_n(z)}{P_n(z)} \qquad (12\text{-}184)$$

Using (12-173), this gives

$$\frac{Z_0(z) + Z_{0^*}(z)}{2} = \frac{1}{P_n(z)P_{n^*}(z)} \tag{12-185}$$

Let

$$P_n(z) = a_0 + a_1 z + \cdots + a_n z^n \qquad a_0 > 0 \tag{12-186}$$

Then, from (12-169) since

$$Z_0(z) = r_0 + 2\sum_{k=1}^{n} r_k z^k + O(z^{n+1}),$$

(12-184) and (12-185) yield

$$\left(r_0 + \sum_{k=1}^{\infty} r_k z^k + \sum_{k=1}^{\infty} r_{-k} z^{-k} \right) \left(a_n + a_{n-1} z + \cdots + a_0 z^n \right) = \frac{z^n}{P_n(z)} \tag{12-187}$$

Since $P_n(z) \neq 0$ in $|z| \leq 1$, $1/P_n(z)$ admits a power series expansion

$$\frac{1}{P_n(z)} = b_0 + b_1 z + b_2 z^2 + \cdots \qquad b_0 = \frac{1}{a_0} > 0 \tag{12-188}$$

with radius of convergence greater than unity. Thus, by comparing coefficients of both sides of (12-187), we obtain

$$\begin{pmatrix} r_0 & r_1 & \cdots & r_n \\ r_{-1} & r_0 & \cdots & r_{n-1} \\ \vdots & \vdots & \ddots & \vdots \\ r_{-n} & r_{-n+1} & \cdots & r_0 \end{pmatrix} \cdot \begin{pmatrix} a_n \\ a_{n-1} \\ \vdots \\ a_0 \end{pmatrix} = \begin{pmatrix} 0 \\ 0 \\ \vdots \\ b_0 \end{pmatrix} \tag{12-189}$$

However, with $\mathbf{a} = [a_n, a_{n-1}, \ldots, a_1, a_0]^t$, from (12-189) [25],

$$P_n(z) = [z^n, z^{n-1}, \ldots, z, 1]\mathbf{a} = \frac{1}{a_0 \Delta_n} \begin{vmatrix} r_0 & r_1 & \cdots & r_{n-1} & r_n \\ r_{-1} & r_0 & \cdots & r_{n-2} & r_{n-1} \\ \vdots & \vdots & & \vdots & \vdots \\ r_{-n+1} & r_{-n+2} & \cdots & r_0 & r_1 \\ z_n & z^{n-1} & \cdots & z & 1 \end{vmatrix}$$

By applying Cramer's rule on (12-189) to solve for b_0 and with the help of (12-188), we get

$$a_0 = \sqrt{\frac{\Delta_{n-1}}{\Delta_n}}$$

which finally gives the compact expression

$$P_n(z) = \frac{1}{\sqrt{\Delta_n \Delta_{n-1}}} \begin{vmatrix} r_0 & r_1 & \cdots & r_{n-1} & r_n \\ r_{-1} & r_1 & \cdots & r_{n-2} & r_{n-1} \\ \vdots & \vdots & & \vdots & \vdots \\ r_{-n+1} & r_{-n+2} & \cdots & r_0 & r_1 \\ z^n & z^{n-1} & \cdots & z & 1 \end{vmatrix} \tag{12-190}$$

Note that an easy determinantal expansion[22] on (12-190) and Δ_n also gives the useful recursion rule (12-177)–(12-178) together with the new formula

$$\frac{\Delta_n}{\Delta_{n-1}} = \frac{\Delta_{n-1}}{\Delta_{n-2}}(1 - |s_n|^2) \tag{12-191}$$

where the junction reflection coefficients also satisfy

$$s_n = (-1)^{n-1}\frac{\Delta_n^{(1)}}{\Delta_{n-1}} \tag{12-192}$$

Here $\Delta_n^{(1)}$ represents the minor of $\mathbf{T}_n$ obtained after deleting its first column and last row. Clearly from (12-191), $\mathbf{T}_n > 0$ implies $|s_k| < 1$, $k = 1 \to n$. Alternatively, (12-191) can also be rewritten in the more convenient form [25]

$$s_n = \left\{ P_{n-1}(z) \sum_{k=1}^{n} r_k z^k \right\}_n P_{n-1}(0) \qquad n \geq 1 \tag{12-193}$$

where $\{\}_n$ denotes the coefficient of z^n in $\{\}$. Notice that expressions (12-177) and (12-178) together with (12-193) can be easily implemented and constitutes the Levinson recursion algorithm for the strict Hurwitz polynomials $P_k(z)$, $k = 1, 2, \ldots, n$ generated from the autocorrelation sequence $r_0, r_1, \ldots, r_n$ that form a positive definite sequence.

This completes the characterization of the class of all power spectral extensions that interpolate the given autocorrelation sequence.

APPENDIX 12A MINIMUM-PHASE FUNCTIONS

A function

$$\mathbf{H}(z) = \sum_{n=0}^{\infty} h_n z^{-n} \tag{12A-1}$$

is called minimum-phase, if it is analytic and its inverse $1/\mathbf{H}(z)$ is also analytic for $|z| > 1$. We shall show that if $\mathbf{H}(z)$ is minimum-phase, then

$$\ln h_0^2 = \frac{1}{2\pi}\int_{-\pi}^{\pi} \ln|\mathbf{H}(e^{j\omega})|^2\, d\omega \tag{12A-2}$$

Proof. Using the identity $|\mathbf{H}(e^{j\omega})|^2 = \mathbf{H}(e^{i\omega})\mathbf{H}(e^{-j\omega})$, we conclude with $e^{j\varphi} = z$, that

$$\int_{-\pi}^{\pi} \ln|\mathbf{H}(e^{j\omega})|^2\, d\omega = \oint \frac{1}{jz}\ln[\mathbf{H}(z)\mathbf{H}(z^{-1})]\, dz$$

where the path of integration is the unit circle. We note further, changing z to $1/z$,

[22] Let A be an $n \times n$ matrix and Δ_{NW}, Δ_{NE}, Δ_{SW}, and Δ_{SE} denote the $(n-1) \times (n-1)$ minors formed from consecutive rows and consecutive columns in the northwest, northeast, southwest, and southeast corners. Further let Δ_C denote the central $(n-2) \times (n-2)$ minor of A. Then, from a special case of an identity due to Jacobi, $\Delta_C|A| = \Delta_{\mathrm{NW}}\Delta_{\mathrm{SE}} - \Delta_{\mathrm{NE}}\Delta_{\mathrm{SW}}$.

that

$$\oint \frac{1}{z}\ln \mathbf{H}(z)\,dz = \oint \frac{1}{z}\ln \mathbf{H}(z^{-1})\,dz$$

To prove (12A-2), it suffices, therefore, to show that

$$\ln|h_0| = \frac{1}{2\pi j}\oint \frac{1}{z}\ln \mathbf{H}(z)\,dz$$

This follows readily because $\mathbf{H}(z)$ tends to h_0 as $z \to \infty$ and the function $\ln \mathbf{H}(z)$ is analytic for $|z| > 1$ by assumption.

Note If z^{-1} is replaced by z in (12A-1), the region of convergence becomes $|z| < 1$ [see (12-134)].

APPENDIX 12B ALL-PASS FUNCTIONS

The unit circle is the locus of points N such that (see Fig. 12-19a)

$$\frac{(NA)}{(NB)} = \frac{|e^{j\omega} - 1/z_i^*|}{|e^{j\omega} - z_i|} = \frac{1}{|z_i|} \qquad |z_i| < 1$$

From this it follows that, if

$$\mathbf{F}(z) = \frac{zz_i^* - 1}{z - z_i} \qquad |z_i| < 1$$

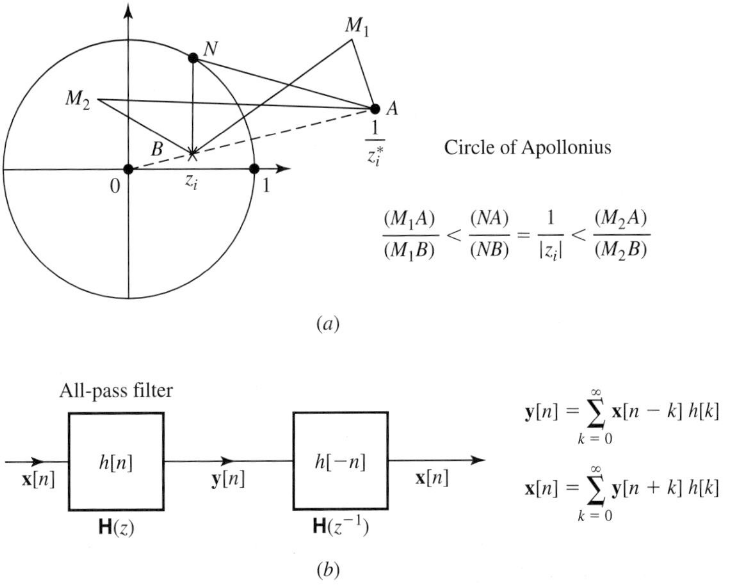

FIGURE 12-19

then $|\mathbf{F}(e^{j\omega})| = 1$. Furthermore, $|\mathbf{F}(z)| > 1$ for $|z| < 1$ and $|\mathbf{F}(z)| < 1$ for $|z| > 1$ because $\mathbf{F}(z)$ is continuous and

$$|\mathbf{F}(0)| = \frac{1}{|z_i|} > 1 \qquad |\mathbf{F}(\infty)| = |z_i^*| < 1$$

Multiplying N bilinear fractions of this form, we conclude that, if

$$\mathbf{H}(z) = \prod_{i=1}^{N} \frac{zz_i^* - 1}{z - z_i} \qquad |z_i| < 1 \tag{12B-1}$$

then

$$|\mathbf{H}(z)| \begin{cases} > 1 & |z| < 1 \\ = 1 & |z| = 1 \\ < 1 & |z| > 1 \end{cases} \tag{12B-2}$$

A system with system function $\mathbf{H}(z)$ as in (12B-1) is called *all-pass*. Thus an all-pass system is stable, causal, and

$$|\mathbf{H}(e^{j\omega T})| = 1$$

Furthermore,

$$\frac{1}{\mathbf{H}(z)} = \prod_{i=1}^{N} \frac{z - z_i}{zz_i^* - 1} = \prod_{i=1}^{N} \frac{1 - z_i/z}{z_i^* - 1/z} = \mathbf{H}\left(\frac{1}{z}\right) \tag{12B-3}$$

because if z_i is a pole of $\mathbf{H}(z)$, then z_i^* is also a pole.

From this it follows that if $h[n]$ is the delta response of an all-pass system, then the delta response of its inverse is $h[-n]$:

$$\mathbf{H}(z) = \sum_{n=0}^{\infty} h[n]z^{-n} \qquad \frac{1}{\mathbf{H}(z)} = \sum_{n=0}^{\infty} h[n]z^{n} \tag{12B-4}$$

where both series converge in a ring containing the unit circle.

PROBLEMS

12-1 Find the mean and variance of the random variable

$$\mathbf{n}_T = \frac{1}{2T}\int_{-T}^{T} \mathbf{x}(t)\,dt \qquad \text{where} \quad \mathbf{x}(t) = 10 + \boldsymbol{\nu}(t)$$

for $T = 5$ and for $T = 100$. Assume that $E\{\boldsymbol{\nu}(t)\} = 0$, $R_\nu(\tau) = 2\delta(\tau)$.

12-2 Show that if a process is normal and distribution-ergodic as in (12-35), then it is also mean-ergodic.

12-3 Show that if $\mathbf{x}(t)$ is normal with $\eta_x = 0$ and $R_x(\tau) = 0$ for $|\tau| > a$, then it is correlation-ergodic.

12-4 Show that the process $\mathbf{a}e^{j(\omega t + \boldsymbol{\varphi})}$ is not correlation-ergodic.

12-5 Show that

$$R_{xy}(\lambda) = \lim_{T\to\infty} \frac{1}{2T}\int_{-T}^{T} \mathbf{x}(t+\lambda)\mathbf{y}(t)\,dt$$

iff

$$\lim_{T\to\infty} \frac{1}{2T}\int_{-2T}^{2T} \left(1-\frac{|\tau|}{2T}\right) E\{\mathbf{x}(t+\lambda+\tau)\mathbf{y}(t+\tau)\mathbf{x}(t+\lambda)\mathbf{y}(t)\}\,d\tau = R_{xy}^2(\lambda)$$

12-6 The process $\mathbf{x}(t)$ is cyclostationary with period T, mean $\eta(t)$, and correlation $R(t_1, t_2)$. Show that if $R(t+\tau, t) \to \eta^2(t)$ as $|\tau| \to \infty$, then

$$\lim_{c\to\infty} \frac{1}{2c}\int_{-c}^{c} \mathbf{x}(t)\,dt = \frac{1}{T}\int_{0}^{T} \eta(t)\,dt$$

Hint: The process $\overline{\mathbf{x}}(t) = \mathbf{x}(t-\boldsymbol{\theta})$ is mean-ergodic.

12-7 Show that if

$$C(t+\tau, t) \xrightarrow[t\to\infty]{} 0$$

uniformly in t; then $\mathbf{x}(t)$ is mean-ergodic.

12-8 The process $\mathbf{x}(t)$ is normal with 0 mean and WSS. (*a*) Show that (Fig. P12-8*a*)

$$E\{\mathbf{x}(t+\lambda) \mid \mathbf{x}(t) = x\} = \frac{R(\lambda)}{R(0)}x$$

(*b*) Show that if D is an arbitrary set of real numbers x_i and $\overline{x} = E\{\mathbf{x}(t) \mid \mathbf{x}(t) \in D\}$, then (Fig. P12-8b)

$$E\{\mathbf{x}(t+\lambda) \mid \mathbf{x}(t) \in D\} = \frac{R(\lambda)}{R(0)}\overline{x}$$

(*c*) Using this, design an analog correlometer for normal processes.

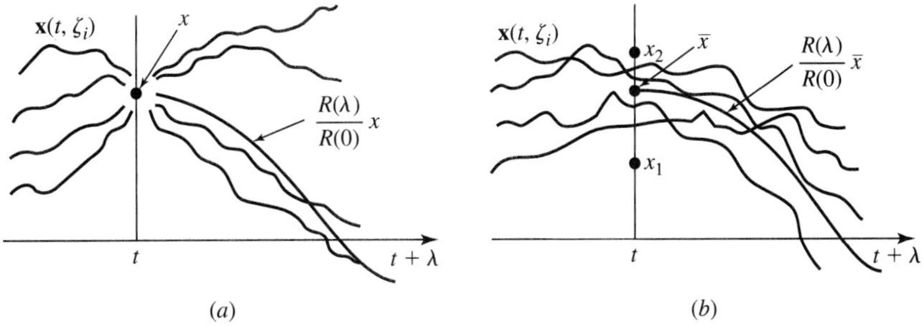

FIGURE P12-8

12-9 The processes $\mathbf{x}(t)$ and $\mathbf{y}(t)$ are jointly normal with zero mean. Show that: (*a*) If $\mathbf{w}(t) = \mathbf{x}(t+\lambda)\mathbf{y}(t)$, then

$$C_{ww}(\tau) = C_{xy}(\lambda+\tau)C_{xy}(\lambda-\tau) + C_{xx}(\tau)C_{yy}(\tau)$$

(*b*) If the functions $C_{xx}(\tau)$, $C_{yy}(\tau)$, and $C_{xy}(\tau)$ tend to 0 as $\tau \to \infty$ then the processes $\mathbf{x}(t)$ and $\mathbf{y}(t)$ are cross-variance ergodic.

12-10 Using Schwarz's inequality (10B-1), show that

$$\left|\int_a^b f(x)\,dx\right|^2 \le (b-a)\int_a^b |f(x)|^2\,dx$$

12-11 We wish to estimate the mean η of a process $\mathbf{x}(t) = \eta + \boldsymbol{\nu}(t)$, where $R_{\nu\nu}(\tau) = 5\delta(\tau)$. (*a*) Using (5-88), find the 0.95 confidence interval of η. (*b*) Improve the estimate if $\boldsymbol{\nu}(t)$ is a normal process.

12-12 (*a*) Show that if we use as estimate of the power spectrum $S(\omega)$ of a discrete-time process $\mathbf{x}[n]$ the function

$$S_w(\omega) = \sum_{m=-N}^{N} w_m R[m] e^{-jm\omega T}$$

then

$$S_w(\omega) = \frac{1}{2\sigma}\int_{-\sigma}^{\sigma} S(y)W(\omega - y)\,dy \qquad W(\omega) = \sum_{-N}^{N} w_n e^{-jn\omega T}$$

(*b*) Find $W(\omega)$ if $N = 10$ and $w_n = 1 - |n|/11$.

12-13 Show that if $\mathbf{x}(t)$ is zero-mean normal process with sample spectrum

$$\mathbf{S}_T(\omega) = \frac{1}{2T}\left|\int_{-T}^{T} \mathbf{x}(t)e^{-j\omega t}\,dt\right|^2$$

and $S(\omega)$ is sufficiently smooth, then

$$E^2\{\mathbf{S}_T(\omega)\} \le \operatorname{Var} \mathbf{S}_T(\omega) \le 2E^2\{\mathbf{S}_T(\omega)\}$$

The right side is an equality if $\omega = 0$. The left side is an approximate equality if $T \gg 1/\omega$. *Hint:* Use (11-74).

12-14 Show that the weighted sample spectrum

$$\mathbf{S}_c(\omega) = \frac{1}{2T}\left|\int_{-T}^{T} c(t)\mathbf{x}(t)e^{-j\omega t}\,dt\right|^2$$

of a process $\mathbf{x}(t)$ is the Fourier transform of the function

$$\mathbf{R}_c(\tau) = \frac{1}{2T}\int_{-T+|\tau|/2}^{T-|\tau|/2} c\left(t+\frac{\tau}{2}\right)c\left(t-\frac{\tau}{2}\right)\mathbf{x}\left(t+\frac{\tau}{2}\right)\mathbf{x}\left(t-\frac{\tau}{2}\right)dt$$

12-15 Given a normal process $\mathbf{x}(t)$ with zero mean and power spectrum $S(\omega)$, we form its sample autocorrelation $\mathbf{R}_T(\tau)$ as in (12-38). Show that for large T,

$$\operatorname{Var} \mathbf{R}_T(\lambda) \simeq \frac{1}{4\pi T}\int_{-\infty}^{\infty} (1 + e^{j2\lambda\omega})S^2(\omega)\,d\omega$$

12-16 Show that if

$$\mathbf{R}_T(\tau) = \frac{1}{2T}\int_{-T+|\tau|/2}^{T-|\tau|/2} \mathbf{x}\left(t+\frac{\tau}{2}\right)\mathbf{x}\left(t-\frac{\tau}{2}\right)dt$$

is the estimate of the autocorrelation $R(\tau)$ of a zero-mean normal process, then

$$\sigma_{R_T}^2 = \frac{1}{2T}\int_{-2T+|\tau|}^{2T-|\tau|} [R^2(\alpha) + R(\alpha+\tau)R(\alpha-\tau)]\left(1 - \frac{|\tau| + |\alpha|}{2T}\right)d\alpha$$

12-17 Show that in Levinson's algorithm,

$$a_k^{N-1} = \frac{a_k^N + K_N a_{N-K}^N}{1 - K_N^2}$$

12-18 Show that if $R[0] = 8$ and $R[1] = 4$, then the MEM estimate of $S(\omega)$ equals

$$S_{\text{MEM}}(\omega) = \frac{6}{|1 - 0.5e^{-j\omega}|^2}$$

12-19 Find the maximum entropy estimate $S_{\text{MEM}}(\omega)$ and the line-spectral estimate (12-111) of a process $\mathbf{x}[n]$ if

$$R[0] = 13 \qquad R[1] = 5 \qquad R[2] = 2$$

12-20 Let $P_n(z)$ represent the Levinson polynomial of the first kind in (12-171). (a) If one of the roots of $P_n(z)$ lie on the unit circle, then show that all other roots of $P_n(z)$ are simple and lie on the unit circle. (b) If the reflection coefficient $s_k \neq 0$, then show that $P_k(z)$ and $P_{k+1}(z)$ have no roots in common.

12-21 If the reflection coefficients satisfy $s_k = \rho^k$, $k = 1 \to \infty$, where $|\rho| < 1$, then show that all zeros of the Levinson polynomial $P_n(z)$, $n = 1 \to \infty$ lie on the circle of radius $1/\rho$.

12-22 Let $P_n(z)$, $n = 0 \to \infty$, represent the Levinson polynomials associated with the reflection coefficients $\{s_k\}_{k=1}^{\infty}$. Define

$$s_k' = \lambda^k s_k, \qquad |\lambda| = 1, \qquad k = 1 \to \infty.$$

Show that the new set of Levinson polynomials are given by $P_n(\lambda z)$, $n = 0 \to \infty$. Thus if $\{s_k\}_{k=1}^{\infty}$ is replaced by $\{(-1)^k s_k\}_{k=1}^{\infty}$, the new set of Levinson polynomials are given by $P_n(-z)$, $n = 0 \to \infty$.

12-23 Consider an MA(1) process with transfer function

$$H(z) = 1 - z.$$

(a) Show that $\Delta_k = k + 2,\ k \geq 0$, and

$$s_k = -\frac{1}{k+1}, \qquad k = 1 \to \infty.$$

(b) Consider the new process with reflection coefficients

$$s_k' = -s_k = \frac{1}{k+1}, \qquad k = 1 \to \infty,$$

and $r_0' = r_0 = 2$. Clearly $\sum_{k=1}^{\infty} |s_k'|^2 < \infty$. Show that the new auto-correlation sequence r_k' is given by

$$r_0' = 2, \qquad r_k' = 1, \qquad k \geq 1,$$

and hence $S(\omega)$ is not integrable since $r_k' \nrightarrow 0$ as $k \to \infty$.

CHAPTER 13

MEAN SQUARE ESTIMATION

13-1 INTRODUCTION[1,2]

In this chapter, we consider the problem of estimating the value of a stochastic process $\mathbf{s}(t)$ at a specific time in terms of the values (data) of another process $\mathbf{x}(\xi)$ specified for every ξ in an interval $a \le \xi \le b$ of finite or infinite length. In the digital case, the solution of this problem is a direct application of the orthogonality principle (see Sec. 7-4). In the analog case, the linear estimator $\hat{\mathbf{s}}(t)$ of $\mathbf{s}(t)$ is not a sum. It is an integral

$$\hat{\mathbf{s}}(t) \triangleq \hat{E}\{\mathbf{s}(t) \mid \mathbf{x}(\xi), a \le \xi \le b\} = \int_a^b h(\alpha)\mathbf{x}(\alpha)\, d\alpha \qquad (13\text{-}1)$$

and our objective is to find $h(\alpha)$ so as to minimize the MS error

$$P = E\{[\mathbf{s}(t) - \hat{\mathbf{s}}(t)]^2\} = E\left\{\left[\mathbf{s}(t) - \int_a^b h(\alpha)\mathbf{x}(\alpha)\, d\alpha\right]^2\right\} \qquad (13\text{-}2)$$

The function $h(\alpha)$ involves a noncountable number of unknowns, namely, its values for every α in the interval (a, b). To determine $h(\alpha)$, we shall use the following extension

[1] N. Wiener: *Extrapolation, Interpolation, and Smoothing of Stationary Time series,* MIT Press, 1950; J. Makhoul: "Linear Prediction: A Tutorial Review," *Proceedings of the IEEE,* vol. 63, 1975.

[2] T. Kailath: "A View of Three Decades of Linear Filtering Theory," *IEEE Transactions Information Theory,* vol. IT-20, 1974.

of the orthogonality principle:

THEOREM 13-1 The MS error P of the estimation of a process $\mathbf{s}(t)$ by the integral in (13-1) is minimum if the data $\mathbf{x}(\xi)$ are orthogonal to the error $\mathbf{s}(t) - \hat{\mathbf{s}}(t)$:

$$E\left\{\left[\mathbf{s}(t) - \int_a^b h(\alpha)\mathbf{x}(\alpha)\,d\alpha\right]\mathbf{x}(\xi)\right\} = 0 \qquad a \le \xi \le b \tag{13-3}$$

or, equivalently, if $h(\alpha)$ is the solution of the integral equation

$$R_{sx}(t, \xi) = \int_a^b h(\alpha) R_{xx}(\alpha, \xi)\,d\alpha \qquad a \le \xi \le b \tag{13-4}$$

Proof. We shall give a formal proof based on the approximation of the integral in (13-1) by its Riemann sum. Dividing the interval (a, b) into m segments $(\alpha_k, \alpha_k + \Delta\alpha)$, we obtain

$$\hat{\mathbf{s}}(t) \simeq \sum_{k=1}^{m} h(\alpha_k)\mathbf{x}(\alpha_k)\,\Delta\alpha \qquad \Delta\alpha = \frac{b-a}{m}$$

Applying (7-82) with $a_k = h(\alpha_k)\,\Delta\alpha$, we conclude that the resulting MS error P is minimum if

$$E\left\{\left[\mathbf{s}(t) - \sum_{k=1}^{m} h(\alpha_k)\mathbf{x}(\alpha_k)\Delta\alpha\right]\mathbf{x}(\xi_j)\right\} = 0 \qquad 1 \le j \le m$$

where ξ_j is a point in the interval $(\alpha_j, \alpha_j + \Delta\alpha)$. This yields the system

$$R_{sx}(t, \xi_j) = \sum_{k=1}^{m} h(\alpha_k) R_{xx}(\alpha_k, \xi_j)\,\Delta\alpha \qquad j = 1, \ldots, m \tag{13-5}$$

The integral equation (13-4) is the limit of (13-5) as $\Delta\alpha \to 0$.

From (7-85) it follows that the LMS error of the estimation of $\mathbf{s}(t)$ by the integral in (13-1) equals

$$P = E\left\{\left[\mathbf{s}(t) - \int_a^b h(\alpha)\mathbf{x}(\alpha)\,d\alpha\right]\mathbf{s}(t)\right\} = R_{ss}(0) - \int_a^b h(\alpha) R_{sx}(t, \alpha)\,d\alpha \tag{13-6}$$

In general, the integral equation (13-4) can only be solved numerically. In fact, if we assign to the variable ξ the values ξ_j and we approximate the integral by a sum, we obtain the system (13-5). In this chapter, we consider various special cases that lead to explicit solutions. Unless stated otherwise, it will be assumed that all processes are WSS and real.

We shall use the following terminology:

If the time t in (13-1) is in the interior of the data interval (a, b), then the estimate $\hat{\mathbf{s}}(t)$ of $\mathbf{s}(t)$ will be called *smoothing*.

If t is outside this interval and $\mathbf{x}(t) = \mathbf{s}(t)$ (no noise), then $\hat{\mathbf{s}}(t)$ is a *predictor* of $\mathbf{s}(t)$. If $t > b$, then $\hat{\mathbf{s}}(t)$ is a "forward predictor"; if $t < a$, it is a "backward predictor."

If t is outside the data interval and $\mathbf{x}(t) \neq \mathbf{s}(t)$, then the estimate is called *filtering and prediction*.

Simple Illustrations

In this section, we present a number of simple estimation problems involving a finite number of data and we conclude with the smoothing problem when the data $\mathbf{x}(\xi)$ are available from $(-\infty, \infty)$. In this case, the solution of the integral equation (13-4) is readily obtained in terms of Fourier transforms.

Prediction. We wish to estimate the future value $\mathbf{s}(t+\lambda)$ of a stationary process $\mathbf{s}(t)$ in terms of its present value

$$\hat{\mathbf{s}}(t+\lambda) = \hat{E}\{\mathbf{s}(t+\lambda) \mid \mathbf{s}(t)\} = a\mathbf{s}(t)$$

From (7-71) and (7-72) it follows with $n = 1$ that

$$E\{[\mathbf{s}(t+\lambda) - a\mathbf{s}(t)]\mathbf{s}(t)\} = 0 \qquad a = \frac{R(\lambda)}{R(0)}$$
$$P = E\{[\mathbf{s}(t+\lambda) - a\mathbf{s}(t)]\mathbf{s}(t+\lambda)\} = R(0) - aR(\lambda)$$

Special case If

$$R(\tau) = Ae^{-\alpha|\tau|} \qquad \text{then} \quad a = e^{-\alpha\lambda}$$

In this case, the difference $\mathbf{s}(t+\lambda) - a\mathbf{s}(t)$ is orthogonal to $\mathbf{s}(t-\xi)$ for every $\xi \ge 0$:

$$E\{[\mathbf{s}(t+\lambda) - a\mathbf{s}(t)]\mathbf{s}(t-\xi)\} = R(\lambda+\xi) - aR(\xi)$$
$$= Ae^{-\alpha(\lambda+\xi)} - Ae^{-\alpha\lambda}e^{-\alpha\xi} = 0$$

This shows that $a\mathbf{s}(t)$ is the estimate of $\mathbf{s}(t+\lambda)$ in terms of its entire past. Such a process is called *wide-sense Markov* of order 1.

We shall now find the estimate of $\mathbf{s}(t+\lambda)$ in terms of $\mathbf{s}(t)$ and $\mathbf{s}'(t)$:

$$\hat{\mathbf{s}}(t+\lambda) = a_1\mathbf{s}(t) + a_2\mathbf{s}'(t)$$

The orthogonality condition (7-82) yields

$$\mathbf{s}(t+\lambda) - \hat{\mathbf{s}}(t+\lambda) \perp \mathbf{s}(t), \mathbf{s}'(t)$$

Using the identities

$$R'(0) = 0 \qquad R_{ss'}(\tau) = -R'(\tau) \qquad R_{s's'}(\tau) = -R''(\tau)$$

we obtain

$$a_1 = R(\lambda)/R(0) \qquad a_2 = R'(\lambda)/R''(0)$$
$$P = E\{[\mathbf{s}(t+\lambda) - a_1\mathbf{s}(t) - a_2\mathbf{s}'(t)]\mathbf{s}(t+\lambda)\} = R(0) - a_1R(\lambda) + a_2R'(\lambda)$$

If λ is small, then

$$R(\lambda) \simeq R(0) \qquad R'(\lambda) \simeq R'(0) + R''(0)\lambda \simeq R''(0)\lambda$$
$$a_1 \simeq 1 \qquad a_2 \simeq \lambda \qquad \hat{\mathbf{s}}(t+\lambda) \simeq \mathbf{s}(t) + \lambda\mathbf{s}'(t)$$

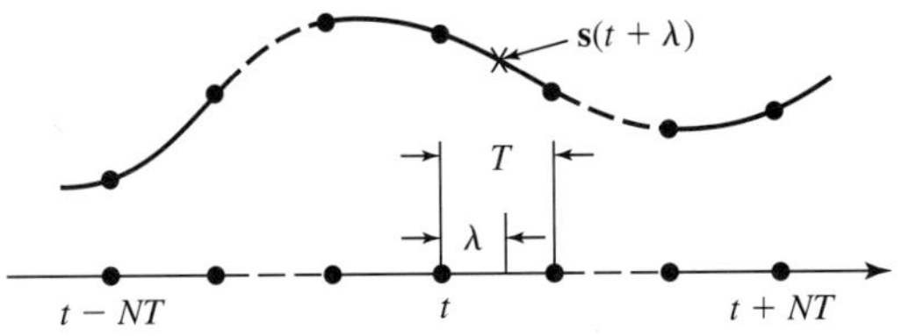

FIGURE 13-1

Filtering. We shall estimate the present value of a process $\mathbf{s}(t)$ in terms of the present value of another process $\mathbf{x}(t)$:

$$\hat{\mathbf{s}}(t) = \hat{E}\{\mathbf{s}(t) \mid \mathbf{x}(t)\} = a\mathbf{x}(t)$$

From (7-71) and (7-72) it follows that

$$E\{[\mathbf{s}(t) - a\mathbf{x}(t)]\mathbf{x}(t)\} = 0 \qquad a = R_{sx}(0)/R_{xx}(0)$$
$$P = E\{[\mathbf{s}(t) - a\mathbf{x}(t)]\mathbf{s}(t)\} = R_{ss}(0) - aR_{sx}(0)$$

Interpolation. We wish to estimate the value $\mathbf{s}(t+\lambda)$ of a process $\mathbf{s}(t)$ at a point $t+\lambda$ in the interval $(t, t+T)$, in terms of its $2N+1$ samples $\mathbf{s}(t+kT)$ that are nearest to t (Fig. 13-1)

$$\hat{\mathbf{s}}(t+\lambda) = \sum_{k=-N}^{N} a_k \mathbf{s}(t+kT) \qquad 0 < \lambda < T \tag{13-7}$$

The orthogonality principle now yields

$$E\left\{\left[\mathbf{s}(t+\lambda) - \sum_{k=-N}^{N} a_k \mathbf{s}(t+kT)\right] \mathbf{s}(t+nT)\right\} = 0 \qquad |n| \le N$$

from which it follows that

$$\sum_{k=-N}^{N} a_k R(kT - nT) = R(\lambda - nT) \qquad -N \le n \le N \tag{13-8}$$

This is a system of $2N+1$ equations and its solution yields the $2N+1$ unknowns a_k. The MS value P of the estimation error

$$\boldsymbol{\varepsilon}_N(t) = \mathbf{s}(t+\lambda) - \sum_{k=-N}^{N} a_k \mathbf{s}(t+kT) \tag{13-9}$$

equals

$$P = E\{\boldsymbol{\varepsilon}_N(t)\mathbf{s}(t+\lambda)\} = R(0) - \sum_{k=-N}^{N} a_k R(\lambda - kT) \tag{13-10}$$

Interpolation as deterministic approximation The error $\boldsymbol{\varepsilon}_N(t)$ can be considered as the output of the system

$$E_N(\omega) = e^{j\omega\lambda} - \sum_{k=-N}^{N} a_k e^{jkT\omega}$$

(error filter) with input $\mathbf{s}(t)$. Denoting by $S(\omega)$ the power spectrum of $\mathbf{s}(t)$, we conclude from (9-152) that

$$P = E\{\boldsymbol{\varepsilon}_N^2(t)\} = \frac{1}{2\pi}\int_{-\infty}^{\infty} S(\omega)\left|e^{j\omega\lambda} - \sum_{k=-N}^{N} a_k e^{jkT\omega}\right|^2 d\omega \qquad \text{(13-11)}$$

This shows that the minimization of P is equivalent to the deterministic problem of minimizing the weighted mean square error of the approximation of the exponential $e^{j\omega\lambda}$ by a trigonometric polynomial (truncated Fourier series).

Quadrature We shall estimate the integral

$$\mathbf{z} = \int_0^b \mathbf{s}(t)\,dt$$

of a process $\mathbf{s}(t)$ in terms of its $N+1$ samples $\mathbf{s}(nT)$:

$$\hat{\mathbf{z}} = a_0\mathbf{s}(0) + a_1\mathbf{s}(T) + \cdots + a_N\mathbf{s}(NT) \qquad T = \frac{b}{N}$$

Applying (7-82), we obtain

$$E\left\{\left[\int_0^b \mathbf{s}(t)\,dt - \hat{\mathbf{z}}\right]\mathbf{s}(kT)\right\} = 0 \qquad 0 \le k \le N$$

Hence

$$\int_0^b R(t-kT)\,dt = a_0R(kT) + \cdots + a_NR(kT-NT) \qquad 0 \le k \le N$$

This is a system of $N+1$ equations and its solution yields the coefficients a_k.

Smoothing

We wish to estimate the present value of a process $\mathbf{s}(t)$ in terms of the values $\mathbf{x}(\xi)$ of the sum

$$\mathbf{x}(t) = \mathbf{s}(t) + \boldsymbol{\nu}(t)$$

available for every ξ from $-\infty$ to ∞. The desirable estimate

$$\hat{\mathbf{s}}(t) = \hat{E}\{\mathbf{s}(t) \mid \mathbf{x}(\xi), -\infty < \xi < \infty\}$$

will be written in the form

$$\hat{\mathbf{s}}(t) = \int_{-\infty}^{\infty} h(\alpha)\mathbf{x}(t-\alpha)\,d\alpha \qquad \text{(13-12)}$$

In this notation, $h(\alpha)$ is independent of t and $\hat{\mathbf{s}}(t)$ can be considered as the output of a linear time-invariant noncausal system with input $\mathbf{x}(t)$ and impulse response $h(t)$. Our problem is to find $h(t)$.

Clearly,

$$\mathbf{s}(t) - \hat{\mathbf{s}}(t) \perp \mathbf{x}(\xi) \qquad \text{all } \xi$$

Setting $\xi = t - \tau$, we obtain

$$E\left\{\left[\mathbf{s}(t) - \int_{-\infty}^{\infty} h(\alpha)\mathbf{x}(t-\alpha)\,d\alpha\right]\mathbf{x}(t-\tau)\right\} = 0 \qquad \text{all} \quad \tau$$

This yields

$$R_{sx}(\tau) = \int_{-\infty}^{\infty} h(\alpha)R_{xx}(\tau-\alpha)\,d\alpha \qquad \text{all} \quad \tau \tag{13-13}$$

Thus, to determine $h(t)$, we must solve the above integral equation. This equation can be solved easily because it holds for all τ and the integral is a convolution of $h(\tau)$ with $R_{xx}(\tau)$. Taking transforms of both sides, we obtain $S_{sx}(\omega) = H(\omega)S_{xx}(\omega)$. Hence

$$H(\omega) = \frac{S_{sx}(\omega)}{S_{xx}(\omega)} \tag{13-14}$$

The resulting system is called the *noncausal Wiener filter*.

The MS estimation error P equals

$$\begin{aligned} P &= E\left\{\left[\mathbf{s}(t) - \int_{-\infty}^{\infty} h(\alpha)\mathbf{x}(t-\alpha)\,d\alpha\right]\mathbf{s}(t)\right\} \\ &= R_{ss}(0) - \int_{-\infty}^{\infty} h(\alpha)R_{sx}(\alpha)\,d\alpha \\ &= \frac{1}{2\pi}\int_{-\infty}^{\infty} [S_{ss}(\omega) - H^*(\omega)S_{sx}(\omega)]\,d\omega \end{aligned} \tag{13-15}$$

If the signal $\mathbf{s}(t)$ and the noise $\boldsymbol{\nu}(t)$ are *orthogonal,* then

$$S_{sx}(\omega) = S_{ss}(\omega) \qquad S_{xx}(\omega) = S_{ss}(\omega) + S_{\nu\nu}(\omega)$$

Hence (Fig. 13-2)

$$H(\omega) = \frac{S_{ss}(\omega)}{S_{ss}(\omega) + S_{\nu\nu}(\omega)} \qquad P = \frac{1}{2\pi}\int_{-\infty}^{\infty} \frac{S_{ss}(\omega)S_{\nu\nu}(\omega)}{S_{ss}(\omega) + S_{\nu\nu}(\omega)}\,d\omega \tag{13-16}$$

If the spectra $S_{ss}(\omega)$ and $S_{\nu\nu}(\omega)$ do not overlap, then $H(\omega) = 1$ in the band of the signal and $H(\omega) = 0$ in the band of the noise. In this case, $P = 0$.

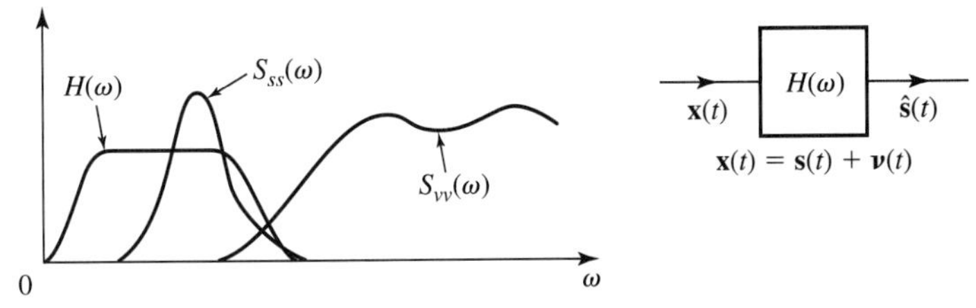

FIGURE 13-2

EXAMPLE 13-1 ▶ If

$$S_{ss}(\omega) = \frac{N_0}{\alpha^2 + \omega^2} \qquad S_{\nu\nu}(\omega) = N \qquad S_{s\nu}(\omega) = 0$$

then (13-16) yields

$$H(\omega) = \frac{N_0}{N_0 + N(\alpha^2 + \omega^2)} \qquad h(t) = \frac{N_0}{2\beta N} e^{-\beta|t|}$$

$$P = \frac{1}{2\pi} \int_{-\infty}^{\infty} \frac{N_0}{\beta^2 + \omega^2}\, d\omega = \frac{N_0}{2\beta} \qquad \beta^2 = \alpha^2 + \frac{N_0}{N}$$ ◀

DISCRETE-TIME PROCESSES. The noncausal estimate $\hat{\mathbf{s}}[n]$ of a discrete-time process in terms of the data

$$\mathbf{x}[n] = \mathbf{s}[n] + \boldsymbol{\nu}[n]$$

is the output

$$\hat{\mathbf{s}}[n] = \sum_{k=-\infty}^{\infty} h[k]\mathbf{x}[n-k]$$

of a linear time-invariant noncausal system with input $\mathbf{x}[n]$ and delta response $h[n]$. The orthogonality principle yields

$$E\left\{\left(\mathbf{s}[n] - \sum_{k=-\infty}^{\infty} h[k]\mathbf{x}[n-k]\right)\mathbf{x}[n-m]\right\} = 0 \qquad \text{all} \quad m$$

Hence

$$R_{sx}[m] = \sum_{k=-\infty}^{\infty} h[k]R_{xx}[m-k] \qquad \text{all} \quad m \tag{13-17}$$

Taking transforms of both sides, we obtain

$$\mathbf{H}(z) = \frac{\mathbf{S}_{sx}(z)}{\mathbf{S}_{xx}(z)} \tag{13-18}$$

The resulting MS error equals

$$P = E\left\{\left[\mathbf{s}[n] - \sum_{k=-\infty}^{\infty} h[k]\mathbf{x}[n-k]\right]\mathbf{s}[n]\right\}$$

$$= R_{ss}(0) - \sum_{k=-\infty}^{\infty} h[k]R_{sx}[k] = \frac{1}{2\pi}\int_{-\pi}^{\pi} [S_{ss}(\omega) - \mathbf{H}(e^{-j\omega T})S_{sx}(\omega)]\, d\omega$$

EXAMPLE 13-2 ▶ Suppose that $\mathbf{s}[n]$ is a first-order AR process and $\boldsymbol{\nu}[n]$ is white noise orthogonal to $\mathbf{s}[n]$:

$$\mathbf{S}_{ss}(z) = \frac{N_0}{(1 - az^{-1})(1 - az)} \qquad \mathbf{S}_{\nu\nu}(z) = N \qquad \mathbf{S}_{s\nu}(z) = 0$$

In this case,

$$\mathbf{S}_{xx}(z) = \mathbf{S}_{ss}(z) + N = \frac{aN(1 - bz^{-1})(1 - bz)}{b(1 - az^{-1})(1 - az)}$$

where

$$0 < b < a < 1 \qquad b + b^{-1} = a + a^{-1} + \frac{N_0}{aN}$$

Hence

$$\mathbf{H}(z) = \frac{bN_0}{aN(1 - bz^{-1})(1 - bz)} \qquad h[n] = cb^{|n|} \qquad c = \frac{bN_0}{aN(1 - b^2)}$$

$$P = \frac{N_0}{1 - a^2}\left[1 - c\sum_{k=-\infty}^{\infty}(ab)^{|k|}\right] = \frac{bN_0}{a(1 - b^2)}$$

◀

13-2 PREDICTION

Prediction is the estimation of the future $\mathbf{s}(t + \lambda)$ of a process $\mathbf{s}(t)$ in terms of its past $\mathbf{s}(t - \tau)$, $\tau > 0$. This problem has three parts: The past (data) is known in the interval $(-\infty, t)$; it is known in the interval $(t - T, t)$ of finite length T; it is known in the interval $(0, t)$ of variable length t. We shall develop all three parts for digital processes only. The discussion of analog predictors will be limited to the first part. In the digital case, we find it more convenient to predict the present $\mathbf{s}[n]$ of the given process in terms of its past $\mathbf{s}[n - k]$, $k \geq r$.

Infinite Past

We start with the estimation of a process $\mathbf{s}[n]$ in terms of its entire past $\mathbf{s}[n - k]$, $k \geq 1$:

$$\hat{\mathbf{s}}[n] = \hat{E}\{\mathbf{s}[n] \mid \mathbf{s}[n - k], k \geq 1\} = \sum_{k=1}^{\infty} h[k]\mathbf{s}[n - k] \tag{13-19}$$

This estimator will be called the one-step predictor of $\mathbf{s}[n]$. Thus $\hat{\mathbf{s}}[n]$ is the response of the *predictor filter*

$$\mathbf{H}(z) = h[1]z^{-1} + \cdots + h[k]z^{-k} + \cdots \tag{13-20}$$

to the input $\mathbf{s}[n]$ and our objective is to find the constants $h[k]$ so as to minimize the MS estimation error. From the orthogonality principle it follows that the error $\boldsymbol{\varepsilon}[n] = \mathbf{s}[n] - \hat{\mathbf{s}}[n]$ must be orthogonal to the data $\mathbf{s}[n - m]$:

$$E\left\{\left(\mathbf{s}[n] - \sum_{k=1}^{\infty} h[k]\mathbf{s}[n - k]\right)\mathbf{s}[n - m]\right\} = 0 \qquad m \geq 1 \tag{13-21}$$

This yields

$$R[m] - \sum_{k=1}^{\infty} h[k]R[m - k] = 0 \qquad m \geq 1 \tag{13-22}$$

We have thus obtained a system of infinitely many equations expressing the unknowns $h[k]$ in terms of the autocorrelation $R[m]$ of $\mathbf{s}[n]$. These equations are called Wiener–Hopf (digital form).

The Wiener–Hopf equations cannot be solved directly with z transforms even though the right side equals the convolution of $h[m]$ with $R[m]$. The reason is that, unlike (13-17), the two sides of (13-22) are not equal for every m. A solution based on the analytic properties of the z transforms of causal and anticausal sequences can be found (see Prob. 13-12); however, the underlying theory is not simple. We shall give presently a very simple solution based on the concept of innovations. We comment first on a basic property of the estimation error $\boldsymbol{\varepsilon}[n]$ and of the error filter

$$\mathbf{E}(z) = 1 - \mathbf{H}(z) = 1 - \sum_{k=1}^{\infty} h[n]z^{-k} \tag{13-23}$$

The error $\boldsymbol{\varepsilon}[n]$ is orthogonal to the data $\mathbf{s}[n-m]$ for every $m \ge 1$; furthermore, $\boldsymbol{\varepsilon}[n-m]$ is a linear function of $\mathbf{s}[n-m]$ and its past because $\boldsymbol{\varepsilon}[n]$ is the response of the causal system $\mathbf{E}(z)$ to the input $\mathbf{s}[n]$. From this it follows that $\boldsymbol{\varepsilon}[n]$ is orthogonal to $\boldsymbol{\varepsilon}[n-m]$ for every $m \ge 1$ and every n. Hence $\boldsymbol{\varepsilon}[n]$ is white noise:

$$R_{\varepsilon\varepsilon}[m] = E\{\boldsymbol{\varepsilon}[n]\boldsymbol{\varepsilon}[n-m]\} = P\delta[m] \tag{13-24}$$

where

$$P = E\{\boldsymbol{\varepsilon}^2[n]\} = E\{(\mathbf{s}[n] - \hat{\mathbf{s}}[n])\mathbf{s}[n]\} = R[0] - \sum_{k=1}^{\infty} h[k]R[k]$$

is the LMS error. This error can be expressed in terms of the power spectrum $S(\omega)$ of $\mathbf{s}[n]$; as we see from (9-152),

$$P = \frac{1}{2\pi}\int_{-\pi}^{\pi} |\mathbf{E}(e^{j\omega})|^2 S(\omega)\, d\omega \tag{13-25}$$

Using this, we shall show that the function $\mathbf{E}(z)$ has no 0's outside the unit circle.

THEOREM 13-2 ▶ If

$$E(z_i) = 0 \qquad \text{then} \quad |z_i| \le 1 \tag{13-26}$$

Proof. We form the function

$$\mathbf{E}_0(z) = \mathbf{E}(z)\frac{1 - z^{-1}/z_i^*}{1 - z_i z^{-1}}$$

This function is an error filter because it is causal and $\mathbf{E}_0(\infty) = \mathbf{E}(\infty) = 1$. Furthermore, if $|z_i| > 1$, then [see (12B-2)]

$$|\mathbf{E}_0(e^{j\omega})| = \frac{1}{|z_i|}|\mathbf{E}(e^{j\omega})| < |\mathbf{E}(e^{j\omega})|$$

Inserting into (13-25), we conclude that if we use as the estimator filter the function $1 - \mathbf{E}_0(z)$, the resulting MS error will be smaller than P. This, however, is impossible because P is minimum; hence $|z_i| \le 1$. ◀

Regular Processes

We shall solve the Wiener–Hopf equations (13-22) under the assumption that the process $\mathbf{s}[n]$ is regular. As we have shown in Sec. 11-1, such a process is linearly equivalent to a white-noise process $\mathbf{i}[n]$ in the sense that

$$\mathbf{s}[n] = \sum_{k=0}^{\infty} l[k]\mathbf{i}[n-k] \tag{13-27}$$

$$\mathbf{i}[n] = \sum_{k=0}^{\infty} \gamma[k]\mathbf{s}[n-k] \tag{13-28}$$

From this it follows that the predictor $\hat{\mathbf{s}}[n]$ of $\mathbf{s}[n]$ can be written as a linear sum involving the past of $\mathbf{i}[n]$:

$$\hat{\mathbf{s}}[n] = \sum_{k=1}^{\infty} h_i[k]\mathbf{i}[n-k] \tag{13-29}$$

To find $\hat{\mathbf{s}}[n]$, it suffices, therefore, to find the constants $h_i[k]$ and to express $\mathbf{i}[n]$ in terms of $\mathbf{s}[n]$ using (13-28). To do so, we shall determine first the cross-correlation of $\mathbf{s}[n]$ and $\mathbf{i}[n]$. We maintain that

$$R_{si}[m] = l[m] \tag{13-30}$$

Proof. We multiply (13-27) by $\mathbf{i}[n-m]$ and take expected values. This gives

$$E\{\mathbf{s}[n]\mathbf{i}[n-m]\} = \sum_{k=0}^{\infty} l[k]E\{\mathbf{i}[n-k]\mathbf{i}[n-m]\} = \sum_{k=0}^{\infty} l[k]\delta[m-k]$$

because $R_{ii}[m] = \delta[m]$, and (13-30) results.

To find $h_i[k]$, we apply the orthogonality principle:

$$E\left\{\left(\mathbf{s}[n] - \sum_{k=1}^{\infty} h_i[k]\mathbf{i}[n-k]\right)\mathbf{i}[n-m]\right\} = 0 \qquad m \ge 1$$

This yields

$$R_{si}[m] - \sum_{k=1}^{\infty} h_i[k]R_{ii}[m-k] = R_{si}[m] - \sum_{k=1}^{\infty} h_i[k]\delta[m-k] = 0$$

and since the last sum equals $h_i[m]$, we conclude that $h_i[m] = R_{si}[m]$. From this and (13-30) it follows that the predictor $\hat{\mathbf{s}}[n]$, expressed in terms of its innovations, equals

$$\hat{\mathbf{s}}[n] = \sum_{k=1}^{\infty} l[k]\mathbf{i}[n-k] \tag{13-31}$$

We shall rederive this important result using (13-27). To do so, it suffices to show that the difference $\mathbf{s}[n] - \hat{\mathbf{s}}[n]$ is orthogonal to $\mathbf{i}[n-m]$ for every $m \ge 1$. This is indeed the case because

$$\boldsymbol{\varepsilon}[n] = \sum_{k=0}^{\infty} l[k]\mathbf{i}[n-k] - \sum_{k=1}^{\infty} l[k]\mathbf{i}[n-k] = l[0]\mathbf{i}[n] \tag{13-32}$$

and $\mathbf{i}[n]$ is white noise.

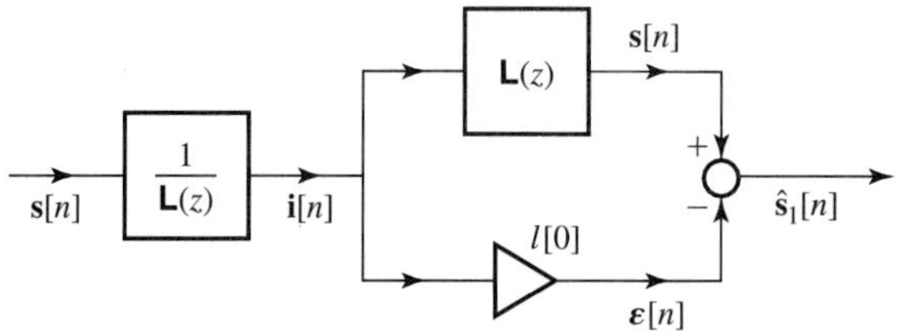

FIGURE 13-3

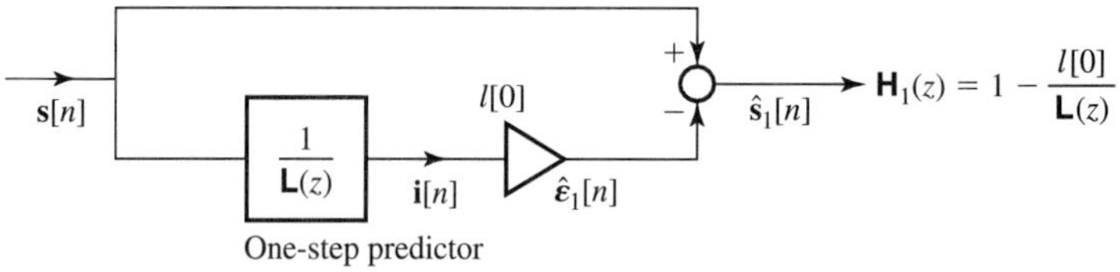

FIGURE 13-4

The sum in (13-31) is the response of the filter

$$\sum_{k=1}^{\infty} l[k]z^{-k} = \mathbf{L}(z) - l[0]$$

to the input $\mathbf{i}[n]$. To complete the specification of $\hat{\mathbf{s}}[n]$, we must express $\mathbf{i}[n]$ in terms of $\mathbf{s}[n]$. Since $\mathbf{i}[n]$ is the response of the filter $1/\mathbf{L}(z)$ to the input $\mathbf{s}[n]$, we conclude, cascading as in Fig. 13-3, that the predictor filter of $\mathbf{s}[n]$ is the product

$$\mathbf{H}(z) = \frac{1}{\mathbf{L}(z)}(\mathbf{L}(z) - l[0]) = 1 - \frac{l[0]}{\mathbf{L}(z)} \tag{13-33}$$

shown in Fig. 13-4. Thus, to obtain $\mathbf{H}(z)$, it suffices to factor $\mathbf{S}(z)$ as in (11-6). The constant $l[0]$ is determined from the initial value theorem:

$$l[0] = \lim_{z\to\infty} \mathbf{L}(z)$$

EXAMPLE 13-3 ▶ Suppose that

$$S(\omega) = \frac{5 - 4\cos\omega}{10 - 6\cos\omega} \qquad \mathbf{L}(z) = \frac{2z-1}{3z-1} \qquad l[0] = \frac{2}{3}$$

as in Example 11-4. In this case, (13-33) yields

$$\mathbf{H}(z) = 1 - \frac{2}{3} \times \frac{3z-1}{2z-1} = \frac{-z^{-1}}{6(1 - z^{-1}/2)}$$

Note that $\hat{\mathbf{s}}[n]$ can be determined recursively:

$$\hat{\mathbf{s}}[n] - \tfrac{1}{2}\hat{\mathbf{s}}[n-1] = -\tfrac{1}{6}\mathbf{s}[n-1]$$ ◀

The Kolmogorov–Szego MS error formula.[3] As we have seen from (13-32), the MS estimation error equals

$$P = E\{\boldsymbol{\varepsilon}^2[n]\} = l^2[0]$$

[3] U. Grenander and G. Szego: *Toeplitz Forms and Their Applications,* Berkeley University Press, 1958 [16].

Furthermore [see (12A-1)]

$$\ln l^2[0] = \frac{1}{2\pi}\int_{-\pi}^{\pi} \ln|\mathbf{L}(e^{i\omega})|^2\, d\omega$$

Since $S(\omega) = |\mathbf{L}(e^{j\omega})|^2$, this yields the identity

$$P = \exp\left\{\frac{1}{2\pi}\int_{-\pi}^{\pi} \ln S(\omega)\, d\omega\right\} \tag{13-34}$$

expressing P directly in terms of $S(\omega)$.

Autoregressive processes. If $\mathbf{s}[n]$ is an AR process as in (11-39), then $l[0] = b_0$ and

$$\begin{aligned} \mathbf{H}(z) &= -a_1 z^{-1} - \cdots - a_N z^{-N} \\ \hat{\mathbf{s}}[n] &= -a_1\mathbf{s}[n-1] - \cdots - a_N\mathbf{s}[n-N] \qquad P = b_0^2 \end{aligned} \tag{13-35}$$

This shows that the predictor $\hat{\mathbf{s}}[n]$ of $\mathbf{s}[n]$ in terms of its entire past is the same as the predictor in terms of the N most recent past values. This result can be established directly: From (11-39) and (13-35) it follows that $\mathbf{s}[n] - \hat{\mathbf{s}}[n] = b_0\mathbf{i}[n]$. This is orthogonal to the past of $\mathbf{s}[n]$; hence

$$\hat{E}\{\mathbf{s}[n]\,|\,\mathbf{s}[n-k],\ 1 \le k \le N\} = \hat{E}\{\mathbf{s}[n]\,|\,\mathbf{s}[n-k],\ k \ge 1\}$$

A process with this property is called wide-sense Markov of order N.

THE r-STEP PREDICTOR. We shall determine the predictor

$$\hat{\mathbf{s}}_r[n] = \hat{E}\{s[n]\,|\,\mathbf{s}[n-k], k \ge r\}$$

of $\mathbf{s}[n]$ in terms of $\mathbf{s}[n-r]$ and its past using innovations. We maintain that

$$\hat{\mathbf{s}}_r[n] = \sum_{k=r}^{\infty} l[k]\mathbf{i}[n-k] \tag{13-36}$$

Proof. It suffices to show that the difference

$$\hat{\boldsymbol{\varepsilon}}_r[n] = \mathbf{s}[n] - \hat{\mathbf{s}}_r[n] = \sum_{k=0}^{r-1} l[k]\mathbf{i}[n-k]$$

is orthogonal to the data $\mathbf{s}[n-k]$, $k \ge r$. This is a consequence of the fact that $\mathbf{s}[n-k]$ is linearly equivalent to $\mathbf{i}[n-k]$ and its past for $k \ge r$; hence it is orthogonal to $\mathbf{i}[n]$, $\mathbf{i}[n-1], \ldots, \mathbf{i}[n-r+1]$.

The prediction error $\hat{\boldsymbol{\varepsilon}}_r[n]$ is the response of the MA filter $l[0] + l[1]z^{-1} + \cdots + l[r-1]z^{-r+1}$ of Fig. 13-5 to the input $\mathbf{i}[n]$. Cascading this filter with $1/\mathbf{L}(z)$ as in Fig. 13-5, we conclude that the process $\hat{\mathbf{s}}_r[n] = \mathbf{s}[n] - \hat{\boldsymbol{\varepsilon}}_r[n]$ is the response of the system

$$\mathbf{H}_r(z) = 1 - \frac{1}{\mathbf{L}(z)}\sum_{k=0}^{r-1} l[k]z^{-k} \tag{13-37}$$

to the input $\mathbf{s}[n]$. This is the *r-step predictor filter* of $\mathbf{s}[n]$. The resulting MS error equals

$$P_r = E\{\boldsymbol{\varepsilon}_r^2[n]\} = \sum_{k=0}^{r-1} l^2[k] \tag{13-38}$$

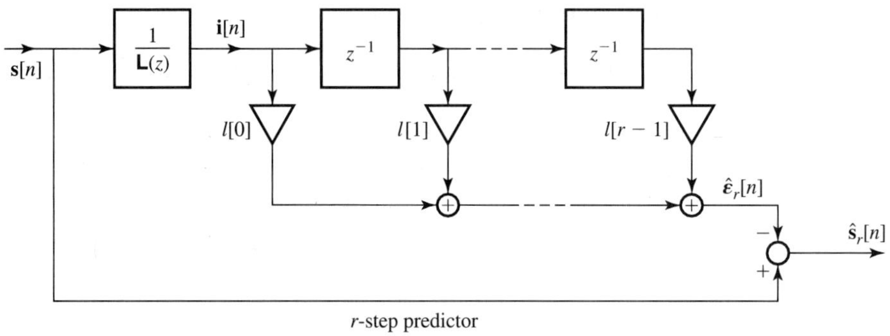

FIGURE 13-5

EXAMPLE 13-4 ▶ We are given a process $\mathbf{s}[n]$ with autocorrelation $R[m]=a^{|m|}$ and we wish to determine its r-step predictor. In this case (see Example 9-30)

$$\mathbf{S}(z)=\frac{a^{-1}-a}{(a^{-1}+a)-(z^{-1}+z)}=\frac{b^2}{(1-az^{-1})(1-az)}\qquad b^2=1-a^2$$

$$\mathbf{L}(z)=\frac{b}{1-az^{-1}}\qquad l[n]=ba^nU[n]$$

Hence

$$\mathbf{H}_r(z)=1-\frac{1-az^{-1}}{b}\sum_{k=0}^{r-1}ba^kz^{-k}=a^rz^{-r}$$

$$\hat{\mathbf{s}}_r[n]=a^r\mathbf{s}[n-r]\qquad P_r=b^2\sum_{k=0}^{r-1}a^{2k}=1-a^{2r}$$ ◀

ANALOG PROCESSES. We consider now the problem of predicting the future value $\mathbf{s}(t+\lambda)$ of a process $\mathbf{s}(t)$ in terms of its entire past $\mathbf{s}(t-\tau)$, $\tau\ge 0$. In this problem, our estimator is an integral:

$$\hat{\mathbf{s}}(t+\lambda)=\hat{E}\{\mathbf{s}(t+\lambda)\,|\,\mathbf{s}(t-\tau),\tau\ge 0\}=\int_0^\infty h(\alpha)\mathbf{s}(t-\alpha)\,d\alpha \qquad (13\text{-}39)$$

and the problem is to find the function $h(\alpha)$. From the analog form (13-4) of the orthogonality principle, it follows that

$$E\left\{\left[\mathbf{s}(t+\lambda)-\int_0^\infty h(\alpha)\mathbf{s}(t-\alpha)\right]\mathbf{s}(t-\tau)\right\}=0\qquad \tau\ge 0$$

This yields the Wiener–Hopf integral equation

$$R(\tau+\lambda)=\int_0^\infty h(\alpha)R(\tau-\alpha)\,d\alpha\qquad \tau\ge 0 \qquad (13\text{-}40)$$

The solution of this equation is the impulse response of the *causal Wiener filter*

$$\mathbf{H}(s)=\int_0^\infty h(t)e^{-st}\,dt$$

The corresponding MS error equals

$$P = E\{[\mathbf{s}(t+\lambda) - \hat{\mathbf{s}}(t+\lambda)]\mathbf{s}(t+\lambda)\} = R(0) - \int_0^\infty h(\alpha)R(\lambda+\alpha)\,d\alpha \qquad (13\text{-}41)$$

Equation (13-40) cannot be solved directly with transforms because the two sides are equal for $\tau \geq 0$ only. A solution based on the analytic properties of Laplace transforms is outlined in Prob. 13-11. We give next a solution using innovations.

As we have shown in (11-8), the process $\mathbf{s}(t)$ is the response of its innovations filter $\mathbf{L}(s)$ to the white-noise process $\mathbf{i}(t)$. From this it follows that

$$\mathbf{s}(t+\lambda) = \int_0^\infty l(\alpha)\mathbf{i}(t+\lambda-\alpha)\,d\alpha \qquad (13\text{-}42)$$

We maintain that $\hat{\mathbf{s}}(t+\lambda)$ is the part of the above integral involving only the past of $\mathbf{i}(t)$:

$$\hat{\mathbf{s}}(t+\lambda) = \int_\lambda^\infty l(\alpha)\mathbf{i}(t+\lambda-\alpha)\,d\alpha = \int_0^\infty l(\beta+\lambda)\mathbf{i}(t-\beta)\,d\beta \qquad (13\text{-}43)$$

Proof. The difference

$$\mathbf{s}(t+\lambda) - \hat{\mathbf{s}}(t+\lambda) = \int_0^\lambda l(\alpha)\mathbf{i}(t+\lambda-\alpha)\,d\alpha \qquad (13\text{-}44)$$

depends only on the values of $\mathbf{i}(t)$ in the interval $(t, t+\lambda)$; hence it is orthogonal to the past of $\mathbf{i}(t)$ and, therefore, it is also orthogonal to the past of $\mathbf{s}(t)$.

The predictor $\hat{\mathbf{s}}(t+\lambda)$ of $\mathbf{s}(t)$ is the response of the system

$$\mathbf{H}_i(s) = \int_0^\infty h_i(t)e^{-st}\,dt \qquad h_i(t) = l(t+\lambda)U(t) \qquad (13\text{-}45)$$

(Fig. 13-6) to the input $\mathbf{i}(t)$. Cascading with $1/\mathbf{L}(s)$, we conclude that $\hat{\mathbf{s}}(t+\lambda)$ is the response of the system

$$\mathbf{H}(s) = \frac{\mathbf{H}_i(s)}{\mathbf{L}(s)} \qquad (13\text{-}46)$$

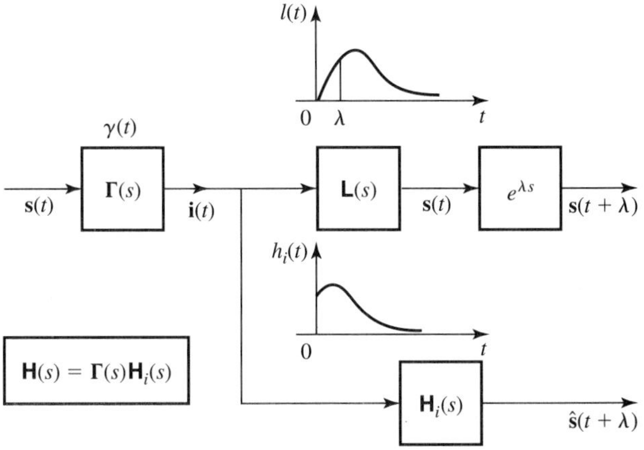

FIGURE 13-6

to the input $\mathbf{s}(t)$. Thus, to determine the predictor filter $\mathbf{H}(s)$ of $\mathbf{s}(t)$, proceed as follows:

Factor the spectrum of $\mathbf{s}(t)$ as in (11-3): $\mathbf{S}(s) = \mathbf{L}(s)\mathbf{L}(-s)$.
Find the inverse transform $l(t)$ of $\mathbf{L}(s)$ and form the function $h_i(t) = l(t+\lambda)U(t)$.
Find the transform $\mathbf{H}_i(s)$ of $h_i(t)$ and determine $\mathbf{H}(s)$ from (13-46).

The MS estimation error is determined from (13-44):

$$P = E\left\{\left|\int_0^{\lambda} l(\alpha)\mathbf{i}(t+\lambda-\alpha)\,d\alpha\right|^2\right\} = \int_0^{\lambda} l^2(\alpha)\,d\alpha \tag{13-47}$$

EXAMPLE 13-5

▶ We are given a process $\mathbf{s}(t)$ with autocorrelation $R(\tau) = 2\alpha e^{-\alpha|\tau|}$ and we wish to determine its predictor. In this problem,

$$\mathbf{S}(s) = \frac{1}{\alpha^2 - s^2} \qquad \mathbf{L}(s) = \frac{1}{\alpha + s} \qquad l(t) = e^{-\alpha t}U(t)$$

$$h_i(t) = e^{-\alpha\lambda}e^{-\alpha t}U(t) \qquad \mathbf{H}_i(s) = \frac{e^{-\alpha\lambda}}{\alpha + s}$$

$$\mathbf{H}(s) = e^{-\alpha\lambda} \qquad \hat{\mathbf{s}}(t+\lambda) = e^{-\alpha\lambda}\mathbf{s}(t)$$

This shows that the predictor of $\mathbf{s}(t+\lambda)$ in terms of its entire past is the same as the predictor in terms of its present $\mathbf{s}(t)$. In other words, if $\mathbf{s}(t)$ is specified, the past has no effect on the linear prediction of the future. ◀

The determination of $\mathbf{H}(s)$ is simple if $\mathbf{s}(t)$ has a rational spectrum. Assuming that the poles of $\mathbf{H}(s)$ are simple, we obtain

$$\begin{aligned} \mathbf{L}(s) &= \frac{N(s)}{D(s)} = \sum_i \frac{c_i}{s - s_i} \qquad l(t) = \sum_i c_i e^{s_i t}U(t) \\ h_i(t) &= \sum_i c_i e^{s_i\lambda}e^{s_i t}U(t) \qquad \mathbf{H}_i(s) = \sum_i \frac{c_i e^{s_i\lambda}}{s - s_i} = \frac{N_1(s)}{D(s)} \end{aligned} \tag{13-48}$$

and (13-46) yields $\mathbf{H}(s) = N_1(s)/N(s)$.

If $N(s) = 1$, then $\mathbf{H}(s)$ is a polynomial:

$$\mathbf{H}(s) = N_1(s) = b_0 + b_1 s + \cdots + b_n s^n$$

and $\hat{\mathbf{s}}(t+\lambda)$ is a linear sum of $\mathbf{s}(t)$ and its first n derivatives:

$$\hat{\mathbf{s}}(t+\lambda) = b_0\mathbf{s}(t) + b_1\mathbf{s}'(t) + \cdots + b_n\mathbf{s}^{(n)}(t)$$

EXAMPLE 13-6

▶ We are given a process $\mathbf{s}(t)$ with

$$\mathbf{S}(s) = \frac{49 - 25s^2}{(1-s^2)(9-s^2)} \qquad \mathbf{L}(s) = \frac{7+5s}{(1+s)(3+s)}$$

and we wish to estimate its future $s(t+\lambda)$ for $\lambda = \log 2$. In this problem, $e^{\lambda} = 2$:

$$\mathbf{L}(s) = \frac{1}{s+1} + \frac{4}{s+3} \qquad \mathbf{H}_i(s) = \frac{e^{-\lambda}}{s+1} + \frac{4e^{-3\lambda}}{s+3} = \frac{s+2}{(s+1)(s+3)}$$

$$\mathbf{H}(s) = \frac{s+2}{5s+7} \qquad h(t) = \frac{1}{5}\delta(t) + \frac{3}{25}e^{-1.4t}U(t)$$

Hence

$$E\{\mathbf{s}(t+\lambda) \mid \mathbf{s}(t-\tau), \tau \geq 0\} = 0.2\mathbf{s}(t) + \hat{E}\{\mathbf{s}(t+\lambda) \mid \mathbf{s}(t-\tau), \tau > 0\}$$ ◀

Notes 1. The integral

$$y(\tau) = \int_0^{\infty} h(\alpha)R(\tau-\alpha)\,d\alpha$$

in (13-40) is the response of the Wiener filter $\mathbf{H}(s)$ to the input $R(\tau)$. From (13-40) and (13-41) it follows that

$$y(\tau) = R(\tau+\lambda) \qquad \text{for} \quad \tau \geq 0 \qquad \text{and} \qquad y(-\lambda) = R(0) - P$$

2. In all MS estimation problems, only second-order moments are used. If, therefore, two processes have the same autocorrelation, then their predictors are identical. This suggests the following derivation of the Wiener–Hopf equation: Suppose that $\boldsymbol{\omega}$ is a random variable with density $f(\omega)$ and $\mathbf{z}(t) = e^{j\boldsymbol{\omega}t}$. Clearly,

$$R_{zz}(\tau) = E\left\{e^{j\boldsymbol{\omega}(t+\tau)}e^{-j\boldsymbol{\omega}t}\right\} = \int_{-\infty}^{\infty} f(\omega)e^{j\omega\tau}\,d\omega$$

From this it follows that the power spectrum of $\mathbf{z}(t)$ equals $2\pi f(\omega)$ [see also (9-140)]. If, therefore, $\mathbf{s}(t)$ is a process with power spectrum $S(\omega) = 2\pi f(\omega)$, then its predictor $h(t)$ will equal the predictor of $\mathbf{z}(t)$:

$$\hat{\mathbf{z}}(t+\lambda) = \hat{E}\left\{e^{j\boldsymbol{\omega}(t+\lambda)} \,\middle|\, e^{j\boldsymbol{\omega}(t-\alpha)}, \alpha \geq 0\right\} = \int_0^{\infty} h(\alpha)e^{j\boldsymbol{\omega}(t-\alpha)}\,d\alpha$$

$$= e^{j\boldsymbol{\omega}t}\int_0^{\infty} h(\alpha)e^{-j\boldsymbol{\omega}\alpha}\,d\alpha = e^{j\boldsymbol{\omega}t}H(\boldsymbol{\omega})$$

And since $\mathbf{z}(t+\lambda) - \hat{\mathbf{z}}(t+\lambda) \perp \mathbf{z}(t-\tau)$, for $\tau \geq 0$, we conclude from the last equation that

$$E\left\{\left[e^{j\boldsymbol{\omega}(t+\lambda)} - e^{j\boldsymbol{\omega}t}H(\boldsymbol{\omega})\right]e^{-j\boldsymbol{\omega}(t-\tau)}\right\} = 0 \qquad \tau \geq 0$$

Hence

$$\int_{-\infty}^{\infty} f(\omega)\left[e^{j\omega(\tau+\lambda)} - e^{j\omega\tau}H(\omega)\right]d\omega = 0 \qquad \tau \geq 0$$

This yields (13-40) because the inverse transform of $f(\omega)e^{j\omega(\tau+\lambda)}$ equals $R(\tau+\lambda)$ and the inverse transform of $f(\omega)e^{j\omega\tau}H(\omega)$ equals the integral in (13-40).

Predictable processes. We shall say that a process $\mathbf{s}[n]$ is *predictable* if it equals its predictor:

$$\mathbf{s}[n] = \sum_{k=1}^{\infty} h[k]\mathbf{s}[n-k] \tag{13-49}$$

In this case [see (13-25)]

$$P = \frac{1}{2\pi}\int_{-\pi}^{\pi} |\mathbf{E}(e^{j\omega})|^2 S(\omega)\,d\omega = 0 \tag{13-50}$$

Since $S(\omega) \geq 0$, the above integral is 0 if $S(\omega) \neq 0$ only in a region R of the ω axis, where $\mathbf{E}(e^{j\omega}) = 0$. It can be shown that this region consists of a countable number of points ω_i—the proof is based on the Paley–Wiener condition (11-9). From this it follows that

$$S(\omega) = 2\pi \sum_{i=1}^{m} \alpha_i \delta(\omega - \omega_i) \qquad \mathbf{E}(e^{j\omega_i}) = 0 \tag{13-51}$$

Thus a process $\mathbf{s}[n]$ is predictable if it is a sum of exponentials as in (11-43):

$$\mathbf{s}[n] = \sum_{i=1}^{m} \mathbf{c}_i e^{j\omega_i n} \qquad E\{\mathbf{c}_i^2\} = \alpha_i > 0 \tag{13-52}$$

We maintain that the converse is also true: If $\mathbf{s}[n]$ is a sum of m exponentials as in (13-52), then it is predictable and its predictor filter equals $1 - D(z)$, where

$$D(z) = (1 - e^{j\omega_1} z^{-1}) \cdots (1 - e^{j\omega_m} z^{-1}) \tag{13-53}$$

Proof. In this case, $\mathbf{E}(z) = D(z)$ and $\mathbf{E}(e^{j\omega_i}) = 0$; hence $\mathbf{E}(e^{j\omega}) S(\omega) = 0$ because $\mathbf{E}(e^{j\omega}) \delta(\omega - \omega_i) = \mathbf{E}(e^{j\omega_i}) \delta(\omega - \omega_i) = 0$. From this it follows that $P = 0$. In that case $|\mathbf{T}_m| = \Delta_m = 0$ and $\Delta_{m-1} > 0$ (see footnote 13, page 564, ch. 12).

Note The preceding result seems to be in conflict with the sampling expansion (10-201) of a BL process $\mathbf{s}(t)$: This expansion shows that $\mathbf{s}(t)$ is predictable in the sense that it can be approximated within an arbitrary error ε by a linear sum involving only its past samples $\mathbf{s}(nT_0)$. From this it follows that the digital process $\mathbf{s}[n] = \mathbf{s}(nT_0)$ is predictable in the same sense. Such an expansion, however, does not violate (13-50). It is only an approximation and its coefficients tend to ∞ as $\varepsilon \to 0$.

GENERAL PROCESSES AND WOLD'S DECOMPOSITION.[4] We show finally that an arbitrary process $\mathbf{s}[n]$ can be written as a sum

$$\mathbf{s}[n] = \mathbf{s}_1[n] + \mathbf{s}_2[n] \tag{13-54}$$

of a regular process $\mathbf{s}_1[n]$ and a predictable process $\mathbf{s}_2[n]$, that these processes are orthogonal, and that they have the same predictor filter. We thus reestablish constructively Wold's decomposition (11-89) in the context of MS estimation.

As we know [see (13-24)], the error $\boldsymbol{\varepsilon}[n]$ of the one-step estimate of $\mathbf{s}[n]$ is a white-noise process. We form the estimator $\mathbf{s}_1[n]$ of $\mathbf{s}[n]$ in terms of $\boldsymbol{\varepsilon}[n]$ and its past:

$$\mathbf{s}_1[n] = \hat{E}\{\mathbf{s}[n] \mid \boldsymbol{\varepsilon}[n-k], k \geq 0\} = \sum_{k=0}^{\infty} w_k \boldsymbol{\varepsilon}[n-k] \tag{13-55}$$

Thus $\mathbf{s}_1[n]$ is the response of the system (Fig. 13-7)

$$\mathbf{W}(z) = \sum_{k=0}^{\infty} w_k z^{-k}$$

[4] A. Papoulis: Predictable Processes and Wold's Decomposition: A Review. *IEEE Transactions on Acoustics, Speech, and Signal Processing,* vol. 22, 1985.

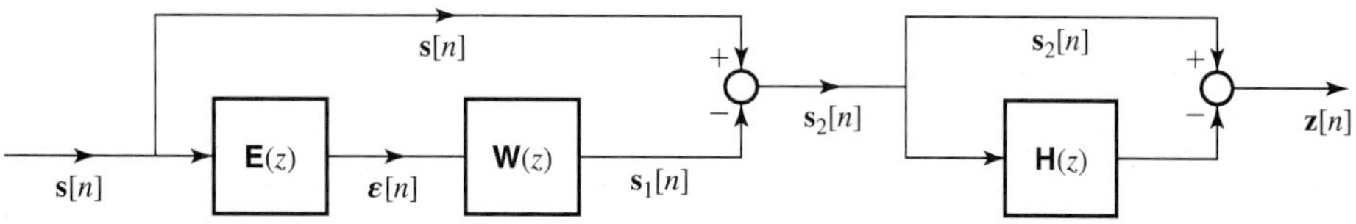

FIGURE 13-7

to the input $\varepsilon[n]$. The difference $s_2[n] = s[n] - s_1[n]$ is the estimation error (Fig. 13-7). Clearly (orthogonality principle)

$$s_2[n] \perp \varepsilon[n-k] \qquad k \geq 0 \tag{13-56}$$

Note that if $s[n]$ is a regular process, then [see (13-32)] $\varepsilon[n] = l[0]i[n]$; in this case, $s_1[n] = s[n]$.

THEOREM 13-3 (*a*) The processes $s_1[n]$ and $s_2[n]$ are orthogonal:

$$s_1[n] \perp s_2[n-k] \qquad \text{all} \quad k \tag{13-57}$$

(*b*) $s_1[n]$ is a regular process.

(*c*) $s_2[n]$ is a predictable process and its predictor filter is the sum in (13-19):

$$s_2[n] = \sum_{k=1}^{\infty} h[k]s_2[n-k] \tag{13-58}$$

Proof. (*a*) The process $\varepsilon[n]$ is orthogonal to $s[n-k]$ for every $k > 0$. Furthermore, $s_2[n-k]$ is a linear function of $s[n-k]$ and its past; hence $s_2[n-k] \perp \varepsilon[n]$ for $k > 0$. Combining with (13-56), we conclude that

$$s_2[n-k] \perp \varepsilon[n] \qquad \text{all} \quad k \tag{13-59}$$

And since $s_1[n]$ depends linearly on $\varepsilon[n]$ and its past, (13-57) follows.

(*b*) The process $s_1[n]$ is the response of the system $W(z)$ to the white noise $\varepsilon[n]$. To prove that it is regular, it suffices to show that

$$\sum_{k=0}^{\infty} w_k^2 < \infty \tag{13-60}$$

From (13-54) and (13-55) it follows that

$$E\{s^2[n]\} = E\{s_1^2[n]\} + E\{s_2^2[n]\} \geq E\{s_1^2[n]\} = \sum_{k=0}^{\infty} w_k^2$$

This yields (13-60) because $E\{s^2[n]\} = R(0) < \infty$.

(*c*) To prove (13-58), it suffices to show that the difference

$$z[n] = s_2[n] - \sum_{k=1}^{\infty} h[k]s_2[n-k]$$

equals 0. From (13-59) it follows that $z[n] \perp \varepsilon[n-k]$ for all k. But $z[n]$ is the response of the

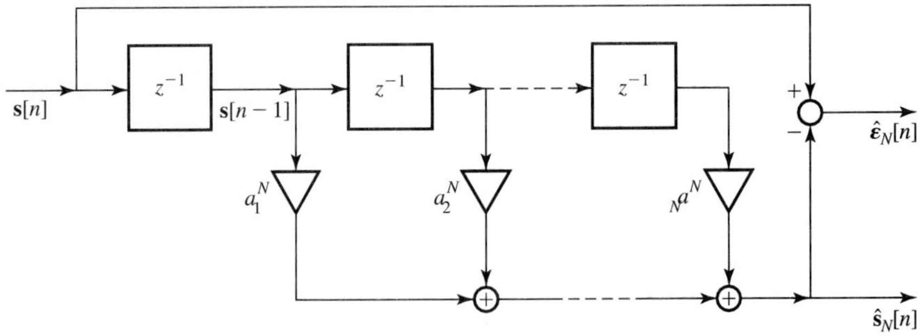

FIGURE 13-8

system $1 - \mathbf{H}(z) = \mathbf{E}(z)$ to the input $\mathbf{s}_2[n] = \mathbf{s}[n] - \mathbf{s}_1[n]$; hence (see Fig. 13-8)

$$\mathbf{z}[n] = \boldsymbol{\varepsilon}[n] - \mathbf{s}_1[n] + \sum_{k=1}^{\infty} h[k]\mathbf{s}_1[n-k] \tag{13-61}$$

This shows that $\mathbf{z}[n]$ is a linear function of $\boldsymbol{\varepsilon}[n]$ and its past. And since it is also orthogonal to $\boldsymbol{\varepsilon}[n]$, we conclude that $\mathbf{z}[n] = 0$.

Note finally that [see (13-61)]

$$\mathbf{s}_1[n] - \sum_{k=1}^{\infty} h[k]\mathbf{s}_1[n-k] = \boldsymbol{\varepsilon}[n] \perp \mathbf{s}_1[n-m] \qquad m \geq 1$$

Hence this sum is the predictor of $\mathbf{s}_1[n]$. We thus conclude that the sum $\mathbf{H}(z)$ in (13-20) is the predictor filter of the processes $\mathbf{s}[n]$, $\mathbf{s}_1[n]$, and $\mathbf{s}_2[n]$. ◀

FIR PREDICTORS. We shall find the estimate $\hat{\mathbf{s}}_N[n]$ of a process $\mathbf{s}[n]$ in terms of its N most recent past values:

$$\hat{\mathbf{s}}_N[n] = \hat{E}\{\mathbf{s}[n] \mid \mathbf{s}[n-k], 1 \leq k \leq N\} = \sum_{k=1}^{N} a_k^N \mathbf{s}[n-k] \tag{13-62}$$

This estimate will be called the *forward* predictor of order N. The superscript in a_k^N identifies the order. The process $\hat{\mathbf{s}}_N[n]$ is the response of the *forward predictor filter*

$$\hat{\mathbf{H}}_N(z) = \sum_{k=1}^{N} a_k^N z^{-k} \tag{13-63}$$

to the input $\mathbf{s}[n]$. Our objective is to determine the constants a_k^N so as to minimize the MS value

$$P_N = E\{\hat{\boldsymbol{\varepsilon}}_N^2[n]\} = E\{(\mathbf{s}[n] - \hat{\mathbf{s}}_N[n])\mathbf{s}[n]\} \tag{13-64}$$

of the forward prediction error $\hat{\boldsymbol{\varepsilon}}_N[n] = \mathbf{s}[n] - \hat{\mathbf{s}}_N[n]$.

The Yule–Walker equations. From the orthogonality principle it follows that

$$E\left\{\left(\mathbf{s}[n] - \sum_{k=1}^{N} a_k^N \mathbf{s}[n-k]\right)\mathbf{s}[n-m]\right\} = 0 \qquad 1 \leq m \leq N$$

This yields the system

$$R[m] - \sum_{k=1}^{N} a_k^N R[m-k] = 0 \qquad 1 \le m \le N \tag{13-65}$$

Solving, we obtain the coefficients a_k^N of the predictor filter $\hat{\mathbf{H}}_N(z)$. The resulting MS error equals [see (12-83)]

$$P_N = R[0] - \sum_{k=1}^{N} a_k^N R[k] = \frac{\Delta_{N+1}}{\Delta_N} \tag{13-66}$$

In Fig. 13-8 we show the ladder realization of $\hat{\mathbf{H}}_N(z)$ and the *forward error filter* $\hat{\mathbf{E}}_N(z) = 1 - \hat{\mathbf{H}}_N(z)$.

As we have shown in Sec. 12-3, the error filter can be realized by the lattice structure of Fig. 13-9. In that figure, the input is $\mathbf{s}[n]$ and the upper output $\hat{\boldsymbol{\varepsilon}}_N[N]$. The lower output $\check{\boldsymbol{\varepsilon}}_N[n]$ is the *backward prediction error* defined as follows: The processes $\mathbf{s}[n]$ and $\mathbf{s}[-n]$ have the same autocorrelation; hence their predictor filters are *identical*. From this it follows that the backward predictor $\check{\mathbf{s}}_N[n]$, that is, the predictor of $\mathbf{s}[n]$ in terms of its N most recent future values, equals

$$\check{\mathbf{s}}_N[n] = \hat{E}\{\mathbf{s}[n] \mid \mathbf{s}[n+k], 1 \le k \le N\} = \sum_{k=1}^{N} a_k^N \mathbf{s}[n+k]$$

The backward error

$$\check{\boldsymbol{\varepsilon}}_N[n] = \mathbf{s}[n-N] - \check{\mathbf{s}}_N[n-N]$$

is the response of the filter

$$\check{\mathbf{E}}_N(z) = z^{-N}\left(1 - a_1^N z - \cdots - a_N^N z^N\right) = z^{-N}\hat{\mathbf{E}}_N(1/z)$$

with input $\mathbf{s}[n]$. From this and (12-94) it follows that the lower output of the lattice of Fig. 13-8 is $\check{\boldsymbol{\varepsilon}}_N[n]$.

In Sec. 12-3, we used the ladder–lattice equivalence to simplify the solution of the Yule–Walker equations. We summarize next the main results in the context of the prediction problem. We note that the lattice realization also has the following advantage. Suppose that we have a predictor of order N and we wish to find the predictor of order $N+1$. In the ladder realization, we must find a new set of $N+1$ coefficients a_k^{N+1}. In the lattice realization, we need only the new reflection coefficient K_{N+1}; the first N reflection coefficients K_k do not change.

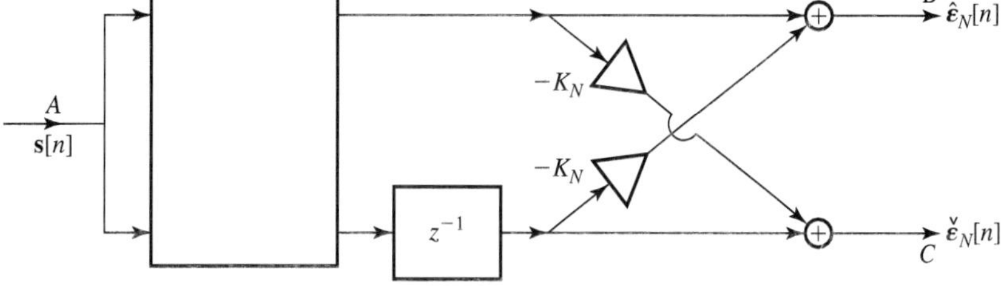

FIGURE 13-9

Levinson's algorithm. We shall determine the constants a_k^N, K_N, and P_N recursively. This involves the following steps: Start with

$$a_1^1 = K_1 = R[1]/R[0] \qquad P_1 = \left(1 - K_1^2\right) R[0]$$

Assume that the $N+1$ constants a_k^{N-1}, K_{N-1}, and P_{N-1} are known. Find K_N and P_N from (12-107) and (12-108):

$$P_{N-1} K_N = R[N] - \sum_{k=1}^{N-1} a_k^{N-1} R[N-k] \qquad P_N = \left(1 - K_N^2\right) P_{N-1} \tag{13-67}$$

Find a_k^N from (12-97)

$$a_N^N = K_N \qquad a_k^N = a_k^{N-1} - K_N a_{N-k}^{N-1} \qquad 1 \le k \le N-1 \tag{13-68}$$

In Levinson's algorithm, the order N of the iteration is finite but it can continue indefinitely. We shall examine the properties of the predictor and of the MS error P_N as $N \to \infty$. It is obvious that P_N is a nonincreasing sequence of positive numbers; hence it tends to a positive limit:

$$P_1 \ge P_2 \cdots \ge P_N \underset{N\to\infty}{\longrightarrow} P \ge 0 \tag{13-69}$$

As we have shown in Sec. 11-3, the zeros z_i of the error filter

$$\hat{\mathbf{E}}_N(z) = 1 - \sum_{k=1}^{N} a_k^N z^{-k}$$

are either all inside the unit circle or they are all on the unit circle:

If $P_N > 0$, then $|K_k| < 1$ for every $k \le N$ and $|z_i| < 1$ for every i [see (12-99)].

If $P_{N-1} > 0$ and $P_N = 0$, then $|K_k| < 1$ for every $k \le N-1$, $|K_N| = 1$, and $|z_i| = 1$ for every i [see (12-101)]. In this case, the process $\mathbf{s}[n]$ is predictable and its spectrum consists of lines.

If $P > 0$, then $|z_i| \le 1$ for every i [see (13-26)]. In this case, the predictor $\hat{\mathbf{s}}_N[n]$ of $\mathbf{s}[n]$ tends to the Wiener predictor $\hat{\mathbf{s}}[n]$ as in (13-19). From this and (13-34) it follows that

$$P = \exp\left\{\frac{1}{2\pi}\int_{-\pi}^{\pi} \ln S(\omega)\, d\omega\right\} = l[0] = \lim_{N\to\infty} \frac{\Delta_{N+1}}{\Delta_N} \tag{13-70}$$

This shows the connection between the LMS error P of the prediction of $\mathbf{s}[n]$ in terms of its entire past, the power spectrum $S(\omega)$ of $\mathbf{s}[n]$, the initial value $l[0]$ of the delta response $l[n]$ of its innovations filter, and the correlation determinant Δ_N.

Suppose, finally, that $P_{M-1} > P_M$ and

$$P_M = P_{M-1} = \cdots = P \tag{13-71}$$

In this case, $K_k = 0$ for $|k| > M$; hence the algorithm terminates at the Mth step. From this it follows that the Mth order predictor $\hat{\mathbf{s}}_M[n]$ of $\mathbf{s}[n]$ equals its Wiener predictor:

$$\hat{\mathbf{s}}_M[n] = \hat{E}\{\mathbf{s}[n] \mid \mathbf{s}[n-k], 1 \le k \le M\} = \hat{E}\{\mathbf{s}[n] \mid \mathbf{s}[n-k], k \ge 1\}$$

In other words, the process $\mathbf{s}[n]$ is *wide-sense Markov* of order M. This leads to the conclusion that the prediction error $\hat{\boldsymbol{\varepsilon}}_M[n] = \mathbf{s}[n] - \hat{\mathbf{s}}_M[n]$ is white noise with average power P [see (13-24)]:

$$\mathbf{s}[n] - \sum_{k=1}^{M} a_k^N \mathbf{s}[n-k] = \hat{\boldsymbol{\varepsilon}}_M[n] \qquad E\{\hat{\boldsymbol{\varepsilon}}_M^2[n]\} = P$$

and it shows that $\mathbf{s}[n]$ is an AR process. Conversely, if $\mathbf{s}[n]$ is AR, then it is also wide-sense Markov.

Autoregressive processes and maximum entropy. Suppose that $\mathbf{s}[n]$ is an AR process of order M with autocorrelation $R[m]$ and $\overline{\mathbf{s}}[n]$ is a general process with autocorrelation $\overline{R}[m]$ such that

$$\overline{R}[m] = R[m] \qquad \text{for} \quad |m| \le M$$

The predictors of these processes of order M are identical because they depend on the values of $R[m]$ for $|m| \le M$ only. From this it follows that the corresponding prediction errors P_M and P_M are equal. As we have noted, $P_M = P$ for the AR process $\mathbf{s}[n]$ and $P_M \ge P$ for the general process $\overline{\mathbf{s}}[n]$.

Consider now the class C_M of processes with identical autocorrelations (data) for $|m| \le M$. Each $R[m]$ is a p.d. extrapolation of the given data. We have shown in Sec. 12-3 that the extrapolating sequence obtained with the maximum entropy (ME) method is the autocorrelation of an AR process [see (12-182)]. This leads to the following relationship between MS estimation and maximum entropy: The ME extrapolation is the autocorrelation of a process $\mathbf{s}[n]$ in the class C_M, the predictor of which *maximizes the minimum MS error P*. In this sense, *the ME method maximizes our uncertainty* about the values of $R[m]$ for $|m| > M$.

Causal Data

We wish to estimate the present value of a regular process $\mathbf{s}[n]$ in terms of its finite past, starting from some origin. The data are now available from 0 to $n-1$ and the desired estimate is given by

$$\hat{\mathbf{s}}_n[n] = \hat{E}\{\mathbf{s}[n] \mid \mathbf{s}[n-k], 1 \le k \le n\} = \sum_{k=1}^{n} a_k^n \mathbf{s}[n-k] \tag{13-72}$$

Unlike the fixed length N of the FIR predictor $\hat{\mathbf{s}}_N[n]$ considered in (13-62), the length n of this estimate is not constant. Furthermore, the values a_k^n of the coefficients of the filter specified by (13-72) depend on n. Thus the estimator of the process $\mathbf{s}[n]$ in terms of its causal past is a linear *time-varying filter*. If it is realized by a tapped-delay line as in Fig. 13-8, the number of the taps increases and the values of the weights change as n increases.

The coefficients a_k^n of $\hat{\mathbf{s}}_n[n]$ can be determined recursively from Levinson's algorithm where now $N = n$. Introducing the backward estimate $\check{\mathbf{s}}[n]$ of $\mathbf{s}[n]$ in terms of its n most recent future values, we conclude from (12-92) that

$$\begin{aligned} \hat{\mathbf{s}}_n[n] &= \hat{\mathbf{s}}_{n-1}[n] + K_n(\mathbf{s}[0] - \check{\mathbf{s}}_{n-1}[0]) \\ \check{\mathbf{s}}_n[0] &= \check{\mathbf{s}}_{n-1}[0] + K_n(\mathbf{s}[n] - \hat{\mathbf{s}}_{n-1}[n]) \end{aligned} \tag{13-73}$$

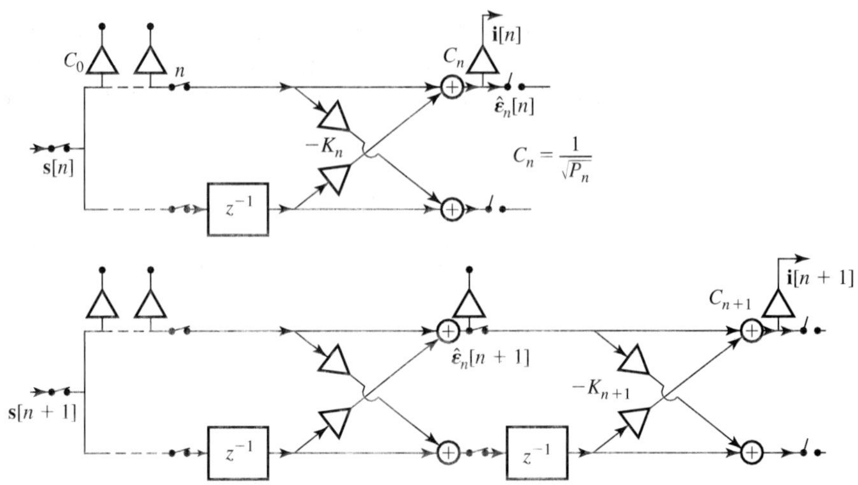

FIGURE 13-10

In Fig. 13-10, we show the *normalized lattice* realization of the error filter $\mathbf{E}_n(z)$, where we use as upper output the process

$$\mathbf{i}[n] = \frac{1}{\sqrt{P_n}}\hat{\boldsymbol{\varepsilon}}_n[n] \qquad E\{\mathbf{i}^2[n]\} = 1 \tag{13-74}$$

The filter is formed by switching "on" successively a new lattice section starting from the left. This filter is again time-varying; however, unlike the tapped-delay line realization, the elements of each section remain unchanged as n increases. We should point out that whereas $\hat{\boldsymbol{\varepsilon}}_k[n]$ is the value of the upper response of the kth section at time n, the process $\mathbf{i}[n]$ does not appear at a fixed position. It is the output of the last section that is switched "on" and as n increases, the point where $\mathbf{i}[n]$ is observed changes.

We conclude with the observation that if the process $\mathbf{s}[n]$ is AR of order M [see (12-81)], then the lattice stops increasing for $n > M$, realizing, thus, the time invariant system $\mathbf{E}_M(z)/\sqrt{P_M}$. The corresponding inverse lattice (see Fig. 12-15) realizes the all-pole system

$$\frac{\sqrt{P_M}}{\mathbf{E}_M(z)}$$

We shall now show that the output $\mathbf{i}[n]$ of the normalized lattice is white noise

$$R_{ii}[m] = \delta[m] \tag{13-75}$$

Indeed, as we know, $\hat{\boldsymbol{\varepsilon}}_n[n] \perp \mathbf{s}[n-k]$ for $1 \le k \le n$. Furthermore, $\hat{\boldsymbol{\varepsilon}}_{n-k}[n-r]$ depends linearly only on $\mathbf{s}[n-r]$ and its past values. Hence

$$\hat{\boldsymbol{\varepsilon}}_n[n] \perp \hat{\boldsymbol{\varepsilon}}_{n-1}[n-1] \tag{13-76}$$

This yields (13-75) because $P_n = E\{\boldsymbol{\varepsilon}_n^2[n]\}$.

Note In a lattice of fixed length, the output $\hat{\boldsymbol{\varepsilon}}_N[n]$ is not white noise and it is not orthogonal to $\hat{\boldsymbol{\varepsilon}}_{N-1}[n]$. However for a specific n, the random variables $\hat{\boldsymbol{\varepsilon}}_N[n]$ and $\hat{\boldsymbol{\varepsilon}}_{N-1}[n-1]$ are orthogonal.

KALMAN INNOVATIONS.[5] The output $\mathbf{i}[n]$ of the time-varying lattice of Fig. 13-10 is an orthonormal process that depends linearly on $\mathbf{s}[n-k]$. Denoting by γ_k^n the response of the lattice at time n to the input $\mathbf{s}[n] = \delta[n-k]$, we obtain

$$\begin{aligned} \mathbf{i}[0] &= \gamma_0^0 \mathbf{s}[0] \\ \mathbf{i}[1] &= \gamma_0^1 \mathbf{s}[0] + \gamma_1^1 \mathbf{s}[1] \\ &\cdots\cdots\cdots\cdots\cdots\cdots \\ \mathbf{i}[n] &= \gamma_0^n \mathbf{s}[0] + \cdots + \gamma_k^n \mathbf{s}[k] + \cdots + \gamma_n^n \mathbf{s}[n] \end{aligned} \tag{13-77}$$

or in vector form

$$\mathbf{I}_{n+1} = \mathbf{S}_{n+1}\Gamma_{n+1} \qquad \Gamma_{n+1} = \begin{bmatrix} \gamma_0^0 & \gamma_0^1 & \cdots & \gamma_0^n \\ & \gamma_1^1 & \cdots & \gamma_1^n \\ & 0 & \cdots & \cdots \\ & & & \gamma_n^n \end{bmatrix}$$

where $\mathbf{S}_{n+1}$ and $\mathbf{I}_{n+1}$ are row vectors with components

$$\mathbf{s}[0], \ldots, \mathbf{s}[n] \qquad \text{and} \qquad \mathbf{i}[0], \ldots, \mathbf{i}[n]$$

respectively.

From this it follows that if

$$\mathbf{s}[n] = \delta[n-k] \qquad \text{then} \quad \mathbf{i}[n] = \gamma_k^n \qquad n \ge k$$

This shows that to determine the delta response of the lattice of Fig. 13-10, we use as input the delta sequence $\delta[n-k]$ and we observe the moving output $\mathbf{i}[n]$ for $n \ge k$.

The elements γ_k^n of the triangular matrix Γ_{n+1} can be expressed in terms of the weights a_k^n of the causal predictor $\hat{\mathbf{s}}_n[n]$. Since

$$\hat{\boldsymbol{\varepsilon}}_n[n] = \mathbf{s}[n] - \hat{\mathbf{s}}_n[n] = \sqrt{P_n}\,\mathbf{i}[n]$$

it follows from (13-72) that

$$\gamma_n^n = \frac{1}{\sqrt{P_n}} \qquad \gamma_{n-k}^n = \frac{-1}{\sqrt{P_n}} a_k^n \qquad k \ge 1$$

The inverse of the lattice of Fig. 13-10 is obtained by reversing the flow direction of the upper line and the sign of the upward weights $-K_n$ as in Fig. 12-15. The turn-on switches close again in succession starting from the left, and the input $\mathbf{i}[n]$ is applied at the terminal of the section that is connected last. The output at A is thus given by

$$\begin{aligned} \mathbf{s}[0] &= l_0^0 \mathbf{i}[0] \\ \mathbf{s}[1] &= l_0^1 \mathbf{i}[0] + l_1^1 \mathbf{i}[1] \qquad & \mathbf{S}_{n+1} = \mathbf{I}_{n+1} L_{n+1} \\ &\cdots\cdots\cdots\cdots\cdots \\ \mathbf{s}[n] &= l_0^n \mathbf{i}[0] + \cdots + l_n^n \mathbf{i}[n] \qquad & L_n = \Gamma_n^{-1} \end{aligned} \tag{13-78}$$

From this it follows that if

$$\mathbf{i}[n] = \delta[n-k] \qquad \text{then} \quad \mathbf{s}[n] = l_k^n \qquad n \ge k$$

---

[5] T. Kailath, A. Vieira, and M. Morf: "Inverses of Toeplitz Operators, Innovations, and Orthogonal Polynomials," *SIAM Review,* vol. 20, no. 1, 1978.

Thus, to determine the delta response l_k^n of the inverse lattice, we use as moving input the delta sequence $\delta[n-k]$ and we observe the left output $\mathbf{s}[n]$ for $n \geq k$.

From the preceding discussion it follows that the random vector $\mathbf{S}_n$ is linearly equivalent to the orthonormal vector $\mathbf{I}_n$. Thus Eqs. (13-77) and (13-78) correspond to the Gram–Schmidt orthonormalization equations (7-100) and (7-103) of Sec. 7-3. Applying the terminology of Sec. 11-1 to causal signals, we shall call the process $\mathbf{i}[n]$ the *Kalman innovations* of $\mathbf{s}[n]$ and the lattice filter and its inverse *Kalman whitening and Kalman innovations* filters, respectively. These filters are *time-varying* and their transition matrices equal Γ_n and L_n, respectively. Their elements can be expressed in terms of the parameters K_n and P_n of Levinson's algorithm because these parameters specify completely the filters.

Cholesky factorization We maintain that the correlation matrix R_n and its inverse can be written as products

$$R_n = L_n^t L_n \qquad R_n^{-1} = \Gamma_n \Gamma_n^t \tag{13-79}$$

where Γ_n and L_n are the triangular matrices introduced earlier. Indeed, from the orthonormality of $\mathbf{I}_n$ and the definition of R_n, it follows that

$$E\{\mathbf{I}_n^t \mathbf{I}_n\} = 1_n \qquad E\{\mathbf{S}_n^t \mathbf{S}_n\} = R_n$$

where 1_n is the identity matrix. Since $\mathbf{I}_n = \mathbf{S}_n \Gamma_n$ and $\mathbf{S}_n = \mathbf{I}_n L_n$, these equations yield

$$\Gamma_n^t R_n \Gamma_n = 1_n \qquad L_n^t 1_n L_n = R_n$$

and (13-79) results.

Autocorrelation as lattice response. We shall determine the autocorrelation $R[m]$ of the process $\mathbf{s}[n]$ in terms of the Levinson parameters K_N and P_N. For this purpose, we form a lattice of order N_0 and we denote by $\hat{q}_N[m]$ and $\check{q}_N[m]$ respectively its upper and lower responses (13-11a) to the input $R[m]$. As we see from the figure

$$\hat{q}_{N-1}[m] = \hat{q}_N[m] + K_N \check{q}_{N-1}[m-1] \tag{13-80a}$$

$$\check{q}_N[m] = \check{q}_{N-1}[m-1] - K_N \hat{q}_{N-1}[m] \tag{13-80b}$$

$$\hat{q}_0[m] = \check{q}_0[m] = R[m] \tag{13-80c}$$

Using this development, we shall show that $R[m]$ can be determined as the response of the inverse lattice of Fig. 13-11b provided that the following boundary and initial conditions are satisfied[6]: The input to the system (point B) is identically 0:

$$\hat{q}_{N_0}[m] = 0 \qquad \text{all} \quad m \tag{13-81}$$

The initial conditions of all delay elements except the first are 0:

$$\check{q}_N[0] = 0 \qquad N > 0 \tag{13-82}$$

[6]E. A. Robinson and S. Treitel: "Maximum Entropy and the Relationship of the Partial Autocorrelation to the Reflection Coefficients of a Layered System," *IEEE Transactions on Acoustics, Speech, and Signal Process,* vol. ASSP-28, no. 2, 1980.

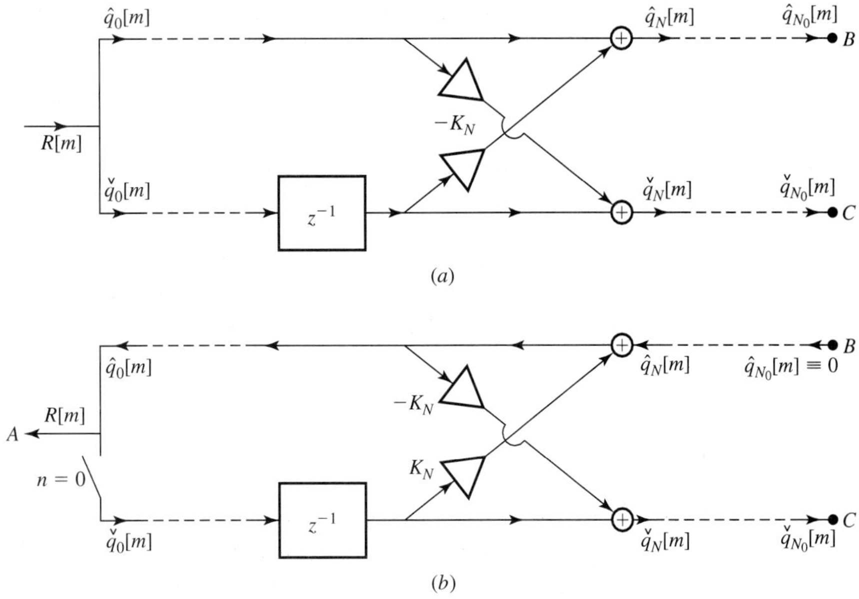

FIGURE 13-11

The first delay element is connected to the system at $m = 0$ and its initial condition equals $R[0]$:

$$\check{q}_0[0] = R[0] \tag{13-83}$$

From the above and (13-81) it follows that

$$\check{q}_N[1] = 0 \qquad N > 1$$

We maintain that under the stated conditions, the left output of the inverse lattice (point A) equals $R[m]$ and the right output of the mth section equals the MS error P_m:

$$\hat{q}_0[m] = R[m] \qquad \check{q}_m[m] = P_m \tag{13-84}$$

Proof. The proof is based on the fact that the responses of the lattice of Fig. 13-10*a* satisfy the equations (see Prob. 13-24)

$$\hat{q}_N[m] = \check{q}_N[m] = 0 \qquad 1 \le m \le N-1 \tag{13-85}$$

$$\check{q}_N[N] = P_N \tag{13-86}$$

From (13-80) it follows that, if we know $\hat{q}_N[m]$ and $\check{q}_{N-1}[m-1]$, then we can find $\hat{q}_{N-1}[m]$ and $\check{q}_N[m]$. By a simple induction, this leads to the conclusion that if $\hat{q}_{N_0}[m]$ is specified for every m (boundary conditions) and $\check{q}_N[1]$ is specified for every N (initial conditions), then all responses of the lattice are determined uniquely. The two systems of Fig. 13-11 satisfy the same equations (13-80) and, as we noted, they have identical initial and boundary conditions. Hence all their responses are identical. This yields (13-84).

13-3 FILTERING AND PREDICTION

In this section, we consider the problem of estimating the future value $\mathbf{s}(t+\lambda)$ of a stochastic process $\mathbf{s}(t)$ (signal) in terms of the present and past values of a regular process $\mathbf{x}(t)$ (signal plus noise)

$$\hat{\mathbf{s}}(t+\lambda) = \hat{E}\{\mathbf{s}(t+\lambda) \mid \mathbf{x}(t-\tau), \tau \geq 0\} = \int_0^\infty h_x(\alpha)\mathbf{x}(t-\alpha)\,d\alpha \qquad (13\text{-}87)$$

Thus $\hat{\mathbf{s}}(t+\lambda)$ is the output of a linear time-invariant causal system $\mathbf{H}_x(s)$ with input $\mathbf{x}(t)$. To determine $\mathbf{H}_x(s)$, we use the orthogonality principle

$$E\left\{\left[\mathbf{s}(t+\lambda) - \int_0^\infty h_x(\alpha)\mathbf{x}(t-\alpha)\,d\alpha\right]\mathbf{x}(t-\tau)\right\} = 0 \qquad \tau \geq 0$$

This yields the *Wiener–Hopf* equation

$$R_{sx}(\tau+\lambda) = \int_0^\infty h_x(\alpha)R_{xx}(\tau-\alpha)\,d\alpha \qquad \tau \geq 0 \qquad (13\text{-}88)$$

The solution $h_x(t)$ of (13-88) is the impulse response of the prediction and filtering system known as the *Wiener filter.* If $\mathbf{x}(t) = \mathbf{s}(t)$, then $h_x(t)$ is a pure predictor as in (13-39). If $\lambda = 0$, then $h_x(t)$ is a pure filter.

To solve (13-88), we express $\mathbf{x}(t)$ in terms of its innovations $\mathbf{i}_x(t)$ (Fig. 13-12)

$$\mathbf{x}(t) = \int_0^\infty l_x(\alpha)\mathbf{i}_x(t-\alpha)\,d\alpha \qquad R_{ii}(\tau) = \delta(\tau) \qquad (13\text{-}89)$$

where $l_x(t)$ is the impulse response of the innovations filter $\mathbf{L}_x(s)$ obtained by factoring the spectrum of $\mathbf{x}(t)$ as in (11-3):

$$\mathbf{S}_{xx}(s) = \mathbf{L}_x(s)\mathbf{L}_x(-s) \qquad (13\text{-}90)$$

As we know, the processes $\mathbf{i}_x(t)$ and $\mathbf{x}(t)$ are linearly equivalent; hence the estimate $\hat{\mathbf{s}}(t+\lambda)$ can be expressed as the output of a causal filter $\mathbf{H}_{i_x}(s)$ with input $\mathbf{i}_x(t)$:

$$\hat{\mathbf{s}}(t+\lambda) = \int_0^\infty h_{i_x}(\alpha)\mathbf{i}_x(t-\alpha)\,d\alpha \qquad (13\text{-}91)$$

To determine $h_{i_x}(t)$, we use the orthogonality principle

$$E\left\{\left[\mathbf{s}(t+\lambda) - \int_0^\infty h_{i_x}(\alpha)\mathbf{i}_x(t-\alpha)\,d\alpha\right]\mathbf{i}_x(t-\tau)\right\} = 0 \qquad \tau \geq 0$$

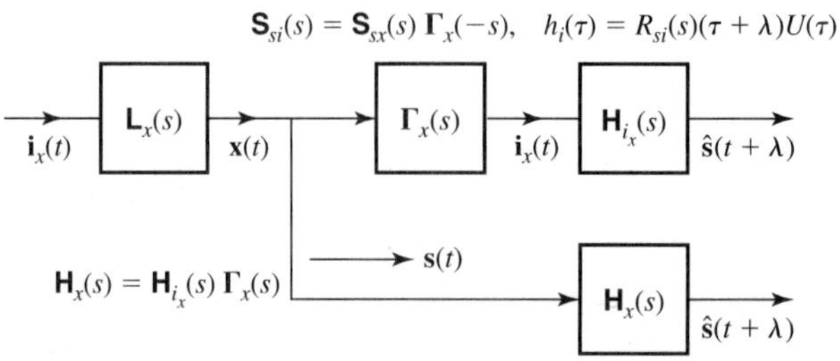

FIGURE 13-12

Since $\mathbf{i}_x(t)$ is white noise, this yields

$$R_{si_x}(\tau+\lambda) = \int_0^{\infty} h_{i_x}(\alpha)\delta(\tau-\alpha)\,d\alpha = h_{i_x}(\tau) \qquad \tau \ge 0 \tag{13-92}$$

This determines $h_{i_x}(\tau)$ for all τ because $h_{i_x}(\tau) = 0$ for $\tau < 0$:

$$h_{i_x}(\tau) = R_{si_x}(\tau+\lambda)U(\tau) \tag{13-93}$$

In (13-92)–(13-93), $R_{si_x}(\tau)$ is the cross-correlation between the signal $\mathbf{s}(t)$ and the process $\mathbf{i}_x(t)$. The function $R_{si_x}(\tau)$ can be expressed in terms of the cross-correlation $R_{sx}(\tau)$ between $\mathbf{s}(t)$ and $\mathbf{x}(t)$. Indeed, since $\mathbf{i}_x(t)$ is the output of the whitening filter $\mathbf{\Gamma}_x(s)$ with input $\mathbf{x}(t)$, we can show as in (9-130) and (9-170) that

$$\mathbf{S}_{si_x}(s) = \mathbf{S}_{sx}(s)\mathbf{\Gamma}_x(-s) \tag{13-94}$$

Thus, since $\mathbf{S}_{sx}(s)$ is assumed known, (13-94) yields $R_{si_x}(\tau)$. Shifting to the left and truncating as in (13-93), we obtain $h_{i_x}(\tau)$.

To complete the specification of $\mathbf{H}_x(s)$, we multiply the transform $\mathbf{H}_{i_x}(s)$ of the function $h_{i_x}(t)$ so obtained with $\mathbf{\Gamma}_x(s)$ (see Fig. 13-12)

$$\mathbf{H}_x(s) = \mathbf{H}_{i_x}(s)\mathbf{\Gamma}_x(s) \tag{13-95}$$

The function $\mathbf{H}_{i_x}(s)$ can be determined directly from (13-94): As we know (shifting theorem) the transform of $R_{si_x}(\tau+\lambda)$ equals

$$\mathbf{S}_{\lambda}(s) = \mathbf{S}_{si_x}(s)e^{\lambda s} = \mathbf{S}_{sx}(s)\mathbf{\Gamma}_x(-s)e^{\lambda s} \tag{13-96}$$

To find $\mathbf{H}_{i_x}(s)$, it suffices to write $\mathbf{S}_{\lambda}(s)$ as a sum

$$\mathbf{S}_{\lambda}(s) = \mathbf{S}_{\lambda}^{+}(s) + \mathbf{S}_{\lambda}^{-}(s) \tag{13-97}$$

where $\mathbf{S}_{\lambda}^{+}(s)$ is analytic in the right-hand s plane and $\mathbf{S}_{\lambda}^{-}(s)$ is analytic in the left-hand s plane. Since the inverse transforms of the function $\mathbf{S}_{\lambda}^{+}(s)$ and $\mathbf{S}_{\lambda}^{-}(s)$ equal $R_{si_x}(\tau+\lambda)U(\tau)$ and $R_{si_x}(\tau+\lambda)U(-\tau)$, respectively, we conclude from (13-93) that (see also, the next Note)

$$\mathbf{H}_{i_x}(s) = \mathbf{S}_{\lambda}^{+}(s) \tag{13-98}$$

To determine the system function $\mathbf{H}_x(s)$ of the Wiener filter, proceed, thus, as follows:

Factor $\mathbf{S}_{xx}(s)$ as in (13-90) and set $\mathbf{\Gamma}_x(s) = 1/\mathbf{L}_x(s)$.
Evaluate $\mathbf{S}_{si_x}(s)$ from (13-94) and form the function $\mathbf{S}_{\lambda}(s)$ using (13-96).
Decompose $\mathbf{S}_{\lambda}(s)$ as in (13-97) and form the function $\mathbf{H}_{i_x}(s)$ using (13-98).
Determine $\mathbf{H}_x(s)$ from (13-95).

If the function $\mathbf{S}_{\lambda}(s)$ is rational, then the decomposition (13-97) can be accomplished by expanding $\mathbf{S}_{si_x}(s)$ into partial fractions. Assuming that $\mathbf{S}_{si_x}(s)$ is a proper fraction with simple poles, we obtain

$$\mathbf{S}_{si_x}(s) = \sum_i \frac{a_i}{s-s_i} + \sum_k \frac{b_k}{s-z_k} \qquad \begin{array}{l}\text{Re } S_i < 0\\ \text{Re } z_k > 0\end{array} \tag{13-99}$$

The inverse of the second sum is 0 for $\tau > 0$. If, therefore, it is shifted to the left, it will remain 0 for $\tau > 0$. This shows that only the first sum will contribute to the term $R_{si_x}(\tau + \lambda)U(\tau)$. In other words,

$$R_{si_x}(\tau + \lambda)U(\tau) = \left[a_1 e^{s_1(\tau+\lambda)} + \cdots + a_n e^{s_n(\tau+\lambda)}\right] U(\tau)$$

The transform of this equation yields

$$\mathbf{S}_\lambda^+(s) = \frac{a_1 e^{s_1\lambda}}{s - s_1} + \cdots + \frac{a_n e^{s_n\lambda}}{s - s_n} \tag{13-100}$$

EXAMPLE 13-7

▶ Suppose that $\mathbf{x}(t) = \mathbf{s}(t) + \boldsymbol{\nu}(t)$ and

$$S_{ss}(\omega) = \frac{N_0}{\alpha^2 + \omega^2} \qquad S_{\nu\nu}(\omega) = N \qquad S_{s\nu}(\omega) = 0 \tag{13-101}$$

as in Example 13-1. In this case, $\mathbf{S}_{sx}(s) = \mathbf{S}_{ss}(s)$ and

$$\mathbf{S}_{xx}(s) = \frac{N_0}{\alpha^2 - s^2} + N = N\frac{\beta^2 - s^2}{\alpha^2 - s^2} \qquad \beta^2 = \alpha^2 + \frac{N_0}{N}$$

Hence

$$\mathbf{L}_x(s) = \sqrt{N}\,\frac{s + \beta}{s + \alpha} \qquad \mathbf{\Gamma}_x(-s) = \frac{1}{\sqrt{N}}\,\frac{\alpha - s}{\beta - s} \tag{13-102}$$

Inserting into (13-94) and expanding into partial fractions, we obtain

$$\mathbf{S}_{si_x}(s) = \frac{N_0}{\alpha^2 - s^2}\,\frac{\alpha - s}{(\beta - s)\sqrt{N}} = \frac{A}{s + \alpha} - \frac{A}{s - \beta} \qquad A = \frac{N_0}{(\alpha + \beta)\sqrt{N}}$$

and with $s_1 = -\alpha$, (13-100) yields

$$\mathbf{S}_\lambda^+(s) = \frac{A}{s + \alpha}e^{-\alpha\lambda}$$

Hence

$$\mathbf{H}_x(s) = \mathbf{S}_\lambda^+(s)\mathbf{\Gamma}_x(s) = \frac{\beta - \alpha}{s + \beta}e^{-\alpha\lambda} \tag{13-103}$$

◀

Note In the decomposition (13-97) of $\mathbf{S}_\lambda(s)$, the functions $\mathbf{S}_\lambda^+(s)$ and $\mathbf{S}_\lambda^-(s)$ are unique within an additive constant. This causes an ambiguity in the determination of $h_{i_x}(t)$. The ambiguity is removed if we impose the condition that

$$\mathbf{S}_\lambda^-(\infty) = 0$$

In the pure filtering case ($\lambda = 0$); the resulting $h_x(t)$ might contain impulses at the origin. This is acceptable because, by assumption the estimate $\hat{\mathbf{s}}(t)$ of $\mathbf{s}(t)$ is a functional of the past *and* the present value of the data $\mathbf{x}(t)$.

Filtering white noise. In the pure filtering problem, the determination of the estimator $\mathbf{H}_x(s)$ can be simplified if $R_{ss}(0) < \infty$ and $\boldsymbol{\nu}(t)$ is white noise orthogonal to the signal as in (13-101). We maintain, in fact, that in this case

$$\mathbf{H}_x(s) = 1 - \sqrt{N}\,\mathbf{\Gamma}_x(s) \tag{13-104}$$

where $\mathbf{\Gamma}_x(s)$ is the whitening filter of $\mathbf{x}(t)$.

Proof. From these assumptions it follows that $\mathbf{S}_{ss}(\infty) = 0$; hence

$$\mathbf{S}_{sx}(s) = \mathbf{S}_{ss}(s) = \mathbf{S}_{xx}(s) - N = \mathbf{L}_x(s)\mathbf{L}_x(-s) - N$$

$$\mathbf{S}_{sx}(\infty) = 0 \qquad \mathbf{S}_{xx}(\infty) = N \qquad \mathbf{L}_x(\pm\infty) = \sqrt{N}$$

Inserting into (13-94), we obtain

$$\mathbf{S}_{si_x}(s) = \mathbf{L}_x(s) - N\mathbf{\Gamma}_x(-s) = \mathbf{L}_x(s) + K - N\mathbf{\Gamma}_x(-s) - K$$

From the preceding note it follows that the constant K must be such that the noncausal component of $\mathbf{S}_{si_x}(s)$ satisfies the infinity condition $-N\mathbf{\Gamma}_x(-\infty) - K = 0$. And since $\mathbf{\Gamma}_x(-\infty) = 1/\mathbf{L}_x(-\infty) = 1/\sqrt{N}$, (13-104) follows from (13-95).

EXAMPLE 13-8

▶ We shall determine the pure filter of the process in Example 13-7. From (13-102) and (13-104) it follows that

$$\mathbf{H}_x(s) = 1 - \frac{\alpha + s}{\beta + s} = \frac{\beta - \alpha}{s + \beta} \qquad h_x(t) = (\beta - \alpha)e^{-\beta t}U(t)$$

in agreement with (13-103). Note that the resulting MS error equals

$$P = E\left\{\left[\mathbf{s}(t) - \int_0^\infty h_x(\alpha)\mathbf{x}(t-\alpha)\,d\alpha\right]\mathbf{s}(t)\right\} = \frac{N_0}{\alpha + \beta}$$ ◀

Discrete-Time Processes

We shall state briefly the discrete-time version of the preceding results. Our problem now is the determination of the future value $\mathbf{s}[n+r]$ of a stochastic process in terms of the present and past values of another process $\mathbf{x}[n]$:

$$\hat{\mathbf{s}}_r[n+r] = \sum_{k=0}^{\infty} h_x^r[k]\mathbf{x}[n-k] \tag{13-105}$$

In this case,

$$\mathbf{s}[n+r] - \hat{\mathbf{s}}_r[n+r] \perp \mathbf{x}[n-m] \qquad m \ge 0$$

hence

$$R_{sx}[m+r] = \sum_{k=0}^{\infty} h_x^r[k]R_{xx}[m-k] \qquad m \ge 0 \tag{13-106}$$

This is the discrete version of the Wiener–Hopf equation (13-88).

To determine $h_x^r[n]$, we proceed as in the analog case: We express $\hat{\mathbf{s}}_r[n+r]$ in terms of the innovations $\mathbf{i}_x[n]$ of $\mathbf{x}[n]$ (Fig. 13-13)

$$\hat{\mathbf{s}}_r[n+r] = \sum_{k=0}^{\infty} h_{i_x}^r[k]\mathbf{i}_x[n-k] \tag{13-107}$$

From this and (7-82) it follows that

$$R_{si_x}[m+r] = \sum_{k=0}^{\infty} h_{i_x}^r[k]\delta[m-k] = h_{i_x}^r[m] \qquad m \ge 0$$

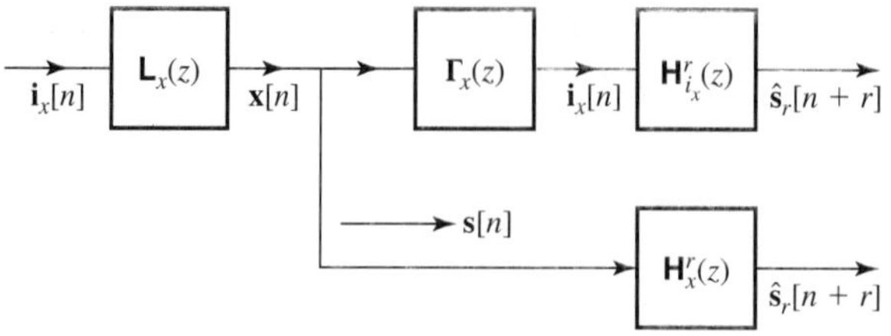

FIGURE 13-13

because $R_{i_x}[m] = \delta[m]$. Hence

$$h^r_{i_x}[m] = R_{si_x}[m+r]U[m] \qquad \text{all} \quad m \tag{13-108}$$

The function $R_{si_x}[m]$ can be expressed in terms of $R_{sx}[m]$ as in (13-94)

$$\mathbf{S}_{si_x}(z) = \mathbf{S}_{sx}(z)\mathbf{\Gamma}_x(z^{-1}) \tag{13-109}$$

Thus the transform of $R_{si_x}[m+r]$ equals

$$\mathbf{S}_r(z) = z^r\mathbf{S}_{si_x}(z) = z^r\mathbf{S}_{sx}(z)\mathbf{\Gamma}_x(z^{-1}) \tag{13-110}$$

The function $\mathbf{S}_r(z)$ is then written as a sum

$$\mathbf{S}_r(z) = \mathbf{S}^+_r(z) + \mathbf{S}^-_r(z) \tag{13-111}$$

where $\mathbf{S}^+_r(z)$ is analytic for $|z| > 1$ and $\mathbf{S}^-_r(z)$ is analytic for $|z| < 1$. Furthermore, the inverse of $\mathbf{S}^-_r(z)$ at the origin is 0. Thus $\mathbf{S}^+_r(z)$ is the transform of the causal function $R_{si_x}[m+r]U[m]$. And since $\mathbf{i}_x[n]$ is the response of the whitening filter $\mathbf{\Gamma}_x(z)$ with input $\mathbf{x}[n]$, we conclude from (12-108) that

$$\mathbf{H}^r_x(z) = \mathbf{H}^r_{i_x}(z)\mathbf{\Gamma}_x(z) = \mathbf{S}^+_r(z)\mathbf{\Gamma}_x(z) \tag{13-112}$$

EXAMPLE 13-9

▶ We shall determine the one-step predictor $\hat{\mathbf{s}}_1[n+1]$ of the process $\mathbf{s}[n]$, where

$$\mathbf{S}_{ss}(z) = \frac{N_0}{(1-az^{-1})(1-az)} \qquad \mathbf{S}_{\nu\nu}(z) = N \qquad \mathbf{S}_{s\nu}(z) = 0$$

In this case (see Example 13-2)

$$\mathbf{L}_x(z) = \sqrt{\frac{Na}{b}}\,\frac{1-bz^{-1}}{1-az^{-1}}$$

From (13-110) it follows with $r = 1$ that

$$z\mathbf{S}_{si_x}(z) = \frac{zN_0\sqrt{b/Na}}{(1-az^{-1})(1-bz)} = \frac{Aaz}{z-a} - \frac{Az/b}{z-1/b} \qquad A = (a-b)\sqrt{\frac{N}{ab}}$$

Since $0 < a < 1$ and $1/b > 1$, we conclude from the above that $\mathbf{S}^+_1(z) = Aaz/(z-a)$ and (13-112) yields

$$\mathbf{H}^1_x(z) = (a-b)\frac{z}{z-b} \qquad h^1_x[n] = (a-b)b^nU[n]$$

We discuss presently a more direct method for determining $\mathbf{H}^r_x(z)$ [see (13-118)]. ◀

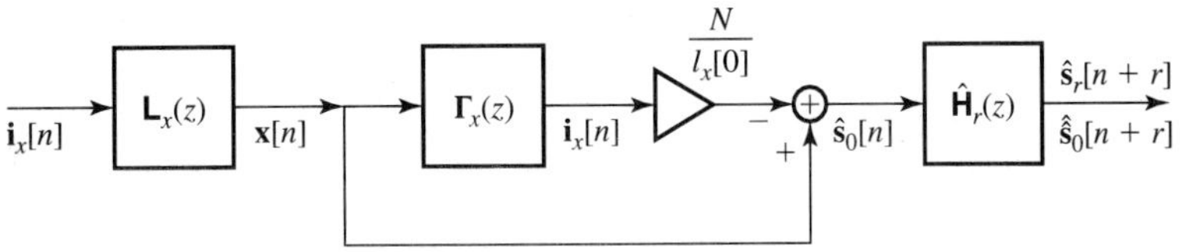

FIGURE 13-14

White noise. We shall examine the nature of the predictor $\mathbf{H}_x^r(z)$ of $\mathbf{s}[n+r]$ under the assumption that the noise is white and orthogonal to the signal

$$R_{\nu\nu}[m] = N\delta[m] \qquad R_{s\nu}[m] = 0 \tag{13-113}$$

Pure filter Suppose first that $r = 0$. In this case, $\mathbf{H}_x^0(z)$ is a pure filter and $\hat{\mathbf{s}}_0[n]$ is the estimate of the signal $\mathbf{s}[n]$ in terms of $\mathbf{x}[n]$ and its past.

We maintain that (Fig. 13-14)

$$\mathbf{H}_x^0(z) = 1 - \frac{D}{\mathbf{L}_x(z)} \qquad D = \frac{N}{l_x[0]} \tag{13-114}$$

Proof. From (13-113) it follows that

$$\mathbf{S}_{sx}(z) = \mathbf{S}_{ss}(z) = \mathbf{S}_{xx}(z) - N = \mathbf{L}_x(z)\mathbf{L}_x(z^{-1}) - N$$

Inserting into (13-109), we obtain

$$\mathbf{S}_{si_x}(z) = \mathbf{L}_x(z) - N\boldsymbol{\Gamma}_x(z^{-1}) \tag{13-115}$$

We wish to find the causal part of the above, including the value of its inverse at $n = 0$. Since the inverse z transform of $\boldsymbol{\Gamma}_x(1/z)$ is 0 for $n > 0$ and for $n = 0$ it equals $\boldsymbol{\Gamma}_x(\infty)$, we conclude that

$$\mathbf{H}_{i_x}^0(z) = \mathbf{L}_x(z) - N\boldsymbol{\Gamma}_x(\infty) \tag{13-116}$$

Multiplying by $\boldsymbol{\Gamma}_x(z)$, we obtain (13-114) because $\boldsymbol{\Gamma}_x(\infty) = 1/l_x[0]$.

Filtering and prediction We shall now show that the estimate $\hat{\mathbf{s}}_r[n+r]$ of $\mathbf{s}[n+r]$ equals the pure predictor $\hat{\hat{\mathbf{s}}}_0[n+r]$ of the estimate $\hat{\mathbf{s}}_0[n]$ of $\mathbf{s}[n]$ (Fig. 13-14)

$$\hat{\mathbf{s}}_r[n+r] = \hat{\hat{\mathbf{s}}}_0[n+r] = \hat{E}\{\hat{\mathbf{s}}_0[n+r] \mid \hat{\mathbf{s}}_0[n-k], k \geq 0\} \tag{13-117}$$

Proof. From (13-110) and (13-115) it follows that

$$\mathbf{S}_r(z) = z^r[\mathbf{L}_x(z) - N\boldsymbol{\Gamma}_x(z^{-1})]$$

But the inverse of $z^r\boldsymbol{\Gamma}_x(1/z)$ is 0 for $n \geq 0$. Hence $\mathbf{S}_r^+(z)$ is the causal part of $z^r\mathbf{L}_x(z)$. Inserting into (13-112), we obtain

$$\mathbf{H}_x^r(z) = z^r\left(\mathbf{L}_x(z) - \sum_{k=0}^{r-1} l_x[k]z^{-k}\right)\boldsymbol{\Gamma}_x(z) = z^r\left(1 - \frac{\sum_{k=0}^{r-1} l_x[k]z^{-k}}{\boldsymbol{\Gamma}_x(z)}\right) \tag{13-118}$$

As we see from Fig. 13-14, the innovations filter of $\hat{\mathbf{s}}_0[n]$ equals $\mathbf{L}_x(z)\mathbf{H}_x^0(z)$. To determine the pure predictor $\hat{\mathbf{H}}_r(z)$ of $\hat{\mathbf{s}}_0[n+r]$, it suffices, therefore, to multiply (13-37) by z^r (we are predicting now the future) and to replace the function $\mathbf{L}(z)$ by $\mathbf{L}_x(z)\mathbf{H}_x^0(z)$. This yields

$$\hat{\mathbf{H}}_r(z) = z^r\left(1 - \frac{\sum_{k=0}^{r-1} l_x[k]z^{-k} - D}{\mathbf{L}_x(z) - D}\right)$$

because the inverse of $\mathbf{L}_x(z) - D$ equals $l_x[n] - D\delta[n]$. Comparing with (13-118), we conclude that

$$\mathbf{H}_x^r(z) = \mathbf{H}_x^0(z)\hat{\mathbf{H}}_r(z)$$

The preceding discussion leads to the following important consequences of the white-noise assumption (13-113):

1. The innovations $\mathbf{i}_x[n]$ of $\mathbf{x}[n]$ are proportional to the difference $\mathbf{x}[n] - \hat{\mathbf{s}}_0[n]$:

$$\mathbf{x}[n] - \hat{\mathbf{s}}_0[n] = D\mathbf{i}_x[n] \qquad D = \frac{N}{l_x[0]} \tag{13-119}$$

Indeed, $\mathbf{x}[n] - \hat{\mathbf{s}}_0[n]$ is the output of the filter

$$\mathbf{L}_x(z) - [\mathbf{L}_x(z) - D] = D$$

with input $\mathbf{i}_x[n]$ (Fig. 13-15*a*). Thus the process $\mathbf{i}_x[n]$ can be realized simply by a feedback system (Fig. 13-15*b*) involving merely the filter $\mathbf{H}_{i_x}^0(z)$.

2. The r-step filtering and prediction estimate $\hat{\mathbf{s}}_0[n+r]$ can be obtained by cascading the pure filter $\mathbf{H}_x^0(z)$ of $\mathbf{s}[n]$ with the pure predictor $\hat{\mathbf{H}}_r(z)$ of $\hat{\mathbf{s}}_0[n+r]$.
3. If the signal $\mathbf{s}[n]$ is an ARMA process, then its estimate $\hat{\mathbf{s}}_0[n]$ is also an ARMA process.

   Indeed, if $\mathbf{L}_x(z) = A(z)/B(z)$ is rational, then [see (13-114)], the filter $\mathbf{H}_x^0(z)$ is also rational. Furthermore, the denominator $B(z)$ of $\mathbf{L}_x(z)$ is the same as the denominator of the forward component $\mathbf{L}_x(z) - D$ of the feedback realization of $\mathbf{H}_x^0(z)$ shown in Fig. 13-15*b*.

As we shall presently see, these results are central in the development of Kalman filters.

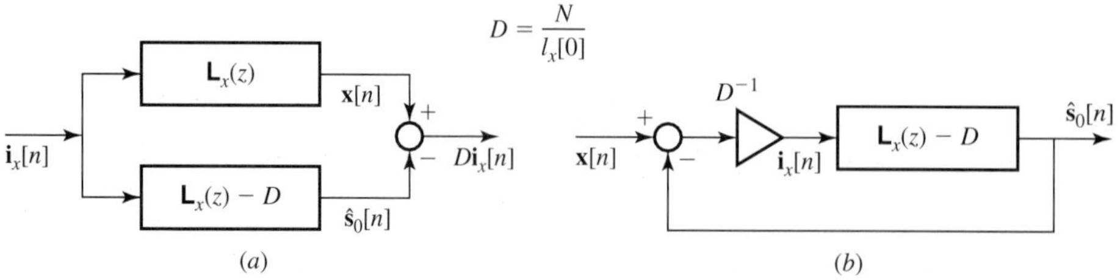

FIGURE 13-15

13-4 KALMAN FILTERS[7]

In this section we extend the preceding results to nonstationary processes with causal data and we show that the results can be simplified if the noise is white and the signal is an ARMA process. The estimate $\hat{\mathbf{s}}_r[n+r]$ of $\mathbf{s}[n+r]$ in terms of the data

$$\mathbf{x}[n] = \mathbf{s}[n] + \boldsymbol{\nu}[n]$$

takes the form

$$\hat{\mathbf{s}}_r[n+r] = \hat{E}\{\mathbf{s}[n+r] \mid \mathbf{x}[k], 0 \le k \le n\} = \sum_{k=0}^{n} h_x^r[n,k]\mathbf{x}[k] \qquad (13\text{-}120)$$

Thus $\hat{\mathbf{s}}_r[n+r]$ is the output of a causal, time-varying system with input $\mathbf{x}[n]U[n]$, and our problem is to find its delta response $h_x^r[n,k]$.

As we know,

$$\mathbf{s}[n+r] - \hat{\mathbf{s}}_r[n+r] \perp \mathbf{x}[m] \qquad 0 \le m \le n$$

This yields

$$R_{sx}[n+r,m] = \sum_{k=0}^{n} h_x^r[n,k]R_{xx}[k,m] \qquad 0 \le m \le n \qquad (13\text{-}121)$$

Thus $h_x^r[n,k]$ must be such that its response to $R_{xx}[n,m]$ (the time variable is n) equals $R_{sx}[n+r,m]$ for every $0 \le m \le n$. For a specific n, this yields $n+1$ equations for the $n+1$ unknowns $h_x^r[n,k]$.

To simplify the determination of $h_x^r[n,k]$, we shall express the desired estimates $\hat{\mathbf{s}}_r[n+r]$ in terms of the *Kalman innovations* [see (13-77)]

$$\mathbf{i}_x[n] = \sum_{k=0}^{n} \gamma_x[n,k]\mathbf{x}[k] \qquad (13\text{-}122)$$

of the process $\mathbf{x}[n]U[n]$, where $\gamma_x[n,k]$ is the Kalman whitening filter. The process $\mathbf{i}_x[n]$ is orthonormal and, if the data are linearly independent, then the processes $\mathbf{x}[n]$ and $\mathbf{i}_x[n]$ are linearly equivalent. This leads to the conclusion that $\hat{\mathbf{s}}_r[n+r]$ can be expressed in terms of $\mathbf{i}_x[n]$ and its past (Fig. 13-16)

$$\hat{\mathbf{s}}_r[n+r] = \sum_{k=0}^{n} h_{i_x}^r[n,k]\mathbf{i}_x[k] \qquad (13\text{-}123)$$

To determine $h_{i_x}^r[n,k]$, we apply the orthogonality principle. Since

$$R_{i_x}[m,n] = \delta[m-n]$$

this yields

$$R_{si_x}[n+r,m] = \sum_{k=0}^{n} h_{i_x}^r[n,k]\delta[k-m]$$

---

[7] R. E. Kalman: "A New Approach to Linear Filtering and Prediction Problems," *ASME Transactions*, vol. 82D, 1960.

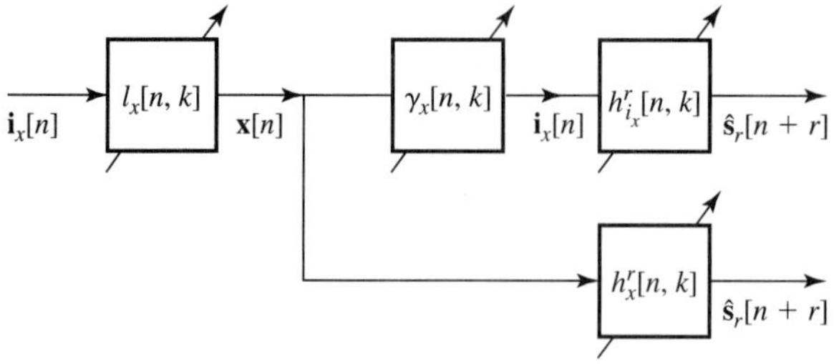

FIGURE 13-16

Hence

$$h^r_{i_x}[n, m] = R_{si_x}[n + r, m] \qquad 0 \le m \le n \tag{13-124}$$

This function can be expressed in terms of the cross-correlation $R_{sx}[m, n]$. Multiplying (13-122) by $\mathbf{s}[m]$, we obtain

$$R_{si_x}[m, n] = \sum_{k=0}^{n} \gamma_k[n, k] R_{sx}[m, k] \tag{13-125}$$

Thus, for a specific m, $R_{si_x}[m, n]$ is the response of the Kalman whitening filter of $\mathbf{x}[n]$ to the function $R_{sx}[m, n]$, where n is the variable. To complete the specification of $\hat{\mathbf{s}}_r[n+r]$, we cascade the filter $h^r_{i_x}[n, m]$ with the whitening filter $\gamma_x[n, k]$ as in Fig. 13-16.

ARMA Signals in White Noise

In the numerical implementation of the above, we are faced with two problems: (1) the realization of the Kalman innovations process $\mathbf{i}_x[n]$: (2) the determination of the sum in (13-123). In general, these problems are complex, involving storage capacity and number of computations proportional to n. However, as we show next, under certain realistic assumptions the problem can be simplified drastically.

ASSUMPTION 1. The noise is white and orthogonal to the signal:

$$R_{\nu\nu}[m, n] = N_n \delta[m - n] \qquad R_{s\nu}[m, n] = 0 \tag{13-126}$$

This leads to the following conclusions.

Property 1 If $\hat{\mathbf{s}}_0[n]$ is the estimate of $\mathbf{s}[n]$ in terms of $\mathbf{x}[n]$ and its past and D_n^2 is the MS estimation error, then the difference $\mathbf{x}[n] - \hat{\mathbf{s}}_0[n]$ is proportional to the Kalman innovations $\mathbf{i}_x[n]$ of the data $\mathbf{x}[n]$:

$$\mathbf{x}[n] - \hat{\mathbf{s}}_0[n] = D_n \mathbf{i}_x[n] \tag{13-127a}$$

$$D_n^2 = E\{|\mathbf{x}[n] - \hat{\mathbf{s}}_0[n]|^2\} \tag{13-127b}$$

Proof. The difference $\mathbf{x}[n] - \hat{\mathbf{s}}_0[n]$ depends linearly on $\mathbf{x}[n]$ and its past. Furthermore, the processes $\boldsymbol{\nu}[n]$ and $\mathbf{s}[n] - \hat{\mathbf{s}}_0[n]$ are orthogonal to the past of $\mathbf{x}[n]$. Hence

$$\mathbf{x}[n] - \hat{\mathbf{s}}_0[n] = \mathbf{s}[n] - \hat{\mathbf{s}}_0[n] + \boldsymbol{\nu}[n] \perp \mathbf{x}[k] \qquad k < n$$

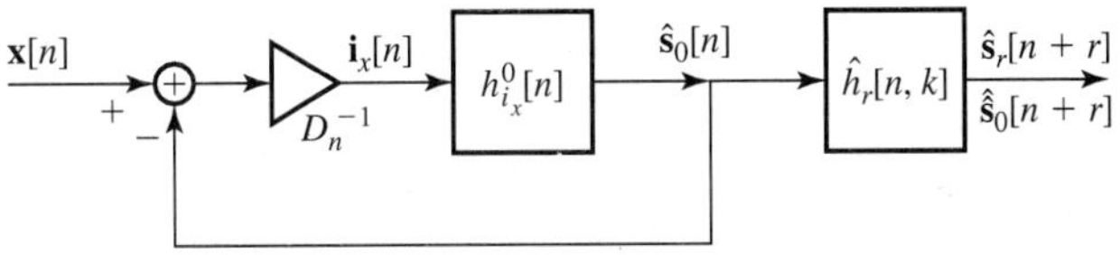

FIGURE 13-17

From this it follows that the process $\mathbf{x}[n] - \hat{\mathbf{s}}_0[n]$ is white noise and

$$\mathbf{x}[n] - \hat{\mathbf{s}}_0[n] \perp \mathbf{i}_x[k] \qquad 0 \le k \le n-1$$

because the processes $\mathbf{x}[k]$ and $\mathbf{i}_x[k]$ are linearly equivalent. And since $\mathbf{x}[n] - \hat{\mathbf{s}}_0[n]$ depends linearly on $\mathbf{i}_x[k]$ for $0 \le k \le n$, (13-127a) results. Equation (13-127b) is a consequence of the requirement that $E\{\mathbf{i}_x^2[n]\} = 1$.

Property 1 shows that the process $\mathbf{i}_x[n]$ can be realized simply by the feedback system of Fig. 13-17. This eliminates the need for designing the whitening filter $\gamma_x[n, k]$.

Property 2 The estimate $\hat{\mathbf{s}}_r[n+r]$ of $\mathbf{s}[n+r]$ equals the pure predictor $\hat{\hat{\mathbf{s}}}_0[n+r]$ of the estimate $\hat{\mathbf{s}}_0[n]$ of $\mathbf{s}[n]$ (Fig. 13-17)

$$\hat{\mathbf{s}}_r[n+r] = \hat{\hat{\mathbf{s}}}_0[n+r] = \sum_{k=0}^{n} \hat{h}_r[n, k]\hat{\mathbf{s}}_0[k] \tag{13-128}$$

provided that, for every $n \ge 0$,

$$E\{\hat{\mathbf{s}}_0[n]\mathbf{i}_x[n]\} = E\{\mathbf{x}[n]\mathbf{i}_x[n]\} - D_n \ne 0 \tag{13-129}$$

Proof. The process $\hat{\mathbf{s}}_0[n]$ is linearly dependent on $\mathbf{x}[n]$ and its past. Condition (13-129) means that the component of $\hat{\mathbf{s}}_0[n]$ in the $\mathbf{i}_x[n]$ direction is not 0. Hence the processes $\hat{\mathbf{s}}_0[n]$ and $\mathbf{x}[n]$ are linearly equivalent. And since

$$\hat{\mathbf{s}}_0[n+r] - \hat{\hat{\mathbf{s}}}_0[n+r] \perp \hat{\mathbf{s}}_0[k] \qquad 0 \le k \le n$$

we conclude that

$$\hat{\hat{\mathbf{s}}}_0[n+r] - \hat{\mathbf{s}}_0[n+r] \perp \mathbf{x}[k] \qquad 0 \le k \le n$$

Furthermore,

$$\mathbf{s}[n+r] - \hat{\mathbf{s}}_0[n+r] \perp \mathbf{x}[k] \qquad 0 \le k \le n+r$$

because $\hat{\mathbf{s}}_0[n+r]$ is the estimate of $\mathbf{s}[n+r]$ in terms of $\mathbf{x}[k]$ for $0 \le k \le n+r$. Finally,

$$\mathbf{s}[n+r] - \hat{\hat{\mathbf{s}}}_0[n+r] = (\mathbf{s}[n+r] - \hat{\mathbf{s}}_0[n+r]) + (\hat{\mathbf{s}}_0[n+r] - \hat{\hat{\mathbf{s}}}_0[n+r])$$

Hence

$$\mathbf{s}[n+r] - \hat{\hat{\mathbf{s}}}_0[n+r] \perp \mathbf{x}[k] \qquad 0 \le k \le n$$

and (13-128) results.

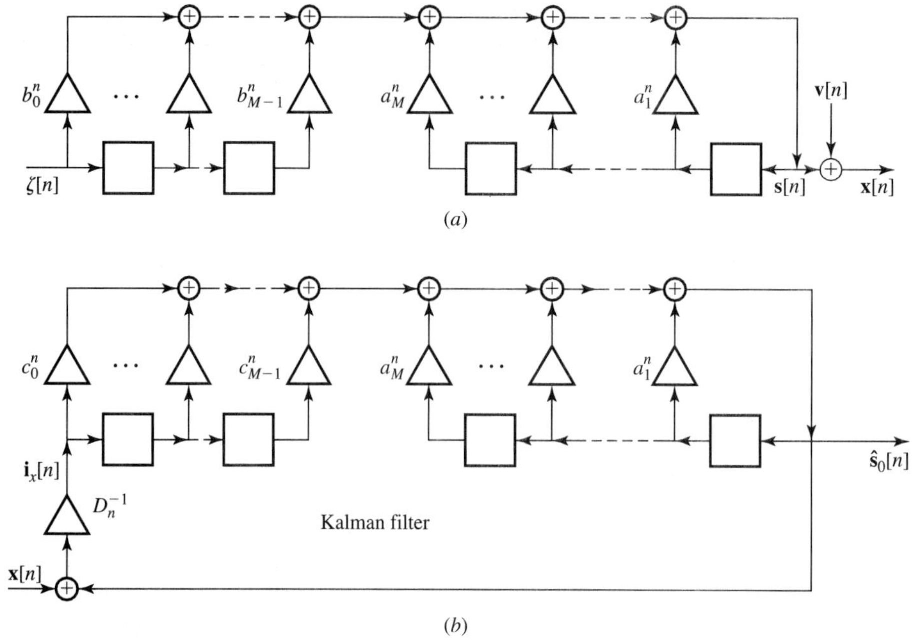

FIGURE 13-18

This property shows that filtering and prediction can be reduced to a cascade of a pure filter and a pure predictor.

ASSUMPTION 2. The signal $\mathbf{s}[n]$ is a time-varying ARMA process (Fig. 13-18a)

$$\mathbf{s}[n] - a_1^n \mathbf{s}[n-1] - \cdots - a_M^n \mathbf{s}[n-M] = \sum_{k=0}^{M-1} b_k^n \boldsymbol{\zeta}[n-k] \qquad (13\text{-}130)$$
$$R_{\zeta\zeta}[m,n] = V_n \delta[m-n]$$

Property 3 The estimate $\hat{\mathbf{s}}_0[n]$ is also an ARMA process

$$\hat{\mathbf{s}}_0[n] - a_1^n \hat{\mathbf{s}}_0[n-1] - \cdots - a_M^n \hat{\mathbf{s}}_0[n-M] = \sum_{k=0}^{M-1} c_k^n \mathbf{i}_x[n-k] \qquad (13\text{-}131)$$

where the coefficients a_k^n are the same as in (13-130) and the coefficients c_k^n are M constants to be determined.

Proof. We assume that (13-131) is true for all past estimates $\hat{\mathbf{s}}_0[n-k]$ and we shall prove that if $\hat{\mathbf{s}}_0[n]$ is given by (13-131), then it is the estimate of $\mathbf{s}[n]$. It suffices to show that if the constants c_k^n are suitably chosen, then the resulting error satisfies the orthogonality principle

$$\boldsymbol{\varepsilon}[n] = \mathbf{s}[n] - \hat{\mathbf{s}}_0[n] \perp \mathbf{x}[r] \qquad 0 \le r \le n \qquad (13\text{-}132)$$

Subtracting (13-131) from (13-130), we obtain

$$\boldsymbol{\varepsilon}[n] = \sum_{k=1}^{M} a_k^n \boldsymbol{\varepsilon}[n-k] + \sum_{k=0}^{M-1} \left(b_k^n \boldsymbol{\zeta}[n-k] - c_k^n \mathbf{i}_x[n-k]\right) \tag{13-133}$$

But

$$\boldsymbol{\zeta}[n_1], \mathbf{i}_x[n_1] \perp \mathbf{x}[r] \qquad \text{for} \quad r < n_1$$

and $\boldsymbol{\varepsilon}[n-k] \perp \mathbf{x}[r]$ for $r \le n-k$ (induction hypothesis). Hence (13-132) is true for $r \le n - M$. It suffices, therefore, to select the M constants c_k^n such that

$$E\{\boldsymbol{\varepsilon}[n]\mathbf{x}[r]\} = 0 \qquad n - M + 1 \le r \le n \tag{13-134}$$

We have thus expressed $\hat{\mathbf{s}}_0[n]$ in terms of $\mathbf{i}_x[n]$. To complete the specification of the filter, we use (13-127a). This yields the feedback system of Fig. 13-18b involving $M+1$ unknown parameters: the constant D_n and the M coefficients c_k^n. These parameters can be determined from (13-127b) and the M equations in (13-134).

The recursion equation (13-131) can be written as a system of M first-order equations (state equations) or, equivalently, as a first-order vector equation (see Sec. 11-2). The unknowns are then the scalar D_n and the coefficients c_k^n. To simplify the analysis, we shall carry out the determination of the unknown parameters for the first-order scalar case only. The results hold also for the vector case mutatis mutandis.

FIRST-ORDER. If

$$\mathbf{s}[n] - A_n \mathbf{s}[n-1] = \boldsymbol{\zeta}[n] \qquad E\{\boldsymbol{\zeta}^2[n]\} = V_n \tag{13-135}$$

then (13-131) yields

$$\hat{\mathbf{s}}_0[n] - A_n \hat{\mathbf{s}}_0[n-1] = K_n(\mathbf{x}[n] - \hat{\mathbf{s}}_0[n]) \tag{13-136}$$

where $K_n = c_0^n / D_n$. This is a first-order system as in Fig. 13-19a. To complete its specification, we must find the constant K_n. We maintain that

$$K_n = \frac{P_n}{N_n - P_n} \qquad P_n = E\{\boldsymbol{\varepsilon}^2[n]\} \tag{13-137}$$

In (13-137), N_n is the average intensity of $\boldsymbol{\nu}[n]$, which we assume known.

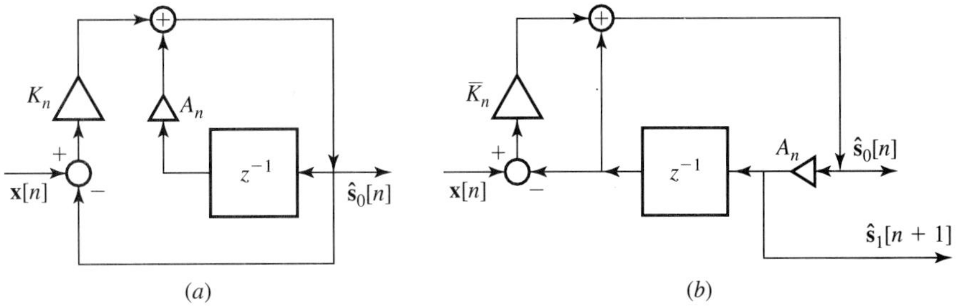

FIGURE 13-19

The MS error P_n can be determined recursively

$$\frac{P_n}{N_n - P_n} = \frac{A_n^2 P_{n-1} + V_n}{N_n} \tag{13-138}$$

Proof. Multiplying the data $\mathbf{x}[n] = \mathbf{s}[n] + \boldsymbol{\nu}[n]$ by the error

$$\boldsymbol{\varepsilon}[n] = \mathbf{s}[n] - \hat{\mathbf{s}}_0[n] = \mathbf{x}[n] - \hat{\mathbf{s}}_0[n] - \boldsymbol{\nu}[n]$$

and using the orthogonality condition (13-132), we obtain

$$E\{\boldsymbol{\varepsilon}[n]\mathbf{x}[n]\} = 0 = P_n + E\{\boldsymbol{\varepsilon}[n]\boldsymbol{\nu}[n]\}$$

From (13-135) and (13-136) it follows that

$$\begin{aligned} \boldsymbol{\varepsilon}[n] &= A_n\boldsymbol{\varepsilon}[n-1] + \boldsymbol{\zeta}[n] - K_n(\boldsymbol{\varepsilon}[n] + \boldsymbol{\nu}[n]) \\ (1 + K_n)\boldsymbol{\varepsilon}[n] &= A_n\boldsymbol{\varepsilon}[n-1] + \boldsymbol{\zeta}[n] - K_n\boldsymbol{\nu}[n] \end{aligned} \tag{13-139}$$

Hence

$$(1 + K_n)E\{\boldsymbol{\varepsilon}[n]\boldsymbol{\nu}[n]\} = -K_n E\{\boldsymbol{\nu}^2[n]\}$$

and (13-137) results.

To prove (13-138), we multiply each side of (13-139) by each side of the identity $\mathbf{s}[n] = A_n\mathbf{s}[n-1] + \boldsymbol{\zeta}[n]$. This yields

$$(1 + K_n)P_n = A_n^2 P_{n-1} + V_n$$

Since $1 + K_N = N_n/(N_n - P_n)$, the above equation yields (13-138).

Note Using (13-135), we can readily show that

$$\hat{\mathbf{s}}_0[n] - A_n\hat{\mathbf{s}}_0[n-1] = \overline{K}_n(\mathbf{x}[n] - A_n\hat{\mathbf{s}}_0[n-1]) \tag{13-140}$$

where

$$\overline{K}_n = \frac{P_n}{N_n} = \frac{A_n^2 P_{n-1} + V_n}{A_n^2 P_{n-1} + V_n + N_n} \tag{13-141}$$

The corresponding system is shown in Fig. 13-19*b*. In the same diagram, we also show the realization of the one-step predictor

$$\hat{\mathbf{s}}_1[n+1] = \hat{\hat{\mathbf{s}}}_0[n+1] = A_n\hat{\mathbf{s}}_0[n]$$

of $\mathbf{s}[n+1]$. This follows readily from (13-128) because the process $\hat{\mathbf{s}}_0[n]$ is AR; hence its pure predictor equals $A_n\hat{\mathbf{s}}_0[n]$.

The iteration The estimate $\hat{\mathbf{s}}_0[n]$ of $\mathbf{s}[n]$ is determined recursively: If $\overline{K}_{n-1}$ and $\hat{\mathbf{s}}_0[n-1]$ are known, then K_n is determined from (13-141) and $\hat{\mathbf{s}}_0[n]$ from (13-140). To start the iteration, we must specify the initial conditions of (13-135). We shall assume that

$$\mathbf{s}[0] = \boldsymbol{\zeta}[0]$$

This leads to the initial estimate

$$\hat{\mathbf{s}}_0[0] = \overline{K}_0\mathbf{x}[0]$$

from which it follows that

$$\overline{K}_0 = \frac{E\{\mathbf{s}[0]\mathbf{x}[0]\}}{E\{\mathbf{x}^2[0]\}} = \frac{E\{\boldsymbol{\zeta}^2[0]\}}{E\{\boldsymbol{\zeta}^2[0]\} + E\{\boldsymbol{\nu}^2[0]\}}$$

Hence

$$\overline{K}_0 = \frac{V_0}{V_0 + N_0} \qquad P_0 = \frac{V_0 N_0}{V_0 + N_0} \tag{13-142}$$

Linearization Equation (13-138) and its equivalent (13-141) are nonlinear. However, each can be replaced by two linear equations. Indeed, if F_n and G_n are two sequences such that

$$\begin{aligned} F_n &= A_n^2 F_{n-1} + V_n G_{n-1} & F_0 &= V_0 N_0 \\ N_n G_n &= A_n^2 F_{n-1} + (V_n + N_n) G_{n-1} & G_0 &= V_0 + N_0 \end{aligned} \tag{13-143}$$

then

$$P_n = \frac{F_n}{G_n}$$

EXAMPLE 13-10 ▶ We shall determine the noncausal, the causal, and the Kalman estimate of a process $\mathbf{s}[n]$ in terms of the data $\mathbf{x}[k] = \mathbf{s}[k] + \boldsymbol{\nu}[k]$, and the corresponding MS error P. We assume that the process $\mathbf{s}[n]$ satisfies the equation

$$\mathbf{s}[n] - 0.8\mathbf{s}[n-1] = \boldsymbol{\zeta}[n]$$

and that

$$R_{\zeta\zeta}[m] = 0.36\delta[m] \qquad R_{\zeta\nu}[m] = 0 \qquad R_{\nu\nu}[m] = \delta[n-m]$$

This is a special case of the process considered in Example 13-2 with

$$a = 0.8 \qquad N = 1 \qquad N_0 = 0.36 \qquad b = 0.5$$

Hence

$$\mathbf{S}_{ss}(z) = \frac{0.36}{(1 - 0.8z^{-1})(1 - 0.8z)} \qquad R_{ss}[m] = 0.8^{|m|}$$

$$\mathbf{S}_{xx}(z) = \mathbf{L}_x(z)\mathbf{L}_x(z^{-1}) \qquad \mathbf{L}_x(z) = \sqrt{1.6}\,\frac{z - 0.5}{z - 0.8}$$

(*a*) *Smoothing:* $\mathbf{x}[k]$ is available for all k. In this case the solution is obtained from Example 13-2 with $b = 0.5$ and $c = 0.375$:

$$h[n] = 0.3 \times 0.5^{|n|} \qquad P = 0.3$$

(*b*) *Causal filter:* $\mathbf{x}[k]$ is available for $k \le n$. The unknown filter is determined from (13-114) where now $l_x[0] = \sqrt{1.6}$:

$$\mathbf{H}_x^0(z) = 1 - \frac{z - 0.8}{1.6(z - 0.5)} = \frac{0.375z}{z - 0.5} \qquad h[n] = 0.375 \times 0.5^n U[n]$$

This shows that the estimate $\hat{\mathbf{s}}[n]$ of $\mathbf{s}[n]$ satisfies the recursion equation

$$\hat{\mathbf{s}}[n] - 0.5\hat{\mathbf{s}}[n-1] = 0.375\mathbf{x}[n] \qquad n \ge 0$$

The resulting MS error equals

$$P = R_{ss}[0] - \sum_{k=0}^{\infty} R_{ss}[k]h[k] = 0.375$$

(c) *Kalman filter:* $\mathbf{x}[k]$ is available for $0 \le k \le n$. Our process is a special case of (13-135) with

$$A_n = 0.8 \qquad V_n = E\{\boldsymbol{\zeta}^2[n]\} = 0.36 \qquad N_n = E\{\mathbf{v}^2[n]\} = 1$$

Inserting into (13-143), we obtain

$$F_n = 0.64F_{n-1} + 0.36G_{n-1} \qquad F_0 = 0.36$$
$$G_n = 0.64F_{n-1} + 1.36G_{n-1} \qquad G_0 = 1.36$$

This is a system of linear recursion equations and can be readily solved with z transforms. Since

$$\overline{K}_n = \frac{P_n}{N} = \frac{F_n}{G_n}$$

and $N = 1$, the solution yields

$$\overline{K}_n = P_n = \frac{0.48z_1^n - 0.12z_2^n}{1.28z_1^n + 0.08z_2^n} \qquad \begin{array}{l} z_1 = 1.6 \\ z_2 = 0.4 \end{array}$$

In particular,

$n =$	0	1	2	3	4
$P_n \simeq$	0.3	0.357	0.371	0.374	0.375

Thus, although the number of the available data increases as n increases, the MS error P_n also increases. The reason is that $\mathbf{s}[n]$ is a nonstationary process with initial second moment $V_0 = 0.36$ because $\mathbf{s}[0] = \boldsymbol{\zeta}[0]$, and, as n increases, $E\{\mathbf{s}^2[n]\}$ approaches the value 1.

We note, finally, that

$$\overline{K}_n = P_n \xrightarrow[n\to\infty]{} \frac{0.48}{1.28} = 0.375$$

and (13-140) yields

$$\hat{\mathbf{s}}_0[n] - 0.8\hat{\mathbf{s}}_0[n-1] = 0.375\mathbf{x}[n] - 0.3\hat{\mathbf{s}}_0[n-1]$$

This shows that, if the process $\mathbf{s}[n]$ is WSS, then its Kalman filter approaches the causal Wiener filter as $n \to \infty$. This is the case for any P_0 because the limit of $F_n G_n$ as $n \to \infty$ equals 0.375 regardless of the initial conditions. ◀

EXAMPLE 13-11 ▶ We wish to estimate the random variable $\mathbf{s}$ in terms of the sum

$$\mathbf{x}[n] = \mathbf{s} + \mathbf{v}[n] \qquad \text{where} \quad E\{\mathbf{s}\mathbf{v}[n]\} = 0 \qquad R_{\nu\nu}[m, n] = N\delta[m-n]$$

The estimate $\hat{\mathbf{s}}_0[n]$ in terms of the data $\mathbf{x}[n]$ can be obtained as the output of a Kalman filter if we consider the random variable $\mathbf{s}$ as a stochastic process satisfying trivially (13-135)

$$\mathbf{s}[n] = \mathbf{s}[n-1] + \boldsymbol{\zeta}[n] \qquad \mathbf{s}[-1] = 0$$

$$\boldsymbol{\zeta}[n] = \begin{cases} \mathbf{s} & n = 0 \\ 0 & n > 0 \end{cases} \qquad V_n = \begin{cases} E\{\mathbf{s}^2\} = M & n = 0 \\ 0 & n > 0 \end{cases}$$

In this case, $A_n = 1$, $N_n = N$, and (13-143) yields

$$F_n = F_{n-1} \qquad NG_n = F_{n-1} + NG_{n-1} \qquad F_0 = MN \qquad G_0 = M + N$$

Solving, we obtain

$$F_n = MN \qquad G_n = M + N + Mn$$

Hence

$$\hat{\mathbf{s}}_0[n] = \frac{N + Mn}{M + N + Mn}\hat{\mathbf{s}}_0[n-1] + \frac{M}{M + N + Mn}\mathbf{x}[n]$$ ◀

Continuous-Time Processes

We wish, finally, to determine the estimate

$$\hat{\mathbf{s}}_0(t) = \hat{E}\{\mathbf{s}(t) \mid \mathbf{x}(\tau), 0 \le \tau \le t\} \tag{13-144}$$

of a continuous-time process $\mathbf{s}(t)$ in terms of the data

$$\mathbf{x}(t) = \mathbf{s}(t) + \boldsymbol{\nu}(t) \tag{13-145}$$

The solution of this problem parallels the discrete-time solution if recursion equations are replaced by differential equations and sums by integrals. It might be instructive, however, to rederive the principal results using a different approach.

To avoid repetition, we start directly with the white-noise assumption

$$R_{\nu\nu}(t, \tau) = N(\tau)\delta(t - \tau) \qquad N(\tau) > 0 \qquad R_{s\nu}(t, \tau) = 0 \tag{13-146}$$

and we show that the process

$$\mathbf{w}(t) = \mathbf{x}(t) - \hat{\mathbf{s}}_0(t)$$

is white noise with autocorrelation

$$R_{ww}(t, \tau) = N(\tau)\delta(t - \tau) \tag{13-147}$$

Proof. As we know

$$\boldsymbol{\varepsilon}(t) = \mathbf{s}(t) - \hat{\mathbf{s}}_0(t) \perp \mathbf{x}(\tau) \qquad \boldsymbol{\nu}(t) \perp \mathbf{x}(\tau)$$

for $\tau < t$. Furthermore, $\mathbf{w}(\tau)$ depends linearly on $\mathbf{x}(\tau)$ and its past. Hence

$$\mathbf{w}(t) = \boldsymbol{\varepsilon}(t) + \boldsymbol{\nu}(t) \perp \mathbf{w}(\tau) \qquad \tau < t \tag{13-148}$$

To complete the proof of (13-147), we shall assume that $\mathbf{s}_0(t)$ is continuous from the left

$$\hat{\mathbf{s}}_0(t^-) = \hat{\mathbf{s}}_0(t)$$

This is not true at the origin if $\mathbf{s}(0) \neq 0$. However, for sufficiently large t, the effect of the initial condition can be neglected. From this it follows that

$$P(t) = E\{\boldsymbol{\varepsilon}^2(t)\} < \infty$$

and since $\boldsymbol{\varepsilon}(t) \perp \boldsymbol{\nu}(\tau)$ for $\tau > t$, we conclude that

$$R_{ww}(t,\tau) = R_{\nu w}(t,\tau) = R_{\nu\nu}(t,\tau) = N(\tau)\delta(t-\tau)$$

Using a limit argument, we can show that, as in the discrete-time case, the normalized process $\mathbf{w}(t)/\sqrt{N(t)}$ is the Kalman innovations of $\mathbf{x}(t)$. The details, however, will be omitted. This leads to the conclusion that $\hat{\mathbf{s}}_0(t)$ can be expressed in terms of $\mathbf{w}(t)$ [see also (13-123)]

$$\hat{\mathbf{s}}_0(t) = \int_0^t h_w(t,\alpha)\mathbf{w}(\alpha)\,d\alpha \tag{13-149}$$

Since $\mathbf{s}(t) - \hat{\mathbf{s}}_0(t) \perp \mathbf{w}(\tau)$ for $\tau \le t$, we conclude from the above and (13-147) that

$$R_{sw}(t,\tau) = \int_0^t h_w(t,\alpha)N(\alpha)\delta(\tau-\alpha)\,d\alpha = h_w(t,\tau)N(\tau)$$

and (13-149) yields

$$\hat{\mathbf{s}}_0(t) = \int_0^t \frac{1}{N(\alpha)} R_{sw}(t,\alpha)\mathbf{w}(\alpha)\,d\alpha \tag{13-150}$$

We note that [see (13-148) and (13-146)]

$$R_{sw}(t,t) = E\{\mathbf{s}(t)[\boldsymbol{\varepsilon}(t) + \boldsymbol{\nu}(t)]\} = P(t)$$

WIDE-SENSE MARKOV PROCESSES. Using the material just discussed, we shall show that, if the signal $\mathbf{s}(t)$ is WS Markov, that is, if it satisfies a differential equation driven by white noise, then its estimate $\hat{\mathbf{s}}_0(t)$ satisfies a similar equation. For simplicity, we consider the first-order case

$$\mathbf{s}'(t) + A(t)\mathbf{s}(t) = \boldsymbol{\zeta}(t) \qquad R_{\zeta\zeta}(t,\tau) = V(\tau)\delta(t-\tau) \tag{13-151}$$

The Kalman–Bucy equations.[8] We maintain that

$$\hat{\mathbf{s}}_0'(t) + A(t)\hat{\mathbf{s}}_0(t) = K(t)[\mathbf{x}(t) - \hat{\mathbf{s}}_0(t)] \tag{13-152}$$

where

$$K(t) = \frac{P(t)}{N(t)} \tag{13-153}$$

Furthermore, the MS error $P(t)$ satisfies the *Riccati equation*

$$P'(t) + 2A(t)P(t) = V(t) - \frac{1}{N(t)}P^2(t) \tag{13-154}$$

---

[8]R. E. Kalman and R. C. Bucy: "New Results in Linear Filtering and Prediction Theory," *ASME Transactions,* vol. 83D, 1961.

Proof. Multiplying the differential equation in (13-151) by $\mathbf{w}(\tau)$, we obtain

$$\frac{\partial}{\partial t} R_{sw}(t,\tau) + A(t) R_{sw}(t,\tau) = 0 \qquad \tau < t \tag{13-155}$$

We next equate the derivatives of both sides of (13-150)

$$\hat{\mathbf{s}}_0'(t) = \frac{1}{N(t)} R_{sw}(t,t)\mathbf{w}(t) + \int_0^t \frac{1}{N(\alpha)} \frac{\partial}{\partial t} R_{sw}(t,\alpha)\,d\alpha$$

Finally, we multiply (13-150) by $A(t)$ and add with the last equation. This yields (13-152) because, as we see from (13-155), the sum of the two integrals is 0.

To prove (13-154), we use the following version of (9-99): If $\mathbf{z}(t)$ is a process with $E\{\mathbf{z}^2(t)\} = I(t)$ and such that

$$\mathbf{z}'(t) + B(t)\mathbf{z}(t) = \boldsymbol{\xi}(t) \qquad R_{\xi\xi}(t,\tau) = Q(\tau)\delta(t-\tau) \tag{13-156}$$

then (see Prob. 9-28*b*)

$$I'(t) + 2B(t)I(t) = Q(t) \tag{13-157}$$

Returning to (13-152), we observe, subtracting from (13-151), that the estimation error $\boldsymbol{\varepsilon}(t)$ satisfies the equation

$$\boldsymbol{\varepsilon}'(t) + [A(t) + K(t)]\boldsymbol{\varepsilon}(t) = \boldsymbol{\zeta}(t) - K(t)\boldsymbol{\nu}(t)$$

In the above, the right side $\boldsymbol{\xi}(t) = \boldsymbol{\zeta}(t) - K(t)\boldsymbol{\nu}(t)$ is white noise as in (13-156) with

$$Q(\tau) = V(\tau) + K^2(\tau)N(\tau)$$

Hence the function $P(t) = E\{\boldsymbol{\varepsilon}^2(t)\}$ satisfies (13-157), where $B(t) = A(t) + K(t)$. This yields

$$P'(t) + 2[A(t) + K(t)]P(t) = V(t) + K^2(t)N(t)$$

and (13-154) results.

Linearization We shall now show that the nonlinear equation (13-154) is equivalent to two linear equations. For this purpose, we introduce the functions $F(t)$ and $G(t)$ such that

$$P(t) = \frac{F(t)}{G(t)} \tag{13-158}$$

Clearly,

$$F'(t) = P'(t)G(t) + P(t)G'(t)$$

and (13-154) yields

$$F'(t) + A(t)F(t) - V(t)G(t) = P(t)\left[G'(t) - A(t)G(t) - \frac{F(t)}{N(t)}\right]$$

This is satisfied if

$$\begin{aligned} F'(t) &= -A(t)F(t) + V(t)G(t) \\ G'(t) &= \frac{F(t)}{N(t)} + A(t)G(t) \end{aligned} \tag{13-159}$$

To solve this system, we must specify $F(0)$ and $G(0)$. Setting arbitrarily $G(0) = 1$, we obtain $F(0) = P(0)$, where

$$P(0) = E\{\mathbf{s}^2(0)\}$$

is the initial value of the MS error $P(t)$. The determination of the Kalman filter thus depends on the second moment of $\mathbf{s}(0)$.

EXAMPLE 13-12 ▶ We shall determine the noncausal, the causal, and the Kalman estimate of a process $\mathbf{s}(t)$ in terms of the data $\mathbf{x}(t) = \mathbf{s}(t) + \boldsymbol{\nu}(t)$, and the corresponding MS error P. We assume that $\mathbf{s}(t)$ satisfies the equation

$$\mathbf{s}'(t) + 2\mathbf{s}(t) = \boldsymbol{\zeta}(t)$$

and that

$$R_{\zeta\zeta}(\tau) = 12\delta(\tau) \qquad R_{s\nu}(\tau) = 0 \qquad R_{\nu\nu}(\tau) = \delta(\tau)$$

This is a special case of the process considered in Example 13-7 with

$$\alpha = 2 \qquad N = 1 \qquad N_0 = 12 \qquad \beta = 4$$

Hence

$$S_{ss}(\omega) = \frac{12}{4+\omega^2} \qquad R_{ss}(\tau) = 3e^{-2|\tau|}$$

$$S_{xx}(\omega) = \frac{16+\omega^2}{4+\omega^2} \qquad \mathbf{L}_x(s) = \frac{s+4}{s+2}$$

(*a*) *Smoothing:* $\mathbf{x}(\xi)$ is available for all ξ. In this case, (13-16) yields

$$H(\omega) = \frac{12}{16+\omega^2} \qquad h(t) = \frac{3}{2}e^{-4|\tau|}$$

The MS error is obtained from (13-15)

$$P = 3 - \frac{9}{2}\int_{-\infty}^{\infty} e^{-4|\tau|}e^{-2|\tau|}\,d\tau = 1.5$$

(*b*) *Causal filter:* $\mathbf{x}(\xi)$ is available for $\xi \le t$. The unknown filter is specified in Example 13-8 with

$$\alpha = 2 \qquad \beta = 4 \qquad N = 12$$

Thus

$$\mathbf{H}_x(s) = \frac{2}{s+4} \qquad h_x(t) = 2e^{-4t}U(t) \qquad P = 2$$

This shows that the estimate $\hat{\mathbf{s}}(t)$ of $\mathbf{s}(t)$ satisfies the differential equation

$$\hat{\mathbf{s}}'(t) + 4\hat{\mathbf{s}}(t) = 2\mathbf{x}(t)$$

(*c*) *Kalman filter:* $\mathbf{x}(\xi)$ is available for $0 \le \xi \le t$. Our problem is a special case of (13-151) with

$$A(t) = 2 \qquad V(t) = 12 \qquad N(t) = 1$$

Hence [see (13-159)]

$$F'(t) = -2F(t) + 12G(t) \qquad G'(t) = F(t) + 2G(t)$$

To solve this system, we must know $P(0)$.

Case 1 If $\mathbf{s}(0) = 0$, then $P(0) = 0$. In this case, $F(0) = 0$, $G(0) = 1$. Inserting the solution of the last system into (13-153), we obtain

$$K(t) = P(t) = \frac{6e^{4t} - 6e^{-4t}}{3e^{4t} + e^{-4t}} \xrightarrow[t\to\infty]{} 2$$

Case 2 We now assume that $\mathbf{s}(t)$ is the stationary solution of the differential equation specifying $\mathbf{s}(t)$. In this case, $E\{\mathbf{s}^2(0)\} = 3$; hence $P(0) = F(0) = 3$ and

$$K(t) = P(t) = \frac{18e^{4t} + 6e^{-4t}}{9e^{4t} - e^{-4t}} \xrightarrow[t\to\infty]{} 2$$

Thus, in both cases, the solution $\hat{\mathbf{s}}_0(t)$ of the Kalman–Bucy equation (13-152) tends to the solution of the causal Wiener filter

$$\hat{\mathbf{s}}'_0(t) + 2\hat{\mathbf{s}}_0(t) = 2\mathbf{x}(t) - 2\hat{\mathbf{s}}_0(t)$$

as $t \to \infty$. ◀

EXAMPLE 13-13 ▶ We wish to estimate the random variable $\mathbf{s}$ in terms of the sum

$$\mathbf{x}(t) = \mathbf{s} + \boldsymbol{\nu}(t) \qquad E\{\mathbf{s}\boldsymbol{\nu}(t)\} = 0 \quad R_{\nu\nu}(\tau) = N\delta(\tau)$$

This is a special case of (13-151) if

$$A(t) = 0 \qquad \mathbf{s}(t) = \mathbf{s} \qquad \zeta(t) = 0 \qquad N(t) = N$$

In this case, $V(t) = 0$, $P(0) = E\{\mathbf{s}^2\} \equiv M$, and (13-159) yields

$$F'(t) = 0 \qquad G'(t) = \frac{F(t)}{N} \qquad F(0) = M \qquad G(0) = 1$$

Hence

$$F(t) = M \qquad G(t) = 1 + \frac{Mt}{N}$$

Inserting into (13-152), we obtain

$$\hat{\mathbf{s}}'_0(t) + \frac{M}{N + Mt}\hat{\mathbf{s}}_0(t) = \frac{M}{N + Mt}\mathbf{x}(t)$$ ◀

PROBLEMS

13-1 If $R_s(\tau) = Ie^{-|\tau|/T}$ and

$$\hat{E}\{\mathbf{s}(t - T/2) \mid \mathbf{s}(t), \mathbf{s}(t - T)\} = a\mathbf{s}(t) + b\mathbf{s}(t - T)$$

find the constants a and b and the MS error.

13-2 Show that if $\hat{\mathbf{z}} = a\mathbf{s}(0) + b\mathbf{s}(T)$ is the MS estimate of

$$\mathbf{z} = \int_0^T \mathbf{s}(t)\,dt \qquad \text{then} \quad a = b = \frac{\int_0^T R_s(\tau)\,d\tau}{R_s(0) + R_s(T)}$$

13-3 Show that if $\mathbf{x}(t) = \mathbf{s}(t) + \boldsymbol{\nu}(t)$, $R_{s\nu}(\tau) = 0$ and

$$\hat{E}\{\mathbf{s}'(t) \mid \mathbf{x}(t), \mathbf{x}(t - \tau)\} = a\mathbf{x}(t) + b\mathbf{x}(t - \tau)$$

then for small τ, $a = -b \simeq R''_{ss}(0)/\tau R''_{xx}(0)$.

13-4 Show that, if $S_x(\omega) = 0$ for $|\omega| > \sigma = \pi/T$, then the linear MS estimate of $\mathbf{x}(t)$ in terms of its samples $\mathbf{x}(nT)$ equals

$$\hat{E}\{\mathbf{x}(t) \mid \mathbf{x}(nT), n = -\infty, \ldots, \infty\} = \sum_{n=-\infty}^{\infty} \frac{\sin(\sigma t - nT)}{\sigma t - n\pi} \mathbf{x}(nT)$$

and the MS error equals 0.

13-5 Show that if

$$\hat{E}\{\mathbf{s}(t+\lambda) \mid \mathbf{s}(t), \mathbf{s}(t-\tau)\} = \hat{E}\{\mathbf{s}(t+\lambda) \mid \mathbf{s}(t)\}$$

then $R_s(\tau) = Ie^{-\alpha|\tau|}$.

13-6 A random sequence $\mathbf{x}_n$ is called a *martingale* if $E\{\mathbf{x}_n = 0\}$ and

$$E\{\mathbf{x}_n \mid \mathbf{x}_{n-1}, \ldots, \mathbf{x}_1\} = \mathbf{x}_{n-1}$$

Show that if the random variables $\mathbf{y}_n$ are *independent,* then their sum $\mathbf{x}_n = \mathbf{y}_1 + \cdots + \mathbf{y}_n$ is a martingale.

13-7 A random sequence $\mathbf{x}_n$ is called *wide-sense martingale* if

$$\hat{E}\{\mathbf{x}_n \mid \mathbf{x}_{n-1}, \ldots, \mathbf{x}_1\} = \mathbf{x}_{n-1}$$

(*a*) Show that a sequence $\mathbf{x}_n$ is WS martingale if it can be written as a sum $\mathbf{x}_n = \mathbf{y}_1 + \cdots + \mathbf{y}_n$, where the random variables $\mathbf{y}_n$ are *orthogonal.*
(*b*) Show that if the sequence $\mathbf{x}_n$ is WS martingale, then

$$E\{\mathbf{x}_n^2\} \geq E\{\mathbf{x}_{n-1}^2\} \geq \cdots \geq E\{\mathbf{x}_1^2\}$$

Hint: $\mathbf{x}_n = \mathbf{x}_n - \mathbf{x}_{n-1} + \mathbf{x}_{n-1}$ and $\mathbf{x}_n - \mathbf{x}_{n-1} \perp \mathbf{x}_{n-1}$.

13-8 Find the noncausal estimators $H_1(\omega)$ and $H_2(\omega)$ respectively of a process $\mathbf{s}(t)$ and its derivative $\mathbf{s}'(t)$ in terms of the data $\mathbf{x}(t) = \mathbf{s}(t) + \boldsymbol{\nu}(t)$, where

$$R_s(\tau) = A\frac{\sin^2 a\tau}{\tau^2} \qquad R_\nu(\tau) = N\delta(\tau) \qquad R_{s\nu}(\tau) = 0$$

13-9 We denote by $H_s(\omega)$ and $H_y(\omega)$, respectively, the noncausal estimators of the input $\mathbf{s}(t)$ and the output $\mathbf{y}(t)$ of the system $T(\omega)$ in terms of the data $\mathbf{x}(t)$ (Fig. P13-9). Show that $H_y(\omega) = H_s(\omega)T(\omega)$.

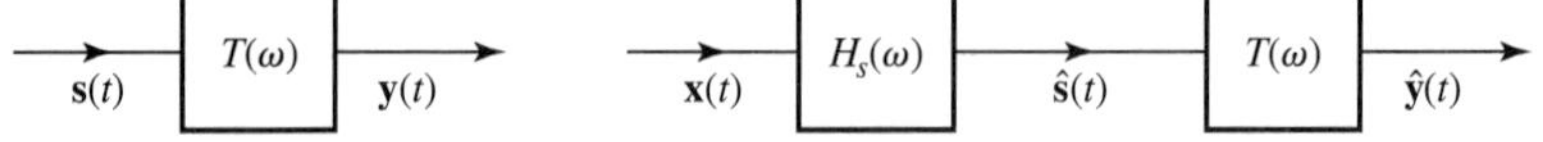

FIGURE P13-9

13-10 Show that if $S(\omega) = 1/(1+\omega^4)$, then the predictor of $\mathbf{s}(t)$ in terms of its entire past equals $\hat{\mathbf{s}}(t+\lambda) = b_0\mathbf{s}(t) + b_1\mathbf{s}'(t)$, where

$$b_0 = e^{-\lambda/\sqrt{2}}\left(\cos\frac{\lambda}{\sqrt{2}} + \sin\frac{\lambda}{\sqrt{2}}\right) \qquad b_1 = \sqrt{2}e^{-\lambda/\sqrt{2}}\sin\frac{\lambda}{\sqrt{2}}$$

13-11 (*a*) Find a function $h(t)$ satisfying the integral equation (Wiener–Hopf)

$$\int_0^\infty h(\alpha)R(\tau-\alpha)\,d\alpha = R(\tau + \ln 2) \qquad t \geq 0 \qquad R(\tau) = \tfrac{3}{2}e^{-\tau} + \tfrac{11}{3}e^{-3\tau}$$

(*b*) The function $\mathbf{H}(s)$ is rational with poles in the left-hand plane. The function $\mathbf{Y}(s)$ is analytic in the left-hand plane. Find $\mathbf{H}(s)$ and $\mathbf{Y}(s)$ if

$$[\mathbf{H}(s) - 2^s]\frac{49 - 25s^2}{9 - 10s^2 + s^4} = \mathbf{Y}(s)$$

(*c*) Discuss the relationship between (*a*) and (*b*).

13-12 (*a*) Find a sequence h_n satisfying the system

$$\sum_{k=0}^{\infty} h_k R_{m-k} = R_{m+1} \qquad m \ge 0 \qquad R_m = \frac{1}{2^m} + \frac{1}{3^m}$$

(*b*) The function $\mathbf{H}(z)$ is rational with poles in the unit circle. The function $\mathbf{Y}(z)$ is rational with poles outside the unit circle. Find $\mathbf{H}(z)$ and $\mathbf{Y}(z)$ if

$$[\mathbf{H}(z) - z]\frac{70 - 25(z + z^{-1})}{6(z + z^{-1})^2 - 35(z + z^{-1}) + 50} = \mathbf{Y}(z)$$

(*c*) Discuss the relationship between (*a*) and (*b*).

13-13 Show that if $\mathbf{H}(z)$ is a predictor of a process $\mathbf{s}[n]$ and $\mathbf{H}_a(z)$ is an all-pass function such that $|\mathbf{H}_a(e^{j\omega})| = 1$, then the function $1 - (1 - \mathbf{H}(z))\mathbf{H}_a(z)$ is also a predictor with the same MS error P.

13-14 We have shown that the one-step predictor $\hat{\mathbf{s}}_1[n]$ of an AR process of order m in terms of its entire past equals [see (13-35)]

$$\hat{E}\{\mathbf{s}[n] \mid \mathbf{s}[n-k], k \ge 1\} = -\sum_{k=1}^{m} a_k \mathbf{s}[n-k]$$

Show that its two-step predictor $\hat{\mathbf{s}}_2[n]$ is given by

$$\hat{E}\{\mathbf{s}[n] \mid \mathbf{s}[n-k], k \ge 2\} = -a_1\hat{\mathbf{s}}_1[n-1] - \sum_{k=2}^{m} a_k \mathbf{s}[n-k]$$

13-15 Using (13-70) show that

$$\lim_{N\to\infty} \log \frac{\Delta_{N+1}}{\Delta_N} = \lim_{N\to\infty} \frac{\log \Delta_N}{N} = \frac{1}{2\pi}\int_{-\pi}^{\pi} \log S_s(\omega)\,d\omega$$

Hint:

$$\frac{1}{N}\sum_{n=1}^{N} \log \frac{\Delta_{n+1}}{\Delta_n} = \frac{1}{N}\log \Delta_{N+1} - \frac{1}{N}\log \Delta_1 \to \lim_{N\to\infty} \log \frac{\Delta_{N+1}}{\Delta_N}$$

13-16 Find the predictor

$$\hat{\mathbf{s}}_N[n] = \hat{E}\{\mathbf{s}[n] \mid \mathbf{s}[n-k], 1 \le k \le N]\}$$

of a process $\mathbf{s}[n]$ and realize the error filter $\mathbf{E}_N(z)$ as an FIR filter (Fig. 13-8) and as a lattice filter (Fig. 12-15) for $N = 1, 2$, and 3 if

$$R_s[m] = \begin{cases} 5(3 - |m|) & |m| < 3 \\ 0 & |m| \ge 3 \end{cases}$$

13-17 The lattice filter of a process $\mathbf{s}[n]$ is shown in Fig. P13-17 for $N = 3$. Find the corresponding FIR filter for $N = 1, 2$, and 3 and the values of $R[m]$ for $|m| \le 3$ if $R[0] = 5$.

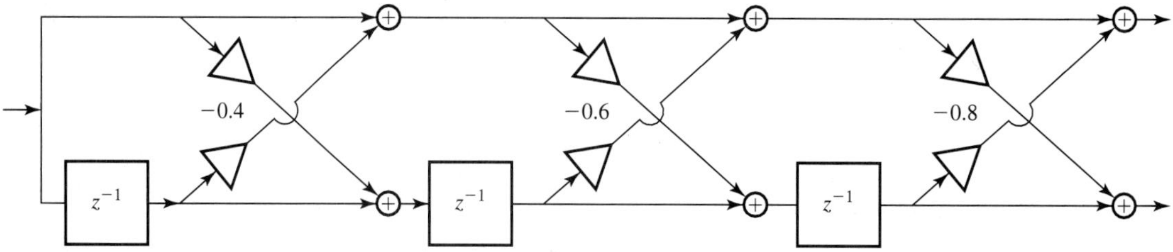

FIGURE P13-17

13-18 We wish to find the estimate $\hat{\mathbf{s}}(t)$ of the random telegraph signal $\mathbf{s}(t)$ in terms of the sum $\mathbf{x}(t) = \mathbf{s}(t) + \boldsymbol{\nu}(t)$ and its past, where

$$R_s(\tau) = e^{-2\lambda|\tau|} \qquad R_\nu(\tau) = N\delta(\tau) \qquad R_{s\nu}(\tau) = 0$$

Show that

$$\hat{\mathbf{s}}(t) = (c - 2\lambda)\int_0^\infty \mathbf{x}(t-\alpha)e^{-c\alpha}\,d\alpha \qquad c = 2\lambda\sqrt{1+\frac{1}{\lambda N}}$$

13-19 Show that if $\hat{\boldsymbol{\varepsilon}}_N[n]$ is the forward prediction error and $\check{\boldsymbol{\varepsilon}}_N[n]$ is the backward prediction error of a process $\mathbf{s}[n]$, then (*a*) $\hat{\boldsymbol{\varepsilon}}_N[n] \perp \hat{\boldsymbol{\varepsilon}}_{N+m}[n+m]$, (*b*) $\check{\boldsymbol{\varepsilon}}_N[n] \perp \check{\boldsymbol{\varepsilon}}_{N+m}[n-m]$, (*c*) $\hat{\boldsymbol{\varepsilon}}_N[n] \perp \check{\boldsymbol{\varepsilon}}_{N+m}[n-N-m]$.

13-20 If $\mathbf{x}(t) = \mathbf{s}(t) + \boldsymbol{\nu}(t)$ and

$$R_s(\tau) = 5e^{-0.2|\tau|} \qquad R_\nu(\tau) = 5\delta(\tau) \qquad R_{s\nu}(\tau) = 0$$

find the following MS estimates and the corresponding MS errors: (*a*) the noncausal filter of $\mathbf{s}(t)$; (*b*) the causal filter of $\mathbf{s}(t)$; (*c*) the estimate of $\mathbf{s}(t+2)$ in terms of $\mathbf{s}(t)$ and its past; (*d*) the estimate of $\mathbf{s}(t+2)$ in terms of $\mathbf{x}(t)$ and its past.

13-21 If $\mathbf{x}[n] = \mathbf{s}[n] + \boldsymbol{\nu}[n]$:

$$R_s[m] = 5 \times 0.8^{|m|} \qquad R_\nu[m] = 5\delta[m] \qquad R_{s\nu}[m] = 0$$

Find the following MS estimates and the corresponding MS errors: (*a*) the noncausal filter of $\mathbf{s}[n]$; (*b*) the causal filter of $\mathbf{s}[n]$; (*c*) the estimate of $\mathbf{s}[n+1]$ in terms of $\mathbf{s}[n]$ and its past; (*d*) the estimate of $\mathbf{s}[n+1]$ in terms of $\mathbf{x}[n]$ and its past.

13-22 Find the Kalman estimate

$$\hat{\mathbf{s}}_0[n] = E\{\mathbf{s}[n] \mid \mathbf{s}[k] + \boldsymbol{\nu}[k], 0 \le k \le n\}$$

of $\mathbf{s}[n]$ and the MS error $P_n = E\{(\mathbf{s}[n] - \hat{\mathbf{s}}_0[n])^2\}$ if

$$R_s[m] = 5 \times 0.8^{|m|} \qquad R_\nu[m] = 5\delta[m] \qquad R_{s\nu}[m] = 0$$

13-23 Find the Kalman estimate

$$\hat{\mathbf{s}}_0(t) = E\{\mathbf{s}(t) \mid \mathbf{s}(\tau) + \boldsymbol{\nu}(\tau), 0 \le \tau \le t\}$$

of $\mathbf{s}(t)$ and the MS error $P(t) = E\{[\mathbf{s}(t) - \hat{\mathbf{s}}_0(t)]^2\}$ if

$$R_s(\tau) = 5e^{-0.2|\tau|} \qquad R_\nu(\tau) = \tfrac{10}{3}\delta(\tau) \qquad R_{s\nu}(\tau) = 0$$

13-24 Show that the sequences $\hat{\mathbf{q}}_N[m]$ and $\check{\mathbf{q}}_N[m]$ of the inverse lattice of Fig. 13-11*b* satisfy (13-85) and (13-86) (see Note 1, page 595).

CHAPTER

14

ENTROPY

14-1 INTRODUCTION

As we have noted in Chap. 1, the probability $P(A)$ of an event A can be interpreted as a measure of our uncertainty about the occurrence or nonoccurrence of A in a single performance of the underlying experiment S. If $P(A) \simeq 0.999$, then we are almost certain that A will occur; if $P(A) = 0.1$, then we are reasonably certain that A will not occur; our uncertainty is maximum if $P(A) = 0.5$. In this chapter, we consider the problem of assigning a measure of uncertainty to the occurrence or nonoccurrence not of a single event of S, but of any event A_i of a partition $\boldsymbol{U}$ of S where, as we recall, a partition is a collection of mutually exclusive events whose union equals S (Fig. 14-1). The measure of uncertainty about $\boldsymbol{U}$ will be denoted by $H(\boldsymbol{U})$ and will be called the *entropy of the partitioning* $\boldsymbol{U}$.

Historically, the functional $H(\boldsymbol{U})$ was derived from a number of postulates based on our heuristic understanding of uncertainty. The following is a typical set of such postulates[1]:

1. $H(\boldsymbol{U})$ is a continuous function of $p_i = P(A_i)$.
2. If $p_1 = \cdots = p_N = 1/N$, then $H(\boldsymbol{U})$ is an increasing function of N.
3. If a new partition $\boldsymbol{B}$ is formed by subdividing one of the sets of $\boldsymbol{U}$, then $H(\boldsymbol{B}) \geq H(\boldsymbol{U})$.

It can be shown that the sum[2]

$$H(\boldsymbol{U}) = -p_1 \log p_1 - \cdots - p_N \log p_N \tag{14-1}$$

[1]C. E. Shannon and W. Weaver: *The Mathematical Theory of Communication,* University of Illinois Press, 1949.

[2]We shall use as logarithmic base either the number 2 or the number *e*. In the first case, the unit of entropy is the *bit.*

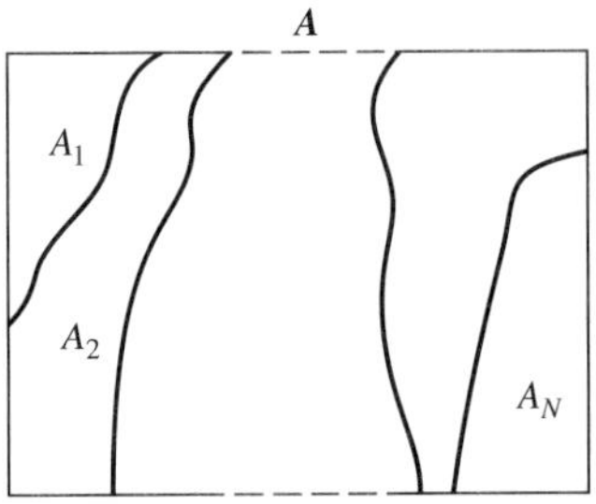

FIGURE 14-1

satisfies these postulates and it is unique within a constant factor. The proof of this assertion is not difficult but we choose not to reproduce it. We propose, instead, to introduce (14-1) as the *definition* of entropy and to develop axiomatically all its properties within the framework of probability. It is true that the introduction of entropy in terms of postulates establishes a link between the sum in (14-1) and our heuristic understanding of uncertainty. However, for our purposes, this is only incidental. In the last analysis, the justification of the concept must ultimately rely on the usefulness of the resulting theory.

The applications of entropy can be divided into two categories. The first deals with problems involving the determination of unknown distributions (Sec. 14-4). The available information is in the form of known expected values or other statistical functionals, and the solution is based on the principle of maximum entropy: We determine the unknown distributions so as to *maximize* the entropy $H(\boldsymbol{U})$ of some partition $\boldsymbol{U}$ subject to the given constraints (statistical mechanics). In the second category (coding theory), we are given $H(\boldsymbol{U})$ (source entropy) and we wish to construct various random variables (code lengths) so as to *minimize* their expected values (Sec. 14-5). The solution involves the construction of optimum mappings (codes) of the random variables under consideration, into the given probability space.

Uncertainty and information In the heuristic interpretation of entropy, the number $H(\boldsymbol{U})$ is a measure of our uncertainty about the events A_i of the partition $\boldsymbol{U}$ prior to the performance of the underlying experiment. If the experiment is performed and the results concerning A_i become known, the uncertainty is removed. We can thus say that the experiment provides *information* about the events A_i equal to the *entropy* of their partition. Thus uncertainty equals information and both are measured by the sum in (14-1).

EXAMPLE 14-1 ▶ (*a*) We shall determine the entropy of the partition $\boldsymbol{U}$ = [even, odd] in the fair-die experiment. Clearly, $P\{\text{even}\} = P\{\text{odd}\} = 1/2$. Hence

$$H(\boldsymbol{U}) = -\tfrac{1}{2}\log\tfrac{1}{2} - \tfrac{1}{2}\log\tfrac{1}{2} = \log 2$$

(*b*) In the same experiment, $\boldsymbol{V}$ is the partition consisting of the elementary events $\{f_i\}$. In this case, $P\{f_i\} = 1/6$; hence

$$H(\boldsymbol{V}) = -\tfrac{1}{6}\log\tfrac{1}{6} - \cdots - \tfrac{1}{6}\log\tfrac{1}{6} = \log 6$$

If the die is rolled and we are told which face showed, then we gain information about the partition $\boldsymbol{V}$ equal to its entropy log 6. If we are told merely that "even" or "odd" showed, then we gain information about the partition $\boldsymbol{U}$ equal to its entropy log 2. In this

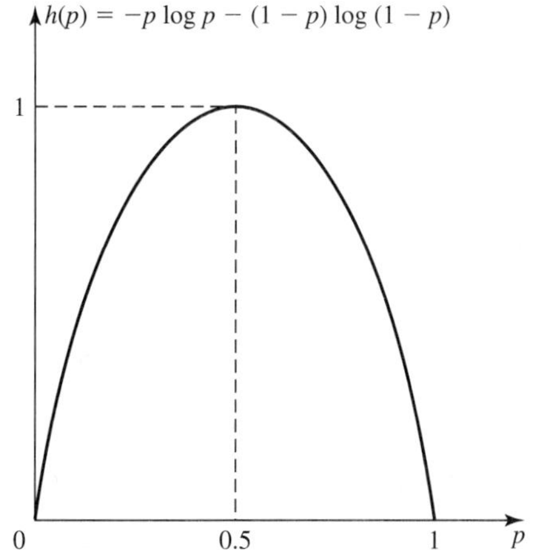

FIGURE 14-2

case, the information gained about the partition $\boldsymbol{V}$ equals again log 2. As we shall see, the difference $\log 6 - \log 2 = \log 3$ is the uncertainty about $\boldsymbol{V}$ assuming $\boldsymbol{U}$ (conditional entropy). ◀

EXAMPLE 14-2

▶ We consider now the coin experiment where $P\{h\} = p$. In this case, the entropy of $\boldsymbol{V}$ equals

$$H(\boldsymbol{V}) = -p \log p - (1 - p) \log(-p) \equiv h(p) \qquad (14\text{-}2)$$

The function $h(p)$ is shown in Fig. 14-2 for $0 \le p \le 1$. This function is symmetrical, convex, even about the point $p = 0.5$, and it reaches its maximum at that point. Furthermore, $h(0) = h(1) = 0$. ◀

Historical note The term *entropy* as a scientific concept was first used in thermodynamics (Clausius 1850). Its probabilistic interpretation in the context of statistical mechanics is attributed to Boltzmann (1877). However, the explicit relationship between entropy and probability was recorded several years later (Planck, 1906). Shannon, in his celebrated paper (1948), used the concept to give an economical description of the properties of long sequences of symbols, and applied the results to a number of basic problems in coding theory and data transmission. His remarkable contributions form the basis of modern information theory. Jaynes[3] (1957) reexamined the method of maximum entropy and applied it to a variety of problems involving the determination of unknown parameters from incomplete data.

Maximum entropy and classical definition. An important application of entropy is the determination of the probabilities p_i of the events of a partition $\boldsymbol{U}$, subject to various constraints, with the method of maximum entropy (MEM). The method states that the unknown p_i's must be so chosen as to maximize the entropy of $\boldsymbol{U}$ subject to the given constraints. This topic is considered in Sec. 14-4. In the following we introduce the

[3]E. T. Jaynes: *Physical Review*, vols. 106–107, 1957.

main idea and we show the equivalence between the MEM and the classical definition of probability (principle of insufficient reason), using as illustration the die experiment.

EXAMPLE 14-3 ▶ (*a*) We wish to determine the probabilities p_i of the six faces of a die, having access to no prior information. The MEM states that the p_i's must be such as to maximize the sum

$$H(V) = -p_1 \log p_1 - \cdots - p_6 \log p_6$$

Since $p_1 + \cdots + p_6 = 1$, this yields

$$p_1 = \cdots = p_6 = \tfrac{1}{6}$$

in agreement with the classical definition.

(*b*) Suppose now that we are given the following information: A player places a bet of one dollar on "odd" and he wins, on the average, 20 cents per game. We wish again to determine the p_i's using the MEM; however, now we must satisfy the constraints

$$p_1 + p_3 + p_5 = 0.6 \qquad p_2 + p_4 + p_6 = 0.4$$

This is a consequence of the available information because an average gain of 20 cents means that $P\{\text{odd}\} - P\{\text{even}\} = 0.2$. Maximizing $H(V)$ subject to the earlier constraints, we obtain

$$p_1 = p_3 = p_5 = 0.2 \qquad p_2 = p_4 = p_6 = 0.133\ldots$$

This agrees again with the classical definition if we apply the principle of insufficient reason to the outcomes of the events {odd} and {even} separately. ◀

Although conceptually the ME principle is equivalent to the principle of insufficient reason, operationally the MEM simplifies the analysis drastically when, as is the case in most applications, the constraints are phrased in terms of probabilities in the space S^n of repeated trials. In such cases the equivalence still holds, although it is less obvious, but the reasoning is involved and rather forced if we derive the unknown probabilities starting from the classical definition.

The MEM is thus a valuable tool in the solution of applied problems. It is used, in fact, even in deterministic problems involving the estimation of unknown parameters from insufficient data. The ME principle is then accepted as a smoothness criterion. We should emphasize, however, that as in the case of the classical definition, the conclusions drawn from the ME principle must be accepted with skepticism particularly when they involve elaborate constraints. This is evident even in the interpretation of the results in Example 14-3: In the absence of prior constraints, we conclude that all p_i's must be equal. This conclusion we accept readily because it is not in conflict with our experience concerning dice. The second conclusion, however, that $p_2 = p_4 = p_6 = 0.133\ldots$ and $p_1 = p_3 = p_5 = 0.2$ is not as convincing, we would think, even though we have no basis for any other conclusion. In our experience, no crooked dice exhibit such symmetries.

One might argue that this apparent conflict between the MEM and our experience is due to the fact that we did not make total use of our prior knowledge. Had we included among the constraints everything we know about dice, there would be no conflict. This might be true; however, it is not always clear how such constraints can be phrased analytically and, even if they can, how complex the required computations might be.

Typical Sequences and Relative Frequency

Suppose that $\boldsymbol{U} = [A_1, \ldots, A_N]$ is an N-element partition of an experiment S. In the space S^n of repeated trials, the elements A_i of $\boldsymbol{U}$ form N^n sequences of the form

$$\{A_i \text{ occurs } n_i \text{ times in a specific order}\} \tag{14-3}$$

and the probability of each sequence equals

$$p_1^{n_1} \cdots p_i^{n_i} \cdots p_N^{n_N} \tag{14-4}$$

where $p_i = P(A_i)$. The numbers n_i are arbitrary subject only to the constraint $n_1 + \cdots + n_N = n$. However, according to the relative frequency interpretation of probability, if n is "sufficiently large," then "almost certainly"

$$n_i \simeq np_i \qquad i = 1, \ldots, N \tag{14-5}$$

This is, of course, only a heuristic statement; hence the resulting consequences must be interpreted accordingly. However, as we know, the approximation (14-5) can be given a precise interpretation in the form of the law of large numbers. Following a similar approach, we prove at the end of the section the main consequence [Eq. (14-10)] of (14-5) in the context of entropy.

Guided by (14-5), we shall separate the N^n sequences of the form (14-3) into two groups: (a) typical and (b) rare. We shall say that a sequence is *typical,* if $n_i \simeq np_i$. All other sequences will be called *rare.* A typical sequence will be identified with the letter $\boldsymbol{t}$:

$$\boldsymbol{t} = \{A_i \text{ occurs } n_i \simeq np_i \text{ times in a specific order}\} \tag{14-6}$$

From the definition it follows that to each set of numbers $n_1, \ldots, n_N$ "close" to the numbers $np_1, \ldots, np_N$ there corresponds one typical sequence. The union of all typical sequences will be denoted by $\boldsymbol{T}$. Thus $\boldsymbol{T}$ is the totality of all sequences of the form (14-3) where $n_i \simeq np$. As we noted, it is almost certain that for large n, each observed sequence is typical. This leads to the conclusion that

$$P(\boldsymbol{T}) \simeq 1 \tag{14-7}$$

The complement $\overline{\boldsymbol{T}}$ of $\boldsymbol{T}$ is the union of all rare sequences and its probability is negligible for large n:

$$P(\overline{\boldsymbol{T}}) \simeq 0 \tag{14-8}$$

Since $n_i \simeq np_i$ for all typical sequences, (14-4) yields

$$P(\boldsymbol{t}) = p_1^{n_1} \ldots p_N^{n_N} \simeq e^{np_1 \log p_1 + \cdots + np_N \log p_N}$$

Hence the probability of each typical sequence equals

$$P(\boldsymbol{t}) = e^{-nH(\boldsymbol{U})} \tag{14-9}$$

where $H(\boldsymbol{U})$ is the entropy of the partition $\boldsymbol{U}$. Denoting by n_T the number of typical sequences, we conclude from (14-7) and (14-9) that

$$n_T = \frac{P(\boldsymbol{T})}{P(\boldsymbol{t})} \simeq e^{nH(\boldsymbol{U})} \tag{14-10}$$

We have thus expressed the number of typical sequences in terms of the entropy of $\boldsymbol{U}$. If all the events of $\boldsymbol{U}$ are equally likely, then $H(\boldsymbol{U}) = \log N$ and $n_T = N^n$. In all other cases, $H(\boldsymbol{U}) < \log N$ [see (14-38)]. Hence

$$n_T \simeq e^{nH(\boldsymbol{U})} \ll N^n \qquad \text{for} \quad n \gg 1 \tag{14-11}$$

This leads to the important conclusion that, if n is sufficiently large, then *most* sequences are rare even though "almost certainly" none will occur.

Notes 1. We should point out that each typical sequence is not more likely than each rare sequence. In fact, the sequence with the largest probability is the rare sequence $\{A_m$ occurs n times$\}$, where A_m is the event with the largest probability. As we presently show, the distinction between typical and rare sequences is best expressed in terms of the events

$$\{A_i \text{ occurs } n_i \text{ times in } \textit{any} \text{ order}\}$$

As we know [see (4-102)], the probability of these events equals

$$\frac{n!}{k_1! \cdots k_N!} p_1^{k_1} \cdots p_N^{k_N}$$

and for large n, it takes significant values only in a small vicinity of the point $(k_1 = n_1 p_1, \ldots, k_N = n_N p_N)$. This follows by repeating the argument leading to (3-17) or, from the DeMoivre–Laplace approximation (3-39).

2. On page 1 of Chap. 1 we noted that the theory of probability applied to averages of mass phenomena leads to useful results only if the ratio k/n approaches a constant as n increases and this constant is the same for any subsequence. This apparently mild requirement results in severe restrictions on the properties of the resulting sequences. It leads to the conclusion that of all possible N^n sequences formed with the N elements of a partition $\boldsymbol{U}$, only the $e^{nH(\boldsymbol{U})}$ typical sequences are likely to occur; all other sequences are nearly impossible.

Typical Sequences and the Law of Large Numbers

We show next that the preceding results can be reestablished rigorously as consequences of the law of large numbers. For simplicity, we consider only two-element partitions and, to be concrete, we assume that A and $\overline{A}$ are the events "heads" and "tails" respectively in the coin experiment. In the space S^n, the probability of the elementary event $\{\zeta_k\} = \{k$ heads in a specific order$\}$ equals

$$P\{\zeta_k\} = p^k q^{n-k}$$

and the probability of the event[4]

$$A_k = \{k \text{ heads in any order}\}$$

equals

$$P(A_k) = \binom{n}{k} p^k q^{n-k} \simeq \frac{1}{\sqrt{2\pi npq}} e^{-(k-np)^2/2npq} \tag{14-12}$$

---

[4]The event A_k is not, of course, an element of the partition $\boldsymbol{U} = [A, \overline{A}]$

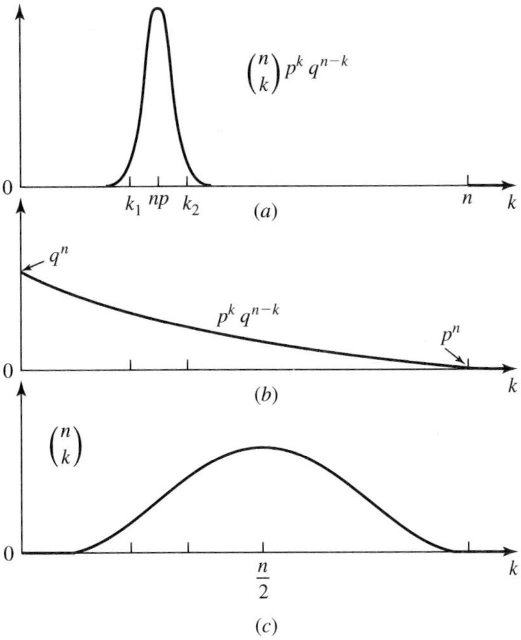

FIGURE 14-3

In Fig. 14-3 we plot the probability $P(A_k)$, the geometric progression $q^n(p/q)^k$, and the binomial coefficients

$$\binom{n}{k} = p^{-k}q^{-(n-k)}P(A_k) \tag{14-13}$$

as functions of k.

α-TYPICAL SEQUENCES. Given a number α between 0 and 1, we form the number ε such that

$$\alpha = 2\text{G}\left(\varepsilon\sqrt{n/pq}\right) - 1 \tag{14-14}$$

where $\text{G}(x)$ is the normal distribution. We shall say that the sequence ζ_k is *α-typical* if k is such that

$$k_1 \le k \le k_2 \qquad \text{where} \quad k_1 = n(p-\varepsilon) \qquad k_2 = n(p+\varepsilon) \tag{14-15}$$

The union of all α-typical sequences is a set $\boldsymbol{T}$ consisting of

$$n_T = \sum_{k=k_1}^{k_2} \binom{n}{k} \tag{14-16}$$

elements and its probability equals α [see (4-92) and (4-96)]

$$P(\boldsymbol{T}) = \sum_{k=k_1}^{k_2} \binom{n}{k} p^k q^{n-k} \simeq 2\text{G}\left(\varepsilon\sqrt{\frac{n}{pq}}\right) - 1 = \alpha \tag{14-17}$$

FUNDAMENTAL THEOREM. For any $\alpha < 1$, the number n_T of α-typical sequences tends to $e^{nH(\boldsymbol{U})}$ in the following sense

$$\frac{\ln n_T}{n} \xrightarrow[n\to\infty]{} H(\boldsymbol{U}) \tag{14-18}$$

Proof. If $p = q = 0.5$, then the DeMoivre–Laplace approximation yields

$$\binom{n}{k} \simeq \frac{2^n}{\sqrt{\pi n/2}} e^{-2(k-n/2)^2/n}$$

for k in the $\sqrt{n}$ vicinity of $n/2$. This approximation cannot be used to evaluate the sum in (14-16) for $p \neq 0.5$ because then the center np of the interval (k_1, k_2) is not $n/2$. We shall bound n_T using (14-13) and (14-16). Clearly,

$$n_T = \sum_{k=k_1}^{k_2} p^{-k} q^{k-n} P(A_k) \tag{14-19}$$

where we assume that $p < q$. As k increases, the term $p^{-k}q^{k-n}$ increases monotonically. Hence

$$q^{-n}\left(\frac{q}{p}\right)^{k_1} \sum_{k=k_1}^{k_2} P(A_k) < n_T < q^{-n}\left(\frac{q}{p}\right)^{k_2} \sum_{k=k_1}^{k_2} P(A_k) \tag{14-20}$$

And since [see (14-17)]

$$\sum_{k=k_1}^{k_2} P(A_k) = P(\boldsymbol{T}) = \alpha$$

(14-20) yields

$$\frac{\alpha}{q^n}\left(\frac{q}{p}\right)^{k_1} < \sum_{k=k_1}^{k_2} \binom{n}{k} < \frac{\alpha}{q^n}\left(\frac{q}{p}\right)^{k_2} \tag{14-21}$$

Setting $k_1 = np - n\varepsilon$ and $k_2 = np + n\varepsilon$ in (14-21) and using the identity

$$p^{-np}q^{-nq} = e^{-n(p\ln p + q\ln q)} = e^{nH(\boldsymbol{U})}$$

we conclude from (14-21) that

$$\alpha e^{nH(\boldsymbol{U})}\left(\frac{q}{p}\right)^{-n\varepsilon} < n_T < \alpha e^{nH(\boldsymbol{U})}\left(\frac{q}{p}\right)^{n\varepsilon}$$

Hence

$$nH(\boldsymbol{U}) + \ln\alpha - n\varepsilon \ln\frac{q}{p} < \ln n_T < nH(\boldsymbol{U}) + \ln\alpha + n\varepsilon\ln\frac{q}{p}$$

Dividing by n, we obtain (14-18) because α is constant and, as we see from (14-14), $\varepsilon \to 0$ as $n \to \infty$.

Important conclusion Theorem (14-18) holds for any $\alpha < 1$; it will be assumed, however, that $\alpha \simeq 1$ and the corresponding sequences will be called typical. With this assumption

$$P(\boldsymbol{T}) = \alpha \simeq 1 \quad P(\overline{\boldsymbol{T}}) = 1 - \alpha \simeq 0 \tag{14-22}$$

The probability of an arbitrary event M equals, therefore, its conditional probability

$$P(M) = P(M \mid \boldsymbol{T})P(\boldsymbol{T}) + P(M \mid \overline{\boldsymbol{T}})P(\overline{\boldsymbol{T}}) \simeq P(M \mid \boldsymbol{T}) \qquad (14\text{-}23)$$

In other words, in any conclusions concerning probabilities in the space S^n, it suffices to consider the subspace of S^n consisting of typical sequences only. This is, of course, only approximately true for finite n. It is, however, exact in the limit as $n \to \infty$.

CONCLUDING REMARKS. In Chap. 1, we presented the following interpretations of the probability $P(A)$ of an event A.

Axiomatic. $P(A)$ is a number assigned to the event A. This number satisfies three axioms but is otherwise arbitrary.

Empirical. For large n,

$$P(A) \simeq \frac{k}{n}$$

where k is the number of times A occurs in n repetitions of the underlying experiment S.

Subjective. $P(A)$ is a measure of our uncertainty about the occurrence of A in a single performance of S.

Principle of insufficient reason. If A_i are N events of a partition $\boldsymbol{U}$ of S and nothing is known about their probabilities, then $P(A_i) = 1/N$.

We give next four related interpretations of the entropy $H(\boldsymbol{U})$ of $\boldsymbol{U}$.

Axiomatic. $H(\boldsymbol{U})$ is a number assigned to each partition of S. This number equals the sum $-\sum p_i \log p_i$, where $p_i = P(A_i)$.

Empirical. This interpretation involves the repeated performance not of the experiment S, but of the experiment S^n of repeated trials. In this experiment, a specific typical sequence $\boldsymbol{t}_j$ is an event with probability $e^{-nH(\boldsymbol{U})}$. Applying the relative frequency interpretation of probability to this event, we conclude that if the experiment S^n is repeated m times and the event $\boldsymbol{t}_j$ occurs m_j times, then for sufficiently large m,

$$P(\boldsymbol{t}_j) = e^{-nH(\boldsymbol{U})} \simeq \frac{m_j}{m} \qquad \text{hence} \quad H(\boldsymbol{U}) \simeq -\frac{1}{n}\ln\frac{m_j}{m}$$

This relates the theoretical quantity $H(\boldsymbol{U})$ to the experimental numbers m_j and m.

Subjective. $H(\boldsymbol{U})$ is a measure of our uncertainty about the occurrence of the events A_i of the partition $\boldsymbol{U}$ in a single performance of S.

Principle of maximum entropy. The probabilities $p_i = P(A_i)$ must be such as to maximize $H(\boldsymbol{U})$ subject to the given constraints. Since $n_t = e^{nH(\boldsymbol{U})}$, the ME principle is equivalent to the principle of maximizing the number of typical sequences. If there are no constraints, that is, if nothing is known about the probabilities p_i, then the ME principle leads to the estimates $p_i = 1/N$, $H(\boldsymbol{U}) = \log N$, and $n_t = N^n$.

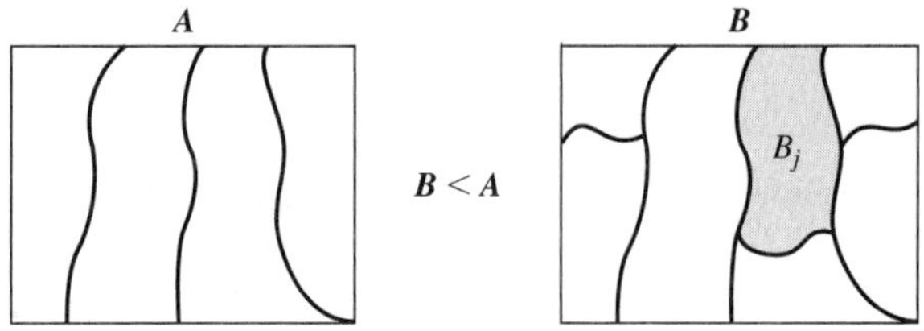

FIGURE 14-4

14-2 BASIC CONCEPTS

In this section, we develop deductively the properties of entropy starting with various notations and set operations. At the end of the section, we reexamine the results in terms of the heuristic notion of entropy as a measure of uncertainty, and we conclude with a typical sequence interpretation of the main theorems.

DEFINITIONS. The notation

$$\boldsymbol{U} = [A_1, \ldots, A_k] \qquad \text{or simply} \qquad \boldsymbol{U} = [A_i]$$

will mean that $\boldsymbol{U}$ is a partition consisting of the events A_i. These events will be called elements[5] of $\boldsymbol{U}$.

I. A partition with only two elements will be called *binary*. Thus

$$\boldsymbol{U} = [A, \overline{A}]$$

is a binary partition consisting of the event A and its complement $\overline{A}$.

II. A partition whose elements are the elementary events $\{\zeta_t\}$ of the space S will be denoted by $\boldsymbol{V}$ and will be called the *element partition*.

III. A *refinement* of a partition $\boldsymbol{U}$ is a partition $\boldsymbol{B}$ such that each element B_j of $\boldsymbol{B}$ is a subset of some element A_i of $\boldsymbol{U}$ (Fig. 14-4). We shall use the notation $\boldsymbol{B} \prec \boldsymbol{U}$ to indicate that $\boldsymbol{B}$ is a refinement of $\boldsymbol{U}$ and we shall say that $\boldsymbol{U}$ is larger[6] than $\boldsymbol{B}$. Thus

$$\boldsymbol{B} \prec \boldsymbol{U} \qquad \text{iff} \quad B_j \subset A_i \tag{14-24}$$

A *common refinement* of two partitions is a refinement of both.

The partition $\boldsymbol{D}$ in Fig. 14-5 is a common refinement of the partitions $\boldsymbol{U}$ and $\boldsymbol{B}$.

IV. The *product*[7] of two partitions $\boldsymbol{U} = [A_i]$ and $\boldsymbol{B} = [B_j]$ is a partition whose elements are all intersections A_iB_j of the elements of $\boldsymbol{U}$ and $\boldsymbol{B}$. This partition will be denoted by

$$\boldsymbol{U} \cdot \boldsymbol{B}$$

Clearly, $\boldsymbol{U} \cdot \boldsymbol{B}$ is the largest common refinement of $\boldsymbol{U}$ and $\boldsymbol{B}$.

---

[5]It will be clear from the context whether the word *element* means an event A_i of a partition $\boldsymbol{U}$ or an element ζ_i of the space S.

[6]The symbol $\prec$ is *not* an ordering of two arbitrary partitions. It has a meaning only if $\boldsymbol{B}$ is a refinement of $\boldsymbol{U}$.

[7]We should emphasize that partition product is *not* a set operation.

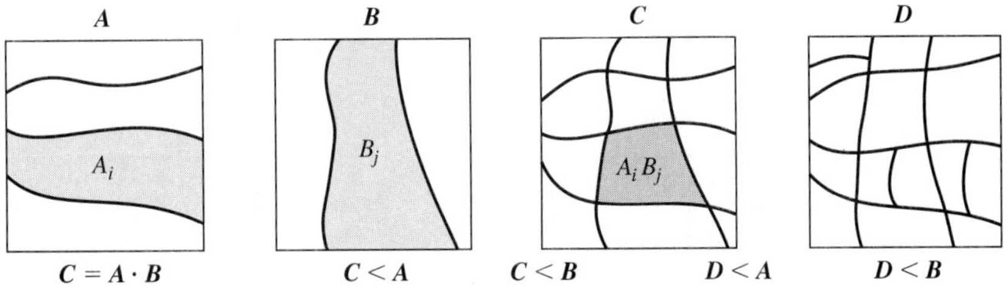

FIGURE 14-5

Properties From the definition it follows that

$$V \prec U \qquad \text{for any} \quad U$$

$$U \cdot B = B \cdot U \qquad U \cdot (B \cdot C) = (U \cdot B) \cdot C$$

$$\text{If} \quad U_1 \prec U_2 \prec U_3 \qquad \text{then} \quad U_1 \prec U_3$$

$$\text{If} \quad B \prec U \qquad \text{then} \quad U \cdot B = B$$

ENTROPY

The entropy of a partition U is by definition the sum

$$H(U) = -(p_1 \log p_1 + \cdots + p_N \log p_N) = \sum_{i=1}^{N} \varphi(p_i) \tag{14-25}$$

where $p_i = P(A_i)$ and $\varphi(p) = -p \log p$.

Since $\varphi(p) \geq 0$ for $0 \leq p \leq 1$, it follows from (14-25) that

$$H(U) \geq 0 \tag{14-26}$$

where $H(U) = 0$ iff one of the p_i's equals 1; all others are then equal to 0.

Binary partitions If $U = [A, \overline{A}]$ and $P(A) = p$, then (Fig. 14-2)

$$H(U) = -p \log p - (1 - p) \log(1 - p) = h(p) \tag{14-27}$$

Equally likely events If

$$p_1 = p_2 = \cdots = p_N$$

then

$$H(U) = -\frac{1}{N} \log \frac{1}{N} - \cdots - \frac{1}{N} \log \frac{1}{N} = \log N \tag{14-28}$$

If, in particular, $N = 2^m$, then $H(U) = m$.

INEQUALITIES. The function $\varphi(p) = -p \log p$ is convex. Therefore (see Fig. 14-6 and Prob. 14-2)

$$\varphi(p_1 + p_2) < \varphi(p_1) + \varphi(p_2) < \varphi(p_1 + \varepsilon) + \varphi(p_2 - \varepsilon) \tag{14-29}$$

where

$$p_1 < p_1 + \varepsilon \leq p_2 - \varepsilon < p_2 \tag{14-30}$$

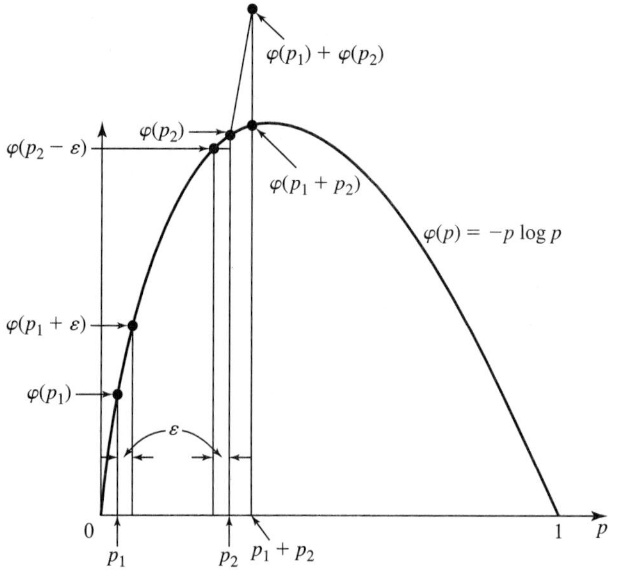

FIGURE 14-6

This leads to the following properties of entropy:

1. Given a partition $\boldsymbol{U} = [A_1, A_2, \ldots, A_N]$, we form the partition $\boldsymbol{B} = [B_a, B_b, A_2, \ldots, A_N]$ obtained by splitting A_1 into the elements B_a and B_b as in Fig. 14-7. We maintain that

$$H(\boldsymbol{U}) \le H(\boldsymbol{B}) \tag{14-31}$$

Proof. Clearly,

$$H(\boldsymbol{U}) - \varphi(p_a + p_b) = H(\boldsymbol{B}) - \varphi(p_a) - \varphi(p_b)$$

because each side equals the contribution to $H(\boldsymbol{U})$ and $H(\boldsymbol{B})$, respectively, due to the common elements of $\boldsymbol{U}$ and $\boldsymbol{B}$. Hence (14-31) follows from the first inequality in (14-29).

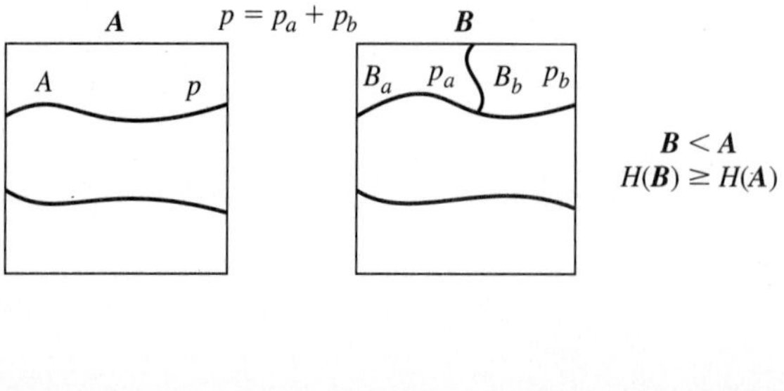

FIGURE 14-7

EXAMPLE 14-4 ▶ In the next table we list the probabilities of the events of a partition $\boldsymbol{U}$ and of its refinement $\boldsymbol{B}$ obtained as above.

$\boldsymbol{U}$	$p = 0.4$		0.35	0.25
$\boldsymbol{B}$	$p_a = 0.22$	$p_b = 0.18$	0.35	0.25

In this case,

$$H(\boldsymbol{U}) = -(0.4 \log 0.4 + 0.35 \log 0.35 + 0.25 \log 0.25) = 1.559$$

$$H(\boldsymbol{B}) = -(0.22 \log 0.22 + 0.18 \log 0.18 + 0.35 \log 0.35 + 0.25 \log 0.25) = 1.956$$

Thus

$$H(\boldsymbol{U}) = 1.559 < 1.956 = H(\boldsymbol{B})$$

in agreement with (14-31). ◀

2. If

$$\boldsymbol{B} \prec \boldsymbol{U} \qquad \text{then} \quad H(\boldsymbol{B}) \geq H(\boldsymbol{U}) \tag{14-32}$$

Proof. Repeating the construction of Fig. 14-7, we form a chain of refinements

$$\boldsymbol{U} = \boldsymbol{U}_1 \prec \cdots \prec \boldsymbol{U}_{m-1} \prec \boldsymbol{U}_m \prec \cdots \prec \boldsymbol{U}_n = \boldsymbol{B}$$

where $\boldsymbol{U}_m$ is obtained by splitting one of the elements of $\boldsymbol{U}_{m-1}$ as in Fig. 14-8. From this and (14-31) it follows that

$$H(\boldsymbol{U}) = H(\boldsymbol{U}_1) \leq \cdots \leq H(\boldsymbol{U}_n) = H(\boldsymbol{B})$$

and (14-32) results.

3. For any $\boldsymbol{U}$:

$$H(\boldsymbol{U}) \leq H(\boldsymbol{V}) \tag{14-33}$$

where $\boldsymbol{V}$ is the element partition.

Proof. It follows from (14-31) because $\boldsymbol{V}$ is a refinement of $\boldsymbol{U}$.

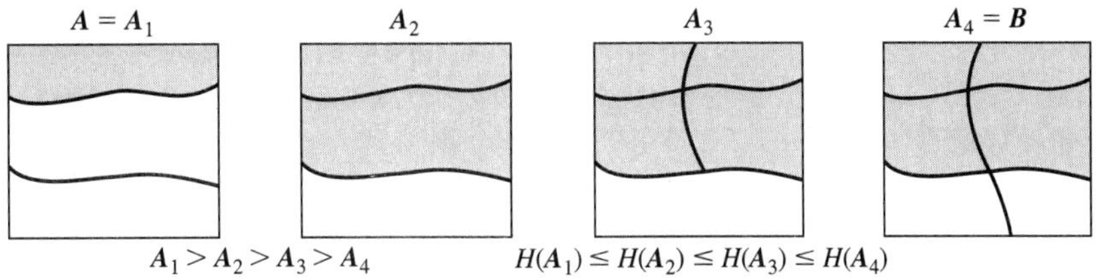

FIGURE 14-8

4. For any $\boldsymbol{U}$ and $\boldsymbol{B}$:

$$H(\boldsymbol{U}\cdot\boldsymbol{B}) \geq H(\boldsymbol{U}) \qquad H(\boldsymbol{U}\cdot\boldsymbol{B}) \geq H(\boldsymbol{B}) \tag{14-34}$$

Proof. It follows from (14-31) because $\boldsymbol{U}\cdot\boldsymbol{B}$ is a refinement of $\boldsymbol{U}$ and of $\boldsymbol{B}$.

EXAMPLE 14-5

▶ In the die experiment, the probabilities of the six events $\{f_1\}, \ldots, \{f_6\}$ equal

$$0.1 \qquad 0.1 \qquad 0.15 \qquad 0.2 \qquad 0.2 \qquad 0.25$$

respectively. The probabilities of the events of the partitions

$$\boldsymbol{U} = [\text{even}, \text{odd}] \qquad \boldsymbol{B} = [i \leq 3, i > 3]$$

are given by

$$P\{\text{even}\} = 0.55 \qquad P\{\text{odd}\} = 0.45 \qquad P\{i \leq 3\} = 0.35 \qquad P\{i > 3\} = 0.65$$

The product $\boldsymbol{U}\cdot B$ is a partition consisting of the four elements

$$\{f_2\} \qquad \{f_1 f_3\} \qquad \{f_4 f_6\} \qquad \{f_5\}$$

with respective probabilities

$$0.1 \qquad 0.25 \qquad 0.45 \qquad 0.2$$

From the above it follows that

$$H(\boldsymbol{U}) = 0.993 \qquad H(\boldsymbol{B}) = 0.934 \qquad H(\boldsymbol{U}\cdot B) = 1.815$$

in agreement with (14-34). ◀

5. Suppose that $\boldsymbol{U}$ and $\boldsymbol{B}$ are two partitions that have the same elements except the first two (Fig. 14-9)

$$\boldsymbol{U} = [A_1, A_2, A_3, \ldots, A_N] \qquad \boldsymbol{B} = [B_1, B_2, A_3, \ldots, A_N]$$

We maintain that if

$$P(A_1) = p_1 \qquad P(A_2) = p_2 \qquad P(B_1) = p_1 + \varepsilon \leq p_2 - \varepsilon = P(B_2)$$

as in (14-30), then

$$H(\boldsymbol{U}) \leq H(\boldsymbol{B}) \tag{14-35}$$

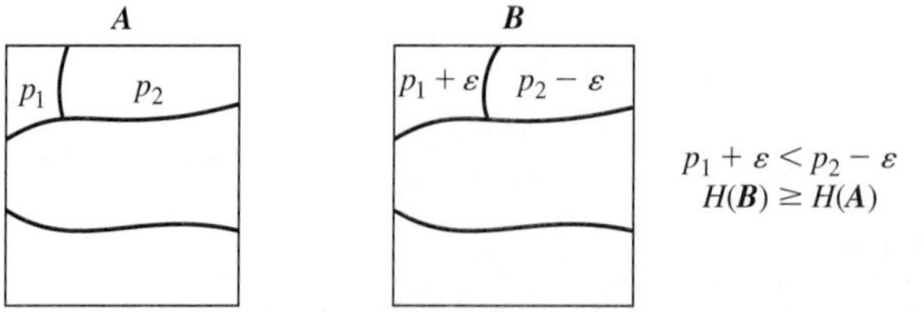

FIGURE 14-9

Proof. Clearly,

$$H(U) - \varphi(p_1) - \varphi(p_2) = H(B) - \varphi(p_1 + \varepsilon) - \varphi(p_2 + \varepsilon)$$

because each side equals the contribution to $H(U)$ and $H(\boldsymbol{B})$, respectively, due to the common elements of $\boldsymbol{U}$ and $\boldsymbol{B}$. Hence (14-35) follows from the second inequality in (14-29).

EXAMPLE 14-6 ▶ In the next table we list the probabilities of the events of the partitions $\boldsymbol{U}$ and $\boldsymbol{B}$.

U	0.1	0.3	0.35	0.25	$p_1 = 0.1$	$\varepsilon = 0.08$
B	0.18	0.22	0.35	0.25	$p_2 = 0.3$	

In this case,

$$H(\boldsymbol{U}) = 1.883 \qquad H(\boldsymbol{B}) = 1.956$$

in agreement with (14-35). ◀

6. If we equalize the entropies of two elements of a partition, leaving all others unchanged, its entropy increases.

Proof. It follows from the above with $\varepsilon = (p_2 - p_1)/2$.

7. The entropy of a partition is maximum if all its elements are equally likely as in (14-28).

Proof. Suppose that $\boldsymbol{U}$ is a partition such that $H(U) = H_m$ is maximum and two of its elements have unequal probabilities. If they are made equal, then (property 6) $H(\boldsymbol{U})$ increases. But this is impossible because H_m is maximum by assumption.

A useful inequality. If a_i and b_i are N positive numbers such that

$$a_1 + \cdots + a_N = 1 \qquad b_1 + \cdots + b_N \le 1 \tag{14-36}$$

then

$$-\sum_i a_i \log a_i \le -\sum_i a_i \log b_i \tag{14-37}$$

with equality iff $a_i = b_i$.

Proof. From the inequality $e^y \ge 1 + y$ it follows that $\ln x \le x - 1$ (Fig. 14-10). With $x = b_i/a_i$, this yields

$$\ln b_i - \ln a_i = \ln \frac{b_i}{a_i} \le \frac{b_i}{a_i} - 1$$

Multiplying by a_i and adding, we obtain

$$\sum_i a_i(\ln b_i - \ln a_i) \le \sum_i a_i\left(\frac{b_i}{a_i} - 1\right) = \sum_i (b_i - a_i) \le 0$$

and (14-37) results.

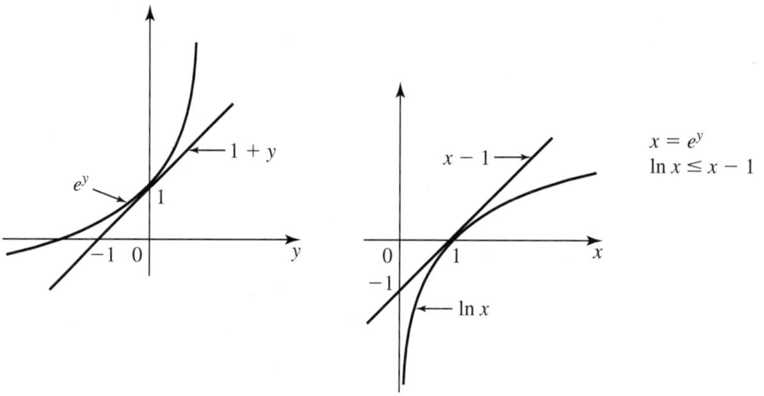

FIGURE 14-10

Maximum entropy. Using (14-37), we shall rederive property 7. It suffices to show that

$$-\sum_i p_i \log p_i \le \log N \tag{14-38}$$

Proof. The numbers $a_i = p_i$ and $b_i = 1/N$ satisfy (14-36). Inserting into (14-37), we conclude that

$$-\sum_i p_i \log p_i \le -\sum_i p_i \log \frac{1}{N} = \log N \sum_i p_i = \log N$$

Conditional Entropy and Mutual Information

The conditional entropy of a partition $\boldsymbol{U}$ assuming M is by definition the sum

$$H(\boldsymbol{U} \mid M) = -\sum_{i=1}^{N_U} P(A_i \mid M) \log P(A_i \mid M) \tag{14-39}$$

where $P(M) \ne 0$, N_U is the number of elements A_i of $\boldsymbol{U}$, and

$$P(A_i \mid M) = \frac{P(A_i M)}{P(M)}$$

As we explain later, $H(\boldsymbol{U} \mid M)$ is the uncertainty about $\boldsymbol{U}$ in the subsequence of trials in which M occurs.

Suppose now that $\boldsymbol{B}$ is a partition consisting of the $N_{\boldsymbol{B}}$ elements B_j. Clearly,

$$H(\boldsymbol{U} \mid B_j) = -\sum_{i=1}^{N_U} P(A_i \mid B_j) \log P(A_i \mid B_j) \tag{14-40}$$

is the conditional entropy of $\boldsymbol{U}$ assuming B_j defined as in (14-39). The conditional entropy of $\boldsymbol{U}$ assuming $\boldsymbol{B}$ is the weighted average of $H(\boldsymbol{U} \mid B_j)$:

$$H(\boldsymbol{U} \mid \boldsymbol{B}) = \sum_{j=1}^{N_B} P(B_j) H(\boldsymbol{U} \mid B_j) \tag{14-41}$$

This equals the uncertainty about $\boldsymbol{U}$ if at each trial we know which of the events B_j of $\boldsymbol{B}$ has occurred.

EXAMPLE 14-7

▶ We shall determine the conditional entropy $H(\boldsymbol{V} \mid \boldsymbol{B})$ of the element partition $\boldsymbol{V}$ in the fair-die experiment where $\boldsymbol{B} = [\text{even, odd}]$.

Clearly, $P\{f_i \mid \text{even}\} = \frac{1}{3}$ if i is even and $P\{f_i \mid \text{even}\} = 0$ if i is odd. Similarly, $P\{f_i \mid \text{odd}\} = \frac{1}{3}$ if i is odd and $P\{f_i \mid \text{odd}\} = 0$ if i is even. Hence

$$H(\boldsymbol{V} \mid \text{even}) = -\left(\tfrac{1}{3}\log\tfrac{1}{3} + \tfrac{1}{3}\log\tfrac{1}{3} + \tfrac{1}{3}\log\tfrac{1}{3}\right) = \log 3 = H(\boldsymbol{V} \mid \text{odd})$$

And since $P\{\text{even}\} = P\{\text{odd}\} = 0.5$, we conclude from (14-41) that

$$H(\boldsymbol{V} \mid \boldsymbol{B}) = 0.5\log 3 + 0.5\log 3 = \log 3$$

Thus, in the absence of any information, our uncertainty about $\boldsymbol{V}$ equals $H(\boldsymbol{V}) = \log 6$. If we know, however, whether at each trial "even" or "odd" showed, then our uncertainty is reduced to $H(\boldsymbol{V} \mid \boldsymbol{B}) = \log 3$. ◀

THEOREM 14-1

▶ If

$$\boldsymbol{B} \prec \boldsymbol{U} \qquad \text{then} \quad H(\boldsymbol{U} \mid \boldsymbol{B}) = 0 \tag{14-42}$$

Proof. Since $\boldsymbol{B}$ is a refinement of $\boldsymbol{U}$, each element B_j of $\boldsymbol{B}$ is a subset of some element A_k of $\boldsymbol{U}$ and, therefore, it is disjoint with all other elements of $\boldsymbol{U}$. Hence $A_i B_j = B_j$ if $i = k$ and $A_i B_j = 0$ otherwise. This leads to the conclusion that

$$P(A_i \mid B_j) = \frac{P(A_i B_j)}{P(B_j)} = \begin{cases} 1 & i = k \\ 0 & i \neq k \end{cases}$$

And since $p \log p = 0$ for $p = 0$ and $p = 1$, we conclude that all terms in (14-40) equal 0; hence

$$H(\boldsymbol{U} \mid B_j) = 0$$

for every j. From this and (14-41) it follows that $H(\boldsymbol{U} \mid \boldsymbol{B}) = 0$.

Independent partitions Two partitions $\boldsymbol{U} = [A_i]$ and $\boldsymbol{B} = [B_j]$ are called independent if the events A_i and B_j are independent for every i and j:

$$P(A_i B_j) = P(A_i)P(B_j) \tag{14-43}$$

◀

THEOREM 14-2

▶ If the partitions $\boldsymbol{U}$ and $\boldsymbol{B}$ are independent, then

$$H(\boldsymbol{U} \mid \boldsymbol{B}) = H(\boldsymbol{U}) \qquad H(\boldsymbol{B} \mid \boldsymbol{U}) = H(\boldsymbol{B}) \tag{14-44}$$

Proof. Clearly, $P(A_i \mid B_j) = P(A_i)$; hence [see (14-40)]

$$H(\boldsymbol{U} \mid B_j) = -\sum_i P(A_i)\log P(A_i) = H(\boldsymbol{U})$$

Inserting into (14-41), we obtain

$$H(\boldsymbol{U} \mid B) = H(\boldsymbol{U}) \sum_j P(B_j) = H(\boldsymbol{U})$$

and (14-43) results. We can show similarly that $H(\boldsymbol{B} \mid \boldsymbol{U}) = H(\boldsymbol{B})$. ◀

THEOREM 14-3 ▶ For any $\boldsymbol{U}$ and $\boldsymbol{B}$:

$$H(\boldsymbol{U} \cdot B) \leq H(\boldsymbol{U}) + H(\boldsymbol{B}) \tag{14-45}$$

Proof. As we know [see (2-41)]

$$P(A_i) = \sum_j P(A_i B_j)$$

Hence

$$H(\boldsymbol{U}) - \sum_i P(A_i) \log P(A_i) = -\sum_{i,j} P(A_i B_j) \log P(A_i)$$

Writing a similar equation for $H(\boldsymbol{B})$ and adding, we obtain

$$H(\boldsymbol{U}) + H(\boldsymbol{B}) = -\sum_{i,j} P(A_i B_j) \log[P(A_i)P(B_j)] \tag{14-46}$$

Clearly, $H(\boldsymbol{U} \cdot \boldsymbol{B})$ is a partition with elements $A_i B_j$. Hence

$$H(\boldsymbol{U} \cdot \boldsymbol{B}) = -\sum_{i,j} P(A_i B_j) \log P(A_i B_j) \tag{14-47}$$

To prove (14-45), we shall apply (14-37) identifying the numbers a_i and b_i with the numbers $P(A_i B_j)$ and $P(A_i)P(B_j)$, respectively. We can do so because

$$\sum_{i,j} P(A_i B_j) = 1 \qquad \sum_{i,j} P(A_i)P(B_j) = 1$$

From (14-37) it follows that the sum in (14-47) cannot exceed the sum in (14-46); hence (14-45) must be true. ◀

COROLLARY ▶

$$H(\boldsymbol{U} \cdot \boldsymbol{B}) = H(\boldsymbol{U}) + H(\boldsymbol{B}) \tag{14-48}$$

iff the partitions $\boldsymbol{U}$ and $\boldsymbol{B}$ are independent.

Proof. This follows from (14-45) because (14-37) is an equality iff $a_i = b_i$ for every i. Hence (14-45) is an equality iff

$$P(A_i B_j) = P(A_i)P(B_j)$$

for every i and j. ◀

THEOREM 14-4 ▶ For any $\boldsymbol{U}$ and $\boldsymbol{B}$:

$$H(\boldsymbol{U} \cdot \boldsymbol{B}) = H(\boldsymbol{B}) + H(\boldsymbol{U} \mid \boldsymbol{B}) = H(\boldsymbol{U}) + H(\boldsymbol{B} \mid \boldsymbol{U}) \tag{14-49}$$

Proof. Since

$$P(A_i B_j) = P(B_j)P(A_i \mid B_j)$$

we conclude from (14-40) that

$$P(B_j)H(\boldsymbol{U} \mid B_j) = -\sum_i P(B_j)P(A_i \mid B_j) \log P(A_i \mid B_j)$$
$$= -\sum_i P(A_i B_j)[\log P(A_i B_j) - \log P(B_j)]$$
$$= -\sum_i P(A_i B_j) \log P(A_i B_j) + P(B_j) \log P(B_j)$$

Summing over all j, we obtain

$$\sum_j P(B_j)H(\boldsymbol{U} \mid B_j) = -\sum_{i,j} P(A_i B_j) \log P(A_i B_j) + \sum_j P(B_j) \log P(B_j)$$

and the first equation in (14-49) follows because the above three sums equal $H(\boldsymbol{U} \mid \boldsymbol{B})$, $H(\boldsymbol{U} \cdot \boldsymbol{B})$, and $-H(\boldsymbol{B})$ respectively. The second equation follows because $\boldsymbol{U} \cdot \boldsymbol{B} = \boldsymbol{B} \cdot \boldsymbol{U}$. ◀

COROLLARIES ▶ The following relationships follow readily from the last two theorems: For any $\boldsymbol{U}$ and $\boldsymbol{B}$:

$$H(\boldsymbol{B}) \le H(\boldsymbol{U} \cdot \boldsymbol{B}) \le H(\boldsymbol{U}) + H(\boldsymbol{B}) \tag{14-50}$$

$$H(\boldsymbol{U} \mid \boldsymbol{B}) \le H(\boldsymbol{U}) \tag{14-51}$$

$$H(\boldsymbol{U}) - H(\boldsymbol{U} \mid \boldsymbol{B}) = H(\boldsymbol{B}) - H(\boldsymbol{B} \mid \boldsymbol{U}) \tag{14-52}$$

◀

Mutual information. The function

$$I(\boldsymbol{U}, \boldsymbol{B}) = H(\boldsymbol{U}) + H(\boldsymbol{B}) - H(\boldsymbol{U} \cdot \boldsymbol{B}) \tag{14-53}$$

is called the mutual information of the partitions $\boldsymbol{U}$ and $\boldsymbol{B}$. From (14-49) it follows that

$$I(\boldsymbol{U}, \boldsymbol{B}) = H(\boldsymbol{U}) - H(\boldsymbol{U} \mid \boldsymbol{B}) = H(\boldsymbol{B}) - H(\boldsymbol{B} \mid \boldsymbol{U}) \tag{14-54}$$

Clearly [see (14-51)]

$$I(\boldsymbol{U}, \boldsymbol{B}) \ge 0 \tag{14-55}$$

As we shall presently see, $I(\boldsymbol{U}, \boldsymbol{B})$ can be interpreted as the "information about $\boldsymbol{U}$ contained in $\boldsymbol{B}$" and it equals the "information about $\boldsymbol{B}$ contained in $\boldsymbol{U}$."

EXAMPLE 14-8 ▶ In the fair-die experiment of Example 14-7,

$$H(\boldsymbol{V}) = \log 6 \qquad H(\boldsymbol{V} \mid \boldsymbol{B}) = \log 3 \qquad H(\boldsymbol{B}) = \log 2 \qquad H(\boldsymbol{B} \mid \boldsymbol{V}) = 0$$

Hence

$$I(\boldsymbol{V}, \boldsymbol{B}) = \log 2$$

Thus the information about the element partition $\boldsymbol{V}$ resulting from the observation of the even–odd partition $\boldsymbol{B}$ equals log 2. ◀

Generalizations. These results can be readily generalized to an arbitrary number of partitions. We list below several special cases leaving the simple proofs as problems:

(*a*) If

$$B \prec V \qquad \text{then} \quad H(U \mid B) \le H(U \mid V) \tag{14-56}$$

(*b*) If the partitions U, B and V are independent, then

$$H(U \cdot B \cdot V) = H(U) + H(B \cdot V) = H(U) + H(B) + H(V) \tag{14-57}$$

(*c*) *Chain rule* For any U, B, and V:

$$H(B \cdot V \mid U) = H(B \mid U) + H(V \mid U \cdot B) \tag{14-58}$$

$$H(U \cdot B \cdot V) = H(U) + H(B \cdot V \mid U) = H(U) + H(B \mid U) + H(V \mid U \cdot B) \tag{14-59}$$

Repeated trials. In the space S^n of repeated trials all outcomes are sequences of the form

$$\zeta_{i_1} \cdots \zeta_{i_k} \cdots \zeta_{i_n} \tag{14-60}$$

where each ζ_{i_k} is an element of S. Consider a partition U of S consisting of N events. At the kth trial, one and only one of these events will occur, namely the event A_{j_k} that contains the element ζ_{i_k}. The cartesian product

$$A_{j,k} = S \times \cdots S \times A_{j_k} \times S \cdots \times S \qquad \zeta_{i_k} \in A_{j_k} \tag{14-61}$$

is an event in S^n with probability

$$P(A_{j,k}) = P(A_{j_k}) \tag{14-62}$$

because it occurs iff the event A_{j_k} occurs at the kth trial. For specific k, the events $A_{j,k}$ form an N element partition of the space S^n. This partition will be denoted by U_k. From (14-62) it follows readily that

$$H(U_k) = H(U) \tag{14-63}$$

We can define similarly the partition B_k of S^n formed with the elements of another partition B of S. Reasoning as in (14-63), we conclude that $H(B_k) = H(B)$ and

$$H(U_k \mid B_k) = H(U \mid B) \qquad I(U_k, B_k) = I(U, B) \tag{14-64}$$

We next form the product of the n partitions U_k:

$$U^n = U_1 \cdot U_2 \cdot \cdots U_n \tag{14-65}$$

The elements of this partition are cartesian products of the form

$$A_{j_1} \times \cdots \times A_{j_k} \times \cdots \times A_{j_n} \tag{14-66}$$

If U is the element partition of S, then U^n is the element partition of S^n. In general, however, the elements of U^n are events consisting of a large number of sequences of the form (14-60). If we picture these sequences as wires, then the elements (14-66) of the partition U^n can be viewed as cables and their union as a collection of such cables (Fig. 14-11).

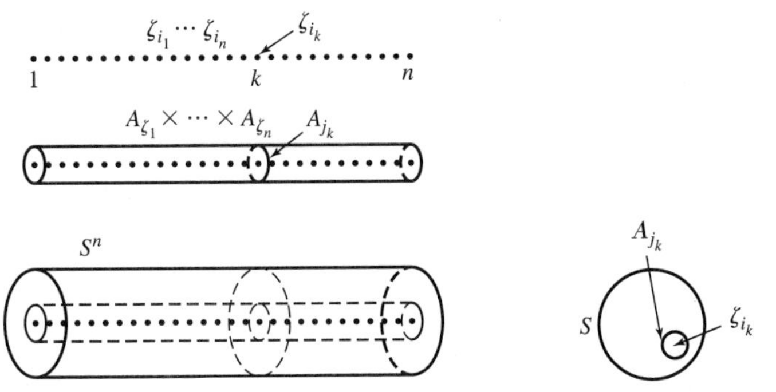

FIGURE 14-11

From the independence of the trials, it follows that the n partitions $\boldsymbol{U}_1, \ldots, \boldsymbol{U}_n$ of S^n are independent. Hence [see (14-57) and (14-63)]

$$H(\boldsymbol{U}^n) = H(\boldsymbol{U}_1) + \cdots + H(\boldsymbol{U}_n) = nH(\boldsymbol{U}) \tag{14-67}$$

Defining similarly the partition $\boldsymbol{B}^n$, we conclude as in (14-64) that

$$H(\boldsymbol{U}^n \mid \boldsymbol{B}^n) = nH(\boldsymbol{B} \mid \boldsymbol{U}) \qquad I(\boldsymbol{U}^n, \boldsymbol{B}^n) = nI(\boldsymbol{U}, \boldsymbol{B}) \tag{14-68}$$

EXAMPLE 14-9

▶ In the coin experiment, the entropy of the element partition equals

$$H(\boldsymbol{V}) = -p \log p - q \log q$$

In the space S^2, the element partition consists of four events with

$$P\{hh\} = p^2 \qquad P\{ht\} = P\{th\} = pq \qquad P\{tt\} = q^2$$

Hence

$$H(\boldsymbol{V}^2) = -p^2 \log p^2 - 2pq \log pq - q^2 \log q^2 = -2p \log p - 2q \log q$$

Thus

$$H(\boldsymbol{V}^2) = 2H(\boldsymbol{V})$$

in agreement with (14-67). ◀

CONDITIONAL ENTROPY AND UNCERTAINTY. As we have noted, the entropy $H(\boldsymbol{U})$ of a partition $\boldsymbol{U} = [A_i]$ gives us a measure of uncertainty about the occurrence of the events A_i at a given trial. Once the trial is performed and the events A_i are observed, the uncertainty is removed. We give next a similar interpretation to the conditional entropy $H(\boldsymbol{U} \mid M)$ of $\boldsymbol{U}$ assuming that the event M has been observed, and of the conditional entropy $H(\boldsymbol{U} \mid \boldsymbol{B})$ of $\boldsymbol{U}$ assuming that the partitioning $\boldsymbol{B}$ has been observed.[8]

If in the definition (14-25) of entropy we replace the probabilities $P(A_i)$ by the conditional probabilities $P(A_i \mid M)$, we obtain the conditional entropy $H(\boldsymbol{U} \mid M)$ of $\boldsymbol{U}$

[8]The expression *a partition* $\boldsymbol{B}$ *is observed* will mean that we know which of the events of $\boldsymbol{B}$ has occurred.

assuming M [see (14-39)]. The relative frequency interpretation of $P(A_i \mid M)$ is the same as that of $P(A_i)$ if we consider not the entire sequence of n trials but only the subsequence of trials in which the event M occurs. From this it follows that $H(\boldsymbol{U} \mid M)$ is the uncertainty about $\boldsymbol{U}$ per trial in that subsequence. In other words, if at a given trial we know that M occurs, then our uncertainty about $\boldsymbol{U}$ equals $H(\boldsymbol{U} \mid M)$; if we know that M occurs, then our uncertainty equals $H(\boldsymbol{U} \mid \overline{M})$. The weighted sum

$$P(M)H(\boldsymbol{U} \mid M) + P(\overline{M})H(\boldsymbol{U} \mid \overline{M})$$

is the uncertainty about $\boldsymbol{U}$ assuming that the binary partition $[M, \overline{M}]$ is observed.

Suppose now that at each trial we observe the partition $\boldsymbol{B} = [B_j]$. We maintain that, under this assumption, the uncertainty per trial about $\boldsymbol{U}$ equals $H(\boldsymbol{U} \mid \boldsymbol{B})$. Indeed, in a sequence of n trials, the number of times the event B_j occurs equals

$$n_j \simeq nP(B_j)$$

In this subsequence, the uncertainty about $\boldsymbol{U}$ equals $H(\boldsymbol{U} \mid B_j)$ per trial. Hence the total uncertainty about $\boldsymbol{U}$ equals

$$\sum_j n_j H(\boldsymbol{U} \mid B_j) \simeq \sum_j nP(B_j)H(\boldsymbol{U} \mid B_j) = nH(\boldsymbol{U} \mid \boldsymbol{B})$$

and the uncertainty per trial equals $H(\boldsymbol{U} \mid \boldsymbol{B})$.

Thus the observation of $\boldsymbol{B}$ reduces the uncertainty about $\boldsymbol{U}$ from $H(\boldsymbol{U})$ to $H(\boldsymbol{U} \mid \boldsymbol{B})$. The difference

$$I(\boldsymbol{U}, \boldsymbol{B}) = H(\boldsymbol{U}) - H(\boldsymbol{U} \mid \boldsymbol{B})$$

is the reduction of the uncertainty about $\boldsymbol{B}$ resulting from the observation of $\boldsymbol{B}$. This justifies the statement that the mutual information $I(\boldsymbol{U}, \boldsymbol{B})$ equals the *information* about $\boldsymbol{U}$ contained in $\boldsymbol{B}$.

We show next the consistency between the properties of entropy developed earlier and the subjective notion of uncertainty.

1. If $\boldsymbol{B}$ is a refinement of $\boldsymbol{U}$ and $\boldsymbol{B}$ is observed, then we know which of the events of $\boldsymbol{U}$ occurred. Hence $H(\boldsymbol{U} \mid \boldsymbol{B}) = 0$ in agreement with (14-42).
2. If the partitions $\boldsymbol{U}$ and $\boldsymbol{B}$ are independent and $\boldsymbol{B}$ is observed, no information about $\boldsymbol{U}$ is gained. Hence $H(\boldsymbol{U} \mid \boldsymbol{B}) = H(\boldsymbol{U})$ in agreement with (14-44).
3. If we observe $\boldsymbol{B}$, our uncertainty about $\boldsymbol{U}$ can only decrease. Hence $H(\boldsymbol{U} \mid \boldsymbol{B}) \le H(\boldsymbol{U})$ in agreement with (14-51).
4. To observe $\boldsymbol{U} \cdot \boldsymbol{B}$, we must observe $\boldsymbol{U}$ and $\boldsymbol{B}$. If only $\boldsymbol{B}$ is observed, the information gained equals $H(\boldsymbol{B})$. The uncertainty about $\boldsymbol{U}$ assuming $\boldsymbol{B}$ equals, therefore, the remaining uncertainty $H(\boldsymbol{U} \mid \boldsymbol{B})$ about $\boldsymbol{B}$. Hence $H(\boldsymbol{U} \cdot \boldsymbol{B}) - H(\boldsymbol{B}) = H(\boldsymbol{U} \mid \boldsymbol{B})$ in agreement with (14-49).
5. Combining 3 and 4, we conclude that $H(\boldsymbol{U} \cdot \boldsymbol{B}) - H(\boldsymbol{B}) \le H(\boldsymbol{U})$ in agreement with (14-45).
6. If $\boldsymbol{B}$ is observed, then the information that is gained about $\boldsymbol{U}$ equals $I(\boldsymbol{U}, \boldsymbol{B})$. If $\boldsymbol{B} \prec \boldsymbol{C}$ and $\boldsymbol{B}$ is observed, then $\boldsymbol{C}$ is known. But knowledge of $\boldsymbol{C}$ yields information about $\boldsymbol{U}$ equal to $I(\boldsymbol{U}, \boldsymbol{C})$. Hence, if $\boldsymbol{B} \prec \boldsymbol{C}$, then $I(\boldsymbol{U}, \boldsymbol{B}) \ge I(\boldsymbol{U}, \boldsymbol{C})$ or, equivalently, $H(\boldsymbol{U} \mid \boldsymbol{B}) \le H(\boldsymbol{U} \mid \boldsymbol{C})$ in agreement with (14-56).

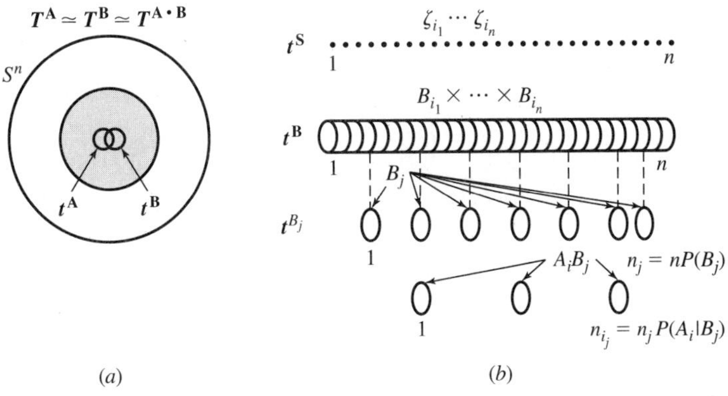

FIGURE 14-12

CONDITIONAL ENTROPY AND TYPICAL SEQUENCES. We give next a typical sequence interpretation of the properties of conditional entropy limiting the discussion to (14-45) and (14-49). The underlying reasoning is used in the proof of the channel capacity theorem (Sec. 14-6).

We denote by $\boldsymbol{t}^{\boldsymbol{U}}$, $\boldsymbol{t}^{\boldsymbol{B}}$, and $\boldsymbol{t}^{\boldsymbol{U}\cdot\boldsymbol{B}}$ the typical sequences of the partition $\boldsymbol{U}$, $\boldsymbol{B}$, and $\boldsymbol{U}\cdot\boldsymbol{B}$, respectively, and by $\boldsymbol{T}^{\boldsymbol{U}}$, $\boldsymbol{T}^{\boldsymbol{B}}$, and $\boldsymbol{T}^{\boldsymbol{U}\cdot\boldsymbol{B}}$ their unions (Fig. 14-12*a*). As we know [see (14-7)]

$$P(\boldsymbol{T}^{\boldsymbol{U}}) \simeq P(\boldsymbol{T}^{\boldsymbol{B}}) \simeq P(\boldsymbol{T}^{\boldsymbol{U}\cdot\boldsymbol{B}}) \simeq 1$$

Furthermore, the number of typical sequences in each of these three sets equals [see (14-10)]

$$n_{T^U} \simeq e^{nH(\boldsymbol{U})} \qquad n_{T^B} \simeq e^{nH(\boldsymbol{B})} \qquad n_{T^{U\cdot B}} \simeq e^{nH(\boldsymbol{U}\cdot\boldsymbol{B})} \tag{14-69}$$

I. We maintain that

$$H(\boldsymbol{U}\cdot\boldsymbol{B}) \le H(\boldsymbol{U}) + H(\boldsymbol{B}) \tag{14-70}$$

Proof. Each $\boldsymbol{t}^{\boldsymbol{U}\cdot\boldsymbol{B}}$ sequence specifies a pair $(\boldsymbol{t}^{\boldsymbol{U}}, \boldsymbol{t}^{\boldsymbol{B}})$. The total number of such pairs formed with all the elements of $\boldsymbol{T}^{\boldsymbol{U}}$ and $\boldsymbol{T}^{\boldsymbol{B}}$ equals $n_{T^U} \cdot n_{T^B}$. However, not all such pairs generate $\boldsymbol{t}^{\boldsymbol{U}\cdot\boldsymbol{B}}$ sequences because, if the partitions $\boldsymbol{U}$ and $\boldsymbol{B}$ are not independent, then not all pairs can occur. For example, if $A_i = B_j$ for some i and j and A_i occurs at the kth trial, then B_j must also occur at this trial. From this it follows that

$$n_{T^{U\cdot B}} \le n_{T^U} \cdot n_{T^B}$$

and (14-70) follows from (14-69).

II. We shall show, finally, that

$$H(\boldsymbol{U}\cdot\boldsymbol{B}) = H(\boldsymbol{B}) + H(\boldsymbol{U} \mid \boldsymbol{B}) \tag{14-71}$$

Proof. There are n_{T^B} sequences in the set $\boldsymbol{T}^{\boldsymbol{B}}$ and $n_{T^{U\cdot B}}$ sequences in the set $\boldsymbol{T}^{\boldsymbol{U}\cdot\boldsymbol{B}}$. The ratio

$$\frac{n_{T^{U\cdot B}}}{n_{T^B}} \simeq e^{n[H(\boldsymbol{U}\cdot\boldsymbol{B}) - H(\boldsymbol{B})]}$$

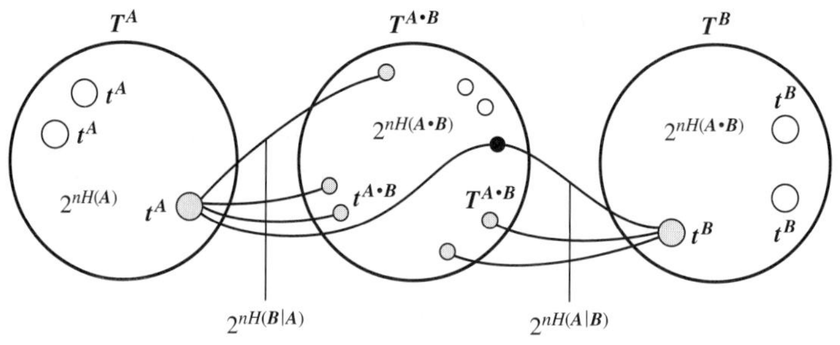

FIGURE 14-13

equals, therefore, the number of $\boldsymbol{t}^{U\cdot B}$ sequences contained in a single $\boldsymbol{t}^B$ sequence *on the average*. To prove (14-71), we must prove, therefore, that this number equals $e^{nH(U|B)}$. We shall prove a stronger statement: The number of $\boldsymbol{t}^{U\cdot B}$ sequences contained in a *single* $\boldsymbol{t}^B$ sequence (Fig. 14-13) equals $e^{nH(U|B)}$.

As we know [see (1-1)], the number of times the event B_j occurs in a $\boldsymbol{t}^B$ sequence "almost certainly" equals

$$n_j \simeq nP(B_j) \tag{14-72}$$

We denote by $\boldsymbol{t}^{B_j}$ subsequence (Fig. 14-12*b*) of $\boldsymbol{t}^B$ in which the number of occurrences of B_j satisfies (14-72). In this subsequence, the relative frequency of the occurrence of an event A_i equals $P(A_i \mid B_j)$ [see page 29].

We shall use (14-10) to show that the number of typical sequences formed with the elements A_i of $\boldsymbol{U}$ that are included in a $\boldsymbol{t}^{B_j}$ sequence equals

$$e^{n_j H(U|B_j)} \simeq e^{nP(B_j)H(U|B_j)} \tag{14-73}$$

Indeed, this follows from (14-10) if we introduce the following changes: We replace $P(A_i)$ by $P(A_i \mid B_j)$, the length n of the original sequences with the length $n_j \simeq nP(B_j)$, and the entropy $H(\boldsymbol{U})$ of $\boldsymbol{U}$ with the conditional entropy $H(\boldsymbol{U} \mid B_j)$.

Returning to the original $\boldsymbol{t}^B$ sequence, we note that it is formed by combining the $\boldsymbol{t}^{B_j}$ sequences that are included in $\boldsymbol{t}^B$. This shows that the total number of $\boldsymbol{t}^U$ sequences that are included in $\boldsymbol{t}^B$ equals the product

$$\prod_j e^{nP(B_j)H(U \mid B_j)} = e^{nH(U|B)} \tag{14-74}$$

But each $\boldsymbol{t}^U$ sequence that is included in $\boldsymbol{t}^B$ is a $\boldsymbol{t}^{U\cdot B}$ sequence. Hence the number of $\boldsymbol{t}^{U\cdot B}$ sequences that is included in $\boldsymbol{t}^B$ equals $e^{nH(U|B)}$.

14-3 RANDOM VARIABLES AND STOCHASTIC PROCESSES

Entropy is a number assigned to a partition. To define the entropy of a random variable we must, therefore, form a suitable partition. This is simple if the random variable is of discrete type. However, for continuous-type random variables we can do so only indirectly.

Discrete type. Suppose that $\mathbf{x}$ is a random variable taking the values x_i with

$$P\{\mathbf{x} = x_i\} = p_i$$

The events $\{\mathbf{x} = x_i\}$ are mutually exclusive and their union is the certain event; hence they form a partition. This partition will be denoted by $\boldsymbol{U}_x$ and will be called the partition of $\mathbf{x}$.

Definition The entropy $H(\mathbf{x})$ of a discrete-type random variable $\mathbf{x}$ is the entropy $H(\boldsymbol{U}_x)$ of its partition $\boldsymbol{U}_x$:

$$H(\mathbf{x}) = H(\boldsymbol{U}_x) = -\sum_i p_i \ln p_i \tag{14-75}$$

Continuous type. The entropy of a continuous-type random variable cannot be so defined because the events $\{\mathbf{x} = x_i\}$ do not form a partition (they are not countable). To define $H(\mathbf{x})$, we form, first, the discrete-type random variable $\mathbf{x}_\delta$ obtained by rounding off $\mathbf{x}$ as in Fig. 14-14:

$$\mathbf{x}_\delta = n\delta \qquad \text{if} \quad n\delta - \delta < \mathbf{x} \le n\delta \tag{14-76}$$

Clearly,

$$P\{\mathbf{x}_\delta = n\delta\} = P\{n\delta - \delta < \mathbf{x} \le \delta\} = \int_{n\delta-\delta}^{n\delta} f(x)\,dx = \delta \bar{f}(n\delta)$$

where $\bar{f}(n\delta)$ is a number between the maximum and the minimum of $f(x)$ in the interval $(n\delta - \delta, n\delta)$. Applying (14-75) to the random variable $\mathbf{x}_\delta$, we obtain

$$H(\mathbf{x}_\delta) = -\sum_{n=-\infty}^{\infty} \delta \bar{f}(n\delta) \ln[\delta \bar{f}(n\delta)]$$

and since

$$\sum_{n=-\infty}^{\infty} \delta \bar{f}(n\delta) = \int_{-\infty}^{\infty} f(x)\,dx = 1$$

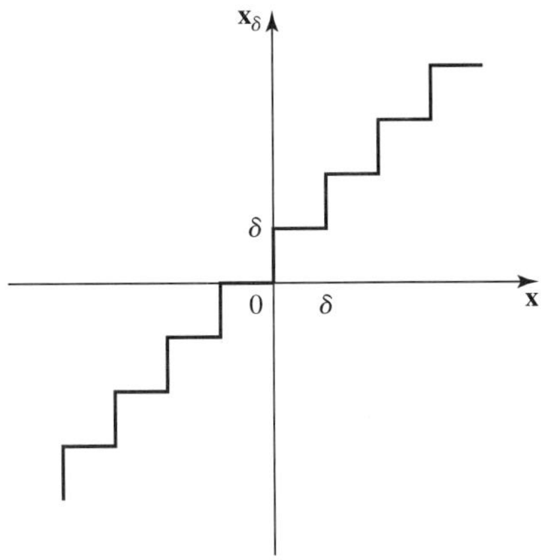

FIGURE 14-14

we conclude that

$$H(\mathbf{x}_\delta) = -\ln\delta - \sum_{n=-\infty}^{\infty} \delta \bar{f}(n\delta) \ln \bar{f}(n\delta) \tag{14-77}$$

As $\delta \to 0$, the random variable $\mathbf{x}_\delta$ tends to $\mathbf{x}$; however, its entropy $H(\mathbf{x}_\delta)$ tends to ∞ because $-\ln\delta \to \infty$. For this reason, we define the entropy $H(\mathbf{x})$ of $\mathbf{x}$ not as the limit of $H(\mathbf{x}_\delta)$ but as the limit of the sum $H(\mathbf{x}_\delta) + \ln\delta$ as $\delta \to 0$. This yields

$$H(\mathbf{x}_\delta) + \ln\delta \underset{\delta\to 0}{\longrightarrow} -\int_{-\infty}^{\infty} f(x)\ln f(x)\,dx \tag{14-78}$$

Definition The entropy of a continuous-type random variable $\mathbf{x}$ is by definition the integral

$$H(\mathbf{x}) = -\int_{-\infty}^{\infty} f(x)\ln f(x)\,dx \tag{14-79}$$

The integration extends only over the region where $f(x) \neq 0$ because $f(x)\ln f(x) = 0$ if $f(x) = 0$.

EXAMPLE 14-10 ▶ If $\mathbf{x}$ is uniform in the interval $(0, a)$, then

$$H(\mathbf{x}) = -\frac{1}{a}\int_0^a \ln\frac{1}{a}\,dx = \ln a \tag{14-80}$$

◀

Notes 1. The entropy $H(\mathbf{x}_\delta)$ of $\mathbf{x}_\delta$ is a measure of our uncertainty about the random variable $\mathbf{x}$ rounded off to the nearest $n\delta$. If δ is small, the resulting uncertainty is large and it tends to ∞ as $\delta \to 0$. This conclusion is based on the assumption that $\mathbf{x}$ can be *observed perfectly;* that is, its various values can be recognized as distinct no matter how close they are. In a physical experiment, however, this assumption is not realistic. Values of $\mathbf{x}$ that differ slightly cannot always be treated as distinct (noise considerations or round-off errors, for example). The presence of the term $\ln\delta$ in (14-78) is, in a sense, a recognition of this ambiguity.

2. As in the case of arbitrary partitions, the entropy of a discrete-type random variable $\mathbf{x}$ is positive and it is used as a measure of uncertainty about $\mathbf{x}$. This is not so, however, for continuous-type random variables. Their entropy can take any value from $-\infty$ to ∞ and it is used to measure only changes in uncertainty. The various properties of partitions also apply to continuous-type random variables if, as is generally the case, they involve only differences of entropies.

Entropy as expected value. The integral in (14-79) is the expected value of the random variable $\mathbf{y} = -\ln f(\mathbf{x})$ obtained through the transformation $g(x) = -\ln f(x)$:

$$H(\mathbf{x}) = E\{-\ln f(\mathbf{x})\} = -\int_{-\infty}^{\infty} f(x)\ln f(x)\,dx \tag{14-81}$$

Similarly, the sum in (14-75) can be written as the expected value of the random variable $-\ln p(\mathbf{x})$:

$$H(\mathbf{x}) = E\{-\ln p(\mathbf{x})\} = -\sum_i p_i \ln p_i \tag{14-82}$$

where now $p(x)$ is a function defined only for $x = x_i$ and such that $p(x_i) = p_i$.

EXAMPLE 14-11 ▶ If

$$f(x) = ce^{-cx}U(x) \qquad \text{then} \qquad E\{-\ln f(\mathbf{x})\} = E\{c\mathbf{x} - \ln c\}$$

Since $E\{c\mathbf{x}\} = 1$, this yields

$$H(\mathbf{x}) = 1 - \ln c = \ln \frac{e}{c} \tag{14-83}$$

◀

EXAMPLE 14-12 ▶ If

$$f(x) = \frac{1}{\sigma\sqrt{2\pi}} e^{-(x-\eta)^2/2\sigma^2}$$

then

$$E\{-\ln f(\mathbf{x})\} = \ln \sigma\sqrt{2\pi} + E\left\{\frac{(\mathbf{x}-\eta)^2}{2\sigma^2}\right\} = \ln \sigma\sqrt{2\pi} + \frac{\sigma^2}{2\sigma^2}$$

Hence the entropy of a *normal* random variable equals

$$H(\mathbf{x}) = \ln \sigma\sqrt{2\pi e} \tag{14-84}$$

◀

Joint entropy. Suppose that $\mathbf{x}$ and $\mathbf{y}$ are two discrete-type random variables taking the values x_i and y_j, respectively, with

$$P\{\mathbf{x} = x_i, \mathbf{y} = y_j\} = p_{ij}$$

Their joint entropy, denoted by $H(\mathbf{x}, \mathbf{y})$, is by definition the entropy of the product of their respective partitions. Clearly, the elements of $\boldsymbol{U}_x \cdot \boldsymbol{U}_y$ are the events $\{\mathbf{x} = x_i, \mathbf{y} = y_i\}$. Hence

$$H(\mathbf{x}, \mathbf{y}) = H(\boldsymbol{U}_x \cdot \boldsymbol{U}_y) = -\sum_{i,j} p_{ij} \ln p_{ij}$$

This can be written as an expected value

$$H(\mathbf{x}, \mathbf{y}) = E\{-\ln p(\mathbf{x}, \mathbf{y})\}$$

where $p(x, y)$ is a function defined only for $x = x_i$ and $y = y_j$ and it is such that $p(x_i, y_j) = p_{ij}$.

The joint entropy $H(\mathbf{x}, \mathbf{y})$ of two continuous-type random variables $\mathbf{x}$ and $\mathbf{y}$ is defined as the limit of the sum

$$H(\mathbf{x}_\delta, \mathbf{y}_\delta) + 2 \ln \delta$$

where $\mathbf{x}_\delta$ and $\mathbf{y}_\delta$ are their staircase approximation. Reasoning as in (14-78), we obtain

$$H(\mathbf{x}, \mathbf{y}) = -\int_{-\infty}^{\infty}\int_{-\infty}^{\infty} f(x, y) \ln f(x, y)\, dx\, dy = E\{-\ln f(\mathbf{x}, \mathbf{y})\} \tag{14-85}$$

EXAMPLE 14-13 ▶ If the random variables **x** and **y** are *jointly normal* as in (6-23), then

$$\ln f(x, y) = \frac{-1}{2(1-r^2)}\left[\frac{(x-\eta_1)^2}{\sigma_1^2} - 2r\frac{(x-\eta_1)(y-\eta_2)}{\sigma_1\sigma_2} + \frac{(y-\eta y_2)^2}{\sigma_2^2}\right]$$
$$- \ln 2\pi\sigma_1\sigma_2\sqrt{1-r^2}$$

In this case,

$$E\left\{\frac{(\mathbf{x}-\eta_1)^2}{\sigma_1^2} - 2r\frac{(\mathbf{x}-\eta_1)(\mathbf{y}-\eta_2)}{\sigma_1\sigma_2} + \frac{(\mathbf{y}-\eta_2)^2}{\sigma_2^2}\right\} = 1 - 2r^2 + 1$$

Hence

$$E\{-\ln f(\mathbf{x}, \mathbf{y})\} = 1 + \ln 2\pi\sigma_1\sigma_2\sqrt{1-r^2}$$

From this and (14-85) it follows that the joint entropy of two jointly normal random variables equals

$$H(\mathbf{x}, \mathbf{y}) = \ln 2\pi e\sqrt{\Delta} \tag{14-86}$$

where

$$\Delta = \mu_{11}\mu_{22} - \mu_{12}^2 \qquad \mu_{11} = \sigma_1^2 \qquad \mu_{22} = \sigma_2^2 \qquad \mu_{12} = r\sigma_1\sigma_2$$ ◀

Conditional entropy. Consider two discrete-type random variables **x** and **y** taking the values x_i and y_j with

$$P\{\mathbf{x} = x_i \,|\, \mathbf{y} = y_j\} = \pi_{ji} = p_{ji}/p_j$$

The conditional entropy $H(\mathbf{x}\,|\,y_j)$ of **x** assuming $\mathbf{y} = y_j$ is by definition the conditional entropy of the partition $\boldsymbol{U}_x$ of **x** assuming $\{\mathbf{y} = y_j\}$. From the above and (14-39) it follows that

$$H(\mathbf{x}\,|\,y_j) = -\sum_i \pi_{ji} \ln \pi_{ji} \tag{14-87}$$

The conditional entropy $H(\mathbf{x}\,|\,\mathbf{y})$ of **x** assuming **y** is the conditional entropy of $\boldsymbol{U}_x$ assuming $\boldsymbol{U}_y$. Thus [see (14-41)]

$$H(\mathbf{x}\,|\,\mathbf{y}) = -\sum_j p_j H(\mathbf{x}\,|\,y_j) = -\sum_{i,j} p_{ji} \ln \pi_{ji} \tag{14-88}$$

For continuous-type random variables the corresponding concepts are defined similarly

$$H(\mathbf{x}\,|\,y) = -\int_{-\infty}^{\infty} f(x\,|\,y) \ln f(x\,|\,y)\,dx \tag{14-89}$$

$$H(\mathbf{x}\,|\,\mathbf{y}) = -\int_{-\infty}^{\infty} f(y)H(\mathbf{x}\,|\,y)\,dy = \int_{-\infty}^{\infty}\int_{-\infty}^{\infty} f(x, y) \ln f(x\,|\,y)\,dx\,dy \tag{14-90}$$

These integrals can be written as expected values [see also (6-240)]

$$H(\mathbf{x}\,|\,y) = E\{-\ln f(\mathbf{x}\,|\,\mathbf{y})\,|\,\mathbf{y} = y\} \tag{14-91}$$

$$H(\mathbf{x}\,|\,\mathbf{y}) = E\{-\ln f(\mathbf{x}\,|\,\mathbf{y})\} = E\{E\{-\ln f(\mathbf{x}\,|\,\mathbf{y})\,|\,\mathbf{y}\}\} \tag{14-92}$$

The discrete case leads to similar expressions.

Mutual information. Guided by (14-53), we shall call the function

$$I(\mathbf{x}, \mathbf{y}) = H(\mathbf{x}) + H(\mathbf{y}) - H(\mathbf{x}, \mathbf{y}) \tag{14-93}$$

the mutual information of the random variables **x** and **y**.

From (14-81) and (14-85) it follows that $I(\mathbf{x}, \mathbf{y})$ can be written as an expected value

$$I(\mathbf{x}, \mathbf{y}) = E\left[\ln \frac{f(\mathbf{x}, \mathbf{y})}{f(\mathbf{x})f(\mathbf{y})}\right] \tag{14-94}$$

Since $f(x, y) = f(x \mid y)f(y)$ it follows from (14-94) and (14-92) that

$$I(\mathbf{x}, \mathbf{y}) = H(\mathbf{x}) - H(\mathbf{x} \mid \mathbf{y}) = H(\mathbf{y}) - H(\mathbf{y} \mid \mathbf{x}) \tag{14-95}$$

EXAMPLE 14-14

▶ If two random variables **x** and **y** are *jointly normal* with zero mean, then [see (6-209)] the conditional density $f(x \mid y)$ is normal with mean $r\sigma_x/\sigma_y$ and variance $\sigma_x^2(1 - r^2)$. From this and (14-84) it follows that

$$H(\mathbf{x} \mid y) = E\{-\ln f(\mathbf{x} \mid y)\} = \ln \sigma_x \sqrt{2\pi e(1 - r^2)} \tag{14-96}$$

Since this is independent of y, it follows from (14-92) that

$$H(\mathbf{x} \mid \mathbf{y}) = H(\mathbf{x} \mid y) \tag{14-97}$$

This yields [see (14-95)]

$$I(\mathbf{x}, \mathbf{y}) = H(\mathbf{x}) - H(\mathbf{x} \mid \mathbf{y}) = -0.5 \ln(1 - r^2) \tag{14-98}$$

We note finally that [see (14-86)]

$$H(\mathbf{x} \mid \mathbf{y}) + H(\mathbf{y}) = \ln 2\pi e\sqrt{\Delta} = H(\mathbf{x}, \mathbf{y})$$

Special Case. Suppose that $\mathbf{y} = \mathbf{x} + \mathbf{n}$, where **n** is independent of **x** and $E\{\mathbf{n}^2\} = N$. In this case,

$$E\{\mathbf{xy}\} = \sigma_x^2 \qquad E\{\mathbf{y}^2\} = \sigma_x^2 + N \qquad r^2 = \frac{\sigma_x^2}{\sigma_x^2 + N}$$

Inserting into (14-98), we obtain

$$I(\mathbf{x}, \mathbf{y}) = 0.5 \ln\left(1 + \frac{\sigma_x^2}{N}\right) \tag{14-99}$$

◀

PROPERTIES. The properties of entropy, developed in Sec. 14-2 for arbitrary partitions, are obviously true for the entropy of discrete-type random variables and can be simply established as appropriate limits for continuous-type random variables. It might be of interest however, to prove directly theorems (14-45) and (14-49) using the representation of entropy as expected value. The proofs are based on the following version of inequality (14-38): If **x** and **y** are two random variables with respective densities $a(x)$ and $b(y)$, then

$$E\{\ln a(\mathbf{x})\} \geq E\{\ln b(\mathbf{x})\} \tag{14-100}$$

Equality holds iff $a(x) = b(x)$.

Proof. Applying the inequality $\ln z \le z - 1$ to the function $z = b(x)/a(x)$, we obtain

$$\ln b(x) - \ln a(x) = \ln \frac{b(x)}{a(x)} \le \frac{b(x)}{a(x)} - 1$$

Multiplying by $a(x)$ and integrating, we obtain

$$\int_{-\infty}^{\infty} a(x)[\ln b(x) - \ln a(x)]\,dx \le \int_{-\infty}^{\infty} [b(x) - a(x)]\,dx = 0$$

and (14-100) results. The right side is 0 because the functions $a(x)$ and $b(x)$ are densities by assumption.

Inequality (14-100) can be readily extended to n-dimensional densities. For example, if $a(x, y)$ and $b(z, w)$ are the joint densities of the random variables $\mathbf{x}, \mathbf{y}$ and $\mathbf{z}, \mathbf{w}$, respectively, then

$$E\{\ln a(\mathbf{x}, \mathbf{y})\} \ge E\{\ln b(\mathbf{x}, \mathbf{y})\} \tag{14-101}$$

THEOREM 14-5

$$H(\mathbf{x}, \mathbf{y}) \le H(\mathbf{x}) + H(\mathbf{y}) \tag{14-102}$$

Proof. Suppose that $f_{xy}(x, y)$ is the joint density of the random variables $\mathbf{x}$ and $\mathbf{y}$ and $f_x(x)$ and $f_y(y)$ their marginal densities. Clearly, the product $f_x(z) f_y(w)$ is the joint density of two independent random variables $\mathbf{z}$ and $\mathbf{w}$. Applying (14-101), we conclude that

$$E\{\ln f_{xy}(\mathbf{x}, \mathbf{y})\} \ge E\{\ln[f_x(\mathbf{x}) f_y(\mathbf{y})]\} = E\{\ln f_x(\mathbf{x})\} + E\{\ln f_y(\mathbf{y})\}$$

and (14-102) results.

THEOREM 14-6

$$H(\mathbf{x}, \mathbf{y}) = H(\mathbf{x} \mid \mathbf{y}) + H(\mathbf{y}) = H(\mathbf{y} \mid \mathbf{x}) + H(\mathbf{x}) \tag{14-103}$$

Proof. Inserting the identity $f(x, y) = f(x \mid y) f(y)$ into (14-85), we obtain

$$H(\mathbf{x}, \mathbf{y}) = E\{-\ln f(\mathbf{x}, \mathbf{y})\} = E\{-\ln f(\mathbf{x} \mid \mathbf{y})\} + E\{-\ln f(\mathbf{y})\}$$

and the first equality in (14-103) results. The second follows because $H(\mathbf{x}, \mathbf{y}) = H(\mathbf{y}, \mathbf{x})$.

COROLLARY

Comparing (14-102) with (14-103), we conclude that

$$H(\mathbf{x} \mid \mathbf{y}) \le H(\mathbf{x}) \tag{14-104}$$

Note If the random variable $\mathbf{y}$ is of discrete type, then $H(\mathbf{y} \mid \mathbf{x}) \ge 0$ and (14-103) yields $H(\mathbf{x}) \le H(\mathbf{x}, \mathbf{y})$. This is not, however, true in general if $\mathbf{y}$ is of continuous type.

Generalizations. The preceding results can be readily generalized to an arbitrary number of random variables: Suppose that $\mathbf{x}_1, \ldots, \mathbf{x}_n$ are n random variables with joint density $f(x_1, \ldots, x_n)$. Extending (14-85), we define their *joint entropy* as an expected value

$$H(\mathbf{x}_1, \ldots, \mathbf{x}_n) = E\{-\ln f(\mathbf{x}_1, \ldots, \mathbf{x}_n)\} \tag{14-105}$$

If the random variables $\mathbf{x}_i$ are *independent,* then

$$f(x_1, \ldots, x_n) = f(x_1) \cdots f(x_n)$$

and (14-105) yields

$$H(\mathbf{x}_1, \ldots, \mathbf{x}_n) = H(\mathbf{x}_1) + \cdots + H(\mathbf{x}_n) \tag{14-106}$$

Conditional entropies are defined similarly. For example [see (14-92)]

$$H(\mathbf{x}_n \mid \mathbf{x}_{n-1}, \ldots, \mathbf{x}_1) = E\{-\ln f(\mathbf{x}_n \mid \mathbf{x}_{n-1}, \ldots, \mathbf{x}_1)\} \tag{14-107}$$

Chain rule From the identity [see (7-37)]

$$f(x_1, \ldots, x_n) = f(x_n \mid x_{n-1}, \ldots, x_1) \cdots f(x_2 \mid x_1) f(x_1)$$

and (14-107) it follows that

$$H(\mathbf{x}_1, \ldots, \mathbf{x}_n) = H(\mathbf{x}_n \mid \mathbf{x}_{n-1}, \ldots, \mathbf{x}_1) + \cdots + H(\mathbf{x}_2 \mid \mathbf{x}_1) + H(\mathbf{x}_1) \tag{14-108}$$

The following relationships are simple extensions of (14-102) and (14-103):

$$\begin{aligned} H(\mathbf{x}, \mathbf{y} \mid \mathbf{z}) &\le H(\mathbf{x} \mid \mathbf{z}) + H(\mathbf{y} \mid \mathbf{z}) \\ H(\mathbf{x}, \mathbf{y} \mid \mathbf{z}) &= H(\mathbf{x} \mid \mathbf{z}) + H(\mathbf{y} \mid \mathbf{x}, \mathbf{z}) \\ H(\mathbf{x}_1, \ldots, \mathbf{x}_n) &\le H(\mathbf{x}_1) + \cdots + H(\mathbf{x}_n) \end{aligned} \tag{14-109}$$

EXAMPLE 14-15 ▶ If the random variables $\mathbf{x}_i$ are jointly normal with covariance matrix C as in (7-58), then

$$E\{-\ln f(\mathbf{x}_1, \ldots, \mathbf{x}_n)\} = \ln \sqrt{(2\pi)^n \Delta} + \tfrac{1}{2} E\{\mathbf{X}C^{-1}\mathbf{X}^t\} \tag{14-110}$$

This yields (see Prob. 7-23)

$$H(\mathbf{x}_1, \ldots, \mathbf{x}_n) = \ln \sqrt{(2\pi e)^n \Delta} \tag{14-111}$$

◀

Transformations of Random Variables

We shall compare the entropy of the random variables $\mathbf{x}$ and $\mathbf{y} = g(\mathbf{x})$.

DISCRETE TYPE. If the random variable $\mathbf{x}$ is of discrete type, then

$$H(\mathbf{y}) \le H(\mathbf{x}) \tag{14-112}$$

with equality iff the transformation $y = g(x)$ has a unique inverse.

Proof. Suppose that $\mathbf{x}$ takes the values x_i with probability p_i and $g(x)$ has a unique inverse. In this case,

$$P\{\mathbf{y} = y_i\} = P\{\mathbf{x} = x_i\} = p_i \qquad y_i = g(x_i)$$

hence $H(\mathbf{y}) = H(\mathbf{x})$. If the transformation is not one-to-one, then $\mathbf{y} = y_i$ for more than one value of $\mathbf{x}$. This results in a reduction of $H(\mathbf{x})$ [see (14-31)].

Continuous type. If the random variable $\mathbf{x}$ is of continuous type, then

$$H(\mathbf{y}) \leq H(\mathbf{x}) + E\{\ln|g'(\mathbf{x})|\} \tag{14-113}$$

with equality iff the transformation $y = g(x)$ has a unique inverse.

Proof. As we know [see (5-16)] if $y = g(x)$ has a unique inverse $x = g^{(-1)}(y)$, then

$$f_y(y) = \frac{f_x(x)}{|g'(x)|} \qquad dy = g'(x)\,dx$$

Hence

$$\begin{aligned} H(\mathbf{y}) &= -\int_{-\infty}^{\infty} f_y(y)\ln f_y(y)\,dy = -\int_{-\infty}^{\infty} f_x(x)\ln\frac{f_x(x)}{|g'(x)|}\,dx \\ &= -\int_{-\infty}^{\infty} f_x(x)\ln f_x(x)\,dx + \int_{-\infty}^{\infty} f_x(x)\ln|g'(x)|\,dx \end{aligned}$$

and (14-113) results.

SEVERAL RANDOM VARIABLES. Reasoning as in (14-113), we can similarly show that if

$$\mathbf{y}_i = g_i(\mathbf{x}_1, \ldots, \mathbf{x}_n) \qquad i = 1, \ldots, n$$

are n functions of the random variables $\mathbf{x}_i$, then

$$H(\mathbf{y}_1, \ldots, \mathbf{y}_n) \leq H(\mathbf{x}_1, \ldots, \mathbf{x}_n) + E\{\ln|J(\mathbf{x}_1, \ldots, \mathbf{x}_n)|\} \tag{14-114}$$

where $J(x_1, \ldots, x_n)$ is the jacobian of the above transformation [see (7-9)]. Equality holds iff the transformation has a unique inverse.

Linear transformations Suppose that

$$\mathbf{y}_i = a_{i1}\mathbf{x}_1 + \cdots + a_{in}\mathbf{x}_n$$

Denoting by Δ the determinant of the coefficients, we conclude from (14-114) that if $\Delta \neq 0$ then

$$H(\mathbf{y}_1, \ldots, \mathbf{y}_n) = H(\mathbf{x}_1, \ldots, \mathbf{x}_n) + \ln|\Delta| \tag{14-115}$$

because the transformation has a unique inverse and Δ does not depend on $\mathbf{x}_i$.

Stochastic Processes and Entropy Rate

As we know, the statistics of most stochastic processes are determined in terms of the joint density $f(x_1, \ldots, x_m)$ of the random variables $\mathbf{x}(t_1), \ldots, \mathbf{x}(t_m)$. The joint entropy

$$H(\mathbf{x}_1, \ldots, \mathbf{x}_m) = E\{-\ln f(\mathbf{x}_1, \ldots, \mathbf{x}_m)\} \tag{14-116}$$

of these random variables is the *mth-order entropy* of the process $\mathbf{x}(t)$. This function equals the uncertainty about these random variables and it equals the information gained when they are observed.

In general, the uncertainty about the values of $\mathbf{x}(t)$ on the entire t axis or even on a finite interval, no matter how small, is infinite. However, if $\mathbf{x}(t)$ can be expressed in terms of its values on a countable set of points, as is the case for bandlimited processes, then a rate of uncertainty can be introduced. It suffices, therefore, to consider only discrete-time processes.

The mth-order entropy of a discrete-time process $\mathbf{x}_n$ is the joint entropy $H(\mathbf{x}_1, \ldots, \mathbf{x}_m)$ of the m random variables

$$\mathbf{x}_n, \mathbf{x}_{n-1}, \ldots, \mathbf{x}_{n-m+1} \tag{14-117}$$

defined as in (14-116). We shall assume throughout that the process $\mathbf{x}_n$ is SSS. In this case, $H(\mathbf{x}_1 \cdots \mathbf{x}_m)$ is the uncertainty about any m consecutive values of the process $\mathbf{x}_n$. The first-order entropy will be denoted by $H(\mathbf{x})$. Thus $H(\mathbf{x})$ equals the uncertainty about $\mathbf{x}_n$ for a specific n.

Clearly [see (14-109)]

$$H(\mathbf{x}_1, \ldots, \mathbf{x}_m) \le H(\mathbf{x}_1) + \cdots + H(\mathbf{x}_m) = mH(\mathbf{x}) \tag{14-118}$$

Special cases (*a*) If the process $\mathbf{x}_n$ is *strictly white,* that is, if the random variables $\mathbf{x}_n, \mathbf{x}_{n-1}, \ldots$ are independent, then [see (14-106)]

$$H(\mathbf{x}_1, \ldots, \mathbf{x}_m) = mH(\mathbf{x}) \tag{14-119}$$

(*b*) If the process $\mathbf{x}_n$ is *Markov,* then [see (15-2)]

$$f(x_1, \ldots, x_m) = f(x_m \mid x_{m-1}) \cdots f(x_2 \mid x_1) f(x_1) \tag{14-120}$$

This yields

$$H(\mathbf{x}_1, \ldots, \mathbf{x}_m) = H(\mathbf{x}_m \mid \mathbf{x}_{m-1}) + \cdots + H(\mathbf{x}_2 \mid \mathbf{x}_1) + H(\mathbf{x}_1) \tag{14-121}$$

From (14-103) and the stationarity of $\mathbf{x}_n$ it follows, therefore, that

$$H(\mathbf{x}_1, \ldots, \mathbf{x}_m) = (m-1)H(\mathbf{x}_1, \mathbf{x}_2) - (m-2)H(\mathbf{x}) \tag{14-122}$$

We have thus expressed the mth-order entropy of a Markov process in terms of its first- and second-order entropies.

CONDITIONAL ENTROPY. The conditional entropy of order m:

$$H(\mathbf{x}_n \mid \mathbf{x}_{n-1}, \ldots, \mathbf{x}_{n-m})$$

of a process $\mathbf{x}_n$ is the uncertainty about its present under the assumption that its m most recent values have been observed. Extending (14-104), we can readily show that

$$H(\mathbf{x}_n \mid \mathbf{x}_{n-1}, \ldots, \mathbf{x}_{n-m}) \le H(\mathbf{x}_n \mid \mathbf{x}_{n-1}, \ldots, \mathbf{x}_{n-m+1}) \tag{14-123}$$

Thus the described conditional entropy is a decreasing function of m. If, therefore, it is bounded from below, it tends to a limit. This is certainly the case if the random variables $\mathbf{x}_n$ are of discrete type because then all entropies are positive. The limit will be denoted by $H_c(\mathbf{x})$ and will be called the *conditional entropy* of the process $\mathbf{x}_n$:

$$H_c(\mathbf{x}) = \lim_{m\to\infty} H(\mathbf{x}_n \mid \mathbf{x}_{n-1}, \ldots, \mathbf{x}_{n-m}) \tag{14-124}$$

The function $H_c(\mathbf{x})$ is a measure of our uncertainty about the present of $\mathbf{x}_n$ under the assumption that its entire past is observed.

Special cases (*a*) If $\mathbf{x}_n$ is *strictly white*, then

$$H_c(\mathbf{x}) = H(\mathbf{x})$$

(*b*) If $\mathbf{x}_n$ is a *Markov* process, then

$$H(\mathbf{x}_n \mid \mathbf{x}_{n-1}, \ldots, \mathbf{x}_{n-m}) = H(\mathbf{x}_n \mid \mathbf{x}_{n-1})$$

Since $\mathbf{x}_n$ is a stationary process, this equation equals $H(\mathbf{x}_2 \mid \mathbf{x}_1)$. Hence

$$H_c(\mathbf{x}) = H(\mathbf{x}_2 \mid \mathbf{x}_1) = H(\mathbf{x}_1, \mathbf{x}_2) - H(\mathbf{x}) \tag{14-125}$$

This shows that if $\mathbf{x}_{n-1}$ is observed, then the past has no effect on the uncertainty of the present.

ENTROPY RATE. The ratio $H(\mathbf{x}_1 \cdots \mathbf{x}_m)/m$ is the average uncertainty per sample in a block of m consecutive samples. The limit of this average as $m \to \infty$ will be denoted by $\overline{H}(\mathbf{x})$ and will be called the *entropy rate* of the process $\mathbf{x}_n$:

$$\overline{H}(\mathbf{x}) = \lim_{m\to\infty} \frac{1}{m} H(\mathbf{x}_1, \ldots, \mathbf{x}_m) \tag{14-126}$$

If $\mathbf{x}_n$ is *strictly white*, then

$$\overline{H}(\mathbf{x}) = H(\mathbf{x}) = H_c(\mathbf{x})$$

If $\mathbf{x}_n$ is *Markov*, then [see (14-122)]

$$\overline{H}(\mathbf{x}) = H(\mathbf{x}_1, \mathbf{x}_2) - H(\mathbf{x}) = H_c(\mathbf{x}) \tag{14-127}$$

Thus, in both cases, the limit in (14-126) exists and it equals $H_c(\mathbf{x})$. We show next that this is true in general.

THEOREM 14-7

▶ The entropy rate of a process $\mathbf{x}_n$ equals its conditional entropy

$$\overline{H}(\mathbf{x}) = H_c(\mathbf{x}) \tag{14-128}$$

Proof. This is a consequence of the following simple property of convergent sequences: If

$$a_k \to a \qquad \text{then} \qquad \frac{1}{m}\sum_{k=1}^{m} a_k \to a \tag{14-129}$$

Since $\mathbf{x}_n$ is stationary we conclude, as in (14-108), that

$$H(\mathbf{x}_1, \ldots, \mathbf{x}_m) = H(\mathbf{x}) + \sum_{k=1}^{m} H(\mathbf{x}_n \mid \mathbf{x}_{n-1}, \ldots, \mathbf{x}_{n-k})$$

Dividing by m and using (14-129), we obtain (14-128) because

$$H(\mathbf{x}_n \mid \mathbf{x}_{n-1}, \ldots, \mathbf{x}_{n-k})$$

tends to $H_c(\mathbf{x})$ as $k \to \infty$. ◀

Note If $\mathbf{x}_n$ equals the samples $\mathbf{x}(nT)$ of $\mathbf{x}(t)$, then the entropy rate is measured in bits per T seconds. If we wish to measure it in bits per second, we must divide by T.

Normal processes. We shall show that if $\mathbf{x}_n$ is a normal process with power spectrum $S(\omega)$, then

$$\overline{H}(\mathbf{x}) = \ln\sqrt{2\pi e} + \frac{1}{4\pi}\int_{-\pi}^{\pi} \ln S(\omega)\,d\omega \qquad (14\text{-}130)$$

Proof. As we know, the function $f(x_{m+1} \mid x_m, \ldots, x_1)$ is a one-dimensional normal density with variance Δ_{m+1}/Δ_m [see (7-97) and (13-66)]. Hence

$$H(\mathbf{x}_n \mid \mathbf{x}_{n-1}, \ldots, \mathbf{x}_{n-m}) = \ln\sqrt{\frac{2\pi e \Delta_{m+1}}{\Delta_m}} \qquad (14\text{-}131)$$

as in (14-84). This leads to the conclusion that

$$H_c(\mathbf{x}) = \ln\sqrt{2\pi e} + \frac{1}{2}\lim_{m\to\infty} \ln\frac{\Delta_{m+1}}{\Delta_m} \qquad (14\text{-}132)$$

and (14-130) follows from (13-70) and Prob. 13-15.

ENTROPY RATE OF SYSTEM RESPONSE. We shall show that the entropy rate $\overline{H}(\mathbf{y})$ of the output $\mathbf{y}_n$ of a linear system $\mathbf{L}(z)$ is given by

$$\overline{H}(\mathbf{y}) = \overline{H}(\mathbf{x}) + \frac{1}{2\pi}\int_{-\pi}^{\pi} \ln|\mathbf{L}(e^{j\omega})|\,d\omega \qquad (14\text{-}133)$$

where $\overline{H}(\mathbf{x})$ is the entropy rate of the input $\mathbf{x}_n$ (Fig. 14-15).

Suppose, first, that $\mathbf{x}_n$ is a normal process. In this case $\mathbf{y}_n$ is also normal and its entropy rate is given by (14-130) where

$$S(\omega) = S_y(\omega) = S_x(\omega)|\mathbf{L}(e^{j\omega})|^2 \qquad (14\text{-}134)$$

This yields

$$\overline{H}(\mathbf{y}) = \ln\sqrt{2\pi e} + \frac{1}{4\pi}\int_{-\pi}^{\pi} [\ln S_x(\omega) + \ln|\mathbf{L}(e^{j\omega})|^2]\,d\omega \qquad (14\text{-}135)$$

and (14-133) follows.

The proof for arbitrary processes is involved. We shall sketch a justification based on (14-115): If the random variables $\mathbf{y}_1, \ldots, \mathbf{y}_m$ depend linearly on the random variables $\mathbf{x}_1, \ldots, \mathbf{x}_m$, then

$$H(\mathbf{y}_1, \ldots, \mathbf{y}_m) = H(\mathbf{x}_1, \ldots, \mathbf{x}_m) + K_o \qquad (14\text{-}136)$$

where $K_o = \log|\Delta|$ is a constant that depends only on the coefficients of the transformation. The process $\mathbf{y}_n$ depends linearly on $\mathbf{x}_n$:

$$\mathbf{y}_n = \sum_{k=0}^{\infty} l_k \mathbf{x}_{n-k} \qquad n = -\infty, \ldots, \infty \qquad (14\text{-}137)$$

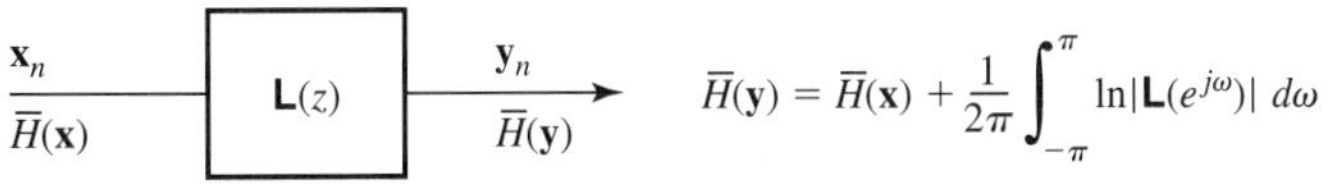

FIGURE 14-15

where now the transformation matrix is of infinity order. Extending (14-136) to infinitely many variables, we conclude with (14-126) that

$$\overline{H}(\mathbf{y}) = \overline{H}(\mathbf{x}) + K \tag{14-138}$$

where again K is a constant that depends only on the parameters of the system $\mathbf{L}(z)$. As we have seen, if $\mathbf{x}_n$ is normal, then K equals the integral in (14-133). And since K is independent of $\mathbf{x}_n$, it must equal that integral for any $\mathbf{x}_n$.

14-4 THE MAXIMUM ENTROPY METHOD

The MEM is used to determine various parameters of a probability space subject to given constraints. The resulting problem can be solved, in general, only numerically and it involves the evaluation of the maximum of a function of several variables. In a number of important cases, however, the solution can be found analytically or it can be reduced to a system of algebraic equations. In this section, we consider certain special cases, concentrating on constraints in the form of expected values. The results can be obtained with the familiar variational techniques involving Lagrange multipliers or Euler's equations. For most problems under consideration, however, it suffices to use the following form of (14-100).

If $f(x)$ and $\varphi(x)$ are two arbitrary densities, then

$$-\int_{-\infty}^{\infty} \varphi(x) \ln \varphi(x)\, dx \le -\int_{-\infty}^{\infty} \varphi(x) \ln f(x)\, dx \tag{14-139}$$

EXAMPLE 14-16 ▶ In the coin experiment, the probability of heads is often viewed as a random variable $\mathbf{p}$ (see bayesian estimation, Sec. 8-2). We shall show that if no prior information about $\mathbf{p}$ is available, then, according to the ME principle, its density $f(p)$ is uniform in the interval $(0, 1)$. In this problem we must maximize $H(\mathbf{p})$ subject to the constraint (dictated by the meaning of $\mathbf{p}$) that $f(p) = 0$ outside the interval $(0, 1)$. The corresponding entropy is, therefore, given by

$$H(\mathbf{p}) = -\int_0^1 f(p) \ln f(p)\, dp$$

and our problem is to find $f(p)$ such as to maximize this integral.

We maintain that $H(\mathbf{p})$ is maximum if

$$f(p) = 1 \qquad H(\mathbf{p}) = 0$$

Indeed, if $\varphi(p)$ is any other density such that $\varphi(p) = 0$ outside the interval $(0, 1)$, then [see (14-139)]

$$-\int_0^1 \varphi(p) \ln \varphi(p) \le -\int_0^1 \varphi(p) \ln f(p)\, dp = 0 = H(\mathbf{p})$$ ◀

EXAMPLE 14-17 ▶ Suppose that $\mathbf{x}$ is a random variable vanishing outside the interval $(-\pi, \pi)$. Using the MEM, we shall determine the density $f(x)$ of $\mathbf{x}$ under the assumption that the coefficients c_n of its Fourier series expansion

$$f(x) = \sum_{n=-\infty}^{\infty} c_n e^{jnx} \qquad -\pi \le x \le \pi$$

are known for $|n| \le N$. Our problem now is to maximize the integral

$$H(\mathbf{x}) = -\int_{-\pi}^{\pi} f(x) \ln f(x)\, dx$$

subject to the constraints

$$c_n = \frac{1}{2\pi}\int_{-\pi}^{\pi} f(x) e^{-jnx}\, dx \qquad |n| \le N \tag{14-140}$$

Clearly, $H(\mathbf{x})$ depends on the unknown coefficients c_n and it is maximum iff

$$\frac{\partial H}{\partial c_n} = \frac{\partial H}{\partial f}\frac{\partial f}{\partial c_n} = -\int_{-\pi}^{\pi} [\ln f(x) + 1] e^{jnx}\, dx = 0 \qquad |n| > N$$

This shows that the coefficients γ_n of the Fourier series expansion of the function $\ln f(x) + 1$ in the interval $(-\pi, \pi)$ are 0 for $|n| > N$. Hence

$$\ln f(x) + 1 = \sum_{k=-N}^{N} \gamma_k e^{jkx}$$

From this it follows that

$$f(x) = \exp\left\{-1 + \sum_{k=-N}^{N} \gamma_k e^{jkx}\right\} \qquad -\pi \le x \le \pi \tag{14-141}$$

We have thus shown that the unknown function is given by an exponential involving the parameters γ_k. These parameters can be determined from (14-140). The resulting system is nonlinear and can only be solved numerically. ◀

Constraints as Expected Values

We shall consider now a class of problems involving constraints in the form of expected values. Such problems are common in statistical mechanics. We start with the one-dimension case.

We wish to determine the density $f(x)$ of a random variable $\mathbf{x}$ subject to the condition that the expected values η_i of n known functions $g_i(\mathbf{x})$ of $\mathbf{x}$ are given

$$E\{g_i(\mathbf{x})\} = \int_{-\infty}^{\infty} g_i(x) f(x)\, dx = \eta_i \qquad i = 1, \ldots, n \tag{14-142}$$

Using (14-139), we shall show that the MEM leads to the conclusion that $f(x)$ must be an exponential

$$f(x) = A \exp\{-\lambda_1 g_1(x) - \cdots - \lambda_n g_n(x)\} \tag{14-143}$$

where λ_i are n constants determined from (14-142) and A is such as to satisfy the density condition

$$A \int_{-\infty}^{\infty} \exp\{-\lambda_1 g_1(x) - \cdots - \lambda_n g_n(x)\}\, dx = 1 \tag{14-144}$$

Proof. Suppose that $f(x)$ is given by (14-143). In this case,

$$\int_{-\infty}^{\infty} f(x)\ln f(x)\,dx = \int_{-\infty}^{\infty} f(x)[\ln A - \lambda_1 g_1(x) - \cdots - \lambda_n g_n(x)]\,dx$$

Hence

$$H(\mathbf{x}) = \lambda_1\eta_1 + \cdots + \lambda_n\eta_n - \log A \tag{14-145}$$

To prove (14-143), it suffices, therefore, to show that, if $\varphi(x)$ is any other density satisfying the constraints (14-142), then its entropy cannot exceed the right side of (14-145). This follows readily from (14-139):

$$\begin{aligned} -\int_{-\infty}^{\infty} \varphi(x)\ln\varphi(x)\,dx &\le -\int_{-\infty}^{\infty} \varphi(x)\ln f(x)\,dx \\ &= \int_{-\infty}^{\infty} \varphi(x)[\lambda_1 g_1(x) + \cdots + \lambda_n g_n(x) - \ln A]\,dx \\ &= \lambda_1\eta_1 + \cdots + \lambda_n\eta_n - \ln A \end{aligned}$$

We note that, if $f(x) = 0$ outside a certain set R, then $f(x)$ is again given by (14-143) for every x in R and the region of integration in (14-144) is the set R.

EXAMPLE 14-18 ▶ We shall determine $f(x)$ assuming that $\mathbf{x}$ is a positive random variable with known mean η. With $g(x) = x$, it follows from (14-143) that

$$f(x) = \begin{cases} Ae^{-\lambda x} & x > 0 \\ 0 & x < 0 \end{cases}$$

We have thus shown that if a random variable is positive with specified mean, then its density obtained with the MEM, is an exponential. ◀

THE PARTITION FUNCTION. In certain problems, it is more convenient to express the given constraints in terms of the partition function (Zustandsumme)

$$Z(\lambda_1, \ldots, \lambda_n) = \frac{1}{A} = \int_{-\infty}^{\infty} \exp\{-\lambda_1 g_1(x) - \cdots - \lambda_n g_n(x)\}\,dx \tag{14-146}$$

Indeed, differentiating with respect to λ_i, we obtain

$$-\frac{\partial Z}{\partial \lambda_i} = \int_{-\infty}^{\infty} g_i(x)\exp\left\{-\sum_{k=1}^{n}\lambda_k g_k(x)\right\}dx = Z\int_{-\infty}^{\infty} g_i(x) f(x)\,dx$$

This yields

$$-\frac{1}{Z}\frac{\partial Z}{\partial \lambda_i} = -\frac{\partial}{\partial \lambda_i}\ln Z = \eta_i \qquad i = 1, \ldots, n \tag{14-147}$$

The above is a system of n equations equivalent to (14-142) and can be used to determine the n parameters λ_i.

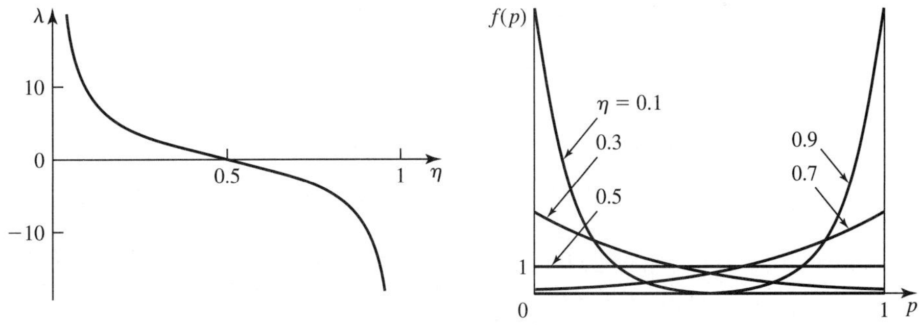

FIGURE 14-16

EXAMPLE 14-19

▶ In the coin experiment of Example 14-16, we assume that **p** is a random variable with known mean η. Since $f(p) = 0$ outside the interval (0, 1), (14-143) yields

$$f(p) = \begin{cases} Ae^{-\lambda p} & 0 \le p \le 1 \\ 0 & \text{otherwise} \end{cases} \qquad Z = \int_0^1 e^{-\lambda p}\, dp = \frac{1 - e^{-\lambda}}{\lambda}$$

The constant λ is determined from (14-147):

$$-\frac{1}{Z}\frac{\partial Z}{\partial \lambda} = \frac{1 - e^{-\lambda} - \lambda e^{-\lambda}}{\lambda(1 - e^{-\lambda})} = \eta$$

In Fig. 14-16, we plot λ and $f(p)$ for various values of η. Note that if $\eta = 0.5$, then $\lambda = 0$ and $f(p) = 1$. ◀

EXAMPLE 14-20

▶ A collection of particles moves in a conservative field whose potential equals $V(x)$. For a specific t, the x component of the position of a particle is a random variable **x** with density $f(x)$ independent of t (stationary state). Thus the probability that the particle is between x and $x + dx$ equals $f(x)\, dx$ and the total energy per unit mass of the ensemble equals

$$I = \int_{-\infty}^{\infty} V(x) f(x)\, dx = E\{V(\mathbf{x})\}$$

We shall find $f(x)$ under the assumption that the function $g(x) = V(x)$ and the mean I of $V(\mathbf{x})$ are given. Inserting into (14-143), we obtain

$$f(x) = \frac{1}{Z} e^{-\lambda V(x)} \tag{14-148}$$

where

$$Z = \int_{-\infty}^{\infty} e^{-\lambda V(x)}\, dx \qquad \frac{1}{Z}\int_{-\infty}^{\infty} V(x) e^{-\lambda V(x)}\, dx = I$$

Special Case. In a gravitational field, the potential $V(x) = Mgx$ is proportional to the distance x from the ground. Since $f(x) = 0$ for $x < 0$, it follows from (14-148) that

$$f(x) = \frac{Mg}{I} e^{-Mgx/I} U(x)$$

The resulting atmospheric pressure is proportional to $1 - F(x)$. ◀

EXAMPLE 14-21 ▶ We shall find $f(x)$ such that $E\{\mathbf{x}^2\} = m_2$. With $g_1(x) = x^2$, (14-143) yields

$$f(x) = Ae^{-\lambda x^2} \tag{14-149}$$

Thus, if the second moment m_2 of a random variable $\mathbf{x}$ is specified, then $\mathbf{x}$ is $N(0, m_2)$. We can show similarly that if the variance σ^2 of $\mathbf{x}$ is specified, then $\mathbf{x}$ is $N(\eta, \sigma^2)$ where η is an arbitrary constant.

Special Case. We consider again a collection of particles in stationary motion and we denote by $\mathbf{v}_x$ the x component of their velocity. We shall determine the density $f(v_x)$ of $\mathbf{v}_x$ under the constraint that the corresponding average kinetic energy $K_x = E\{M\mathbf{v}_x^2/2\}$ is specified. This is a special case of (14-149) with $m_2 = 2K_x/M$. Hence

$$f(v_x) = \sqrt{\frac{M}{4\pi K_x}}e^{-Mv^2/4K_x}$$ ◀

Discrete type random variables. Suppose that a random variable $\mathbf{x}$ takes the values x_k with probability p_k. We shall use the MEM to determine p_k under the assumption that the expected values

$$E\{g_i(\mathbf{x})\} = \sum_k g_i(x_k)p_k = \eta_i \tag{14-150}$$

of the n known functions $g_i(\mathbf{x})$ are given.

Using (14-37), we can show as in (14-143) that the unknown probabilities equal

$$p_k = A\exp\{-\lambda_1 g_1(x_k) - \cdots - \lambda_n g_n(x_k)\} \tag{14-151}$$

where

$$\frac{1}{A} = Z = \sum_k \exp\{-[\lambda_1 g_1(x_k) + \cdots + \lambda_n g_n(x_k)]\} \tag{14-152}$$

The n constants λ_i are determined either from (14-150) or from the equivalent system

$$-\frac{1}{Z}\frac{\partial Z}{\partial \lambda_i} = \eta_i \qquad i = 1, \ldots, n \tag{14-153}$$

EXAMPLE 14-22 ▶ A die is rolled a large number of times and the average number of dots up equals η. Assuming that η is known, we shall determine the probabilities p_k of the six faces f_k using the MEM. For this purpose, we form a random variable $\mathbf{x}$ such that $\mathbf{x}(f_k) = k$. Clearly,

$$E\{\mathbf{x}\} = p_1 + 2p_2 + \cdots + 6p_6 = \eta$$

With $g(x) = x$, it follows from (14-151) that

$$p_k = \frac{1}{Z}e^{-k\lambda} \qquad Z = w + w^2 + \cdots + w^6$$

where $w = e^{-\lambda}$. Hence

$$p_k = \frac{w^k}{w + w^2 + \cdots + w^6} \qquad \frac{w + 2w^2 + \cdots + 6w^6}{w + w^2 + \cdots + w^6} = \eta$$

as in Fig. 14-17. We note that if $\eta = 3.5$, then $p_k = \frac{1}{6}$. ◀

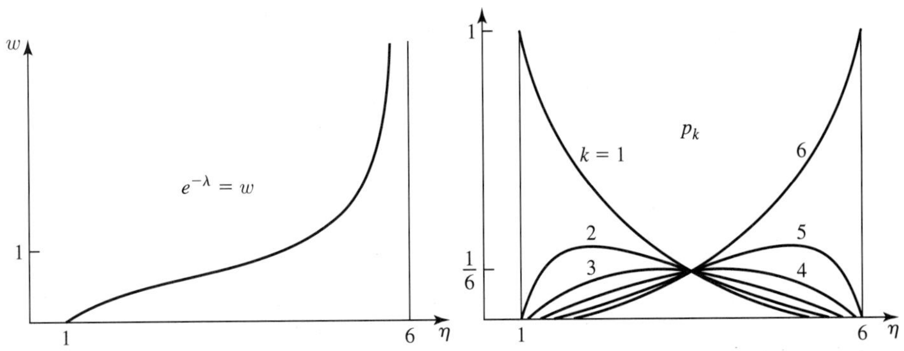

FIGURE 14-17

Joint density. The MEM can be used to determine the density $f(X)$ of the random vector $\mathbf{X}$: $[\mathbf{x}_1, \ldots, \mathbf{x}_M]$ subject to the n constraints

$$E\{g_i(\mathbf{X})\} = \eta_i \qquad i = 1, \ldots, n \tag{14-154}$$

Reasoning as in the scalar case, we conclude that

$$f(X) = A \exp\{-\lambda_1 g_1(X) - \cdots - \lambda_n g_n(X)\} \tag{14-155}$$

Second-Order Moments and Normality

We are given the correlation matrix

$$R = E\{\mathbf{X}^t \mathbf{X}\} \tag{14-156}$$

of the random vector $\mathbf{X}$ and we wish to find its density using the MEM. We maintain that $f(X)$ is normal with zero mean as in (7-58)

$$f(X) = \frac{1}{\sqrt{(2\pi)^M \Delta}} \exp\left\{ -\tfrac{1}{2} X R^{-1} X^t \right\} \tag{14-157}$$

Proof. The elements $R_{jk} = E\{\mathbf{x}_j \mathbf{x}_k\}$ of R are the expected values of the M^2 random variables $g_{jk}(\mathbf{X}) = \mathbf{x}_j \mathbf{x}_k$. Changing the subscript i in (14-154) to double subscript, we conclude from (14-155) that

$$f(X) = A \exp\left\{ -\sum_{j,k} \lambda_{jk} x_j x_k \right\} \tag{14-158}$$

This shows that $f(X)$ is normal. The M^2 coefficients λ_{jk} can be determined from the M^2 constraints in (14-156). As we know [see (8-58)], these coefficients equal the elements of the matrix $R^{-1}/2$ as in (14-157).

These results are acceptable only if the matrix R is positive definite. Otherwise, the function $f(X)$ in (14-157) is not a density. The p.d. condition is, of course, satisfied if the given R is a true correlation matrix. However, even then (14-157) might not be acceptable if only a subset of the elements of R is specified. In such cases, it might be necessary, as we shall presently see, to introduce the unspecified elements of R as auxiliary constraints.

Suppose, first, that we are given only the diagonal elements of R:

$$E\{\mathbf{x}_i^2\} = R_{ii} \qquad i = 1, \ldots, M \tag{14-159}$$

Inserting the functions $g_{ii}(x) = x_i^2$ into (14-155), we obtain

$$f(X) = A \exp\{-\lambda_{11}x_1^2 - \cdots - \lambda_{MM}x_M^2\} \tag{14-160}$$

This shows that the random variables $\mathbf{x}_i$ are normal, independent, with zero mean and variance $R_{ii} = 1/2\lambda_{ii}$.

This solution is acceptable because $R_{ii} > 0$. If, however, we are given $N < M^2$ arbitrary joint moments, then the corresponding quadratic in (14-158) will contain only the terms $x_j x_k$ corresponding to the given moments. The resulting $f(X)$ might not then be a density. To find the ME solution for this case, we proceed as follows: We introduce as constraints the M^2 joint moments R_{jk}, where now only N of these moments are given and the other $M^2 - N$ moments are unknown parameters. Applying the MEM, we obtain (14-157). The corresponding entropy equals [see (14-111)]

$$H(\mathbf{x}_1, \ldots, \mathbf{x}_M) = \ln\sqrt{(2\pi e)^M \Delta} \qquad \Delta = |R| \tag{14-161}$$

This entropy depends on the unspecified parameters of R and it is maximum if its determinant Δ is maximum. Thus the random variables $\mathbf{x}_i$ are again normal with density as in (14-157) where the unspecified parameters of R are such as to maximize Δ.

Note From the developments just discussed it follows that the determinant Δ of a correlation matrix R is such that

$$\Delta \le R_{11} \cdots R_{MM}$$

with equality iff R is diagonal. Indeed, (14-159) is a restricted moment set; hence the ME solution (14-160) maximizes Δ.

Stochastic processes. The MEM can be used to determine the statistics of a stochastic process subject to given constraints. We shall discuss the following case.

Suppose that $\mathbf{x}_n$ is a WSS process with autocorrelation

$$R[m] = E\{\mathbf{x}_{n+m}\mathbf{x}_n\}$$

We wish to find its various densities assuming that $R[m]$ is specified either for some or for all values of m. As we know [see (14-158)] the MEM leads to the conclusion that, in both cases, $\mathbf{x}_n$ must be a normal process with zero mean. This completes the statistical description of $\mathbf{x}_n$ if $R[m]$ is known for all m. If, however, we know $R[m]$ only partially, then we must find its unspecified values. For finite-order densities, this involves the maximization of the corresponding entropy with respect to the unknown values of $R[m]$ and it is equivalent to the maximization of the correlation determinant Δ [see (14-161)]. An important special case is the MEM solution to the extrapolation problem considered in Sec. 12-3. We shall reexamine this problem in the context of the entropy rate.

We start with the simplest case: Given the average power $E\{\mathbf{x}_n^2\} = R[0]$ of $\mathbf{x}_n$, we wish to find its power spectrum. In this case, the entropy of the random variables

$$\mathbf{x}_n, \ldots, \mathbf{x}_{n+M}$$

is maximum if these random variables are normal and independent for any M [see (14-160)], that is, if the process $\mathbf{x}_n$ is normal white noise with $R[m] = R[0]\delta[m]$.

Suppose now that we are given the $N + 1$ values (data)

$$R[0], \ldots, R[N]$$

of $R[m]$ and we wish to find the density $f(X)$ of the $M+1$ random variables $\mathbf{x}_n, \ldots, \mathbf{x}_{n+M}$. If $M \leq N$, then the correlation matrix of $\mathbf{X}$ is specified in terms of the data and $f(X)$ is given by (14-157). This is not the case, however, if $M > N$ because then only the center diagonal and the N upper and lower diagonals of the correlation matrix

$R_{M+1} =$

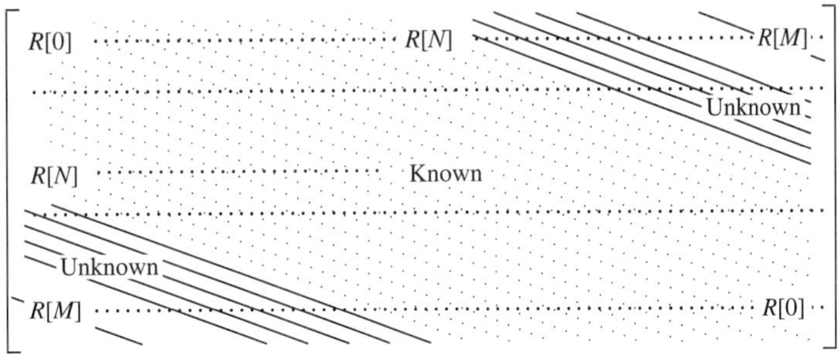

are known. To complete the specification of R_{M+1}, we maximize the determinant Δ_{M+1} with respect to the unknown values of $R[m]$.

EXAMPLE 14-23

▶ Given $R[0]$ and $R[1]$, we shall find $R[2]$ using the maximum determinant method. In this case,

$$\Delta = \begin{vmatrix} R[0] & R[1] & R[2] \\ R[1] & R[0] & R[1] \\ R[2] & R[1] & R[0] \end{vmatrix}$$

Hence

$$\frac{\partial \Delta}{\partial R[2]} = -2R[0]R[2] + 2R^2[1] = 0 \qquad R[2] = \frac{R^2[1]}{R[0]}$$ ◀

THE MEM IN SPECTRAL ESTIMATION. We are given again $R[m]$ for $|m| \leq N$. The power spectrum

$$S(\omega) = R[0] + 2\sum_{m=1}^{\infty} R[m] \cos m\omega$$

of $\mathbf{x}_n$ involves the values of $R[m]$ for every m. To find its unspecified values, we maximize the correlation determinant Δ_M and examine the form of the resulting $R[m]$ as $M \to \infty$. This is equivalent to the maximization of the *entropy rate* $\overline{H}(\mathbf{x})$ of the process $\mathbf{x}_n$. Using this equivalence, we shall develop a more direct method for determining $S(\omega)$.

As we know, the MEM leads to the conclusion that under the given constraints (second-order moments), the process $\mathbf{x}_n$ must be normal with zero mean. From this and

(14-130) it follows that

$$\overline{H}(\mathbf{x}) = \ln\sqrt{2\pi e} + \frac{1}{4\pi}\int_{-\pi}^{\pi} \ln S(\omega)\, d\omega$$

The entropy rate $\overline{H}(\mathbf{x})$ depends on the unspecified values of $R[m]$ and it is maximum if

$$\frac{\partial \overline{H}}{\partial R[m]} = \frac{1}{2\pi}\int_{-\pi}^{\pi} \frac{1}{S(\omega)} e^{-jm\omega}\, d\omega = 0 \qquad |m| > N \tag{14-162}$$

This shows that the coefficients of the Fourier series expansion of $1/S(\omega)$ are 0 for $|m| > N$. Hence

$$\frac{1}{S(\omega)} = \sum_{k=-N}^{N} c_k e^{-jk\omega}$$

Factoring the resulting $\mathbf{S}(z)$ as in (11-6), we obtain

$$S(\omega) = \frac{1}{|b_0 + b_1 e^{-j\omega} + \cdots + b_N e^{-jN\omega}|^2} \tag{14-163}$$

This is the spectrum obtained in Sec. 12-4 [see (12-182)] and it shows that the MEM leads to an AR model. The coefficients b_k can be obtained either from the Yule–Walker equations or from Levinson's algorithm.

Note The MEM also has applications in nonprobabilistic problems involving the determination of unknown parameters from insufficient data. In such cases, probabilistic models are created where the unknown parameters take the form of statistical variables that are determined with the MEM. We should point out, however, that the results obtained are not unique because more than one model can be used in the same problem. In the following, we illustrate this approach using as an example the one-dimensional form of an important problem in *crystallography*.

A deterministic application of the MEM. We wish to find a nonnegative periodic function $f(x)$ with period 2π:

$$0 < f(x) = \sum_{n=-\infty}^{\infty} c_n e^{jnx}$$

having access only to partial information about its Fourier series coefficients

$$c_n = r_n e^{j\varphi_n}$$

The truncation problem We assume that c_n is known only for $|n| \le N$.

Solution 1. We create the following probabilistic model: In the interval $(-\pi, \pi)$, the unknown function $f(x)$ is the density of a random variable $\mathbf{x}$ taking values between $-\pi$ and π. We determine $f(x)$ so as to maximize the entropy

$$I = -\int_{-\pi}^{\pi} f(x) \ln f(x)\, dx$$

of $\mathbf{x}$. This yields [see (14-141)]

$$f(x) = \exp\left\{-1 + \sum_{n=-N}^{N} \gamma_n e^{jnx}\right\}$$

The constants γ_n are determined in terms of the known values of c_n.

Solution 2. We assume that $f(x)$ is the power spectrum of a stochastic process $\mathbf{x}_n$ and we determine $f(x)$ so as to maximize the entropy rate (we omit incidental constants)

$$I = \int_{-\pi}^{\pi} \ln f(x)\, dx$$

of $\mathbf{x}_n$. In this case, $f(x)$ is given by [see (14-163)]

$$f(x) = \frac{1}{\sum_{n=-N}^{N} d_n e^{jnx}}$$

The constants d_n are again determined in terms of the known values of c_n (Levinson's algorithm).

The phase problem We assume that we know only the amplitudes r_n of c_n for $|n| \le N$.

To solve the problem, we form again the integral I, either as the entropy or as the entropy rate, and we maximize it with respect to the unknown parameters that are now the coefficients c_n (amplitudes and phases) for $|n| > N$, and the phase φ_n for $|n| \le N$. An equivalent approach involves the determination of $f(x)$ as in the truncation problem, treating the phases φ_n as parameters, and the maximization of the resulting I with respect to these parameters. In either case, the required computations are not simple.

14-5 CODING

Coding belongs to a class of problems involving the efficient search and identification of an object ζ_i from a set S of N objects. This topic is extensive and it has many applications. We shall present here merely certain aspects related to entropy and probability, limiting the discussion to binary instantly decodable codes. The underlying ideas can be readily generalized.

Binary coding can be also described in terms of the familiar game of 20 questions: A person selects an object ζ_i from a set S. Another person wants to identify the object by asking "yes" or "no" questions. The purpose of the game is to find ζ_i using the smallest possible number of questions.

The various search techniques can be described in three equivalent forms: *(a)* as chains of dichotomies of the set S; *(b)* in the form of a binary tree; *(c)* as binary codes (Fig. 14-18). We start with an explanation of these approaches, ignoring for the moment optimality considerations. The criteria for selecting the "best" search method will be developed later.

Set dichotomies. We subdivide the set S into two nonempty sets A_0 and A_1 (first-generation sets). We subdivide each of the sets A_0 and A_1 into two nonempty sets A_{00}, A_{01} and A_{10}, A_{11} (second-generation sets). We continue with such dichotomies until the final sets consist of a single element each.

The indices of the sets of each generation are binary numbers formed by attaching 0 or 1 to the indices of the preceding generation sets.

In Fig. 14-18, we illustrate the above with a set consisting of nine elements. We shall use the chain of sets so formed to identify the element ζ_7 by a sequence of appropriate questions (set dichotomies): Is it in A_0? No. Is it in A_{10}? No. Is it in A_{110}? Yes. Is it in A_{1100}? Yes. Hence the unknown element is ζ_7 because $A_{1100} = \{\zeta_7\}$.

Binary trees. A tree is a simply connected graph consisting of line segments called *branches*. In a binary tree, each branch splits into *two* other branches or it terminates. The points of termination are the *endpoints* of the tree and the starting point R is its *root* (Fig. 14-18). A *path* is a part of the tree from R to an endpoint. The two branches closest

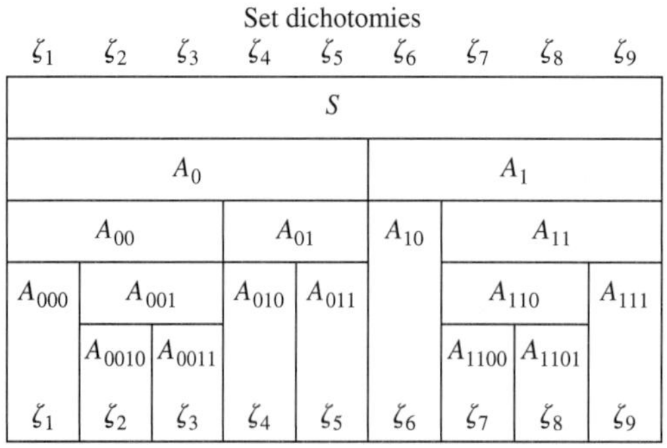

Binary code

x_1	000	0010	0011	010	011	10	1100	1101	111
l_i	3	4	4	3	3	2	4	4	3

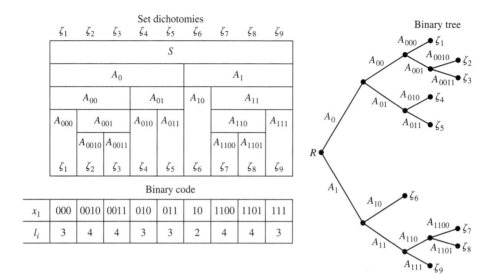

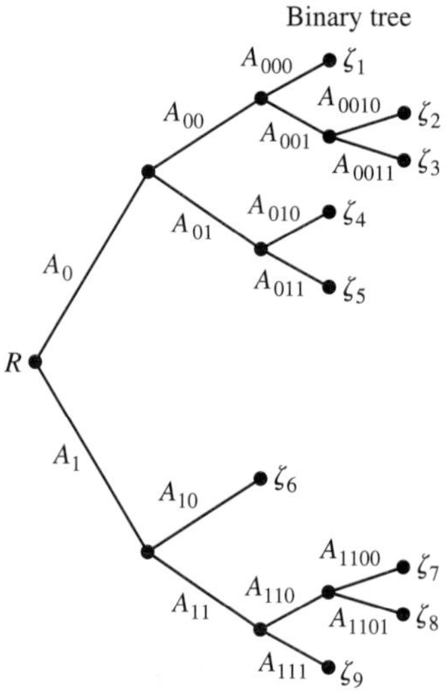

FIGURE 14-18

to the root are the first-generation branches. They split into two branches each, forming the second generation. Since each branch splits into two or it terminates, the number of branches in each generation is always *even.* The *length* of a path is the total number of its branches.

There is one-to-one correspondence between set dichotomies and trees. The kth-generation sets correspond to the kth-generation branches and each set dichotomy to the splitting of the corresponding branch. The terminal sets $\{\zeta_i\}$ correspond to the terminal branches and the elements ζ_i to the endpoints of the tree. The indices of the sets are also used to identify the corresponding branches where we use the following convention: When a branch splits, 0 is assigned to the left new branch and 1 to the right. The index of a terminal branch is also used to identify the corresponding endpoint ζ_i. Thus each element ζ_i of S is identified by a binary number x_i (Fig. 14-18). The number of digits l_i of x_i equals the length of the path ending at ζ_i. This number also equals the number of questions (dichotomies) required to identify ζ_i.

Binary codes. A binary code is a one-to-one correspondence between the elements ζ_i of a set S and the elements x_i of a set $X = \{x_1, x_2, \ldots\}$ of binary numbers. *Encoding* is the process of constructing such a correspondence.

The set S will be called the *source* and its elements ζ_i the *source words*. The corresponding binary numbers x_i will be called the *code words.* The binary digits 0 and 1 form the *code alphabet.* The *length* l_i of a code word x_i is the total number of its binary digits.

A *message* is a sequence of source words

$$\zeta_{i_1} \cdots \zeta_{i_k} \cdots \zeta_{i_n} \qquad \zeta_{i_k} \in S \tag{14-164}$$

The sequence of the corresponding code words

$$x_{i_1} \cdots x_{i_k} \cdots x_{i_n} \tag{14-165}$$

is a *coded message.*

Tree contraction

$$\frac{1}{2^4} + \frac{1}{2^4} + \sum_{i=1}^{5} \frac{1}{2^{l_i}} \qquad\qquad \frac{1}{2^3} + \sum_{i=1}^{5} \frac{1}{2^{l_i}}$$

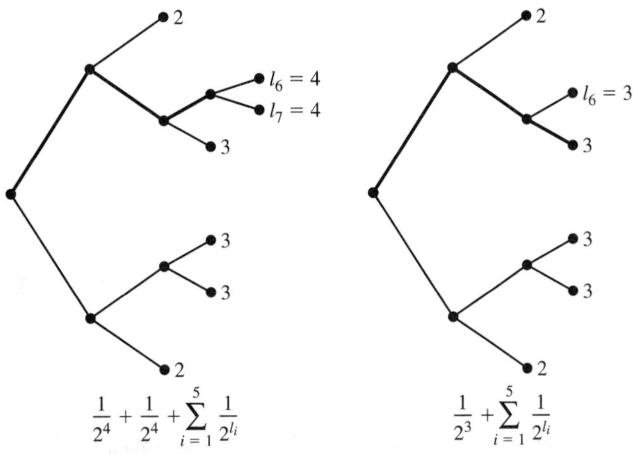

FIGURE 14-19

The indices of the terminal elements of a tree, or, equivalently, of a chain of set dichotomies, specify a code. Codes can, of course, be formed in other ways; however, other codes will not be considered here. The term *code* will mean a *binary code* specified by a tree as above.

In Fig. 14-18, we show the code words x_i of a source S consisting of $N = 9$ elements, and the corresponding word lengths l_i.

THEOREM 14-8 ▶ If a source S has N words and the lengths of the corresponding code words equal l_i, then

$$\sum_{i=1}^{N} \frac{1}{2^{l_i}} = 1 \qquad\qquad (14\text{-}166)$$

Proof. The last-generation branches of the tree are terminal and they form pairs. The two branches of one such pair are the ends of two paths of length l_r (Fig. 14-19). If they are removed, the tree *contracts* into a tree with $N - 1$ endpoints. In this operation, the two paths are replaced with one path of length $l_r - 1$ and the two terms 2^{-l_r} in (14-166) are replaced with the term $2^{-(l_r-1)}$. Since

$$2^{-l_r} + 2^{-l_r} = 2^{-(l_r-1)} \qquad\qquad (14\text{-}167)$$

the sum does not change. Thus the binary length sum in (14-166) is invariant to a contraction. Repeating the process until we are left with only two first-generation branches, we obtain (14-166) because $2^{-1} + 2^{-1} = 1$. ◀

THEOREM 14-9 ▶ Given N integers l_i satisfying (14-166), we can construct a code with lengths l_i.

CONVERSE THEOREM

Proof. It suffices to construct a binary tree with path lengths l_i. From (14-166) it follows that if l_r is the largest of the integers l_i, then the number n of lengths that equal l_r is even. Using $n = 2m$ segments, we form the rth (last) generation branches of our tree. If each of the m pairs of integers l_r is replaced by a single integer $l_r - 1$ and all others are not changed, the resulting set of numbers

Tree construction

l	1	2	3	4	5	6	7	8
l_i	2	2	3	3	4	4	4	4
	2	2	3	3	3		3	
	2	2	2		2			
	1		1					
x_i	11	10	011	010	0011	0010	0001	0000

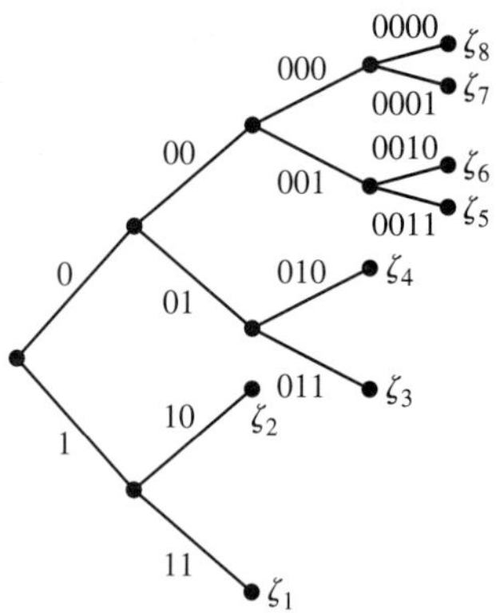

FIGURE 14-20

will satisfy (14-166) [see (14-167)]. We can, therefore, continue this process until we are left with only two terms. These terms yield the two first-generation branches. This procedure is illustrated in Fig. 14-20 for $N = 8$. ◀

Decoding. In the earlier discussion, we presented a method for encoding the words ζ_i of a source S. Encoding of an entire message of the form (14-164) can be obtained by encoding each word successively. The result is a coded message as in (14-165). Decoding is the reverse process: Given a coded message, find the corresponding source message.

Since word coding is a one-to-one correspondence between ζ_i and x_i, the decoding of each word of a message is unique. However, an entire message cannot always be so decoded because there is no space separating the code words (this would require an additional letter in the code alphabet). The problem of separation does not exist for codes constructed through dichotomies (they are, we repeat, the only codes considered here) because such codes have the following property: *No code word is the beginning of another code word.* This property is a consequence of the fact that in any tree, each path terminates at an endpoint; therefore, it cannot be part of another path. Codes with this property are called "instantaneous" because they are instantly decodable; that is, if we start from the beginning of a message, we can identify in real time the end of each word without any reference to the future.

EXAMPLE 14-24 ▶ We wish to decode the message

$$1011010000101000101111000000010$$

formed with the code shown in Fig. 14-18. Starting from the beginning, we identify the code words by underlying them with the help of the table of Fig. 14-18:

$$\underline{10}\quad\underline{1101}\quad\underline{000}\quad\underline{010}\quad\underline{10}\quad\underline{0010}\quad\underline{111}\quad\underline{1100}\quad\underline{000}\quad\underline{0010}$$

The corresponding source message is the sequence

$$\zeta_6\zeta_8\zeta_1\zeta_4\zeta_6\zeta_2\zeta_9\zeta_7\zeta_1\zeta_2$$

◀

Note We have identified each source word with a single symbol ζ_i. It is possible, however, that ζ_i might be a grouping of other symbols. For example, the source S might consist of: All the *letters* of the English alphabet, certain frequently used words (for instance, the word *the*) and even a number of common phrases like *happy birthday*. Such sources are equivalent to single-symbol sources if each word is viewed as a single element.

Optimum Codes

In the absence of prior information, the two subsets of each set dichotomy are so chosen as to have nearly equal elements. The resulting code lengths are then nearly equal to $\log N$. If, however, prior information is available, then more efficient codes can be constructed. The information is usually given in terms of relative frequencies and it is used to form codes with minimum average length. Since relative frequencies are best described in terms of probabilities, we shall assume from now on that the source S is a probability space.

DEFINITIONS. A *random code* is a process of assigning to every source word ζ_i a binary number x_i.

Since ζ_i is an element of the probability space S, a random code defines a random variable $\mathbf{x}$ such that

$$\mathbf{x}(\zeta_i) = x_i$$

The *length* of a random code is a random variable $\mathbf{L}$ such that

$$\mathbf{L}(\zeta_i) = l_i \tag{14-168}$$

where l_i is the length of the code word x_i assigned to the element ζ_i.

The expected value of $\mathbf{L}$ is denoted by L and it is called the *average length* of the random code $\mathbf{x}$. Thus

$$L = E\{\mathbf{L}\} = \sum_i p_i l_i \tag{14-169}$$

where $p_i = P\{\mathbf{x} = x_i\} = P\{\zeta_i\}$.

Optimum code. An *optimum code* is a code whose average length does not exceed the average length of any other code. A basic objective of coding theory is the determination of such a code. Optimum codes have these properties:

1. Suppose that ζ_a and ζ_b are two elements of S such that

$$p_a = P\{\zeta_a\} \qquad p_b = P\{\zeta_b\} \qquad l(\zeta_a) = l_a \qquad l(\zeta_b) = l_b$$

We maintain that if the code is optimum and

$$p_a > p_b \qquad \text{then} \quad l_a \le l_b \tag{14-170}$$

Proof. Suppose that $l_a > l_b$. Interchanging the codes assigned to the elements ζ_a and ζ_b, we obtain a new code with average length

$$L_1 = L - (p_a l_a + p_b l_b) + (p_a l_b + p_b l_a) = L - (p_a - p_b)(l_a - l_b)$$

And since $(p_a - p_b)(l_a - l_b) > 0$, we conclude that $L_1 < L$. This, however, is impossible because L is the optimum code length; hence $l_a \le l_b$.

Repeated application of (14-170) leads to the conclusion that if

$$p_1 \geq p_2 \geq \cdots \geq p_N \qquad \text{then} \quad l_1 \leq l_2 \leq \cdots \leq l_N \tag{14-171}$$

2. The elements (source words) with the two smallest probabilities p_{N-1} and p_N are in the last generation of the tree; that is, their code lengths are l_{N-1} and l_N.

Proof. This is a consequence of (14-171) and the fact that the number of branches in each generation is even.

The following basic theorem shows the relationship between the entropy

$$H(V) = -\sum_{i=1}^{N} p_i \log p_i$$

of the source word partition V and the average length L of an arbitrary random code **x**.

THEOREM 14-10

$$H(V) \leq L \tag{14-172}$$

Proof. As we have seen from (14-166), if l_i are the lengths of the code words of **x** and $q_i = 1/2^{l_i}$, then the sum of the q_i's equals 1. With $a_i = p_i$ and $b_i = q_i$ it follows, therefore, from (14-37) that

$$-\sum_i p_i \log p_i \leq -\sum_i p_i \log q_i = \sum_i p_i l_i = L \tag{14-173}$$

and (14-172) results.

In general, $H(V) < L$. We maintain, however, that $H(V) = L$ iff the probabilities p_i are binary decimals, that is, iff $p_i = 1/2^{n_i}$.

Proof. If $H(V) = L$, then (14-173) is an equality; hence $p_i = q_i = 1/2^{l_i}$ [see (14-37)] and our assertion is true because the lengths l_i are integers.

Conversely, if $p_i = 1/2^{n_i}$ and n_i are integers, then we can construct a code with lengths $l_i = n_i$ because the sum of the p_i's equals 1. The length L of this code equals $H(V)$. In other words, if all p_i's are binary decimals, then the code with lengths $l_i = n_i$ is optimum. ◀

Shannon, Fano, and Huffman Codes

The last theorem gives us a low bound for the average code length L but it does not say how close we can come to this bound. At the end of the section we show that, if we encode not each word but an entire message, then we can construct codes with average length per word less than $H(V) + \varepsilon$ for any $\varepsilon > 0$.

Now, we present three well-known codes including the optimum code (Huffman). The description of these codes is clarified in Example 14-25.

THE SHANNON CODE. As we noted, if all probabilities p_i are binary decimals, then the code with lengths $l_i = -\log p_i$ is optimum. Guided by this, we shall construct a code for all other cases.

Each p_i specifies an integer n_i such that

$$\frac{1}{2^{n_i}} \leq p_i < \frac{1}{2^{n_i - 1}} \tag{14-174}$$

where $p_i > 1/2^{n_i}$ for at least one p_i (assumption). With n_m the largest of the integers n_i, it follows from (14-174) that

$$\sum_{i=1}^{N} \frac{1}{2^{n_i}} \le 1 - \frac{1}{2^{n_m}} \tag{14-175}$$

because the left side is a binary integer smaller than 1. If, therefore, n_m is changed to $n_m - 1$, the resulting value of the sum in (14-175) will not exceed 1. We continue the process of reducing the largest integer by 1 until we reach a set of integers l_i such that

$$\sum_{i=1}^{N} \frac{1}{2^{l_i}} = 1 \qquad l_i \le n_i \tag{14-176}$$

With this set of integers we construct a code and we denote by L^a its average length. Thus

$$L^a = \sum_{i=1}^{N} p_i l_i \le \sum_{i=1}^{N} p_i n_i$$

We maintain that

$$H(\mathbf{V}) \le L^a < H(\mathbf{V}) + 1 \tag{14-177}$$

Proof. From (14-174) it follows that $n_i < -\log p_i + 1$. Multiplying by p_i and adding, we obtain

$$\sum_{i=1}^{N} p_i n_i < \sum_{i=1}^{N} p_i(-\log p_i + 1) = H(\mathbf{V}) + 1$$

and (14-177) results [see (14-172)].

THE FANO CODE. We shall describe this code in terms of set dichotomies based on the following rule of subdivision. We number the probabilities p_i in descending order

$$p_1 \ge p_2 \ge \cdots \ge p_N \tag{14-178}$$

and we select the sets A_0 and A_1 of the first generation so as to have equal or nearly equal probabilities. To do so, we determine k such that

$$p_1 + \cdots + p_k \le 0.5 \le p_{k+1} + \cdots + p_N$$

and we set A_0 equal to $\{\zeta_1, \ldots, \zeta_k\}$ or to $\{\zeta_1, \ldots, \zeta_{k+1}\}$. The same rule is used in all subsequent subdivisions. As we see in Example 14-25, the length L^b of the resulting code is close to the Shannon code length L^a.

We note that, since there is an ambiguity in the choice of the subsets in each dichotomy, the Fano code is not unique.

THE HUFFMAN CODE. We denote by $\mathbf{x}_N^0$ the optimum N-element code and by L_N^0 its average length. We shall determine $\mathbf{x}_N^0$ using the following operation: We arrange the probabilities p_i of the elements ζ_i of S in descending order as in (14-178) and we number the corresponding elements ζ_i accordingly. We then replace the last two elements ζ_{N-1} and ζ_N with a new element and we assign to this element the probability

$p_{N-1} + p_N$. A new source results with $N - 1$ elements. This operation will be called *Huffman contraction.*

In the table of Example 14-25, the new element is identified by a box in which the replaced elements are shown.

Rearranging the probabilities of the new source in descending order, we repeat the Huffman contraction operation until we reach a set with only two elements.

To each element ζ_i of the source S, we shall assign a code word x_i starting from the last digit: We assign the numbers 0 and 1, respectively, to the last digits of the code words of the elements ζ_{N-1} and ζ_N. At each subsequent contraction, we assign the numbers 0 and 1 to the *left* of the partially completed code words of all elements that are included in the last two boxes.

The code so formed (Huffman) will be denoted by $\mathbf{x}_N^c$ and its average length by L_N^c. We shall show that this code is optimal.

Proof. The proof of the optimality is based on the following observation. We can readily see that the last two code words x_{N-1} and x_N have the same length l_r. In Example 14-25,

$$N = 9 \qquad x_8 = 00000 \qquad x_9 = 00001 \qquad l_r = 5$$

If we replace these two words with a single word consisting of their common part, we obtain the Huffman code $\mathbf{x}_{N-1}^c$ for the set of $N - 1$ elements and the code length of the new element equals l_{r-1}. This leads to the conclusion that

$$L_N^c - (p_{N-1} + p_N)l_r = L_{N-1}^c - (p_{N-1} + p_N)(l_r - 1)$$

Hence

$$L_N^c = L_{N-1}^c + p_{N-1} + p_N \tag{14-179}$$

In the example

$$L_9^c = \sum_{i=1}^{7} p_i l_i + 5p_8 + 5p_9 \qquad L_8^c = \sum_{i=1}^{7} p_i l_i + 4(p_8 + p_9)$$

Induction The Huffman code is optimum for $N = 2$ because there is only one code with two words. We assume that it is optimum for every source with $k \le N - 1$ elements and we shall show that it is optimum for $k = N$. Suppose that there is an N-element source S for which this is not true, that is, suppose that

$$L_N^0 < L_N^c \tag{14-180}$$

As we know, the two elements ζ_{N-1} and ζ_N with the smallest probabilities are in the last-generation branches of the optimum code tree. If they are removed, the contracted tree specifies a new code with length L_{N-1}. Reasoning as in (14-179), we conclude with (14-180) that

$$L_{N-1} + p_{N-1} + p_N = L_N^0 < L_N^c = L_{N-1}^c + p_{N-1} + p_N$$

hence $L_{N-1} < L_{N-1}^c$. But this is impossible because the Huffman code of order $N - 1$ is optimum by assumption.

EXAMPLE 14-25 ▶ We shall describe the above codes using as source a set S with nine elements. Their probabilities are shown in the table below:

i	1	2	3	4	5	6	7	8	9
p_i	0.22	0.19	0.15	0.12	0.08	0.07	0.07	0.06	0.04

The resulting entropy equals

$$H(V) = -\sum_{i=1}^{9} p_i \log p_i = 2.703$$

Arbitrary code We form a code using a chain of dichotomies chosen arbitrarily as in Fig. 14-19. In the table below we show the code words and their lengths.

i	1	2	3	4	5	6	7	8	9
x_i	000	0010	0011	010	011	10	1100	1101	111
l_i	3	4	4	3	3	2	4	4	3

$$L = \sum_{i=1}^{9} p_i l_i = 3.40$$

Shannon code In the table below we show the integers n_i determined from (14-174) and the required reductions until the final lengths l_i are reached. The corresponding code tree is shown in Fig. 14-20.

p_i	0.22	0.19	0.15	0.12	0.08	0.07	0.07	0.06	0.04	
	$\frac{1}{2^3} \le p_i < \frac{1}{2^2}$			$\frac{1}{2^4} \le p_i < \frac{1}{2^3}$				$\frac{1}{2^5} \le p_i < \frac{1}{2^4}$		$\sum_{i=1}^{N} \frac{1}{2^{n_i}}$
n_i	3	3	3	4	4	4	4	5	5	12/16
	3	3	3	3	3	4	4	4	4	14/16
l_i	3	3	3	3	3	3	3	4	4	1
x_i	000	001	010	011	100	101	110	1110	1111	$L^a = 3.1$

Fano code In the table below we show the subsets obtained with the Fano dichotomies, and their probabilities. The last-generation sets are the elements ζ_i of S; their probabilities are shown on the first row of the table. The dichotomies start with

$$A_0 = \{\zeta_1, \zeta_2, \zeta_3\} \qquad P(A_0) = 0.22 + 0.19 + 0.15 = 0.56$$

p_i	0.22	0.19	0.15	0.12	0.08	0.07	0.07	0.06	0.04	
	A_0	0.56			A_1		0.44			
	A_{00}	A_{01}	0.34	A_{10}	0.20		A_{11}	0.24		
		A_{010}	A_{011}	A_{100}	A_{101}	A_{110}	0.14	A_{111}	0.10	
						A_{1100}	A_{1101}	A_{1110}	A_{1111}	
	ζ_1	ζ_2	ζ_3	ζ_4	ζ_5	ζ_6	ζ_7	ζ_8	ζ_9	
x_i	00	010	011	100	101	1100	1101	1110	1111	
l_i	2	3	3	3	3	4	4	4	4	$L^b = 3.02$

Optimum code In the table below we show the original set S_9 consisting of nine elements and the sets obtained with each Huffman contraction. The elements ζ_i are identified by their indices and the combined elements by boxes. Each box contains all elements ζ_i of the original source involved in each contraction, and the evolution of their code words x_i starting with the last digit. The rows below each S_i line show the probabilities of the various elements of S_i. For example, the number 0.10 in the line below S_7 is the probability of the box (element of S_7) that contains the elements ζ_8 and ζ_9.

The column at the extreme right shows the sum of the two smallest probabilities of the elements in S_i. This number is used to form the row S_{i+1} by reordering the elements of S_i.

Evolution of Huffman code

S_9	1	2	3	4	5	6	7	8	9	
$p_{i,9}$	0.22	0.19	0.15	0.12	0.08	0.07	0.07	0.06	0.04	0.10
S_8	1	2	3	4	8	9	5	6	7	
					0	1				
$p_{i,8}$	0.22	0.19	0.15	0.12		0.10	0.08	0.07	0.07	0.14
S_7	1	2	3	6	7	4	8	9	5	
				0	1		0	1		
$p_{i,7}$	0.22	0.19	0.15	0.14		0.12		0.10	0.08	0.18
S_6	1	2	8	9	5	3	6	7	4	
			00	01	1		0	1		
$p_{i,6}$	0.22	0.19		0.18		0.15		0.14	0.12	0.26
S_5	6	7	4	1	2	8	9	5	3	
	00	01	1			00	01	1		
$p_{i,5}$		0.26		0.22	0.19		0.18		0.15	0.33
S_4	8	9	5	3	6	7	4	1	2	
	000	001	01	1	100	01	1			
$p_{i,4}$		0.33				0.26		0.22	0.19	0.41
S_3	1	2	8	9	5	3	6	7	4	
	0	1	000	001	01	1	00	01	1	
$p_{i,3}$		0.41			0.33			0.26		0.59
S_2	8	9	5	3	6	7	4	1	2	
	0000	001	001	01	100	101	11	0	1	
$p_{i,2}$				0.59					0.41	1
S_1	8	9	5	3	6	7	4	1	2	
	00000	00001	0001	001	0100	0101	011	10	11	

The completed code words x_i taken from the last line of the table and their code lengths l_i are listed below.

	1	2	3	4	5	6	7	8	9	
x_i	10	11	001	011	0001	0100	0101	00000	00001	$L^0 = 3.01$
l_i	2	2	3	3	4	4	4	5	5	

◀

The Shannon Coding Theorem

In the earlier discussion, we considered only codes of the elements ζ_i of a set S and we showed that the optimum code is between $H(V)$ and $H(V)+1$:

$$H(V) \le L^0 \le H(V) + 1 \tag{14-181}$$

This follows from (14-172) and (14-177). We show next that if we encode not merely single words but entire messages, then the code length per word can be reduced to less than $H(V) + \varepsilon$ for any $\varepsilon > 0$.

A message of length n is any element of the product space S^n. The number of such messages is N^n and a code of the space S^n is a correspondence between its elements and a set of N^n binary numbers. This correspondence defines the random variable $\mathbf{x}_n$ (random code) on the space S^n and the lengths of the code words form another random variable $\mathbf{L}_n$ (random code length). The expected value L_n of $\mathbf{L}_n$ is the average code length. From the definition it follows that L_n is the average number of digits required to encode the elements of S^n. The ratio

$$\overline{L} = \frac{L_n}{n} \tag{14-182}$$

is the *average code length* per word. The term *word* means, of course, an element of S.

We shall assume that S^n is the space of n *independent* trials.

THEOREM 14-11 ▶ We can construct a code of the space S^n such that

$$H(V) \le \overline{L} \le H(V) + \frac{1}{n} \tag{14-183}$$

Proof. We shall give two proofs. The first is a direct consequence of (14-181). The second is based on the concept of typical sequences.

1. Applying the earlier results to the source S^n, we construct a code L_n such that

$$H(V^n) \le L_n < H(V^n) + 1 \tag{14-184}$$

This yields (14-183) because $L_n = n\overline{L}$ and $H(V^n) = nH(V)$ [see (14-67)].

2. As we know the space S^n can be divided into two sets: the set $\boldsymbol{T}$ of all typical sequences and the set $\overline{\boldsymbol{T}}$ of all rare sequences. To prove (14-183), we construct a code tree consisting of $2^{nH(V)} - 1$ short paths of length $l_t = nH(V)$ and 2^l paths of length $I_t + l$. The short paths are used as the code words of the typical sequences and the long paths for the long sequences (Fig. 14-21). Since $P(\boldsymbol{T}) \simeq 1$ and $P(\boldsymbol{T}) \simeq 0$, we conclude that the average length of the resulting code equals

$$L_n = l_t P(\boldsymbol{T}) + (l + l_t) P(\overline{\boldsymbol{T}}) \simeq l_t = nH(V)$$

Thus $\overline{L} \simeq H(V)$ and (14-183) results.

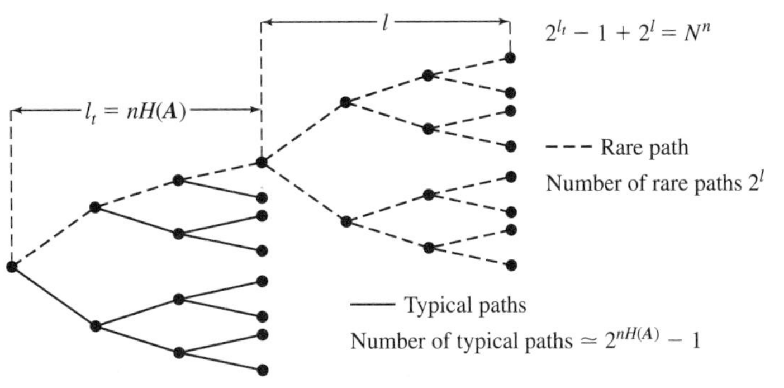

FIGURE 14-21

We note that (14-184) holds even if the trials are not independent. In this case, the theorem is true if $H(V)$ is replaced by $H(V^n)/n$. ◀

14-6 CHANNEL CAPACITY

We wish to transmit a message from point A to point B by means of a communications channel (a telephone cable, for example). The message to be transmitted is a stationary process $\mathbf{x}_n$ generating at the receiving end another process $\mathbf{y}_n$. The output $\mathbf{y}_n$ depends not only on the input $\mathbf{x}_n$ but also on the nature of the channel. Our objective is to determine the maximum rate of information that can be transmitted through the channel. To simplify the discussion, we make the following assumptions:

1. The channel is *binary;* that is, the input $\mathbf{x}_n$ and the output $\mathbf{y}_n$ take only the values 0 and 1.
2. The channel is *memoryless;* that is, the present value of $\mathbf{y}_n$ depends only on the present value of $\mathbf{x}_n$.
3. The input $\mathbf{x}_n$ is *strictly white noise.*

From assumptions 2 and 3 it follows that $\mathbf{y}_n$ is also white noise.

4. The messages are transmitted at the rate of *one word per second.*

This is a mere normalization stating that the duration T of each transmitted state equals one second.

EXAMPLE 14-26 ▶ In Fig. 14-22 we show a simple realization of a channel as a system with input $\mathbf{x}_n$ and output $\mathbf{y}_n$. The input to the physical channel is a time signal $\mathbf{x}(t)$ taking the values E and $-E$ (binary transmission). These values correspond to the two states 1 and 0 of $\mathbf{x}_n$. The received signal $\mathbf{y}(t)$ is a distorted version of $\mathbf{x}(t)$ contaminated possibly by noise. The system output $\mathbf{y}_n$ is obtained by some decision rule (detector) translating the time signal $\mathbf{y}(t)$ into a discrete-time signal consisting of 0's and 1's. ◀

Noiseless Channel

We shall say that a channel is noiseless[9] if there is a one-to-one correspondence between the input $\mathbf{x}_n$ and the output $\mathbf{y}_n$. For a binary channel this means that if $\mathbf{x}_n = 0$, then $\mathbf{y}_n = 0$; if $\mathbf{x}_n = 1$, then $\mathbf{y}_n = 1$.

In a given channel, the uncertainty per transmitted word equals the entropy rate $\overline{H}(\mathbf{x}) = H(\mathbf{x})$ of the input $\mathbf{x}_n$. If the channel is noiseless, then the observed output $\mathbf{y}_n$ determines $\mathbf{x}_n$ uniquely; hence it removes this uncertainty. Thus the rate of transmitted information equals $H(\mathbf{x})$.

[9]This definition does not lead to any conclusion about the actual presence of noise in the channel.

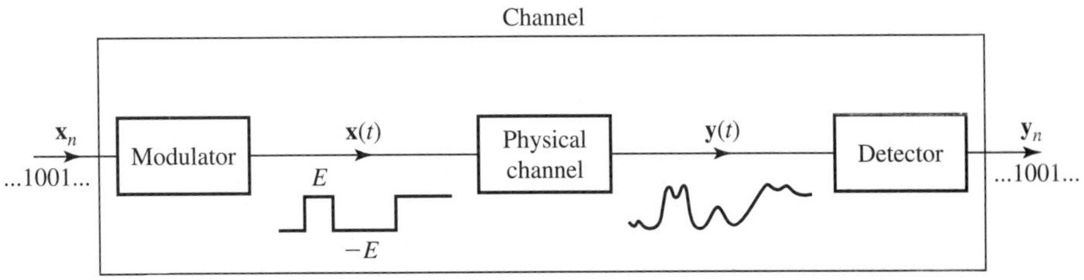

FIGURE 14-22

DEFINITION OF CHANNEL CAPACITY. The maximum value of $H(\mathbf{x})$, as $\mathbf{x}$ ranges over all possible inputs, is denoted by C and is called the *channel capacity*

$$C = \max_{\mathbf{x}_n} H(\mathbf{x}) \tag{14-185}$$

It appears that C does not depend on the channel but that is not so because the channel determines the number of the input states. If it is binary, then $\mathbf{x}_n$ has two possible states with probabilities p and $q = 1 - p$, respectively; hence

$$H(\mathbf{x}) = -p \log p - (1 - p) \log(1 - p) = h(p) \tag{14-186}$$

where $h(p)$ is the function of Fig. 14-2. Since $h(p)$ is maximum for $p = 0.5$ and $h(0.5) = 1$, we conclude that the capacity of a binary noiseless channel equals 1 bit/s.

Similarly, if the channel accepts N input states, then its capacity equals $\log N$ bit/s.

RATE OF INFORMATION TRANSMISSION. We repeat: The channel transmits messages at the rate of 1 word/s. It transmits information at the rate $H(\mathbf{x})$ bits/s. This rate depends on the source and it is maximum if the two states of the source are equally likely.

THEOREM 14-12

▶ The maximum rate of 1 bit/s can be reached even if the input $\mathbf{x}_n$ is arbitrary, provided that it is properly encoded prior to transmission.

Proof. 1. An m-word message is a binary number with m digits. There are 2^m such messages forming the space S_x^m and every realization of the input $\mathbf{x}_n$ is a sequence of such messages. We encode optimally the space S_x^m into a set of binary numbers $\overline{\mathbf{x}}_n$ using the techniques of the last section (Fig. 14-23). The number of digits (code length) of each $\overline{\mathbf{x}}_n$ is a random variable $\mathbf{L}_m$ with

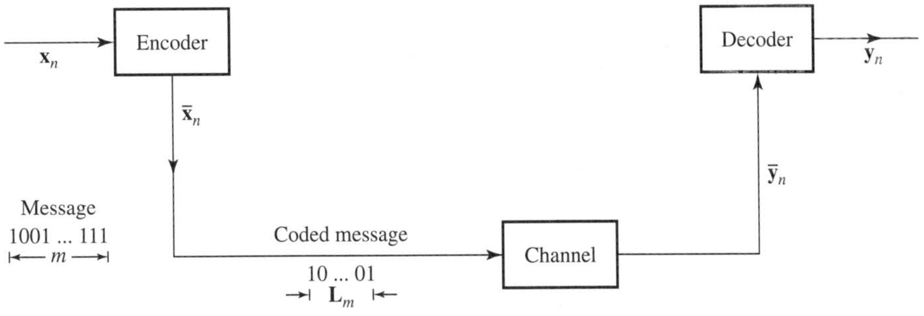

FIGURE 14-23

mean $L_m = E\{\mathbf{L}_m\}$. As we know,

$$mH(\mathbf{x}) \le L_m < mH(\mathbf{x}) + 1 \tag{14-187}$$

Hence $\overline{L}_m \simeq H(\mathbf{x})$ for large m. A code word $\overline{\mathbf{x}}_n$ requires $\mathbf{L}_m$ seconds to be transmitted because it consists of $\mathbf{L}_m$ binary digits. Hence the average time required to transmit the m-word messages of $\mathbf{x}_n$ in code form equals $\mathbf{L}_m \simeq mH(\mathbf{x})$ seconds. And since the information contained in each message equals $mH(\mathbf{x})$ bits, we conclude that the average rate of information transmission equals $mH(\mathbf{x})/mH(\mathbf{x}) = 1$ bit/s.

Proof. 2. We have 2^m messages of length m. In a direct transmission (not encoded), each message requires the same transmission time: m seconds. However, of all these messages, only $2^{mH(\mathbf{x})}$ are likely to occur (typical sequences). To reduce the time of transmission, we encode all typical sequences into words of length $l_t \simeq mH(\mathbf{x})$ as in Fig. 14-21. The rare sequences require longer codes; however, the probability of their occurrence is negligible. Hence the average time of transmission of each message is reduced from m seconds to $mH(\mathbf{x})$ seconds. ◀

Noisy Channel

Due to a variety of factors, a physical channel establishes not a functional but a statistical relationship between the input $\mathbf{x}_n$ and the output $\mathbf{y}_n$. For a binary channel, this relationship is completely specified in terms of the probabilities

$$P\{\mathbf{x}_n = 0\} = p \qquad P\{\mathbf{x}_n = 1\} = q$$

of the two states of the input, and the conditional probabilities

$$P\{\mathbf{y}_n = j \mid \mathbf{x}_n = i\} = \pi_{ij} \qquad i, j = 0, 1 \tag{14-188}$$

The probabilities of the output states are given by

$$P\{\mathbf{y}_n = 0\} = \pi_{00}p + \pi_{10}q \qquad P\{\mathbf{y}_n = 1\} = \pi_{01}p + \pi_{11}q \tag{14-189}$$

DEFINITION ▶ A noisy channel is a random system establishing a statistical relationship between the input $\mathbf{x}_n$ and the output $\mathbf{y}_n$.

For a memoryless channel, this relationship is completely specified in terms of the *channel matrix* Π whose elements π_{ij} are the conditional probabilities between the input states and the output states. For a binary channel

$$\Pi = \begin{bmatrix} \pi_{00} & \pi_{01} \\ \pi_{10} & \pi_{11} \end{bmatrix} \qquad \text{where} \qquad \begin{aligned} \pi_{00} + \pi_{01} &= 1 \\ \pi_{10} + \pi_{11} &= 1 \end{aligned} \tag{14-190}$$

The channel is called *symmetrical* if $\pi_{10} = \pi_{01} = \beta$. In a symmetrical channel, $\pi_{00} = \pi_{11} = 1 - \beta$ and

$$\Pi = \begin{bmatrix} 1-\beta & \beta \\ \beta & 1-\beta \end{bmatrix} \tag{14-191}$$

◀

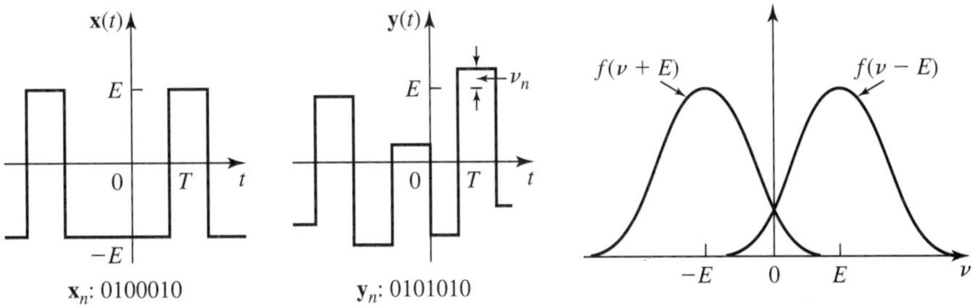

FIGURE 14-24

EXAMPLE 14-27 ▶ To give some idea of the nature of the channel matrix, we show in Fig. 14-24 a simple version of a symmetrical channel. The input $\mathbf{x}(t)$ is a time signal as in Example 14-26, and the resulting output $\mathbf{y}(t)$ is the sum

$$\mathbf{y}(t) = \mathbf{x}(t) + \boldsymbol{\nu}_n \qquad nT \le t < nT + T \tag{14-192}$$

where $\boldsymbol{\nu}_n$ is a sequence of independent random variables with density the even function $f(\nu)$. The output states are determined as follows:

$$\mathbf{y}_n = \begin{cases} 1 & \text{if} \quad \mathbf{y}(t) \ge 0 \\ 0 & \text{if} \quad \mathbf{y}(t) < 0 \end{cases}$$

From this we conclude that the channel is symmetrical and

$$\beta = P\{\mathbf{y}_n = 1 \mid \mathbf{x}_n = 0\} = \int_0^\infty f(\nu + E)\,d\boldsymbol{\nu} = P\{\boldsymbol{\nu} > E\}$$ ◀

CHANNEL CAPACITY. Prior to transmission, the uncertainty about the input $\mathbf{x}_n$ equals $H(\mathbf{x})$ per word. In a noiseless channel, the observed output $\mathbf{y}_n$ reduces the uncertainty to 0. This is not so, however, for a noisy channel because $\mathbf{y}_n$ does not determine $\mathbf{x}_n$ uniquely. Knowledge of $\mathbf{y}_n$ reduces the uncertainty about $\mathbf{x}_n$ from $H(\mathbf{x})$ to $H(\mathbf{x} \mid \mathbf{y})$ and the difference

$$I(\mathbf{x}, \mathbf{y}) = H(\mathbf{x}) - H(\mathbf{x} \mid \mathbf{y}) \tag{14-193}$$

is the *rate of information transmission.*[10]

If the channel is noiseless, then $H(\mathbf{x} \mid \mathbf{y}) = 0$; hence $I(\mathbf{x}, \mathbf{y}) = H(\mathbf{x})$. If the output $\mathbf{y}_n$ is independent of the input, then $H(\mathbf{x} \mid \mathbf{y}) = H(\mathbf{x})$; hence $I(\mathbf{x}, \mathbf{y}) = 0$. In other words, such a channel is useless (it does not transmit any information).

DEFINITION ▶ The function $I(\mathbf{x}, \mathbf{y})$ depends on the matrix Π and on the input $\mathbf{x}_n$. The capacity C of a noisy channel is the maximum value of $I(\mathbf{x}, \mathbf{y})$ as $\mathbf{x}_n$ ranges over all possible inputs

$$C = \max_{\mathbf{x}_n} I(\mathbf{x}, \mathbf{y}) \tag{14-194}$$

This is consistent with (14-185) because, for noiseless channels, $I(\mathbf{x}, \mathbf{y}) = H(\mathbf{x})$. ◀

[10]The conditional entropy $H(\mathbf{x} \mid \mathbf{y})$ is Shannon's *equivocation*.

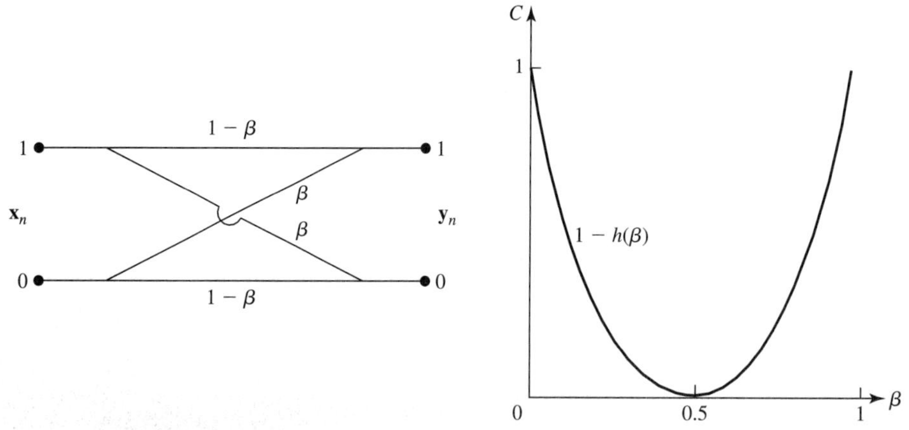

FIGURE 14-25
Binary symmetric channel.

EXAMPLE 14-28 ▶ We shall show that the capacity of a *binary symmetrical channel* with channel matrix as in (14-191) (Fig. 14-25) equals

$$C = 1 - h(\beta) \qquad \text{where} \quad h(p) = -p \log p - q \log q \tag{14-195}$$

Proof. The entropy of a two-state partition equals $h(p)$, where p is the probability of one of the states. Thus the entropy $H(\mathbf{x})$ of the input to the channel equals $h(p)$ and the entropy of the output equals

$$H(\mathbf{y}) = h(\gamma) \qquad \gamma = (1 - 2\beta)p + \beta \tag{14-196}$$

because [see (14-189)]

$$P\{\mathbf{y}_n = 0\} = (1 - \beta)p + \beta(1 - p) = \gamma$$

This holds also for conditional entropies. Thus, since

$$P\{\mathbf{y}_n = 0 \mid \mathbf{x}_n = 0\} = P\{\mathbf{y}_n = 1 \mid \mathbf{x}_n = 1\} = 1 - \beta$$

we conclude that

$$H(\mathbf{y} \mid \mathbf{x}_n = 0) = H(\mathbf{y} \mid \mathbf{x}_n = 1) = h(1 - \beta)$$

Inserting into (14-41) and using the fact that $h(\beta) = h(1 - \beta)$, we obtain

$$H(\mathbf{x} \mid \mathbf{y}) = H(\mathbf{y} \mid \mathbf{x}) = ph(\beta) + qh(\beta) = h(\beta)$$

From the above it follows that $I(\mathbf{x}, \mathbf{y}) = h(\gamma) - h(\beta)$. This yields (14-195) because $h(\beta)$ does not depend on p and $h(\gamma)$ is maximum if $\gamma = 0.5$. ◀

Redundant and random codes Consider a set A (source) with N elements and a set B (code) with M elements where $N < M$. A redundant code is a one-to-one correspondence between the elements of A and the elements of a subset B_1 of B.

The subset B_1 consists of N elements that can be selected in many ways. If the elements of B_1 are chosen at random from the M elements of B, the resulting code is

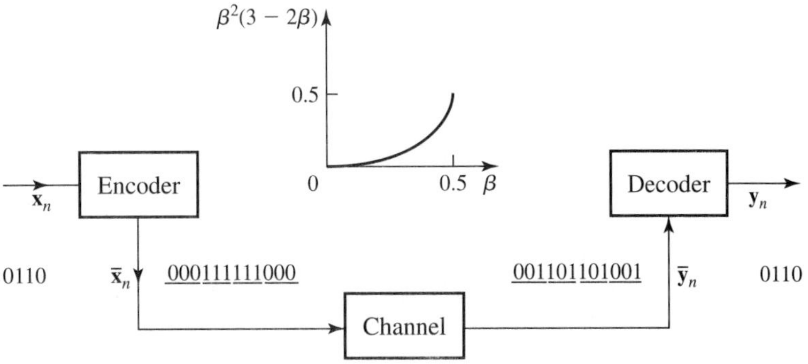

FIGURE 14-26

called *random*.[11] From the definition it follows that the probability that a specific element of B is in the randomly selected set B_1 equals N/M.

In Example 14-29 we show that redundant encoding can be used to reduce the probability of error in transmission.

EXAMPLE 14-29 ▶ In a symmetrical channel, the probability of error equals β. To reduce this error, we encode the input set $A = \{0, 1\}$ into the subset $B_1 = \{000, 111\}$ of the set B of all three-digit binary numbers. In the earlier notation, $N = 2$ and $M = 8$.

The input $\mathbf{x}_n$ is thus encoded into a signal $\bar{\mathbf{x}}_n$ consisting of triplets of 0's and 1's yielding as output a signal $\bar{\mathbf{y}}_n$ (Fig. 14-26). The decoding scheme is the *majority rule:* If the received triplet consists of at least two 0's, then $\mathbf{y}_n = 0$, otherwise $\mathbf{y}_n = 1$.

It can be readily seen that (Prob. 14-23) the probability that a transmitted word will be detected incorrectly equals $\beta^2(3 - 2\beta)$. This is less than β if $\beta < 0.5$. However, the rate of transmission is also reduced from 1 word per second to 1 word per three seconds. ◀

It appears from the above that reduction of the probability of error by redundant encoding must result in transmission rates that tend to 0 as the error tends to 0. This, however, is not so. As the following remarkable theorem shows, it is possible to achieve arbitrarily small error probabilities while maintaining the rate of information transmission close to the channel capacity.

The Channel Capacity Theorem

Information can be transmitted through a noisy channel at a rate nearly equal to the channel capacity C with negligible probability of error.

Proof. *Preliminary remarks* From the definition of channel capacity, it follows that the maximum of $H(\mathbf{x})$ is at least equal to C because

$$H(\mathbf{x}) = I(\mathbf{x}, \mathbf{y}) + H(\mathbf{x} \mid \mathbf{y}) \geq I(\mathbf{x}, \mathbf{y}) \tag{14-197}$$

[11]This definition of a random code is not the definition given on page 677.

This shows that we can find a source with entropy rate as close to C as we want. We shall show that if $\mathbf{x}_n$ is a source with entropy rate

$$H(\mathbf{x}) < C \tag{14-198}$$

then it can be transmitted at the rate of 1 word per second with probability of error less than α for any $\alpha > 0$. This will prove the theorem because the information per word equals $H(\mathbf{x})$.

As in the noiseless case, the proof is based on proper encoding of the space S_x^m consisting of all possible segments of $\mathbf{x}_n$ of length m. However, as the following remarks show, the objectives are different.

Noiseless channel The code set consists of two groups of binary numbers (Fig. 14-27*a*). The first group has 2^{m_1} elements of length $m_1 = mH(\mathbf{x})$ and it is used to encode the 2^{m_1} typical sequences of the input space S_x^m. The second group is used to encode the rare sequences of S_x^m. Since the set of all rare sequences has negligible probability, the average length of the code equals m_1.

Thus, in the noiseless case, the purpose of coding is reduction of the time of transmission of m-word messages from m seconds to m_1 seconds. This results in an increase of the rate of information transmission from $mH(\mathbf{x})$ bits per m seconds to $mH(\mathbf{x})$ bits per $m_1 = mH(\mathbf{x})$ seconds.

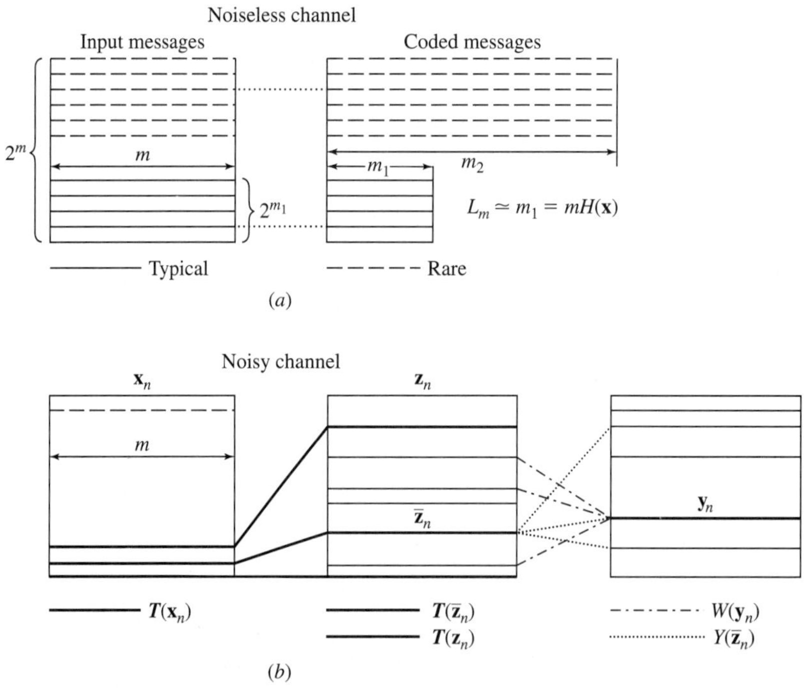

FIGURE 14-27

Noisy channel Reasoning as in (14-197), we conclude that, given $\varepsilon > 0$, we can find a process $\mathbf{z}_n$ such that

$$H(\mathbf{z}) - H(\mathbf{z} \mid \mathbf{y}) > C - \varepsilon \tag{14-199}$$

Choosing $\varepsilon < C - H(\mathbf{x})$, we obtain

$$H(\mathbf{z}) > H(\mathbf{x}) + H(\mathbf{z} \mid \mathbf{y}) \geq H(\mathbf{x}) \tag{14-200}$$

because $H(\mathbf{z} \mid \mathbf{y}) > 0$.

All sequences of $\mathbf{z}_n$ of length m form a space S_z^m consisting of 2^m elements. We can, therefore, encode the input set S_x^m into the set S_z^m. The resulting code is one-to-one (Fig. 14-27*b*). The code can, however, be viewed as redundant if we consider only the mapping of the subset $\boldsymbol{T}(\mathbf{x}_n)$ of all typical sequences of S_x^m into the subset $\boldsymbol{T}(\mathbf{z}_m)$ of all typical sequences of S_z^m. Indeed, $\boldsymbol{T}(\mathbf{x}_n)$ has $N = 2^{mH(\mathbf{x})}$ elements and $\boldsymbol{T}(\mathbf{z}_n)$ has $M = 2^{mH(\mathbf{z})}$ elements where

$$N = 2^{mH(\mathbf{x})} \ll 2^{mH(\mathbf{z})} = M \tag{14-201}$$

because $H(\mathbf{x}) < H(\mathbf{z})$ and $m \gg 1$. We denote by $\bar{\mathbf{z}}_n$ the code word of a typical $\mathbf{x}_n$ message and by $\boldsymbol{T}(\bar{\mathbf{z}}_n)$ the set of all such code words. Clearly, $\boldsymbol{T}(\bar{\mathbf{z}}_n)$ is a subset of the set $\boldsymbol{T}(\mathbf{z}_n)$ consisting of $N \ll M$ elements.

The purpose of the coding is to select the set $\boldsymbol{T}(\bar{\mathbf{z}}_n)$ such that its elements are at a "large distance" from each other in the following sense: Since the channel is noisy, the output due to a specific element $\bar{\mathbf{z}}_n$ is not unique. We denote by $Y(\bar{\mathbf{z}}_n)$ the set of all output sequences due to this element, and we attempt to design the code such that the probability of the intersection of the output sets $Y(\bar{\mathbf{z}}_n)$ as $\bar{\mathbf{z}}_n$ ranges over every element of the set $\boldsymbol{T}(\bar{\mathbf{z}}_n)$ is negligible. This will ensure the unique determination of $\bar{\mathbf{z}}_n$ in terms of the observed output $\mathbf{y}_n$.

Random code To complete the proof, we shall show that among all N-element subsets of the set $\boldsymbol{T}(\mathbf{z}_n)$ there exists at least one that meets our requirements. In fact, we shall prove a stronger statement: If we select *at random* N elements $\bar{\mathbf{z}}_n$ from the M elements of $\boldsymbol{T}(\mathbf{z}_n)$ and use the resulting set $\boldsymbol{T}(\bar{\mathbf{z}}_n)$ to encode the set $\boldsymbol{T}(\mathbf{x}_n)$ then, almost certainly, the probability of error in transmission will be negligible.

We note that, once the *code set* $\boldsymbol{T}(\bar{\mathbf{z}}_n)$ has been selected, the probability that an element of $\boldsymbol{T}(\mathbf{z}_n)$ is in $\boldsymbol{T}(\bar{\mathbf{z}}_n)$ equals N/M. From this it follows that, if W is a randomly selected subset of $\boldsymbol{T}(\mathbf{z}_n)$ consisting of N_w elements, then the probability P_w that it will intersect the set $\boldsymbol{T}(\bar{\mathbf{z}}_n)$ equals

$$P_w = -1 - \left(1 - \frac{N}{M}\right)^{N_w} \simeq \frac{NN_w}{M} \tag{14-202}$$

because $N \ll M$.

Suppose that we transmit the selected m-word message $\mathbf{z}_n$ through the channel and we observe at the output the m-word message $\mathbf{y}_n$. Since the channel is noisy, the same $\mathbf{y}_n$ might result from many other input messages. We denote by $W(\mathbf{y}_n)$ the set consisting of all elements of $\boldsymbol{T}(\mathbf{z}_n)$ that will produce the same output $\mathbf{y}_n$, excluding the actually transmitted message $\bar{\mathbf{z}}_n$ (Fig. 14-27*b*). If the set $W(\mathbf{y}_n)$ does not intersect the code set $\boldsymbol{T}(\bar{\mathbf{z}}_n)$, there is no error because the observed signal $\mathbf{y}_n$ determines uniquely the transmitted signal $\bar{\mathbf{z}}_n$. The error probability equals, therefore, the probability P_w that the sets $W(\mathbf{y}_n)$ and $\boldsymbol{T}(\bar{\mathbf{z}}_n)$ intersect. As we know [see (14-74)] the number N_w of typical elements in $W(\mathbf{y}_n)$ equals $2^{mH(\mathbf{z}\,|\,\mathbf{y})}$. Neglecting all others, we conclude from (14-202) that

$$P_w \simeq \frac{NN_w}{M} = 2^{mH(\mathbf{z}\,|\,\mathbf{y})}2^{m[H(\mathbf{x})-H(\mathbf{z})]}$$

This shows that

$$P_w \to 0 \qquad \text{as} \quad m \to \infty$$

because $H(\mathbf{z}\,|\,\mathbf{y}) + H(\mathbf{x}) - H(\mathbf{z}) < 0$, and the proof is complete.

We note, finally, that the maximum rate of information transmission cannot exceed C bits per second.

Indeed, to achieve a rate higher than C, we would need to transmit a signal $\mathbf{z}_n$ such that $H(\mathbf{z}) - H(\mathbf{z}\,|\,\mathbf{y}) > C$. This, however, is impossible [see (14-194)].

PROBLEMS

14-1 Show that $H(\boldsymbol{U}\cdot\boldsymbol{B}\,|\,\boldsymbol{B}) = H(\boldsymbol{U}\,|\,\boldsymbol{B})$.

14-2 Show that if $\varphi(p) = -p\log p$ and $p_1 < p_1 + \varepsilon < p_2 - \varepsilon < p_2$, then

$$\varphi(p_1 + p_2) < \varphi(p_1) + \varphi(p_2) < \varphi(p_1 + \varepsilon) + \varphi(p_2 - \varepsilon)$$

14-3 In Fig. P14-3*a*, we give a schematic representation of the identities

$$H(\boldsymbol{U}\cdot\boldsymbol{B}) = H(\boldsymbol{U}) + H(\boldsymbol{B}\,|\,\boldsymbol{U}) = H(\boldsymbol{U}) + H(\boldsymbol{B}) - I(\boldsymbol{U},\boldsymbol{B})$$

where each quantity equals the area of the corresponding region. Extending formally this representation to three partitions (Fig. P14-3*b*), we obtain the identities

$$H(\boldsymbol{U}\cdot\boldsymbol{B}\cdot\boldsymbol{C}) = H(\boldsymbol{U}) + H(\boldsymbol{B}\cdot\boldsymbol{C}|\boldsymbol{U}) = H(\boldsymbol{U}\cdot\boldsymbol{B}) + H(\boldsymbol{C}|\boldsymbol{U}\cdot\boldsymbol{B})$$

$$H(\boldsymbol{U}\cdot\boldsymbol{B}\cdot\boldsymbol{C}) = H(\boldsymbol{U}) + H(\boldsymbol{B}|\boldsymbol{U}) + H(\boldsymbol{C}|\boldsymbol{U}\cdot\boldsymbol{B})$$

$$H(\boldsymbol{B}\cdot\boldsymbol{C}|\boldsymbol{U}) = H(\boldsymbol{B}|\boldsymbol{U}) + H(\boldsymbol{C}|\boldsymbol{U}\cdot\boldsymbol{B})$$

Show that these identities are correct.

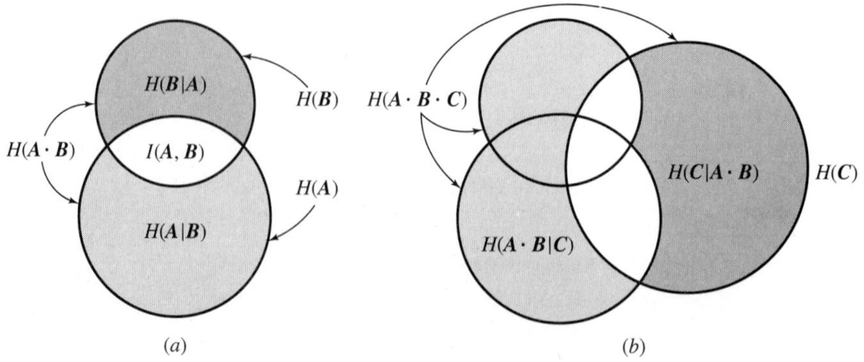

FIGURE P14-3

14-4 Show that

$$I(\boldsymbol{U}\cdot\boldsymbol{B},\boldsymbol{C})+I(\boldsymbol{U},\boldsymbol{B})=I(\boldsymbol{U}\cdot\boldsymbol{C},\boldsymbol{B})+I(\boldsymbol{U},\boldsymbol{C})$$

and identify each quantity in the representation of Fig. P14-3*b*.

14-5 The conditional mutual information of two partitions $\boldsymbol{U}$ and $\boldsymbol{B}$ assuming $\boldsymbol{C}$ is by definition

$$I(\boldsymbol{U},\boldsymbol{B}|\boldsymbol{C})=H(\boldsymbol{U}|\boldsymbol{C})+H(\boldsymbol{B}|\boldsymbol{C})-H(\boldsymbol{U}\cdot\boldsymbol{B}|\boldsymbol{C})$$

(*a*) Show that

$$I(\boldsymbol{U},\boldsymbol{B}|\boldsymbol{C})=I(\boldsymbol{U},\boldsymbol{B}\cdot\boldsymbol{C})-I(\boldsymbol{U},\boldsymbol{C}) \qquad \text{(i)}$$

and identify each quantity in the representation of Fig. P14-3*b*.
(*b*) From (i) it follows that $I(\boldsymbol{U},\boldsymbol{B}\cdot\boldsymbol{C})\geq I(\boldsymbol{U},\boldsymbol{C})$. Interpret this inequality in terms of the subjective notion of mutual information.

14-6 In an experiment S, the entropy of the binary partition $\boldsymbol{U}=[A,\overline{A}]$ equals $h(p)$ where $p=P(A)$. Show that in the experiment $S^3=S\times S\times S$, the entropy of the eight-element partition $\boldsymbol{U}^3=\boldsymbol{U}\cdot\boldsymbol{U}\cdot\boldsymbol{U}$ equals $3h(p)$ as in (14-67).

14-7 Show that

$$H(\mathbf{x}+a)=H(\mathbf{x}) \qquad H(\mathbf{x}+\mathbf{y}\,|\,\mathbf{x})=H(\mathbf{y}\,|\,\mathbf{x})$$

In this, $H(\mathbf{x}+a)$ is the entropy of the random variable $\mathbf{s}+a$ and $H(\mathbf{x}+\mathbf{y}\,|\,\mathbf{x})$ is the conditional entropy of the random variable $\mathbf{x}+\mathbf{y}$.

14-8 The random variables $\mathbf{x}$ and $\mathbf{y}$ are of discrete type and independent. Show that if $\mathbf{z}=\mathbf{x}+\mathbf{y}$ and the line $x+y=z_i$ contains no more than one mass point, then

$$H(\mathbf{z}\,|\,\mathbf{x})=H(\mathbf{y})\leq H(\mathbf{z})$$

Hint: Show that $\boldsymbol{U}_z=\boldsymbol{U}_x\cdot\boldsymbol{U}_y$.

14-9 The random variable $\mathbf{x}$ is uniform in the interval $(0,a)$ and the random variable $\mathbf{y}$ equals the value of $\mathbf{x}$ rounded off to the nearest multiple of δ. Show that $I(\mathbf{x},\mathbf{y})=\log a/\delta$.

14-10 Show that, if the transformation $\mathbf{y}=g(\mathbf{x})$ is one-to-one and $\mathbf{x}$ is of discrete type, then

$$H(\mathbf{x},\mathbf{y})=H(\mathbf{x})$$

Hint: $p_{ij}=P\{\mathbf{x}=x_i\}\delta[i-j]$.

14-11 Show that for discrete-type random variables

$$H(\mathbf{x},\mathbf{x})=H(\mathbf{x}) \qquad H(\mathbf{x}\,|\,\mathbf{x})=0 \qquad H(\mathbf{y}\,|\,\mathbf{x})=H(\mathbf{y},\mathbf{x}\,|\,\mathbf{x})$$

$$H(\mathbf{y}\,|\,\mathbf{x}_1,\ldots,\mathbf{x}_n)=H\left(\mathbf{y},\sum_{k=1}^{n}a_k\mathbf{x}_k\,|\,\mathbf{x}_1,\ldots,\mathbf{x}_n\right)$$

For continuous-type random variables, the relevant densities are singular. This holds, however, if we set $H(\mathbf{x},\mathbf{x})=H(\mathbf{x})$ and use theorem (14-103) and its extensions to several variables to define recursively all conditional entropies.

14-12 The process $\mathbf{x}_n$ is normal white noise with $E\{\mathbf{x}_n^2\}=5$, and

$$\mathbf{y}_n=\sum_{k=0}^{\infty}2^{-k}\mathbf{x}_{n-k}$$

(*a*) Find the mutual information of random variables $\mathbf{x}_n$ and $\mathbf{y}_n$. (*b*) Find the entropy rate of the process $\mathbf{y}_n$.

14-13 The random variables $\mathbf{x}_n$ are independent and each is uniform in the interval (4, 6). Find the entropy rate of the process

$$\mathbf{y}_n = 5 \sum_{k=0}^{\infty} 2^{-k} \mathbf{x}_{n-k}$$

14-14 Find the ME density of a random variable $\mathbf{x}$ if $f(x) = 0$ for $|x| > 1$ and $E\{\mathbf{x}\} = 0.31$.

14-15 It is observed that the duration of the telephone calls is a number $\mathbf{x}$ between 1 and 5 minutes and its mean is 3 min 37 sec. Find its ME density.

14-16 We are given a die with $P\{\text{even}\} = 0.5$ and are told that the mean of the number $\mathbf{x}$ of faces up equals 4.44. Find the ME values of $p_i = P\{\mathbf{x} = i\}$.

14-17 Suppose that $\mathbf{x}$ is a random variable with entropy $H(\mathbf{x})$ and $\mathbf{y} = 3\mathbf{x}$. Express the entropy $H(\mathbf{y})$ of $\mathbf{y}$ in terms of $H(\mathbf{x})$ (*a*) if $\mathbf{x}$ is of discrete type, (*b*) if $\mathbf{x}$ is of continuous type.

14-18 In the experiment of two fair dice. $\boldsymbol{U}$ is a partition consisting of the events $A_1 = \{\text{seven}\}$, $A_2 = \{\text{eleven}\}$, and $A_3 = \overline{A_1 \cup A_2}$. (*a*) Find its entropy. (*b*) The dice were rolled 100 times. Find the number of typical and atypical sequences formed with the events A_1, A_2, and A_3.

14-19 The process $\mathbf{x}[n]$ is SSS with entropy rate $\overline{H}(\mathbf{x})$. Show that, if

$$\mathbf{w}_n = \sum_{k=0}^{n} \mathbf{x}_{n-k} h_k$$

then

$$\lim_{n\to\infty} \frac{1}{n+1} H(\mathbf{w}_0, \ldots, \mathbf{w}_n) = \overline{H}(\mathbf{x}) + \log h_0$$

14-20 In the coin experiment, the probability of "heads" is a random variable $\mathbf{p}$ with $E\{\mathbf{p}\} = 0.6$. Using the MEM, find its density $f(p)$.

14-21 (The Brandeis dice problem[12]) In a die experiment, the average number of dots up equals 4.5. Using the MEM, find $p_i = P\{f_i\}$.

14-22 Using the MEM, find the joint density $f(x_1, x_2, x_3)$ of the random variables $\mathbf{x}_1$, $\mathbf{x}_2$, and $\mathbf{x}_3$ if

$$E\{\mathbf{x}_1^2\} = E\{\mathbf{x}_2^2\} = E\{\mathbf{x}_3^2\} = 4 \qquad E\{\mathbf{x}_1\mathbf{x}_2\} = E\{\mathbf{x}_1\mathbf{x}_3\} = 1$$

14-23 A source has seven elements with probabilities

$$0.3 \qquad 0.2 \qquad 0.15 \qquad 0.15 \qquad 0.1 \qquad 0.06 \qquad 0.04$$

respectively. Construct a Shannon, a Fano, and a Huffman code and find their average code lengths.

14-24 Show that in the redundant coding of Example 14-29, the probability of error equals $\beta^2(3 - 2\beta)$.

Hint: $P\{\mathbf{y}_n = 1 \mid \mathbf{x}_n = 0\} = \beta^3 + 3\beta^2(1 - \beta)$.

14-25 Find the channel capacity of a symmetrical binary channel if the received information is always wrong.

---

[12]E. T. Jaynes: Brandeis lectures, 1962.

CHAPTER 15

MARKOV CHAINS

15-1 INTRODUCTION

Markov processes represent the simplest generalization of independent processes by permitting the outcome at any instant to depend only on the outcome that preceeds it and none before that. Thus in a Markov process $\mathbf{x}(t)$, the past has no influence on the future if the present is specified. This means that if $t_{n-1} < t_n$, then

$$P[\mathbf{x}(t_n) \le x_n \mid \mathbf{x}(t), t \le t_{n-1}] = P[\mathbf{x}(t_n) \le x_n \mid \mathbf{x}(t_{n-1})] \qquad (15\text{-}1)$$

From (15-1) it follows that if $t_1 < t_2 < \cdots < t_n$, then

$$P[\mathbf{x}(t_n) \le x_n \mid \mathbf{x}(t_{n-1}), \ldots, \mathbf{x}(t_1)] = P[\mathbf{x}(t_n) \le x_n \mid \mathbf{x}(t_{n-1})] \qquad (15\text{-}2)$$

A special kind of Markov process is a Markov chain where the system can occupy a finite or countably infinite number of states $e_1, e_2, \ldots, e_j, \ldots$ such that the future evolution of the process, once it is in a given state, depends only on the present state and not on how it arrived at that state. Both Markov chains and Markov processes can be discrete-time or continuous-time, depending on whether the time index set is discrete or continuous. This chapter is mostly concerned with the transient and steady state limiting behavior of discrete-time Markov chains. In addition, the behavior of various occupation times, first passage times, state occupancy times, and their probability distributions are of special interest. Examples 15-1 to 15-15 illustrate the abundance of Markov processes in nature and day-to-day problems.

Markov processes are named after A. A. Markov (1856–1922), who introduced this concept for discrete parameter systems with a finite number of states (1907). The theory for denumerable (countably infinite) chains was initiated by Kolmogorov (1936) followed by Doeblin (1937), Doob (1942), Levy (1951), and many others.

EXAMPLE 15-1

RANDOM WALKS

▶ The one-dimensional random walk model considered in Sec. 10-1 is a special case of a Markov chain. The sequence of Bernoulli trials $\mathbf{x}_1, \mathbf{x}_2, \ldots, \mathbf{x}_n, \ldots$, at each stage are independent, and the accumulated partial sum $\mathbf{s}_n$ in (10-1) that represents the relative position of the particle satisfies the recursion $\mathbf{s}_{n+1} = \mathbf{s}_n + \mathbf{x}_{n+1}$. Given $\mathbf{s}_n = j$, for $j = 0, \pm 1, \pm 2, \ldots, \pm n, \ldots$, the random variable $\mathbf{s}_{n+1}$ can assume only two values: $\mathbf{s}_{n+1} = j+1$ with probability p, and $\mathbf{s}_{n+1} = j-1$ with probability q. Thus

$$
\begin{aligned}
P(\mathbf{s}_{n+1} = j+1 \mid \mathbf{s}_n = j) &= p \\
P(\mathbf{s}_{n+1} = j-1 \mid \mathbf{s}_n = j) &= q
\end{aligned}
\tag{15-3}
$$

These conditional probabilities for $\mathbf{s}_{n+1}$ depend only on the values of $\mathbf{s}_n$ and are not affected by the values of $\mathbf{s}_1, \mathbf{s}_2, \ldots, \mathbf{s}_{n-1}$. ◀

EXAMPLE 15-2

BRANCHING PROCESSES

▶ Consider a population that is able to produce new offspring of like kind. For each member let p_k, $k = 0, 1, 2, \ldots$ represent the probability of creating k new members. The direct descendents of the nth generation form the $(n+1)$st generation. The members of each generation are independent of each other. Suppose $\mathbf{x}_n$ represents the size of the nth generation. It is clear that $\mathbf{x}_n$ depends only on $\mathbf{x}_{n-1}$ since $\mathbf{x}_n = \sum_{i=1}^{\mathbf{x}_{n-1}} \mathbf{y}_i$, where $\mathbf{y}_i$ represents the number of offspring of the ith member of the $(n-1)$ generation, and the manner in which the value of $\mathbf{x}_{n-1}$ was reached is of no consequence. Thus $\mathbf{x}_n$ represents a Markov chain. Nuclear chain reactions, survival of family surnames, gene mutations, and waiting lines in a queueing system are all examples of branching processes. In a nuclear chain reaction, a particle such as a neutron scores a hit with probability p, creating m new particles, and $q = 1-p$ represents the probability that it remains inactive with no descendants. In that case, the only possible number of descendants is zero and m with probabilities q and p. If p is close to one, the number of particles is likely to increase indefinitely, leading to an explosion, whereas if p is close to zero the process may never start. ◀

EXAMPLE 15-3

SURVIVAL OF FAMILY SURNAMES

▶ In the family surname survival scene, let p_k represent the probability of a newborn to become the progenitor of exactly k descendents. If p_k, $k = 0, 1, 2, \ldots$ are assumed to be constants over generations, the probability of finding a total of m carriers of the family name in the nth generation is of interest; in particular the probability of extinction of the line ($m = 0$) and the conditions under which that is possible are of special interest. This problem was first treated by Francis Galton (1873) and the first solution was given by Galton and Watson in 1874. In 1930, Steffensen gave the complete solution to the problem. In the gene mutation problem, every gene has a chance to reappear in its k direct descendants with probability p_k, $k = 1, 2, \ldots$, and a spontaneous mutation produces a single gene that plays the role of a zero-generation particle. The spread of the mutant gene through later generations follows a Markov process, and the probability of a mutant gene being present in k new offspring is of interest. ◀

EXAMPLE 15-4

WAITING LINES (QUEUES)

▶ In any type of queue or a waiting line, customers (jobs) arrive randomly and wait for service. A customer arriving when the server is free receives immediate service; otherwise the customer joins the queue and waits for service. The server continues service according to some schedule such as first in, first out, as long as there are customers in the queue waiting for service. The total duration of uninterrupted service beginning at $t = 0$ is known as the busy period. Under the assumption that various arrivals and service times are mutually independent random variables, the total number of customers during the busy period, the duration of the busy period, and the probability of its termination are of special interest. As we shall see, queues are universally characterized depending on the type of interarrival time distribution as well as the service time distribution and the total number of channels or servers employed. For example, the interarrivals and service times can be independent exponential distributions, while the number of servers may be just one or several of them in parallel.

Let $\mathbf{x}_n$ denote the number of customers (jobs) waiting in line for service at the instant t_n when the nth customer departs after completing service. If we consider the first customer arriving at an empty counter and receiving immediate service as representing the zeroth generation, then the direct descendents are the $\mathbf{x}_1$ customers arriving during the service time of the first customer and forming a waiting line. The process continues as long as the queue lasts. Referring to the family surname survival problem, it is clear that the number of customers $\mathbf{x}_n$ waiting for service at the departure instant t_n of the nth customer form a Markov chain. The probability of busy time termination corresponds to the extinction probability of a family line. Note that the Markov chain $\{\mathbf{x}_n\}$ defined here is known as an *imbedded Markov chain,* since it corresponds to observing the underlying stochastic process $\mathbf{x}(t)$, that represents the total number of customers at time t, at a sequence of random time instants $\{t_n\}$ corresponding to the instants when successive customers depart after completing service from the system. The underlying stochastic process $\mathbf{x}(t)$ in general requires additional information regarding the actual service time of each customer to predict its future bahavior and hence $\mathbf{x}(t)$ need *not be* markovian. ◀

TRANSITION PROBABILITIES. In a discrete-time Markov chain $\{\mathbf{x}_n\}$ with a finite or infinite set of states $e_1, e_2, \ldots, e_i, \ldots$, let $\mathbf{x}_n = \mathbf{x}(t_n)$ represent the state of the system at $t = t_n$. If $t_n = nT$, then for $n \geq m \geq 0$, the sequence $\mathbf{x}_m \to \mathbf{x}_{m+1} \to \cdots \mathbf{x}_n, \ldots$ represents the evolution of the system. Let

$$p_i(m) = P\{\mathbf{x}_m = e_i\} \tag{15-4}$$

represent the probability that at time $t = t_m$ the system occupies the state e_i, and

$$p_{ij}(m,n) \triangleq P\{\mathbf{x}_n = e_j \mid \mathbf{x}_m = e_i\} \tag{15-5}$$

the probability that the system goes into state e_j at $t = t_n$ given that it was in state e_i at $t = t_m$ (regardless of its behavior prior to t_m). The numbers $p_{ij}(m,n)$ represent the transition probabilities of the Markov chain from state e_i at t_m to e_j at t_n. Notice that (15-4)–(15-5) completely determine the system, since for $m < n < r$,

$$\begin{aligned} P\{\mathbf{x}_r = e_i, \mathbf{x}_n = e_j, \mathbf{x}_m = e_k\} &\\ = P\{\mathbf{x}_r = e_i \mid \mathbf{x}_n = e_j\}P\{\mathbf{x}_n = e_j \mid \mathbf{x}_m = e_k\}P\{\mathbf{x}_m = e_k\} &\\ = p_{ji}(n,r)p_{kj}(m,n)p_k(m) & \end{aligned} \tag{15-6}$$

HOMOGENEOUS CHAIN

▶ A Markov chain is said to be *homogeneous* in time if $p_{ij}(m,n)$ depends only on the difference $n - m$. In that case, the transition probabilities are said to be *stationary* and

$$P\{\mathbf{x}_{m+n} = e_j \mid \mathbf{x}_m = e_i\} \triangleq p_{ij}(n) = p_{ij}^{(n)} \tag{15-7}$$

represents the conditional probability that a homogeneous Markov chain will move from state e_i to state e_j in n steps. The one-step transition probabilities are usually denoted simply as p_{ij}. Thus

$$p_{ij} = P\{\mathbf{x}_{n+1} = e_j \mid \mathbf{x}_n = e_i\} \tag{15-8}$$

The time duration $\mathbf{y}$ that a homogeneous Markov process spends in a given state (interarrival time) must be memoryless, since the present state is sufficient to determine the future. Thus in the discrete case if the time instants t_n are uniformly placed at $t_n = nT$, then $\mathbf{y}$ satisfies the relation

$$P(\mathbf{y} > m + n \mid \mathbf{y} > m) = P(\mathbf{y} > n) \tag{15-9}$$

which shows that $\mathbf{y}$ is a geometric random variable. Thus the duration that a homogeneous discrete-time (uniform) Markov chain spends in *any* state has a geometric distribution. ◀

STOCHASTIC MATRIX

▶ It is convenient to arrange the transition probabilities $p_{ij}(m,n)$ in a matrix form $P(m,n)$ as

$$P(m,n) = \begin{pmatrix} p_{11}(m,n) & p_{12}(m,n) & \cdots & p_{1j}(m,n) & \cdots \\ p_{21}(m,n) & p_{22}(m,n) & \cdots & \cdots & \cdots \\ \vdots & \vdots & \vdots & \vdots & \vdots \\ p_{i1}(m,n) & \cdots & \cdots & p_{ij}(m,n) & \cdots \\ \cdots & \cdots & \cdots & \cdots & \cdots \end{pmatrix} \tag{15-10}$$

Clearly $P(m,n)$ is a matrix whose entries are all nonnegative, and elements in each row add to unity, since

$$\sum_j p_{ij}(m,n) = \sum_j P\{\mathbf{x}_n = e_j \mid \mathbf{x}_m = e_i\} = 1 \tag{15-11}$$

Such a matrix represents a *stochastic matrix*. As we show later, together with the initial distribution in (15-4), the transition probability matrices completely define the Markov chain. In the special case of a homogeneous Markov chain, the one step transition matrix P is given by

$$P = \begin{pmatrix} p_{11} & p_{12} & p_{13} & \cdots & p_{1j} & \cdots \\ p_{21} & p_{22} & p_{23} & \cdots & \cdots & \cdots \\ p_{31} & p_{32} & p_{33} & \cdots & \cdots & \cdots \\ \vdots & \vdots & \vdots & \vdots & \vdots & \vdots \\ p_{i1} & p_{i2} & p_{i3} & \cdots & p_{ij} & \cdots \\ \cdots & \cdots & \cdots & \cdots & \cdots & \cdots \end{pmatrix} \tag{15-12}$$

and along with the initial probability distribution

$$p_k(0) \triangleq P\{\mathbf{x}_0 = e_k\} \tag{15-13}$$

it completely defines the process. ◀

The one-step probability transition matrices for several interesting problems are given in Examples 15-5 to 15-15. Starting from some initial distribution, our immediate goal is to study the evolutionary behavior of these Markov processes.

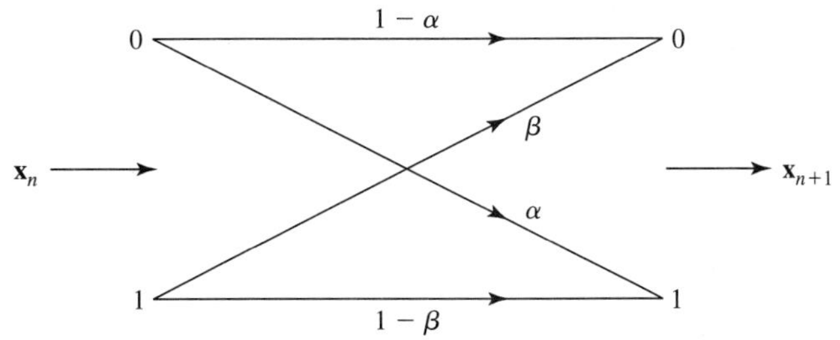

FIGURE 15-1
Binary communication channel.

EXAMPLE 15-5

BINARY COMMUNICATION CHANNEL

▶ Figure 15-1 represents a time-invariant binary communication channel, where $\mathbf{x}_n$ denotes the input and $\mathbf{x}_{n+1}$ the output. The input and the output each possess two states e_0 and e_1 that represent the two binary symbols "0" and "1," respectively. The channel delivers the input symbol to the output with a certain error probability that may depend on the symbol being transmitted. As Fig. 15-1 shows, let $\alpha < 1/2$ and $\beta < 1/2$ represent the two kinds of channel error probabilities. In a time-invariant channel, these error probabilities remain constant over various transmitted symbols so that

$$P\{\mathbf{x}_{n+1} = 1 \mid \mathbf{x}_n = 0\} = p_{01} = \alpha \qquad P\{\mathbf{x}_{n+1} = 0 \mid \mathbf{x}_n = 1\} = p_{10} = \beta \tag{15-14}$$

and the corresponding Markov chain is homogeneous.

The 2×2 homogeneous probability transition matrix P in this case is given by

$$P = \begin{pmatrix} p_{00} & p_{01} \\ p_{10} & p_{11} \end{pmatrix} = \begin{pmatrix} 1-\alpha & \alpha \\ \beta & 1-\beta \end{pmatrix} \tag{15-15}$$

In a binary symmetric channel, the two kinds of error probabilities are equal so that $\alpha = \beta = p$. ◀

EXAMPLE 15-6

RANDOM WALKS

▶ Consider a general one-dimensional random walk on the possible states $e_0, e_1, e_2, \ldots$. Let $\mathbf{s}_n$ represent the location of the particle at time n on a straight line such that at each interior state e_j, the particle either moves to the right to e_{j+1} with probability p_j, or to the left to e_{j-1} with probability q_j or remains where it is at e_j with probability r_j (see Fig. 15-2). Obviously when at state e_0, it can either stay there with probability r_0 or move to the right to e_1 with probability p_1. This gives the corresponding transition matrix P to be

$$P = \begin{pmatrix} r_0 & p_0 & 0 & 0 & 0 & \cdots \\ q_1 & r_1 & p_1 & 0 & 0 & \cdots \\ 0 & q_2 & r_2 & p_2 & 0 & \cdots \\ 0 & 0 & q_3 & r_3 & p_3 & \cdots \\ \vdots & \vdots & \vdots & \vdots & \vdots & \vdots \end{pmatrix} \tag{15-16}$$

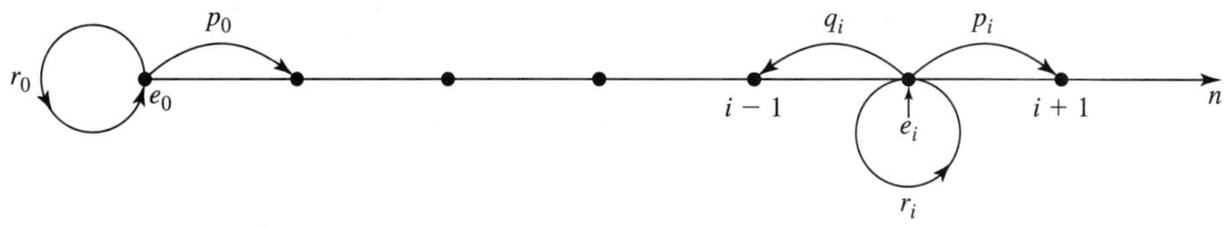

FIGURE 15-2
Random walk on a line.

with

$$r_0 + p_0 = 1 \qquad q_i + r_i + p_i = 1 \qquad i = 1, 2, \ldots \tag{15-17}$$

Thus

$$p_{00} = r_0 \qquad p_{01} = p_0 \qquad p_{0j} = 0 \qquad j > 1 \tag{15-18}$$

and for $i \geq 1$

$$p_{ij} = \begin{cases} p_i > 0 & j = i+1 \\ r_i \geq 0 & j = i \\ q_i > 0 & j = i-1 \\ 0 & \text{otherwise} \end{cases} \tag{15-19}$$

This model with $p_i = p, q_i = 1-p, r_i = 0$ for $i > 1$, and $r_0 = 1$ corresponds to the gambler's ruin problem discussed in Chap. 3 (Ex. 3-21) where one of the players is infinitely wealthy (e.g., a gambling casino). For the unrestricted one-dimensional random walk considered in Sec. 10-1, we have $p_i = p, q_i = q$, and $r_i = 0$ for all integers i (positive and negative), and the matrix P is infinitely large in all four directions there. The following special cases of the random walk are also of interest. ◀

EXAMPLE 15-7

RANDOM WALK WITH ABSORBING BARRIERS

▶ Let the number of states in a random walk be finite $(e_0, e_1, e_2, \ldots, e_N)$, and consider the special case of (15-16) given by

$$P = \begin{pmatrix} 1 & 0 & 0 & 0 & \cdot & \cdot & \cdot & 0 \\ q & 0 & p & 0 & 0 & \cdot & \cdot & 0 \\ 0 & q & 0 & p & 0 & \cdot & \cdot & 0 \\ \vdots & \vdots & \vdots & \vdots & \vdots & \vdots & \vdots & \vdots \\ 0 & 0 & \cdot & \cdot & \cdot & q & 0 & p \\ 0 & 0 & \cdot & \cdot & \cdot & 0 & 0 & 1 \end{pmatrix} \tag{15-20}$$

Thus from the interior states $e_1, e_2, \ldots, e_{N-1}$, transitions to the left and right neighbors are possible with probabilities q and p, respectively, while no transition is possible from e_0 and e_N to any other state. The system may move from one interior state to the other, but once it reaches a boundary it stays there forever (the particle gets absorbed). It is easy to see that the gambler's ruin problem discussed in Example 3-21, where both players have finite wealth, corresponds to this case with $N = a + b$. In that case the game starts from the fixed point a (state e_a) of the interval $(0, a+b)$, which corresponds to the initial distribution $P\{\mathbf{x} = e_a\} = 1$ and zero otherwise in (15-13). On the other hand, if the initial state was randomly chosen, that would correspond to the distribution $P\{\mathbf{x}_0 = e_k\} = 1/(N+1)$. In Example 15-26 (page 747) we shall make use of this model to analyze the game of tennis in detail. ◀

EXAMPLE 15-8

RANDOM WALK WITH REFLECTING BARRIERS

▶ Suppose the two boundaries in Example 15-7 reflect the particle back to the adjacent state instead of absorbing it. With $e_1, e_2, \ldots, e_N$ representing the N states, the end reflection probabilities to the right and left are given by

$$p_{1,2} = p \qquad \text{and} \qquad p_{N,N-1} = q \tag{15-21}$$

and this gives the $N \times N$ transition matrix to be

$$P = \begin{pmatrix} q & p & 0 & 0 & \cdot & \cdot & \cdot & 0 \\ q & 0 & p & 0 & 0 & \cdot & \cdot & \cdot \\ 0 & q & 0 & p & 0 & \cdot & \cdot & \cdot \\ \vdots & & & & & \vdots & \vdots & \vdots \\ 0 & 0 & 0 & \cdot & \cdot & q & 0 & p \\ 0 & 0 & 0 & \cdot & \cdot & 0 & q & p \end{pmatrix} \tag{15-22}$$

In gambling, this corresponds to a *fun* game where every time a player loses the game, his adversary returns just the stake amount so that the game is kept alive and it continues forever. ◀

EXAMPLE 15-9

CYCLIC RANDOM WALKS

▶ Here the two end-boundary states e_0 and e_{N-1} loop together to form a circle so that e_{N-1} has neighbors e_0 and e_{N-2} (Fig. 15-3). The random walk continues on this circular boundary by passing from one state either to the right or left neighbor and this corresponds to the following $N \times N$ transition matrix

$$P = \begin{pmatrix} 0 & p & 0 & 0 & \cdot & \cdot & \cdot & 0 & q \\ q & 0 & p & 0 & \cdot & \cdot & \cdot & \cdot & 0 \\ 0 & q & 0 & p & 0 & \cdot & \cdot & \cdot & 0 \\ \vdots & & & & & & & & \vdots \\ 0 & 0 & \cdot & \cdot & \cdot & \cdot & q & 0 & p \\ p & 0 & \cdot & \cdot & \cdot & \cdot & 0 & q & 0 \end{pmatrix} \tag{15-23}$$

More generally, if we permit transition between any two states $e_0, e_1, \ldots, e_{N-1}$, then since moving k steps to the right on a circle is the same as moving $N-k$ to the left (Fig 15-3), we obtain the following circulant transition matrix

$$P = \begin{pmatrix} q_0 & q_1 & q_2 & \cdots & q_{N-1} \\ q_{N-1} & q_0 & q_1 & \cdots & q_{N-2} \\ q_{N-2} & q_{N-1} & q_0 & \cdots & q_{N-3} \\ \vdots & & & & \vdots \\ q_1 & q_2 & \cdots & q_{N-1} & q_0 \end{pmatrix} \tag{15-24}$$

Here

$$q_k = P\{\mathbf{x}_{n+1} = e_{i+k} \mid \mathbf{x}_n = e_i\} = P\{\mathbf{x}_{n+1} = e_{i-(N-k)} \mid \mathbf{x}_n = e_i\} \tag{15-25}$$

◀

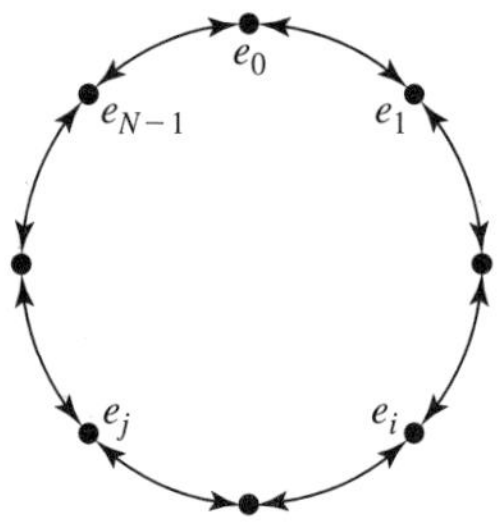

FIGURE 15-3
Cyclic Random Walk in (15-23).

EXAMPLE 15-10

EHRENFEST'S DIFFUSION MODEL (NON-UNIFORM RANDOM WALK)

▶ Let N represent the combined population of two cities A and B. Suppose migration occurs between the cities one at a time with probability proportional to the population of the city, and let the population of A determine the state of the system. Then $e_0, e_1, \ldots, e_N$ represent the possible states, and from state e_k, at the next step, A can move into either e_{k-1} or e_{k+1} with probabilities k/N or $1 - k/N$, respectively. Thus

$$P = \begin{pmatrix} 0 & 1 & 0 & \cdot & \cdot & \cdot & \cdot & 0 \\ p & 0 & 1-p & 0 & \cdot & \cdot & \cdot & 0 \\ 0 & 2p & 0 & 1-2p & \cdot & \cdot & \cdot & 0 \\ 0 & 0 & 3p & 0 & 1-3p & 0 & \cdot & 0 \\ \cdot & \cdot & \cdot & \cdot & \cdot & \cdot & \cdot & \cdot \\ 0 & 0 & \cdot & \cdot & \cdot & 1-p & 0 & p \\ 0 & 0 & \cdot & \cdot & \cdot & 0 & 1 & 0 \end{pmatrix} \tag{15-26}$$

where $p = 1/N$. We can also think of this model as a random walk with totally reflective barriers where the probability of a step varies with the position or state. From (15-26), if $k < N/2$, the particle is more likely to move to the right, while if $k > N/2$, it is more likely to move to the left. Thus the particle has a tendency to move toward the center, which should correspond to an equilibrium distribution. ◀

EXAMPLE 15-11

SUCCESS RUNS (RANDOM WALK)

▶ Consider a type of one-dimensional random walk over $0, 1, 2, \ldots,$ where the particle moves from i to $i+1$ with probability p or moves back to the origin with probability q. This gives

$$p_{ij} = \begin{cases} p & j = i+1 \\ q & j = 0 \\ 0 & \text{otherwise} \end{cases} \tag{15-27}$$

Thus at the nth trial the system is in state e_i only if the previous failure occurred at $n - i$, and the index i represents the number of uninterrupted successes up to the nth trial. More generally, we can let

$$p_{ij} = \begin{cases} p_i & j = i+1 \\ q_i & j = 0 \\ 0 & \text{otherwise} \end{cases} \tag{15-28}$$

where $p_i + q_i = 1$. In this case the probability that the time between two successive returns to zero equals k is given by the product $p_1 p_2 \cdots p_{k-1} q_k$. ◀

EXAMPLE 15-12

RANDOM OCCUPANCY

▶ Consider N cells (compartments) and a sequence of independent trials where a new ball is placed at random into one of the cells at each trial. Each cell can hold multiple balls. Let e_k, $k = 0, 1, \ldots, N$, represent the state where k cells are occupied (and $N - k$ cells are empty). At the next trial, the next ball can go into one of the occupied cells ($e_k \rightarrow e_k$) with probability k/N or an empty cell ($e_k \rightarrow e_{k+1}$) with probability $(N-k)/N$. This gives the transition probabilities to be

$$p_{kk} = \frac{k}{N} \qquad p_{k,k+1} = 1 - \frac{k}{N} \qquad p_{kj} = 0 \qquad j \neq k \quad j \neq k+1 \tag{15-29}$$

An all empty initial distribution in this case corresponds to $P\{\mathbf{x}_0 = e_0\} = 1$, $P\{\mathbf{x}_0 = e_k\} = 0$, $k \neq 0$ in (15-13).

This model can be used to study an interesting birthday statistics problem: What is the minimum number of people required in a random group so that every day is a birthday for someone in that group (with probability p)? (See Example 15-18 for solution.) ◀

EXAMPLE 15-13

GENETICS

▶ Suppose each cell of an organism contains two types of genes A and B, where the total number of genes in each cell adds up to N. The cell is in state e_j, $j = 0, 1, 2, \ldots, N$, if it contains exactly j genes of type A and $N - j$ genes of type B. Prior to cell division each gene duplicates itself so that the two new cells in the next generation each inherit N genes chosen at random from the pool of $2j$ genes of type A and $2N - 2j$ genes of type B. The probability that a new cell has moved into state e_k is given by the hypergeometric distribution

$$p_{jk} = \frac{\binom{2j}{k}\binom{2N-2j}{N-k}}{\binom{2N}{N}} \qquad j, k = 0, 1, \ldots, \quad \max(0, 2j - N) \leq k \leq \min(2j, N) \tag{15-30}$$

In another genetic model, let e_j represent the present state as defined above so that the probability of selecting a gene of type A in the next generation is simply $p = j/N$. Suppose the N genes in the next generation are determined by random selection resulting from N Bernoulli trails with the A-gene probability equal to p. In that case, the transition probability that the next generation has moved into state e_k (k genes of type A and $N - k$ genes of type B) from state e_j is given by the binomial distribution with

$$p_{jk} = \binom{N}{k}\left(\frac{j}{N}\right)^k \left(1 - \frac{j}{N}\right)^{N-k} \qquad j, k = 0, 1, \ldots, N \tag{15-31}$$

It will be interesting to study the limiting behavior of the population based on these models after several generations[1]. Notice that in both models, the states e_0 and e_N contain genes of the same type, and no exit from these states is possible. (For further generalization and insight into these models, see Appendix 15A.) ◀

[1]This problem was originally studied by R. A. Fisher and S. Wright in connection with the evolution of cornfield population.

EXAMPLE 15-14

IMBEDDED MARKOV CHAINS

▶ Consider an imbedded Markov chain $\{\mathbf{x}_n\}$ that represents the number of customers or jobs waiting for service in a queue at the departure of the nth customer, as defined in Example 15-4. To determine the transition probabilities, let $\mathbf{y}_n$ denote the number of customers arriving at the queue during the service time of the nth customer. Then the number of customers waiting for service at the departure of the $(n+1)$st customer is given by

$$\mathbf{x}_{n+1} = \begin{cases} \mathbf{x}_n + \mathbf{y}_{n+1} - 1 & \mathbf{x}_n \neq 0 \\ \mathbf{y}_{n+1} & \mathbf{x}_n = 0 \end{cases} \tag{15-32}$$

If the queue was empty at the departure of the nth customer, then $\mathbf{x}_n = 0$, and the next customer to arrive is the $(n+1)$st one. During that service, $\mathbf{y}_{n+1}$ customers arrive, so that $\mathbf{x}_{n+1} = \mathbf{y}_{n+1}$, otherwise the number of customers left behind by the $(n+1)$st customer is $\mathbf{x}_n - 1 + \mathbf{y}_{n+1}$.

We can use (15-32) to compute the transition probabilities of an imbedded Markov chain. Thus

$$p_{ij} = P\{\mathbf{x}_{n+1} = j \mid \mathbf{x}_n = i\} = \begin{cases} P\{\mathbf{y}_{n+1} = j\} \stackrel{\Delta}{=} a_j & i = 0 \\ P\{\mathbf{y}_{n+1} = j - i + 1\} \stackrel{\Delta}{=} a_{j-i+1} & i \geq 1 \end{cases} \tag{15-33}$$

and the probability transition matrix is given by

$$P = \begin{pmatrix} a_0 & a_1 & a_2 & a_3 & \cdot & \cdot & \cdot \\ a_0 & a_1 & a_2 & a_3 & \cdot & \cdot & \cdot \\ 0 & a_0 & a_1 & a_2 & \cdot & \cdot & \cdot \\ 0 & 0 & a_0 & a_1 & \cdot & \cdot & \cdot \\ 0 & 0 & 0 & a_0 & \cdot & \cdot & \cdot \\ \vdots & \vdots & \vdots & \vdots & \vdots & \vdots & \vdots \end{pmatrix} \tag{15-34}$$

Recall that the underlying stochastic process $\mathbf{x}(t)$ need not be markovian, whereas the sequence $\mathbf{x}_n = \mathbf{x}(t_n)$ generated at the random departure time instants t_n is markovian. Thus the method of imbedded Markov chains converts a non-markovian problem into markovian, and the limiting behavior of the imbedded chain can be used to study the underlying process $\mathbf{x}(t)$ since it has been shown that when such limiting behavior exists (Cohen [38], Khintchin [39]) (see also page 810)

$$\lim_{t\to\infty} P\{\mathbf{x}(t) = k\} = \lim_{n\to\infty} P\{\mathbf{x}_n = k\} \tag{15-35}$$

that is, for certain queues after a long time, the limiting behavior at an arbitrary time instant t is the same as those at the random departure instants. ◀

EXAMPLE 15-15

DOUBLY STOCHASTIC MATRICES

▶ A transition probability matrix P is called *doubly stochastic* if in addition to the row sums, the column sums are also unity. For example, a symmetric binary communication channel as in (15-15) with $\alpha = \beta$ corresponds to a doubly stochastic matrix. ◀

15-2 HIGHER TRANSITION PROBABILITIES AND THE CHAPMAN–KOLMOGOROV EQUATION

The transition probability function of any Markov chain $\{\mathbf{x}_n\}$ satisfies the Chapman–Kolmogorov equation. We can use the basic Markov relation in (15-6) to derive this fundamental equation that governs the evolution of all chains. For $n > r > m$, we have

$$\begin{aligned} P\{\mathbf{x}_n = e_j, \mathbf{x}_m = e_i\} &= \sum_k P\{\mathbf{x}_n = e_j, \mathbf{x}_r = e_k, \mathbf{x}_m = e_i\} \\ &= \sum_k P\{\mathbf{x}_n = e_j \mid \mathbf{x}_r = e_k, \mathbf{x}_m = e_i\} P\{\mathbf{x}_r = e_k, \mathbf{x}_m = e_i\} \\ &= \sum_k P\{\mathbf{x}_n = e_j \mid \mathbf{x}_r = e_k\} P\{\mathbf{x}_r = e_k, \mathbf{x}_m = e_i\} \end{aligned}$$

Thus

$$\begin{aligned} p_{ij}(m, n) &= P\{\mathbf{x}_n = e_j \mid \mathbf{x}_m = e_i\} \\ &= \sum_k P\{\mathbf{x}_n = e_j \mid \mathbf{x}_r = e_k\} P\{\mathbf{x}_r = e_k \mid \mathbf{x}_m = e_i\} \end{aligned} \tag{15-36}$$

or

$$p_{ij}(m, n) = \sum_k p_{ik}(m, r) p_{kj}(r, n) \tag{15-37}$$

In terms of the probability transition matrices in (15-10), this relation reduces to

$$P(m, n) = P(m, r) P(r, n) \tag{15-38}$$

where $m < r < n$, and by letting $r = m + 1, m + 2, \ldots$ we get

$$P(m, n) = P(m, m + 1) P(m + 1, m + 2) \cdots P(n - 1, n) \tag{15-39}$$

Thus to obtain $P(m, n)$ for all $n \geq m$, it is sufficient to know the one-step transition probability matrices

$$P(0, 1), P(1, 2), P(2, 3), \ldots, P(n, n + 1), \ldots \tag{15-40}$$

For a homogeneous Markov chain, all transition probability matrices in (15-39) and (15-40) are equal to P in (15-12), so that (15-39) reduces to

$$P(m, n) = P^{n-m} \tag{15-41}$$

n-STEP TRANSITION PROBABILITIES

▶ From (15-7), for a homogeneous chain $p_{ij}^{(n)}$ represents the (i, j)th entry of $P(0, n) = P^n$. Thus

$$P^n \triangleq \left(p_{ij}^{(n)}\right) \tag{15-42}$$

and since $P^{n+m} = P^m P^n = P^n P^m$, we obtain the useful relation

$$p_{ij}^{(m+n)} = \sum_k p_{ik}^{(m)} p_{kj}^{(n)} = \sum p_{ik}^{(n)} p_{kj}^{(m)} \tag{15-43}$$

In particular, the one-step recursion relation is given by

$$p_{ij}^{(n+1)} = \sum_k p_{ik} p_{kj}^{(n)} = \sum_k p_{ik}^{(n)} p_{kj} \tag{15-44}$$

Finally, the unconditional probability distribution at $t = nT$ is given by

$$\begin{aligned} p_j(n) = P\{\mathbf{x}_n = e_j\} &= \sum_i P\{\mathbf{x}_n = e_j \mid \mathbf{x}_m = e_i\} P\{\mathbf{x}_m = e_i\} \\ &= \sum_i p_{ij}(m, n) p_i(m) \end{aligned} \tag{15-45}$$

or

$$p(n) = p(m) P(m, n) \tag{15-46}$$

where

$$p(n) \triangleq [p_1(n), p_2(n), \ldots, p_j(n), \ldots] \tag{15-47}$$

For a homogeneous chain, (15-46) and (15-47) reduce to

$$p(n) = p(0) P^n \tag{15-48}$$

◀

In general, it is difficult to obtain explicit formulas for the n-step transition probabilities $p_{ij}^{(n)}$. However, for a homogeneous Markov chain with finitely many states $e_1, e_2, \ldots, e_N$, the transition matrix P is $N \times N$ and certain simplifications are possible.

Because the matrix P is $N \times N$, it has N eigenvalues $\lambda_1, \lambda_2, \ldots, \lambda_N$. For simplicity, we shall slightly restrict the generality and begin by focusing on the case where the eigenvalues are simple (distinct) and nonzero. The assumption that the eigenvalues are distinct is satisfied in many practical cases, except for decomposable (reducible) and periodic chains, and they require only minor changes to modify. However, zero could be among the eigenvalues, and if it is of multiplicity one, as we show later, it is easy to modify that case also. Under these assumptions, let (λ_i, u_i), $i = 1, 2, \ldots, N$ represent the N eigenvalue–eigenvector pairs for P. Thus

$$P u_i = \lambda_i u_i \qquad i = 1 \to N \tag{15-49}$$

or

$$PU = U\Lambda \tag{15-50}$$

where the square matrix U is given by

$$U \triangleq [u_1, u_2, \ldots, u_N] \tag{15-51}$$

and

$$\Lambda \triangleq \begin{pmatrix} \lambda_1 & & & 0 \\ & \lambda_2 & & \\ & & \ddots & \\ 0 & & & \lambda_N \end{pmatrix} \tag{15-52}$$

Since u_i, $i = 1, 2, \ldots, N$ are $N \times 1$ linearly independent column vectors, U is an $N \times N$ nonsingular matrix. Hence from (15-50),

$$P = U \Lambda U^{-1} \triangleq U \Lambda V \tag{15-53}$$

or

$$VP = \Lambda V \tag{15-54}$$

where

$$V \triangleq U^{-1} = \begin{bmatrix} v_1 \\ v_2 \\ \vdots \\ v_N \end{bmatrix} \tag{15-55}$$

with v_k, $k = 1, 2, \ldots, N$, representing the kth row vector of V. Thus

$$VU = I \quad \text{or} \quad v_k u_k = 1, \qquad v_i u_k = 0 \quad i \neq k \quad i, k = 1, 2, \ldots, N \tag{15-56}$$

From (15-53) we also obtain

$$P^n = U \Lambda^n V = \sum_{k=1}^{N} \lambda_k^n u_k v_k \tag{15-57}$$

or

$$p_{ij}^{(n)} = \sum_{k=1}^{N} \lambda_k^n u_{ik} v_{kj} \tag{15-58}$$

To summarize, from (15-50) and (15-54) for each eigenvalue λ_k, $k = 1, 2, \ldots, N$, the vectors u_k and v_k satisfy two sets of N linear equations given by

$$\sum_{j=1}^{N} p_{ij} x_j^{(k)} = \lambda_k x_i^{(k)} \tag{15-59}$$

and

$$\sum_{i=1}^{N} y_i^{(k)} p_{ij} = \lambda_k y_j^{(k)} \tag{15-60}$$

respectively. To start with, one can obtain the eigenvalues λ_k for $k = 1, 2, \ldots, N$ by solving the characteristic equation $\det(P - \lambda I) = 0$. For each λ_k, obtain the $\{x_i^{(k)}\}$ and $\{y_i^{(k)}\}$ vector components from (15-59) and (15-60). The normalization condition in (15-56) gives

$$v_k u_k = c_k \sum_{i=1}^{N} x_i^{(k)} y_i^{(k)} = 1$$

or

$$c_k = \frac{1}{\sum_{i=1}^{N} x_i^{(k)} y_i^{(k)}} \tag{15-61}$$

Finally in terms of $x^{(k)}$, $y^{(k)}$, and c_k, we may rewrite (15-58) as

$$p_{ij}^{(n)} = \sum_{k=1}^{N} c_k \, \lambda_k^n \, x_i^{(k)} y_j^{(k)} \tag{15-62}$$

If one of the eigenvalues of P is zero (with multiplicity one), then by letting $\lambda_N = 0$, the representation in (15-62) is seen to be valid for all $n \geq 1$, and for $n = 0$ the zero eigenvalue contributes an additional constant term to $p_{ij}^{(n)}$. Thus $p_{ij}^{(n)}$ is the sum of $N - 1$ terms in (15-62) for $n \geq 1$ if P is singular with a simple zero eigenvalue. Next, we illustrate this procedure for obtaining the higher transition probabilities $p_{ij}^{(n)}$ through several examples.

EXAMPLE 15-16

BINARY COMMUNICATION CHANNEL

▶ Referring to the nonsymmetric binary communication channel model in Example 15-5, we get the characteristic equation of the transition matrix in (15-15) to be

$$\det(P-\lambda I) = \begin{vmatrix} 1-\alpha-\lambda & \alpha \\ \beta & 1-\beta-\lambda \end{vmatrix}$$
$$= \lambda^2 - \lambda(2-\alpha-\beta) + (1-\alpha-\beta) = 0 \qquad (15\text{-}63)$$

By inspection, $\lambda_1 = 1$ and $\lambda_2 = 1-\alpha-\beta < 1$ are the two eigenvalues of P. By solving (15-59) and (15-60), after normalization, we get

$$U = \begin{pmatrix} 1 & -\alpha \\ 1 & \beta \end{pmatrix} \qquad V = \frac{1}{\alpha+\beta}\begin{pmatrix} \beta & \alpha \\ -1 & 1 \end{pmatrix}$$

This gives the n-step transition probability matrix as

$$P^n = \frac{1}{\alpha+\beta}\begin{pmatrix} 1 \\ 1 \end{pmatrix}(\beta, \alpha) + \frac{(1-\alpha-\beta)^n}{\alpha+\beta}\begin{pmatrix} -\alpha \\ \beta \end{pmatrix}(-1, 1)$$
$$= \frac{1}{\alpha+\beta}\begin{pmatrix} \beta & \alpha \\ \beta & \alpha \end{pmatrix} + \frac{(1-\alpha-\beta)^n}{\alpha+\beta}\begin{pmatrix} \alpha & -\alpha \\ -\beta & \beta \end{pmatrix} \qquad (15\text{-}64)$$

Notice that physically P^n corresponds to the transition probability matrix of a cascade of n binary channels each given by (15-15).

We can use (15-64) to compute the probability that a digit arriving as 1 through this cascaded channel was in fact transmitted as 1. From Bayes' theorem, this is given by

$$P\{\mathbf{x}_0 = 1 \mid \mathbf{x}_n = 1\} = \frac{P\{\mathbf{x}_n = 1 \mid \mathbf{x}_0 = 1\}\, P\{\mathbf{x}_0 = 1\}}{P\{\mathbf{x}_n = 1\}}$$
$$= \frac{p_{11}^{(n)}\, p_1(0)}{p_1(n)} = \frac{p_{11}^{(n)}\, p}{p_{11}^{(n)}\, p + p_{01}^{(n)}\, q}$$
$$= \frac{[\alpha + (1-\alpha-\beta)^n \beta]\, p}{\alpha + (1-\alpha-\beta)^n(\beta p - \alpha q)} \qquad (15\text{-}65)$$

where $p \triangleq P\{\mathbf{x}_0 = 1\}$ and $q \triangleq P\{\mathbf{x}_0 = 0\}$. Notice that as $n \to \infty$, $P\{\mathbf{x}_0 = 1 \mid \mathbf{x}_n = 1\}$ as well as $P\{\mathbf{x}_0 = 1 \mid \mathbf{x}_n = 0\}$ tend to its unconditional value p implying that even if the individual error probabilities α and β are negligibly small, too many sections in cascade tend to increase the overall unreliability of the channel making the final output practically useless in terms of containing any useful information. ◀

EXAMPLE 15-17

CYCLIC RANDOM WALK

▶ The transition matrix in (15-23) corresponding to a cyclic random walk is a special case of the more general circulant transition matrix in (15-24). For an $N \times N$ circulant matrix with first row equal to $q_0, q_1, \ldots, q_{N-1}$, its eigenvalues and eigenvectors are given by[2]

$$\lambda_m = \sum_{i=0}^{N-1} q_i e^{j2\pi i m/N} \qquad m = 0, 1, \ldots, N-1$$

---

[2] "Matrix computations," by G. H. Golub and C. F. Van Loan, Johns Hopkins Press, 3rd Ed., 1996.

and

$$x_i^{(m)} = e^{j2\pi im/N} \qquad y_i^{(m)} = e^{-j2\pi im/N}$$

Notice that the eigenvalues represent the discrete Fourier transform (DFT) of the sequence $\{q_i\}$, and $x_i^{(m)}$ and $y_i^{(m)}$ represent the DFT vectors, so that $c_m = 1/N$, and using (15-62)

$$p_{ik}^{(n)} = \frac{1}{N}\left\{1 + \sum_{m=1}^{N-1} \lambda_m^n \, e^{j2\pi m(i-k)/N}\right\} \tag{15-66}$$

For the circulant random walk model in (15-23), the eigenvalues are given by

$$\lambda_0 = 1, \quad \lambda_m = \theta^m\left(p + q\theta^{m(N-2)}\right) \qquad m = 1, 2, \ldots, N-1 \tag{15-67}$$

where $\theta = e^{j2\pi/N}$. If N is even, we have $N = 2K$ so that $\theta = e^{j\pi/K}$, which gives $\lambda_K = \theta^K(p + qe^{j2\pi(K-1)}) = \theta^K = -1$. In particular for $N = 4$, we get $\lambda_0 = 1$, $\lambda_1 = j(p-q)$, $\lambda_2 = -1$, and $\lambda_3 = -j(p-q)$. ◀

EXAMPLE 15-18

RANDOM OCCUPANCY

▶ Referring to Example 15-12, the system is in state e_i if there are i occupied cells and $N - i$ empty cells. If n additional balls are placed at random into this situation, then $p_{ij}^{(n)}$ represents the probability that there will be j occupied cells and $N - j$ empty cells. Clearly $j \geq i$ and hence $p_{ij}^{(n)} = 0$ if $j < i$.

From (15-29), we have $p_{ii} = i/N$ and $p_{i,i+1} = (N-i)/N$ so that (15-59) reduces to (suppressing the superscripts)

$$(N\lambda - i)x_i = (N - i)x_{i+1} \tag{15-68}$$

For $\lambda = 1$, this gives $x_i = 1$ for all i. When $\lambda \neq 1$, with $i = N$ in (15-68) we obtain $x_N = 0$, and by direct substitution of $i = N - 1$ in (15-68) we get $x_{N-1} = 0$, and so on. Since the eigenvectors are not identically zero vectors, for each such eigenvalue there must exist an integer k such that $x_{k+1} = 0$ but $x_k \neq 0$. In that case from (15-68) we get $N\lambda - k = 0$, and hence the eigenvalues are given by

$$\lambda_k = \frac{k}{N} \qquad k = 1, 2, \ldots, N \tag{15-69}$$

The corresponding solutions for (15-68) are given by

$$(k - i)x_i^{(k)} = (N - i)x_{i+1}^{(k)} \qquad i \leq k$$

or

$$x_i^{(k)} = \frac{(k-i+1)}{(N-i+1)}x_{i-1}^{(k)} = \frac{(N-i)!}{N!}\,\frac{k!}{(k-i)!}$$

$$= \begin{cases} \dfrac{\binom{k}{i}}{\binom{N}{i}} & i \leq k \\ 0 & i > k \end{cases} \tag{15-70}$$

Similarly for $\lambda_k = k/N$, the system of equations in (15-60) reduces to

$$y_{j-1}^{(k)} p_{j-1,j} + y_j^{(k)} p_{jj} = \lambda_k y_j^{(k)} \tag{15-71}$$

or

$$(k-j)y_j^{(k)} = (N-j+1)y_{j-1}^{(k)}$$

which gives $y_{k-1}^{(k)} = 0$ and so on. Thus

$$y_j^{(k)} = \begin{cases} 0 & j < k \\ (-1)^{j-k}\binom{N-k}{j-k} & j \geq k \end{cases} \tag{15-72}$$

Since $x_i^{(k)} = 0$ for $i > k$ and $y_i^{(k)} = 0$ for $i < k$, from (15-61) we get $c_k = \binom{N}{k}$. Substituting (15-70) and (15-72) into (15-62), we get

$$\begin{aligned} p_{ij}^{(n)} &= \sum_{k=i}^{j} \lambda_k^n c_k x_i^{(k)} y_j^{(k)} \\ &= \sum_{k=i}^{j} \left(\frac{k}{N}\right)^n (-1)^{j-k} \binom{N}{k}\binom{k}{i}\binom{N-k}{j-k} \Big/ \binom{N}{i} \\ &= \sum_{k=i}^{j} \left(\frac{k}{N}\right)^n \frac{(N-i)!}{(N-j)!} \frac{(-1)^{j-k}}{(j-k)!(k-i)!} \\ &= \binom{N-i}{N-j} \sum_{k=i}^{j} (-1)^{j-k} \left(\frac{k}{N}\right)^n \binom{j-i}{k-i} \\ &= \binom{N-i}{N-j} \sum_{r=0}^{j-i} (-1)^{j-i-r} \left(\frac{r+i}{N}\right)^n \binom{j-i}{r} \qquad j \geq i \end{aligned} \tag{15-73}$$

and $p_{ij}^{(n)} = 0$ for $j < i$. In particular, for an all empty initial state ($i = 0$), Eq. (15-73) reduces to

$$p_{0,j}^{(n)} = \binom{N}{N-j} \sum_{r=0}^{j} \left(\frac{r}{N}\right)^n \binom{j}{r} (-1)^{j-r} \tag{15-74}$$

and it represents the probability of finding j cells occupied (or $N - j$ cells empty) when n balls are distributed randomly among N initially empty cells. We can use this formula to answer the particular birthday problem raised in Example 15-12. To simplify (15-74) further, let $m = N - j$ represent the number of empty cells at stage n, and define the new variable $\nu = N - m - r$ so that

$$\begin{aligned} p_{0,j}^{(n)} &= \binom{N}{m} \sum_{\nu=0}^{N-m} (-1)^{\nu} \binom{N-m}{\nu} \left(1 - \frac{m+\nu}{N}\right)^n \\ &= \frac{1}{m!} \sum_{\nu=0}^{N-m} \frac{(-1)^{\nu}}{\nu!} \frac{N!}{(N-m-\nu)!} \left(1 - \frac{m+\nu}{N}\right)^n \end{aligned} \tag{15-75}$$

where $j = N - m$. To derive the limiting form of (15-75) observe that when $N \to \infty$, and $n \to \infty$,

$$\frac{N!}{(N-m-\nu)!} = \prod_{k=0}^{m+\nu-1} (N-k)$$

$$= N^{m+\nu} \prod_{k=0}^{m+\nu-1} \left(1 - \frac{k}{N}\right) \to N^{m+\nu} \tag{15-76}$$

and

$$\left(1 - \frac{m+\nu}{N}\right)^n \to e^{-(m+\nu)n/N} \tag{15-77}$$

Substituting these into (15-75), we obtain

$$\lim_{N,n\to\infty} p_{0,N-m}^{(n)} = \lim_{N,n\to\infty} \frac{1}{m!} \sum_{\nu=0}^{N-m} \frac{(-1)^\nu}{\nu!} N^{m+\nu} e^{-(m+\nu)n/N}$$

$$= \frac{(Ne^{-n/N})^m}{m!} \sum_{\nu=0}^{\infty} \frac{(-Ne^{-n/N})^\nu}{\nu!}$$

$$= \frac{\lambda^m}{m!} e^{-\lambda} \tag{15-78}$$

where

$$\lambda \triangleq Ne^{-n/N} \tag{15-79}$$

a formula originally derived by R. von Mises ([3], Vol. I). To summarize, if N and n increase such that λ in (15-79) remains *constrained,* then the probability of finding m empty cells when n balls are distributed randomly among N (initially empty) cells is given by the Poisson distribution in (15-78). For example, the probability that all 365 days in a year correspond to birthdays in a population of size 2000 is given by $e^{-\lambda} = e^{-1.5226} = 0.2181$ [with $m = 0$ in (15-78)], where $\lambda = 365e^{-2000/365} = 1.5226$. However, the probability of finding 3 days in a year that are not birthdays in that group is only 0.128. For fixed λ, from (15-79),

$$n = N \log N + N \log(1/\lambda) \tag{15-80}$$

and in that case, when n balls are distributed among N initially empty cells, the probability of finding all N cells occupied is given by $e^{-\lambda}$. It follows that for all 365 days in a year to be birthdays with 98% probability for at least someone in a crowd, we get $\lambda = \log(1/0.98) = 0.0202$, and from (15-80) the size of that crowd should be around 3500.

By the same token, a bank with an average daily volume of 500 customers is guaranteed to be busy (with 60% probability) throughout the day with a customer arriving every 5 minutes (here "a day" has 8 hours with 96 slots of 5 minutes duration so that $N = 96$, $n = 500$ gives $\lambda = 0.525$ and $p = e^{-\lambda} = 0.591$). If it takes on the average 10 minutes of service time/customer, then at least two employees must be dedicated for customer service. ◀

EXAMPLE 15-19

RANDOM WALK WITH REFLECTING BARRIERS

▶ Referring to Example 15-8, let $e_1, e_2, \ldots, e_N$ represent the states with two end-reflecting barriers and transition matrix as in (15-22). Thus $p_{i,i+1} = p$, $p_{i,i-1} = q$ for $2 \le i \le N-1$, $p_{11} = q$, $p_{12} = p$, and $p_{N,N-1} = q$, $p_{NN} = p$. Substituting these into (15-59) we get

$$x_1 = s(qx_1 + px_2) \tag{15-81}$$

$$x_i = s(qx_{i-1} + px_{i+1}) \qquad i = 2, 3, \ldots, N-1 \tag{15-82}$$

$$x_N = s(qx_{N-1} + px_N) \tag{15-83}$$

where we have used $s = 1/\lambda$. Clearly $\lambda = 1$ corresponds to the specific solution $x_i = 1$. To find all other solutions, notice that (15-82) satisfies the particular solution

$$x_i = \xi^i$$

provided ξ is a root of the quadratic equation

$$\xi = qs + ps\xi^2 \tag{15-84}$$

The two roots of this equation are

$$\xi_1(s) = \frac{1 + \sqrt{1 - 4pqs^2}}{2ps} \qquad \xi_2(s) = \frac{1 - \sqrt{1 - 4pqs^2}}{2ps} \tag{15-85}$$

and the general solution to (15-82) is given by

$$x_i = a(s)\xi_1^i(s) + b(s)\xi_2^i(s) \qquad i = 2, \ldots, N-1 \tag{15-86}$$

where $a(s)$ and $b(s)$ are yet to be determined. For (15-81) to satisfy (15-86), it must have the same form as (15-82), and hence we must have $x_1 = x_0$. Similarly for (15-83) to satisfy (15-86), on comparing it with (15-82), we must have $x_N = x_{N+1}$. But

$$x_1 = x_0 \Rightarrow a(s)[1 - \xi_1(s)] = -b(s)[1 - \xi_2(s)] \tag{15-87}$$

and

$$x_N = x_{N+1} \Rightarrow a(s)[1 - \xi_1(s)]\xi_1^N(s) = -b(s)[1 - \xi_2(s)]\xi_2^N(s) \tag{15-88}$$

and from (15-87) and (15-88) we must have

$$\xi_1^N(s) = \xi_2^N(s) \qquad \text{with} \quad \xi_1(s) \ne \xi_2(s) \tag{15-89}$$

But from (15-84), we get $\xi_1(s)\xi_2(s) = q/p$ so that (15-89) reduces to

$$\left[\sqrt{p/q}\,\xi_1(s)\right]^{2N} = 1$$

Thus $\sqrt{p/q}\,\xi_1(s)$ is a $2N$th root of unity, and hence

$$\xi_1(s_k) = \sqrt{q/p}\, e^{jk\pi/N} \qquad 0 \ge k \ge 2N-1 \tag{15-90}$$

and from (15-84) this corresponds to

$$s_k = \frac{\xi_1(s_k)}{q + p\xi_1^2(s_k)} = \frac{1}{2\sqrt{pq}\cos(\pi k/N)} \tag{15-91}$$

or

$$\lambda_k = \frac{1}{s_k} = 2\sqrt{pq}\cos(\pi k/N) \qquad k = 1, 2, \ldots, N-1 \tag{15-92}$$

For $\lambda_0 = 1$ we obtain directly

$$x_i^{(0)} = 1 \tag{15-93}$$

as before, and solving (15-87) and (15-88) using (15-91) and substituting into (15-86) we get

$$x_i^{(k)} = \left(\frac{q}{p}\right)^{i/2} \sin\frac{\pi k i}{N} - \left(\frac{q}{p}\right)^{(i+1)/2} \sin\frac{\pi k(i-1)}{N} \qquad k = 1, 2, \ldots, N-1 \tag{15-94}$$

Proceeding in a similar manner, the equations in (15-60) reduce to

$$y_1 = sq(y_1 + y_2) \tag{15-95}$$

$$y_i = s(py_{k-1} + qy_{k+1}) \qquad k = 2, 3, \ldots, N-1 \tag{15-96}$$

$$y_N = sp(y_{N-1} + y_N). \tag{15-97}$$

Notice that (15-96) is the same as (15-82) provided p and q are interchanged, and hence its general solution is given by (15-86) with p and q interchanged. Equations (15-95) and (15-97) are satisfied if $qy_1 = py_0$ and $py_N = qy_{N+1}$. After some calculations, the solutions to (15-95) and (15-97) turn out to be

$$y_j^{(0)} = \left(\frac{p}{q}\right)^j \tag{15-98}$$

and

$$y_j^{(k)} = \left(\frac{p}{q}\right)^{j/2} \sin\frac{\pi k j}{N} - \left(\frac{p}{q}\right)^{(j-1)/2} \sin\frac{\pi k(j-1)}{N} \qquad k = 1, 2, \ldots, N-1 \tag{15-99}$$

Finally, using (15-61) we get

$$c_0 = \frac{q}{p}\frac{1-(p/q)}{1-(p/q)^N} \tag{15-100}$$

and

$$c_k = \frac{2p/N}{1 - 2\sqrt{pq}\cos(\pi k/N)} \qquad k \geq 1 \tag{15-101}$$

Using (15-92)–(15-94) and (15-98)–(15-101) in (15-62) we get the higher transition probabilities for the random walk model with two reflecting barriers to be

$$p_{ij}^{(n)} = \frac{1-(p/q)}{1-(p/q)^N}\left(\frac{p}{q}\right)^{j-1} + \frac{2p}{N}\sum_{k=1}^{N-1} \frac{x_i^{(k)} y_j^{(k)} [2\sqrt{pq}\cos(\pi k/N)]^n}{1 - 2\sqrt{pq}\cos(\pi k/N)} \tag{15-102}$$

◀

Repeating Eigenvalues

Equation (15-62) has been derived under the assumption that all eigenvalues of P are distinct. However, some of the eigenvalues of P can repeat with multiplicity greater than one. For example, if a stochastic matrix P is of the form

$$P = \begin{pmatrix} P_1 & 0 & 0 \\ 0 & P_2 & 0 \\ & & P_3 \end{pmatrix} \tag{15-103}$$

where $P_i, i \geq 1$, are themselves stochastic matrices, then $\lambda_1 = 1$ is a multiple eigenvalue of P irrespective of whether some of the other eigenvalues are multiple or otherwise.

In general, suppose that the eigenvalue λ_i occurs with multiplicity $r_i \geq 1, i = 1, 2, \ldots, k$ so that $\sum_{i=1}^{k} r_i = N$, the size of the matrix P. In that case, P has the Jordan canonical representation

$$P = U \Lambda U^{-1} \tag{15-104}$$

where Λ is given by

$$\Lambda = \begin{pmatrix} \Lambda_1 & 0 & 0 & 0 \\ 0 & \Lambda_2 & 0 & 0 \\ 0 & 0 & \ddots & 0 \\ & & & \Lambda_k \end{pmatrix} \tag{15-105}$$

Here Λ_i is an $r_i \times r_i$ square matrix that is no longer diagonal and given by

$$\Lambda_i = \begin{pmatrix} \lambda_i & 1 & 0 & 0 \\ 0 & \lambda_i & 1 & 0 \\ 0 & 0 & \ddots & 1 \\ & & & \lambda_i \end{pmatrix} \tag{15-106}$$

From (15-104),

$$P^n = U \Lambda^n U^{-1} \tag{15-107}$$

where

$$\Lambda^n = \begin{pmatrix} \Lambda_1^n & 0 & \cdots & 0 \\ 0 & \Lambda_2^n & 0 & 0 \\ & & \ddots & \\ 0 & 0 & & \Lambda_k^n \end{pmatrix} \tag{15-108}$$

and

$$\Lambda_i^n = \begin{pmatrix} \lambda_i^n, & \binom{n}{1} \lambda_i^{n-1}, & \ldots, & \binom{n}{r_i - 1} \lambda_i^{n-r_i+1} \\ 0 & \lambda_i^n & \cdots & \binom{n}{r_i - 2} \lambda_i^{n-r_i+2} \\ & & \ddots & \\ 0 & 0 & \cdots & \lambda_i^n \end{pmatrix} \tag{15-109}$$

Thus, in the general case of multiple roots, we have (15-107)–(15-109), where the columns of U represent the generalized eigenvectors of P as in (15-104).

In examples 15-16, 15-17, and 15-19, we notice that the first term in $p_{ij}^{(n)}$ converges to a limit independent of the starting state e_i, and the remaining terms converge to zero as $n \to \infty$, indicating that P^n converges to a matrix with identical rows. We shall see that this is usually the case, implying that the influence of the initial state should gradually wear off, and for large n, $p_j(n) = P\{\mathbf{x}_n = e_j\}$ in (15-47) should be independent of the initial distribution. In other words, regardless of the initial state, the Markov chain (under some restrictions) reaches a steady or stable limiting distribution after a large number of transitions. When such limits exist, the system settles down and becomes stable. However, there are exceptions, and to investigate the conditions for stability, we begin with a classification of states and chains.

15-3 CLASSIFICATION OF STATES

Given any two states e_i and e_j, if the probability $p_{ij}^{(n)} > 0$ for some n, then there is a positive probability of reaching the state e_j starting from e_i in n steps. In that case the state e_j is said to be *accessible* from the state e_i. If e_i and e_j are accessible from each other (*i.e.*, if either state can be reached from the other one), then we say e_i *communicates* with e_j. If every state in a Markov chain is accessible from every other state (possibly in different number of transitions), then the chain and the corresponding transition matrix is said to be *irreducible* (*communicating chain*). For example, in the random walk model in Example 15-1, every state can be reached from every other state, and hence it represents an irreducible chain. Evidently, the same is true for the binary communication model in Example 15-5, as well as the cyclic model in Example 15-9.

Closed sets. If C is a set of states such that no state outside C can be reached from any state in C, then C is said to be *closed*. Thus if C is a closed set and if $e_i \in C$, $e_j \notin C$, then $p_{ij} = 0$. In that case $p_{ij}^{(2)} = \sum_k p_{ik} p_{kj} = 0$, since one term in the product is always zero, and more generally $p_{ij}^{(n)} = 0$, $n \geq 1$, so that no state outside C can be reached from any state inside C in any number of transitions. A closed set may contain one or more states. If a closed set contains only one state, then it is called an *absorbing state*. If e_i is an absorbing state, then $p_{ii} = 1$ and $p_{ij} = 0$, $i \neq j$. Once the system enters an absorbing state, it gets trapped there. Nothing ever escapes from an absorbing state.

In Examples 15-7 and 15-13, the states e_0 and e_N are absorbing states. It follows that a Markov chain and the corresponding transition matrix are irreducible it there exists no closed set other than the set of all states. Thus a chain is irreducible if and only if every state can be reached from every other state (if and only if all states communicate with each other).

In a chain with states $e_1, e_2, \ldots, e_n, \ldots$ suppose a subset of states $e_1, e_2, \ldots, e_r$, form a closed set C. Then the $r \times r$ upper left-hand matrix in P is itself stochastic, and we can exhibit P in the form

$$P = \begin{pmatrix} U & 0 \\ V & W \end{pmatrix} \tag{15-110}$$

where U and W are square matrices, and $p_{ij} = 0$ whenever $e_i \in C$, and e_j belongs to its complement. This gives

$$P^n = \begin{pmatrix} U^n & 0 \\ V_n & W^n \end{pmatrix} \tag{15-111}$$

which shows that $p_{ij}^{(n)} = 0$ if $e_i \in C$, but $e_j \notin C$. Moreover U^n in (16-157) indicates that when both e_i and e_j are in C, the transition probabilities $p_{ij}^{(n)}$ are obtained by restricting the summation over the closed set C only. Similarly, W^n indicates that the same is true if both e_i and e_j belong to the complement of C, in which case summation is only over the complement of C.

As an example, consider the following 7×7 transition matrix

$$P = \begin{pmatrix} 0 & 0 & 0 & 0 & a_{15} & 0 & a_{17} \\ 0 & 0 & 0 & a_{24} & 0 & 0 & 0 \\ 0 & 0 & a_{33} & 0 & 0 & 0 & 0 \\ 0 & a_{42} & 0 & a_{44} & 0 & 0 & 0 \\ a_{51} & 0 & 0 & 0 & 0 & 0 & a_{57} \\ 0 & a_{62} & a_{63} & 0 & 0 & a_{66} & a_{67} \\ a_{71} & 0 & 0 & 0 & a_{75} & 0 & a_{77} \end{pmatrix} \tag{15-112}$$

where $a_{ij} > 0$ represent positive probabilities. Since a_{24} and a_{33} are the only nonzero entries in rows 2 and 3, we have $a_{24} = 1$, $a_{33} = 1$, and hence the state e_3 is *absorbing*. From e_2, transition takes place to e_4 and from there to e_2 or itself. Hence e_2 and e_4 form a *closed* set. Similarly from e_1, transitions are possible to e_5 and e_7 and from there to e_1, e_5, and e_7 only. As a result e_1, e_5, and e_7 form another *closed* set. From e_6 transitions are possible to all seven states. On rearranging the states as e_3, e_2, e_4, e_1, e_5, e_7, and e_6, the new transition matrix has the general structure shown in (15-103), with block square matrices P_1, P_2 and P_3 along the main diagonal followed by P_4. Notice that P_1 corresponds to e_3, P_2 to e_2 and e_4, and P_3 to e_1, e_5 and e_7, while the 1×7 matrix $P_4 \triangleq [V, W]$ is as in (15-110), where W is 1×1 and it equals a_{66}.

PERSISTENT OR RECURRENT AND TRANSIENT STATES. Starting from any state e_i, whether the system ever returns with certainty to the same state is an important question. If so, one may ask how long does it take on the average for that event to happen? To analyze these questions, we first generalize the event *"the first return to origin"* introduced in Sec. 10-1 in connection with the random walk model, and define $f_{ij}^{(n)}$ to be the probability that starting from state e_i, the chain reaches the state e_j for the *first time* in n steps. Thus [37]

$$f_{ij}^{(n)} = P[\mathbf{x}_n = e_j, \mathbf{x}_m \neq e_j, 0 < m < n \mid \mathbf{x}_0 = e_i] \tag{15-113}$$

and $f_{ij}^{(n)}$ represents the *first passage probability* from e_i to e_j in n steps. Notice that $p_{ij}^{(n)}$ represents the probability of reaching e_j starting from e_i in n steps, but *not necessarily* for the first time.

It is easy to establish a relation between $f_{ij}^{(n)}$ and $p_{ij}^{(n)}$ by arguing as in (10-10). Starting from e_i, the state e_j can be reached for the first time at the rth step with probability $f_{ij}^{(r)}$, $r \leq n$, and again in the remaining $n-r$ steps with probability $p_{jj}^{(n-r)}$ for $1 \leq r \leq n$. Summing over all these mutually exclusive possibilities, we obtain a key relation

$$p_{ij}^{(n)} = \sum_{r=1}^{n} f_{ij}^{(r)} p_{jj}^{(n-r)} \qquad n \geq 1 \tag{15-114}$$

with

$$f_{ij}^{(0)} = 0 \qquad p_{jj}^{(0)} = 1 \qquad p_{ij}^{(0)} = 0 \qquad i \neq j \quad \text{and} \quad f_{ij}^{(1)} = p_{ij} \tag{15-115}$$

Let $P_{ij}(z)$ and $F_{ij}(z)$ represent the moment generating functions of the sequences

$\{p_{ij}^{(n)}\}$ and $\{f_{ij}^{(n)}\}$, respectively. Then proceeding as in (10-11)

$$P_{ij}(z) = \sum_{n=0}^{\infty} p_{ij}^{(n)} z^n = p_{ij}^{(0)} + \sum_{n=1}^{\infty} \sum_{r=1}^{n} f_{ij}^{(r)} p_{jj}^{(n-r)} z^n$$

$$= p_{ij}^{(0)} + \sum_{r=1}^{\infty} f_{ij}^{(r)} z^r \sum_{k=0}^{\infty} p_{jj}^{(k)} z^k = p_{ij}^{(0)} + F_{ij}(z) P_{jj}(z) \qquad (15\text{-}116)$$

where

$$F_{ij}(z) = \sum_{n=1}^{\infty} f_{ij}^{(n)} z^n \qquad (15\text{-}117)$$

In particular for $i = j$, we obtain the useful relation

$$P_{ii}(z) = 1 + F_{ii}(z) P_{ii}(z) \qquad (15\text{-}118)$$

or

$$P_{ii}(z) = \frac{1}{1 - F_{ii}(z)} \qquad (15\text{-}119)$$

Clearly,

$$f_{ij} \triangleq \sum_{n=1}^{\infty} f_{ij}^{(n)} = F_{ij}(1) \qquad (15\text{-}120)$$

represents the *first passage* probability that starting from state e_i, the system will sooner or later ever pass through state e_j. Thus $f_{ij} \leq 1$ always, and when $f_{ij} = 1$, the sequence $\{f_{ij}^{(n)}\}$ represents a proper probability distribution, and we refer to it as the *first passage distribution* for the state e_j, with $f_{ij}^{(n)}$ defined as in (15-113). In particular, if $f_{jj} = 1$, then $\{f_{jj}^{(n)}\}$ represents the distribution for the *recurrence times* of e_j, and in that case if $\mathbf{y}_j$ represents the recurrence time random variable for the state e_j, then

$$P\{\mathbf{y}_j = n\} = f_{jj}^{(n)} \qquad (15\text{-}121)$$

and

$$\mu_j = E\{\mathbf{y}_j\} = \sum_{n=1}^{\infty} n f_{jj}^{(n)} \qquad (15\text{-}122)$$

represents the *mean recurrence time* for the state e_j.

PERSISTENT AND TRANSIENT STATES

▶ The state e_j is said to be *persistent*[3] if $f_{jj} = 1$ (i.e., starting from state e_j, return to the state e_j is certain). If $f_{jj} < 1$, then e_j is said to be *transient* (return to e_j is not absolutely certain).

For example, in (15-112) the state e_6 represents a transient state while all other states are persistent states.

A persistent state e_j is called *a null state* if its mean recurrence time in (15-122) $\mu_j = \infty$, and *nonnull* if $\mu_j < \infty$. ◀

PERIODIC STATES

▶ A state e_j is said to be *periodic* with period T, if return to that state is possible only at instants $T, 2T, 3T, \ldots$ (multiples of T), that is, $p_{jj}^{(n)} = 0$ unless $n = kT$, where T is

[3]Sometimes the term *recurrent* is also used to identify *persistent* states, and *nonrecurrent* to identify *transient* states. See Chung [37] and Parzen [46].

the greatest common divisor of all n for which $p_{ij}^{(n)} > 0$ (see also Appendix 15-B). The state e_j is *aperiodic* if no such T (>1) exits.

For example, the $m \times m$ transition matrix

$$P = \begin{pmatrix} 0 & 1 & 0 & 0 & \cdots & 0 \\ 0 & 0 & 1 & 0 & \cdots & 0 \\ 0 & 0 & 0 & 1 & \cdots & 0 \\ 0 & 0 & 0 & 0 & \cdots & 1 \\ 1 & 0 & 0 & 0 & \cdots & 0 \end{pmatrix} \tag{15-123}$$

represents a periodic chain with period $T = m$.

In an unrestricted random walk (Sec. 10-1) and the cyclic random walk model with an *even* number of states in (15-23) (Example 15-9) all states have period 2. In the random walk model with absorbing barriers in Example 15-7, the absorbing states e_0 and e_N are aperiodic, whereas the internal states are periodic with period 2. In the random walk model with reflecting barriers inExample 15-8, all states are aperiodic. ◀

Finally, a *persistent, nonnull,* and *aperiodic* state is said to be an *ergodic state*. A Markov chain, all of whose states are *ergodic,* is said to be an *ergodic chain.*

Theorem 15-1 expresses the conditions for the various type of states in terms of the transition probabilities $p_{ij}^{(n)}$.

THEOREM 15-1

PERSISTENT AND TRANSIENT STATES

▶ (*i*) The state e_i is *persistent* if and only if

$$\sum_{n=0}^{\infty} p_{ii}^{(n)} = \infty \tag{15-124}$$

and *transient* if and only if

$$\sum_{n=0}^{\infty} p_{ii}^{(n)} < \infty \tag{15-125}$$

If state e_j is transient, then for all i

$$\sum_{n=0}^{\infty} p_{ij}^{(n)} < \infty \tag{15-126}$$

(*ii*) The state e_j is a *persistent null* state if and only if

$$\sum_{n=0}^{\infty} p_{jj}^{(n)} = \infty \qquad \text{and} \qquad p_{jj}^{(n)} \to 0 \qquad \text{as} \quad n \to \infty \tag{15-127}$$

In this case, for all i

$$p_{ij}^{(n)} \to 0 \qquad \text{as} \quad n \to \infty \tag{15-128}$$

(*iii*) An aperiodic persistent state e_j is *ergodic* if and only if $\mu_j < \infty$. In that case as $n \to \infty$,

$$p_{ij}^{(n)} \to \frac{f_{ij}}{\mu_j} \tag{15-129}$$

(*iv*) If state e_j is *persistent and periodic* with period T, then

$$p_{jj}^{(nT)} \to \frac{T}{\mu_j} \tag{15-130}$$

Proof. (i) To say that e_i is *persistent* means that

$$f_{ii} = \sum_{n=1}^{\infty} f_{ii}^{(n)} = \lim_{z \to 1} F_{ii}(z) = 1$$

or equivalently from (15-119)

$$\sum_{n=0}^{\infty} p_{ii}^{(n)} = \lim_{z \to 1} P_{ii}(z) = \lim_{z \to 1} \frac{1}{1 - F_{ii}(z)} = \infty$$

Conversely, suppose $\sum_{n=0}^{\infty} p_{ii}^{(n)} < \infty$. Then, since $p_{ii}^{(n)}$ are all nonnegative, $P_{ii}(z)$ increases monotonically as $z \to 1$, and

$$\sum_{n=0}^{N} p_{ii}^{(n)} \le \lim_{z \to 1} P_{ii}(z) \le \sum_{n=0}^{\infty} p_{ii}^{(n)}$$

for every N, and hence, taking the limit as $N \to \infty$, we get

$$\lim_{z \to 1} P_{ii}(z) = \sum_{n=0}^{\infty} p_{ii}^{(n)} < \infty$$

Thus $P_{ii}(z)$ approaches a finite limit as $z \to 1$ if and only if (16-166) holds. In that case from (15-119) and (16-163), we have $f_{ii} < 1$, that is, the state e_i is transient. Equivalently, $P_{ii}(z) \to \infty$ as $z \to 1$, that is, the state e_i is persistent if and only if (16-165) holds.

To prove (16-167), we can make use of (16-159). From there

$$\sum_{n=0}^{\infty} p_{ij}^{(n)} = \sum_{n=0}^{\infty} \sum_{r=0}^{n} f_{ij}^{(r)} p_{jj}^{(n-r)} = \sum_{m=0}^{\infty} p_{jj}^{(m)} \sum_{r=0}^{\infty} f_{ij}^{(r)} \le \sum_{m=0}^{\infty} p_{jj}^{(m)} < \infty \tag{15-131}$$

if e_j is *transient* [use (16-166)]. This proves (16-167). In particular, if state e_j is transient, then from (16-173) we obtain

$$p_{ij}^{(n)} \to 0 \tag{15-132}$$

To prove (ii) and (iii) formally, define

$$v_n = p_{jj}^{(n)} - p_{jj}^{(n-1)} \qquad n \ge 1, \qquad v_0 = p_{jj}^{(0)}$$

so that

$$\sum_{k=0}^{n} v_k = p_{jj}^{(n)} \tag{15-133}$$

and using (15-119)

$$V(z) \triangleq \sum_{n=0}^{\infty} v_n z^n = P_{jj}(z) - zP_{jj}(z) = \frac{1 - z}{1 - F_{jj}(z)} \tag{15-134}$$

Thus

$$\lim_{z \to 1} V(z) = \lim_{z \to 1} \frac{1}{(1 - F_{jj}(z))/(1 - z)} = \frac{1}{F'_{jj}(1)} = \frac{1}{\mu_j} \tag{15-135}$$

But

$$\lim_{z \to 1} V(z) = \lim_{z \to 1} \sum_{k=0}^{\infty} v_k z^k = \lim_{n \to \infty} \sum_{k=0}^{n} v_k = \lim_{n \to \infty} p_{jj}^{(n)} \tag{15-136}$$

where we have used (16-176). From (16-182) and (16-185) we get

$$\lim_{n\to\infty} p_{jj}^{(n)} = \frac{1}{\mu_j} \tag{15-137}$$

Together with (16-165), equation (16-186) shows that e_j is a persistent null state ($\mu_j = \infty$) if and only if (16-168) holds. In that case $p_{jj}^{(n)} \to 0$.

From (16-159) and (16-186), we obtain

$$\lim_{n\to\infty} p_{ij}^{(n)} = \lim_{n\to\infty} \sum_{k=1}^{n} f_{ij}^{(k)} p_{jj}^{(n-k)} = \sum_{k=1}^{\infty} \frac{f_{ij}^{(k)}}{\mu_j} = \frac{f_{ij}}{\mu_j} \tag{15-138}$$

If e_j is a persistent null state, then $\mu_j = \infty$, so that

$$\lim_{n\to\infty} p_{ij}^{(n)} \to 0$$

which proves (16-169).

Finally if e_j is aperiodic, persistent, and ergodic, then by definition $\mu_j < \infty$. Conversely, if $\mu_j < \infty$, from (16-186), $p_{jj}^{(n)}$ tends to a nonzero constant as $n \to \infty$, and hence $\lim_{n\to\infty} \sum_n p_{jj}^{(n)} = \infty$ so that e_j is persistent and ergodic. In that case, we also have (16-187), proving (16-170).

To prove (*iv*), note that if e_j is periodic with period T, then $f_{jj}^{(n)} = 0$ unless n is a multiple of T, and hence $F_{jj}(z)$ only contains powers of z^T. Let

$$F_{jj}(z) = \varphi(z^T)$$

so that from (15-119)

$$P_{jj}(z) = \frac{1}{1 - \varphi(z^T)}$$

or

$$P_{jj}(z^{1/T}) = \frac{1}{1 - \varphi(z)} = \sum_{n=0}^{\infty} p_{jj}^{(nT)} z^n$$

and arguing as in (16-181)–(16-186), we get

$$p_{jj}^{(nT)} \to \frac{1}{\varphi'(1)} = \frac{T}{F'_{jj}(1)} = \frac{T}{\mu_j}$$

which proves (16-171) and $p_{jj}^{(m)} \to 0$ if $m \neq kT$. This completes the proof of Theorem 15-1.

Thus returns to persistent states are bound to happen with probability 1. In fact, as Theorem 15-2 states, it is possible to specify the total number of visits to both persistent and transient states.

THEOREM 15-2

If an initial state e_i is persistent, then with probability 1 the system returns *infinitely* often to e_i as $n \to \infty$. If e_i is transient, the system returns to e_i only finitely often, and after certain number of visits, the system *never* returns to e_i again.

Proof. Suppose the system first returns to e_i after N_1 steps, returns for a second time after N_2 steps, etc. In that case the event $\{N_k < \infty\}$ represents that there are at least k returns to e_i, and

$$P\{N_1 < \infty\} = f_{ii} = \sum_{m=1}^{\infty} f_{ii}^{(m)} \tag{15-139}$$

represents the probability of the system returning to e_i sooner or later at least once. After every return to e_i, the system starts all over again, and hence after the first return the behavior gets repeated till the second return occurs. Thus,

$$P\{N_2 < \infty \mid N_1 < \infty\} = f_{ii} \tag{15-140}$$

Moreover $\{N_1 = \infty\}$ implies $\{N_2 = \infty\}$ and hence $\{N_2 < \infty\}$ implies $\{N_1 < \infty\}$. Thus

$$P\{N_2 < \infty\} = P\{N_2 < \infty \mid N_1 < \infty\}P\{N_1 < \infty\} = f_{ii}^2 \tag{15-141}$$

and in general,

$$P\{N_k < \infty\} = f_{ii}^k \tag{15-142}$$

If e_i is transient, then $f_{ii} < 1$ and hence

$$\sum_{k=1}^{\infty} P\{N_k < \infty\} = \sum_{k=1}^{\infty} f_{ii}^k = \frac{1}{1 - f_{ii}} < \infty \tag{15-143}$$

In that case, by the first part of Borel–Cantelli lemma [see (2-69)], with probability 1 only finitely many of the events $\{N_k < \infty\}$ do occur. Thus the system returns to a transient state only for a finite number of times with probability one, and after a certain number of steps the system never returns to a transient state. If on the other hand e_i is persistent, then $f_{ii} = 1$ and hence for every k

$$P\{N_k < \infty\} = 1 \tag{15-144}$$

Let N be the number of times the system returns to e_i as $n \to \infty$. Since the events $\{N_k < \infty\}$ and $\{N \geq k\}$ are equivalent, (16-200) implies that N exceeds any preassigned number k with probability one, that is, for every k,

$$P\{N > k\} = 1$$

or

$$P\{N = \infty\} = 1 \tag{15-145}$$

Thus the system returns to a persistent state infinitely often as $n \to \infty$, and this completes the proof. ◀

Theorem 15-3 shows that all states that are accessible from a persistent state are themselves persistent.

THEOREM 15-3

▶ If a state e_j is accessible from a persistent state e_i, then e_i is also accessible from e_j, and moreover e_j is persistent.

Proof. Suppose a state e_j is accessible from a persistent state e_i, but e_i is not accessible from e_j. Thus the system goes from e_i to e_j in a certain number of steps with positive probability $p_{ij}^{(m)} = a > 0$, and after that it does not return to e_i. Consequently starting from e_i the probability of the system not returning to e_i is at least a, or the probability of the system eventually returning to e_i cannot exceed $1 - a$. Thus $f_{ii} \leq 1 - a$. But $1 - a$ is strictly less than 1, contradicting the assumption that e_i is persistent. Hence e_i must be accessible from e_j, that is, $p_{ji}^{(r)} = b > 0$ for some r. From (15-43), we have

$$p_{ij}^{(n+m)} \geq p_{ik}^{(m)} p_{kj}^{(n)} \qquad \text{for any} \quad k \tag{15-146}$$

and hence

$$p_{ii}^{(n+m+r)} \geq p_{ij}^{(m)} p_{ji}^{(n+r)} \geq p_{ij}^{(m)} p_{jj}^{(n)} p_{ji}^{(r)} = abp_{jj}^{(n)} \tag{15-147}$$

Similarly

$$p_{jj}^{(n+m+r)} \geq p_{ji}^{(r)} p_{ii}^{(n)} p_{ij}^{(m)} = abp_{ii}^{(n)} \tag{15-148}$$

Thus the two series $\sum_{n=0}^{\infty} p_{ii}^{(n)}$ and $\sum_{n=0}^{\infty} p_{jj}^{(n)}$ converge or diverge together. But $\sum_n p_{ii}^{(n)} = \infty$, since e_i is persistent, and it now follows that e_j is also persistent. This completes the proof of the theorem. ◀

If the Markov chain is also irreducible, then all states are accessible from each other, and the proof of Theorem 15-3 shows that in that case all states are of the same type, that is, they are all either transient or persistent. Theorem 15-4 summarizes this observation.

THEOREM 15-4

▶ In an irreducible Markov chain, all states are of the same type. They are either all transient, all persistent null or all persistent nonnull. All the states are either aperiodic or periodic with the same period.

Proof. The chain is irreducible, and hence *every* state is accessible from *every other state*. In that case, from (16-213) and (16-214) for any two states, the series $\sum_n p_{ii}^{(n)}$ and $\sum_n p_{jj}^{(n)}$ converge or diverge together, and hence all states are either transient or persistent. If e_i is persistent null, then $p_{ii}^{(n)} \to 0$ as $n \to \infty$ and from (16-213), $p_{jj}^{(n)} \to 0$ as $n \to \infty$ so that e_j and all other states are also persistent null. Finally if e_i is persistent nonnull and has period T, then $p_{ii}^{(n)} > 0$ whenever n is a multiple of T only. From (16-212)

$$p_{ii}^{(m+r)} \geq p_{ij}^{(m)} p_{ji}^{(r)} = ab > 0 \tag{15-149}$$

since e_i and e_j are mutually accessible. Hence from (16-216), $(m + r)$ must be a multiple of T. Finally from (16-214),

$$p_{jj}^{(n+m+r)} \geq abp_{ii}^{(n)} > 0$$

where n and hence $(n + m + r)$ are multiples of T. Thus T is also the period of the state e_j, and this proves the theorem. ◀

One way to show that an irreducible chain is aperiodic is to exhibit a state e_k for which $p_{kk} > 0$. Such a state is clearly aperiodic.

Next, we shall use Theorems 15-1–15-4 to analyze the limiting behavior of various random walk models in one and higher dimensions.

EXAMPLE 15-20

RANDOM WALK IN ONE AND HIGHER DIMENSIONS

▶ **One-dimensional random walk.** Consider the one-dimensional unrestricted random walk model introduced in Sec. 10-1. Every state is accessible there from every other state (see Fig. 10-1*a*) and hence all of them are of the same type. The one-step transition probabilities in that case are given by

$$p_{ij} = \begin{cases} p & j = i+1 \\ q & j = i-1 \\ 0 & \text{otherwise} \end{cases} \tag{15-150}$$

and moreover from (10-5) for any state e_i

$$p_{ii}^{(n)} = \begin{cases} u_{2k} & n = 2k \\ 0 & n = 2k+1 \end{cases} \tag{15-151}$$

Thus all states are periodic with period 2, and

$$p_{ii}^{(2n)} = u_{2n} = \binom{2n}{n}(pq)^n = \frac{2n!}{n!n!}(pq)^n$$

$$\simeq \frac{\sqrt{4\pi n}(2n)^{2n}e^{-2n}}{(\sqrt{2\pi n}\, n^n e^{-n})^2}(pq)^n = \frac{(4pq)^n}{\sqrt{\pi n}} \qquad n \geq 1 \tag{15-152}$$

where we have made use of *Stirling's approximation formula* given by

$$n! = \sqrt{2\pi n} n^n e^{-n} \qquad n \to \infty \tag{15-153}$$

From (16-219), both the series

$$\sum_{n=0}^{\infty} p_{ii}^{(2n)} \qquad \text{and} \qquad 1 + \sum_{n=1}^{\infty} \frac{(4pq)^n}{\sqrt{\pi n}} \tag{15-154}$$

either converge or diverge together. Since

$$4pq = (p+q)^2 - (p-q)^2 = 1 - (p-q)^2 \leq 1,$$

for $p \neq q$, we have $4pq < 1$, and the later series in (16-221) converges since it is bounded by $1/(1-4pq)$, and hence

$$\sum_{n=0}^{\infty} p_{ii}^{(2n)} < \infty \qquad \text{for every} \quad e_i \tag{15-155}$$

Thus if $p \neq q$, then every state is *transient* in a one-dimensional random walk. If $p > q$, the particle will gradually work its way out to the right and eventually will permanently abandon any state e_i. However, if $p = q = 1/2$, we have $4pq = 1$, and the later series in (16-221) diverges. In that case $\sum_{n=0}^{\infty} p_{ii}^{(n)}$ diverges, and every state is *persistent* and from Theorem 15-2 the particle will return to each state infinitely often. Note that these conclusions also have been obtained more directly in (10-15) by making use of the generating function $F_{ii}(z) = V(z)$ in (10-13). From (10-16), the mean recurrence time $\mu_i = V'(1) = \infty$ if $p = q$, and from Theorem 15-1 it now follows that all states in a symmetric random walk are persistent null with periodicity 2. From (16-169), we also have $p_{ij}^{(n)} \to 0$ as $n \to \infty$.

Two-dimensional random walk. In a two-dimensional random walk, the particle moves in unit steps in both the x and y directions independently, starting at the origin, so that its path contains all points on the plane with integer-valued coordinates. Each position of the particle has four neighbors. Once again every state is accessible from every other state, and a return to origin is possible only if the number of steps in the positive x and y directions equal those in the negative x and y directions. Hence (16-218) holds in this case also and all states are periodic with period 2. If n represents the total number of steps in the positive x and y directions, and k those in the positive x direction, then for a symmetric random walk using the multinomial distribution we get

$$
\begin{aligned}
p_{ii}^{(2n)} &= \sum_{k=0}^{n} \frac{2n!}{k!k!(n-k)!(n-k)!} \left(\frac{1}{2}\right)^{2n} \left(\frac{1}{2}\right)^{2n} \\
&= \frac{1}{4^{2n}} \binom{2n}{n} \sum_{k=0}^{n} \binom{n}{k}^2 = \left\{ \binom{2n}{n} 2^{-2n} \right\}^2 \\
&\simeq \frac{1}{\pi n} \qquad n \to \infty
\end{aligned}
\tag{15-156}
$$

Since the series $\sum_{n=1}^{\infty} 1/n$ diverges, $\sum_{n=0}^{\infty} p_{ii}^{(2n)}$ also diverges, and all states are persistent. In addition, from (16-239) $p_{ii}^{(2n)} \to 0$ as $n \to \infty$, and hence all states are persistent null with periodicity 2. It is interesting to note that in a two-dimensional *symmetric* random walk, in spite of the greater freedom present (compared to one dimension) to wander away, the particle does return to each state infinitely often. Figure 15-4 illustrates four separate runs of a two-dimensional random walk each consisting of 1250 trials.

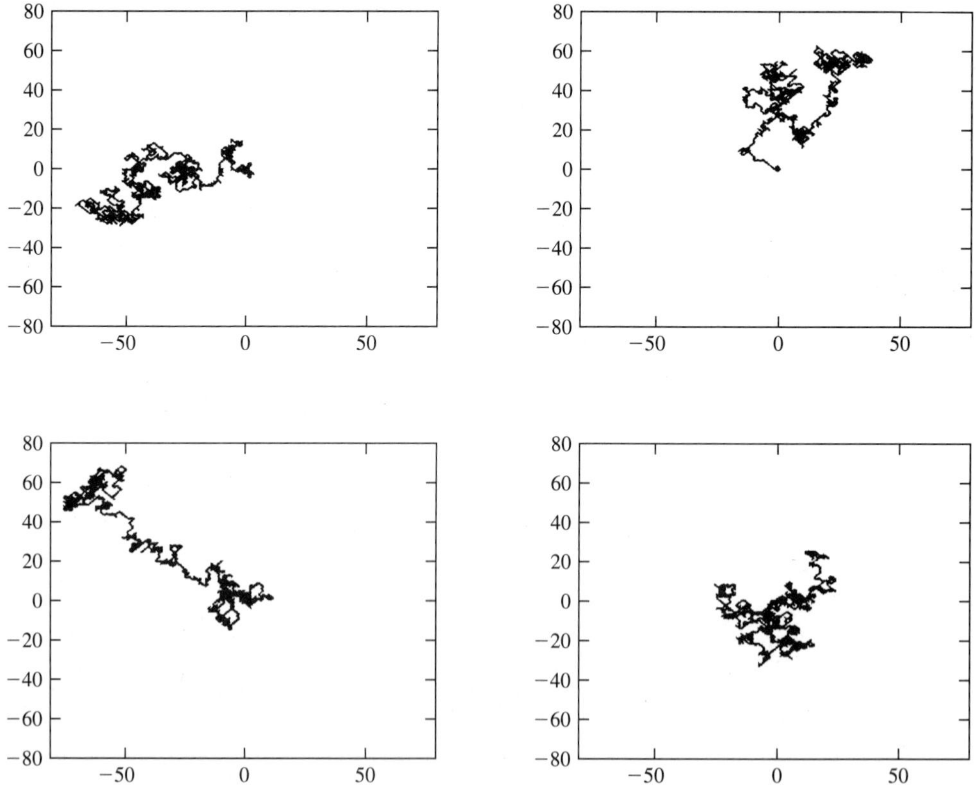

FIGURE 15-4
Two-dimensional random walk. Four different runs each consisting of 1250 trials.

Three-dimensional random walk. In a general three-dimensional random walk, the particle undergoes a one-dimensional random walk at every step along any one of the three axes chosen at random. Thus the particle moves independently in one of the six directions generated by the x, y, or z axes at every step. In that case, each position has six neighbors. In a slightly restricted model, the particle undergoes three independent one-dimensional random walks at every step along the three axes simultaneously, and that model gives rise to eight neighbors for each position (corners of the cube as opposed to centers of the faces of the cube). In this case also, every state is accessible from every other state and all states are periodic with period 2. A return to the origin is possible in the later case if and only if we have returns to origin in all three directions. By the independent assumption, this gives

$$p_{ii}^{(2n)} \simeq P\{\mathbf{s}_{2n}^{(x)} = 0\} P\{\mathbf{s}_{2n}^{(y)} = 0\} P\{\mathbf{s}_{2n}^{(z)} = 0\} = \left\{ \binom{2n}{n} 2^{-2n} \right\}^3 \tag{15-157}$$

so that for large n,

$$p_{ii}^{(2n)} \simeq \frac{1}{\pi^{3/2} n^{3/2}} \tag{15-158}$$

In this case, however, the series $\sum_{n=1}^{\infty} 1/n^{3/2} < 3$ converges, and hence $\sum_{n=0}^{\infty} p_{ii}^{(2n)}$ also converges. Using (16-166), it follows that all states in a three-dimensional random walk are *transient*. Thus after a certain number of visits the particle never returns to the initial position. By direct computation using (16-240), we obtain

$$\sum_{n=0}^{100} p_{ii}^{(2n)} = 1.3574 \simeq P_{ii}(1)$$

and from (15-119) and (16-163), this gives the probability of ever returning to the origin (or any other state) to be

$$f_{ii} = F_{ii}(1) = 1 - \frac{1}{P_{ii}(1)} \simeq 0.2633 \tag{15-159}$$

for a three-dimensional symmetric random walk with eight neighbors for each position. In the general model referred to earlier with six neighbors for each state, the probability of ever revisiting the origin is slightly higher due to the lower number of neighbors present there, and it is given by $0.2633 \times 8/6 \simeq 0.351$. The expected number of returns in that case equals $0.649 \sum k(0.351)^k = 0.351/0.649 \simeq 0.54$, an old result due to Polya ([3], Vol. I).

To summarize, in a symmetric random walk in one and two dimensions, with probability one, the particle will sooner or later return to the origin. However, in three dimensions this probability is only about 0.26 (or 0.35). Thus in a three dimensional symmetric random walk, all states are *transient,* and after a certain number of visits the particle never again returns to the initial position.[4] The extra degrees of freedom do seem to have made a difference in the three-dimensional case. It follows that two independent random walks on a plane will meet infinitely often, whereas in three dimensions there

[4]Polya had remarked the dimensional breakdown $n \le 2$ versus $n \ge 3$ as "newsworthy" and not intuitively obvious.

is a positive probability that they will never meet. In a diffusion process containing a large number of particles, it goes without saying that the probability of even a small number, let alone all particles, regrouping to origin at a later point in time is essentially zero. Similar conclusions hold in higher dimensional random walks also. In the case of a four-dimensional symmetric random walk, the probability that a particle will return to the origin is approximately 0.105 or 0.2 depending on whether the model assumes that each state has eight neighbors (faces of the hypercube) or 16 neighbors (corners of the hypercube). ◀

More on classification. For every persistent state e_i interestingly there exists a subset C of states that are accessible from e_i. From Theorem 15-3, all such states are persistent and accessible from each other, forming an irreducible closed set C. Further, for every pair of states e_i and e_j in C

$$f_{ij} = 1 \qquad f_{ji} = 1 \tag{15-160}$$

To prove (15-160), we can argue as follows: Let α denote the probability of the event "starting from the state e_i the system reaches e_j without returning to e_i," and on reaching e_j the probability of never returning to e_i is given by $1 - f_{ji}$. Hence the probability of the event, "starting from e_i, the system never returns to e_i," is at least $\alpha(1 - f_{ji})$. But since e_i is persistent $f_{ii} = 1$, and hence this no-return probability is zero. Hence $f_{ji} = 1$ for every e_j that can be reached from e_i that is persistent. Since e_j is also persistent, we get $f_{ij} = 1$. If C denotes the set of all states that can be reached from e_i, then all of them are persistent and accessible from each other. Using this argument it follows that (15-160) is true for every pair in C. According to (15-160), starting from any state in C, the system is certain to pass through every other state in C, and no exit from C is ever possible.

Thus every persistent state is contained in an irreducible closed set containing other persistent states that are accessible from each other. As a result no transient state can be ever reached from a persistent state. Hence, if e_i is persistent and e_j is transient, then

$$f_{ij} = 0 \tag{15-161}$$

In summary, if a chain contains both persistent and transient states, then the transition matrix P can be partitioned as in (16-156), where U corresponds to persistent states. U can consist of several irreducible subsets such as $P_1, P_2, \ldots,$ as in (15-103) in which case no transition between such subsets is possible. Together with Theorem 15-4, this leads us to Theorem 15-5.

THEOREM 15-5

▶ A Markov chain can be partitioned in a unique manner into nonoverlapping sets $T, C_1, C_2, \ldots$ where T consists of all transient states and $C_1, C_2, \ldots$ are irreducible closed sets containing persistent states of the same type. Further if e_i belongs to some C_r, then

$$f_{ij} = 1 \quad \text{for all} \quad e_j \in C_r \qquad f_{ik} = 0 \qquad e_k \notin C_r \tag{15-162}$$

◀

Example 15-20 shows that all states in a chain can be transient ($p \neq q$), in which case, as Theorem 15-6 shows, the chain must be necessarily *infinite*.

THEOREM 15-6

In a finite chain it is impossible that all states are transient. Further, if all states are accessible from each other, then all of them are persistent nonnull.

Proof. Since there are only a finite number of states, the system must return to at least one of them infinitely often as $n \to \infty$. From Theorem 15-2, all states cannot be transient in that case, and at least one of them must be persistent. In addition, if all states are accessible from each other, then since one of them is persistent, from Theorem 15-3 all of them are persistent and form an irreducible closed set. Since the rows for P^n contain a finite number of elements and each row adds to unity, it is impossible that $p_{jk}^{(n)} \to 0$ for every pair j, k. Hence all states cannot be persistent null, and consequently at least one of them must be persistent nonnull. But then by Theorem 15-4, all states are persistent nonnull. This completes the proof.

Of course, a finite chain can have more than one closed set containing persistent states (e.g., random walk with absorbing barriers). In that case, obviously all states are not accessible from each other.

15-4 STATIONARY DISTRIBUTIONS AND LIMITING PROBABILITIES

So far we have seen that every persistent state belongs to an irreducible closed set C containing similar states. Since no exit from C is ever possible, its asymptotic behavior can be studied independently of the remaining states.

An important question in this context is whether in the long run a Markov chain can reach a limiting distribution *irrespective* of the initial distribution? Thus under what conditions, if any, does

$$p_{jk}^{(n)} \to q_k \qquad \text{as} \quad n \to \infty \tag{15-163}$$

regardless of the initial starting state e_j? When such limits do exist, the system shows long-run regularity or stationary behavior. Interestingly, for irreducible chains containing aperiodic, persistent nonnull states (ergodic chain), from Theorem 15-1, (16-170), and (15-160), it at once follows that

$$\lim_{n\to\infty} p_{ij}^{(n)} \to \frac{f_{ij}}{\mu_j} = \frac{1}{\mu_j} \stackrel{\Delta}{=} q_j > 0 \tag{15-164}$$

As Theorem 15-7 shows, the converse is also true.

THEOREM 15-7

STEADY STATE PROBABILITIES

In an irreducible ergodic chain, the limits

$$q_k \stackrel{\Delta}{=} \lim_{n\to\infty} p_{jk}^{(n)} > 0 \tag{15-165}$$

exist *independent* of the initial state e_j. Further

$$\sum_k q_k = 1 \tag{15-166}$$

and these limiting probabilities satisfy the equations

$$q_j = \sum_i q_i p_{ij} \tag{15-167}$$

Conversely, if a chain is irreducible and aperiodic and there exist numbers $q_k \geq 0$ satisfying (15-166) and (15-167), then the chain is ergodic, and

$$q_k = \frac{1}{\mu_k} > 0 \tag{15-168}$$

where μ_k is the mean recurrence time for the persistent nonnull state e_k.

Proof. Let the chain be irreducible and ergodic. Then (15-164) establishes (15-165). Further, since

$$p_{kj}^{(n+1)} = \sum_i p_{ki}^{(n)} p_{ij} \tag{15-169}$$

as $n \to \infty$, we have $p_{kj}^{(n+1)} \to q_j$ and $p_{ki}^{(n)} p_{ij} \to q_i p_{ij}$. Taking only finitely many terms in the sum of (15-169) and letting $n \to \infty$, we get

$$q_j \geq \sum_i q_i p_{ij} \tag{15-170}$$

Since $\sum_j p_{kj}^{(n+1)} = 1$, we have $\sum_j q_j = 1$, and summing over j in (15-170) we get

$$\sum_j q_j \geq \sum_i q_i$$

in which the inequality is impossible. Hence (15-170) holds with equality for all j, and this proves (15-166) and (15-167). Conversely, let $q_k \geq 0$ and (15-166) and (15-167) be true. Then by repeated use of (15-167) we get

$$q_k = \sum_i q_i p_{ik}^{(n)} \qquad n \geq 1 \tag{15-171}$$

Since the chain is irreducible, all states are of the same type. But they cannot be transient or null states since in that case $p_{ik}^{(n)} \to 0$, and $\sum q_k = 1$ cannot be satisfied. Thus all states are persistent non-null, and the chain is ergodic. Then from (15-164) we get

$$q_k = \sum_i q_i \frac{1}{\mu_k} = \frac{1}{\mu_k} \tag{15-172}$$

and this completes the proof of the theorem. ◀

Since the probability of the state e_j at the nth step is given by

$$P\{\mathbf{x}_n = e_j\} = \sum_i P\{\mathbf{x}_n = e_j \mid \mathbf{x}_0 = e_i\} P\{\mathbf{x}_0 = e_i\} = \sum_i p_{ij}^{(n)} p_i(0) \tag{15-173}$$

where $p_i(0)$ represents the initial distribution, we have

$$\lim_{n\to\infty} P\{\mathbf{x}_n = e_j\} = \sum_i q_j p_i(0) = q_j \sum_i p_i(0) = q_j > 0 \tag{15-174}$$

Thus when the chain exhibits steady state or stationary behavior, then the system in the long run settles down to the invariant probability distribution in (15-174) *irrespective* of the initial distribution. If there are N independent activities involved, then after a long time, Nq_k among them will be in state e_k, reaching a steady state or equilibrium distribution.

Interestingly, (15-171) also shows that if e_k is either a transient state or a persistent null state then $q_k = 0$. To summarize, an irreducible chain possesses an invariant positive probability distribution if and only if it is ergodic (consists of aperiodic, persistent nonnull states). In that case $q_k > 0$ and $\lim_{n\to\infty} P\{\mathbf{x}_n = e_k\} \to q_k$.

Finite Chains and Perron's Theorem

If a Markov chain has only a finite number of states $e_1, e_2, \ldots, e_N$, then Eq. (15-163) can be expressed in matrix form as

$$P^n \to Q \tag{15-175}$$

where P is the probability transition matrix in (15-12), and Q is a matrix with identical rows equal to q given by

$$q = [q_1, q_2, \ldots, q_N] \tag{15-176}$$

Moreover, in that case (15-167) can be rewritten as

$$q = qP \tag{15-177}$$

with

$$\sum_{i=1}^{N} q_i = 1 \tag{15-178}$$

Equation (15-177) represents the left-eigenvector equation of P corresponding to the eigenvalue $\lambda = 1$. This raises an interesting question: Is $\lambda = 1$ an eigenvalue of any probability transition matrix P, and if so, why should the left eigenvector q associated with $\lambda = 1$ always consist of all strictly positive entries?

Since P is a stochastic matrix, it is easy to establish that $\lambda = 1$ is indeed an eigenvalue of P. In fact, since each row of P sums to unity we have

$$Px_1 = x_1 \qquad \text{with} \quad x_1 = [1, 1, 1, \ldots, 1]^T \tag{15-179}$$

which proves the claim. To establish the strictly positive nature of the left eigenvector q, we can make use of a theorem by Perron (1907) in connection with positive matrices and later generalized by Frobenius (1912) for nonnegative irreducible matrices.

A matrix is said to be positive (*not* positive-definite) if its entries are *all* strictly positive. Let $\rho(A)$ represent the spectral radius of A (i.e., maximum of the absolute value of the eigenvalues of A). Thus

$$\rho(A) = \max_i |\lambda_i(A)| \tag{15-180}$$

It is easy to show that if r_i, c_i, represent respectively the ith row sum and the ith column sum of A for $i = 1, 2, \ldots, N$, then

$$\min_i r_i \le \rho(A) \le \max_i r_i \tag{15-181}$$

and

$$\min_i c_i \le \rho(A) \le \max_i c_i \tag{15-182}$$

Using (15-181), it follows that the spectral radius of a probability transition matrix is unity.

THEOREM 15-8

PERRON'S THEOREM[5]

For a positive matrix A,

(*i*) $\rho(A) > 0$ and $\rho(A)$ is an eigenvalue of A with multiplicity one.

(*ii*) There exists an eigenvector with all positive entries corresponding to the eigenvalue $\rho(A)$.

(*iii*) If λ is any other eigenvalue of A, then $|\lambda| < \rho(A)$. In particular, there are no other eigenvalues λ such that $|\lambda| = \rho(A)$.

A probability transition matrix with all positive entries would correspond to all one-step transition probabilities being positive, and as we have seen that is not the case in most of the Markov chains. Fortunately Perron's theorem can be shown to be true even if some of the entries of A are zeros, *provided* A^n is a positive matrix for some integer n. In that case A is said to be a *primitive* matrix. Thus, if the probability transition matrix P is *primitive,* then for some n

$$\min_n p_{ij}^{(n)} > 0 \tag{15-183}$$

implying that *all* states are accessible from each other at the *same* stage n. This need not be the case with all irreducible chains (see definition on page 715). For example, the matrix

$$P = \begin{pmatrix} 0 & 1 \\ 1 & 0 \end{pmatrix} \tag{15-184}$$

is irreducible[6] since $p_{12}^{(1)} = p_{21}^{(1)} = 1$, and $P^2 = I$ gives $p_{11}^{(2)} = p_{22}^{(2)} = 1$. But P is not primitive since $p_{11}^{(n)} = 0$ if n is odd and $p_{12}^{(n)} = 0$ if n is even. Note that P is periodic here with period 2. Primitive matrices exclude periodicity. More generally a nonnegative matrix A is primitive if and only if it is irreducible and aperiodic. The $m \times m$ matrix in (15-123) is also not primitive, since all of its eigenvalues have absolute value equal to unity. However, the $m \times m$ matrix P given by

$$P = \begin{pmatrix} 0 & 1 & 0 & \cdot & \cdots & 0 & 0 \\ 0 & 0 & 1 & 0 & \cdots & \cdot & 0 \\ \vdots & \vdots & \vdots & \vdots & \vdots & \vdots & \vdots \\ 0 & 0 & 0 & 0 & \cdots & 0 & 1 \\ p & q & 0 & 0 & \cdots & \cdot & 0 \end{pmatrix} \tag{15-185}$$

is primitive for any $p > 0$ and $q > 0$ since P^n is positive for $n = m^2 - 2m + 2$. Similarly,

---

[5]*Matrix Algebra and Its Applications to Statistics and Econometrics,* C. R. Rao and M. B. Rao, World Scientific, 1998; *The Theory of Matrices,* P. Lancaster and M. Tismenetsky, Academic Press, 1985.

[6]If P is an irreducible matrix, then for any states e_i, e_j we have $p_{ij}^{(m)} > 0$ for some $m = m_{ij}$. Note that for a primitive matrix there exists an n in (15-183) that is independent of i, j there, whereas for an irreducible matrix no such n may exist, although every state e_i is still accessible from every other state e_j at some stage m_{ij} (not necessarily at the same stage).

If A is a nonnegative irreducible matrix with period T, then A has exactly T eigenvalues given by $\lambda_i = \rho(A)e^{2\pi i j/T}, i = 1, 2, \ldots, T$, that are related through the T-th roots of unity, and all other eigenvalues are strictly less than $\rho(A)$ in magnitude (compare this with (iii) in Perron's theorem). See (15-67) with N even for an example and Appendix 15-B for a proof.

it follows that

$$P = \begin{pmatrix} p_0 & p_1 & p_2 & \cdots & p_m \\ 1 & 0 & 0 & \cdots & 0 \\ 1 & 0 & 0 & \cdots & 0 \\ \vdots & \vdots & \vdots & \vdots & \vdots \\ 1 & 0 & 0 & \cdots & 0 \end{pmatrix} \qquad (15\text{-}186)$$

with $\sum_i p_i = 1$ also forms a primitive stochastic matrix since P^2 is a positive matrix.

If A is primitive, then A^N is positive for some N, and hence Perron's theorem must be true for A^N. Thus $\rho(A)^N$ is a *simple* eigenvalue of A^N with all positive entries in its corresponding eigenvector. Moreover, all other eigenvalues of A^N are *strictly* less than $\rho(A)^N$ in magnitude. Since A and A^N have the same set of eigenvectors, it follows that $\rho(A)$ is a simple eigenvalue of A with the *same* positive entries in its eigenvector, and all other eigenvalues of A are strictly less than $\rho(A)$ in magnitude. Thus Perron's theorem is true if and only if A is a *primitive* matrix.

If P is a stochastic matrix, then $\rho(P) = 1$ and it follows that the eigenvector associated with $\lambda = 1$ has all positive entries, with all other eigenvalues being strictly less than unity in magnitude, if and only if P is also a primitive matrix. We can apply Perron's theorem to its transpose P' to obtain the desired result. Since $\rho(P') = \rho(P) = 1$, in that case also there exists a vector y_1 with *positive entries* (left-Perron vector of P) such that

$$P' y_1 = y_1 \qquad \text{or} \qquad y_1' P = y_1' \qquad (15\text{-}187)$$

On comparing (15-187) with (15-177), we obtain $q = y_1'$. Thus, the desired invariant distribution, if it exists, is given by the left-Perron vector of P subject to the normalization condition in (15-178). Finally to examine whether such an invariant distribution exists, we can make use of the eigen decomposition for P with x_1 and y_1, as defined in (15-179) and (15-187) and $x_k, y_k, k = 2, \ldots, N$ representing the remaining eigenvectors of P and P' respectively. Thus

$$P = \lambda_1 x_1 y_1' + \sum_{k=2}^{N} \lambda_k x_k y_k' = x_1 q + \sum_{k=2}^{N} \lambda_k x_k y_k' \qquad (15\text{-}188)$$

Since $|\lambda_k| < 1$ for $k \geq 2$ and q is a positive row vector if and only if P is a primitive stochastic matrix, we obtain

$$P^n = x_1 q + \sum_{k=2}^{N} \lambda_k^n x_k y_k' \to x_1 q = Q \qquad (15\text{-}189)$$

To summarize, *for finite Markov chains $P^n \to Q$, a positive matrix with identical rows equal to q, if and only if the transition matrix P is a primitive stochastic matrix. In that case the invariant distribution q is given by the left-Perron vector of P.* ◀

It may be remarked that (15-177) and (15-178) may have a positive solution *even if* P is not primitive. For example, the random walk with two absorbing barriers (Example 15-7) represents a nonprimitive transition matrix P in (15-20). Hence there exists no invariant distribution. However, every probability distribution of the form $q = (a, 0, 0, \ldots, 0, 1 - a), 0 < a \leq 1$ assigning positive weights to the absorbing states satisfies the equation $q = qP$. Similarly for the transition matrix example in (15-123), it is easy to see that

$$q = \left[\frac{1}{m}, \frac{1}{m}, \ldots, \frac{1}{m}\right] \tag{15-190}$$

is an obvious solution to $qP = q$, but nevertheless P^n does not converge to a positive matrix Q with identical rows, since P there is not a primitive matrix. The solution to (15-177) is meaningful as the unique invariant distribution to the Markov chain if and only if P is a primitive stochastic matrix. It follows that a doubly stochastic matrix (as in Example 15-15) has an equally likely invariant distribution given in (15-190) if and only if it is also a primitive matrix.

EXAMPLE 15-21

▶ (*i*) The $m \times m$ probability transition matrix given in (15-185) is primitive, and hence an invariant distribution exists. In that case solving $qP = q$, we obtain the stable distribution

$$q_1 = \frac{p}{p+m-1} \qquad q_2 = q_3 = \cdots = q_m = \frac{1}{p+m-1} \tag{15-191}$$

Figure 15-5 shows the state transition probabilities and the steady state matrix Q for $m = 5$ with probabilities as in (15-191). In steady state, every state communicates with every other state—even states that do not initially communicate directly, end up communicating with each other with positive probability.

(*ii*) As another example, consider the primitive transition matrix in (15-186). In that case, the matrix equation $qP = q$ leads to

$$q_0 = q_0 p_0 + \sum_{k=1}^{m} q_k \qquad q_k = q_0 p_k \qquad k \geq 1$$

The normalization in (15-178) gives

$$\sum_{k=0}^{m} q_k = q_0 + q_0 \sum_{k=1}^{m} p_k = q_0 + q_0(1 - p_0) = q_0(2 - p_0) = 1$$

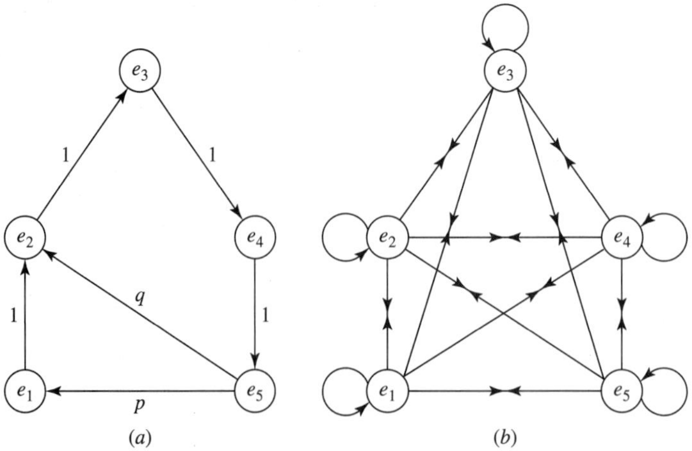

FIGURE 15-5
State transition probabilities and steady state distribution.

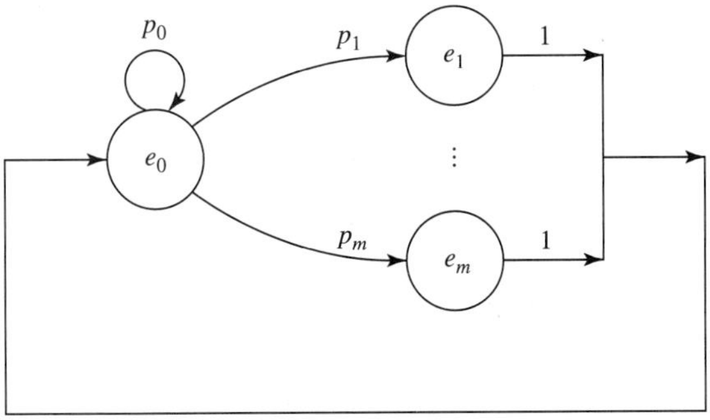

FIGURE 15-6
State transition probabilities.

Thus

$$q_0 = \frac{1}{2 - p_0} \qquad q_k = \frac{p_k}{2 - p_0} \qquad k = 1, 2, \ldots, m \tag{15-192}$$

represents the unique steady state distribution. Once again although the state transition probabilities are as in Fig. 15-6, in the long run, every state communicates with every other state with probabilities given by (15-192).

For the binary communication channel in Example 15-5, clearly P is primitive if $\alpha > 0$ and $\beta > 0$, and in that case the steady state distribution is given by

$$q_0 = \frac{\beta}{\alpha + \beta} \qquad q_1 = \frac{\alpha}{\alpha + \beta} \tag{15-193}$$

Notice that (15-193) agrees with the steady state part of the solution for $p_{jk}^{(n)}$ in (15-64). ◀

EXAMPLE 15-22

RANDOM WALKS (CONTINUED)

▶ **One reflecting barrier.** The general random walk model in Example 15-6 (page 699) with possible states $e_0, e_1, e_2, \ldots$ represents an irreducible aperiodic chain since $p_{00} = r_0 < 1$ (reflecting barrier). To examine the conditions under which the chain exhibits a steady state solution, let $u_0, u_1, u_2, \ldots$ represent the invariant probabilities if they exist. Thus

$$p_{jk}^{(n)} \to u_k \qquad k \geq 0 \tag{15-194}$$

and

$$\sum_{k=0}^{\infty} u_k = 1 \tag{15-195}$$

The sequence $\{u_k\}$ satisfies the equation [see (15-166)–(15-167)]

$$u_k = \sum_i u_i p_{ik} \tag{15-196}$$

and using (15-18)–(15-19), this reduces to

$$u_0 = u_0 r_0 + u_1 q_1$$

$$u_j = u_{j-1} p_{j-1} + u_j r_j + u_{j+1} q_{j+1} \qquad j \geq 1$$

Since $r_0 = 1 - p_0$, and $r_j = 1 - p_j - q_j$, $j \geq 1$, we get

$$u_1 q_1 = u_0 p_0 \qquad \text{and} \qquad u_{j+1} q_{j+1} - u_j p_j = u_j q_j - u_{j-1} p_{j-1} \qquad j \geq 1 \tag{15-197}$$

which gives iteratively

$$u_{k+1}q_{k+1} = u_k p_k$$

Thus

$$u_k = \frac{p_0 p_1 \cdots p_{k-1}}{q_1 q_2 \cdots q_k} u_0 \stackrel{\Delta}{=} \rho_k u_0 \tag{15-198}$$

where

$$\rho_k = \frac{p_0 p_1 \cdots p_{k-1}}{q_1 q_2 \cdots q_k} \tag{15-199}$$

For $\{u_k\}$ to satisfy (15-195), from (15-198) it is necessary and sufficient that

$$\sum_{k=1}^{\infty} \rho_k = \sum_{k=1}^{\infty} \frac{p_0 p_1 \cdots p_{k-1}}{q_1 q_2 \cdots q_k} < \infty \tag{15-200}$$

In that case the general random walk in (15-16) with one reflecting barrier possesses a steady state distribution. Using the normalization condition (15-195) we get

$$u_0 = \frac{1}{1 + \sum_{j=1}^{\infty} \rho_j} \qquad u_k = \frac{\rho_k}{1 + \sum_{j=1}^{\infty} \rho_j} \qquad k \geq 1 \tag{15-201}$$

In the special case when all steps in the random walk are identical, we have $p_k = p$, $q_k = q$, and $\rho_k = (p/q)^k$, so that for $p < q$,

$$u_0 = 1 - \frac{p}{q} \qquad u_k = \left(1 - \frac{p}{q}\right)\left(\frac{p}{q}\right)^k \qquad k \geq 1 \tag{15-202}$$

The steady state distribution in (15-202) represents a geometric random variable with expected value equal to $p/(q-p)$. From the general random walk model in Example 15-20 (see page 723) it is clear that the states are transient when $p > q$, and persistent null when $p = q$ in this case. For $p < q$, the states are all persistent nonnull, although from (15-202) the probability of the system occupying higher and higher states becomes more and more unlikely.

For the unrestricted random walk model in Sec. 10-1, since all states are either transient ($p = q$) or persistent null ($p \neq q$), as shown in Example 15-20, there is no stationary distribution in the limiting case.

Two reflecting barriers. For the random walk model with two reflecting barriers over a finite number of states $e_1, e_2, \ldots, e_N$ (Example 15-8, page 701), from (15-198) the steady state distribution satisfies

$$u_k = \left(\frac{p}{q}\right)^{k-1} u_1 \qquad k = 1, 2, \ldots, N$$

and

$$\sum_{k=1}^{N} u_k = u_1 \sum_{k=1}^{N} \left(\frac{p}{q}\right)^{k-1} = u_1 \frac{1 - (p/q)^N}{1 - p/q} = 1$$

Thus the steady state distribution is given by ($p \neq q$)

$$u_j = \frac{1 - (p/q)}{1 - (p/q)^N} \left(\frac{p}{q}\right)^{j-1} \qquad j = 1, 2, \ldots, N \tag{15-203}$$

For this finite chain there is no restriction on the size of p relative to q, and the states are all persistent nonnull. Referring to the exact expression for $p_{ij}^{(n)}$ in (15-102) for the reflecting barrier model (Example 15-19), as $n \to \infty$, the terms inside the summation there tend to zero (since $pq < 1$), and the remaining first term agrees with (15-203).

Cyclic random walk. The steady state behavior of the cyclic random walk in Example 15-9 (page 701) is easy to analyze. In that case, since P is doubly stochastic, the equally likely distribution in (15-190) holds good in the steady state provided P is primitive. For the special case in (15-23) (cyclic random walk), P is primitive only if the number of states are odd. (In the even case, P is not primitive since both 1 and -1 are eigenvalues for P; see also (15-67) with $N = 2K$).

Nonuniform random walk. For the nonuniform random walk model with finite number of states (Ehrenfest's diffusion model) discussed in Example 15-10 (page 702), the transition matrix P in (15-26) is primitive and with p_k as defined in (15-16), we have $p_k = 1 - k/N$ so that from (15-199)

$$\rho_k = \frac{\left(1-\frac{1}{N}\right)\left(1-\frac{2}{N}\right)\cdots\left(1-\frac{k-1}{N}\right)}{\frac{1}{N}\frac{2}{N}\cdots\frac{k}{N}} = \binom{N}{k}$$

and hence from (15-198), $u_k = \rho_k u_0 = \binom{N}{k} u_0$. Using the normalization condition $\sum_{k=0}^{N} u_k = 1$, we get

$$u_k = \binom{N}{k} 2^{-N} \qquad k = 0, 1, 2, \ldots, N \tag{15-204}$$

since $\sum_{k=0}^{N} \binom{N}{k} = (1+1)^N = 2^N$. Notice that for large N, the binomial distribution in (15-204) can be replaced by the normal approximation

$$u_k \simeq \sqrt{\frac{2}{\pi N}} e^{-2(k-N/2)^2/N} \tag{15-205}$$

which shows that the states in a very large system have Gaussian distribution, and the maximum of their probabilities correspond to $k = N/2$. Thus, in steady state, the system will most likely stay in those states for which $k = N/2$ or very close to it, and from (15-205) they are equiprobable states. For example, if $N = 10^6$, the probability that k is off by 0.5% from its equilibrium value is about 10^{-24}. The initial state of the system has no bearing on the final equilibrium distribution. ◀

Note Referring to Example 15-10, if A and B represent any imaginary partition of equal size in a container, then in steady state from (15-205) it is certain that one-half the molecules will be on each side of every such partition. Since the partition is arbitrary, it follows that at a macroscopic level, the molecules must be uniformly distributed inside the container in steady state. All other states, although not improbable, have a negligible probability of occurrence. It is possible that the system may once in a while drift into one of these improbable states, but then as time progresses the system is more likely to drift back toward equilibrium during the next transition than away from it, simply because of the probability distribution in (15-205). Technically all states are accessible from each other, however, the *large number* of particles involved makes it practically certain that only the "most likely states" are ever visited by the system in equilibrium. Since equiprobable states correspond to maximum disorder, the arrow of time does seem to indicate transitions from order to disorder.

EXAMPLE 15-23

SUCCESS RUNS

▶ Referring to the return to zero random walk model of (15-27) and (15-28) in Example 15-11 (page 702), every state is accessible from every other state there, since for $p_i = p$ and $q_i = q$ by expanding P^n sequentially we get

$$p_{jk}^{(n)} = \begin{cases} qp^k & k = 0, 1, 2, \ldots, n-1 \\ p^n & k = j + n \\ 0 & \text{otherwise} \end{cases}$$

Thus $p_{jk}^{(n)} > 0$ for all j and k and the chain is irreducible so that all states are either transient or persistent. If the system is initially at state e_0, then the product $p_0 p_1 \cdots p_{k-1}$ represents the probability of k consecutive transitions from e_0 to e_k. Hence, $\prod_{k=0}^{\infty} p_k$ represents the probability of the system never returning to zero. Let

$$\lim_{k\to\infty} \prod_{i=0}^{k} p_i = v_0$$

so that $f_{00} = 1 - v_0$ represents the probability of the system returning to e_0. If $v_0 > 0$, then $f_{00} < 1$ and the state e_0 is transient and hence so are all other states. Then as $n \to \infty$, the particle moves off to infinity in the positive direction.

But if $v_0 = 0$, then $f_{00} = 1$, and in that case e_0 and all other states are persistent. To determine the conditions under which a steady state distribution exists, using (15-28) in (15-196) we get for $k \geq 1$

$$u_k = p_{k-1} u_{k-1} = p_0 p_1 p_2 \cdots p_{k-1} u_0$$

and

$$\sum_{k=0}^{\infty} u_k = \left(1 + \sum_{k=1}^{\infty} p_0 p_1 \cdots p_{k-1}\right) u_0 = 1 \tag{15-206}$$

From (15-206), a stationary solution exists if and only if the series

$$1 + \sum_{k=1}^{\infty} p_0 p_1 \cdots p_{k-1} < \infty \tag{15-207}$$

In that case

$$u_0 = \frac{1}{1 + \sum_{k=1}^{\infty} p_0 p_1 \cdots p_{k-1}} \qquad u_j = \frac{p_0 p_1 \cdots p_{j-1}}{1 + \sum_{k=1}^{\infty} p_0 p_1 \cdots p_{k-1}} \qquad j \geq 1 \tag{15-208}$$

In particular, if all steps are identical, then $p_i = p < 1$ as in (15-27) and

$$\sum_{k=0}^{\infty} u_k = \sum_{k=0}^{\infty} p^k = \frac{1}{1-p} = \frac{1}{q}$$

so that

$$u_k = qp^k \qquad k \geq 0 \tag{15-209}$$

Thus the steady state distribution in the case of uniform success runs is also geometric [compare this result with (15-202)] with mean value equal to p/q. ◀

EXAMPLE 15-24

IMBEDDED MARKOV CHAINS

▶ Referring to the imbedded Markov chain formulation in Example 15-14 (page 704), using the transition matrix in (15-33)–(15-34) the system of equations in (15-167) reduces to

$$q_j = q_0 a_j + \sum_{i=1}^{j+1} q_i a_{j-i+1} \qquad j \geq 0 \tag{15-210}$$

Define the moment generating functions

$$Q(z) = \sum_{j=0}^{\infty} q_j z^j \qquad A(z) = \sum_{k=0}^{\infty} a_k z^k \tag{15-211}$$

so that using (15-210) we get

$$\begin{aligned} Q(z) &= q_0 \sum_{j=0}^{\infty} a_j z^j + \sum_{j=0}^{\infty}\sum_{i=1}^{j+1} q_i a_{j-i+1} z^j \\ &= q_0 A(z) + \sum_{i=1}^{\infty} q_i z^i \sum_{m=0}^{\infty} a_m z^m \cdot z^{-1} \\ &= q_0 A(z) + (Q(z) - q_0)A(z)/z \end{aligned} \tag{15-212}$$

or

$$Q(z) = \frac{q_0(1-z)A(z)}{A(z) - z} \tag{15-213}$$

Since $Q(1) = A(1) = 1$, using l'Hôpital's rule above, we obtain

$$Q(1) = \lim_{z \to 1} q_0 \frac{(1-z)A'(z) - A(z)}{A'(z) - 1} = \frac{q_0}{1-\rho} = 1$$

where

$$\rho = A'(1) = \sum_{k=0}^{\infty} k a_k = \sum_{k=0}^{\infty} k P\{\mathbf{y}_n = k\} = E\{\mathbf{y}_n\} \tag{15-214}$$

represents the average number of new customers (jobs) arriving per service period, or the traffic rate. Thus $q_0 = 1 - \rho > 0$, and

$$Q(z) = \frac{(1-\rho)(1-z)A(z)}{A(z) - z} \qquad \rho < 1 \tag{15-215}$$

From (15-35), Eq. (15-215) can be used to determine the long-term behavior of the underlying queue. For example, from (15-174) $\lim_{n\to\infty} P\{\mathbf{x}_n = k\} = q_k$, and the average number of customers (jobs) in the system in steady state is given by

$$\lim_{n\to\infty} E\{\mathbf{x}_n\} = Q'(1) = \sum_{k=0}^{\infty} k q_k \tag{15-216}$$

and for various appropriate input and service time distributions this quantity can be computed [see (16-211)].

For example, if $\mathbf{s}$ denotes the general service time random variable with probabiliy density function $f_s(t)$, under the Poisson arrival assumption at the input, we get (k new arrivals in duration $\mathbf{s} = t$)

$$P\{\mathbf{y}_{n+1} = k \mid \mathbf{s} = t\} = e^{-\lambda t}\frac{(\lambda t)^k}{k!} \tag{15-217}$$

and from (15-33) this gives

$$\begin{aligned} a_k = P\{\mathbf{y}_{n+1} = k\} &= \int_0^\infty P\{\mathbf{y}_{n+1} = k \mid \mathbf{s} = t\} f_s(t)\,dt \\ &= \int_0^\infty e^{-\lambda t}\frac{(\lambda t)^k}{k!} f_s(t)\,dt \end{aligned} \tag{15-218}$$

From (15-211),

$$\begin{aligned} A(z) &= \int_0^\infty e^{-\lambda t}\left(\sum_{k=0}^\infty \frac{(z\lambda t)^k}{k!}\right) f_s(t)\,dt \\ &= \int_0^\infty e^{-\lambda(1-z)t} f_s(t)\,dt \triangleq \Phi_s(\lambda(1-z)) \end{aligned} \tag{15-219}$$

where Φ_s represents the Laplace transform of the service time probability density function $f_s(t)$. With (15-219) in (15-215), we obtain the well-known *Pollaczek–Khinchin* formula.

In particular if we assume the service times to be also exponentially distributed, then $f_s(t) = \mu e^{-\mu t}$, $t \ge 0$, and (15-219) reduces to

$$A(z) = \frac{\mu}{\mu + \lambda(1-z)} = \frac{\mu}{\mu+\lambda}\sum_{k=0}^\infty \left(\frac{\lambda}{\mu+\lambda}\right)^k z^k \tag{15-220}$$

and $A'(1) = \rho = \lambda/\mu$, so that

$$a_k = \frac{\mu}{\mu+\lambda}\left(\frac{\lambda}{\mu+\lambda}\right)^k = \frac{1}{1+\rho}\left(\frac{\rho}{1+\rho}\right)^k \qquad k = 0, 1, 2, \ldots \tag{15-221}$$

Further

$$Q(z) = \frac{1-\rho}{1-\rho z} = (1-\rho)\sum_{k=0}^\infty \rho^k z^k \qquad \rho < 1 \tag{15-222}$$

Thus under the Poisson input arrivals and exponential service time assumptions, for $\rho < 1$ from (15-222) we get

$$\lim_{t\to\infty} P\{\mathbf{x}(t) = k\} = \lim_{n\to\infty} P\{\mathbf{x}_n = k\} = q_k = (1-\rho)\rho^k \qquad k \ge 0 \tag{15-223}$$

and $E[\mathbf{x}_n] \to \rho/(1-\rho)$.

From (15-223), as time goes on the number of customers waiting for service approaches a geometric distribution in such a queue, provided the average number of customers arriving per service period is strictly less than one. Thus, if the service rate is faster than the arrival rate, then the queue attains a steady state geometric distribution. From (15-221), the arrivals during the interservice periods are also geometrically distributed. Interestingly, unlike the queue, the arrival distribution in (15-221) is valid for all values of ρ in that case [see also (10-89)]. ◀

15-5 TRANSIENT STATES AND ABSORPTION PROBABILITIES

In general, as we have seen a chain can contain both persistent and transient states. From Theorem 15-5, every state is either persistent and is contained in one of the closed irreducible sets $C_1, C_2, C_3, \dots$, or it is in the set T containing all transient states. From (15-162), if the system is in a persistent state, it stays forever in the irreducible set C_r to which that state belongs. What if the system started from a transient state? As far as the evolution from a transient state is concerned, only two possibilities can occur. The system can either stay among the transient states forever, or it moves over (gets absorbed) into one of the closed sets C and stays there from then onwards. Notice that the first possibility is impossible in a finite state system since from Theorem 15-2 the system visits a transient state only a finite number of times, and hence ultimately the system must get absorbed into a closed set of persistent states (s) that are guaranteed to exist by Theorem 15-6. Of course, if the states are infinitely many, then both possibilities have finite probability, and our immediate goal is to determine these probabilities.

First, we start with an easy situation involving special chains known as *martingales*.

MARTINGALES

A Markov chain is said to be a *martingale* if for every i the expectation of the probability distribution $\{p_{ij}\}$ equals i. Thus in a martingale, we have

$$\sum_j j p_{ij} = i \tag{15-224}$$

In the genetics models in Example 15-13, both probability transition matrices (15-30) and (15-31) satisfy this definition. Hence they represent finite-chain martingales.

It is easy to compute the absorption probabilities for such chains. Let $e_0, e_1, \dots, e_N$ represent the states in a martingale. With $i = 0$ and $i = N$ in (15-224), we obtain $p_{00} = p_{NN} = 1$, and hence e_0 and e_N are absorbing states. If we assume that these are the only persistent states in the chain, then $e_1, e_2, \dots, e_{N-1}$ are transient states, and hence the system ultimately gets absorbed into either e_0 or e_N. From (15-224), by induction we also obtain[7]

$$\sum_{k=0}^{N} k p_{jk}^{(n)} = j \tag{15-225}$$

for all n. But $p_{jk}^{(n)} \to 0$ for every transient state $e_k, k = 1, 2, \dots, N-1$, and hence for $j > 0$ (15-225) gives the only solution

$$p_{jN}^{(n)} \to \frac{j}{N} \tag{15-226}$$

Since there are only two absorbing states, we also obtain

$$p_{j,0}^{(n)} \to 1 - \frac{j}{N} \tag{15-227}$$

[7]From (15-225), we have $E\{\mathbf{x}_{n+m} \mid \mathbf{x}_m\} = \mathbf{x}_m$ for all n, m, which is usually the definition for a *martingale*. The concept of martingales originally introduced by Levy was developed by Doob [12] who recognized its potential usefulness in probability theory.

Thus if the system starts from e_j, the probabilities of ultimate absorption into e_0 and e_N are $1 - j/N$ and j/N, respectively. If all states are equally likely to start with, then the probability of ultimate absorption into e_N is given by

$$\lim_{n\to\infty} \sum_{j=0}^{N} p_j^{(0)} p_{j,N}^{(n)} = \sum_{j=1}^{N} \frac{1}{N+1}\frac{j}{N} = \frac{1}{2} \tag{15-228}$$

Hence for a randomly chosen initial distribution, ultimate absorption into either e_0 or e_N are both equally likely events for a finite state martingale. ◀

Returning to the martingale models in Example 15-13, it follows that irrespective of the actual mechanism of the model, starting from an initial state e_j, the ultimate absorption probabilities into e_0 (all A genes) and e_N (all B genes) are $1 - j/N$ and j/N, respectively. In the long run, only "pure breeds" are allowed to survive in this case, and "mixed breeds" become gradually extinct.

Transient State Probabilities

To study the evolution of the system among transient states, we can make use of the general partition of the probability transition matrix P as in (15-156), where U represents the transition probabilities within the irreducible sets $C_1, C_2, \ldots$ containing all persistent states, and W represents the transition probabilities among all transient states[8]. Notice that

$$w_{ij}^{(n)} = p_{ij}^{(n)} \qquad e_i, e_j \in T \tag{15-229}$$

represents the transition probability from transient state e_i into transient state e_j in n steps. Thus W^n represents the evolutionary behavior among transient states, where W is the *substochastic* matrix obtained from P by deleting all rows and columns associated with persistent states. From (15-229), starting from a transient state e_i, the ith row sum of W^n given by

$$\sigma_i^{(n)} = \sum_{j\in T} w_{ij}^{(n)} \qquad e_i \in T \tag{15-230}$$

represents the probability of the system staying among the transient states after n transitions. To study its limiting behavior we can make use of the relation

$$w_{ij}^{(n+1)} = \sum_{k\in T} w_{ik} w_{kj}^{(n)} \qquad e_i, e_j \in T \tag{15-231}$$

that follows from the identity $W^{n+1} = WW^n$. Using (15-231) in (15-230), we get

$$\sigma_i^{(n+1)} = \sum_{k\in T} w_{ik} \sigma_k^{(n)} \qquad e_i \in T \tag{15-232}$$

Since $\sigma_i^{(1)} = \sum_{k\in T} w_{ik} \le 1$, from (15-232) we get

$$\sigma_i^{(2)} = \sum_{k\in T} w_{ik}\sigma_k^{(1)} \le \sum_{k\in T} w_{ik} = \sigma_i^{(1)}$$

---

[8]Recall that U is a block diagonal stochastic matrix with block entries $P_1, P_2 \cdots$, that correspond to the irreducible sets $C_1, C_2, \cdots$ [see (15-103)].

and by induction $\sigma_i^{(n+1)} \leq \sigma_i^{(n)}$. Thus the sequence $\{\sigma_i^{(n)}\}$ decreases monotonically to a limit

$$\sigma_i = \lim_{n\to\infty} \sigma_i^{(n)} = \lim_{n\to\infty} \sum_{j\in T} w_{ij}^{(n)} \qquad e_i \in T \tag{15-233}$$

and these limiting values satisfy equation (15-232) given by

$$\sigma_i = \sum_{j\in T} p_{ij}\sigma_j \qquad e_i \in T \tag{15-234}$$

Notice that these equations are different from the steady state equations derived in (15-167). When infinitely many states are involved, nonzero solutions can exist to the set of equations in (15-234) that satisfy $0 \leq \sigma_i \leq 1$. However, when the chain consists of *finitely* many states, as we show below $\sigma_i = 0$, for $e_i \in T$ is the only solution.

If e_j is transient, then from Theorem 15-1, (16-174), we have $p_{ij}^{(n)} \to 0$ for all i, and hence in particular $w_{ij}^{(n)} \to 0$ for all $e_i \in T$. In the finite state situation, σ_i is a sum of finitely many terms in (15-233) and it follows that

$$\sigma_i = \sum_{j\in T} \lim_{n\to\infty} w_{ij}^{(n)} \to 0$$

We can also arrive at the same conclusion by making use of the condition for equality in (15-181). In either case, equality is attained there if and only if all the row sums $r_1, r_2, \ldots, r_N$ are equal. In a finite substochastic matrix since at least one row sum is strictly less than unity (why?), it follows that $\rho(W) < 1$, and hence all eigenvalues λ_i of W are strictly less than unity in magnitude. In the case of a finite chain, W is also finite and it yields a Jordan canonical representation as in (15-104)–(15-106) with all $|\lambda_i| < 1$. As a result $W^n = U\Lambda^n U^{-1} \to 0$ so that $w_{ij}^{(n)} \to 0$, and all $\sigma_i \to 0$ in (15-234). Equivalently, (15-234) reads $Wx = \lambda x$, with x representing a column eigenvector corresponding to $\lambda = 1$. But all eigenvalues of W are strictly less than 1 in magnitude, and hence $x \equiv 0$. This shows that in a finite chain, starting from any transient state, the probability of the system remaining forever among the set of transient states is zero. In other words, in a finite chain, the system will ultimately settle down among the persistent states.

On the other hand, in a chain involving infinitely many transient states, starting from a transient state e_i, the system can remain *forever* among the transient states with probability $\sigma_i > 0$, and these probabilities satisfy the set of equations in (15-234). These equations may yield no solution, or give one or several independent valid solutions $\{x_i\}$ that satisfy $0 \leq x_i \leq 1$. Which among these solutions should one choose? To identify the desired solution, notice that if $\{x_i\}$ represents such a solution set, then $x_i \leq 1$, and since they satisfy $x_i = \sum_{j\in T} p_{ij}x_j \leq \sum_{j\in T} p_{ij} = \sigma_i^{(1)}$, by induction $x_i \leq \sigma_i^{(m)}$ implies

$$x_i = \sum_{j\in T} p_{ij}x_j \leq \sum_{j\in T} p_{ij}\sigma_j^{(m)} = \sigma_i^{(m+1)}$$

Thus $x_i \leq \sigma_i^{(n)}$ for all n and hence

$$0 \leq x_i \leq 1 \qquad \text{implies} \quad 1 \geq \sigma_i \geq x_i \tag{15-235}$$

From Eq. (15-235) it follows that the desired solution $\{\sigma_i\}$ in (15-234) satisfies the maximal property in (15-235) among all solutions, and they represent the probability that starting from any transient state, the system *never* moves into a persistent state. Thus starting from a transient state, the probability that the system stays forever among the transient states is given by the *maximal solution* of (15-234).

As Feller [3] has shown, we can use this result to derive the necessary and sufficient condition for an irreducible chain with states $e_0, e_1, \ldots$ to be persistent. In that case the solution to the system of equations

$$x_i = \sum_{j=1}^{\infty} p_{ij} x_j \qquad i \geq 1 \tag{15-236}$$

represent the probability of the system staying forever among the states $e_1, e_2, e_3, \ldots$, and never entering the state e_0. But since the states are persistent and the chain is irreducible, $f_{i,0} = 1$ and hence $x_i = 0$ for all i. Hence e_0 and all other states in an irreducible chain are persistent if and only if the system of equations in (15-236) admits no solution except the identically zero solution.

Equation (15-236) can be used to determine the conditions for a general random walk model in Example 15-6 with possible states $e_0, e_1, e_2, \ldots$ to be persistent. In that case, using (15-16)–(15-19), (15-236) reduces to

$$x_1 = r_1 x_1 + p_1 x_2$$

and

$$x_j = q_j x_{j-1} + r_j x_j + p_j x_{j+1} \qquad j \geq 2$$

Since $r_j = 1 - p_j - q_j$, these equations simplify to

$$x_2 - x_1 = \frac{q_1}{p_1} x_1$$

and

$$x_{j+1} - x_j = \frac{q_j}{p_j}(x_j - x_{j-1}) = \frac{q_j q_{j-1} \cdots q_1}{p_j p_{j-1} \cdots p_1} x_1 = \sigma_j x_1 \tag{15-237}$$

where we define

$$\sigma_j = \frac{q_j q_{j-1} \cdots q_1}{p_j p_{j-1} \cdots p_1} \qquad j \geq 1 \tag{15-238}$$

From (15-237)

$$x_{k+1} = \left(1 + \sum_{j=1}^{k} \sigma_j\right) x_1 = \left(\sum_{j=0}^{k} \sigma_j\right) x_1$$

with $\sigma_0 = 1$. From the above equation, clearly a bounded nonnegative solution $\{x_k\}$ exists if and only if the series

$$\sum_{k=0}^{\infty} \sigma_k < \infty \tag{15-239}$$

If the states are persistent, such a solution does not exist, and hence this series must diverge. Hence the necessary and sufficient condition for the general random walk model to be persistent is that the series in (15-239) diverge. In the special case if $p_j = p, q_j = q$, then $\sigma_k = (q/p)^k$ and the series in (15-239) diverges if and only if $q/p \geq 1$ or $p \leq q$. Referring to Example 15-22 (page 734), this condition agrees with the conclusions reached there for the states of a uniform random walk with one reflecting barrier to be persistent. In that case the steady state distribution in (15-202) corresponds to the set of persistent states $\{e_k\}$.

Mean Time to Absorption

Interestingly, we can use $\sigma_i^{(n)}$ in (15-230) to compute the mean time m_i that, starting from a transient state e_i, the system spends among the transient states before absorption into a persistent state. Thus, starting at $e_i \in T$, let $m_i^{(n)}$ represent the mean time that the chain spends among the transient states at the completion of n transitions. Then after one more transition, since the chain occupies one more unit of time among the transient states with probability $\sigma_i^{(n+1)}$, we have

$$m_i^{(n+1)} = m_i^{(n)} + \sigma_i^{(n+1)} = \sum_{k=0}^{n+1} \sigma_i^{(k)}.$$

Hence the mean time to absorption m_i (starting at $e_i \in T$, the mean time that the chain spends among the transient states) is given by

$$\begin{aligned} m_i &= \lim_{n\to\infty} m_i^{(n)} = \sum_{n=0}^{\infty} \sigma_i^{(n)} = \sum_{n=0}^{\infty} \sum_{j\in T} p_{ij}^{(n)} \\ &= 1 + \sum_{n=1}^{\infty} \sum_{j\in T} p_{ij}^{(n)} = 1 + \sum_{n=1}^{\infty} \sum_{j\in T} \sum_{k\in T} p_{ik} p_{kj}^{(n-1)} \\ &= 1 + \sum_{k\in T} p_{ik} \sum_{n=0}^{\infty} \sum_{j\in T} p_{kj}^{(n)} = 1 + \sum_{k\in T} p_{ik} m_k. \end{aligned} \tag{15-240}$$

For finite chains the above equation simplifies to [with W as in (15-110)]

$$m = (I - W)^{-1} E = ME$$

where

$$m \triangleq [m_1, m_2, \cdots, m_i, \cdots]^t \qquad \text{for all} \qquad e_i \in T \qquad E \triangleq [1, 1, \cdots, 1, \cdots]^t$$

and

$$M \triangleq (I - W)^{-1}$$

represents the *fundamental matrix* of the absorbing chain. The above equation has a unique positive (finite) solution for finite chains. Notice that $m_i = \sum_{j\in T} M_{ij}$, so that M_{ij} represents the mean time that the chain spends in the transient state e_j starting from the transient state e_i. Finally if v_i denotes the initial probability of the system belonging to the transient state e_i, then

$$m_a = \sum_{e_i \in T} v_i m_i$$

represents the mean time to absorption for the chain.

As an example, referring back to the random walk with two absorbing barriers in Example 15-7 (page 700), there are $(N - 1)$ transient states in (15-20), and the fundamental matrix M is the inverse of a tridiagonal Toeplitz matrix. With

$$r = p/q$$

the elements of M in that case are given by [42]

$$M_{ij} = \begin{cases} \dfrac{(r^j - 1)(r^{N-i} - 1)}{(p-q)(r^N - 1)} & j \le i \\ \dfrac{(r^i - 1)(r^{N-i} - r^{j-i})}{(p-q)(r^N - 1)} & j \ge i \end{cases}$$

for $p \neq q$, and for $p = q = 1/2$

$$M_{ij} = \begin{cases} 2j(1 - i/N) & j \le i \\ 2i(1 - j/N) & j \ge i \end{cases}$$

These expressions can be verified also by direct substitution into the identify $M(I - W) = I$.

Using these results, starting from $e_i \in T$, the mean time to absorption m_i for the chain in (15-20) is given by

$$m_i = \sum_{j=1}^{N-1} M_{ij} = \begin{cases} \dfrac{1}{p-q}\left(n\dfrac{1 - r^{-i}}{1 - r^{-n}} - i\right) & p \neq 1/2 \\ i(n-i) & p \neq 1/2 \end{cases}$$

In particular for a random walk over three transient states [$N = 4$ in (15-20)], we obtain

$$W = \begin{pmatrix} 0 & p & 0 \\ q & 0 & p \\ 0 & q & 0 \end{pmatrix} \qquad M = \frac{1}{p^2 + q^2}\begin{pmatrix} p + q^2 & p & p^2 \\ q & 1 & p \\ q^2 & q & p^2 + q \end{pmatrix}$$

and the mean time to absorption vector equals

$$m = \begin{pmatrix} m_1 \\ m_2 \\ m_3 \end{pmatrix} = \frac{1}{p^2 + q^2}\begin{pmatrix} 1 + 2p^2 \\ 2 \\ 1 + 2q^2 \end{pmatrix}$$

For a symmetric random walk this gives the mean time to absorption vector to be $m = (3, 4, 3)^t$.

Absorption Probabilities

Starting from a transient state e_i, the other alternative is for the system to visit a persistent state e_j and stay in the closed set C_r containing that state e_j. In fact, for finite chains this is the only choice, and hence such systems ultimately get absorbed into a closed set of persistent state(s). Since f_{ij} represents the probability of the system ever visiting the state e_j starting from e_i, in the present situation for an $e_i \in T$ and $e_j \in C_r$, an irreducible closed set of persistent nonnull states, f_{ij} represents the probability of absorption to state e_j starting from the transient state e_i. To compute these absorption probabilities we can make use of the fundamental relation (15-44) and Theorem 15-1. Using (16-170) in (15-44) we obtain the useful relation

$$f_{ij} = \sum_k p_{ik} f_{kj} \qquad e_j \in C_r \tag{15-241}$$

where C_r denotes as before the closed set containing the persistent state e_j. In particular, if $e_i \in T$, then this relation simplifies into

$$f_{ij} = \sum_{k \in T} p_{ik} f_{kj} + \sum_{k \in C_r} p_{ik} \qquad e_j \in C_r \tag{15-242}$$

since

$$f_{kj} = \begin{cases} 1 & e_j \in C_r, e_k \in C_r \\ 0 & e_j \in C_r, e_k \notin C_r, e_k \notin T \end{cases}$$

In fact for any two persistent states e_j and e_m that belong to the same C_r, Eq. (15-242) yields the *same* solution. Thus for any $e_i \in T$

$$f_{ij} = f_{im} \triangleq \beta_i^{(r)} \qquad e_j, e_m \in C_r$$

and hence, starting from any transient state e_i, the probability of absorption into any persistent state in a closed set C_r is the same, and is given by $f_{ij} = \beta_i^{(r)}$ for all $e_j \in C_r$. In that case from (15-242), the probability of absorption into any closed set C_r satisfies the relation

$$\beta_i^{(r)} = \sum_{k\in T} p_{ik}\beta_k^{(r)} + a_i^{(r)}, \qquad a_i^{(r)} \triangleq \sum_{k\in C_r} p_{ik} \qquad e_i \in T \tag{15-243}$$

which for finite chains translates into

$$\beta_r = (I - W)^{-1} a_r = M a_r \tag{15-244}$$

where M represents the fundamental matrix of the absorbing chain and

$$\beta_r \triangleq [\beta_1^{(r)}, \beta_2^{(r)}, \cdots, \beta_i^{(r)} \cdots]^t, \quad a_r \triangleq [a_1^{(r)}, a_2^{(r)}, \cdots, a_i^{(r)} \cdots]^t \qquad \text{for all} \qquad e_i \in T.$$

Note that the set of Eq. (15-243) possesses a unique solution if and only if its homogeneous part given by

$$\beta_i^{(r)} = \sum_{k\in T} p_{ik}\beta_k^{(r)} \qquad e_i \in T \tag{15-245}$$

possess the null solution as its only bounded solution. Notice that (15-245) is identical to the transient probability Eq. (15-234), and, in the finite state situation from the discussion there, it is guaranteed to have the null solution as its only solution. Hence in a finite Markov chain the absorption probabilities $\beta_i^{(r)} = f_{ir}$ for $e_i \in T$ always exist as a unique solution to (15-243).

Next we illustrate the computation of the absorption probabilities for the random walk model in Example 15-7 and compare them with certain results obtained earlier in Chap. 3.

EXAMPLE 15-25

ABSORPTION PROBABILITIES FOR RANDOM WALK

► For the random walk model with two absorbing barriers in Example 15-7 (page 700), consider a slight generalization of the transition probability matrix in (15-20) with p_{ij} as in (15-19) for $i \geq 1$. In that case states e_0 and e_N are absorbing states, and $e_1, e_2, \ldots, e_{N-1}$ are transient states. Hence

$$f_{0,0} = 1 \qquad f_{N,N} = 1 \qquad f_{N,0} = 0 \tag{15-246}$$

and Eq. (15-241) directly yields for $j = 0$

$$f_{i,0} = q_i f_{i-1,0} + r_i f_{i,0} + p_i f_{i+1,0} \qquad i \geq 1 \tag{15-247}$$

or

$$(f_{i+1,0} - f_{i,0})p_i = q_i(f_{i,0} - f_{i-1,0})$$

Thus

$$f_{i+1,0} - f_{i,0} = \frac{q_i}{p_i}(f_{i,0} - f_{i-1,0}) = \frac{q_i q_{i-1}\cdots q_1}{p_i p_{i-1}\cdots p_1}(f_{1,0} - 1)$$

$$= \sigma_i(f_{1,0} - 1) \tag{15-248}$$

where

$$\sigma_i \triangleq \frac{q_i q_{i-1}\cdots q_1}{p_i p_{i-1}\cdots p_1} \qquad \sigma_0 = 1 \tag{15-249}$$

Hence

$$f_{k,0} - 1 = \sum_{i=0}^{k-1}(f_{i+1,0} - f_{i,0}) = \sum_{i=0}^{k-1}\sigma_i(f_{1,0} - 1) \tag{15-250}$$

With $k = N$ in (15-250) we get $f_{1,0} - 1 = -1/\sum_{i=0}^{N-1}\sigma_i$, and hence starting from any transient state e_k the desired probability of absorption into state e_0 is given by

$$f_{k,0} = 1 - \frac{\sum_{i=0}^{k-1}\sigma_i}{\sum_{i=0}^{N-1}\sigma_i} \qquad k = 1, 2, \ldots, N-1 \tag{15-251}$$

In the special case of a uniform random walk $p_i = p$, $q_i = q$, $r_i = 0$, as in (15-20), we have

$$f_{k,0} = 1 - \frac{1 - \left(\frac{q}{p}\right)^k}{1 - \left(\frac{q}{p}\right)^N} = \frac{\left(\frac{q}{p}\right)^k - \left(\frac{q}{p}\right)^N}{1 - \left(\frac{q}{p}\right)^N}$$

$$= \frac{1 - \left(\frac{p}{q}\right)^{N-k}}{1 - \left(\frac{p}{q}\right)^N} \qquad k = 1, 2, \ldots, N-1 \tag{15-252}$$

Referring to the gambler's ruin problem in Example 3-15, since the total wealth $\$(a+b)$ of the two,players corresponds to N here, we have the absorption probability starting at state e_a (probability of ruin for player A starting with $\$a$) is given by

$$f_{a,0} = \frac{1 - \left(\frac{p}{q}\right)^b}{1 - \left(\frac{p}{q}\right)^{a+b}} \tag{15-253}$$

which agrees with the expression in (3-47) for the probability of ruin P_a for player A. Here p corresponds to the probability of winning for A at each play. As remarked earlier [see also (3-52)]

$$f_{a,0} \to \begin{cases} (q/p)^a & p > q \\ 1 & p \le q \end{cases} \tag{15-254}$$

Thus A is sure to be ruined while playing against a rich adversary who is also more skillful ($q \ge p$), such as a casino. From (15-254), the only situation where it makes sense to play against a rich adversary is when the game is advantageous to A as in that case the probability of ruin $f_{a,0} \to 0$ as a increases. The expected return (gain/loss) is given by

$$\eta = (N-k)(1 - f_{k,0}) - kf_{k,0} = b(1 - f_{a,0}) - af_{a,0} \tag{15-255}$$

and as expected it is negative when $p < q$. ◀

We conclude this section with a detailed analysis of the game of tennis as a Markov chain, and its absorption probabilities (win/loss).

EXAMPLE 15-26

THE GAME OF TENNIS[9]

▶ Tennis is played between two players—the server and the receiver. The scoring system in tennis is 15, 30, 40, and 60, so that if the server wins the first point, the score is $15:0$, otherwise it is $0:15$ in favor of the receiver. If the server also wins the second point, the score becomes $30:0$. After winning the third point the score is $40:0$, and after winning the fourth point the score is $60:0$ and the server wins the *game*. When a player loses the second point after winning the first point, 15 is scored by the opponent so that the score is $15:15$. *Deuce* is a tie in points starting with the sixth point. *Advantage in/advantage out* (Adv. in/Adv. out) starts with the seventh point if the server scores/loses the point after *deuce*. If the server scores the next point after advantage in, the server wins *the game*. If the receiver scores the point after advantage out, the receiver wins *the game*. Thus the score in a game can be only one of the following (server's score is always the first number): $15:0$, $0:15$, $15:15$, $30:0$, $30:15$, $30:30$, $15:30$, $0:30$, $40:0$, $40:15$, $40:30$, $30:40$, $15:40$, $0:40$, *deuce, advantage in, advantage out, the game.*

Once the first game is over, the second game starts with the server and receiver alternating their roles until one side wins at least six games with a margin of at least two games. This completes a set. Thus the score in a completed set can be only one of the following: $6:0$, $6:1$, $6:2$, $6:3$, $6:4$, $7:5$, $8:6$, or in its reverse order and so on. One set is followed by another set until one player wins the match by taking two sets out of three (or three-sets out of five), depending on the rules of the match.

As we shall see, the requirement that there be a two-point margin in every game as well as every set enables us to model the later portion of every game (and set) as a *random walk* over five states with two absorbing barriers. In principle, the match could continue for a very long time, and to conserve time and players' energy, sets are not continued indefinitely until a two-game margin is realized by one side. Instead, at the score of $6:6$, a *tie-breaker* gameis played in which the player who scores the first seven points with a two-point lead wins the game and the set. Otherwise the game is continued till a two-point margin is achieved.

Game Figure 15-7 shows the state diagram for a *game* where states are identified by scores. Transition from one state to the next depends only on the present state and the corresponding transition probabilities, and not on the previous history, and hence the game can be modeled as a Markov chain. Let p denote the probability of the server winning a point, and $q = 1 - p$ that of the receiver winning a point. Thus the first point is scored with probabilities

$$P\{15:0\} = p \qquad P\{0:15\} = q$$

The second point is scored with probabilities $P\{30:0\} = p^2$, $P\{15:15\} = 2pq$, $P\{0:30\} = q^2$. Similarly the third point is scored with probabilities $P\{40:0\} = p^3$, $P\{30:15\} = 3p^2q$, $P\{15:30\} = 3pq^2$, $P\{0:40\} = q^3$, and the fourth point is scored

[9]Patented in 1874 by a retired British major, W. Wingfield, tennis probably goes back to thirteenth-century France, where a similar game was played by throwing the ball to each other. The present-day scoring system is derived from stake values used in those olden day French games.

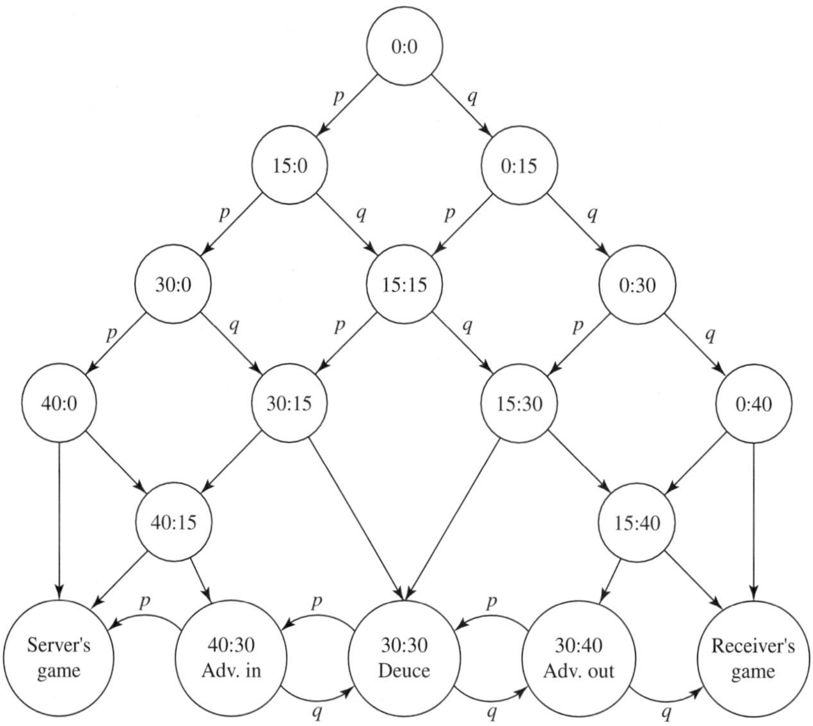

FIGURE 15-7
State diagram for a game in tennis. Each game results in a random walk among five states that are initialized by the probability distribution in (15-256)–(15-258).

with probabilities

$$P\{server\ wins\} = p^4 \qquad P\{40:15\} = 4p^3q \qquad P\{deuce\} = 6p^2q^2$$
$$P\{15:40\} = 4pq^3 \qquad P\{receiver\ wins\} = q^4$$

Finally, the fifth point is scored with probabilities

$$p_0 = P\{server\ wins\} = p^4(1+4q) \qquad p_1 = P\{adv.\ in\} = 4p^3q^2 \tag{15-256}$$

$$p_2 = P\{deuce\} = 6p^2q^2 \tag{15-257}$$

$$p_3 = P\{adv.\ out\} = 4p^2q^3 \qquad p_4 = P = \{receiver\ wins\} = q^4(1+4p) \tag{15-258}$$

The rest of the game resembles a random walk over five states with two absorption states at the two ends $(e_0, e_4) \equiv$ (*server wins; receiver wins*) and three transient states (*adv. in, deuce, adv. out*). The transition probability matrix for this random walk is given by

$$P = \begin{pmatrix} 1 & 0 & 0 & 0 & 0 \\ p & 0 & q & 0 & 0 \\ 0 & p & 0 & q & 0 \\ 0 & 0 & p & 0 & q \\ 0 & 0 & 0 & 0 & 1 \end{pmatrix} \tag{15-259}$$

and from (15-256) and (15-258) this random walk starts with the initial distribution

$$p(0) = [p_0, p_1, p_2, p_3, p_4] \tag{15-260}$$

For the three transient states e_j, $j = 1, 2, 3$, from (16-174) we have $p_{ij}^{(n)} \to 0$, and since the chain has only a finite number of states, from Theorem 15-6 the system must get absorbed into one of the two end barriers in this case. We can use the results from Example 15-25 to compute the absorption probabilities $f_{k,0}$ and $f_{k,4}$ from the transient states e_k, $k = 1, 2, 3$ into the absorbing states e_0 and e_4, respectively. Thus in the long run

$$P^n \to Q = \begin{pmatrix} 1 & 0 & 0 & 0 & 0 \\ f_{1,0} & 0 & 0 & 0 & f_{1,4} \\ f_{2,0} & 0 & 0 & 0 & f_{2,4} \\ f_{3,0} & 0 & 0 & 0 & f_{3,4} \\ 0 & 0 & 0 & 0 & 1 \end{pmatrix} \tag{15-261}$$

where $f_{k,0}$ are given by (15-252) as [on comparing (15-20) and (15-259) the roles of p and q should be reversed in (15-252)]

$$f_{k,0} = \frac{1 - (q/p)^{N-k}}{1 - (q/p)^N} \tag{15-262}$$

with $N = 4$, and

$$f_{k,4} = 1 - f_{k,0} \tag{15-263}$$

Using (15-260) in (15-48) we obtain the long-run distribution for the game to be

$$\lim_{n\to\infty} p(n) = \lim_{n\to\infty} p(0)P^n = p(0)Q \triangleq [p_g, 0, 0, 0, 1 - p_g] \tag{15-264}$$

where

$$p_g = P\{\textit{server wins the game}\} = \sum_{k=0}^{4} p_k f_{k,0} = \frac{1 - \sum_{k=0}^{4} p_k (q/p)^{4-k}}{1 - (q/p)^4} \tag{15-265}$$

with p_k, $k = 0, 1, \ldots, 4$, as in (15-260). For example, if the server plays twice as well as the receiver ($p = 2/3$, $q = 1/3$), from (15-265) we get the probability of the server winning a game to be 0.856 and the receiver winning the game to be 0.144. On the other hand, if the players are of about the same strength, with the server having a slight advantage so that $p = 0.52$ and $q = 0.48$, then the probabilities for winning the game for the server and receiver are 0.55 and 0.45, respectively. Notice that while the probability of winning a point differs only by 0.04, the probability of winning a game differs by 0.1. As we shall see, this amplification of even the slightest advantage of the stronger player is brought out in an even more pronounced manner in a set by the underlying random walk there.

Set To complete a set, games are played sequentially until one side wins at least six games with a margin of at least two games. Figure 15-8*a* shows the state diagram for a set where the states are once again identified by scores. Notice that the probability of the server winning a game is given by p_g in (15-265) and that of the receiver winning a game by $q_g = 1 - p_g$. Thus $P(6:0) = p_g^6$, and from Fig. 15-8*a* at the 11th or 12th game a new random walk phenomenon similar to (15-259) takes place with p and q replaced by p_g and q_g, respectively. Proceeding as is (15-256)–(15-258), it is easy to show that the 11th game is scored with probabilities

$$\begin{aligned} v_0 &\triangleq P\{\textit{server wins}\} = P\{6:0) \cup (6:1) \cup (6:2) \cup (6:3) \cup (6:4\} \\ &= p_g^6 + 6p_g^6 q_g + 21p_g^6 q_g^2 + 56p_g^6 q_g^3 + 126p_g^6 q_g^4 \end{aligned} \tag{15-266}$$

$$v_1 \triangleq P\{\textit{one game server adv.}\} = P\{6:5\} = 252p_g^6 q_g^5 \tag{15-267}$$

$$v_2 \triangleq P\{\textit{equal score after five all}\} = 0 \tag{15-268}$$

$$v_3 \triangleq P\{\textit{one game receiver adv.}\} = P\{5:6\} = 252q_g^6 p_g^5 \tag{15-269}$$

and

$$v_4 \triangleq P\{\textit{receiver wins}\} = q_g^6 + 6q_g^6 p_g + 21q_g^6 p_g^2 + 56q_g^6 p_g^3 + 126q_g^6 p_g^4 \tag{15-270}$$

These probabilities act as the initial distribution for the random walk in Fig. 15-8*b*. Proceeding as in (15-261)–(15-265) we obtain the long run probability distribution for a set of games to be

$$\lim_{n\to\infty} p(n) \to [v_0, v_1, v_2, v_3, v_4]\, Q_g \triangleq [p_s, 0, 0, 0, 1 - p_s] \tag{15-271}$$

where (Q_g represents the counter part of Q in (15-261) for sets)

$$p_s = \frac{1 - \sum_{k=0}^{4} v_k (q_g/p_g)^{4-k}}{1 - (q_g/p_g)^4} \tag{15-272}$$

with $v_k, k = 0, 1, \ldots, 4$ as in (15-266)–(15-270) and p_g as in (15-265). In summary, p_s represents the probability of winning a set for the server.

Table 15-1 shows the probability of winning a set for various levels of player skills. For example, an opponent with twice the skills will win each set with probability 0.9987, whereas among two equally seeded players, the one with a slight advantage ($p = 0.51$) will win each set only with probability 0.5734. In the later case, the odds in favor of the stronger player are not very significant in any one set, and hence several sets must be played to bring out the better among the two top seeded players.

Match Usually three or five sets are played to complete the match. To win a three-set match, a player needs to score either a (2 : 0) or (2 : 1), and hence the probability of

TABLE 15-1
Game of tennis

Player skills		Prob. of winning a game		Prob. of winning a set		Prob. of winning the match: 3 sets		Prob. of winning the match: 5 sets	
p	q	p_g	$1 - p_g$	p_s	$1 - p_s$	p_m	$1 - p_m$	p_m	$1 - p_m$
0.75	0.25	0.949	0.051	1.000	0	1	0	1	0
0.66	0.34	0.856	0.144	0.9987	0.0013	1	0	1	0
0.60	0.40	0.736	0.264	0.9661	0.0339	0.9966	0.0034	0.9996	0.0004
0.55	0.45	0.623	0.377	0.8215	0.1785	0.9158	0.0842	0.9573	0.0427
0.52	0.48	0.550	0.450	0.6446	0.3554	0.7109	0.2891	0.7564	0.2436
0.51	0.49	0.525	0.475	0.5734	0.4266	0.6093	0.3907	0.6357	0.3643

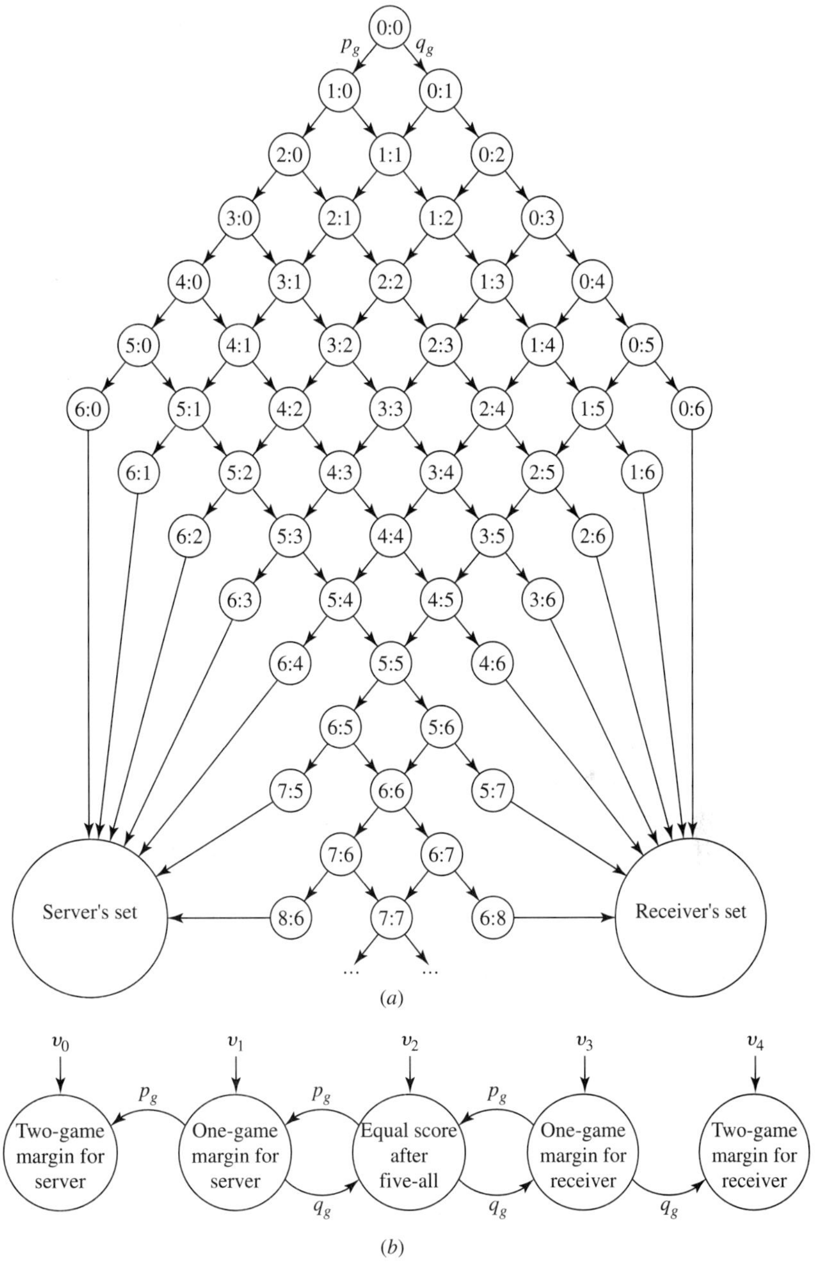

FIGURE 15-8
State diagram for a set in tennis (*a*) Set initialization. Each circle represents a game. (*b*) Each set results in a random walk among five states that are initialized by the distribution in (15-266)–(15-270).

winning a three-set match is given by

$$p_m = P\{2:0\} + P\{2:1\} = p_s^2 + 2p_s^2 q_s \tag{15-273}$$

where p_s represents the probability of winning a set for the player as given in (15-272), and $q_s = 1 - p_s$. Similarly, the probability of winning a five-set match for the same player is given by (Fig. 15-9)

$$p_m = P\{3:0\} + P\{3:1\} + P\{3:2\} = p_s^3 + 3p_s^3 q_s + 6p_s^3 q_s^2 \tag{15-274}$$

Referring to Table 15-1, top seeds and their low-ranked opponents ($p = 0.66, q = 0.34$) should be able to settle the match in three sets in favor of the top seed with probability one, which is almost always the case in the early part of any tournament. For closely seeded players of approximately equal skills ($p = 0.51, q = 0.49$), the probability of winning a three-set match is 0.609, and winning a five set match is 0.636 for the player with the slight advantage. Thus to bring out the contrast between two closely seeded players (0.51 vs. 0.49), it is necessary to play at least a five-set match (0.636 vs. 0.364), or even a seven-set match, the later of course being physically much more strenuous on the players in addition to being far too long. Recall that a game is usually 5 to 10 minutes long and an average set consists of about 10 games. Hence a three-set match lasts about 3 to 4 hours, and a five-set match about 4 to 5 hours long [10].

The game of tennis has two random walk models imbedded in it at two levels—one at the game level and the other at the set level—and they are designed to bring out the better among two players of approximately equal skill. Using the 5×5 transition matrix for the random walk in a set, it is easy to show that the total games in a set can continue to a considerable number (beyond 12) before an absorption takes place especially between top seeded players, and to conserve time and players' energy, *tie-breakers* are introduced into sets.

Tie-breakers [49] At the score of 6 : 6 in every set, *tie-breakers* are played, and the player whose turn it is to serve starts the game. The opponent serves the next two points and the server is alternated after every two points until the player who scores the first seven points with a two-point lead wins the game and the set. Notice that the two-point lead requirement once again introduces yet another random walk model towards the later part of the tie-breaker game.

The players' strategy in a tie-breaker game is quite different from that in regular games, since it is a decisive game for the set. It is quite natural that after losing a point,

---

[10] Both the initialization portion as well as the random walk part of a game (and set) contribute to its average duration (mean absorption time). Thus the average duration of a game m_g is given by $m_g = m_i + m_r$, where the (4 or 5 point) initialization part contributes [use (15-256)–(15-258)]

$$m_i = 4(p^4 + 6p^2q^2 + q^4) + 5(4p^4q + 4p^3q^2 + 4p^2q^3 + 4pq^4)$$

and the random walk part with two absorbing states and three transient states as in (15-259) contributes (see page 744)

$$m_r = p_1m_1 + p_2m_2 + p_3m_3 = \frac{1}{p^2+q^2}\{(1+2q^2)4p^3q^2 + 2(6p^2q^2) + (1+2p^2)4p^2q^3\}$$

Between two equally skilled players ($p = q = 1/2$) the above expressions give the average duration of a game to be $6\frac{3}{4}$ points. If each point is played in about 1–2 minute duration, then each game lasts for approximately 10 minutes. Similarly the average duration of a set between two equally skilled players is about 10.03 games. (Show this!).

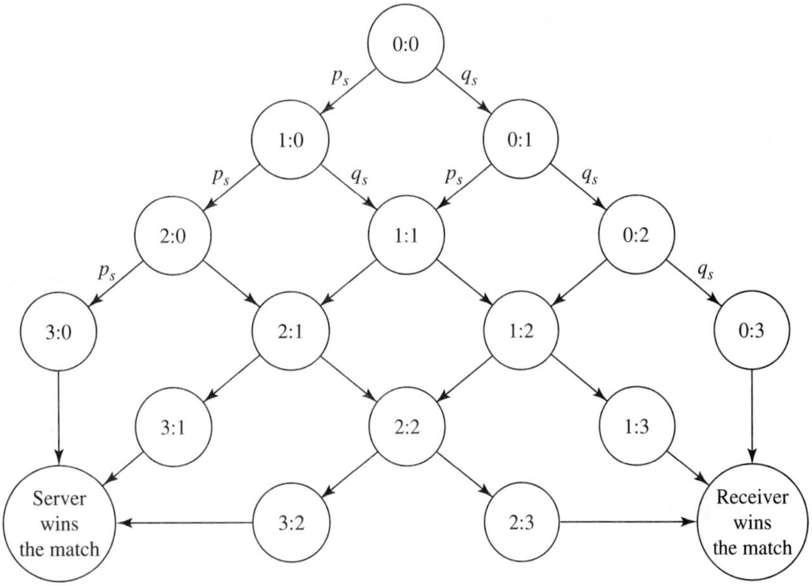

FIGURE 15-9
State diagram for the "match" in tennis. Each circle here represents a set.

each player puts in a little more effort and determination to win the next point. After winning a point the player may not be under that much pressure to win the next point. Thus with

$$P\{\textit{server wins the next point} \mid \textit{receiver won the last point}\} = \alpha \qquad (15\text{-}275)$$

and

$$P\{\textit{receiver wins the next point} \mid \textit{server won the last point}\} = \beta \qquad (15\text{-}276)$$

we can assume both $\alpha > 0.5$ and $\beta > 0.5$, and the 2×2 transition matrix for the tie-breaker points has the form

$$P_t = \begin{array}{l} \textit{server won last} \\ \textit{receiver won last} \end{array} \overset{\begin{array}{cc} \textit{server wins next} & \textit{receiver wins next} \end{array}}{\begin{pmatrix} 1-\beta & \beta \\ \alpha & 1-\alpha \end{pmatrix}} \qquad (15\text{-}277)$$

as in (15-15) for a nonsymmetric binary communication channel! The chain in (15-277) is ergodic, and its steady state probability distribution is given by (see (15-64) with α and β reversed)

$$P_t^n \to \frac{1}{\alpha+\beta}\begin{pmatrix} \alpha & \beta \\ \alpha & \beta \end{pmatrix}$$

Thus *irrespective* of which player wins the first point, after four or five points, the probability of the server winning the next point in a tie-breaker game settles down to

$$\rho = \frac{\alpha}{\alpha + \beta} \tag{15-278}$$

and the receiver winning the next point settles down to

$$1 - \rho = \frac{\beta}{\alpha + \beta} \tag{15-279}$$

and they remain as steady state values for the rest of the tie-breaker game. For example, if $\alpha = 0.7$ and $\beta = 0.6$, then after four or five points are scored, the server wins the next point with probability $7/13$ and the receiver wins it with probability $6/13$. From (15-278) and (15-279), if $\alpha > \beta$, then $\rho > 1/2$, and hence in a tie-breaker game, after losing a point, to score the next one it is advantageous to yourself to exert more than your opponent in the same situation.

The state diagram for a tie-breaker game is quite similar to that of a set in Fig. 15-8 with p_g and q_g there are replaced by ρ and $1 - \rho$ respectively. Once again as in Fig. 15-8*b* at the 11th or 12th point, the game enters a random walk with transition probability matrix as in (15-259), where p and q are replaced this time by ρ and $1-\rho$, respectively. However, the initial probability distribution for this random walk is somewhat different from those in (15-266)–(15-270) because of the slightly different requirement that to win a player must score seven points (compared to six in a set) with a two-point margin. Proceeding as before, under this rule, the 11th game is scored with probabilities

$$\begin{aligned} u_0 &= P\{7:0\} + P\{7:1\} + P\{7:2\} + P\{7:3\} + P\{7:4\} \\ &= \rho^7 \left\{1 + 7(1-\rho) + 28(1-\rho)^2 + 84(1-\rho)^3 + 210(1-\rho)^4\right\} \end{aligned} \tag{15-280}$$

$$u_1 = P\{6:5\} = 462\rho^6(1-\rho)^5 \qquad u_2 = 0 \qquad u_3 = P\{5:6\} = 462\rho^5(1-\rho)^6 \tag{15-281}$$

and

$$u_4 = (1-\rho)^7 \left\{1 + 7\rho + 28\rho^2 + 84\rho^3 + 210\rho^4\right\} \tag{15-282}$$

These probabilities act as initial distribution for the random walk ahead. Finally, using these quantities and proceeding as in (15-271)–(15-272), we obtain the probability of winning a tie-breaker game for the server to be

$$p_t = \frac{1 - \sum_{k=0}^{4} u_k[(1-\rho)/\rho]^{4-k}}{1 - [(1-\rho)/\rho]^4} \tag{15-283}$$

For $\rho = 7/13$, the probability of winning a tie-breaker game is 0.6197. On comparing (15-283) with the probability of winning a set in (15-272), if we let $p_s = \psi(p_g)$, then[11]

$$p_t \simeq \psi(\rho) \tag{15-284}$$

and it shows that the tie-breaker game is played essentially in the same spirit of an entire set. The tie-breaker is a set played rapidly within a set at an accelerated pace.

[11] The relation (15-284) is only approximate because of the difference in the initial distributions $\{u_i\}$ and $\{v_i\}$. For example, $\rho = 7/13$ gives $p_t = 0.6197$ but $\psi(\rho) = 0.612$.

Note that the initial probability distribution in (15-280)–(15-282) is somewhat idealized since it assumes that the chain represented in (15-277) has attained steady state from the first point onward. The chain reaches steady state only after four or five points, and the probability distribution during those initial points is slightly different from the steady state values in (15-278) and (15-279). To compute these probabilities exactly, let p_0 represent the probability of the server winning the first point and $q_0 = 1 - p_0$ that of the receiver winning the first point in a tie-breaker game. Then with (p_1, q_1) representing the probabilities of the server/receiver winning the second point in a tie-breaker, we get

$$[p_1, q_1] = [p_0, q_0]P_t = [p_0(1-\beta) + q_0\alpha,\ p_0\beta + q_0(1-\alpha)] \tag{15-285}$$

Similarly with $[p_k, q_k]$ representing the probabilities of the server/receiver winning the $(k+1)$st point in a tie-breaker, we have

$$[p_k, q_k] = \begin{cases} [p_{k-1}, q_{k-1}]P_t & k = 1, 2, 3, 4 \\ [\rho, 1-\rho] & k \geq 5 \end{cases} \tag{15-286}$$

These probabilities can be used in (15-280)–(15-282) to recompute the initial distribution $\{u_i\}$. For example, in that case the first term in (15-280) becomes

$$P(7:0) = p_0 p_1 p_2 p_3 p_4 \rho^2$$

Other terms can be obtained similarly; however, the computations become more involved. ◀

Note A closer examination of Fig. 15-9 reveals that each circle there represents the *set configuration* in Fig. 15-8, each with its own random walk in it. Looking further into the circles in each set diagram, one notices the *match configuration* in Fig. 15-7 embedded in every one of them. If we include also *tie-breakers,* then sometimes sets are played within sets. Thus the game of tennis represents a *self-similar process* that exhibits similar behavior for three layers deep into its segments.

15-6 BRANCHING PROCESSES

Consider a population that is able to reproduce, and let $\mathbf{x}_n$ represent the size of the nth generation (total number of offspring of the $(n-1)$th generation). If $\mathbf{y}_i$ represents the number of offspring of the ith member of the nth generation, then

$$\mathbf{x}_{n+1} = \sum_{i=1}^{\mathbf{x}_n} \mathbf{y}_i \tag{15-287}$$

Let us assume that the various offspring of different individuals are independent, identically distributed, random variables with common distribution given by (over all generations)

$$p_k = P\{\mathbf{y} = k\} = P\{\textit{an individual has k offspring}\} \geq 0 \tag{15-288}$$

and common moment generating function

$$P(z) = E\{z^{\mathbf{y}}\} = \sum_{k=0}^{\infty} p_k z^k \tag{15-289}$$

with $p_0 > 0$, $p_0 + p_1 < 1$, $p_i \neq 1$, for all i. To compute the transition probabilities

$$p_{jk} = P\{\mathbf{x}_{n+1} = k \mid \mathbf{x}_n = j\} \tag{15-290}$$

in this case, we can use the conditional moment generating function

$$\sum_{k=0}^{\infty} p_{jk} z^k = \sum_{k=0}^{\infty} z^k P\{\mathbf{x}_{n+1} = k \mid \mathbf{x}_n = j\} \triangleq E[z^{\mathbf{x}_{n+1}} \mid \mathbf{x}_n = j]$$

$$= E\left[z^{\sum_{i=1}^{j} \mathbf{y}_i} \mid \mathbf{x}_n = j\right] = [E\{z^{\mathbf{y}_i}\}]^j = P^j(z) \tag{15-291}$$

Thus the one-step transition probability p_{jk} is given by the coefficient of z^k in the expansion of $P^j(z)$ i.e., using the notation developed earlier [see (12-193)]

$$p_{jk} = \{P^j(z)\}_k$$

From (15-291), we also get the (unconditional) moment generating function of $\mathbf{x}_{n+1}$ to be

$$P_{n+1}(z) \triangleq E\{z^{\mathbf{x}_{n+1}}\} = E[E\{z^{\mathbf{x}_{n+1}} \mid \mathbf{x}_n = j\}]$$

$$= E\left\{P^j(z) \mid \mathbf{x}_n = j\right\} = \sum_{j=0}^{\infty} [P(z)]^j P\{\mathbf{x}_n = j\}$$

$$= P_n(P(z))$$

since

$$P_n(z) = E\{z^{\mathbf{x}_n}\} = \sum_{j=0}^{\infty} p_j(n) z^j \tag{15-292}$$

where

$$p_j(n) = P\{\mathbf{x}_n = j\} \tag{15-293}$$

Thus

$$P_n(z) = P_{n-1}(P(z)) \tag{15-294}$$

which gives $P_2(z) = P(P(z))$, $P_3(z) = P_2(P(z))$, and so on. Iterating (15-294), we obtain

$$P_n(z) = P_{n-1}(P(z)) = P_{n-2}(P(P(z)) = P_{n-2}(P_2(z)) \tag{15-295}$$

For $n = 3$, this gives $P_3(z) = P(P_2(z))$, and once again iterating the above equation, we get

$$P_n(z) = P_{n-3}(P(P_2(z))) = P_{n-3}(P_3(z))$$

and in general

$$P_n(z) = P_{n-k}(P_k(z)) \qquad k = 0, 1, 2, \ldots, n$$

which for $k = n - 1$ gives

$$P_n(z) = P_1(P_{n-1}(z)) = P(P_{n-1}(z)) \tag{15-296}$$

Together with (15-295), we obtain the useful relation

$$P_n(z) = P_{n-1}(P(z)) = P(P_{n-1}(z)) \triangleq \sum_{k=0}^{\infty} p_k(n) z^k \tag{15-297}$$

For example, if we assume that the direct descendants follow a geometric distribution[12] given by $p_k = qp^k$ in (15-288), then $P(z) = q/(1 - pz)$ in (15-289), and an explicit calculation for $P_2(z)$, $P_3(z)$ leads to the general formula

$$P_n(z) = q\frac{p^n - q^n - (p^{n-1} - q^{n-1})pz}{p^{n+1} - q^{n+1} - (p^n - q^n)pz} \qquad p \neq q \tag{15-298}$$

and for $p = q$, we get

$$P_n(z) = \frac{n - (n-1)z}{n + 1 - nz} \tag{15-299}$$

In a slightly different model, if we assume the direct descendent distribution to be

$$p_k = \begin{cases} cp^k & k \geq 1 \\ p_0 = 1 - \dfrac{cp}{1-p} & k = 0 \end{cases} \tag{15-300}$$

then

$$P(z) = p_0 + \sum_{k=1}^{\infty} p_k z^k = p_0 + \frac{cpz}{1 - pz} \tag{15-301}$$

According to Lotka (1931), the statistics for the average American family ($\simeq$ 1920s) satisfy (15-300) with $p = 0.7358$, $p_0 = 0.4823$ (roughly 48% families have no children), so that $c = 0.1859$, and the moment-generating function in (15-301) simplifies to

$$P(z) = \frac{0.4823 - 0.2181z}{1 - 0.7358z} \tag{15-302}$$

In a similar manner, to determine the higher order transition probabilities $p_{jk}^{(n)}$, we can proceed as in (15-291). Thus,

$$\begin{aligned}
\sum_{k=0}^{\infty} p_{jk}^{(n)} z^k &= \sum_{k=0}^{\infty} z^k P\{\mathbf{x}_n = k \mid \mathbf{x}_0 = j\} = E\{z^{\mathbf{x}_n} \mid \mathbf{x}_0 = j\} \\
&= \sum_{k=0}^{\infty}\sum_{i=0}^{\infty} z^k P\{\mathbf{x}_n = k \mid \mathbf{x}_{n-1} = i, \mathbf{x}_0 = j\} P\{\mathbf{x}_{n-1} = i \mid \mathbf{x}_0 = j\} \\
&= \sum_{i=0}^{\infty} E\{z^{\mathbf{x}_n} \mid \mathbf{x}_{n-1} = i\} P\{\mathbf{x}_{n-1} = i \mid \mathbf{x}_0 = j\} \\
&= \sum_{i=0}^{\infty} [P(z)]^i \, P\{\mathbf{x}_{n-1} = i \mid \mathbf{x}_0 = j\} \\
&= E\{[P(z)]^{\mathbf{x}_{n-1}} \mid \mathbf{x}_0 = j\} \\
&= E\{[P(P(z))]^{\mathbf{x}_{n-2}} \mid \mathbf{x}_0 = j\} = E\{[P_{n-1}(z)]^{\mathbf{x}_1} \mid \mathbf{x}_0 = j\} \\
&= \{E\,[P_{n-1}(z)]^{\mathbf{y}_1}\}^j = [P(P_{n-1}(z))]^j = [P_n(z)]^j
\end{aligned} \tag{15-303}$$

[12]Refer to Example 15A-3 in Appendix 15A for an interesting physical justification for the geometric model.

since $\mathbf{x}_1 = \sum_{i=1}^{j} \mathbf{y}_i$. Equation (15-303) represents the moment generating function of the nth generation given that the process starts with j ancestors. Thus $p_{jk}^{(n)}$ is given by the coefficient of z^k in the power series expansion of $[P_n(z)]^j$, that is,

$$p_{jk}^{(n)} = \{[P_n(z)]^j\}_k$$

Notice that $p_{1,k}^{(n)}$ is the same as $p_k(n)$ in (15-297).

Extinction Probability

An important question, first raised by Galton (1873) in connection with extinction of family surnames, is to determine the extinction probability

$$\pi_0 = \lim_{n\to\infty} p_0(n) = \lim_{n\to\infty} p_{1,0}^{(n)} = P_n(0) \tag{15-304}$$

the limit of the probability $p_{1,0}^{(n)} = P\{\mathbf{x}_n = 0 \mid \mathbf{x}_0 = 1\}$ of *zero individuals* in the nth generation, given that $\mathbf{x}_0 = 1$. For the geometric distribution model in (15-298), we have

$$p_0(n) \to \begin{cases} q/p & p > q \\ 1 & p \le q \end{cases} \tag{15-305}$$

To find its analogue for *any* general distribution in (15-289), let

$$z_n \triangleq p_{1,0}^{(n)} = P_n(0)$$

so that $z_1 = P(0) = p_0$ and

$$z_n = P(P_{n-1}(0)) = P(z_{n-1}) \tag{15-306}$$

If $p_0 = 0$, then $z_1 = 0, z_2 = 0, \ldots, z_n = 0$, and so on. Similarly if $p_0 = 1$, then $z_1 = P(1) = 1, z_2 = 1, \ldots, z_n = 1, \ldots,$ that is, if the probability of no offspring is one, then extinction is bound to occur after the zeroth generation. Excluding these extreme cases, we have $0 < p_0 < 1$. Since $P(z)$ is a strictly increasing (convex) function of z, we have $z_2 = P(z_1) > P(0) = p_0 = z_1$, and by induction $z_1 < z_2 < \cdots < z_n < z_{n+1} \le 1$. Thus $p_0(n)$ is a bounded increasing sequence, and a limit $\pi_0 \le 1$ exists. From (15-306), it is clear that the above limit satisfies the equation[13]

$$z = P(z) \tag{15-307}$$

Referring to Fig. 15-10, in the interval $0 \le z \le 1$, the convex curve $P(z)$ starts at the point $(0, p_0)$ above the bisector and ends at the point $(1, 1)$ on the bisector. As a result, two situations are possible, as shown in Fig. 15-10*a* and 15-10*b*.

In Fig. 15-10*a*, the graph $P(z)$ is entirely above the bisector line. In this case, $z = 1$ is the unique root of the equation $z = P(z)$, and hence $z_n \to 1$. Since $P(z) \ge z$ in $0 \le z \le 1$, we have $1 - P(z) \le 1 - z$ or $(1 - P(z))/(1 - z) \le 1$, and letting $z \to 1$, we also obtain in that case the mean value $\mu = P'(1) \le 1$ (see also Fig. 15-10*a*). The slope at $z = 1$ is less than or equal to one.

In Fig. 15-10*b*, the graph $P(z)$ intersects the bisector line at some point $\pi_0 < 1$, in addition to that at $z = 1$. Since a convex curve can intersect a straight line at most

---

[13]Starting with $z_1 = P(0)$, the recursion in (15-306) can be used to determine the extinction probability numerically (alternating projections onto convex sets). The condition $p_0 + p_1 < 1$ guarantees strict convexity for $P(z)$.

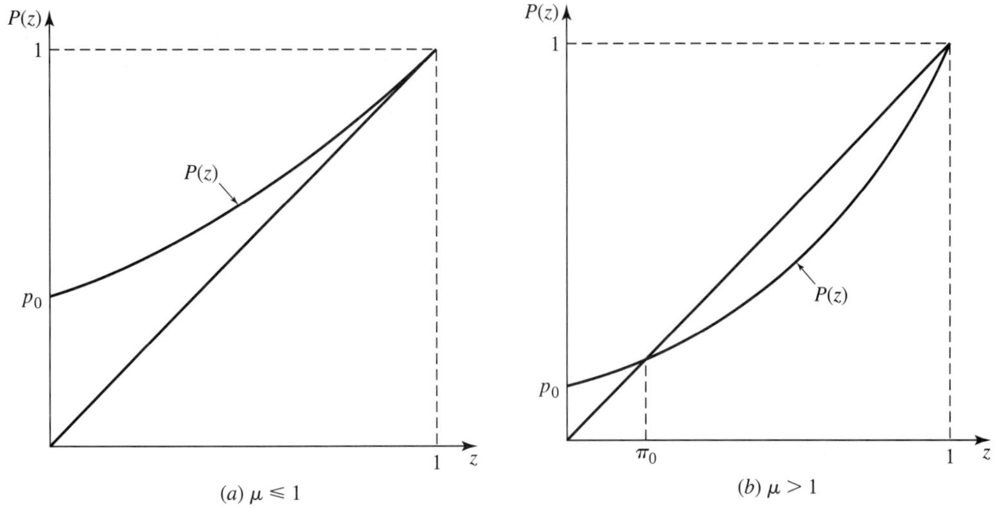

FIGURE 15-10
Probability of extinction for branching processes.

at two points, we have $P(z) > z$ for $z < \pi_0$and $P(z) < z$ for $\pi_0 < z < 1$. To start with, since $0 < \pi_0$, we get $z_1 = P(0) = p_0 < P(\pi_0) = \pi_0$, and by induction $z_n = P(z_{n-1}) < P(\pi_0) = \pi_0$. In that case, $z_n \to \pi_0 < 1$, and the graph $P(z)$ crosses over the bisector at $z = 1$. Hence we must have $\mu = P'(1) > 1$ here. This also follows from the mean value theorem by which there exists a point between π_0 and 1 at which the derivative equals $(P(1) - P(\pi_0))/(1 - \pi_0) = 1$. Since the derivative $P'(z)$ is monotone, we have $P'(1) > 1$. Thus the two cases are characterized by the mean value μ of the descendant's distribution being greater than unity or otherwise. We summarize these observations in Theorem 15-8.

THEOREM 15-9

EXTINCTION PROBABILITY

Let $\{p_k\}$ represent the common descendant distribution of a branching process, and $P(z) = \sum_{k=0}^{\infty} p_k z^k$ its moment-generating function. If the mean value

$$\mu = \sum_{k=0}^{\infty} k p_k = P'(1) \leq 1 \tag{15-308}$$

then the process dies out eventually with probability one, and if $\mu > 1$, the probability that the process terminates on or before the nth generation tends to the unique positive root $\pi_0 < 1$ of the equation $P(z) = z$.

As an example, consider the moment generating function given by (15-301). In that case the identity $P(z) = z$ leads to a quadratic equation whose roots are given by unity and

$$\pi_0 = \frac{1 - p(1 + c)}{p(1 - p)} < 1 \tag{15-309}$$

In particular, for the simplified American population model in (15-302), we get $\pi_0 = 0.6554$. Thus in a broad sense, the probability of extinction is about 0.65 for the American population represented by (15-302). However, immigration into the population makes the matters more interesting.

In general, the average size of the population at the nth stage is given by

$$\mu_n = E\{\mathbf{x}_n\} = P_n'(1) \tag{15-310}$$

But from (15-294), $P_n'(z) = P_{n-1}'(P(z))P'(z)$, so that

$$\mu_n = P_n'(1) = P'(1)P_{n-1}'(1) = \mu\mu_{n-1} = \mu^n \to \begin{cases} 0 & \mu < 1 \\ \infty & \mu > 1 \end{cases} \tag{15-311}$$

Thus it is not surprising that the process is bound for extinction when $\mu < 1$, but that a stable solution is still impossible for $\mu = 1$ is somewhat surprising. Finally $\mu > 1$, corresponds to a geometric growth in population, with a probability of extinction equal to π_0. For $\mu \le 1$, the probability of extinction is unity, implying that *almost surely* there will be no descendants in the long run. On the other hand, for $\mu > 1$, after a sufficient number of generations it is quite likely that either there are no descendants with probability π_0, or a great (infinitely) many descendants with probability $1 - \pi_0$. Thus the two extreme situations (zero population and infinite population) correspond to *absorbing states,* and all intermediate states with finite population are *transient states*. To summarize, *in the long run, irrespective of the mean value of the descendant distribution, every species either dies out completely, or its population explodes without bound,* both unpleasant conclusions either way.

We can also arrive at this conclusion by observing that

$$\lim_{n\to\infty} P_n(z) = \lim_{n\to\infty} P_{n-1}(P(z)) = \pi_0 \tag{15-312}$$

since the limit satisfies the equation $P(z) = z$, *irrespective* of the mean value μ of the descendents' distribution. From (15-312), the coefficients of $z, z^2, z^3, \ldots$ all tend to zero in $P_n(z)$. Thus using (15-292) and (15-293), we get

$$\lim_{n\to\infty} P\{\mathbf{x}_n = 0\} = \pi_0 \qquad \lim_{n\to\infty} P\{\mathbf{x}_n = k\} = 0 \tag{15-313}$$

for any *finite* positive k, and hence

$$\lim_{n\to\infty} P\{\mathbf{x}_n = \infty\} = 1 - \pi_0 \tag{15-314}$$

Total Number of Descendants' Distribution

Let $\mathbf{s}_n$ represent the total number of descendants up to and including the nth generation. Then

$$\mathbf{s}_n = 1 + \mathbf{x}_1 + \mathbf{x}_2 + \cdots + \mathbf{x}_n \tag{15-315}$$

To determine the long-term behavior of the total population size and its distribution function, let $H_n(z)$ represent the moment generation function of the total population $\mathbf{s}_n$. Then

$$H_n(z) = \sum_{k=0}^{\infty} P\{\mathbf{s}_n = k\}z^k = zG_n(z) \tag{15-316}$$

where $G_n(z)$ represents the moment-generating function of the random variable

$$\mathbf{u}_n = \mathbf{x}_1 + \mathbf{x}_2 + \cdots + \mathbf{x}_n$$

that is, the total population up to the nth generation *without* the ancestor $\mathbf{x}_0 = 1$. Thus

$$\begin{aligned}
G_n(z) &= \sum_{k=0}^{\infty} P\{\mathbf{u}_n = k\} z^k = E\{z^{\mathbf{u}_n}\} \\
&= E\{E(z^{\mathbf{u}_n} \mid \mathbf{x}_1 = j)\} = E\left\{E\left(z^{(j+\mathbf{x}_2+\cdots+\mathbf{x}_n)} \,\middle|\, \mathbf{x}_1 = j\right)\right\} \\
&= E\left\{z^j E\left(z^{(\mathbf{x}_2+\cdots+\mathbf{x}_n)} \,\middle|\, \mathbf{x}_1 = j\right)\right\} \\
&= E\left[z^j \left\{E\left(z^{(\mathbf{x}_1+\mathbf{x}_2+\cdots+\mathbf{x}_{n-1})} \,\middle|\, \mathbf{x}_0 = j\right)\right\}\right] \\
&= E[z^j \{E(z^{\mathbf{u}_{n-1}} \mid \mathbf{x}_0 = j)\}] = E\{[zG_{n-1}(z)]^j\} \\
&= \sum p_j [zG_{n-1}(z)]^j = P(zG_{n-1}(z)) = P(H_{n-1}(z)) \qquad (15\text{-}317)
\end{aligned}$$

where we have made use of the fact that if the process starts with one ancestor, the moment generating function of the succeeding n generations is given by $G_n(z)$, and if starts with j ancestors, then the corresponding moment generating function is given by $(G_n(z))^j$ (see also (15-303)). Substituting (15-317) into (15-316) we obtain the desired recursion formula

$$H_n(z) = zP(H_{n-1}(z)) \qquad (15\text{-}318)$$

for the total population size. If $H_n(z) = \sum_{k=0}^{\infty} h_k^{(n)} z^k$, from (15-316) $h_k^{(n)}$ represents the probability that the total population size up to and including the nth generation equals k. Thus

$$h_k^{(n)} = P\{\mathbf{s}_n = k\} \qquad (15\text{-}319)$$

From (15-315)–(15-316), and (15-287)–(15-289) for $0 < z < 1$, since $H_1(z) = zP(z) < z$, from the convexity of $P(z)$ we have $P(zP(z)) < P(z)$, and hence

$$H_1(z) = zP(z) > zP(H_1(z)) = H_2(z)$$

and by induction assuming that $H_m(z) < H_{m-1}(z)$, we get

$$H_{m+1}(z) = zP(H_m(z)) < zP(H_{m-1}(z)) = H_m(z) \qquad (15\text{-}320)$$

where we have made use of the convexity property of $P(z)$. Hence $H_n(z) < H_{n-1}(z)$ for all $n > 0$ for every $0 < z < 1$, and $H_n(z)$ represents a monotone decreasing sequence that is bounded from below. Let

$$\lim_{n\to\infty} H_n(z) = H(z) = \sum_{k=0}^{\infty} h_k z^k \qquad (15\text{-}321)$$

represent this limit. Then h_k are nonnegative numbers such that $H(1) = \sum_{k=0}^{\infty} h_k \le 1$ and taking the limit in (15-318), we obtain that the limiting function in (15-321) satisfies the equation

$$H(z) = zP(H(z)) \qquad 0 < z < 1 \qquad (15\text{-}322)$$

THEOREM 15-10

DESCENDANT'S DISTRIBUTION

▶ The function $H(z)$ in (15-322) is the unique root of the equation

$$x = zP(x) \tag{15-323}$$

such that $H(z) \leq \pi_0$, where π_0 is the smallest positive root of the equation

$$x = P(x) \tag{15-324}$$

and $H(1) = \sum_{k=0}^{\infty} h_k = \pi_0 \leq 1$.

Proof. For $0 < z \leq 1$ let $x = H(z)$ in (15-322) so that the desired $H(z)$ is the solution of (15-323) that is bounded by unity. Let π_0 represent the smallest positive root of (15-324). From Theorem 15-8, clearly $\pi_0 \leq 1$. For every fixed $z < 1$, it is easy to see that the convex function $y = zP(x)$ lies entirely below the function $y = P(x)$ for $0 \leq x \leq 1$ in Fig. 15-10, and hence the function $zP(x)$ intersects the line $y = x$ at a unique point that is strictly less than π_0. This unique point $H(z) < \pi_0$ for $z < 1$. At $z = 1$, $H(1)$ is the smallest root π_0 of the Eq. (15-324) and hence for $0 < z < 1$ we have a unique function $H(z) \leq 1$, and this completes the proof. ◀

From Theorem 15-9, if the mean value $\mu \leq 1$, then $H(1) = \sum_{k=0}^{\infty} h_k = \pi_0 = 1$, and the limiting function in (15-321) represents a proper moment-generating function. However, if $\mu > 1$, then $\pi_0 < 1$, and

$$\lim_{n\to\infty} P\{\mathbf{s}_n = \infty\} = 1 - \pi_0 > 0$$

that is, the total population explodes with probability $1 - \pi_0$, a conclusion that agrees with (15-314). Interestingly, as we shall see in Chap. 16, the limiting distribution $H(z)$ also represents the total number of customers served during the busy periods in certain type of queues [see (16-230)–(16-236)].

Immigration

Several species (plants, animals) have been eliminated from the face of this planet, and total extinction is certainly possible. Thus left to themselves, populations either die out completely or grow without bound. However, immigration from *outside* into an otherwise unstable population ($\mu \leq 1$) can have stabilizing effects on the population. To see this, consider a population model with descendant distribution $\{p_k\}$ and moment generating function $P(z)$ as in (15-288) and (15-289). Suppose $\mathbf{m}_n$ immigrants enter the nth generation independently with probability distribution function

$$P\{\mathbf{m}_n = j\} = b_j \qquad j = 0, 1, 2, \ldots \tag{15-325}$$

The totality of immigrants $\mathbf{m}_n$, $n = 0, 1, 2, \ldots$ entering into successive generations are independent, identically distributed, random variables with common moment generating function

$$B(z) = E\{z^{\mathbf{m}_n}\} = \sum_{k=0}^{\infty} b_k z^k \tag{15-326}$$

and they contribute to the next generations in the same way as those already present in the population. Thus with $\mathbf{x}_n$ representing the size of the population prior to immigration at the nth stage, let

$$\mathbf{w}_n = \mathbf{x}_n + \mathbf{m}_n \tag{15-327}$$

represent the total population size at that stage, so that similar to (15-287), we get

$$\mathbf{x}_{n+1} = \sum_{i=1}^{\mathbf{w}_n} \mathbf{y}_i \tag{15-328}$$

In this case the transition probabilities p_{jk} for the total population satisfy

$$\begin{aligned}
\sum_{k=0}^{\infty} p_{jk} z^k &= \sum_{k=0}^{\infty} z^k P\{\mathbf{w}_{n+1} = k \mid \mathbf{w}_n = j\} \\
&= \sum_{k=0}^{\infty} z^k P\{\mathbf{x}_{n+1} + \mathbf{m}_{n+1} = k \mid \mathbf{w}_n = j\} \\
&= \sum_{k=0}^{\infty} \sum_{i=0}^{\infty} z^{k-i} P\{\mathbf{x}_{n+1} = k - i \mid \mathbf{w}_n = j\} z^i P\{\mathbf{m}_{n+1} = i\} \\
&= E\{z^{\mathbf{x}_{n+1}} \mid \mathbf{w}_n = j\} E\{z^{\mathbf{m}_{n+1}}\} = E\left(z^{\sum_{i=1}^{j} \mathbf{y}_i}\right) B(z) \\
&= [P(z)]^j B(z)
\end{aligned} \tag{15-329}$$

Thus p_{jk} is given by the coefficient of z^k in $B(z)[P(z)]^j$. Further, let $Q_n(z)$ denote the moment generating function of the total population $\mathbf{w}_n$ at stage n. Thus

$$\begin{aligned}
Q_n(z) &= \sum_k z^k P\{\mathbf{w}_n = k\} = \sum_k z^k P\{\mathbf{x}_n + \mathbf{m}_n = k\} \\
&= \sum_k \sum_i z^{k-i} P\{\mathbf{x}_n = k - i\} z^i P\{\mathbf{m}_n = i\} \\
&= \sum_{m=0}^{\infty} E\{z^{\mathbf{x}_n} \mid \mathbf{w}_{n-1} = m\} P\{\mathbf{w}_{n-1} = m\} E\{z^{\mathbf{m}_n}\} \\
&= B(z) \sum_{m=0}^{\infty} E\left\{z^{\sum_{i=1}^{m} \mathbf{y}_i}\right\} P\{\mathbf{w}_{n-1} = m\}
\end{aligned}$$

where we have made use of (15-326)–(15-328). Hence

$$\begin{aligned}
Q_n(z) &= B(z) \sum_{m=0}^{\infty} [E\{z^{\mathbf{y}_1}\}]^m P\{\mathbf{w}_{n-1} = m\} \\
&= B(z) \sum_m [P(z)]^m P\{\mathbf{w}_{n-1} = m\} \\
&= B(z) Q_{n-1}(P(z))
\end{aligned} \tag{15-330}$$

Thus if $\lim_{n\to\infty} Q_n(z) = G(z) = \sum_{k=0}^{\infty} g_k z^k$, then it satisfies the equation

$$G(z) = B(z) G(P(z)) \tag{15-331}$$

and in that case, if $G(z)$ represents a proper probability-generating function, then the limiting distribution $\lim_{n\to\infty} P\{\mathbf{w}_n = k\} = g_k$ exists for all k. It has been shown by Heathcote that if the descendants' mean value $\mu < 1$ and $B'(1) < \infty$, then $G(z)$ satisfying (15-331) exists as a proper probability generating function if and only if $\sum_{k=1}^{\infty} b_k \log k < \infty$. In that case

$$\lim_{n\to\infty} P\{\mathbf{w}_n = k\} \to g_k \qquad k = 0, 1, 2, \ldots \tag{15-332}$$

Finally if $\mu = 1$, and if the descendants' distribution has finite variance, then Seneta has shown that $\mathbf{w}_n/n$ converges in distribution to a gamma random variable. Thus under *immigration,* the *transient states* become *persistent nonnull states.* In summary, it is possible to avoid population extinction and achieve stabilization through immigration.

The assumption that the descendants' distribution $\{p_k\}$ in (15-288) remains the same throughout all generations may be an oversimplification, and it has been replaced by a time-dependent offspring distribution $\{p_{n,k}\}$ for the nth generation by Jagers and others. In yet another generalization by Wilkinson and others, the offspring distribution for each generation is selected randomly from a class of distributions, and interesting results in random environments have been obtained by Wilkinson, Atheya, Karlin, Kaplan, and many others [34–35, 42, 44, 48].

APPENDIX 15A
MIXED TYPE POPULATION OF CONSTANT SIZE

Consider two populations of types A and B each multiplying independently according to branching processes $\{\mathbf{x}_n\}$ and $\{\mathbf{y}_n\}$ given by

$$\mathbf{x}_{n+1} = \sum_{i=1}^{\mathbf{x}_n} \boldsymbol{\xi}_i \qquad \mathbf{y}_{n+1} = \sum_{j=1}^{\mathbf{y}_n} \boldsymbol{\eta}_j \tag{15A-1}$$

Let

$$P\{\boldsymbol{\xi}_i = k\} = a_k \geq 0 \qquad P\{\boldsymbol{\eta}_j = k\} = b_k \geq 0 \qquad k = 0, 1, 2, \ldots \tag{15A-2}$$

represent the respective progency distributions for single individuals in each population. Then

$$A(z) = \sum_{k=0}^{\infty} a_k z^k \qquad B(z) = \sum_{k=0}^{\infty} b_k z^k \tag{15A-3}$$

represent their respective moment generating functions, and from (15-290) and (15-291), $A^i(z)$ gives the generating function for the number of offspring of i individuals for the type-A population[14], that is,

$$P\{\mathbf{x}_{n+1} = j \mid \mathbf{x}_n = i\} = \{A^i(z)\}_j \tag{15A-4}$$

[14]As before, the notation $\{P(z)\}_j$ represents the coefficient of z^j in $P(z)$.

The two-dimensional process evolves as a sequence of pairs of random variables $(\mathbf{x}_n, \mathbf{y}_n)$ composed of the independent branching processes $\{\mathbf{x}_n\}$ and $\{\mathbf{y}_n\}$ so that

$$\begin{aligned} P\{\mathbf{x}_{n+1} = j_1, \mathbf{y}_{n+1} = j_2 \mid \mathbf{x}_n = i_1, \mathbf{y}_n = i_2\} &= P\{\mathbf{x}_{n+1} = j_1 \mid \mathbf{x}_n = i_1\} P\{\mathbf{y}_{n+1} = j_2 \mid \mathbf{y}_n = i_2\} \\ &= \{A^{i_1}(z)\}_{j_1} \{B^{i_2}(z)\}_{j_2} \end{aligned} \tag{15A-5}$$

Consider the special situation, where the combined population remains fixed over all generations. Thus

$$\mathbf{x}_n + \mathbf{y}_n = N \qquad n = 0, 1, 2, \ldots \tag{15A-6}$$

In that case if $\{\mathbf{x}_n = i\}$, then necessarily $\{\mathbf{y}_n = N - i\}$, so that the one-step transition probability for event $\{\mathbf{x}_{n+1} = j\}$ given $\{\mathbf{x}_n = i\}$ simplifies as [42]

$$\begin{aligned} p_{ij} &= P\{\mathbf{x}_{n+1} = j \mid \mathbf{x}_n = i, \mathbf{x}_n + \mathbf{y}_n = \mathbf{x}_{n+1} + \mathbf{y}_{n+1} = N\} \\ &= \frac{P\{\mathbf{x}_{n+1} = j, \mathbf{x}_n = i, \mathbf{x}_n + \mathbf{y}_n = \mathbf{x}_{n+1} + \mathbf{y}_{n+1} = N\}}{P\{\mathbf{x}_{n+1} + \mathbf{y}_{n+1} = N, \mathbf{x}_n = i, \mathbf{x}_n + \mathbf{y}_n = N\}} \\ &= \frac{P\{\mathbf{x}_{n+1} = j, \mathbf{x}_{n+1} + \mathbf{y}_{n+1} = N \mid \mathbf{x}_n = i, \mathbf{x}_n + \mathbf{y}_n = N\}}{P\{\mathbf{x}_{n+1} + \mathbf{y}_{n+1} = N \mid \mathbf{x}_n = i, \mathbf{x}_n + \mathbf{y}_n = N\}} \\ &= \frac{P\{\mathbf{x}_{n+1} = j, \mathbf{y}_{n+1} = N - j \mid \mathbf{x}_n = i, \mathbf{y}_n = N - i\}}{P\{\mathbf{x}_{n+1} + \mathbf{y}_{n+1} = N \mid \mathbf{x}_n = i, \mathbf{y}_n = N - i\}} \\ &= \frac{\{A^i(z)\}_j \{B^{N-i}(z)\}_{N-j}}{\{A^i(z) B^{N-i}(z)\}_N} \qquad i, j = 0, 1, \ldots, N \end{aligned} \tag{15A-7}$$

Here we have used (15A-5) in simplifying the numerator, and the denominator expression follows, since the moment generating function for the sum random variable

$$\mathbf{z}_{n+1} = \mathbf{x}_{n+1} + \mathbf{y}_{n+1} = \sum_{m=1}^{\mathbf{x}_n} \xi_m + \sum_{m=1}^{\mathbf{y}_n} \eta_m \tag{15A-8}$$

under the condition $\mathbf{x}_n = i$, $\mathbf{y}_n = N - i$, is given by $A^i(z) B^{N-i}(z)$. Interestingly (15A-7) represents the transition probability matrix for a finite Markov chain with state space $\{0, 1, 2, \ldots, N\}$.

As Examples 15A-1 and 15A-2 show the genetic models in Example 15-13 can be derived as special cases of this model.

EXAMPLE 15A-1

SECOND ORDER BINOMIAL MODEL

▶ Suppose the individuals in either population A or B can have at most two progeny with common probabilities (for ξ_i or η_i)

$$P\{\xi_i = 0\} = q^2 \qquad P\{\xi_i = 1\} = 2pq \qquad P\{\xi_i = 2\} = p^2 \qquad q = 1 - p \qquad 0 < p < 1 \tag{15A-9}$$

so that their common moment generating function is given by

$$A(z) = B(z) = (q + pz)^2$$

In this case, (15A-7) reduces to

$$p_{ij} = \frac{\binom{2i}{j}\binom{2(N-i)}{N-j}}{\binom{2N}{N}} \qquad i, j = 0, 1, \ldots, N \tag{15A-10}$$

which coincides with the *hypergeometric* genetic model in (15-30). ◀

EXAMPLE 15A-2

POISSON POPULATION MODEL

▶ As another example, suppose the two branching processes A and B follow independent Poisson progency distributions with mean values λ and μ respectively. Then,

$$A(z) = e^{\lambda(z-1)} \qquad B(z) = e^{\mu(z-1)} \tag{15A-11}$$

and hence from (15A-7) we obtain

$$\begin{aligned} p_{ij} &= \frac{\left(e^{-i\lambda}(i\lambda)^j/j!\right)\left(e^{-(N-i)\mu}[(N-i)\mu]^{N-j}/(N-j)!\right)}{e^{-(i\lambda+(N-i)\mu)}\left([i\lambda+(N-i)\mu]^N/N!\right)} \\ &= \binom{N}{j}\left(\frac{i\lambda}{i\lambda+(N-i)\mu}\right)^j\left(\frac{(N-i)\mu}{i\lambda+(N-i)\mu}\right)^{N-j} \qquad i, j = 0, 1, 2, \ldots, N \end{aligned} \tag{15A-12}$$

which represents a *binomial* model. Notice that in the special case when $\lambda = \mu$, (15A-12) simflifies to

$$p_{ij} = \binom{N}{j}\left(\frac{i}{N}\right)^j\left(1-\frac{i}{N}\right)^{N-j} \qquad i, j = 0, 1, \ldots, N \tag{15A-13}$$

and it coincides with the *binomial* sampling model in (15-31). Interestingly, $\lambda > \mu$ expresses certain bias in terms of advantage of type-A over type-B individuals, and the general distribution in (15A-12) can be used to analyze the natural selection phenomenon in that case. ◀

EXAMPLE 15A-3

LOCALLY POISSON POPULATION MODEL

▶ Consider the population model over a large area where the progeny distribution is locally Poisson with parameter λ. Suppose that λ is a random variable depending on the sub area as shown in Fig 15-11, and the distribution of λ over the whole area is exponential with parameter $a > 0$. Thus

$$P\{\xi_i = k \mid \lambda\} = e^{-\lambda}\frac{\lambda^k}{k!} \qquad k = 0, 1, 2, \ldots \qquad f(\lambda) = ae^{-a\lambda} \qquad \lambda \ge 0 \tag{15A-14}$$

and hence from (15A-2)

$$\begin{aligned} a_k = P\{\xi_i = k\} &= \int_0^\infty P(\xi_i = k \mid \lambda) f(\lambda)\, d\lambda = \int_0^\infty e^{-\lambda}\frac{\lambda^k}{k!} ae^{-\lambda a}\, d\lambda \\ &= \frac{a}{1+a}\left(\frac{1}{1+a}\right)^k \triangleq qp^k \qquad k = 0, 1, 2, \ldots \end{aligned} \tag{15A-15}$$

with $p = 1/(1+a)$ and $q = 1 - p$. Thus the progency distribution is *geometric* in this case with moment generating function

$$A(z) = \frac{q}{1-pz} \tag{15A-16}$$

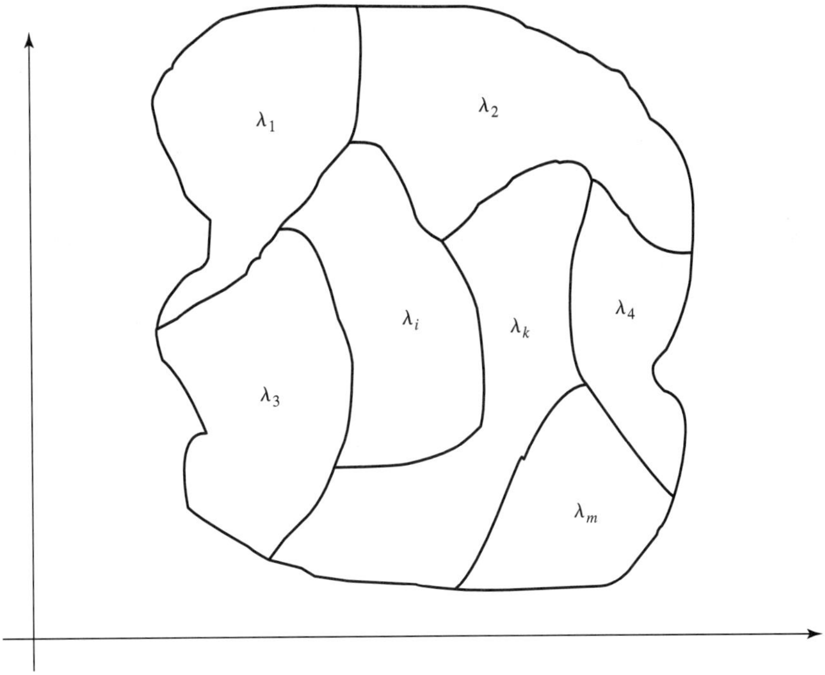

FIGURE 15-11
Locally Poisson population with exponentially distributed mean value.

and if we assume the two independent populations to have this common progency distribution, then the corresponding transition probabilities for the finite Markov chain (15A-7) turn out to be

$$p_{ij} = \frac{\binom{i+j-1}{j}\binom{2N-i-j-1}{N-j}}{\binom{2N-1}{N}} \qquad i, j = 0, 1, \ldots, N \tag{15A-17}$$

An obvious generalization in this case is to relax the exponential assumption in (15A-14) to a gamma random variable with m degrees of freedom. In that case (15A-15) corresponds to a *negative binomial* distribution and the generalization to (15A-16)–(15A-17) is immediate. ◀

APPENDIX 15B STRUCTURE OF PERIODIC CHAINS

For an irreducible Markov chain, if (see pages 715, 730)

$$p_{ii}^{(n)} = 0 \qquad \text{if} \quad n \neq kT \tag{15B-1}$$

then the chain is said to be periodic with period T. By Theorem 15-4 all states in a chain have the same period, and since the chain is irreducible, for any two states e_i and e_j there exist integers m and n such that $p_{ij}^{(m)} > 0$ and $p_{ji}^{(n)} > 0$. But

$$p_{ii}^{(m+n)} \geq p_{ij}^{(m)} p_{ji}^{(n)} > 0 \tag{15B-2}$$

and hence from (15B-1), $m + n = kT$ in (15B-2), or simplifying we get $m = r + sT$, where $1 \leq r \leq T$. Here r is a fixed integer that is characteristic of the states e_i and e_j. Thus starting with any state e_{i_0}, let C_r represent the set of states $\{e_j\}$ for which $p_{i_0,j}^{(m_0)} > 0$, where m_0 is of the form

$$m_0 = r + kT \qquad 1 \leq r \leq T \tag{15B-3}$$

Continuing this procedure over all states in the chain, the reminder r exhausts all integer values in (15B-3) (if not, the period will be less than T).

Thus the set of states can be divided into T mutually exclusive classes $C_1, C_2, \cdots, C_T$ such that

$$p_{ij} = 0 \qquad \text{if} \quad e_i \in C_k \qquad e_j \notin C_{k+1} \tag{15B-4}$$

and hence

$$\sum_{j \in C_{k+1}} p_{ij} = 1 \qquad e_i \in C_k \tag{15B-5}$$

These T classes can be cyclically ordered so that one-step transitions are possible only to a state in a neighboring class to the right (C_k to C_{k+1} and finally C_T to C_1), and T such steps always lead back to a state of the same class.[15] In this sense the chain has a periodic behavior. As a result, the transition matrix for a periodic chain has the following block structure (see (15-123) for an example):

$$P = \begin{pmatrix} 0 & P_1 & 0 & 0 & \cdots & 0 \\ 0 & 0 & P_2 & 0 & \cdots & 0 \\ \vdots & \vdots & \vdots & \vdots & \vdots & \vdots \\ 0 & 0 & 0 & \cdot & \cdots & P_{T-1} \\ P_T & 0 & \cdot & \cdot & \cdots & 0 \end{pmatrix} \tag{15B-6}$$

By direct computation

$$P^2 = \begin{pmatrix} 0 & 0 & A_1 & 0 & \cdots & 0 \\ 0 & 0 & 0 & A_2 & \cdots & 0 \\ \vdots & \vdots & \vdots & \vdots & \vdots & \vdots \\ A_{T-1} & 0 & 0 & \cdot & \cdots & 0 \\ 0 & A_T & 0 & \cdot & \cdots & 0 \end{pmatrix} \tag{15B-7}$$

[15]Note that it may not be possible to reach all states in the next class after one transition. Similarly to get back to the same state it may take several rounds of T transitions.

Finally, the Tth power of P gives the block diagonal stochastic matrix

$$P^T = \begin{pmatrix} B_1 & 0 & 0 & \cdot & \cdot & 0 \\ 0 & B_2 & 0 & \cdot & \cdot & 0 \\ \cdot & \cdot & \cdot & \cdot & \cdot & \cdot \\ \cdot & \cdot & \cdot & \cdot & \cdot & \cdot \\ 0 & 0 & \cdot & \cdot & B_{T-1} & 0 \\ 0 & 0 & \cdot & \cdot & \cdot & B_T \end{pmatrix} \tag{15B-8}$$

where the block entries $B_1, B_2, \ldots, B_T$ correspond to the T-step transition matrices for the set of states in classes $C_1, C_2, \ldots, C_T$ respectively. Thus each class C_k forms an irreducible closed set with respect to a chain with transition matrix B_k. From Theorem 15-5, since every state can be reached with certainty within the same irreducible closed set, we have $f_{ij} = 1$ if $e_i, e_j \in C_k$ and together with (15-130), from (15-114) we obtain

$$p_{ij}^{(nT)} \to \begin{cases} \dfrac{T}{\mu_j} & e_i, e_j \in C_k, k = 1, 2, \ldots, T \\ 0 & \text{otherwise} \end{cases} \tag{15B-9}$$

For finite chains, these steady state probabilities in (15B-9) also can be computed directly from the uncoupled set of equations

$$x_k = x_k B_k \qquad k = 1, 2, \ldots, T \tag{15B-10}$$

that follow from (15-177) and (15B-8) with x_k representing the steady state probability row vector for the states in the class C_k. Note that the largest eigenvalue of each stochastic matrix B_k equals unity, and hence P^T in (15B-8) possesses T repeated eigenvalues that equal unity. It follows that for a chain with period T, the T roots of unity are among the eigenvalues of the original transition matrix P (see also footnote 6, page 730).

PROBLEMS

15-1 Classify the states of the Markov chains with the following transition probabilities

$$P = \begin{pmatrix} 0 & 1/2 & 1/2 \\ 1/2 & 0 & 1/2 \\ 1/2 & 1/2 & 0 \end{pmatrix} \qquad P = \begin{pmatrix} 0 & 0 & 1/3 & 2/3 \\ 1 & 0 & 0 & 0 \\ 0 & 1 & 0 & 0 \\ 0 & 0 & 1 & 0 \end{pmatrix} \qquad P = \begin{pmatrix} 1/2 & 1/2 & 0 & 0 & 0 \\ 1/2 & 1/2 & 0 & 0 & 0 \\ 0 & 0 & 1/3 & 2/3 & 0 \\ 0 & 0 & 2/3 & 1/3 & 0 \\ 1/3 & 1/3 & 0 & 0 & 1/3 \end{pmatrix}$$

15-2 Consider a Markov chain $\{\mathbf{x}_n\}$ with states $e_0, e_1, \ldots, e_m$ and transition probability matrix

$$P = \begin{pmatrix} q & p & 0 & \cdot & \cdot & \cdot & 0 \\ 0 & q & p & 0 & \cdot & \cdot & 0 \\ \cdot & \cdot & \cdot & \cdot & \cdot & \cdot & \cdot \\ 0 & 0 & \cdot & \cdot & \cdot & q & p \\ p & 0 & \cdot & \cdot & \cdot & 0 & q \end{pmatrix}$$

Determine P^n, and the limiting distribution

$$\lim_{n\to\infty} P\{\mathbf{x}_n = e_k\} \qquad k = 0, 1, 2, \ldots, m$$

15-3 Find the stationary distribution $q_0, q_1, \dots$ for the Markov chain whose only nonzero stationary probabilities are

$$p_{i,1} = \frac{i}{i+1} \qquad p_{i,i+1} = \frac{1}{i+1} \qquad i = 1, 2, \dots$$

15-4 Show that the probability of extinction of a population given that the zeroth generation has size m is given by π_0^m, where π_0 is the smallest positive root in Theorem 15-8. Show that the probability that the population grows indefinitely in that case is $1 - \pi_0^m$.

15-5 Consider a population in which the number of offspring for any individual is at most two. Show that if the probability of occurrence of two offspring/individual is less than the probability of occurrence of zero offspring/individual, then the entire population is bound to extinct with probability one.

15-6 Let $\mathbf{x}_n$ denote the size of the nth generation in a branching process with probability generating function $P(z)$ and mean value $\mu = P'(1)$. Define $\mathbf{w}_n = \mathbf{x}_n/\mu^n$. Show that

$$E\{\mathbf{w}_{n+m} \mid \mathbf{w}_n\} = \mathbf{w}_n$$

15-7 Show that the sums $\mathbf{s}_n = \mathbf{x}_1 + \mathbf{x}_2 + \cdots + \mathbf{x}_n$ of independent zero mean random variables form a *martingale*.

15-8 **Time Reversible Markov Chains**. Consider a stationary Markov chain $\cdots \mathbf{x}_n, \mathbf{x}_{n+1}, \mathbf{x}_{n+2}, \dots$ with transition probabilities $\{p_{ij}\}$ and steady state probabilities $\{q_i\}$. (a) Show that the reversed sequence $\dots \mathbf{x}_n, \mathbf{x}_{n-1}, \mathbf{x}_{n-2}, \dots$ is also a stationaty Markov process with transition probabilities

$$P\{\mathbf{x}_n = j \mid \mathbf{x}_{n+1} = i\} \triangleq p_{ij}^* = \frac{q_j p_{ji}}{q_i}$$

and steady state probabilities $\{q_i\}$.

A Markov chain is said to be *time reversible* if $p_{ij}^* = p_{ij}$ for all i, j. (b) Show that a necessary condition for time reversibility is that

$$p_{ij}\, p_{jk}\, p_{ki} = p_{ik}\, p_{kj}\, p_{ji} \qquad \text{for all} \quad i, j, k$$

which states that the transition $e_i \to e_j \to e_k \to e_i$ has the same probability as the reversed transition $e_i \to e_k \to e_j \to e_i$. In fact, for a reversible chain starting at any state e_i, any path back to e_i has the same probability as the reversed path.

15-9 Let $A = (a_{ij})$ represent a symmetric matrix with positive entries, and consider an associated probability transition matrix P generated by

$$p_{ij} = \frac{a_{ij}}{\sum_k a_{ik}}$$

(a) Show that this transition matrix represents a time-reversible Markov chain.
(b) Show that the stationary probabilities of this chain is given by

$$q_i = c \sum_j a_{ij} = \frac{\sum_j a_{ij}}{\sum_i \sum_j a_{ij}}$$

Note: In a connected graph, if a_{ij} represents the weight associated with the segment (i, j), then p_{ij} represents the probability of transition from node i to j.

15-10 For transient states e_i, e_j in a Markov chain, starting from state e_i, let m_{ij} represent the average time spent by the chain in state e_j. Show that [see also (15-240)]

$$m_{ij} = \delta_{ij} + \sum_{e_k \in T} p_{ik} m_{kj} \qquad e_i, e_j \in T$$

or

$$M = (I - W)^{-1}$$

where $M = (m_{ij})$, and W represents the substochastic matrix associated with the transient states [see (15-110)]. Determine M for

$$W = \begin{pmatrix} 0 & p & 0 & 0 & 0 \\ q & 0 & p & 0 & 0 \\ 0 & q & 0 & p & 0 \\ 0 & 0 & q & 0 & p \\ 0 & 0 & 0 & q & 0 \end{pmatrix}$$

15-11 Every Stochastic matrix corresponds to a Markov chain for which it is the one-step transition matrix. However, show that not every stochastic matrix can correspond to the two-step transition matrix of a Markov chain. In particular, a 2×2 stochastic matrix is the two-step transition matrix of a Markov chain if and only if the sum of its diagonal elements is greater than or equal to unity.

15-12 **Genetic model with mutation.** In the genetic model (15-31), consider the possibility that prior to the formation of a new generation each gene can spontaneously mutate into a gene of the other kind with probabilities

$$P\{A \to B\} = \alpha > 0 \qquad \text{and} \qquad P\{B \to A\} = \beta > 0$$

Thus for a system in state e_j, after mutation there are $N_A = j(1-\alpha) + (N-j)\beta$ genes of type A and $N_B = j\alpha + (N-j)(1-\beta)$ genes of type B. Hence the modified probabilities prior to forming a new generation are

$$p_j = \frac{N_A}{N} = \frac{j}{N}(1-\alpha) + \left(1 - \frac{j}{N}\right)\beta$$

and

$$q_j = \frac{N_B}{N} = \frac{j}{N}\alpha + \left(1 - \frac{j}{N}\right)(1-\beta)$$

for the A and B genes, respectively. This gives

$$p_{jk} = \binom{N}{k} p_j^k q_j^{N-k} \qquad j, k = 0, 1, 2, \ldots, N$$

to be the modified transition probabilities for the Morkov chain with mutation. Derive the steady state distribution for this model, and show that, unlike the models in (15-30) and (15-31), fixation to "the pure gene states" does not occur in this case.

15-13 [41] (*a*) Show that the eigenvalues for the finite state Markov chain with probability transition matrix as in (15-30) are given by

$$\lambda_0 = 1 \qquad \lambda_1 = 1 \qquad \lambda_r = 2^r \frac{\binom{2N-r}{N-r}}{\binom{2N}{N}} < 1 \qquad r = 2, 3, \ldots, N$$

(*b*) Show that the eigenvalues for the finite state Markov chain with probability transition matrix as in (15-31) are given by

$$\lambda_0 = 1 \qquad \lambda_r = \left(1 - \frac{1}{N}\right)\left(1 - \frac{2}{N}\right)\cdots\left(1 - \frac{r-1}{N}\right) \le 1 \qquad r = 1, 2, \ldots, N$$

(c) Consider a finite state Markov chain with transition probabilities (see Example 15A-3, in Appendix 15A)

$$p_{ij} = \frac{\binom{i+j-1}{j}\binom{2N-i-j-1}{N-j}}{\binom{2N-1}{N}} \qquad i, j = 0, 1, 2, \ldots, N$$

Show that the eigenvalues of the corresponding probability transition matrix are given by

$$\lambda_0 = 1 \qquad \lambda_1 = 1 \qquad \lambda_r = \frac{\binom{2N-1}{N-r}}{\binom{2N-1}{N}} < 1 \qquad r = 2, 3, \ldots, N$$

Note: The eigenvalues $\lambda_0 = 1$ and $\lambda_1 = 1$ correspond to the two absorbing "fixed" states in all these Markov chains, and λ_2 measures the rate of approach to absorption for the system.

15-14 Determine the mean time to absorption for the genetic models in Example 15-13. [Hint: Use (15-240).]

15-15 Determine the mean time to absorption for the random walk model in Example 15-25. In the context of the gambler's ruin problem discussed there, show that the mean time to absorption for player A (starting with \$$a$) reduces to Eq. (3-53). (See page 65.)

CHAPTER 16

MARKOV PROCESSES AND QUEUEING THEORY

16-1 INTRODUCTION

In this chapter we shall study Markov processes that represent the continuous analogue of Markov chains discussed in Chap. 15. Thus in a Markov process, the time index t varies continuously, and the process can occupy either a finite or infinite number of states $e_0, e_1, e_2, e_3, \ldots$, as before. In general, for a Markov process the state space can vary continuously, and the time index can be discrete or continuous. In addition, starting from some initial state at $t = 0$, the process changes its state randomly as time goes on. Once again information about the past has no effect on the future if the present state of the process is specified. As we shall see, the evolution of the Markov processes is governed by the Kolmogorov equations, and their transient and steady state analysis will characterize the near-term and long-term (steady state) behavior of the processes.

A wide variety of queueing phenomena can be modeled as Markov processes. Recall that a queue, or a waiting line, involves arriving items (customers, jobs) that demand service at a service station, such as incoming telephone calls at a trunk station or inoperative machines that wait for a repairman for service. If the server is busy with another item, the newly arrived items form a waiting line until the server is free, or they may get impatient and leave the system with or without waiting for service. In the meantime, other items may arrive for service. The queue so formed can be described by the *arrival (input) process,* the *queue discipline,* and the *service mechanism.* The queue discipline determines the manner in which arriving items form a queue and behave while waiting. The input process and service mechanism are specified by the characteristics of the *interarrival times* and *service times,* respectively. It is reasonable to assume that the successive service times are independent of each other and also independent of the sequence of interarrival times. In addition, if one or both of the associated processes are assumed to have specific markovian characteristics, then the Kolmogorov equations can be used to analyze their behavior for better understanding of the queues in terms of their *waiting time distributions* and other useful features. As we shall see, the specific form of the queueing parameters distinguishes various queueing phenomena.

The first major contribution to queueing theory dates back to the work of A. K. Erlang[1] (1908) on telephone traffic problems. Erlang's primary interest was with the equilibrium behavior of traffic at telephone exchanges, and he derived the equilibrium form of Kolmogorov equations for Markov processes along with results for the probability of the different number of calls waiting, equilibrium waiting time for calls and the probability of a call loss. Erlang's work stimulated further research in this area (Fry, Molina, O'Dell), and new mathematical ideas such as link systems, where a set of sources may have limited access to a set of destinations, were introduced. Among other things, Pollaczek developed results for the single-channel non-markovian queue with various types of input, service times, and arbitrary queue disciplines. The waiting time distribution in the transient case for an ordered queue with Poisson input with time dependent parameter and for arbitrary service time distribution was developed by L. Takacs (1955) [35, 39, 48, 52].

The concept of imbedded Markov chains was first introduced by D. G. Kendall (1951) based on the point of regeneration concept due to Palm, and it was followed by a queue classification paper (1953), both of which have been in wide use since that time. O'Brien followed by Jackson studied the first "network of queues" (1954) by investigating two and three queues in series and giving expressions for the length distribution and waiting time for Poisson input and exponential service time. Burke, Reich, and Cohen have independently established that the output from a Poisson queue is also Poisson [39, 43, 48].

The theory of queues has been applied to a wide variety of problems that provide service for randomly arising demands—telephone traffic (Erlang, O'Dell, Vaulot, Pollaczek, Kendall, Takacs etc.), machine breakdown and repair (Khinchin, 1943, Kronig, Mondria, Palm, Takacs, Ashcroft, Cox), air-traffic control (Pollaczek, Pearcey), inventory control (Arrow, Karlin, Scarf), insurance risk theory (Lundberg, Seal), data communications networks (Jackson, Burke, Sondhi), and dams and storage systems (Downton, Gani, Moran, Prabhu). By examining the input process, the service mechanism and the queue discipline, it is possible to develop a unified approach to analyze these seemingly diverse problems.

16-2 MARKOV PROCESSES

A continuous-time Markov process $\mathbf{x}(t)$ can occupy randomly a finite or infinite number of states $e_0, e_1, e_2, e_3, \ldots$ at time t. The status of the process at time t is described by $\mathbf{x}(t)$ and it equals the state e_j that the process occupies at that time. Suppose that the process $\mathbf{x}(t)$ is in state e_i at time t_0. For a Markov process, from (15-2) the probability

[1]Danish scientist who for many years (1908–1922) worked for the Copenhagen Telephone Company.

that the process goes into the state e_j at time $t_0 + t$ is given by

$$P\{\mathbf{x}(t_0 + t) = e_j \mid \mathbf{x}(t_0) = e_i\} \tag{16-1}$$

and this probability is *independent* of the behavior of the process $\mathbf{x}(t)$ *prior to* the instant t_0. If $\mathbf{x}(t)$ is a *homogeneous Markov process,* then this transition probability from state e_i to state e_j does not depend on the initial epoch t_0 but depends only on the elapsed time t between the transitions. Thus in the case of a homogeneous Markov chain (16-1) reduces to

$$p_{ij}(t) = P\{\mathbf{x}(t_0 + t) = e_j \mid \mathbf{x}(t_0) = e_i\} \tag{16-2}$$

In particular, we have

$$p_{ij}(t) = P\{\mathbf{x}(t) = e_j \mid \mathbf{x}(0) = e_i\} \tag{16-3}$$

where

$$p_i(0) = P\{\mathbf{x}(0) = e_i\} \tag{16-4}$$

represents the initial probability distribution of the states. For all states e_i, e_j we have

$$0 \le p_{ij}(t) \le 1 \qquad \sum_j p_{ij}(t) = 1 \tag{16-5}$$

and the unconditional probability of the event "$\mathbf{x}(t)$ is in state e_j" is given by

$$\begin{aligned} p_j(t) = P\{\mathbf{x}(t) = e_j\} &= \sum_i P\{\mathbf{x}(t) = e_j \mid \mathbf{x}(0) = e_i\}P\{\mathbf{x}(0) = e_i\} \\ &= \sum_i p_i(0)p_{ij}(t) \end{aligned} \tag{16-6}$$

More generally for arbitrary t and s, we have

$$\begin{aligned} p_{ij}(t+s) &= P\{\mathbf{x}((t+s) = e_j \mid \mathbf{x}(0) = e_i\} \\ &= \sum_k P\{\mathbf{x}(t+s) = e_j \mid \mathbf{x}(t) = e_k, \mathbf{x}(0) = e_i\}P\{\mathbf{x}(t) = e_k \mid \mathbf{x}(0) = e_i\} \\ &= \sum_k P\{\mathbf{x}(t) = e_k \mid \mathbf{x}(0) = e_i\}P\{\mathbf{x}(t+s) = e_j \mid \mathbf{x}(t) = e_k\} \\ &= \sum_k p_{ik}(t)p_{kj}(s) \end{aligned} \tag{16-7}$$

and it represents the continuous version of the Chapman–Kolmogorov equation in (15-43).

SOJOURN TIME

All Markov processes share the interesting property that the time it takes for a change of state (*sojourn time*) is an *exponentially* distributed random variable. To see this, let $\boldsymbol{\tau}_i$ represent the waiting time for a change of state for a Markov process $\mathbf{x}(t)$, given that it is in state e_i at time t_0. If $\boldsymbol{\tau}_i > s$, then the process will be in the same state e_i at time $t_0 + s$ as at t_0, and (being a Markov process) its subsequent behavior is *independent* of s. Hence

$$P\{\boldsymbol{\tau}_i > t + s \mid \boldsymbol{\tau}_i > s\} = P\{\boldsymbol{\tau}_i > t\} \triangleq \varphi_i(t) \tag{16-8}$$

represents the probability that the event $\{\boldsymbol{\tau}_i > t+s\}$ given that $\{\boldsymbol{\tau}_i > s\}$. But

$$\begin{aligned}\varphi_i(t+s) &= P\{\boldsymbol{\tau}_i > t+s\} = P\{\boldsymbol{\tau}_i > t+s, \boldsymbol{\tau}_i > s\} \\ &= P\{\boldsymbol{\tau}_i > t+s \mid \boldsymbol{\tau}_i > s\}P\{\boldsymbol{\tau}_i > s\} = \varphi_i(t)\varphi_i(s)\end{aligned}$$

or

$$\log \varphi_i(t+s) = \log \varphi_i(t) + \log \varphi_i(s) \tag{16-9}$$

Notice that the only function that satisfies (16-9) for arbitrary t and s is either of the form ct, where c is a constant or unbounded in every interval. Thus

$$\log \varphi_i(t) = -\lambda_i t \qquad \varphi_i(t) = P\{\boldsymbol{\tau}_i > t\} = e^{-\lambda_i t} \qquad t \geq 0$$

or

$$F_{\tau_i}(t) = P\{\boldsymbol{\tau}_i \leq t\} = 1 - e^{-\lambda_i t} \qquad t \geq 0 \tag{16-10}$$

which shows that the *sojourn time* (waiting time in any state) has an exponential distribution for all Markov processes. The parameter λ_i represents the density of transition out of the state e_i and in general it can depend on the final state e_j also. If $\lambda_i > 0$ the probability of the process undergoing a change of state from e_i in a small interval Δt is given by

$$P\{\boldsymbol{\tau} \leq \Delta t\} = 1 - e^{-\lambda_i \Delta t} = \lambda_i \Delta t + o(\Delta t) \tag{16-11}$$

and the probability that there is no change of state from e_i in the same interval is given by

$$P\{\boldsymbol{\tau} > \Delta t\} = 1 - \lambda_i \Delta t + o(\Delta t) \tag{16-12}$$

where $o(\Delta t)$ represents an infinitesimal of higher order than Δt. ◀

The Kolmogorov Equations

We can make use of Eq. (16-7) to study the evolution of a Markov process. Using (16-7) we obtain

$$p_{ij}(t+\Delta t) = \sum_k p_{ik}(t)p_{kj}(\Delta t) = \sum_k p_{ik}(\Delta t)p_{kj}(t) \tag{16-13}$$

But from (16-11) and (16-12) for a Markov process

$$p_{kj}(\Delta t) = \begin{cases} P\{\boldsymbol{\tau}_{kj} \leq \Delta t\} = \lambda_{kj}\Delta t + o(\Delta t) & k \neq j \\ P\{\boldsymbol{\tau}_j > \Delta t\} = 1 - \lambda_j \Delta t + o(\Delta t) & k = j \end{cases} \tag{16-14}$$

and substituting this into (16-13) we obtain

$$\frac{p_{ij}(t+\Delta t) - p_{ij}(t)}{\Delta t} = \sum_{k \neq j} p_{ik}(t)\lambda_{kj} - p_{ij}(t)\lambda_j + \frac{o(\Delta t)}{\Delta t} \tag{16-15}$$

and

$$\frac{p_{ij}(t+\Delta t) - p_{ij}(t)}{\Delta t} = \sum_{k \neq i} \lambda_{ik} p_{kj}(t) - \lambda_i p_{ij}(t) + \frac{o(\Delta t)}{\Delta t} \tag{16-16}$$

Define

$$\lambda_{ii} = -\lambda_i \qquad i = 0, 1, 2, \ldots \tag{16-17}$$

so that right sides of (16-15) and (16-16) become $\sum_k p_{ik}(t)\lambda_{kj} + o(\Delta t)/\Delta t$ and $\sum_k \lambda_{ik} p_{kj}(t) + o(\Delta t)/\Delta t$, respectively. Both sums have definite limits as $\Delta t \to 0$ in the case of finite chains, since $o(\Delta t)/\Delta t \to 0$ in that case. As a result the left sides of (16-15)–(16-16) tend to the derivative $p'_{ij}(t)$, and it gives rise to the differential equations

$$p'_{ij}(t) = \sum_k p_{ik}(t)\lambda_{kj} \qquad i, j = 0, 1, 2, \ldots \tag{16-18}$$

and

$$p'_{ij}(t) = \sum_k \lambda_{ik} p_{kj}(t) \qquad i, j = 0, 1, 2, \ldots \tag{16-19}$$

under the initial conditions

$$p_{ij}(0) = 0 \qquad i \neq j \qquad p_{ii}(0) = 1 \tag{16-20}$$

Thus the transition probabilities satisfy the two systems of linear differential equations given by (16-18) and (16-19), and they are known as the *forward* and *backward Kolmogorov equations,* respectively. Using (16-14) and (16-17), the condition $\sum_j p_{ij}(\Delta t) = 1$ reduces to

$$\sum_j p_{ij}(\Delta t) = 1 + \sum_j \lambda_{ij}\Delta t = 1$$

or we obtain

$$\sum_j \lambda_{ij} = 0 \qquad \Rightarrow \qquad \lambda_{ii} = -\sum_{j\neq i} \lambda_{ij} \tag{16-21}$$

The Kolmogorov equations also hold in the case of a countably infinite number of states, provided the error term $o(\Delta t)/\Delta t$ tends to zero uniformly for all i, j.

Using (16-14) and (16-20), we also get

$$\lambda_{ij} = \begin{cases} \dfrac{p_{ij}(\Delta t)}{\Delta t} = \dfrac{p_{ij}(\Delta t) - p_{ij}(0)}{\Delta t} & i \neq j \\[2ex] \dfrac{p_{ij}(\Delta t) - 1}{\Delta t} = \dfrac{p_{ii}(\Delta t) - p_{ii}(0)}{\Delta t} & i = j \end{cases} \tag{16-22}$$

and hence

$$\lambda_{ij} = \left.\frac{dp_{ij}(t)}{dt}\right|_{t=0} \tag{16-23}$$

are known as the *transition densities* of the process. Let

$$A \triangleq (\lambda_{ij}) \qquad i, j = 0, 1, 2, \ldots \tag{16-24}$$

represent the matrix consisting of the transition densities λ_{ij}. From (16-21), all diagonal entries of A are negative, the off-diagonal entries are all positive, and row elements in each row sum to zero. Let

$$P(t) \triangleq (p_{ij}(t)) \qquad i, j = 0, 1, 2, \ldots \tag{16-25}$$

represent the matrix of transition probabilities. In this notation, the forward and backward

Kolmogorov equations simplify to

$$P'(t) = P(t)A = AP(t) \tag{16-26}$$

under the initial condition $P(0) = I$.

For a finite state process $e_0, e_1, \ldots, e_m$, the transient solution of (16-26) takes the form

$$P(t) = e^{At} \tag{16-27}$$

where

$$e^{At} = I + \sum_{n=1}^{\infty} \frac{A^n t^n}{n!} \tag{16-28}$$

Explicit solutions for $P(t)$ in terms of the λ_{ij}s are often difficult except in simple situations. In the event when the transition density matrix A has distinct eigenvalues, (16-27) can be expressed in a rather compact form. Since zero is always an eigenvalue of A, let $d_1, d_2, \ldots, d_m$ represent the remaining distinct nonzero eigenvalues of A. Then from (15-53), $A = UDU^{-1}$ so that $A^n = UD^nU^{-1}$ and (16-27) and (16-28) simplify to

$$P(t) = Ue^{Dt}U^{-1} \tag{16-29}$$

where

$$e^{Dt} = \begin{pmatrix} 1 & & & & 0 \\ & e^{d_1 t} & & & \\ & & e^{d_2 t} & & \\ & & & \ddots & \\ 0 & & & & e^{d_m t} \end{pmatrix} \tag{16-30}$$

The forward Kolmogorov equations are concerned with ways of reaching a state e_j from other states; the backward equations consider ways of getting out of a state e_j to other states. In general, their solutions with same initial conditions are identical.

The structure of the transition density matrix A characterizes various Markov processes, and the class of processes for which

$$\lambda_{ij} = 0 \qquad |i - j| > 1 \tag{16-31}$$

are known as the *birth and death processes*. Thus for birth and death processes transitions occur only between adjacent neighbors. Specific values of λ_{ij} for $|i - j| \leq 1$ in (16-31) give rise to various birth-death processes, the simplest among them being the *Poisson process*.

EXAMPLE 16-1

THE POISSON PROCESS

▶ Consider a Markov process $\mathbf{x}(t)$ with states $e_0, e_1, e_2, \ldots$ that can only change from state e_i by going into the state e_{i+1} with probability that is independent of the state. Therefore the transition densities are

$$\lambda_{kj} = \begin{cases} \lambda & j = k+1 \\ 0 & j \neq k, k+1 \end{cases} \tag{16-32}$$

and from (16-21), we obtain

$$\lambda_{kk} = -\lambda \tag{16-33}$$

The forward Kolmogorov equations in (16-18) become

$$p'_{ii}(t) = -\lambda p_{ii}(t) \tag{16-34}$$

$$p'_{ij}(t) = \lambda p_{i,j-1} - \lambda p_{ij}(t) \qquad j = i+1, i+2, \ldots \tag{16-35}$$

Let $p_j(t) = P\{\mathbf{x}(t) = e_j\}$ and $p_i(0) = 0$ for all $i \neq 0$ in (16-4). Then $p_0(0) = 1$ and using (16-6) we get $p_j(t) = p_{0j}(t)$, and hence (16-34) and (16-35) reduce to

$$p'_0(t) = -\lambda p_0(t) \tag{16-36}$$

and

$$p'_n(t) = \lambda p_{n-1}(t) - \lambda p_n(t) \qquad n = 1, 2, \ldots \tag{16-37}$$

under the initial conditions $p_0(0) = 1$, $p_n(0) = 0$, $n \neq 1$. To solve (16-36) and (16-37), define

$$q_n(t) = e^{\lambda t} p_n(t) \qquad n = 0, 1, 2, \ldots \tag{16-38}$$

Then

$$q'_0(t) = e^{\lambda t} p'_0(t) + \lambda q_0(t) = -\lambda e^{\lambda t} p_0(t) + \lambda q_0(t) = 0 \tag{16-39}$$

and

$$\begin{aligned} q'_n(t) &= e^{\lambda t} p'_n(t) + \lambda q_n(t) \\ &= e^{\lambda t}\{\lambda p_{n-1}(t) - \lambda p_n(t)\} + \lambda q_n(t) = \lambda q_{n-1}(t) \end{aligned} \tag{16-40}$$

with $q_0(0) = 1$, $q_n(0) = 0$, $n \neq 1$. Under these initial conditions (16-39) gives $q_0(t) = 1$, and (16-40) iteratively yield

$$q_1(t) = \lambda t \qquad q_2(t) = \frac{(\lambda t)^2}{2!}, \ldots, q_n(t) = \frac{(\lambda t)^n}{n!}$$

and hence from (16-38) we obtain

$$p_n(t) = P\{\mathbf{x}(t) = n\} = e^{-\lambda t}\frac{(\lambda t)^n}{n!} \qquad n = 0, 1, 2, \ldots \tag{16-41}$$

and it represents a valid probability density function to be the desired solution. Notice that, for a Poisson process, from (16-32) the transition probabilities are independent of the current state, and at any time the process can either remain in the current state or move over to the next state with constant probability. ◀

Historically, Poisson processes were initially observed to form in telephone traffic, where calls originated by a Poisson process, and the duration of calls was experimentally verified to have an exponential distribution as well. Poisson distributions are characterized by the property that in a small interval chances are very small that more than a single arrival occurs, and together with the "memoryless" property of the exponential distribution [see (4-32)], they have wide applicability. Since transitions occur only in one direction ($e_i \to e_{i+1}$), Poisson processes represent a special case of the pure birth processes.

EXAMPLE 16-2

THE PURE BIRTH PROCESS

▶ If the constant transition probability assumption is relaxed in the Poisson case in (16-32) so that

$$\lambda_{kj} = \begin{cases} \lambda_k & j = k+1 \\ 0 & j \neq k, k+1 \end{cases} \tag{16-42}$$

then we get the *pure birth process*. In that case, from (16-21)

$$\lambda_{i,i+1} = \lambda_i \qquad \lambda_{ii} = -\sum_{j \neq i} \lambda_{ij} = -\lambda_{i,i+1} = -\lambda_i$$

and the forward Kolmogorov equations take the form

$$p'_{ii}(t) = -\lambda_i p_{ii}(t) \tag{16-43}$$

and

$$p'_{ij}(t) = \lambda_{j-1} p_{i,j-1}(t) - \lambda_j p_{ij}(t) \tag{16-44}$$

Thus in a pure birth process, the transition probability (birth rate) is a function of the state that the system is in at time t. Once again, transitions take place only in the forward direction (see (16-42)) so that if e_j represents the population size, then the population is a strictly increasing function of time. If we assume that the birth rate is proportional to the "current population size," then $\lambda_j = j\lambda$ in (16-43) and (16-44), and it gives size to a *linear birth process* whose explicit solution has been shown to be

$$p_{ij}(t) = \begin{cases} \binom{j-1}{j-i} e^{-i\lambda t}(1 - e^{-\lambda t})^{j-i} & j \geq i \\ 0 & \text{otherwise} \end{cases} \tag{16-45}$$

In the general birth process, since the birth rate depends on the current state, it is possible that a rapid increase in the birth rate can lead to the degenerate condition $\sum_{j=0}^{\infty} p_j(t) < 1$, that corresponds to a "population explosion" in a finite time. It has been shown by Feller and Lundberg that for "nondegenerate behavior" of a birth process ($\sum_{j=0}^{\infty} p_j(t) = 1$), the necessary and sufficient (Feller–Lundberg) condition is given by

$$\sum_{k=0}^{\infty} \frac{1}{\lambda_k} = \infty \tag{16-46}$$

◀

EXAMPLE 16-3

THE DEATH PROCESS

▶ In this case the process $\mathbf{x}(t)$ is a strictly decreasing function of time so that if the process is in state e_i at time t, it can only go into state e_{i-1} at $t + \Delta t$ with probability $\mu_i \Delta t$. Thus

$$\lambda_{kj} = \begin{cases} \mu_k & j = k-1 \\ 0 & j \neq k, k-1 \end{cases} \tag{16-47}$$

which gives

$$\lambda_{k,k-1} = \mu_k \qquad \lambda_{kk} = -\sum_{j \neq k} \lambda_{kj} = -\lambda_{k,k-1} = -\mu_k \tag{16-48}$$

and hence the forward Kolmogorov equations become

$$p'_{ij}(t) = \mu_{j+1} p_{i,j+1}(t) - \mu_j p_{ij}(t) \tag{16-49}$$

◀

More generally, we have the *birth-death processes,* where transitions to/from adjacent neighboring states are allowed.

EXAMPLE 16-4

THE BIRTH-DEATH PROCESSES

▶ Consider a process $\mathbf{x}(t)$ that combines the features of a pure birth process as well as the simple death process, that is, if the process is in state e_i at t, it can go into state e_{i+1} at time $t + \Delta t$ with probability $\lambda_i \Delta t$ or to state e_{i-1} with probability $\mu_i \Delta t$. This gives

$$\lambda_{kj} = \begin{cases} \lambda_k & j = k+1 \\ \mu_k & j = k-1 \\ 0 & j \neq k, k-1, k+1 \end{cases} \tag{16-50}$$

so that

$$\lambda_{k,k+1} = \lambda_k \qquad \lambda_{k,k-1} = \mu_k$$

and

$$\lambda_{kk} = -\sum_{j \neq k} \lambda_{kj} = -(\lambda_{k,k+1} + \lambda_{k,k-1}) = -(\lambda_k + \mu_k) \tag{16-51}$$

In this case the forward Kolmogorov equations reduce to (Fig. 16-1)

$$p'_{ij}(t) = \lambda_{j-1} p_{i,j-1}(t) - (\lambda_j + \mu_j) p_{ij}(t) + \mu_{j+1} p_{i,j+1}(t) \tag{16-52}$$

$$p'_{i0}(t) = -\lambda_0 p_{i0}(t) + \mu_1 p_{i,1}(t) \tag{16-53}$$

The transition density matrix A for the general birth-death process has the form

$$A = \begin{pmatrix} -\lambda_0 & \lambda_0 & 0 & 0 & & & 0 & \cdots \\ \mu_1 & -(\lambda_1 + \mu_1) & \lambda_1 & 0 & & & & \cdots \\ 0 & \mu_2 & -(\lambda_2 + \mu_2) & \lambda_2 & & & & \\ \cdot & \cdot & \cdot & \cdot & & & & \\ \cdot & \cdot & \cdot & \cdot & & & & \\ \cdot & \cdot & \cdot & \cdot & \mu_i & -(\lambda_i + \mu_i) & \lambda_i & \cdots \\ \cdot & \cdot & \cdot & \cdot & \cdot & \cdot & \cdot & \cdots \\ \cdot & \cdot & \cdot & \cdot & \cdot & \cdot & \cdot & \cdots \end{pmatrix} \tag{16-54}$$

where $\lambda_{ij} = 0$ if $|i - j| > 1$, $\lambda_i > 0$ for $i \geq 0$, $\mu_i > 0$ for $i \geq 1$, and $\mu_0 \geq 0$.

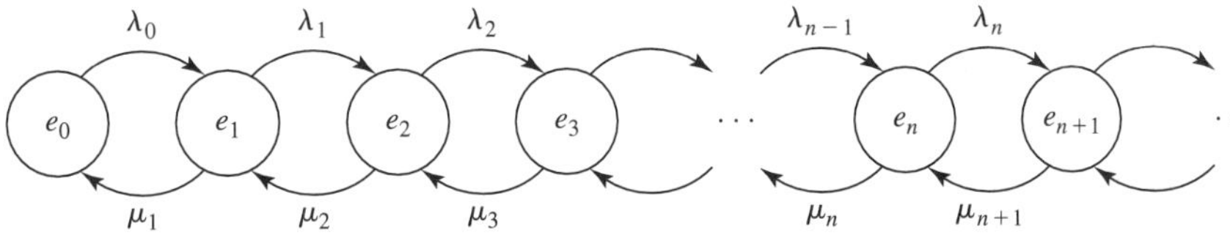

FIGURE 16-1
State diagram for the birth-death process.

If $\lambda_n = \lambda$, $\mu_n = \mu$, the birth-death equation in (16-52) and (16-53) describes a single channel process, since in that case in a small interval Δt the process either remains in the current state with no arrivals and departures with probability $1 - (\lambda + \mu)\Delta t$, or moves over to the next state (single arrival) with probability $\lambda \Delta t$, or moves back to the previous state (one departure) with probability $\mu \Delta t$. Similarly, the backward Kolmogorov equation has the form

$$p'_{ij}(t) = \lambda_i p_{i+1,j}(t) - (\lambda_i + \mu_i) p_{ij}(t) + \mu_i p_{i-1,j}(t) \tag{16-55}$$

The birth-death process is of considerable interest, as this model is encountered in many fields of application including queueing theory, where "births" correspond to arriving customers and "deaths" correspond to customers departing after completing service at the server. Recall that these processes are characterized by the property that the interval of time between state transitions of the same type (births or deaths) is a random variable with exponential distribution. ◀

The general solution of (16-52) for arbitrary time t is quite complicated. However, a special case with two states (e_0 and e_1) and constant birth (arrival) and death (departure) rates ($\lambda_k = \lambda$, $\mu_k = \mu$) can be readily solved using the method in (16-26)–(16-30).

EXAMPLE 16-5

TWO-STATE PROCESS WITH EXPONENTIAL HOLDING TIMES

▶ Suppose a system is either free (state e_0) or remains busy (state e_1) and the lengths of the free period as well as the busy period are independent exponential random variables with parameters λ and μ, respectively. Hence the probability $p_{01}(\Delta t)$ of the system going from e_0 to e_1 in Δt is $\lambda \Delta t + o(\Delta t)$ and similarly $p_{10}(\Delta t) = \mu \Delta t + o(\Delta t)$. This gives the probability transition matrix in (16-24) to be

$$A = \begin{pmatrix} -\lambda & \lambda \\ \mu & -\mu \end{pmatrix} \tag{16-56}$$

where [see (16-25)]

$$P(t) = \begin{pmatrix} p_{00}(t) & p_{01}(t) \\ p_{10}(t) & p_{11}(t) \end{pmatrix} \tag{16-57}$$

The eigenvalues of A can be readily verified to be 0 and $-(\lambda + \mu)$, and hence

$$A = U \begin{pmatrix} 0 & 0 \\ 0 & -(\lambda + \mu) \end{pmatrix} U^{-1} \tag{16-58}$$

where

$$U = \begin{pmatrix} 1 & \lambda \\ 1 & -\mu \end{pmatrix} \qquad U^{-1} = \frac{1}{\lambda + \mu} \begin{pmatrix} \mu & \lambda \\ 1 & -1 \end{pmatrix} \tag{16-59}$$

and using (16-29) and (16-30) we obtain

$$P(t) = U \begin{pmatrix} 1 & 0 \\ 0 & e^{-(\lambda+\mu)t} \end{pmatrix} U^{-1} = \frac{1}{\lambda + \mu} \begin{pmatrix} \mu + \lambda e^{-(\lambda+\mu)t} & \lambda - \lambda e^{-(\lambda+\mu)t} \\ \mu - \mu e^{-(\lambda+\mu)t} & \lambda + \mu e^{-(\lambda+\mu)t} \end{pmatrix} \tag{16-60}$$

◀

Equilibrium Behavior and Limiting Probabilities

The equilibrium behavior of the process is governed by the limiting probabilities $p_j = \lim_{t\to\infty} p_j(t)$ in (16-6). An important problem is to determine conditions under which the above limiting probabilities p_j exist.

For a Markov process $\mathbf{x}(t)$ that is irreducible and ergodic, with states $e_0, e_1, e_2, \ldots,$ the limiting probabilities

$$p_j = \lim_{t\to\infty} p_{ij}(t) \geq 0 \qquad \sum_j p_j = 1 \tag{16-61}$$

do exist, and they *do not* depend on the initial state e_i. The proof is essentially the same as that of Theorem 15-7 for Markov chains, and similar definitions for classification of states hold here also. Moreover, for irreducible finite chains, the continuous analogue of the conditions (15-183) is automatically satisfied here.

In particular, taking the limit as $t \to \infty$ in (16-7) and using (16-61) we get $p_j = \sum_k p_k p_{kj}(s)$, or

$$p_j = \sum_i p_i p_{ij}(t) \tag{16-62}$$

Suppose the transition probabilities satisfy (16-18)–(16-23). Differentiating (16-62) and setting $t = 0$, we obtain

$$\sum_i p_i \lambda_{ij} = 0 \qquad j = 0, 1, 2, \ldots \tag{16-63}$$

where λ_{ij} represents the transition density from state e_i to e_j as defined in (16-22)–(16-23). In matrix form (16-63) has the representation

$$pA = 0 \tag{16-64}$$

where

$$p = [p_0, p_1, p_2, \ldots, p_j, \ldots] \qquad \sum_i p_i = 1 \tag{16-65}$$

Notice that (16-64) is similar in structure to its discrete counterpart in (15-177). The matrices A and $(P - I)$ both have nonnegative off-diagonal elements, zero row sums, and a unique positive eigenvector corresponding to the simple zero eigenvalue. Equation (16-64) can also be obtained directly from the forward-Kolmogorov equations in (16-26) by putting

$$p_j = \lim_{t\to\infty} p_{ij}(t) \qquad \lim_{t\to\infty} p'_{ij}(t) = 0 \tag{16-66}$$

EXAMPLE 16-6

LIMITING PROBABILITIES FOR THE BIRTH-DEATH PROCESS

▶ Using (16-50) and (16-51) in (16-63), the (forward) steady state equations for the general birth-death process are given by

$$0 = \lambda_{j-1} p_{j-1} - (\lambda_j + \mu_j) p_j + \mu_{j+1} p_{j+1} \tag{16-67}$$

and

$$0 = -\lambda_0 p_0 + \mu_1 p_1 \tag{16-68}$$

Rewriting these equations, we obtain the iterative identity

$$\mu_{j+1}p_{j+1} - \lambda_j p_j = \mu_j p_j - \lambda_{j-1}p_{j-1} = \mu_{j-1}p_{j-1} - \lambda_{j-2}p_{j-2}$$
$$= \mu_1 p_1 - \lambda_0 p_0 = 0 \tag{16-69}$$

which gives

$$p_{j+1} = \frac{\lambda_j}{\mu_{j+1}}p_j = \frac{\lambda_j\lambda_{j-1}}{\mu_{j+1}\mu_j}p_{j-1}$$
$$= \frac{\lambda_j\lambda_{j-1}\cdots\lambda_0}{\mu_{j+1}\mu_j\cdots\mu_1}p_0 \tag{16-70}$$

or

$$p_n = \prod_{k=1}^{n}\frac{\lambda_{k-1}}{\mu_k}p_0 \qquad n = 1, 2, \ldots \tag{16-71}$$

The condition $\sum_{n=0}^{\infty} p_n = 1$ gives

$$\left(1 + \sum_{n=1}^{\infty}\prod_{k=1}^{n}\frac{\lambda_{k-1}}{\mu_k}\right)p_0 = 1 \tag{16-72}$$

and hence the necessary and sufficient condition for the existence of a steady state solution in (16-52) and (16-53) is the convergence of the infinite series $\sum_{n=1}^{\infty}\prod_{k=1}^{n}(\lambda_{k-1}/\mu_k)$ in (16-72) (Karlin and McGregor). When that series converges, the steady state probabilities for the birth-death process is given by

$$p_n = \lim_{t\to\infty} P\{\mathbf{x}(t) = n\} = \prod_{k=1}^{n}\frac{\lambda_{k-1}}{\mu_k}\cdot p_0 \qquad n = 1, 2, \ldots \tag{16-73}$$

where

$$p_0 = \lim_{t\to\infty} P\{\mathbf{x}(t) = 0\} = \frac{1}{1 + \sum_{n=1}^{\infty}\prod_{k=1}^{n}(\lambda_{k-1}/\mu_k)} \tag{16-74}$$

In particular, if $\lambda_n = \lambda$, and $\mu_n = \mu$, $n = 0, 1, 2, \ldots$, then we obtain the steady state solutions

$$p_n = \left(1 - \frac{\lambda}{\mu}\right)\left(\frac{\lambda}{\mu}\right)^n \qquad n = 0, 1, 2, \ldots \tag{16-75}$$

provided $\lambda/\mu < 1$. ◀

We shall use several variations of this birth-death model in Sec. 16-3 to study various markovian queues that are popular models in queueing theory.

16-3 QUEUEING THEORY

Queueing theory dates back to A. K. Erlang's (1878–1929) fundamental work on the study of congestion in telephone traffic, and since then it has been applied to a wide variety of applications such as inventory control, road traffic congestion, aviation traffic control, machine interference problem, biology, astronomy, nuclear cascade theory and, of course, voice and data communication networks. Simple queues collectively form a chain of queues, where queues, in turn, feed other queues, and this process can go on for several layers forming complex networks of queues. The mathematical characterization and study of these phenomena constitute queueing theory.

A queue, or a waiting line, is formed by arriving customers/jobs requiring service from a service station. If service is not immediately available, the arriving units may join the queue and wait for service and leave the system after being served, or may leave sooner without being served for various reasons. In the meantime, other units may arrive for service. The source from which the arriving units come may be finite or infinite. An arrival may consist of a single unit or in bulk (several units in a group). The service system may have either a limited or unlimited capacity for holding units (waiting room capacity), and depending on that, an arriving unit may join or leave the system. Service may be rendered either singly or in bulk (in batches). The basic features of a queue are: (*i*) the *input process,* (*ii*) the *service mechanism,* (*iii*) the *queue discipline,* and (*iv*) the *server's capacity.*

The input process specifies the probability law governing the arrival statistics of the customers at the server at times $\mathbf{t}_1, \mathbf{t}_2, \ldots, \mathbf{t}_n$, where $\mathbf{t}_i < \mathbf{t}_{i+1}$ (Fig. 16-2). Let $\boldsymbol{\tau}_n = \mathbf{t}_{n+1} - \mathbf{t}_n$ represent the *interarrival time* between the $(n+1)$st and nth customers. Then the input process is specified by the probability distribution of the *sequence of arrival instants* $\{\mathbf{t}_n\}$ and the *sequence of interarrival times* $\{\boldsymbol{\tau}_n\}$. The simplest model for the input process is one in which the arrival times follow a Poisson process with parameter λ (see Examples 9-5, 16-1 and Sec. 10-2 for Poisson process). In that case, as we have seen the interarrival times (sojourn time) $\boldsymbol{\tau}_n$ are independent exponential random variables with common parameter λ (see (16-10)) and the input process is markovian or memoryless. A strong argument in favor of the Poisson arrivals is that the limiting form of a binomial distribution is Poisson [see (4-107)]. Thus if a phenomenon is the collective sum of several bernoulli-type events, all of which are independent and each has a small probability of occurrence, then as we have seen, the overall phenomenon tends to be Poisson. The exponential assumption may be relaxed to include an arbitrary distribution $A(\tau)$ for the interarrival times while maintaining their independence assumption, in which case the input process is no longer markovian. (Not all queues are markovian!)

The service mechanism is specified by the *sequence of service times* $\{\mathbf{s}_n\}$, where $\mathbf{s}_n$ denotes the time required to serve the nth customer (Fig. 16-2). It is reasonable to

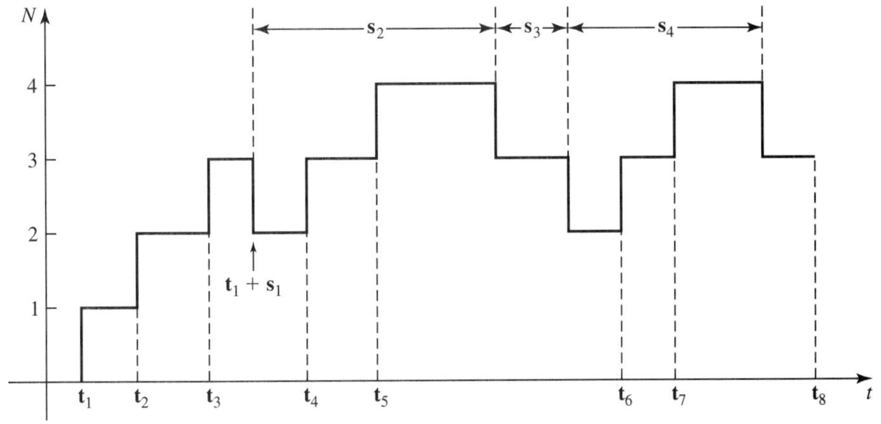

FIGURE 16-2
Arrivals and departures at a queue. Here $\{t_i\}$ refer to the arrival instants, $\{s_i\}$ refer to the service times, and N represents the number of customers in the system.

assume that the successive durations $\{\mathbf{s}_n\}$ are statistically independent of one another and also of the sequence of interarrival times $\{\boldsymbol{\tau}_n\}$. The simplest models in this case are either a constant service duration ($\mathbf{s}_n = T$), or an exponential distribution with parameter μ. Recall that both these models can be represented as special cases of the Erlang-n density function (see also (4-38))

$$f(\tau) = \begin{cases} \dfrac{(n\mu)^n}{n!}\tau^{n-1}e^{-n\mu\tau} & \tau \geq 0 \\ 0 & \text{otherwise} \end{cases} \tag{16-76}$$

Since (16-76) represents the p.d.f. of the sum of n independent exponential random variables with parameter $n\mu$, if the service duration satisfies the above model, the input unit must pass through n "phases" of service before a new unit is admitted for service. Although the Erlang model can be given a phase-type interpretation, it is obviously not restricted to modeling situations where there are only phases of service. As (16-76) shows, the Erlang model has greater flexibility than the exponential model and it gives a better fitting in many practical situations. In general, let $B(\tau)$ represent the common service duration distribution.

The queue discipline specifies the rule by which the arriving units form a queue, the manner in which they behave while waiting (patient vs. impatient customers) and the type of service offered at the server. The usual discipline is to process the units in the order of their arrival, that is, "first come, first served" (FIFO or first in, first out). However, other forms such as "last in, first served," "random selection for service," and "priority servicing" (emergency rooms) also can be adopted. The behavior of the customers that do not receive immediate service can vary widely. An arriving unit may choose to wait for service, or may immediately decide not to join the queue (*balking*), perhaps because of the length of the queue. A unit may join the queue, but may become impatient and leave the queue (*renege*), if the wait becomes longer than expected. The units may arrive later than scheduled, and when there are several queues, impatient units may *jockey* back and forth among them. The present discussion will assume the most common *first in, first out* procedure.

The service system may have one or several channels that provide service at the same or different rates to the arriving units, and in addition the system may have either a limited or unlimited capacity for holding waiting units. In a single channel case, the ratio

$$\rho = \frac{\lambda}{\mu} = \frac{\text{mean arrival rate (number of arrivals/unit time)}}{\text{mean service rate (number served/unit time)}} \tag{16-77}$$

denotes the traffic intensity, and it can be modified appropriately in other situations.

Description of queues. A notational system proposed by Kendall (1951) is universally used to specify queues. In this description, a three-part symbol (sometimes four-part) is used, where the first symbol specifies the *input process* (interarrival distribution), the second symbol specifies the *service mechanism* (service time distribution), and the third symbol denotes the *number of channels* or servers in use. If the system has a limited holding capacity for waiting items, then a fourth symbol is used to specify this information. The following symbols are usually used to specify the input process and the service mechanism:

QUEUE NOTATION

M: Poisson or exponential (markovian or memoryless)
D: deterministic or regular
E_n: Erlangian distribution
G: arbitrary service time distribution function $B(\tau)$
GI: arbitrary independent inter-arrival distribution function $A(\tau)$

In this notation, $M/G/r$ stands for a queue with Poisson arrivals, no special assumption about service-time distribution $B(\tau)$, and r number of servers. Notice that only for $M/M/r$ queues, the associated stochastic processes are markovian.

Characterization of queues. To quantify the queueing systems and to determine their performance, the following parameters are generally used:

The number of *waiting units* in the system at time t, *including* the one being served, if any.

The *waiting time distribution* for the queue, that is, the distribution of the duration of the time $\mathbf{w}_q(t)$ that a unit has to spend in the queue, and $\mathbf{w}_s(t)$ that it has to spend in the system, and the waiting time distribution for the nth arrival.

The *busy period distribution,* that is, the interval from the instant a unit arrives at an empty counter to the instant the server becomes free for the first time.

A complete characterization of the queueing system is given by their time-dependent solutions, which are usually difficult to obtain in general. Fortunately, often one is more interested in the steady state behavior resulting from the system being in operation for a long time. If such limiting behavior as in (16-66) exists, then the system goes into equilibrium and the steady state solutions can be used to determine the long-term properties of the system.

WAITING TIME DISTRIBUTIONS. An arriving item may or may not have to wait in the queue, and if the queue is empty it directly goes for service into the system. Let $\mathbf{w}_q$ represent the random waiting time duration in the queue, and if $\mathbf{s}$ denotes the service duration of an item, then the waiting time duration $\mathbf{w}_s$ in the system is given by

$$\mathbf{w}_s = \mathbf{w}_q + \mathbf{s} \tag{16-78}$$

Notice that unlike $\mathbf{w}_q$, the waiting time in the system is always nonzero for all units, since the service time of each item is always nonzero. We can make use of the conditional probability law

$$f_w(t) = \sum_n p_n f_w(t \mid n) \tag{16-79}$$

where p_n denotes the probability that there are n items waiting in the queue, to determine the p.d.f.s of these waiting times. If the queue has r channels in parallel, then the waiting time is zero if the number of items in the system n is less than r. In that case

$$f_w(t) = P\{n \le r-1\}\delta(t) + \sum_{n=r}^{\infty} p_n f_w(t \mid n) \tag{16-80}$$

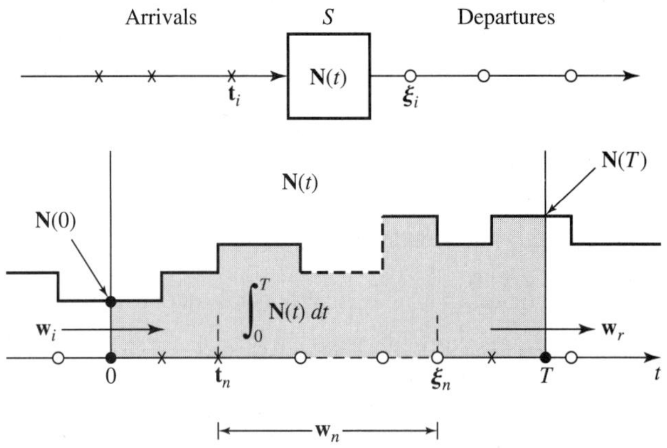

FIGURE 16-3

A general result that does not rely on any special conditions about the input and the nature of the system can be derived assuming that all processes are strict sense stationary with finite second order moments. Let $\mathbf{N}(t)$ represent the number of units in the system and $\{\mathbf{t}_i\}$ the input arrival instants and $\{\boldsymbol{\xi}_i\}$ the output departure instants. If $\mathbf{w}_i$ represents the total time spend by the ith unit in the system (waiting time and service time), then (Fig. 16-3)

$$\boldsymbol{\xi}_i = \mathbf{t}_i + \mathbf{w}_i \tag{16-81}$$

Thus $\mathbf{N}(t)$ increases by 1 at $\mathbf{t}_i$ and decreases by 1 at $\boldsymbol{\xi}_i$.

THEOREM 16-1

LITTLE'S THEOREM

▶ Suppose that the processes $\mathbf{t}_i$ and $\mathbf{w}_i$ are mean-ergodic

$$\frac{\mathbf{n}_T}{T} \to \lambda \quad \text{as} \quad T \to \infty; \qquad \frac{1}{n}\sum_{k=1}^{n} \mathbf{w}_k \to E\{\mathbf{w}_n\} \quad \text{as} \quad n \to \infty \tag{16-82}$$

In (16-82), $\mathbf{n}_T$ is the number of points $\mathbf{t}_i$ in the interval $(0, T)$ and $\lambda = E\{\mathbf{n}_T\}/T$ is the mean density of these points.

In that case[2]

$$E\{\mathbf{N}(t)\} = \lambda E\{\mathbf{w}_i\} \qquad \text{or} \qquad L = \lambda W \tag{16-83}$$

where L is the expected number of units in the system, and W is the expected waiting time in the system in the steady state. In fact, we shall establish the stronger statement that $\mathbf{N}(t)$ is also mean-ergodic:

$$\lim_{T\to\infty} \frac{1}{T}\int_0^T \mathbf{N}(t)\,dt = \lambda E\{\mathbf{w}_i\} = E\{\mathbf{N}(t)\} \tag{16-84}$$

Equations (16-83) seem reasonable: The mean $E\{\mathbf{N}(t)\}$ of the number of units in the system equals the mean number λ of arrivals per second multiplied by the mean time

[2] F. J. Beutler: "Mean Sojourn Times . . . ," *IEEE Transactions Information Theory,* March 1983.

$E\{\mathbf{w}_i\}$ that each unit remains in the system. It is not, however, always true, although it holds under general conditions.

Proof. We start with the observation that

$$-\sum_{r=1}^{\mathbf{N}(T)} \mathbf{w}_r \le \int_0^T \mathbf{N}(t)\,dt - \sum_{n=1}^{\mathbf{n}_T} \mathbf{w}_n \le \sum_{i=1}^{\mathbf{N}(0)} \mathbf{w}_i \tag{16-85}$$

In (16-85), the terms $\mathbf{w}_n$ of the second sum are due to the $\mathbf{n}_T$ units that arrived in the interval $(0, T)$; the terms $\mathbf{w}_i$ of the last sum are due to the $\mathbf{N}(0)$ units that are in the system at $t = 0$; the terms $\mathbf{w}_r$ of the first sum are due to the $\mathbf{N}(T)$ units that are still in the system at $t = T$. The details of the reasoning that establishes (16-85) are omitted. As we know (see Prob. 7-9)

$$E\left\{\left(\sum_{k=1}^{\mathbf{N}(t)} \mathbf{w}_k\right)^2\right\} \le E\{\mathbf{N}(t)^2\}E\left\{\mathbf{w}_k^2\right\} < \infty \tag{16-86}$$

Dividing (16-85) by T, we conclude that if T is sufficiently large, then

$$\frac{1}{T}\int_0^T \mathbf{N}(t)\,dt \simeq \frac{1}{T}\sum_{n=1}^{\mathbf{n}_T} \mathbf{w}_n \tag{16-87}$$

because the left and right sides of (16-85) tend to 0 after the division by T (see (16-86)). Furthermore, assumption (16-82) yields $\mathbf{n}_T \simeq \lambda T$ and

$$\frac{1}{T}\sum_{n=1}^{\mathbf{n}_T} \mathbf{w}_n \simeq \frac{\lambda}{\mathbf{n}_T}\sum_{n=1}^{\mathbf{n}_T} \mathbf{w}_n \simeq \lambda E\{\mathbf{w}_n\}$$

Inserting into (16-87), we obtain the first equality in (16-84). The second follows because the mean of the left side equals $E\{\mathbf{N}(t)\}$. ◀

Next we shall examine the steady state behavior of some of the specific queueing systems starting with the classic markovian queue with a single server.

Markovian Queues

$M/M/1$ QUEUE

▶ In this case, as Fig. 16-4 shows, the arrivals occur according to a Poisson process with parameter λ, so that from (16-11) and (16-32) the probability that a single arrival occurs in Δt is $\lambda\Delta t + o(\Delta t)$ while that of more than one arrival is $o(\Delta t)$. The interarrival durations $\boldsymbol{\tau}_n$ are independent exponential random variables with p.d.f. given by $a(\tau) = \lambda e^{-\lambda\tau}$, $\tau > 0$, and the service time durations $\mathbf{s}_n$ are also independent exponential with p.d.f. given by $b(\tau) = \mu e^{-\mu\tau}$. Thus the probability that service for one unit is completed in an interval Δt is given by $\mu\Delta t + o(\Delta t)$, and that of more than one completion there is $o(\Delta t)$. Let $\mathbf{N}(t)$ denote the number of items n in the system (those in the queue and the one being served, if any) at $t \ge 0$. Then $\mathbf{N}(t)$ is a continuous-time Markov process of the

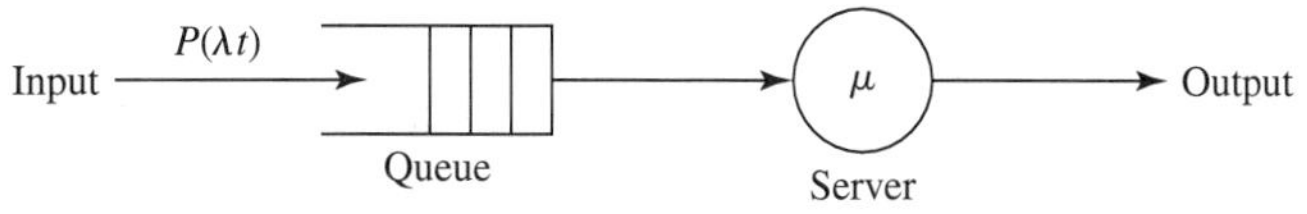

FIGURE 16-4
$M/M/1$ queue.

birth-death type discussed in Example 16-4, with $\lambda_n = \lambda$, $\mu_n = \mu$, and from Example 16-6 its limiting probabilities are given by (16-75). Thus the probability that there are n items in the system is given by

$$p_n = \lim_{t\to\infty} P\{\mathbf{N}(t) = n\} = (1-\rho)\rho^n \qquad n \geq 0 \tag{16-88}$$

provided the traffic intensity $\rho = \lambda/\mu < 1$. Notice that $1-\rho$ represents the probability that the system is empty, and the probability that the system is not empty is given by $P(\mathbf{N}(t) \geq 1) = \rho$. Since (16-88) represents a geometric distribution, the expected number in the system is given by

$$L = \lim_{t\to\infty} E\{\mathbf{N}(t)\} = \frac{\rho}{1-\rho} = \frac{\lambda}{\mu-\lambda} \tag{16-89}$$

and

$$\lim_{t\to\infty} \text{Var}\{\mathbf{N}(t)\} = \frac{\rho}{(1-\rho)^2} = \frac{\lambda\mu}{(\lambda-\mu)^2} = L + L^2 \tag{16-90}$$

Clearly $\text{Var}\{\mathbf{N}(t)\}$ is quite large compared to L and it increases rapidly as $\rho \to 1$. Hence the mean value in (16-89) has a great amount of uncertainly in the immediate neighborhood of $\rho = 1$.

WAITING TIME DISTRIBUTIONS. We can make use of (16-78)–(16-80) to determine the waiting time distributions in the queue as well as in the system. Given that there are n units in the system, the waiting time for the nth item in the queue is given by

$$\mathbf{w}_q = \mathbf{s}_1' + \mathbf{s}_2 + \cdots + \mathbf{s}_n \tag{16-91}$$

where $\mathbf{s}_1'$ represents the residual service time of the item being served, and $\mathbf{s}_2, \mathbf{s}_3, \ldots, \mathbf{s}_n$, the service times of the $n-1$ units ahead in the queue. Since $\mathbf{s}_1'$ is the residual of an exponential random variable with mean $1/\mu$, it is also an exponential distribution with the same mean, and $\mathbf{s}_2, \mathbf{s}_3, \ldots, \mathbf{s}_n$ represent independent exponential distributions with mean $1/\mu$. Hence $f_{w_q}(t \mid n)$ in (16-91) is a gamma random variable as in (4-37) or (10-87) with λ replaced by μ, and substituting this into (16-80) with $r = 1$, we obtain the probability density function for the waiting time in the queue to be

$$\begin{aligned} f_{w_q}(t) &= (1-\rho)\delta(t) + \sum_{n=1}^{\infty} (1-\rho)\rho^n \frac{\mu^n t^{n-1}}{(n-1)!} e^{-\mu t} \\ &= (1-\rho)\delta(t) + \mu(1-\rho)\rho e^{-\mu t} \sum_{n=0}^{\infty} \frac{(\mu\rho t)^n}{n!} \\ &= (1-\rho)\delta(t) + \mu(1-\rho)\rho e^{-\mu(1-\rho)t} \qquad t > 0 \end{aligned} \tag{16-92}$$

where the first term represents the probability that the waiting time is zero in the queue. From (16-92)

$$\begin{aligned} E\{\mathbf{w}_q\} &= (1-\rho)\cdot 0 + \int_{0^+}^{\infty} t\mu(1-\rho)\rho e^{-\mu(1-\rho)t}\, dt \\ &= \frac{\rho}{\mu(1-\rho)} = \frac{\lambda}{\mu(\mu-\lambda)} \end{aligned} \tag{16-93}$$

The probability that the waiting time in an $M/M/1$ queue is no more than t is given by

$$P\{\mathbf{w}_q \le t\} = 1 - P(\mathbf{w}_q > t) = 1 - \int_t^\infty f_{w_q}(x)\,dx$$
$$= 1 - \rho e^{-\mu(1-\rho)t} \tag{16-94}$$

and there is a finite probability equal to $(1-\rho)$ that the waiting time in the queue is in fact zero.

To determine the total waiting time distribution in the system, we can make use of relations (16-78)–(16-79) and (16-91). From there, for the nth item in the queue, the total waiting time in the system equals the waiting time in the queue plus its own service time $\mathbf{s}_{n+1}$. Thus given that there are n units in the system

$$\mathbf{w}_s = \mathbf{w}_q + \mathbf{s}_{n+1} = \mathbf{s}_1' + \mathbf{s}_2 + \mathbf{s}_3 + \cdots + \mathbf{s}_n + \mathbf{s}_{n+1} \tag{16-95}$$

and its conditional distribution is given by the gamma distribution

$$f_{w_s}(t \mid n) = \frac{\mu^{n+1}t^n}{n!}e^{-\mu t} \tag{16-96}$$

Hence using (16-79) the distribution of the waiting time in the $M/M/1$ queueing system is given by

$$f_{w_s}(t) = \sum_{n=0}^{\infty} p_n f_{w_s}(t \mid n) = \sum_{n=0}^{\infty}(1-\rho)\rho^n \frac{\mu^{n+1}t^n}{n!}e^{-\mu t}$$
$$= \mu(1-\rho)e^{-\mu t}\sum_{n=0}^{\infty}\frac{(\mu\rho t)^n}{n!} = \mu(1-\rho)e^{-\mu(1-\rho)t} \qquad t \ge 0 \tag{16-97}$$

and it represents an exponential p.d.f. with mean

$$E\{\mathbf{w}_s\} = \frac{1}{\mu(1-\rho)} = \frac{1}{\mu-\lambda} \tag{16-98}$$

Clearly Eqs. (16-89) and (16-98) agree with Little's formula in (16-83). ◀

***M/M/r* QUEUE**

▶ Consider a queueing model where a Poisson input with parameter λ feeds r identical servers (channels) that operate in parallel as shown in Fig. 16-5. Each server has an independent, identically distributed exponential service time holding distribution with parameter μ. If $n < r$ channels are busy, then the system is in state e_n, and the total number of services completed form a Poisson process with parameter $n\mu$, and the time between two successive service completions is exponential with parameter $n\mu$. On the other hand, if $n > r$, the time between two successive service completions is exponential with parameter $r\mu$ for all values of n. If the number $\mathbf{N}(t)$ of items present in the system is in state e_n at time t, then transition from e_n to e_{n+1} takes place in a small interval Δt with probability $\lambda\Delta t + o(\Delta t)$, and the probability that any one of the busy channels becomes free is $\mu\Delta t + o(\Delta t)$. Hence for $n < r$, the probability that none of the n busy channels become free equals $[1-\mu\Delta t + o(\Delta t)]^n$, since the channels are independent. Thus the probability that at least one server becomes free in the interval Δt is given by

$$1 - [1-\mu\Delta t + o(\Delta t)]^n = \begin{cases} n\mu\Delta t + o(\Delta t) & n < r \\ r\mu\Delta t + o(\Delta t) & n \ge r \end{cases} \tag{16-99}$$

For small intervals Δt, the probability that one or more servers become free is the same as the probability that one server becomes free, and hence (16-99) in fact represents

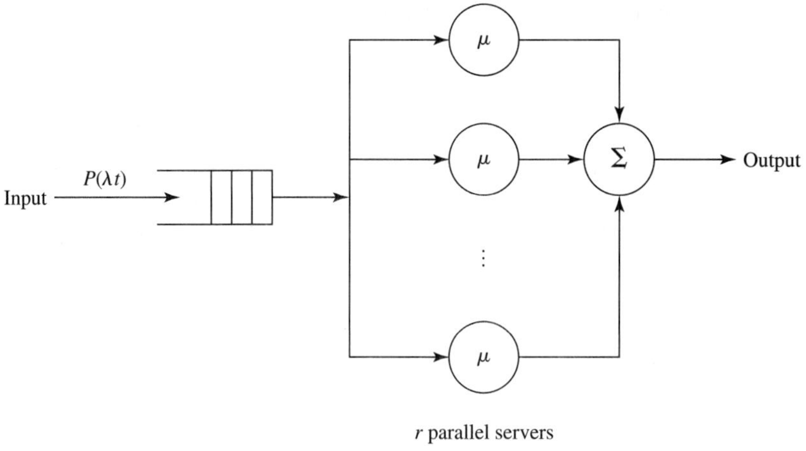

FIGURE 16-5
$M/M/r$ queue.

the probability of transition from e_n to e_{n-1}. The transitions from state e_j to states other than e_{j-1} or e_{j+1} have probability of order $o(\Delta t)$. This gives the nonzero probability transition densities in (16-50) and (16-51) for this specific "birth-death" process to be

$$\lambda_{j,j+1} = \lambda_j = \lambda \qquad \lambda_{j,j-1} = \mu_j = \begin{cases} j\mu & j < r \\ r\mu & j \geq r \end{cases} \tag{16-100}$$

and

$$\lambda_{jj} = -(\lambda_j + \mu_j) = \begin{cases} -(\lambda + j\mu) & j < r \\ -(\lambda + r\mu) & j \geq r \end{cases} \tag{16-101}$$

and hence the steady state probabilities p_j satisfy the equations (16-67) and (16-68) with parameters as in (16-100) and (16-101). Thus $N(t)$ is a birth-death process with constant birth (arrival) rate and state dependent service rates. Substituting these values into (16-71), we obtain the steady state probabilities to be

$$p_n = \begin{cases} \dfrac{\lambda \cdot \lambda \cdots \lambda}{\mu \cdot 2\mu \cdots n\mu} p_0 = \dfrac{(\lambda/\mu)^n}{n!} p_0 & n < r \\ \dfrac{\lambda \cdot \lambda \cdots \lambda}{\mu \cdot 2\mu \cdots r\mu \cdot r\mu \cdots r\mu} p_0 = \dfrac{r^r}{r!} (\lambda/r\mu)^n p_0 & n \geq r \end{cases} \tag{16-102}$$

The condition $\sum_{n=0}^{\infty} p_n = 1$ gives

$$\left\{ 1 + \sum_{n=1}^{r-1} \frac{(\lambda/\mu)^n}{n!} + \frac{r^r}{r!} \sum_{n=r}^{\infty} (\lambda/r\mu)^n \right\} p_0 = 1 \tag{16-103}$$

and the series in (16-103) is guaranteed to converge, provided $\lambda/r\mu < 1$. In the r channel case,

$$\rho = \lambda/r\mu \tag{16-104}$$

represents the traffic intensity, and

$$p_0 = \frac{1}{\displaystyle\sum_{n=0}^{r-1} \frac{(\lambda/\mu)^n}{n!} + \frac{(\lambda/\mu)^r}{r!(1-\rho)}} \tag{16-105}$$

We can use the above steady state probability distribution to compute various parameters that measure the effectiveness of a parallel queueing system.

The average number of items waiting in an $M/M/r$ queue is given by

$$L = \sum_{n=r+1}^{\infty} (n-r)p_n = p_r \frac{\rho}{(1-\rho)^2} \tag{16-106}$$

where

$$p_r = \frac{(\lambda/\mu)^r}{r!} p_0 \tag{16-107}$$

CALL WAITING. The probability that an arriving item has to wait in an $M/M/r$ queue is given by the probability that there are at least r items in the queue. Thus

$$p_{\text{cw}} = P\{\mathbf{N}(t) \geq r\} = \sum_{n=r}^{\infty} p_n = \frac{(\lambda/\mu)^r}{r!(1-\rho)} p_0 = \frac{p_r}{1-\rho} \tag{16-108}$$

This is also the probability that all r channels are occupied in an $M/M/r$ queue. This formula has wide application in telephone traffic theory, and it gives the *probability of call waiting* in an exchange with r trunk lines, where no server is available to handle

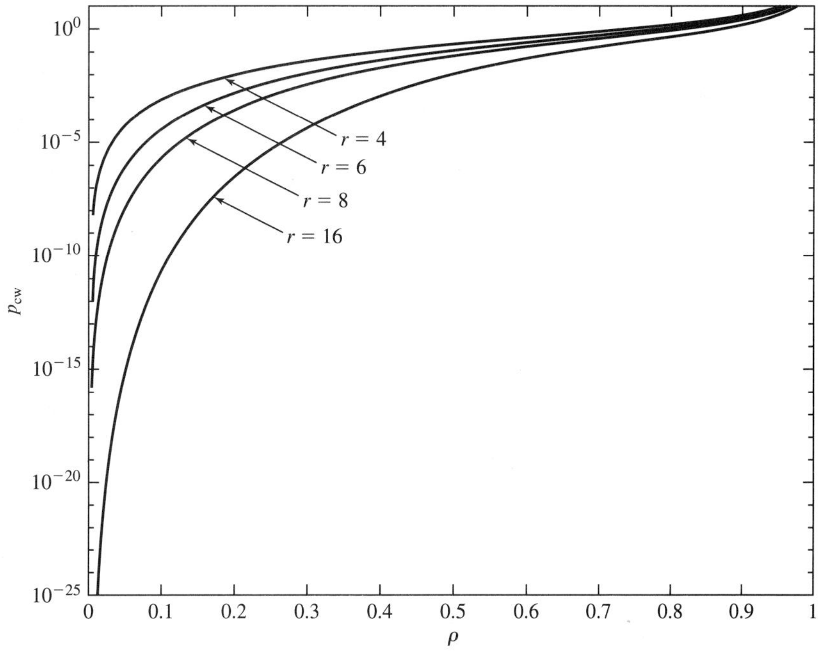

FIGURE 16-6
Probability of call waiting p_{cw} in an $M/M/r$ queue.

an incoming call. It is referred to as *Erlang's delayed-call* formula. Figure 16-6 shows the probability of call waiting given in (16-108) as a function of the load factor ρ. From there it is clear that the probability that all channels are occupied (a call has to wait) is negligible when the servers are lightly loaded and the wait is almost certain when they are heavily loaded. This well-known queueing phenomenon—high sensitivity to increase in the load factor when the system is already substantially loaded—should be taken into account in system design [39].

SINGLE QUEUE VS. SEVERAL QUEUES

The probability that there will be some item waiting in the queue is given by

$$\sum_{n=r+1}^{\infty} p_n = p_r \frac{\rho}{1-\rho} \tag{16-109}$$

From (16-106) and (16-109), the average number of waiting items, for those who actually wait in an $M/M/r$ queue, is given by

$$\frac{\sum_{n=r+1}^{\infty}(n-r)p_n}{\sum_{n=r+1}^{\infty} p_n} = \frac{1}{1-\rho} = \frac{r\mu}{r\mu-\lambda} \tag{16-110}$$

and it represents the average number of waiting items in front of a waiting customer. These items must go into service before the waiting customer can actually obtain service. Hence the mean waiting time in the queue for those who actually wait is given by

$$T_r = \frac{1}{r\mu}\frac{1}{1-\rho} = \frac{1}{r\mu-\lambda} \tag{16-111}$$

since $1/r\mu$ represents the average time between two successive service completions in a busy queue. Interestingly, we can use the above results to show the superiority of an $M/M/r$ configuration compared to r distinct $M/M/1$ queues in parallel that operate independently, each with its own waiting line.

For an $M/M/1$ channel, the average waiting time for those who actually wait is given by $(r=1)$ in (16-111)

$$T_1(\lambda) = \frac{1}{\mu-\lambda} \tag{16-112}$$

If the same Poisson process with arrival rate λ feeds r parallel queues of the $M/M/1$ type randomly, then from (9-25) each such input to these $M/M/1$ queues is Poisson with parameter $\lambda' = \lambda/r$ and replacing λ by λ/r in (16-112), we obtain the average waiting time for those who actually wait in such r separate $M/M/1$ parallel queues to be

$$T_1(\lambda/r) = \frac{r}{r\mu-\lambda} = rT_r \tag{16-113}$$

From (16-111) and (16-113), clearly the average wait is smaller by a factor of r in an $M/M/r$ queue compared to r separate $M/M/1$ parallel queues, and hence when r servers are available it is much more efficient to operate with a *single queue* rather than several independent queues.

WAITING TIME DISTRIBUTION. Let $\mathbf{w}_q$ denote the random variable representing the waiting time in the queue as before. From (16-108)

$$P\{\mathbf{w}_q = 0\} = P(N(t) \le r-1) = 1 - P(N(t) \ge r) = 1 - \frac{p_r}{1-\rho} \tag{16-114}$$

and $\mathbf{w}_q > 0$, if the number of items in the system is $n \geq r$, out of which r of them are in servers, and $n - r$ items are in the queue. Hence given that there are n items in the system, the waiting time in the queue is given by

$$\mathbf{w}_q = \min(\mathbf{s}_1', \mathbf{s}_2', \ldots, \mathbf{s}_r') + \mathbf{s}_1 + \mathbf{s}_2 + \cdots + \mathbf{s}_{n-r} \tag{16-115}$$

where the first term represents the least residual service time among the r items in the servers. Since each residual service time $\mathbf{s}_i'$ is independent and exponentially distributed with parameter μ, $\min(\mathbf{s}_1', \mathbf{s}_2', \ldots, \mathbf{s}_r')$ is an exponentially distributed random variable with parameter $r\mu$. From (16-99), the remaining $n - r$ independent service times $\mathbf{s}_i$ are also exponentially distributed with parameter $r\mu$ (the time between two successive service completions is $r\mu$), since all the channels are busy, and hence (16-115) represents a gamma random variable with parameters $(n - r + 1)$ and $n\mu$. Thus

$$f_{w_q}(t \mid n) = \frac{(r\mu)^{n-r+1}}{(n-r)!} t^{n-r} e^{-r\mu t} \qquad t \geq 0 \tag{16-116}$$

and from (16-79)–(16-80), the waiting time distribution $f_{w_q}(t)$ for $t > 0$ is given by

$$\begin{aligned} f_{w_q}(t) &= \sum_{n=r}^{\infty} \frac{r^r}{r!} \left(\frac{\lambda}{r\mu}\right)^n p_0 \frac{(r\mu)^{n-r+1}}{(n-r)!} t^{n-r} e^{-r\mu t} \\ &= r\mu \frac{(\lambda/\mu)^r}{r!} p_0 e^{-r\mu t} \sum_{n=r}^{\infty} \frac{(\lambda t)^{n-r}}{(n-r)!} \\ &= r\mu p_r e^{-(r\mu-\lambda)t} \qquad t > 0 \end{aligned} \tag{16-117}$$

and $P\{\mathbf{w}_q = 0\}$ is given by (16-114). Notice that the probability that an arriving item has to wait in an $M/M/r$ queue is given by

$$P\{\mathbf{w}_q > 0\} = \int_{0^+}^{\infty} f_{w_q}(t)\,dt = \frac{p_r}{1-\rho} \tag{16-118}$$

which is the same as the probability in (16-108) that there are r or more items in the system. Similarly the probability that the waiting time in an $M/M/r$ queue is longer than T is given by

$$P\{\mathbf{w}_q > T\} = \int_{T}^{\infty} f_{w_q}(t) = \frac{p_r}{1-\rho} e^{-r\mu(1-\rho)T} = P\{\mathbf{w}_q > 0\} e^{-r\mu(1-\rho)T} \tag{16-119}$$

From (16-117) the average waiting time in the queue for all arrivals turns out to be

$$\begin{aligned} E\{\mathbf{w}_q\} &= \int_{0^+}^{\infty} t f_{w_q}(t)\,dt = \frac{r\mu p_r}{(r\mu-\lambda)^2} \\ &= \frac{p_r/r\mu}{(1-\rho)^2} = \frac{p_r}{\lambda} \frac{\rho}{(1-\rho)^2} \end{aligned} \tag{16-120}$$

From Little's formula, the expected number of items in the queue is given by

$$L = \lambda E\{\mathbf{w}_q\} = p_r \frac{\rho}{(1-\rho)^2}$$

and it agrees with (16-106).

From (16-108) and (16-120), the average waiting time in an $M/M/r$ queue, given that the arriving unit has to wait, equals

$$\frac{E\{\mathbf{w}_q\}}{\sum_{n=r}^{\infty} p_n} = \frac{p_r/r\mu(1-\rho)^2}{p_r/(1-\rho)} = \frac{1}{r\mu(1-\rho)} = \frac{1}{r\mu-\lambda} \tag{16-121}$$

Similarly, using (16-109) the average waiting time in an $M/M/r$ queue, for those who actually wait is given by

$$\eta_r = \frac{E\{\mathbf{w}_q\}}{\sum_{n=r+1}^{\infty} p_n} = \frac{p_r\rho/\lambda(1-\rho)^2}{p_r\rho/(1-\rho)} = \frac{1}{\lambda(1-\rho)} = \frac{r\mu}{\lambda(r\mu-\lambda)} \tag{16-122}$$

Once again similar conclusions as in (16-113) follow, showing the superiority of the $M/M/r$ configuration over r distinct $M/M/1$ queues operating in parallel.

From (16-77), the average waiting time in the system is given by

$$E\{\mathbf{w}_s\} = E\{\mathbf{s}\} + E\{\mathbf{w}_q\} = \frac{1}{\mu} + \frac{p_r}{r\mu}\frac{1}{(1-\rho)^2} \tag{16-123}$$

and the average number of items in the system equals

$$L_s = \lambda E\{\mathbf{w}_s\} = \frac{\lambda}{\mu} + p_r\frac{\rho}{(1-\rho)^2} \tag{16-124}$$

◀

$M/M/r/r$ QUEUE (ERLANG'S MODEL)

▶ In this case, the number of servers are the same as before, but there is no facility in the system to wait and form a queue. If an arriving item finds all channels busy, it leaves the system without waiting for service (impatient customer). Erlang had originally used this loss model to investigate the distribution of busy channels in telephone systems. Such a system can handle up to r incoming calls at once. An incoming call "goes through" if at least one server is free, otherwise it is rejected (the call is lost) if all servers are busy (i.e., $\lambda_j = 0,\ j > r$). If the arrivals are assumed to be Poisson with parameter λ, and the service durations (call holding time) are also exponential with parameter μ, then Erlang's model represents a birth-death process with

$$\lambda_j = \begin{cases} \lambda & j \le r \\ 0 & j > r \end{cases} \qquad \mu_j = \begin{cases} j\mu & j < r \\ r\mu & j \ge r \end{cases} \tag{16-125}$$

From (16-71), this gives the steady state probabilities to be

$$p_n = \begin{cases} \dfrac{(\lambda/\mu)^n}{n!}p_0 & n = 0, 1, 2, \ldots, r \\ 0 & \text{otherwise} \end{cases} \tag{16-126}$$

The normalization condition $\sum_{k=0}^{r} p_k = 1$ gives

$$p_0 = \frac{1}{\displaystyle\sum_{k=0}^{r} \frac{(\lambda/\mu)^k}{k!}} \tag{16-127}$$

and hence the probability that n channels are busy in the steady state is given by (*Erlang's first formula*)

$$p_n = \frac{\dfrac{(\lambda/\mu)^n}{n!}}{\displaystyle\sum_{k=0}^{r} \frac{(\lambda/\mu)^k}{k!}} \qquad n = 0, 1, 2, \ldots, r \tag{16-128}$$

Notice that (16-128) represents a truncated Poisson distribution with parameter $\rho = \lambda/\mu$, and it is defined for *all values* of λ and μ. The main characteristic of the quality of service of a system with refusals (zero waiting room capacity), is measured by the probability of refusal or loss of a customer. In the Erlang model, from (16-128), the probability that an arriving item is lost (call congestion) is given by

$$p_r = \frac{\dfrac{(\lambda/\mu)^r}{r!}}{\displaystyle\sum_{k=0}^{r} \frac{(\lambda/\mu)^k}{k!}} \tag{16-129}$$

and it is the same as the probability that all channels are busy (time congestion). Equation (16-129) represents *Erlang's loss formula or blocking formula*. It has been shown by Pollaczek, Palm, Vaulot, and others that Erlang's loss formula holds for *any* distribution of service time, provided the input is Poisson with parameter λ (i.e., $M/G/r$ queue). (See also Prob. 16-16 for a related result.) The average number of busy servers in an $M/M/r/r$ queue is given by

$$N_s = \sum_{n=0}^{r} n p_n = \rho(1 - p_r) \tag{16-130}$$

A slight generalization of the $M/M/r/r$ queue (Erlang's model) can be obtained by considering a model with a limited waiting room capacity. ◀

$M/M/r/m$ QUEUE

▶ In this case $m > r$, and the number of servers are the same as in an $M/M/r$ model, but a limited number of items can wait on a first-come basis if they find all r channels busy. This gives the probability transition rates to be

$$\lambda_j = \begin{cases} \lambda & j \le m \\ 0 & j > m \end{cases} \qquad \mu_j = \begin{cases} j\mu & j \le r \\ r\mu & j > r \end{cases} \tag{16-131}$$

Here m represents the holding capacity of the system and $m - r$ the waiting room capacity. From (16-71) and (16-131), we get

$$p_n = \begin{cases} \dfrac{(\lambda/\mu)^n}{n!} p_0 & n \le r \\[2ex] \dfrac{(\lambda/\mu)^n}{r!\,r^{n-r}} p_0 & r < n \le m \end{cases} \tag{16-132}$$

where

$$p_0 = \frac{1}{\displaystyle\sum_{n=0}^{r} \frac{(\lambda/\mu)^n}{n!} + \frac{1}{r!} \sum_{n=r+1}^{m} \frac{(\lambda/\mu)^n}{r^{n-r}}} \tag{16-133}$$

Notice that when $m = r$, (16-132) coincides with Erlang's model. The probability of a call loss in this case is given by

$$p_m = \frac{\dfrac{(\lambda/\mu)^m}{r!r^{m-r}}}{\displaystyle\sum_{n=0}^{r} \frac{(\lambda/\mu)^n}{n!} + \frac{(\lambda/\mu)^r}{r!} \sum_{n=1}^{m-r} (\lambda/r\mu)^n} \tag{16-134}$$

The mean number of busy servers in steady state is given by

$$N_s = \sum_{n=1}^{r} np_n + r \sum_{n=r+1}^{m} p_n = \frac{\displaystyle\sum_{n=0}^{r-1} \frac{(\lambda/\mu)^n}{n!} + \frac{(\lambda/\mu)^r}{(r-1)!} \sum_{n=1}^{m-r} (\lambda/r\mu)^n}{\displaystyle\sum_{n=0}^{r} \frac{(\lambda/\mu)^n}{n!} + \frac{(\lambda/\mu)^r}{r!} \sum_{n=1}^{m-r} (\lambda/r\mu)^n} \tag{16-135}$$

Table 16-1 and Fig. 16-7 show the probability of call loss for various values of the load factor $\rho = \lambda/\mu$, for the Erlang model *without* and *with* waiting room facility. From Fig. 16-7 it is clear that when the servers are lightly loaded, the probability of losses decrease substantially with an increase in the number of servers. However, the situation is almost the same in all cases if the load is large. Interestingly, even one additional waiting facility reduces the probability of call loss, provided the load factor is not very large. ◀

$M/M/\infty$ QUEUE

▶ Here the number of servers is assumed to be unlimited so that there is no queue, and an arriving item is instantly served. In this can $\lambda_n = \lambda$, $\mu_n = n\mu$, $n = 0, 1, 2, \ldots$ so that the steady state solutions can be obtained by letting $r \to \infty$ in the $M/M/r$ or

TABLE 16-1
Probability of call loss for an Erlang queue without and with waiting room capacity ($M/M/r/r$ vs. $M/M/r/r+1$)

	Probability of call loss p_m					
	Waiting room capacity = 0, ($m = r$), r			Waiting room capacity = 1, ($m = r + 1$), r		
ρ	2	4	8	2	4	8
0.1	0.0045	0	0	0.0002	0	0
0.2	0.0164	0	0	0.0016	0	0
0.4	0.0541	0.0007	0	0.0107	0	0
0.8	0.1509	0.0077	0	0.0569	0.0015	0
1	0.2000	0.0154	0	0.0909	0.0038	0
2	0.4000	0.0952	0.0009	0.2857	0.0455	0.0002
4	0.6154	0.3107	0.0304	0.5517	0.2370	0.0150
8	0.7805	0.5746	0.2356	0.7574	0.5347	0.1907
12	0.8471	0.6985	0.4227	0.8356	0.6769	0.3880
16	0.8828	0.7674	0.5452	0.8760	0.7543	0.5216
20	0.9050	0.8109	0.6270	0.9005	0.8020	0.6105

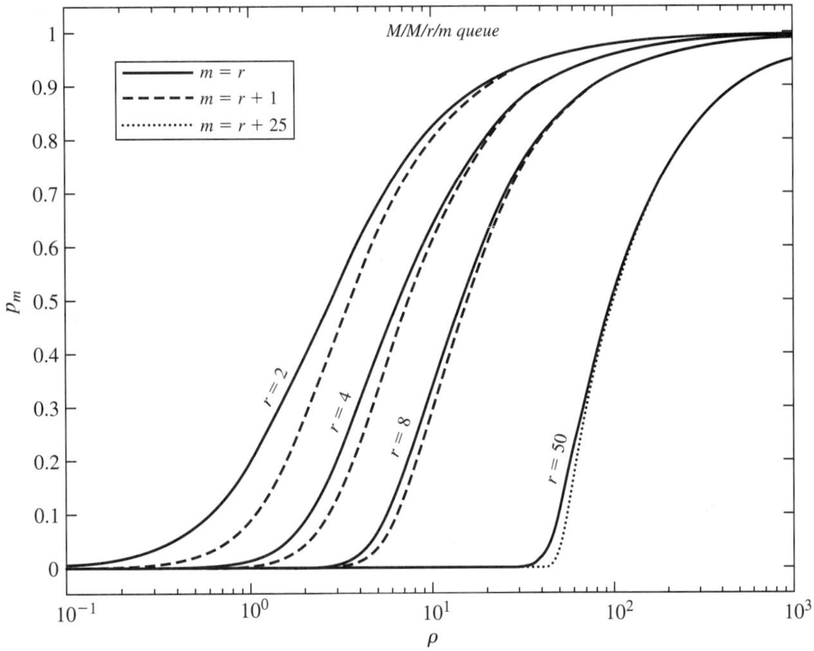

FIGURE 16-7
Probability of call loss for an Erlang queue without and with waiting room capacity. For $r = 2, 4, 8$, the waiting room capacity is either zero or one. For $r = 50$, the waiting room capacity is zero or 25.

$M/M/r/r$ cases. From (16-127),

$$p_0 = \frac{1}{\displaystyle\sum_{k=0}^{\infty} \frac{(\lambda/\mu)^k}{k!}} = e^{-(\lambda/\mu)} \tag{16-136}$$

and

$$p_n = p_0 \frac{(\lambda/\mu)^n}{n!} = e^{-(\lambda/\mu)} \frac{(\lambda/\mu)^n}{n!} \qquad n = 0, 1, 2, \ldots \tag{16-137}$$

Thus in steady state, the number of items in an $M/M/\infty$ queue is Poisson distributed with parameter λ/μ (see also Prob. 16-16 for a similar result on $M/G/\infty$ queues). When the number of servers is large, (16-137) can be used in an $M/M/r/r$ model to compute p_n reasonably accurately. ◀

IMPATIENT CUSTOMERS (RENEGING)

▶ In some systems, an arriving item that finds all servers busy joins the queue but waits only for a limited duration, and after that it departs the queue (reneging). The item may also decide to leave the queue depending on the existing size of the queue. Suppose the system has Poisson input arrivals with rate λ, and r independent identical servers with exponential service rate μ. Further, there is no limitation regarding the waiting space. In that case $\lambda_j = \lambda$, $j = 0, 1, 2, \ldots$, and $\mu_j = j\mu$ for $j \le r$. For $j > r$, the transition from state e_j to state e_{j-1} in an interval Δt can occur in two mutually exclusive ways: either one of the r busy servers becomes free with probability $r\mu\Delta t + o(\Delta t)$ or one of the waiting units in the queue departs with probability $\nu_j \Delta t + o(\Delta t)$. Thus the resulting markovian process has the probability transition densities given by

$$\lambda_j = \lambda \qquad \mu_j = \begin{cases} j\mu & j \le r \\ r\mu + \nu_j & j > r \end{cases} \tag{16-138}$$

Substituting these values into (16-73) and (16-74), we obtain the long-run probability that there are n items in the system to be

$$p_n = \begin{cases} \dfrac{(\lambda/\mu)^n}{n!} p_0 & n \le r \\ \dfrac{\lambda^n}{r!\mu^r \prod_{j=r+1}^{n}(r\mu + \nu_j)} p_0 & n > r \end{cases} \tag{16-139}$$

where

$$p_0 = \frac{1}{\displaystyle\sum_{n=0}^{r} \frac{(\lambda/\mu)^n}{n!} + \frac{1}{r!\mu^r} \sum_{n=r+1}^{\infty} \frac{\lambda^n}{\prod_{j=r+1}^{n}(r\mu + \nu_j)}} \tag{16-140}$$

In particular, if we assume the waiting time of a unit is an independent exponentially distributed random variable with parameter ν, then arguing as before, the probability that one of the $(n - r)$ waiting items departs the queue prior to service is given by

$$1 - (1 - \nu \Delta t)^{n-r} = (n - r)\nu \Delta t$$

so that

$$\nu_j = (j - r)\nu \tag{16-141}$$

and this value can be used in (16-139) and (16-140) to determine the steady state probabilities of such a queue. ◀

So far, we have analyzed queues with an infinite source population. That need not be the case always, and in many situations sources may have only finite resources. Next we shall examine finite source models with m items that are served by r servers ($r < m$).

Finite Input Sources and Machine Servicing Problem

Finite input situations occur in machine servicing problems, where a set of m machines that break down from time to time are serviced by r repairmen ($r < m$). Each time a machine breaks down, a repairman works on it to put it in operational form. If the number of inoperative machines at any time exceeds the total number of repairmen, then the excess machines will have to wait till repairmen are available for servicing them. In this context, the rate of machine breakdown, the time taken to repair an inoperative machine, and an optimum strategy that minimizes the overall cost in terms of repairmen and inoperative machine loss are of special interest.

In the queueing theory terminology, the machines that breakdown are the customers, the repairmen are the servers, and customers waiting time corresponds to the idle time of an inoperative machine before repair is performed. However, unlike the queueing models studied earlier, the arrival of customers is finite here, since when all the machines are inoperative, another machine cannot break down and no further customer can join the queue. Thus the probability of an additional customer joining the queue depends on how many customers are already waiting in the queue at that time.

Here, we first review the classical $M/M/r$ model due to Palm (1947), and a more general $M/G/1$ model due to Takacs (1957) in connection with the machine servicing problem.

PALM'S FINITE INPUT $M/M/r$ MODEL

Suppose the input population consists of m machines that break down independently and call for service, by an exponential distribution with parameter λ. Thus the probability density function of the operating time of a working machine is given by

$$f(t) = \begin{cases} \lambda e^{-\lambda t} & t \geq 0 \\ 0 & \text{otherwise} \end{cases} \tag{16-142}$$

Hence from (16-11) the probability that a working machine at time t breaks down in the interval $(t, t+\Delta t)$ is given by $\lambda \Delta t + o(\Delta t)$, and the average working time of a machine is $1/\lambda$. Let e_n represent the state that n out of m machines are *not working* at time t. Thus the probability of transition from state e_n to e_{n+1} in the interval $(t, t+\Delta t)$ equals the probability that at least one of the $m-n$ working machines break down in that interval, and it is given by $1 - [1 - \lambda\Delta t + o(\Delta t)]^{m-n} = (m-n)\lambda\Delta t + o(\Delta t)$. Thus

$$\lambda_{n,n+1} = \lambda_n = \begin{cases} (m-n)\lambda & n = 0, 1, 2, \ldots, m-1 \\ 0 & n \geq m \end{cases} \tag{16-143}$$

Service is provided by r repairmen, $r < m$, and duration of each service is an exponential random variable with parameter μ. Thus when a machine is being serviced at time t, the probability that the servicing will be completed in $(t, t+\Delta t)$ is given by $\mu\Delta t + o(\Delta t)$. As before, when n machines are being serviced, the probability that one service gets completed in $(t, t+\Delta t)$ equals $1 - [1 - \mu\Delta t + o(\Delta t)]^n = n\mu\Delta t + o(\Delta t)$ for $n \leq r$, and equals $r\mu\Delta t + o(\Delta t)$ for $n > r$. This gives

$$\mu_n = \begin{cases} n\mu & n \leq r \\ r\mu & n \geq r \end{cases} \tag{16-144}$$

When a machine breaks down, it is at once serviced if one of the r repairmen is available, otherwise it forms a queue and waits for service. *Machine interference time* corresponds to the duration when a machine breaks down and waits for a repairman (idle time), who in turn may be busy repairing other machines or doing related work. Thus the machine servicing problem is equivalent to a birth-death Markov process (Examples 16-4 and 16-6) $\mathbf{x}(t)$ with parameters as in (16-143) and (16-144), where $\mathbf{x}(t)$ represents the number of *nonworking machines* at time t. Using (16-73) and (16-74), the steady state probabilities that n out of m machines are *not working* turn out to be

$$p_n = \begin{cases} \dfrac{1}{n!}\displaystyle\prod_{i=0}^{n-1}(m-i)\left(\frac{\lambda}{\mu}\right)^n p_0 = \binom{m}{n}\left(\frac{\lambda}{\mu}\right)^n p_0 & n = 0, 1, \ldots, r \\ \dfrac{1}{r!r^{n-r}}\displaystyle\prod_{i=0}^{n-1}(m-i)\left(\frac{\lambda}{\mu}\right)^n p_0 = \dfrac{m!r^r}{(m-n)!r!}\left(\frac{\lambda}{r\mu}\right)^n p_0 & r \le n \le m \end{cases} \tag{16-145}$$

where

$$p_0 = \frac{1}{\displaystyle\sum_{n=0}^{r}\binom{m}{n}(\lambda/\mu)^n + \frac{r^{r-1}}{(r-1)!}\sum_{n=r+1}^{m}\frac{m!}{(m-n)!}(\lambda/r\mu)^n} \tag{16-146}$$

This solution was first obtained by Palm (1947). Following Naor (1956), if we use the notation [48]

$$\rho \triangleq \frac{\lambda}{r\mu} \tag{16-147}$$

$$p(k, \rho) \triangleq e^{-\rho}\frac{\rho^k}{k!} \tag{16-148}$$

$$P(n, \rho) \triangleq \sum_{k=n}^{\infty} p(k, \rho) \tag{16-149}$$

and

$$S(m, r, \rho) \triangleq \sum_{n=0}^{r-1}\frac{r^n}{n!}p(m-n, \rho) + \frac{r^{r-1}}{(r-1)!}[1 - P(m-r+1, \rho)] \tag{16-150}$$

then (16-145) and (16-146) reduce to the compact form (show this)

$$p_n = \begin{cases} \dfrac{r^n}{n!}\dfrac{p(m-n, 1/\rho)}{S(m, r, 1/\rho)} & n < r \\ \dfrac{r^{r-1}}{(r-1)!}\dfrac{p(m-n, 1/\rho)}{S(m, r, 1/\rho)} & r \le n \le m \end{cases} \tag{16-151}$$

and it can be verified by direct substitution. For a single repairman case, (16-151) simplifies to

$$p_n = \frac{\dfrac{(1/\rho)^{m-n}}{(m-n)!}}{\displaystyle\sum_{k=0}^{m}\frac{(1/\rho)^k}{k!}} \qquad n = 0, 1, 2, \ldots, m \tag{16-152}$$

and it represents the probability that n out of m machines are *not* working. We can use (16-151) to deduce several interesting conclusions. First, the long-term probability that s machines are in *working order* is given by p_{m-s}. Notice the close similarity of p_{m-s} in the single repairman case with Erlang's first formula in (16-128).

Let the random variables $\mathbf{x}$, $\mathbf{y}$, and $\mathbf{z}$ represent, respectively, the number of machines in working order, the number of machines being serviced, and the number of machines waiting in line for service at any time. Then

$$\mathbf{x} + \mathbf{y} + \mathbf{z} = m \tag{16-153}$$

and using (16-151) their average values are given by (show this)

$$a = E\{\mathbf{x}\} = \sum_{n=0}^{m}(m-n)p_n = \frac{\mu r}{\lambda}\frac{S(m-1,r,1/\rho)}{S(m,r,1/\rho)} \tag{16-154}$$

$$b = E\{\mathbf{y}\} = \sum_{n=0}^{r-1} np_n + r\sum_{n=r}^{m} p_n$$

$$= r - \sum_{n=0}^{r-1}(r-n)p_n = r\frac{S(m-1,r,1/\rho)}{S(m,r,1/\rho)} \tag{16-155}$$

and

$$c = E\{\mathbf{z}\} = \sum_{n=r+1}^{m}(n-r)p_n \tag{16-156}$$

so that

$$a + b + c = m \tag{16-157}$$

Equations (16-153) and (16-157) state that a machine has to be in one of these three states. From (16-154) and (16-155) we also get

$$\frac{a}{b} = \frac{\mu}{\lambda} \tag{16-158}$$

which states that the ratio of the average number of working machines to those being serviced equals the ratio of the average working time of a machine $(1/\lambda)$ to the average servicing time $(1/\mu)$, an obvious equality. We can define a/m to be the *machine efficiency* factor, so that from (16-157), it is given by $1-(b/m-c/m)$, where b/m represents the loss factor due to repairs, and c/m the loss factor due to interference that occurs because of other machines being repaired. Interestingly, b also represents the average number of occupied (busy) repairmen so that

$$\left.\begin{matrix}\textit{Average number of}\\ \textit{unoccupied repairmen}\end{matrix}\right\} = r - b = r\left(1 - \frac{S(m-1,r,1/\rho)}{S(m,r,1/\rho)}\right) \tag{16-159}$$

and

$$\frac{b(r)}{r} = \frac{S(m-1,r,1/\rho)}{S(m,r,1/\rho)} \tag{16-160}$$

represents *repairmen efficiency*.

For maximum repairmen efficiency the optimization criterion is $b(r) = r$, or $b(r)/r$ needs to be maximized, and for maximum machine efficiency, the quantity $a(r)/m = (1/\rho m)b(r)/r$ needs to be maximized. These two requirements are conflicting, since to keep up servicemen efficiency, r should be as small as possible, whereas to keep up machine efficiency r should be as large as possible. Equating $b(r)/r$ and $a(r)/m$, we get the optimum number of servicemen to be $r_0 = m\lambda/\mu$. Figure 16-8 shows $b(r)/r$ as well as the machine efficiency $a(r)/m$ for $m = 100$ and $\lambda/\mu = 0.15$. From Fig. 16-8, $r \le 10$ keeps the servicemen completely busy, whereas at least 30% of the machines will be unproductive in that case. On the other hand, $r_0 = 15$ gives about 85% machine efficiency and also keeps the servicemen occupied 85% of the time. Increasing the servicemen beyond 15 does not result in any significant gain in machine efficiency in this case. ◀

Several variations to the above model are possible. Different machines can have different stoppage parameters λ_i and different service parameters μ_i. Further, each

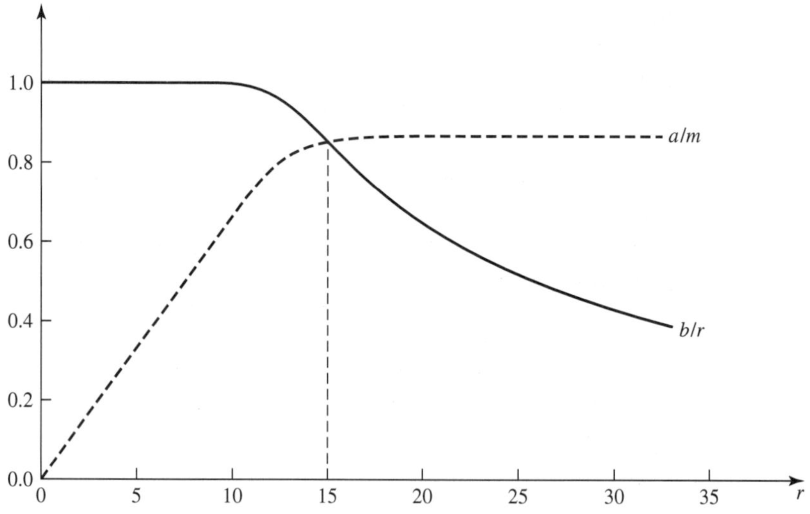

FIGURE 16-8
Machine Servicing Problem: Servicemen efficiency (b/r) and Machine efficiency (a/m). Here $m = 100$ and $\lambda/\mu = 0.15$.

repairman may specialize in one or several (not all) types of repairs. The repairmen may also spend part of their working time attending related work such as collecting materials or traveling between inoperative machines. Benson and Cox have investigated this ancillary work problem.

The general machine interference problem involving finite Poisson input population and arbitrary service-time distribution has been studied for a single repairman case $(M/G/1)$ by Khinchin, Kronig, Palm, Ashcroft, and Takacs. Next we examine an elegant technique due to Takacs that is based on a method given by Ashcroft.

TAKACS' FINITE INPUT $M/G/1$ MODEL

In the Takacs' model, the queueing system consists of m machines that work independently and a single repairman. The operating time of each machine is exponential as in (16-142), so that the probability that a working machine at time t breaks down in the interval $(t, t + \Delta t)$ is given by $\lambda \Delta t + o(\Delta t)$. When a breakdown occurs, the machine will be serviced immediately unless the repairman is servicing another machine, in which case it joins a queue of machines waiting to be repaired. The service time is assumed to be a positive random variable with *arbitrary* density function $f_s(t)$, and hence the underlying queueing system is of type $M/G/1$.

Unlike the Palm model, Takacs defines $\mathbf{x}(t)$ to represent the number of *working machines* at time t. Since the process $\mathbf{x}(t)$ is in general non-markovian, Takacs first examines the imbedded chain obtained by observing the number of working machines just before the termination of service of machines under repair. Thus if $t_1, t_2, \ldots, t_n, \ldots$ denote the end points of consecutive service-time periods, then $\mathbf{x}_n = \mathbf{x}(t_n - 0)$ represents an imbedded Markov chain that is aperiodic with a finite number of states $j = 0, 1, 2, \ldots, m - 1$. Notice that it does not make sense to talk about m working machines just before the termination of a service, since there is a total of only m machines, and at the time of observation, at least one of them is still under repair. Since

$t_1, t_2, \ldots$ correspond to the departure instants of customers in a regular queue, on comparing with Examples 15-14 and 15-24, the present method is very similar to that of generating imbedded Markov chains in those cases.

To compute the limiting probabilities π_j, where

$$\pi_j = \lim_{n\to\infty} P\{\mathbf{x}_n = j\} \qquad j = 0, 1, 2, \ldots, m-1 \tag{16-161}$$

define p_{ij} to be the probability that there are j working machines just before the end of the current service, given that there were i working machines just before the end of the previous service, that is,

$$p_{ij} = P\{\mathbf{x}_n = j \mid \mathbf{x}_{n-1} = i\} \tag{16-162}$$

To evaluate these transition probabilities we argue as follows: During the service time t of a machine, from (16-142) the probability that a machine is in working order is given by $p = P(\tau > t) = e^{-\lambda t}$, and the probability that it calls for service equals $q = 1 - e^{-\lambda t}$. After the previous machine under repair has finished repair work, there are $i+1$ working machines, and to have j of them work during the current service time t is given by the binomial expression $\binom{i+1}{j} e^{-j\lambda t}(1-e^{-\lambda t})^{i+1-j}$. Since the probability density of the service time duration t is $f_s(t)dt$, we have [53]

$$p_{ij} = \int_0^\infty \binom{i+1}{j} e^{-j\lambda t}(1-e^{-\lambda t})^{i+1-j} f_s(t)\,dt \qquad i = 1, 2, \ldots, m-2 \quad j \le i+1 \tag{16-163}$$

and

$$p_{m-1,j} = p_{m-2,j} \qquad i = m-1 \tag{16-164}$$

At least one machine is under repair during a service time, so that always $i \le m-1$. The ergodic nature of the Markov chain gives the desired steady state probabilities π_j in (16-161) as the solution of the linear equations (15-167) given by

$$\pi_j = \sum_{i=j-1}^{m-1} \pi_i p_{ij} \qquad 1 \le j \le m-1, \qquad \sum_{j=0}^{m-1} \pi_j = 1 \tag{16-165}$$

To obtain explicit expressions for π_j consider its moment generating function

$$\begin{aligned} P(z) &= \sum_{j=0}^{m-1} \pi_j z^j = \sum_{j=0}^{m-1}\sum_{i=j-1}^{m-1} \pi_i p_{ij} z^j \\ &= \int_0^\infty \left\{ \sum_{j=0}^{m-1}\sum_{i=j-1}^{m-2} \pi_i \binom{i+1}{j} (zp)^j (1-p)^{i+1-j} \right. \\ &\quad \left. + \pi_{m-1} \sum_{j=0}^{m-1} \binom{m-1}{j} (zp)^j (1-p)^{m-1-j} \right\} f_s(t)\,dt \end{aligned} \tag{16-166}$$

where we have used (16-163) and (16-164) with $p \overset{\Delta}{=} e^{-\lambda t}$. But the terms inside the parenthesis in (16-166) simplify to

$$\begin{aligned}\sum_{i=0}^{m-2} \pi_i \sum_{j=0}^{i+1} \binom{i+1}{j} (zp)^j (1-p)^{i+1-j} + \pi_{m-1}(1-p+zp)^{m-1} \\ = \sum_{i=0}^{m-2} \pi_i (1-p+zp)^{i+1} + \pi_{m-1}(1-p+zp)^{m-1} \\ = (1-p+zp)P(1-p+zp) - \pi_{m-1}(z-1)p(1-p+zp)^{m-1}\end{aligned} \tag{16-167}$$

where $P(z)$ is as defined in (16-166). Substituting (16-167) into (16-166) we get

$$\begin{aligned}P(z) = \int_0^\infty [1+(z-1)e^{-\lambda t}] P(1+(z-1)e^{-\lambda t}) f_s(t)\,dt \\ - \pi_{m-1}(z-1) \int_0^\infty e^{-\lambda t}[1+(z-1)e^{-\lambda t}]^{m-1} f_s(t)\,dt\end{aligned} \tag{16-168}$$

Write

$$P(z) \triangleq \sum_{j=0}^{m-1} B_j (z-1)^j \tag{16-169}$$

where B_j represents the binomial moments of π_j, and they are related as

$$B_j = \sum_{i=j}^{m-1} \binom{i}{j} \pi_i \qquad \pi_j = \sum_{i=j}^{m-1} (-1)^{i-j} \binom{i}{j} B_i \tag{16-170}$$

Notice that $B_0 = 1$ and $B_{m-1} = \pi_{m-1}$. Using (16-169) we get

$$P(1+(z-1)e^{-\lambda t}) = \sum_{j=0}^{m-1} B_j (z-1)^j e^{-j\lambda t} \tag{16-171}$$

Finally, substituting (16-171) into (16-168), we obtain

$$\begin{aligned}P(z) = \sum_{j=0}^{m-1} B_j [\phi_j (z-1)^j + \phi_{j+1}(z-1)^{j+1}] \\ - \pi_{m-1} \sum_{j=0}^{m-1} \binom{m-1}{j} \phi_{j+1}(z-1)^{j+1}\end{aligned} \tag{16-172}$$

where the constant

$$\phi_k \triangleq \int_0^\infty f_s(t) e^{-k\lambda t}\,dt \tag{16-173}$$

represents the Laplace transform of the service-time distribution evaluated at $k\lambda$. On comparing (16-169) and (16-172), we get

$$B_j(1-\phi_j) - B_{j-1}\phi_j = -\pi_{m-1} \binom{m-1}{j-1} \phi_j \qquad 1 \le j \le m-1 \tag{16-174}$$

Since $B_{m-1} = \pi_{m-1}$, define

$$\beta_j = \frac{B_j}{\pi_{m-1}} \qquad 0 \le j \le m-1 \tag{16-175}$$

so that $\beta_{m-1} = 1$, and (16-174) can be expressed as

$$\beta_j = \frac{\phi_j}{1-\phi_j}\left[\beta_{j-1} - \binom{m-1}{j-1}\right] \triangleq M_j[\beta_{j-1} - N_{j-1}] \qquad 1 \le j \le m-1 \tag{16-176}$$

with $M_j = \phi_j/(1-\phi_j)$ and $N_{j-1} = \binom{m-1}{j-1}$. Iterating the above equation we get

$$\begin{aligned}\beta_j &= \left(\prod_{i=1}^{j} M_i\right)\beta_0 - \sum_{i=0}^{j-1}\left(\prod_{k=j-i}^{j} M_k\right) N_{j-i-1} \\ &= C_j\left(\beta_0 - \sum_{i=0}^{j-1}\frac{N_{j-i-1}}{C_{j-i-1}}\right) \\ &= C_j\left[\beta_0 - \sum_{k=0}^{j-1}\binom{m-1}{k}\frac{1}{C_k}\right]\end{aligned} \tag{16-177}$$

where we define

$$C_0 = 1 \qquad C_j = \prod_{k=1}^{j}\frac{\phi_k}{1-\phi_k} \qquad 1 \le k \le m-1 \tag{16-178}$$

With $j = m-1$ in (16-177) we obtain

$$\beta_0 = \frac{1}{C_{m-1}} + \sum_{k=0}^{m-2}\binom{m-1}{k}\frac{1}{C_k} = \sum_{k=0}^{m-1}\binom{m-1}{k}\frac{1}{C_k} \tag{16-179}$$

so that from (16-177)

$$\beta_j = C_j \sum_{k=j}^{m-1}\binom{m-1}{k}\frac{1}{C_k} \tag{16-180}$$

and $j = 0$ in (16-175) gives

$$\pi_{m-1} = \frac{B_0}{\beta_0} = \frac{1}{\sum_{k=0}^{m-1}\binom{m-1}{k}(1/C_k)} \tag{16-181}$$

Finally, using (16-180) and (16-181) in (16-175), we obtain

$$B_j = C_j\frac{\sum_{i=j}^{m-1}\binom{m-1}{i}(1/C_i)}{\sum_{i=0}^{m-1}\binom{m-1}{i}(1/C_i)} \qquad 0 \le j \le m-1 \tag{16-182}$$

Using (16-181) in (16-170), the desired steady state probabilities π_k, that there are k machines in working order just prior to the completion of a service, simplify to

$$\pi_k = \frac{\sum_{j=k}^{m-1}(-1)^{j-k}\binom{j}{k}C_j\sum_{i=j}^{m-1}\binom{m-1}{i}(1/C_i)}{\sum_{i=0}^{m-1}\binom{m-1}{i}(1/C_i)} \qquad 0 \le k \le m-1 \tag{16-183}$$

If, in particular, the service times are exponentially distributed, then

$$f_s(t) = \mu e^{-\mu t} \qquad t \geq 0$$

and

$$\phi_j = \frac{\mu}{\mu + j\lambda} \qquad C_j = \frac{(\mu/\lambda)^j}{j!}$$

so that (16-183) simplifies to

$$\pi_k = \frac{\dfrac{(\mu/\lambda)^k}{k!}}{\displaystyle\sum_{j=0}^{m-1} \frac{(\mu/\lambda)^j}{j!}} \qquad 0 \leq k \leq m-1 \tag{16-184}$$

The absolute probabilities

$$p_j = \lim_{t\to\infty} P\{\mathbf{x}(t) = j\} \qquad j = 0, 1, \ldots, m \tag{16-185}$$

can be similarly derived by using the transition probabilities r_{ij} instead of p_{ij} in (16-163) and (16-164), obtained by replacing $f_s(t)$ in (16-163) with the remaining service time distribution $P\{s > t\} = [1 - F_s(t)]$. This gives [35, 48]

$$p_n = \frac{m\pi_{n-1}}{n(\pi_{n-1} + m\lambda/\mu)} \qquad n = 1, 2, \ldots, m \tag{16-186}$$

and

$$p_0 = 1 - \sum_{n=1}^{m} p_n \tag{16-187}$$

◀

The above non-markovian queue assumes finite input source models. Next we examine general non-markovian queues of the $M/G/1$ and $GI/M/1$ type, where either the service times or the interarrival times do not possess the memoryless property. Thus the process $\mathbf{x}(t)$ that represents the state of the system will no longer be Markovian; however, by examining the discrete-time process generated just immediately after each departure instant t_n (or just before each arrival instant t_n), an imbedded Markov chain $\mathbf{x}_n = \mathbf{x}(t_n + 0)$ (or $\mathbf{x}_n = \mathbf{x}(t_n - 0)$) can be extracted, and its steady state behavior can be studied using techniques developed in Chap. 15. When such steady state probabilities exist for the chain, Khinchin has also shown that they represent the steady–state behavior of the non-markovian process at arbitrary times (see (15-35)) [40].

Interestingly, this approach of Takacs can be used to analyze a generalized Erlang model for telephone traffic, where the input arrivals are no longer Poisson, but possess an arbitrary distribution.

$GI/M/r/r$ QUEUE (GENERALIZED ERLANG MODEL)

▶ Let the interarrival durations of the input process be independent random variables with common distribution $A(\tau)$ and mean value $1/\lambda$. The service times in all r servers in the system are assumed to be independent exponential distributions with common parameter μ. Once again the system has no facility to wait, and if an arriving unit finds all channels busy, it simply leaves the system (call is lost). Hence the possible states of the system are $e_0, e_1, e_2, \ldots, e_r$, and let $\mathbf{x}(t)$ represent the number of busy channels at time t. If $t_1, t_2, \ldots, t_n, \ldots$ denote the call arrival instants, then $\mathbf{x}_n = \mathbf{x}(t_n - 0)$ denotes the number of busy channels just prior to the nth arrival, and

$$\pi_j = \lim_{n\to\infty} P\{\mathbf{x}_n = j\} \qquad j = 0, 1, 2, \ldots, r \tag{16-188}$$

their limiting probabilities. In this case, following the arguments for computing the transition probabilities p_{ij} in (16-163) that there are j busy channels (each with probability $p = e^{-\mu t}$) just before a new arrival, given that there were i busy channels just before the previous arrival, we obtain (with service period replaced by inter-arrival duration t with distribution $A(t)$)

$$p_{ij} = \int_0^\infty \binom{i+1}{j} e^{-j\mu t}(1-e^{-\mu t})^{i+1-j}\, dA(t) \qquad i = 0, 1, 2, \ldots, r-1 \quad j \le i+1 \tag{16-189}$$

and for $i = r$, we have

$$p_{r,j} = p_{r-1,j} \tag{16-190}$$

Proceeding as in (16-164)–(16-182) with $m - 1$ replaced by r, we find the desired probability that n channels are occupied prior to the next arrival equals

$$\pi_n = \frac{\sum_{j=n}^{r}(-1)^{j-n}\binom{j}{n}C_j\sum_{i=j}^{r}\binom{r}{i}(1/C_i)}{\sum_{i=0}^{r}\binom{r}{i}(1/C_i)} \qquad n = 0, 1, 2, \ldots, r \tag{16-191}$$

where

$$C_0 = 1 \qquad C_j = \prod_{k=1}^{j}\frac{\psi_k}{1-\psi_k} \tag{16-192}$$

with

$$\psi_k = \int_0^\infty e^{-k\mu\tau}\, dA(\tau) \tag{16-193}$$

representing the Laplace transform of the interarrival distribution $A(\tau)$ evaluated at $k\mu$. Notice that π_r in (16-191) represents the *probability of call congestion* in a $GI/M/r/r$ *queue,* and it generalizes the formula in (16-128) for an arbitrary interarrival distribution $A(\tau)$. In particular, if the interarrival distribution is also assumed to be exponential with parameter λ, then π_n in (16-191) reduces to Erlang's first formula in (16-128). ◀

SINGLE SERVER QUEUES WITH POISSON INPUT AND GENERAL SERVICE

$M/G/1$ QUEUE

▶ The single server queue with a homogeneous Poisson arrival input and independent identically distributed (general) service times is known as the $M/G/1$ queue. Let $B(\tau)$ represent the common service time distribution in this case.

Examples 15-4, 15-14, and 15-24 in fact discuss the $M/G/1$ queue in detail. From there, $\mathbf{x}_n$ represents the number of customers waiting for service just immediately after the departure of the nth customer, and $\mathbf{y}_n$ the number of customers that arrive during the service time of the nth customer. These quantities are related as in (15-32), and the transition matrix for the associated Markov chain is given by (15-34), where

$$P\{\mathbf{y}_n = k\} = a_k \tag{16-194}$$

represents the probability of k arrivals during the service time of any customer. Let

$$q_j = \lim_{n\to\infty} P\{\mathbf{x}_n = j\} \qquad j = 0, 1, 2, \ldots \tag{16-195}$$

represent the steady state probabilities of the Markov chain $\{\mathbf{x}_n\}$. When the chain is ergodic, these probabilities satisfy (15-167) and their characteristic function is given by (see (15-211)–(15-215))

$$Q(z) = \sum_{k=0}^{\infty} q_k z^k = \frac{(1-\rho)(1-z)A(z)}{A(z)-z} \qquad \rho < 1 \tag{16-196}$$

where

$$A(z) = \sum_{k=0}^{\infty} P\{\mathbf{y}_n = k\} z^k = \sum_{k=0}^{\infty} a_k z^k \tag{16-197}$$

and

$$\rho = A'(1) = E\{\mathbf{y}_n\} \tag{16-198}$$

represents the average number of customers arriving per service period.

Interestingly, we can use Theorem 15-8 to investigate the conditions under which $Q(z)$ defines a proper probability generating function, thereby making the Markov chain $\{\mathbf{x}_n\}$ ergodic. From Theorem 15-8, if the mean value $\rho = A'(1) \le 1$ in (16-198), then in the interval $0 \le z \le 1$ the equation $A(z) = z$ has only one root at $z = 1$. Hence

$$A(z) - z = (1-z)B(z) \tag{16-199}$$

and

$$Q(z) = (1-\rho)\frac{A(z)}{B(z)} \tag{16-200}$$

If $Q(z)$ represents a proper probability generating function, then it must be analytic in $|z| \le 1$, or $B(z)$ must be free of zeros within the closed unit circle. To examine this, from (16-199)

$$B(z) = \frac{A(z)-z}{1-z} = 1 - \frac{1-A(z)}{1-z} = 1 - C(z) \tag{16-201}$$

where

$$C(z) = \frac{1-A(z)}{1-z} \triangleq \sum_{k=0}^{\infty} c_k z^k \tag{16-202}$$

Notice that $c_0 = 1 - a_0 > 0$, $-a_k = c_k - c_{k-1}$, $k \ge 1$, so that $c_n = 1 - \sum_{k=1}^{n} a_k \ge 0$ and $c_n \to 0$, since $\sum_{k=0}^{\infty} a_k = 1$, $a_k \ge 0$. Moreover, $C(1) = \sum_{k=0}^{\infty} c_k = A'(1) = \rho < 1$, and hence $|C(z)| \le \rho < 1$, inside the closed unit circle. From (16-201) and (16-202) it now follows that when $\rho < 1$, the function $B(z) = 1 - C(z)$ is in fact free of zeros inside the closed unit circle (Rouche's theorem; see hint for Problem 16-8. Use $f(z) \equiv 1$ and $g(z) = -C(z)$.), and $Q(z)$ in (16-200) is analytic in $|z| \le 1$ with $q_k \ge 0$. Since $\sum_{k=0}^{\infty} q_k = 1$, $Q(z)$ in (16-196) represents a probability generating function provided

$\rho = A'(1) < 1$. In that case, from (16-200)–(16-202), we get

$$Q(z) = (1-\rho)A(z)\sum_{m=0}^{\infty} C^m(z)$$

and if we let

$$C^m(z) \triangleq \sum_{k=0}^{\infty} c_k^{(m)} z^k$$

then we obtain the steady state probabilities for a general single server queue (under the imbedded Markov chain assumption) to be

$$q_n = (1-\rho)\sum_{m=0}^{\infty}\sum_{k=0}^{n} a_k c_{n-k}^{(m)} \tag{16-203}$$

where

$$c_0 = 1 - a_0 \qquad c_n = 1 - \sum_{k=1}^{n} a_k \tag{16-204}$$

Here the sequence $\{c_k^{(m)}\}$ represents the m-fold convolution of the sequence $\{c_k\}$.

In particular, if the input arrivals are Poisson with rate λ, then (see (15-218))

$$a_k = P\{\mathbf{y}_n = k\} = \int_0^{\infty} e^{-\lambda\tau}\frac{(\lambda\tau)^k}{k!}dB(\tau) \tag{16-205}$$

and from (15-219), $A(z) = \Phi_s(\lambda(1-z))$, where $\Phi_s(s)$ represents the Laplace transform of the service time distribution $B(\tau)$. Substituting this into (16-196), we obtain the well known *Pollaczek–Khinchin* formula

$$Q(z) = \frac{(1-\rho)(1-z)\Phi_s(\lambda(1-z))}{\Phi_s(\lambda(1-z)) - z} \tag{16-206}$$

for the steady state distribution $\{q_j\}$ of $M/G/1$ queues that exists under the condition $\rho < 1$.

WAITING TIME DISTRIBUTION. We can use (16-206) to obtain the limiting distribution of the waiting time random variable $\mathbf{w}$ for an $M/G/1$ queue. Let $F_w(t)$ represent the waiting time distribution in the long run and $\Psi_w(s)$ its Laplace transform. A departing customer at $t = t_n$ leaves behind $\mathbf{x}_n$ customers, all of whom must have arrived during the total time spent by the departing customer (waiting time plus service time). Thus if $\mathbf{s}$ denotes the service time of the departing customer, then we also have [51]

$$q_j = P\{\mathbf{x}_n = j\} = \int_0^{\infty} e^{-\lambda t}\frac{(\lambda t)^j}{j!}P\{\mathbf{w}+\mathbf{s} \le t\}\,dt$$

and hence

$$\begin{aligned} Q(z) = \sum_{j=0}^{\infty} q_j z^j &= \int_0^{\infty} e^{-(1-z)\lambda t}P\{\mathbf{w}+\mathbf{s} \le t\}\,dt \\ &= \Phi_s((1-z)\lambda)\,\Psi_w((1-z)\lambda) \end{aligned} \tag{16-207}$$

Equating (16-206) and (16-207), we get

$$\Psi_w((1-z)\lambda) = \frac{(1-\rho)(1-z)}{\Phi_s((1-z)\lambda) - z} \tag{16-208}$$

or the Laplace transform of the waiting time distribution is given by [replace $(1-z)\lambda$ with s in (16-208)]

$$\Psi_w(s) = \frac{(1-\rho)s}{s - \lambda + \lambda\Phi_s(s)} \tag{16-209}$$

Using the relation $\rho = A'(1) = -\lambda\Phi_s'(0) = \lambda E\{\mathbf{s}\}$, and from (16-209), the mean waiting time for an $M/G/1$ queue equals

$$E\{\mathbf{w}\} = -\Psi_w'(0) = \frac{\lambda E\{\mathbf{s}^2\}}{2(1-\rho)} = \frac{\lambda^2 \operatorname{Var}\{\mathbf{s}\} + \rho^2}{2\lambda(1-\rho)} \tag{16-210}$$

and from (16-207) the mean queue length is given by

$$L = Q'(1) = \rho - \lambda\Psi_w'(0) = \rho + \frac{\lambda^2 E\{\mathbf{s}^2\}}{2(1-\rho)} = \rho + \frac{\lambda^2 \operatorname{Var}\{\mathbf{s}\} + \rho^2}{2(1-\rho)} \tag{16-211}$$

Equations (16-210) and (16-211) indicate that for given average arrival and service times we can decrease the expected queue length and expected waiting time by decreasing the variance of service time. Hence it follows that the constant service time ($M/D/1$) queue is optimum in that sense. ◀

Examples 16-7 and 16-8 illustrate two specific $M/G/1$ queues: $M/E_m/1$ and $M/D/1$.

EXAMPLE 16-7

$M/E_m/1$ QUEUE

▶ In this case, the service time is Erlang-m with p.d.f. as in (16-76) (with n replaced by m), and from the discussion there the service-time consists of m independent identical phases of exponential type, each with parameter $m\mu$. Thus a customer enters phase 1 of the service, progresses through the remaining phases sequentially, and must complete the last phase before the next customer is allowed into the first phase of the service. In that case, we have

$$\Phi_s(u) = \int_0^\infty e^{-u\tau}\, dB(\tau) = \left(\frac{\mu m}{u + \mu m}\right)^m \tag{16-212}$$

and

$$A(z) = \Phi_s(\lambda(1-z)) = \frac{1}{(1 + \rho(1-z)/m)^m} \tag{16-213}$$

where

$$\rho = A'(1) = -\lambda\Phi_s'(0) = \lambda E\{\mathbf{s}\} = \frac{\lambda}{\mu}$$

Substituting (16-213) into (16-206), we get

$$Q(z) = \frac{(1-\rho)(1-z)}{1 - z(1 + \rho(1-z)/m)^m} \tag{16-214}$$

from which the steady state queue distribution $\{q_n\}$ as well as their mean and variances can be determined. Note that $m = 1$ gives the $M/M/1$ queue. ◀

EXAMPLE 16-8

***M/D/1* QUEUE**

▶ In this case, the service time is of constant duration and it follows as a special case of Erlang-m, since from (4-37) with $\lambda = m\mu$, as $m \to \infty$ the distribution tends to concentrate at $t = 1/\mu$, that is,

$$F_x(t) = \begin{cases} 1 & t > 1/\mu \\ 0 & \text{otherwise} \end{cases} \tag{16-215}$$

Thus the service time is of constant duration $1/\mu$. From (16-213), for constant service time we get ($m \to \infty$)

$$\Phi_s\left(\lambda(1-z)\right) \to e^{-\rho(1-z)} \tag{16-216}$$

so that

$$\begin{aligned} Q(z) &= \frac{(1-\rho)(1-z)}{1-ze^{\rho(1-z)}} = (1-\rho)(1-z)\sum_{k=0}^{\infty} z^k e^{k\rho} e^{-k\rho z} \\ &= (1-\rho)(1-z)\sum_{k=0}^{\infty} z^k e^{k\rho} \sum_{j=0}^{\infty} \frac{(-k\rho z)^j}{j!} \\ &= (1-\rho)(1-z)\sum_{n=0}^{\infty}\sum_{k=0}^{n} e^{k\rho}\frac{(-k\rho)^{n-k}}{(n-k)!} z^n \end{aligned} \tag{16-217}$$

which gives the steady state probabilities to be

$$q_0 = 1-\rho$$

and

$$q_n = (1-\rho)\left(e^{n\rho} + \sum_{k=1}^{n-1}(-1)^{n-k} e^{k\rho}\frac{(k\rho)^{n-k-1}}{(n-k)!}\left[n-k(1-\rho)\right]\right) \qquad n \ge 1 \tag{16-218}$$

◀

Next we examine the busy period distribution for $M/G/1$ type queues.

BUSY PERIOD DISTRIBUTION FOR $M/G/1$ QUEUES (KENDALL, TAKACS). A busy period starts when an item goes into service (at the end of an idle period), and it ends when the last item in the queue formed during that uninterrupted service operation has completed service, with no items arriving immediately thereafter (see Fig. 16-9). Thus the busy periods correspond to the durations of uninterrupted service periods. The duration of a busy period may be thought of as consisting of the duration of the service period for the first item followed by the durations of the busy periods for subsequent items arriving during the first service period.

As Fig. 16-9 shows, let $\mathbf{z}_1, \mathbf{z}_2, \mathbf{z}_3, \ldots, \mathbf{z}_n, \ldots$ represent independent identically distributed busy period durations with common probability distribution $G(t) = P(\mathbf{z}_n \le t)$ for a single channel $M/G/1$ queue with Poisson input with parameter λ, and arbitrary service time distribution $B(\tau)$. The length of the busy period $\mathbf{z}$ does not exceed t, if the service time for the first item lasts $\tau\,(0 < \tau \le t)$, and if the service times of *all* items arriving during that τ does not exceed the remaining time $t - \tau$. The probability that the service time for first item lasts τ is given by $B(\tau)$, and to compute the probability

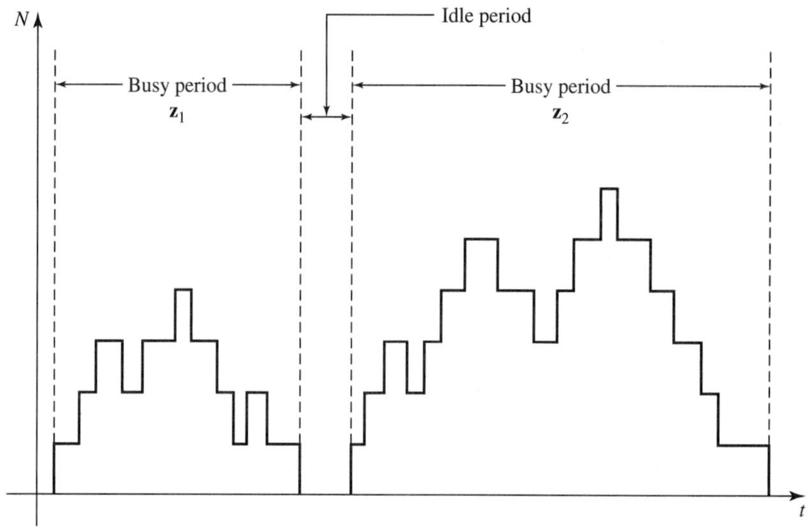

FIGURE 16-9
Busy periods in a queue.

of the second event we argue as follows: The probability that n customers arrive during the service time τ of the initial customer is given by

$$\frac{(\lambda\tau)^n}{n!}e^{-\lambda\tau} \qquad n = 0, 1, 2, \ldots \tag{16-219}$$

Notice that as far as the computations of the busy period are concerned, the particular order in which arriving items are served is irrelevant. This affects the customers only, but the distribution function of the busy period remains the same. Hence suppose that the first among these newly arrived n items goes into service immediately after the completion of service of the initial item, and if any further arrivals occur during its service, they are all served first. After completion of all such items associated with the first item (which corresponds to a busy period), another item from the remaining $n-1$ items is admitted for service. Hence the later portion $(t-\tau)$ of the busy period consists of the sum of n independent busy periods, each with distribution $G(x)$, and their cumulative distribution $G_n(x)$ is given by the n–fold convolution of $G(x)$. Thus the probability that the service time of the n customers (that have arrived during the service time of the initial customer) does not exceed $t-\tau$ is given by $G_n(t-\tau)$. Hence

$$G(t) = \int_0^\infty \sum_{n=0}^{\infty} \frac{(\lambda\tau)^n}{n!} e^{-\lambda\tau} G_n(t-\tau)\, dB(\tau) \tag{16-220}$$

To simplify this expression, let

$$\Gamma(s) = \int_0^\infty e^{-st} dG(t) \qquad \Phi(s) = \int_0^\infty e^{-st} dB(t)$$

represent the Laplace transforms of the unknown busy-time distribution $G(t)$ and the service-time distribution $B(t)$. Then $\Gamma^n(s)$ represents the Laplace transform of the n-fold convolution $G_n(x)$, and the Laplace transform of (16-220) is given by

$$\Gamma(s) = \sum_{n=0}^{\infty} \frac{(\lambda\Gamma(s))^n}{n!} \int_0^\infty \tau^n e^{-(s+\lambda)\tau} dB(\tau)$$
$$= \int_0^\infty e^{-[s+\lambda-\lambda\Gamma(s)]\tau} dB(\tau)$$
$$= \Phi(s+\lambda-\lambda\Gamma(s)) \tag{16-221}$$

The functional equation in (16-221) was first obtained by Kendall. Takacs gives a proof that the busy time distribution function $G(t)$ can be uniquely determined from (16-221), and it represents a proper distribution function provided $\rho = \lambda/\mu \leq 1$, where $1/\mu$ represents the mean service time. Otherwise, the busy-time period can be infinite with probability equal to $[1 - \lim_{t\to\infty} G(t)]$. This result can be stated as follows:

THEOREM 16-2

BUSY PERIODS FOR $M/G/1$ QUEUE

The busy period distribution transform $\Gamma(s)$ for an $M/G/1$ queue is the unique solution of Eq. (16-221) for Re $s > 0$ subject to the condition $|\Gamma(s)| \leq 1$. Further, if π_0 denotes the smallest root of the equation

$$\Phi(\lambda(1-z)) = z \tag{16-222}$$

then

$$G(\infty) = \pi_0 \tag{16-223}$$

If $\rho = \lambda/\mu \leq 1$, then $\pi_0 = 1$ and $G(t)$ is a proper distribution function. Otherwise π_0 is strictly less than one, and $1 - \pi_0$ represents the probability that the busy period is infinite.

Proof. If $G(t)$ represents a proper probability distribution function, then for Re $s > 0$, $|\Gamma(s)| \leq 1$ and $\Gamma(0) = 1$. Let $z = \Gamma(s)$ so that (16-221) reads

$$z = \Phi(s + \lambda(1-z)) \tag{16-224}$$

and we have $|z| < 1$ for Re $s > 0$. Note that for every s that is real and positive, Eq. (16-224) yields a positive solution $\pi_0(s)$ that is bounded by unity (see Fig. 16-10*b*). Since $\Phi(s)$ is continuous

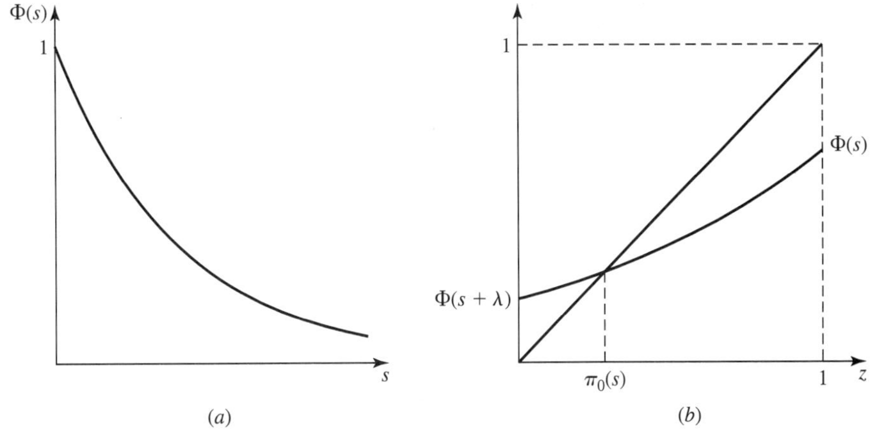

FIGURE 16-10
Busy period distribution $\Gamma(s) = \pi_0(s)$.

and positive for Re $s > 0$, the two sides of (16-224) intersect at a point $\pi_0(s) < 1$. This root is also unique since $\Phi(s)$ is a convex function (Fig. 16-10a). As a result, $\Gamma(s) = \pi_0(s)$ is uniquely determined on the positive real semiaxis. By analytic continuation, $\Gamma(s)$ can be extended uniquely over the entire right half plane.

As $s \to 0$, the function $\Phi(s + \lambda(1 - z))$ tends to $\Phi(\lambda) < 1$ at $z = 0$, and it equals unity at $z = 1$. This situation refers to Fig. 15-10, and there are two cases to be distinguished depending on the value of $\Phi'(\lambda(1 - z))|_{z=1} = -\lambda\Phi'(0) = \lambda/\mu = \rho$. If $\rho \le 1$, the situation corresponds to Fig 15-10a, and $z = 1$ is the only solution of the equation

$$z = \Phi(\lambda(1 - z)) \tag{16-225}$$

However, if $\rho > 1$, the solution corresponds to Fig. 15-10b and from there $\pi_0 < 1$ is the unique smallest root of (16-225). Hence

$$\lim_{s\to 0} \pi_0(s) = \pi_0 = \lim_{s\to 0} \Gamma(s) = \Gamma(0) = G(\infty)$$

so that the probability that the busy period $\mathbf{z}_n$ equals infinity is given by

$$P\{\mathbf{z}_n = \infty\} = 1 - P\{\mathbf{z}_n \le \infty\} = 1 - G(\infty) = 1 - \pi_0 \tag{16-226}$$

In summary, if $\rho \le 1$, the busy periods for an $M/G/1$ queue will always end with probability 1. Otherwise they will explode with probability given by (16-226).

This completes the proof of the theorem. ◀

From (16-221), the mean value of the busy time distribution simplifies to

$$E\{\mathbf{z}\} = \int_0^\infty t\,dG(t) = -\Gamma'(0) = \frac{E\{\mathbf{s}\}}{1 - \lambda E\{\mathbf{s}\}} = \frac{1/\mu}{1 - \lambda/\mu} = \frac{1}{\mu - \lambda}$$

where $E\{\mathbf{s}\} = -\Phi'(0) = 1/\mu$, and with $\rho = \lambda/\mu$ we also get

$$E\{\mathbf{z}^2\} = \frac{E\{\mathbf{s}^2\}}{(1 - \rho)^3}$$

This gives the variance of the busy time distribution to be

$$\text{Var}\{\mathbf{z}\} = \frac{\text{Var}\{\mathbf{s}\} + \rho(E\{\mathbf{s}\})^2}{(1 - \rho)^3} \tag{16-227}$$

and for given average arrival and service times, once again an $M/D/1$ queue attains minimum variance for the busy time distribution as well [see also (16-210)–(16-211)].

Takacs has also derived the waiting time distribution for $M/G/1$ queue, for the probability $P(w, t)$ that for an item arriving at t the waiting time $\{\mathbf{w}(t) \le w\}$, as an integrodifferential equation involving the service time distribution $B(t)$.

Next we illustrate Theorem 16-1 by computing the busy period distribution for an $M/M/1$ queue.

EXAMPLE 16-9

▶ Determine the busy period distribution for an $M/M/1$ queue.

For an $M/M/1$ queue, the service time transform $\Phi(s) = \mu/(s + \mu)$, so that the functional equation (16-221) becomes

$$\Gamma(s) = \frac{\mu}{s + \lambda + \mu - \lambda\Gamma(s)} \tag{16-228}$$

which simplifies to

$$\lambda\Gamma^2(s) - (s + \lambda + \mu)\Gamma(s) + \mu = 0$$

This quadratic equation in Γ has two roots, one with magnitude greater than unity in

$\operatorname{Re} s > 0$, and another with magnitude less than unity. Since $|\Gamma(s)| \le 1$ in $\operatorname{Re} s > 0$, the desired solution is given by the smaller root

$$\Gamma(s) = \frac{(s+\lambda+\mu) - \sqrt{(s+\lambda+\mu)^2 - 4\lambda\mu}}{2\lambda} \tag{16-229}$$

and its inverse transform represents the busy time distribution $G(t)$.

If $\rho = \lambda/\mu > 1$, the unique smallest root of (16-225) in this case is given by ($s = 0$ in (16-229))

$$\pi_0 = \frac{\mu}{\lambda} < 1$$

and from (16-226) an $M/M/1$ queue becomes a never-ending one with probability $1 - \mu/\lambda$, when the load factor ρ is greater than unity. ◀

DISTRIBUTION OF THE NUMBER OF CUSTOMERS SERVED DURING BUSY PERIODS. A busy period is initiated by a single customer whose service begins instantly, and let $\mathbf{y}_1$ represent the number of customers that arrive during the first customer's service period, $\mathbf{y}_2$ those who arrive during the service periods of the $\mathbf{y}_1$ customers prior to them and so on. Then with

$$\mathbf{s}_n = 1 + \mathbf{y}_1 + \mathbf{y}_2 + \cdots + \mathbf{y}_n \tag{16-230}$$

we have

$$\lim_{n\to\infty} P\{\mathbf{s}_n = k\} \stackrel{\Delta}{=} h_k \qquad k \ge 0 \tag{16-231}$$

represents the limiting distribution of the total number of customers served during any busy period. Further, let

$$H(z) = \sum_{k=0}^{\infty} h_k z^k \tag{16-232}$$

represent the corresponding moment generating function. From Sec. 15-6 and Theorem 15-9, this situation is identical to the total number of descendents' distribution in a population, and $H(z)$ in (16-232) satisfies the functional equation in (15-322), which translates in this case into

$$H(z) = zA(H(z)) = z\Phi(\lambda - \lambda H(z)) \qquad 0 < z < 1 \tag{16-233}$$

Here $A(z) = \sum_{k=0}^{\infty} P\{\mathbf{y}_n = k\} z^k$ represents the moment generating function of the random variable representing the number of customers arriving per service period as in (16-197), and $\Phi(s)$ represents the Laplace transform of the service time distribution in an $M/G/1$ queue. From Theorem 15-9, the solution to (16-233) is given by the unique root $x(z)$ of the equation

$$x = z\Phi(\lambda(1-x)) \tag{16-234}$$

such that $H(z) = x(z) \le \pi_0$, where π_0 is the smallest positive root of the equation

$$x = \Phi(\lambda(1-x)) \tag{16-235}$$

and $H(1) = \sum_{k=0}^{\infty} h_k = \pi_0 \le 1$. From Theorem 15-9, if $\rho \le 1$, then $H(1) = 1$, and $H(z)$ represents a proper moment generating function. However if $\rho > 1$, then $\pi_0 < 1$ and

$$\lim_{n\to\infty} P\{\mathbf{s}_n = \infty\} = 1 - \pi_0 \tag{16-236}$$

EXAMPLE 16-10 ▶ Returning back to the $M/M/1$ queue in Example 16-9, the moment generating function for the number of customers served during a busy period satisfies the equation

$$\lambda H^2(z) - (\lambda + \mu)H(z) + \mu z = 0$$

Here $|H(z)| \leq 1$ in $|z| < 1$ gives the unique solution

$$\begin{aligned} H(z) &= \frac{(\lambda+\mu) - \sqrt{(\lambda+\mu)^2 - 4\lambda\mu z}}{2\lambda} \\ &= \frac{(\lambda+\mu)}{2\lambda}\sum_{k=1}^{\infty}(-1)^{k-1}\binom{1/2}{k}\left(\frac{4\lambda\mu}{(\lambda+\mu)^2}\right)^k z^k \\ &= \frac{(\lambda+\mu)}{2\lambda}\sum_{k=1}^{\infty}\binom{2k}{k}2^{-2k}\left(\frac{4\lambda\mu}{(\lambda+\mu)^2}\right)^k z^k \\ &\simeq \frac{(\lambda+\mu)}{2\lambda}\sum_{k=1}^{\infty}\frac{1}{\sqrt{\pi k}}\left(\frac{4\lambda\mu}{(\lambda+\mu)^2}\right)^k z^k \end{aligned} \qquad (16\text{-}237)$$

Thus

$$h_k \simeq \frac{(\mu+\lambda)}{2\lambda\sqrt{\pi k}}\left(\frac{4\lambda\mu}{(\lambda+\mu)^2}\right)^k = \frac{(1+\rho)}{2\rho\sqrt{\pi k}}\left(\frac{4\rho}{(1+\rho)^2}\right)^k \qquad k \geq 1$$

represents the probability that k customers are served during a busy period in an $M/M/1$ queue. If $\rho \leq 1$, then $\sum_{k=0}^{\infty} h_k = 1$, whereas it equals $1/\rho < 1$ if $\rho > 1$. In that case from (16-236), the quantity $(1 - 1/\rho) < 1$ represents the probability that the number of customers served during a busy period becomes infinite. ◀

GENERAL INPUT AND EXPONENTIAL SERVICE

***GI/M/1* QUEUE**

▶ A single server queue with an arbitrary inter-arrival distribution $A(\tau)$ and exponential service time with parameter μ gives rises to an $GI/M/1$ queue. Compared to the $M/G/1$ queue, since the roles of the exponential and arbitrary distributions are reversed here, let $\mathbf{t}_1, \mathbf{t}_2, \ldots, \mathbf{t}_n, \ldots$ represent the arrival instants of the items (rather than the departure instants), and define $\mathbf{x}_n = \mathbf{x}(\mathbf{t}_n - 0)$ to represent the number of customers in the system just before the *arrival* of the nth customer. Further, let $\mathbf{z}_n$ denote the number of customers served during the interarrival time $(\mathbf{t}_n, \mathbf{t}_{n+1})$ between the nth and $(n+1)$st customers. Then as in (15-32)

$$\mathbf{x}_{n+1} = \mathbf{x}_n + 1 - \mathbf{z}_n \qquad \text{if} \quad \mathbf{x}_n \geq 0 \qquad (16\text{-}238)$$

and $\mathbf{z}_n \leq \mathbf{x}_n + 1$. The sequence $\{\mathbf{x}_n\}$ represents a Markov chain and the transition probabilities $p_{ij} = P\{\mathbf{x}_{n+1} = j \mid \mathbf{x}_n = i\}$ are given by

$$p_{ij} = \begin{cases} P\{\mathbf{z}_n = i - j + 1\} = b_{i-j+1} & i+1 \geq j \geq 1 \qquad i \geq 0 \\ 0 & j > i+1 \end{cases} \qquad (16\text{-}239)$$

where

$$P\{\mathbf{z}_n = j\} = b_j \qquad j = 0, 1, 2, \ldots \qquad (16\text{-}240)$$

represents the probability that j items were served during the interarrival time $\boldsymbol{\tau}$ between the nth and $(n+1)$st item. Since the distribution of $\boldsymbol{\tau}$ is $A(\tau)$ and the service times are exponential, proceeding as in (15-218) we get

$$P\{\mathbf{z}_n = j\} = b_j = \int_0^\infty e^{-\mu\tau} \frac{(\mu\tau)^j}{j!}\, dA(\tau) \tag{16-241}$$

and this gives the moment generating function of the random variable $\mathbf{z}_n$ to be

$$\begin{aligned} B(z) &= \sum_{j=0}^\infty b_j z^j = \int_0^\infty \sum_{j=0}^\infty \frac{(z\mu\tau)^j}{j!} e^{-\mu\tau}\, dA(\tau) \\ &= \int_0^\infty e^{-\mu(1-z)\tau} dA(\tau) = \Psi_A\left(\mu(1-z)\right) \end{aligned} \tag{16-242}$$

where $\Psi_A(s)$ represents the Laplace transform of the interarrival distribution $A(\tau)$.

Since (16-239) is only valid for $j \geq 1$, to obtain $p_{i,0}$, we can make use of the identity $\sum_{j=0}^{i+1} p_{ij} = 1$ for $i = 0, 1, 2, \ldots$, which gives

$$p_{i,0} = 1 - \sum_{j=1}^{i+1} p_{ij} = 1 - \sum_{j=1}^{i+1} b_{i-j+1} = 1 - \sum_{k=0}^{i} b_k \stackrel{\Delta}{=} c_i \qquad i \geq 0 \tag{16-243}$$

From (16-239) and (16-243), we get the probability transition matrix to be

$$P = \begin{pmatrix} c_0 & b_0 & 0 & \cdot & \cdot & 0 \\ c_1 & b_1 & b_0 & 0 & \cdot & \cdots \\ c_2 & b_2 & b_1 & b_0 & 0 & \cdots \\ \vdots & \vdots & \vdots & \vdots & \vdots & \vdots \\ c_k & b_k & b_{k-1} & \cdots & b_0 & \cdots \\ \cdot & \cdot & \cdot & \cdot & \cdot & \cdots \end{pmatrix} \tag{16-244}$$

As in (16-195), let q_j, $j = 0, 1, 2, \ldots$ represent the steady state probabilities of the Markov chain $\{\mathbf{x}_n\}$. When the chain is ergodic, once again these probabilities satisfy the matrix equation $q = qP$ with P as in (16-244), and it can be rewritten as a set of linear equations

$$q_0 = \sum_{k=0}^\infty q_k c_k \tag{16-245}$$

and

$$q_j = \sum_{k=0}^\infty q_{k+j-1} b_k \qquad j \geq 1 \tag{16-246}$$

As before let $Q(z)$ represent the moment generating function of these steady state probabilities $\{q_j\}$. Then

$$\begin{aligned} Q(z) &= \sum_{j=0}^\infty q_j z^j = q_0 + \sum_{j=1}^\infty \sum_{k=0}^\infty q_{k+j-1} b_k z^j \\ &= q_0 + \sum_{m=0}^\infty q_m \sum_{k=0}^\infty b_k z^{m-k+1} = q_0 + zQ(z)B(1/z) \end{aligned} \tag{16-247}$$

or

$$Q(z) = \frac{q_0}{1 - zB(1/z)} \tag{16-248}$$

where $Q(z)$ must be analytic in $|z| \leq 1$ for it to represent a valid probability generating function. Equation (16-248) can be formally rewritten as

$$Q(1/z) = \frac{q_0 z}{z - B(z)} \tag{16-249}$$

where $Q(1/z)$ must be analytic in $|z| \geq 1$. Once again from Theorem 15-8, if $B'(1) > 1$, then the equation $B(z) - z$ has a real positive root π_0 with magnitude strictly less than unity, and arguing as in (16-201)–(16-202), it follows that $B(z) - z$ is free of any other zeros in $|z| < 1$. Using this in (16-248) and (16-249), after some simplifications we get

$$Q(z) = \frac{q_0}{1 - \pi_0 z} + R(z) \qquad 0 < \pi_0 < 1$$

where $R(z)$ must be analytic in the entire z plane. Thus $R(z)$ is a constant, and $z = 0$ in the above identity gives $R(z)$ to be zero. Finally, using the condition $\sum_{n=0}^{\infty} q_n = 1$, we obtain

$$q_n = (1 - \pi_0)\,\pi_0^n \qquad n = 0, 1, 2, \ldots \qquad 0 < \pi_0 < 1 \tag{16-250}$$

Thus the steady state distribution in an $GI/M/1$ queue exists under the condition

$$B'(1) = -\mu\Psi_A'(0) = \mu/\lambda = 1/\rho > 1 \tag{16-251}$$

where $1/\lambda$ represents the mean arrival time of the interarrival distribution $A(\tau)$. To summarize, when the traffic rate $\rho = \lambda/\mu < 1$, the steady state distribution in an $GI/M/1$ queue is *geometric* as in an $M/M/1$ queue. However, unlike the $M/M/1$ case, the queue parameter here is not easily related to ρ, and it is given by the unique positive root $\pi_0 < 1$ of the equation

$$B(z) - z = \Psi_A(\mu(1 - z)) - z = 0 \tag{16-252}$$

that exists whenever $\rho < 1$.

The waiting time distribution for the $GI/M/1$ is the same that as in (16-92) with ρ replaced by π_0 there. ◀

Next we examine the queue structure resulting from interconnection of various queues.

16-4 NETWORKS OF QUEUES

Networks of queues arise when a set of resources are shared by a set of customers. Each resource represents a service center that may have multiple servers operating in parallel. If an incoming item finds a particular center busy, it will join the queue at that center and wait for service (or may leave that queue and may go for another type of service). After completion of service at one station, the item may move to another service center, or reenter the same center, or leave the system.

In a chain of queues that are connected in series as in Fig. 16-11, once an item is in the system, it stays on for service through all phases of the system. In general,

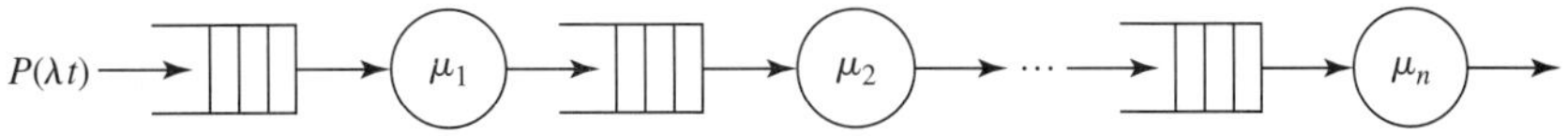

FIGURE 16-11
Network of queues.

waiting is allowed before each server. Note that the phase type service discussed earlier in connection with Erlang-n models is a special case of this series model, where *no* waiting is allowed before the servers except the first one. In that case a new item is admitted into the system for service only after the previous item has completed service through *all* of the n identical phases.

The behavior of networks of queues is characterized by the output distributions and the service time distributions of the servers in addition to the input distribution and the various service disciplines. In a series network, since the output from one server forms the input to the next server, the steady state properties of the network are dictated by the queue output distributions.

In this context, Burke has shown that in an $M/M/r$ queue, the interdeparture time intervals are independent random variables in steady state. Moreover, the outputs from such a queue form a Poisson process with the same parameter λ as that of the input. It follows from Burke's result that when a Poisson input process with parameter λ feeds a series network of $M/M/r$ queues, all subsequent input and output processes are also Poisson with the same parameter λ in the steady state. We next prove this important result.

THEOREM 16-3

NYQUIST THEOREM

The steady state output of an $M/M/r$ queue, with (Poisson) input parameter λ, is also Poisson with parameter λ.

Proof. Let $\boldsymbol{\tau}$ denote the length of interval between any two consecutive departures, and $\mathbf{n}(t)$ the number of items in the system after the previous departure. The joint distribution of these two random variables is given by

$$F_n(t) \triangleq P\{\boldsymbol{\tau} > t, \mathbf{n}(t) = n\} \tag{16-253}$$

where

$$F_n(0) = p_n \tag{16-254}$$

represents the steady state probability of n units in the system as in (16-102). From (16-253), we get

$$F(t) \triangleq P\{\boldsymbol{\tau} > t\} = \sum_{n=0}^{\infty} F_n(t) \tag{16-255}$$

and $1 - F(t) = P(\tau \le t)$ represents the marginal distribution of the length of time between departures. Since the interarrival distributions are independent exponential random variables with parameter λ, a new arrival occurs in any interval of length Δt with probability $\lambda \Delta t + o(\Delta t)$, and a new departure occurs with probability $\mu_n \Delta t + o(\Delta t)$, where n represents the number of items in the system. Thus

$$(1 - \lambda \Delta t + o(\Delta t))(1 - \mu_n \Delta t + o(\Delta t))$$

represents the probability that there are no arrivals or departures in an interval of length Δt. Now

$$F_n(t+\Delta t) = P\{\boldsymbol{\tau} > t+\Delta t, \mathbf{n}(t+\Delta t) = n\}$$

and the system is in state e_n at $t+\Delta t$ either because the system was at state e_{n-1} at t and one arrival occurred in $(t, t+\Delta t)$ with probability $\lambda\Delta t$ so that the system moved over to state e_n, or the system was at e_n at t and it remained in e_n with probability $(1-\lambda\Delta t)(1-\mu_n\Delta t)$. Notice that since the inter-departure duration $\tau > t+\Delta t$, there is no departure in the interval $(0, t+\Delta t)$. This gives

$$F_n(t+\Delta t) = (1-\lambda\Delta t)(1-\mu_n\Delta t)F_n(t) + \lambda F_{n-1}(t) + o(\Delta t) \tag{16-256}$$

where μ_n is as given in (16-100). Proceeding as in (16-15) and (16-16), the above equations simplify to

$$F_0'(t) = -\lambda F_0(t) \tag{16-257}$$

and

$$F_n'(t) = \begin{cases} -(\lambda+n\mu)F_n(t) + \lambda F_{n-1}(t) & n < r \\ -(\lambda+r\mu)F_n(t) + \lambda F_{n-1}(t) & n \geq r \end{cases} \tag{16-258}$$

Under the initial condition (16-254), the Laplace transforms of (16-257) and (16-258) yield

$$P_n(s) = \frac{1}{s+\lambda+\mu_n}[p_n + \lambda P_{n-1}(s)] \qquad n \geq 0 \tag{16-259}$$

where $P_n(s)$ represents the Laplace transform of $F_n(t)$. Direct substitution using (16-102) for p_n shows that (16-259) satisfies the solution

$$P_n(s) = \frac{p_n}{s+\lambda} \tag{16-260}$$

or

$$F_n(t) = p_n e^{-\lambda t} \qquad t > 0 \tag{16-261}$$

Using this in (16-255) we obtain

$$F(t) = P\{\boldsymbol{\tau} > t\} = e^{-\lambda t} \tag{16-262}$$

or

$$P\{\boldsymbol{\tau} \leq t\} = 1 - F(t) = 1 - e^{-\lambda t} \qquad t \geq 0 \tag{16-263}$$

that is, in the case of Poisson arrivals, the marginal distribution of the intervals between departures is the same as the distribution between arrivals. Using the markovian property of the system, it follows that $\boldsymbol{\tau}$ is also independent of the set of lengths of all subsequent intervals between departures and hence the output stream is also Poisson with parameter λ. This proves the theorem.

It is easy to show that $\mathbf{n}(t)$ and $\boldsymbol{\tau}$ are in fact independent random variables since

$$P\{t < \boldsymbol{\tau} < t+\Delta t, \mathbf{n}(t) = n\} = F_{n+1}(t)\mu_{n+1}\Delta t + o(\Delta t) \tag{16-264}$$

and using (16-261), this factors into the probability distribution functions of $\mathbf{n}(t)$ and $\boldsymbol{\tau}$, establishing their independence. ◀

A partial converse of this result, due to Reich, can be stated as follows: If the arrivals and departures of a single server queue are Poisson distributed, then the service time distribution is either exponential or a step function at zero.

One may argue on intuitive grounds that in steady state "what goes in must come out," thus justifying Burke's result. However, as the following result due to Reich shows, such conclusions are often erroneous.

REICH'S RESULT. The distribution of the interdeparture time from a single-server queue, where interarrival time variables and service-time variables are each the sum of two identically distributed exponential random variables with parameters λ and μ, respectively, is *not* the sum of two exponentially distributed random variables.

Jackson has considered the general problem where every server in a network has independent Poisson arrivals in addition to feedbacks from other server outputs. In steady state, such a complex network essentially reduces to a series network with independent servers where each server has an equivalent input rate and service rate. As a prelude to the study of arbitrary networks of queues with feedback and Poisson arrivals at various servers, we first examine the simplest network of two single-server $M/M/1$ queues in series.

TWO QUEUES IN SERIES

Consider a two-stage series (tandem) network consisting of two servers with service rates μ_1 and μ_2, respectively, as in Fig. 16-12. The input to the first server is Poisson with parameter λ, and the output of the first server becomes the input to the second server. The system can be modeled as a stochastic process whose states are specified by the pair (n_1, n_2), where $n_i \geq 0$ represents the number of items in the ith phase (queue plus server) of the system. A change of state occurs either on completion of service at one of the two servers (first server to second $(n_1+1, n_2-1) \to (n_1, n_2)$ or second server to output $(n_1, n_2+1) \to (n_1, n_2)$), or on an external arrival $((n_1-1, n_2) \to (n_1, n_2))$ as shown in Fig. 16-13. Let $p(n_1, n_2, t)$ represent the probability that there are n_1 items in the first phase, and n_2 items in the second phase. From Fig. 16-13, the transient equations are given by

$$p'(0,0,t) = -\lambda p(0,0,t) + \mu_2 p(0,1,t) \tag{16-265}$$

$$p'(n_1,0,t) = -(\lambda+\mu_1)p(n_1,0,t) + \mu_2 p(n_1,1,t) + \lambda p(n_1-1,0,t) \tag{16-266}$$

$$p'(0,n_2,t) = -(\lambda+\mu_2)p(0,n_2,t) + \mu_1 p(1,n_2-1,t) + \mu_2 p(0,n_2+1,t) \tag{16-267}$$

and

$$\begin{aligned} p'(n_1,n_2,t) = {} & -(\lambda+\mu_1+\mu_2)p(n_1,n_2,t) + \mu_1 p(n_1+1,n_2-1,t) \\ & +\mu_2 p(n_1,n_2+1,t) + \lambda p(n_1-1,n_2,t) \end{aligned} \tag{16-268}$$

From Burke's theorem, interdeparture time intervals at both phases are exponential in steady state, and the stochastic process is markovian. The steady state equations corresponding to (16-267) and (16-268) have the solution

$$p(n_1,n_2) = (1-\rho_1)\rho_1^{n_1}(1-\rho_2)\rho_2^{n_2} \qquad n_1 \geq 0 \qquad n_2 \geq 0 \tag{16-269}$$

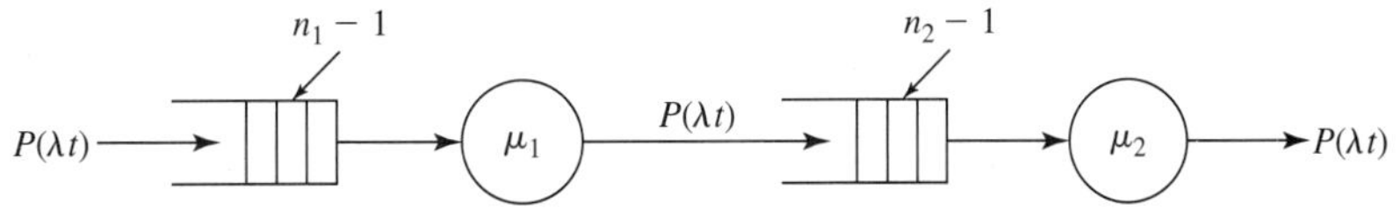

FIGURE 16-12
Two queues in series.

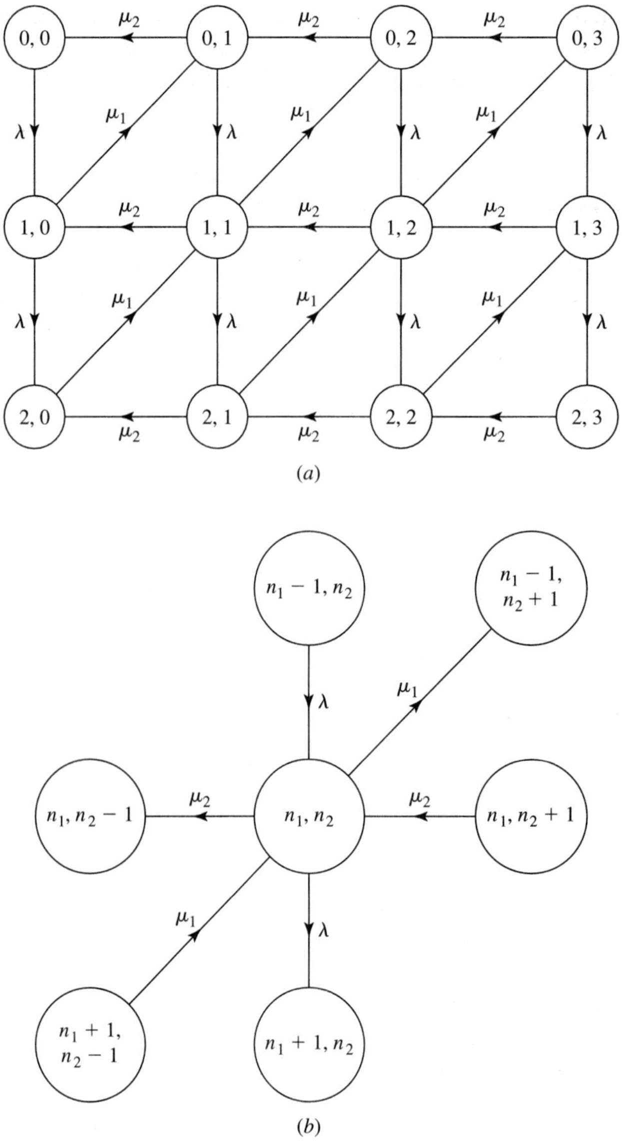

FIGURE 16-13
State diagram for a two-stage series network.

where

$$\rho_1 = \frac{\lambda}{\mu_1} \qquad \rho_2 = \frac{\lambda}{\mu_2} \tag{16-270}$$

and it can be verified by direct substitution. From (16-269), the probability of n_i items in phase i in steady state is given by

$$p_i(n_i) = (1 - \rho_i)\rho_i^{n_i} \qquad n_i \geq 0 \qquad i = 1, 2 \tag{16-271}$$

which represents solution similar to an $M/M/1$ queue. Since both phases have Poisson arrivals with rate λ in steady state and exponential service times, they both represent $M/M/1$ queues. Further, from (16-269) and (16-271), we have

$$p(n_1, n_2) = p_1(n_1)p_2(n_2) \tag{16-272}$$

which shows that the number of items in each phase are independent random variables. Thus the two queues are independent $M/M/1$ queues. ◀

Generalizing this argument to a cascade connection of m phases it follows that when the input to the first phase is Poisson, all intermediate inputs and outputs to subsequent phases become Poisson with the same rate, and each phase behaves like an independent $M/M/1$ queue in steady state. The expected number of items in such a series interconnection of m phases is given by

$$\begin{aligned} L &= \sum_{n_1, n_2, \ldots, n_m} (n_1 + n_2 + \cdots + n_m)p(n_1, n_2, \ldots, n_m) \\ &= \sum_{i=1}^{m} \sum_{n_i=0}^{\infty} n_i p_i(n_i) = \sum_{i=1}^{m} (1-\rho_i) \sum_{n_i=0}^{\infty} n_i \rho_i^{n_i} \\ &= \sum_{i=1}^{m} \frac{\rho_i}{1-\rho_i} \end{aligned} \tag{16-273}$$

Using Little's formula, this gives the average waiting time in the system to be

$$W = \frac{L}{\lambda} = \sum_{i=1}^{m} \frac{1}{\mu_i - \lambda} = \sum_{i=1}^{m} W_i \tag{16-274}$$

which is the sum of the waiting times in each $M/M/1$ queue.

MULTIPLE SERVERS IN SERIES AND PARALLEL. R. R. P. Jackson has generalized the series interconnection of single server queues to a series interconnection of m phases, where the ith phase consists of r_i parallel channels, all with exponential service-rate μ_i, as shown in Fig. 16-14. The input to the first phase is an unlimited Poisson input with parameter λ, and queueing is allowed before each phase. With n_i units in the ith phase, the probability that an item finishes service in Δt is given by $\mu_{n_i}\Delta t + o(\Delta t)$, where

$$\mu_{n_i} = \begin{cases} n_i \mu_i & n_i < r_i \\ r_i \mu_i & n_i \ge r_i \end{cases} \tag{16-275}$$

In steady state, repeated use of Burke's theorem shows that all inputs and outputs are Poisson with rate λ, and proceeding as in (16-265)–(16-268) the steady state equations, for example, become

$$\begin{aligned} \left(\lambda + \sum_{i=1}^{m} \mu_i\right) p(n_1, n_2, \ldots, n_m) &= \sum_{i=1}^{m} \mu_{n_i} p(n_1, n_2, \ldots, n_i + 1, n_{i+1} - 1, \ldots, n_m) \\ &\quad + \lambda p(n_1 - 1, n_2, \ldots, n_m) \qquad n_i > 0 \end{aligned} \tag{16-276}$$

with μ_{n_i} as in (16-275). Here $p(n_1, n_2, \ldots, n_m)$ represents the probability that there are n_1 items in the first phase, n_2 items in the second phase, and so on. R. R. P. Jackson has shown that the unique solution to (16-276) is given by the product form

$$p(n_1, n_2, \ldots, n_m) = p_1(n_1)p_2(n_2)\cdots p_i(n_i)\cdots p_m(n_m) \tag{16-277}$$

where

$$p_i(n_i) = \begin{cases} \dfrac{(\lambda/\mu_i)^{n_i}}{n_i!}p_{i,0} & n_i < r_i \\ \dfrac{r_i^{r_i}\rho_i^{n_i}}{r_i!}p_{i,0} & n_i \geq r_i \end{cases} \tag{16-278}$$

Here

$$p_{i,0} = \frac{1}{\displaystyle\sum_{n=0}^{r_i-1}\frac{(\lambda/\mu_i)^n}{n!} + \frac{(\lambda/\mu_i)^{r_i}}{r_i!(1-\rho_i)}} \tag{16-279}$$

and

$$\rho_i = \lambda/r_i\mu_i \tag{16-280}$$

Equation (16-278) gives the probability of n_i items in the ith phase. Notice that (16-278)–(16-280) represents an $M/M/r_i$ queue with n_i items, and from (16-277) it follows that in steady state a series-parallel network as in Fig. 16-14 will behave like a cascade of independent $M/M/r_i$ queues, provided all servers in each parallel configuration have identical service rates. Using (16-106) and (16-107), the average number of items waiting in such a network is given by

$$L = \sum_{i=1}^{m} p_{r_i}\frac{\rho_i}{(1-\rho_i)^2} = \sum_{i=1}^{m} L_i \tag{16-281}$$

where

$$p_{r_i} = \frac{(\lambda/\mu_i)^{r_i}}{r_i!}p_{i,0} \tag{16-282}$$

and it equals the sum of the average number of items waiting in each phase.

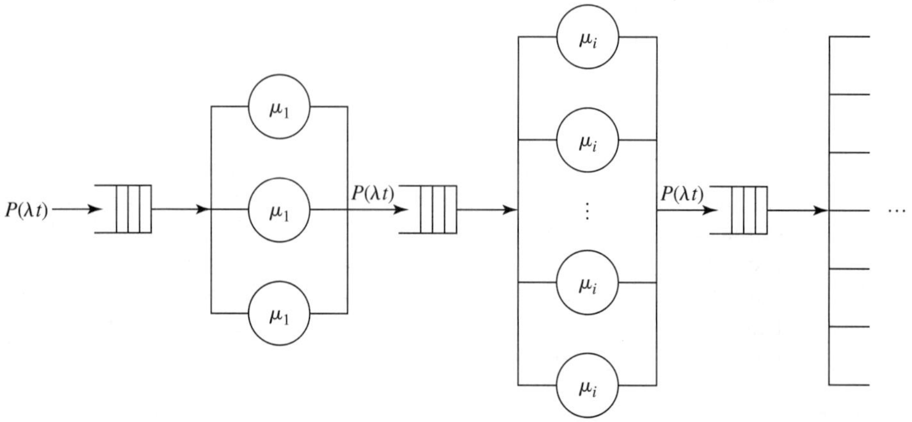

FIGURE 16-14
Servers in series and parallel.

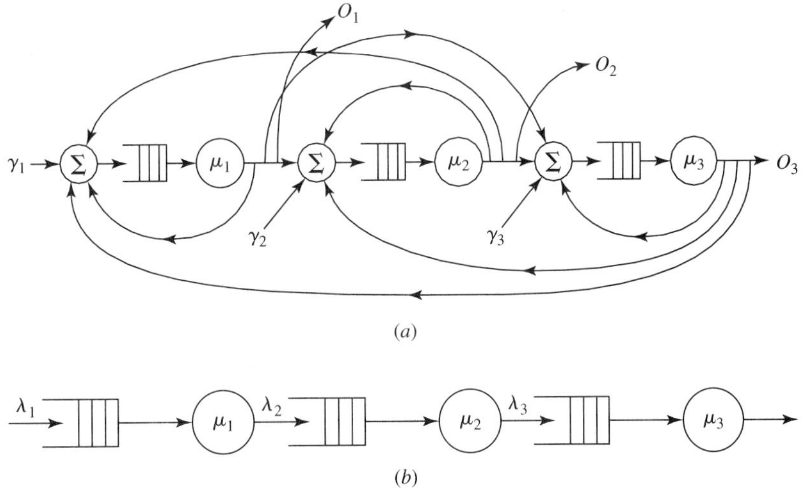

FIGURE 16-15
(*a*) Three queues with feedback and feed-forward. (*b*) Equivalent network in steady state.

R. R. Jackson has generalized this result by permitting additional Poisson arrivals to each phase from outside the system, and feedbacks from various phases within the system (Fig. 16-15). Thus a unit arrives at a phase with different probabilities. The service distributions are exponential, with the ith phase consisting of r_i parallel channels with identical service rates μ_i. Poisson arrivals from outside the system occur at the ith phase with rate γ_i, and after finishing service at the ith phase, an item either leaves for the jth phase with probability q_{ij}, where it is served in the order of their arrival along with Poisson arrivals from outside, or it leaves the system with probability

$$q_{i,0} = 1 - \sum_{j=1}^{m} q_{ij} \tag{16-283}$$

Figure 16-15a shows such an interconnected network for $m = 3$. Let λ_i represent the average arrival rate at the ith phase. Then λ_is satisfy

$$\lambda_i = \gamma_i + \sum_{k=1}^{m} q_{ki}\lambda_k \qquad i = 1, 2, \ldots, m \tag{16-284}$$

R. R. Jackson has shown that in steady state, the distribution of the number of items in each phase of such an interconnected network is independent of the distribution of the number of items in any other phase, and they satisfy (16-277). Proceeding as in (16-278) with λ replaced by $\lambda_1, \lambda_2, \ldots, \lambda_m$ for each state, Jackson has shown that an interconnected feedback/feed-forward network with Poisson arrivals at various phases behaves like a cascade connection of independent queues with input rate λ_i and service rate μ_i at the ith phase.

THEOREM 16-4

JACKSON'S THEOREM

▶ Consider a network of m phases with the ith phase consisting of r_i parallel servers, all with identical service rate μ_i. The network allows feedback and feed forward from phase i to j with probability q_{ij}, in addition to Poisson arrivals from outside to each phase at rate γ_i. Then the probability that there are n_i items in phase i, $i = 1, 2, \ldots, m$ is given by

$$p(n_1, n_2, \ldots, n_m) = \prod_{i=1}^{m} p_i(n_i) \tag{16-285}$$

where $p_i(n_i)$ is as in (16-278) and (16-279) with λ replaced by λ_i given by (16-284). ◀

From (16-283) we also obtain

$$\sum_{i=1}^{m} q_{i,0}\, \lambda_i = \sum_{i=1}^{m} \gamma_i$$

and the total output from the system equals the total input into the system.

Thus any complex network with external Poisson feeds behaves like a cascade connection of $M/M/r_i$ queues in steady state. Jackson's theorem is noteworthy considering that the combined input to each phase in presence of feedback is *no longer* Poisson, and consequently the server outputs are no longer Poisson. Nevertheless from (16-285) the phases are independent and they behave like $M/M/r_i$ queues with input rate λ_i and service rate μ_i, $i = 1, 2, \ldots, m$.

PROBLEMS

16-1 ***M/M/1/m* queue.** Consider a single server Poisson queue with limited system capacity m. Write down the steady state equations and show that the steady state probability that there are n items in the system is given by

$$p_n = \begin{cases} \dfrac{1-\rho}{1-\rho^{m+1}}\rho^n & \rho \neq 1 \\ \dfrac{1}{r+1} & \rho = 1 \end{cases}$$

where $\rho = \lambda/\mu$. (*Hint:* Refer to (16-132) with $r = 1$.)

16-2 (*a*) Let $\mathbf{n}_1(t)$ represent the total number of items in two identical $M/M/1$ queues, each operating independently with input rate λ and service rate μ. Show that in the steady state

$$P\{\mathbf{n}_1(t) = n\} = (n+1)(1-\rho)^2\rho^n \qquad n \geq 0$$

when $\rho = \lambda/\mu < 1$.

(*b*) If the two queues are merged to form the input to a single $M/M/2$ queue, show that the number of items in steady state in the system is given by

$$P\{\mathbf{n}_2(t) = n\} = \begin{cases} 2(1-\rho)\rho^n/(1+\rho) & n \geq 1 \\ (1-\rho)/(1+\rho) & n = 0 \end{cases}$$

(*c*) Let L_1 and L_2 represent the average number of items waiting in the two configurations above. Prove that

$$L_1 = \frac{2\rho^2}{1-\rho} \qquad L_2 = \frac{2\rho^3}{1-\rho^2} < L_1$$

which shows a single queue configuration is more efficient than separate queues. (*Hint:* Use (16-106) and (16-107).)

16-3 A system has m components that become ON and OFF independently. Suppose the ON and OFF processes are independent Poisson processes with parameter λ and μ, respectively. Determine the steady state probability of exactly k components being ON, $k = 0, 1, 2, \ldots, m$.

16-4 State-dependent service. Consider an $M/M/1$ queue where the mean service rate depends on the state of the system. Suppose the server has two rates, the slow rate μ_1 and the fast rate μ_2. The server works at the slow rate till there are m customers in the system, after which it switches over to the fast rate. (a) Show that the steady state probabilities are given by

$$p_n = \begin{cases} \rho_1^n p_0 & 0 \geq n < m \\ \rho_1^{m-1} \rho_2^{n-m+1} p_0 & n \geq m \end{cases}$$

where

$$p_0 = \left(\frac{1-\rho_1^m}{1-\rho_1} + \frac{\rho_2 \rho_1^{m-1}}{1-\rho_2} \right)^{-1}$$

with $\rho_1 = \lambda/\mu_1$, and $\rho_2 = \lambda/\mu_2$.

(b) Determine the mean system size $L = \sum_{n=0}^{\infty} n p_n$.

16-5 Patient and impatient customers. Consider an $M/M/r$ queue with arrival rate λ and service rate μ that contains both patient and impatient customers at its input. If all servers are busy, patient customers join the queue and wait for service, while impatient customers leave the system instantly. If p represents the probability of an arriving customer to be patient, show that when $p\lambda/r\mu < 1$ the steady state distribution in the system is given by

$$p_n = \begin{cases} \dfrac{(\lambda/\mu)^n}{n!} p_0 & n < r \\[2mm] \dfrac{(\lambda/\mu)^r}{r!} (p\lambda/r\mu)^{n-r} p_0 & n \geq r \end{cases}$$

where

$$p_0 = \frac{1}{\displaystyle\sum_{n=0}^{r-1} \frac{(\lambda/\mu)^n}{n!} + \frac{(\lambda/\mu)^r}{r!(1-p\lambda/r\mu)}}$$

(Hospitals, restaurants, barber shops, department stores, and telephone exchanges all lose customers who are either inherently impatient or cannot afford to wait.)

16-6 Let $\mathbf{w}$ represent the waiting time in an $M/M/r/m$ queue ($m > r$). Show that

$$P\{\mathbf{w} > t\} = \frac{p_r e^{-r\mu t}}{1-\rho} \sum_{k=0}^{m-r-1} \frac{(r\mu t)^k}{k!} (\rho^k - \rho^{m-k})$$

where $\rho = \lambda/r\mu$, and

$$p_r = \frac{(\lambda/\mu)^r/r!}{\displaystyle\sum_{n=0}^{r} \frac{(\lambda/\mu)^n}{n!} + \frac{(\lambda/\mu)^r \rho(1-\rho^{m-r})}{r!(1-\rho)}}$$

(*Hint:* $P(\mathbf{w} > t) = \int_t^\infty f_w(\tau)\, d\tau = \sum_{n=r}^{m-1} p_n \int_t^\infty f_w(\tau \mid n)\, d\tau$.)

16-7 Bulk arrivals ($M^{[x]}/M/1$) queue. In certain situations, arrivals and/or service can occur in groups (in bulk or batches). The simplest such generalization with respect to arrival process is to assume that the arrival instants are markovian as in an $M/M/1$ queue, but each arrival is a random variable of size $\mathbf{x}$ with

$$P\{\mathbf{x} = k\} = c_k \qquad k = 1, 2, \ldots, \infty$$

that is, arrivals occur in batches of random size $\mathbf{x}$ with the above probability distribution. Referring to (10-113)–(10-115), the input process to the system represents a compound Poisson process.

(*a*) Show that the steady probabilities $\{p_n\}$ for a single-server queue with compound Poisson arrivals and exponential service times $(M^{[\mathbf{x}]}/M/1)$ satisfy the system of equations [40]

$$0 = -(\lambda + \mu)p_n + \mu p_{n+1} + \lambda \sum_{k=1}^{n} p_{n-k}c_k \qquad n \ge 1$$

$$0 = -\lambda p_0 + \mu p_1$$

Hint: For the forward Kolmogorov equations in (16-18), the transition densities in this case are given by

$$\lambda_{kj} = \begin{cases} \lambda c_i & j = k+i, \quad i = 1, 2, \dots \\ \mu & j = k-1 \\ 0 & \text{otherwise} \end{cases}$$

Notice that although the process $M^{[\mathbf{x}]}/M/1$ is markovian, it represents a non–birth/death process.

(*b*) Let $P(z) = \sum_{n=0}^{\infty} p_n z^n$ and $C(z) = \sum_{k=1}^{\infty} c_k z^k$ represent the moment generating functions of the desired steady state probabilities $\{p_n\}$ and the bulk arrival probabilities $\{c_k\}$, respectively. Show that

$$P(z) = \frac{(1-\rho_0)(1-z)}{1 - z - \rho z(1 - C(z))} = \frac{1-\rho_0}{1 - \rho z D(z)}$$

where $\rho = \lambda/\mu$, $\rho_0 = \rho E\{\mathbf{x}\} = \rho C'(1) < 1$ and

$$D(z) = \frac{1 - C(z)}{1 - z} = \sum_{k=0}^{\infty} d_k z^k$$

with $d_0 = 1 - c_0$, $d_k = 1 - \sum_{i=0}^{k} c_i \ge 0$, $k \ge 1$. Hence for $\rho_0 < 1$, show that

$$p_n = \lim_{t\to\infty} P\{\mathbf{x}(t) = n\} = (1-\rho_0)\sum_{k=0}^{n} \rho^k d_{n-k}^{(k)}$$

where $\{d_n^{(k)}\}$ represents the k-fold convolution of the sequence $\{d_n\}$ above with itself.

(*c*) Determine the mean system size L in terms of $E\{\mathbf{x}\}$ and $E\{\mathbf{x}^2\}$.

(*d*) $M^{[m]}/M/1$ queue: Suppose the bulk arrivals are of constant size m. Determine $P(z)$ and the mean value L.

(*e*) Suppose the bulk arrivals are geometric random variables with parameter p. Determine $P(z)$ and the mean value L.

(*f*) Suppose the bulk arrivals are binomial random variables with parameters m and p. Determine $P(z)$ and the mean value L.

16-8 Bulk service ($M/M^{[\mathbf{y}]}/1$). Consider a single-server queue with Poisson arrivals, where items are served on a first-come first-served basis in batches not exceeding a certain number m. Thus m items are served together, if the length of the queue is greater than m, otherwise, the entire queue is served in a batch, and new arrivals immediately enter into service up to the limit m and finish along with the others. The amount of time required for the server is exponential, irrespective of the batch size ($\le m$).

(*a*) Show that the steady state probabilities $\{p_n\}$ for the bulk service mechanism described above satisfy the equations [40]

$$0 = -(\lambda + \mu)p_n + \mu p_{n+m} + \lambda p_{n-1} \qquad n \ge 1$$

$$0 = -\lambda p_0 + \mu p_1 + \mu p_2 + \cdots + \mu p_m$$

Hint: The transition densities for this non–birth/death process are given by

$$\lambda_{kj} = \begin{cases} \lambda & j = k+1 \\ \mu & j = k-m \end{cases}$$

(*b*) Let $P(z)$ represent the moment-generating function of the steady state probabilities in part (a). Show that

$$P(z) = \frac{\sum_{k=0}^{m-1} p_k z^k - p_0(1+\rho)z^m}{\rho z^{m+1} - (\rho+1)z^m + 1}$$

where $\rho = \lambda/\mu$.

(*c*) Using Rouche's theorem show that $P(z)$ in part (*b*) can be simplified as

$$P(z) = \frac{z_0 - 1}{z_0 - z}$$

where z_0 represents the unique positive root that is greater than unity of the denominator polynomial $\rho z^{m+1} - (\rho+1)z^m + 1$. Hence

$$p_n = (1 - r_0)r_0^n \qquad n \geq 0$$

where $r_0 = 1/z_0$, and the bulk service model presented here behaves like an $M/M/1$ queue.

(*Hint:* **Rouche's theorem:** If $f(z)$ and $g(z)$ are analytic inside and on a closed contour C and if $|f(z)| > |g(z)|$ on the contour C, then $f(z)$ and $f(z)+g(z)$ have the same number of zeros inside C.

Try $f(z) = (\rho+1)z^m$, $g(z) = \rho z^{m+1} + 1$ on a contour C defined by the circle $|z| = 1 + \epsilon$ with $\epsilon > 0$.)

(*d*) Derive the average system size for this model.

16-9 Bulk service $M/M^{[m]}/1$. In Prob. 16-8, assume that the batch size must be exactly m for the server to commence service, and if not the server waits until such a time to start service.

(*a*) Show that the steady state probabilities $\{p_n\}$ satisfy the equations [40]

$$0 = -(\lambda+\mu)p_n + \mu p_{n+m} + \lambda p_{n-1} \qquad n \geq m$$

$$0 = -\lambda p_n + \mu p_{n+m} + \lambda p_{n-1} \qquad 1 \leq n < m$$

$$0 = -\lambda p_0 + \mu p_m$$

(*b*) Show that the moment generating function in this case is given by

$$P(z) = \frac{(1-z^m)\sum_{k=0}^{m-1} p_k z^k}{\rho z^{m+1} - (\rho+1)z^m + 1} = \frac{(z_0 - 1)\sum_{k=0}^{m-1} z^k}{m(z_0 - z)}$$

where the latter form is obtained by applying Rouche's theorem with z_0 and ρ as defined in Prob. 16-8.

(*c*) Determine the steady state probabilities $\{p_n\}$ and the mean system size in this case.

16-10 $E_m/M/1$ queue versus bulk service. Consider an Erlang-m input queue with arrival rate λ and exponential service rate μ. From the $GI/M/1$ analysis it follows that such a queue behaves like an $M/M/1$ queue with parameter π_0, where $\pi_0 < 1$ is the unique solution of the equation $B(z) = z$, provided $\lambda/\mu < 1$. Here $B(z) = \Psi_A(\mu(1-z))$, and $\Psi_A(s)$ equals the Laplace transform of the interarrival probability density function. Show that the parameter π_0 satisfies the characteristic equation

$$\rho x^{m+1} - (\rho+1)x^m + 1 = 0$$

with $\rho = m\lambda/\mu$.

(*Hint:* Use the transformation $z = x^{-m}$ in $B(z) = z$, which corresponds to a phase approach to the Erlang arrivals.)

On comparing the last equation with the bulk service model in Prob. 16-9, conclude that an Erlang arrival model with exponential service is essentially equivalent to a bulk service model with exponential arrivals. This analogy, however, is true only in a broad sense in terms of the steady state probabilities, since in bulk service several items leave together after completing service, whereas in the $E_m/M/1$ model items leave one by one after service.

16-11 ***$M/E_m/1$* versus bulk arrivals.** Consider an Erlang-m service model with service rate μ and Poisson arrivals with parameter λ. Using the $M/G/1$ analysis in Example 16-4 show by proper transformation, the characteristic equation in that case can be rewritten as

$$\rho z^{m+1} - (1+\rho)z + 1 = 0$$

Hence conclude that $M/E_m/1$ queue exhibits similar characteristics in terms of steady state probabilities as the constant the bulk arrival model $M^{[m]}/M/1$ discussed in Prob. 16-7.

16-12 ***$M/G/1$* queue.** (*a*) Suppose the service mechanism is described by a k-point distribution where

$$\frac{dB(t)}{dt} = \sum_{i=1}^{k} c_i \delta(t - T_i) \qquad \sum_{i=1}^{k} c_i = 1$$

Determine the steady state probabilities $\{p_n\}$.

(*b*) Find the steady state probabilities for a state-dependent $M/G/1$ queue where

$$B_i(t) = \begin{cases} 1 - e^{-\mu_1 t} & i = 1 \\ 1 - e^{-\mu_2 t} & i > 1 \end{cases}$$

Thus service is provided at the rate of μ_1 if there is no queue, and at the rate of μ_2 if there is a queue.

16-13 Show that the waiting time distribution $F_w(t)$ for an $M/G/1$ queue can be expressed as

$$F_w(t) = (1-\rho)\sum_{n=0}^{\infty} \rho^n F_r^{(n)}(t)$$

where $F_r(t)$ represents the residual service time distribution given by

$$F_r(t) = \mu \int_0^t [1 - B(\tau)]\, d\tau$$

and $F_r^{(n)}(t)$ is the n-fold convolution of $F_r(t)$ with itself.

(*Hint:* Use (16-209).)

16-14 **Transient *$M/G/1$* queue.** Show that an $M/G/1$ queueing system is transient if the mean value of the number of customers that arrive during a service period is greater than one.

(*Hint:* In (16-197) and (16-198) if $\rho > 1$, then the equation $A(z) = z$ has a unique root $\pi_0 < 1$. Show that $\sigma_j = \pi_0^j$ satisfies the transient state probability equation in (15-236).)

16-15 ***$M/G/1/m$* queue.** Consider a finite capacity $M/G/1$ queue where atmost $m-1$ items are allowed to wait. (*a*) Show that the $m \times m$ probability transition matrix for such a finite $M/G/1$ queue is given by the $m \times (m-1)$ upper left hand block matrix in (15-34) followed by an mth column that makes each row sum equal to unity.

(*b*) Show that the steady state probabilities $\{q_j^*\}_{j=0}^{m-1}$ satisty the first $m-1$ equations in (15-210) together with the normalization condition $\sum_{j=0}^{m-1} q_j^* = 1$. Hence conclude that $q_j^* = cq_j,\ j = 0, 1, \ldots, m-1$, where $\{q_j\}$ correspond to the $M/G/1/\infty$ queue (see (16-203)).

16-16 ***$M/G/\infty$* queues.** Consider Poisson arrivals with parameter λ and general service times $\mathbf{s}$ with distribution $B(\tau)$ at a counter with infinitely many servers. Let $\mathbf{x}(t)$ denote the number of busy servers at time t. Show that [52] (a)

$$P\{\mathbf{x}(t) = k\} = e^{-\lambda\alpha(t)} \frac{[\lambda\alpha(t)]^k}{k!} \qquad k = 0, 1, 2, \ldots$$

where

$$\alpha(t) \triangleq \int_0^t [1 - B(\tau)]\, d\tau$$

and (b)

$$\lim_{t\to\infty} P\{\mathbf{x}(t) = k\} = e^{-\rho} \frac{\rho^k}{k!} \qquad k = 0, 1, 2, \ldots$$

where $\rho = \lambda E\{\mathbf{s}\}$. Thus in the long run, all $M/G/\infty$ queues behave like $M/M/\infty$ queues.

Hint: The event $\{\mathbf{x}(t) = k\}$ can occur in several mutually exclusive ways, namely, in the interval $(0, t)$, n customers arrive and k of them continue their service beyond t. Let A_n = "n arrivals in $(0, t)$," and $B_{k,n}$ = "exactly k services among the n arrivals continue beyond t," then by the theorem of total probability

$$P\{\mathbf{x}(t) = k\} = \sum_{n=k}^{\infty} P\{A_n \cap B_{k,n}\} = \sum_{n=k}^{\infty} P\{B_{k,n} \mid A_n\} P(A_n)$$

But $P(A_n) = e^{-\lambda t}(\lambda t)^n/n!$, and to evaluate $P\{B_{k,n} \mid A_n\}$, we argue as follows: From (9-28), under the condition that there are n arrivals in $(0, t)$, the joint distribution of the arrival instants agrees with the joint distribution of n independent random variables arranged in increasing order and distributed uniformly in $(0, t)$. Hence the probability that a service time $\mathbf{s}$ does not terminate by t, given that its starting time $\mathbf{x}$ has a uniform distribution in $(0, t)$ is given by

$$\begin{aligned} p_t &= \int_0^t P\{\mathbf{s} > t - x \mid \mathbf{x} = x\} f_x(x)\, dx \\ &= \int_0^t [1 - B(t - x)] \frac{1}{t}\, dx = \frac{1}{t} \int_0^t [1 - B(\tau)]\, d\tau = \frac{\alpha(t)}{t} \end{aligned}$$

It follows that $B_{k,n}$ given A_n has a binomial distribution, so that

$$P\{B_{k,n} \mid A_n\} = \binom{n}{k} p_t^k (1 - p_t)^{n-k} \qquad k = 0, 1, 2, \ldots, n$$

and

$$P\{\mathbf{x}(t) = k\} = \sum_{n=k}^{\infty} e^{-\lambda t} \frac{(\lambda t)^n}{n!} \binom{n}{k} \left(\frac{\alpha(t)}{t}\right)^k \left(\frac{1}{t} \int_0^t B(\tau)\, d\tau\right)^{n-k} \qquad k = 0, 1, 2, \ldots$$

BIBLIOGRAPHY

PROBABILITY THEORY

[1] Chung, Kai Lai, *A Course in Probability Theory* (3rd edition), Academic Press, San Diego, 2001.

[2] Cramer, H., *Mathematical Methods to Statistics,* Princeton University Press, Princeton, NJ, 1946.

[3] Feller, William, *An Introduction to Probability Theory and Its Applications,* Volume I (3rd edition) and Volume II (2nd edition), John Wiley and Sons, New York, 1968 and 1971.

[4] Gnedenko, B., *The Theory of Probability,* trans. by G. Yankovsky, MIR Publishers, Moscow, 1978.

[5] Loeve, Michel, *Probability Theory* (3rd edition), Van Nostrand, Princeton, NJ, 1963.

[6] Parzen, E., *Modern Probability Theory and Its Applications,* Wiley, New York, 1960.

[7] Rao, C. Radhakrisna, *Linear Statistical Inference and Its Applications* (2nd edition), John Wiley and Sons, New York, 1973.

[8] Rohatgi, Vijay K., and A. K. Saleh, *An Introduction to Probability and Statistics,* John Wiley and Sons, New York, 2001.

[9] Uspensky, J. V., *Introduction to Mathematical Probability,* McGraw-Hill, New York, 1965.

STOCHASTIC PROCESSES

[10] Childers, D. G., *Modern Spectrum Analysis,* Wiley, New York, 1978.

[11] Davenport, W. B., Jr., and W. L. Root, *An Introduction to the Theory of Random Signals and Noise,* McGraw-Hill, New York, 1958.

[12] Doob, J. L., *Stochastic Processes,* John Wiley and Sons, New York, 1953.

[13] Franks, L. E., *Signal Theory,* Prentice-Hall, Englewood Cliffs, NJ, 1979.

[14] Geronimus, Y. L., *Polynomials Orthogonal on a Circle and System Identification,* Springer-Verlag, New York, 1954.

[15] Gikhman, I. I., and A. V. Skorokhod, *Introduction to the Theory of Random Processes,* Dover Publications, New York, 1996.

[16] Grenander, U., and G. Szegö, *Toeplitz Forms and Their Applications,* Chelsea, New York, 1984.

[17] Helstom, C. W., *Statistical Theory of Signal Detection* (2nd edition), Pergamon Press, New York, 1968.

[18] Leon-Garcia, A., *Probability and Random Processes for Electrical Engineering,* Addison Wesley, New York, 1994.

[19] Oppenheim, A. V., and R. W. Schafer, *Digital Signal Processing,* Prentice-Hall, Englewood Cliffs, NJ, 1975.

[20] Papoulis, A., *The Fourier Integral and Applications,* McGraw-Hill, New York, 1962.

[21] Papoulis, A., *Systems and Transforms with Applications in Optics,* McGraw-Hill, New York, 1968.

[22] Papoulis, A., *Signal Analysis,* McGraw-Hill, New York, 1977.

[23] Papoulis, A., *Circuits and Systems: A Modern Approach,* Holt, Rinehart and Winston, New York, 1980.

[24] Papoulis, J., *Probability and Statistics,* Prentice-Hall, Englewood Cliffs, NJ, 1990.

[25] Pillai, S. U., and T. I. Shim, *Spectrum Estimation and System Identification,* Springer-Verlag, New York, 1993.
[26] Proakis, J., *Introduction to Digital Communications,* McGraw-Hill, New York, 1977.
[27] Schwartz, M., and L. Shaw, *Signal Processing,* McGraw-Hill, New York, 1975.
[28] Wainstein, L. A., and V. D. Zubakov, *Extraction of Signals from Noise* (trans. from Russian), Prentice-Hall, Englewood Cliffs, NJ, 1962.
[29] Wiener, N., *Extrapolation, Interpolation, and Smoothing of Stationary Time Series,* MIT Press, Cambridge, MA, 1949.
[30] Woodward, P., *Probability and Information Theory with Applications to Radar,* Pergamon, New York, 1953.
[31] Yaglom, A. M., *Stationary Random Functions* (trans. from Russian), Prentice-Hall, Englewood Cliffs, NJ, 1962.
[32] Yaglom, A. M., *Correlation Theory of Stationary and Related Random Functions,* 2 Vols., Springer, New York, 1987.
[33] Youla, D. C., *Lecture Notes on Network Theory,* Polytechnic University, Farmingdale, NY, 2000.

QUEUEING THEORY

[34] Atreya, K. B., and P. Tagers (Ed.), *Classical and Modern Branching Processes,* Springer-Verlag, New York, 1991.
[35] Bharucha-Reid, A. T., *Elements of the Theory of Markov Processes and Their Applications,* Dover Publications, New York, 1988.
[36] Bremaud, Pierre, *Markov Chains,* Springer, New York, 2001.
[37] Chung, Kai Lai, *Markov Chains* (2nd edition), Springer-Verlag, New York, 1967.
[38] Cohen, J. W., *The Single Server Queue,* North Holland, Amsterdam, Amsterdam, 1969.
[39] Gnedenko, B. V., and I. N. Kovalenko, *Introduction to Queueing Theory,* trans. by S. Kotz (2nd edition), Birkhanuser, Boston, 1989.
[40] Gross, D., and C. M. Harris, *Fundamentals of Queueing Theory* (3rd edition), John Wiley and Sons, New York, 1998.
[41] Karlin, Samuel, *A First Course in Stochastic Processes,* Academic Press, New York, 1966.
[42] Kemeny, John, G., and Snell, Laurie, J., *Finite Markov Chains*, Van Nostrand, Princeton, NJ, 1960.
[43] Kleinrock, L., *Queueing Systems,* Volumes I and II, John Wiley and Sons, New York, 1975.
[44] Medhi, J., *Stochastic Processes,* Wiley Eastern Ltd., New Delhi, 1991.
[45] Neuts, Marcel F., *Structured Stochastic Matrices of M/G/1 Type and Their Applications,* Marcal Dekker, New York, 1989.
[46] Parzen, Emanuel, *Stochastic Processes,* Classics in Applied Mathematics Series, No. 24, SIAM, Philadelphia, PA, 1999.
[47] Prabhu, N. U., *Stochastic Storage Processes* (2nd edition), Springer, New York, 1998.
[48] Saaty, Thomas, *Elements of Queueing Theory with Applications,* Dover Publications, New York, 1983.
[49] Sadovskii, L. E., and A. L. Sadovski, *Mathematics and Sports,* trans. by S. Maker-Limanov, University Press (India), Hyderabad, 1998.
[50] Schwartz, M., *Computer-Communication Network Design and Analysis,* Prentice-Hall, Englewood Cliffs, NJ, 1977.
[51] Srinivasan, S. K., and K. M. Mehata, *Stochastic Processes,* McGraw-Hill, New York, 1978.
[52] Takacs, Lajos, *Introduction to the Theory of Queues,* Oxford University Press, New York, 1962.
[53] Takacs, Lajos, *Stochastic Processes, Problems and Solutions,* trans. by P. Zador, Methuen and Co. and Science Paperbacks, London, 1974.
[54] Takacs, Lajos, *Combinatorial Methods in the Theory of Stochastic Processes,* John Wiley and Sons, New York, 1967.
[55] Trivedi, Kishore, S., *Probability and Statistics with Reliability, Queueing and Computer Applications,* Prentice-Hall, Englewood Cliffs, NJ, 1982.

INDEX

Interrelationship among Random Variables

$\mathbf{x}_i \sim N(0, \sigma^2), i = 1, 2, \ldots, n$ $\mathbf{x}_i$ independent	$\sum_{a=1}^{n} a_i \mathbf{x}_i \sim N\left(0, a^2\sigma^2\right), \quad a^2 = \sum_{i=1}^{n} a_i^2$ $\mathbf{x}_i^2 + \mathbf{x}_k^2 \sim$ exponential $\mathbf{a} = \sqrt{\mathbf{x}_i^2 + \mathbf{x}_k^2} \sim$ Rayleigh $\sum_{i=1}^{n} \mathbf{x}_i^2/\sigma^2 \sim$ Chi-square-n $(\chi^2(n))$ $\mathbf{r} = \sqrt{(\mathbf{x}_i - a)^2 + (\mathbf{x}_k - b)^2} \sim$ Rician $\boldsymbol{\psi} = \tan^{-1}\left(\frac{\mathbf{x}_i}{\mathbf{x}_k}\right) \sim$ uniform $\mathbf{x}_i + j\mathbf{x}_k = \mathbf{a}e^{j\boldsymbol{\psi}} \sim$ complex Gaussian random variable. $\mathbf{a}$, $\boldsymbol{\psi}$ independent Rayleigh and uniform respectively.
$\mathbf{x} \sim G(\alpha, \beta)$ $\mathbf{y} \sim G(\alpha_0, \beta)$ $\mathbf{x}, \mathbf{y}$ independent	$\alpha = 1 \Rightarrow \mathbf{x} \sim$ exponential $\alpha = \frac{n}{2}, \beta = 2 \Rightarrow \mathbf{x} \sim \chi^2(n)$ $\alpha = n, \beta = \frac{1}{n\lambda} \Rightarrow \mathbf{x} \sim$ Erlang-n $\alpha = m, \beta = \Omega/m \Rightarrow \sqrt{\mathbf{x}} \sim$ Nakagami-m $\mathbf{x} + \mathbf{y} \sim G(\alpha + \alpha_0, \beta)$ $\frac{\mathbf{x}}{\mathbf{x} + \mathbf{y}} \sim \beta(\alpha, \alpha_0)$
$\mathbf{x} \sim$ exponential (λ) $\mathbf{y} \sim$ exponential (λ) $\mathbf{x}, \mathbf{y}$ independent	$\mathbf{x}^{1/\beta} \sim$ Weibull $\sqrt{\mathbf{x}} \sim$ Rayleigh ($\beta = 2$ in Weibull) $\mathbf{x} + \mathbf{y} \sim$ gamma $G(2, \lambda)$ $\min(\mathbf{x}, \mathbf{y}) \sim$ exponential(2λ)
$\mathbf{x} \sim$ Nakagami-m	$\mathbf{x}^2 \sim$ gamma $m = 1 \Rightarrow \mathbf{x} \sim$ Rayleigh
$\mathbf{x} \sim P(\lambda_1), \mathbf{y} \sim P(\lambda_2)$ $\mathbf{x}, \mathbf{y}$ independent	$\mathbf{x} + \mathbf{y} \sim P(\lambda_1 + \lambda_2)$
$\mathbf{x} \sim B(n, p), \mathbf{y} \sim B(m, p)$ $\mathbf{x}, \mathbf{y}$ independent	$\mathbf{x} + \mathbf{y} \sim B(n + m, p)$
$\mathbf{x}_i \sim$ Geometric $(p), i = 1, 2, \ldots, r$ $\mathbf{x}_i$ independent	$\sum_{i=1}^{r} \mathbf{x}_i \sim NB(r, p)$ $\min(\mathbf{x}_1, \mathbf{x}_2, \ldots, \mathbf{x}_r) \sim$ Geometric $[1 - (1 - P)^r]$
$\mathbf{x} \sim NB(m, p), \mathbf{y} \sim NB(n, p)$ $\mathbf{x}, \mathbf{y}$ independent	$\mathbf{x} + \mathbf{y} \sim NB(m + n, p)$
$\mathbf{x} \sim C(\alpha_1, \mu_1), \mathbf{y} \sim C(\alpha_2, \mu_2)$ $\mathbf{x}, \mathbf{y}$ independent	$\mu_1 = 0 \Rightarrow \frac{1}{\mathbf{x}} \sim C(1/\alpha_1, 0)$ $\mathbf{x} + \mathbf{y} \sim C(\alpha_1 + \alpha_2, \mu_1 + \mu_2)$
$\mathbf{x}, \mathbf{y} \sim U(0, 1)$ $\mathbf{x}, \mathbf{y}$ independent	$\mathbf{z} = \sqrt{-2\log \mathbf{x}}\ \cos(2\pi \mathbf{y}) \sim N(0, 1)$